Basic Concepts of String Theory

Theoretical and Mathematical Physics

The series founded in 1975 and formerly (until 2005) entitled *Texts and Monographs in Physics* (TMP) publishes high-level monographs in theoretical and mathematical physics. The change of title to *Theoretical and Mathematical Physics* (TMP) signals that the series is a suitable publication platform for both the mathematical and the theoretical physicist. The wider scope of the series is reflected by the composition of the editorial board, comprising both physicists and mathematicians.

The books, written in a didactic style and containing a certain amount of elementary background material, bridge the gap between advanced textbooks and research monographs. They can thus serve as basis for advanced studies, not only for lectures and seminars at graduate level, but also for scientists entering a field of research.

Editorial Board

W. Beiglböck, Institute of Applied Mathematics, University of Heidelberg, Germany
P. Chrusciel, Hertford College, University of Oxford, UK
J.-P. Eckmann, Department of Theoretical Physics, University of Geneva, Switzerland
H. Grosse, Institute of Theoretical Physics, University of Vienna, Austria
A. Kupiainen, University of Helsinki, Finland
H. Löwen, Heinrich-Heine-University, Düsseldorf, Germany
M. Loss, School of Mathematics, Georgia Institute of Technology, Atlanta, GA, USA
N. A. Nekrasov, Institut des Hautes Études Scientifiques, Bures-sur-Yvette, France
M. Ohya, Tokyo University of Science, Noda, Japan
M. Salmhofer, Institute for Theoretical Physics, University of Heidelberg, Germany
S. Smirnov, Mathematics Section, University of Geneva, Switzerland
L. Takhtajan, Department of Mathematics, Stony Brook University, NY, USA
J. Yngvason, Institute of Theoretical Physics, University of Vienna, Austria

For further volumes:
http://www.springer.com/series/720

Ralph Blumenhagen
Dieter Lüst
Stefan Theisen

Basic Concepts
of String Theory

Springer

Ralph Blumenhagen
Werner-Heisenberg-Institut
Max-Planck-Institut für Physik
München
Germany

Stefan Theisen
Albert-Einstein-Institut
Max-Planck-Institut für Gravitationsphysik
Golm
Germany

Dieter Lüst
Ludwig-Maximilians Universität München
Arnold-Sommerfeld Zentrum für
Theoretische Physik
München
Germany

ISSN 1864-5879　　　　　　　ISSN 1864-5887 (electronic)
ISBN 978-3-642-42999-6　　　ISBN 978-3-642-29497-6 (eBook)
DOI 10.1007/978-3-642-29497-6
Springer Heidelberg New York Dordrecht London

© Springer-Verlag Berlin Heidelberg 2013
Softcover reprint of the hardcover 1st edition 2013
This work is subject to copyright. All rights are reserved by the Publisher, whether the whole or part of the material is concerned, specifically the rights of translation, reprinting, reuse of illustrations, recitation, broadcasting, reproduction on microfilms or in any other physical way, and transmission or information storage and retrieval, electronic adaptation, computer software, or by similar or dissimilar methodology now known or hereafter developed. Exempted from this legal reservation are brief excerpts in connection with reviews or scholarly analysis or material supplied specifically for the purpose of being entered and executed on a computer system, for exclusive use by the purchaser of the work. Duplication of this publication or parts thereof is permitted only under the provisions of the Copyright Law of the Publisher's location, in its current version, and permission for use must always be obtained from Springer. Permissions for use may be obtained through RightsLink at the Copyright Clearance Center. Violations are liable to prosecution under the respective Copyright Law.
The use of general descriptive names, registered names, trademarks, service marks, etc. in this publication does not imply, even in the absence of a specific statement, that such names are exempt from the relevant protective laws and regulations and therefore free for general use.
While the advice and information in this book are believed to be true and accurate at the date of publication, neither the authors nor the editors nor the publisher can accept any legal responsibility for any errors or omissions that may be made. The publisher makes no warranty, express or implied, with respect to the material contained herein.

Printed on acid-free paper

Springer is part of Springer Science+Business Media (www.springer.com)

Preface

This is a revision of Vol. 346 of Springer Lecture Notes in Physics written by D. Lüst and S. Theisen in 1989. The notes have long been out of print, but over the years we had continuous positive feedback from students, who found the book helpful in their attempt to enter string theory. To make the book useful for a new generation of string theorists required a revision and a substantial extension. Unfortunately this means that it became intimidatingly voluminous.

The purpose of this new edition is the same as of the old one: to prepare the reader for research in string theory. It is not a compendium of results but is intended to be a textbook in the sense that, at least in most parts, the reader is not referred to the original literature for derivations. We try to be pedagogical, avoiding excessive use of phrases such as "it is well known," "one can show," which are often frustrating (not only) for the beginner.

Aiming at a pedagogical introduction to a vast subject such as string theory also means that we had to make a selection of topics. Major omissions are black holes in string theory, strings at finite temperature, string cosmology, anomalies in string theory, model building, matrix model description of M-theory, string field theory, to name a few.

We give references at the end of each chapter. We restrict ourselves to some basic papers and reviews from which we have profited and also some additional references which cover material which goes beyond what we could cover. Almost all references are easily available. With few exceptions they are either published in journals, available as preprints on the arXiv or scanned at: http://www-lib.kek.jp/KISS/kiss_prepri.html.

The influence of the classic string monographs by Green, Schwarz, Witten, and Polchinski can be felt throughout most chapters.

München, Germany	Ralph Blumenhagen
München, Germany	Dieter Lüst
Golm, Germany	Stefan Theisen

Acknowledgement

R.B. thanks the Albert Einstein Institute in Potsdam, the KITP in Santa Barbara and the KITPC in Beijing for hospitality during part of this work. D.L. acknowledges the hospitality of the theory group at CERN. S.T. is grateful to the Heisenberg Institute in Munich, to the Departamento de Física of the Universidad Central de Venezuela and to the Institute for Theoretical Physics at the University of Heidelberg for extended hospitality while writing parts of this book. I. Adam, A. Font, S. Förste, S. Fredenhagen, M. Gaberdiel, D. Ghoshal, M. Haack, A. Hebecker, S. Hosono, A. Kleinschmidt, O. Lechtenfeld, I. Melnikov, R. Minasian, O. Schlotterer, A. Schwimmer, S. Stieberger, D. Tsimpis, T. Weigand were very patient answering questions on various topics of string theory. We also thank N. Akerblom, S. Moster and E. Plauschinn for their help in recovering the 1989 edition of the book and for preparing part of the figures. Moreover, we thank A. Deser, X. Gao, B. Jurke, T. Rahn, F. Rennecke and H. Roschy for comments on the manuscript. All their help is very much appreciated.

Contents

1	**Introduction**	1
2	**The Classical Bosonic String**	7
	2.1 The Relativistic Particle	7
	2.2 The Nambu-Goto Action	10
	2.3 The Polyakov Action and Its Symmetries	12
	2.4 Oscillator Expansions	23
	2.5 Examples of Classical String Solutions	31
3	**The Quantized Bosonic String**	35
	3.1 Canonical Quantization of the Bosonic String	35
	3.2 Light-Cone Quantization of the Bosonic String	42
	3.3 Spectrum of the Bosonic String	46
	3.4 Covariant Path Integral Quantization	53
	3.5 Appendix: The Virasoro Algebra	59
4	**Introduction to Conformal Field Theory**	63
	4.1 General Introduction	63
	4.2 Application to Closed String Theory	85
	4.3 Boundary Conformal Field Theory	93
	4.4 Free Boson Boundary States	101
	4.5 Crosscap States for the Free Boson	103
5	**Parametrization Ghosts and BRST Quantization**	107
	5.1 The Ghost System as a Conformal Field Theory	107
	5.2 BRST Quantization	110
6	**String Perturbation Theory and One-Loop Amplitudes**	121
	6.1 String Perturbation Expansion	121
	6.2 The Polyakov Path Integral for the Closed Bosonic String	126
	6.3 The Torus Partition Function	146
	6.4 Torus Partition Functions for Rational CFTs	152
	6.5 The Cylinder Partition Function	157

	6.6	Boundary States and Cylinder Amplitude for RCFTs	163
	6.7	Crosscap States, Klein Bottle and Möbius Strip Amplitudes	166
	6.8	Appendix: D-brane Tension	171

7 The Classical Fermionic String ... 175
- 7.1 Motivation for the Fermionic String ... 175
- 7.2 Superstring Action and Its Symmetries ... 176
- 7.3 Superconformal Gauge ... 180
- 7.4 Oscillator Expansions ... 189
- 7.5 Appendix: Spinor Algebra in Two Dimensions ... 192

8 The Quantized Fermionic String ... 195
- 8.1 Canonical Quantization ... 195
- 8.2 Light-Cone Quantization ... 200
- 8.3 Spectrum of the Fermionic String, GSO Projection ... 202
- 8.4 Path Integral Quantization ... 208
- 8.5 Appendix: Dirac Matrices and Spinors in d Dimensions ... 210

9 Superstrings ... 223
- 9.1 Spin Structures and Superstring Partition Function ... 223
- 9.2 Boundary States for Fermions ... 232
- 9.3 D-branes ... 235
- 9.4 The Type I String ... 241
- 9.5 Stable Non-BPS Branes ... 251
- 9.6 Appendix: Theta-Functions and Twisted Fermionic Partition Functions ... 254

10 Toroidal Compactifications: 10-Dimensional Heterotic String ... 263
- 10.1 Motivation ... 263
- 10.2 Toroidal Compactification of the Closed Bosonic String ... 264
- 10.3 Toroidal Partition Functions ... 280
- 10.4 The $E_8 \times E_8$ and $SO(32)$ Heterotic String Theories ... 284
- 10.5 Toroidal Orbifolds ... 294
- 10.6 D-branes on Toroidal Compactifications ... 308

11 Conformal Field Theory II: Lattices and Kač-Moody Algebras ... 321
- 11.1 Kač-Moody Algebras ... 321
- 11.2 Lattices and Lie Algebras ... 327
- 11.3 Frenkel-Kač-Segal Construction ... 337
- 11.4 Fermionic Construction of the Current Algebra: Bosonization ... 340
- 11.5 Unitary Representations and Characters of Kač-Moody Algebras ... 343
- 11.6 Highest Weight Representations of $\widehat{su}(2)_k$ ... 349

12 Conformal Field Theory III: Superconformal Field Theory ... 355
- 12.1 $N = 1$ Superconformal Symmetry ... 355
- 12.2 $N = 2$ Superconformal Symmetry ... 370
- 12.3 Chiral Ring and Topological Conformal Field Theory ... 382

Contents   xi

13 Covariant Vertex Operators, BRST and Covariant Lattices 391
     13.1 Bosonization and First Order Systems 391
     13.2 Covariant Vertex Operators, BRST and Picture Changing 403
     13.3 D-branes and Space-Time Supersymmetry 411
     13.4 The Covariant Lattice 415
     13.5 Heterotic Strings in Covariant Lattice Description 422

14 String Compactifications 427
     14.1 Conformal Invariance and Space-Time Geometry 427
     14.2 T-Duality and Buscher Rules 432
     14.3 Compactification 439
     14.4 Mathematical Preliminaries.. 450
     14.5 Calabi-Yau Manifolds.. 473
     14.6 Compactification of the Type II String on CY Threefolds 487
     14.7 Compactification of the Heterotic String on CY Threefolds 497
     14.8 Appendix: Some Riemannian Geometry, Hodge Duals, etc. 508

15 CFTs for Type II and Heterotic String Vacua 521
     15.1 Motivation 521
     15.2 Supersymmetric Orbifold Compactifications 522
     15.3 Orientifold Compactifications 534
     15.4 World-Sheet Aspects of Supersymmetric Compactifications 544
     15.5 Gepner Models 567
     15.6 Four-Dimensional Heterotic Strings via Covariant Lattices....... 578

16 String Scattering Amplitudes and Low Energy Effective Field Theory 585
     16.1 Generalities 585
     16.2 String Scattering Amplitudes 588
     16.3 From Amplitudes to the Low Energy Effective Field Theory 615
     16.4 Effective Supergravities in Ten and Eleven Dimensions 622
     16.5 Effective Action for D-branes 630
     16.6 Appendix: Integrals in Veneziano and Virasoro-Shapiro Amplitudes 636

17 Compactifications of the Type II Superstring with D-branes and Fluxes 641
     17.1 Brane Worlds and Fluxes 641
     17.2 Supersymmetry, Tadpoles and Massless Spectra 644
     17.3 Flux Compactifications 657
     17.4 Fluxes and SU(3) Structure Group 663
     17.5 Appendix 670

18 String Dualities and M-Theory 675
     18.1 General Remarks 675
     18.2 Simple Examples: Modular Invariance and T-Duality............. 677
     18.3 Extended Objects: Some Generalities 680

18.4	Central Charges and BPS Bound	684
18.5	Brane Solutions in Supergravity	688
18.6	Non-perturbative Dualities	705
18.7	M-Theory	716
18.8	F-Theory	724
18.9	AdS/CFT Correspondence	733

Index 775

Chapter 1
Introduction

String theory started in the late 1960s as an attempt to organize and to explain the observed spectrum of hadrons and their interactions. It was then discarded as a candidate theory of strong interactions, a development which was mainly triggered by the rapid success of quantum chromodynamics. One problem was the existence of a critical dimension, which is 26 for the bosonic string and 10 for the fermionic string. Another obstacle for the interpretation of string theory as a theory of strong interactions was the existence of a massless spin two particle which is not present in the hadronic world. In 1974 Scherk and Schwarz suggested to turn the existence of this mysterious massless spin two particle into an advantage by interpreting it as the graviton, i.e. the field quantum of gravity. This means that the fundamental mass scale, i.e. the tension of a string, is related to the characteristic mass scale of gravity, namely the Planck mass $M_P = \sqrt{\hbar c/G_N} \simeq 10^{19} \,\text{GeV}/c^2$. They also realized that at low energies this stringy graviton interacts according to the covariance laws of general relativity. With this insight, string theory became a candidate for a quantum theory of gravity which could, at least in principle, achieve a unification of all particles and their interactions.

There are open and closed strings. The massless spin two particle appears in the spectrum of the closed string. Since any open string theory with local interactions, which consist of splitting and joining of strings, automatically contains closed strings, gravity is unavoidable in string theory. At that time it was only known how to incorporate non-Abelian gauge symmetries with open strings. However, these open string theories are plagued by gravitational and gauge anomalies which were believed to be fatal. Renewed interest in string theory started in 1984 when M. Green and J. Schwarz showed that the open superstring is anomaly free if and only if the gauge group is $SO(32)$. But they also realized that the ten-dimensional supersymmetric Einstein-Yang-Mills field theory is anomaly free not only for the gauge group $SO(32)$ but also for $E_8 \times E_8$. So far this gauge group had not appeared in any of the known string theories. This soon changed with the formulation of the heterotic string by D. Gross, J. Harvey, E. Martinec and R. Rohm. It is a theory of closed strings only and represents the most economical way of incorporating both gravitational and gauge interactions. The allowed gauge symmetries are precisely

$E_8 \times E_8$ and $SO(32)$, and they originate from affine Lie algebras, also called Kač-Moody algebras, which are infinite-dimensional extensions of ordinary Lie algebras. Phenomenologically, the group E_8 was considered to be very promising as it contains all simple GUT groups like $SU(5)$, $SO(10)$ or E_6 as subgroups.

Until 1994 the situation was that there were five string theories in ten-dimensions: the type I theory, the two heterotic theories and the type IIA and type IIB theory. Only the type I theory appeared to contain open strings. The only known link between these theories was the so-called T-duality, an inherently stringy symmetry relating the type IIA to the type IIB theory and the heterotic $E_8 \times E_8$ to the heterotic $SO(32)$ theory upon compactifying on a circle.

In a stunning talk at the "Strings 1995" conference at the University of Southern California, E. Witten provided theoretical evidence for an intricate web of dualities among all five superstring theory, where an essential role was reserved for a new theory in eleven dimensions, whose low energy limit should be the well known eleven-dimensional supergravity theory. In this picture, the five known superstring theories are various perturbatively defined limits of this eleven dimensional theory, which Witten called M-theory. A qualitative boost to our understanding of string theory occurred when J. Polchinski realized that string theories have additional degrees of freedom, so-called D-branes. From the point of view of the weakly coupled string theories, these are non-perturbative objects which turned out to be the key ingredients for establishing the non-perturbative dualities between all five string theories.

String theory in its critical dimension obviously fails to reproduce an important experimentally established fact, namely that we live in four dimensions. This led to the concept of compactification or, more generally, to the choice of an internal conformal field theory. This choice determines the low-energy physics as seen by a four-dimensional observer. For a purely geometric compactification the requirement of conformal symmetry implies equations of motion for the metric which, at leading order in a derivative expansion, are the familiar source free Einstein equations. Non-trivial solutions were known to exist: Ricci-flat metrics on Calabi-Yau manifolds. The relation between string compactifications and Calabi-Yau manifolds had already been realized in 1985 by P. Candelas, G. Horowitz, A. Strominger and E. Witten.

One of the short-comings of these compactifications is the existence of many massless scalar fields, also called moduli, which cause phenomenological and cosmological problems. To remedy this, one can consider more general backgrounds, where not only the metric but also fields corresponding to massless bosonic string excitations, the dilaton or the various anti-symmetric tensor fields, have non-trivial background values. These backgrounds are called flux-compactifications and in general lead to isolated string vacua with no or only a small number of massless scalars. Gauge fields can arise on lower dimensional D-branes leading to the general idea of a brane world scenario, where the gauge interactions are confined to the brane and only gravity propagates into the ten-dimensional bulk. The exploration of the possible compactifications or, in other words, of the space of ground states of string theory, has led to the development of the concept of the string landscape.

1 Introduction

Even though this general concept dates back to the mid 1980s, it has received a major boost with the construction of flux vacua and the realization that their number is enormously large.

These developments introduced one concept into string theory which is a blatant blow to the early hope that string theory might have an (almost) unique ground state that correctly describes our world: the anthropic principle. From this perspective, string theory together with a mechanism to populate the landscape, like eternal inflation, might be considered as a physical framework for anthropic reasoning. There is still an ongoing controversy about the apparent loss of predictivity, but the pure fact that string theory is so rich that it describes not just a single universe but a multiverse is not in contradiction with the uniqueness of the fundamental ten or eleven dimensional theory. Even though there might exist many vacua, it is far from true that 'anything goes'. From a bottom-up perspective, not all possible four-dimensional anomaly free quantum field theories can be consistently extended to include quantum gravity.

Another major development in string theory occurred in 1997 with J. Maldacena's formulation of his conjecture, which is also known as the AdS/CFT correspondence or the gauge-gravity duality. One central feature of this duality is its holographic nature, i.e. it relates theories in different dimensions, namely a gravity or string theory in the bulk to a field theory on its boundary. In some sense, it has brought string theory closer to its beginnings: it offers a very concrete realization of the QCD string of strong interactions as the fundamental type IIB string. In the AdS/CFT correspondence, string theory is not so much considered the fundamental theory of quantum gravity but rather serves as a theoretical tool with the potential to furnish dual descriptions of interesting physical systems of various sorts, e.g. Yang-Mills theories in the infrared, at finite temperature, fluids, superconductors, etc. In all these cases, string theory is believed to be the consistent ultraviolet completion. After all, the challenging question remains, whether the theories realized in our universe, e.g. QCD with $SU(3)$ color symmetry and dynamical quarks, real fluids and superconductors are among the theories allowing for such a dual description.

One of the criticism string theory is often confronted with is that it does not make any—hopefully testable—predictions. As a matter of fact, it makes plenty of general predictions and some of them are quite generic. For instance, there are computable corrections to general relativity in the form of higher derivative corrections to the Einstein-Hilbert action. The problem is that generically these corrections are very difficult to test because they are very small. But this is not an intrinsic problem of string theory but of any theory of quantum gravity which is simply related to its tiny length scale, which is the Planck length $\ell_p = \sqrt{G_N \hbar / c^3} \simeq 10^{-33}$ cm. In string theory, the Planck scale is a derived quantity, whereas the fundamental scale is the string scale which can be parametrically smaller than the Planck scale. This leads to the logical possibility of having a low quantum gravity scale which can be detected experimentally, e.g. in the form of Kaluza-Klein gravitons, in collider experiments. Other general predictions are the existence of high scale supersymmetry and of axions, whose four-dimensional masses however depend on the details of the compactification and the process of supersymmetry breaking.

For specific predictions, e.g. the precise values of coupling constants and masses in the Standard Model, ones needs to know the string vacuum which correctly describes our world. In view of the landscape paradigm, this might indeed be difficult to find.

It is fair to say that string theory provides a framework of ultimate unification. All particles, matter and interactions have a common origin: they are excitations of the string. The theory seems to be consistent and much richer than originally thought. It has led to new developments in mathematics and has led particle physicists, both theorists and experimentalists, to explore, literally speaking, new dimensions. While it is not clear whether string theory is the correct theory to describe nature, it is fair to say that over the last 25 years, it has ignited many interesting developments in theoretical higher energy physics, in mathematical physics, in various branches of mathematics and, more recently, also in other areas of theoretical physics. It is therefore worth studying.

General References on String Theory and Background Material

Text books on string theory

- M.B. Green, J. Schwarz, E. Witten, *Superstring Theory*, vols. 1, 2 (Cambridge University Press, Cambridge, 1987)
- J. Polchinksi, *String Theory*, vols. 1, 2 (Cambridge University Press, Cambridge, 1998)
- C.V. Johnson, *D-Branes* (Cambridge University Press, Cambridge, 2003)
- B. Zwiebach, *A First Course in String Theory* (Cambridge University Press, Cambridge, 2009)
- K. Becker, M. Becker, J. Schwarz, *String Theory and M-Theory* (Cambridge University Press, Cambridge, 2007)
- E. Kiritsis, *String Theory in a Nutshell* (Princeton University Press, Princeton, 2007)
- A. Uranga, L. Ibáñez, *String Theory and Particle Physics: An Introduction to String Phenomenology* (Cambridge University Press, Cambridge, 2012)

Early reviews of string theory are

- P.H. Frampton, *Dual Resonance Models and Superstrings* (World Scientific, Singapore, 1986)
- M. Jacob (ed.), *Dual Models*, a reprint collection of several Physics Reports by the fathers of string theory (Elsevier, Amsterdam, 1974)
- J. Scherk, An introduction to the theory of dual models and strings. Rev. Mod. Phys. **47**, 123 (1975)
- J.H. Schwarz, Superstring theory. Phys. Rep. **89**, 223–322 (1982)

The history of the early days of string theory is recalled in

- P. Di Vecchia, A. Schwimmer, The beginning of string theory: a historical sketch. Lect. Notes Phys. **737**, 119 (2008) [arXiv:0708.3940 [physics.hist-ph]]

1 Introduction

For the necessary background in Quantum Field Theory, see e.g.

- M. Peskin, D. Schroeder, *An Introduction to Quantum Field Theory* (Addison-Wesley, Reading, 1995)
- S. Weinberg, *The Quantum Theory of Fields. Vol. 1: Foundations; Vol. 2: Modern Applications* (Cambridge University Press, Cambridge, 1995 and 1996)
- M. Srednicki, *Quantum Field Theory* (Cambridge University Press, Cambridge, 2007)

For General Relativity consult

- R.M. Wald, *General Relativity* (Chicago University Press, Chicago, 1984)
- S. Weinberg, *Gravitation and Cosmology* (Wiley, London, 1972)
- H. Stephani, *Relativity: An Introduction to Special and General Relativity* (Cambridge University Press, Cambridge, 2004)

and for Supersymmetry and Supergravity

- J. Wess, J. Bagger, *Supersymmetry and Supergravity* (Princeton University Press, Princeton, 1992)
- I. Buchbinder, S. Kuzenko, *Ideas and Methods of Supersymmetry and Supergravity: Or a Walk Through Superspace* (IOP, Bristol, 1998)
- S. Weinberg, *The Quantum Theory of Fields. Vol. 3: Supersymmetry* (Cambridge University Press, Cambridge, 2000)
- D. Freedman, A. van Proeyen, *Supergravity* (Cambridge University Press, Cambridge, 2012)
- A. Salam, E. Sezgin (eds.), *Supergravities in Diverse Dimensions. Vol. 1, 2* (North-Holland and World Scientific, NY, 1989)
- H.P. Nilles, Supersymmetry, supergravity and particle physics. Phys. Rep. **110**, 1–162 (1984)

Good and useful introductions to group theory, addressed to physicists, are

- H. Georgi, *Lie Algebras in Particle Physics*, 2nd edn. (Westview, Boulder, 1999)
- J. Fuchs, C. Schweigert, *Symmetries, Lie Algebras and Representations: A Graduate Course for Physicists* (Cambridge University Press, Cambridge, 1997)
- P. Ramond, *Group Theory* (Cambridge University Press, Cambridge, 2010)

Chapter 2
The Classical Bosonic String

Abstract Even though we will eventually be interested in a quantum theory of interacting strings, it will turn out to be useful to start two steps back and treat the free classical string. We will set up the Lagrangian formalism which is essential for the path integral quantization which we will treat in Chap. 3. We will then solve the classical equations of motion for single free closed and open strings. These solutions will be used for the canonical quantization which we will discuss in detail in the next chapter.

2.1 The Relativistic Particle

Before treating the relativistic string we will, as a warm up exercise, first study the free relativistic particle of mass m moving in a d-dimensional Minkowski space-time. Its action is simply the length of its world-line[1]

$$S = -m \int_{s_0}^{s_1} dx = -m \int_{\tau_0}^{\tau_1} d\tau \left[-\frac{dx^\mu}{d\tau} \frac{dx^\nu}{d\tau} \eta_{\mu\nu} \right]^{1/2}, \qquad (2.1)$$

where τ is an arbitrary parametrization along the world-line, whose embedding in d-dimensional Minkowski space is described by d real functions $x^\mu(\tau)$, $\mu = 0, \ldots, d-1$. We use the metric $\eta_{\mu\nu} = \mathrm{diag}(-1, +1, \ldots, +1)$. The action (2.1) is invariant under τ-reparametrizations $\tau \to \tilde{\tau}(\tau)$. Under infinitesimal reparametrizations $\tau \to \tau + \xi(\tau)$, x^μ transforms like

$$\delta x^\mu(\tau) = -\xi(\tau)\,\partial_\tau x^\mu(\tau). \qquad (2.2)$$

---

[1] It is easy to generalize the action to the case of a particle moving in a curved background by simply replacing the Minkowski metric $\eta_{\mu\nu}$ by a general metric $G_{\mu\nu}(x)$.

The action is invariant as long as $\xi(\tau_0) = \xi(\tau_1) = 0$. The momentum conjugate to $x^\mu(\tau)$ is

$$p_\mu = \frac{\partial L}{\partial \dot{x}^\mu} = m \frac{\dot{x}^\mu}{\sqrt{-\dot{x}^2}}, \qquad (2.3)$$

where $\dot{x} = \partial_\tau x$ and $\dot{x}^2 = \eta_{\mu\nu} \dot{x}^\mu \dot{x}^\nu$. Equation (2.3) immediately leads to the following constraint equation

$$\phi \equiv p^2 + m^2 = 0. \qquad (2.4)$$

Constraints which, as the one above, follow from the definition of the conjugate momenta without the use of the equations of motion are called primary constraints. Their number equals the number of zero eigenvalues of the Hessian matrix $\frac{\partial p_\mu}{\partial \dot{x}^\nu} = \frac{\partial^2 L}{\partial \dot{x}^\mu \partial \dot{x}^\nu}$ which, in the case of the free relativistic particle, is one, the corresponding eigenvector being $\dot{x}^\mu$. The absence of zero eigenvalues is necessary (via the inverse function theorem) to express the 'velocities' $\dot{x}^\mu$ uniquely in terms of the 'momenta' and 'coordinates', p_μ and x^μ. Systems where the rank of $\frac{\partial^2 L}{\partial \dot{x}^\mu \partial \dot{x}^\nu}$ is not maximal, thus implying the existence of primary constraints, are called singular. For singular systems the τ-evolution is governed by the Hamiltonian $H = H_{\text{can}} + \sum c_k \phi_k$, where H_{can} is the canonical Hamiltonian, the ϕ_k an irreducible set of primary constraints and the c_k are constants in the coordinates and momenta. This is so since the Hamiltonian is well defined only on the submanifold of phase space defined by the primary constraints and can be arbitrarily extended off that submanifold. For the free relativistic particle we find that $H_{\text{can}} = \frac{\partial L}{\partial \dot{x}^\mu} \dot{x}^\mu - L$ vanishes identically and the dynamics is completely determined by the constraint Eq. (2.4). The condition $H_{\text{can}} \equiv 0$ implies the existence of a zero eigenvalue of the Hessian: $\frac{\partial^2 L}{\partial \dot{x}^\mu \partial \dot{x}^\nu} \dot{x}^\nu = \frac{\partial}{\partial \dot{x}^\mu} H_{\text{can}} = 0$. This is always the case for systems with 'time' reparametrization invariance and follows from the fact that the 'time' evolution of an arbitrary phase-space function $f(x, p)$, given by $\frac{df}{d\tau} = \frac{\partial f}{\partial \tau} + \{f, H\}_{\text{P.B.}}$, should also be valid for $\tilde{\tau} = \tilde{\tau}(\tau)$ on the constrained phase-space; here $\{\,,\,\}_{\text{P.B.}}$ is the usual Poisson bracket, $\{f, g\}_{\text{P.B.}} = \left(\frac{\partial f}{\partial x} \frac{\partial g}{\partial p} - \frac{\partial f}{\partial p} \frac{\partial g}{\partial x} \right)$. From this we also see that a particular choice of the constants c_n corresponds to a particular gauge choice which, for the relativistic particle, means a choice of the 'time' variable τ. We write

$$H = \frac{N}{2m}(p^2 + m^2) \qquad (2.5)$$

and find that

$$\frac{dx^\mu}{d\tau} = \{x^\mu, H\}_{\text{P.B.}} = \frac{N}{m} p^\mu = \frac{N \dot{x}^\mu}{\sqrt{-\dot{x}^2}}, \qquad (2.6)$$

from which $\dot{x}^2 = -N^2$ follows. For the choice $N = 1$ the parameter τ is the proper time of the particle.

2.1 The Relativistic Particle

At this point it is appropriate to introduce the concept of first and second class constraints. If $\{\phi_k\}$ is the collection of all constraints and if $\{\phi_a, \phi_k\}_{P.B.} = 0, \forall k$ upon application of the constraints, we say that ϕ_a is first class. Otherwise it is called second class. First class constraints are associated with gauge conditions.

For the relativistic particle the constraint given in Eq. (2.4) is trivially first class and reflects τ reparametrization invariance.

Classically, we can describe the free relativistic particle by an alternative action which has two advantages over Eq. (2.1): (1) it does not contain a square root, thus leading to simpler equations of motion and (2) it allows the generalization to the massless case. This is achieved by introducing an auxiliary variable $e(\tau)$, which should, however, not introduce new dynamical degrees of freedom. The action containing x^μ and e is

$$S = \frac{1}{2} \int_{\tau_0}^{\tau_1} e\left(e^{-2}\dot{x}^2 - m^2\right) d\tau. \tag{2.7}$$

e plays the role of an ein-bein on the world-line. To see that (2.7) is equivalent to (2.1), we derive the equations of motion

$$\frac{\delta S}{\delta e} = 0 \quad \Rightarrow \quad \dot{x}^2 + e^2 m^2 = 0,$$

$$\frac{\delta S}{\delta x^\mu} = 0 \quad \Rightarrow \quad \frac{d}{d\tau}\left(e^{-1}\dot{x}^\mu\right) = 0. \tag{2.8}$$

Since the equation of motion for e is purely algebraic, e does not represent a new dynamical degree of freedom. We can solve for e and substitute it back into the action (2.7) to obtain (2.1), thus showing their classical equivalence.[2] We note that since $\frac{\partial^2 L}{\partial \dot{x}^\mu \partial \dot{x}^\nu} = e^{-1}\eta_{\mu\nu}$ has maximal rank, we now do not have primary constraints. The constraint equation $p^2 + m^2 = 0$ does not follow from the definition of the conjugate momenta alone; in addition one has to use the equations of motion. Constraints of this kind are called secondary constraints. But since it is first class, it implies a symmetry. Indeed, the action Eq. (2.7) is invariant under τ reparametrizations under which $x'^\mu(\tau') = x^\mu(\tau)$, $e'(\tau') = (\partial \tau'/\partial \tau)^{-1} e(\tau)$ or, in infinitesimal form with $\tau' = \tau + \xi$

$$\delta x^\mu = -\xi \partial_\tau x^\mu,$$

$$\delta e = -\partial_\tau(\xi e) \tag{2.9}$$

and we can make a τ-reparametrization to go to the gauge $e = 1/m$. If we then naively used the gauge fixed action to find the equations of motion, we would find

---

[2] It is important to point out that classical equivalence does not necessarily imply quantum equivalence.

$\ddot{x}^\mu = 0$, whose solutions are all straight lines in Minkowski space, which we know to be incorrect. This simply means that we cannot use the reparametrization freedom to fix e and then forget about it. We rather have to use the gauge fixed equation of motion for e, $T \equiv \dot{x}^2 + 1 = 0$, as a constraint. This excludes all time-like and light-like lines and identifies the parameter τ in this particular gauge as the proper time of the particle. In the massless case we set $e = 1$ and have to supplement the equation $\ddot{x}^\mu = 0$ by the constraint $T \equiv \dot{x}^2 = 0$, which leaves only the light-like world-lines. Note that the equation of motion, $\ddot{x}^\mu = 0$, does not imply $T = 0$, but it implies that $\frac{dT}{d\tau} = 0$, i.e. $T = 0$ is a constraint on the initial data and is conserved.

2.2 The Nambu-Goto Action

Let us now turn to the string. The generalization of Eq. (2.1) to a one-dimensional object is to take as its action the area of the world-sheet Σ swept out by the string, i.e.

$$\begin{aligned}
S_{\text{NG}} &= -T \int_\Sigma dA \\
&= -T \int_\Sigma d^2\sigma \left[-\det{}_{\alpha\beta} \left(\frac{\partial X^\mu}{\partial \sigma^\alpha} \frac{\partial X^\nu}{\partial \sigma^\beta} \eta_{\mu\nu} \right) \right]^{1/2} \\
&= -T \int_\Sigma d^2\sigma \left[(\dot{X} \cdot X')^2 - \dot{X}^2 X'^2 \right]^{1/2} \\
&\equiv -T \int_\Sigma d^2\sigma \sqrt{-\Gamma},
\end{aligned} \qquad (2.10)$$

where $\sigma^\alpha = (\sigma, \tau)$ are the two coordinates on the world-sheet; we choose them such that $\tau_i < \tau < \tau_f$ and $0 \leq \sigma < \ell$. The dot denotes derivative with respect to τ and the prime derivative with respect to σ. $X^\mu(\sigma, \tau)$, $\mu = 0, \ldots, d-1$ are maps of the world-sheet into d-dimensional Minkowski space and T is a constant of mass dimension two (mass/length), the string tension. Our conventions are such that X^μ has dimensions of length and so do σ and τ. $\Gamma_{\alpha\beta} = \frac{\partial X^\mu}{\partial \sigma^\alpha} \frac{\partial X^\nu}{\partial \sigma^\beta} \eta_{\mu\nu}$ is the induced metric on the world-sheet, inherited from the ambient d-dimensional Minkowski space through which the string moves and $\Gamma < 0$ is its determinant. The requirement that Γ be negative means that at each point of the world-sheet there is one time-like or light-like and one space-like tangent vector. This is necessary for causal propagation of the string. Requiring $\dot{X}^\mu + \lambda X'^\mu$ to be time-like and space-like when λ is varied gives $\Gamma < 0$. The action Eq. (2.10) was first considered by Nambu and Goto, hence the subscript NG.

One distinguishes between open and closed strings. The world-sheet of a free open string has the topology of a strip while the world-sheet of a closed string has

2.2 The Nambu-Goto Action

that of a cylinder. The string tension T is the only dimensionful quantity in string theory. Instead of the tension, one also uses the parameter

$$\alpha' = \frac{1}{2\pi T} \qquad (2.11)$$

also called the Regge slope. α' has dimension (length)^2. The open and closed string tensions are the same because in the interacting theory an open string can close and become a closed string and vice versa.

It is also common to introduce the string length scale

$$\ell_s = 2\pi \sqrt{\alpha'} \qquad (2.12)$$

and the string mass scale

$$M_s = (\alpha')^{-1/2}. \qquad (2.13)$$

Being the area of the world-sheet, the Nambu-Goto action is invariant under reparametrizations under which X^μ transforms as a scalar[3]

$$\delta X^\mu(\sigma, \tau) = -\xi^\alpha \partial_\alpha X^\mu(\sigma, \tau), \qquad (2.14)$$

as long as $\xi^a = 0$ on the boundary of the world-sheet. In addition to local coordinate transformations, global Poincaré transformations of the space-time coordinates, $X^\mu \to X^\mu + a^\mu$, are also a symmetry of the action.

To derive the equations of motion for the string we vary its trajectory, keeping initial and final positions fixed, i.e. $\delta X^\mu(\sigma, \tau_i) = 0 = \delta X^\mu(\sigma, \tau_f)$. This gives

$$\frac{\partial}{\partial \tau} \frac{\partial \mathscr{L}}{\partial \dot{X}^\mu} + \frac{\partial}{\partial \sigma} \frac{\partial \mathscr{L}}{\partial X'^\mu} = 0 \qquad (2.15)$$

together with the boundary conditions for the open string

---

[3] A general tensor density of rank, say (1,1), and weight w transforms under reparametrizations $\sigma^\alpha \to \tilde{\sigma}^\alpha(\sigma, \tau)$ of the world-sheet as

$$t_\alpha^\beta(\sigma, \tau) \to \tilde{t}_\alpha^\beta(\tilde{\sigma}, \tilde{\tau}) = \left| \frac{\partial(\tilde{\sigma}, \tilde{\tau})}{\partial(\sigma, \tau)} \right|^w \frac{\partial \sigma^\gamma}{\partial \tilde{\sigma}^\alpha} \frac{\partial \tilde{\sigma}^\beta}{\partial \sigma^\delta} t_\gamma^\delta(\sigma, \tau),$$

where the first factor is the Jacobian of the transformation. For infinitesimal transformations $\tilde{\sigma}^\alpha(\sigma, \tau) \to \sigma^\alpha + \xi^\alpha(\sigma, \tau)$, this gives

$$\delta t_\alpha^\beta(\sigma, \tau) \equiv \tilde{t}_\alpha^\beta(\sigma, \tau) - t_\alpha^\beta(\sigma, \tau) = -(\xi^\gamma \partial_\gamma - w \partial_\gamma \xi^\gamma) t_\alpha^\beta - t_\gamma^\beta \partial_\alpha \xi^\gamma + t_\alpha^\delta \partial_\delta \xi^\beta.$$

The generalization to tensors of arbitrary rank is obvious.

$$\frac{\partial \mathcal{L}}{\partial X'^{\mu}} \delta X^{\mu} = 0 \quad \text{at} \quad \sigma = 0, \ell \tag{2.16}$$

and the periodicity condition for the closed string

$$X^{\mu}(\sigma + \ell, \tau) = X^{\mu}(\sigma, \tau). \tag{2.17}$$

For each coordinate direction μ and at each of the two ends of the open string there are two ways to satisfy the boundary condition of the open string:

1. we may impose Neumann boundary conditions which amounts to requiring δX^{μ} to be arbitrary at the boundary. This requires $\frac{\partial \mathcal{L}}{\partial X'^{\mu}} = 0$. Physically this conditions means that no momentum flows off the end of the string. This will be become clear below.
2. Alternatively, we may impose Dirichlet boundary conditions where we set $\delta X^{\mu} = 0$ at the boundary. In other words, we fix the position of the boundary of the string. Thus Dirichlet boundary condition breaks space-time translational invariance. We will discuss the consequences in Sect. 2.4

Due to the square root in the Lagrangian, the equations of motion are rather complicated. The canonical momentum is

$$\Pi_{\mu} = \frac{\partial \mathcal{L}}{\partial \dot{X}^{\mu}} = -T \frac{(\dot{X} \cdot X') X'_{\mu} - (X')^2 \dot{X}_{\mu}}{\left[(X' \cdot \dot{X})^2 - \dot{X}^2 X'^2\right]^{1/2}}. \tag{2.18}$$

The Hessian $\frac{\partial^2 \mathcal{L}}{\partial \dot{X}^{\nu} \partial \dot{X}^{\mu}} = \frac{\partial}{\partial \dot{X}^{\nu}} \Pi_{\mu}$ has, for each value of σ, two zero eigenvalues with eigenvectors $\dot{X}^{\mu}$ and X'^{μ}. The resulting primary constraints are

$$\Pi_{\mu} X'^{\mu} = 0 \tag{2.19}$$

and

$$\Pi^2 + T^2 X'^2 = 0. \tag{2.20}$$

After gauge fixing they become non-trivial constraints on the dynamics and play an important role in string theory, as we will see later. The canonical Hamiltonian, $H_{\text{can}} = \int_0^L d\sigma (\dot{X} \cdot \Pi - \mathcal{L})$ is easily seen to vanish identically and hence the dynamics is completely governed by the constraints.

2.3 The Polyakov Action and Its Symmetries

Due to the occurrence of the square root, the Nambu-Goto action is difficult to deal with. As in the case of the relativistic particle, one can remove the square root at the expense of introducing an additional (auxiliary) field on the world-sheet. This

2.3 The Polyakov Action and Its Symmetries

field is a metric $h_{\alpha\beta}(\sigma,\tau)$ on the world-sheet with signature $(-,+)$. In the resulting action the d massless world-sheet scalars X^μ are coupled to two-dimensional gravity $h_{\alpha\beta}$:

$$\begin{aligned}S_P &= -\frac{T}{2}\int_\Sigma d^2\sigma\,\sqrt{-h}\,h^{\alpha\beta}\,\partial_\alpha X^\mu\,\partial_\beta X^\nu \eta_{\mu\nu} \\ &= -\frac{T}{2}\int_\Sigma d^2\sigma\,\sqrt{-h}\,h^{\alpha\beta}\,\Gamma_{\alpha\beta}\,,\end{aligned} \qquad (2.21)$$

where $h = \det h_{\alpha\beta}$. This form of the string action is the starting point for the path integral quantization of Polyakov, hence the subscript P. Note that the world-sheet metric does not appear with derivatives, in accord with our requirement that it is not dynamical. The components of the metric play the role of Lagrange multipliers which impose the Virasoro constraints which are now no longer primary constraints.

The action is easy to generalize to a string moving in a curved background: one replaces the Minkowski metric $\eta_{\mu\nu}$ by a general metric $G_{\mu\nu}(X)$. In this general form, the action is that of a non-trivial, interacting field theory: a non-linear sigma-model. Choosing $G_{\mu\nu} = \eta_{\mu\nu}$ can be considered as the zeroth order term in a perturbative expansion around a flat background. This is of course a limitation and a complete theory should determine its own background in which the string propagates, much in the same way as in general relativity where the metric of space-time is determined by the matter content according to Einstein's equations. However, at this point this is simply a consequence of how the theory is formulated and it is not an inherent problem. We will discuss strings in non-trivial backgrounds in Chap. 14. For now we use (2.21).

We now define the energy-momentum tensor of the world-sheet theory in the usual way as the response of the system to changes in the metric under which $\delta S_P = \frac{1}{4\pi}\int d^2\sigma\,\sqrt{-h}\,T_{\alpha\beta}\,\delta h^{\alpha\beta}$ ($\delta h^{\alpha\beta} = -h^{\alpha\gamma}h^{\beta\delta}\delta h_{\gamma\delta}$), i.e.

$$T_{\alpha\beta} = \frac{4\pi}{\sqrt{-h}}\frac{\delta S_P}{\delta h^{\alpha\beta}} \qquad (2.22)$$

is the world-sheet energy-momentum tensor. Using $\delta h = -h_{\alpha\beta}(\delta h^{\alpha\beta})\,h$ we find

$$T_{\alpha\beta} = -\frac{1}{\alpha'}\left(\partial_\alpha X^\mu \partial_\beta X_\mu - \frac{1}{2}h_{\alpha\beta}\,h^{\gamma\delta}\,\partial_\gamma X^\mu\,\partial_\delta X_\mu\right) \qquad (2.23)$$

and the equations of motion are

$$T_{\alpha\beta} = 0\,, \qquad (2.24a)$$

$$\Box X^\mu = \frac{1}{\sqrt{-h}}\partial_\alpha(\sqrt{-h}\,h^{\alpha\beta}\,\partial_\beta X^\mu) = 0 \qquad (2.24b)$$

with the appropriate boundary and periodicity conditions:

$$X^\mu(\tau, \sigma + \ell) = X^\mu(\tau, \sigma) \tag{2.25}$$

for the closed string and

$$n^\alpha \partial_\alpha X^\mu \delta X_\mu|_{\sigma=0,\ell} = 0 \tag{2.26}$$

for the open string. Here n^α is a normal vector at the boundary. We require the boundary condition at each end of the string separately, since locality demands that we take δX^μ independently at the two ends.

Energy-momentum conservation, $\nabla^\alpha T_{\alpha\beta} = 0$, which is a consequence of the diffeomorphism invariance of the Polyakov action, is easily verified with the help of the equations of motion for X^μ. ∇_α is a covariant derivative with the usual Christoffel connection $\Gamma^\gamma_{\alpha\beta} = \frac{1}{2} h^{\gamma\delta}(\partial_\alpha h_{\delta\beta} + \partial_\beta h_{\alpha\delta} - \partial_\delta h_{\alpha\beta})$. From the vanishing of the energy-momentum tensor we derive $\det_{\alpha\beta}(\partial_\alpha X^\mu \partial_\beta X_\mu) = \frac{1}{4} h(h^{\gamma\delta} \partial_\gamma X_\mu \partial_\delta X^\mu)^2$ which, when inserted into S_P, shows the classical equivalence of the Polyakov and Nambu-Goto actions.

One checks that the constraints, Eqs. (2.19), (2.20), which were primary in the Nambu-Goto formulation, follow here only if we use the equation of motion $T_{\alpha\beta} = 0$, i.e. they are secondary. This is the same situation which we encountered in the case of the relativistic particle.

Note that we have introduced two metrics on the world-sheet, namely the metric inherited from the ambient space, i.e. the induced metric, $\Gamma_{\alpha\beta} = \partial_\alpha X^\mu \partial_\beta X^\nu \eta_{\mu\nu}$ which enters the Nambu-Goto action and the intrinsic metric $h_{\alpha\beta}$ which appears in the Polyakov action. They are, a priori, unrelated. The Polyakov action is not the area of the world-sheet measured with the intrinsic metric, which would simply be $\int d^2\sigma \sqrt{-h}$ and could be added to S_P as a cosmological term (see below). However, for any real symmetric 2×2 matrix A we have the inequality $(\text{tr } A)^2 \geq 4 \det A$ with equality for $A \propto \mathbf{1}$. With the choice $A^\alpha{}_\beta = h^{\alpha\gamma} \Gamma_{\gamma\beta}$ it follows that $S_P \geq S_{NG}$. Equality holds if and only if $h_{\alpha\beta} \propto \Gamma_{\alpha\beta}$, i.e. if the two metrics are conformally related. This is the case if the equation of motion for $h_{\alpha\beta}$, Eq. (2.24a), is satisfied.

We can now ask whether there are other terms one could add to S_P. If we restrict ourselves to closed strings moving in Minkowski space-time without any other background fields, the only possibilities compatible with d-dimensional Poincaré invariance and power counting renormalizability (at most two derivatives) of the two-dimensional theory are[4]

$$S_1 = \lambda_1 \int_\Sigma d^2\sigma \sqrt{-h} \tag{2.27}$$

---

[4] For the open string with boundary $\partial\Sigma$ there are further possible terms besides S_1 and S_2, which are defined on the boundary of the world-sheet: $S_3 = \lambda_3 \int_{\partial\Sigma} ds$ and $S_4 = \lambda_4 \int_{\partial\Sigma} k\, ds$. Here k is the extrinsic curvature of the boundary. It turns out that these terms can also be discarded.

2.3 The Polyakov Action and Its Symmetries

which is the cosmological term mentioned above, and

$$S_2 = \frac{\lambda_2}{4\pi} \int_\Sigma d^2\sigma \sqrt{-h} R = \lambda_2 \chi(\Sigma) \tag{2.28}$$

where R is the curvature scalar for the metric $h_{\alpha\beta}$. S_2 is the two-dimensional Gauss-Bonnet term and χ the Euler number of the world-sheet, which is a topological invariant. The integrand is (locally) a total derivative and consequently does not contribute to the classical equations of motion. S_2 does, however, play a role in the organization of string perturbation theory. λ_2 turns out to be the constant background value of the dilaton field Φ, which is one of the massless excitations of the closed string and which couples to the world-sheet via $\frac{1}{4\pi} \int d^2\sigma \sqrt{-h} \Phi R$. Inclusion of the cosmological term S_1 would lead to the equation of motion $T_{\alpha\beta} = -\frac{\lambda_1}{2T} h_{\alpha\beta}$ from which we conclude that $\lambda_1 h^{\alpha\beta} h_{\alpha\beta} = 0$. This is unacceptable unless $\lambda_1 = 0$.[5] We will thus consider the action S_P, Eq. (2.21), which is the action of a collection of d massless real scalar fields (X^μ) coupled to gravity ($h_{\alpha\beta}$) in two dimensions.

Let us now discuss the symmetries of the Polyakov action.

1. Global symmetries:

- Space-time Poincaré invariance:

$$\delta X^\mu = a^\mu{}_\nu X^\nu + b^\mu \qquad (a_{\mu\nu} = -a_{\nu\mu}),$$
$$\delta h_{\alpha\beta} = 0 \tag{2.29}$$

2. Local symmetries:

- Reparametrization invariance

$$\delta X^\mu = -\xi^\alpha \partial_\alpha X^\mu,$$
$$\delta h_{\alpha\beta} = -(\xi^\gamma \partial_\gamma h_{\alpha\beta} + \partial_\alpha \xi^\gamma h_{\gamma\beta} + \partial_\beta \xi^\gamma h_{\alpha\gamma})$$
$$= -(\nabla_\alpha \xi_\beta + \nabla_\beta \xi_\alpha),$$
$$\delta \sqrt{-h} = -\partial_\alpha(\xi^\alpha \sqrt{-h}). \tag{2.30}$$

- Weyl rescaling

$$\delta X^\mu = 0,$$
$$\delta h_{\alpha\beta} = 2\Lambda h_{\alpha\beta}. \tag{2.31}$$

---

[5] Note that inclusion of $S_{1,2}$ breaks classical Weyl invariance. In the quantum theory the regularization procedure leads to an explicit breakdown of Weyl invariance and divergent counter-terms associated with S_1 and S_2 are generated.

Here ξ^α and Λ are arbitrary (infinitesimal) functions of (σ,τ) and $a_{\mu\nu}$ and b_μ are constants. From Eq. (2.29) we see that X^μ is a Minkowski space vector whereas $h_{\alpha\beta}$ is a scalar. Under reparametrizations of the world-sheet, Eq. (2.30), the X^μ are world-sheet scalars, $h_{\alpha\beta}$ a world-sheet tensor and $\sqrt{-h}$ a scalar density of weight -1. The scale transformations of the world-sheet metric, Eq. (2.31), is the infinitesimal version of $h_{\alpha\beta}(\sigma,\tau) \to \Omega^2(\sigma,\tau)h_{\alpha\beta}(\sigma,\tau)$ for $\Omega^2(\sigma,\tau) = e^{2\Lambda(\sigma,\tau)} \sim 1 + 2\Lambda(\sigma,\tau)$.

One immediate important consequence of Weyl invariance of the action is the tracelessness of the energy-momentum tensor:

$$T^\alpha{}_\alpha = h^{\alpha\beta} T_{\alpha\beta} = 0 \qquad (2.32)$$

which is satisfied by the expression Eq. (2.23) without invoking the equations of motion. It is not difficult to see that this has to be so. Consider an action which depends on a metric and a collection of fields ϕ_i which transform under Weyl rescaling as $h_{\alpha\beta} \to e^{2\Lambda} h_{\alpha\beta}$ and $\phi_i \to e^{d_i \Lambda} \phi_i$. If the action is scale invariant, i.e. if $S[e^{2\Lambda} h_{\alpha\beta}, e^{d_i \Lambda}\phi_i] = S[h_{\alpha\beta}, \phi_i]$, then

$$0 = \delta S = \int d^2\sigma \left\{ -2 \frac{\delta S}{\delta h^{\alpha\beta}} h^{\alpha\beta} + \sum_i d_i \frac{\delta S}{\delta \phi_i} \phi_i \right\} \delta\Lambda. \qquad (2.33)$$

If we now use the equations of motion for ϕ_i, $\frac{\delta S}{\delta \phi_i} = 0$ and the definition $T_{\alpha\beta} \propto \frac{\delta S}{\delta h^{\alpha\beta}}$, tracelessness of the energy-momentum tensor is immediate. We note that it follows without the use of the equations of motion if and only if $d_i = 0$, $\forall i$. This is, for instance, the case for the Polyakov action of the bosonic string (where $\{\phi_i\} = \{X^\mu\}$) but will not be satisfied for the fermionic string in Chap. 7.

The local invariances allow for a convenient gauge choice for the world-sheet metric $h_{\alpha\beta}$, called conformal or orthonormal gauge. Reparametrization invariance is used to choose coordinates such that locally $h_{\alpha\beta} = \Omega^2(\sigma,\tau)\eta_{\alpha\beta}$ with $\eta_{\alpha\beta}$ being the two-dimensional Minkowski metric defined by $ds^2 = -d\tau^2 + d\sigma^2$. It is not hard to show that this can always be done. Indeed, for any two-dimensional Lorentzian metric $h_{\alpha\beta}$, consider two null vectors at each point. In this way we get two vector fields and their integral curves which we label by σ^+ and σ^-. Then $ds^2 = -\Omega^2 d\sigma^+ d\sigma^-$; $h_{++} = h_{--} = 0$ since the curves are null. Now let

$$\sigma^\pm = \tau \pm \sigma, \qquad (2.34)$$

from which it follows that $ds^2 = \Omega^2(-d\tau^2 + d\sigma^2)$. A choice of coordinate system in which the two-dimensional metric is conformally flat, i.e. in which

$$ds^2 = \Omega^2(-d\tau^2 + d\sigma^2) = -\Omega^2 d\sigma^+ d\sigma^- \qquad (2.35)$$

is called a conformal gauge. The world-sheet coordinates $\sigma^\pm$ introduced above are called light-cone, isothermal or conformal coordinates. In these coordinates $\gamma_{\alpha\beta} \equiv \frac{h_{\alpha\beta}}{\sqrt{-h}} = \eta_{\alpha\beta}$. We can now use Weyl invariance to set $h_{\alpha\beta} = \eta_{\alpha\beta}$.

2.3 The Polyakov Action and Its Symmetries

We collect some results about the world-sheet light-cone coordinates (2.34) which we will frequently use below. The components of the Minkowski metric in light-cone coordinates are

$$\eta_{+-} = \eta_{-+} = -\frac{1}{2}, \quad \eta^{+-} = \eta^{-+} = -2,$$
$$\eta_{++} = \eta_{--} = \eta^{++} = \eta^{--} = 0. \tag{2.36}$$

We will also need

$$\partial_{\pm} = \frac{1}{2}(\partial_{\tau} \pm \partial_{\sigma}) \tag{2.37}$$

and indices are raised and lowered according to

$$\xi^+ = -2\xi_- \quad \text{and} \quad \xi^- = -2\xi_+. \tag{2.38}$$

It is important to realize that reparametrizations which satisfy $\mathcal{L}_\xi h_{\alpha\beta} = -(\nabla_\alpha \xi_\beta + \nabla_\beta \xi_\alpha) \propto h_{\alpha\beta}$ can be compensated by a Weyl rescaling. Expressed in light-cone coordinates the conformal gauge preserving diffeomorphisms are those which satisfy $\partial_+ \xi^- = \partial_- \xi^+ = 0$, i.e. $\xi^\pm = \xi^\pm(\sigma^\pm)$.[6] (Here we have used that $\nabla_+ \xi_+ = h_{+-}\nabla_+ \xi^- = h_{+-}\partial_+ \xi^-$ since the only non-vanishing Christoffel symbols in conformal gauge with $\Omega = e^\Lambda$ are $\Gamma^+_{++} = 2\partial_+ \Lambda$ and $\Gamma^-_{--} = 2\partial_- \Lambda$.) Indeed, instead of $\sigma^\pm$ we could as well have chosen $\tilde{\sigma}^\pm = \tilde{\sigma}^\pm(\sigma^\pm)$ or, in infinitesimal form, $\tilde{\sigma}^\pm = \sigma^\pm + \xi^\pm(\sigma^\pm)$. Note that the transformation $\sigma^\pm \to \tilde{\sigma}^\pm(\sigma^\pm)$ corresponds to $\begin{pmatrix}\tau\\\sigma\end{pmatrix} \to \begin{pmatrix}\tilde{\tau}\\\tilde{\sigma}\end{pmatrix} = \frac{1}{2}[\tilde{\sigma}^+(\tau+\sigma) \pm \tilde{\sigma}^-(\tau-\sigma)]$; i.e. any $\tilde{\tau}$ and $\tilde{\sigma}$ satisfying the two-dimensional wave equation will do the job.

Conformal gauge is unique to two dimensions. In $d > 0$ dimensions a metric $h_{\alpha\beta}$, being symmetric, has $\frac{1}{2}d(d+1)$ independent components. Reparametrization invariance allows to fix d of them, leaving $\frac{1}{2}d(d-1)$ components. In two dimensions this suffices to go to conformal gauge. The Polyakov action then still has one extra local symmetry, namely Weyl transformations, which allow us to eliminate the remaining metric component. This also shows that gravity in two dimensions is trivial in the sense that the graviton can be gauged away completely. For $d > 2$ Weyl invariance, even if present as for instance in conformal gravity, won't suffice to gauge away all metric degrees of freedom.[7]

The argument given above that conformal gauge is always possible was a local statement. We will now set up a global criterion and consider the general case with gauge condition

---

[6] After Wick rotation to Euclidean signature on the world-sheet these are conformal transformations. More about this later.

[7] Note that the action for the relativistic particle was not Weyl invariant; there reparametrization invariance was sufficient to eliminate the one metric degree of freedom.

$$h_{\alpha\beta} = e^{2\phi} \hat{h}_{\alpha\beta} .\qquad(2.39)$$

In conformal gauge $\hat{h}_{\alpha\beta} = \eta_{\alpha\beta}$. Under reparametrizations and Weyl rescaling the metric changes as

$$\begin{aligned}\delta h_{\alpha\beta} &= -(\nabla_\alpha \xi_\beta + \nabla_\beta \xi_\alpha) + 2\Lambda\, h_{\alpha\beta}\\ &\equiv -(P\xi)_{\alpha\beta} + 2\tilde{\Lambda}\, h_{\alpha\beta} ,\end{aligned}\qquad(2.40)$$

where the operator P maps vectors into symmetric traceless tensors according to

$$(P\xi)_{\alpha\beta} = \nabla_\alpha \xi_\beta + \nabla_\beta \xi_\alpha - (\nabla_\gamma \xi^\gamma)\, h_{\alpha\beta} ,\qquad(2.41)$$

and we have defined $2\tilde{\Lambda} = 2\Lambda - \nabla_\gamma \xi^\gamma$. The decomposition into symmetric traceless and trace part is orthogonal with respect to the inner product $(\delta h^{(1)} | \delta h^{(2)}) = \int d^2\sigma \sqrt{-h}\, h^{\alpha\gamma} h^{\beta\delta} \delta h^{(1)}_{\alpha\beta} \delta h^{(2)}_{\gamma\delta}$. The trace part of $\delta h_{\alpha\beta}$ can always be cancelled by a suitable choice of Λ. It then follows that for the gauge Eq. (2.39) to be possible globally, there must exist a globally defined vector field ξ^α such that

$$(P\xi)_{\alpha\beta} = t_{\alpha\beta}\qquad(2.42)$$

for arbitrary symmetric traceless $t_{\alpha\beta}$. If the operator P has zero modes, i.e. if there exist vector fields ξ_0 such that $P\xi_0 = 0$, then for any solution ξ we also have the solution $\xi + \xi_0$. In this case the gauge fixing is not complete and those reparametrizations which can be absorbed by a Weyl rescaling are still allowed, as we have already seen above.

The adjoint of P, $P^\dagger$, maps traceless symmetric tensors to vectors via

$$(P^\dagger t)_\alpha = -2\nabla^\beta\, t_{\alpha\beta} .\qquad(2.43)$$

Zero modes of $P^\dagger$ are symmetric traceless tensors which cannot be written as $(P\xi)_{\alpha\beta}$ for any vector field ξ. Indeed, if $(P^\dagger t_0)_\alpha = 0$, then for all ξ^σ, $(\xi, P^\dagger t_0) = (P\xi, t_0) = 0$. This means that zero modes of $P^\dagger$ are metric deformations which cannot be absorbed by reparametrization and Weyl rescaling. If they do not exist, the gauge is possible globally. This applies in particular to the conformal gauge; there the condition is that the equations $\partial_- t_{++} = 0$ and $\partial_+ t_{--} = 0$ have no globally defined solutions. We will further discuss the solutions to these equations in Chap. 6. The equation

$$(P\xi)_{\alpha\beta} = 0\qquad(2.44)$$

is the conformal Killing equation and its solutions are called conformal Killing vectors. In contrast to Killing vectors which generate isometries, conformal Killing vectors generate Weyl rescalings of the metric; in particular, they preserve the conformal gauge.

2.3 The Polyakov Action and Its Symmetries

In conformal gauge the Polyakov action simplifies to

$$\begin{aligned} S_P &= -\frac{T}{2} \int d^2\sigma \, \eta^{\alpha\beta} \, \partial_\alpha X^\mu \, \partial_\beta X_\mu \\ &= \frac{T}{2} \int d^2\sigma \, (\dot{X}^2 - X'^2) \\ &= 2T \int d^2\sigma \, \partial_+ X \cdot \partial_- X \, . \end{aligned} \qquad (2.45)$$

Varying with respect to X^μ such that $\delta X^\mu(\tau_0) = 0 = \delta X^\mu(\tau_1)$ we obtain[8]

$$\delta S_p = T \int d^2\sigma \, \delta X^\mu \, (\partial_\sigma^2 - \partial_\tau^2) \, X_\mu - T \int_{\tau_0}^{\tau_1} d\tau \, X'_\mu \, \delta X^\mu \bigg|_{\sigma=0}^{\sigma=\ell} . \qquad (2.46)$$

The surface term is absent for the closed string for which we impose the periodicity condition[9]

$$\text{(closed string)} \qquad X^\mu(\sigma + \ell) = X^\mu(\sigma) \, . \qquad (2.47)$$

To achieve the vanishing of the boundary term for the open string we have to impose either Dirichlet or Neumann boundary conditions for each X^μ and at each of the two ends of the string:

$$\partial_\sigma X^\mu|_{\sigma=0,\ell} = 0 \qquad \text{(Neumann)} \qquad (2.48)$$

or   (open string)

$$\delta X^\mu|_{\sigma=0,\ell} = 0 \qquad \text{(Dirichlet)} \, . \qquad (2.49)$$

The Dirichlet boundary condition means that the end-point of the open string is fixed in space-time. This boundary condition thus breaks space-time Poincaré invariance. As we will discuss below, these boundary conditions have important implications.

The vanishing of (2.46) leads to the following equations of motion

$$(\partial_\sigma^2 - \partial_\tau^2) \, X^\mu = 4\partial_+ \partial_- X^\mu = 0 \qquad (2.50)$$

which have to be solved subject to (2.47) or (2.48), (2.49).

---

[8] One can show that on the strip and the cylinder one can always go to conformal gauge and preserve $0 \leq \sigma \leq \ell$.
[9] More general periodicity conditions $X^\mu(\sigma + \ell) = M^\mu{}_\nu X^\nu(\sigma)$ for any constant $O(1, d-1)$ matrix M also leave the action invariant. If we want to interpret X^μ as coordinates in Minkowski space, only (2.47) is allowed, i.e. they are the only periodicity conditions which are invariant under d-dimensional Poincaré transformations. When we consider compactifications of the string we will consider so-called twisted boundary conditions for which M is non-trivial.

Eq. (2.50) is the two-dimensional massless wave equation with the general solution

$$X^\mu(\sigma, \tau) = X_L^\mu(\sigma^+) + X_R^\mu(\sigma^-). \tag{2.51}$$

Here $X_{L,R}^\mu$ are arbitrary functions of their respective arguments, subject only to periodicity or boundary conditions. They describe the "left"- and "right"-moving modes of the string, respectively. In the case of the closed string the left- and right-moving components are completely independent for the unconstrained system, an observation which is crucial for the formulation of the heterotic string. This is however not the case for the open string where the boundary condition mixes left- with right-movers through reflection at the ends of the string. We will present explicit Fourier series solutions for all possible boundary conditions in the next two subsection.

On a solution of the equations of motion we still have to impose the constraints resulting from the gauge fixed equations of motion for the metric: we have to require that the energy-momentum tensor vanishes; i.e.

$$T_{01} = T_{10} = -2\pi T (\dot X \cdot X') = 0, \tag{2.52a}$$

$$T_{00} = T_{11} = -\pi T (\dot X^2 + X'^2) = 0 \tag{2.52b}$$

which can be alternatively expressed as

$$(\dot X \pm X')^2 = 0. \tag{2.53}$$

In light-cone coordinates they become

$$T_{++} = -2\pi T(\partial_+ X \cdot \partial_+ X) = 0, \tag{2.54a}$$

$$T_{--} = -2\pi T(\partial_- X \cdot \partial_- X) = 0, \tag{2.54b}$$

$$T_{+-} = T_{-+} = 0, \tag{2.54c}$$

where $T_{++} = \frac{1}{2}(T_{00} + T_{01})$, $T_{--} = \frac{1}{2}(T_{00} - T_{01})$; Eq. (2.54c) expresses the tracelessness of the energy-momentum tensor. In terms of the left- and right-movers the constraints Eqs. (2.54a), (2.54b) become $\dot X_R^2 = \dot X_L^2 = 0$. Energy-momentum conservation, i.e. $\nabla^\alpha T_{\alpha\beta} = 0$ becomes

$$\partial_- T_{++} + \partial_+ T_{-+} = 0, \tag{2.55a}$$

$$\partial_+ T_{--} + \partial_- T_{+-} = 0 \tag{2.55b}$$

which, using Eq. (2.54c), simply states that

$$\partial_- T_{++} = 0, \tag{2.56a}$$

$$\partial_+ T_{--} = 0 \tag{2.56b}$$

2.3 The Polyakov Action and Its Symmetries

i.e.
$$T_{++} = T_{++}(\sigma^+) \quad \text{and} \quad T_{--} = T_{--}(\sigma^-). \tag{2.57}$$

The conservation equations (2.56) imply the existence of an infinite number of conserved charges. In fact, for any function $f(\sigma^+)$ we have $\partial_-(f(\sigma^+)T_{++}) = 0$ and the corresponding conserved charges are

$$L_f = 2T \int_0^\ell d\sigma \, f(\sigma^+) T_{++}(\sigma^+) \tag{2.58}$$

and likewise for the right-movers.

The Hamiltonian for the string in conformal gauge is

$$\begin{aligned} H &= \int_0^\ell d\sigma \, (\dot{X} \cdot \Pi - \mathscr{L}) \\ &= \frac{T}{2} \int_0^\ell d\sigma \, (\dot{X}^2 + X'^2) \\ &= T \int_0^\ell d\sigma \, ((\partial_+ X)^2 + (\partial_- X)^2), \end{aligned} \tag{2.59}$$

where, as before, the canonical momentum is $\Pi^\mu = \partial \mathscr{L}/\partial \dot{X}_\mu = T\dot{X}^\mu$. We note that the Hamiltonian is just one of the constraints. This was to be expected from our discussion of constrained systems in the context of the relativistic particle. Indeed, we saw that the canonical Hamiltonian derived from the Nambu-Goto action vanishes identically and the τ-evolution is completely governed by the constraints, i.e.

$$H = \int_0^\ell d\sigma \left\{ N_1(\sigma, \tau) \Pi \cdot X' + N_2(\sigma, \tau) (\Pi^2 + T^2 X'^2) \right\}, \tag{2.60}$$

where N_1 and N_2 are arbitrary functions of σ and τ. Using the basic equal τ Poisson brackets

$$\begin{aligned} \{X^\mu(\sigma, \tau), X^\nu(\sigma', \tau)\}_{\text{P.B.}} &= \{\Pi^\mu(\sigma, \tau), \Pi^\nu(\sigma', \tau)\}_{\text{P.B.}} = 0, \\ \{X^\mu(\sigma, \tau), \Pi^\nu(\sigma', \tau)\}_{\text{P.B.}} &= \eta^{\mu\nu} \delta(\sigma - \sigma') \end{aligned} \tag{2.61}$$

we find

$$\dot{X}^\mu = N_1 X'^\mu + 2 N_2 \Pi^\mu \tag{2.62}$$

and

$$\dot{\Pi}^\mu = \partial_\sigma (N_1 \Pi^\mu + 2T^2 N_2 X'^\mu). \tag{2.63}$$

If we choose $N_1 = 0$ and $N_2 = \frac{1}{2T}$, Eqs. (2.62) and (2.63) lead to the equation of motion $(\partial_\sigma^2 - \partial_\tau^2)X^\mu = 0$ which we have obtained previously from the action in conformal gauge. This means that choosing $N_1 = 0$ and $N_2 = \frac{1}{2T}$ is equivalent to fixing the conformal gauge. With this choice for the functions N_1 and N_2 we also get the Hamiltonian (2.59).

In conformal gauge the Poisson brackets are

$$\{X^\mu(\sigma,\tau), X^\nu(\sigma',\tau)\}_{\text{P.B.}} = \{\dot{X}^\mu(\sigma,\tau), \dot{X}^\nu(\sigma',\tau)\}_{\text{P.B.}} = 0,$$

$$\{X^\mu(\sigma,\tau), \dot{X}^\nu(\sigma',\tau)\}_{\text{P.B.}} = \frac{1}{T}\eta^{\mu\nu}\delta(\sigma-\sigma'). \tag{2.64}$$

With their help one readily shows that $-T\int \dot{X}\cdot X' d\sigma$ and $\frac{1}{2}T\int(\dot{X}^2 + X'^2)d\sigma$ generate constant σ- and τ-translations, respectively. More generally, using the explicit expression for T_{++} one finds that the charges L_f of Eq. (2.58) generate transformations $\sigma^+ \to \sigma^+ + f(\sigma^+)$, i.e. those reparametrizations which do not lead out of conformal gauge:

$$\{L_f, X(\sigma)\}_{\text{P.B.}} = -f(\sigma^+)\partial_+ X(\sigma). \tag{2.65}$$

So far we have only discussed issues connected with world-sheet symmetries. However, invariance under d-dimensional global Poincaré transformations, Eq. (2.29), leads, via the Noether theorem, to two conserved currents; invariance under translations gives the energy-momentum current

$$P^\alpha_\mu = -T\sqrt{h}\, h^{\alpha\beta}\partial_\beta X_\mu, \tag{2.66}$$

whereas invariance under Lorentz transformations gives the angular momentum current

$$J^\alpha_{\mu\nu} = -T\sqrt{h}\, h^{\alpha\beta}(X_\mu \partial_\beta X_\nu - X_\nu \partial_\beta X_\mu) = X_\mu P^\alpha_\nu - X_\nu P^\alpha_\mu. \tag{2.67}$$

Using the equations of motion, it is easy to check conservation of P^α_μ and $J^\alpha_{\mu\nu}$. The total conserved charges (momentum and angular momentum) are obtained by integrating the currents over a space-like section of the world-sheet, say $\tau = 0$. Then the total momentum in conformal gauge is

$$P_\mu = \int_0^\ell d\sigma\, P^\tau_\mu = T\int_0^\ell d\sigma\, \partial_\tau X_\mu(\sigma) \tag{2.68}$$

and the total angular momentum is

$$J_{\mu\nu} = \int_0^\ell d\sigma\, J^\tau_{\mu\nu} = T\int_0^\ell d\sigma\, (X_\mu \partial_\tau X_\nu - X_\nu \partial_\tau X_\mu). \tag{2.69}$$

2.4 Oscillator Expansions

It is straightforward to see that P_μ and $J_{\mu\nu}$ are conserved for the closed string. Indeed, $\frac{\partial P_\mu}{\partial \tau} = \int_0^\ell d\sigma\, \partial_\tau^2 X_\mu = \int_0^\ell d\sigma\, \partial_\sigma^2 X_\mu = \partial_\sigma X_\mu(\sigma = \ell) - \partial_\sigma X_\mu(\sigma = 0)$ which vanishes for the closed string by periodicity. For the open string it only vanishes if we impose Neumann boundary conditions at both ends. Hence our earlier statement that Neumann boundary conditions have the physical interpretation that no momentum flows off the ends of the string. This is not the case, however, for Dirichlet boundary conditions. They break Poincaré invariance and consequently space-time momentum is not conserved. Conservation of the total angular momentum is also easy to check for closed strings and open strings with Neumann boundary conditions at both ends.

With the help of the Poisson brackets Eq. (2.64) it is straightforward to verify that P^μ and $J^{\mu\nu}$ generate the Poincaré algebra:

$$\{P^\mu, P^\nu\}_{\text{P.B.}} = 0,$$

$$\{P^\mu, J^{\rho\sigma}\}_{\text{P.B.}} = \eta^{\mu\sigma} P^\rho - \eta^{\mu\rho} P^\sigma,$$

$$\{J^{\mu\nu}, J^{\rho\sigma}\}_{\text{P.B.}} = \eta^{\mu\rho} J^{\nu\sigma} + \eta^{\nu\sigma} J^{\mu\rho} - \eta^{\nu\rho} J^{\mu\sigma} - \eta^{\mu\sigma} J^{\nu\rho}. \tag{2.70}$$

2.4 Oscillator Expansions

Let us now solve the classical equations of motion of the string in conformal gauge, taking into account the boundary conditions. We will do this for the unconstrained system. The constraints then have to be imposed on the solutions. We have to distinguish between the closed and the open string and will treat them in turn.

Closed Strings

The general solution of the two-dimensional wave equation compatible with the periodicity condition $X^\mu(\sigma, \tau) = X^\mu(\sigma + \ell, \tau)$ is[10]

$$X^\mu(\sigma, \tau) = X_R^\mu(\tau - \sigma) + X_L^\mu(\tau + \sigma) \tag{2.71}$$

where

$$X_R^\mu(\tau - \sigma) = \frac{1}{2}(x^\mu - c^\mu) + \frac{\pi \alpha'}{\ell} p^\mu(\tau - \sigma) + i\sqrt{\frac{\alpha'}{2}} \sum_{n \neq 0} \frac{1}{n} \alpha_n^\mu e^{-\frac{2\pi}{\ell} i n (\tau - \sigma)},$$

$$\tag{2.72a}$$

---

[10] A more general condition which guarantees that the boundary term in (2.46) vanishes is $X^\mu(\tau, \sigma) = X^\mu(\tau + \Delta, \sigma + \ell)$. In fact, the periodicity of the solution is not preserved under a world-sheet Lorentz transformation. We can always find a Lorentz frame in which the more general periodicity condition reduces to the usual one.

$$X_L^\mu(\tau+\sigma) = \frac{1}{2}(x^\mu + c^\mu) + \frac{\pi\alpha'}{\ell}p^\mu(\tau+\sigma) + i\sqrt{\frac{\alpha'}{2}}\sum_{n\neq 0}\frac{1}{n}\overline{\alpha}_n^\mu e^{-\frac{2\pi}{\ell}in(\tau+\sigma)} \tag{2.72b}$$

with $n \in \mathbb{Z}$ and arbitrary Fourier modes α_n^μ and $\overline{\alpha}_n^\mu$. The normalizations have been chosen for later convenience, and we have also introduced the parameters c^μ, which will become relevant in Chap. 10 when we discuss toroidal compactifications. Here we can choose the zero mode part of the expansion (2.72a) and (2.72b) left-right symmetric and set $c^\mu = 0$. Our notation is such that the α_n^μ are positive frequency modes for $n < 0$ and negative frequency modes for $n > 0$. Note that the left- and right-moving parts are only coupled through the zero modes x^μ and p^μ. The requirement that $X^\mu(\sigma,\tau)$ be a real function implies that x^μ and p^μ are real and that

$$\alpha_{-n}^\mu = (\alpha_n^\mu)^* \quad \text{and} \quad \overline{\alpha}_{-n}^\mu = (\overline{\alpha}_n^\mu)^* . \tag{2.73}$$

If we define

$$\alpha_0^\mu = \overline{\alpha}_0^\mu = \sqrt{\frac{\alpha'}{2}}p^\mu , \tag{2.74}$$

we can write

$$\partial_- X^\mu = \dot{X}_R^\mu = \frac{2\pi}{\ell}\sqrt{\frac{\alpha'}{2}}\sum_{n=-\infty}^{+\infty}\alpha_n^\mu e^{-\frac{2\pi}{\ell}in(\tau-\sigma)} , \tag{2.75a}$$

$$\partial_+ X^\mu = \dot{X}_L^\mu = \frac{2\pi}{\ell}\sqrt{\frac{\alpha'}{2}}\sum_{n=-\infty}^{+\infty}\overline{\alpha}_n^\mu e^{-\frac{2\pi}{\ell}in(\tau+\sigma)} . \tag{2.75b}$$

From

$$P^\mu = \int_0^\ell d\sigma\, \Pi^\mu = \frac{1}{2\pi\alpha'}\int_0^\ell d\sigma\, \dot{X}^\mu = p^\mu , \tag{2.76}$$

we conclude that p^μ is the total space-time momentum of the string. From

$$q^\mu(\tau) \equiv \frac{1}{\ell}\int_0^\ell d\sigma\, X^\mu = x^\mu + \frac{2\pi\alpha'}{\ell}p^\mu\tau , \tag{2.77}$$

we learn that x^μ is the 'center of mass' position of the string at $\tau = 0$. Using the expression for the total angular momentum, we find

$$J^{\mu\nu} = \int_0^\ell d\sigma\,(X^\mu\Pi^\nu - X^\nu\Pi^\mu) = \frac{1}{2\pi\alpha'}\int_0^\ell d\sigma\,(X^\mu\dot{X}^\nu - X^\nu\dot{X}^\mu)$$

$$= l^{\mu\nu} + E^{\mu\nu} + \overline{E}^{\mu\nu} \tag{2.78}$$

2.4 Oscillator Expansions

with

$$l^{\mu\nu} = x^\mu p^\nu - x^\nu p^\mu \tag{2.79}$$

and

$$E^{\mu\nu} = -i \sum_{n=1}^{\infty} \frac{1}{n}(\alpha_{-n}^\mu \alpha_n^\nu - \alpha_{-n}^\nu \alpha_n^\mu) \tag{2.80}$$

with a similar expression for $\overline{E}^{\mu\nu}$.

From the Poisson brackets Eq. (2.64) we derive the brackets for the α_n^μ, $\overline{\alpha}_n^\mu$, x^μ and p^μ:

$$\{\alpha_m^\mu, \alpha_n^\nu\}_{\text{P.B.}} = \{\overline{\alpha}_m^\mu, \overline{\alpha}_n^\nu\}_{\text{P.B.}} = -im\delta_{m+n}\,\eta^{\mu\nu}\,, \tag{2.81a}$$

$$\{\overline{\alpha}_m^\mu, \alpha_n^\nu\}_{\text{P.B.}} = 0\,, \tag{2.81b}$$

$$\{x^\mu, p^\nu\}_{\text{P.B.}} = \eta^{\mu\nu}\,. \tag{2.81c}$$

We have introduced the notation $\delta_m = \delta_{m,0}$. x^μ and p^μ, the center of mass position and momentum, are canonically conjugate. The Hamiltonian (2.59), expressed in terms of oscillators, is

$$H = \frac{\pi}{\ell} \sum_{n=-\infty}^{+\infty} (\alpha_{-n} \cdot \alpha_n + \overline{\alpha}_{-n} \cdot \overline{\alpha}_n)\,. \tag{2.82}$$

We have seen above that the constraints (2.54a) and (2.54b), together with energy-momentum conservation, give rise to an infinite number of conserved charges Eq. (2.58), with a similar expression for the right-movers. We now choose for the functions $f(\sigma^\pm)$ a complete set satisfying the periodicity condition appropriate for the closed string: $f_m(\sigma^\pm) = \exp(\frac{2\pi i}{\ell} m\sigma^\pm)$ for all integers m. We then define the Virasoro generators as the corresponding charges at $\tau = 0$ [11]

$$L_n = -\frac{\ell}{4\pi^2} \int_0^\ell d\sigma\, e^{-\frac{2\pi i}{\ell} n\sigma}\, T_{--} = \frac{1}{2} \sum_m \alpha_{n-m} \cdot \alpha_m\,,$$

$$\overline{L}_n = -\frac{\ell}{4\pi^2} \int_0^\ell d\sigma\, e^{+\frac{2\pi i}{\ell} n\sigma}\, T_{++} = \frac{1}{2} \sum_m \overline{\alpha}_{n-m} \cdot \overline{\alpha}_m\,. \tag{2.83}$$

With the help of the representation of the periodic δ-function

---

[11] Since the Hamiltonian is one of the constraints and the constraints form a closed algebra under Poisson brackets (i.e. they are first class; cf. below), it is clear that the L_n are constant in τ (modulo the constraints); this is indeed easily verified.

$$\frac{1}{\ell} \sum_{n \in \mathbb{Z}} e^{\frac{2\pi i}{\ell} n(\sigma - \sigma')} = \delta(\sigma - \sigma'), \tag{2.84}$$

one can invert the above definitions:

$$T_{--}(\sigma) = -\left(\frac{2\pi}{\ell}\right)^2 \sum_n L_n \, e^{\frac{2\pi i}{\ell} n\sigma} \tag{2.85}$$

and likewise for T_{++}.

The L_m's satisfy the reality condition

$$L_n = L_{-n}^*, \quad \overline{L}_n = \overline{L}_{-n}^*. \tag{2.86}$$

Comparing with Eq. (2.83), we find that the Hamiltonian is simply

$$H = \frac{2\pi}{\ell}(L_0 + \overline{L}_0). \tag{2.87}$$

The general τ evolution operator would have been $H = \sum_n (c_n L_n + \overline{c}_n \overline{L}_n)$; the choice implied by Eq. (2.87), $c_n = \overline{c}_n = \delta_n$ is the conformal gauge. Since the constraint $T \int_0^\ell d\sigma \dot{X} \cdot X' = \frac{2\pi}{\ell}(L_0 - \overline{L}_0)$ generates rigid σ-translations and since on a closed string no point is special, we need to require that $L_0 - \overline{L}_0 = 0$. It is through this condition that the left-movers know about the right-movers. The Virasoro generators satisfy an algebra, called the (centerless) Virasoro algebra:

$$\{L_m, L_n\}_{\text{P.B.}} = -i(m-n)L_{m+n},$$

$$\{\overline{L}_m, \overline{L}_n\}_{\text{P.B.}} = -i(m-n)\overline{L}_{m+n},$$

$$\{\overline{L}_m, L_n\}_{\text{P.B.}} = 0. \tag{2.88}$$

In the mathematical literature this algebra is called Witt algebra. Equation (2.88) is straightforward to verify. It is nothing but the Fourier decomposition of the (equal τ) algebra of the Virasoro constraints:

$$\{T_{--}(\sigma), T_{--}(\sigma')\}_{\text{P.B.}} = +2\pi[T_{--}(\sigma) + T_{--}(\sigma')]\partial_\sigma \delta(\sigma - \sigma'),$$

$$\{T_{++}(\sigma), T_{++}(\sigma')\}_{\text{P.B.}} = -2\pi[T_{++}(\sigma) + T_{++}(\sigma')]\partial_\sigma \delta(\sigma - \sigma'),$$

$$\{T_{++}(\sigma), T_{--}(\sigma')\}_{\text{P.B.}} = 0. \tag{2.89}$$

It is useful to recognize that if we replace the Poisson brackets by Lie brackets, a realization of the Virasoro algebra is furnished by the vector fields $\overline{L}_n = e^{\frac{2\pi i}{\ell} n\sigma^+} \partial_+$ and $L_n = e^{\frac{2\pi i}{\ell} n\sigma^-} \partial_-$. They are the generators of the reparametrizations $\sigma^\pm \to \sigma^\pm +$

2.4 Oscillator Expansions

$f_n(\sigma^\pm)$. If we define the variable $z = e^{\frac{2\pi i}{\ell}\sigma^-} \in S^1$, we get $L_n = iz^{n+1}\partial_z$, which are reparametrizations of the circle S^1. The algebra (2.89) expresses the conformal invariance of the classical string theory. Its quantum version will be one of the central themes in the following chapters.

Open Strings

Next we discuss open strings, where we have to distinguish between Neumann and Dirichlet boundary conditions. Because the boundary reflects left- into right-movers, and vice versa, the open string solutions have only one set of oscillator modes.

For Neumann boundary conditions at both ends, we have to require $X'^\mu = 0$ at $\sigma = 0$ and $\sigma = \ell$. The general solution of the wave equation subject to these boundary conditions is

$$\text{(NN)} \quad X^\mu(\sigma,\tau) = x^\mu + \frac{2\pi\alpha'}{\ell}p^\mu\tau + i\sqrt{2\alpha'}\sum_{n\neq 0}\frac{1}{n}\alpha_n^\mu e^{-i\frac{\pi}{\ell}n\tau}\cos\left(\frac{n\pi\sigma}{\ell}\right) \tag{2.90}$$

from which we get

$$\partial_\pm X^\mu = \frac{1}{2}(\dot{X}^\mu \pm X'^\mu) = \frac{\pi}{\ell}\sqrt{\frac{\alpha'}{2}}\sum_{n=-\infty}^{+\infty}\alpha_n^\mu e^{-\frac{i\pi n}{\ell}(\tau\pm\sigma)}. \tag{2.91}$$

We have defined

$$\alpha_0^\mu = \sqrt{2\alpha'}p^\mu. \tag{2.92}$$

As in the case of the closed string we easily show that x^μ and p^μ are the center of mass position and total space-time momentum of the open string. The total angular momentum is

$$J^{\mu\nu} = \frac{1}{2\pi\alpha'}\int_0^\pi d\sigma\,(X^\mu\dot{X}^\nu - X^\nu\dot{X}^\mu) = l^{\mu\nu} + E^{\mu\nu} \tag{2.93}$$

with $l^{\mu\nu}$ and $E^{\mu\nu}$ as in Eqs. (2.79) and (2.80). We again find

$$\{\alpha_m^\mu, \alpha_n^\nu\}_{\text{P.B.}} = -im\delta_{m+n}\eta^{\mu\nu}, \tag{2.94a}$$

$$\{x^\mu, p^\nu\}_{\text{P.B.}} = \eta^{\mu\nu}. \tag{2.94b}$$

In terms of the oscillators the Hamiltonian for the open string is

$$H = \frac{\pi}{2\ell}\sum_{n=-\infty}^{+\infty}\alpha_{-n}\cdot\alpha_n. \tag{2.95}$$

The easiest way to derive this is to use the doubling trick and to write

$$H = \frac{1}{2\pi\alpha'} \int_0^\ell d\sigma \left((\partial_+ X)^2 + (\partial_- X)^2\right) = \frac{1}{2\pi\alpha'} \int_{-\ell}^\ell d\sigma (\partial_+ X)^2 \quad (2.96)$$

which is possible because of $X'(\sigma) = -X'(-\sigma)$. On the interval $-\ell \leq \sigma \leq \ell$ the functions $e^{i\pi m\sigma/\ell}$ are periodic.

The open string boundary conditions mix left- with right-movers and consequently T_{++} with T_{--}. We define the Virasoro generators for the open string as (again at $\tau = 0$)

$$\begin{aligned}
L_m &= -\frac{\ell}{2\pi^2} \int_0^\ell d\sigma \left(e^{\frac{i\pi}{\ell} m\sigma} T_{++} + e^{-\frac{i\pi}{\ell} m\sigma} T_{--}\right) \\
&= \frac{\ell}{2\pi^2 \alpha'} \int_0^\ell d\sigma \left(e^{\frac{i\pi}{\ell} m\sigma} (\partial_+ X)^2 + e^{-\frac{i\pi}{\ell} m\sigma} (\partial_- X)^2\right) \\
&= \frac{\ell}{2\pi^2 \alpha'} \int_{-\ell}^{+\ell} d\sigma \, e^{\frac{i\pi}{\ell} m\sigma} (\partial_+ X)^2 \\
&= \frac{1}{2} \sum_{n=-\infty}^{+\infty} \alpha_{m-n} \cdot \alpha_n .
\end{aligned} \quad (2.97)$$

The L_m are a complete set of conserved charges respecting the open string boundary conditions. Comparison with Eq. (2.96) gives

$$H = \frac{\pi}{\ell} L_0 , \quad (2.98)$$

which, as in the closed string case, reflects the fact that we are in conformal gauge. The L_m satisfy the Virasoro algebra

$$\{L_m, L_n\}_{\text{P.B.}} = -i(m-n) L_{m+n} . \quad (2.99)$$

The second choice of boundary conditions are Dirichlet conditions at both ends of the string. We impose them by requiring $\dot{X}^\mu = 0$ at $\sigma = 0$ and at $\sigma = \ell$. The positions of the ends are fixed at $X^\mu(\sigma = 0, \tau) = x_0^\mu$, $X^\mu(\sigma = \ell, \tau) = x_1^\mu$. The general solution of the wave equation, subject to these boundary conditions, is

$$\begin{aligned}
\text{(DD)} \quad X^\mu(\sigma, \tau) &= x_0^\mu + \frac{1}{\ell}(x_1^\mu - x_0^\mu) \sigma \\
&+ \sqrt{2\alpha'} \sum_{n \neq 0} \frac{1}{n} \alpha_n^\mu \, e^{-\frac{i\pi}{\ell} n\tau} \sin\left(\frac{\pi n \sigma}{\ell}\right)
\end{aligned} \quad (2.100)$$

2.4 Oscillator Expansions

with $n \in \mathbb{Z}$. There is no center of mass momentum. From (2.100) we derive

$$\partial_\pm X^\mu = \pm \frac{\pi}{\ell} \sqrt{\frac{\alpha'}{2}} \sum_{n=-\infty}^{\infty} \alpha_n^\mu e^{-i\frac{\pi n}{\ell}(\tau \pm \sigma)}, \qquad (2.101)$$

where

$$\alpha_0^\mu = \frac{1}{\sqrt{2\alpha'}} \frac{1}{\pi}(x_1^\mu - x_0^\mu). \qquad (2.102)$$

The oscillator modes satisfy Eq. (2.94a). There is no center of mass momentum. The center of mass position is

$$q^\mu = \frac{1}{\ell} \int_0^\ell d\sigma\, X^\mu(\sigma, \tau) = \left(\frac{x_0^\mu + x_1^\mu}{2}\right). \qquad (2.103)$$

For the Virasoro generators L_m for $m \neq 0$ one gets the same expressions as for Neumann boundary conditions, Eq. (2.97). The L_m also satisfy the algebra (2.99). With $H = \frac{\pi}{\ell} L_0$ we find for the Hamiltonian

$$H = \frac{T}{2\ell}(x_1^\mu - x_0^\mu)^2 + \frac{\pi}{2\ell} \sum_{n \neq 0} \alpha_{-n} \cdot \alpha_n. \qquad (2.104)$$

The first term is the potential energy of the stretched string.

We can also impose mixed boundary conditions, i.e. different boundary conditions at the two ends of the open string. For Neumann boundary conditions at $\sigma = 0$ and Dirichlet boundary condition at $\sigma = \ell$ the general solution reads

$$\text{(ND)} \quad X^\mu(\sigma, \tau) = x^\mu + i\sqrt{2\alpha'} \sum_{r \in \mathbb{Z}+\frac{1}{2}} \frac{1}{r} \alpha_r^\mu e^{-i\frac{\pi}{\ell}r\tau} \cos\left(\frac{\pi r \sigma}{\ell}\right), \qquad (2.105)$$

where x^μ is the position of the $\sigma = \ell$ end of the open string. Note that the center of mass momentum vanishes and that the oscillators carry half-integer modes. They also satisfy Eq. (2.94a). For completeness we also give the last possible combination of boundary conditions

$$\text{(DN)} \quad X^\mu(\sigma, \tau) = x^\mu + \sqrt{2\alpha'} \sum_{r \in \mathbb{Z}+\frac{1}{2}} \frac{1}{r} \alpha_r^\mu e^{-i\frac{\pi}{\ell}r\tau} \sin\left(\frac{\pi r \sigma}{\ell}\right), \qquad (2.106)$$

where x^μ is the position of the $\sigma = 0$ end of the open string. We have $(\alpha_n^\mu)^* = \alpha_{-n}^\mu$ for all possible boundary conditions. From (2.105) and (2.106) one derives

$$\partial_\pm X^\mu = \begin{cases} \dfrac{\pi}{\ell}\sqrt{\dfrac{\alpha'}{2}}\sum_r \alpha_r^\mu e^{-\frac{i\pi r}{\ell}(\tau\pm\sigma)} & \text{(ND)} \\ \pm\dfrac{\pi}{\ell}\sqrt{\dfrac{\alpha'}{2}}\sum_r \alpha_r^\mu e^{-\frac{i\pi r}{\ell}(\tau\pm\sigma)} & \text{(DN)}. \end{cases} \quad (2.107)$$

For all four boundary conditions one can use the doubling trick and combine $\partial_\pm X$ into one field, say left moving which is defined on the doubled interval $0 \leq \sigma \leq 2\ell$:

$$\begin{aligned}\partial_+ X^\mu &= \begin{cases} \partial_+ X^\mu(\sigma), & 0\leq\sigma\leq\ell \\ \pm\partial_- X^\mu(2\ell-\sigma), & \ell\leq\sigma\leq 2\ell \end{cases} \quad \begin{cases} +\text{sign for (NN) and (ND)},\\ -\text{sign for (DD) and (DN)}, \end{cases} \\ &= \dfrac{\pi}{\ell}\sqrt{\dfrac{\alpha'}{2}}\sum \alpha_n^\mu e^{-\frac{i\pi n}{\ell}(\tau+\sigma)}, \quad 0\leq\sigma\leq 2\ell \quad \begin{cases} n\in\mathbb{Z} & \text{for NN and DD},\\ n\in\mathbb{Z}+\tfrac{1}{2} & \text{for DN and ND}. \end{cases}\end{aligned}$$
$$(2.108)$$

The signs are chosen to have continuity at $\sigma = \ell$. Clearly

$$\begin{aligned}\partial_+ X^\mu(\sigma+2\ell) &= \partial_+ X^\mu(\sigma) & \text{for (NN) and (DD)}. \\ \partial_+ X^\mu(\sigma+2\ell) &= -\partial_+ X^\mu(\sigma) & \text{for (DN) and (ND)}.\end{aligned} \quad (2.109)$$

If the string moves in d space-time dimensions, one can combine various boundary conditions. For instance, one can have open strings with $(p+1)$ Neumann directions and $(d-p-1)$ Dirichlet directions. The end points of the open string are then confined to $(p+1)$ dimensional subspaces of the d dimensional target space. Space-time translation symmetry along the $(d-p-1)$ transverse directions is broken by these solutions. As we have already remarked, this means that the space-time momentum in the Dirichlet directions, which is carried by the open string, is not conserved; it can flow off the ends of the string. Since the translation invariance is spontaneously broken, momentum must be conserved. One is thus forced to consider the subspaces, to which the endpoints are attached, as dynamical objects which exchange momentum with the open strings ending on them. These objects are called Dp-branes. The world-volume of a Dp-brane is $(p+1)$-dimensional. String endpoints can move along them (these are the Neumann directions), but cannot leave them. In other words, open string cannot simply end in free space. They are always attached to D-branes.

We will further analyze D-branes in Chaps. 6 and 9. In particular, we will show that they also have tension and therefore a mass density. However, as we will see, the tension scales like $1/g_s$ with the string coupling constant, which indicates that these objects are not visible in string perturbation theory, but should be considered as non-perturbative objects. That means they are string theory analogues of monopoles or instantons.

2.5 Examples of Classical String Solutions

By choosing N boundary conditions in some and D boundary conditions in the remaining coordinate directions, we obtained static D-branes at fixed transverse position and of infinite extent. Since branes carry a finite tension, which we will compute in Chap. 6, branes of infinite extent are therefore infinitely heavy and can absorb any amount of space-time momentum. But in general they are dynamical objects and their dynamics is governed by world-volume actions which are $(p+1)$-dimensional generalizations of the Nambu-Goto action, to be discussed in Chap. 16. The transverse fluctuations of a D-brane correspond to massless scalar fields in this $(p+1)$-dimensional field theory. These are the Goldstone bosons of the spontaneously broken translation invariance.

If we impose N boundary conditions in all space-time directions, we obtain space-time filling D-branes. The other extreme is a D-instanton, which only exists at one space-time point. A D1-brane is also called a D-string, to be distinguished from the fundamental string that we have studied so far and which might end on a D-string. D-branes play a central role in all recent developments of string theory and we will learn more about them as we go along.

2.5 Examples of Classical String Solutions

The solutions to the wave equations satisfying various periodicity and boundary conditions which we have found in the previous section are still subject to the Virasoro constraints: $T_{00} = T_{01} = 0$. We will now construct simple explicit solutions of the classical equations of motion which satisfy the constraints.

Since in conformal gauge the coordinate functions X^μ are solutions of the wave equation, we can use the remaining gauge freedom to set $X^0 = t = \kappa\tau$ for some constant κ. The X^i, $i = 1, \ldots, d-1$ then satisfy

$$(\partial_\sigma^2 - \partial_\tau^2)X^i = 0 \tag{2.110}$$

with solution

$$X^i(\sigma,\tau) = \frac{1}{2}a^i(\sigma+\tau) + \frac{1}{2}b^i(\sigma-\tau). \tag{2.111}$$

The constraint $\dot{X}\cdot X' = -\dot{X}^0 X'^0 + \dot{X}^i X'^i = 0$ leads to $a'^2 = b'^2$ and $\dot{X}^2 + X'^2 = 0$ to $\frac{1}{2}(a'^2 + b'^2) = \kappa^2$. Combined this gives

$$a'^2 = b'^2 = \kappa^2. \tag{2.112}$$

The simplest example of an open string with N boundary conditions is

$$X^1 = L\cos\left(\frac{\pi\sigma}{\ell}\right)\cos\left(\frac{\pi\tau}{\ell}\right) \qquad X^0 = t = \frac{\pi L}{\ell}\tau,$$

$$X^2 = L \cos\left(\frac{\pi\sigma}{\ell}\right) \sin\left(\frac{\pi\tau}{\ell}\right),$$

$$X^i = 0, \quad i = 3,\ldots,d-1. \tag{2.113}$$

It clearly satisfies the constraints. It is a straight string of length $2L$ rotating around its midpoint in the (X^1, X^2)-plane. Its total (spatial) momentum vanishes and its energy is $E = L\pi T$ from which we derive the mass $M^2 = -P^\mu P_\mu = (L\pi T)^2$. The angular momentum is $J = J_{12} = \frac{1}{2}L^2\pi T$ and we find that $J = \frac{1}{2\pi T}M^2 = \alpha' M^2$. This is a straight line in the (M^2, J) plane with slope $\alpha' = (2\pi T)^{-1}$, called a Regge trajectory. It can actually be shown that for any classical open string solution $J < \alpha' M^2$. (In the gauge chosen here and in the center of mass frame $J^2 = \frac{1}{2}J_{ij}J^{ij}$, $i,j = 1,\ldots,d-1$.) The velocity of the string is $v^2 = \cos^2(\frac{\pi\sigma}{\ell})$. It is one at both endpoints. This is an immediate consequence of the constraint $\dot{X}^2 + X'^2 = 0$ and holds for any open string with Neumann boundary conditions ($X' = 0$ at the ends).

A second simple open string solution is

$$X^0 = t = \tau \quad X^1 = vt, \quad X^2 = \frac{1}{\ell}L\sigma, \quad X^3 = \cdots = 0. \tag{2.114}$$

This describes a string which is spanned between D-branes which are a distance L apart. The string moves rigidly along the X^1 direction with a velocity v. It satisfies N b.c. along X^0 and X^1 and Dirichlet boundary conditions along X^2. The constraints are satisfied if $L = \ell\sqrt{1-v^2}$. This is simply the relativistic length contraction.

For the closed string the periodicity requirement leads to $a(\sigma + \ell) = a(\sigma)$ and $b(\sigma + \ell) = b(\sigma)$. From

$$\begin{aligned}X\left(\sigma + \frac{\ell}{2}, \tau + \frac{\ell}{2}\right) &= \frac{1}{2}a(\sigma + \tau + \ell) + \frac{1}{2}b(\sigma - \tau) \\ &= \frac{1}{2}a(\sigma + \tau) + \frac{1}{2}b(\sigma - \tau), \end{aligned} \tag{2.115}$$

we find that the period of a classical closed string is $\ell/2$. For an initially static closed string configuration, i.e. one that satisfies $\dot{X}(\sigma, \tau = 0) = 0$, we find $X(\sigma, \tau) = \frac{1}{2}(a(\sigma + \tau) + a(\sigma - \tau))$. After half a period, i.e. at $\tau = \frac{\ell}{4}$, $X(\sigma, \frac{\ell}{4}) = X(\sigma + \frac{\ell}{2}, \frac{\ell}{4})$, the loop doubles up and goes around itself twice: $X(\sigma) = X(\sigma + \frac{\ell}{2})$. A simple closed string configuration is

$$X^0 = t = \frac{2\pi}{\ell}R\tau,$$

$$X^1 = \frac{1}{2}R\left[\cos\left(\frac{2\pi}{\ell}(\sigma + \tau)\right) + \cos\left(\frac{2\pi}{\ell}(\sigma - \tau)\right)\right] = R\cos\left(\frac{2\pi\sigma}{\ell}\right)\cos\left(\frac{2\pi\tau}{\ell}\right),$$

$$X^2 = \frac{1}{2}R\left[\sin\left(\frac{2\pi}{\ell}(\sigma+\tau)\right) + \sin\left(\frac{2\pi}{\ell}(\sigma-\tau)\right)\right] = R\sin\left(\frac{2\pi\sigma}{\ell}\right)\cos\left(\frac{2\pi\tau}{\ell}\right).$$
(2.116)

At $t = 0$ it represents a circular string of radius R in the (X^1, X^2)-plane, centered around the origin. Its energy is $E = 2\pi R T$. Linear and angular momentum vanish. At $\tau = \frac{\ell}{4}$ ($t = \frac{\pi}{2}R$) it has collapsed to a point and at $\tau = \frac{\ell}{2}$ ($t = \pi R$) it has expanded again to its original size. Similar to the open string case, one can show that a general classical closed string configuration satisfies $J \leq \frac{1}{2}\alpha' M^2$.

Further Reading

Constrained systems with applications to string theory are discussed in

- K. Sundermeyer, *Constrained Dynamics*. Lecture Notes in Physics, vol. 169 (Springer, Heidelberg, 1982)
- A. Hanson, T. Regge, C. Teitelboim, *Constrained Hamiltonian Systems* (Accademia Nazionale de Lincei, Roma, 1976), also available at https://scholarworks.iu.edu/dspace/handle/2022/3108
- J. Govaerts, *Halitonian Quantization and Constrained Dynamics* (Leuven University Press, Leuven, 1991)

Discussion of various terms which can be added to the Polyakov action:

- O. Alvarez, Theory of strings with boundaries. Nucl. Phys. B **216**, 125 (1983)

Classical string solutions:

- P. Shellard, A. Vilenkin, *Cosmic Strings and Other Topological Defects* (Cambridge University Press, Cambridge, 1994)

Chapter 3
The Quantized Bosonic String

Abstract In this chapter the quantization of the bosonic string is discussed. This leads to the notion of a critical dimension ($d = 26$) in which the bosonic string can consistently propagate. Its discovery was of great importance for the further development of string theory. We will discuss both the quantization in so-called light-cone gauge and the covariant path integral quantization, which leads to the introduction of ghost fields.

3.1 Canonical Quantization of the Bosonic String

In this section we will discuss first the quantization of the bosonic string in terms of operators, i.e. we will consider the functions $X^\mu(\sigma,\tau)$ as quantum mechanical operators. This is equivalent to the transition from classical mechanics to quantum mechanics via canonical commutation relations for the coordinates and their canonically conjugate momenta. We replace Poisson brackets by commutators according to

$$\{\,,\,\}_{\text{P.B.}} \to \frac{1}{i}[\,,\,]. \qquad (3.1)$$

In this way we obtain for the equal time commutators[1]

$$[X^\mu(\sigma,\tau), \dot{X}^\nu(\sigma',\tau)] = 2\pi i\, \alpha'\eta^{\mu\nu}\delta(\sigma-\sigma'),$$
$$[X^\mu(\sigma,\tau), X^\nu(\sigma',\tau)] = [\dot{X}^\mu(\sigma,\tau), \dot{X}^\nu(\sigma',\tau)] = 0. \qquad (3.2)$$

---

[1] Our notation does not distinguish between classical and quantum quantities and between operators and their eigenvalues. Only when confusion is possible we will denote operators by hatted symbols.

The Fourier expansion coefficients in Eqs. (2.72a), (2.72b), etc. are now operators for which the following commutation relations hold:

$$[x^\mu, p^\nu] = i\eta^{\mu\nu},$$
$$[\alpha_m^\mu, \alpha_n^\nu] = [\bar{\alpha}_m^\mu, \bar{\alpha}_n^\nu] = m\,\delta_{m+n}\eta^{\mu\nu,0},$$
$$[\bar{\alpha}_m^\mu, \alpha_n^\nu] = 0. \tag{3.3}$$

For the open string the $\bar{\alpha}_m^\mu$ are, of course, absent. The reality conditions (cf. (2.73)) become hermiticity conditions

$$(\alpha_m^\mu)^\dagger = \alpha_{-m}^\mu, \qquad (\bar{\alpha}_m^\mu)^\dagger = \bar{\alpha}_m^\mu \tag{3.4}$$

which follow from requiring hermiticity of the operators $X^\mu(\sigma, \tau)$. If we rescale the α_m^μ's and define, for $m > 0$, $a_m^\mu = \frac{1}{\sqrt{m}}\alpha_m^\mu$, $(a_m^\mu)^\dagger = \frac{1}{\sqrt{m}}\alpha_{-m}^\mu$, then the a_m^μ satisfy the familiar harmonic oscillator commutation relations $[a_m^\mu, (a_n^\nu)^\dagger] = \delta_{m,n}\eta^{\mu\nu}$. One defines the oscillator ground state as the state which is annihilated by all positive modes α_m^μ, $m > 0$. They are annihilation operators while the negative modes α_{-m}^μ $m > 0$ are creation operators. This does not yet completely specify the state; we can choose it to be an eigenstate of the center of mass momentum operator with eigenvalue p^μ. If we denote this state by $|0; p^\mu\rangle$, we have

$$\alpha_m^\mu |0; p^\mu\rangle = 0 \qquad \text{for} \quad m > 0,$$
$$\hat{p}^\mu |0; p^\mu\rangle = p^\mu |0; p^\mu\rangle. \tag{3.5}$$

The number operator for the m'th mode ($m > 0$) is $\hat{N}_m =: \alpha_m \cdot \alpha_{-m} := \alpha_{-m} \cdot \alpha_m$, where the normal ordering symbol which, like the number operator, is defined w.r.t. the vacuum $|0\rangle$, instructs us to put annihilation operators to the right of creation operators. The number operator satisfies $\hat{N}_n \alpha_{\pm m}^\mu = \alpha_{\pm m}^\mu (N_n \mp \delta_{n,m})$. From now on we drop the hat on $\hat{N}$ and denote by N both the operator and its eigenvalue and likewise for p.

We now face the following problem: Since the Minkowski metric $\eta^{\mu\nu}$ has $\eta^{00} = -1$, we get $[\alpha_m^0, \alpha_{-m}^0] = [\alpha_m^0, \alpha_m^{0\,\dagger}] = -m$ and states of the form $\alpha_{-m}^0|0\rangle$ with $m > 0$ satisfy $\langle 0|\alpha_m^0 \alpha_{-m}^0|0\rangle = -m\langle 0|0\rangle < 0$; i.e. these states have negative norm. They are called ghosts.[2] Negative norm states are bad news since they are in conflict with the probabilistic interpretation of quantum mechanics. However, just as we had to impose the constraints on the solutions of the classical equations of motion, we have to impose them, now as operators, as subsidiary conditions on the states. We then can hope that the ghosts decouple from the physical Hilbert space. Indeed, one can prove a no-ghost theorem which states that the ghosts decouple in 26 dimensions (i.e. $d = 26$) if the normal ordering constant to be discussed below

---

[2] These ghosts are not to be confused with the Faddeev-Popov ghosts of Sect. 3.4.

3.1 Canonical Quantization of the Bosonic String

is -1. We will not prove this theorem here, but instead arrive at the same consistency condition by different means (c.f. below and Chap. 5).

Let us now determine the propagators for the fields $X^\mu(\sigma, \tau)$. As usual, we define them as

$$\langle X^\mu(\sigma, \tau) X^\nu(\sigma', \tau')\rangle = T[X^\mu(\sigma, \tau) X^\nu(\sigma', \tau')] - :[X^\mu(\sigma, \tau) X^\nu(\sigma', \tau')]:,\tag{3.6}$$

where T denotes time ordering and $:\cdots:$ normal ordering. Zero modes need special care. We define $:p^\nu x^\mu:=x^\mu p^\nu$. This corresponds to the choice of a translationally invariant in-vacuum $p^\mu|0\rangle = 0$.

We start with the closed string, i.e. with the propagator on the cylinder. Expressed in terms of the variables $(z, \bar z) = (e^{2\pi i(\tau-\sigma)/\ell}, e^{2\pi i(\tau+\sigma)/\ell}) \in S^1 \times S^1$ we find[3]

$$\langle X_L^\mu(\bar z) X_L^\nu(\bar w)\rangle = \frac{1}{4}\alpha' \eta^{\mu\nu} \ln \bar z - \frac{1}{2}\alpha' \eta^{\mu\nu} \ln(\bar z - \bar w),\tag{3.7a}$$

$$\langle X_R^\mu(z) X_R^\nu(w)\rangle = \frac{1}{4}\alpha' \eta^{\mu\nu} \ln z - \frac{1}{2}\alpha' \eta^{\mu\nu} \ln(z - w),\tag{3.7b}$$

$$\langle X_R^\mu(z) X_L^\nu(\bar w)\rangle = -\frac{1}{4}\alpha' \eta^{\mu\nu} \ln z,\tag{3.7c}$$

$$\langle X_L^\mu(\bar z) X_R^\nu(w)\rangle = -\frac{1}{4}\alpha' \eta^{\mu\nu} \ln \bar z \tag{3.7d}$$

and

$$\langle X^\mu(z, \bar z) X^\nu(w, \bar w)\rangle = -\frac{\alpha'}{2}\eta^{\mu\nu} \ln\left((z-w)(\bar z - \bar w)\right).\tag{3.8}$$

For $z \in \mathbb{C}$ the expression on the r.h.s. of (3.8) is familiar from two-dimensional electrostatics where $-q \ln|z - w|^2$ is the potential at w due to a charge q at z and vice versa. The right hand sides of (3.7c, 3.7d) do not vanish because X_R abd X_L have common zero mode operators. If we define the fields

$$X_R^\mu(z) = x_R^\mu + \frac{\pi}{\ell}\alpha' p_R^\mu(\tau - \sigma) + \text{oscillators},\tag{3.9a}$$

$$X_L^\mu(\bar z) = x_L^\mu + \frac{\pi}{\ell}\alpha' p_L^\mu(\tau + \sigma) + \text{oscillators}\tag{3.9b}$$

with

$$[x_R^\mu, p_R^\nu] = [x_L^\mu, p_L^\nu] = i\eta^{\mu\nu}\tag{3.10}$$

---

[3]In Chap. 4 we will perform a Wick rotation to a Euclidean world-sheet and $z \in \mathbb{C}$ with $\bar z$ its complex conjugate.

and

$$[x_L^\mu, p_R^\nu] = [x_R^\mu, p_L^\nu] = 0 \tag{3.11}$$

we find

$$\langle X_R^\mu(z) X_R^\nu(w) \rangle = -\frac{1}{2}\alpha' \ln(z-w)\eta^{\mu\nu}, \tag{3.12a}$$

$$\langle X_L^\mu(\bar{z}) X_L^\nu(\bar{w}) \rangle = -\frac{1}{2}\alpha' \ln(\bar{z}-\bar{w})\eta^{\mu\nu} \tag{3.12b}$$

with vanishing cross terms. We note that the propagators for $X_L + X_R$ are the same in both cases. Treating left- and right-movers as completely independent fields will be the key ingredient for the construction of the heterotic string, which we will discuss in Chap. 10.

For the open string we compute the propagator on the strip of width ℓ. In terms of the variables $(z, \bar{z}) = (e^{i\pi(\tau-\sigma)/\ell}, e^{i\pi(\tau+\sigma)/\ell})$ we find

$$\langle X^\mu(z,\bar{z}) X^\nu(w,\bar{w}) \rangle_{\substack{NN \\ DD}} = -\frac{\alpha'}{2}\left\{\ln|z-w|^2 \pm \ln|z-\bar{w}|^2\right\}\eta^{\mu\nu}, \tag{3.13a}$$

$$\langle X^\mu(z,\bar{z}) X^\nu(w,\bar{w}) \rangle_{\substack{ND \\ DN}} = -\frac{\alpha'}{2}\left\{\ln\left|\frac{\sqrt{z}-\sqrt{w}}{\sqrt{z}+\sqrt{w}}\right|^2 \pm \ln\left|\frac{\sqrt{\bar{z}}-\sqrt{w}}{\sqrt{\bar{z}}+\sqrt{w}}\right|^2\right\}\eta^{\mu\nu}. \tag{3.13b}$$

They are manifestly symmetric under $(z, \bar{z}) \leftrightarrow (w, \bar{w})$ and they satisfy the necessary boundary conditions. For instance, the DD propagator vanishes at the two ends of the string where $z = \bar{z}$ while for the NN propagator $\partial_\sigma = -\frac{i\pi}{\ell}(z\partial_z - \bar{z}\partial_{\bar{z}})$ vanishes there. When checking the boundary conditions of the DN and ND propagators one has to use that at the $\sigma = 0$ boundary, $\sqrt{\bar{z}} = \sqrt{z}$ while at $\sigma = \ell$, $\sqrt{\bar{z}} = -\sqrt{z}$. The electrostatic analogy is again clear; e.g. Eq. (3.13a) follows from Eq. (3.8) by the method of images. We either have a charge of the same sign at $\bar{z}$ (NN) or of opposite sign (DD).

We now turn to the constraints. In the classical theory they were shown to correspond to $T_{++} = T_{--} = 0$ or, expressed through the Fourier components, $L_n = \bar{L}_n = 0$ (in the open string case the $\bar{L}_n$ are absent). However, in the quantum theory any expression containing non-commuting operators is ill-defined without specifying an operator ordering prescription. This applies in particular to L_0 (the other L_n's are safe). Classically it was given by $L_0 = \frac{1}{2}\sum_{n=-\infty}^{+\infty} \alpha_{-n} \cdot \alpha_n$. In the quantum theory we define the L_n's by their normal ordered expressions, i.e.

$$L_n = \frac{1}{2}\sum_{m=-\infty}^{+\infty} :\alpha_{n-m} \cdot \alpha_m: \tag{3.14}$$

3.1 Canonical Quantization of the Bosonic String

and, in particular,

$$L_0 = \frac{1}{2}\alpha_0^2 + \sum_{m=1}^{+\infty} \alpha_{-m} \cdot \alpha_m . \tag{3.15}$$

For L_0 there is an ordering ambiguity, i.e. L_0 is not completely determined by its classical expression. This ambiguity is taken into account by including a normal ordering constant in expressions containing L_0, i.e. by replacing $L_0 \to L_0 + a$.

We now have to determine the algebra of the L_n's. Due to normal ordering the calculation has to be done with great care; the details can be found in the appendix of this chapter. We find the Virasoro algebra $\hat{v}$

$$[L_m, L_n] = (m-n)\, L_{m+n} + \frac{c}{12} m(m^2 - 1)\, \delta_{m+n} , \tag{3.16}$$

where c is called the central charge. The Virasoro algebra is the central extension of the Witt algebra.

Let us state the mathematical definition of a central extension $\hat{g} = g \oplus \mathbb{C}c$ of a Lie algebra g by c. It is characterized by the commutation relations

$$[x, y]_{\hat{g}} = [x, y]_g + c\, p(x, y) , \quad x, y \in g ,$$

$$[x, c]_{\hat{g}} = 0 ,$$

$$[c, c]_{\hat{g}} = 0 , \tag{3.17}$$

where c belongs to the center $\hat{g}$, i.e. it commutes with all generators. By Schur's lemma, it is constant in any irreducible representation of the algebra. We use the same symbol c for this constant. $p : g \times g \to \mathbb{C}$, called a 2-cocycle, is bilinear and anti-symmetric. As a consequence of the Jacobi-identity it satisfies the cocycle condition $p([x, y], z) + \text{cyclic} = 0$. A cocycle is called trivial if we can redefine the generators $x \to \tilde{x}$ s.t. $[\tilde{x}, \tilde{y}] = \tilde{z}$. For the Virasoro algebra this is the case for the term linear term in m. Indeed, $\tilde{L}_m = L_m - \alpha \delta_m$ satisfies the Virasoro algebra with an anomaly $\left(\frac{c}{12}m^3 + (2\alpha - \frac{c}{12})m\right)\delta_{m+n}$ and for $\alpha = c/24$ we can eliminate the linear term. This shift also changes the normal ordering constant in L_0 to $a \to a - \alpha$. Only the term in the anomaly proportional to m^3 has an invariant meaning. Central extensions of algebras are closely related to projective representations which are common in quantum mechanics.

Physically the term proportional to c in (3.16) arises as a quantum effect. It is an anomaly due to the breaking of Weyl invariance of the quantum theory. In the representation through d free bosons, $c = \eta_\mu^\mu = d$. In the present context, d is the dimension of the embedding space-time, the number of free scalar fields on the world-sheet. This means that each free scalar field contributes one unit to the central charge. In Sect. 3.4 we derive the contribution of the Faddeev-Popov ghosts to the central charge and in later chapters of other fields such as free world-sheet fermions.

The quantum version of the Virasoro algebra in the form of Eq. (2.89) is

$$[T_{++}(\sigma), T_{++}(\sigma')] = -2\pi i\{T_{++}(\sigma) + T_{++}(\sigma')\}\partial_\sigma \delta(\sigma - \sigma') - \frac{ic\pi}{6}\partial_\sigma^3 \delta(\sigma - \sigma'),$$

$$[T_{--}(\sigma), T_{--}(\sigma')] = +2\pi i\{T_{--}(\sigma) + T_{--}(\sigma')\}\partial_\sigma \delta(\sigma - \sigma') + \frac{ic\pi}{6}\partial_\sigma^3 \delta(\sigma - \sigma'),$$

$$[T_{++}(\sigma), T_{--}(\sigma')] = 0 \tag{3.18}$$

which corresponds to the choice $\alpha = \frac{c}{24}$.

It is now easy to see that even though in the classical theory the constraints are $L_n = 0$, $\forall n$, this cannot be implemented on quantum mechanical states $|\phi\rangle$ since

$$\langle \phi | [L_n, L_{-n}] | \phi \rangle = \langle \Phi | 2nL_0 | \phi \rangle + \frac{c}{12} n(n^2 - 1) \langle \phi | \phi \rangle. \tag{3.19}$$

i.e. we cannot impose $L_n|\phi\rangle = 0$, $\forall n$. The most we can do is to demand that on physical states

$$L_n|\text{phys}\rangle = 0, \quad n > 0, \tag{3.20a}$$

$$(L_0 + a)|\text{phys}\rangle = 0, \tag{3.20b}$$

i.e. the positive modes annihilate physical states. The conditions (3.20) are called Virasoro constraints or physical state conditions. This is consistent since the L_n for $n > 0$ form a closed subalgebra and the requirement $L_n|\text{phys}\rangle = 0$ for $n > 0$ only, effectively incorporates all constraints since with $L_n = (L_{-n})^\dagger$ we find that[4]

$$\langle \text{phys}' | L_n | \text{phys} \rangle = 0 \quad \forall n \neq 0. \tag{3.21}$$

For the closed string we have in addition the $\bar{L}_n$'s. They also satisfy a Virasoro algebra and commute with the L_n's. We impose the condition Eq. (3.20b) also for the $\bar{L}_n$'s and the level matching condition

$$(L_0 - \bar{L}_0)|\text{phys}\rangle = 0. \tag{3.22}$$

The reason for this constraint is that the unitary operator

$$U_\delta = e^{2\pi i \frac{\delta}{\ell}(L_0 - \bar{L}_0)} \tag{3.23}$$

satisfies

---

[4] Note that the situation is very similar to the one in the quantization of electromagnetism. There we can only impose the positive frequency part of the gauge condition $\partial \cdot A = 0$ on physical states. This suffices to get $\langle \text{phys}' | \partial \cdot A | \text{phys} \rangle = 0$. In this restricted Hilbert space longitudinal and scalar photons decouple.

3.1 Canonical Quantization of the Bosonic String

$$U_\delta^\dagger X^\mu(\sigma,\tau) U_\delta = X^\mu(\sigma+\delta,\tau), \tag{3.24}$$

i.e. U_δ generates rigid σ translations. This already follows from our discussion in Chap. 2 of the motions generated by the constraints $L_0 - \overline{L}_0 = -\int d\sigma X' \cdot \Pi$. Since no point on a closed string should be distinct, we have to impose Eq. (3.22). (This also follows from Eq. (3.20b) and the equivalent condition for $\overline{L}_0$ if $a = \overline{a}$ which is the case for the bosonic string moving through Minkowski space-time.)

We define the level number operator for the open string as

$$N = \sum_{n=1}^{\infty}(\alpha^\mu_{-n}\alpha_{\mu,n} + \alpha^i_{-n}\alpha_{i,n}) + \sum_{r\in\mathbb{N}_0+\frac{1}{2}} \alpha^a_{-r}\alpha_{a,r}, \tag{3.25}$$

where μ enumerates the NN directions, i the DD directions and a the DN and ND directions. Using the condition (3.20b) we find that physical open string excitations have to satisfy

$$\alpha' m^2 = N + \alpha'(T\Delta x)^2 + a \tag{3.26}$$

where $m^2 = -p^\mu p_\mu$ is the mass of the excitation and $(\Delta x)^2 = \Delta x^i \Delta x_i$ the distance between the two ends of the open string in the DD directions. N is the total level number of the string state, i.e. the eigenvalue of the level number operator. Note that the mass of the ground state of the open string is determined by the normal ordering constant and a piece which grows with the minimal length of the string. Due to its tension it costs energy to stretch the string a distance $|\Delta x|$.

For closed strings we find from $L_0 = \sum_{n=1}^{\infty} \alpha_{-n}\cdot\alpha_n + \frac{1}{2}\alpha_0^2 = N + \frac{\alpha'}{4}p^2$ and the analogous expression for $\overline{L}_0$

$$m^2 = -p^\mu p_\mu = m_L^2 + m_R^2, \tag{3.27}$$

where

$$\alpha' m_L^2 = 2(\overline{N} + a),$$
$$\alpha' m_R^2 = 2(N + a) \tag{3.28}$$

and

$$m_L^2 = m_R^2 \tag{3.29}$$

as a consequence of $L_0 - \overline{L}_0 = 0$. The mass of the ground state ($N = \overline{N} = 0$) is again determined by the normal ordering constant.

We want to remark that the normal ordering constants drop out from the expressions for the angular momentum operators and one easily verifies the Poincaré algebra (2.70) as an operator algebra, after replacing the Poisson brackets by commutators.

3.2 Light-Cone Quantization of the Bosonic String

It is possible to choose a gauge, called light-cone gauge, in which the Virasoro constraints can be solved explicitly and the theory can be formulated in terms of physical degrees of freedom only. It is therefore a unitary gauge but it is a non-covariant gauge. As the formulation in light-cone gauge is obtained from a manifestly Lorentz invariant theory via gauge fixing, one might expect that d-dimensional Lorentz invariance is automatic, though not manifest. However, as we will see shortly, this is true in the quantum theory only if $d = 26$, which is the critical dimension of the bosonic string.

Quantization in light-cone gauge is the fastest way to get the excitation spectrum of the string. For the computation of scattering amplitudes, however, a covariant quantization procedure is more convenient. We will discuss covariant quantization in Sect. 3.4 and in Chap. 5. It leads to the same result for the critical dimension.

Space-time light-cone coordinates are defined as (X^+, X^-, X^i), $i = 2, \ldots, d-1$, with

$$X^\pm = \frac{1}{\sqrt{2}}(X^0 \pm X^1). \tag{3.30}$$

X^i are the transverse coordinates. The scalar product in terms of light-cone components is $V \cdot W = V^i W^i - V^+ W^- - V^- W^+$ and indices are raised and lowered according to $V^+ = -V_-$, $V^- = -V_+$ and $V^i = V_i$.

Light-cone gauge is a conformal gauge where the residual gauge freedom $\tau \to \tilde{\tau} = f(\sigma^+) + g(\sigma^-)$, $\sigma \to \tilde{\sigma} = f(\sigma^+) - g(\sigma^-)$ is fixed by choosing[5] $X^+ \propto \tau$. On-shell this is possible since X^+ satisfies the two-dimensional wave equation. The constant of proportionality is determined by (2.68):

$$X^+ = \frac{2\pi\alpha'}{\ell} p^+ \tau. \tag{3.31}$$

For the open string we need that $X^\pm$ satisfy Neumann boundary conditions. The oscillators in the X^+ direction all vanish, except for the zero mode for which

$$\alpha_0^+ = \bar{\alpha}_0^+ = \sqrt{\frac{\alpha'}{2}} p^+ \quad \text{(closed string)}, \quad \alpha_0^+ = \sqrt{2\alpha'} p^+ \quad \text{(open string)}. \tag{3.32}$$

The action in light-cone gauge is[6]

---

[5]The proper way to go to light-cone gauge would be to use the local symmetries on the world-sheet to fix components of the world-sheet metric and X^+. One then has to show that no propagating ghosts are introduced in this process of gauge fixing. Rather than going through these steps we take this a posteriori justifiable short-cut.

[6]Here and below a summation over $i = 2, \ldots, d-1$ is implied.

3.2 Light-Cone Quantization of the Bosonic String

$$S = \frac{1}{4\pi\alpha'} \int d\tau d\sigma \, \{(\dot{X}^i)^2 - (X'^i)^2\} - \int d\tau \, p^+ \dot{q}^- = \int d\tau \, L, \quad (3.33)$$

where $q^- = \frac{1}{\ell} \int_0^\ell d\sigma X^-$. The canonical momenta are

$$p_- = -p^+ = \frac{\partial L}{\partial \dot{q}^-}$$

$$\Pi^i = \frac{\partial L}{\partial \dot{X}^i} = \frac{1}{2\pi\alpha'} \dot{X}^i \quad (3.34)$$

and the canonical Hamiltonian in light-cone gauge becomes

$$H_{\text{l.c.}} = p_- \dot{q}^- + \int_0^\ell d\sigma \, \Pi_i \dot{X}^i - L,$$

$$= \frac{1}{4\pi\alpha'} \int_0^\ell d\sigma \, \{(\dot{X}^i)^2 + (X'^i)^2\}. \quad (3.35)$$

With the help of the constraints $(\dot{X}^\mu \pm X'^\mu)^2 = 0$ we can express X^- through the transverse coordinates X^i via

$$\partial_\pm X^- = \frac{\ell}{2\pi\alpha' p^+} (\partial_\pm X^i)^2. \quad (3.36)$$

Adding the two constrains we find

$$\partial_\tau X^- = \frac{\ell}{2\pi\alpha' p^+} \{(\dot{X}^i)^2 + (X'^i)^2\} \quad (3.37)$$

or, with $p^\mu = \frac{1}{2\pi\alpha'} \int d\sigma \, \dot{X}^\mu$,

$$p^- = \frac{\ell}{2\pi\alpha' p^+} H_{\text{l.c.}}. \quad (3.38)$$

Subtracting the constraints we find, after integrating over the length of the closed string and using periodicity,

$$\text{(closed string)} \qquad \int_0^\ell d\sigma X'^i \Pi^i = \frac{1}{2\pi\alpha'} \int_0^\ell d\sigma X'^i \dot{X}^i = 0. \quad (3.39)$$

This is the σ-translation operator and we have to impose this constraint on physical closed string states.

We now turn to the light-cone quantization of the bosonic string. The dynamical variables are p_-, q^-, X^i, Π^i. We impose the following canonical commutation relations

$$[q^-, p^+] = -i \,,$$

$$[q^i, p^j] = i\delta^{ij} \,,$$

$$[\alpha_n^i, \alpha_m^j] = n\delta^{ij}\,\delta_{n+m,0}\,, \qquad [\overline{\alpha}_n^i, \overline{\alpha}_m^j] = n\delta^{ij}\,\delta_{m+n,0}\,, \qquad (3.40)$$

where we have used the mode expansions of Chap. 2. For the open string the $\overline{\alpha}$ oscillators are, of course, absent.

When we express e.g. the light-cone Hamiltonian in terms of oscillators, we have to worry about ordering ambiguities of the oscillators. As before, we define the Hamiltonian as a normal ordered expression and include a normal ordering constant which we determine momentarily. For the closed string we find

$$H_{\text{l.c.}} = \frac{2\pi}{\ell}\left(\sum_{n>0}(\alpha_{-n}^i\alpha_n^i + \overline{\alpha}_{-n}^i\overline{\alpha}_n^i) + a + \overline{a}\right) + \frac{\pi\alpha'}{\ell}p^i p^i \qquad (3.41)$$

and for the open string

$$H_{\text{l.c.}} = \frac{\pi}{\ell}\sum_{n>0}\left(\alpha_{-n}^i\alpha_n^i + a\right) + \frac{\pi\alpha'}{\ell}\sum_{\text{NN}} p^i p^i + \frac{1}{4\pi\alpha'\ell}\sum_{\text{DD}}(x_1^i - x_2^i)^2\,, \qquad (3.42)$$

where the mode numbers are integer for NN and DD boundary conditions and half-integer for DN and ND.

We now determine the normal ordering constants. We first give a formal derivation which we will then justify by a physically motivated one. Recall how the normal ordering constant arose:

$$\sum_{n\neq 0}\alpha_{-n}\alpha_n = \sum_{n\neq 0}:\alpha_{-n}\alpha_n: + \sum_{n=1}^{\infty}n = 2\left(\sum_{n=1}^{\infty}\alpha_{-n}\alpha_n + \frac{1}{2}\sum_{n=1}^{\infty}n\right)\,. \qquad (3.43)$$

For simplicity we have assumed $n \in \mathbb{Z}$. The last sum in this expression is undefined and must be regularized. We do this in two ways. The first method uses so-called ζ-function regularization. Consider the sum $\sum_{n=1}^{\infty} n^{-s} \equiv \zeta(s)$ which is the Riemann zeta-function. The sum converges for $\text{Re}(s) > 1$ and has a unique analytic continuation at $s = -1$ where it has the value $\zeta(-1) = -\frac{1}{12}$. This determines the contribution to a from one coordinate of the closed string and from a NN or DD direction of the open string as $-\frac{1}{24}$. For half-integer modes we have to use the generalized ζ-function $\zeta(s,q) = \sum_{n=0}^{\infty}(n+q)^{-s}$ with

$$\sum_{n=0}^{\infty}(n+q) = \zeta(-1,q) = -\frac{1}{12}(6q^2 - 6q + 1)\,. \qquad (3.44)$$

3.2 Light-Cone Quantization of the Bosonic String

Let us now justify this 'mathematical trick'. Consider the open string light-cone Hamiltonian

$$H_{\text{l.c.}} = \frac{\pi}{\ell} \sum_{n>0} \alpha^i_{-n} \alpha^i_n + \frac{\pi}{2\ell}(d-2) \sum_{n=1}^{\infty} n + \frac{\pi \alpha'}{\ell} \sum_i p^i p^i. \tag{3.45}$$

For simplicity we consider only the case where all directions are NN. The generalization to other boundary conditions is straightforward. We regularize the sum

$$\frac{\pi}{\ell} \sum_{n=1}^{\infty} n \to \frac{\pi}{\ell} \sum_{n=1}^{\infty} n \exp\left(-\frac{\pi n}{\ell \Lambda}\right) = \frac{\pi}{\ell} \frac{e^{-\pi/\ell\Lambda}}{(1 - e^{-\pi/\ell\Lambda})^2}$$

$$= \frac{\ell}{\pi} \Lambda^2 - \frac{1}{12} \frac{\pi}{\ell} + \mathcal{O}(1/\Lambda), \tag{3.46}$$

where we have introduced the UV energy cut-off Λ. The divergent part, being proportional to the length of the string, can be cancelled by adding a cosmological constant $\propto \Lambda^2 \int d^2\sigma \sqrt{-h}$ to the Polyakov action. The remaining finite part reproduces what we had obtained via ζ-function regularization.

We can now write down the mass operator $m^2 = 2p^+ p^- - p^i p^i$. Using (3.38) and the expressions for the light-cone Hamiltonian, we find for the closed string

$$m^2 = m_L^2 + m_R^2 \tag{3.47}$$

with

$$\alpha' m_L^2 = 2\left(\overline{N}_{\text{tr}} - \frac{1}{24}(d-2)\right), \qquad \alpha' m_R^2 = 2\left(N_{\text{tr}} - \frac{1}{24}(d-2)\right) \tag{3.48}$$

and

$$m_L^2 = m_R^2 \tag{3.49}$$

as a result of (3.39). N_{tr} is the level number in the transverse directions and (3.49) is the level-matching condition

$$N_{\text{tr}} = \overline{N}_{\text{tr}} \tag{3.50}$$

in light-cone gauge, which follows from (3.39).

In light-cone gauge there are no non-zero oscillator modes in the $\pm$-directions. Therefore $L_0 = N_{\text{tr}} + \frac{\alpha'}{4} p^2$. With $p^2 = p^i p^i - 2p^+ p^-$ and (3.38) we find $L_0 + \overline{L}_0 + a + \bar{a} = 0$. Combining this with $L_0 - \overline{L}_0 = N - \overline{N} = 0$, we recover the physical state condition (3.20b) with $a = \bar{a}$ the same constant as in (3.41).

For the open string the mass operator is

$$\alpha' m^2 = N_{\mathrm{tr}} - \frac{d-2}{24} + \frac{\nu}{16} + \alpha'(T\Delta X)^2, \qquad (3.51)$$

where ν is the number of directions with mixed boundary conditions, i.e. DN or ND. It remains to determine d, the number of space-time dimensions in which the string can be consistently quantized. To find d we look at the mass spectrum of the string and require Lorentz invariance, which is not manifest in the light-cone gauge.

3.3 Spectrum of the Bosonic String

In light-cone gauge the excited states of the string are generated by acting with the transverse oscillators on the ground state. They must fall into representations of the little group. Consider first a massive particle moving through d-dimensional Minkowski space. Since any massive particle necessarily moves with a speed less than the speed of light, we can make a Lorentz boost and go to its rest frame. In this frame the particle's momentum is $p^\mu = (m, 0, \ldots, 0)$ with $p^2 = -m^2$ whose invariance subgroup (isotropy group, little group) is $SO(d-1)$. This means that massive particles, and by the same token, massive string excitations, can be classified by representations of $SO(d-1)$. For a massless particle the situation is different. Since they necessarily move at the speed of light and satisfy $p^2 = 0$, we can choose a frame in which its momentum is $p^\mu = (E, 0, \ldots, 0, E)$. The invariance group of this vector is $E(d-2)$, the group of motions in $(d-2)$-dimensional Euclidean space, whose generators are rotations and translations in $(d-2)$-dimensional Euclidean space. Massless string states form, however, representations of its connected component $SO(d-2) \subset E(d-2)$. We now discuss open and closed strings in turn.

Open String Spectrum

We start with NN boundary conditions in all directions, i.e. unbroken d-dimensional Poincaré symmetry. This corresponds to a space-time filling D-brane. The ground state $|0; p\rangle$ is unique up to Lorentz boosts. Its mass is given by its eigenvalue of the mass operator $\alpha' m^2 |0; p\rangle = a |0; p\rangle$. The first excited state is $\alpha^i_{-1} |0; p\rangle$. It is a $(d-2)$-dimensional vector of the transverse rotation group $SO(d-2)$. Lorentz invariance requires that this state is massless. Acting on it with the mass operator we get

$$0 = \alpha' m^2 \left(\alpha^i_{-1} |0; p\rangle \right) = (1+a) \alpha^i_{-1} |0; p\rangle \qquad (3.52)$$

3.3 Spectrum of the Bosonic String

Table 3.1 The five lowest mass levels of the oriented open bosonic string

Level	$\alpha'(m)^2$	States and their $SO(24)$ representation contents	Little group	Representation contents with respect to the little group
0	-1	$\|0\rangle$ • (1)	$SO(25)$	• (1)
1	0	$\alpha^i_{-1}\|0\rangle$ □ (24)	$SO(24)$	□ (24)
2	$+1$	$\alpha^i_{-2}\|0\rangle$ $\alpha^i_{-1}\alpha^j_{-1}\|0\rangle$ □ □□ + • (24) (299) + (1)	$SO(25)$	□□ (324)
3	$+2$	$\alpha^i_{-3}\|0\rangle$ $\alpha^i_{-2}\alpha^j_{-1}\|0\rangle$ $\alpha^i_{-1}\alpha^j_{-1}\alpha^k_{-1}\|0\rangle$ □ ⊟ + □ + • □□□ + □ (24) (276) + (299) + (1) (2576) + (24)	$SO(25)$	□□□ + ⊟ (2900) (300)
4	$+3$	$\alpha^i_{-4}\|0\rangle$ $\alpha^i_{-3}\alpha^j_{-1}\|0\rangle$ $\alpha^i_{-2}\alpha^j_{-2}\|0\rangle$ □ ⊟ + □ + • □□ + • (24) (276) + (299) + (1) (299) + (1) $\alpha^i_{-2}\alpha^j_{-1}\alpha^k_{-1}\|0\rangle$ $\alpha^i_{-1}\alpha^j_{-1}\alpha^k_{-1}\alpha^l_{-1}\|0\rangle$ $2\times$□ + □□□ + ⊞ □□□□ + □□ + • $2\times$(24) + (2576) + (4576) (17250) + (299) + (1)	$SO(25)$	□□□□ + ⊞ (20150) (5175) + □□ + • (324) (1)

i.e. we have to require $a = -1$. Comparison with (3.51) immediately leads to the important result $d = 26$. This value is called the critical dimension of the bosonic string. To summarize, space-time Lorentz invariance of the quantized bosonic string theory in Minkowski space requires $a = -1$ and $d = 26$. A more rigorous argument, which also relies on Lorentz invariance, is to check the closure of the Lorentz algebra; the commutator $[M^{i-}, M^{j-}]$ is crucial. In contrast to the Lorentz generators in covariant gauge, M^{i-} contains cubic normal ordered expressions due to the appearance of α_n^- which is a quadratic expression of the transverse oscillators (cf. Eq. (3.36)). The actual calculation of this commutator is quite tedious and will not be presented here.

An immediate consequence of $a = -1$ is that the ground state satisfies $\alpha' m^2 = -1$, i.e. it is a tachyon. The presence of a tachyon is not necessarily a fatal problem for the theory. It means that the 'ground state' is unstable and some other, stable ground state might exist. Another way to get rid of the tachyon is to introduce anti-commuting degrees of freedom whose normal ordering constant cancels that of the commuting degrees of freedom of the bosonic string. This is indeed what is done in the superstring theory.

In Table 3.1 we have collected the light-cone states of the open bosonic string with NN boundary conditions up to the fourth level. It is demonstrated how, for the massive states, the light-cone states, which are tensors of $SO(24)$, combine uniquely into representations of $SO(25)$. It can be shown that this occurs at all mass levels and depends crucially on the choices $a = -1$ and $d = 26$. Since at level n with mass $\alpha' m^2 = (n-1)$ we always have a state described by a symmetric tensor of rank n, we find that the maximal spin at each level is $j_{\max} = n = \alpha' m^2 + 1$. All states satisfy $j \leq \alpha' m^2 + 1$ and, since j and m^2 are quantized, all states lie on Regge trajectories, with the tachyon lying on the leading trajectory.

Open String Spectrum for Dp-branes

Consider an open string with $(p+1)$ Neumann boundary conditions and $(25-p)$ Dirichlet conditions. This is an open string with end points on a single Dp-brane. The story is very similar to the previous case, except for the difference between excitations parallel to the brane and normal to the brane which breaks $SO(24)$ to $SO(p-1) \times SO(25-p)$. Let us denote the parallel (N) directions as X^i and the normal (D) ones as X^a.

Indeed the ground state only carries momentum along X^i, i.e. $|0, p^i\rangle$. The first excited massless states are $\alpha_{-1}^i |0, p^i\rangle$ and $\alpha_{-1}^a |0, p^i\rangle$, which transform as a vector and $(25-p)$ scalars of $SO(p-1)$, respectively. Therefore, on the Dp-branes one gets still a massless vector field in addition to a number of scalars which can be considered as the Goldstone modes for the broken translational symmetry in the directions orthogonal to the Dp-branes. As is well known from quantum field theory, massless vector fields are gauge fields and, indeed, by analysing the couplings of these open string massless modes one finds that they precisely couple like gauge fields among themselves and to other charged matter fields. We will study this more closely in Chap. 16. We will not discuss higher excited (massive) states, though it is straightforward.

One can also consider an open string, which is stretched between two parallel Dp-branes located at, in general, different positions x_0^a and x_1^a. The oscillator excitations are completely analogous to the previous case, the only difference being that all mass levels experience a shift by the energy of the stretched non-excited open string

$$m_0^2 = \frac{1}{(\alpha')^2} \sum_a \left(\frac{x_1^a - x_0^a}{2\pi} \right)^2 . \tag{3.53}$$

This implies that the modes $\alpha_{-1}^i |\Delta x^a, p^i\rangle$ are not any longer massless.

If one places N such Dp branes on top of each other, one can write for instance the massless vector excitations as

$$\alpha_{-1}^i |k, l; p^i\rangle, \qquad k, l \in \{1, \ldots N\} . \tag{3.54}$$

The discrete degrees of freedom k, l are labels which encode the information on which brane the open string starts and ends. These are called Chan-Paton labels.

3.3 Spectrum of the Bosonic String

We can represent the N^2 states labelled by k, l in terms of a matrix which we expand into a complete set of $N \times N$ matrices $\lambda_{kl}^a, a = 1, \ldots, N^2$ and write

$$|k, l; p^i\rangle = \lambda_{kl}^a |a; p^i\rangle. \tag{3.55}$$

The λ_{kl}^a are called Chan-Paton (CP) factors. Clearly, this gives N^2 massless vector fields with momentum along the world-volume of the Dp-branes. We will later argue that these are the gauge bosons of an $U(N)$ Yang-Mills theory on the world-volume of the D-branes, i.e. a natural choice for the λ_{kl}^a are the Hermitian generators of $U(N)$ satisfying $\lambda^\dagger = \lambda$. The transverse scalar modes

$$\alpha_{-1}^a |k, l; p^i\rangle, \qquad k, l \in \{1, \ldots N\} \tag{3.56}$$

also carry the N^2 Chan-Paton factors and they also transform in the adjoint representation of the gauge group $U(N)$.

If we separate the branes in the transverse directions, due to (3.53) some of the open strings become massive and the gauge symmetry is spontaneously broken to a subgroup. In the $(p + 1)$-dimensional $U(N)$ gauge theory on the world-volume of the Dp-brane this is described by giving a non-zero vacuum expectation value to some of the adjoint scalars (3.56) proportional to the separation of the branes.

The fact that the massless modes of open strings ending on Dp-branes lead to non-abelian gauge fields in $p + 1$ dimensions is important for building realistic particle physics models within string theory.

Closed String Spectrum

On a closed string we can excite both left- and right-moving oscillators. Its states are simply tensor products of the open string states, subject to the level matching condition which means that the excitation level in both sectors has to be the same. The ground state is again a scalar tachyon $|0\rangle$ with mass $\alpha' m^2 = -4a$. The first excited state is $\alpha_{-1}^i \bar{\alpha}_{-1}^j |0\rangle$ with mass $\alpha' m^2 = 4(1 - a)$. We can decompose this state into irreducible representations of the transverse rotation group $SO(d - 2)$ as follows:

$$\alpha_{-1}^i \bar{\alpha}_{-1}^j |0, p\rangle = \left(\alpha_{-1}^{(i} \bar{\alpha}_{-1}^{j)} - \frac{1}{d-2} \delta^{ij} \alpha_{-1}^k \bar{\alpha}_{-1}^k \right) |0, p\rangle + \alpha_{-1}^{[i} \bar{\alpha}_{-1}^{j]} |0, p\rangle$$

$$+ \frac{1}{d-2} \delta^{ij} \alpha_{-1}^k \bar{\alpha}_{-1}^k |0, p\rangle, \tag{3.57}$$

where indices in parentheses and brackets are symmetrized and anti-symmetrized, respectively. As for the open string we conclude that $a = 1$ and $d = 26$. The states (3.57) represent a massless spin two particle, an antisymmetric tensor and a massless scalar which we will later identify as the graviton, an antisymmetric tensor

Table 3.2 The three lowest mass levels of the oriented closed bosonic string

Level	$\alpha'(m)^2$	States and their $SO(24)$ representation contents	Little group	Representation contents with respect to the little group
0	-4	$\|0\rangle$ • (1)	$SO(25)$	• (1)
1	0	$\alpha^i_{-1}\bar{\alpha}^j_{-1}\|0\rangle$ $\square \times \square$ (24) (24)	$SO(24)$	$\square\square$ + $\begin{array}{c}\square\\\square\end{array}$ + • (299) (276) (1)
2	$+4$	$\alpha^i_{-2}\bar{\alpha}^j_{-2}\|0\rangle$ $\alpha^i_{-1}\alpha^j_{-1}\bar{\alpha}^k_{-1}\bar{\alpha}^l_{-1}\|0\rangle$ $\square\times\square$ $(\square\square+\bullet)\times(\square\square+\bullet)$ (24) (24) (299)+(1) (299)+(1) $\alpha^i_{-2}\bar{\alpha}^j_{-1}\bar{\alpha}^k_{-1}\|0\rangle$ $\alpha^i_{-1}\alpha^j_{-1}\bar{\alpha}^k_{-2}\|0\rangle$ $\square\times(\square\square+\bullet)$ $(\square\square+\bullet)\times\square$ (24) (299)+(1) (299)+(1) (24)	$SO(25)$	$\square\square\times\square\square = \square\square\square\square + \begin{array}{c}\square\square\\\square\square\end{array}$ (324) (324) (20150) (32175) + $\begin{array}{c}\square\square\\\square\square\\\square\end{array}$ + $\square\square$ + $\begin{array}{c}\square\\\square\end{array}$ + • (52026) (324) (300) (1)

field and the dilaton. The spectrum for the first three levels is displayed in Table 3.2. Again, the massive states combine into representations of the little group $SO(25)$. The relation between the maximal spin and the mass is now $j_{\max} = \frac{1}{2}\alpha'm^2 + 2$.

Oriented vs. Non-oriented Strings

Even after going to light-cone gauge the original diffeomorphism invariance of the Polyakov action has not yet been completely fixed. The discrete world-sheet parity transformation, which possesses no infinitesimal version, $\sigma \to \tilde{\sigma} = \ell - \sigma$, $\tau \to \tilde{\tau} = \tau$, is still possible. It respects the periodicity of the closed string and maps the two end points of the open string to each other but it reverses the orientation $d\sigma \wedge d\tau$ of the world-sheet.

The concept of orientability can be made precise by defining a unitary operator Ω which implements the above discrete diffeomorphism and thus reverses the orientation of the string, i.e.

$$\Omega X^\mu(\sigma, \tau) \Omega^{-1} = X^\mu(\ell - \sigma, \tau). \tag{3.58}$$

Expressed in terms of oscillators this means for the modes of the closed string

$$\Omega \alpha_n^\mu \Omega^{-1} = \bar{\alpha}_n^\mu, \qquad \Omega \bar{\alpha}_n^\mu \Omega^{-1} = \alpha_n^\mu \tag{3.59}$$

and for the open string

$$\Omega \alpha_n^\mu \Omega^{-1} = (-1)^n \alpha_n^\mu \qquad \text{(NN)} \tag{3.60a}$$

3.3 Spectrum of the Bosonic String

$$\left. \begin{array}{l} \Omega \, \alpha_n^\mu \, \Omega^{-1} = (-1)^{n+1} \, \alpha_n^\mu \\ \Omega \, x_{0,1}^\mu \, \Omega^{-1} = \quad x_{1,0}^\mu \end{array} \right\} \quad \text{(DD)} . \tag{3.60b}$$

For mixed boundary conditions, we get that the ND-sector is mapped to the DN-sector and vice versa

$$\Omega \, \alpha_{n+\frac{1}{2}}^{\mu,\text{ND}} \, \Omega^{-1} = i \, (-1)^n \, \alpha_{n+\frac{1}{2}}^{\mu,\text{DN}} . \tag{3.61}$$

The action of Ω on a state is fixed only after we specify its action on the ground state. Requiring $\Omega^2 = 1$ on all states, which makes $\frac{1}{2}(1 + \Omega)$ a projection operator, leaves a sign choice. As we will see in Chap. 9, it is constrained by so-called tadpole cancellation conditions. Once it is fixed, the action of Ω on all oscillator excitations of the ground state is also fixed.

There are now two possibilities. If we gauge the discrete symmetry we arrive at unoriented strings whose physical states must be invariant under Ω. Otherwise we have the oriented string without such a restriction on the spectrum of physical states. The spectrum of unoriented closed strings consists of states symmetric under the interchange of left- and right-moving oscillators. Of the massless states only the symmetric (graviton) and singlet piece (dilaton) survive. For historical reason the closed oriented bosonic string theory is referred to as the extended Shapiro-Virasoro model, whereas the unoriented theory is called the restricted Shapiro-Virasoro model.

For the case that Ω acts on the ground state with a plus sign, the spectrum of an unoriented open string with NN (DD) boundary conditions consists only of states with even (odd) level number. For the case of a single D9-brane this projects out the massless vector. However, for $2N$ branes on top of each other, one has to take also the action of Ω on the CP factors into account. Since Ω changes the orientation

$$\Omega |k, l; p^i\rangle = |l, k; p^i\rangle , \tag{3.62}$$

at the massless level, $N(2N - 1)$ of the $(2N)^2$ CP-factors survive. This gives a massless vector of a $SO(2N) \subset U(2N)$ gauge theory. Contrarily, if Ω acts with a minus sign on the open string vacuum, the CP labels are symmetrized and we get a massless vector of a $Sp(2N) \subset U(2N)$ gauge theory.[7] The symplectic groups are non-compact and the compact gauge group is $USp(2N) = Sp(2N; \mathbb{C}) \cap U(2N)$.

One can dress the world-sheet parity transformation by additional $\mathbb{Z}_2$ symmetries of the string theory. In this way one can define more general gaugings. Such constructions are also called orientifolds. For instance, the combination ΩI_n, where I_n reflects n of the 26 space-time coordinates, leads to orientifolds, where the $26 - n$

---

[7] The group $Sp(2N)$ is defined as the group generated by $2N \times 2N$-matrices with $MJM^{\mathrm{t}} = J$ where $J = \begin{pmatrix} 0 & \mathbb{1} \\ -\mathbb{1} & 0 \end{pmatrix}$ and $\mathbb{1}$ is the $N \times N$ unit matrix. In our conventions $Sp(2N)$ has rank N.

dimensional fixed locus of I_n defines a so-called orientifold $O(25-n)$-plane. One can think of these as defects in space-time, which carry a mass-density. We will discuss such quotient theories in more detail in Chap. 9, where we will see that in contrast to D-branes the O-planes are not dynamical.

Let us now interpret the mass spectra. Any interacting string theory with local interactions has to contain closed strings. String interactions consist of strings splitting and joining. How this is incorporated in string theory will be discussed in Chap. 6. The assumption of locality implies that two joining ends of strings can be either two ends of two different open strings or the two ends of one open sting, joining to form a closed string.

Looking at the closed string spectrum we see that a massless spin two particle is always present. It is very suggestive to identify it with the graviton, i.e. the gauge particle of the ubiquitous gravitational interaction. If we do this we have to relate the string scale set by the slope parameter to the Planck scale, i.e. $\alpha' \sim G$ where $G = M_P^{-2}$ is Newtons constant and M_P the Planck mass. It is of course one of the attractive and encouraging features of string theory that it necessarily contains gravity.[8] This however also means that the massive states, since their mass is an integer multiple of the string mass M_s, cannot be identified with known particles or hadronic resonances as was the original motivation for string theory. But this is all right, as the higher mass and spin resonances of hadronic physics have by now found an adequate description by QCD and the prospect of having a consistent quantum theory including gravity has led to a shift in the interpretation of string theory from the hadronic scale (100 MeV) to the Planck scale (10^{19} GeV). The other massless states of the closed string, the singlet and the antisymmetric tensor part of Eq. (3.57) can be interpreted as a dilaton and an antisymmetric tensor particle.

As we will discuss in Chap. 17, within the so-called brane-world scenario, there exists the possibility to have a low string scale which could bring the masses of low string excitations within the reach of particle accelerators. But even with a high string scale one can describe hadronic physics, or at least a theory with similar features, within string theory. This is in the context of the AdS/CFT correspondence which we will introduce in Chap. 18.

What about the states of the open string? As already mentioned, the massless vector could be interpreted as a gauge boson, if we associate a non-abelian charge in the adjoint representation of the gauge group to the open string. This can indeed be done by attaching to one end of the string the charge of the fundamental representation and to the other end the charge of its complex conjugate representation. This is the method of Chan and Paton. For a Dp-brane the momentum of the vector is confined to the $p+1$ longitudinal Neumann directions so that the gauge boson cannot leave the brane. It is confined to it. For space-time filling D25-branes, consistency of the interacting theory restricts the possible gauge groups to

---

[8]The presence of a massless spin two particle is a priori not sufficient to have gravity. We will show in the last chapter that at low energies it couples to matter and to itself like the graviton of general relativity.

$SO(8192)$ for the bosonic string. We will go into details of the Chan-Paton method in Chap. 9 where we also explain why in the superstring with space-time filling D9 branes, the gauge group $SO(32)$ is singled out. In Chap. 9 we will also learn how to get non-abelian gauge symmetries in a theory of only closed strings.

At the end of this discussion about a possible connection of string theory with known physics we have to say a word of caution. The suggested interpretation of the massless particles can, of course, only hold if they have the interactions appropriate to gravitons, gauge bosons etc. This means that the theory has to be gauge invariant and especially generally coordinate invariant as a 26 dimensional theory. This is not a priori obvious and will be demonstrated in Chap. 16. Also, we still have to find a way to go from the critical dimension to 4 dimensions. We will address this important question in Chap. 14.

3.4 Covariant Path Integral Quantization

Path integral quantization à la Faddeev-Popov has proven useful for theories with local symmetries, e.g. Yang-Mills theories. As such it is also applicable in string theory and an alternative to the non-covariant light-cone gauge quantization. The starting point is the Polyakov action (2.21). As discussed in Chap. 2, the induced metric $\Gamma_{\alpha\beta} = \partial_\alpha X^\mu \partial_\beta X^\nu \eta_{\mu\nu}$ and the intrinsic world-sheet metric $h_{\alpha\beta}$ are related only on-shell, i.e. through the classical equations of motion $T_{\alpha\beta} = 0$. Quantum mechanically this does not need to be so. In fact, just as we integrate in the Feynman path integral approach to quantum mechanics over all paths and not only classical ones, we have to integrate over $h_{\alpha\beta}$ and the embeddings X^μ independently. One must, however, find a measure for the functional integrations which respects the (local) symmetries of the classical theory, which are reparametrizations and Weyl rescalings. In the classical theory the symmetries allowed us to gauge away the three degrees of freedom of $h_{\alpha\beta}$. If these symmetries cannot be maintained in the quantum theory, one has an anomaly and the quantum theory has additional degrees of freedom. Indeed, the measures for the integrations over the metrics and the embeddings are both not Weyl invariant. However, Polyakov has shown that their Weyl variations cancel in 26 dimensions in which case there is no Weyl anomaly. This is the same value which we found from Lorentz invariance in the light-cone gauge.

One can also consider so-called non-critical string theories where $d < d_{\text{crit}}$. For these non-critical string theories not all world-sheet fields can be interpreted as coordinates of Minkowski space. If we want to quantize string theory in Minkowski space without any additional degrees of freedom, we need $d = d_{\text{crit}}$. We will not discuss non-critical strings.

Polyakov's analysis will not be presented here but we will rederive the critical dimension by the requirement that the central term in the Virasoro algebra cancels when the contributions from the Faddeev-Popov ghosts are included. We will show in Chap. 4 that this is equivalent to requiring a vanishing Weyl (trace) anomaly.

We will assume in the following discussion that we are in the critical dimension in which the combined integration measure is Weyl and diffeomorphism invariant.

Consider the vacuum to vacuum amplitude or partition function[9]

$$Z = \int \mathcal{D}h \mathcal{D}X \, e^{iS_P[h,X]} . \quad (3.63)$$

The integration measures in Eq. (3.63) are defined by means of the norms

$$||\delta h|| = \int d^2\sigma \sqrt{-h} h^{\alpha\beta} h^{\gamma\delta} \delta h_{\alpha\gamma} \delta h_{\beta\delta} ,$$

$$||\delta X|| = \int d^2\sigma \sqrt{-h} \delta X^\mu \delta X_\mu . \quad (3.64)$$

For finite dimensional spaces the norm $||\delta X||^2 = g_{ij}\delta X^i \delta X^j$ leads to the volume element $\sqrt{g}d^n x$. In the infinite dimensional case the measure can be implicitly defined via the Gaussian integral

$$1 = \int \mathcal{D}(\delta X) \, e^{-\frac{1}{4\pi\alpha'}||\delta X||^2} \quad (3.65)$$

and likewise for the other variables. The measures are invariant under reparametrizations but neither measure is invariant under Weyl rescaling of $h_{\alpha\beta}$. The measure in Eq. (3.63) is not complete. Factors involving the volume of the symmetry group will be discussed below and in Chap. 6.

The initial step in the Faddeev-Popov quantization procedure is to fix the gauge. As described in Chap. 2, we can use reparametrizations to go to a gauge in which the metric is equivalent to a fixed reference metric $\hat{h}_{\alpha\beta}$; i.e. our gauge condition is

$$h_{\alpha\beta} = e^{2\phi} \hat{h}_{\alpha\beta} . \quad (3.66)$$

We have also seen that under reparametrizations and Weyl rescalings the changes of the metric can be decomposed as

$$\delta h_{\alpha\beta} = -(P\xi)_{\alpha\beta} + 2\tilde{\Lambda} h_{\alpha\beta} , \quad (3.67)$$

where the operator P maps vectors to symmetric traceless tensors. The covariant derivatives in above expressions are with respect to the metric $h_{\alpha\beta}$, which is also used to raise and lower indices. The integration measure can now be written as

$$\mathcal{D}h = \mathcal{D}(P\xi)\mathcal{D}\tilde{\Lambda} = \mathcal{D}\xi \mathcal{D}\Lambda \left| \frac{\partial(P\xi, \tilde{\Lambda})}{\partial(\xi, \Lambda)} \right| . \quad (3.68)$$

---

[9] Later we will compute scattering amplitudes as correlation functions with this partition function.

3.4 Covariant Path Integral Quantization

The Jacobian is easy to evaluate formally:

$$\left| \frac{\partial(P\xi, \tilde{\Lambda})}{\partial(\xi, \Lambda)} \right| = \left| \det \begin{pmatrix} P & 0 \\ * & 1 \end{pmatrix} \right| = |\det P| = (\det PP^\dagger)^{1/2}, \qquad (3.69)$$

where $*$ is some operator which does not enter the determinant. The integral over reparametrizations simply gives the volume of the diffeomorphism group (more precisely, the volume of the component connected to the identity). This volume depends on the Weyl degree of freedom as the measure $\mathscr{D}\xi$ does. We do, however, assume that all dependence on the conformal factor will eventually drop out in the critical dimension. We thus ignore it and drop the integral over Λ. We then have

$$Z = \int \mathscr{D}X^\mu \, (\det PP^\dagger)^{1/2} \, e^{iS_P[e^{2\phi}\hat{h}_{\alpha\beta}, X^\mu]}. \qquad (3.70)$$

The last step of the Faddeev-Popov procedure is to rewrite the determinant by introducing anti-commuting ghost fields c^α and $b_{\alpha\beta}$, where $b_{\alpha\beta}$ is symmetric and traceless. We then obtain

$$(\det PP^\dagger)^{1/2} = \int \mathscr{D}c\mathscr{D}b \, \exp\left(\frac{1}{2\pi} \int d^2\sigma \sqrt{-h} \, h^{\alpha\beta} \, b_{\beta\gamma} \nabla_\alpha c^\gamma \right), \qquad (3.71)$$

where $h_{\alpha\beta} = e^{2\phi}\hat{h}_{\alpha\beta}$ is the gauge fixed metric. The normalization of the ghost fields are chosen for later convenience. Note that c^α corresponds to infinitesimal reparametrizations and $b_{\alpha\beta}$ to variations perpendicular to the gauge slice. One often refers to $b_{\alpha\beta}$ as the anti-ghost. If we insert Eq. (3.71) into the partition function we get

$$Z = \int \mathscr{D}X^\mu \, \mathscr{D}c \, \mathscr{D}b \, e^{iS[X, \hat{h}, b, c]}, \qquad (3.72)$$

where

$$S = -\frac{1}{4\pi\alpha'} \int d^2\sigma \sqrt{-\hat{h}} \hat{h}^{\alpha\beta} \left\{ \partial_\alpha X^\mu \partial_\beta X_\mu + 2i\alpha' b_{\beta\gamma} \hat{\nabla}_\alpha c^\gamma \right\}. \qquad (3.73)$$

The action is independent of the Weyl factor. If we choose b and c Hermitian, the action is real. For the conformal gauge choice $\hat{h}_{\alpha\beta} = \eta_{\alpha\beta}$ we find

$$S = S[X] + S_{\text{ghost}}[b, c] \qquad (3.74)$$

with

$$S[X] = \frac{1}{\pi\alpha'} \int d^2\sigma \partial_+ X^\mu \partial_- X_\mu,$$

$$S_{\text{ghost}}[b, c] = \frac{i}{\pi} \int d^2\sigma (c^+ \partial_- b_{++} + c^- \partial_+ b_{--}). \qquad (3.75)$$

Since $b_{\alpha\beta}$ is traceless symmetric its only non-vanishing components are b_{++} and b_{--}.

We have to point out that our treatment above left unmentioned several subtle points. One, having to do with the conformal anomaly, was already touched upon and will be taken up shortly. Another issue has to do with reparametrizations which satisfy $(P\xi)_{\alpha\beta} = 0$, i.e. with the possible existence of conformal Killing vector fields. The equation of motion for the c ghosts is the conformal Killing equation, i.e. the c zero modes correspond to diffeomorphisms which can be absorbed by a Weyl rescaling. In the functional integration we are to integrate over each metric deviation only once. Since the ones generated by conformal Killing vectors are already taken care of by the integration over the conformal factor, we have to omit the zero modes from the integration over c.

Another problem concerns the question whether all symmetric traceless metric deformations can be generated by reparametrizations. As we know from Chap. 2, this is not the case if $P^\dagger$ has zero modes; they correspond to zero modes of the b ghosts. If present, they have to be treated separately to get a non-vanishing result, since $\int d\theta 1 = 0$ for θ a Grassmann variable. We will come back to the issue of ghost zero modes in Chap. 6.

The energy momentum tensor of the ghosts fields can be derived from Eq. (3.73) using Eq. (2.22). Dropping hats from now on, we find

$$T_{\alpha\beta} = -i(b_{\alpha\gamma}\nabla_\beta c^\gamma + b_{\beta\gamma}\nabla_\alpha c^\gamma - c^\gamma \nabla_\gamma b_{\alpha\beta} - h_{\alpha\beta} b_{\gamma\delta} \nabla^\gamma c^\delta), \qquad (3.76)$$

which is traceless as a consequence of the Weyl invariance of the ghost action. In the derivation one also has to vary the metric dependence of the covariant derivatives and has to take into account the tracelessness of $b_{\alpha\beta}$, e.g. via a Lagrange multiplier. One can verify that $\nabla^\alpha T_{\alpha\beta} = 0$ if one uses the equations of motion. In light-cone gauge, the non-vanishing components are

$$T_{++} = -i\left(2b_{++}\partial_+ c^+ + (\partial_+ b_{++})c^+\right),$$
$$T_{--} = -i\left(2b_{--}\partial_- c^- + (\partial_- b_{--})c^-\right) \qquad (3.77)$$

and energy-momentum conservation is

$$\partial_- T_{++} = \partial_+ T_{--} = 0. \qquad (3.78)$$

The equations of motion for the ghost fields are

$$\partial_- b_{++} = \partial_+ b_{--} = 0,$$
$$\partial_+ c^- = \partial_- c^+ = 0. \qquad (3.79)$$

They have to be supplemented by periodicity (closed string) and boundary conditions (open string). The periodicity condition is simply $b(\sigma + \ell) = b(\sigma)$ and

3.4 Covariant Path Integral Quantization

likewise for c. In the closed string case the equations of motion imply that b_{++} and c^+ are purely left-moving whereas b_{--} and c^- are purely right-moving. For the open string, the boundary terms which arise in the derivation of the equations of motion vanish if we require

$$b_{01} = c^1 = 0 \tag{3.80}$$

at the two ends of the open string. This is equivalent to $b_{++} = b_{--}$ and $c^+ = c^-$. This choice, which is not the only possible one, has the following interpretation: through the relation between c^α and the parameter of reparametrizations it means that we are restricting to variations which do not move the boundary. The covariant form of these boundary conditions is $n^\alpha \delta c^\beta b_{\alpha\beta}|_{\partial \Sigma} = 0$ with $\delta c^\alpha \sim t^\alpha$, where n^α and t^α are normal and tangential to the boundary of the world-sheet.

The ghost system, consisting of Grassmann odd fields, is quantized by the following canonical anti-commutation relations:

$$\{b_{++}(\sigma, \tau), c^+(\sigma', \tau)\} = 2\pi \delta(\sigma - \sigma'),$$
$$\{b_{--}(\sigma, \tau), c^-(\sigma', \tau)\} = 2\pi \delta(\sigma - \sigma') \tag{3.81}$$

with all others vanishing. We can solve the equations of motion and express the canonical brackets in terms of Fourier modes. We then define the Virasoro operators of the b, c system as the moments of the constraints $T_{++} = T_{--} = 0$.

For the closed string the solutions of the equations of motion, periodic in σ with period ℓ are

$$c^+(\sigma, \tau) = \frac{\ell}{2\pi} \sum_{n=-\infty}^{+\infty} \bar{c}_n e^{-2\pi i n(\tau+\sigma)/\ell},$$

$$c^-(\sigma, \tau) = \frac{\ell}{2\pi} \sum_{n=-\infty}^{+\infty} c_n e^{-2\pi i n(\tau-\sigma)/\ell} \tag{3.82}$$

and

$$b_{++}(\sigma, \tau) = \left(\frac{2\pi}{\ell}\right)^2 \sum_{n=-\infty}^{+\infty} \bar{b}_n e^{-2\pi i n(\tau+\sigma)/\ell},$$

$$b_{--}(\sigma, \tau) = \left(\frac{2\pi}{\ell}\right)^2 \sum_{n=-\infty}^{+\infty} b_n e^{-2\pi i n(\tau-\sigma)/\ell} \tag{3.83}$$

and the canonical anti-commutators become

$$\{b_m, c_n\} = \delta_{m+n},$$
$$\{b_m, b_n\} = \{c_m, c_n\} = 0 \tag{3.84}$$

and likewise for the barred oscillators. Left-moving modes anticommute with right-moving modes. The Virasoro operators are defined as in Eq. (2.83) and we get

$$L_m = \sum_{n=-\infty}^{+\infty} (m-n) \, {}^*_* b_{m+n} \, c_{-n} \, {}^*_* ,$$

$$\bar{L}_m = \sum_{n=-\infty}^{+\infty} (m-n) \, {}^*_* \bar{b}_{m+n} \, \bar{c}_{-n} \, {}^*_* \quad (3.85)$$

Here the normal ordering symbol means that we move the modes c_n, b_n with $n > 0$ to the right, taking due care of minus sign arising from the anticommutativity of the modes. The modes c_0 and b_0 do not appear in L_0, which is the only one of the L_n's which is effected by normal ordering ambiguities. We distinguish this normal ordering symbol from the more familiar one used e.g. in (3.14) and more generally introduced in Chap. 4. Normal ordering is always defined with respect to a choice of vacuum state which defines which oscillator modes are creation and which are annihilation operators. We will encounter two choices of vacua for the ghost system in Chap. 5, one corresponding to the : and one to the ${}^*_*$ prescription.

Hermiticity of b and c entails

$$c_n = c^\dagger_{-n}, \quad b_n = b^\dagger_{-n} \quad (3.86)$$

from which we get

$$L^\dagger_m = L_{-m} \quad (3.87)$$

with identical relations for the left-movers.

We can now compute the commutator of the Virasoro operators to find the algebra they satisfy. Following the same steps as the ones outlined in the appendix, we obtain

$$[L_m, L_n] = (m-n) L_{m+n} + A(m) \delta_{m+n}, \quad (3.88)$$

where the anomaly is

$$A(m) = \frac{1}{12}(-26 m^3 + 2 m), \quad (3.89)$$

i.e. the central charge of the ghost system is $c = -26$.

Let us now look at the combined matter-ghost system. We define the Virasoro operators as the sum of the Virasoro operators for the X^μ fields and the conformal ghost system

$$L_m = L^X_m + L^{gh}_m + a\delta_m, \quad (3.90)$$

where the last term accounts for a normal ordering ambiguity in $L_0^X + L_0^{gh}$. We then get

$$[L_m, L_n] = (m-n) L_{m+n} + A(m) \delta_{m+n} \qquad (3.91)$$

with

$$A(m) = \frac{d}{12} m(m^2 - 1) + \frac{1}{6}(m - 13m^3) - 2am. \qquad (3.92)$$

The first term is due to the X^μ fields ($\mu = 0, \ldots, d-1$), the second is due to the ghosts and the last arises from the shift in L_0. $A(m)$ vanishes if and only if $d = 26$ and $a = -1$. These are precisely the values we found from requiring Lorentz invariance of the theory quantized in light-cone gauge. Here they arose from requiring that the total (ghost plus matter) anomaly of the Virasoro algebra vanishes. This in turn is the condition for conformal symmetry to be preserved in the transition from the classical theory to the quantum theory. We will show at the end of Sect. 4.1 that $c \neq 0$ implies a non-zero trace of the energy-momentum tensor. If $T_{\alpha\beta}$ is no longer traceless, the theory is not Weyl invariant if it is coupled to gravity. Weyl invariance is, however, the condition that all three components of the world-sheet metric are unphysical which is an essential feature of the Polyakov formulation of the string.

The discussion for the open string is almost identical. We will not present it here and only give the mode expansion for the ghost fields:

$$b_{\pm\pm}(\sigma, \tau) = \left(\frac{\pi}{\ell}\right)^2 \sum_{n \in \mathbb{Z}} b_n\, e^{-i\pi n(\tau \pm \sigma)/\ell},$$

$$c^{\pm}(\sigma, \tau) = \frac{\ell}{\pi} \sum_{n \in \mathbb{Z}} c_n\, e^{-i\pi n(\tau \pm \sigma)/\ell}. \qquad (3.93)$$

The Virasoro generators are the same as for one sector of the closed string.

The anomaly in the Virasoro algebra in light-cone gauge is 24, the number of transverse dimensions. Even though in the critical dimension light-cone quantization is completely consistent, we cannot expect the anomaly to vanish since in making this gauge choice we have completely fixed the gauge so that the light-cone action is no longer invariant under the transformations generated by T_{++} and T_{--}, i.e. they are not a symmetry.

3.5 Appendix: The Virasoro Algebra

In this appendix we derive the algebra satisfied by the Virasoro operators. The fact that in the quantum theory the L_n's are normal ordered expressions, requires some care. For purpose of illustration we present some details of the calculation. For notational reasons we use a Euclidean metric so that we do not have to distinguish between upper and lower indices.

The following commutator will be useful:

$$[\alpha_m^i, L_n] = \frac{1}{2} \sum_{p=-\infty}^{+\infty} [\alpha_m^i, :\alpha_p^j \alpha_{n-p}^j :]. \tag{3.94}$$

Here we can drop the normal ordering symbol since α_m^i commutes with c-numbers. Using $[A, BC] = [A, B]C + B[A, C]$ we get

$$[\alpha_m^i, L_n] = \frac{1}{2} \sum_{p=-\infty}^{+\infty} \left\{ [\alpha_m^i, \alpha_p^j]\alpha_{n-p}^j + \alpha_p^j [\alpha_m^i, \alpha_{n-p}^j] \right\}$$

$$= \frac{1}{2} \sum_{p=-\infty}^{+\infty} \left\{ \delta_{m+p} \alpha_{n-p}^j + \delta_{m+n-p} \alpha_p^j \right\} m\delta^{ij} = m\alpha_{m+n}^i. \tag{3.95}$$

Next we write

$$[L_m, L_n] = \frac{1}{2} \sum_{p=-\infty}^{+\infty} [:\alpha_p^i \alpha_{m-p}^i :, L_n] \tag{3.96}$$

and break up the sum to eliminate normal ordering and use (3.95):

$$[L_m, L_n] = \frac{1}{2} \sum_{p=-\infty}^{0} [\alpha_p^i \alpha_{m-p}^i, L_n] + \frac{1}{2} \sum_{p=1}^{+\infty} [\alpha_{m-p}^i \alpha_p^i, L_n]$$

$$= \frac{1}{2} \sum_{p=-\infty}^{0} \left\{ (m-p)\alpha_p^i \alpha_{m+n-p}^i + p\alpha_{n+p}^i \alpha_{m-p}^i \right\}$$

$$+ \frac{1}{2} \sum_{p=1}^{+\infty} \left\{ (m-p)\alpha_{m+n-p}^i \alpha_p^i + p\alpha_{m-p}^i \alpha_{n+p}^i \right\}. \tag{3.97}$$

Now we change the summation variable in the second and forth term to $q = p + n$ and get

$$[L_m, L_n] = \frac{1}{2} \left\{ \sum_{p=-\infty}^{0} (m-p)\alpha_p^i \alpha_{m+n-p}^i + \sum_{q=-\infty}^{n} (q-n)\alpha_q^i \alpha_{m+n-q}^i \right.$$

$$\left. + \sum_{p=1}^{+\infty} (m-p)\alpha_{m+n-p}^i \alpha_p^i + \sum_{q=n+1}^{+\infty} (q-n)\alpha_{m+n-q}^i \alpha_q^i \right\}. \tag{3.98}$$

Let us now assume that $n > 0$ (the case $n \leq 0$ is treated similarly). We then get

3.5 Appendix: The Virasoro Algebra

$$[L_m, L_n] = \frac{1}{2}\Bigg\{\sum_{q=-\infty}^{0}(m-n)\alpha_q^i\alpha_{m+n-q}^i + \sum_{q=1}^{n}(q-n)\alpha_q^i\alpha_{m+n-q}^i$$
$$+ \sum_{q=n+1}^{+\infty}(m-n)\alpha_{m+n-q}^i\alpha_q^i + \sum_{q=1}^{n}(m-q)\alpha_{m+n-q}^i\alpha_q^i\Bigg\}. \quad (3.99)$$

We now notice that except for the second term all terms are already normal ordered (the only critical case is when $m + n = 0$). The second term can be rewritten as

$$\sum_{q=1}^{n}(q-n)\alpha_q^i\alpha_{m+n-q}^i = \sum_{q=1}^{n}(q-n)\alpha_{m+n-q}^i\alpha_q^i + \sum_{q=1}^{n}(q-n)qd\,\delta_{m+n}, \quad (3.100)$$

where $d = \delta_i^i$. Using this we get

$$[L_m, L_n] = \frac{1}{2}\sum_{q=-\infty}^{+\infty}(m-n) : \alpha_q^i\alpha_{m+n-q}^i : + \frac{1}{2}d\sum_{q=1}^{n}(q^2 - nq)\delta_{m+n}. \quad (3.101)$$

If we now use

$$\sum_{q=1}^{n}q^2 = \frac{1}{6}n(n+1)(2n+1) \quad \text{and} \quad \sum_{q=1}^{n}q = \frac{1}{2}n(n+1), \quad (3.102)$$

we finally obtain the Virasoro algebra

$$[L_m, L_n] = (m-n)L_{m+n} + \frac{d}{12}m(m^2-1)\delta_{m+n}. \quad (3.103)$$

In Chap. 4 we will see how conformal field theory provides a simple tool to rederive this algebra. But if we are only interested in the structure of the central term, we can proceed as follows. Since only L_0 is affected by normal ordering subtleties, we know that the central term is of the form $A(m)\delta_{m+n}$. One first shows that $A(-m) = -A(m)$. The Jacobi-identity can then be used to derive $A(m) = A_1 m + A_2 m^3$. One finally determines A_1 and A_2 by evaluating e.g. the $[L_2, L_{-2}]$ commutator on the vacuum state.

Further Reading

The light-cone gauge for the NG string was introduced by

- P. Goddard, J. Goldstone, C. Rebbi, C.B. Thorn, Quantum dynamics of a massless relativistic string. Nucl. Phys. **B56**, 109 (1973)

For the light-cone gauge in the Polyakov formulation see Polchinski, vol. 1 and

- R.E. Rudd, Light cone gauge quantization of 2-D sigma models. Nucl. Phys. **B427**, 81 (1994) [hep-th/9402106]

The no-ghost theorem was proven in

- R.C. Brower, Spectrum generating algebra and no ghost theorem for the dual model. Phys. Rev. **D6**, 1655 (1972)
- P. Goddard, C.B. Thorn, Compatibility of the dual pomeron with unitarity and the absence of ghosts in the dual resonance model. Phys. Lett. **B40**, 235 (1972)

The path integral quantization of the Polyakov string was initiated in

- A.M. Polyakov, Quantum geometry of bosonic strings. Phys. Lett. **B103**, 207 (1981)

and elaborated in

- D. Friedan, in *Introduction to Polyakov's String Theory*, ed. by J.B. Zuber, R. Stora. Recent Advances in Field Theory and Statistical Mechanics, Les Houches (North Holland, Amsterdam, 1982)
- O. Alvarez, Theory of strings with boundaries. Nucl. Phys. **B216**, 125 (1983)

Chapter 4
Introduction to Conformal Field Theory

Abstract This chapter is an introduction to conformal field theory (CFT) which is the basic tool for studying world-sheet properties of string theory. First, we discuss CFT defined on the complex plane, which is relevant for closed strings at tree level. As an application, we then discuss the CFT of a free boson, i.e. string theory in flat space. Finally, we discuss boundary conformal field theory relevant for describing open strings ending on D-branes and non-oriented strings in orientifolds.

4.1 General Introduction

Conformal field theory deals with quantum field theories having conformal symmetry. The conformal group is the subgroup of those general coordinate transformations which preserve the angle between any two vectors. They leave the metric invariant up to a scale transformation. In distinction to higher dimensions the conformal algebra in two dimensions is infinite dimensional, as it is generated by analytic and anti-analytic vector fields. Associated with the infinity of generators is an infinity of conserved charges. That imposes important restrictions on the structure of two-dimensional conformally invariant theories.

One class of physical systems which is described by conformal field theory (CFT) are two dimensional statistical systems showing the behavior of a second order phase transition at a critical temperature T_C. At this particular point, the system has long range correlations, i.e. it has no particular length scale and is in fact conformally invariant. The representation theory of the conformal algebra places constraints on the critical exponents at T_C. A simple such system is the two-dimensional Ising model. We will, however, not cover this application of CFT.

The second important application of conformal field theory is to string theory. We have seen in Chap. 2 that the string action in conformal gauge is invariant under conformal transformations with the infinite dimensional Virasoro algebra as symmetry algebra. The classical solutions of string theory are conformally invariant two dimensional field theories. A particular choice corresponds to a particular vacuum which determines e.g. the number of space-time dimensions, the gauge group, etc. There are of course constraints that a conformal field theory has to satisfy in order to be an acceptable string vacuum. One obvious condition that we have already encountered is the vanishing of the conformal anomaly. Others, coming from modular invariance, spin-statistics etc. will be discussed in subsequent chapters. We can then use methods of conformal field theory to determine the string spectrum and to compute string scattering amplitudes.

In a more general context, e.g. two-dimensional systems at the critical point, there is no need for the CFT to have $c = 0$. In fact, there c has a physical (and measurable) interpretation as Casimir energy. In string theory, which is a theory of gravity on the world-sheet, coupled to matter fields, one needs this condition to maintain in the quantum theory the local world-sheet symmetries, which were needed to gauge away the metric degrees of freedom.

In order for the full power of conformal field theory to be applicable to string theory, we have to continue the signature of the world-sheet metric from Minkowskian to Euclidean. Consider the world-sheet of a free closed string—the cylinder—parametrized by $\sigma \in [0, \ell)$ and $\tau \in (-\infty, +\infty)$. We now make a Wick rotation, i.e. go to imaginary $\tau : \tau \to -i\tau$ or

$$\sigma^{\pm} = \tau \pm \sigma \to -i(\tau \pm i\sigma). \tag{4.1}$$

Define complex coordinates on the cylinder

$$w = \tau - i\sigma,$$
$$\bar{w} = \tau + i\sigma. \tag{4.2}$$

We can then map the cylinder to the complex plane via the conformal transformation

$$z = e^{\frac{2\pi}{\ell}w} = e^{\frac{2\pi}{\ell}(\tau - i\sigma)},$$
$$\bar{z} = e^{\frac{2\pi}{\ell}\bar{w}} = e^{\frac{2\pi}{\ell}(\tau + i\sigma)}. \tag{4.3}$$

This is illustrated in Fig. 4.1. The conformal map from the cylinder to the (punctured) plane will not change the theory if it is conformally invariant. This will be the case for string theory in the critical dimension.

Having defined the theory on the complex plane, we can use powerful techniques of complex analysis. Lines of equal time τ are mapped into circles around the origin. Integrals over σ will be replaced by contour integrals around the origin. The infinite past becomes $z = 0$ and the infinite future $z = \infty$. In addition, σ-translations become rotations: $\sigma \to \sigma + \frac{\ell}{2\pi}\theta \Rightarrow z \to e^{-i\theta}z$ and time translations become scale transformations, which are also called dilations: $\tau \to \tau + \frac{\ell}{2\pi}a \Rightarrow z \to e^a z$. In the quantized theory the generator of dilations will take the role of the Hamiltonian and time ordering will be replaced by radial ordering. Equal time commutators will be equal radius commutators. This is known as radial quantization. Products of fields

4.1 General Introduction

Fig. 4.1 Conformal map from the cylinder to the complex plane

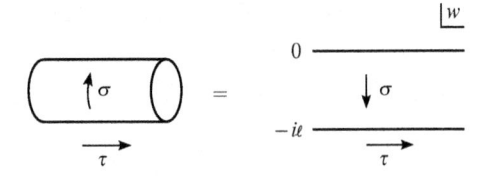

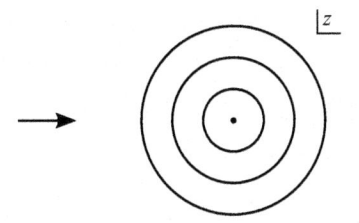

are only defined if we put them in radial order. Radial order is defined in analogy with time order in ordinary quantum field theory:

$$R\left(\phi_1(z)\,\phi_1(w)\right) = \begin{cases} \phi_1(z)\,\phi_2(w) \text{ for } |z| > |w| \\ \phi_2(w)\,\phi_1(z) \text{ for } |w| > |z|. \end{cases} \quad (4.4)$$

There will be a relative minus sign for the case of two anti-commuting fields. Products of operators will always be assumed to be R-ordered and we drop the ordering symbol. The necessity to put operators in radial order will be illustrated below. The equal radius commutator is then defined by

$$[\phi_1(z), \phi_2(w)]_{|z|=|w|} = \lim_{\delta \to 0} \left\{ (\phi_1(z)\,\phi_2(w))\,|_{|z|=|w|+\delta} - (\phi_2(w)\,\phi_1(z))\,|_{|z|=|w|-\delta} \right\}. \quad (4.5)$$

After Wick rotation Eq. (4.1) and application of the map (4.3), right- and left-moving is replaced by holomorphic in z and $\bar{z}$, respectively. We will use both terminologies interchangeably. Also, we will call fields holomorphic in $\bar{z}$ anti-holomorphic or anti-analytic. Most expressions of Chaps. 2 and 3 which were expressed in isothermal coordinates are unchanged if we replace σ^- by z and σ^+ by $\bar{z}$ and include the Jacobian factors from the map (4.3). For instance, the non-vanishing components of the energy-momentum tensor are now T_{zz} and $T_{\bar{z}\bar{z}}$.

The basic objects of a conformal field theory are the conformal fields, also called primary fields, $\phi(z, \bar{z})$. Consider a conformal transformation $z \to z' = f(z)$, $\bar{z} \to \bar{z}' = \bar{f}(\bar{z})$. Primary fields transform as tensors under conformal transformations[1]:

$$\phi(z, \bar{z}) \to \phi'(z', \bar{z}') = \left(\frac{\partial z'}{\partial z}\right)^{-h} \left(\frac{\partial \bar{z}'}{\partial \bar{z}}\right)^{-\bar{h}} \phi(z, \bar{z}). \quad (4.6)$$

---
[1] $\phi'(z', \bar{z}')\,(dz')^h\,(d\bar{z}')^{\bar{h}} = \phi(z, \bar{z})\,(dz)^h\,(d\bar{z})^{\bar{h}}$

All other fields are called secondary fields. Under infinitesimal transformations[2]

$$z' = z + \xi(z), \quad \bar{z}' \to \bar{z} + \bar{\xi}(\bar{z}) \tag{4.7}$$

primary fields transform as

$$\phi(z, \bar{z}) \to \phi'(z, \bar{z}) = \phi(z, \bar{z}) + \delta_{\xi, \bar{\xi}} \phi(z, \bar{z}) \tag{4.8}$$

with

$$\delta_{\xi, \bar{\xi}} \phi(z, \bar{z}) = -\left(h\, \partial \xi + \bar{h}\, \bar{\partial} \bar{\xi} + \xi\, \partial + \bar{\xi}\, \bar{\partial}\right) \phi(z, \bar{z}). \tag{4.9}$$

We have introduced the notation $\partial = \frac{\partial}{\partial z}$ and $\bar{\partial} = \frac{\partial}{\partial \bar{z}}$. h and $\bar{h}$ are called the conformal weights of ϕ under analytic and anti-analytic transformations. Tensors with $\bar{h} = 0$ ($h = 0$) are called holomorphic (anti-holomorphic) tensors. Note that $\bar{h}$ does not denote the complex conjugate of h. Purely left- or right-moving fields are called chiral. If we want to make a distinction between left- and right-moving chiral fields we call the latter anti-chiral. (Anti)chiral fields necessarily have $\bar{h} = 0$ ($h = 0$). Under a rescaling (dilation) $z \to \lambda z$, $\lambda \in \mathbb{R}$, $\phi'(\lambda z, \lambda \bar{z}) = \lambda^{-(h+\bar{h})} \phi(z, z)$ and $h + \bar{h}$ is called the scaling dimension of ϕ. In radial quantization the generator of dilations plays the role of the Hamiltonian and the scaling dimension is related to the energy. Under rotations $z \to e^{i\theta} z$, $\theta \in \mathbb{R}$ one finds $\phi'(e^{i\theta} z, e^{-i\theta} \bar{z}) = e^{-i(h-\bar{h})\theta} \phi(z, \bar{z})$ and $h - \bar{h}$ is referred to as the conformal spin of ϕ.

Consider the map (4.3) from the cylinder to the complex plane. Applying Eq. (4.6), chiral fields on the cylinder and the plane are related as follows:

$$\phi_{\text{plane}}(z) = \left(\frac{\ell}{2\pi}\right)^h z^{-h} \phi_{\text{cyl}}(w). \tag{4.10}$$

The generalization to arbitrary primary fields is obvious. If $\phi_{\text{cyl}}(w)$ has a mode expansion

$$\phi_{\text{cyl}}(w) = \left(\frac{2\pi}{\ell}\right)^h \sum_{n \in \mathbb{Z}} \phi_n\, e^{-\frac{2\pi}{\ell} nw} \tag{4.11}$$

then the mode expansion on the complex plane is

$$\phi_{\text{plane}}(z) = \sum_{n \in \mathbb{Z}} \phi_n\, z^{-n-h} \tag{4.12}$$

---

[2]On the Riemann sphere $\mathbb{C} \cup \infty$ the only analytic functions are the constants. When we speak about infinitesimal conformal transformations, we mean that $\xi(z)$ is analytic and small in the region where the operators are inserted. Outside this region, they are also small but not necessarily analytic.

4.1 General Introduction

with the same coefficients ϕ_n. From now on, unless otherwise stated, all fields will be on the complex plane. The inverse of Eq. (4.12) is

$$\phi_n = \oint_{C_0} \frac{dz}{2\pi i} \, \phi(z) \, z^{n+h-1} \,, \tag{4.13}$$

where the integration is counterclockwise around the origin. The value of the integral is independent of the specific contour as long as it is running around the origin. For fields to be single valued on the complex plane the mode numbers n have to be such that $n + h \in \mathbf{Z}$. They will, however, always be integer spaced. Note that due to the Jacobian factor of the map (4.3) single valuedness on the plane does not mean single valuedness on the cylinder and vice versa.

We already know from Chap. 2 that in a conformally invariant theory the energy-momentum tensor is traceless, i.e. $T_\alpha^\alpha = 0$. Expressed in conformal coordinates this reads

$$T_{z\bar{z}} = 0 \,. \tag{4.14}$$

This, together with energy-momentum conservation

$$\partial_{\bar{z}} T_{zz} + \partial_z T_{\bar{z}z} = 0 \,, \qquad \partial_z T_{\bar{z}\bar{z}} + \partial_{\bar{z}} T_{z\bar{z}} = 0 \tag{4.15}$$

shows that in a conformally invariant theory we have

$$\partial_{\bar{z}} T_{zz} = 0 \,, \qquad \partial_z T_{\bar{z}\bar{z}} = 0 \,. \tag{4.16}$$

The two non-vanishing components of the energy-momentum tensor of a conformal field theory are analytic and anti-analytic functions, respectively. We will use the notation $T(z) = T_{zz}(z)$ and $\overline{T}(\bar{z}) = T_{\bar{z}\bar{z}}(\bar{z})$. T and $\overline{T}$ are therefore chiral and anti-chiral fields, respectively. From the conservation law Eq. (4.16) we immediately find that, if $T(z)$ is conserved, so is $\xi(z) T(z)$ as long as ξ depends only analytically on its argument.[3] This infinity of conserved currents is equivalent to the statement at the beginning of this chapter that the conformal algebra in two dimensions is infinite dimensional. With each current we associate a conserved charge

$$T_\xi = \oint_{C_0} \frac{dz}{2\pi i} \, \xi(z) \, T(z) \tag{4.17}$$

which generates infinitesimal conformal transformations

$$z \to z' = z + \xi(z) \tag{4.18}$$

---

[3] With indices made explicit, the conserved currents are $\xi^z(z) T_{zz}(z)$.

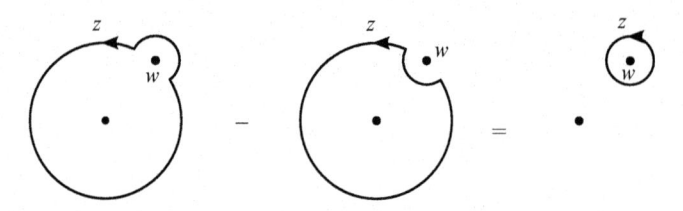

Fig. 4.2 Integration contours in Eq. (4.21)

with a similar expression for the anti-analytic component $\overline{T}$. From now on we will restrict our attention to chiral fields, say right-moving ones.

It is important to keep in mind that, due to the holomorphicity of $T(z)$, we obtain an infinite number of conserved charges, here parametrized by analytic functions $\xi(z)$. In general, if we have a conserved current $\partial_\alpha j^\alpha = 0$ or, in conformal coordinates, $\partial_{\bar{z}} j_z + \partial_z j_{\bar{z}} = 0$, the only conserved charge is the combination

$$Q = \frac{1}{2\pi i} \oint dz\, j_z + \frac{1}{2\pi i} \oint d\bar{z}\, j_{\bar{z}} \tag{4.19}$$

with the convention that both contour integrals are taken in the counter clockwise orientation, in the z- and $\bar{z}$-plane, respectively.

Returning to the energy-momentum tensor, the transformation Eq. (4.9) is implemented by the commutator of $\phi(z)$ and T_ξ

$$\delta_\xi \phi(w) = -[T_\xi, \phi(w)]. \tag{4.20}$$

Using the prescription of radial ordering in Eq. (4.5), this gives

$$\delta_\xi \phi(w) = -\oint_{\substack{C_0 \\ |z|>|w|}} \frac{dz}{2\pi i}\, \xi(z)\, T(z)\, \phi(w) + \oint_{\substack{C_0 \\ |z|<|w|}} \frac{dz}{2\pi i}\, \xi(z)\, T(z)\, \phi(w)$$

$$= -\oint_{C_w} \frac{dz}{2\pi i}\, \xi(z)\, T(z)\, \phi(w). \tag{4.21}$$

The contours are shown in Fig. 4.2.

For the anti-holomorphic part one similarly gets

$$\delta_{\bar{\xi}} \phi(\bar{w}) = \oint_{C_w} \frac{d\bar{z}}{2\pi i}\, \bar{\xi}(\bar{z})\, \overline{T}(\bar{z})\, \phi(\bar{w}). \tag{4.22}$$

where the path C_w is counter-clockwise in the z-plane, i.e. clockwise in the $\bar{z}$-plane.

Recall that all operator products are assumed to be radially ordered. Comparing Eq. (4.21) with Eq. (4.9) for $\bar{h} = 0$ and $\bar{\partial}\phi = 0$ we find with the help of the Cauchy-Riemann formula

4.1 General Introduction

$$\oint_{C_z} \frac{dw}{2\pi i} \frac{f(w)}{(w-z)^n} = \frac{1}{(n-1)!} f^{(n-1)}(z) \tag{4.23}$$

that any conformal field must have the following (*R*-ordered) operator product with $T(z)$:

$$T(z)\phi(w) = \frac{h\phi(w)}{(z-w)^2} + \frac{\partial \phi(w)}{(z-w)} + \text{finite terms}. \tag{4.24}$$

Therefore, instead of Eq. (4.6), the operator product with the energy-momentum tensor can serve as the definition of a primary field of weight h.

Equation (4.24) is our first example of an operator product expansion (OPE) of two fields. The basic idea is that if $\{O_i\}$ is a complete set of local operators with definite scaling dimensions, then the product of two operators can be expanded as

$$O_i(z,\bar{z}) O_j(w,\bar{w}) = \sum_k C_{ij}{}^k\big((z-w)\big) O_k(w,\bar{w}). \tag{4.25}$$

A simple example is the product of any operator with the identity operator

$$\phi(z) \mathrm{I} = \sum_{n=0}^{\infty} \frac{(z-w)^n}{n!} \partial_w^n \phi(w). \tag{4.26}$$

Only the first term in the sum is a primary field, i.e. ϕ itself, all others are called descendants of ϕ.

The operator product is associative and for Grassmann even fields commutative. Covariance under rescaling specifies the structure functions up to numerical constants:

$$C_{ij}{}^k\big((z-w)\big) = (z-w)^{h_k - h_i - h_j} (\bar{z}-\bar{w})^{\bar{h}_k - \bar{h}_i - \bar{h}_j} C_{ij}{}^k \tag{4.27}$$

where h_i are the scaling dimensions of the fields which are not necessarily primary. Operator products should always be thought as inserted into correlation functions (cf. below). The radius of convergence of the operator product expansion is restricted by the positions of the other operators in the correlation function. Completeness of the set of operators $\{O_i\}$ means that any state can be generated by their linear action. Note that the set of chiral operators closes under the operator product expansion, i.e. the chiral fields form a subalgebra.

We now examine the conformal transformation properties of the energy-momentum tensor. Using the commutation properties of infinitesimal conformal transformations

$$[\delta_{\xi_1}, \delta_{\xi_2}] = \delta_{(\xi_2 \partial \xi_1 - \xi_1 \partial \xi_2)} \tag{4.28}$$

we find

$$T(z)\,T(w) = \frac{c/2}{(z-w)^4} + \frac{2\,T(w)}{(z-w)^2} + \frac{\partial T(w)}{(z-w)} + \text{finite terms}. \qquad (4.29)$$

The first term is allowed by Eq. (4.28) and is consistent with Bose symmetry and scale invariance. We can rewrite Eq. (4.29) in an equivalent way as

$$\delta_\xi T(z) = -2\partial\xi(z)\,T(z) - \xi(z)\,\partial T(z) - \frac{c}{12}\,\partial^3\xi(z). \qquad (4.30)$$

We see that $T(z)$ transforms as a tensor of weight two under those transformations for which $\partial^3\xi(z) = 0$ but fails to do so for general conformal transformations if $c \neq 0$. Classically c is zero and $c \neq 0$ represents a conformal anomaly, a purely quantum mechanical effect. We will elaborate on the relation between the central charge of the Virasoro algebra and the breakdown of Weyl invariance at the end of this section. Note that the scaling dimension of $T(z)$ does not get modified by quantum effects. The reason is that for a rescaling $\xi \propto z$ the last term in (4.30) vanishes. One should compare Eq. (4.29) with Eq. (3.18).

Let us now show that the OPE $T(z)\,T(w)$ is equivalent to the Virasoro algebra with central charge c. For this purpose, we expand $T(z)$ in modes

$$T(z) = \sum_n z^{-n-2}\,L_n \qquad (4.31)$$

which in turn gives

$$L_n = \oint \frac{dz}{2\pi i}\,z^{n+1}\,T(z), \qquad (4.32)$$

where L_n's are the Virasoro generators. They satisfy the Hermiticity relation

$$(L_n)^\dagger = L_{-n} \qquad (4.33)$$

which follows from the reality of the energy-momentum tensor in Minkowski space (c.f. Chap. 2). We remark that the Hermitian conjugate of a chiral field of weight h with mode expansion (4.12) is generally defined by

$$[\phi(z)]^\dagger = \phi^\dagger\left(\frac{1}{\bar z}\right)\frac{1}{\bar z^{2h}}. \qquad (4.34)$$

For the modes this means

$$(\phi^\dagger)_{-n} = (\phi_n)^\dagger. \qquad (4.35)$$

The generalization to non-chiral fields $\phi(z,\bar z)$ is straightforward. The conformal properties of a field and its Hermitian conjugate are identical and we can, if we

4.1 General Introduction

wish, choose Hermitian combinations. Note that a real field in Minkowski space corresponds to a Hermitian field on the complex plane, i.e. it satisfies $\phi^\dagger = \phi$ and therefore $(\phi_n)^\dagger = \phi_{-n}$. As a consequence, also for real fields there is no relation between the holomorphic and anti-holomorphic sector of the CFT on the complex plane. Consider the continuation back to the Minkowski space cylinder. The missing factors of i in Euclidean time-evolution, $\phi(\sigma, \tau) = e^{H\tau}\phi(\sigma, 0)e^{-H\tau}$, must be compensated in the definition of the adjoint by an explicit time reversal $\tau \to -\tau$. On the complex plane this is $z \to 1/\bar{z}$.

The Virasoro algebra is now easily obtained, using Eq. (4.29) as the difference of a double contour integral:

$$[L_m, L_n] = \oint_{C_0} \frac{dw}{2\pi i} \oint_{C_w} \frac{dz}{2\pi i} z^{m+1} w^{n+1} \left[\frac{c/2}{(z-w)^4} + \frac{2T(w)}{(z-w)^2} + \frac{\partial T(w)}{(z-w)}\right]$$

$$= \frac{c}{12} m(m^2 - 1)\delta_{n+m,0} + (m-n) L_{m+n}. \quad (4.36)$$

$\bar{T}(\bar{z})$ or, equivalently, the $\bar{L}_n$ satisfy identical algebras as $T(z)$ and the L_n. The two algebras commute, i.e. $[L_n, \bar{L}_m] = 0$. Thus we have shown that the singular terms in the OPE of the energy-momentum tensor are equivalent to the Virasoro algebra. As we have mentioned, the set of chiral fields closes under OPE so that the singular terms give rise to a chiral algebra. In fact such extensions of the Virasoro algebra by other chiral primaries describe extended symmetry algebras of CFTs and are denoted as $\mathscr{A} \otimes \bar{\mathscr{A}}$. They are also called $\mathscr{W}$-algebras.

The L_n's act as the generators of all possible conformal transformations. A primary field is defined via the L_n's as

$$[L_n, \phi(z)] = z^n [z\partial + (n+1)h] \phi(z) \quad (4.37)$$

or, in terms of the modes of ϕ:

$$[L_n, \phi_m] = (n(h-1) - m) \phi_{n+m}. \quad (4.38)$$

Comparing Eq. (4.58b) with Eq. (4.9), we see that L_n is associated with the infinitesimal transformation $\epsilon(z) = -z^{n+1}$. In particular, L_0, L_1, L_{-1} generate infinitesimal transformations $\delta z = \alpha + \beta z + \gamma z^2$; they are the generators of $SL(2, \mathbb{C})$ (for $\alpha, \beta, \gamma \in \mathbb{C}$), the maximal closed finite dimensional subalgebra of the Virasoro algebra. The finite transformations are

$$z \to z' = \frac{az+b}{cz+d}$$

$$\begin{pmatrix} a & b \\ c & d \end{pmatrix} \in SL(2, \mathbb{C}) \quad \text{i.e.} \quad \begin{matrix} a,b,c,d \in \mathbb{C}, \\ ad - bc = 1, \end{matrix} \quad (4.39)$$

from which we infer that a $\mathbb{Z}_2$ reflection $(a, b, c, d) \to (-a, -b, -c, -d)$ does not change the conformal map so that the actual conformal group is $PSL(2, \mathbb{C}) = SL(2, \mathbb{C})/\mathbb{Z}_2$. Indeed, if we expand (4.39) around $a = d = 1, b = c = 0$ we obtain $\delta z = \delta b + 2\delta a\, z - \delta c\, z^2$. The anti-holomorphic modes $\overline{L}_0, \overline{L}_1, \overline{L}_{-1}$ generate a second $PSL(2, \mathbb{C})$. $PSL(2, \mathbb{C})$ transformations are the only globally defined invertible one-to-one conformal mappings of the Riemann sphere $\mathbb{C} \cup \infty$ onto itself. This is in fact easy to see. The transformations δ_ϵ are generated by the vector fields $\epsilon(z)\,\partial_z$; only those for which $\partial^3 \epsilon(z) = 0$ are defined at $z = 0$ and at $z = \infty$ (for $z = \infty$ one uses the map $w = 1/z$). The transformations (4.39) are called fractional linear or Möbius transformations. If we treat z and $\bar{z}$ as independent variables, as we often do, the globally defined conformal transformation act independently on z and $\bar{z}$ and they form the group $PSL(2, \mathbb{C}) \times PSL(2, \mathbb{C})$. If we treat them as complex conjugates, it is $PSL(2, \mathbb{C}) \simeq SO(3, 1; \mathbb{R})/\mathbb{Z}_2$ and $\bar{z} = z^*$ defines the real surface.

A special class of secondary fields are those which transform as tensors under Möbius transformations. They are called quasi-primary fields. The energy-momentum tensor is an example of a quasi-primary field as the last term in (4.30) vanishes for $sl(2, \mathbb{C})$ transformations. As usual, we denote the group by capital and the algebra by small letters.

The behavior of $T(z)$ under finite conformal transformations $z \to w(z)$ can be obtained as follows. From Eq. (4.30) we know that it must be of the form

$$T(z) \to T'(w) = \left(\frac{dw}{dz}\right)^{-2} \left\{ T(z) - \frac{c}{12}\{w, z\} \right\}, \qquad (4.40)$$

where $\{z, w\}$ must satisfy the following properties:

(i) $\{z + \epsilon, z\} = -\epsilon'''(z) + \mathcal{O}(\epsilon^2)$

(ii) $\{w, z\} = 0 \iff w = \dfrac{az + b}{cz + d}, \quad ad - bc \neq 0$

(iii) $\{u, z\} = \left(\dfrac{dw}{dz}\right)^2 \{u, w\} + \{w, z\}$

(iv) $\{w, z\} = -\left(\dfrac{dw}{dz}\right)^2 \{z, w\}.$ \hfill (4.41)

With the help of (iv) we can rewrite (4.40) in the form

$$T'(w) = \left(\frac{dz}{dw}\right)^2 T(z) + \frac{c}{12}\{z, w\}. \qquad (4.42)$$

Property (i) follows from comparison with Eq. (4.30); property (ii) is a consequence of the fact that T is a quasi-primary field; (iii) follows from considering two consecutive transformations $z \to w \to u$ and property (iv) from (iii) with $u = z$. Taking (iii) with $u = w + \delta w$ one finds

4.1 General Introduction

$$\delta\{w, z\} \equiv \{w + \delta w, z\} - \{w, z\} = (w')^2 \partial_w^3 \delta w, \tag{4.43}$$

where $w' = \partial_z w$. Using the chain rule to write $\partial_w = \frac{1}{w'}\partial_z$ we find

$$\delta\{w, z\} = \delta\left\{\frac{w'''}{w'} - \frac{3}{2}\left(\frac{w''}{w'}\right)^2\right\} \tag{4.44}$$

and, upon integration,

$$\{w, z\} = \frac{w'''}{w'} - \frac{3}{2}\left(\frac{w''}{w'}\right)^2. \tag{4.45}$$

Consistency with (i) means that the integration constant vanishes. $\{w, z\}$ is the Schwarzian derivative. It is the only object with properties (i)–(iv). For the map from the cylinder to the plane Eq. (4.40) gives

$$T_{\text{cyl}}(w) = \left(\frac{2\pi}{\ell}\right)^2 \left(z^2 T_{\text{plane}}(z) - \frac{c}{24}\right). \tag{4.46}$$

In particular for L_0 we find

$$(L_0)_{\text{cyl}} = (L_0)_{\text{plane}} - \frac{c}{24}. \tag{4.47}$$

Let us now consider the Hilbert space and some representation theory of conformal field theories. Denote the in-vacuum by $|0\rangle$. Regularity of the energy-momentum tensor at $z = 0$ ($\tau = -\infty$), i.e.

$$\lim_{z \to 0} \sum_{n \in \mathbb{Z}} L_n z^{-n-2} |0\rangle = \text{reg.} \tag{4.48}$$

requires

$$L_n |0\rangle = 0 \quad \text{for} \quad n > -2. \tag{4.49}$$

The L_n's with $n > -2$ generate conformal transformations which are regular at the origin. To get the conditions on the out-vacuum $\langle 0|$ following from regularity at $z = \infty$ ($\tau = +\infty$) we map the point at infinity to the origin via $w = 1/z$. The mode expansion of T is then $T'(w) = \sum w^{n-2} L_n$ and we find

$$\langle 0| L_n = 0 \quad \text{for} \quad n < 2. \tag{4.50}$$

Equations (4.49) and (4.50) are Hermitian conjugates of each other. Identical relations also hold for the $\overline{L}_n$. The generators of $SL(2, \mathbb{C})$ annihilate both the in- and the out-vacuum. We refer to this vacuum as the $SL(2, \mathbb{C})$ invariant vacuum. Similarly, the requirement of regularity at $z = 0$ and $z = \infty$ leads for a primary field of weight h with mode expansion as in Eq. (4.12) to

$$\phi_n |0\rangle = 0 \quad \text{for } n > -h,$$
$$\langle 0| \phi_n = 0 \quad \text{for } n < h. \tag{4.51}$$

Remarkably, for $h < 0$ there are modes of ϕ which annihilate neither the in- nor the out-vacuum. We will see below that unitarity restricts the conformal weights to $h \geq 0$. This restriction is avoided by the non-unitary ghost system. The c-ghost has $h = -1$ and indeed the three zero modes c_{-1}, c_0 and c_1 do not annihilate the $SL(2, \mathbb{C})$ invariant vacuum. We will come back to this in the next chapter.

Let us now construct the asymptotic in- and out-states of the conformal field theory. Since the time $\tau \to -\infty$ on the cylinder is mapped to the origin on the z-plane, it is natural to define in-states as

$$|\phi_{\text{in}}\rangle \equiv |\phi\rangle = \lim_{z \to 0} \phi(z)|0\rangle = \phi(0)|0\rangle = \phi_{-h}|0\rangle, \tag{4.52}$$

where

$$\phi_{-h} = \oint_{C_0} \frac{dz}{2\pi i} \frac{1}{z} \phi(z). \tag{4.53}$$

This 1-1 correspondence between states $|\phi\rangle$ and operators $\phi(z)$ is called the state-operator correspondence. With

$$\phi(z) = e^{zL_{-1}} \phi(0) e^{-zL_{-1}} \tag{4.54}$$

one obtains $\phi(z)|0\rangle = e^{zL_{-1}}|\phi\rangle$.

To define the out-states we have to construct the analogous objects for $z \to \infty$. If we require $\langle \phi_{\text{out}}| = |\phi_{\text{in}}\rangle^\dagger$, then Eqs. (4.34) and (4.35) lead to the following definition for $\langle \phi_{\text{out}}|$:

$$\langle \phi_{\text{out}}| = \lim_{z \to \infty} \langle 0| \phi^\dagger(z) z^{2h} = \langle 0|(\phi^\dagger)_h. \tag{4.55}$$

We define the inner product

$$\langle \phi_i | \phi_j \rangle = \lim_{\substack{z \to 0 \\ w \to \infty}} w^{2h_j} \langle 0| \phi_j^\dagger(w) \phi_i(z)|0\rangle \tag{4.56}$$

and normalize $\langle 0|0\rangle = 1$.

As we will see below, when we consider correlation functions, another notion of an out-state, the so called BPZ-conjugate state to $|\phi\rangle$, is relevant. It is defined as

$$\langle \phi| = \lim_{z \to \infty} \langle 0| \phi(z) z^{2h} \tag{4.57}$$

4.1 General Introduction

so that no Hermitian conjugation is involved. While our discussion was for chiral fields, the state-operator correspondence also holds for non-chiral fields according to $|\phi\rangle = \lim_{z,\bar{z} \to 0} \phi(z,\bar{z})|0\rangle$.

Since ϕ is primary, one derives from Eqs. (4.38) and (4.52)

$$L_0|\phi\rangle = h|\phi\rangle, \tag{4.58a}$$

$$L_n|\phi\rangle = 0 \quad \forall\, n > 0. \tag{4.58b}$$

In addition one gets

$$L_0\Big(L_{-n}|\phi\rangle\Big) = (n+h)\Big(L_{-n}|\phi\rangle\Big) \qquad \text{for } n \geq 0, \tag{4.59}$$

i.e. the L_{-n} ($n \geq 0$) raise the eigenvalue of L_0. For quasi-primary fields (4.58b) is replaced by the single condition $L_1|\phi\rangle = 0$. If one wants to check whether a given field is primary, it suffices to verify (4.58a) and $L_1|\phi\rangle = L_2|\phi\rangle = 0$. The Virasoro algebra then guarantees that (4.58b) holds for all $n \geq 1$.

States which satisfy Eq. (4.58) should be called 'lowest weight states'; however, in analogy to the terminology used in the representation theory of Lie algebras they are called highest weight states of the Virasoro algebra. Via (4.52) we have thus established a correspondence between conformal fields and highest weight states. In fact, as we will see below successive action of the modes L_n, $n < 0$ on the state $|\phi\rangle$ amounts to taking derivatives and normal ordered products with the energy-momentum tensor. Moreover, the vacuum $|0\rangle$ is itself a highest weight state; in a unitary theory ($h \geq 0$, cf. below) it has the lowest eigenvalue of the 'Hamiltonian' L_0. Highest weight states with different L_0 eigenvalues are orthogonal.

The complete Hilbert space is obtained by acting with the raising operators L_{-n} ($n > 0$) on highest weight states. The new states obtained in this way are called descendant states. On each highest weight state $|\phi\rangle$ we can act with a whole tower of L_{-n} modes. Using the Virasoro algebra, we can order the modes and get the linear independent states in the Hilbert space

$$|\phi^{k_1\ldots k_m}\rangle = L_{-k_1}\cdots L_{-k_m}|\phi\rangle, \qquad k_1 \geq k_2 \geq \ldots \geq k_m > 0 \tag{4.60}$$

with L_0 eigenvalue $h + \sum_i k_i$. The collection of these states is called the Verma module $V(c,h)$ of the highest weight state $|\phi_j\rangle$. If we choose a basis of highest weight states such that $\langle \phi_i|\phi_j\rangle = 0$ for $\phi_j \neq \phi_i^\dagger$, states in different Verma modules are orthogonal to each other as are states in the same Verma module but at different levels.

Descendant states are created from the vacuum by descendant fields which are not primary but rather secondary operators. They are contained in the operator product of the primary fields with the energy-momentum tensor:

$$T(z)\,\phi(w) = \sum_{k=0}^{\infty}(z-w)^{-2+k}\,\phi^{(k)}(w) \tag{4.61}$$

i.e.
$$\phi^{(k)}(w) = \oint \frac{dz}{2\pi i} (z-w)^{1-k} T(z) \phi(w) \equiv \hat{L}_{-k} \phi(w). \quad (4.62)$$

In particular
$$\phi^{(0)}(z) = \hat{L}_0 \phi(z) = h \phi(z),$$
$$\phi^{(1)}(z) = \hat{L}_{-1} \phi(z) = \partial \phi(z). \quad (4.63)$$

The descendants for $k \geq 2$ appear in the regular terms of the operator products Eq. (4.61) and can be understood as normal ordered products of $T(z)$ and $\phi(w)$. The fields $\phi^{(k)}$ do not exhaust the descendants of the primary field ϕ. The operator product $T(z)\phi^{(k_1)}(w)$ contains the fields $\phi^{(k_1,k_2)}(w)$ and so on. For the descendant field that creates the state (4.60) we get

$$\phi^{\{k\}}(z) = \hat{L}_{-k_1} \ldots \hat{L}_{-k_m} \phi(z). \quad (4.64)$$

The fields constitute the conformal family $[\phi]$. We have already encountered one example of a secondary field, namely the energy-momentum tensor $T(z)$. It is in the conformal family of the identity operator $[I]$ which is present in any conformal field theory. Indeed

$$I^{(2)}(z) = \hat{L}_{-2} I(z) = \oint_{C_z} \frac{dw}{2\pi i} \frac{T(w)}{w-z} I(z) = T(z). \quad (4.65)$$

A state is defined to be in the n'th level of the Virasoro algebra if its L_0 eigenvalue is $h + n$. Thus the n'th level is spanned by the vectors of (4.60) with $\sum k_i = n$. There are $P(n)$ such states, where $P(n)$ is the number of ways to write n as a sum of positive integers, i.e. the number of partitions of n. One easily convinces oneself that the generating function for the partitions $P(n)$ is

$$\sum_{n=0}^{\infty} P(n) q^n = (1 + q + q^2 + \ldots)(1 + q^2 + q^4 + \ldots)(1 + q^3 + \ldots) \cdots$$
$$= \prod_{n=1}^{\infty} \frac{1}{(1-q^n)}, \quad (4.66)$$

where we have defined $P(0) = 1$.

The asymptotic growth of the number of partitions of n is extremely fast. It is given by the Hardy-Ramanujan formula

$$P(n) \stackrel{n \to \infty}{\sim} \frac{1}{4\sqrt{3}n} e^{\pi \sqrt{\frac{2n}{3}}} \left(1 + \mathcal{O}\left(\frac{\ln n}{n^{1/4}}\right)\right). \quad (4.67)$$

4.1 General Introduction

The character of a representation of the Virasoro algebra is a generating function for the degeneracy $a(n)$ of states at all levels:

$$\chi_j(\tau) = \text{Tr}_j\, q^{L_0 - \frac{c}{24}} = q^{h_j - \frac{c}{24}} \sum_{n=0}^{\infty} a(n)\, q^n, \qquad q = e^{2\pi i \tau}$$

$$= q^{h_j - \frac{c}{24}} \prod_{n=1}^{\infty} (1-q^n)^{-1}, \qquad (4.68)$$

where the trace is over all members of the conformal family $[\phi_j]$. In Chap. 6 the identification of τ with the complex structure modulus of the world-sheet torus, relevant for one-loop string calculations, will be explained. Considering the torus as a cylinder whose two ends have been identified, the relation (4.47) explains the shift by $-\frac{c}{24}$.

The character of a generic Verma module $V(c,h)$ has $a(n) = P(n)$. However, it might happen that for certain values of the central charge and highest weight h_j there appear states $|\chi\rangle$ in $V(c,h_j)$ which are themselves annihilated by all L_n, $n > 0$. Such states, called singular vectors or null vectors, are orthogonal to all states in $V(c,h_j)$, in particular they have zero norm. Furthermore, they create their own Verma module which is contained in $V(c,h_j)$ and all states in these submodules have zero norm as well. If a Verma module contains singular vectors it is therefore not irreducible. One constructs an irreducible representation of the Virasoro algebra by taking equivalence classes of states, where states which differ by a state of zero norm (at the same level) are identified. The characters of irreducible representations of the Virasoro algebra therefore satisfy $a(n) \leq P(n)$.

One can show that for $c > 1$ there are never null vectors with the exception of $L_{-1}|0\rangle$ in the Verma module of the identity operator. The irreducible character is therefore

$$\chi_0(\tau) = q^{-\frac{c}{24}} \prod_{n=2}^{\infty} (1-q^n)^{-1}. \qquad (4.69)$$

Let us now take up the discussion of operator product expansions, Eq. (4.25). In general, the product of two operators contains primary fields and descendant fields. In particular, if ϕ_i and ϕ_j are two primaries, one finds

$$\phi_i(z,\bar{z})\,\phi_j(w,\bar{w}) = \sum_k \sum_{\{l,\bar{l}\}} C_{ij}^{\{l\bar{l}\}k}\, (z-w)^{h_k + \sum_n l_n - h_i - h_j}$$
$$(\bar{z}-\bar{w})^{\bar{h}_k + \sum_n \bar{l}_n - \bar{h}_i - \bar{h}_j}\, \phi_k^{\{l\bar{l}\}}(w,\bar{w}). \qquad (4.70)$$

It can be shown that $C_{ij}^{\{l\bar{l}\}k} = C_{ij}{}^k \beta_{ij}^{\{l\}k} \beta_{ij}^{\{\bar{l}\}k}$ where the $\beta_{ij}^{\{l\}k}$ are uniquely determined by conformal invariance in terms of the dimensions h_i, h_j, h_k and the central charge c. Then the spectrum of primary fields, their operator product coefficients $C_{ij}{}^k$ and the central charge of the Virasoro algebra completely specify a conformal field theory. These parameters cannot be determined from conformal symmetry. One needs additional, dynamical principles such as associativity of the operator algebra. Not any set of parameters $\{c, h_i, C_{ij}{}^k\}$ defines a conformal field theory. A classification is possible only under certain additional restrictions.[4]

The information, which conformal families are contained in the operator product of two primary fields is encoded in the fusion rules

$$[\phi_i] \times [\phi_j] = \sum_k N_{ij}{}^k [\phi_k], \qquad N_{ij}{}^k \in \mathbb{N}_0. \tag{4.71}$$

$N_{ij}{}^k = N_{ji}{}^k > 1$ means that there is more than one way the primary field ϕ_k is contained in the product of ϕ_i and ϕ_j. This is similar to the situation in the theory of finite dimensional groups where an irreducible representation can appear more than once in the product of two representations. Due to the associativity of the OPE, the fusion algebra is also associative, which leads to the relation

$$N_{ij}{}^k N_{kl}{}^m = N_{ik}{}^m N_{jl}{}^k. \tag{4.72}$$

Defining the matrices $N_i = (N_i)^k_j$, Eq. (4.72) can be written in the form

$$[N_i, N_j] = 0. \tag{4.73}$$

We will now investigate the constraints of unitarity for representations of the Virasoro algebra. Unitarity means that the inner product in the Hilbert space is positive definite. The inner product of any two states can be computed from Eqs. (4.36) and (4.58)

$$\langle \phi_j | L_n L_{-n} | \phi_j \rangle = \left[2n\, h_j + \frac{c}{12}(n^3 - n) \right] \langle \phi_j | \phi_j \rangle. \tag{4.74}$$

Taking n sufficiently large implies $c > 0$,[5] while for $n = 1$ we find that $h_j \geq 0$; i.e. the vacuum is the state of lowest energy (L_0 eigenvalue) and this state is unique.

Indeed, consider a field ϕ^0 with scaling dimension $h = 0$. The associated state satisfies $L_0|\phi^0\rangle = 0$. If we assume that the CFT is unitary, we also have to require $L_{+1}|\phi^0\rangle = 0$. Otherwise this state would have negative energy. (The same argument gives $L_n|\phi^0\rangle = 0$ for all $n > 0$.) Using the Virasoro algebra, it then follows that

---

[4]In analogy, Lie algebras cannot be classified, but finite dimensional semi-simple Lie algebras allow for the so-called Cartan classification.

[5]Alternatively, take the vacuum state and $n = 2$, then $\langle 0|L_2 L_{-2}|0\rangle = \frac{c}{2}$, which is the two-point function of the energy-momentum tensor.

4.1 General Introduction

$||L_{-1}|\phi^0\rangle||^2 = 0$ and from (4.63) that ϕ^0 must be a constant.[6] Reversing the argument shows that in unitary theories chiral (anti-chiral) fields necessarily have $\bar{h} = 0$ ($h = 0$). In particular, the only field with $h = \bar{h} = 0$ is the unit operator. If there were another field with zero weight, the state created by it would also be $SL(2, \mathbb{C})$ invariant. This implies that the vacuum state in a unitary CFT is unique.

Further analysis shows that unitarity places no additional constraints for $c \geq 1$. Then one has in general an infinite number of primary fields and c and h can take continuous values.[7] On the other hand, if $c < 1$, one encounters the aforementioned null-states. A detailed analysis shows that only for special quantized values of h and c unitary representations can exist. For this so-called unitary series one finds that c is given by

$$c = 1 - \frac{6}{m(m+1)}, \qquad m = 3, 4, \ldots \qquad (4.75)$$

and for a given s there is a finite set of allowed values for h

$$h_{p,q} = \frac{[(m+1)p - mq]^2 - 1}{4m(m+1)}$$

$$p = 1, 2, \ldots, m-1, \qquad q = 1, 2, \ldots, p. \qquad (4.76)$$

The conformal theories with c and h given by Eqs. (4.75) and (4.76) are called minimal models. c and h are rational and there is only a finite number of primary fields. Conformal field theories with these properties are also called Rational Conformal Field Theories (RCFT). For the first few values of m the minimal models have been identified with two-dimensional statistical models. The first non-trivial one is obtained for $m = 3$ and describes the continuum limit of the two-dimensional Ising model at the critical point.

Rational Conformal Field Theories often have extended chiral and anti-chiral symmetry algebras $\mathcal{A}$ and $\overline{\mathcal{A}}$, respectively. Besides the two commuting Virasoro algebras there are additional symmetry generators which also generate infinite dimensional symmetry algebras. We will encounter explicit examples in Chap. 12. By definition of a RCFT the Hilbert space splits into a finite number of irreducible representations of the symmetry algebra

$$\mathcal{H} = \bigoplus_{i,\bar{j}} M_{i\bar{j}}\, \mathcal{H}_i \otimes \overline{\mathcal{H}}_{\bar{j}} \qquad (4.77)$$

So far we have only defined the character of the, say, right-moving representation. We still need to combine the left- and the right moving sectors. The information

---

[6] In quantum field theory, locality and causality demand that operators at space-like separation commute. It follows that an operator which does not depend on position commutes with all other operators and is therefore a c-number.
[7] We will see examples of this later.

about the entire Hilbert space can then be encoded into a partition function where the trace is over the left- and the right moving sectors. This partition function is a combination of left- and right-moving characters as follows

$$Z(\tau, \bar{\tau}) = \sum_{i,j} M_{i\bar{j}}\, \chi_i(\tau)\, \bar{\chi}_{\bar{j}}(\bar{\tau}) \,. \tag{4.78}$$

One obvious constraint on the coefficients $M_{i\bar{j}}$ is that they are non-negative. Modular invariance, which will be discussed in Chap. 6, imposes additional less trivial restrictions on the coefficients.

Let us now make some general remarks about correlation functions. For simplicity we will only exhibit the chiral part. For the anti-chiral one completely analogous results hold. Recall (from quantum field theory) that correlation functions are vacuum expectations values of R-ordered products of fields. In a CFT their general structure is severely restricted, as we will now demonstrate. Since the vacuum is invariant under $SL(2, \mathbb{C})$ (however not under all generators of the infinite dimensional conformal algebra) correlation functions have to satisfy[8]

$$\langle \phi_1'(z_1) \cdots \phi_n'(z_n) \rangle = \langle \phi_1(z_1) \cdots \phi_n(z_n) \rangle \,, \tag{4.79}$$

where $\phi'(z) = U\,\phi(z)\,U^{-1}$ is the $SL(2, \mathbb{C})$ transformed of $\phi(z)$ and U leaves the in and out vacua invariant. For primary fields we know how they transform under conformal transformations (see Eq. (4.6)). The following discussion is, however, valid for quasi-primary fields, which transform as tensors under $SL(2, \mathbb{C})$ but not necessarily under the full conformal algebra. Many secondary fields, e.g. the energy-momentum tensor, are of this type. The generators of $SL(2, \mathbb{C})$ act as

L_{-1} : translations $\quad U = e^{bL_{-1}} \quad \phi'(z) = \phi(z+b)\,,$

L_0 : dilations and rotations $\quad U = e^{\ln a\, L_0} \quad \phi'(z) = a^{h_i}\,\phi(a\,z)\,,$

L_{+1} : special conformal transformations $\quad U = e^{cL_{+1}} \quad \phi'(z) = \left(\frac{1}{1-cz}\right)^{2h} \phi\left(\frac{z}{1-cz}\right).$
$$\tag{4.80}$$

This leads to the three conditions on the correlation functions of quasi-primary operators

$$\sum_i \partial_i \langle \phi_1(z_1) \cdots \phi_n(z_n) \rangle = 0\,, \tag{4.81a}$$

$$\sum_i (z_i \partial_i + h_i) \langle \phi_1(z_1) \cdots \phi_n(z_n) \rangle = 0\,, \tag{4.81b}$$

---

[8] As we remarked above, $SL(2, \mathbb{C})$ transformations map the Riemann sphere to itself. More general conformal transformations change the geometry.

4.1 General Introduction

$$\sum_i (z_i^2 \partial_i + 2z_i h_i)\langle \phi_1(z_1) \cdots \phi_n(z_n)\rangle = 0 \qquad (4.81c)$$

with similar equations satisfied by the anti-holomorphic part. Invariance under translations tell us that a general n-point function of quasi-primary fields can only depend on the coordinate differences $z_{ij} = z_i - z_j$. This means in particular that one-point functions must be constant. From dilation invariance, it then follows that they vanish

$$\langle \phi_i(z) \rangle = \delta_{i0} \qquad (4.82)$$

for all quasi-primary fields; the label 0 refers to the unit operator. This is no longer true in other geometries, e.g. on the cylinder. From (4.46) we find e.g.

$$\langle T(w) \rangle_{\text{cyl}} = -\frac{c}{24}\left(\frac{2\pi}{\ell}\right)^2. \qquad (4.83)$$

This is possible because the geometry provides a scale, the radius of the cylinder ℓ. Equation (4.83) also provides a physical interpretation for the central charge as a Casimir energy in a finite geometry.

Returning to the complex plane, for two-point functions invariance under dilations and rotations means that they can only be $\langle \phi_i(z)\phi_j(w)\rangle = G_{ij}(z-w)^{-(h_i+h_j)}$ where G_{ij} is a constant which cannot be determined from $SL(2,\mathbb{C})$ invariance. Invariance under special conformal transformations restricts this further to $h_i = h_j$ so that finally

$$\langle \phi_i(z)\phi_j(w)\rangle = \begin{cases} \frac{G_{ij}}{(z-w)^{2h_i}} & h_i = h_j, \\ 0 & \text{otherwise}. \end{cases} \qquad (4.84)$$

We assume that G_{ij} is non-degenerate.

We have already seen that chiral fields generate a subalgebra of the operator product algebra. From (4.84) we learn that locality, i.e. absence of branch cuts, restricts the conformal dimension of chiral quasi-primary fields to be integer or half-integer. For general (i.e. non-chiral fields) the restriction is that $h - \bar{h}$ must be integer or half-integer.

We can always find a basis $\{\phi_i\}$ of primary fields of a given theory such that the only non-vanishing two-point functions are

$$\langle \phi_i(z)\phi_j(w)\rangle = \frac{d_{ij+}}{(z-w)^{2h_i}}, \qquad (4.85)$$

where $d_{ij+} = 0$ if $\phi_i \neq \phi_j^c$ where ϕ_j^c is the unique field which is conjugate to ϕ_j. For real fields $\phi_j^c = \phi_j$. Clearly $h_{j+} = h_j$. Note that $d_{ij+} = C_{ij}{}^0 = \langle \phi_i | \phi_j \rangle$. Therefore $C_{ij}{}^0$ is not a free parameter of the CFT but merely a normalization factor.

Three-point functions are constrained by dilations and rotations to be of the form $f(z_{12}, z_{13}, z_{23})$ where f is a homogeneous function of degree $h_1 + h_2 + h_3$. This function is completely determined, up to a constant, by invariance under special conformal transformations. We find

$$\langle \phi_i(z_1)\phi_j(z_2)\phi_k(z_3)\rangle = \frac{C_{ijk}}{z_{12}^{h_i+h_j-h_k} z_{13}^{h_i+h_k-h_j} z_{23}^{h_j+h_k-h_i}}, \tag{4.86}$$

where $C_{ijk} = C_{ij}{}^{k^+} d_{k^+k}$ are related to the operator product coefficients, cf. Eq. (4.25). In the presentation we restricted ourselves to chiral fields, but the same results also hold for non-chiral fields $\phi(z, \bar{z})$.

Using this equation we can also evaluate the expectation value of a field $\phi_j(z)$ between an asymptotic in-state $|\phi_{k\,\text{in}}\rangle$ and an asymptotic BPZ out-state $\langle\phi_{i\,\text{out}}|$. This simply amounts to multiplying Eq. (4.86) by $z_1^{2h_i}$ and taking the limit $z_1 \to \infty$ and $z_3 \to 0$:

$$\langle \phi_{i\,\text{out}}|\phi_j(z)|\phi_{k\,\text{in}}\rangle = \frac{C_{ijk}}{z^{h_j+h_k-h_i}}. \tag{4.87}$$

For correlation functions of four and more quasi-primary fields the situation is more complicated. They are no longer determined up to a constant. The reason is that out of four points z_i we can form anharmonic quotients or cross ratios

$$X_{ij}^{kl} = \frac{z_{ij} z_{kl}}{z_{il} z_{kj}} \tag{4.88}$$

which are invariant under $SL(2, \mathbb{C})$ transformations of the z_i. For four points there is only one independent cross ratio. For n-point functions there are $n-3$. Then, repeating above reasoning, we find the following general structure for an arbitrary correlation function of n quasi-primary fields:

$$\langle \phi_1(z_1) \cdots \phi_n(z_n)\rangle = \prod_{i<j} z_{ij}^{-\gamma_{ij}} f(X_{ij}^{kl}), \tag{4.89}$$

where the $\gamma_{ij} = \gamma_{ji}$ are any solutions of the set of $\frac{1}{2}n(n-1)$ equations

$$\sum_{j \neq i} \gamma_{ij} = 2h_i \tag{4.90}$$

and f is an undetermined function of the $(n-3)$ independent cross ratios; it cannot be determined from $SL(2, \mathbb{C})$ invariance. (Note that, if γ_{ij} and $\tilde{\gamma}_{ij}$ are two different solutions of Eq. (4.90), $\prod_{i<j} z_{ij}^{-(\gamma_{ij}-\tilde{\gamma}_{ij})}$ is always a function of the X_{ij}^{kl}.) One solution for $n=4$ is

$$\langle \phi_1(z_1)\phi_2(z_2)\phi_3(z_3)\phi_4(z_4)\rangle = \frac{z_{13}^{h_2+h_4} z_{24}^{h_1+h_3}}{z_{12}^{h_1+h_2} z_{23}^{h_2+h_3} z_{34}^{h_3+h_4} z_{14}^{h_1+h_4}} f\left(\frac{z_{12} z_{34}}{z_{13} z_{24}}\right). \tag{4.91}$$

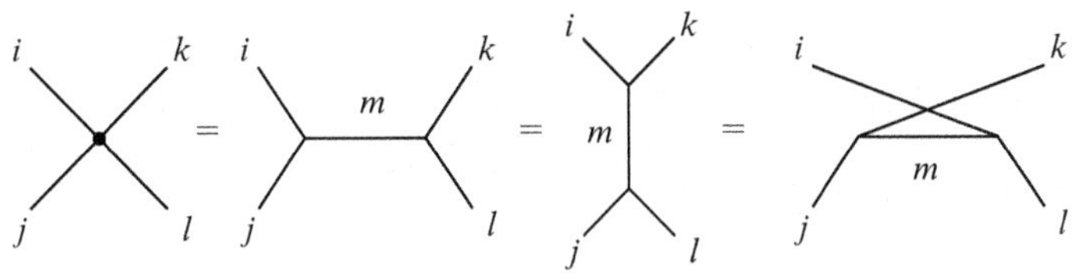

4.1 General Introduction

Fig. 4.3 Crossing symmetry of the four point amplitude

The four-point amplitudes can be used to obtain some constraints on the operator product coefficients C_{ijk}. One evaluates the four-point function in two ways as shown schematically in Fig. 4.3. Associativity of the operator product algebra implies that the two ways give the same result. This is known as crossing symmetry or duality of the four-point amplitude. In this way we obtain a number of equations that the C_{ijk} have to satisfy. The procedure of solving these relations is known as conformal bootstrap; in general this is very difficult to do in practice. We note that one can also use this bootstrap approach to construct chiral symmetry algebras, where this method has been worked out for many cases.

We complete our discussion of general properties of correlation functions in conformal field theory by writing down the conformal Ward identities satisfied by correlation functions of primary fields $\phi_i(z)$. Ward identities among correlation functions generally reflect the symmetries of a theory. We want to investigate the constraints of the local conformal algebra on the correlation functions of the primary fields. Therefore consider the action of the generator of arbitrary infinitesimal conformal transformations, $\oint_{C_0} \frac{dz}{2\pi i} \xi(z) T(z)$, on the correlation function of n primary fields $\phi_j(w_j, \overline{w}_j)$ ($j = 1, \ldots, n$) where the z-contour surrounds all points w_j. Analyticity allows to deform the contour to a sum over contours encircling each of the points w_j (we take the w_i and $\overline{w}_j$ as independent variables):

$$\begin{aligned}
&\delta \langle \phi_1(w_1, \overline{w}_1) \ldots \phi_n(w_n, \overline{w}_n) \rangle \\
&= -\left\langle \oint \frac{dz}{2\pi i} \xi(z) T(z) \phi_1(w_1, \overline{w}_1) \ldots \phi_n(w_n, \overline{w}_n) \right\rangle \\
&= -\sum_{j=1}^n \left\langle \phi_1(w_1, \overline{w}_1) \ldots \left(\oint_{C(w_j)} \frac{dz}{2\pi i} \xi(z) T(z) \phi_j(w_j, \overline{w}_j) \right) \ldots \phi_n(w_n, \overline{w}_n) \right\rangle \\
&= -\sum_{j=1}^n \left\langle \phi_1(w_1, \overline{w}_1) \ldots \left[h_j \, \partial \xi(w_j) + \xi(w_j) \, \partial_{w_j} \right] \phi_j(w_j, \overline{w}_j) \ldots \phi_n(w_n, \overline{w}_n) \right\rangle.
\end{aligned}$$

(4.92)

Using (4.23) we can write

$$0 = \oint \frac{dz}{2\pi i} \xi(z) \left[\langle T(z) \phi_1(w_1, \overline{w}_1) \ldots \phi_n(w_n, \overline{w}_n) \rangle \right.$$
$$\left. - \sum_{j=1}^{n} \left(\frac{h_j}{(z-w_j)^2} + \frac{1}{z-w_j} \partial_{w_j} \right) \langle \phi_1(w_1, \overline{w}_1) \ldots \phi_n(w_n, \overline{w}_n) \rangle \right]. \tag{4.93}$$

Since this must hold for all $\xi(z)$, the integrand must vanish and we conclude

$$\langle T(z) \phi_1(w_1, \overline{w}_1) \ldots \phi_n(w_n, \overline{w}_n) \rangle$$
$$= \sum_{j=1}^{n} \left(\frac{h_j}{(z-w_j)^2} + \frac{1}{z-w_j} \partial_{w_j} \right) \langle \phi_1(w_1, \overline{w}_1) \ldots \phi_n(w_n, \overline{w}_n) \rangle. \tag{4.94}$$

This is the unintegrated form of the conformal Ward identity for correlation functions of primary fields. Note that the l.h.s. is a meromorphic function of z with simple and double poles at w_i, the insertion points of the primary fields. Expanding (4.94) for large z and using that the first three terms have to vanish on account of (4.81), we conclude that

$$\langle T(z)\phi_1(w_1, \overline{w}_1) \ldots \phi_n(w_n, \overline{w}_n) \rangle \stackrel{z \to \infty}{\sim} \frac{1}{z^4} \Gamma(w_1, \ldots w_n) \tag{4.95}$$

or simply $T(z) \stackrel{z \to \infty}{\sim} \frac{1}{z^4}$, valid inside correlation functions. This is, of course, an immediate consequence of $\langle 0 | L_n = 0$ for $n < 2$. Similarly, if we view the correlation function of n primary fields as a function of w, i.e. the insertion point of one of the primaries, it follows from (4.52) that it is regular at small w and from (4.57) that the leading large w behavior is $\sim w^{-2h} \overline{w}^{-2\overline{h}}$. For $h, \overline{h} \neq 0$ this excludes non-vanishing one-point functions, which translational invariance requires to be constant.

We now show that the central charge in the Virasoro algebra is related to the trace anomaly of the energy-momentum tensor. One way the anomaly manifests itself is that, after coupling a CFT to an external c-number metric $h_{\alpha\beta}$, the vacuum expectation value (vev) of the trace of the energy-momentum tensor no longer vanishes. In two dimensions, the only possible trace anomaly is of the form[9]

$$\langle T^{\alpha}_{\alpha} \rangle = -\frac{c}{12} R, \tag{4.96}$$

---

[9] The r.h.s. can be written as $\frac{4\pi}{\sqrt{h}} h^{\alpha\beta} \frac{\delta}{\delta h^{\alpha\beta}} \left(\frac{c}{96\pi} \int \sqrt{h} R \frac{1}{\Box} R \, d^2\sigma \right)$, where the action in parenthesis is the non-local effective action for the external metric, also called Polyakov action, which one obtains if one integrates out the CFT. The fact that the trace is the variation of a non-local action means that it cannot be removed by the addition of local counter terms, i.e. we have a true anomaly.

4.2 Application to Closed String Theory

where R is the Ricci scalar of the metric. An infinitesimal Weyl variation of this expression followed by setting $h_{\alpha\beta} = \delta_{\alpha\beta}$ gives the two-point function in flat space. Expressed in complex coordinates it is

$$\langle T_{z\bar{z}}(z) T_{w\bar{w}}(w) \rangle = -c \frac{\pi}{6} \partial_z \bar{\partial}_{\bar{z}} \delta^{(2)}(z-w), \tag{4.97}$$

where we used $\delta_\sigma R = -2\sigma R - 2(d-1)\Box\sigma$. Here we see another manifestation of the trace anomaly as a contact term in special correlator functions. On the other hand, from the Virasoro algebra we know that

$$\langle T_{zz}(z) T_{ww}(w) \rangle = \frac{\frac{c}{2}}{(z-w)^4}. \tag{4.98}$$

Applying $\bar{\partial}_{\bar{z}} \bar{\partial}_{\bar{w}}$ to this expression, using Eq. (4.15) and[10]

$$\partial\bar{\partial} \log |z|^2 = \bar{\partial}\frac{1}{z} = 2\pi\delta^{(2)}(z), \tag{4.99}$$

we reproduce (4.97).

Here we have assumed that energy-momentum conservation is non-anomalous. We could instead choose to quantize the theory in such a way that it is violated but the trace vanishes. There is, however, no way to quantize such as to respect all symmetries unless $c = 0$.

4.2 Application to Closed String Theory

So far we have considered abstract CFTs and everything we have learned about them are consequences of the symmetry. In particular, there was no need for a Lagrangian description.

We have seen in Chap. 2 that string theory becomes a conformal field theory after fixing the conformal gauge. Prior to gauge fixing, we had a two dimensional field theory coupled to two-dimensional gravity with full diffeomorphism invariance. In conformal gauge this huge symmetry was reduced to conformal invariance which is still infinite dimensional. String theory, when viewed as a CFT, remembers its origin as a diffeomorphism invariant theory in several ways. First, there are the reparametrization ghosts whose CFT we treat in the next chapter. Second, the

---

[10]One way to see this is as follows: $\partial\bar{\partial}\log|z| = \frac{1}{2}\partial\frac{1}{\bar{z}} = \frac{1}{2}\partial\frac{z}{|z|^2} = \frac{1}{2}\lim_{\epsilon\to 0}\partial\frac{z}{|z|^2+\epsilon^2} = \frac{1}{2}\lim_{\epsilon\to 0}\frac{\epsilon^2}{(|z|^2+\epsilon^2)^2} = \pi\delta^{(2)}(z)$. An alternative way is to integrate with a test function, e.g. $e^{-\alpha z\bar{z}}$. The normalization is such that $\int d^2z\,\delta^{(2)}(z) = 1$. For $z = \sigma^1 + i\sigma^2, \bar{z} = \sigma^1 - i\sigma^2$, i.e. $d^2z = 2\,d\sigma^1 d\sigma^2$, $\delta^{(2)}(z) = \frac{1}{2}\delta^{(2)}(\sigma)$ with $\int d^2\sigma\,\delta^{(2)}(\sigma) = 1$.

conformal anomaly has to vanish, i.e. we have to work in the critical dimension where the anomaly due to 26 free scalar fields cancels that of the ghost system. Thirdly, we have to impose a physical state condition which is the remnant of the equation of motion for the world-sheet metric which is invisible in conformal gauge.

The matter fields on the world-sheet are free massless scalars $X(z,\bar{z})$. They furnish the simplest example of a conformal field theory. This is a Lagrangian field theory where all the abstract notions become very explicit. In Euclidean space the action for a free scalar field is

$$S = \frac{1}{2\pi\alpha'} \int d^2z\, \partial X(z,\bar{z})\, \bar{\partial} X(z,\bar{z}) \qquad (4.100)$$

which is, up to the index μ, the Euclidean action of the bosonic string. It leads to the equation of motion

$$\partial\bar{\partial} X(z,\bar{z}) = 0 \qquad (4.101)$$

with general solution

$$X(z,\bar{z}) = X(z) + \overline{X}(\bar{z}) \qquad (4.102)$$

and mode expansion

$$X(z,\bar{z}) = q - i\frac{\alpha'}{2}p \log|z|^2 + i\sqrt{\frac{\alpha'}{2}} \sum_{n\neq 0} \frac{1}{n}(\alpha_n z^{-n} + \bar{\alpha}_n \bar{z}^{-n}). \qquad (4.103)$$

The fields $X(z)$ and $\overline{X}(\bar{z})$ are the right- and left-moving coordinates of the closed bosonic string, respectively. The two-point function for the free boson $X(z,\bar{z})$ following from the action (4.100) is

$$\langle X(z,\bar{z}) X(w,\bar{w}) \rangle = -\frac{\alpha'}{2} \log|(z-w)/R|^2. \qquad (4.104)$$

In order to make the argument of the logarithm well-defined, we have introduced an arbitrary scale R which also serves as an IR regulator and can be thought of as the size of the system. We will demonstrate later that this is not in conflict with scale invariance by showing that the dependence on R cancels in all correlation functions. The two-point function (4.104) satisfies the equation

$$\partial_z \bar{\partial}_{\bar{z}} \langle X(z,\bar{z}) X(w,\bar{w}) \rangle = -\pi\alpha'\, \delta^{(2)}(z-w). \qquad (4.105)$$

An easy way to derive (4.104) and (4.105) is via the path integral identity

$$0 = \int [DX] \frac{\delta}{\delta X(z,\bar{z})} \left(X(w,\bar{w})\, e^{-S[X]} \right)$$

$$= \int [DX] \left(\delta^{(2)}(z-w) + \frac{1}{\pi\alpha'} \partial_z \bar{\partial}_{\bar{z}} X(z,\bar{z}) X(w,\bar{w}) \right) e^{-S[X]} \qquad (4.106)$$

4.2 Application to Closed String Theory

To arrive at (4.105) use (4.99). From (4.106) we also derive (4.101) as an operator equation which is valid in any correlation function where all other insertions have no support at $(z, \bar{z})$.

From (4.102) it follows that the derivative fields $\partial X(z)$ and $\bar{\partial} X(\bar{z})$ are holomorphic and anti-holomorphic fields, respectively. The two-point function for these chiral fields are

$$\langle \partial X(z) \, \partial X(w) \rangle = -\frac{\alpha'}{2} \frac{1}{(z-w)^2},$$

$$\langle \bar{\partial} X(\bar{z}) \, \bar{\partial} X(\bar{w}) \rangle = -\frac{\alpha'}{2} \frac{1}{(\bar{z}-\bar{w})^2}. \qquad (4.107)$$

From its two-point function we see that the field X does not have a definite scaling dimension. However, we will only need its derivatives and exponential which, as we will see below, have definite scaling dimensions and are, in fact, primary fields of the CFT of a free boson.

The energy momentum tensor following from the action Eq. (4.100) is

$$T(z) = -\frac{1}{\alpha'} : \partial X(z) \, \partial X(z) : \qquad (4.108)$$

and likewise for $\bar{T}(\bar{z})$. This can be found with reference to Eq. (2.54a). An alternative way to derive the energy-momentum tensor, which does not require the coupling to a world-sheet metric, is to compute the change of the action under infinitesimal non-holomorphic coordinate transformations $\delta z = \xi(z, \bar{z})$ and $\delta \bar{z} = \bar{\xi}(z, \bar{z})$ (the action is invariant under holomorphic reparametrizations):

$$\delta S = \frac{1}{2\pi} \int d^2z \left(\partial \bar{\xi} \, \bar{T} + \bar{\partial} \xi \, T \right). \qquad (4.109)$$

Equation (4.108) then follows with $\delta X = -(\xi \, \partial X + \bar{\xi} \, \bar{\partial} X)$. In Eq. (4.108) normal ordering is defined by

$$:\phi_i(z) \, \phi_j(z): = \lim_{w \to z} \left(\phi_i(w) \, \phi_j(z) - \text{poles} \right) = \frac{1}{2\pi i} \oint dw \, \frac{\phi_i(w) \, \phi_j(z)}{w - z}, \qquad (4.110)$$

where the pole terms to be subtracted are those arising in the operator product expansion of $\phi_i(w) \, \phi_j(z)$. This assumes that we are dealing with analytic and mutually local fields, i.e. the only singularities in the operator product are poles. Using contour deformations shown in Fig. 4.2, one can express the modes of the normal ordered product as

$$\left(:\phi_i \, \phi_j:\right)_m = \sum_{n \leq -h_i} \phi_{i,n} \, \phi_{j,m-n} + \sum_{n > -h_i} \phi_{j,m-n} \, \phi_{i,n}, \qquad (4.111)$$

i.e. normal ordering : ** : is defined with respect to the $SL(2,\mathbb{Z})$ invariant vacuum (4.51). Sometimes it is convenient to define normal ordering such that all positive modes are moved to the right. This normal ordering will be denoted by ${}^*_*\cdots{}^*_*$. More generally, the normal ordered product of two fields at non-coincident points, $:\phi_i(z)\phi_j(w):$ is defined as the regular part of their operator product. The singular part is called the contraction, often denoted by $\underline{\phi_i(z)\phi_j(w)}$ so that the operator product is

$$\phi_i(z)\,\phi_j(w) = \underline{\phi_i(z)\phi_j(w)} + :\phi_i(z)\,\phi_j(w): \,. \tag{4.112}$$

For the free field X we have

$$\partial X(z)\,\partial X(w) = -\frac{\alpha'}{2}\frac{1}{(z-w)^2} + :\partial X(z)\,\partial X(w): \tag{4.113}$$

and we can define the energy momentum tensor (4.108) as

$$T(z) = \lim_{\epsilon \to 0}\left\{-\frac{1}{\alpha'}\left(\partial X(z+\epsilon)\,\partial X(z) + \frac{\alpha'}{2\epsilon^2}\right)\right\}\,. \tag{4.114}$$

In terms of the oscillator modes the Virasoro generators are

$$T(z) = \sum_n L_n\, z^{-n-2} \tag{4.115}$$

with

$$L_n = \frac{1}{2}\sum_m :\alpha_{n-m}\,\alpha_m: \,. \tag{4.116}$$

Note that free fields have the characteristic property that there is only one singular term in the operator product, but this is not true in general, as the following example shows.[11]

It is straightforward to compute the operator product of the energy-momentum tensor with itself. Since we are dealing with free fields, we can use Wick's theorem. Remembering that no contractions are to be made within normal ordered expressions, we find with the help of the basic contraction Eq. (4.113)

---

[11] Great care is required when dealing with normal ordered products of more than two fields. If they are not free fields, the product is not associative and Wick's theorem cannot be applied straightforwardly. An example will be the Sugawara form of the energy-momentum tensor which we will discuss in Chap. 11.

4.2 Application to Closed String Theory

$$T(z)\,T(w) = \frac{\frac{1}{2}}{(z-w)^4} + \frac{2\,T(w)}{(z-w)^2} + \frac{\partial T(w)}{(z-w)} + \dots \quad (4.117)$$

This shows that we have a conformal field theory with $c = 1$.

Finally, we have to specify the conformal fields of this model. As already mentioned, $X(z)$ is not a conformal field due to the logarithmic z-dependence of its propagator. However, we have already seen that $\partial X(z)$ is a chiral field. Computing the operator product of $T(z)$ with $\partial X(z)$,

$$T(z)\,\partial X(w) = \frac{\partial X(w)}{(z-w)^2} + \frac{\partial(\partial X(w))}{(z-w)} + \dots \quad (4.118)$$

shows that $\partial X(z)$ is a primary field with dimension $h = 1$. Higher derivatives $\partial^n X(z)$ are not primary but descendant fields with $h = n$. For example, $\partial^2 X(z) = L_{-1}\partial X(z)$. The only other conformal fields in the free scalar model are normal ordered exponentials of $X(z)$:

$$T(z)\,:e^{ikX(w,\bar{w})}: \;=\; \left[\frac{\frac{\alpha'}{4}k^2}{(z-w)^2} + \frac{\partial_w}{(z-w)}\right]\,:e^{ikX(w,\bar{w})}: \quad (4.119)$$

and a similar result for the anti-holomorphic OPE. This is again shown using Wick expansion. We find that the operator $:e^{ikX(z,\bar{z})}:$ has conformal dimension $(h,\bar{h}) = \left(\frac{\alpha'}{4}k^2, \frac{\alpha'}{4}k^2\right)$ with k being a continuous variable.

The complete operator product algebra among the conformal fields has the following form[12]:

$$:e^{ipX(z,\bar{z})}:\;:e^{iqX(w,\bar{w})}: \;=\; |z-w|^{\alpha' pq}\;:e^{i(p\,X(z,\bar{z}) + q\,X(w,\bar{w}))}:$$

$$= |z-w|^{\alpha' pq}\;:e^{i(p+q)X(w,\bar{w})}:$$

$$+ ip|z-w|^{\alpha' pq}\Big((z-w)\,:\partial X(w)\,e^{i(p+q)X(w,\bar{w})}:$$

$$+ (\bar{z}-\bar{w})\,:\bar\partial X(\bar{w})\,e^{i(p+q)X(w,\bar{w})}:\Big) + \dots, \quad (4.120a)$$

$$\partial X(z)\,:e^{ipX(w,\bar{w})}: \;=\; -\frac{\alpha'}{2}\frac{ip}{z-w}\,:e^{ipX(w,\bar{w})}:\;+\text{finite}\,. \quad (4.120b)$$

From Eq. (4.107) we find

$$[\alpha_m, \alpha_n] = m\,\delta_{m+n}\,. \quad (4.121)$$

The relation $\alpha^\dagger_{-n} = \alpha_n$ follows from Hermiticity of $i\partial X$. As a check, let us use this algebra to compute the two-point function on the plane

---

[12]Note that the field conjugate to $e^{ipX(z)}$ is $e^{-ipX(z)}$.

$$\langle \partial X(z) \, \partial X(w) \rangle = -\frac{\alpha'}{2} \sum_{m,n} \langle 0|\alpha_n \alpha_m|0\rangle \, z^{-n-1} \, w^{-m-1}$$

$$= -\frac{\alpha'}{2} \sum_{n>0} n \, z^{-n-1} \, w^{n+1}$$

$$= -\frac{\alpha'}{2} \frac{1}{(z-w)^2} \quad \text{for} \quad |z| > |w| \,, \tag{4.122}$$

where we have used Eq. (4.51). The above derivation demonstrates that the operators have to be radially ordered for the sum to converge.

In the bosonic string theory one deals with d ($d = 26$) identical free bosonic fields X^μ ($\mu = 0, \ldots, d-1$) and their contribution to the central charge is $c = \bar{c} = d$. Physical string states must satisfy the following conditions:

$$L_n|\phi\rangle = \bar{L}_n|\phi\rangle = 0, \quad n > 0,$$
$$(L_0 - 1)|\phi\rangle = (\bar{L}_0 - 1)|\phi\rangle = 0,$$
$$(L_0 - \bar{L}_0)|\phi\rangle = 0 \,. \tag{4.123}$$

This means that physical string states correspond to primary fields of the conformal field theory:

$$|\phi\rangle = \phi(0)|0\rangle = \lim_{z,\bar{z}\to 0} \phi(z,\bar{z})|0\rangle \,. \tag{4.124}$$

$\phi(z,\bar{z})$ are conformal fields and create asymptotic states. They are called vertex operators. String scattering amplitudes are then simply correlation functions of vertex operators (cf. Chaps. 6 and 16). Equation (4.123) also implies that the conformal dimension of vertex operators is $(h,\bar{h}) = (1,1)$ and that expressions such as $\int d^2z \, \phi(z,\bar{z})$ are well-defined.

Let us briefly discuss the spectrum of the closed bosonic string in the context of conformal field theory. The lowest state is the tachyon; it is a space-time scalar with momentum k

$$|k\rangle = \lim_{z,\bar{z}\to 0} :e^{ik \cdot X(z,\bar{z})}: |0\rangle \,. \tag{4.125}$$

The physical state condition Eq. (4.123) requires that $:e^{ik \cdot X(z,\bar{z})}:$ has conformal dimension $h = \bar{h} = 1$. With Eq. (4.119) this leads to $\alpha' k^2 = 4$ which is the mass shell condition for a tachyon: $m^2 = -k^2 = -\frac{4}{\alpha'}$. Let us verify that the state $|k\rangle$ carries momentum k_μ. We restrict our attention to one, say the right-moving sector. The momentum operator is $p^\mu = \sqrt{\frac{2}{\alpha'}} \alpha_0^\mu = \frac{2}{\alpha'} \oint \frac{dz}{2\pi i} i \partial X^\mu$ and

$$p^\mu |k\rangle = \frac{2}{\alpha'} \oint \frac{dz}{2\pi i} i \partial X^\mu(z) :e^{ik \cdot X(z,\bar{z})}: |0\rangle = k^\mu |k\rangle \,. \tag{4.126}$$

4.2 Application to Closed String Theory

We have used the operator product Eq. (4.120b). The states at the next level have the form:

$$|k, \epsilon\rangle = -\frac{2}{\alpha'}\epsilon_{\mu\nu}(k) \lim_{z,\bar{z}\to 0} :\partial X^{\mu}(z)\bar{\partial}X^{\nu}(\bar{z}) e^{ik\cdot X(z,\bar{z})}: |0\rangle$$
$$= \alpha^{\mu}_{-1}\bar{\alpha}^{\nu}_{-1}|k\rangle \epsilon_{\mu\nu}. \tag{4.127}$$

Here $\epsilon_{\mu\nu}$ is a polarization tensor. These states are those discussed in Eq. (3.57). We still have to check whether this vertex operator is a conformal field with dimensions $(h, \bar{h}) = (1, 1)$. We therefore take its operator product with the energy-momentum tensor $T(z)$; we can do this for the holomorphic and anti-holomorphic parts separately:

$$T(z) :\epsilon_{\mu\nu}\partial X^{\mu}(w)\bar{\partial}X^{\nu}(\bar{w})e^{ik\cdot X(w,\bar{w})}: = -\frac{i\alpha'}{2}\frac{k^{\mu}\epsilon_{\mu\nu}}{(z-w)^3} :\bar{\partial}X^{\nu}(\bar{w})e^{ik\cdot X(w,\bar{w})}:$$
$$+ \left[\frac{\frac{\alpha'}{4}k^2 + 1}{(z-w)^2} + \frac{\partial_w}{(z-w)}\right]\epsilon_{\mu\nu} :\partial X^{\mu}(w)\bar{\partial}X^{\nu}(\bar{w}): e^{ik\cdot X(w,\bar{w})}: + \dots \tag{4.128}$$

In order to get rid of the unwanted cubic singularity one has to demand that

$$k^{\mu}\epsilon_{\mu\nu} = \epsilon_{\mu\nu}k^{\nu} = 0, \tag{4.129}$$

where the second condition comes from the anti-holomorphic sector. Together with $k^2 = 0$, to obtain $h = 1$, this is the on-shell condition for a massless tensor particle which we identify with the graviton, antisymmetric tensor and dilaton, depending on whether $\epsilon_{\mu\nu}$ is symmetric traceless, antisymmetric or pure trace. In terms of the state (4.127) the conditions (4.129) are a consequence of the requirement $L_1|k, \epsilon\rangle = \bar{L}_1|k, \epsilon\rangle = 0$. Higher L_n's do not lead to non-trivial conditions. At the first massive level, third and fourth order singularities can appear and forcing them to be absent leads to conditions on the polarization tensors. Equivalently, L_1 and L_2 have to annihilate the state, etc.

Scattering amplitudes of asymptotic string states are correlation functions of the corresponding vertex operators.[13] We have already said that the vertex operators are primary fields of the Virasoro algebra of weight $(h, \bar{h}) = (1, 1)$. It involves, however, also an integration over the positions of the vertex operators. (This will be discussed in Chap. 16, where we explicitly compute string scattering amplitudes.) We then get expressions of the form

$$\int \prod_{i=1}^{n} d^2 z_i \, \langle V(z_1, \bar{z}_1) \dots V(z_n, \bar{z}_n)\rangle. \tag{4.130}$$

---

[13] The discussion here is limited to string tree level amplitudes since there the world-sheet is the (Riemann) sphere which is conformally equivalent to $\mathbb{C} \cup \infty$.

Since the V's have weight $(h, \bar{h}) = (1, 1)$ and the integration measure transforms under $SL(2, \mathbb{C})$ as $d^2z \to \frac{d^2z}{|cz+d|^4}$ we find, using Eq. (4.6), that string amplitudes are $SL(2, \mathbb{C})$ invariant.

Let us now demonstrate that secondary states decouple from string scattering amplitudes. Consider the correlation function

$$A = \langle \phi^{(k)}(z) \phi_1(z_1) \ldots \phi_n(z_n) \rangle, \qquad (4.131)$$

where the $\phi_i, i = 1, \ldots, n$ are primary and

$$\phi^{(k)}(z) = \hat{L}_{-k} \phi(z) = \oint_{C_z} \frac{dw}{2\pi i} \frac{T(w)}{(w-z)^{k-1}} \phi(z) \qquad (4.132)$$

is secondary. We now insert Eq. (4.132) into (4.131) and deform the contour to enclose the points $z_1, \ldots, z_n$ instead of z. (This is easy to visualize for the sphere which is conformally equivalent to $\mathbb{C} \cup \infty$.) Then, expanding the operator products $T(w) \phi_i(z_i)$, we get

$$A = \sum_{i=1}^{n} \left\{ -\frac{(k-1)h_i}{(z_i - z)^k} + \frac{\partial_{z_i}}{(z_i - z)^{k-1}} \right\} \langle \phi(z) \phi_1(z_1) \ldots \phi_n(z_n) \rangle. \qquad (4.133)$$

If the ϕ_i are vertex operators, we have $h_i = 1, \forall i$ and this can be written as

$$A = \sum_{i=1}^{n} \partial_{z_i} \left\{ \frac{1}{(z_i - z)^{k-1}} \langle \phi(z) \phi_1(z_1) \ldots \phi_n(z_n) \rangle \right\}, \qquad (4.134)$$

which vanishes upon integration over the positions of the vertex operators. This argument can be generalized to the case of several descendant fields.

We still need to comment on the dependence on the infrared mass scale R appearing in the two-point function (4.104). Let us first notice that the two-point function diverges for $z \to w$. To regulate this divergence we introduce an UV cut-off a and define $\langle X(z, \bar{z}) X(w, \bar{w}) \rangle = -\alpha' \log\left(\frac{a}{R}\right)$ for $|z - w| < a$. Using Wick's theorem[14] it is straightforward to show that

$$e^{ikX} = \left(\frac{a}{R}\right)^{\frac{1}{2}\alpha' k^2} : e^{ikX} : . \qquad (4.136)$$

---

[14] Recall Wick's theorem which reads schematically:

$$\underbrace{\phi \cdots \phi}_{n} = : \phi \cdots \phi : + \sum_{\substack{\text{single} \\ \text{contractions}}} \underbrace{\phi \phi}_{} : \underbrace{\phi \cdots \phi}_{n-2} : + \sum_{\substack{\text{double} \\ \text{contractions}}} \underbrace{\phi \phi}_{} \underbrace{\phi \phi}_{} : \underbrace{\phi \cdots \phi}_{n-4} : + \ldots \qquad (4.135)$$

and use $\underbrace{\phi \phi}_{} = -\alpha' \log(a/R)$.

The R-dependence of n-point functions $\langle \prod_i e^{i k_i X(z_i, \bar{z}_i)} \rangle$ is $R^{-\frac{\alpha'}{2}(\sum k_i)^2}$ and disappears if $\sum k_i = 0$. In Euclidean signature the correlation functions vanish in the limit $R \to \infty$, unless this condition is satisfied. In the Polyakov path integral momentum conservation is due to the integration over the zero modes x^μ. The dependence of (4.136) on the UV cut-off shows that the anomalous dimension of the tachyon vertex operator is $\frac{\alpha'}{2} k^2$. The a-dependence can be absorbed in a rescaling of the vertex operator. A similar argument holds, if we replace the tachyon by a vertex operator of, for instance, the form $:\partial X e^{ikX}:$.

We want to close this section with an interesting remark. The generic form of an excited string state is $\alpha^{i_1}_{-n_1} \cdots \alpha^{i_k}_{-n_k} |0\rangle$ with a similar expression for the left-movers. The number of states at level $n = \sum_i n_i$ for one oscillator direction was quoted in (4.67). The generalization of the density of states to $d - 2 = c$ directions is

$$\rho(L_0) \sim e^{2\pi \sqrt{\frac{c L_0}{6}}}. \tag{4.137}$$

We have written the density of states at level L_0 in a form which can be shown to be valid for a general CFT with central charge c. This is known as the Cardy formula. For a bosonic string excitation at level n with $m^2 \sim E^2 \sim n/\alpha'$ we find $\rho(E) \sim \exp(\tilde{c} E \sqrt{\alpha'})$ ($\tilde{c}$ is a constant) and therefore for the entropy $S \sim E$. The temperature $T = (\partial S/\partial E)^{-1}$ approaches a limiting value $T_H \sim 1/\sqrt{\alpha'}$ which is called the Hagedorn temperature. No matter how much energy we pump into the system, we cannot increase its temperature beyond T_H. This is due to the exponential increase of the density of states with energy, which is a characteristic feature of string theory.

Strings at finite temperature are an interesting subject, e.g. within the context of string cosmology, but we will not elaborate on it any further.

4.3 Boundary Conformal Field Theory

So far we have discussed CFT on the complex plane which is directly applicable to closed string theory. When working with open strings, we have boundaries. The presence of boundaries forced us to impose boundary conditions for the free boson fields X^μ. We distinguished between Dirichlet and Neumann boundary conditions and were led to introduce the concept of D-branes. However, in an abstract CFT usually there is no Lagrangian formulation available and no boundary terms will arise from a variational principle. Therefore, to proceed, we need a more inherent formulation of a boundary and boundary conditions. In the following, we give an introduction to boundary CFT and to the so-called boundary state formalism, where we illustrate the main features for the example of the free boson. As an application we compute the boundary state of a bosonic Dp-brane in the next section.

We have seen in the previous section how the (Euclidean) world-sheet of a free closed string, the infinitely long cylinder, can be conformally mapped to the Riemann sphere, cf. (4.3). For a freely propagating open string the world-sheet is an infinite strip of width ℓ. In the following we parametrize the strip as $-\ell \leq \sigma \leq 0$ and $-\infty < \tau < \infty$. After a Wick rotation $\tau \to -i\tau$ and with $w = \tau - i\sigma$ we can map the strip to the upper half plane $\mathbb{H}_+$ via

$$z = e^{\pi(\tau - i\sigma)/\ell} = e^{\pi w/\ell} \,. \tag{4.138}$$

The lower and upper edges of the strip are mapped to the positive and negative real axis, respectively.

We now impose a physical boundary condition on the energy-momentum tensor: no (world-sheet) energy-momentum flows across the real axis, i.e. $T_{01} = 0$ at the boundary. Expressed in complex coordinates this is

$$T(z) = \overline{T}(\bar{z})\Big|_{z=\bar{z}} \,. \tag{4.139}$$

Via this condition, the boundary breaks the conformal symmetry from two independent Virasoro algebras to a single (diagonal) Virasoro algebra.

The conformal transformations which leave the real axis $z = \bar{z}$ fixed, are the real analytic functions, i.e. functions which satisfy $\overline{\xi(z)} = \xi(\bar{z})$. The invertible one-to-one maps of the upper half plane to itself are generated by $L_{0,\pm 1}$ with real coefficient. The group is $PSL(2, \mathbb{R})$. It maps the real axis to itself.

In passing we mention another common representation of the open string worldsheet, the unit disk. The map from the upper half-plane z to the unit disk z' is[15]

$$z' = \frac{1 + iz}{1 - iz} \,. \tag{4.140}$$

The real axis is mapped to the unit circle. We will use the expressions disk and upper half plane interchangeably. These two maps are shown in Fig. 4.4.

In Sect. 4.1 we used the independence of the holomorphic and anti-holomorphic transformations to get a holomorphic and an anti-holomorphic Ward identity. Since these two transformations are now linked, there is only one Ward identity

$$\langle \delta_{\xi,\bar{\xi}} X \rangle = -\oint_C \frac{dz}{2\pi i} \, \xi(z) \, \langle T(z) \, X \rangle + \oint_C \frac{d\bar{z}}{2\pi i} \, \bar{\xi}(\bar{z}) \, \langle \overline{T}(\bar{z}) \, X \rangle \,. \tag{4.141}$$

$X = \phi_1(w_1, \bar{w}_1) \cdots \phi_n(w_n, \bar{w}_n)$ is the insertion of a product of fields (as in (4.92)) with scaling dimensions $(h_i, \bar{h}_i)$ inserted in the upper half plane. The counterclockwise contour C encloses all insertion points.

We now rewrite (4.141) in the form (4.92) by using the following doubling trick (cf. Eq. (2.97)). We consider the dependence of correlators on the anti-holomorphic coordinates $\bar{w}_i$, with w_i in the upper half-plane, as a dependence on holomorphic

---

[15]More generally, $z' = e^{i\alpha} \frac{z - z_0}{z - \bar{z}_0}$, $\alpha \in \mathbb{R}$, $\text{Im}(z_0) > 0$.

4.3 Boundary Conformal Field Theory

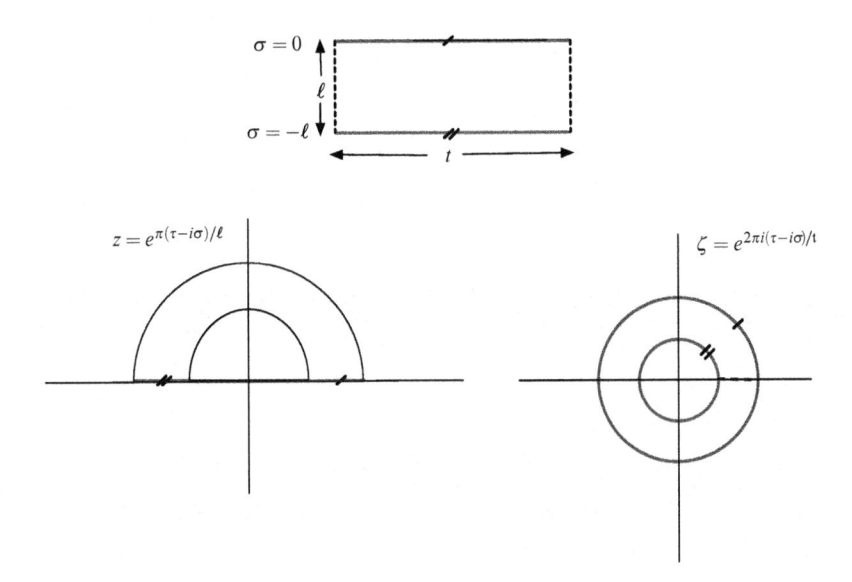

Fig. 4.4 The doubling trick

coordinates $w'_i = \bar{w}_i$ in the lower half-plane. We use the boundary condition (4.139) to analytically continue T to the lower half plane via $T(z') = \bar{T}(z)$ for $\text{Im}\, z' \leq 0$ (with $z' = \bar{z}$). We now have a holomorphic energy-momentum tensor $T(z)$ on the whole complex plane. The contour of the second integral in (4.141) is now reflected into the lower half z-plane. To make it again counter-clockwise we have to reverse the direction, which cancels the relative sign in (4.141). Using (4.139), the two contours combine into one which encloses the w_i in the upper half-plane and the w'_i in the lower half plane; the two horizontal pieces cancel (Fig. 4.4). The conformal Ward-identity is (cf. (4.94))

$$\langle T(z) \prod_{i=1}^{n} \phi_i(w_i, w'_i) \rangle$$
$$= \sum_{j=1}^{n} \left(\frac{h_j}{(z-w_j)^2} + \frac{1}{z-w_j}\partial_{w_j} + \frac{h_j}{(z-w'_j)^2} + \frac{1}{z-w'_j}\partial_{w'_j} \right) \langle \prod_{i=1}^{n} \phi_i(w_i, w'_i) \rangle \quad (4.142)$$

As in the case without boundary, the Virasoro generators are

$$L_n = \oint \frac{dz}{2\pi i} z^{n+1} T(z). \quad (4.143)$$

The difference is that there are no $\bar{L}_n$'s.

One important consequence of the presence of a boundary is that one-point functions (tadpoles) on the upper half-plane are no longer required to vanish. On the plane their vanishing is a consequence of translational invariance, but this is broken by the boundary. The distance from the boundary being the only scale, one point functions must be of the form

$$\langle \phi_i(z, \bar{z}) \rangle = \frac{d_{C(i)i+}}{|z - \bar{z}|^{2h_i}} \delta_{h_i, \bar{h}_i} \tag{4.144}$$

which has the structure of a two-point function on the plane. The two indices on $d_{C(i)i+}$ refer to the holomorphic and anti-holomorophic parts of ϕ_i. $C(i)$ contains the information about the boundary condition that we have imposed along the real axis and d_{ij+} are the constants appearing in the bulk two-point functions (4.85). Recall that for these not to vanish i and j must be conjugate to each other. More generally, n-point functions on the upper half-plane have the structure of $2n$-point functions on the plane. This is obvious in the case of factorized fields but is also true in general.

The discussion which led, in the presence of a boundary, to the breaking of the direct sum of a left-moving and a right-moving Virasoro algebra to the diagonal subalgebra, can be generalized. Given a holomorphic field $W(z)$ of dimension h, we can define conserved charges

$$W_n = \oint \frac{dz}{2\pi i} \, W(z) \, z^{n+h-1} \, . \tag{4.145}$$

On the cylinder the charges are integrals at constant τ and conservation means that they are time independent. On the strip conservation is not guaranteed because the current can have a component normal to the boundary and charge can flow off the boundaries of the strip. This will happen unless particular boundary conditions are imposed. On the upper half-plane we define the charges as in (4.145) but the integral is over a semi-circle in the upper half-plane with the origin as its center. This integral is not independent of the radius of the semi-circle because of the contributions from the real axis; cf. the lower left picture in Fig. 4.5. However, if we also have an antiholomorphic current $\overline{W}(\bar{z})$ such that we can impose (cf. Eq. (4.139)

$$W(z) = \overline{W}(\bar{z}) \Big|_{\text{Im} z = 0} \tag{4.146}$$

the charges

$$W_n = \int \frac{dz}{2\pi i} \, W(z) \, z^{n+h-1} - \int \frac{d\bar{z}}{2\pi i} \, \overline{W}(\bar{z}) \, \bar{z}^{n+h-1}$$

$$= \oint \frac{dz}{2\pi i} \, W(z) \, z^{n+h-1} \tag{4.147}$$

are conserved. In the first line the integrals are over semi-circles. In the second line we have used the doubling trick and the integral is over a circle enclosing the

4.3 Boundary Conformal Field Theory

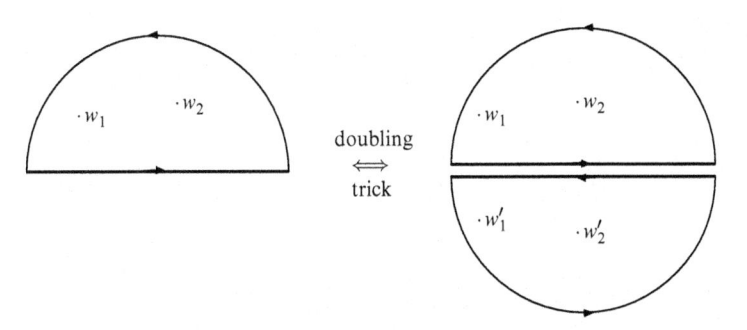

Fig. 4.5 Two maps of a finite strip to regions in (*left*) the upper half-plane and (*right*) the complex plane

origin. While there were two sets of charges, W_n and $\overline{W}_n$ in the complex plane, the boundary has broken them to a diagonal subset.

We now consider the free scalar field on the upper half plane. Its mode expansion is

$$X(z,\bar{z}) = \begin{cases} q - i\alpha' p \log |z|^2 + i\sqrt{\dfrac{\alpha'}{2}} \sum_{n\neq 0} \dfrac{1}{n} \alpha_n \left(z^{-n} + \bar{z}^{-n}\right) & \text{(NN)} \\[2mm] \dfrac{1}{2\pi i}(q_1 - q_0) \log \left|\dfrac{z}{\bar{z}}\right| + i\sqrt{\dfrac{\alpha'}{2}} \sum_{n\neq 0} \dfrac{1}{n} \alpha_n \left(z^{-n} - \bar{z}^{-n}\right) & \text{(DD)} \\[2mm] q + i\sqrt{\dfrac{\alpha'}{2}} \sum_{r\in\mathbb{Z}+\frac{1}{2}} \dfrac{1}{r} \alpha_r \left(z^{-r} \pm \bar{z}^{-r}\right) & \begin{matrix}\text{(ND)}\\ \text{(DN),}\end{matrix} \end{cases}$$

(4.148)

where the boundary conditions are imposed along the real axis. For the mixed boundary conditions they change at the origin.

Using the doubling trick one can compute the two-point functions on the upper half-plane which, in general, is already a four-point function on the complex plane. One finds (cf. (3.13a))

$$\langle X(z,\bar{z}) X(w,\bar{w}) \rangle_{\substack{\text{NN}\\\text{DD}}} = -\frac{\alpha'}{2}\left(\log|z-w|^2 \pm \log|z-\bar{w}|^2\right). \qquad (4.149)$$

For mixed boundary conditions we find from (3.13b)

$$\langle X(z,\bar{z}) X(w,\bar{w}) \rangle_{\substack{\text{ND}\\\text{DN}}} = -\frac{\alpha'}{2}\left(\log\left|\frac{\sqrt{z}-\sqrt{w}}{\sqrt{z}+\sqrt{w}}\right|^2 \pm \log\left|\frac{\sqrt{z}-\sqrt{\bar{w}}}{\sqrt{z}+\sqrt{\bar{w}}}\right|^2\right). \qquad (4.150)$$

We can use (4.149) to compute the one-point functions

$$\langle(\partial X \bar\partial X)(z,\bar z)\rangle_{\substack{NN\\DD}} = \pm\frac{\alpha'}{2}\frac{1}{|z-\bar z|^2} \qquad (4.151)$$

which should be compared to (4.144). On the other hand, from (4.150) we find

$$\langle(\partial X \bar\partial X)(z,\bar z)\rangle_{\substack{ND\\DN}} = \pm\frac{\alpha'}{4}\frac{1}{|z-\bar z|^2}\left(\sqrt{\frac{\bar z}{z}} + \sqrt{\frac{z}{\bar z}}\right) \qquad (4.152)$$

which is not of the form (4.144). The reason is that it is a four-point function with so-called boundary changing operators inserted at the origin and at infinity.

So far we have considered only bulk fields. In the string theory context they are closed string vertex operators which are inserted in the bulk of open string world-sheets. Examples are the vertex operators which we discussed in the previous section. In addition there are boundary operators which have to be inserted on the boundary of the world-sheet. They are the open-string vertex operators and one has to distinguish two types. The first one arises from quantizing an open string with both end-points on the same type of boundary (D-brane). The corresponding open string fields $\lambda_{ij}^a \phi_a(x)$ (no sum on a implied) can be inserted on the boundary, i.e. on the real axis of the complex plane $x \in \mathbb{R}$. Here λ_{ij}^a is the Chan-Paton factor introduced in Chap. 3. The n-point functions of such operators are very similar to the closed string correlation functions. Using the $PSL(2,\mathbb{R})$ symmetry of the real axis, the two- and three-point functions are fixed to be of the form ($x_{ij} = x_i - x_j$)

$$\langle\phi_a(x_1)\phi_b(x_2)\rangle = \frac{d_{ab}}{x_{12}^{2h_a}}\delta_{h_a,h_b}$$

$$\langle\phi_1(x_a)\phi_b(x_2)\phi_c(x_3)\rangle = \frac{C_{abc}}{x_{12}^{h_a+h_b-h_c} x_{23}^{h_b+h_c-h_a} x_{13}^{h_a+h_c-h_b}} . \qquad (4.153)$$

The second type of boundary operators originate from open strings with the two end-points on different types of boundaries (D-branes). Examples are the open strings with mixed Neumann-Dirichlet boundary conditions. The corresponding fields $\lambda_{ij}^{ab} \phi_{ab}(x)$ can also be inserted on the real axis, but they must have different boundary conditions a and b to the left and to the right, i.e. at their position the type of boundary changes. For this reason, such operators are also called boundary changing operators. Their correlators are, in general, much harder to compute.

For the free boson we can again be more specific. The open string propagators can be obtained from (4.149) and (4.150). For NN boundary conditions we find from (4.149) with $z = \bar z = x$, $w = \bar w = y$

$$\langle X(x)X(y)\rangle = -2\alpha' \log|x-y|, \qquad (4.154)$$

4.3 Boundary Conformal Field Theory

while it vanishes for DD boundary conditions, but

$$\langle \partial_n X(x) \, \partial_n X(y) \rangle = \frac{2\alpha'}{(x-y)^2} \,. \tag{4.155}$$

For completeness we also give the vertex operators for the open string tachyon and gauge boson, where we also indicate the on-shell conditions, i.e. the conditions that the vertex operators are primary fields of weight $h = 1$:

$$V(x) = :e^{ik \cdot X(x)}: \,, \qquad k^2 = \frac{1}{\alpha'},$$

$$V(x) = \epsilon_\mu :\dot{X}^\mu e^{ik \cdot X(x)}: \,, \qquad k^2 = 0, \ k \cdot \epsilon = 0, \tag{4.156}$$

where $\dot{X} = \partial_x X$. The notation indicates that the real axis is the image of the end of the open string which is parametrized by world-sheet time under the map of the strip to the upper half-plane. In the presence of several D-branes the vertex operators still have to be multiplied by CP factors. The on-shell conditions are easily verified.

We now come to an essential point in boundary conformal field theory, which is the so-called closed-open string duality, also known as tree-channel–loop-channel equivalence. Under the map (4.138) lines of constant world-sheet time τ are concentric semi-circles around the origin. A finite strip is mapped to a semi-annulus in the z-plane, as indicated in Fig. 4.5. Also shown in the figure is the image of the strip under the map

$$\zeta = e^{2\pi i(\tau - i\sigma)/t} = e^{2\pi i w/t} \tag{4.157}$$

which is an annulus, cut open along the positive real axis.

Using the transformation rule of the energy-momentum tensor, Eq. (4.40), the condition (4.139) becomes

$$\zeta^2 T^{(\zeta)}(\zeta) = \bar{\zeta}^2 \overline{T}^{(\zeta)}(\bar{\zeta}) \qquad \text{at} \qquad |\zeta| = 1 \text{ and } e^{-2\pi \ell/t} \tag{4.158}$$

(the contributions from the Schwarzian derivative cancel). For a field $W(z)$ of conformal weight h, for which we impose (4.146), we obtain

$$\zeta^h W^{(\zeta)}(\zeta) = (-1)^h \bar{\zeta}^h \overline{W}^{(\zeta)}(\bar{\zeta}) \,. \tag{4.159}$$

The factor $(-1)^h$ is due to the Jacobian $\frac{\partial \zeta}{\partial z} = 2i \cdot \frac{\ell}{t} \cdot \frac{\zeta}{z}$ and will be relevant later when we discuss fields with non-integer weight.

If we perform radial quantization on the ζ-plane, which amounts to considering the radial direction as world-sheet time and expand $T^{(\zeta)}(\zeta) = \sum L_n^\zeta \zeta^{-n-2}$, we find that the boundary condition on T translates to[16]

$$L_n^{(\zeta)} = \overline{L}_{-n}^{(\zeta)} \quad \text{at} \quad |\zeta| = 1 \tag{4.160}$$

or, more generally,

$$W_n^{(\zeta)} = (-1)^h \overline{W}_{-n}^{(\zeta)} \quad \text{at} \quad |\zeta| = 1. \tag{4.161}$$

The map from the upper half z-plane to the ζ-plane essentially exchanges the roles of σ and τ and the two maps of the strip lead to quite different interpretations of the open string diagram. If we identify the two vertical ends of the strip, the world-sheet becomes a cylinder of circumference t and height ℓ. This can be viewed as an open string one loop diagram[17] where the string propagates for a time t. The fact that we identify the two ends means that we take a trace. The resulting open string amplitude is

$$\text{Tr } e^{-t H_{\text{op}}} = \text{Tr } e^{-\frac{\pi t}{\ell}(L_0 - \frac{c}{24})}, \tag{4.162}$$

where we have used (2.98) and (4.47). But we can exchange the roles of σ and τ and view σ as the world-sheet time and the cylinder as a tree-level closed string diagram, where the closed string propagates for a time ℓ. This is precisely what the coordinate transformation $z \leftrightarrow \zeta$ accomplishes.

On the ζ-plane the interpretation of the strip is as a tree-level closed string diagram: a closed string of length t propagates for a time ℓ. Using (2.87) with $\ell \to t$ and (4.47) one obtains the closed-string tree-level amplitude

$$\langle \beta | e^{-\ell H_{\text{cl}}} | \alpha \rangle = \langle \beta | e^{-\frac{2\pi \ell}{t}(L_0 + \overline{L}_0 - \frac{c}{12})} | \alpha \rangle, \tag{4.163}$$

where the information about the boundary conditions α, β is encoded in the boundary states between which the closed string propagates. This is an essential difference between the open and closed string pictures. In the former the boundary conditions are encoded in the Hamiltonian which propagates the open string along the cylinder. In the latter they enter through the choice of boundary states while the Hamiltonian is blind to the choices for α and β. The fact that (4.162) and (4.163) are equal is the statement of closed-open string duality. We will return to this issue in Chap. 6.

---

[16] The mode expansion on the z-plane would be $T(z) = \sum L_n z^{-n-2}$, where the L_n's are not the same as the $L_n^{(\zeta)}$. Below we will drop the superscript ζ.

[17] We will discuss one-loop diagrams in detail in Chap. 6.

4.4 Free Boson Boundary States

To compute the amplitude in the tree-channel one needs a way to describe the coupling of the closed string to the boundary/D-brane. This information is encoded in a coherent state in the closed string Hilbert space

$$|B\rangle = \sum_{i,\bar{j}\in \mathcal{H}\otimes\overline{\mathcal{H}}} \alpha_{i\bar{j}} |i,\bar{j}\rangle . \tag{4.164}$$

Here $i, \bar{j}$ label the states in the left- and right-moving sectors of $\mathcal{H} \otimes \overline{\mathcal{H}}$, and the coefficients $\alpha_{i\bar{j}}$ encode the strength of how the closed string mode $|i, \bar{j}\rangle$ couples to the boundary $|B\rangle$. Such a coherent state, which inserts a boundary on the world-sheet, is called a boundary state and provides the world-sheet/CFT description of a D-brane in String Theory. The coefficients $\alpha_{i\bar{j}}$ are nothing else than one-point functions (4.144) of closed strings on a disk world-sheet with the boundary lying on the D-brane

$$\alpha_{i\bar{j}} = d_{C(i),\bar{j}^+} . \tag{4.165}$$

Therefore the boundary state of a D-brane contains a sum over all closed string tadpoles on the disk with boundary on the D-brane. As mentioned, $C(i)$ encodes the information about the type of the boundary. Instead of computing all these infinitely many one-point functions directly, one formulates the type of boundary as conditions on the boundary states, which can then be solved directly. As we will see for the free boson in the next section, this approach turns out to be very powerful and as a by-product gives all closed string one-point functions on the disk.

4.4 Free Boson Boundary States

Let us now translate the Neumann and Dirichlet boundary conditions for open strings, $\partial_\sigma X^\mu|_{\sigma=0} = 0$ and $X^\mu|_{\sigma=0} = x_0^\mu$ into the picture of boundary states. Clearly, we have to exchange $(\sigma, \tau)_{\text{op}} \to (\tau, \sigma)_{\text{cl}}$ and readily obtain

$$\partial_\tau X^\mu_{\text{closed}}|_{\tau=0} |B_N\rangle = 0 \qquad \text{Neumann condition,}$$

$$X^\mu_{\text{closed}}|_{\tau=0} |B_D\rangle = x_0^\mu |B_D\rangle \quad \text{Dirichlet condition.} \tag{4.166}$$

Using the mode expansion for the closed string, these conditions give rise to conditions for the modes acting on the boundary state

$$(\alpha_n^\mu + \bar{\alpha}_{-n}^\mu) |B_N\rangle = 0 , \qquad p^\mu |B_N\rangle = 0 \qquad \text{Neumann,}$$

$$(\alpha_n^\mu - \bar{\alpha}_{-n}^\mu) |B_D\rangle = 0 , \qquad x^\mu |B_D\rangle = x_0^\mu |B_D\rangle \quad \text{Dirichlet,} \tag{4.167}$$

for each n. Here x_0^μ is the position of the D-brane. Such conditions relating the left- and right-moving modes acting on the boundary state are called gluing conditions.

One can give an explicit form of the solution to these gluing conditions. If we split the coordinates into Neumann and Dirichlet directions according to $\mu = (\alpha, i)$ with $\alpha = 0, \ldots, p$ and $i = p+1, \ldots, d-1$, then the solution for the boundary state for an infinite, static flat Dp-brane is, up to a normalization,

$$|B_{\text{Dp}}\rangle = \exp\left(-\sum_{n=1}^{\infty} \frac{1}{n} \alpha_{-n}^\mu D_{\mu\nu} \bar{\alpha}_{-n}^\nu\right)|x_0^i\rangle, \qquad (4.168)$$

where

$$D_{\mu\nu} = (\eta_{\alpha\beta}, -\delta_{ij}) \qquad (4.169)$$

and $|x_0^i\rangle$ is an eigenstate of the position operator in the Dirichlet directions. Its eigenvalues are the position of the D-brane. Furthermore, it is annihilated by all the positive oscillator modes and by the oscillator zero modes, i.e, the momentum operator in the directions along the brane. One can verify this form of the boundary states by straightforwardly evaluating the gluing conditions.

Note that the boundary state also satisfies the gluing condition (cf. (4.160))

$$(L_n - \bar{L}_{-n})|B_{\text{N,D}}\rangle = 0 \qquad (4.170)$$

for the generators of conformal transformations. In this sense it preserves a diagonal subalgebra of the left- and the right-moving conformal symmetries.

For later generalizations to more abstract CFTs we provide an alternative notation for the boundary states. We consider one single direction with Neumann boundary conditions. Applying the identity

$$\exp\left(\sum_{n=1}^{\infty} c_n\right) = \prod_{n=1}^{\infty} \sum_{m_n=0}^{\infty} \frac{1}{m_n!} c_n^{m_n} = \sum_{m_1=0}^{\infty} \sum_{m_2=0}^{\infty} \cdots \prod_{n=1}^{\infty} \frac{1}{m_n!} c_n^{m_n} \qquad (4.171)$$

to $c_n = -\frac{1}{n}\alpha_{-n}\bar{\alpha}_{-n}$ we obtain

$$|B_{\text{N}}\rangle = \exp\left(-\sum_{n=1}^{n} \frac{1}{n} \alpha_{-n} \bar{\alpha}_{-n}\right)|0\rangle$$

$$= \sum_{m_1=0}^{\infty} \sum_{m_2=0}^{\infty} \cdots \prod_{n=1}^{\infty} \frac{1}{\sqrt{m_n!}} \left(\frac{\alpha_{-n}}{\sqrt{n}}\right)^{m_n}|0\rangle \otimes \left(\frac{-\bar{\alpha}_{-n}}{\sqrt{n}}\right)^{m_n}|\bar{0}\rangle. \qquad (4.172)$$

4.5 Crosscap States for the Free Boson

Here $|0\rangle$ is the oscillator ground state with zero center of mass momentum. Next, we note that the following states form a complete orthonormal basis

$$|\mathbf{m}\rangle = |m_1, m_2, \ldots\rangle = \prod_{n=1}^{\infty} \frac{1}{\sqrt{m_n!}} \left(\frac{\alpha_{-n}}{\sqrt{n}}\right)^{m_n} |0\rangle. \quad (4.173)$$

We now introduce an operator acting as

$$U \bar{\alpha}_n U^{-1} = -\bar{\alpha}_n = -\left(\bar{\alpha}_{-n}\right)^{\dagger}. \quad (4.174)$$

With the orthonormal basis (4.173) and the operator U we can express (4.172) in a more general way as

$$|B_\mathrm{N}\rangle = \sum_{\mathbf{m}} |\mathbf{m}\rangle \otimes |U\,\bar{\mathbf{m}}\rangle, \quad (4.175)$$

which, as we will see in Chap. 6, generalizes to more complicated CFTs. From (4.175) we can read off the non-vanishing one-point functions of closed string fields on the disc with Neumann boundary condition as

$$\langle \phi_{\mathbf{m},\bar{\mathbf{m}}} \rangle_\mathrm{N} = \prod_n (-1)^{m_n}. \quad (4.176)$$

The expression for a Dirichlet direction is almost identical, with U replaced by the unit operator so that the one-point functions are simply $\langle \phi_{\mathbf{m},\bar{\mathbf{m}}} \rangle_\mathrm{D} = 1$.

The full boundary state of the Dp-brane is then simply the tensor product

$$|B_\mathrm{Dp}\rangle = \bigotimes_{\alpha=0}^{p} |B_\mathrm{N}^\alpha\rangle \otimes \bigotimes_{i=p+1}^{d-1} |B_\mathrm{D}^i\rangle. \quad (4.177)$$

In Chap. 6 we will use these boundary states to compute the closed string tree amplitude and compare it to the open string one-loop amplitude. This will also enable us to fix the normalizations of the boundary states.

4.5 Crosscap States for the Free Boson

In the case of non-oriented strings, one can introduce a very similar structure, which we briefly introduce in this section. Recall that for the closed string the parity transformation acts as $\Omega : (\tau, \sigma) \to (\tau, 2\pi - \sigma)$. We take the length of the closed string to be 2ℓ. This reverses the orientation of the world-sheet and forces us to also consider non-orientable world-sheets. Note that in general this symmetry can

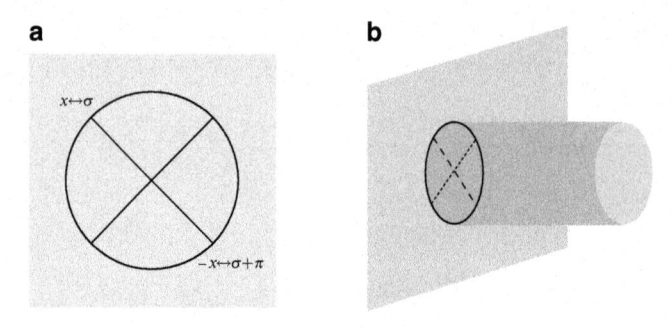

Fig. 4.6 Illustration of how points are identified on a crosscap, and how a closed string couples to a crosscap. (**a**) Identification of points on a crosscap, (**b**) Closed string at a crosscap

be dressed with space-time symmetries R. We will give a simple example at the end of this section.

The simplest Euclidean non-orientable surface is $\mathbb{RP}^2$, which can be viewed as S^2 with anti-podal points identified or, equivalently, as a hemisphere with opposite points on the equator identified. As shown on the left of Fig. 4.6, it looks like a disk with a special identification of points on the boundary. Such a world-sheet is also called a crosscap. As shown on the right in Fig. 4.6, in the same way as the boundary of the disk is attached to a D-brane, the crosscap is attached to an orientifold plane. As mentioned already in Chap. 3, the latter can be regarded as a defect in space-time, which is invariant under the geometric part R of the orientifold projection. We will see in Chap. 9 how these orientifold planes automatically arise when performing a quotient by ΩR.

One can also consider closed string n-point functions on such non-oriented world-sheets. Like for the disk the simplest ones are the one-point functions. Similarly to a boundary state which describe the coupling of closed string states to a boundary, for orientifold theories there should also exist a coherent state encoding all these one-point functions.

We now discuss such crosscap states for the example of the free bosons. Similarly to boundary states, we do not infer them from direct computations of the one-point functions, but by first finding gluing conditions and then solving them. From Fig. 4.6 it is clear that the field X at (σ, τ) should be identified with the field X at $(\sigma + \pi, \tau)$. More concretely, we have to impose

$$X(\sigma, \tau)\big|_{\tau=0} |C\rangle = X(\sigma + \pi, \tau)\big|_{\tau=0} |C\rangle ,$$
$$\partial_\tau X(\sigma, \tau)\big|_{\tau=0} |C\rangle = -\partial_\tau X(\sigma + \pi, \tau)\big|_{\tau=0} |C\rangle . \quad (4.178)$$

We have located the crosscap at $\tau = 0$. The minus sign in the second line reflects the change of the orientation. It guarantees that the boundary term at fixed τ, $\int_0^{2\pi} \dot{X} \delta X \, d\sigma$, which arises in the variation of the Polyakov action, vanishes. From (4.178) we easily derive the gluing conditions for the crosscap states

4.5 Crosscap States for the Free Boson

$$\left(\alpha_n^\mu + (-1)^n \bar{\alpha}_{-n}^\mu\right) |C\rangle = 0, \qquad p^\mu |C\rangle = 0. \tag{4.179}$$

The center of mass coordinate x_0 of the closed string is unconstrained and for the center of mass momentum one gets the same gluing condition as for a Neumann boundary condition for a brane. Therefore, the object which satisfies these constraints can be considered as being extended in all directions X^μ, $\mu = 0, \ldots, d-1$. It might be called an orientifold plane. To indicate its dimensionality we refer to it as an $O(d-1)$-plane.

Apart from the factor $(-1)^n$, the gluing conditions are those of a boundary state with Neumann conditions. The solution to the gluing conditions is therefore similar and reads, up to normalization,

$$\left|C_{O(d-1)}\right\rangle = \exp\left(-\sum_{n=1}^\infty \frac{(-1)^n}{n} \alpha_{-n}^\mu \bar{\alpha}_{-n}^\mu\right)|0\rangle. \tag{4.180}$$

The crosscap state can also be written as

$$\left|C_{O(d-1)}\right\rangle = \sum_\mathbf{m} |\mathbf{m}\rangle \otimes |U\overline{\mathbf{m}}\rangle \tag{4.181}$$

where U acts as

$$U \bar{\alpha}_n^\mu U^{-1} = -(-1)^n \bar{\alpha}_n^\mu = -(-1)^n \left(\bar{\alpha}_{-n}^\mu\right)^\dagger. \tag{4.182}$$

Analogous to Dirichlet boundary conditions, we can also get an overall sign change in the crosscap gluing conditions. This is achieved by dressing the orientifold projection Ω by an additional reflection $R: X^i(\tau, \sigma) \mapsto -X^i(\tau, \sigma)$ for $i = p+1, \ldots, d-1$. The combined action then reads

$$\Omega R: \quad X^i(\tau, \sigma) \mapsto -X^i(\tau, \pi + \sigma). \tag{4.183}$$

This replaces (4.178) for the directions X^i by

$$\begin{aligned}
X^i(\tau, \sigma)\big|_{\tau=0}|C\rangle &= -X^i(\tau, \sigma + \pi)\big|_{\tau=0}|C\rangle, \\
\partial_\tau X^i(\tau, \sigma)\big|_{\tau=0}|C\rangle &= \partial_\tau X^i(\tau, \sigma + \pi)\big|_{\tau=0}|C\rangle,
\end{aligned} \tag{4.184}$$

and leads to the gluing conditions

$$\left(\alpha_n^i - (-1)^n \bar{\alpha}_{-n}^i\right)|C\rangle = 0, \qquad x^i |C\rangle = x_0^i |C\rangle, \tag{4.185}$$

where in general x_0^i is a fixed point of R, i.e. $R x_0^i = x_0^i$. In our situation we just get $x_0^i = 0$. Combining the crosscap states for the unreflected and reflected directions we get the complete crosscap states for a Op-plane in d-dimensions

$$|C_{Op}\rangle = \exp\left(-\sum_{n=1}^{\infty} \frac{(-1)^n}{n} \alpha_{-n}^{\mu} D_{\mu\nu} \bar{\alpha}_{-n}^{\nu}\right)|x_0^i\rangle \quad (4.186)$$

with $D_{\mu\nu} = (\eta_{\alpha\beta}, -\delta_{ij})$, as before.

We close this chapter with the following observation: the conditions (4.167) for the boundary state and (4.179) and (4.185) for the crosscap state are related via a unitary transformation generated by $e^{i\pi L_0}$. Therefore $|B\rangle$ and $|C\rangle$ are related via $|C\rangle = e^{i\pi L_0}|B\rangle = i^{L_0+\bar{L}_0}|B\rangle$, where we used (4.170).

Further Reading

The basic references of two-dimensional CFT are:

- A.A. Belavin, A.M. Polyakov, A.B. Zamolodchikov, Infinite conformal symmetry in two-dimensional quantum field theory. Nucl. Phys. **B241**, 333 (1984)
- J.L. Cardy, Boundary conditions, fusion rules and the Verlinde formula. Nucl. Phys. **B324**, 581 (1989)

Some useful reviews are:

- P. Ginsparg, *Applied conformal field theory*, in *Fields, Strings and Critical Phenomena, Les Houches 1988*, E. Brezin, J. Zinn-Justin (eds.), North Holland, p. 1; hep-th/9108028
- P. Di Francesco, P. Matthieu, D. Sénéchal, *Conformal Field Theory* (Springer, Berlin, 1996)
- C. Itzykson, L.-M. Drouffe, *Statistic Field Theory*, vol. 2, Chap. 9 (CUP, Cambridge, 1989)
- R. Blumenhagen, E. Plauschinn, *Introduction to Conformal Field Theory*. LNP, vol. 779 (Springer, Berlin, 2009)
- M. Henkel, *Conformal Invariance and Critical Phenomena* (Springer, Berlin, 1999)
- M. Gaberdiel, An introduction to conformal field theory. Rep. Prog. Phys. **63**, 607 (2000)
- J. Cardy, *Conformal Invariance and Statistical Mechanics*, Les Houches lectures 1988 and 2008

Good discussions of boundary states can be found in

- J. Polchinski, Y. Cai, Consistency of open superstring theories. Nucl. Phys. **B296**, 91 (1988)
- M. Gaberdiel, Lectures on non-BPS dirichlet branes. Class. Quant. Grav. **17**, 3483 (2000)

Chapter 5
Parametrization Ghosts and BRST Quantization

Abstract As another application of conformal field theory, we want to examine the reparametrization ghosts which we introduced within the path integral quantization of the bosonic string in Chap. 3. In the second part of this chapter we briefly study the very much related issue of BRST quantization of the bosonic string, where we encounter another characterization of physical string states, namely as states in the cohomology of a nilpotent BRST charge.

5.1 The Ghost System as a Conformal Field Theory

In conformal coordinates the ghost action is

$$S = \frac{1}{2\pi} \int d^2z \, (b_{zz} \, \partial_{\bar{z}} c^z + b_{\bar{z}\bar{z}} \, \partial_z c^{\bar{z}}) \tag{5.1}$$

and the solutions of the equations of motion are

$$c^z = c(z), \qquad b_{zz} = b(z),$$
$$c^{\bar{z}} = \bar{c}(\bar{z}), \qquad b_{\bar{z}\bar{z}} = \bar{b}(\bar{z}). \tag{5.2}$$

The ghost fields are effectively free fermions with integer spin. $b(z)$ is a holomorphic conformal field of dimension (spin) $(h, \bar{h}) = (2, 0)$, $c(z)$ is a field of dimension $(h, \bar{h}) = (-1, 0)$; $\bar{b}(\bar{z})$ and $\bar{c}(\bar{z})$ are anti-holomorphic conformal tensors with $(h, \bar{h}) = (0, 2)$ and $(0, -1)$ respectively. In the following we will restrict the discussion to the holomorphic fields $b(z)$ and $c(z)$, i.e. to the holomorphic sector of the closed string. Following from the action Eq. (5.1) (cf. the derivation of the X propagator in Sect. 4.2), their propagator is

$$\langle b(z) \, c(w) \rangle = \langle c(z) \, b(w) \rangle = \frac{1}{z - w}. \tag{5.3}$$

It satisfies
$$\partial_{\bar{z}} \langle b(z) c(w) \rangle = 2\pi \delta^{(2)}(z-w), \qquad (5.4)$$

where we have used the relation
$$\partial_{\bar{z}} \frac{1}{z-w} = 2\pi \delta^{(2)}(z-w) \qquad (5.5)$$

which directly follows from (4.99) on page 85. From the propagator we deduce the following operator products
$$c(z) b(w) = b(z) c(w) = \frac{1}{z-w} + \ldots . \qquad (5.6)$$

Next we expand the ghost fields in modes:
$$c(z) = \sum_n c_n z^{-n+1},$$
$$b(z) = \sum_n b_n z^{-n-2} \qquad (5.7)$$

with Hermiticity conditions $b_n^\dagger = b_{-n}$ and $c_n^\dagger = c_{-n}$. On the $SL(2,\mathbb{Z})$ invariant ghost vacuum $|0\rangle_{b,c}$ the oscillators act as
$$b_n |0\rangle_{b,c} = 0 \quad \text{for } n \geq -1,$$
$$c_n |0\rangle_{b,c} = 0 \quad \text{for } n \geq 2. \qquad (5.8)$$

From the operator products Eq. (5.6) we derive the following anticommutation relations:
$$\{b_m, c_n\} = \delta_{n+m},$$
$$\{c_n, c_m\} = \{b_n, b_m\} = 0. \qquad (5.9)$$

Note that since we are dealing with anticommuting fields, the contour trick of Chap. 4 leads to anticommutators. The energy momentum tensor of the b, c system is (3.77)
$$T^{b,c}(z) = \lim_{w \to z} \left\{ -2b(w) \partial c(z) - \partial b(w) c(z) + \frac{1}{(z-w)^2} \right\}$$
$$= -2 :b \partial c(z): - :\partial b c(z): . \qquad (5.10)$$

5.1 The Ghost System as a Conformal Field Theory

Normal ordering is defined as in (4.110).[1] One can check that this corresponds to the ordering such that the modes which annihilate $|0\rangle_{b,c}$ are moved to the right. The mode expansions of the ghost fields lead to the following expression for the Virasoro generators (cf. Eq. (3.85) on page 58):

$$L_n^{b,c} = \sum_{m=-\infty}^{\infty} (n-m) : b_{n+m} c_{-m} : . \qquad (5.12)$$

Normal ordering is only relevant for L_0, for which one finds

$$L_0^{b,c} = \sum_{m \geq 1} m (b_{-m} c_m + c_{-m} b_m) - 1 = \sum_m m \, {}_\ast^\ast b_{-m} c_m {}_\ast^\ast - 1, \qquad (5.13)$$

after taking due care of minus signs arising due to the Grassmann property of the ghosts. The normal ordering prescription ${}_\ast^\ast \ldots {}_\ast^\ast$ was already introduced in Chap. 4. It is w.r.t. a state which is annihilated by all positive modes of b and c. We will define such a state later in this chapter. The constant -1 arises from the need to reorder expressions containing c_1 and b_{-1} and it represents the expectation value of L_0 in the chosen vacuum. It is zero in the $SL(2,\mathbb{Z})$ invariant vacuum $|0\rangle_{b,c}$.

It is straightforward to work out the operator product of the stress tensors with itself:

$$T^{b,c}(z) T^{b,c}(w) = \frac{-26/2}{(z-w)^4} + \frac{2T^{b,c}(w)}{(z-w)^2} + \frac{\partial T^{b,c}(w)}{(z-w)} + \text{finite terms} . \qquad (5.14)$$

This shows that the central charge of the b, c system is $c^{b,c} = -26$. With $c < 0$ the conformal field theory of the ghost system is not unitary, as expected. This already follows from the negative conformal weight of $c(z)$. Adding the contribution to the central charge from d bosonic fields and from the ghost fields one gets

$$c^{\text{tot}} = c^X + c^{b,c} = d - 26 \qquad (5.15)$$

so that again the conformal anomaly vanishes if $d = 26$. The operator product of $T^{b,c}$ with the ghost fields can be easily worked out and shows that b and c are primary fields.

Another important operator of the b, c system is the $U(1)$ ghost number current $j(z)$,

$$j(z) = - : b(z) c(z) := \sum_n z^{-n-1} j_n , \qquad (5.16)$$

---

[1] For two anti-commuting fields one gets an extra minus for the modes of the normal ordered product

$$: \phi_i \phi_j :_m = \sum_{n \leq -h_i} \phi_{i,n} \phi_{j,m-n} - \sum_{n > -h_i} \phi_{j,m-n} \phi_{i,n} . \qquad (5.11)$$

where

$$j_n = \sum_m : c_{n-m} b_m : \,. \qquad (5.17)$$

$j(z)$ is a conformal field of dimension $h = 1$. Classically the ghost number current is conserved. In the quantum theory there is however an anomaly. This will be discussed in Chap. 13. The ghost charge is given by the contour integral of $j(z)$:

$$N_g = \oint_{C_0} \frac{dz}{2\pi} j(z) = j_0 = \sum_m : c_{-m} b_m : \,. \qquad (5.18)$$

Thus, the ghost charge N_g of a particular conformal field $\phi(z)$ is given by the singular part of its operator product with j.

$$j(z)\,\phi(w) = \frac{N_g\,\phi(w)}{(z-w)} + \text{finite terms}\,. \qquad (5.19)$$

It follows that $c(z)$ and $b(z)$ have $N_g = +1$ and -1 respectively. In fact, the ghost number current is the Noether current for the continuous symmetry $c \to e^{i\alpha} c$, $b \to e^{-i\alpha} b$.

5.2 BRST Quantization

Let us now turn to the question of how to identify physical states. We have seen in Chap. 3 that in light-cone quantization the longitudinal and time-like components of X^μ are not independent degrees of freedom and can be eliminated. All states can be built as excitations of the transverse oscillators. In covariant Polyakov quantization we keep all components of X^μ and, in addition, have the ghost fields b and c. The excitation of the longitudinal and time-like components of X^μ and the ghosts will now become part of the spectrum of string theory and we need some way to distinguish physical from unphysical states. The tool to do this is the BRST charge. So let us briefly review some general aspects of BRST quantization and then apply it to the bosonic string.

BRST quantization was introduced to quantize systems with a local gauge symmetry. After gauge fixing, BRST symmetry is a remnant of the local gauge symmetry. Let us first review the general strategy. Consider a system with gauge invariances generated by charges K_i which form a closed finite dimensional Lie algebra g[2]

$$[K_i, K_j] = f_{ij}{}^k K_k, \quad i,j,k = 1,\ldots,\dim g\,; \qquad (5.20)$$

---

[2] For instance, in a Yang-Mills theory the K_i are the non-abelian generalizations of Gauss law.

5.2 BRST Quantization

$f_{ij}{}^k$ are the structure constants of g. One now defines a Hermitian nilpotent operator which commutes with the Hamiltonian and which acts on all fields like a fermionic gauge transformation. The gauge parameter is replaced by an anticommuting variable c^i, called a ghost. This operator is the BRST charge Q. An explicit expression for the Hermitian BRST charge is[3]

$$Q = c^i \left(K_i - \frac{1}{2} f_{ij}{}^k c^j b_k \right)$$
$$= c^i \left(K_i + \frac{1}{2} K_i^{\text{ghost}} \right), \qquad (5.21)$$

where we have introduced the anti-ghosts b_i which obey the following commutation relations with the c^i:

$$\{c^i, b_j\} = \delta^i_j . \qquad (5.22)$$

In the following, both b_i and c^i will be collectively referred to as ghosts. The first part of Q clearly acts in the required way on the fields. The second part is needed to make Q nilpotent. It acts like a gauge transformation on the ghost fields. Nilpotency of Q is easy to verify. Using the symmetry algebra Eq. (5.20), the anticommutation relations Eq. (5.22) and antisymmetry of the structure constants, we find without difficulty

$$Q^2 = \frac{1}{4} f_{[ij}{}^k f_{l]k}{}^m (c^j c^i c^l b_m) = 0 . \qquad (5.23)$$

The Jacobi identity was used in the last step. The BRST transformation acts on the ghosts as

$$\delta c^i = \{Q, c^i\} = -\frac{1}{2} f_{kl}{}^i c^k c^l ,$$
$$\delta b_i = \{Q, b_i\} = K_i - f_{ij}{}^k c^j b_k = K_i + K_i^{\text{ghosts}} \equiv \tilde{K}_i . \qquad (5.24)$$

One now shows that

$$[\tilde{K}_i, \tilde{K}_j] = f_{ij}{}^k \tilde{K}_k \qquad (5.25)$$

i.e. the $\tilde{K}_i$ also satisfy the symmetry algebra but, in contrast to the K_i, also incorporate the ghost degrees of freedom. Finally, one also introduces a ghost number operator

---

[3] For $(c^i)^\dagger = c^i$, $(b_i)^\dagger = b_i$ and $K_i^\dagger = K_i$, Q is Hermitian. Note that (5.20) implies that the structure constants are purely imaginary.

$$N_g = -\sum_{i=1}^{\dim g} b_i c^i \qquad (5.26)$$

so that c^i and b_i have ghost number $+1$ and -1 respectively and Q has ghost number $+1$.

Let us now consider the Hilbert space of the theory. Eigenstates of the Hamiltonian are said to be BRST invariant if they are annihilated by Q:

$$Q|\phi\rangle = 0 . \qquad (5.27)$$

States in the zero ghost sector which satisfy Eq. (5.27) are obviously gauge invariant. There are two types of BRST invariant states. First, any state of the form

$$|\phi\rangle = Q|\lambda\rangle \qquad (5.28)$$

is trivially BRST invariant due to the nilpotency of Q. The states $|\phi\rangle$ and $|\lambda\rangle$ form a BRST doublet. They differ in ghost charge by one unit. (We work in a basis of states with definite ghost number.) $|\phi\rangle$ has zero norm, due to the Hermiticity and nilpotency of the BRST charge: $\langle\lambda|QQ|\lambda\rangle = 0$. These states decouple in S-matrix elements. (Recall that Q commutes with the Hamiltonian.) Therefore we have to look for states of the form

$$Q|\phi\rangle = 0, \quad |\phi\rangle \neq Q|\lambda\rangle . \qquad (5.29)$$

They are BRST singlets. These states will henceforth be referred to as physical states. Two states $|\phi\rangle$ and $|\phi'\rangle$ are said to be equivalent if

$$|\phi\rangle - |\phi'\rangle = Q|\lambda\rangle . \qquad (5.30)$$

The equivalence classes are called BRST cohomology classes. Clearly, all states within one given cohomology class have the same ghost number. S-matrix elements are independent of which representative of a cohomology class one uses: $\langle\phi_1|S|\phi_2\rangle = \langle\phi_1'|S|\phi_2'\rangle$, for ϕ and ϕ' related as in Eq. (5.30).

If $|\phi\rangle = \lim_{z,\bar{z}\to 0} \phi(z,\bar{z})|0\rangle$ is a physical BRST singlet, then $[Q,\phi] = 0$. For states without ghost excitation this implies $[K_i,\phi] = 0$. Those are the states identified with physical particles.

Let us now apply the BRST formalism to the bosonic string. In distinction to the case of gauge theories with finite dimensional symmetry groups, we are now dealing with the infinite dimensional Virasoro algebra. Expressions such as the BRST charge and the ghost number must be normal ordered and a normal ordering constant will appear. Finally, nilpotency of the BRST charge, which contains the symmetry generators, might be anomalous. Let us start with the BRST charge. The generalization of Eq. (5.21) to the case of the Virasoro algebra is

5.2 BRST Quantization

$$\begin{aligned}Q &= \sum_{m=-\infty}^{\infty}\left(c_{-m}L_m^X - \frac{1}{2}\sum_{n=-\infty}^{\infty}(m-n):c_{-m}c_{-n}b_{m+n}:\right)\\ &= \sum_{m}:c_{-m}\left[L_m^X + \frac{1}{2}L_m^{b,c}\right]:\\ &= \sum_{m=-\infty}^{\infty}\left(c_{-m}L_m^X - \frac{1}{2}\sum_{n=-\infty}^{\infty}(m-n)\colon\!c_{-m}c_{-n}b_{m+n}\!\colon\right) - c_0\,.\end{aligned} \quad (5.31)$$

In the first line we have used the explicit form for the structure constants of the Virasoro algebra, $f_{mn}{}^p = (m-n)\delta_{p,m+n}$ and the identification $c^m \equiv c_{-m}$. The $L_m^{(b,c)}$ were given in Eq. (5.12). One easily verifies that Q is Hermitian. Note that $L_m^{\text{tot}} = L_m^X + L_m^{b,c}$ corresponds to $\tilde{K}_i$ in Eq. (5.24).

Q can be equivalently written as a contour integral:

$$\begin{aligned}Q &= \oint_{C_0}\frac{dz}{2\pi i}:c(z)\left[T^X(z) + \frac{1}{2}T^{b,c}(z)\right]:\\ &= \oint_{C_0}\frac{dz}{2\pi i}\left[:c(z)\left[T^X(z) + \frac{1}{2}T^{b,c}(z)\right]: - \frac{1}{z^2}c(z)\right]\\ &= \oint_{C_0}\frac{dz}{2\pi i}\, j_{\text{BRST}}(z)\,.\end{aligned} \quad (5.32)$$

The operator j_{BRST} is the BRST current. Equation (5.32) defines it only up to a total derivative which must however be of dimension one and ghost number one. The most general form is then

$$j_{\text{BRST}} = c\,T^{(X)} + \frac{1}{2}c\,T^{(b,c)} + \kappa\,\partial^2 c\,. \quad (5.33)$$

Requiring the BRST current to be a conformal field of weight one gives $\kappa = \frac{3}{2}$.

We need to check the nilpotency of the BRST charge Q which is crucial for the identification of physical states. This is most easily done by computing the $j(z)\,j(w)$ operator product. One finds

$$j_{\text{BRST}}(z)\,j_{\text{BRST}}(w) = \cdots + \frac{D-26}{12}\frac{1}{z-w}(\partial^3 c\,c)(w) + \ldots\,. \quad (5.34)$$

More singular terms and total derivatives do not contribute to Q^2. We thus find once again $D = 26$, this time as the condition for nilpotency of the BRST charge. Alternatively we can compute

$$Q^2 = \frac{1}{2}\{Q,Q\} = \frac{1}{2}\sum_{m,n=-\infty}^{\infty}\left([L_m^{\text{tot}},L_n^{\text{tot}}] - (m-n)L_{m+n}^{\text{tot}}\right)c_{-m}c_{-n} \quad (5.35)$$

which implies that $Q^2 = 0$ if the total anomaly of the matter-plus-ghost system vanishes.

For the BRST transformation properties of the various fields we find

$$[Q, X^\mu(z)] = c\, \partial X^\mu(z),$$

$$[Q, T^{\text{tot}}(z)] = \frac{1}{12}(D - 26)\, \partial^3 c(z),$$

$$\{Q, c(z)\} = c\, \partial c(z),$$

$$\{Q, b(z)\} = T^{\text{tot}}(z), \tag{5.36}$$

where $T^{\text{tot}} = T^X + T^{b,c}$. One also verifies

$$[Q, j_{\text{gh}}(z)] = -j_{\text{BRST}}(z) \tag{5.37}$$

provided one chooses $\kappa = \frac{3}{2}$ and j_{gh} as in (5.16). j_{BRST} and T^{tot} are therefore BRST exact.

Expressed in terms of modes Eqs. (5.36) are

$$[Q, \alpha_n^\mu] = -\sum_m n\, c_m\, \alpha_{n-m}^\mu,$$

$$[Q, L_n^{\text{tot}}] = -\frac{1}{12}(D - 26)\, n\, (n^2 - 1)\, c_n,$$

$$\{Q, c_n\} = -\sum_m (2n + m)\, c_{-m}\, c_{m+n},$$

$$\{Q, b_n\} = L_n^{\text{tot}}, \tag{5.38}$$

where

$$L_n^{\text{tot}} = \sum_{m=-\infty}^{\infty} \left(\frac{1}{2} : \alpha_{n-m} \cdot \alpha_m : + (n + m) : b_{n-m} c_m : \right) \tag{5.39}$$

and, in particular

$$L_0^{\text{tot}} = \sum_{m>0} (\alpha_{-m} \cdot \alpha_m + m\, (b_{-m} c_m + c_{-m} b_m)) + \frac{\alpha'}{4} p^2 - 1$$

$$= N^{\text{tot}} + \frac{\alpha'}{4} p^2 - 1. \tag{5.40}$$

Here, N^{tot} is the total level number, including the two light-cone directions and the ghost contribution. Note that the normal ordering constant -1 is due to the ghost sector. In the critical dimension, i.e. when $Q^2 = 0$, Eq. (5.38) are nothing but the general results (5.24). For X one simply has the behavior under holomorphic

5.2 BRST Quantization

coordinate transformations with the symmetry parameter replaced by the ghost. This is according to the general BRST procedure.

It is now straightforward to verify that the total action

$$S = S^X + S^{b,c}$$
$$= \frac{1}{2\pi\alpha'} \int d^2z \, (\partial X \cdot \bar{\partial} X + \alpha' b \bar{\partial} c + \alpha' \bar{b} \partial \bar{c}) \quad (5.41)$$

is invariant under BRST transformations. In fact, one can derive the BRST current as the Noether current of this symmetry.

With the help of (5.38) it is possible to show that $Q^2 = 0$ implies that the conformal anomaly is zero. Indeed,

$$[L_m, L_n] = [L_m, \{Q, b_n\}] = -\{Q, [b_n, L_m]\} + \{b_n, [L_m, Q]\}$$
$$= \{Q, [L_m, b_n]\} + \{b_n, [\{Q, b_m\}, Q]\}$$
$$= (m-n)\{Q, b_{m+n}\} = (m-n)L_{m+n}, \quad (5.42)$$

where we have repeatedly used the graded Jacobi identity, $Q^2 = 0$ and Eq. (4.38).

We want to comment on the relation of the BRST procedure presented here to the one familiar from gauge theories. There the ghost contribution to the action can be written as $\mathcal{L}^{\text{ghost}} \sim \delta(b^i F_i)$ where F_i is called the gauge fixing function. This guarantees the BRST invariance of the total action. One introduces the auxiliary field B via $\delta b^i = B^i$ and $\delta B^i = 0$ as a consequence of $Q^2 = 0$. B^i serves as a Lagrange multiplier for the gauge fixing. One then eliminates B^i via the equations of motion (for the non-ghost fields). The gauge fixing function leading to the action Eq. (5.41) would have been $F^{\alpha\beta} = \sqrt{h}(h^{\alpha\beta} - \eta^{\alpha\beta})$.

Let us now turn to the problem of identifying physical states. We again demand that they are BRST singlet states, i.e. $Q|\phi\rangle = 0$ but $|\phi\rangle \neq Q|\lambda\rangle$. According to our discussion in Chap. 4, a state $|\phi\rangle$ is created from the $SL(2, \mathbb{C})$ invariant vacuum by a local vertex operator: $|\phi\rangle = \phi(0)|0\rangle$. BRST invariance then implies that

$$[Q, \phi(z)] = \oint_{C_z} \frac{dw}{2\pi i} \, j_{\text{BRST}}(w) \, \phi(z) = \text{total derivative} \quad (5.43)$$

i.e. the operator product of j_{BRST} and ϕ must not have a pole of order one, unless the residue is a total derivative, in which case it vanishes upon integration over the insertion point of the vertex operator. Then correlation functions will be BRST invariant. Consider states without ghost excitations. Then

$$\oint \frac{dw}{2\pi i} \, j_{\text{BRST}}(w) \, \phi(z) = \oint \frac{dw}{2\pi i} \, c(w) \, T^\phi(w) \, \phi(z)$$
$$= \oint \frac{dw}{2\pi i} \, c(w) \left[\frac{h_\phi \phi(z)}{(w-z)^2} + \frac{\partial \phi(z)}{w-z} + \ldots \right]$$
$$= h_\phi \, (\partial c) \, \phi(z) + c \, \partial \phi(z) \, , \quad (5.44)$$

which is a total derivative whenever the conformal weight of ϕ is $h_\phi = 1$. We have thus found that primary fields of dimension one create asymptotic BRST invariant states.

Let us now look at the ghost sector of the theory. Both b and c have zero frequency components which satisfy the anti-commutation relations $b_0^2 = c_0^2 = 0$ and $\{c_0, b_0\} = 1$. b_0 and c_0 commute with the Hamiltonian L_0 in Eq. (5.13). There are then two degenerate states. One, denoted $|\uparrow\rangle$, is annihilated by c_0. The second is then defined as $|\downarrow\rangle = b_0|\uparrow\rangle$:

$$c_0|\uparrow\rangle = 0, \qquad b_0|\downarrow\rangle = 0$$
$$b_0|\uparrow\rangle = |\downarrow\rangle, \qquad c_0|\downarrow\rangle = |\uparrow\rangle . \qquad (5.45)$$

These states clearly have zero norm. They differ in ghost number by one unit. We can add a constant to j_0 in (5.18) such that the ghost number assignments of the two degenerate states are symmetric:

$$N_g = \frac{1}{2}(c_0 b_0 - b_0 c_0) + \sum_{n>0}(c_{-n} b_n - b_{-n} c_n) . \qquad (5.46)$$

This corresponds to the definition $j(z) = -:b(z) c(z): -\frac{3}{2z}$. With this convention the states $|\uparrow\rangle$ and $|\downarrow\rangle$ have ghost charge $+\frac{1}{2}$ and $-\frac{1}{2}$ respectively.

On the other hand we notice that the $SL(2, \mathbb{C})$ invariant ghost vacuum $|0\rangle_{b,c}$ obeys Eq. (5.8). This means that while being a highest weight state of the Virasoro algebra, it is not a highest weight state of the b, c algebra since it is not annihilated by all positive frequency modes:

$$c_1|0\rangle_{b,c} = c(0)|0\rangle_{b,c} \neq 0 . \qquad (5.47)$$

Since $[L_0, c_1] = -c_1$, $|0\rangle_{b,c}$ is not the ground state of the ghost system. The state

$$c_0 c_1 |0\rangle_{b,c} = -c\, \partial c(0)|0\rangle_{b,c} \neq 0 \qquad (5.48)$$

has also L_0 eigenvalue -1; i.e. the two states $c_1|0\rangle_{b,c}$ and $c_0 c_1|0\rangle_{b,c}$ are degenerate. It is easy to see that if we identify $c_1|0\rangle_{b,c} = |\downarrow\rangle$ and $c_0 c_1|0\rangle_{b,c} = |\uparrow\rangle$, we can verify the relations Eq. (5.45) and their ghost number assignments. In addition, we find that $\langle\uparrow|\uparrow\rangle = \langle\downarrow|\downarrow\rangle = 0$ and $\langle\uparrow|\downarrow\rangle = \langle\downarrow|\uparrow\rangle = {}_{b,c}\langle 0|c_{-1} c_0 c_1|0\rangle_{b,c} \neq 0$ and that $|\uparrow\rangle$ and $|\downarrow\rangle$ are annihilated by all positive ghost modes. We choose the normalization such that

$${}_{b,c}\langle 0|c_{-1} c_0 c_1|0\rangle_{b,c} = 1 . \qquad (5.49)$$

This shows that the $SL(2, \mathbb{C})$ invariant vacuum carries three units of ghost number. They correspond to the three global diffeomorphisms of the sphere (compactified plane), generated by $L_0, L_{\pm 1}$. Since correlation functions are invariant under

5.2 BRST Quantization

$SL(2,\mathbb{C})$ (cf. Chap. 4), the gauge fixing is not complete which is reflected by the presence of ghost zero modes. For details we refer to the following chapter where we learn how to deal with the ghost zero modes in the computation of scattering amplitudes. In the operator language, the two-point functions (5.3) are ${}_{b,c}\langle 0|c_{-1} c_0 c_1 b(z) c(w)|0\rangle_{b,c}$ and ${}_{b,c}\langle 0|c_{-1} c_0 c_1 c(z) b(w)|0\rangle_{b,c}$.

Physical states are characterized by BRST cohomology classes of some definite ghost number. We can build states on either of the two ghost ground states $|\uparrow\rangle$ and $|\downarrow\rangle$. Let us consider states of the form

$$|\psi\rangle = |\phi\rangle_X \otimes |\downarrow\rangle_{b,c} . \tag{5.50}$$

BRST invariance of this state then requires

$$Q|\psi\rangle = \left(c_0 (L_0^X - 1) + \sum_{n>0} c_{-n} L_n^X \right) |\psi\rangle = 0 \tag{5.51}$$

which is equivalent to the physical state conditions

$$(L_0^X - 1)|\phi\rangle_X = 0 \quad \text{and} \quad L_n^X |\phi\rangle_X = 0, \quad \text{for } n > 0 . \tag{5.52}$$

Had we instead taken the state $|\uparrow\rangle$, we could not have obtained the condition $(L_0^X - 1) = 0$ since $c_0|\uparrow\rangle = 0$. We therefore require $b_0|\psi\rangle = 0$ and for the closed string also $\bar{b}_0|\psi\rangle = 0$. This eliminates all states built upon the $|\uparrow\rangle_{b,c}$ vacuum. For the closed string it also implies $\{Q, b_0 - \bar{b}_0\}|\psi\rangle = (L_0 - \bar{L}_0)|\psi\rangle = 0$, i.e. level matching.

From $\{Q, b_0\} = L_0^{\text{tot}}$ it follows that any state of definite level number N^{tot} and definite momentum p^μ which satisfies $Q|\psi\rangle = 0$ and $b_0|\psi\rangle = 0$, satisfies $\alpha' m^2 = 4(N^{\text{tot}} - 1) = 0$, cf. Eq. (5.40). Furthermore, any Q-closed state of definite level number and momentum, which does not satisfy $\alpha' m^2 = 4(N^{\text{tot}} - 1)$, is Q-exact. Indeed, denote such a state by $|N, p\rangle$. Then $\{Q, b_0\}|N, p\rangle = Q\, b_0|N, p\rangle = L_0|N, p\rangle$ with L_0-eigenvalue $\frac{\alpha'}{4} p^2 + N - 1$. It follows that if this does not vanish, $|N, p\rangle = Q\left(\frac{1}{L_0} b_0|N, p\rangle\right)$. It takes more effort to show that every non-trivial Q-cohomology class has a representative of the form

$$|\psi\rangle = |\phi\rangle_X \otimes |\downarrow\rangle = |\phi\rangle_X \otimes (c_1|0\rangle_{b,c}) , \tag{5.53}$$

where $|\phi\rangle_X$ is a highest weight state of the Virasoro algebra with L_0 eigenvalue $+1$. From the oscillator expression of Q it follows that any state (5.53) which is Q-exact must be of the form $\sum_{m>0} L_m|\psi_m\rangle \otimes |\downarrow\rangle_{b,c}$. We have shown in Chap. 4 that such states decouple from string scattering amplitudes.

The vertex operators of physical states (5.53) are of the form

$$\psi(z) = c(z)\,\phi(z) . \tag{5.54}$$

One easily shows that

$$[Q, \psi(z)] = (h_\phi - 1) \; :\partial c \, c \, \phi(z): \qquad (5.55)$$

which vanishes for $h_\phi = 1$; i.e. if $\phi(z)$ with $h_\phi = 1$ satisfies Eq. (5.44), $c\phi(z)$ commutes with Q without any derivative terms. The fields $c\phi(z)$ then have zero conformal weight.

To summarize, there are two versions for each vertex operator of a physical state: an unintegrated, $\psi(z,\bar{z})$, and an integrated, $\int d^2z \, \phi(z,\bar{z})$ s.t. $[Q,\phi] = \partial(c\phi)$, $[\overline{Q},\phi] = \bar{\partial}(\bar{c}\phi)$ and $[Q,\psi] = [\overline{Q},\psi] = 0$ with $\psi = c\bar{c}\,\phi$. Here we have combined holomorphic and anti-holomorphic parts. For open string vertex operators we have $\psi = c\phi$ and the integrated vertex operator is $\int dx \, \phi(x)$ where the integral is over the real axis.

As an example, consider the closed string tachyon with

$$|\phi\rangle_X = |k\rangle = \lim_{z,\bar{z}\to 0} :e^{ik\cdot X(z,\bar{z})}: |0\rangle_X. \qquad (5.56)$$

Its mass is given by its $L_0^{b,c}$ eigenvalue. Since the ghost ground state gives $L_0^{b,c}|\downarrow\rangle = -|\downarrow\rangle$, we have $\frac{1}{4}\alpha' p^2 = -1$. Therefore we can attribute the negative (mass)^2 of the tachyon to the ghost contribution to the string spectrum.

So far most of our discussion was restricted to the holomorphic sector of the closed string. The anti-holomorphic sector is completely isomorphic. For the open string, left and right movers are again coupled through the boundary conditions.

Further Reading

A very readable introduction to the general BRST method is:

- T. Kugo, S. Uehara, General procedure of gauge fixing based on BRS invariance principle. Nucl. Phys. **B197**, 378 (1982)

The application to string theory can be found in:

- M. Kato, K. Ogawa, Covariant quantization of string based on BRS invariance. Nucl. Phys. **B212**, 443 (1983)

The explicit construction of the BRST charge in terms of the generators and structure constants of the gauge algebra is due to Batalin and Vilkovisky. It was applied to the bosonic string theory in:

- S. Hwang, Covariant quantization of the string in dimensions $D \leq 26$ using a Becchi-Rouet-Stora formulation. Phys. Rev. **D28**, 2614 (1983)

Our discussion follows:

- D. Friedan, E. Martinec, S. Shenker, Covariant quantization, supersymmetry and stringtheory. Nucl. Phys. **B271**, 93 (1986)

Useful reviews are:

- M. Peskin, *Introduction to String and Superstring Theory II*, Santa Cruz TASI proceedings *From the Planck scale to the weak scale, TASI 1986*, H. Haber (ed.),World Scientific 1987, p.277
- J.H. Schwarz, Faddeev-Popov ghosts and BRS symmetry in string theories. Prog. Theor. Phys. **S86**, 70 (1986)
- P. West, An introduction to string theory. Acta Phys. Polon. **B20**, 471 (1989)

The Q-cohomology is discussed in

- M.D. Freeman, D.I. Olive, BRS cohomology in string theory and the no ghost theorem. Phys. Lett. **B175**, 151 (1986)

Chapter 6
String Perturbation Theory and One-Loop Amplitudes

Abstract In this chapter we study issues of relevance for the perturbation theory of oriented bosonic strings. After giving a general description of world-sheets of higher genus, we discuss in some detail string one loop diagrams. We first do this for the closed string leading to torus diagrams, which we discuss both for the bosonic string and, continuing our presentation from Chap. 4, also for abstract conformal field theories. In this context we also present the simple current method, which provides a powerful tool for generating modular invariant partition functions. We also discuss the one-loop amplitude for open strings. From the one-loop amplitude of an open string stretched between two bosonic Dp-branes we extract the D-brane tension.

6.1 String Perturbation Expansion

String world-sheets are two dimensional surfaces. Clearly, fixing the number of in- and out-going strings does not yet specify the world-sheet. More complicated world-sheets intuitively correspond to higher orders in perturbation theory. To illustrate this, we look at the closed oriented string whose world-sheets are orientable surfaces. Consider as an example the tree level scattering amplitude of four strings shown in Fig. 6.1. The interactions of strings result from their splitting and joining. The corresponding world-sheet has tubes extending into the past and the future for incoming and outgoing closed strings, respectively. In the Polyakov formulation[1] a scattering amplitude is a functional integral over oriented surfaces bounded by the position curves of the initial and final string configurations, weighted with the exponential of the free action (Polyakov action) and integrated with the string wave functions. The key observation is now that conformal invariance allows to consider punctured world-sheets instead of surfaces with boundaries

---

[1] The alternative operator approach leads to the same results.

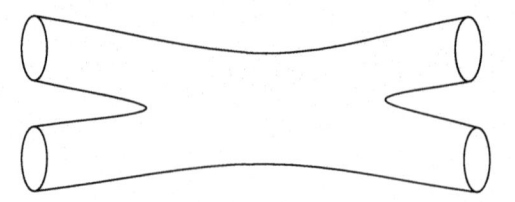

Fig. 6.1 Tree level scattering of four closed strings

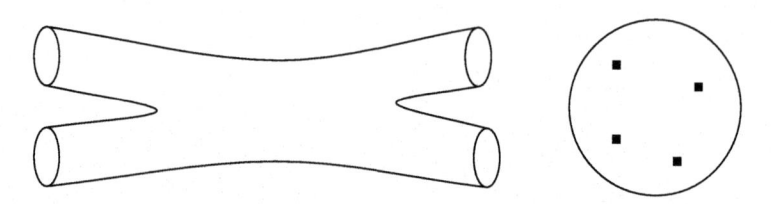

Fig. 6.2 Map of asymptotic string states to points on the sphere

corresponding to incoming and outgoing strings. The incoming and outgoing strings are conformally mapped to points (the punctures) of the two-dimensional surface (see Fig. 6.2).

Consider, for example, the case of a world-sheet with only one incoming and one outgoing string, described by a cylinder with metric $ds^2 = d\tau^2 + d\sigma^2$, $-\infty < \tau < \infty$, $0 \leq \sigma < 2\pi$. From here on, unless specified otherwise, we will always work with world-sheets with Euclidean signature metrics. There are several reasons for this. First, it allows us to use techniques of two-dimensional CFT, like the ones we have developed in Chap. 4 and the mathematics of Riemann surfaces. Second, Riemann surfaces generally do not admit non-singular Lorentzian metrics. The only exception is the torus.[2] Taking $\tau = \ln r$ this becomes $ds^2 = r^{-2}(dr^2 + r^2 d\sigma^2)$. The incoming string ($\tau = -\infty$) has been mapped to the point $r = 0$ and the outgoing string ($\tau = +\infty$) to $r = \infty$. The string world-sheet has been mapped to the plane. A suitable choice of a conformal factor maps the plane to the sphere. We rescale the metric by $4r^2(1 + r^2)^{-2}$ and get $d\tilde{s}^2 = 4\frac{dr^2 + r^2 d\sigma^2}{(1+r^2)^2} = \frac{4 dz d\bar{z}}{(1+|z|^2)^2}$ where $z =$

---

[2] To have a Lorentzian metric one needs a globally defined vector field which defines the direction of time. Given such a vector field t^α and a Riemannian metric $g_{\alpha\beta}$ which always exists on the surfaces of interest, the metric

$$h_{\alpha\beta} = g_{\alpha\beta} - 2\frac{g_{\alpha\gamma}g_{\beta\delta}t^\gamma t^\delta}{(g_{\gamma\delta}t^\gamma t^\delta)}$$

is Lorentzian. For a surface without boundary such a vector field only exists on the torus. For surfaces with boundaries a Lorentzian metric exists on the cylinder but not e.g. for the surface shown on the left hand side of Fig. 6.2, if we require that the boundary components are space-like, e.g. time flowing in from the left and out to the right boundary circles. It is clear that the 'arrow of time' must merge and split somewhere where it is not well defined.

6.1 String Perturbation Expansion

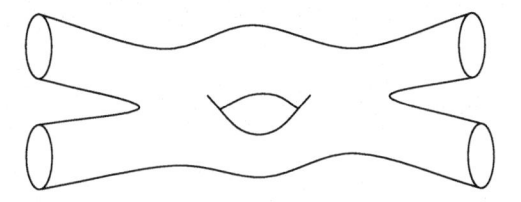

Fig. 6.3 One loop scattering of four closed strings

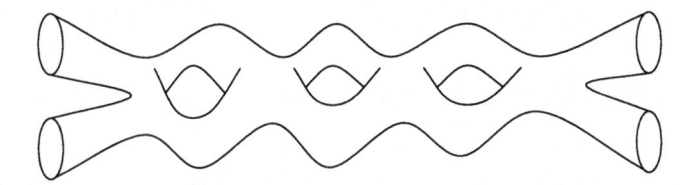

Fig. 6.4 Multiloop scattering of four closed strings

$re^{i\sigma}$. This is the standard round metric of the sphere, stereographically projected onto the plane. Indeed, with $z = \cot(\frac{\theta}{2})\, e^{i\phi}$ we find $d\tilde{s}^2 = d\theta^2 + \sin^2\theta\, d\phi^2$. The incoming and outgoing strings are now finite points, namely the south and north pole of the sphere. For more complicated string diagrams with several incoming and outgoing strings, the conformal factor can always be chosen to map all of them to points on the sphere. This remains also true for loop diagrams (cf. below) where the external strings are mapped to points on spheres with g handles, where g denotes the number of loops. The quantum numbers of the external string states are generated by local operators inserted at these points. These are the vertex operators introduced in Chap. 4. In summary, performing the conformal mapping, the world-sheet becomes a two-dimensional surface with the incoming and outgoing particles inserted by local vertex operators. In this sense vertex operators can be viewed as conformal projections of asymptotic states. It is however only known how to construct vertex operators for on-shell states.[3] Then string scattering amplitudes of on-shell particles are correlation functions of vertex operators.

Analogously, the one loop scattering amplitudes are described by world-sheets with one 'hole' (handle) as shown in Fig. 6.3. A four-point multi-loop diagram is drawn in Fig. 6.4.

Two dimensional oriented surfaces without boundaries are topologically completely characterized as spheres with g handles. The Euler number, which is a topological invariant, is $\chi = \frac{1}{4\pi} \int d^2\sigma \sqrt{h} R = 2(1-g)$. The number of handles is called the genus of the surface. In summary, an n-point, g-loop amplitude of the closed oriented string is described by a two-dimensional surface with g handles and

---

[3]In previous chapters we have seen how the requirement of conformal or BRST invariance puts them on shell.

n vertex operator insertions. No interaction terms need to be added to the Polyakov action. Interactions are encoded in the topology of the world-sheet.

In perturbation theory, general closed string n-point amplitudes can then be computed as the following (formal) path integral:

$$A_n = \sum_{g=0}^{\infty} A_n^{(g)}$$

$$= \sum_{g=0}^{\infty} C_{\Sigma_g} \int \mathcal{D}h \, \mathcal{D}X^\mu \int d^2z_1 \ldots d^2z_n \, V_1(z_1,\bar{z}_1) \ldots V_n(z_n,\bar{z}_n) \, e^{-S[X,h]},$$
(6.1)

where we sum over all topologies of the world-sheet and integrate over the insertion points of the vertex operators. C_{Σ_g} is a weight factor which depends only on the topology of the world-sheet. It can be computed from first principles, but one often determines it in an indirect way by imposing unitarity. This will be done for C_{S^2} and C_{D_2} in Chap. 16. $A_0^{(g)}$ is called the genus g partition function and is commonly denoted by $Z^{(g)}$. To shed some light on the meaning of this formal expression is the purpose of the next section. However, the only amplitudes we will compute explicitly are tree-level N-point amplitudes in Chap. 16 and one-loop vacuum amplitudes (this chapter). In these cases, most of the rather heavy machinery developed here can be avoided by resorting to simpler, somewhat heuristic, arguments.[4]

The generalization to the open oriented string is straightforward: we simply have to allow for surfaces with boundaries. Two-dimensional surfaces with boundary can be obtained from surfaces without boundary by removing disks. The scattering amplitude of open and closed strings is associated with Riemann surfaces with boundaries where the asymptotic open strings are realized as vertex operator insertions at a boundary component and closed strings are realized as vertex operator insertions in the bulk of the surface. The Euler number of a Riemann surface with g handles and b boundary components, each of which is isomorphic to a circle S^1, is $\chi = 2 - 2g - b$.

For the perturbation theory of unoriented strings we also need to consider non-orientable world-sheets. On a non-orientable world-sheet there are closed non-contractible paths such that if one parallel transports a pair of vectors around they change their relative orientation. A simple example with boundary is the Möbius strip. It arises if we consider the propagation of an open string. We can glue the two ends of the strip to form a cylinder or we can glue up to a Ω transformation which results in a Möbius strip. Non-orientable world-sheets can be obtained from orientable world-sheets by adding crosscaps. A crosscap is obtained if one

---

[4] For higher loop amplitudes, e.g. when discussing questions of finiteness of string amplitudes or to prove unitarity of the theory, these tools are necessary. They are fully under control for the bosonic string. For the superstring they are much more subtle and not yet fully worked out in all detail.

6.1 String Perturbation Expansion

removes a disk and identifies opposite points on the boundary. For instance, the real projective plane can be defined as the space of all lines through the origin in $\mathbb{R}^2$, or, equivalently, as $S^2/\mathbb{Z}_2$ where the action of $\mathbb{Z}_2$ is the identification of antipodal points. It is topologically the same as S^2 with a crosscap. The Möbius strip is a hemisphere with a disk removed (i.e. a cylinder) with a crosscap glued in. If one glues in a second crosscap one obtains the Klein bottle. The sphere, the disk and the projective plane are tree level and torus, cylinder, Möbius strip and Klein bottle are one-loop world-sheets. We will encounter the former again in Chap. 16 and the latter later in this chapter.

The Euler number of a non-orientable surface with g handles, b holes and c crosscaps is

$$\chi = 2 - 2g - b - c. \tag{6.2}$$

The Euler number is a topological invariant. The Euler number, the number of boundaries and orientability completely specify the topology of a two-dimensional (connected) manifold. For instance, attaching two crosscaps to a non-orientable surface is equivalent to attaching a handle.

The Gauss-Bonnet theorem states that[5]

$$\chi(\Sigma) = \frac{1}{4\pi} \int_\Sigma \sqrt{h} R d^2\sigma + \frac{1}{2\pi} \int_{\partial\Sigma} k ds, \tag{6.3}$$

where R is the curvature scalar of h and k the trace of the extrinsic curvature on the boundary which, in general, consists of several components.[6] But these are precisely the terms which we can add to the Polyakov action without changing the equations of motion. If we include in S_P the term $-\lambda\chi$ and define $g_s = e^\lambda$, then each term in the perturbation series (6.1) will be weighted by

$$g_s^{-2+2g+b+c+n_c+\frac{1}{2}n_o} = g_s^{-\chi+n_c+\frac{1}{2}n_o}, \tag{6.4}$$

where $n_c(n_o)$ is the number of closed (open) string vertex operator insertions. The dependence on χ is clear. The factors arising from the vertex operator insertions are also easily understood as follows. Consider any world-sheet and add a handle to it, changing $g \to g + 1$. This corresponds to the emission and reabsorption of a closed string, i.e. to the insertion of two closed string vertex operators, each of which contributes g_s. Consider now a world-sheet with boundary and add a 'handle' to the boundary. This changes $b \to b + 1$ and is equivalent to the insertion of two open string vertex operators. In other words, we attach to each closed string vertex operator a factor g_s and to each open string vertex operator a factor $g_s^{1/2}$.

---

[5]If the boundary has corners, there is a correction term $\frac{1}{2\pi}\sum_i (\pi - \theta_i)$, where the sum is over all interior angles.

[6]For non-orientable surfaces one goes to the orientable double cover (which always exists) with twice as many boundary components; (6.3) then computes twice the Euler number of the non-orientable surface.

The simplest interacting closed string diagram, namely one ingoing and two outgoing closed strings is proportional to g_s. Therefore g_s is called the string coupling

$$g_s = e^\lambda = e^{\langle\Phi\rangle}$$

constant. However, as we have already mentioned in Chap. 2 and will elaborate further in Chap. 14, λ is the constant background value of the dilaton Φ and g_s is not a free dimensionless parameter. In fact, there is no free dimensionless parameter in string theory. Any rescaling of g_s can be reabsorbed in the action and the normalization of the vertex operators.

6.2 The Polyakov Path Integral for the Closed Bosonic String

Since the action S is invariant under conformal transformations and diffeomorphisms of the world-sheet, the Polyakov path integral is highly divergent. One integrates infinitely many times over gauge equivalent metric configurations, and in the absence of anomalies, i.e. in the critical dimension, any two metrics which are equivalent under diffeomorphisms and rescalings give the same contribution to the path integral. To compensate for this overcounting one has to divide the measure in Eq. (6.1) by the volume of the symmetry group which is generated by diffeomorphisms and Weyl rescaling; i.e. we need to consider

$$\int \frac{\mathscr{D}h\,\mathscr{D}X}{\mathrm{Vol}(\mathrm{Diff})\,\mathrm{Vol}(\mathrm{Weyl})}. \tag{6.5}$$

We thus have to examine the measure of the integration over metrics, $\mathscr{D}h$. First, fix a particular metric $h_{\alpha\beta}$. We know how it changes under diffeomorphisms and Weyl rescaling. However, on a general world-sheet not all metrics can be generated from $h_{\alpha\beta}$ by such a symmetry transformation. Clearly, if the operator $P^\dagger$ which we defined in Chap. 2 has zero modes, this is not possible. A point in the moduli space of metrics is an equivalence class of metrics where two metrics are in the same class if they can be transformed into each other by a diffeomorphism and a Weyl rescaling:

$$\mathscr{M}_g = \frac{\{\text{metrics}\}}{\{\text{Weyl rescalings}\} \times \{\text{diffeomorphisms}\}}. \tag{6.6}$$

In the absence of anomalies the Polyakov path integral thus reduces to an integral over the moduli space. Its dimension is the number of zero modes of $P^\dagger$. Our aim is to disentangle the integral over metrics into an integral over diffeomorphisms, an

6.2 The Polyakov Path Integral for the Closed Bosonic String

integral over Weyl transformations and an integral over moduli in such a way that the first two factors cancel the infinite volume factors and what is left is a finite dimensional integral over the moduli. The functional integral over metrics is then reduced to a (finite dimensional) integral over a gauge slice[7] such that each metric can be reached from a unique metric on the slice by a gauge transformation. We will say more about the moduli spaces of string world-sheets below. It is the separation of the measure that we now turn to. A word of warning: we will not keep track of numerical factors.

Let $h_{\alpha\beta}$ be the reference metric, i.e. a point on the chosen gauge slice. An arbitrary infinitesimal change of $h_{\alpha\beta}$ can be written in the form

$$\delta h_{\alpha\beta} = \delta \Lambda\, h_{\alpha\beta} + (\nabla_\alpha \xi_\beta + \nabla_\beta \xi_\alpha) + \sum_i \delta\tau_i \frac{\partial}{\partial \tau_i} h_{\alpha\beta}, \qquad (6.7)$$

which is the sum of a Weyl rescaling, a reparametrization and the rest. τ_i are the moduli parameters, i.e. local coordinates on moduli space at a point in the vicinity of the reference metric $h_{\alpha\beta}$ which, by definition, is at the origin of a local coordinate system on $\mathcal{M}_g$. We can absorb the trace parts of the latter two terms in (6.7) into the first term by defining

$$\delta\tilde{\Lambda} = \delta\Lambda + \frac{1}{2}\nabla\cdot\xi + \frac{1}{2}\sum_i \delta\tau^i\, h^{\alpha\beta}\, \partial_i h_{\alpha\beta}. \qquad (6.8)$$

Defining ($\partial_i = \frac{\partial}{\partial \tau_i}$)

$$\mu^i_{\alpha\beta} = \partial_i h_{\alpha\beta} - \frac{1}{2} h_{\alpha\beta}\, h^{\gamma\delta}\, \partial_i h_{\gamma\delta}, \qquad (6.9)$$

we can rewrite (6.7) as

$$\delta h_{\alpha\beta} = \delta\tilde{\Lambda}\, h_{\alpha\beta} + (P\,\xi)_{\alpha\beta} + \sum \delta\tau^i\, \mu^i_{\alpha\beta}. \qquad (6.10)$$

This is a decomposition of the tangent space of the space of metrics at the point $h_{\alpha\beta}$. The space of metrics has a natural structure of a fiber bundle with a finite dimensional base space (gauge slice) and infinite dimensional fibers. The fiber over a point in the base consists of all metrics which can be obtained by applying all diffeomorphisms and Weyl rescalings to the metric at that point. The first two terms in (6.10) are along the fiber at $h_{\alpha\beta}$ while the last term is along the gauge slice. The Eq. (6.10) is not yet what we want, because it is not an orthogonal decomposition.

---

[7] The choice of a good gauge slice for the world-sheet gravitino of the RNS string is considerably more difficult than for the metric for the bosonic string. We will not discuss these issues.

Orthogonality is defined with respect to the inner product

$$(\delta h^{(1)}|\delta h^{(2)}) = \int \sqrt{h}\, h^{\alpha\gamma} h^{\beta\delta}\, \delta h^{(1)}_{\alpha\beta}\, \delta h^{(2)}_{\gamma\delta}, \qquad ||\delta h||^2 = (\delta h|\delta h) \qquad (6.11)$$

which we had already defined in Chap. 3. Analogous definitions will be used for the inner product of other tensors. Note that neither $||\delta h||^2$ nor $||\delta X||^2$ are Weyl invariant, but $||\delta \xi||$ is. All norms are invariant under diffeomorphisms. The total measure is only Weyl-invariant in the critical dimension.

The last two terms in (6.10) are not yet orthogonal with respect to (6.11). The space of symmetric traceless tensors orthogonal to the image of P is spanned by the quadratic differentials $\phi_{\alpha\beta}$, i.e. by solutions of $P^\dagger \phi = 0$. Let $\{\phi^i\}$ be a basis of quadratic differentials on the world-sheet Σ_g and denote by $\Pi = \sum_{ij} |\phi_i\rangle M_{ij}^{-1} \langle \phi_j|$ with $M_{ij} = (\phi_i|\phi_j)$ the projection operator on the space spanned by the ϕ_i. We then have the decomposition

$$\sum_i \delta\tau_i\, \mu^i = \sum_i \delta\tau^i\, (\Pi\mu)^i + \sum_i \delta\tau^i\, ((1-\Pi)\mu)^i$$
$$= \sum_i \delta\tau^i\, (\Pi\mu)^i + \sum_i \delta\tau^i\, (P\zeta^i) \qquad (6.12)$$

for some vector fields ζ^i (because $(1-\Pi)\mu$ is in the range of P). We then find the orthogonal decomposition of the tangent space of the space of metrics at $h_{\alpha\beta}$

$$\delta h_{\alpha\beta} = \delta\Lambda\, h_{\alpha\beta} + (P\xi)_{\alpha\beta} + \sum_{ijk} \phi^j_{\alpha\beta}\, M_{jk}^{-1}\, (\phi^k|\mu^i)\, \delta\tau^i, \qquad (6.13)$$

where we have dropped the tilde on Λ and we have redefined $\xi + \sum \delta\tau^i\, \zeta^i \to \xi$. The decomposition of the norm is therefore

$$||\delta h||^2 = 2||\delta\Lambda||^2 + (\xi, P^\dagger P \xi) + \sum_{ij} \delta\tau^i\, \delta\tau^j\, (\mu^i|\phi^k)\, M_{kl}^{-1}\, (\phi^l|\mu^j), \qquad (6.14)$$

where $||\delta\Lambda||^2 = \int_\Sigma \sqrt{h}(\delta\Lambda)^2$. Given the decomposition (6.14) we can finally write the measure over the space of metrics as

$$\mathcal{D}h = \mathcal{J}\mathcal{D}\Lambda \mathcal{D}'\xi \prod_i d\tau^i \qquad (6.15)$$

with Jacobian

$$\mathcal{J} = {\det}'^{1/2}(P^\dagger P)\, \frac{\det(\phi_i|\mu_j)}{\det^{1/2}(\phi_k|\phi_l)}. \qquad (6.16)$$

6.2 The Polyakov Path Integral for the Closed Bosonic String

The prime means that we do not integrate over the zero modes of P: they are the conformal Killing vectors (CKV) whose effect on $\delta h_{\alpha\beta}$ is already accounted for in the integral over the Weyl factor. We will come back to the subtlety related to CKV's later. Before we complete our discussion of the integration measure, we introduce some useful concepts and notation.

Let us consider two-dimensional compact orientable manifolds without boundary, Σ. Topologically Σ is completely specified by its genus, i.e. it is a sphere with g handles. We can now add additional structure. A Riemannian structure is the choice of a Riemannian metric on Σ. This is always possible on a compact manifold (in contrast to a Lorentzian structure, as we have discussed above). We showed in Chap. 2 that locally on a world-sheet of Minkowski signature we can introduce coordinates in which the metric has the simple form $ds^2 = 2e^{2\varphi}((d\sigma^1)^2 - (d\sigma^0)^2)$. After Wick rotation $\sigma^0 = -i\sigma^2$ and introducing complex coordinates $z = \sigma^1 + i\sigma^2$, $\bar{z} = \sigma^1 - i\sigma^2$ the metric becomes $ds^2 = 2e^{2\varphi}dzd\bar{z} \equiv 2h_{z\bar{z}}dzd\bar{z}$.[8] Coordinates in which the metric takes this simple form are called conformal coordinates.[9] Conformal transformations $z \to w = f(z)$ and $\bar{z} \to \bar{w} = \bar{f}(\bar{z})$ only change the conformal factor $e^{2\varphi}$ of the metric. However, we know that if the operator $P^\dagger$ defined in Chap. 2 has zero modes, then there exist metrics on the worldsheet which are not conformally related, i.e. they cannot be obtained from each other by reparametrization and Weyl rescaling. They are said to have different conformal structures. All metrics which are conformaly related will take the form $ds^2 = 2e^{2\varphi}dzd\bar{z}$ in some fixed coordinate system. Conformally unrelated metrics will take the form $ds^2 = 2e^{2\tilde{\varphi}}|dz + \mu d\bar{z}|^2$ in that same coordinate system where $\mu = \mu_{\bar{z}}^z(z,\bar{z})$ is called a Beltrami differential. If we make an infinitesimal change of coordinates to $z \to z + \delta z$ where $(\delta z, \delta \bar{z})$ is not a globally defined vector field, we find to first order in δz, $\delta h_{\bar{z}\bar{z}} = h_{z\bar{z}} \partial_{\bar{z}} \delta z = h_{z\bar{z}} \mu_{\bar{z}}^z$ with

$$\mu_{\bar{z}}^z = h^{\bar{z}z} \delta h_{\bar{z}\bar{z}}. \tag{6.18}$$

The conformal structures are distinguished by a finite number of parameters, the moduli τ^i. We can then write the change of metric as $\delta h_{\bar{z}\bar{z}} = 2\nabla_{\bar{z}} \xi_{\bar{z}} + \sum \delta\tau^i \mu_{\bar{z}\bar{z}}^i$ where $\mu_{\bar{z}}^{i\,z} = h^{\bar{z}z} \partial_i h_{\bar{z}\bar{z}}$. The first part in $\delta h_{\bar{z}\bar{z}}$ is generated by diffeomorphisms and

---

[8] The transformation $z \to f(\bar{z})$, $\bar{z} \to \bar{w} = \bar{f}(z)$ reverses the orientation. The determinant of the Jacobian is $-|\partial_z \bar{f}|^2 < 0$.

[9] The existence of conformal coordinates in Euclidean signature is guaranteed by the following result: Consider a metric $ds^2 = g_{ij}(x)dx^i dx^j$ written in a local coordinate system. We can define local complex coordinates $z = x^1 + ix^2$ so that the metric has the general form

$$ds^2 = 2e^{2\varphi}|dz + \mu d\bar{z}|^2. \tag{6.17}$$

Via a non-holomorphic coordinate transformation $z \to w(z,\bar{z})$ one can bring the metric (6.17) to the form $ds^2 = 2e^{2\tilde{\varphi}}|dw|^2$ with $\tilde{\varphi} = \varphi - \ln|\partial_z w|$ provided w is a solution of the Beltrami equation $\partial_{\bar{z}} w = \mu \partial_z w$. The Jacobian of this coordinate transformation is $|\partial w|^2 - |\bar{\partial} w|^2$ which must be positive to preserve the orientation. Then $|\mu|^2 < 1$. One can show that a solution of the Beltrami equation always exists locally.

the second, more interesting part, by changes of the complex structure, a concept which we will now explain.

We know from Footnote 9 that we can choose local complex coordinates, i.e. we can cover the world-sheet by conformal coordinate patches U_a and the metric has the form $ds^2 = 2e^{\varphi_a}|dz_a|^2$. On overlaps $U_a \cap U_b \neq \emptyset$ the transition functions are then necessarily analytic. This defines a global system of complex coordinates. A system of analytic coordinate patches is called a complex structure and a manifold with a complex structure is called a complex manifold. A two-dimensional connected manifold with a complex structure is called a Riemann surface. From the above discussion it is clear that for Riemann surfaces a complex structure is the same as a conformal structure and one may view the study of Riemann surfaces as the study of conformally invariant properties of two-dimensional Riemannian manifolds. This is why they play an important role in string perturbation theory and in conformal field theory.[10] The equivalence between complex and conformal structures only holds in two dimensions. We will discuss complex manifolds in arbitrary (even) dimension in Sect. 14.4.

Having globally defined complex coordinates we can define on any Riemann surface vectors $V^z \partial_z$ and $V^{\bar z} \partial_{\bar z}$ and 1-forms $V_z \, dz$ and $V_{\bar z} \, d\bar z$ and in general tensors with components $V^{z \ldots z\bar z \ldots \bar z}{}_{z \ldots z\bar z \ldots \bar z}$. Since the indices z and $\bar z$ range only over one value all tensors are one component objects. Denote by $T^{(n,\bar n)}_{(m,\bar m)}$ a tensor with $n(\bar n)$ upper $z(\bar z)$ and $m(\bar m)$ lower $z(\bar z)$ indices. Under conformal transformations it transforms with weight $(h = m - n, \bar h = \bar m - \bar n)$. Its conformal spin is $h - \bar h = (m - n) - (\bar m - \bar n)$. The non-vanishing metric component $h_{z\bar z}$ and its inverse $h^{z\bar z}$ allow one to convert upper $\bar z$ indices to lower z indices and vice versa. It follows that any tensor can be written with one type of indices only, say z. Such a tensor is called holomorphic. A holomorphic tensor with p lower and q upper indices is said to have rank $n = p - q$. Its rank is also equal to the conformal weight h and to the conformal spin (since $\bar h = 0$). A rank n holomorphic tensor transforms under analytic coordinate transformations as $T(z, \bar z) \to (\partial_z f(z))^n \, T(f(z), \bar f(\bar z))$. From now on we will only consider holomorphic tensors. We will call the space of holomorphic rank n tensors $\mathcal{T}^{(n)}$. Note that elements of $\mathcal{T}^{(n)}$ are in general functions of z and $\bar z$. An analytic tensor is a holomorphic tensor whose components depend only analytically on the coordinates in each local coordinate chart.

We now define a scalar product and norm on $\mathcal{T}^{(n)}$ by ($\sqrt{h} = h_{z\bar z}$)

$$(V^{(n)}|W^{(n)}) = \int d^2z \, \sqrt{h} \, (h^{z\bar z})^n \, V^{(n)*} W^{(n)} \qquad (6.19)$$

and

$$||V^{(n)}||^2 = (V^{(n)}|V^{(n)}), \qquad (6.20)$$

---

[10] This is even true for non-orientable world-sheets as each such surface has a double cover which is a Riemann surface.

6.2 The Polyakov Path Integral for the Closed Bosonic String

where $V^{(n)}, W^{(n)} \in \mathcal{T}^{(n)}$. This is the only possible covariant ultralocal norm (contains no derivatives). We note that it is invariant under Weyl rescalings of the metric only for $n = 1$.

We can define covariant derivatives. As connection we take the Levi-Civita connection with the Christoffel symbols as connection coefficients. In conformal coordinates only two of them are non-zero (recall $h_{z\bar{z}} = e^{2\varphi}$, $\partial = \partial_z$)

$$\Gamma^z_{zz} = 2\partial\varphi, \quad \Gamma^{\bar{z}}_{\bar{z}\bar{z}} = 2\bar{\partial}\varphi \tag{6.21}$$

and the Riemann tensor for a conformally flat metric is (recall that in two dimensions the curvature tensor has only one independent component)

$$\begin{aligned} R_{z\bar{z}z\bar{z}} = -h_{z\bar{z}} R_{z\bar{z}} &= -\frac{1}{2}(h_{z\bar{z}})^2 R \\ &= \partial\bar{\partial} h_{z\bar{z}} - h^{z\bar{z}} \partial h_{z\bar{z}} \bar{\partial} h_{z\bar{z}} \\ &= 2 e^{2\varphi} \partial\bar{\partial}\varphi, \end{aligned} \tag{6.22}$$

where $R_{z\bar{z}}$ is the Ricci tensor and R the Ricci curvature scalar

$$\begin{aligned} R &= -2 h^{z\bar{z}} \partial\bar{\partial} \ln h_{z\bar{z}} \\ &= -4 e^{-2\varphi} \partial\bar{\partial}\varphi. \end{aligned} \tag{6.23}$$

Our conventions are such that for the sphere with $h_{z\bar{z}} = \frac{2}{(1+|z|^2)^2}$ one gets $R = 2$. We now have the following covariant derivatives:

$$\nabla_z^{(n)} : \mathcal{T}^{(n)} \to \mathcal{T}^{(n+1)} : \nabla_z^{(n)} T^{(n)}(z,\bar{z}) = (\partial - 2n\, \partial\varphi)\, T^{(n)}(z,\bar{z}),$$

$$\nabla^z_{(n)} : \mathcal{T}^{(n)} \to \mathcal{T}^{(n-1)} : \nabla^z_{(n)} T^{(n)}(z,\bar{z}) = h^{z\bar{z}} \nabla_{\bar{z}} T^{(n)} = h^{z\bar{z}} \bar{\partial} T^{(n)}(z,\bar{z}) \tag{6.24}$$

which are nothing but the ordinary covariant derivatives with connection coefficients Eq. (6.21). They commute with holomorphic coordinate changes. Note that $(PV)_{zz} = 2\nabla_z^{(1)} V_z$ and $(PV)^{zz} = 2\nabla^z_{(-1)} V^z$ where P is the operator defined in (2.41). The adjoint of $\nabla_z^{(n)}$ is defined as $(W^{(n+1)}|\nabla_z^{(n)} V^{(n)}) = (\nabla_z^{(n)\dagger} W^{(n+1)}|V^{(n)})$. We find

$$(\nabla_z^{(n)})^\dagger = -\nabla^z_{(n+1)}. \tag{6.25}$$

The Ricci identity is also easy to derive:

$$[\nabla^z_{(n+1)} \nabla_z^{(n)} - \nabla_z^{(n-1)} \nabla^z_{(n)}] = \frac{1}{2} n R. \tag{6.26}$$

The zero modes of $(\nabla_z^{(+1)})^\dagger$ are the quadratic differentials. They satisfy

$$h_{z\bar{z}} \nabla^z_{(+2)} \phi^i_{zz} = \partial_{\bar{z}} \phi^i_{zz} = 0 \tag{6.27}$$

i.e. they are global analytic tensors of rank 2. They have a natural pairing with Beltrami differentials

$$(\mu^i|\phi^j) = \int d^2z \, \mu^{iz}_{\bar{z}} \, \phi^j_{zz} \tag{6.28}$$

which does not require a metric. The quadratic differentials thus span the cotangent space to the gauge slice. The kernel of $\nabla_z^{(+1)}$ is spanned by tensors $\in \mathcal{T}^{(1)}$ which satisfy

$$\nabla_z^{(+1)} V_z = h_{z\bar{z}} \, \partial_z V^{\bar{z}} = 0 \tag{6.29}$$

which defines conformal Killing vectors $V^{\bar{z}}$. They are globally defined vector fields which span the kernel of $\nabla_z^{(+1)}$. They generate the conformal Killing group (CKG), the group of conformal isometries. The diffeomorphisms generated by them can be completely absorbed by Weyl rescaling the metric.

We can now rewrite the variation of the metric in complex coordinates:

$$\begin{aligned} \delta h_{z\bar{z}} &= \Lambda h_{z\bar{z}}, \\ \delta h_{zz} &= \nabla_z^{(+1)} \xi_z + \sum_{ijk} \phi^j_{zz} M^{-1}_{jk}(\phi^k|\mu^i)\delta\tau^i \end{aligned} \tag{6.30}$$

in place of (6.13) and the orthogonal decomposition of the tangent space $T_{(h)}$ of the space of metrics at $h_{\alpha\beta}$ is

$$T_{(h)} = \{\Lambda h_{z\bar{z}}\} \oplus \{\text{image } \nabla_z^{(+1)}\} \oplus \{\text{ker } \nabla^z_{(+2)}\} \oplus \text{c.c.} \tag{6.31}$$

The question, how many moduli parameters τ_i exist for a compact Riemann surface of genus g without boundary, can be answered with the help on an index theorem, the Riemann-Roch theorem, which we will state without proof. If we define the index of $\nabla_z^{(n)}$ to be the number of its zero modes minus the number of zero modes of its adjoint $\nabla^z_{(n+1)}$, then the theorem states that

$$\text{ind } \nabla_z^{(n)} = \dim_\mathbb{C} \text{ker } \nabla_z^{(n)} - \dim_\mathbb{C} \text{ker } \nabla^z_{(n+1)} = -(2n+1)(g-1). \tag{6.32}$$

For $n = 1$ this tells us that the number of conformal Killing vectors minus the number of complex moduli parameters is $-3g + 3 = \frac{3}{2}\chi$ with $\chi = 2(1-g)$ denoting the Euler number of Σ_g.

6.2 The Polyakov Path Integral for the Closed Bosonic String

It is not hard to find the number of conformal Killing vectors for any compact Riemann surface without boundary. They have to be globally defined analytic vector fields whose norm is finite:

$$\int_{\Sigma_g} d^2 z \sqrt{h} \, h_{z\bar{z}} \, V^z \, V^{\bar{z}} = \text{finite}, \qquad (6.33)$$

where $V^z = \sum_n V_n z^n$. On the sphere ($g = 0$) the metric is $ds^2 = \frac{4 dz d\bar{z}}{(1+|z|^2)^2}$. It then follows that there are three independent conformal Killing vectors: ∂_z, $z\partial_z$ and $z^2 \partial_z$. To show that these fields are also well behaved at ∞, we study their behavior at $w \to 0$ where $w = 1/z$. Then the three conformal Killing vectors become $-w^2 \partial_w$, $-w \partial_w$ and $-\partial_w$. They are the only holomorphic vector fields which are well-behaved at the origin and at infinity. They correspond to the transformations generated by L_0 and $L_{\pm 1}$. The conformal Killing group is thus $SL(2, \mathbb{C})$, as follows from our discussion in Chap. 4. From the Riemann-Roch theorem we get for the dimension of moduli space $\dim \mathcal{M}_0 = 0$; i.e. there are no moduli parameters on the sphere which therefore admits a unique conformal and complex structure. All metrics on the sphere are conformally equivalent and there is a unique Riemann surface at genus zero. In the same way that we have shown $\dim \ker \nabla_z^{(+1)} = 3$, we can show that $\dim \ker \nabla_{(2)}^z = 0$. In fact we easily find that $\dim \ker \nabla_z^{(n)} = 2n + 1$ and $\dim \ker \nabla_{(n)}^z = 0$ (for $n > 0$) thus verifying the Riemann-Roch theorem explicitly for the case $g = 0$. For $g > 0$ we use the Ricci identity (6.26). Then for $V^{(n)} \in \ker \nabla_z^{(n)}$

$$\begin{aligned} 0 &= \left(\nabla_z^{(n)} V^{(n)} | \nabla_z^{(n)} V^{(n)} \right) \\ &= -\left(V^{(n)} | \nabla_{(n+1)}^z \nabla_z^{(n)} V^{(n)} \right) \\ &= -\frac{1}{2} \left(V^{(n)} | \left(\nabla_{(n+1)}^z \nabla_z^{(n)} + \nabla_z^{(n-1)} \nabla_{(n)}^z + \frac{1}{2} nR \right) V^{(n)} \right) \\ &= +\frac{1}{2} \left\{ \left(\nabla_z^{(n)} V^{(n)} | \nabla_z^{(n)} V^{(n)} \right) + \left(\nabla_{(n)}^z V^{(n)} | \nabla_{(n)}^z V^{(n)} \right) - \frac{1}{2} n \left(V^{(n)} | R V^{(n)} \right) \right\}, \end{aligned} \qquad (6.34)$$

where the first two terms are non-negative. The torus ($g = 1$) admits globally a flat metric $ds^2 = dz d\bar{z}$, i.e. $R = 0$ and we find that $\partial_z V^{(n)} = \partial_{\bar{z}} V^{(n)} = 0$, i.e. $V^{(n)} = \text{const.}$ and $\dim \ker \nabla_z^{(n)} = 1$. For $n = 1$ this is the generator of complex translations generating the conformal Killing group $U(1) \times U(1)$ of the torus. This group has one complex generator and therefore, by the Riemann-Roch theorem, the torus is described by one complex modulus τ. To get information about higher genus surfaces we use a result from the theory of Riemann surfaces which states that any Riemann surface with $g > 1$ admits a metric with constant negative curvature. We conclude that $\dim \ker \nabla_z^{(n)} = 0$ for $g > 1$, $n > 0$ and $\dim \ker \nabla_{(n+1)}^z = (2n+1)(g-1)$. For $n = 0$ $\dim \ker \nabla_z^{(0)}$ is spanned by constant functions. We can then complete

Table 6.1 Number of zero modes of $\nabla_z^{(n)}$ and $\nabla_{(n+1)}^z$

g	dim ker $\nabla_z^{(n)}$	dim ker $\nabla_{(n+1)}^z$
0	$2n+1$	0
1	1	1
>1	1 for $n=0$	g
	0 for $n>0$	$(2n+1)(g-1)$

Table 6.1, valid for $n \geq 0$. To get the results for $n < 0$ we use dim ker $\nabla_z^{(-n)}$ = dim ker $\nabla_{(n)}^z$.

Below we will compute the one-loop partition function of the closed string for which the relevant Riemann surface is the torus. Let us therefore investigate the difference between two conformally inequivalent tori more carefully. Heuristically, the fat and the thin torus, depicted in Fig. 6.5 are conformally inequivalent; roughly speaking, the modulus is given by the ratio of the two radii of the torus. More precisely, consider the complex z-plane and pick two complex numbers λ_1 and λ_2 such that $\text{Im}(\lambda_2/\lambda_1) \neq 0$, as shown in Fig. 6.6. $\lambda_{1,2}$ generate a lattice $\Lambda = \{n\lambda_1 + m\lambda_2 | n, m \in \mathbb{Z}\}$. The torus is defined by making the following identifications on the complex plane:

$$z \approx z + n\lambda_1 + m\lambda_2, \quad n, m \in \mathbb{Z}, \quad \lambda_1, \lambda_2 \in \mathbb{C} \quad \text{Im}(\lambda_2/\lambda_1) \neq 0, \qquad (6.35)$$

i.e. the torus is $\mathbb{C}/\Lambda$ and $\mathbb{C}$ is its universal covering space. Since λ_1, λ_2 are rescaled and rotated by the conformal transformation $z' = \alpha z$, $\alpha \in \mathbb{C}$, it is clear that only their ratio

$$\tau = \frac{\lambda_2}{\lambda_1} \equiv \tau_1 + i\tau_2 \qquad (6.36)$$

can be a conformal invariant. We can therefore set $\lambda_1 = 1$ and due to the freedom of interchanging λ_2 and λ_1, we may restrict $\text{Im}\,\tau > 0$. The tori are thus characterized by points τ in the upper half plane as illustrated in Fig. 6.7, where opposite sides of the parallelograms are identified

$$z \approx z + n + m\tau \quad n, m \in \mathbb{Z}. \qquad (6.37)$$

Here τ is called Teichmüller parameter and describes points in Teichmüller space which for the case of the torus is the upper-half plane $\mathbb{H}_+$. The Teichmüller space has the same dimension as the moduli space. It is the universal covering space of the moduli space.

It is the latter space that we are interested in. To understand the relation between these two spaces, the following fact is crucial: it is not quite true that τ is a conformal invariant that cannot be changed by rescalings and diffeomorphisms. The reason is that we must also consider global diffeomorphisms which are not smoothly connected to the identity. They leave the torus invariant but change the Teichmüller parameter τ. Images of the Teichmüller parameter under global diffeomorphisms represent the same point in moduli space.

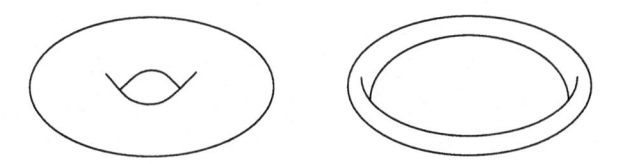

6.2 The Polyakov Path Integral for the Closed Bosonic String

Fig. 6.5 Two conformally inequivalent tori

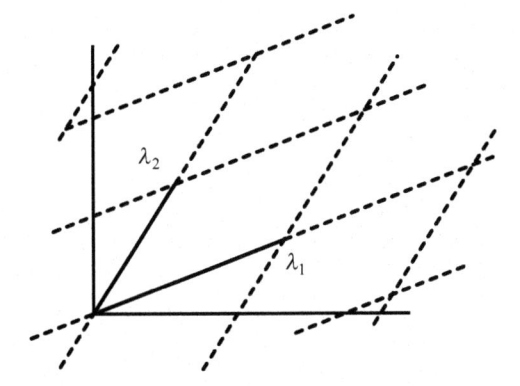

Fig. 6.6 Definition of the two-dimensional torus by the complex numbers λ_1 and λ_2

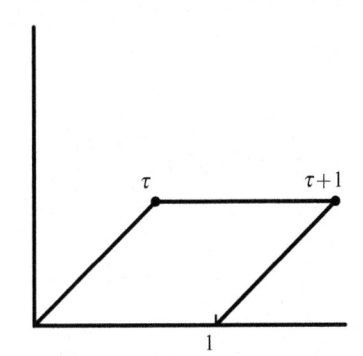

Fig. 6.7 Definition of the two-dimensional torus by the complex number τ

On the torus, the global diffeomorphisms are the following operations. Cut the torus along the cycle a indicated in Fig. 6.8, twist one of its ends by 2π and glue them back together. Points that were in a neighborhood of each other before the twist will be so after the twist. Yet this twist is not connected to the identity transformation. The same can now also be done along the cycle b. These operations are called Dehn twists and generate all global diffeomorphisms of the torus. The action on τ of a Dehn twist around the a cycle is shown in Fig. 6.9. In terms of λ_1, λ_2 it corresponds to $\lambda_1 \to \lambda_1$, $\lambda_2 \to \lambda_1 + \lambda_2$, which means $\tau \to \tau + 1$. A Dehn twist around the b cycle is shown in Fig. 6.10. To bring the transformed parallelogram into standard form, we have to rotate and rescale it. Under the

Fig. 6.8 The two independent cycles on the torus

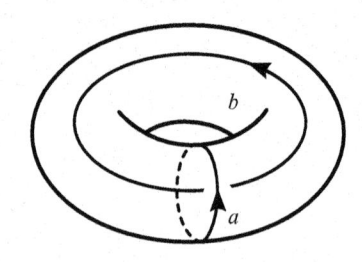

Fig. 6.9 Action on τ of the Dehn twist around the a cycle

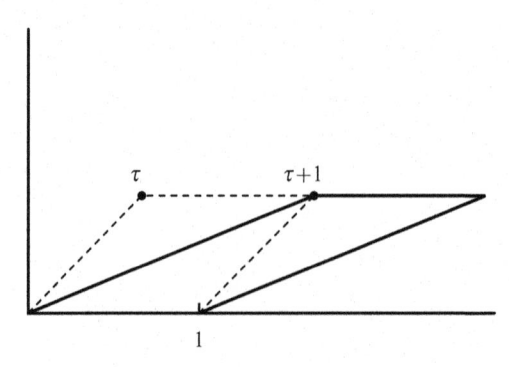

Fig. 6.10 Action on τ of the Dehn twist around the b cycle

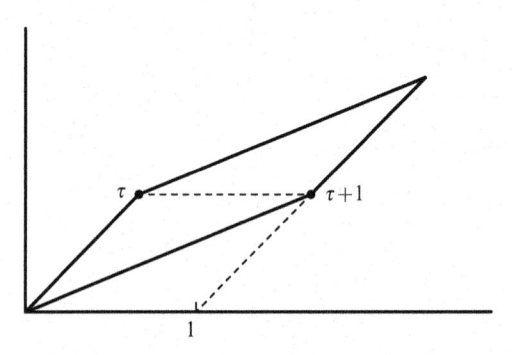

combined transformation we have $\tau \to \frac{\tau}{\tau+1}$. This again follows easily from the action on λ_1, λ_2: $\lambda_1 \to \lambda_1 + \lambda_2$, $\lambda_2 \to \lambda_2$. The two transformations $\tau \to \tau + 1$ and $\tau \to \frac{\tau}{\tau+1}$ generate the group $SL(2, \mathbb{Z})$:

$$\tau \to \tau' = \frac{a\tau + b}{c\tau + d}, \quad a, b, c, d \in \mathbb{Z}, \quad ad - bc = 1. \tag{6.38}$$

Indeed, a general transformation is $\lambda_1 \to d\lambda_1 + c\lambda_2$, $\lambda_2 \to b\lambda_1 + a\lambda_2$ and the condition $ad - bc = 1$ preserves the area of the parallelogram. Since the two $SL(2, \mathbb{Z})$ matrices $\pm \begin{pmatrix} a & b \\ c & d \end{pmatrix}$ generate the same transformation of τ, the group of global diffeomorphisms, called the modular group of the torus, is $SL(2, \mathbb{Z})/\mathbb{Z}_2 = PSL(2, \mathbb{Z})$. We have thus learned that the parameter τ, subject to the equivalence

6.2 The Polyakov Path Integral for the Closed Bosonic String

Fig. 6.11 Fundamental region $\mathscr{F}$ in Teichmüller space H^+ and its images under S and T

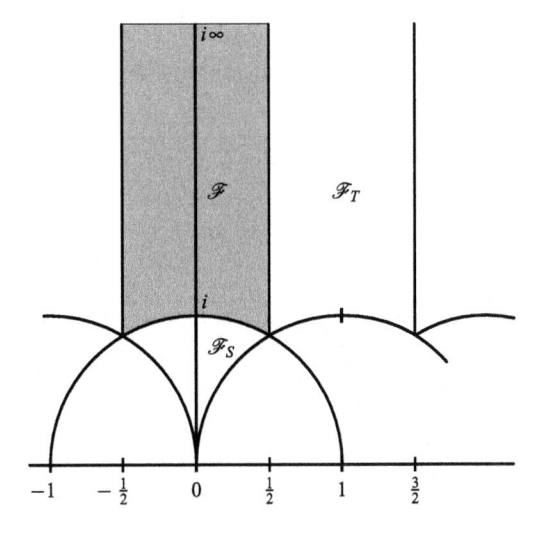

relation Eq. (6.38), describes conformally inequivalent tori. Therefore, the moduli space of the torus is the quotient of the Teichmüller space and the modular group:

$$\mathscr{M}_1 = \frac{\text{Teichmüller space}}{\text{modular group}}. \tag{6.39}$$

The Dehn twists correspond to the following $SL(2,\mathbb{Z})$ matrices: $D_a = \begin{pmatrix} 1 & 1 \\ 0 & 1 \end{pmatrix}$ and $D_b = \begin{pmatrix} 1 & 0 \\ 1 & 1 \end{pmatrix}$. Instead of the two Dehn twists, one often uses the following transformations as the generators of the modular group:

$$T: \quad \tau \to \tau + 1,$$
$$S: \quad \tau \to -\frac{1}{\tau}. \tag{6.40}$$

We note that $TST: \tau \to \frac{\tau}{\tau+1}$. Any element of $SL(2,\mathbb{Z})$ can then be composed of S and T transformations. Any point in the upper half plane is, via a $PSL(2,\mathbb{Z})$ transformation, related to a point in the so-called fundamental region $\mathscr{F}$ of the modular group. The fundamental region of the torus is

$$\mathscr{F} = \{-\frac{1}{2} \leq \operatorname{Re}\tau \leq 0, |\tau|^2 \geq 1 \cup 0 < \operatorname{Re}\tau < \frac{1}{2}, |\tau|^2 > 1\}. \tag{6.41}$$

$\mathscr{F}$ is shown in Fig. 6.11. It is the moduli space of the torus and points in $\mathscr{F}$ describe inequivalent tori. Any non-trivial modular transformation takes τ out of the fundamental region. The transformation S maps the fundamental region onto $\mathscr{F}_S$ shown in Fig. 6.11. T maps $\mathscr{F}$ onto $\mathscr{F}_T$. Of course, any image of $\mathscr{F}$ can serve equally well to parametrize the moduli space $\mathscr{M}_1$. Note that the modular group does not act freely on modular space. Indeed $\tau = i$ is a fixed point of $S: \tau \to -\frac{1}{\tau}$,

$S^2 = 1$ and $\tau = e^{2i\pi/3}$ of $ST : \tau \to -\frac{1}{\tau+1}$, $(ST)^3 = 1$. Because of these fixed points $\mathcal{M}_1$ is not a smooth manifold but rather a so-called orbifold with singularities at the fixed points.

It is instructive to show explicitly that the moduli space parametrizes different complex structures on the torus. Recall that the torus is defined as the quotient $\mathbb{C}/\Lambda$. Its complex structure is the one inherited from $\mathbb{C}$. Consider two lattices $\Lambda(\lambda_1, \lambda_2)$ and $\Lambda'(\lambda'_1, \lambda'_2)$ and the two tori $T = \mathbb{C}/\Lambda$ and $T' = \mathbb{C}/\Lambda'$. For T and T' to define the same complex manifold (as real manifolds all tori are diffeomorphic) there must be a holomorphic 1-1 map $f : T \to T'$. f induces a holomorphic map $F : \mathbb{C} \to \mathbb{C}$ between the universal covering spaces of the two tori which has to satisfy

$$F(z + \lambda_1) = F(z) + a\lambda'_1 - c\lambda'_2,$$
$$F(z + \lambda_2) = F(z) - b\lambda'_1 + d\lambda'_2 \qquad (6.42)$$

with a, b, c, d integers such that $ad - bc = \pm 1$. Differentiating these equations w.r.t. z shows that F' is a doubly periodic function with periods $\lambda_{1,2}$ and is therefore holomorphic on $T = \mathbb{C}/\Lambda$. With T compact, F' must be a constant (Liouville's theorem), i.e. $F(z) = \alpha z + \beta$. We set $\beta = 0$ by shifting z. Inserting into (6.42) gives

$$\alpha\lambda_1 = a\lambda'_1 - c\lambda'_2,$$
$$\alpha\lambda_2 = -b\lambda'_1 + d\lambda'_2, \qquad (6.43)$$

and we conclude that the moduli of two tori with the same complex structure must be related as in (6.38). Conversely, if τ and τ' are related via (6.38), there exists a non-vanishing complex number α such that (6.43) is satisfied and there is a holomorphic map $F = \alpha z : T \to T'$. The restriction to $ad - bc = +1$ arises if we require $\operatorname{Im} \tau > 0$. The space of complex structures of the torus is therefore $\mathscr{F}$ or any copy under $SL(2, \mathbb{Z})$.

Since it is irrelevant which fundamental integration region we choose, the integrand of one-loop string amplitudes must be invariant under modular transformations Eq. (6.38). This requirement of modular invariance plays an important role in string theory and has far reaching implications. We will encounter it for instance in the construction of the heterotic string where it leads to strong restrictions on the possible gauge groups and is deeply related to anomalies in the space-time theory.

The requirement of modular invariance, in particular invariance under the S-transformation, merely reflects the arbitrariness of which of the two (Euclidean) world-sheet coordinates we call time and which we call space. This observation also plays an important role for the cylinder amplitude which can be viewed either as a tree level closed string or an one-loop open string diagram. We will discuss this in detail later in this chapter.

Let us now turn to the higher genus Riemann surfaces. We will do this mainly to introduce some commonly used language and to point out some of the difficulties one encounters when going to higher loops. We have seen that for $g \geq 2$ Riemann surfaces have no conformal isometries but $3g - 3$ complex moduli parameters, their number being identical to the complex dimension of moduli space. Let us choose

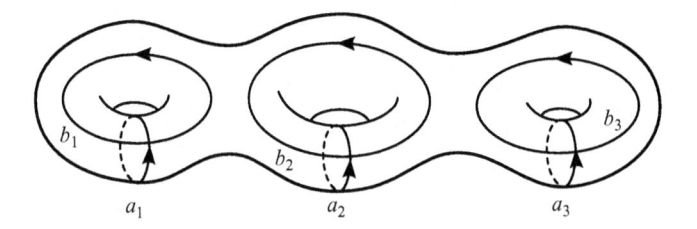

6.2 The Polyakov Path Integral for the Closed Bosonic String

Fig. 6.12 Cycles a_i, b_i as basis of first homology group $H_1(\Sigma_g, \mathbb{Z})$

$2g$ linear independent cycles a_i, b_i ($i = 1, \ldots, g$) on Σ_g forming a basis of the first homology group $H_1(\Sigma_g, \mathbb{Z}) = \mathbb{Z}^{2g}$. They are shown in Fig. 6.12. This basis has the property that the intersection pairings of cycles satisfy (including orientation)

$$(a_i, a_j) = (b_i, b_j) = 0,$$
$$(a_i, b_j) = -(b_i, a_j) = \delta_{ij}. \tag{6.44}$$

Any such basis is called canonical. Now one can also find a set of g holomorphic and g antiholomorphic closed one forms $\omega_i, \bar{\omega}_i$ which are called Abelian differentials. A standard way of normalizing the ω_i's is to require:

$$\int_{a_i} \omega_j = \delta_{ij}. \tag{6.45}$$

Then the periods over the b cycles are completely determined as

$$\int_{b_i} \omega_j = \Omega_{ij}. \tag{6.46}$$

It is not hard to see that the integrals over ω_j only depend on the homology class of the cycle; cf. the discussion in Chap. 14, page 461 *ff*. Ω_{ij} is called the period matrix of the Riemann surface; it can be shown to be a symmetric matrix with positive definite imaginary part, i.e. $(\text{Im}\,\Omega)_{ij}\, u^i \bar{u}^j > 0$. The space of all period matrices is a complex $\frac{1}{2}g(g+1)$-dimensional space known as Siegel's upper-half plane $\mathbb{H}_+^g$. In fact, the Ω_{ij} can be used to parametrize conformally inequivalent Riemann surfaces. However it is a highly redundant description, since the same surface will have in general many different matrices Ω corresponding to different canonical bases (cf. below). Remember that the dimension of moduli space is 0 for $g = 0$, 1 for $g = 1$ and $3g - 3$ for $g \geq 2$. On the other hand, the dimension of Siegel's upper half plane is $\frac{1}{2}g(g+1)$. The dimensions coincide only for $g = 0, 1, 2, 3$. For instance for $g = 1$ the period matrix Ω_{ij} is the Teichmüller parameter τ—Siegel's upper half plane and Teichmüller space are identical. The Abelian differentials are constant one-forms. However, for $g \geq 4$ Teichmüller space T_g is embedded in a complicated way in Siegel's upper half plane $\mathbb{H}_+^g$—not every symmetric $g \times g$ matrix with positive definite imaginary part corresponds to a point in Teichmüller space.

This embedding problem and its formal solution can be phrased as the solutions to complicated differential equations (the so-called KP equations). We will not discuss this any further.

Analogous to the one loop case, the second source of redundancy has to do with the reduction of Teichmüller space $\mathscr{T}_g$ to the moduli space $\mathscr{M}_g$, i.e. to find a fundamental region in $\mathscr{T}_g$. In general, the moduli space is obtained by dividing the Teichmüller space by the group of disconnected diffeomorphisms of Σ_g. This group is known as the mapping class group (MCG). We then have the following relations:

$$\mathscr{T}_g = \frac{M_h}{\text{Weyl} \times \text{Diff}_0},$$

$$\mathscr{M}_g = \frac{M_h}{\text{Weyl} \times \text{Diff}} = \frac{\mathscr{T}_g}{\text{MCG}},$$

$$\text{MCG} = \frac{\text{Diff}}{\text{Diff}_0}. \tag{6.47}$$

Here M_h is the space of all metrics on Σ_g and Diff_0 the diffeomorphisms connected to the identity.

A subclass of the mapping class group is the group of modular transformations which act non-trivially on a given homology basis. Suppose two canonical bases of the same Riemann surface are related by

$$\begin{pmatrix} a' \\ b' \end{pmatrix} = \begin{pmatrix} D & C \\ B & A \end{pmatrix} \begin{pmatrix} a \\ b \end{pmatrix}, \tag{6.48}$$

where A, B, C, D are $g \times g$ matrices. To preserve Eq. (6.44), the matrix in Eq. (6.48) must be a symplectic modular matrix with integer coefficients, i.e. an element of $Sp(2g, \mathbb{Z})$. These transformations are the analogue of the one loop modular transformation; indeed, for $g = 1$ $Sp(2, \mathbb{Z}) = SL(2, \mathbb{Z})$. We can now compute the transformation of Ω_{ij} induced by the change of homology basis such that

$$\int_{a'_i} \omega'_j = \delta_{ij}. \tag{6.49}$$

It follows that $\omega'_j = \omega_k (C\Omega + D)^{-1}_{kj}$ and the new period matrix is

$$\Omega' = (A\Omega + B)(C\Omega + D)^{-1}. \tag{6.50}$$

The generators of modular transformations are the Dehn twists along the homologically non-trivial curves of Fig. 6.13. We have two generators for each handle and one generator for each curve linking the holes of two consecutive handles.

As in the torus case, there is a way of representing Dehn twists in terms of matrices. A Dehn twist around a non-trivial curve acts non-trivially on the homology basis. For example, a Dehn twist around a_1 induces the following transformation

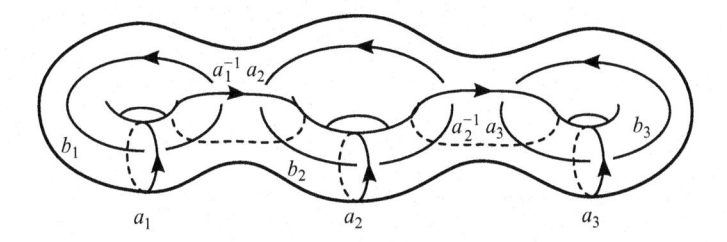

6.2 The Polyakov Path Integral for the Closed Bosonic String

Fig. 6.13 Cycles a_i, b_i as basis of first homology group $H_1(\Sigma_g, Z)$

on the homology basis: $a_1 \to a_1, b_1 \to b_1 + a_1$. Let D_γ be the modular transformation defined by a twist around γ. Then one can show that the matrices $D_{a_1}, D_{b_1}, D_{a_1^{-1}a_2}, \ldots, D_{a_g}, D_{b_g}$ generate all matrices of $Sp(2g, \mathbb{Z})$. For the same reason as for the one loop case the integrand of the higher loop string amplitudes must be invariant under these modular transformations. For instance for $g = 2$, the generators of $Sp(4, \mathbb{Z})$ are given by the following 4×4 matrices:

$$D_{a_1} = \begin{pmatrix} 1 & 0 & 0 & 0 \\ 0 & 1 & 0 & 0 \\ 1 & 0 & 1 & 0 \\ 0 & 0 & 0 & 1 \end{pmatrix}, \quad D_{b_1} = \begin{pmatrix} 1 & 0 & 1 & 0 \\ 0 & 1 & 0 & 0 \\ 0 & 0 & 1 & 0 \\ 0 & 0 & 0 & 1 \end{pmatrix},$$

$$D_{a_2} = \begin{pmatrix} 1 & 0 & 0 & 0 \\ 0 & 1 & 0 & 0 \\ 0 & 0 & 1 & 0 \\ 0 & 1 & 0 & 1 \end{pmatrix}, \quad D_{b_2} = \begin{pmatrix} 1 & 0 & 0 & 0 \\ 0 & 1 & 0 & 1 \\ 0 & 0 & 1 & 0 \\ 0 & 0 & 0 & 1 \end{pmatrix}, \quad D_{a_1^{-1}a_2} = \begin{pmatrix} 1 & 0 & 0 & 0 \\ 0 & 1 & 0 & 0 \\ -1 & 1 & 1 & 0 \\ 1 & -1 & 0 & 1 \end{pmatrix}. \quad (6.51)$$

However, it is important to note that the modular transformations, i.e. the Dehn twists around the homologically non-trivial cycles do not generate the whole mapping class group. There are also twists around trivial cycles so that they do not affect the homology basis but nevertheless correspond to non-trivial diffeomorphisms. These form a subgroup of the mapping class group called the Torelli group. The quotient of the mapping class group and the Torelli group is precisely the symplectic modular group $Sp(2g, \mathbb{Z})$. We will however not consider this subtlety since for $g = 1$ (which we will be mainly interested in) the Torelli group is trivial so that the mapping class group is identical to the modular group $SL(2, \mathbb{Z})$. This concludes our considerations about the modular transformations.

We are now ready to resume our discussion of the integration measure. If we use complex bases for the Beltrami differentials and the quadratic differentials, we can write the integration over the metrics as

$$\int_{\mathcal{M}_g} \prod_i d\tau_i^2 \int \frac{\mathcal{D}X \mathcal{D}'\xi \mathcal{D}\Lambda}{\text{Vol(Diff)Vol(Weyl)}} \frac{\det(\phi|\mu)\det(\mu|\phi)}{\det(\phi|\phi)} \det{}'\nabla^z_{(-1)} \det{}'\nabla^{(+1)}_z. \quad (6.52)$$

The prime on $\mathscr{D}'\xi$ indicates that we do not integrate over the conformal Killing vectors. Their effect on δh is already taken care of by the integral over Λ. The prime on the determinants means that one has to drop the zero modes (which are the conformal Killing vectors). The τ_i are complex parameters. If we define $|\text{MCG}| = \frac{\text{Vol}(\text{Diff})}{\text{Vol}(\text{Diff}_0)}$ and make an orthogonal decomposition of Diff_0 with

$$\text{Vol}(\text{Diff}_0) = \text{Vol}(\text{Diff}_0^\perp)\text{Vol}(\text{CKG}), \tag{6.53}$$

we can write the measure as

$$\frac{1}{|\text{MCG}|} \int_{\mathcal{M}_g} \prod_i d\tau_i^2 \int \frac{\mathscr{D}X \mathscr{D}'\xi \mathscr{D}\Lambda}{\text{Vol}(\text{Diff}_0^\perp) \, \text{Vol}(\text{CKG}) \, \text{Vol}(\text{Weyl})}$$
$$\times \frac{\det(\phi|\mu) \det(\mu|\phi)}{\det(\phi|\phi)} \det{}'\nabla_z^{(+1)} \det{}'\nabla_{(-1)}^z. \tag{6.54}$$

In the critical dimension, i.e. in the absence of a conformal anomaly, we can cancel $\int \mathscr{D}'\xi \mathscr{D}\Lambda$ against $\text{Vol}(\text{Diff}_0^\perp)\text{Vol}(\text{Weyl})$. Furthermore, we can replace $\frac{1}{|\text{MCG}|} \int_{\mathcal{M}_g} \prod_i d\tau_i$ by an integral over a fundamental region. We then get the following expression for scattering amplitudes of the closed bosonic string

$$A_n = \sum_{g=0}^{\infty} \int d^2 z_1 \ldots d^2 z_n \int_{\mathcal{F}_g} \prod_i d\tau_i \int \frac{\mathscr{D}X}{\text{Vol}(\text{CKG})} \frac{\det(\phi|\mu) \det(\mu|\phi)}{\det(\phi|\phi)}$$
$$\times \det{}'\nabla_z^{(+1)} \det{}'\nabla_{(-1)}^z \, V(z_1, \bar{z}_1, \tau_i) \cdots V(z_n, \bar{z}_n, \tau_i) \, e^{-S[X,\tau_i]}. \tag{6.55}$$

At tree level there are no quadratic differentials and the corresponding factor in the measure is absent. At two and higher loop order there are no conformal Killing vectors. At one loop there is one of each, both being constants on the torus.

As we did in Chap. 3, we can replace the Jacobian determinant by an integral over anti-commuting Faddeev-Popov ghosts:

$$\det{}'\nabla_z^{(+1)} \det{}'\nabla_{(-1)}^z = \int \mathscr{D}'(b\bar{b}c\bar{c}) \, e^{-S[b,c,\tau_i]}, \tag{6.56}$$

where the ghost action is (cf. Chaps. 3 and 5)

$$S[b,c] = \frac{1}{2\pi} \int d^2 z \, (b \bar{\partial} c + \bar{b} \partial \bar{c}). \tag{6.57}$$

In Eq. (6.56) we have excluded the integration over the ghost zero modes. The integral would vanish otherwise. We will show in the following that the remaining factors $\frac{1}{\text{Vol}(\text{CKG})}$ and $\frac{\det(\phi|\mu) \det(\mu|\phi)}{\det(\phi|\phi)}$ can be attributed to the c- and b-ghost zero modes respectively. This is in fact easy to see. The ghost zero modes satisfy the equations

6.2 The Polyakov Path Integral for the Closed Bosonic String

$$\bar{\partial} c = h_{\bar{z}\bar{z}} \nabla^z_{(-1)} c = 0,$$
$$\bar{\partial} b = h_{\bar{z}\bar{z}} \nabla^z_{(+2)} b = 0 \quad (6.58)$$

which tell us that the c zero modes correspond to the conformal Killing vectors and the b zero modes to the quadratic differentials. By the Riemann-Roch theorem, Eq. (6.32), we then get

$$N_b - N_c = 3g - 3 \quad (6.59)$$

where $N_{b,c}$ denotes the number of zero modes. (Note that $\dim \ker \nabla^z_{(-1)} = \dim \ker \nabla^{(+1)}_z$ as $\nabla^{(n)}_z$ is the complex conjugate of $\nabla^z_{(-n)}$.) The presence of ghost zero modes means that the ghost number current is not conserved. We will give a more detailed and general description in Chap. 13.

Due to the anti-commutativity of the ghosts, integration over their zero modes will give a vanishing answer if we do not insert them into the integrand. To illustrate this, consider an anti-commuting variable ψ and split it into its zero-mode part ψ_0 and the remainder ψ':

$$\psi(z,\bar{z}) = \psi_0(z) + \psi'(z,\bar{z}) = \sum_{i=1}^{N} \psi_0^i \phi^i(z) + \psi'(z,\bar{z}), \quad (6.60)$$

where the zero mode wave functions ϕ^i satisfy $\bar{\partial}\phi^i = 0$. (E.g. for the b ghosts they are the quadratic differentials.) The ψ_0^i are constant anti-commuting parameters satisfying $\int d\psi_0^i \psi_0^i = 1, \int d\psi_0^i = 0$. Since the action for ψ does not depend on ψ_0, $\int \mathcal{D}(\psi\bar{\psi}) e^{-S}$ will vanish unless we restrict the integration to the non-zero modes or absorb them by inserting $\prod_{i=1}^{N} \psi(z_i)$ into the integrand:

$$\int \mathcal{D}(\psi\bar{\psi}) \prod_{i=1}^{N} \psi(z_i)\bar{\psi}(\bar{z}_i) e^{-S[\psi']} = \frac{|\det(\phi^i(z_j))|^2}{\det(\phi^i|\phi^j)} \int \mathcal{D}'(\psi\bar{\psi}) e^{-S[\psi']}, \quad (6.61)$$

where the factor $\det(\phi^i|\phi^j)$ is relevant if the ϕ^i do not form an orthonormal basis. It renders the zero mode contribution basis independent.[11]

Using this it is now easy to rewrite the string measure including the integration over the ghost zero modes. Let us first, for simplicity, restrict ourselves to the case $g \geq 2$ where there are no conformal Killing vectors. Using Eq. (6.61) we find the following simple expression for the partition function:

$$Z_{g\geq 2} = \int_{\mathcal{F}_g} d^2 \prod_i \tau_i \int \mathcal{D}X \mathcal{D}(b\bar{b}c\bar{c}) \prod_{i=1}^{3g-3} |(\mu^i|b)|^2 e^{-S[X,b,c,\tau_i]}. \quad (6.62)$$

---
[11] Note that $||\psi_0||^2 = (\phi^i|\phi^j)\psi_0^i \bar{\psi}_0^j$.

Let us now turn to the cases $g = 0$ and $g = 1$. At tree level we have no b zero modes (no moduli) but instead three complex c zero modes, corresponding to the conformal Killing vectors that generate the group $PSL(2, \mathbb{C})$. The zero mode wave functions were found to be $1, z$ and z^2. $PSL(2, \mathbb{C})$ acts freely on the insertion points z_i of the vertex operators and one can fix three of them, say z_1, z_2 and z_3, at arbitrary points with a unique $PSL(2, \mathbb{C})$ transformation. They are generated by vector fields $(\alpha + \beta z + \gamma z^2)\partial_z$ and we can trade the integrations over z_1, z_2, z_3 for an integration over α, β, γ. From $\delta z_i = \alpha + \beta z_i + \gamma z_i^2$ we find the Jacobian

$$\left|\frac{\partial(z_1, z_2, z_3)}{\partial(\alpha, \beta, \gamma)}\right|^2 = |(z_1 - z_2)(z_2 - z_3)(z_1 - z_3)|^2. \tag{6.63}$$

The integration over α, β, γ then cancels the Vol(CKG) factor. Note that the Jacobian is also $|\det(V^i(z_j))|^2$ where $\{V^i\} = \{1, z, z^2\}$ are the conformal Killing vectors (which are orthogonal to each other). More importantly, it can be written as

$$|\langle 0|c(z_1)c(z_2)c(z_3)|0\rangle|^2 = |\langle 0|c_{-1}c_0c_1|0\rangle|^2 \det\begin{vmatrix} 1 & 1 & 1 \\ z_3 & z_2 & z_1 \\ z_3^2 & z_2^2 & z_1^2 \end{vmatrix}^2$$

$$= |(z_1 - z_2)(z_1 - z_3)(z_2 - z_3)|^2, \tag{6.64}$$

where we have used results from Chap. 5. For tree level amplitudes this means that the presence of conformal Killing vectors is taken care of if we drop the integration over the positions of three of the vertex operators and multiply each of them by $c(z_i) \bar{c}(\bar{z}_i)$. We know from Chap. 5 that if $\int V$ is BRST invariant, then so is $c\bar{c}V$. Tree level scattering amplitudes of the closed bosonic string then take the form:

$$A_n^{g=0} = \int \mathcal{D}X\mathcal{D}(b\bar{b}c\bar{c}) \, c\bar{c} \, V(z_1, \bar{z}_1) \, c\bar{c} \, V(z_2, \bar{z}_2) \, c\bar{c} \, V(z_3, \bar{z}_3)$$

$$\times \prod_{i=4}^{n} \int d^2z_i \, V(z_i, \bar{z}_i) \, e^{-S[X,b,c]}$$

$$= \left\langle c\bar{c} \, V(z_1, \bar{z}_1) \, c\bar{c} \, V(z_2, \bar{z}_2) \, c\bar{c} \, V(z_3, \bar{z}_3) \prod_{i=4}^{n} \int d^2z_i \, V(z_i, \bar{z}_i) \right\rangle, \tag{6.65}$$

where the last line is a CFT correlation function of the vertex operators. If we have less than three vertex operator insertions we cannot completely factor out the $PSL(2, \mathbb{C})$ volume and the correlation functions vanish upon dividing by the infinite factor Vol(CKG). So 0, 1 and 2 point functions vanish at tree level.[12] There

---

[12] If we have less then three vertex operator insertions the subgroup of $PSL(2, \mathbb{C})$ which leaves their positions fixed is non-compact and has infinite volume.

6.2 The Polyakov Path Integral for the Closed Bosonic String

Table 6.2 Number of conformal Killing vectors and quadratic differentials for surfaces relevant for tree-level and one-loop string amplitudes

World-sheet	Euler number	dim ker P	dim ker $P^\dagger$
Sphere	2	6	0
Disk	1	3	0
Projective plane	1	3	0
Torus	0	2	2
Cylinder	0	1	1
Möbius strip	0	1	1
Klein bottle	0	1	1

is no tree level cosmological constant, no tree level tadpoles and no tree level mass or wave-function renormalization. In CFT there are, of course, two point functions. But in the CFT's which arise from string theory, i.e. from a theory with gauge fixed reparametrization invariance there are ghosts with zero modes and we have to divide by the infinite volume of the conformal Killing group. To get a non-vanishing answer it has to be cancelled. This happens by trading the integration over the positions of three vertex operators for the integration over the parameters of the group.

Although we do not present a general discussion of the measure for open strings and for non-orientable strings, we will give some results which are useful for tree level and one-loop amplitudes. As we have just seen, we need to know the number of conformal Killing vectors and of quadratic differentials, i.e. the number of zero modes of the operators P and $P^\dagger$. For the various tree-level and one-loop amplitudes we have collected them in Table 6.2 which counts real dimensions. The index theorem $\dim \ker P - \dim \ker P^\dagger = 3\chi(\Sigma)$ can be checked in all cases.

The results in the table can be understood as follows. The disk and the projective plane can be realized as $\mathbb{Z}_2$ quotients of the sphere, cf. Chap. 16.2. Half of the CKVs survive the projection. The real modulus of the cylinder is the ratio of its height to its circumference. The CKV is translation around the cylinder at constant height. The same is true for the Möbius strip which is simply a cylinder with a crosscap on one of its boundary and also for the Klein bottle where both ends of the cylinder are closed off with a crosscap.

None of the tree-level world-sheets have moduli but they all have conformal Killing vectors. The sphere was discussed above. The CKG of the disk and the projective plane is $SU(1,1)$ and $SU(2)$, respectively; this will be derived in Chap. 16.2 when we discuss open string tree level amplitudes. Therefore the disk and the projective plane have three real CKVs. The volume of the non-compact group $SU(1,1) \simeq PSL(2,\mathbb{R})$ is infinite and zero-point functions on the disk vanish. The volume of $SU(2)$, which is compact, is finite and zero-point functions on the projective plane do not necessarily vanish. For the one-point functions and two-point functions on the disk the story is more subtle. One- and two point functions of open string vertex operators, which have to be inserted on the boundary of the disk, vanish because the residual symmetry which keeps their position fixed is $\mathbb{R}^2$ and $\mathbb{R}$ and one has to divide by their infinite volume. However, for the one-point function of a closed string vertex operator, which is inserted in the interior of the disk, the residual symmetry is $U(1)$ with finite volume 2π. Therefore, one-point functions of closed strings on the disk are not necessarily zero.

On the disk we may use $PSL(2,\mathbb{R})$ transformations to either fix the positions of three open string vertex operators on the real axis or the position of one open string and one closed string vertex operator. There are some restrictions. For instance, the cyclic order of the positions of vertex operators on the real axis (compactified by adding the point at infinity) cannot be changed by a $PSL(2,\mathbb{R})$ transformation. The positions of the remaining vertex operators have to be integrated over. In analogy to (6.63), there will be a Jacobian due to the change of variables from the integration over the positions of the vertex operators to the parameters of the CKG. This will be discussed further in Chap. 16 for tree level amplitudes. In the remainder if this chapter we discuss the one-loop vacuum amplitudes of the closed and the open bosonic string. We restrict ourselves to orientable world-sheets, i.e. to the torus and the cylinder/annulus. Amplitudes on non-orientable world-sheets will be discussed for the fermionic string in Chap. 9.

6.3 The Torus Partition Function

We start with the torus. Here the conformal Killing vector V and the quadratic differential ϕ are (complex) constants, which we set to one for simplicity. We can parametrize the torus by two real variables ξ^1, ξ^2 with $0 \le \xi^1, \xi^2 \le 1$, in terms of which the complex coordinates become $z = \xi^1 + \tau \xi^2$ and we can use Weyl invariance to set $ds^2 = |dz|^2$. The area of the torus is $\int d^2\xi \sqrt{h} = \text{Im}\,\tau$. If we change τ to $\tau \to \tau + \delta\tau$, we find (up to a rescaling) $ds^2 \to |dz + \delta\tau \frac{i}{2\text{Im}\tau} d\bar z|^2$ and therefore $\mu_{\bar z}^z = \frac{i}{2\text{Im}\tau}$. Neglecting constant factors independent of τ, we compute $(\phi|\phi) = \text{Im}\,\tau$, $(\mu|\phi) = 1$ and $\text{Vol}(CKG) = (\text{Im}\,\tau)^2$. This latter result needs an explanation. The two conformal Killing vectors on the torus are ∂_{ξ^1} and ∂_{ξ^2}. They are in fact Killing vectors. They generate shifts $\xi^\alpha \to \xi^\alpha + a^\alpha$ with $0 \le a^\alpha \le 1$. Defining the metric on the space of Killing vectors V_i^α as $g_{ij} = \int \sqrt{h}\, h_{\alpha\beta} V_i^\alpha V_j^\beta$, $\text{Vol}(CKG)$ is $\int \sqrt{\det g}\, da^1 da^2 = \tau_2^2$. Consequently, the total contribution from the ghost zero modes is $(\frac{1}{\text{Im}\tau})^3$. For the vacuum amplitude the integral over X^μ can be easily performed. Again, the zero mode of the scalar Laplace operator needs special treatment. We expand X^μ into eigenmodes of the Laplace operator, $X^\mu = \sum_{n=0}^\infty c_n^\mu \phi_n \equiv X_0^\mu + X'^\mu$ where $\Box \phi_n = -\lambda_n \phi_n$ with $(\phi_n|\phi_m) = \delta_{m,n}$ and $\lambda_n \ge 0$. In particular the normalized zero mode is $\phi_0 = (\int \sqrt{h})^{-1/2} = \tau_2^{-1/2}$.

The functional integral reduces to an integral over the expansion coefficients. We will do the calculation for one space-time direction. With the normalization (3.65) the measure in the functional integral is $\mathcal{D}X = \prod_n \frac{dc_n}{2\pi\sqrt{\alpha'}}$ and therefore

$$\int \mathcal{D}X\, e^{-\frac{1}{4\pi\alpha'}\int d^2\sigma \sqrt{h}\, X\Box X} = \int \frac{dc_0}{2\pi\sqrt{\alpha'}} \prod_{n=1}^\infty \frac{dc_n}{2\pi\sqrt{\alpha'}}\, e^{-\frac{1}{4\pi\alpha'}\lambda_n c_n^2}. \tag{6.66}$$

6.3 The Torus Partition Function

The Gaussian integral over the non-zero modes produces

$$\prod_{n>0} \frac{1}{\lambda^{1/2}} \equiv (\det' \Box)^{-1/2}, \tag{6.67}$$

where the product over the eigenvalues has to be regularised. The integration over the zero modes gives $\int dc_0 = \int \left(\frac{dX_0}{\phi_0}\right) = L\sqrt{\tau_2}$ where L is the size of the system. Combining the contribution from all 26 space-time directions we finally find for the torus vacuum amplitude, up to numerical constants,

$$\begin{aligned} A_0^{(g=1)} &\sim \frac{V_{26}}{\alpha'^{13}} \int_{\mathscr{F}_1} \frac{d^2\tau}{(\mathrm{Im}\tau)^2} \frac{1}{\mathrm{Im}\tau} (\mathrm{Im}\tau)^{13} (\det' \Box)^{-13} \det' \nabla_z^{(+1)} \det' \nabla_{(-1)}^z \\ &\sim \frac{V_{26}}{\alpha'^{13}} \int_{\mathscr{F}_1} \frac{d^2\tau}{(\mathrm{Im}\tau)^2} (\mathrm{Im}\tau)^{12} \int \mathscr{D}'X \, \mathscr{D}'c \, \mathscr{D}'b \, e^{-S[X,\tau]-S[b,c,\tau]}, \end{aligned} \tag{6.68}$$

where $V_{26} = L^{26}$ is the volume of space-time. The primes on the measures in the second line mean that we integrate only over the non-zero modes.

The next step would be the computation of the determinants. We will not present this computation which can be done, e.g. using ζ-function or other methods of regularization of the infinite product over the eigenvalues of the Laplacian. Infinite products which appear in the process can be absorbed in the Polyakov action; cf. the brief discussion in Footnote 5 on page 15. For instance, a different normalization in (3.65) would have produced an infinite factor in the above calculation which we could absorb in a bare cosmological constant on the world-sheet. We will not present any of the details of these calculations, which lead to a well-defined one-loop partition function with a definite normalization. Instead, we will simply state the result. We will then rewrite the partition function in a way which exhibits the relation with the Hamiltonian formalism in light-cone gauge and which gives a heuristic understanding of the result of the careful evaluation of the Euclidean path integral. The Hamiltonian formalism is an easy way to compute the torus partition function and it generalizes to other conformal field theories and to other one-loop amplitudes. We will also use it when we discuss the partition function of the fermionic string in Chap. 9.

At this point we simply state the result for the determinants:

$$\det' \nabla_z^{(+1)} \det' \nabla_{(-1)}^z \sim \det' \Box = (\mathrm{Im}\tau)^2 |\eta(\tau)|^4, \tag{6.69}$$

where $\eta(\tau)$ is the Dedekind eta-function defined as

$$\eta(\tau) = q^{1/24} \prod_{n=1}^{\infty} (1 - q^n), \quad q = e^{2\pi i \tau}. \tag{6.70}$$

The final, correctly normalized expression for the one-loop vacuum amplitude is

$$A_0^{(g=1)} \equiv \mathcal{T} = \frac{V_{26}}{\ell_s^{26}} \int_{\mathcal{F}} \frac{d^2\tau}{4(\mathrm{Im}\tau)^2} \, Z(\tau, \bar{\tau}), \tag{6.71}$$

where

$$Z(\tau, \bar{\tau}) = (\mathrm{Im}\tau)^{-12} |\eta(\tau)|^{-48} \tag{6.72}$$

is the closed string partition function. V_{26} is the volume of the space-time. We will refer to $\mathcal{T}$ as the torus amplitude and to $Z(\tau, \bar{\tau})$ as the partition function. The equation (6.69) indicates that the effect of the ghosts is to cancel the contribution of two coordinate degrees of freedom which correspond to the longitudinal and time-like string excitations. Therefore, the partition function counts only the physical transverse string excitations.

Let us now check modular invariance of the one loop partition function. First note that the measure $\frac{d^2\tau}{(\mathrm{Im}\tau)^2}$ is invariant by itself. This follows from

$$d^2\tau \to |c\tau + d|^{-4} d^2\tau,$$
$$\mathrm{Im}\tau \to |c\tau + d|^{-2} \mathrm{Im}\tau. \tag{6.73}$$

To check modular invariance of $Z(\tau, \bar{\tau})$, it suffices to do so for the two generators S and T of the modular group. With the transformation properties of the eta-function

$$\eta(\tau + 1) = e^{i\pi/12} \eta(\tau),$$
$$\eta\left(-\frac{1}{\tau}\right) = \sqrt{-i\tau}\, \eta(\tau) \tag{6.74}$$

modular invariance follows straightforwardly.

We will now rewrite the torus amplitude in a way which exhibits the connection between the Euclidean path integral and the Hamiltonian formalism. Recall from (3.41) the definition of the light-cone Hamiltonian $H_{\mathrm{l.c.}}$, the generator of τ translations, and also that the generator of σ translations, which we denote as $P_{\mathrm{l.c}}$:

$$H_{\mathrm{l.c.}} = \sum_{n>0} \left(\alpha^i_{-n}\alpha^i_n + \bar{\alpha}^i_{-n}\bar{\alpha}^i_n\right) - 2 + \tfrac{\alpha'}{2} p^i p^i,$$

$$P_{\mathrm{l.c.}} = \sum_{n>0} \left(\bar{\alpha}_{-n}\bar{\alpha}_n - \alpha_{-n}\alpha_n\right). \tag{6.75}$$

Using

$$\mathrm{tr}_{\mathrm{osc.}}\!\left[\bar{q}^{N_L-1} q^{N_R-1}\right] = (q\bar{q})^{-1} \prod_{n=1}^{\infty} (1-\bar{q}^n)^{-24} (1-q^n)^{-24}$$

$$= \frac{1}{|\eta(\tau)|^{-48}} \tag{6.76}$$

6.3 The Torus Partition Function

and

$$\int \frac{d^{24}p}{(2\pi)^{24}} e^{-\pi\alpha\tau_2 p^2} = \frac{1}{(2\pi\sqrt{\alpha'})^{24}} \cdot \frac{1}{\tau_2^{12}}, \tag{6.77}$$

it is not difficult to see that we can then write (6.71) as

$$\begin{aligned}\mathcal{T} &= \frac{V_{26}}{\ell_s^{26}} \int_{\mathcal{F}} \frac{d^2\tau}{4\tau_2} \frac{1}{\tau_2^{13}} \mathrm{tr}_{\mathrm{osc.}}\left(e^{-2\pi\tau_2(N+\overline{N}-2)} e^{2\pi i\tau_1(N-\overline{N})}\right) \\ &= V_{26} \int_{\mathcal{F}} \frac{d^2\tau}{16\pi^2 \alpha' \tau_2^2} \int \frac{d^{24}p}{(2\pi)^{24}} \mathrm{tr}_{\mathrm{osc.}}\left(e^{-2\pi\tau_2 H_{\mathrm{l.c.}}} e^{-2\pi i\tau_1 P_{\mathrm{l.c.}}}\right)\end{aligned} \tag{6.78}$$

with $q = e^{2\pi i\tau}$. A more succinct way to write the torus amplitude is

$$\mathcal{T} = \int_{\mathcal{F}} \frac{d^2\tau}{4\tau_2} \, \mathrm{tr}_{\mathcal{H}_{\mathrm{cl}}}\left(q^{L_0 - \frac{c}{24}} \bar{q}^{\overline{L}_0 - \frac{\bar{c}}{24}}\right), \tag{6.79}$$

where $L_0 = N_{\mathrm{tr}} + \frac{1}{2} p^\mu p_\mu$ and tr includes the trace over momenta:

$$\int \frac{d^{26}p}{(2\pi)^{26}} \langle p|e^{-\pi\alpha'\tau_2 p^2}|p\rangle = \int \frac{d^{26}p}{(2\pi)^{26}} e^{-\pi\alpha'\tau_2 p^2} V_{26}. \tag{6.80}$$

We have used the normalization

$$\langle p|p'\rangle = (2\pi)^{26} \delta^{26}(p-p') \quad \text{with} \quad \langle p|p\rangle = (2\pi)^{26} \delta(0) = V_{26}. \tag{6.81}$$

In (6.79) we have the trace over the closed string Hilbert space. As far as the oscillators are concerned, we can either only consider the transverse directions or include all directions, but then we also have to include the ghost sector, which cancels the contribution of two bosonic oscillators. Either way, the net contribution is that of 24 oscillator directions. Contrary to that, all 26 momentum directions contribute.

The interpretation of (6.78) is as follows: the imaginary part of the modulus of the torus plays the role of an Euclidean time variable or, in statistical mechanics language, of the inverse temperature. If $\mathrm{Re}\tau = 0$, we obtain the functional integral as $\mathrm{tr}\,\exp(-2\pi\tau_2 H_{\mathrm{l.c.}})$, where the time evolution operator in the τ_2 direction is the light-cone Hamiltonian. (We have rescaled Fig. 6.7 by a factor 2π in order to obtain a string of length $\ell = 2\pi$. This leads to the 2π in the exponent of the trace.) The partition function counts the number of states propagating around the torus in the τ_2 direction and weights them with a factor $\exp(-2\pi\tau_2 H_{\mathrm{l.c.}})$. If one thinks of the torus as a cylinder of length τ_2 whose ends are identified, one can twist the two ends relatively to each other by an angle $2\pi\tau_1$ before joining them. The operator generating this twist is $P_{\mathrm{l.c.}}$.

Writing the torus amplitude in Hamiltonian form, it is evident that $Z(\tau,\bar\tau)$ contains the information about the level density of string states, i.e. the number of states of each mass level. Expanding $Z(\tau,\bar\tau)$ in powers of $q = e^{2\pi i\tau}$ one obtains a power series of the form $\sum d_{mn}\bar q^m q^n$. Here d_{mn} is simply the number of states with $\frac{\alpha'}{2}m_L^2 = m$ and $\frac{\alpha'}{2}m_R^2 = n$. The first few terms of the expansion are

$$Z(\tau,\bar\tau) \sim |\eta(\tau)|^{-48} = \frac{1}{|q|^2} + 24\,q^{-1} + 24\,\bar q^{-1} + 576 + \dots \qquad (6.82)$$

The first term corresponds to the negative (mass)^2 tachyon and the constant term to the massless string states, namely to the on-shell graviton, antisymmetric tensor field and dilaton. The second and third term do not satisfy level-matching and are projected out by the τ_2 integration. Due to the tachyon pole, one finds that the one-loop cosmological constant of the closed bosonic string is infinite. The normalization of the partition function can be fixed by requiring that it correctly reproduces the ground state degeneracy, which is one in this simple case.

Let us now comment on the connection between these explicit results for the one-loop partition function of the bosonic string and the general discussion in Chap. 4 where we defined the partition function (4.78) as a sum over characters of the Virasoro algebra (cf. also Sect. 6.4). One obvious difference is that in CFT there is no integration over the modulus τ. In the string context it is a remnant of the integral over world-sheet metrics. But for the rest, the string partition function should be a special case of the more general situation. Note first that in light-cone gauge the constraints which lead to the physical state conditions have already been taken care of, i.e. they do not impose restrictions on the trace. So we should be able to write $Z(\tau,\bar\tau)$ as a sum over Virasoro characters. We expect that the partition function factorizes into holomorphic and anti-holomorphic parts, apart from the zero mode contribution, i.e. the integral over p which is easily done and which we will ignore. Then, in the absence of null states, which is the case for $c > 1$,[13] the chiral partition function has the general form

$$Z(\tau) = \frac{\sum_{m=0}^{\infty} p(m)\, q^{m-\frac{c}{24}}}{\prod_{n>0}(1-q^n)}, \qquad (6.83)$$

where $p(m)$ is the number of highest weight states with weight m and the denominator is due to the sum over the descendants of each highest weight state. Here we use the fact that for the light-cone bosonic string CFT all states are created from the states $|p\rangle$ by the action of transverse oscillators with integer (negative) mode number. It is easy to extract the numbers $p(m)$ from our explicit results above:

$$\prod_{n>0}(1-q^n)\frac{1}{\eta^{24}(q)} = \frac{1}{q} + 23 + 299q + 2852q^2 + \dots \qquad (6.84)$$

---

[13] With the exception of $L_{-1}|0\rangle$, but this is irrelevant here since we are considering excitations of states $|p\rangle$ and $L_{-1}|p\rangle \neq 0$.

6.3 The Torus Partition Function

The interpretation of the coefficients goes as follows. The first one (1) is the highest weight state corresponding to the tachyon. The second, 23, are massless states $\alpha_{-1}^i|p\rangle$ where there is one linear combination, namely $p^i\alpha_{-1}^i|p\rangle \propto L_{-1}|p\rangle$, which is a descendant of the tachyon, i.e. it is already accounted for. At the next level there are 324 states but 25 of them are descendants: $L_{-2}|p\rangle$ and $L_{-1}\alpha_{-1}^i|p\rangle$ (the state $L_{-1}L_{-1}|p\rangle \propto p^i L_{-1}\alpha_{-1}^i|p\rangle$ is not independent), etc. So we see that the light-cone partition function does indeed decompose as a sum over Verma modules of the $c = 24$ CFT.

The last point to clarify is the physical interpretation of the torus amplitude as the one-loop cosmological constant of the bosonic string. Recall that the effective action for a scalar field is defined as

$$e^{-\Gamma_E} = \int \mathcal{D}\phi \, e^{-S_E[\phi]}, \tag{6.85}$$

where S_E is the Euclidean action. For a free (massive) scalar field $\Gamma_E = VE_{\text{vac}}$ where V is the volume of space-time and E_{vac} is the vacuum energy, i.e. the cosmological constant. In this case $S_E = \frac{1}{2}\int d^d x \, \phi(-\Box + m^2)\phi$ and the Gaussian integral gives

$$\int \mathcal{D}\phi \, e^{-S_E} = \left(\det(-\Box + m^2)\right)^{-1/2}. \tag{6.86}$$

From this we obtain

$$VE_{\text{vac}} = -\log \det{}^{1/2}(-\Box + m^2) = -\frac{1}{2}\log \det \Delta^{-1} = -\frac{1}{2}\text{tr} \log \Delta^{-1}$$

$$= -\frac{V}{2}\int \frac{d^d k}{(2\pi)^d} \log(k^2 + m^2), \tag{6.87}$$

where $\Delta = \frac{1}{-\Box + m^2}$ is the Euclidean propagator and we have used

$$\text{tr} \log \Delta^{-1} = \int d^d x \, \langle x|\log \Delta^{-1}|x\rangle$$

$$= \int d^d x \int d^d k \int d^d k' \, \langle x|k\rangle \langle k|\log \Delta^{-1}|k'\rangle \langle k'|x\rangle$$

$$= V \int \frac{d^d k}{(2\pi)^d} \log(k^2 + m^2). \tag{6.88}$$

This is the vacuum energy for a single bosonic particle of mass m. Therefore the vacuum energy density for all string excitations is

$$\rho = -\frac{1}{2}\sum_I \int \frac{d^d k}{(2\pi)^d} \log(k^2 + m_I^2), \tag{6.89}$$

where the sum is over all physical excitations of the string. One can rewrite this using Schwinger's proper time representation

$$\begin{aligned}\rho &= \frac{1}{2}\sum_I \int \frac{d^d k}{(2\pi)^d} \int_0^\infty \frac{dt}{t} e^{-t(k^2+m_I^2)} \\ &= \int_0^\infty \frac{dt}{2t} \int_{-\pi}^{+\pi} \frac{d\theta}{2\pi} \left(\frac{\pi}{t}\right)^{d/2} \text{tr}_{\text{osc.}} \left(e^{-\frac{2t}{\alpha'}(k^2+(N+\overline{N}-2))} e^{i(N-\overline{N})\theta}\right),\end{aligned} \quad (6.90)$$

where we have performed the integration over the momenta and we have used $\frac{\alpha'}{2}m^2 = N + \overline{N} - 2$. Here N, $\overline{N}$ are the transverse oscillator excitation numbers; cf. (3.47). We have also implemented the level matching condition $N = \overline{N}$ via

$$\delta_{N,\overline{N}} = \int_{-\pi}^{+\pi} \frac{d\theta}{2\pi} e^{i\theta(N-\overline{N})}. \quad (6.91)$$

The t-integral diverges at the lower end and has to be regularized. This divergence reflects the UV divergence of the momentum integral in (6.89). We introduce

$$2\pi\tau = \theta + i\frac{2t}{\alpha'} \equiv 2\pi(\tau_1 + i\tau_2) \quad (6.92)$$

and perform the trace over the momenta to obtain ($d^2\tau = 2d\tau_1 d\tau_2$, $d = 26$)

$$VE_{\text{vac}} = \frac{V}{\ell_s^{26}} \int \frac{d^2\tau}{4\tau_2} \frac{1}{\tau_2^{13}} \text{tr}_{\text{osc.}} \left(e^{-2\pi\tau_2(N+\overline{N}-2)} e^{2\pi i \tau_1(N-\overline{N})}\right). \quad (6.93)$$

If we interpret τ as the modular parameter of the torus and restrict the τ-integration to a fundamental domain—this also cuts off the divergence at $\tau_2 = 0$—Eq. (6.93) agrees, including the normalization, with (6.71). The latter was the result of a careful evaluation of the Polyakov path integral. From this 'derivation' we learn that the torus amplitude has the interpretation as the one-loop cosmological constant of the closed string. It is positive and of the order of $1/\ell_s^d$. When we compute the partition function of the superstring in Chap. 9, we will find that it vanishes on account of space-time supersymmetry.

6.4 Torus Partition Functions for Rational CFTs

The torus partition functions, which we derived for the bosonic string in the previous section, is an important object in general CFT's. One interesting aspect is the relation between the fusion rules (4.71) and the properties of the partition function. We begin by recapitulating a few facts from Chap. 4.

6.4 Torus Partition Functions for Rational CFTs

Consider a rational CFT with central charge c, i.e. there exist only a finite number of highest weight representations of the chiral algebra. The prototype example are the minimal models of the Virasoro algebra (4.75). The character of an irreducible representation $|h_i\rangle$ with highest weight h_i is defined as

$$\chi_i(\tau) = \mathrm{tr}_{\mathcal{H}_i}\left(q^{L_0 - \frac{c}{24}}\right), \tag{6.94}$$

where $\mathcal{H}_i$ denotes the Hilbert space built upon the (irreducible) highest weight state $|h_i\rangle$. The complete torus partition function can then be written in terms of these characters as

$$Z(\tau, \bar{\tau}) = \sum_{i\bar{j}} M_{i\bar{j}}\, \chi_i(\tau) \bar{\chi}_{\bar{j}}(\bar{\tau}). \tag{6.95}$$

Requiring that this partition function is invariant under global diffeomorphisms of the torus, i.e. under modular transformations, provides very strong constraints on the non-negative integers $M_{i\bar{j}}$.

On the space of characters there must exist an action of the two generators $T: \tau \to \tau + 1$ and $S: \tau \to -\frac{1}{\tau}$. In particular, there exists a matrix T_{ij} such that a modular T-transformation acts as

$$\chi_i(\tau + 1) = \sum_j T_{ij}\, \chi_j(\tau). \tag{6.96}$$

If the levels in the character differ by an integer (this excludes characters in fermionic theories to be introduced in Chap. 9), the matrix T_{ij} is diagonal with

$$T_{ij} = \delta_{ij}\, e^{2\pi i \left(h_i - \frac{c}{24}\right)}. \tag{6.97}$$

For $M_{i\bar{j}} \neq 0$, T-invariance requires $h_i - h_j \in \mathbb{Z}$ (if $c = \bar{c}$).

Similarly, the action of S is described by the modular S-matrix

$$\chi_i\left(-\frac{1}{\tau}\right) = \sum_j S_{ij}\, \chi_j(\tau), \tag{6.98}$$

which can be shown to satisfy $S^2 = C$. The so-called charge conjugation matrix C satisfies $C^2 = 1$, i.e. $S^4 = 1$. Furthermore, the S-matrix is unitary and symmetric. $S^2 \neq 1$ is only possible if the characters form an unfaithful representation of the modular group. This happens when there are complex primary fields. The characters of ϕ and $\phi^\dagger$ are identical and get interchanged under S^2. Modular invariance of the partition function $Z(\tau, \bar{\tau})$ requires

$$[M, S] = [M, T] = 0. \tag{6.99}$$

One consequence of (6.99) is that $M_{i\bar{j}} = \delta_{i\bar{j}}$ always leads to a modular invariant partition function. This is called the diagonal invariant.

It is one of the deepest results in CFT that there exists an intricate relation between the modular S-matrix (torus partition function) and the fusion algebra for the OPE on the sphere (tree level). Namely, the fusion coefficients $N_{ij}{}^k \in \mathbb{Z}_0^+$ can be computed from the S-matrix via the Verlinde formula

$$N_{ij}{}^k = \sum_m \frac{S_{im} S_{jm} S_{mk}^*}{S_{0m}}, \qquad (6.100)$$

where S^* denotes the complex conjugate of S and the index 0 labels the vacuum representation. It is quite remarkable that the above combination of $S_{ij} \in \mathbb{C}$ always gives non-negative integers.

We know from (4.73) that the fusion matrices N_i can be diagonalized simultaneously. If we write the Verlinde formula, using matrix notation,

$$N_i = S\, D_i\, S^{-1}, \qquad (D_i)_{mn} = \left(\frac{S_{im}}{S_{0m}}\right) \delta_{mn}, \qquad (6.101)$$

we can state it by saying that the modular S-matrix diagonalizes the fusion matrices.

Even with the explicit form of the modular T and S matrices known, it is highly non-trivial to find all solutions for the coefficients $M_{i\bar{j}}$ in the torus partition function (6.95). For the minimal models (4.75) of the Virasoro algebra, such a classification has been achieved. Intriguingly, it has turned out that these modular invariant partition functions obey an ADE classification, very similar to the ADE classification of simply laced Lie-algebras, the discrete subgroups of $SU(2)$ and of a certain class of singularities of complex surfaces. However, in general, to find modular invariant partition functions is quite a challenge. It would thus be helpful to have a method which allows to generate modular invariant partition functions without explicitly classifying all matrices M satisfying (6.99).

Such a method is the simple current construction. Since it has found extensive application in string theory, cf. the construction of Gepner models in Chap. 15, we will briefly present it here. For a unitary RCFT, the definition of a simple current is in terms of the fusion rules of the chiral algebra. Given such a theory with highest weight representations $[\phi_i]$ and fusion algebra (4.71), a highest weight representation J is called a simple current if its fusion with any other highest weight takes the simple form

$$[J] \times [\phi_i] = [\phi_{J(i)}]. \qquad (6.102)$$

The notation $J(i)$ means that J permutes the indices of the fields ϕ_i. The fact that only one field appears on the right hand side is the reason why the currents are called simple. In general the action of J can have fixed points, i.e. $J(i) = i$. Every CFT has at least one simple current, namely the identity. But we are interested in less

6.4 Torus Partition Functions for Rational CFTs

trivial examples. Also, it follows from the associativity of the fusion algebra that J^n for any n is also a simple current.

Since we are considering a RCFT with only a finite number of highest weight representations, there must exist an integer $L \in \mathbb{Z}$, called the order of J, such that

$$[J]^L = [\mathbf{1}]. \tag{6.103}$$

It follows that $J^c = J^{L-1}$ and an equivalent characterization of a simple current J is that in the OPE with J^c only the identity appears.[14] Similarly, for any primary field ϕ_i there is an integer l_i such that $[J]^{l_i} \times [\phi_i] = [\phi_i]$. Therefore, J organizes all highest weight representations into orbits of length $l_i = L/p$ where p is a divisor of L

$$(\phi_i, J\phi_i, J^2\phi, \ldots, J^{l_i-1}\phi_i). \tag{6.104}$$

Consider now the general form of the OPE of a simple current J and a primary field ϕ_i

$$J(z)\phi_i(w) \sim \frac{1}{(z-w)^{Q(\phi_i)}} \Big([J] \times [\phi_i] \Big)(w), \tag{6.105}$$

which defines the so-called monodromy charge of the field ϕ_l

$$Q(\phi_i) = h(J) + h(\phi_i) - h(J\phi_i) \quad \text{mod } 1. \tag{6.106}$$

The monodromy charge measures the phase that is accumulated if one moves J around ϕ_i. Note that this notion only makes sense because the currents are simple. If one moves it around the product of fields $\phi_i(z)\phi_j(w)$, one finds that the phases simply add, i.e.

$$Q(\phi_i\phi_j) = Q(\phi_i) + Q(\phi_j). \tag{6.107}$$

Applying this to $J^{n-1}\phi_i$ one finds that

$$h(J^n\phi_i) = h(\phi_i) + h(J^n) - nQ(\phi_i) \quad \text{mod } 1. \tag{6.108}$$

For $n = L$ this leads to

$$Q(\phi_i) = \frac{t(\phi_i)}{L} \quad \text{with} \quad t(\phi_i) \in \mathbb{Z}. \tag{6.109}$$

---

[14]Proof: assume $J \times \phi_1 = \phi_2 + \phi_3 + \ldots$, multiply both sides with J^c and obtain $\mathbb{1} \times \phi_1 = J^c \times \phi_2 + J^c \times \phi_3 + \ldots$ which is in contradiction with the fusion rule of the identity operator.

The integers $t(\phi_i)$ are in general different for different primary fields. Their explicit form is not needed in the following.

Let us now consider the modular S-transformation of characters in theories with simple currents. To do so we decompose the space of primary fields into J-orbits and in each orbit we choose a representative ϕ_a. We also introduce the short-hand notation: $J^\alpha \phi_a =: (\alpha a)$. Let us further restrict ourselves to the case with $h(J) \in \mathbb{Z}$. In this case it is consistent with locality of the CFT to set $\bar{h}(J) = 0$ and the simple current can be included in the symmetry algebra of the CFT. This is called a simple current extension and simple currents with integer conformal dimension are called orbit simple currents. Integrality of $h(J)$ also implies that $Q(J) = 0$ and that $Q(J^\alpha \phi_a)$ is independent of α. With these assumptions, the requirement that the fusion algebra respects the monodromy charge Q restricts the modular S-matrix to have the following form:

$$S_{(\alpha a)(\beta b)} = \exp\left(2\pi i \left(Q(\phi_a)\beta + Q(\phi_b)\alpha\right)\right) S_{ab}. \tag{6.110}$$

The proof, which heavily relies on the Verlinde formula, will not be presented here.

With the help of (6.109) one verifies that orbits

$$\widehat{\chi}_a(\tau) = \sum_{\alpha=0}^{l_a-1} \chi_{(\alpha a)}(\tau) \tag{6.111}$$

of integer monodromy charge $Q(a)$ transform among themselves under modular transformations. Using this fact one can construct for each simple current of integer conformal dimension a new non-diagonal modular invariant partition function in the following way

$$Z_J(\tau, \bar{\tau}) = \frac{1}{\mathcal{N}} \sum_{\substack{(\alpha a) \\ Q(a) \in \mathbb{Z}}} \sum_{\beta=0}^{L-1} \chi_{((\alpha+\beta)a)}(\tau)\, \bar{\chi}_{(\alpha a)}(\bar{\tau}). \tag{6.112}$$

The normalization constant $\mathcal{N}$ is fixed by the requirement that the vacuum should appear only once in $Z_J(\tau, \bar{\tau})$. Note that the sum over α extends over the length l_a of the orbit a, while the sum over β runs over L, the order of J. One can verify that $M_{(\alpha a)(\beta b)} = \sum_{\gamma=0}^{L-1} \delta_{ab}\, \delta_{\beta,\alpha+\gamma}\, \delta(Q(a))$ indeed satisfies (6.99). Here we have defined

$$\delta^{(n)}(q) = 1 \quad \text{for} \quad q = 0 \bmod n, \quad \text{and} \quad \delta(q) \equiv \delta^{(1)}(q). \tag{6.113}$$

An important feature of Z_J is that only states with integer monodromy charge contribute, the others are projected out.

6.5 The Cylinder Partition Function

So far, we have considered only one-loop partition functions for closed strings, respectively CFTs on the torus. Let us now discuss the one-loop partition function for open strings, i.e. for CFTs with boundaries.

To do so, recall that we defined the one-loop partition function for closed strings as follows. We started from a theory defined on the infinite cylinder parametrized by (τ, σ) where $\sigma \in [0, 2\pi]$ was periodic and $\tau \in (-\infty, +\infty)$. We then imposed periodicity conditions also on the time coordinate τ yielding the topology of a torus. In the present case, the space coordinate σ is not periodic and thus we start from a theory defined on the infinite strip with $\sigma \in [0, \pi]$ and $\tau \in (-\infty, +\infty)$. For the definition of the one-loop partition function, we again make the time coordinate τ periodic. This leaves us with the topology of a cylinder instead of a torus. This is illustrated in Fig. 6.14.

As is evident form Table 6.2, similarly to the complex modular parameter τ of the torus, there is a real modular parameter t with $0 \leq t < \infty$ which parametrises different cylinders. The inequivalent cylinders are described by $\{(\tau, \sigma) : 0 \leq \sigma \leq \pi, 0 \leq \tau \leq 2\pi t\}$.[15] Recall from Eq. (4.162) that the cylinder/annulus partition function is defined as $Z^A(t) = \text{tr}\left[\exp(-2\pi t\, H_{\text{op}})\right]$ with $H_{\text{op}} = (L_{\text{cyl.}})_0 = L_0 - \frac{c}{24}$. The complete cylinder one-loop diagram can be written as

$$\mathscr{A} = \int_0^\infty \frac{dt}{2t}\, \text{tr}_{\mathscr{H}_{\text{op}}}\left(q^{L_0 - \frac{c}{24}}\right) \qquad (q = e^{-2\pi t}). \tag{6.114}$$

The derivation is as for the closed string torus amplitude. We have also introduced the notation $\mathscr{A} = A_0^{(g=1, b=2)}$.

Consider this amplitude for an open string stretched between two parallel bosonic Dp-branes. In this case the longitudinal center of mass position of the string introduces a regularized world-volume factor V_{p+1}. Moreover, as for the torus partition function the trace for the longitudinal center of mass momentum is written as an integral

$$\mathscr{A} = V_{p+1} \int_0^\infty \frac{dt}{2t} \frac{d^{p+1}p}{(2\pi)^{p+1}}\, \text{tr}_{\text{osc}}\left(q^{L_0 - \frac{c}{24}}\right). \tag{6.115}$$

The trace is now only over oscillator excitations of the open string. We have not yet included the trace over the Chan-Paton indices. In the presence of N parallel Dp-branes which lie on top of each other this is simply $N^2 = N + 2 \times \frac{1}{2}N(N-1)$, where the first contribution is due to the strings whose two end-points lie on the same brane and the second due to the strings which stretch between two different

---

[15] If the cylinder is mapped to the plane its image is an annulus. The map is $z = \exp(\frac{i}{t}(\tau - i\sigma))$. The modular parameter t' of the annulus is the ratio of the radii of the two boundary circles. Two fundamental regions are $0 < t' \leq 1$ and $1 \leq t' < \infty$ which can be mapped to each other by an S transformation $t' \to 1/t'$.

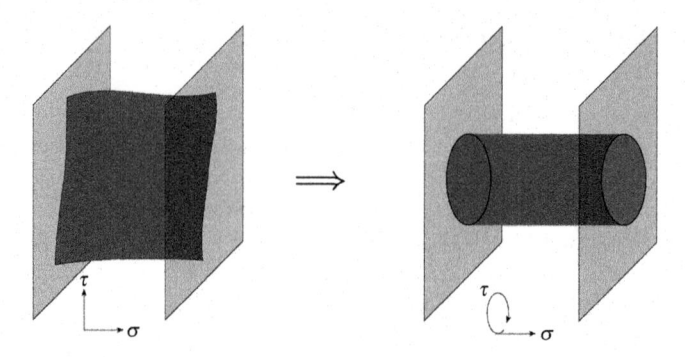

Fig. 6.14 Illustration how the cylinder partition function is obtained from the infinite strip by cutting out a finite piece and identifying the ends

branes. The factor two in the second term is for the two possible orientations of the string. If the branes are separated along their common transverse directions one has to take this into account in the expression for L_0.

For each single free boson X^μ the trace over the oscillators takes the familiar form

$$Z^A(t) = \text{tr}_{\mathcal{H}_{\text{osc}}}\left(q^{L_0 - \frac{c}{24}}\right) = \frac{1}{\eta(it)}. \tag{6.116}$$

But we also have to take into account the effect of the zero modes, i.e. the center of mass positions and center of mass momentum modes of the open string. Here we have to distinguish between directions with NN and DD boundary conditions. We denote the longitudinal directions to the Dp-brane as X^μ and the transverse directions as X^a.

For the case of NN boundary conditions for coordinate X^μ, the momentum mode p^μ is unconstrained and contributes to the trace. Since it is a continuous variable, the sum is replaced by an integral

$$\int_{-\infty}^{\infty} dp^\mu \, e^{-2\pi t \alpha' (p^\mu)^2} = \frac{1}{\sqrt{2\alpha' t}}. \tag{6.117}$$

For the DD case, we have seen in Eq. (3.42) that we get an extra contribution from the stretched string. Therefore, we get a contribution to the partition function of the form

$$\exp\left(-\frac{t}{2\pi\alpha'} \sum_i (x_B^a - x_A^a)^2\right) \equiv \exp\left(-\frac{t}{2\pi\alpha'} Y^2\right). \tag{6.118}$$

Combining the oscillator and zero mode contributions, we can compute the one-loop string partition function for an open string stretched between a bosonic Dp-brane at transverse position x_A^a and a second one at position x_B^a. For the total one-loop diagram we obtain

6.5 The Cylinder Partition Function

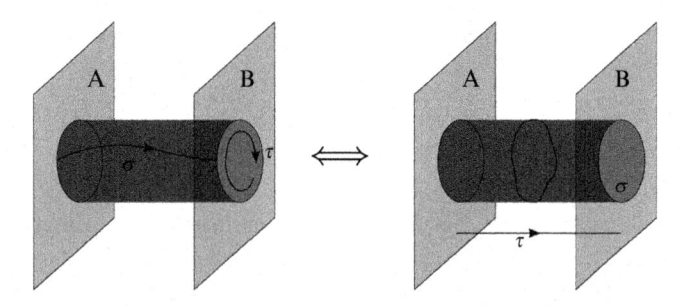

Fig. 6.15 Illustration of world-sheet duality relating the cylinder amplitude in the open and closed sector

$$\mathscr{A}_{Dp,Dp} = V_{p+1} \int_0^\infty \frac{dt}{t} \frac{1}{(8\pi^2\alpha't)^{\frac{p+1}{2}}} \exp\left(-\frac{t}{2\pi\alpha'}Y^2\right) \frac{1}{\eta^{24}(it)}. \quad (6.119)$$

The overall factor of two as compared to (6.114) is $N = 2$, as explained after (6.115).

So far, we have defined the Dp-branes indirectly via the target-space locus where the end points of the open strings are confined to. The question now is whether one can find a more intrinsic description of these boundaries. This is in fact possible within the boundary state formalism where the D-brane is represented as a coherent state in the closed string Hilbert space.

Let us repeat the observation about the cylinder diagram which we made in Chap. 4. It is illustrated more explicitly in Fig. 6.15. Upon interchanging the roles of τ and σ we can reinterpret the open string cylinder partition function shown on the left-hand side as the closed string tree-level diagram shown on the right hand side. The tree level amplitude describes the emission of a closed string at boundary A which propagates to boundary B where it is absorbed. In both views the boundaries are D-branes. In the open string picture they are objects on which open strings end. In the closed string picture they are objects which emit and absorb closed strings. The fact that the same amplitude can be considered either as an open string one loop-amplitude (loop-channel) or as a tree-level closed string exchange (tree-channel) is known as world-sheet duality between open and closed strings.

We now use the boundary states constructed in Chap. 4 to compute the cylinder diagram (6.114) directly in the closed string tree-channel, i.e. by the exchange of closed strings between two boundaries. Referring again to Fig. 6.15, we interpret this diagram as a closed string which is emitted at the boundary A, propagating via the closed sector Hamiltonian $H_{\text{closed}} = L_0 + \bar{L}_0 - \frac{c+\bar{c}}{24}$ for a (Euclidean) time l until it reaches the boundary B where it is absorbed. Here we have chosen $\ell = 2\pi$ for the length of the closed string. We work in light-cone gauge which means that we have to impose Neumann boundary conditions along $X^\pm$. The zero mode integrations have to be performed over all Neumann directions. An alternative would be to consider the contribution from all 26 oscillators and of the ghosts and then verify directly that the ghost contribution precisely cancels that of two oscillators.

In analogy to Quantum Mechanics, such an amplitude is given by the overlap

$$\widetilde{\mathscr{A}} = \int_0^\infty dl\, \widetilde{Z}^A(l) = \int_{-\pi}^{+\pi} \frac{d\theta}{2\pi} \int_0^\infty dl\, \langle B|\, e^{-2\pi l\left(L_0+\bar{L}_0-\frac{c+\bar{c}}{24}\right)}\, e^{i\theta(L_0-\bar{L}_0)}\,|B\rangle, \tag{6.120}$$

where the tilde indicates that the computation is performed in the closed sector (or at tree-level) and $2\pi l$ is the length of the cylinder connecting the two boundaries. The integral over θ has again been included to impose level matching. As we will see momentarily, the boundary states have $(L_0-\bar{L}_0)$-eigenvalue zero, so the integral over θ gives a trivial contribution. We observe that

$$\int_{-\pi}^{+\pi} \frac{d\theta}{2\pi} \int_0^\infty dl\, e^{-2\pi l\left(L_0+\bar{L}_0-\frac{c+\bar{c}}{24}\right)}\, e^{i\theta(L_0-\bar{L}_0)} = \frac{\delta_{m_L^2, m_R^2}}{\alpha'\pi(p^2 + m_L^2 + m_R^2)} \tag{6.121}$$

is the string propagator.

We will now evaluate (6.120). We define $\widetilde{Z} = \mathcal{N}_{Dp}^{-2} \widetilde{Z}^A_{NN} \widetilde{Z}^A_{DD}$ where $\mathcal{N}_{Dp}$ is a normalization constant to be determined and the $\widetilde{Z}^A$ are evaluated with the unnormalized boundary states of Chap. 4. Recall from there the orthonormal basis states (4.173). For each Neumann direction we obtain, using that $L_0|\mathbf{m}\rangle = \sum_n n m_n |\mathbf{m}\rangle$ and likewise for the $\bar{L}_0$ eigenvalue of $|U\mathbf{m}\rangle$, and the orthonormality of these states,

$$\widetilde{Z}^A_{NN}(l) = \sum_{m_1=0}^{\infty} \sum_{m_2=0}^{\infty} \cdots e^{\frac{\pi l}{6}} \exp\left(-4\pi l \sum_{n>0} n m_n\right) \langle 0|0\rangle$$

$$= e^{\frac{\pi l}{6}} \prod_{n=1}^{\infty} \left(\sum_{m_n=0}^{\infty} e^{-4\pi l m_n}\right) \langle 0|0\rangle$$

$$= e^{\frac{\pi l}{6}} \prod_{n=1}^{\infty} \left(1 - e^{-4\pi l n}\right)^{-1} \langle 0|0\rangle = \frac{1}{\eta(2il)} \langle 0|0\rangle. \tag{6.122}$$

We normalize the momentum eigenstates as in (6.81)

$$\langle 0|0\rangle = 2\pi\delta(0) = L, \tag{6.123}$$

where L is the (regularized) volume of the Neumann direction. We thus find for the contribution of a Dp brane

$$\widetilde{Z}^A_{NN}(l) = \frac{1}{\eta(2il)^{p-1}} V_{p+1}, \tag{6.124}$$

where V_{p+1} is the regularized world-volume of the brane.

6.5 The Cylinder Partition Function

Next, we consider the case of a coordinate with DD boundary conditions. For the oscillator modes, the calculation is completely analogous to the case with NN conditions. However, instead of (6.123) we now have to compute the contribution from the zero modes of $(L_0 + \bar{L}_0)$:

$$\langle x_B | e^{-\pi l \alpha' p^2} | x_A \rangle = \int \frac{dp}{2\pi} e^{ip(x_B - x_A) - \pi l \alpha' p^2} = \frac{1}{\sqrt{4\pi^2 l \alpha'}} e^{-\frac{1}{4\pi l \alpha'}(x_B - x_A)^2}, \tag{6.125}$$

where we have inserted a complete set of momentum eigenstates $1 = \frac{1}{2\pi} \int dp |p\rangle\langle p|$. The contribution from $d_\perp = d - p - 1$ Dirichlet directions is then

$$\widetilde{Z}^A_{\text{DD}} = \frac{1}{\eta(2il)^{d_\perp}} \left(\frac{1}{\sqrt{4\pi^2 l \alpha'}} \right)^{d_\perp} e^{-\frac{1}{4\pi l \alpha'}(x_B - x_A)^2}. \tag{6.126}$$

Combining the NN and DD amplitudes we find the overlap between the boundary states for two bosonic Dp-branes

$$\widetilde{\mathcal{A}}_{\text{Dp,Dp}} = \frac{V_{p+1}}{\mathcal{N}_{\text{Dp}}^2} \int_0^\infty dl \, e^{-\frac{1}{4\pi l \alpha'} Y^2} \frac{1}{(4\pi^2 \alpha' l)^{\frac{25-p}{2}}} \frac{1}{\eta(2il)^{24}}, \tag{6.127}$$

where $Y^2 = \sum_i (x_B^i - x_A^i)^2$ is the distance between the two branes.

For different boundary conditions at the two ends of the open string one has to compute the overlap of a Neumann and a Dirichlet boundary state. There is no zero mode contribution and the oscillator part can be expressed as

$$\widetilde{Z}^A_{\text{ND}}(l) = e^{\frac{\pi l}{6}} \prod_{k=1}^\infty \frac{1}{1 + e^{-4\pi l k}} = \sqrt{2} \sqrt{\frac{\eta(2il)}{\vartheta_2(2il)}}. \tag{6.128}$$

where we have written the infinite product in terms of a ϑ-function. We will introduce them more systematically in Chap. 9.

Comparison of the tree level closed string amplitude with the one-loop open string amplitude derived above allows us to fix the normalization $\mathcal{N}_{\text{Dp}}$ of the boundary state. For this we have to establish the relation between the parameters l and t. Recall that the modulus of a cylinder is the ratio of its circumference to its height. In the tree diagram this is $\frac{2\pi}{2\pi l}$ whereas in the loop diagram it is $\frac{2\pi t}{\pi}$. Comparison leads to the relation

$$t = \frac{1}{2l}. \tag{6.129}$$

This is the formal expression for the pictorial loop-channel–tree-channel equivalence of the cylinder diagram illustrated in Fig. 6.15.

With this relation we can express the cylinder amplitudes in terms of the closed string parameter l. All we need is the modular transformation of the η-function

$$\eta(it) = \sqrt{2l}\,\eta(2il) \tag{6.130}$$

so that we find

$$\mathcal{N}_{Dp}^{-1} = \frac{1}{2^6}(4\pi^2\alpha')^{\frac{1}{2}(12-p)}. \tag{6.131}$$

From the closed string point of view the amplitude is a sum over closed string states which are exchanged by the D-branes. If we expand the η-function we can separate the contributions from the different closed string mass levels. The first two terms in this expansion are

$$\frac{1}{\eta^{24}(2il)} = e^{4\pi l} + 24 + O(e^{-4\pi l}). \tag{6.132}$$

The first term is the tachyon contribution. It apparently leads to a divergence in the annulus amplitude which will be absent in the supersymmetric case to be discussed later. The constant term in (6.132) is due to the exchange of massless closed string modes. In particular the coupling to the graviton means that D-branes carry an energy density in the form of a brane tension and as such they are a source for the gravitational field.

From (6.127) with the expansion (6.132) and the substitution $\bar{l} = \alpha' l$ we get

$$\widetilde{\mathcal{A}}_{Dp,Dp} = \frac{24 V_{p+1}}{2^{p+13}\,\pi^{p+1}(\alpha')^{p-11}} \int_0^\infty d\bar{l}\,\exp\!\left(-\frac{|Y|^2}{4\pi\bar{l}}\right) \frac{1}{(\bar{l})^{\frac{25-p}{2}}} + \cdots$$

$$= \frac{3\pi}{2^7}(4\pi^2\alpha')^{11-p}\,V_{p+1}\,G_{25-p}(|Y|) + \cdots, \tag{6.133}$$

where

$$G_D(|Y|) = \int \frac{d^D p}{(2\pi)^D} \frac{e^{ip\cdot Y}}{p^2}$$

$$= \int_0^\infty ds \int \frac{d^D p}{(2\pi)^D} e^{-sp^2 - ip\cdot Y} = \frac{1}{\pi^{D/2} 2^D} \int_0^\infty ds\,\frac{1}{s^{D/2}} e^{-\frac{|Y|^2}{4s^2}}$$

$$= \frac{1}{4\pi^{\frac{D}{2}}} \Gamma\!\left(\frac{D-2}{2}\right) |Y|^{2-D} = \frac{1}{(D-2)\text{Vol}(S^{D-1})} \frac{1}{|Y|^{D-2}} \tag{6.134}$$

is the scalar Green function in the space transverse to the two D-branes, i.e. it is a solution to the equation $\Box G_D(|Y|) = -\delta^{(D)}(Y)$. Here S^{D-1} is the $(D-1)$-dimensional unit sphere. Note that the set of points a unit distance away from the brane in the transverse directions is a S^{d-p-2}.

The field theoretical interpretation of this result is quite clear: the Y-dependence is that of the interaction potential created by an extended object, the Dp-brane, with $25 - p$ transverse spatial dimensions, where the interaction is due to the exchange of massless particles. The familiar example is, of course, the Newton potential of a point mass which is proportional to the mass and, in d space-time dimensions, has a $1/|Y|^{d-3}$ fall-off. The equivalent of the mass for an extended object is its tension and the fall-off is determined by the number of transverse directions. One determines the brane tension T_p by comparing the string calculation with an effective field theory calculation of the interaction of two branes due to the tree level exchange of massless particles. It turns out that only the graviton and dilaton contribute. This calculation is relegated to Appendix 6.8. The result of the comparison is

$$T_p = \frac{\sqrt{\pi}}{16\tilde{\kappa}_{26}}(4\pi^2\alpha')^{\frac{1}{2}(11-p)}, \tag{6.135}$$

where $\tilde{\kappa}_d$ is the gravitational coupling constant in d dimensions. If the dilaton has a non-zero expectation (background) value Φ_0, this is not the coupling constant which governs Newton's law which is derived in a weak field approximation around a Minkowski metric. In this case, the relevant coupling constant is

$$\kappa_d = g_s\,\tilde{\kappa}_d, \qquad g_s = e^{\Phi_0}. \tag{6.136}$$

This will be explained in more detail in Chaps. 16 and 18. Likewise, the string tension, as measured in an asymptotic Minkowski space-time is

$$\tau_p = \frac{T_p}{g_s}. \tag{6.137}$$

In Chap. 16 we will also encounter two different gauge couplings $\tilde{g}_d$ and g_d, whose relation will depend on the type of string theory.

6.6 Boundary States and Cylinder Amplitude for RCFTs

In this section we provide the formalism how the notion of boundary states and the computation of annulus diagrams can be generalized to more abstract CFTs without a Lagrangian description and without an apparent space-time interpretation. Due to the lack of the latter it is not directly clear what the generalization of Neumann and Dirichlet boundary conditions is.

However, one can generalize the concept of boundary states and the gluing conditions they have to satisfy. We have already introduced in Eq. (4.164) the concept of a coherent state collecting all one-point functions of closed string states on the disk. Generalizing the results from the free boson theory, a boundary state $|B\rangle$ in the RCFT preserving the symmetry algebra $\mathcal{A} = \overline{\mathcal{A}}$ has to satisfy the following gluing conditions

$$(L_n - \overline{L}_{-n})|B\rangle = 0 \quad \text{conformal symmetry,}$$

$$(W_n^i - (-1)^{h_i} \overline{W}_{-n}^i)|B\rangle = 0 \quad \text{extended symmetries,} \quad (6.138)$$

where W_n^i is the Laurent mode of the extended symmetry generator W^i with conformal weight $h^i = h(W^i)$, and $\overline{W}^i$ denotes the generator in the anti-holomorphic sector. The derivation is completely analogous to the one of (4.160).

The condition for the extended symmetries can be relaxed, so that also Dirichlet boundary conditions similar to the example of a free boson are covered

$$\left(W_n^i - (-1)^{h^i} \Omega(\overline{W}_{-n}^i)\right)|B\rangle = 0, \quad (6.139)$$

where $\Omega : \mathcal{A} \to \mathcal{A}$ is an automorphism of the chiral algebra $\mathcal{A}$. Such an automorphism Ω is also called a gluing automorphism and for the example of the free boson with Dirichlet boundary conditions, it is simply $\Omega : \overline{W}_n \mapsto -\overline{W}_n$.

The next step is to find solutions to these gluing conditions. In general this is a formidable task but for the simple case of the so-called charge conjugate modular invariant partition function $\sum_i \chi_i \overline{\chi}_{i^+}$ a general solution has been found by Ishibashi. Here, the charge conjugation matrix is defined by $C = S^2$ and maps highest weight representations i to their charge conjugate i^+. Denoting the Hilbert space built upon the charge conjugate representation by $\mathcal{H}_i^+$, we can state the result: For $\overline{\mathcal{A}} = \mathcal{A}$ and $\overline{\mathcal{H}}_i = \mathcal{H}_i^+$, to each highest weight representation ϕ_i of $\mathcal{A}$ one can associate an (up to a multiplicative constant) unique state $|\mathcal{B}_i\rangle\rangle$ such that the gluing conditions are satisfied.

Since we are considering a RCFT, there is only a finite number of such so-called Ishibashi states $|\mathcal{B}_i\rangle\rangle$. We now construct the Ishibashi states in analogy to the boundary states of the free boson (4.172). Denoting by $|\phi_i, \mathbf{m}\rangle$ an orthonormal basis for $\mathcal{H}_i$, the Ishibashi states are written as

$$|\mathcal{B}_i\rangle\rangle = \sum_{\mathbf{m}} |\phi_i, \mathbf{m}\rangle \otimes U|\overline{\phi}_i, \overline{\mathbf{m}}\rangle, \quad (6.140)$$

where $U : \overline{\mathcal{H}} \to \overline{\mathcal{H}}^+$ is an anti-unitary operator acting on the symmetry generators $\overline{W}^i$ as follows

$$U\overline{W}_n^i U^{-1} = (-1)^{h^i} (\overline{W}_{-n}^i)^\dagger. \quad (6.141)$$

To show that the Ishibashi states satisfy the gluing conditions, it suffices to show that it holds on a complete set of states $\langle\phi_k, \mathbf{n}_1| \otimes \langle\overline{U(\phi_l, \mathbf{n}_2)}|$. This is done in the following calculation:

6.6 Boundary States and Cylinder Amplitude for RCFTs

$$\sum_{\mathbf{m}} \langle \phi_k, \mathbf{n}_1 | \otimes \langle \overline{U\phi_l, \mathbf{n}_2} | (W_n - (-1)^h \overline{W}_{-n}) | \phi_i, \mathbf{m} \rangle \otimes | \overline{U(\phi_i, \mathbf{m})} \rangle$$

$$= \sum_{\mathbf{m}} \Big\{ \langle \phi_k, \mathbf{n}_1 | W_n | \phi_i, \mathbf{m} \rangle \langle \overline{U(\phi_l, \mathbf{n}_2)} | \overline{U(\phi_i, \mathbf{m})} \rangle$$

$$- (-1)^{h_i} \langle \phi_k, \mathbf{n}_1 | \phi_i, \mathbf{m} \rangle \langle \overline{U(\phi_l, \mathbf{n}_2)} | \overline{W}_{-n} | \overline{U(\phi_i, \mathbf{m})} \rangle \Big\}$$

$$= \sum_{\mathbf{m}} \Big\{ \langle \phi_k, \mathbf{n}_1 | W_n | \phi_i, \mathbf{m} \rangle \langle \phi_l, \mathbf{n}_2 | \phi_i, \mathbf{m} \rangle^* - \langle \phi_k, \mathbf{n}_1 | \phi_i, \mathbf{m} \rangle \langle \phi_l, \mathbf{n}_2 | W_{-n} | \phi_i, \mathbf{m} \rangle^* \Big\}$$

$$= \delta_{ki} \delta_{li} \Big\{ \langle \phi_k, \mathbf{n}_1 | W_n | \phi_k, \mathbf{n}_2 \rangle - \langle \phi_k, \mathbf{n}_2 | W_{-n} | \phi_k, \mathbf{n}_1 \rangle^* \Big\} = 0. \tag{6.142}$$

In going from the second to the third line we have dropped the overbar as they are only a mnemonic to distinguish left- from right-movers and we have used the property of anti-unitary operators $\langle Ua|b\rangle = \langle a|U^{-1}b\rangle^*$ together with (6.141).

Given the form of the Ishibashi states (6.140) and using the convention of the out-states (4.55) the tree-channel overlap of two such states is

$$\langle\!\langle \mathscr{B}_j | e^{-2\pi l \left(L_0 + \bar{L}_0 - \frac{c+\bar{c}}{24}\right)} | \mathscr{B}_i \rangle\!\rangle = \delta_{ij}\, \chi_i(2il), \tag{6.143}$$

where the sum over all states $|\mathbf{m}\rangle$ in (6.140) amounts to the appearance of the characters χ_i of the highest weight ϕ_i. Performing a modular S-transformation for this overlap, we expect to obtain an open string partition function. However, because the modular S-transform of a character $\chi_i(2il)$ in general does not give non-negative integer coefficients in the loop-channel, the Ishibashi states themselves are not bona fide boundary states. However, they are their building blocks guaranteed to satisfy the gluing conditions. A true boundary state can be found by taking an appropriate linear combination of Ishibashi states

$$|B_\alpha\rangle = \sum_i B_\alpha^i\, |\mathscr{B}_i\rangle\!\rangle. \tag{6.144}$$

The complex coefficients B_α^i in (6.144) are called reflection coefficients and are further constrained by loop-channel–tree-channel equivalence. For RCFTs this condition is also called the Cardy condition.

Indeed, using relation (6.143) the cylinder amplitude between two boundary states can be expressed as

$$\widetilde{Z}_{\alpha\beta}^A(l) = \langle B_\alpha | e^{-2\pi l \left(L_0 + \bar{L}_0 - \frac{c+\bar{c}}{24}\right)} | B_\beta \rangle$$

$$= \sum_{i,j} (B_\alpha^j)^* B_\beta^i \langle\!\langle \mathscr{B}_j | e^{-2\pi l \left(L_0 + \bar{L}_0 - \frac{c+\bar{c}}{24}\right)} | \mathscr{B}_i \rangle\!\rangle$$

$$= \sum_i (B_\alpha^i)^* B_\beta^i\, \chi_i(2il). \tag{6.145}$$

Performing a modular S-transformation $l \mapsto \frac{1}{2t}$ on the characters χ_i, this closed sector cylinder diagram is transformed to the following expression in the open sector

$$Z^A_{\alpha\beta}(t) = \sum_{i,j} (B^i_\alpha)^* B^i_\beta S_{ij} \chi_j(it) \equiv \sum_j n^j_{\alpha\beta} \chi_j(it). \qquad (6.146)$$

Now, the Cardy condition is the requirement that this expression can be interpreted as a partition function in the open sector. That is, for all pairs of boundary states $|B_\alpha\rangle$ and $|B_\beta\rangle$ in a RCFT the following combinations have to be non-negative integers

$$n^j_{\alpha\beta} = \sum_i (B^i_\alpha)^* B^i_\beta S_{ij} \quad \in \mathbb{Z}^+_0. \qquad (6.147)$$

In general it is a challenge to find solutions to these equations. However, by observing that this condition is very reminiscent of the Verlinde formula (6.100), in certain cases a solution can be found. More precisely, for the case of a charge conjugate modular invariant partition function we can construct a generic solution to the Cardy condition by choosing the reflection coefficients in the following way

$$B^i_\alpha = \frac{S_{\alpha i}}{\sqrt{S_{0i}}}. \qquad (6.148)$$

Note, for each highest weight representation ϕ_i in the RCFT, there not only exists an Ishibashi state but also a boundary state, i.e. the index α in $|B_\alpha\rangle$ also runs from one to the number of highest weight representations. Employing then the Verlinde formula (6.100), the Cardy condition for the coefficients $n^j_{\alpha\beta}$ is always satisfied

$$n^j_{\alpha\beta} = \sum_i \frac{(S_{\alpha i})^* S_{\beta i} S_{ji}}{S_{0i}} = \sum_i \frac{S_{ji} S_{\beta i} (S_{\alpha i})^*}{S_{0i}} = N_{j\beta}{}^\alpha \quad \in \mathbb{Z}^+_0. \qquad (6.149)$$

Building upon this formula boundary states for various RCFT models have been constructed. Here to mention is in particular the construction of boundary states for the so-called Gepner models. Gepner models will be the subject of Sect. 15.5, albeit only in the closed string context.

6.7 Crosscap States, Klein Bottle and Möbius Strip Amplitudes

Having discussed the boundary states and how they are related to the computation of open string one-loop diagrams, in this section we introduce an analogous notion for non-oriented strings. This continues Sect. 4.5 and generalizes the concept of crosscap states beyond the free boson case. We will also discuss how these crosscap states are related to non-oriented one-loop diagrams, i.e. the Klein bottle and Möbius

6.7 Crosscap States, Klein Bottle and Möbius Strip Amplitudes

strip amplitudes. Since we will have an elaborate computation of the Klein bottle and Möbius strip amplitudes for the ten-dimensional type I superstring in Chap. 9, here instead we introduce these concepts for rational conformal field theories. The special case of a theory of free bosons is essentially included, where only the zero modes require some extra treatment.

Analogous to boundary states one can also define crosscap states for RCFTs. The crosscap gluing conditions for the generators of a symmetry algebra $\mathcal{A} \otimes \overline{\mathcal{A}}$ are in analogy to the conditions (4.179) for the example of the free boson

$$\left(L_n - (-1)^n \overline{L}_{-n} \right) |C\rangle = 0 \qquad \text{conformal symmetry,}$$

$$\left(W_n^i - (-1)^n (-1)^{h^i} \overline{W}_{-n}^i \right) |C\rangle = 0 \qquad \text{extended symmetries} \quad (6.150)$$

with again $h^i = h(W^i)$. For $\mathcal{A} = \overline{\mathcal{A}}$ and $\overline{\mathcal{H}}_j = \mathcal{H}_i^+$ we can define crosscap Ishibashi states $|\mathscr{C}_i\rangle\rangle$ satisfying the crosscap gluing conditions. A crosscap state $|C\rangle$ can then be expressed as a linear combination of the crosscap Ishibashi states

$$|C\rangle = \sum_i \Gamma^i |\mathscr{C}_i\rangle\rangle. \qquad (6.151)$$

The crosscap Ishibashi states and the boundary Ishibashi states are related via

$$|\mathscr{C}_i\rangle\rangle = e^{\pi i (L_0 - h^i)} |\mathscr{B}_i\rangle\rangle. \qquad (6.152)$$

Using (6.152) and the result for the boundary Ishibashi states (6.143), we can compute the overlap of two crosscap Ishibashi states as

$$\langle\langle \mathscr{C}_j | e^{-2\pi l \left(L_0 + \overline{L}_0 - \frac{c+\bar{c}}{24} \right)} |\mathscr{C}_i\rangle\rangle = \delta_{ij}\, \chi_i(2il). \qquad (6.153)$$

Next we can compute the overlap of two crosscap states (6.151)

$$\widetilde{Z}^K(l) = \langle C | e^{-2\pi l \left(L_0 + \overline{L}_0 - \frac{c+\bar{c}}{24} \right)} |C\rangle = \sum_i |\Gamma^i|^2\, \chi_i(2il) \qquad (6.154)$$

which has the interpretation of a closed string exchange between two crosscap states. From our experience with boundary states, we expect that this tree channel amplitude also has an interpretation as a loop diagram. Indeed, as shown in Fig. 6.16, by a number of cuts and shifts the cylinder with a crosscap at each end is related to the Klein bottle, which is a one-loop diagram of non-oriented strings. Therefore, in this channel we would write the same amplitude as

$$Z^K(t) = \text{tr}_{\mathcal{H}_{\text{cl}}} \left(\Omega e^{-2\pi t (L_0 + \overline{L}_0 - \frac{c}{12})} \right)$$

$$= \text{tr}_{\mathcal{H}_{\text{cl}}} \left(\Omega e^{-4\pi t (L_0 - \frac{c}{24})} \right). \qquad (6.155)$$

(a) Klein bottle

(b) Divide

(c) Shift

(d) Flip

(e) Tree-channel diagram

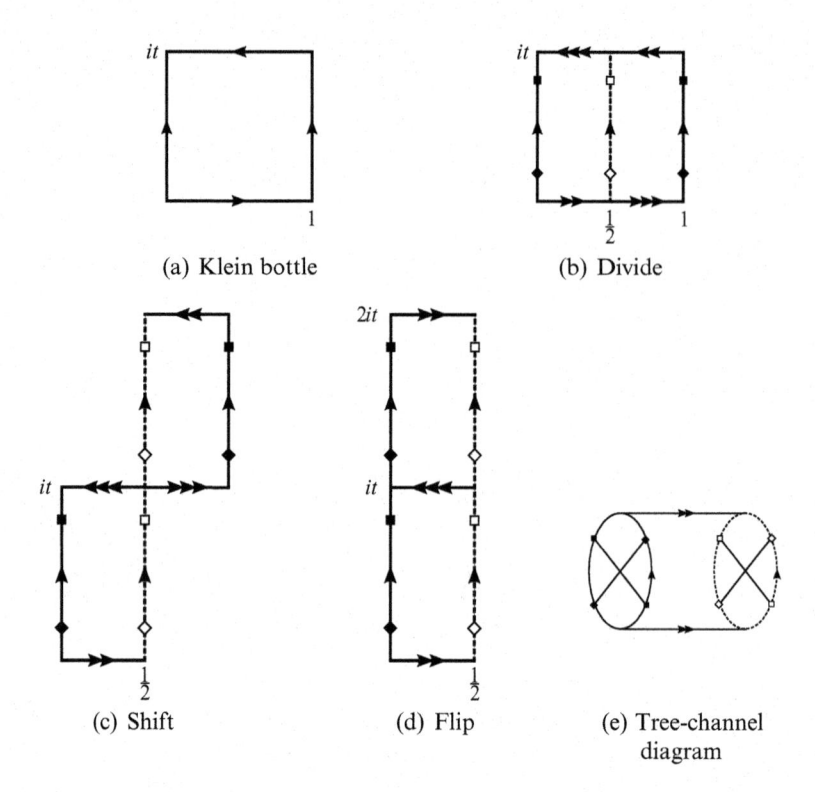

Fig. 6.16 Transformation of the fundamental domain of the Klein bottle to a tree-channel diagram between two crosscaps

As before, Ω is the orientation reversal operator on the world-sheet. In the second line we have used that only left-right symmetric states contribute to the trace. Moreover, the parameter t is the single real modulus of the Klein bottle (see Table 6.2). From Fig. 6.16 we can deduce that it is related to the parameter l from tree channel via

$$\frac{4\pi t}{\pi} = \frac{2\pi}{2\pi l}, \quad \rightarrow \quad l = \frac{1}{4t} \qquad (6.156)$$

Performing now a modular S-transformation of (6.154) we arrive at the loop-channel Klein bottle amplitude

$$Z^K(t) = \sum_{i,j} |\Gamma^i|^2 S_{ij}\, \chi_j(2it). \qquad (6.157)$$

From the definition of the loop-channel Klein bottle amplitude as a trace over the closed string Hilbert space, one has to require that

$$\sum_i |\Gamma^i|^2 S_{ij} = \kappa_j \in \mathbb{Z}. \qquad (6.158)$$

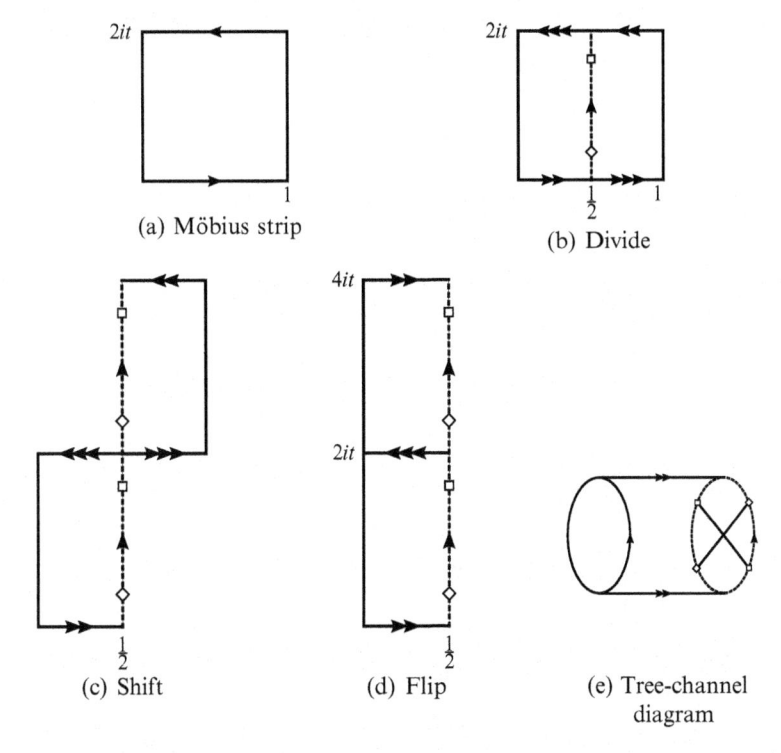

6.7 Crosscap States, Klein Bottle and Möbius Strip Amplitudes

Fig. 6.17 Transformation of the fundamental domain of the Möbius strip to a tree-channel diagram between an ordinary boundary and a crosscap

Since Ω can act with a minus sign, the κ_j are not necessarily non-negative, while the expansion coefficients in the torus + Klein bottle amplitude have to be non-negative integers.

Very similarly one can also compute the overlap of a crosscap state and a boundary state. This leads to a closed string exchange between a D-brane and an orientifold plane. Again by the series of cutting and shifting operations shown in Fig. 6.17, this is related to the non-oriented open string one-loop diagram, i.e. a Möbius strip diagram. The latter loop-channel amplitude is defined as

$$Z^M(t) = \text{tr}_{\mathcal{H}_{\text{op}}}\left(\Omega e^{-2\pi t(L_0 - \frac{c}{24})}\right). \quad (6.159)$$

Since from the action of Ω we get extra signs in the expressions, it is useful to introduce so-called hatted characters as

$$\hat{\chi}(\tau) = e^{-\pi i(h - \frac{c}{24})} \chi(\tau + \tfrac{1}{2}). \quad (6.160)$$

For the Möbius strip tree channel amplitude we readily compute

$$\widetilde{Z}^M(l) = \langle C | e^{-2\pi l\left(L_0 + \overline{L}_0 - \frac{c + \bar{c}}{24}\right)} | B_\alpha \rangle = \sum_i B_\alpha^i \left(\Gamma^i\right)^* \hat{\chi}_i(2il). \quad (6.161)$$

From Fig. 6.17 we infer the relation between the loop-channel and tree-channel parameter

$$\frac{4\pi t}{\pi/2} = \frac{2\pi}{2\pi l}, \quad \rightarrow \quad l = \frac{1}{8t}. \tag{6.162}$$

Since these hatted characters contain this shift by $1/2$ in their argument, we cannot simply apply a modular S-transformation to relate the tree- and the loop-channel. However, from the mapping of the modular parameter $2il$ under the combination of S- and T-transformations

$$2il \xrightarrow{T^{\frac{1}{2}}} 2il + \tfrac{1}{2} \xrightarrow{TST^2S} \tfrac{i}{8l} + \tfrac{1}{2} \xrightarrow{T^{-\frac{1}{2}}} \tfrac{i}{8l}, \tag{6.163}$$

we can infer the transformation of the hatted characters $\widehat{\chi}(\tau)$ as

$$\widehat{\chi}_i\left(\tfrac{i}{8l}\right) = \sum_j P_{ij}\,\widehat{\chi}_j(2il) \quad \text{with} \quad P = T^{\frac{1}{2}} S T^2 S T^{\frac{1}{2}}, \tag{6.164}$$

where $T^{\frac{1}{2}}$ is defined as the square root of the entries in the diagonal matrix T_{ij} shown in equation (6.97). One can show that P is symmetric and like S satisfies $P^2 = C$.

Using the P-transformation one can formally compute the Möbius strip amplitude as

$$Z^M(t) = \sum_{i,j} B^i_\alpha \left(\Gamma^i\right)^* P_{ij}\,\widehat{\chi}_j(it). \tag{6.165}$$

Interpreting this expression as a loop-channel partition function, we again have to require that the coefficients have to be integer

$$\sum_i B^i_\alpha \left(\Gamma^i\right)^* P_{ij} = m_{\alpha j} \in \mathbb{Z}. \tag{6.166}$$

What we discussed so far is the general structure of the crosscap states and non-oriented one-loop diagrams, which also feature a tree-channel loop-channel equivalence. What remains is to find solutions to the resulting Cardy conditions (6.158) and (6.166). Similar to the boundary states, for the charge conjugated modular invariant partition function, one can show that these integer conditions are indeed satisfied for the reflection coefficients of the form

$$\Gamma^i = \frac{P_{0i}}{\sqrt{S_{0i}}}, \qquad B^i_\alpha = \frac{S_{\alpha i}}{\sqrt{S_{0i}}}. \tag{6.167}$$

The Klein bottle and Möbius coefficients can be expressed by two Verlinde type formulas of the form

$$\kappa_j = \sum_i \frac{P_{0i}^* P_{0i} S_{ij}}{S_{0i}} = Y_{j0}^0, \qquad m_{\alpha j} = \sum_i \frac{S_{\alpha i} P_{0i}^* P_{ij}}{S_{0i}} = Y_{\alpha j}^0 \qquad (6.168)$$

with the definition

$$Y_{ij}^k = \sum_l \frac{S_{il} P_{jl} P_{kl}^*}{S_{0l}}. \qquad (6.169)$$

One can show that the combination Y_{ij}^k is always an integer, guaranteeing that the loop channel Klein bottle and Möbius strip amplitudes contain only integer coefficients.

6.8 Appendix: D-brane Tension

Consider two parallel Dp-branes in a d-dimensional space-time. In order to compare the string computation with a field theory computation, one needs to study the exchange of massless modes between the two branes. D-branes are dynamical objects which are expected to have a mass-density (tension) and they therefore couple to gravity and possibly other massless excitations of the closed string, such as the dilaton and the anti-symmetric tensor. It turns out that the latter does not contribute to lowest order (there is no field-theoretical one-particle exchange involving this field) and we will therefore ignore it. In the following we compute the one graviton and one dilaton exchange amplitude between two Dp-branes from the effective space-time action, cf. Chap. 16,

$$S = S_{\text{bulk}} + S_{\text{brane}}$$
$$= \frac{1}{2\tilde{\kappa}_d^2} \int d^d x \sqrt{-G} \, e^{-2\Phi} \left[R + 4\partial_\mu \Phi \partial^\mu \Phi \right] - T_p \int_{\mathcal{W}} d^{p+1}\xi \, e^{-\Phi} \sqrt{-\Gamma}. \qquad (6.170)$$

The first part, the bulk action, governs the space-time dynamics of the closed string modes. The dilaton dependence signals that it is derived from tree-level closed string theory, i.e. from a world-sheet with Euler number $\chi = 2$. The second part is the world-volume action of the Dp-brane. ξ^α are local coordinates and $\Gamma_{\alpha\beta} = \partial_\alpha X^\mu \partial_\beta X^\nu G_{\mu\nu}$ is the induced metric on the world-volume $\mathcal{W}$ of the brane. $X^\mu(\xi)$ describe the embedding of the brane into space-time. The gauge fields, i.e. the massless excitations of the open string, also appear in the D-brane action but they also do not contribute to the interaction between the two branes because they are confined to the world-volume and cannot be exchanged between the two branes. The dilaton dependence indicates that this is the effective action extracted from looking at string scattering amplitudes on the disk with Euler number $\chi = 1$. This world-volume action is a straightforward generalization of the Nambu-Goto-action to higher dimensional world-volumes. It was first considered by Dirac in the context of membranes.

For this bulk action the propagator will not be diagonal between the graviton and the dilaton which are the fluctuations of $g_{\mu\nu}$ and Φ around their background values. In order to decouple them at the quadratic level we first separate off the constant background value of the dilaton: $\Phi = \Phi_0 + \phi$ and then perform a dilaton dependent Weyl rescaling of the space-time metric g such that the Ricci scalar appears without dilaton prefactor:

$$G_{\mu\nu} = e^{\frac{4\phi}{d-2}} g_{\mu\nu} \qquad (6.171)$$

leading to

$$S = \frac{1}{2\kappa_d^2} \int d^dx \sqrt{-g} \left[R - \frac{4}{d-2} \partial_\mu \phi \partial^\mu \phi \right] - \tau_p \int d^{p+1}\xi \, e^{\frac{4-d+2p}{d-2}\phi} \sqrt{-\Gamma}, \qquad (6.172)$$

where now $\Gamma_{\alpha\beta} = \partial_\alpha X^\mu \partial_\beta X^\nu g_{\mu\nu}$. The action (6.170) is said to be in the string frame whereas (6.172) is in the (modified) Einstein frame, cf. Chap. 18. The field redefinitions have, of course, no physical consequences.

We now expand the metric around Minkowski space and choose a convenient normalization for the dilaton

$$g_{\mu\nu} = \eta_{\mu\nu} + 2\kappa_d h_{\mu\nu}, \qquad \phi = \frac{1}{2}\kappa_d \sqrt{d-2}\, D. \qquad (6.173)$$

To proceed we need to gauge fix the symmetries. For the world-volume diffeomorphisms we choose the static gauge, i.e. we identify

$$X^\alpha = \xi^\alpha, \qquad X^i = \text{const}, \qquad \alpha = 0, \ldots, p, \quad i = p+1, \ldots, d-1. \qquad (6.174)$$

The condition on the X^i means that we fix the position of the branes. We neglect fluctuations which do not contribute, for the same reason as the gauge fields do not contribute. In the static gauge the induced metric is $\Gamma_{\alpha\beta} = g_{\alpha\beta}$ and to first order in the quantum fields the world-volume action is

$$S_{\text{brane}} = T_p \tilde{\kappa}_d \int d^{p+1}\xi \left(D^{\mu\nu} h_{\mu\nu} - \left(\frac{4-d+2p}{2\sqrt{d-2}} \right) D \right), \qquad (6.175)$$

where $D^{\mu\nu} = \begin{pmatrix} \eta^{\alpha\beta} & 0 \\ 0 & 0 \end{pmatrix}$ and we used $\tau_p \kappa_d = T_p \tilde{\kappa}_d$. If we had included the antisymmetric tensor in the world-volume action it would not appear at linear order. That is why it does not contribute.

6.8 Appendix: D-brane Tension

To fix the space-time diffeomorphisms we add a gauge fixing term

$$-\left(\partial_\nu h^\nu{}_\mu - \frac{1}{2}\partial_\mu h^\nu{}_\nu\right)^2 \tag{6.176}$$

to S_{bulk} and expand to second order in the quantum fields. This yields ($h = \eta^{\mu\nu} h_{\mu\nu}$)

$$S_{\text{bulk}} = \int d^d x \left(\frac{1}{2} h^{\mu\nu} \Box h_{\mu\nu} - \frac{1}{4} h \Box h + \frac{1}{2} D \Box D\right), \tag{6.177}$$

from which we get the propagators

$$\Delta_{\mu\nu,\rho\sigma}(q) = \frac{1}{2q^2}\left(\eta_{\mu\rho}\eta_{\nu\sigma} + \eta_{\mu\sigma}\eta_{\nu\rho} - \frac{2}{d-2}\eta_{\mu\nu}\eta_{\rho\sigma}\right) \equiv P_{\mu\nu,\rho\sigma}\frac{1}{q^2},$$

$$\Delta(q) = \frac{1}{q^2} = \int d^d x\, e^{-i q \cdot x} G(|\mathbf{x}|). \tag{6.178}$$

If we write the brane action, i.e. the interaction term, as an integral over the d-dimensional bulk, the one-particle exchange amplitude becomes

$$\widetilde{\mathscr{A}}^{\text{eff}}_{\text{Dp,Dp}} = (T_p \tilde{\kappa}_d)^2 \int d^d x \int d^d x'\, \delta(x^\perp - x_A^\perp)\delta(x'^\perp - x_B^\perp)$$

$$\times \left(D^{\mu\nu} P_{\mu\nu,\rho\sigma} D^{\rho\sigma} + \frac{1}{4}\frac{(4-d+2p)^2}{(d-2)}\right) G(|x-x'|)$$

$$= \frac{d-2}{4}(T_p \tilde{\kappa}_d)^2 \int d^{p+1} x^{\|} \int d^{p+1} x'^{\|} \int \frac{d^d q}{(2\pi)^d} \frac{e^{i q^{\|}\cdot(x^{\|}-x'^{\|})}}{q^2} e^{i q^\perp \cdot (x_A^\perp - x_B^\perp)}$$

$$= \frac{d-2}{4}(T_p \tilde{\kappa}_d)^2 V_{p+1} \int \frac{dq^\perp}{(2\pi)^{d-p-1}} \frac{1}{(q^\perp)^2} e^{i q^\perp \cdot (x_A^\perp - x_B^\perp)}$$

$$= \frac{d-2}{4}(T_p \tilde{\kappa}_d)^2 V_{p+1}\, G^\perp(|x_B^\perp - x_A^\perp|). \tag{6.179}$$

Comparison of (6.179) with (6.133) immediately leads to (6.135).

Further Reading

Global aspects of the string world-sheet were addressed in:

- D. Friedan, *Introduction to Polyakov's String Theory*. Recent Advances in Field Theory and Statistical Mechanics, Les Houches, France, Aug 2–Sep 10, 1982
- O. Alvarez, Theory of strings with boundaries: Fluctuations, topology and quantum geometry. Nucl. Phys. **B216**, 125 (1983)

The role of modular invariance for closed string loop calculations was first discussed in:

- J.A. Shapiro, Loop graph in the dual tube model. Phys. Rev. **D5**, 1945 (1972)

Some reviews which influenced the presentation in this chapter and which contain further extensions of these topics are:

- E. D'Hoker, D.H. Phong, The geometry of string perturbation theory. Rev. Mod. Phys. **60**, 917 (1988)
- L. Alvarez-Gaumé, P.C. Nelson, *Riemann Surfaces and String theories*, Proceedings of the 4th Trieste Spring School on Supersymmetry, Supergravity and Superstrings, Trieste, p. 419
- M. Nakahara, *Geometry, Topology and Physics*, Chap. 14 (IOP, Bristol, 2003)

For a careful treatment of the Polyakov path integral of the bosonic string consult

- J. Polchinski, Evaluation of the one loop string path integral. Comm. Math. Phys. **104**, 37 (1986)
- S. Weinberg, *Covariant Path Integral Approach to String Theory*, Lectures given at 4th Jerusalem Winter School for Theoretical Physics, 1986; published in the proceedings

All one-loop vacuum amplitudes for the bosonic string were computed in

- C.P. Burgess, T.R. Morris, Open and unoriented strings à la Polyakov. Nucl. Phys. **B291**, 256 (1987)

Expansion of curvature to second order and derivation of graviton propagator:

- G. 't Hooft, M.J.G. Veltman, One loop divergencies in the theory of gravitation. Annales Poincaré Phys. Theor. **A20**, 69 (1974)
- M.J.G. Veltman, *Quantum Theory of Gravitation*, In 'Les Houches 1975, Proceedings, Methods In Field Theory', p. 265

The D-Brane tension was first computed in

- J. Polchinski, Dirichlet-Branes and Ramond-Ramond charges. Phys. Rev. Lett. **75**, 4724 (1995) [arXiv:hep-th/9510017]

References for the boundary state formalism are given in Chaps. 4 and 9.

Chapter 7
The Classical Fermionic String

Abstract The fermionic string theory presented in this and the following chapter is the Neveu-Schwarz-Ramond spinning string. We present the world-sheet action and discuss its symmetries, most notably the local $N = 1$ world-sheet supersymmetry. The admissible periodicity and boundary conditions lead to the distinction between Neveu-Schwarz and Ramond sectors. The oscillator expansions of the world-sheet fermions differ in the two sectors. We close with an appendix on spinors in two dimensions. Quantisation of the fermionic string will be the subject of the following chapter.

7.1 Motivation for the Fermionic String

So far we have only discussed bosonic strings. That means that all physical world-sheet degrees of freedom had been described by bosonic variables. We have treated the classical and the quantum theory, the algebra of the constraints (the Virasoro algebra) and we have found that at the quantum level the theory makes sense only in the critical dimension which was found to be $d = 26$. The spectra of the open and closed, oriented and unoriented theories were found to contain a tachyon, a fact which is at least alarming. Let us recall that its negative (mass)^2 arose from the (regularized) zero point energy of an infinite set of bosonic harmonic oscillators. The problem with the tachyon may be cured if we introduce, on the world sheet, fermionic degrees of freedom which are quantized with anti-commutators. Then there is a chance that the zero point energies cancel and the tachyon is absent.

One basic symmetry principle that guarantees the absence of a tachyon in the string spectrum is space-time supersymmetry. It is important to keep in mind the distinction between world-sheet and space-time supersymmetry. The fermionic string theories that we will discuss all possess world-sheet supersymmetry but not necessarily space-time supersymmetry and are not all tachyon-free. Whether a particular string theory is space-time supersymmetric or not will manifest itself, for instance, in the spectrum. Especially the existence of one or more massless gravitinos will signal space-time supersymmetry.

Here we will present the Neveu-Schwarz-Ramond (RNS) superstring, which features manifest world-sheet supersymmetry but lacks manifest space-time supersymmetry. We should mention that there exists also the so-called Green-Schwarz (GS) formalism in which space-time supersymmetry is manifest at the cost of manifest world-sheet supersymmetry. It uses in a crucial way the triality property of $SO(8)$, the transverse Lorentz group in ten dimensions, which is, as we will see later in this chapter, the critical dimension for the fermionic string. The fermionic degrees of freedom are world-sheet scalars which carry an $SO(8)$ spinor index. We will not discuss the GS formulation of the superstring but it will appear in the duality between type I and heterotic theories in Sect. 18.6. A covariant extension of the GS formalism is the pure-spinor formulation of the superstring, which we won't discuss either.

7.2 Superstring Action and Its Symmetries

In analogy to our treatment of the bosonic string, we start with a discussion of the classical fermonic string action and its symmetries (but we skip the discussion of the superparticle). In particular, we want to find the requirements on the field content coming from world-sheet supersymmetry and to set up the supersymmetric extension of the Polyakov action. The bosonic string theory was described by the action for a collection of d scalar fields $X^\mu(\sigma,\tau)$ coupled to gravity $h_{\alpha\beta}$ in two dimensions. The purely gravitational part of the action was trivial, being a total derivative. This left us with

$$S_1 = -\frac{1}{4\pi\alpha'} \int d^2\sigma \sqrt{-h}\, h^{\alpha\beta}\, \partial_\alpha X^\mu \partial_\beta X_\mu \tag{7.1}$$

which is the covariant kinetic energy for the "matter fields" X^μ. The supersymmetric extension of S_1 should be the coupling of supersymmetric "matter" to two-dimensional supergravity. With respect to the d-dimensional target space, which can be considered as an internal space from the world-sheet point of view, the fields $X^\mu(\sigma,\tau)$ transform as a vector. Hence, their supersymmetric partners should be world-sheet spinors with a target space vector index. We will denote them by $\psi^\mu(\sigma,\tau)$. We do not exhibit the spinor indices. We refer to the appendix of this chapter for further details on spinors in two dimensions. Let us now see how the balance between bosonic and fermionic degrees of freedom works out. The fields X^μ representing d real scalars, provide d bosonic degrees of freedom. If we impose on the d world-sheet fermions ψ^μ a Majorana condition, they provide $2d$ fermionic degrees of freedom. We have to introduce d real auxiliary scalar fields F^μ. Together (X^μ, ψ^μ, F^μ) form an off-shell scalar multiplet of two-dimensional $N=1$ supersymmetry.[1] On-shell (X^μ, ψ^μ) suffice.

---

[1] We denote the amount of world-sheet supersymmetry by N and reserve $\mathcal{N}$ for space-time supersymmetry.

7.2 Superstring Action and Its Symmetries

Let us now turn to the gravity sector. The supergravity multiplet consists of the zwei-bein e_α^a (n-bein in n dimensions) and the gravitino χ_α. The zwei-bein has two different kinds of indices. a is a Lorentz index and takes part in local Lorentz transformations whereas α is called an Einstein index and takes part in coordinate transformations (reparametrizations). Einstein indices are raised and lowered with the world-sheet metric $h_{\alpha\beta}$ and Lorentz indices with the Lorentz metric η_{ab}. The zwei-bein allows to transform Lorentz into Einstein indices and vice versa. The introduction of the zwei-bein is necessary to describe spinors on a curved manifold, as the group $GL(n, \mathbb{R})$ does not have spinor representations[2] whereas the tangent space group, $SO(n-1, 1)$, does. The inverse of e_α^a, denoted by e_a^α, is defined by

$$e_\alpha^a e_b^\alpha = \delta_b^a. \tag{7.2}$$

e_a^α defines an orthonormal set of basis vectors at each point, i.e. it satisfies

$$e_a^\alpha e_b^\beta h_{\alpha\beta} = \eta_{ab}, \tag{7.3}$$

from which we derive

$$e_a^\alpha e_b^\beta \eta^{ab} = h^{\alpha\beta}. \tag{7.4}$$

The gravitino is a world-sheet vector and a world-sheet Majorana spinor.

In n dimensions the n-bein e_α^a has n^2 components. There are n reparametrizations and $\frac{1}{2}n(n-1)$ local Lorentz transformations as gauge symmetries, leaving $\frac{1}{2}n(n-1)$ degrees of freedom. The gravitino, being a Majorana spinor-vector, has $2^{[\frac{n}{2}]}n$ components, where $[\frac{n}{2}]$ denotes the integer part of $\frac{n}{2}$. For $N = 1$ supersymmetry there are $2^{[\frac{n}{2}]}$ supersymmetry parameters leaving $(n-1)2^{[\frac{n}{2}]}$ degrees of freedom. For the case of interest, namely $n = 2$, we find one bosonic and two fermionic degrees of freedom. To get a complete off-shell supergravity multiplet, we have to introduce one auxiliary real scalar field A. The complete off-shell supergravity multiplet is then $(e_\alpha{}^a, \chi_\alpha, A)$. The on-shell supergravity multiplet is $(e_\alpha{}^a, \chi_\alpha)$.

As an aside, we list the number of on-shell degrees of freedom of various massless fields in n dimensions

$$\begin{aligned} n-\text{bein} \quad & e_\alpha{}^a \\ \text{metric} \quad & g_{\alpha\beta} \end{aligned} \quad : \quad \frac{1}{2}n(n-3),$$

$$\text{Weyl fermion} \quad \psi \quad : \quad 2^{[\frac{n}{2}]-\epsilon},$$

$$\text{Gravitino} \quad \chi_\alpha \quad : \quad 2^{[\frac{n}{2}]-\epsilon}(n-3),$$

$$p-\text{form} \quad A^{(p)} \quad : \quad \binom{n-2}{p}, \tag{7.5}$$

---

[2]Under general coordinate transformations, tensor indices are acted on with elements of $GL(n, \mathbb{R})$.

where $[m]$ is the integer part of m. For fermions we count the number of real components. The parameter ϵ assumes values $0, 1, 2$ for Dirac (0), Weyl (1), Majorana (1) and Majorana-Weyl (2) fermions (provided they exist in n dimensions). The gravitino satisfies the condition $\Gamma^\alpha \chi_\alpha = 0$, where Γ^α are Dirac matrices in n dimensions. p-form fields play an important role when we discuss the spectrum of the fermionic string. If its field strength satisfies a self-duality constraint, which can only happen for a p-form in $2(p+1)$ dimensions, the number of physical components is $\frac{1}{2}\binom{2p-2}{p}$. Note that the number of components of the various fields are the dimensions of irreducible representations of the little group $SO(n-2)$.

So far the discussion was independent of any particular action and only a statement of the field content of the two-dimensional supersymmetry multiplets. Let us now complete the string action. The kinetic energy term for the gravitino vanishes identically in two dimensions.[3] The kinetic energy for the matter fermions ψ^μ and the contribution of the auxiliary fields F^μ is

$$S_2 = \frac{1}{4\pi}\int d^2\sigma \, e \left(-i\overline{\psi}^\mu \rho^\alpha \partial_\alpha \psi_\mu + F^\mu F_\mu\right), \tag{7.6}$$

where $e = |\det e_\alpha^a| = \sqrt{-h}$. Our notation is summarized in the appendix to this chapter. The fact that the derivative in S_2 is an ordinary derivative rather than a covariant derivative containing the spin connection is a consequence of the Majorana spin-flip property Eq. (7.76). The action $S_1 + S_2$ is not yet locally supersymmetric. It is simply the covariantized form of the action of a scalar multiplet. Local supersymmetry requires the additional term

$$S_3 = \frac{i}{8\pi}\int d^2\sigma \, e \, \overline{\chi}_\alpha \rho^\beta \rho^\alpha \psi^\mu \left(\sqrt{\tfrac{2}{\alpha'}}\, \partial_\beta X_\mu - \tfrac{i}{4}\overline{\chi}_\beta \psi_\mu\right). \tag{7.7}$$

The auxiliary field A does not appear and the auxiliary matter scalars F^μ can be eliminated via their equations of motion. This will be assumed to be done from now on.

The complete action[4]

$$S = -\frac{1}{8\pi}\int d^2\sigma \, e \bigg(\tfrac{2}{\alpha'} h^{\alpha\beta} \partial_\alpha X^\mu \partial_\beta X_\mu + 2i\,\overline{\psi}^\mu \rho^\alpha \partial_\alpha \psi_\mu$$
$$-i\overline{\chi}_\alpha \rho^\beta \rho^\alpha \psi^\mu \left(\sqrt{\tfrac{2}{\alpha'}}\, \partial_\beta X_\mu - \tfrac{i}{4}\overline{\chi}_\beta \psi_\mu\right)\bigg) \tag{7.8}$$

is invariant under the following local world-sheet symmetries:

---

[3] In any number of dimensions it is $\overline{\chi}_\alpha \Gamma^{\alpha\beta\gamma} D_\beta \chi_\gamma$ where $\Gamma^{\alpha\beta\gamma}$ is the anti-symmetrized product of three Dirac matrices which vanishes in two dimensions.

[4] The fields and parameters have mass dimensions $[\psi] = [\chi] = 1/2$, $[\epsilon] = -1/2$, $[\xi] = [X] = -1$. All others are dimensionless. The fact that we choose X to have dimension of length, rather than being dimensionless, as is more common for a scalar field in two dimensions, is the origin of the various powers of $\sqrt{\tfrac{2}{\alpha'}}$.

7.2 Superstring Action and Its Symmetries

1. Supersymmetry

$$\sqrt{\frac{2}{\alpha'}}\delta_\epsilon X^\mu = i\bar{\epsilon}\psi^\mu,$$

$$\delta_\epsilon \psi^\mu = \frac{1}{2}\rho^\alpha \left(\sqrt{\frac{2}{\alpha'}}\partial_\alpha X^\mu - \frac{i}{2}\bar{\chi}_\alpha\psi^\mu\right)\epsilon,$$

$$\delta_\epsilon e_\alpha{}^a = \frac{i}{2}\bar{\epsilon}\rho^a \chi_\alpha,$$

$$\delta_\epsilon \chi_\alpha = 2D_\alpha \epsilon, \tag{7.9}$$

where $\epsilon(\sigma, \tau)$ is a Majorana spinor which parametrizes supersymmetry transformations and D_α a covariant derivative with torsion:

$$D_\alpha \epsilon = \partial_\alpha \epsilon - \frac{1}{2}\omega_\alpha \bar{\rho}\epsilon,$$

$$\omega_\alpha = -\frac{1}{2}\epsilon^{ab}\omega_{\alpha a b} = \omega_\alpha(e) + \frac{i}{4}\bar{\chi}_\alpha \bar{\rho}\rho^\beta \chi_\beta,$$

$$\omega_\alpha(e) = -\frac{1}{e}e_{\alpha a}\epsilon^{\beta\gamma}\partial_\beta e_\gamma{}^a. \tag{7.10}$$

$\omega_\alpha(e)$ is the spin connection without torsion.

2. Weyl transformations

$$\delta_\Lambda X^\mu = 0,$$

$$\delta_\Lambda \psi^\mu = -\frac{1}{2}\Lambda\psi^\mu,$$

$$\delta_\Lambda e_\alpha{}^a = \Lambda e_\alpha{}^a,$$

$$\delta_\Lambda \chi_\alpha = \frac{1}{2}\Lambda\chi_\alpha. \tag{7.11}$$

3. Super-Weyl transformations

$$\delta_\eta \chi_\alpha = \rho_\alpha \eta,$$

$$\delta_\eta(\text{others}) = 0 \tag{7.12}$$

with $\eta(\sigma, \tau)$ a Majorana spinor parameter.

4. Two-dimensional Lorentz transformations

$$\delta_l X^\mu = 0,$$

$$\delta_l \psi^\mu = -\frac{1}{2} l \, \bar{\rho} \psi^\mu,$$

$$\delta_l e_\alpha{}^a = l \, \epsilon^a{}_b \, e_\alpha{}^b,$$

$$\delta_l \chi_\alpha = -\frac{1}{2} l \, \bar{\rho} \chi_\alpha. \tag{7.13}$$

5. Reparametrizations

$$\delta_\xi X^\mu = -\xi^\beta \partial_\beta X^\mu,$$

$$\delta_\xi \psi^\mu = -\xi^\beta \partial_\beta \psi^\mu,$$

$$\delta_\xi e_\alpha{}^a = -\xi^\beta \partial_\beta e_\alpha{}^a - e_\beta{}^a \partial_\alpha \xi^\beta,$$

$$\delta_\xi \chi_\alpha = -\xi^\beta \partial_\beta \chi_\alpha - \chi_\beta \partial_\alpha \xi^\beta. \tag{7.14}$$

If we combine reparametrizations with a Lorentz transformation with parameter $l = \xi^\alpha \omega_\alpha(e)$, they can be written in covariant form

$$e_{\beta a} \delta_\xi e_\alpha^a = -\nabla_\alpha \xi_\beta = -\frac{1}{2}(P\xi)_{\alpha\beta} + \frac{1}{2e^2}\epsilon_{\alpha\beta}(\epsilon^{\gamma\delta}\nabla_\gamma \xi_\delta) - \frac{1}{2}h_{\alpha\beta}\nabla \cdot \xi,$$

$$\delta_\xi \chi_\alpha = -\xi^\beta \nabla_\beta \chi_\alpha - \chi_\beta \nabla_\alpha \xi^\beta,$$

$$\delta_\xi \psi^\mu = -\xi^\alpha \nabla_\alpha \psi^\mu, \tag{7.15}$$

where ∇_α is a covariant derivative without torsion and the operator P has been defined in Chap. 2.

Note that in Eqs. (7.11)–(7.15) Λ, l and ξ are infinitesimal functions of (σ, τ).

There are several ways to get the complete action and the symmetry transformation rules. One possibility is to use the Noether method; another is to go to superspace. Either way, the procedure is analogous to the four dimensional case. In addition to these local world-sheet symmetries, (7.8) is also invariant under global space-time Poincaré transformations (2.29) together with $\delta \psi^\mu = a^\mu{}_\nu \psi^\nu$, $\delta \chi_\alpha = 0$.

7.3 Superconformal Gauge

We can now use local supersymmetry, reparametrizations and Lorentz transformations to gauge away two degrees of freedom of the zwei-bein and two degrees of freedom of the gravitino. To do this we decompose the gravitino as

7.3 Superconformal Gauge

$$\chi_\alpha = \left(h_\alpha{}^\beta - \frac{1}{2}\rho_\alpha\rho^\beta\right)\chi_\beta + \frac{1}{2}\rho_\alpha\rho^\beta\chi_\beta$$

$$= \frac{1}{2}\rho^\beta\rho_\alpha\chi_\beta + \frac{1}{2}\rho_\alpha\rho^\beta\chi_\beta$$

$$\equiv \tilde{\chi}_\alpha + \rho_\alpha\lambda, \qquad (7.16)$$

where $\tilde{\chi}_\alpha = \frac{1}{2}\rho^\beta\rho_\alpha\chi_\beta$ is ρ-traceless, i.e. $\rho\cdot\tilde{\chi} = 0$ and $\lambda = \frac{1}{2}\rho^\alpha\chi_\alpha$. This corresponds to a decomposition of the spin 3/2 gravitino into helicity $\pm 3/2$ and helicity $\pm 1/2$ components. It is orthogonal with respect to the inner product $(\phi|\psi) = \int d^2\sigma \overline{\phi}^\alpha \psi_\alpha$. We can make the same decomposition for the supersymmetry transformation of the gravitino

$$\delta_\epsilon \chi_\alpha = 2D_\alpha\epsilon$$

$$\equiv 2(\Pi\epsilon)_\alpha + \rho_\alpha\rho^\beta D_\beta\epsilon, \qquad (7.17)$$

where we have defined the operator

$$(\Pi\epsilon)_\alpha = \left(h_\alpha{}^\beta - \frac{1}{2}\rho_\alpha\rho^\beta\right)D_\beta\epsilon = \frac{1}{2}\rho^\beta\rho_\alpha D_\beta\epsilon \qquad (7.18)$$

which maps spin 1/2 fields to ρ-traceless spin 3/2 fields. We can now write, at least locally, $\tilde{\chi}_\alpha = \rho^\beta\rho_\alpha D_\beta\kappa$ for some spinor κ where we have used the identity Eq. (7.78). Comparing this with Eq. (7.17) we see that κ can be eliminated by a supersymmetry transformation. We then use reparametrizations and local Lorentz transformations to transform the zwei-bein into the form $e_\alpha{}^a = e^\phi \delta_\alpha^a$ which is always possible locally, cf. Chap. 2. These transformations do not reintroduce traceless parts into the gravitino, as under reparametrizations it transforms as $\rho_\alpha(\sigma)\lambda(\sigma) \to \tilde{\rho}_\alpha(\tilde{\sigma})\tilde{\lambda}(\tilde{\sigma}) = (\frac{\partial\sigma^\beta}{\partial\tilde{\sigma}^\alpha})\rho_\beta(\sigma)\lambda(\sigma)$. In this way we arrive at the so-called superconformal gauge which is a generalization of the conformal gauge to the supersymmetric case:

$$e_\alpha{}^a = e^\phi \delta_\alpha^a, \quad \chi_\alpha = \rho_\alpha\lambda. \qquad (7.19)$$

In the classical theory we can still use a Weyl rescaling and super-Weyl transformation to gauge away the remaining metric and gravitino degrees of freedom ϕ and λ, leaving only $e_\alpha{}^a = \delta_\alpha^a$ and $\chi_\alpha = 0$. In analogy to the bosonic case, these symmetries will be broken in the quantum theory except in the critical dimension.

The above arguments, used to go to superconformal gauge, are only true locally and one has to check under what conditions the superconformal gauge can be reached globally. From our foregoing discussion it is clear that the condition is that there exists a globally defined spinor ϵ and a vector field ξ^α such that

$$(\Pi\epsilon)_\alpha = \tau_\alpha \quad \text{and} \quad (P\xi)_{\alpha\beta} = t_{\alpha\beta} \qquad (7.20)$$

for arbitrary τ_α which satisfies $\rho\cdot\tau = 0$ and arbitrary symmetric traceless tensor $t_{\alpha\beta}$.

In Chap. 2 we have seen that the second condition is equivalent to the absence of zero modes of the operator $P^\dagger$. In the same way we can show that the absence of zero modes of the operator $\Pi^\dagger$, the adjoint of Π, allows to gauge away the trace part of the gravitino. $\Pi^\dagger$ maps ρ-traceless spin $3/2$ fields to spin $1/2$ fields via

$$(\Pi^\dagger \tau) = -2D^\alpha \tau_\alpha. \tag{7.21}$$

The zero modes of $P^\dagger$ were called moduli. In analogy we call the zero modes of $\Pi^\dagger$ supermoduli. We thus have

$$\text{\# of moduli} = \dim \ker P^\dagger,$$
$$\text{\# of supermoduli} = \dim \ker \Pi^\dagger. \tag{7.22}$$

Zero modes of the operators P and Π mean that the gauge fixing is not complete. The zero modes of P are the conformal Killing vectors (CKV) (cf. Chap. 2); the zero modes of Π will be referred to as conformal Killing spinors (CKS); i.e.

$$\text{\# of CKV} = \dim \ker P,$$
$$\text{\# of CKS} = \dim \ker \Pi. \tag{7.23}$$

We will compute the dimensions of the kernels of Π and $\Pi^\dagger$ in Chap. 9, whereas P and $P^\dagger$ have already been treated in Chap. 6.

In superconformal gauge the action simplifies to

$$S = -\frac{1}{8\pi} \int d^2\sigma \left(\tfrac{2}{\alpha'} \partial_\alpha X^\mu \partial^\alpha X_\mu + 2i\, \overline{\psi}^\mu \rho^\alpha \partial_\alpha \psi_\mu \right) \tag{7.24}$$

which is nothing else than the action of a free scalar superfield in two dimensions. To arrive at Eq. (7.24), we have rescaled the matter fermions by $e^{\phi/2}\psi \to \psi$. Worldsheet indices are now raised with the flat metric $\eta^{\alpha\beta}$ and $\rho^\alpha = \delta^\alpha_a \rho^a$. Also, the torsion piece in the spin connection now vanishes due to the identity Eq. (7.78) and we have $\omega_\alpha = -\epsilon_\alpha{}^\beta \partial_\beta \phi$. The action is still invariant under those local reparametrizations and supersymmetry transformations which satisfy $P\xi = 0$ and $\Pi \epsilon = 0$. Under the supersymmetry transformations the fields transform as

$$\sqrt{\tfrac{2}{\alpha'}}\, \delta_\epsilon X^\mu = i\overline{\epsilon}\psi^\mu$$
$$\delta_\epsilon \psi^\mu = \sqrt{\tfrac{2}{\alpha'}}\, \tfrac{1}{2} \rho^\alpha \partial_\alpha X^\mu \epsilon\,. \tag{7.25}$$

To see this, we note that the zwei-bein is not taken out of superconformal gauge, if a supersymmetry transformation with parameter ϵ is accompanied by a Lorentz transformation with parameter $l = \tfrac{i}{2} \overline{\epsilon} \rho \lambda$. The Weyl degree of freedom ϕ then changes according to $\delta\phi = \tfrac{i}{2} \overline{\epsilon} \lambda$. Likewise, the gravitino stays in the gauge

7.3 Superconformal Gauge

Eq. (7.19), if the supersymmetry parameter satisfies $\rho^\beta \rho_\alpha D_\beta \epsilon = 0$ which is the condition $(\Pi \epsilon) = 0$ found above. If we now redefine $e^{\phi/2}\psi = \tilde{\psi}$ and $e^{-\phi/2}\epsilon = \tilde{\epsilon}$ we find, after dropping tildes, Eqs. (7.25). It is also easy to show that the condition $\rho^\beta \rho_\alpha D_\beta \epsilon = 0$ reduces in superconformal gauge to $\rho^\beta \rho_\alpha \partial_\beta \tilde{\epsilon} = 0$. It is of course also straightforward to verify directly that the action Eq. (7.24) is invariant under the transformations (7.25) with ϵ satisfying $\rho^\beta \rho_\alpha \partial_\beta \epsilon = 0$.

Under diffeomorphisms the fields transform as

$$\delta_\xi X^\mu = \xi^\alpha \partial_\alpha X^\mu$$

$$\delta_\xi \psi^\mu = \xi^\alpha \partial_\alpha \psi^\mu + \frac{1}{4}(\partial \cdot \xi)\psi^\mu - \frac{1}{4}(\epsilon^{\alpha\beta}\partial_\alpha \xi_\beta)\bar{\rho}\psi^\mu, \qquad (7.26)$$

where $\delta_\xi \psi$ is that combination of diffeomorphisms, Weyl and Lorentz transformations which is needed to preserve $e_\alpha{}^a \propto \delta_\alpha^a$.

The equations of motion derived from the action Eq. (7.24) are

$$\partial_\alpha \partial^\alpha X^\mu = 0,$$

$$\rho^\alpha \partial_\alpha \psi^\mu = 0. \qquad (7.27)$$

As in the bosonic theory they have to be supplemented by boundary conditions. We will discuss them at the end of this section.

For theories with fermions the energy momentum tensor is defined as

$$T_{\alpha\beta} = \frac{2\pi}{e}\frac{\delta S}{\delta e_a^\beta}e_{\alpha a}. \qquad (7.28)$$

We can analogously define the supercurrent as the response to variations of the gravitino; we will denote it by T_F, indicating that it is a fermionic object related to the energy-momentum tensor T by supersymmetry. This means that $\delta S = \frac{1}{2\pi}\int d^2\sigma\, e\, i\delta\bar{\chi}^\alpha T_{F\alpha}$ with

$$T_{F\alpha} = \frac{2\pi}{e}\frac{\delta S}{i\delta\bar{\chi}^\alpha}. \qquad (7.29)$$

The normalization is such that

$$\delta_\epsilon T_{F\alpha} = \frac{1}{2}T_{\alpha\beta}\rho^\beta \epsilon. \qquad (7.30)$$

The equations of motion for the metric and gravitino are

$$T_{\alpha\beta} = 0, \quad T_{F\alpha} = 0. \qquad (7.31)$$

They are constraints on the system and generate symmetries, analogous to the bosonic case. We will say more about this below. After going to superconformal gauge we find

$$T_{\alpha\beta} = -\frac{1}{\alpha'}\left(\partial_\alpha X^\mu \partial_\beta X_\mu - \frac{1}{2}\eta_{\alpha\beta}\partial^\gamma X^\mu \partial_\gamma X_\mu\right)$$
$$- \frac{i}{4}\left(\overline{\psi}^\mu \rho_\alpha \partial_\beta \psi_\mu + \overline{\psi}^\mu \rho_\beta \partial_\alpha \psi_\mu\right) = 0,$$
$$T_{F\alpha} = -\frac{1}{4}\sqrt{\frac{2}{\alpha'}}\rho^\beta \rho_\alpha \psi^\mu \partial_\beta X_\mu = 0. \tag{7.32}$$

Here we have used the equations of motion for ψ^μ to cast $T_{\alpha\beta}$ into its symmetric form. Tracelessness follows also upon using the equations of motion. Note that

$$\rho^\alpha T_{F\alpha} = 0 \tag{7.33}$$

which is the analogue of $T^\alpha_\alpha = 0$. It is a consequence of super-Weyl invariance. Again, with the help of the equations of motion, it is easy to show that the energy-momentum tensor and the supercurrent are conserved:

$$\partial^\alpha T_{\alpha\beta} = 0,$$
$$\partial^\alpha T_{F\alpha} = 0. \tag{7.34}$$

These conservation laws lead, as in the bosonic theory, to an infinite number of conserved charges. This is most easily analyzed in light-cone coordinates on the world-sheet. In terms of these, Eqs. (7.24) and (7.27) become

$$S = \frac{1}{2\pi}\int d^2\sigma \left(\tfrac{2}{\alpha'}\partial_+ X \cdot \partial_- X + i\left(\psi_+ \cdot \partial_- \psi_+ + \psi_- \cdot \partial_+ \psi_-\right)\right) \tag{7.35}$$

and

$$\partial_+ \partial_- X^\mu = 0,$$
$$\partial_- \psi^\mu_+ = \partial_+ \psi^\mu_- = 0. \tag{7.36}$$

The conditions on the allowed reparametrizations and supersymmetry transformations take the simple form

$$\partial_+ \xi^- = \partial_- \xi^+ = 0,$$
$$\partial_+ \epsilon^- = \partial_- \epsilon^+ = 0. \tag{7.37}$$

We have defined[5] $\psi_A = \begin{pmatrix}\psi_+ \\ \psi_-\end{pmatrix}$ and $\epsilon^A = \begin{pmatrix}\epsilon^+ \\ \epsilon^-\end{pmatrix}$ following the conventions of the appendix. $\psi_\pm$ and $\epsilon^\pm$ are real. Note that for spinors $\pm$ denote their spinor

---

[5]In later chapters we will also use the notations ψ_L and $\overline{\psi}$ for left-moving fermions and ψ_R and ψ for right-moving fermions.

7.3 Superconformal Gauge

components whereas for vectors they denote vector components in conformal coordinates. The symmetries of (7.35) are

$$\delta_\xi X^\mu = -(\xi^+ \partial_+ X^\mu + \xi^- \partial_- X^\mu),$$

$$\delta_\xi \psi_+^\mu = -(\xi^+ \partial_+ \psi_+^\mu + \xi^- \partial_- \psi_+^\mu + \frac{1}{2}\partial_+ \xi^+ \, \psi_+^\mu),$$

$$\delta_\xi \psi_-^\mu = -(\xi^+ \partial_+ \psi_-^\mu + \xi^- \partial_- \psi_-^\mu + \frac{1}{2}\partial_- \xi^- \, \psi_-^\mu) \tag{7.38}$$

and

$$\sqrt{\frac{2}{\alpha'}}\delta_\epsilon X^\mu = i(\epsilon^+ \psi_+^\mu + \epsilon^- \psi_-^\mu)$$

$$\delta_\epsilon \psi_+^\mu = -\sqrt{\frac{2}{\alpha'}}\epsilon^+ \partial_+ X^\mu,$$

$$\delta_\epsilon \psi_-^\mu = -\sqrt{\frac{2}{\alpha'}}\epsilon^- \partial_- X^\mu \tag{7.39}$$

with ϵ and ξ subject to (7.37).

The energy-momentum tensor in world-sheet light-cone coordinates is

$$T_{++} = -\frac{1}{\alpha'}\partial_+ X \cdot \partial_+ X - \frac{i}{2}\psi_+ \cdot \partial_+ \psi_+,$$

$$T_{--} = -\frac{1}{\alpha'}\partial_- X \cdot \partial_- X - \frac{i}{2}\psi_- \cdot \partial_- \psi_-,$$

$$T_{+-} = T_{-+} = 0 \tag{7.40}$$

with

$$\partial_- T_{++} = \partial_+ T_{--} = 0. \tag{7.41}$$

Due to Eq. (7.33), two of the four components of $T_{F\alpha}$ vanish identically. The two non-vanishing components, which we denote by $T_{F\pm}$, are

$$T_{F\pm} = -\frac{1}{2}\sqrt{\frac{2}{\alpha'}}\psi_\pm \cdot \partial_\pm X \tag{7.42}$$

with

$$\partial_- T_{F+} = \partial_+ T_{F-} = 0. \tag{7.43}$$

The expressions (7.40) and (7.42) can also be derived as the Noether currents associated to the symmetries (7.39) and (7.38), respectively. From the equations of motion we learn that X^μ can again be split into left- and right-movers and that $\psi_+^\mu = \psi_+^\mu(\sigma^+)$ and $\psi_-^\mu = \psi_-^\mu(\sigma^-)$. The conservation laws tell us that T_{++} and T_{F+} are functions of σ^+ only whereas T_{--} and T_{F-} only depend on σ^-.

We have discussed in Chap. 2 how energy-momentum conservation results in an infinite number of conserved charges which generate the transformations $\sigma^\pm \to \sigma^\pm + f(\sigma^\pm)$ under which the action is invariant after going to conformal gauge. These are precisely the transformations which do not lead out of conformal gauge. This carries over to the fermionic string. But now we also have the conserved supercharges $\int d\sigma\, \epsilon^+(\sigma^+) T_{F+}(\sigma^+)$ reflecting the fact that the action and the superconformal gauge condition are invariant under supersymmetry transformations with parameters satisfying Eq. (7.37).

Next, let us find the algebra of T and T_F, which is the supersymmetric extension of the algebra Eq. (2.89). To do this we need the basic equal τ Poisson brackets[6]

$$\{\psi_A^\mu(\sigma,\tau), \Pi_\nu^B(\sigma',\tau)\}_{\text{P.B.}} = -\delta_A^B \delta(\sigma - \sigma') \delta_\nu^\mu. \tag{7.45}$$

The bracket of ψ with itself vanishes. For the momentum conjugate to ψ_A^μ we find

$$\Pi_\mu^A = \frac{i}{4\pi} \psi_{\mu B} (C\rho^0)^{BA}. \tag{7.46}$$

If, however, we use this definition of Π in Eq. (7.45), we find a contradiction. Let us explain the way out of this. We notice that Eq. (7.46) constitutes a (primary) constraint. We define

$$\phi_\mu^A = \Pi_\mu^A - \frac{i}{4\pi} \psi_{\mu B} (C\rho^0)^{BA} \tag{7.47}$$

and find

$$\{\phi_\mu^A(\sigma,\tau), \phi_\nu^B(\sigma',\tau)\}_{\text{P.B.}} = \frac{i}{2\pi} \delta(\sigma - \sigma')(C\rho^0)^{AB} \eta_{\mu\nu}. \tag{7.48}$$

In contrast to the constraints we have encountered so far (primary and secondary), the Poisson bracket of this constraint with itself does not vanish on the constrained hypersurface of phase space. Constraints of this kind are called second class constraints; the constraints we have encountered so far have all been first class. If second class constraints are present, Poisson brackets have to be replaced by Dirac brackets. If ϕ_i are a complete set of second class constraints, we define $\{\phi_i, \phi_j\}_{\text{P.B.}} = C_{ij}$. The Dirac bracket is defined as

---

[6]For anticommuting variables they are defined as

$$\{F, G\}_{\text{P.B.}} = \left(\frac{\partial F}{\partial q^i} \frac{\partial G}{\partial p_i} - \frac{\partial F}{\partial p_i} \frac{\partial G}{\partial q^i}\right) + (-1)^{\varepsilon_F} \left(\frac{\partial F}{\partial \theta^\alpha} \frac{\partial G}{\partial \pi_\alpha} + \frac{\partial F}{\partial \pi_\alpha} \frac{\partial G}{\partial \theta^\alpha}\right) \tag{7.44}$$

where (q, p) are the usual Grassmann even phase-space variables and θ^α and $\pi_\alpha = \frac{\partial \mathcal{L}}{\partial \dot\theta^\alpha}$ are Grassmann odd phase-space variables. F and G are functions on phase-space and ε_F is the Grassmann parity of F, i.e. $\varepsilon_F = 0(1)$ for F even (odd). All derivatives are left-derivatives. The canonical Hamiltonian is defined as $H = \dot q p + \dot\theta \pi - L$ where the order in the second term matters.

7.3 Superconformal Gauge

$$\{A, B\}_{\text{D.B.}} = \{A, B\}_{\text{P.B.}} - \{A, \phi_i\}_{\text{P.B.}} C_{ij}^{-1} \{\phi_j, B\}_{\text{P.B.}}. \tag{7.49}$$

This leads to

$$\{\psi_\pm^\mu(\sigma, \tau), \psi_\pm^\nu(\sigma', \tau)\}_{\text{D.B.}} = -2\pi i\, \delta(\sigma - \sigma')\eta^{\mu\nu},$$
$$\{\psi_+^\mu(\sigma, \tau), \psi_-^\nu(\sigma', \tau)\}_{\text{D.B.}} = 0. \tag{7.50}$$

Using this and the basic brackets for $X^\mu(\sigma, \tau)$, we find

$$\{T_{\pm\pm}(\sigma), T_{\pm\pm}(\sigma')\}_{\text{D.B.}} = \pm\Big(2\, T_{\pm\pm}(\sigma')\, \partial' + \partial' T_{\pm\pm}(\sigma')\Big)\, 2\pi\delta(\sigma - \sigma'),$$

$$\{T_{\pm\pm}(\sigma), T_{F\pm}(\sigma')\}_{\text{D.B.}} = \pm\Big(\frac{3}{2}\, T_{F\pm}(\sigma')\, \partial' + \partial' T_{F\pm}(\sigma')\Big)\, 2\pi\delta(\sigma - \sigma'),$$

$$\{T_{F\pm}(\sigma), T_{F\pm}(\sigma')\}_{\text{D.B.}} = \frac{i}{2}\, T_{\pm\pm}(\sigma')\, 2\pi\delta(\sigma - \sigma'). \tag{7.51}$$

It is also easily verified that

$$\Big\{T_{F+}(\sigma), \sqrt{\tfrac{2}{\alpha'}}\, X^\mu(\sigma')\Big\}_{\text{D.B.}} = \frac{1}{2}\psi_+^\mu(\sigma')\, 2\pi\delta(\sigma - \sigma'),$$

$$\{T_{F+}(\sigma), \psi_+^\mu(\sigma')\}_{\text{D.B.}} = \frac{i}{2}\sqrt{\tfrac{2}{\alpha'}}\, \partial_+ X^\mu(\sigma')\, 2\pi\delta(\sigma - \sigma'), \tag{7.52}$$

which are the supersymmetry transformations. Under the transformations generated by T_{++} the field ψ transforms as

$$\{T_{++}(\sigma), \psi_+(\sigma')\}_{\text{D.B.}} = \Big(\frac{1}{2}\psi_+(\sigma')\, \partial' + \partial' \psi_+(\sigma')\Big)\, 2\pi\delta(\sigma - \sigma'). \tag{7.53}$$

This last equation tells us that the world-sheet fermions transform under conformal transformations with weight $\frac{1}{2}$.

We now turn to the discussion of the boundary conditions. To derive the equations of motion (7.27), we impose $\delta\psi^\mu(\tau_0) = \delta\psi^\mu(\tau_1) = 0$. Further, we have to impose boundary conditions such that the boundary term

$$\delta S = \frac{1}{2\pi}\int_{\tau_0}^{\tau_1} d\tau\, (\psi_+ \cdot \delta\psi_+ - \psi_- \cdot \delta\psi_-)\Big|_{\sigma=0}^{\sigma=l} \tag{7.54}$$

vanishes. For the closed string this requires $(\psi_+ \cdot \delta\psi_+ - \psi_- \cdot \delta\psi_-)(\sigma) = (\psi_+ \cdot \delta\psi_+ - \psi_- \cdot \delta\psi_-)(\sigma + \ell)$ which is solved by

$$\psi_+^\mu(\sigma) = \pm\psi_+^\mu(\sigma + \ell),$$
$$\psi_-^\mu(\sigma) = \pm\psi_-^\mu(\sigma + \ell) \tag{7.55}$$

with the same conditions on $\delta\psi_\pm$. Anti-periodicity of $\psi_\pm$ is possible as they are fermions on the world-sheet. As the action is quadratic in ψ, it is periodic for either sign. Periodic boundary conditions in (7.55) are referred to as Ramond (R) boundary conditions, whereas anti-periodic boundary conditions are called Neveu-Schwarz (NS) boundary conditions. This means that fermions on the world-sheet satisfy $\psi(\sigma+\ell) = e^{2\pi i\phi}\psi(\sigma)$, where $\phi = 0$ for the R-sector and $\phi = \frac{1}{2}$ for the NS-sector. More general phases are not allowed for real ψ. Space-time Poincaré invariance requires that we impose the same boundary conditions in all directions μ. This also guarantees that the supercurrents $T_{F\pm}$ have definite periodicity. However, the conditions for the two spinor components ψ_+ and ψ_- can be chosen independently, leading to a total of four possibilities: (R,R), (NS,NS), (NS,R) and (R,NS). Obviously, the two components of the supersymmetry parameter ϵ have to be chosen such that $\delta X^\mu \propto \overline{\epsilon}\psi^\mu$ is periodic. We will show in the next chapter that string states in the R sector are space-time fermions and states in the NS sector are space-time bosons. Therefore, the two sectors (R,R) and (NS,NS) lead to space-time bosons and the remaining sectors, (NS,R) and (R,NS) to space-time fermions.

For the open string the variation (7.54) has to be canceled on each boundary, i.e. at $\sigma = 0$ and $\sigma = \ell$, separately. Given that ψ_+ and ψ_- are real, this leads to the following admissible boundary conditions

$$\psi_+^\mu(0) = \pm\psi_-^\mu(0), \qquad \psi_+^\mu(\ell) = \pm\psi_-^\mu(\ell), \tag{7.56}$$

which can be imposed for each direction μ independently. As for the open bosonic string, we can impose Neumann or Dirichlet boundary conditions.

If we want to preserve space-time Poincaré invariance along $(p+1)$ directions, we have to impose the same boundary conditions. In analogy to the bosonic string we call these Neumann directions and without loss of generality we specify NN boundary conditions for the world-sheet fermions as

(NN) $$\psi_+^\alpha(0) = \psi_-^\alpha(0), \qquad \psi_+^\alpha(\ell) = \eta\psi_-^\alpha(\ell), \tag{7.57}$$

where $\eta = \pm 1$. Only the relative sign in the boundary conditions at $\sigma = 0$ and $\sigma = \ell$ is relevant and by a redefinition $\psi_- \to \pm\psi_-$, which leaves the action invariant, we can always move the sign η to the $\sigma = \ell$ boundary. We therefore have to distinguish between two sectors, $\eta = +1$ and $\eta = -1$. The choice $\eta = +1$ is called Ramond sector and the choice $\eta = -1$ is called Neveu-Schwarz sector of the open string. States in the R (NS) sector will turn out to be space-time fermions (bosons).

We will now demonstrate that the boundary conditions (7.57) break half of the world-sheet supersymmetry. Indeed, in order to preserve the boundary conditions under (7.39) we need to require

$$\epsilon^+(0) = \epsilon^-(0), \qquad \epsilon^+(\ell) = \eta\epsilon^-(\ell). \tag{7.58}$$

Here we use the fact that for NN b.c.'s $\partial_\sigma X$ vanishes at the two ends while $\partial_\tau X$ is free.

7.4 Oscillator Expansions

If we want to impose DD boundary conditions along the remaining directions we require

(DD) $$\psi_+^i(0) = -\psi_-^i(0), \qquad \psi_+^i(\ell) = -\eta\psi_-^i(\ell). \tag{7.59}$$

and for mixed boundary conditions

(ND) $$\psi_+^a(0) = +\psi_-^a(0), \qquad \psi_+^a(\ell) = -\eta\psi_-^a(\ell), \tag{7.60}$$

(DN) $$\psi_+^a(0) = -\psi_-^a(0), \qquad \psi_+^a(\ell) = +\eta\psi_-^a(\ell). \tag{7.61}$$

In all four cases $\eta = +1$ is the R-sector and $\eta = -1$ the NS sector and all boundary conditions impose (7.58) on the supersymmetry parameters. The requirement that they preserve the same world-sheet supersymmetry fixes them.

We can summarize the boundary conditions as

$$\begin{aligned}\psi_+^\mu(0) &= D^\mu{}_\nu(0)\psi_-^\nu(0) \\ \psi_+^\mu(\ell) &= \eta D^\mu{}_\nu(\ell)\psi_-^\nu(\ell)\end{aligned} \quad \text{with} \quad \eta = \begin{cases} +1 & \text{R} \\ -1 & \text{NS}, \end{cases} \tag{7.62}$$

where, as in (4.169), $D^\mu{}_\nu = (\delta^\alpha{}_\beta, -\delta^i{}_j)$.

7.4 Oscillator Expansions

We now proceed as in the bosonic string and solve the equations of motion for the unconstrained system. The treatment for the bosonic coordinates is identical to the one in Chap. 2 and will not be repeated here. The fermionic fields require some care.

Closed Strings

We have to distinguish between two choices of boundary conditions for each chirality. The general solutions of the two-dimensional Dirac equation with periodic (R) and antiperiodic (NS) boundary conditions are

$$\psi_+^\mu(\sigma,\tau) = \sqrt{\frac{2\pi}{\ell}} \sum_{r\in\mathbb{Z}+\phi} \bar{b}_r^\mu e^{-2\pi i r(\tau+\sigma)/\ell}$$

$$\psi_-^\mu(\sigma,\tau) = \sqrt{\frac{2\pi}{\ell}} \sum_{r\in\mathbb{Z}+\phi} b_r^\mu e^{-2\pi i r(\tau-\sigma)/\ell} \quad \text{where} \quad \begin{cases} \phi = 0 & \text{(R)} \\ \phi = \frac{1}{2} & \text{(NS)} \end{cases}. \tag{7.63}$$

ϕ can be chosen independently for the left- and right-movers. The reality of the Majorana spinors translates into the following conditions for the modes:

$$(b_r^\mu)^* = b_{-r}^\mu, \qquad (\bar{b}_r^\mu)^* = \bar{b}_{-r}^\mu. \tag{7.64}$$

In terms of the fermionic oscillator modes the basic Dirac bracket Eq. (7.50) translates to

$$\{b_r^\mu, b_s^\nu\}_{\text{D.B.}} = -i\,\eta^{\mu\nu}\delta_{r+s},$$
$$\{\bar{b}_r^\mu, \bar{b}_s^\nu\}_{\text{D.B.}} = -i\,\eta^{\mu\nu}\delta_{r+s},$$
$$\{b_r^\mu, \bar{b}_s^\nu\}_{\text{D.B.}} = 0. \tag{7.65}$$

Next we decompose the generators of conformal and superconformal transformations into modes. We restrict ourselves to one sector, say the right-moving one. The results for the left-moving sector are obtained by simply placing a bar over all oscillators. We define

$$L_n = -\frac{\ell}{4\pi^2}\int_0^\ell d\sigma\, e^{-2\pi in\sigma/\ell}\, T_{--}(\sigma),$$
$$G_r = -\frac{1}{\pi}\sqrt{\frac{\ell}{2\pi}}\int_0^\ell d\sigma\, e^{-2\pi ir\sigma/\ell}\, T_{F-}(\sigma). \tag{7.66}$$

From Eq. (7.42) it is clear that T_{F-} satisfies the same periodicity condition as the ψ^μ: periodic in the R-sector and antiperiodic in the NS-sector. Consequently, the mode numbers are integer and half-integer respectively. In terms of oscillators we find $L_m = L_m^{(\alpha)} + L_m^{(b)}$ where

$$L_n^{(\alpha)} = \frac{1}{2}\sum_{m\in\mathbb{Z}} \alpha_{-m}\cdot\alpha_{m+n} \quad \text{(as before)},$$
$$L_n^{(b)} = \frac{1}{2}\sum_r \left(r + \frac{n}{2}\right) b_{-r}\cdot b_{n+r},$$
$$G_r = \sum_m \alpha_{-m}\cdot b_{r+m}. \tag{7.67}$$

Note that $\sum_r b_{-r}\cdot b_{n+r} = 0$. This term has been included to make the expression for $L_n^{(b)}$ look more symmetric. It corresponds to the mode expansion of $\partial_-(\psi_-\psi_-)$ which vanishes by the Grassmann property of ψ_-. The generators L_m and G_r satisfy the following reality conditions

$$L_m^* = L_{-m}, \qquad G_r^* = G_{-r}. \tag{7.68}$$

7.4 Oscillator Expansions

One now verifies, using the basic brackets Eqs. (2.81a) and (7.65), the following (classical) algebra:

$$\{L_m, L_n\}_{\text{D.B.}} = -i(m-n)L_{m+n},$$
$$\{L_m, G_r\}_{\text{D.B.}} = -i\left(\frac{1}{2}m - r\right)G_{m+r},$$
$$\{G_r, G_s\}_{\text{D.B.}} = -2i L_{r+s}. \tag{7.69}$$

It can also be derived from Eq. (7.51) and the definitions Eq. (7.66). This algebra is called the centerless super-Virasoro algebra. In the next chapter we will show how it is modified in the quantum theory. For the closed string there are two copies of this algebra, one for the left- and one for the right-movers.

Open Strings

For the open string we also expand the fermionic fields in modes and implement the boundary conditions (7.56). As expected, the boundary conditions relate the left- and right-movers and there is only one set of oscillators. We find

(NN) $\quad \psi_\pm^\alpha(\sigma, \tau) = \sqrt{\dfrac{\pi}{\ell}} \sum_r b_r^\alpha e^{-i\pi r(\tau\pm\sigma)/\ell} \quad r \in \begin{cases} \mathbb{Z} + \frac{1}{2} & \text{NS} \\ \mathbb{Z} & \text{R,} \end{cases}$ (7.70a)

(DD) $\quad \psi_\pm^i(\sigma, \tau) = \pm\sqrt{\dfrac{\pi}{\ell}} \sum_r b_r^i e^{-i\pi r(\tau\pm\sigma)/\ell} \quad r \in \begin{cases} \mathbb{Z} + \frac{1}{2} & \text{NS} \\ \mathbb{Z} & \text{R,} \end{cases}$ (7.70b)

(ND) $\quad \psi_\pm^a(\sigma, \tau) = \sqrt{\dfrac{\pi}{\ell}} \sum_r b_r^a e^{-i\pi r(\tau\pm\sigma)/\ell} \quad r \in \begin{cases} \mathbb{Z} & \text{NS} \\ \mathbb{Z} + \frac{1}{2} & \text{R,} \end{cases}$ (7.70c)

(DN) $\quad \psi_\pm^a(\sigma, \tau) = \pm\sqrt{\dfrac{\pi}{\ell}} \sum_r b_r^a e^{-i\pi r(\tau\pm\sigma)/\ell} \quad r \in \begin{cases} \mathbb{Z} & \text{NS} \\ \mathbb{Z} + \frac{1}{2} & \text{R.} \end{cases}$ (7.70d)

In all four cases the surviving supersymmetry parameter has an expansion in integer (half-integer) modes in the R (NS) sector. This is consistent with $\delta X \sim \epsilon \psi$.

As in the bosonic case we can use the doubling trick and define, for each of the four boundary conditions purely, say left-moving fields ψ_+ on the interval $0 \leq \sigma \leq 2\ell$ by combining $\psi_+(\sigma)$ with $\pm\psi_-(2\ell-\sigma)$. The sign is chosen to have continuity at $\sigma = \ell$. In the R-sector these fields are periodic (with period 2ℓ) for (NN) and (DD) boundary conditions and anti-periodic for (DN) and (ND) boundary conditions. In the NS sector the assignments are opposite.

To find the oscillator expressions for the symmetry generators, we also use the doubling trick, i.e. L_n and G_r are defined as

$$L_n = -\frac{\ell}{2\pi^2} \int_{-\ell}^{+\ell} d\sigma\, e^{i\pi n\sigma/\ell}\, T_{++}(\sigma).$$

$$G_r = -\frac{1}{\pi}\sqrt{\frac{\ell}{\pi}} \int_{-\ell}^{+\ell} d\sigma\, e^{i\pi r\sigma/\ell}\, T_{F+}(\sigma). \qquad (7.71)$$

One finds

$$L_n = \frac{1}{2}\sum_m \alpha_{-m} \cdot \alpha_{m+n} + \frac{1}{2}\sum_r \left(r + \frac{n}{2}\right) b_{-r} \cdot b_{n+r},$$

$$G_r = \sum_m \alpha_{-m} \cdot b_{r+m}. \qquad (7.72)$$

Notice that for all boundary conditions the modes of G_r are integer (half-integer) in the R (NS) sector. This is required in order for the supersymmetry generators $\int \bar{\epsilon} T_F$ to be well-defined and can be considered the defining property of the two sectors.

The Dirac brackets for the oscillators and the modes of the symmetry generators are the same as those for, say the right-movers, of the closed string.

7.5 Appendix: Spinor Algebra in Two Dimensions

In this appendix we summarize our notation for spinors in two dimensions and provide some identities which will prove useful in this and the following chapter. The two-dimensional Dirac matrices satisfy

$$\{\rho^\alpha, \rho^\beta\} = 2h^{\alpha\beta}. \qquad (7.73)$$

They transform under coordinate transformations and are related to the constant Dirac matrices ρ^a through the zweibein: $\rho^\alpha = e^\alpha_a \rho^a$ from which $\{\rho^a, \rho^b\} = 2\eta^{ab}$ with $\eta^{ab} = \begin{pmatrix} -1 & 0 \\ 0 & +1 \end{pmatrix}$ follows. A convenient basis for the ρ^a is

$$\rho^0 = i\sigma^2 = \begin{pmatrix} 0 & 1 \\ -1 & 0 \end{pmatrix}, \quad \rho^1 = \sigma^1 = \begin{pmatrix} 0 & 1 \\ 1 & 0 \end{pmatrix}, \qquad (7.74)$$

and we define $\bar{\rho} = \rho^0 \rho^1 = \sigma^3 = \begin{pmatrix} 1 & 0 \\ 0 & -1 \end{pmatrix}$, which is the analogue of γ^5 in four dimensions. We define the charge conjugation matrix as $C = \rho^0$. Then $(\rho^a)^T = -C\rho^a C^{-1}$. A Majorana spinor satisfies $\bar{\lambda} = \lambda^\dagger \rho^0 = \lambda^T C$. This means that Majorana spinors are real. Using spinor indices, an expression of the form $\bar{\chi}\Gamma\psi$, where Γ is some combination of Dirac matrices, can be alternatively written as $\chi^A \Gamma_A{}^B \psi_B$, where $\chi^A = \chi_B C^{BA}$. Two-dimensional spinor indices take values

7.5 Appendix: Spinor Algebra in Two Dimensions

$A = \pm$; i.e. $\psi^+ = -\psi_-$, $\psi^- = \psi_+$. The index structures of the Dirac matrices and of the charge conjugation matrix are $(\rho^\alpha)_A{}^B$, C^{AB} and $(C^{-1})_{AB}$. With these results we can convert covariant expressions into component expressions; e.g.

$$\overline{\psi}\rho^\alpha \partial_\alpha \psi = -2(\psi_+ \partial_- \psi_+ + \psi_- \partial_+ \psi_-). \tag{7.75}$$

It is easy to prove the following spin-flip property, valid for anti-commuting Majorana spinors:

$$\overline{\lambda}_1 \rho^{\alpha_1} \cdots \rho^{\alpha_n} \lambda_2 = (-1)^n \overline{\lambda}_2 \rho^{\alpha_n} \cdots \rho^{\alpha_1} \lambda_1. \tag{7.76}$$

This and the following Fierz identity, again valid for anticommuting spinors, are needed to show the invariance of the action under supersymmetry transformations:

$$(\overline{\psi}\lambda)(\overline{\phi}\chi) = -\frac{1}{2}\left\{(\overline{\psi}\chi)(\overline{\phi}\lambda) + (\overline{\psi}\overline{\rho}\chi)(\overline{\phi}\overline{\rho}\lambda) + (\overline{\psi}\rho^\alpha \chi)(\overline{\phi}\rho_\alpha \lambda)\right\}. \tag{7.77}$$

The identity

$$\rho^\alpha \rho_\beta \rho_\alpha = 0 \tag{7.78}$$

follows from the Dirac algebra in two dimensions. Another useful relation is

$$\rho^\alpha \rho^\beta = h^{\alpha\beta} + \frac{1}{e}\epsilon^{\alpha\beta}\overline{\rho}, \quad (\epsilon^{01} = 1). \tag{7.79}$$

A purely helicity $\pm 3/2$ fermion $\tilde{\psi}$, i.e. $\rho \cdot \tilde{\psi} = 0$, satisfies $\rho_\alpha \tilde{\psi}_\beta = \rho_\beta \tilde{\psi}_\alpha$. Our convention for the two-dimensional ϵ-tensor is $\epsilon^{\alpha\beta} = \pm 1$ ($\epsilon^{01} = 1$) and $\epsilon_{\alpha\beta} = \pm h$ ($\epsilon_{01} = h = -e^2$). Then $e_{\alpha a} e_{\beta b} \epsilon^{ab} = \frac{1}{e}\epsilon_{\alpha\beta}$.

Further Reading

The formulation of fermionic string theories is due to

- P. Ramond, Dual theory for free fermions. Phys. Rev. **D3**, 2415 (1971)
- A. Neveu, J.H. Schwarz, Factorizable dual model of pions. Nucl. Phys. **B31**, 86 (1971)

The world-sheet action and its symmetries were found in

- L. Brink, P. Di Vecchia, P. Howe, A locally super symmetric and reparametrization invariant action for the spinning string. Phys. Lett. **65B**, 471 (1976)
- P.S. Howe, Super Weyl transformations in two dimensions. J. Phys. **A12**, 393 (1979)
- S. Deser, B. Zumino, A complete action for the spinning string. Phys. Lett. **65B**, 369 (1976)

The Green-Schwarz formalism is reviewed in

- J. Schwarz, Superstring theory. Phys. Rep. **89**, 223 (1982)

An introduction to the pure spinor formalism is

- O.A. Bedoya, N. Berkovits, *GGI Lectures on the Pure Spinor Formalism of the Superstring* [arXiv:0910.2254 [hep-th]]

The canonical formalism for Grassmann odd variables is extensively treated in

- M. Henneaux, C. Teitelboim, *Quantization of Gauge Systems* (Princeton University Press, Princeton, 1992)

Chapter 8
The Quantized Fermionic String

Abstract The fermionic string is quantized analogously to the bosonic string, though this time leading to a critical dimension $d = 10$. We first quantize in light-cone gauge and construct the spectrum. To remove the tachyon one has to perform the so-called GSO projection, which guarantees space-time supersymmetry of the ten-dimensional theory. There are two possible space-time supersymmetric GSO projections which result in the type IIA and the type IIB superstring. We also present the covariant path integral quantization. The chapter closes with an appendix on spinors in d dimensions.

8.1 Canonical Quantization

We proceed in the same way as in the bosonic theory by making the replacement (3.1) and, in addition, by replacing the Dirac bracket for the anticommuting world-sheet fermions by an anti-commutator:

$$\{\,,\,\}_{\text{D.B.}} \to \frac{1}{i}\{\,,\,\}. \tag{8.1}$$

We find

$$\{\psi_+^\mu(\sigma,\tau), \psi_+^\nu(\sigma',\tau)\} = 2\pi\, \eta^{\mu\nu}\delta(\sigma-\sigma'),$$
$$\{\psi_-^\mu(\sigma,\tau), \psi_-^\nu(\sigma',\tau)\} = 2\pi\, \eta^{\mu\nu}\delta(\sigma-\sigma'),$$
$$\{\psi_+^\mu(\sigma,\tau), \psi_-^\nu(\sigma',\tau)\} = 0 \tag{8.2}$$

or, in terms of oscillators[1]

---

[1] Again, we will only write down the expressions for the right-moving sector of the closed string. The left-moving expressions are easily obtained by simply putting bars over all mode operators.

$${\{b_r^\mu, b_s^\nu\} = \eta^{\mu\nu}\delta_{r+s}.} \tag{8.3}$$

We define oscillators with positive mode numbers as annihilation operators and oscillators with negative mode numbers are creation operators. We have seen in Chap. 3 that α_0^μ and $\bar{\alpha}_0^\mu$ correspond to the center of mass momentum of the string. We will see below how the fermionic zero mode operators b_0^μ and $\bar{b}_0^\mu$ in the R-sector are to be interpreted. But we observe already here that they satisfy a Clifford algebra:

$$\{b_0^\mu, b_0^\nu\} = \eta^{\mu\nu}. \tag{8.4}$$

The level number operator is

$$N = N^{(\alpha)} + N^{(b)}, \tag{8.5}$$

where

$$N^{(\alpha)} = \sum_{m=1}^{\infty} \alpha_{-m} \cdot \alpha_m,$$

$$N^{(b)} = \sum_{r \in \mathbb{Z}+\phi > 0} r\, b_{-r} \cdot b_r. \tag{8.6}$$

The oscillator expressions of the super-Virasoro generators are again undefined without giving an operator ordering prescription. As in Chap. 3 we define them by their normal ordered expressions, i.e.

$$L_n = L_n^{(\alpha)} + L_n^{(b)} \tag{8.7}$$

with

$$L_n^{(\alpha)} = \frac{1}{2} \sum_{m \in \mathbb{Z}} :\alpha_{-m} \cdot \alpha_{m+n}:,$$

$$L_n^{(b)} = \frac{1}{2} \sum_{r \in \mathbb{Z}+\phi} \left(r + \frac{n}{2}\right) :b_{-r} \cdot b_{n+r}: \tag{8.8}$$

and

$$G_r = \sum_{m \in \mathbb{Z}} \alpha_{-m} \cdot b_{r+m}. \tag{8.9}$$

Obviously, normal ordering is only required for L_0 and we include again an as yet undetermined normal ordering constant a in all formulas containing L_0.

Unless stated otherwise, the expressions for the open string coincide with the ones for the right-moving sector of the closed string.

8.1 Canonical Quantization

The algebra satisfied by the L_n and G_r can now be determined. Great care is again required due to normal ordering. One finds

$$[L_m, L_n] = (m-n) L_{m+n} + \frac{d}{8} m (m^2 - 2\phi) \delta_{m+n},$$

$$[L_m, G_r] = \left(\frac{m}{2} - r\right) G_{m+r},$$

$$\{G_r, G_s\} = 2 L_{r+s} + \frac{d}{2}\left(r^2 - \frac{\phi}{2}\right) \delta_{r+s}. \tag{8.10}$$

This is the super-Virasoro algebra. A straightforward and save way to derive it is to evaluate the (anti)commutators between two states which are annihilated by all annihilation operators or to use techniques of superconformal field theory. These will be presented in Chap. 12. We note that the R ($\phi = 0$) and NS ($\phi = \frac{1}{2}$) algebras agree formally except for the terms in the anomalies which can be removed by shifting L_0 by a constant. Indeed, if we define $L_0^R \to L_0^R + \frac{d}{16}$, it takes the form Eq. (8.10) with $\phi = \frac{1}{2}$ in both sectors.

The algebra in the form of Eq. (7.51) is modified by quantum effects as follows:

$$[T_{++}(\sigma), T_{++}(\sigma')] = \cdots - \frac{i\pi d}{4} \partial_{\sigma'}^3 \delta(\sigma - \sigma'),$$

$$\{T_{F+}(\sigma), T_{F+}(\sigma')\} = \cdots - \frac{\pi d}{4} \partial_{\sigma'}^2 \delta(\sigma - \sigma'), \tag{8.11}$$

where we have only written the non-trivial quantum corrections. The other terms follow from (7.51) and (8.1).

Let us now examine the states in the Hilbert space of the theory. In doing so, we have to distinguish between two sectors, R and NS. The oscillator ground states in the two sectors are defined by

$$\alpha_m^\mu |0\rangle_{\text{NS}} = b_r^\mu |0\rangle_{\text{NS}} = 0, \quad m = 1, 2, \ldots, \quad r = \frac{1}{2}, \frac{3}{2}, \ldots \tag{8.12}$$

and

$$\alpha_m^\mu |a\rangle_R = b_m^\mu |a\rangle_R = 0, \quad m = 1, 2, \ldots \tag{8.13}$$

(we suppress the dependence on the center of mass momentum). The fact that the R sector ground state carries a label is because in this sector there are the b_0^μ zero modes. They do not change the mass of the ground state or any other state. The mass operators for the fermionic string are given by the same expressions as in the bosonic case but with the level numbers as in Eq. (8.5) and as yet to be determined normal ordering constants a_{NS} and a_R which are a priori different. It is then easy to check that $[b_0^\mu, m^2] = 0$, i.e. the states $|0\rangle$ and $b_0^\mu |0\rangle$ are degenerate in mass. But α_n^μ, b_r^μ for $n, r < 0$ increase $\alpha' m^2$ by $2n$ and $2r$ units respectively (for the closed string;

for the open string the increase is by n and r units). This means that in the NS sector there is a unique ground state which must therefore be spin zero. In the R sector the ground state is degenerate. These degenerate R ground states form a representation of the Clifford algebra (8.4) and transform as a spinor of $SO(d-1,1)$. We will see this explicitly below.

Since the degenerate ground states transform in a spinor representation of the space-time rotation group $SO(d-1,1)$, they are space-time fermions, whereas states in the NS sector are space-time bosons. The oscillators, all being space-time vectors, cannot change bosons into fermions or vice versa. Whether a state belongs to the R or the NS sector depends on the ground state it is built on. We will come back to this important point in Chap. 12.

We write the R ground state as $|a\rangle$, where a is a $SO(d-1,1)$ spinor index. We know from the zero mode algebra (8.4) that the b_0^μ satisfy the Dirac algebra, and it is very suggestive to represent them as Dirac matrices.[2] We identify $b_0^\mu|a\rangle = \frac{1}{\sqrt{2}}(\gamma^\mu)_b^a|b\rangle$ where γ^μ is a Dirac matrix in d dimensions, satisfying $\{\gamma^\mu, \gamma^\nu\} = 2\eta^{\mu\nu}$.

As announced above, we will now give an explicit description of the Ramond ground states for d-even. One first defines the fermionic raising and lowering operators[3]

$$b_0^\pm = \frac{i}{\sqrt{2}}\left(b_0^0 \pm b_0^1\right),$$

$$b_i^\pm = \frac{1}{\sqrt{2}}\left(b_0^{2i} \pm i b_0^{2i+1}\right) \qquad i = 1, \ldots, \frac{d-2}{2} \qquad (8.14)$$

so that in this basis the Clifford algebra reads ($i = 0, \ldots, \frac{d-2}{2}$)

$$\{b_i^+, b_j^-\} = \delta_{ij}. \qquad (8.15)$$

Define a highest weight state $|0\rangle_R$ which satisfies $b_i^+|0\rangle_R = 0$. Successive application of the lowering operators b_i^- generates the $2^{\frac{d}{2}}$ dimensional spinor representation of $SO(d-1,1)$. These states are denoted as[4]

$$|s\rangle = |s_0, s_1, s_2, s_3, s_4\rangle, \qquad \text{with} \quad s_i = \pm\frac{1}{2}, \qquad (8.16)$$

---

[2] In fact, the zero modes generate a finite group of order 2^{d+1} under multiplication. Using results from the representation theory of finite groups, one can show that for d even there is only one inequivalent irreducible representation of this group which is not one-dimensional. This representation must therefore be the one in terms of Dirac matrices. The one-dimensional representations of the group clearly violate the Dirac algebra. For d odd, there are two representations but they only differ by a sign of one of the Dirac matrices.

[3] We anticipate that the critical dimension will be even.

[4] In this way the states are presented by their weight vector; cf. Chap. 13.

8.1 Canonical Quantization

where $|0\rangle_R = |+\frac{1}{2},+\frac{1}{2},+\frac{1}{2},+\frac{1}{2},+\frac{1}{2}\rangle$ and the operator b_i^- lowers s_i from $+\frac{1}{2}$ to $-\frac{1}{2}$. This leads to a $2^{\frac{n}{2}}$-dimensional representation of the b_0^μ and it is obvious from our discussion that the representation matrices satisfy the Dirac algebra, i.e. they are Dirac-matrices. Further properties are collected in the appendix of this chapter.

The discussion above was for the right-moving part of the closed string and the results immediately carry over to the left-moving sector of the closed string. However, we have to be careful when using them for the open string. For NN and DD boundary conditions they are valid. But for ND and DN boundary conditions the results for the R and NS sectors are interchanged, because in these mixed sectors the moding of the fermionic oscillators is reversed. In other words, for ND and DN boundary conditions the oscillator ground state in the NS is degenerate while it is unique in the R sector. We postpone a discussion of the propagators for the world-sheet fermions until Chap. 12.

We still have to implement the constraints on the states of the theory. Due to the anomalies in the super-Virasoro algebra, it is again impossible to impose $L_m|\text{phys}\rangle = G_r|\text{phys}\rangle = 0$ for all m and r. The best we can do is to demand that

$$G_r|\text{phys}\rangle = 0 \quad r > 0,$$
$$L_m|\text{phys}\rangle = 0 \quad m > 0 \quad \text{(NS)},$$
$$(L_0 + a)|\text{phys}\rangle = 0 \tag{8.17a}$$

in the NS sector, and

$$G_r|\text{phys}\rangle = 0 \quad r \geq 0,$$
$$L_m|\text{phys}\rangle = 0 \quad m > 0 \quad \text{(R)},$$
$$L_0|\text{phys}\rangle = 0 \tag{8.17b}$$

in the R sector. Note that we do not have included a normal ordering constant in the last equation. There are several reasons for this. From the super-Virasoro algebra we find that $G_0^2 = L_0$, i.e. if we had $(L_0 - \mu^2)|\text{phys}\rangle = 0$ we also needed $(G_0 - \mu)|\text{phys}\rangle = 0$. However, G_0 has no normal ordering ambiguity and the normal ordering constants arising from the bosonic and the fermionic oscillators cancel in L_0 in the Ramond sector. Also, G_0 is anti-commuting whereas the normal ordering constant is a commuting c-number. When we discuss the spectrum we will find that setting $\mu = 0$ is indeed correct.

For the closed string there is, of course, a second set of conditions for the left movers and we also have to demand that

$$(L_0 - \overline{L}_0)|\text{phys}\rangle = 0, \tag{8.18}$$

again expressing the fact that no point on a closed string is distinguished.

Consider the ground state in the Ramond sector. For it to be a physical state, it has to satisfy $G_0|a\rangle = 0$. With the mode expansion of G_0 and (8.13) we find

$$G_0|a\rangle = \alpha_{0\mu}b_0^\mu|a\rangle \propto p_\mu(\gamma^\mu)_b^a|b\rangle = 0 \,. \tag{8.19}$$

If we introduce the polarization spinor u_a and define the state $|u\rangle = u_a|a\rangle$ this means that $|u\rangle$ is a physical state if $p^\mu\gamma_\mu u = 0$, i.e. if u satisfies the massless Dirac equation. In Chap. 13 we will find the same condition by requiring BRST invariance of the vertex operator of a space-time fermion.

So far the quantization has been canonical and covariant. We know, however, that due to the negative eigenvalue of $\eta^{\mu\nu}$ there are negative norm states (ghosts). As in the bosonic theory one can prove a no-ghost theorem which states that the negative norm states decouple in the critical dimension d for a particular value of the normal ordering constant a. It turns out that for the fermionic string $d = 10$ and $a = -1/2$. As it was for the conformal symmetry in the bosonic case, in the fermionic theory the superconformal symmetry is just big enough to allow for the ghost decoupling. We will not prove the no-ghost theorem here but instead follow our treatment of the bosonic theory and discuss the non-covariant light-cone quantization where the constraints are solved explicitly. At the end of this chapter we will discuss the covariant path integral quantization. Both approaches will also lead to the above values for the critical dimension and the normal ordering constant.

8.2 Light-Cone Quantization

In the bosonic theory the light-cone gauge was obtained by the choice

$$X^+ = \frac{2\pi\alpha'}{\ell}p^+\tau \tag{8.20}$$

(cf. Eq. (3.31) on page 42) which fixed the gauge completely. This choice is again possible in the fermionic theory and also completely eliminates the reparametrization invariance. But now we still have local supersymmetry transformations. In going to super-conformal gauge we have fixed it partially leaving only transformations satisfying $\partial_+\epsilon^- = \partial_-\epsilon^+ = 0$. This freedom can now be used (cf. (7.39)) to transform ψ^+ away; i.e. in addition to Eq. (8.20) the light-cone gauge condition in the fermionic theory is

$$\psi^+ = 0 \tag{8.21}$$

or, equivalently, $b_r^+ = 0$, $\forall r$. (Here and below the superscript denotes the light-cone component; i.e. $\psi^\pm = \frac{1}{\sqrt{2}}(\psi^0 \pm \psi^1)$.) The canonical Hamiltonian in light-cone gauge is

$$H_{\text{l.c.}} = \frac{\pi\alpha'}{\ell}p^i p^i + \frac{2\pi}{\ell}\left(N_{\text{tr}}^{(\alpha)} + \bar{N}_{\text{tr}}^{(\tilde\alpha)} + N_{\text{tr}}^{(b)} + \bar{N}_{\text{tr}}^{(\tilde b)} + a + \bar{a}\right) \tag{8.22}$$

8.2 Light-Cone Quantization

for the closed string and

$$H_{\text{l.c.}} = \frac{\pi}{\ell}\left(N_{\text{tr}}^{(\alpha)} + N_{\text{tr}}^{(b)} + a\right) + \frac{\pi\alpha'}{\ell}\sum_{\text{NN}} p^i p^i + \frac{1}{4\pi\alpha'\ell}\sum_{\text{DD}}(x_2^i - x_1^i)^2 \quad (8.23)$$

for the open string. Here we have used the number operators for the transverse degrees of freedom

$$N_{\text{tr}}^{(\alpha)} = \sum_{n>0} \alpha_{-n}^i \alpha_n^i \quad \text{and} \quad N_{\text{tr}}^{(b)} = \sum_{r>0} r\, b_{-r}^i b_r^i \quad (8.24)$$

and we have also introduced normal ordering constants which we will determine later. They will turn out to be different for the NS and the R sectors.

After going to light-cone gauge we can solve the constraints. The bosonic constraints $T_{\pm\pm} = 0$ lead to

$$\partial_\pm X^- = \frac{1}{2p^+}\frac{\ell}{2\pi}\left(\frac{2}{\alpha'}\partial_\pm X^i \partial_\pm X^i + i\psi_\pm^i \partial_\pm \psi_\pm^i\right) \quad (8.25)$$

and solving the fermionic constraints $T_{F\pm} = 0$ we find

$$\psi_\pm^- = \frac{2}{\alpha' p^+}\frac{\ell}{2\pi}\psi_\pm^i \partial_\pm X^i, \quad (8.26)$$

which leaves only the transverse components X^i and ψ^i as independent degrees of freedom. The corresponding oscillator expressions are

$$\alpha_m^- = \frac{1}{\sqrt{2\alpha'}\,p^+}\left\{\sum_n :\alpha_n^i \alpha_{m-n}^i: + \sum_r \left(\frac{m}{2} - r\right):b_r^i b_{m-r}^i: + a\,\delta_m\right\} \quad (8.27)$$

and

$$b_r^- = \sqrt{\frac{2}{\alpha'}}\frac{1}{p^+}\sum_q \alpha_{r-q}^i b_q^i. \quad (8.28)$$

These expressions are valid for the right-moving part of the closed string and have to be supplemented by their left-moving counterpart. For the open string we find (8.27) and (8.28) but with an additional multiplicative factor of $\frac{1}{2}$ on the right hand side.

From (8.27) with $m = 0$ and the expressions for the light-cone Hamiltonians we find

$$p^- = \frac{\ell}{2\pi\alpha' p^+} H_{\text{l.c.}}. \quad (8.29)$$

For the closed string, the expression $\int d\sigma (\partial_+ - \partial_-) X^- = \int d\sigma \partial_\sigma X^- = 0$ leads to the level matching condition

$$N_{\mathrm{tr}}^{(\alpha)} + N_{\mathrm{tr}}^{(b)} = \bar{N}_{\mathrm{tr}}^{(\bar{\alpha})} + \bar{N}_{\mathrm{tr}}^{(\bar{b})}. \tag{8.30}$$

The mass operators are now easily obtained from (8.29). For the closed string we find

$$\alpha' m_L^2 = 2(\bar{N}_{\mathrm{tr}}^{(\bar{\alpha})} + \bar{N}_{\mathrm{tr}}^{(\bar{b})} + \bar{a}) \qquad \alpha' m_R^2 = 2(N_{\mathrm{tr}}^{(\alpha)} + N_{\mathrm{tr}}^{(b)} + a) \tag{8.31}$$

and

$$m_L^2 = m_R^2 \tag{8.32}$$

as a consequence of level matching and assuming $a = \bar{a}$. For the open string the mass operator reads

$$\alpha' m^2 = N_{\mathrm{tr}}^{(\alpha)} + N_{\mathrm{tr}}^{(b)} + \frac{1}{4\pi^2 \alpha'}(\Delta X)^2 + a. \tag{8.33}$$

We still need to determine the normal ordering constants. Here we have to distinguish between the two sectors. In the NS sector of the closed string it is $a_{\mathrm{NS}} = \bar{a}_{\mathrm{NS}} = \frac{1}{2}(d-2)(\sum_{n=0}^{\infty} n - \sum_{r=1/2}^{\infty} r) = \frac{1}{2}(d-2)(-\frac{1}{12} - \frac{1}{24}) = -\frac{1}{16}(d-2)$, where again we have used ζ-function regularization (3.44). In the R sector of the closed string the sum in the fermionic sector is over integer modes and it cancels the contribution from the bosons. Therefore one simply gets $a_R = \bar{a}_R = 0$.

For the open string, in addition to the distinction between the NS and R sector, we also have to distinguish between various boundary conditions. For NN and DD boundary conditions one obtains the same results as for the closed string except that d is replaced by the number $d_{\mathrm{NN/DD}}$ of NN plus DD boundary conditions. In the DN and ND sectors, the bosons are half-integer moded. Following our discussion in Sect. 7.3, the fermions in the R sector have the same moding as the bosons while in the NS sector they have the opposite (integer vs. half-integer) moding. We find $a_{\mathrm{NS}} = -\frac{1}{16}(d_{\mathrm{NN/DD}} - 2) + \frac{m}{16} = -\frac{1}{16}(d-2) + \frac{m}{8}$ and $a_R = 0$, where m is the number of directions with mixed boundary conditions. Note that in the R-sector the contributions from bosons and fermions always cancel.

8.3 Spectrum of the Fermionic String, GSO Projection

Let us now look at the spectrum of the fermionic string, where we have to distinguish between NS and R sectors. We first discuss the open fermionic string spectrum with NN boundary conditions. The spectrum of the closed string then follows easily as the tensor product of two open string spectra, one for the left and one for the right movers, subject to level matching.

8.3 Spectrum of the Fermionic String, GSO Projection

1. NS-sector: The ground state is the oscillator vacuum $|0\rangle$ with $\alpha' m^2 = a$. The first excited state is $b^i_{-1/2}|0\rangle$ with $\alpha' m^2 = \frac{1}{2} + a$. This is a vector of $SO(d-2)$ and, following the argument of Chap. 3, must be massless, leading to a value $a = -\frac{1}{2}$ for the normal ordering constant. Using $a = -\frac{1}{16}(d-2)$ we derive $d = 10$ for the critical dimension of the fermionic string. At the next excitation level we have the states $\alpha^i_{-1}|0\rangle$ and $b^i_{-1/2}b^j_{-1/2}|0\rangle$ with $\alpha' m^2 = \frac{1}{2}$, comprising $8 + 28$ bosonic states. It can again be shown that these and all other massive light-cone states, which are tensors of $SO(8)$, combine uniquely to tensors of $SO(9)$, the little group for massive states in ten dimensions.

2. R-sector: We already know that the R ground state is a spinor of $SO(9,1)$. A Dirac spinor in ten space-time dimensions has 2^5 independent complex or 64 real components. On shell this reduces to 32 components since the Dirac equation $\gamma^\mu \partial_\mu \psi = 0$ relates half of the components to the other half. We can now still impose a Weyl or a Majorana condition, each of which reduces the number of independent components further by a factor of two. In ten dimensional space-time it is even possible to impose both simultaneously[5] leaving 8 independent on-shell components. They can also be viewed as the components of a Majorana-Weyl spinor of $SO(8)$. It is easy to see that the Ramond ground state is indeed massless. In light-cone gauge, using the description given in (8.14), one only has the raising and lowering zero modes $b_i^\pm$ with $i = 1, \ldots, 4$. Therefore, the degenerate R-ground can be described by the 16 states $|s_1, s_2, s_3, s_4\rangle$ with $s_i = \pm\frac{1}{2}$. We can now choose the ground state to have either one of two possible chiralities, which we will denote by $|a\rangle$ and $|\dot{a}\rangle$ respectively. The ground state $|a\rangle$ contains all states with $\sum_i s_i \in 2\mathbb{Z}$ and $|\dot{a}\rangle$ all states with $\sum_i s_i \in 2\mathbb{Z} + 1$. The first excitation level consists of states $\alpha^i_{-1}|a\rangle$ and $b^i_{-1}|a\rangle$ plus their chiral partners with $\alpha' m^2 = 1$. Again, for $d = 10$, all the massive light-cone states can be uniquely assembled into representations of $SO(9)$. In Table 8.1 this is demonstrated for the first few mass levels.

It can be shown that the fermionic string theory with all the states in both the NS and R sectors is inconsistent. This can be reconciled by making a truncation of the spectrum, called the GSO (Gliozzi-Scherk-Olive) projection, which renders the spectrum tachyon free and space-time supersymmetric. We will prove both statements, the necessity of the truncation and the space-time supersymmetry of the resulting spectrum, in Chap. 9 where the first assertion follows from the requirement of modular invariance and the second from the vanishing of the one-loop partition function. Another argument would be the locality of the operator algebra generated by the vertex operators, cf. Chap. 15.

---

[5]The general statement is that we can impose Majorana and Weyl conditions simultaneously on spinors of $SO(p,q)$ if and only if $p - q = 0 \mod 8$. For Minkowski space-times ($q = 1$) this is the case for $d = 2 + 8n$ and for Euclidean spaces ($q = 0$) for $d = 8n$. More details can be found in the appendix of this chapter.

Table 8.1 Open fermionic string spectrum

α'(mass)^2	States and their $SO(8)$ representation contents	Little group	$(-1)^F$	Representation contents with respect to the little group
\multicolumn{5}{c}{NS-sector}				
$-\frac{1}{2}$	$\|0\rangle$ (1)	$SO(9)$	-1	(1)
0	$b^i_{-1/2}\|0\rangle$ $(8)_v$	$SO(8)$	$+1$	$(8)_v$
$+\frac{1}{2}$	$\alpha^i_{-1}\|0\rangle \quad b^i_{-1/2}b^j_{-1/2}\|0\rangle$ $(8)_v \qquad\qquad (28)$	$SO(9)$	-1	(36)
$+1$	$b^i_{-1/2}b^j_{-1/2}b^k_{-1/2}\|0\rangle$ $(56)_v$ $\alpha^i_{-1}b^j_{-1/2}\|0\rangle \qquad b^i_{-3/2}\|0\rangle$ $(1)+(28)+(35)_v \qquad (8)_v$	$SO(9)$	$+1$	$(84)+(44)$
\multicolumn{5}{c}{R-sector}				
0	$\|a\rangle$ $(8)_s$ $\|\dot a\rangle$ $(8)_c$	$SO(8)$	$+1$ -1	$(8)_s$ $(8)_c$
$+1$	$\alpha^i_{-1}\|a\rangle \qquad b^i_{-1}\|\dot a\rangle$ $(8)_c+(56)_c \quad (8)_s+(56)_s$ $\alpha^i_{-1}\|\dot a\rangle \qquad b^i_{-1}\|a\rangle$ $(8)_s+(56)_s \quad (8)_c+(56)_c$	$SO(9)$	$+1$ -1	(128) (128)

Here we will turn the argument around and motivate the GSO projection by requiring a space-time supersymmetric spectrum. By inspection of Table 8.1 we see that at the massless level this can be achieved by projecting out one of the two possible chiralities of the R ground state. This leaves us with the on-shell degrees of freedom of $\mathcal{N}=1$, $d=10$ Super-Yang-Mills theory: a massless spinor and a massless vector. Obviously, we also have to get rid of the tachyon. Let us define a quantum number which is the eigenvalue of the operator $(-1)^F$ where F is the world-sheet fermion number. If we assign the NS vacuum $(-1)^F|0\rangle = -|0\rangle$, we can write in the NS sector $F = \sum_{r>0} b^i_{-r} b^i_r - 1$. If we then require that all states satisfy $(-1)^F = 1$, we remove all states with half-integer $\alpha' m^2$ (for which there are

8.3 Spectrum of the Fermionic String, GSO Projection

no space-time fermions). In particular the tachyon disappears. A general state in the NS sector, $\alpha^{i_1}_{-n_1} \cdots \alpha^{i_N}_{-n_N} b^{j_1}_{-r_1} \cdots b^{j_M}_{-r_M}|0\rangle$ has $(-1)^F = (-1)^{M+1}$ and all states with M even are projected out.

In the R sector the equivalent of $(-1)^{F_L}$ is a generalized chirality operator

$$(-1)^F = 16\, b_0^2 \cdots b_0^9 \, (-1)^{\sum_{n>0} b^i_{-n} b^i_n}, \qquad (8.34)$$

where $\gamma = 16\, b_0^2 \cdots b_0^9$ is the chirality operator in the eight transverse dimensions and $\sum_{n>0} b^i_{-n} b^i_n$ the world-sheet fermion number operator. It is easy to see that $\{(-1)^F, \psi^\mu\} = 0$ and the eigenvalues of the R ground states are ± 1, depending on their chirality, if we define $(-1)^F|a\rangle = 16 \prod_{i=2}^9 b_0^i|a\rangle = +1|a\rangle$ and $(-1)^F|\dot a\rangle = -1|\dot a\rangle$. Then a general state in the R-sector $\alpha^{i_1}_{-n_1} \cdots \alpha^{i_N}_{-n_N} b^{j_1}_{-m_1} \cdots b^{j_M}_{-m_M}|a\rangle$ has $(-1)^F = (-1)^M (-1)^{\sum_i \delta_{m_i,0}}$ and the analogous state built on the $|\bar a\rangle$ ground state has $(-1)^F = -(-1)^M (-1)^{\sum_i \delta_{m_i,0}}$. The GSO projection then amounts to demanding that all states have either chirality $(-1)^F = 1$ or $(-1)^F = -1$. We see from Table 8.1 that making the GSO projection we arrive at a supersymmetric spectrum (at least up to the level displayed there).

Of course, the consistency of the truncation requires that in the interacting theory we do not produce any of the projected-out states. This will follow from demanding locality of the operator product algebra of the vertex operators for all allowed states. We will discuss this further in Chap. 15.

To obtain the closed string spectrum, we take the tensor product of two open string spectra, one for the left- and one for the right-movers, obeying the constraint Eq. (8.18). We have to distinguish between four sectors, two of which, namely (NS,NS) and (R,R), lead to space-time bosons and the two sectors (NS,R) and (R,NS) to space-time fermions. An additional complication arises because we can choose between two possible chiralities for the left and right R ground state. Since in each sector we have to satisfy the constraint $L_0 - \bar L_0 = 0$, or, equivalently $m_R^2 = m_L^2$, the closed string states are tensor products of open string states at the same mass level. The possible states up to the massless level are shown in Table 8.2. There are too many states at the massive level to display, but it is straightforward to work out the continuation of the table. Again, we have to make the GSO projection. One way to perform it is for the right- and left-movers separately. For the NS states we require $(-1)^F = +1$ and $(-1)^{\bar F} = +1$ and for the R-sector states $(-1)^F = +1$ or $(-1)^F = -1$ and likewise for $(-1)^{\bar F}$. This leads to two inequivalent possibilities, $(-1)^F = (-1)^{\bar F}$ and $(-1)^F = -(-1)^{\bar F}$. For instance, the theory with $(-1)^F = (-1)^{\bar F} = +1$ has no tachyon and the following massless states:

(IIB)
Bos: $[(1) + (28) + (35)_v]_{(NS,NS)} + [(1) + (28) + (35)_s]_{(R,R)}$,

Fermi: $[(8)_c + (56)_c]_{(NS,R)} + [(8)_c + (56)_c]_{(R,NS)}$, $\qquad (8.35)$

i.e. we find a total of 128 bosonic and 128 fermionic states, indicating a supersymmetric spectrum.

Table 8.2 Closed fermionic string spectrum

$\alpha'(\text{mass})^2$	States and their $SO(8)$ representation contents	Little group	$(-1)^{\overline{F}}$	$(-1)^F$	Representation contents with respect to the little group
\multicolumn{6}{c}{(NS,NS)-sector}
-2	$\begin{array}{c}\|0\rangle_L \times \|0\rangle_R \\ (1) \qquad (1)\end{array}$	$SO(9)$	-1	-1	(1)
0	$\begin{array}{c}\overline{b}^{\,i}_{-1/2}\|0\rangle_L \times b^{\,j}_{-1/2}\|0\rangle_R \\ (8)_v \qquad\qquad (1)\end{array}$	$SO(8)$	$+1$	$+1$	$(1) + (28) + (35)_v$
\multicolumn{6}{c}{(R,R)-sector}
0	$\begin{array}{c}\|a\rangle_L \times \|b\rangle_R \\ (8)_s \qquad (8)_s\end{array}$	$SO(8)$	$+1$	$+1$	$(1) + (28) + (35)_s$
	$\begin{array}{c}\|\dot{a}\rangle_L \times \|\dot{b}\rangle_R \\ (8)_c \qquad (8)_c\end{array}$		-1	-1	$(1) + (28) + (35)_c$
	$\begin{array}{c}\|\dot{a}\rangle_L \times \|b\rangle_R \\ (8)_c \qquad (8)_s\end{array}$		-1	$+1$	$(8)_v + (56)_v$
	$\begin{array}{c}\|a\rangle_L \times \|\dot{b}\rangle_R \\ (8)_s \qquad (8)_c\end{array}$		$+1$	-1	$(8)_v + (56)_v$
\multicolumn{6}{c}{(R,NS)-sector}
0	$\begin{array}{c}\|a\rangle_L \times b^{\,i}_{-1/2}\|0\rangle_R \\ (8)_s \qquad\qquad (8)_v\end{array}$	$SO(8)$	$+1$	$+1$	$(8)_c + (56)_c$
	$\begin{array}{c}\|\dot{a}\rangle_L \times b^{\,i}_{-1/2}\|0\rangle_R \\ (8)_c \qquad\qquad (8)_v\end{array}$		-1	$+1$	$(8)_s + (56)_s$
\multicolumn{6}{c}{(NS,R)-sector}
0	$\begin{array}{c}\overline{b}^{\,i}_{-1/2}\|0\rangle_L \times \|a\rangle_R \\ (8)_v \qquad\qquad (8)_s\end{array}$	$SO(8)$	$+1$	$+1$	$(8)_c + (56)_c$
	$\begin{array}{c}\overline{b}^{\,i}_{-1/2}\|0\rangle_L \times \|\dot{a}\rangle_R \\ (8)_v \qquad\qquad (8)_c\end{array}$		$+1$	-1	$(8)_s + (56)_s$

The projection as given above defines the type IIB superstring theory whose massless spectrum is that of type IIB supergravity (SUGRA) in ten dimensions. The $(35)_v$ represents the on-shell degrees of freedom of a graviton, the two (28)'s represent two antisymmetric tensor fields and the $(35)_s$ a rank four self-dual antisymmetric tensor. In addition there are two real scalars. The fermionic degrees of freedom are those of two on-shell gravitinos (the $(56)_c$) with spin 3/2 and of two spin 1/2 fermions, called dilatinos. The number of components for the massless fields follows from (7.5). The presence of two gravitinos means that this theory has $\mathcal{N} = 2$ supersymmetry. Since both gravitinos are of the same handedness, it is a chiral theory. It is worth mentioning that the handedness of the dilatinos is opposite to that of the gravitini. Together a gravitino and a dilatino form a reducible vector spinor ψ_μ whose traceless part $\gamma^\mu \psi_\mu = 0$ is the gravitino and whose trace part is the dilatino.

The choice $(-1)^F = -(-1)^{\overline{F}} = 1$ leads to the following massless spectrum:

(IIA)
$$\text{Bos:} \quad [(1) + (28) + (35)_v]_{(\text{NS,NS})} + [(8)_v + (56)_v]_{(\text{R,R})},$$
$$\text{Fermi:} \quad [(8)_c + (56)_c]_{(\text{NS,R})} + [(8)_s + (56)_s]_{(\text{R,NS})}, \quad (8.36)$$

representing the degrees of freedom of a graviton $(35)_v$, an antisymmetric rank three tensor $(56)_v$, an antisymmetric rank two tensor (28), one vector $((8)_v)$ and one real scalar, the dilaton. The fermions can be interpreted as two gravitinos of spin $3/2$ and two dilatinos of spin $1/2$, one of each handedness. Again, we have $\mathcal{N} = 2$ supersymmetry but non-chiral. This theory is called the type IIA superstring and its massless sector type IIA supergravity. We already note here and will discuss it in more detail in Chap. 16, that the massless spectrum of the type IIA theory can be obtained by dimensional reduction of eleven-dimensional supergravity down to ten dimensions. The fermionic degrees of freedom combine into those of one eleven-dimensional Majorana spinor with $8 \cdot 2^4 = 128$ degrees of freedom, cf. (7.5) on page 177. As for the bosons, $[(1) + (8) + (35)]_{SO(8)} = 44_{SO(9)}$ which are the on-shell degrees of freedom of an eleven-dimensional graviton and $[(28) + (56)]_{SO(8)} = 84_{SO(9)}$ which is a three-form.

Both supergravity theories, type IIA and type IIB, have $\mathcal{N} = 2$ supersymmetry in ten-dimensions. This means that there are two fermionic generators Q^I, $I = 1, 2$ which are Majorana-Weyl spinors of $SO(1, 9)$. Together they have 32 real components. Often this is expressed by saying that the type II supergravities have 32 supercharges. When one compactifies on a torus to four dimensions, the resulting four-dimensional theory has $\mathcal{N} = 8$ supersymmetry with eight fermionic generators Q^I, $I = 1, \ldots, 8$ which are Majorana spinors of $SO(1, 3)$. Again, the number of supercharges is 32. In other words, denoting the amount of supersymmetry by the number of supercharges is independent of space-time dimension and is invariant under compactification on a torus. Compactification on manifolds with curvature will, on the other hand, break some or all supersymmetries and reduces the number of supercharges. For instance, in Chap. 14 we consider compactification on Calabi-Yau manifolds which preserves one quarter, i.e. eight, supercharges, that is $\mathcal{N} = 2$ supersymmetry in $D = 4$. The type I supergravity theory which we consider next has 16 supercharges.

Let us finally look at the unoriented closed string. It is clear that its states are a subset of those of the left-right symmetric type IIB theory, namely those which are symmetric under world-sheet parity Ω which interchanges left- and right-movers

$$\Omega X^\mu(\sigma, \tau) \Omega^{-1} = X^\mu(\ell - \sigma, \tau), \qquad \Omega \psi^\mu_\pm(\sigma, \tau) \Omega^{-1} = \psi^\mu_\mp(\ell - \sigma, \tau). \quad (8.37)$$

In terms of the oscillator modes this becomes Eq. (3.59) for the bosons and for the fermions one obtains

$$\Omega\, b_r^\mu\, \Omega^{-1} = e^{-2\pi i r}\, \overline{b}_r^\mu, \qquad \Omega\, \overline{b}_r^\mu\, \Omega^{-1} = e^{-2\pi i r}\, b_r^\mu \qquad (8.38)$$

with $r \in \mathbb{Z}$ ($r \in \mathbb{Z} + \tfrac{1}{2}$) for the R (NS) sector. From this it is clear that Ω exchanges the (NS,R) sector with the (R,NS) sector. We also have to define how Ω acts on the closed string ground sates. Taking into account that the Ramond ground states are space-time fermions with odd Grassmann parity, one defines

$$\begin{aligned}
(\mathrm{NS,NS}): &\quad \Omega\big(|0\rangle_L \times |0\rangle_R\big) = |0\rangle_L \times |0\rangle_R, \\
(\mathrm{R,R}): &\quad \Omega\big(|a\rangle_L \times |b\rangle_R\big) = -|b\rangle_L \times |a\rangle_R, \\
(\mathrm{NS,R}): &\quad \Omega\big(|0\rangle_L \times |a\rangle_R\big) = |a\rangle_L \times |0\rangle_R, \\
(\mathrm{R,NS}): &\quad \Omega\big(|a\rangle_L \times |0\rangle_R\big) = |0\rangle_L \times |a\rangle_R.
\end{aligned} \qquad (8.39)$$

Thus, among the massless (NS,NS) sector states the $(1) + (35)$ survive, and among the (R,R) states the (28). Among the fermions a diagonal combination of the $(8)_c + (56)_c$ survives so that the massless closed string spectrum of the non-orientable theory is

$$\begin{aligned}
\text{Bosons:} &\quad [(1) + (35)_v]_{(\mathrm{NS,NS})} + [(28)]_{(\mathrm{R,R})}, \\
\text{Fermions:} &\quad [(8)_c + (56)_c]_{(\mathrm{NS,R})+(\mathrm{R,NS})}.
\end{aligned} \qquad (8.40)$$

These are the states of $\mathcal{N} = 1$ supergravity in ten dimensions which is the massless closed string sector of the type I superstring theory. Both the field theory and the string theory obtained from the type IIB theory by retaining only states invariant under reversal of the world-sheet orientation are inconsistent. We will come back to this in Chap. 9 when we discuss the full type I superstring theory. We will see that consistency requires the addition of so-called twisted sectors which, in this case, are open strings which give rise to massless gauge bosons.

8.4 Path Integral Quantization

Let us now turn to the path integral quantization of Polyakov. In Chap. 3 we have seen that to obtain the ghost action we had to re-express the Faddeev-Popov determinant as an integral over anti-commuting ghost fields. The ghost action was than simply $S_{\mathrm{gh}} \sim \int b^{\alpha\beta} \frac{\delta h_{\alpha\beta}}{\delta \xi^\gamma} c^\gamma$ where $b_{\alpha\beta}$ was symmetric and traceless, the tracelessness following from the Weyl-invariance of the theory (at least in the critical dimension). In other words, the ghost Lagrangian followed immediately from the traceless symmetric variation of the metric by replacing the gauge parameter ξ^α by a ghost field c^α of opposite statistics and introducing the anti-ghost $b_{\alpha\beta}$ with the

8.4 Path Integral Quantization

same tensor structure as the gauge field $h_{\alpha\beta}$, but opposite statistics. We will now apply the same procedure to the fermionic string.

The transformations of the zwei-bein and the gravitino under reparametrizations and supersymmetry transformations were discussed in the previous chapter. The starting point is (7.15). We make compensating Weyl, Lorentz and super-Weyl transformations to eliminate the trace and antisymmetric part in $e_{\alpha a}\delta e_\beta{}^a$ and to achieve $\rho^\alpha \delta \chi_\alpha = 0$. The parameters of these transformations are

$$\Lambda = \frac{1}{2}\nabla^\alpha \xi_\alpha,$$

$$l = \frac{1}{2e}\epsilon^{\alpha\beta}\nabla_\alpha \xi_\beta,$$

$$\eta = -\rho^\alpha \nabla_\alpha \epsilon + \frac{1}{4}\rho^\alpha \chi^\beta (P\xi)_{\alpha\beta}. \tag{8.41}$$

In the critical dimension these are good symmetries and they can be used to make χ ρ-traceless. This is what we have done. In this case the torsion piece in the spin-connection vanishes and all covariant derivatives will be without torsion. We obtain

$$e_{\alpha a}\delta e_\beta^a = -\frac{1}{2}(P\xi)_{\alpha\beta} - \frac{i}{2}(\overline{\chi}_\beta \rho_\alpha \epsilon),$$

$$\delta \chi_\alpha = 2(\Pi\epsilon)_\alpha - \frac{3}{4}\rho^\beta \rho^\gamma \chi_\alpha(\nabla_\beta \xi_\gamma) + \frac{1}{2}\chi_\alpha(\nabla \cdot \xi) - (\nabla_\beta \chi_\alpha)\xi^\beta. \tag{8.42}$$

The ghost action is then

$$S_{\text{ghost}} = -\frac{i}{2\pi}\int d^2\sigma \sqrt{-h}\left\{ b^{\alpha\beta}\left[e_{\alpha a}\frac{\delta e_\beta^a}{\delta \xi^\gamma}c^\gamma + \frac{i}{2}e_{\alpha a}\frac{\delta e_\beta^a}{\delta \epsilon}\gamma \right] \right.$$

$$\left. + i\overline{\beta}^\alpha\left[\frac{\delta \chi_\alpha}{\delta \xi^\beta}c^\beta + \frac{i}{2}\frac{\delta \chi_\alpha}{\delta \epsilon}\gamma\right]\right\}$$

$$= -\frac{i}{2\pi}\int d^2\sigma \sqrt{-h}\left\{ b^{\alpha\beta}\nabla_\alpha c_\beta + \overline{\beta}^\alpha \nabla_\alpha \gamma \right.$$

$$\left. - i\overline{\chi}_\alpha\left[c^\beta \nabla_\beta \beta^\alpha + \frac{3}{2}\beta^\beta \nabla^\alpha c_\beta - \frac{i}{4}b^{\alpha\beta}\rho_\beta \gamma \right]\right\}. \tag{8.43}$$

Here, β_α and γ are commuting spin 3/2 and spin 1/2 ghosts with $\rho \cdot \beta = 0$ and $b_{\alpha\beta}$ and c^γ are as in Chap. 5. The factors of i have been included to make γ real and β imaginary. Note that the term in brackets which multiplies $\overline{\chi}_\alpha$ will be absent in superconformal gauge. We could also have introduced ghosts for Weyl, Lorentz and Super-Weyl transformations, but they would have been integrated out, giving constraints on $b_{\alpha\beta}$ and β_α, which imply them to be symmetric-traceless and

ρ-traceless, respectively. Also, we would find that only the helicity $\pm 3/2$ components of the gravitino couple, reflecting super-Weyl invariance.

From the ghost action we can now derive the ghost energy-momentum tensor and the ghost supercurrent. Using the definitions (7.28) and (7.29), the equations of motion and a gravitino in superconformal gauge, we find

$$T_{\alpha\beta} = -i\left\{b_{\alpha\gamma}\nabla_\beta c^\gamma + b_{\beta\gamma}\nabla_\beta c^\gamma - c^\gamma \nabla_\gamma b_{\alpha\beta}\right.$$
$$\left. + \frac{3}{4}(\overline{\beta}_\alpha \nabla_\beta + \overline{\beta}_\beta \nabla_\alpha)\gamma + \frac{1}{4}(\nabla_\alpha \overline{\beta}_\beta + \nabla_\beta \overline{\beta}_\alpha)\gamma\right\},$$

$$T_{F\gamma} = i\left\{-\frac{3}{2}\beta^\alpha \nabla_\gamma c_\alpha - (\nabla_\alpha \beta_\gamma)c^\alpha + \frac{i}{4}b_{\gamma\delta}\rho^\delta \gamma\right\}. \qquad (8.44)$$

We could now proceed as in Chap. 3, expand the ghost fields in modes, find their contribution to the super Virasoro operators and show that the conformal anomaly vanishes in the critical dimension. We will however not do this here but rather postpone the discussion until Chap. 12, where we will be able to derive the same results in a much easier way using superconformal field theory. There we will also discuss the supersymmetry of the ghost system.

8.5 Appendix: Dirac Matrices and Spinors in d Dimensions

Clifford Algebra and Dirac Matrices

The Clifford algebra in d dimensions is

$$\{\gamma^\mu, \gamma^\nu\} = 2\eta^{\mu\nu} \quad \text{(Minkowskian)}, \qquad \{\gamma^\mu, \gamma^\nu\} = 2\delta^{\mu\nu} \quad \text{(Euclidean)}, \qquad (8.45)$$

where $\mu, \nu = 0, \ldots, d-1$.[6] The extension to arbitrary signatures is possible. We will restrict the discussion to Minkowskian and Euclidean signatures. All indices are flat tangent space indices. One can convert them to curved indices with the help of a d-bein. Often, in particular when we work in the critical dimension $d = 10$ of the superstring, we also use the notation Γ^μ for the Dirac matrices.

Define the anti-symmetrized products

$$\gamma^{\mu_1 \ldots \mu_p} = \gamma^{[\mu_1} \cdots \gamma^{\mu_p]} = \frac{1}{p!}\left(\gamma^{\mu_1}\gamma^{\mu_2}\cdots\gamma^{\mu_p} \pm \text{permutations}\right). \qquad (8.46)$$

---

[6] In Euclidean signature the labelling is usually $\gamma^1, \ldots, \gamma^d$, but for unity of notation we use $\mu = 0, \ldots, d-1$ in both cases.

8.5 Appendix: Dirac Matrices and Spinors in d Dimensions

The 2^{d+1} matrices $\pm\mathbb{1}, \pm\gamma^\mu, \pm\gamma^{\mu\nu}, \ldots$ generate a finite group. Every representation of a finite group can be made unitary by a similarity transformation. If we assume that all γ^μ are unitary, then it follows from the Clifford algebra (8.45) that all γ's are Hermitian, except γ^0, which is anti-Hermitian in Minkowski signature; i.e.

$$(\gamma^\mu)^\dagger = \gamma^\mu \quad \text{(Euclidean)}, \qquad (\gamma^\mu)^\dagger = \gamma^0 \gamma^\mu \gamma^0 \quad \text{(Minkowskian)}. \tag{8.47}$$

The Hermiticity property is preserved under unitary transformations.

Consider the product $\gamma_{d+1} \propto \gamma^0 \gamma^1 \cdots \gamma^{d-1}$. For odd d it commutes with all γ^μ and therefore, by Schur's lemma, it is a multiple of $\mathbb{1}$. For even $d = 2n$ it anticommutes with all γ^μ and one defines the chirality operator

$$\gamma_{d+1} = \alpha \gamma^0 \gamma^1 \cdots \gamma^{d-1} \quad \text{with} \quad \alpha = \begin{cases} i^{n-1} & \text{Minkowskian} \\ i^n & \text{Euclidean} \end{cases}. \tag{8.48}$$

The phase α has been chosen such that $\gamma_{d+1}^2 = \mathbb{1}$. Furthermore $\gamma_{d+1}^\dagger = \gamma_{d+1}$.

If γ^μ is a Hermitian representation of (8.45), so is $-\gamma^\mu$ and $\pm(\gamma^\mu)^T$. But they are not all inequivalent. In fact, with the help of the representation theory of finite groups one can prove (independent of the signature) the fundamental theorem of Dirac matrices: (1) for even d all irreducible representations of the Clifford algebra are equivalent and are $n \times n$ (Dirac) matrices with $n = 2^{d/2}$; (2) for odd d there are two inequivalent irreducible representations. They are in terms of $n \times n$ (Dirac) matrices with $n = 2^{(d-1)/2}$. For $d = 2n$ and $d = 2n+1$ the dimensions of the Dirac matrices agree. The two inequivalent representations in $d = 2n+1$ are obtained by taking the Dirac matrices in $d = 2n$ and adding $+\gamma_{2n+1}$ or $-\gamma_{2n+1}$.

It follows from the algebra that in Euclidean signature all matrices γ^μ are traceless and that they square to $+\mathbb{1}$. Therefore, their eigenvalues are ± 1 in equal number; in Minkowski signature γ^0 squares to $-\mathbb{1}$ and its eigenvalues are $\pm i$.

For any two equivalent representations γ^μ and γ'^μ, there exists an intertwiner A such that $A\gamma^\mu A^{-1} = \gamma'^\mu$. If such an intertwiner exists, it is essentially unique. Indeed, assume A and $\tilde{A}$ are two intertwiners. Then $\tilde{A}^{-1} A$ must commute with all γ^μ and must therefore be proportional to $\mathbb{1}$.

The fact that for d odd not all representations of (8.45) are equivalent can be demonstrated as follows. From $A\gamma^\mu A^{-1} = -\gamma^\mu$ we derive $A\gamma_{d+1}A^{-1} = (-1)^d \gamma_{d+1}$ which is in contradiction with the previous result that $\gamma_{d+1} \propto \mathbb{1}$ for d odd. For d even, $A = \gamma_{d+1}$.

One defines the 'charge conjugation' matrices $C_\pm$ as the intertwiners between γ^μ and $\mp(\gamma^\mu)^T$:

$$(\gamma^\mu)^T = \mp C_\pm \gamma^\mu C_\pm^{-1}. \tag{8.49}$$

The existence of $C_\pm$ in even dimensions and of one of them in odd dimensions (cf. below) is guaranteed by the fundamental theorem of Dirac matrices. Iterating (8.49) we find that $C^{-1}C^T$ commutes with all γ^μ and must therefore be proportional to the identity matrix, i.e. $C^T = \eta C$. Iterating this leads to $\eta^2 = 1$ or

$$C^T = \pm C. \tag{8.50}$$

Using (8.49) and its complex conjugate together with (8.50) one shows that $C^\dagger C = \alpha \mathbb{1}$. Since the l.h.s. is a positive matrix, $\alpha > 0$ and we can rescale C such that it is unitary.

Before we proceed we give an explicit representation for the γ-matrices and the charge conjugation matrices.[7] Let σ_i be the Pauli matrices. The set of $d = 2n$ n-fold direct products

$$\begin{aligned}
\gamma^0 &= \sigma_1 \otimes \mathbb{1} \otimes \mathbb{1} \otimes \cdots \otimes \mathbb{1} \\
\gamma^1 &= \sigma_2 \otimes \mathbb{1} \otimes \mathbb{1} \otimes \cdots \otimes \mathbb{1} \\
\gamma^2 &= \sigma_3 \otimes \sigma_1 \otimes \mathbb{1} \otimes \cdots \otimes \mathbb{1} \\
\gamma^3 &= \sigma_3 \otimes \sigma_2 \otimes \mathbb{1} \otimes \cdots \otimes \mathbb{1} \\
\gamma^4 &= \sigma_3 \otimes \sigma_3 \otimes \sigma_1 \otimes \cdots \otimes \mathbb{1} \\
&\vdots \quad \vdots \quad \vdots \\
\gamma^{2n-1} &= \sigma_3 \otimes \sigma_3 \otimes \sigma_3 \otimes \cdots \otimes \sigma_2
\end{aligned} \tag{8.51}$$

satisfy the Euclidean signature Clifford algebra. For Minkowski signature one replaces γ^0 by $-i\sigma_1 \otimes \mathbb{1} \otimes \cdots$. In odd dimensions ($d = 2n+1$) one adds

$$\gamma^{2n} = \sigma_3 \otimes \sigma_3 \otimes \sigma_3 \otimes \cdots \propto \gamma^0 \gamma^1 \cdots \gamma^{2n-1}. \tag{8.52}$$

In $d = 2n$, $\gamma^{2n} = \gamma_{d+1}$ is the chirality matrix with eigenvalues ± 1.

In even dimensions we have the following two explicit representations of $C_\pm$.

$$\begin{aligned}
C_+ &= \sigma_2 \otimes \sigma_1 \otimes \sigma_2 \otimes \sigma_1 \otimes \cdots \\
C_- &= \sigma_1 \otimes \sigma_2 \otimes \sigma_1 \otimes \sigma_2 \otimes \cdots \propto C_+ \gamma_{2n+1}
\end{aligned} \tag{8.53}$$

which satisfy $C_\pm = C_\pm^\dagger = C_\pm^{-1}$ and

$$C_\pm^T = (-1)^{\frac{1}{2}n(n\pm 1)} C_\pm. \tag{8.54}$$

One might wonder how much of these and other properties derived below depend on a particular representation of the Dirac matrices. If we make a unitary basis

---

[7] In Chap. 7 we used a different representation for $d = 2$.

8.5 Appendix: Dirac Matrices and Spinors in d Dimensions

transformation of the Dirac matrices, which preserves their Hermiticity properties, and define $\gamma'^{\mu} = U^{-1}\gamma^{\mu}U$ then (8.49) is preserved with C replaced by $C' = U^{\mathrm{T}}CU$. C' has the same symmetry as C, i.e. (8.54) and also unitarity of $C_{\pm}$ is representation independent, but the property $C^2 = \mathbb{1}$, which is satisfied by (8.53), is not.

If we denote by $\gamma^{[p]}$ an antisymmetrized product of p γ-matrices, we find, with the help of (8.49) and (8.54), that it enjoys the following symmetry properties:

$$\left(C_{\pm}\gamma^{[p]}\right)^{\mathrm{T}} = (-1)^{\frac{1}{2}n(n\pm 1)+\frac{1}{2}p(p\pm 1)} C_{\pm}\gamma^{[p]}. \tag{8.55}$$

In odd dimensions $\gamma^{[2n]}$ is one of the Dirac matrices and we have to obtain the same sign in (8.55) for $p = 1$ and for $p = 2n$. This means that

$$d = 2n+1: \qquad C = \begin{cases} C_{+} & n \text{ odd} \\ C_{-} & n \text{ even.} \end{cases} \tag{8.56}$$

For the appropriate C Eq. (8.55) is then still valid for $d = 2n+1$. In all cases $(\gamma^{[2]})^{\mathrm{T}} = -C\gamma^{[2]}C^{-1}$.

Let us remark that $\pm(\gamma^{\mu})^{*}$ also satisfy the Clifford algebra and have the same Hermiticity properties as γ^{μ}. Therefore we can define intertwiners $B_{\pm}$ such that $(\gamma^{\mu})^{*} = \pm B_{\pm}\gamma^{\mu}B_{\pm}^{-1}$. Using the above results one easily shows that $B_{\pm} = C_{\pm}\gamma^{0}$.

Spinor Representations

We now discuss spinor representations of $SO(d)$ and $SO(1, d-1)$. The matrices

$$T^{\mu\nu} = -\frac{i}{2}\gamma^{\mu\nu} \tag{8.57}$$

satisfy the algebra

$$[T^{\mu\nu}, T^{\rho\sigma}] = i\,(\eta^{\mu\rho}T^{\nu\sigma} + \eta^{\nu\sigma}T^{\mu\rho} - \eta^{\mu\sigma}T^{\nu\rho} - \eta^{\nu\rho}T^{\mu\sigma}), \tag{8.58}$$

as one easily verifies using (8.45). They are generators of infinitesimal $SO(d)$ and $SO(1, d-1)$ transformations in the spinor representation. A spinor ψ transform under an infinitesimal transformation with parameters $\omega_{\mu\nu}$ as $\delta\psi = \frac{i}{2}\omega_{\mu\nu}T^{\mu\nu}\psi$. In Euclidean signature the generators are Hermitian. In Minkowski signature T^{0i} are anti-Hermitian and the others are Hermitian ($i = 1, \ldots, d-1$). In $d = 2n$ and $d = 2n+1$ dimensions the representation space is spanned by the 2^n dimensional Dirac spinors.

Given any irreducible representation $T^{\mu\nu}$ of the algebra (8.58), $-(T^{\mu\nu})^{*}$ is also a representation. If the two representations are equivalent, i.e. if there exists a matrix R such that

$$-(T^{\mu\nu})^* = R T^{\mu\nu} R^{-1}, \tag{8.59}$$

the representation is called real. Otherwise it is complex. One can show that R is essentially unique and using that $T^* = T^T$ for Hermitian and $T^* = -T^T$ for anti-Hermitian generators, one can show that $R^T = \pm R$. A more refined characterization is

$$R = +R^T \quad \text{(real)}, \qquad R = -R^T \quad \text{(pseudo-real)}. \tag{8.60}$$

Let us consider odd and even dimensions in turn.

- $d = 2n + 1$

  We easily show that

  $$(T^{\mu\nu})^* = -(CD) T^{\mu\nu} (CD)^{-1}. \tag{8.61}$$

  With $D = \mathbb{1}$ and $D = i\gamma^0$ for Euclidean and Lorentzian signature, respectively and $C = C_+$ for n odd and $C = C_-$ for n even. Examining the symmetry properties of CD we find

	$SO(2n+1)$	$SO(1,2n)$
Real	$n = 0, 3 \bmod 4$	$n = 0, 1 \bmod 4$
Pseudo-real	$n = 1, 2 \bmod 4$	$n = 2, 3 \bmod 4$

- $d = 2n$

  Since γ_{d+1} commutes with all the $T^{\mu\nu}$, they cannot generate an irreducible representation of the Lorentz algebra. We therefore project the generators to the irreducible subspaces of chiral spinors (more on them below) and define $T_\pm^{\mu\nu} = T^{\mu\nu} \frac{1}{2}(1 \pm \gamma_{d+1})$. They satisfy the Lorentz algebra and

  $$\text{(Euclidean)} \quad (T_\pm^{\mu\nu})^* = \begin{cases} -(C_\pm) T_\mp^{\mu\nu} (C_\pm)^{-1} & \text{for } n \text{ odd} \\ -(C_\pm) T_\pm^{\mu\nu} (C_\pm)^{-1} & \text{for } n \text{ even} \end{cases} \tag{8.62}$$

  and

  $$\text{(Minkowskian)} \quad (T_\pm^{\mu\nu})^* = \begin{cases} -(C_\pm \gamma^0) T_\mp^{\mu\nu} (C_\pm \gamma^0)^{-1} & \text{for } n \text{ odd} \\ -(C_\pm \gamma^0) T_\pm^{\mu\nu} (C_\pm \gamma^0)^{-1} & \text{for } n \text{ even.} \end{cases} \tag{8.63}$$

  Analyzing the symmetry of $(C_\pm D)$ we find

	$SO(2n)$	$SO(1, 2n-1)$
Real	$n = 0 \bmod 4$	$n = 1 \bmod 4$
Pseudo-real	$n = 2 \bmod 4$	$n = 3 \bmod 4$
Complex	$n = 1, 3 \bmod 4$	$n = 0, 2 \bmod 4$

8.5 Appendix: Dirac Matrices and Spinors in d Dimensions

For the complex representations the two chiralities are obviously related by complex conjugation.

Define (here we take d even)[8]

$$b^i = \frac{1}{2}(\gamma^{2i} + i\gamma^{2i+1}), \quad b_i = \frac{1}{2}(\gamma^{2i} - i\gamma^{2i+1}), \quad i = 0, \ldots, \frac{d}{2} - 1. \quad (8.64)$$

They satisfy $b_i = (b^i)^\dagger$ and

$$\{b^i, b_j\} = \delta^i_j \quad \{b_i, b_j\} = \{b^i, b^j\} = 0, \quad (8.65)$$

i.e. they satisfy the algebra of fermionic oscillators. In Minkowski signature we have to replace $b^0 = \frac{i}{2}(\gamma^0 + \gamma^1)$ and $b_0 = \frac{i}{2}(\gamma^0 - \gamma^1)$. The oscillator algebra (8.65) remains unchanged.

Define a highest weight state $|\Omega\rangle$ by the condition $b^i|\Omega\rangle = 0$. Altogether there are 2^n states in this representation which we can label as $|\pm\frac{1}{2}, \ldots, \pm\frac{1}{2}\rangle$ where the highest weight state is $|\Omega\rangle = |+\frac{1}{2}, \ldots, +\frac{1}{2}\rangle$. These 2^n states span the (reducible) spinor representation of $SO(2n)$. The components of the two irreducible representations of definite chirality are the states with an even (odd) number of $-\frac{1}{2}$ entries. They are obtained from $|\Omega\rangle$ by the action with an even (odd) number of creation operators.

In this notation, the $2n$-dimensional vector representation of $SO(2n)$ consists of the states $|\pm e_i\rangle$, $i = 1, \ldots, n$ where e_i is the unit vector with components $(e_i)_j = \delta_{ij}$. More generally, given a weight vector $\lambda = \lambda_i e_i$ of $SO(2n)$, the corresponding state is $|\lambda_1, \ldots, \lambda_n\rangle$. This will be discussed further in Chap. 11.

Let $(t_a)^j_i$ be the (traceless) generators in the fundamental representation **n** of $SU(n)$. Using (8.65) one shows that

$$t_a = b^i \, (t_a)^j_i \, b_j \quad (8.66)$$

satisfy the $SU(n)$ algebra. Because of

$$b^i b_j = \frac{1}{2}\{b^i, b_j\} + \frac{1}{2}[b^i, b_j]$$
$$= \frac{1}{2}\delta^i_j + \frac{i}{2}T^{2i,2j} + \frac{i}{2}T^{2i+1,2j+1} + \frac{1}{2}T^{2i,2j+1} - \frac{1}{2}T^{2i+1,2j}, \quad (8.67)$$

the t_a can be written as linear combinations of the $SO(2n)$ generators (8.57). They therefore generate a subalgebra of $SO(2n)$. It also follows from (8.65) that

$$[t_a, b^i] = b^j \, (t^a)^i_j \quad [t_a, b_i] = -b_j \, (t^a)^j_i, \quad (8.68)$$

---

[8] Another common notation is $b^i = \gamma^i$ and $b_i = \gamma^{\bar{i}}$.

i.e. the b^i and b_i transform in the **n** and $\bar{\mathbf{n}}$ of $SU(n)$, respectively. The states $b_{i_1} \cdots b_{i_k} |\Omega\rangle$ transform in the antisymmetric product of k $\bar{\mathbf{n}}$'s of $SU(n)$. In particular $|\Omega\rangle$ and $|\overline{\Omega}\rangle$ are singlets. Here $|\overline{\Omega}\rangle = |-\frac{1}{2}, \ldots, -\frac{1}{2}\rangle$ is the lowest weight state; it is annihilated by all b_i. From this one can work out the decomposition of the two spinor representations of $SO(2n)$ into irreducible $SU(n)$ representations.

$SU(n)$ is not a maximal subgroup of $SO(2n)$, but $SU(n) \times U(1)$ is. The $U(1)$ generator, which commutes with all t_a, can be straightforwardly constructed. It is essentially the number operator. A convenient normalization is

$$Q = 2\sum_{i=1}^{n} b_i b^i - n \qquad (8.69)$$

such that the two singlets have opposite $U(1)$ charge $\mp n$.

Dirac, Weyl, Majorana, Majorana-Weyl Spinors

Given a complex (Dirac) spinor ψ we can define its Dirac conjugate $\bar{\psi}_D$, or simply $\bar{\psi}$, whose components are linear combinations of the components of ψ^*, i.e. we define $\bar{\psi}_D = \psi^\dagger D$ where the matrix D is chosen such that $\bar{\psi}\psi$ transforms as a scalar under rotations. In Euclidean signature $D = 1$ while in Lorentzian signature $D = \gamma^0$.[9] Under Lorentz transformations it transforms as $\delta \bar{\psi}_D = -\frac{i}{2}\bar{\psi}_D T^{\mu\nu} \omega_{\mu\nu}$.

One defines the Majorana conjugate $\bar{\psi}_M$ of ψ to be that linear combination of the components of ψ which transform under Lorentz transformations in the same way as $\bar{\psi}_D$, i.e. $\bar{\psi}_M = \psi^T M$ where M has to satisfy $(T^{\mu\nu})^T = -M T^{\mu\nu} M^{-1}$. We can therefore identify $M = C$, i.e. $\bar{\psi}_M = \psi^T C$. If we further require that ψ_M and $\bar{\psi}$ satisfy the same Dirac equation we find for massless spinors the condition $C\gamma^\mu C^{-1} = \pm(\gamma^\mu)^T$ and for massive spinors the stronger condition $C\gamma^\mu C^{-1} = -(\gamma^\mu)^T$ which leaves only C_+.

A Majorana spinor is a spinor whose Majorana conjugate is proportional to its Dirac conjugate, i.e. $\bar{\psi}_M = \alpha \bar{\psi}_D$. This means that a Majorana spinor is essentially real. While a Dirac spinor has 2^n complex components a Majorana spinor has 2^n real components. Iterating the Majorana condition one finds the condition

$$1 = |\alpha|^2 D^*(C^*)^{-1} D C^{-1}. \qquad (8.70)$$

From (8.70), using the unitarity of $C_\pm$ and the explicit forms of D we derive $|\alpha| = 1$ and

$$C_\pm^T = C_\pm \quad \text{(Euclidean)}, \qquad C_\pm^T = \mp C_\pm \quad \text{(Minkowskian)} \qquad (8.71)$$

---

[9] Another common choice is $D = i\gamma^0$ to make $\bar{\psi}_D \psi$ Hermitian.

8.5 Appendix: Dirac Matrices and Spinors in d Dimensions

Table 8.3 Majorana spinors in various dimensions

d	2	3	4	5	6	7	8	9	10	11
Euclidean	M_				M_+	M_-	MW	M_-	M_-	
Minkowski	MW	M_+	M_+				M_-	M_-	MW	M_+

A subscript indicates if only C_+ or C_- is allowed. A Dirac spinor in $d = 2n$ and $d = 2n+1$ dimensions has 2^{n+1} real components. Majorana and Weyl conditions reduce them by a factor 2. The table continues d mod 8

as the conditions for the existence of Majorana spinors. Redefining the phase of ψ we can set $\alpha = 1$, i.e. Majorana spinors satisfy $\overline{\psi}_M = \overline{\psi}_D$.

For d odd, Dirac spinors furnish an irreducible representation of the Lorentz group. For d even it is reducible. Indeed, in even dimensions γ_{d+1} commutes with the Lorentz generators and by Schur's lemma its value on an irreducible representation is constant. The two irreducible representations are those with $\gamma_{d+1} = \pm 1$. They are the Weyl spinors of positive and negative chirality. The projection operators which project a Dirac spinor on its two irreducible components are $\gamma_\pm = \frac{1}{2}(1 \pm \gamma_{d+1})$. Note that $\gamma_\pm$ commute with the massless Dirac operator $i\gamma^\mu \partial_\mu$, i.e. chirality is a conserved quantum number of massless fermions. A Weyl spinor has 2^{n-1} complex components.

If one imposes in addition a Majorana condition one arrives at a Majorana-Weyl spinor with 2^{n-1} real components. Majorana-Weyl spinors do not exist in all even dimensions in which Majorana spinors exist. We have to satisfy the additional requirement that $\gamma_{d+1}\psi$ also satisfies the Majorana condition. This leads to the condition $D^{-1}\gamma_{d+1}D = C^{-1}(\gamma_{d+1})^T C$ where we have used that γ_{d+1} is Hermitian. If we now use that γ_{d+1} is the product of an even number of Dirac matrices which obey (8.49), this conditions leads to the conclusion that Majorana-Weyl spinors exist in those even dimensions $d = 2n$ where Majorana spinors exist and if $n = 0$ (mod 2) for Euclidean and $n = 1$ (mod 2) for Minkowski signature.

We have collected the results of this analysis about the existence of various types of spinors in Table 8.3. For both signatures it continues mod 8 to higher dimensions.

If Majorana spinors exist, we can always find a representation of the Dirac matrices, the Majorana representation, where $C = D$, i.e. $C = \mathbb{1}$ for Euclidean signature and $C = \gamma^0$ for Minkowski signature. To prove this we have to show that there exists a unitary matrix U such that $C'(D')^{-1} = U^T C D^{-1} U = \mathbb{1}$. One first checks that under the condition that (8.71) holds, CD^{-1} is both symmetric and unitary. Then, by a theorem of linear algebra,[10] a unitary matrix with the required properties exists. In a Majorana representation, $\psi = \psi^*$ and the Dirac matrices are either purely real (Minkowskian with C_+ and Euclidean with C_-) or purely imaginary (Minkowskian with C_- and Euclidean with C_+). If it exists we will always use C_+, e.g. in ten-dimensional Minkowski space or eight-dimensional Euclidean space.

In those dimensions where no Majorana spinors exist one can define modified Majorana spinors. For this one needs to consider several spinors ψ_i for which one

---

[10] Proven e.g. in B. Zumino, Normal forms of complex matrices. J. Math. Phys. **3**, 1055 (1962).

defines
$$\overline{\psi}_{iM} = (\psi^j)^T \Omega_{ji}. \tag{8.72}$$

If we now require $\overline{\psi}_{iM} = \alpha \overline{\psi}_{iD}$, the left hand side of (8.70) changes to $\Omega^*\Omega$. We therefore find

$$\Omega^*\Omega = -1 \tag{8.73}$$

as the condition for the existence of modified Majorana spinors in those dimensions in which Majorana spinors do not exist, i.e. where Eq. (8.70) cannot be satisfied. Due to the positivity of $\Omega^\dagger \Omega$, (8.73) implies $\Omega^T = -\Omega$, and for Ω to be non-generate we need the range of i to be even. The choice $\Omega = \begin{pmatrix} 0 & 1 \\ -1 & 0 \end{pmatrix}$ leads to so-called symplectic Majorana spinors. If we further impose the Weyl condition, we arrive at symplectic Majorana-Weyl spinors. In Euclidean signature they exist in 4 (mod 8) dimensions and in Minkowski signature in 6 (mod 8) dimensions.

Spinor Indices, etc.

So far we have suppressed spinor indices. Sometimes it is useful to exhibit them explicitly. By convention, a spinor has a lower index α, i.e. ψ_α, Dirac matrices have index structure $\gamma_\alpha{}^\beta$, the charge conjugation matrix is $C = (C^{\alpha\beta})$ and its inverse $C^{-1} = (C_{\alpha\beta})$ such that $C^{\alpha\beta} C_{\beta\gamma} = \delta^\alpha_\gamma$. The Hermitian conjugate of a spinor is $\psi^{\dagger\alpha} = (\psi_\alpha)^\dagger$ and its Majorana conjugate $\psi_\beta C^{\beta\alpha}$. The symmetry properties (8.55) are those of $(\gamma^{[p]}C^{-1})_{\alpha\beta}$ and of $(C\gamma^{[p]})^{\alpha\beta}$. The charge conjugation matrix and its inverse can thus be used to raise and lower spinor indices.

In even dimensions the spinors ψ_α do not transform irreducibly under Lorentz transformations. One splits the spinor index $\alpha = (a, \dot{a})$ where ψ_a ($\psi_{\dot{a}}$) are the components of a positive (negative) chirality spinor.[11]

Since Dirac matrices change the chirality, they decompose into two blocks $(\gamma^\mu)_a{}^{\dot{b}}$ and $(\gamma^\mu)_{\dot{a}}{}^b$. More generally, the matrices (8.46) have the following block form

$$\gamma^{[p]}{}_\alpha{}^\beta = \begin{pmatrix} 0 & \gamma^{[p]}{}_a{}^{\dot{b}} \\ \gamma^{[p]}{}_{\dot{a}}{}^b & 0 \end{pmatrix} \quad (p \text{ odd}),$$

$$\gamma^{[p]}{}_\alpha{}^\beta = \begin{pmatrix} \gamma^{[p]}{}_a{}^b & 0 \\ 0 & \gamma^{[p]}{}_{\dot{a}}{}^{\dot{b}} \end{pmatrix} \quad (p \text{ even}). \tag{8.74}$$

In this basis the chirality matrix is

---

[11] In later chapters we will also use $(\alpha, \dot{\alpha})$ for chiral spinor indices.

8.5 Appendix: Dirac Matrices and Spinors in d Dimensions

$$\gamma_{d+1} = \begin{pmatrix} \mathbb{1} & 0 \\ 0 & -\mathbb{1} \end{pmatrix}, \tag{8.75}$$

where $\mathbb{1}$ is the 2^{n-1} dimensional unit matrix. The charge conjugation matrix has the index structure

$$C^{\alpha\beta} = \begin{pmatrix} C^{ab} & 0 \\ 0 & C^{\dot{a}\dot{b}} \end{pmatrix} \text{ for } d = 4n, \quad C^{\alpha\beta} = \begin{pmatrix} 0 & C^{a\dot{b}} \\ C^{\dot{a}b} & 0 \end{pmatrix} \text{ for } d = 4n+2 \tag{8.76}$$

and likewise for its inverse, but with lower indices. The index structures of $\gamma^{[p]}C^{-1}$ are

- $d = 4n$

$$(\gamma^{[p]}C^{-1})_{\alpha\beta} = \begin{pmatrix} 0 & \gamma^{[p]}_{a\dot{b}} \\ \gamma^{[p]}_{\dot{a}b} & 0 \end{pmatrix} \quad (p \text{ odd}),$$

$$(\gamma^{[p]}C^{-1})_{\alpha\beta} = \begin{pmatrix} \gamma^{[p]}_{ab} & 0 \\ 0 & \gamma^{[p]}_{\dot{a}\dot{b}} \end{pmatrix} \quad (p \text{ even}) \tag{8.77}$$

- $d = 4n+2$

$$(\gamma^{[p]}C^{-1})_{\alpha\beta} = \begin{pmatrix} \gamma^{[p]}_{ab} & 0 \\ 0 & \gamma^{[p]}_{\dot{a}\dot{b}} \end{pmatrix} \quad (p \text{ odd}),$$

$$(\gamma^{[p]}C^{-1})_{\alpha\beta} = \begin{pmatrix} 0 & \gamma^{[p]}_{a\dot{b}} \\ \gamma^{[p]}_{\dot{a}b} & 0 \end{pmatrix} \quad (p \text{ even}). \tag{8.78}$$

The matrices $C\gamma^{[p]}$ with two upper indices have the same block structure as those with lower indices.

Group theoretically the Dirac matrices $\gamma^{[p]}$ are Clebsch-Gordan coefficients for the coupling of spinors to anti-symmetric tensors. For instance, the absence of C_{ab} components in $d = 10$ means that the singlet does not appear in the tensor product of two positive chirality spinors. This immediately follows from the form of the spinor weights which we discuss in Chap. 11. As another application of the above results, in $d = 8$ we have the expansions

$$\psi_\alpha \lambda_\beta = \sum_{p \text{ even}} A^{(p)}_{i_1 \dots i_p} (\gamma^{i_1 \dots i_p} C^{-1})_{\alpha\beta},$$

$$\psi_\alpha \lambda_{\dot{\beta}} = \sum_{p \text{ odd}} A^{(p)}_{i_1 \dots i_p} (\gamma^{i_1 \dots i_p} C^{-1})_{\alpha\dot{\beta}} \tag{8.79}$$

with $A^{(p)}_{i_1...i_p} \propto \psi C \gamma_{i_1...i_p} \lambda$, which expresses the fact that in the massless spectrum of type IIB (IIA) there are p forms with p even (odd). If we did the analogous expansion in $d = 10$, we would get the same information, but now expressed in terms of the field strengths $F_{(p+1)} = dA_{(p)}$. In the above expansions not all coefficients are independent. Indeed, for $d = 8$ there are only 64 independent matrices with the given index structure, say $(\alpha\beta)$, e.g. C, $\gamma^{[2]}$ and $\gamma^{[4]}_+$ where $\gamma^{[4]}_+$ is the self-dual part of $\gamma^{[4]}$, i.e. it satisfies

$$\gamma^{i_1...i_4}_\pm = \pm \frac{1}{4!} \gamma^{j_1...j_4}_\pm \epsilon_{j_1...j_4}{}^{i_1...i_4}. \tag{8.80}$$

The other terms in the expansions (8.79) are related by Hodge duality where the Hodge dual of $\gamma^{[d-p]}$ is defined as

$$(*\gamma)^{i_1...i_p} = \frac{1}{(d-p)!} \gamma^{j_1...j_{d-p}} \epsilon_{j_1...j_{d-p}}{}^{i_1...i_p}, \tag{8.81}$$

where we define $\epsilon_{0...d-1} = +1$. This is because in the chiral basis $\gamma^0 \cdots \gamma^{d-1} = \pm \mathbb{1}$. In fact, in the first line of (8.79) only the self-dual part of $\gamma^{[4]}$ contributes due to the property, valid in Euclidean signature,

$$\frac{1}{2}(1 \pm \gamma) \gamma^{[4]}_\pm = \gamma^{[4]}_\pm, \qquad \frac{1}{2}(1 \pm \gamma) \gamma^{[4]}_\mp = 0, \qquad (\gamma = \gamma^0 \ldots \gamma^7). \tag{8.82}$$

The analogous relation in Minkowski signature is

$$\frac{1}{2}(1 + \gamma) \gamma^{[5]}_+ = \gamma^{[5]}_+, \qquad \frac{1}{2}(1 + \gamma) \gamma^{[5]}_- = 0, \qquad (\gamma = \gamma^0 \ldots \gamma^9). \tag{8.83}$$

These relations show that the 4-form potential of the type IIB theory has a self-dual field strength.

Further Reading

The proof of the no-ghost theorem and the determination of $d = 10$ first appeared in the work of

- J.H. Schwarz, Physical states and pomeron poles in the dual pion model. Nucl. Phys. **B46**, 61 (1972)
- P. Goddard, C.B. Thorn, Compatibility of the dual Pomeron with unitarity and the absence of ghosts in the dual resonance model. Phys. Lett. **40B**, 235 (1972)
- R.C. Brower, K.A. Friedman, Spectrum-generating algebra and no-ghost theorem for the Neveu-Schwarz model. Phys. Rev. **D7**, 535 (1973)

The path integral quantization of the fermionic string was initiated in

8.5 Appendix: Dirac Matrices and Spinors in d Dimensions

- A.M. Polyakov, Quantum geometry of fermionic strings. Phys. Lett. **103B**, 211 (1981)

The ghost sector is examined in detail in

- D.Z. Freedman, N.P. Warner, Locally supersymmetric string Jacobian. Phys. Rev. **D34**, 3084 (1986)

The GSO projection was invented in

- F. Gliozzi, J. Scherk, D.I. Olive, Supergravity and the spinor dual model. Phys. Lett. **B65**, 282 (1976)
- F. Gliozzi, J. Scherk, D.I. Olive, Supersymmetry, supergravity theories and the dual spinor model. Nucl. Phys. **B122**, 253 (1977)

Useful references for the appendix are the GSO paper and

- P. van Nieuwenhuizen, in *Six Lectures on Supergravity*, ed. by S. Ferrara, J.G. Taylor. Supergravity '81 (CUP, Cambridge, 1982)
- P. van Nieuwenhuizen, in *An Introduction to Simple Supergravity and the Kaluza-Klein Program*, ed. by B.S. deWitt, R. Stora. Les Houches Lectures 1983, published in Relativity, Groups and Topology II (Elsevier, Amsterdam, 1984)
- T. Van Proeyen, *Tools for Supersymmetry* Spring School on Quantum Field Theory, Supersymmetry and Superstrings, Calimanesti, Romania [arXiv:hep-th/9910030]

Chapter 9
Superstrings

Abstract In the first part of this chapter we compute the one-loop partition function of the closed fermionic string. We will do this in light cone gauge. The possibility to assign to the world-sheet fermions periodic or anti-periodic boundary conditions leads to the concept of spin structures. The requirement of modular invariance is then shown to result in the GSO projection. We also generalize some of the results of Chap. 6 to the case of fermions. We then consider open superstrings, i.e. we extend the formalism of conformal field theories with boundaries to include free fermionic fields. This gives rise to D-branes in superstring theories. We also discuss non-oriented superstrings, which result form performing a quotient of the type IIB superstring by the world-sheet parity transformation. We show that one-loop diagrams are divergent unless D-branes are present in the model. This defines the type I superstring, whose construction we discuss in some detail.

9.1 Spin Structures and Superstring Partition Function

The RNS formulation of the superstring, discussed in the previous two chapters, possesses world-sheet spinor variables. Given a topological manifold we can always endow it with a Riemannian structure, i.e. we can always define a metric on it but there are topological conditions which have to be met in order to define spinors. On a d-dimensional orientable manifold, the transition functions of the frame bundle are elements of $SO(d)$. $SO(d)$ has no spinor representations, but its double cover $Spin(d)$ does. Spinors can be defined if the transition functions of the frame bundle can be consistently lifted to $Spin(d)$. The set of transition functions defines a spin structure, which does not have to be unique. Manifolds which admit a spin structure are called spin manifolds. On an orientable manifold the necessary and sufficient condition to be spin is that the second Stiefel-Whitney class of the tangent bundle vanishes.[1]

In string theory the question of spin structures poses itself at two levels: first at the level of the world-sheet and second at the level of the background space-time through which the string propagates. The latter occurs when we discuss the low-energy effective action and solutions to its equations of motion. In this chapter we will only be concerned with world-sheet properties. (The space-time is always Minkowski space which certainly does allow spinors.)

Except for a few comments at the end, the discussion in this section will be concerned with the closed string whose (Euclidean) world-sheet is a genus g Riemann surface Σ_g. Riemann surfaces are spin and, as we will now discuss, except for $g = 0$, they admit more than one spin-structure. Let us begin by characterizing spin structures on Σ_g. We know from Chap. 6 that there are two non-contractible loops associated with each of the g holes. All other non-contractible loops can be generated by deforming and joining elements of this basis. When we have spinors defined over Σ_g we can assign to them either periodic or anti-periodic boundary conditions around each of the $2g$ loops. Each of these 2^{2g} possible assignments is called a spin structure on Σ_g. An important distinction is that of even and odd spin structures which is connected to the zero mode structure of the chiral Dirac operator which, on Σ_g, is simply ∇_z and $\nabla_{\bar{z}}$ for the two chiralities. We will treat these operators in more detail below. We call a spin structure even if the number of zero modes of the chiral Dirac operator is even and we call it odd otherwise. We will first study the situation for the torus and then generalize to arbitrary genus.

We can put a flat metric on the torus for which the chiral Dirac operator is ∂_z. It is then clear that the only global zero mode is the constant spinor. Obviously, only $(+, +)$ boundary conditions allow for a constant spinor where the two entries refer to the boundary conditions along the two non-contractible loops. This means that there are three even and one odd spin structure on the torus. For the generalization to arbitrary genus and the properties under modular transformations the following facts are important. We state without proof:

- For a given spin structure, the number of chiral Dirac zero modes modulo two is a topological invariant;
- The number of chiral Dirac zero modes modulo two is additive when we glue together two Riemann surfaces.

The second fact together with our result for the torus can be used to find the number of even and odd spin structures for arbitrary Riemann surfaces. It is not hard to see that there are $\sum_{m \text{ odd}} \binom{g}{m} 3^{g-m} = 2^{g-1}(2^g - 1)$ odd and $\sum_{m \text{ even}} \binom{g}{m} 3^{g-m} = 2^{g-1}(2^g + 1)$ even spin structures. Since the number of zero modes of the Dirac operator mod 2 is a modular invariant (this follows from the first item above), the two classes of spin structures transform separately under modular

---

[1] Even if this condition is not satisfied, one might still have spinors. But this requires the existence of a suitable complex line bundle to which they couple. This leads to the notion of Spin$_C$-structures.

9.1 Spin Structures and Superstring Partition Function

transformations. In fact it can be shown that they transform irreducibly. This means that in the computation of the partition function or any correlation function we have to sum over all boundary conditions leading to even or odd spin structures. The relative phases between the different contributions are then determined by modular invariance.

Let us illustrate the above and work out the details for the vacuum amplitude on the torus, the one-loop partition function. As already discussed in Chap. 6 we parametrize the torus by two coordinates $\xi^1, \xi^2 \in [0, 1]$. We have the following possible boundary conditions for fermions, leading to four spin structures:

$$\psi(\xi^1 + 1, \xi^2) = \pm \psi(\xi^1, \xi^2),$$
$$\psi(\xi^1, \xi^2 + 1) = \pm \psi(\xi^1, \xi^2). \tag{9.1}$$

Periodic boundary conditions in ξ^1 correspond to the Ramond sector and anti-periodic boundary conditions to the Neveu-Schwarz sector.

We now want to determine how the boundary conditions transform under modular transformations of the torus. To this end we look at the more general boundary conditions[2]

$$\psi(\xi^1 + 1, \xi^2) = h\psi(\xi^1, \xi^2),$$
$$\psi(\xi^1, \xi^2 + 1) = g\psi(\xi^1, \xi^2). \tag{9.2}$$

Recall from Chap. 6 that the torus is generated by two complex numbers $\lambda_{1,2}$, such that $\text{Im}(\tau) > 0$ where $\tau = \frac{\lambda_2}{\lambda_1}$. If ψ satisfies (9.2) on a torus with modular parameter τ, then the boundary conditions on a torus with complex structure $\frac{a\tau+b}{c\tau+d}$, generated by $\lambda'_2 = a\lambda_2 + b\lambda_1$, $\lambda'_1 = c\lambda_2 + d\lambda_1$, will be

$$(h, g, \tau) \leftrightarrow \left(h^d g^c, h^b g^a; \frac{a\tau + b}{c\tau + d} \right) \tag{9.3}$$

or, equivalently

$$\left(h, g, \frac{a\tau + b}{c\tau + d} \right) \leftrightarrow (h^a g^{-c}, h^{-b} g^d; \tau). \tag{9.4}$$

From this we find that under an S-transformation $\tau \to -1/\tau$ with modular matrix $\begin{pmatrix} 0 & 1 \\ -1 & 0 \end{pmatrix}$ the spin structures transform as

---

[2] For a real fermion the only choices for g, h are ± 1. For complex fermions they can be arbitrary phases and if we consider several fermions they can be orthogonal matrices; see also the appendix to this chapter.

$$S: \quad \begin{aligned} (--) &\to (--) \\ (++) &\to (++), \quad (+-) \to (-+) \\ (-+) &\to (+-) \end{aligned} \qquad (9.5)$$

For a modular T-transformation $\tau \to \tau + 1$ with modular matrix $\begin{pmatrix} 1 & 1 \\ 0 & 1 \end{pmatrix}$ we find

$$T: \quad \begin{aligned} (--) &\to (-+) \\ (++) &\to (++), \quad (+-) \to (+-) \\ (-+) &\to (--) \end{aligned} \qquad (9.6)$$

This demonstrates our general statement that even and odd spin structures transform irreducibly under modular transformations.

In string theory one of the basic principles is invariance under diffeomorphisms of the world-sheet including also global ones. Since, as we have seen, they change the spin structure, we have, in order to find modular invariant expressions, to sum over all different spin structures in each class (even and odd). At the one loop level the $(++)$ spin structure is invariant by itself, being the only odd one and so is its contribution to the partition function. The other three must all be included in a modular invariant way. This means in particular that we must include both the R and the NS sectors.

It is now important to note that due to the world-sheet supersymmetry algebra, world-sheet fermions ψ^μ as well as the gravitino χ_α and consequently also the superconformal ghosts β, γ, all have the same spin structure. Left- and right-movers however can have different spin structures. We then denote the contribution to the partition function of the right-moving fermions with spin structure $(++)$ by $\chi_{\text{ferm}}^{(++)}(\tau)$ and likewise for the other three cases and the left-movers. In light-cone gauge we get the following expressions which are trivial generalizations of the corresponding expression for the bosonic string:

$$\chi_{\text{ferm}}^{(++)}(\tau) = \eta_{(++)} \operatorname{tr} e^{2\pi i \tau H_R} (-1)^F,$$

$$\chi_{\text{ferm}}^{(+-)}(\tau) = \eta_{(+-)} \operatorname{tr} e^{2\pi i \tau H_R},$$

$$\chi_{\text{ferm}}^{(--)}(\tau) = \eta_{(--)} \operatorname{tr} e^{2\pi i \tau H_{\text{NS}}},$$

$$\chi_{\text{ferm}}^{(-+)}(\tau) = \eta_{(-+)} \operatorname{tr} e^{2\pi i \tau H_{\text{NS}}} (-1)^F, \qquad (9.7)$$

where the η are phases to be determined by modular invariance. Let us comment on the $(-1)^F$ factors (F is the world-sheet fermion number). For anticommuting variables the trace automatically implies that the fermions satisfy anti-periodic boundary conditions along ξ^2. If we want to have periodic boundary conditions, we

9.1 Spin Structures and Superstring Partition Function

have to insert the operator $(-1)^F$.[3] The light-cone Hamiltonians in the two sectors for a closed string of length $\ell = 2\pi$ are (cf. Chap. 7):

$$H_R = \sum_{i=2}^{9} \sum_{m=1}^{\infty} m b^i_{-m} b^i_m + \frac{1}{3},$$

$$H_{NS} = \sum_{i=2}^{9} \sum_{r=\frac{1}{2}}^{\infty} r b^i_{-r} b^i_r - \frac{1}{6}. \tag{9.8}$$

The normal ordering constants follow directly by subtracting the bosonic contribution $-\frac{d-2}{24}$ from the total normal ordering constant in each sector, namely 0 (R) and $-\frac{1}{2}$ (NS). It is now easy to evaluate the different contributions to the partition function. For instance, for $\chi^{(--)}_{\text{ferm}}(\tau)$ we find ($q = e^{2\pi i \tau}$)

$$\chi^{(--)}_{\text{ferm}}(\tau) = \eta_{(--)} \operatorname{tr} e^{2\pi i \tau H_{NS}}$$

$$= \eta_{(--)} q^{-\frac{1}{6}} \operatorname{tr}\left(q^{\sum_{r=\frac{1}{2}}^{\infty} r b^i_{-r} b^i_r}\right)$$

$$= \eta_{(--)} q^{-\frac{1}{6}} \prod_r \left(\sum_{N_r=0,1} q^{r N_r}\right)^8$$

$$= \eta_{(--)} q^{-\frac{1}{6}} \left(\prod_{n=1}^{\infty} (1+q^{n-\frac{1}{2}})\right)^8. \tag{9.9}$$

The calculation is analogous to the bosonic case only that the occupation numbers are now restricted by the Pauli principle to $N_r = 0$ and 1. (This is just the grand partition function of an ideal Fermi gas with energy levels $E_r = r$.) We can now write

$$q^{-\frac{1}{6}} \prod_{n=1}^{\infty}\left(1+q^{n-\frac{1}{2}}\right)^8 = \left\{q^{-1/24}\prod_{n=1}^{\infty}(1-q^n)^{-1}\right\}^4 \left\{\prod_{n=1}^{\infty}(1-q^n)^4 (1+q^{n-\frac{1}{2}})^8\right\}$$

$$= \frac{\vartheta_3^4(0|\tau)}{\eta^4(\tau)}, \tag{9.10}$$

where $\eta(\tau)$ is the previously encountered Dedekind η-function and ϑ_3 is one of the four Jacobi theta-functions. In general, one defines the ϑ-functions

---

[3] For instance, if one computes the partition function of a fermionic oscillator in the path integral formulation, one obtains the correct result only if one uses anti-periodic boundary conditions for the fermion along the compact Euclidean time (temperature) direction.

$$\vartheta\begin{bmatrix}\alpha\\\beta\end{bmatrix}(0|\tau) = \eta(\tau)\, e^{2\pi i\alpha\beta} q^{\frac{\alpha^2}{2}-\frac{1}{24}} \prod_{n=1}^{\infty}\left(1+q^{n+\alpha-\frac{1}{2}}e^{2\pi i\beta}\right)\left(1+q^{n-\alpha-\frac{1}{2}}e^{-2\pi i\beta}\right)$$

$$= \sum_{n=-\infty}^{\infty} \exp\left[i\pi(n+\alpha)^2\tau + 2\pi i(n+\alpha)\beta\right]. \tag{9.11}$$

Through the one-loop partition function the ϑ-functions for arbitrary α and β are in correspondence to the generalized fermion boundary conditions[4] as

$$\psi(\xi^1+1,\xi^2) = -e^{2\pi i\alpha}\psi(\xi^1,\xi^2),$$
$$\psi(\xi^1,\xi^2+1) = -e^{2\pi i\beta}\psi(\xi^1,\xi^2). \tag{9.12}$$

The different spin structures then correspond to

(R)
$$(++) \quad \alpha = 1/2,\ \beta = 1/2 \quad \vartheta\begin{bmatrix}1/2\\1/2\end{bmatrix} = \vartheta_1,$$

$$(+-) \quad \alpha = 1/2,\ \beta = 0 \quad \vartheta\begin{bmatrix}1/2\\0\end{bmatrix} = \vartheta_2,$$

$$(--) \quad \alpha = 0,\ \beta = 0 \quad \vartheta\begin{bmatrix}0\\0\end{bmatrix} = \vartheta_3,$$

(NS)
$$(-+) \quad \alpha = 0,\ \beta = 1/2 \quad \vartheta\begin{bmatrix}0\\1/2\end{bmatrix} = \vartheta_4. \tag{9.13}$$

The Jacobi ϑ-functions and their generalizations to higher genus Riemann surfaces (the Riemann theta-functions) play an important role in string theory and conformal field theory. They satisfy many intriguing identities. At one loop, some of the most useful ones are

$$\vartheta_2^4(\tau) - \vartheta_3^4(\tau) + \vartheta_4^4(\tau) = 0 \quad \text{(Riemann identity)}, \tag{9.14a}$$

$$\vartheta_2(\tau)\,\vartheta_3(\tau)\,\vartheta_4(\tau) = 2\,\eta^3(\tau) \quad \text{(Jacobi triple product identity)}, \tag{9.14b}$$

$$\vartheta'_1(\tau) = 2\pi\,\eta^3(\tau). \tag{9.14c}$$

The notation here and below is $\vartheta_i(\tau) = \vartheta_i(0|\tau)$ and $\vartheta'_i(\tau) = \partial_z\vartheta_i(z|\tau)|_{z=0}$ (cf. appendix). Also, it is easy to see that $\vartheta_1(\tau) = 0$. In the same way that we have derived the partition function for the $(--)$ spin structure, we compute

$$\chi_{\text{ferm}}^{(--)}(\tau) = \eta_{(--)}\frac{\vartheta_3^4(\tau)}{\eta^4(\tau)},$$

---

[4] This will be derived in the appendix to this chapter.

9.1 Spin Structures and Superstring Partition Function

$$\chi_{\text{ferm}}^{(-+)}(\tau) = \eta_{(-+)} \frac{\vartheta_4^4(\tau)}{\eta^4(\tau)},$$

$$\chi_{\text{ferm}}^{(+-)}(\tau) = \eta_{(+-)} \frac{\vartheta_2^4(\tau)}{\eta^4(\tau)},$$

$$\chi_{\text{ferm}}^{(++)}(\tau) = \eta_{(++)} \frac{\vartheta_1^4(\tau)}{\eta^4(\tau)}. \tag{9.15}$$

Obviously, $\chi_{\text{ferm}}^{(++)}(\tau) = 0$, i.e. the partition function for the odd spin structure vanishes. This is not surprising since we know that the Dirac operator has a zero mode and $\chi_{\text{ferm}}^{(++)} \sim \int D\psi \exp(-\psi \slashed{D}\psi) = 0$. We remark that ϑ_2 and ϑ_1' both have an overall factor of 2. In the partition function this is the degeneracy of the R ground state. The $\chi_{\text{ferm}}^{(\alpha\beta)}(\tau)$ are the determinants of the Dirac operator on the torus with modulus τ with periodicity conditions (α, β) along its two cycles. We have computed them using, as we did in Chap. 6, the connection between the Euclidean path integral and the Hamiltonian formulation.

We have stated above that odd and even spin structures transform irreducibly under modular transformations. This should be reflected in the transformation properties of the ϑ-functions. Indeed, from their representation as infinite sums it is not hard to show that

$$\vartheta_1'(\tau+1) = e^{i\pi/4}\,\vartheta_1'(\tau),$$

$$\vartheta_2(\tau+1) = e^{i\pi/4}\,\vartheta_2(\tau),$$

$$\vartheta_3(\tau+1) = \vartheta_4(\tau),$$

$$\vartheta_4(\tau+1) = \vartheta_3(\tau),$$

$$\eta(\tau+1) = e^{i\pi/12}\,\eta(\tau), \tag{9.16}$$

which reflects the transformation properties of the spin structures under T (cf. Eq. (9.6)). Under S they transform as

$$\vartheta_1'(-1/\tau) = (-i\tau)^{\frac{3}{2}}\,\vartheta_1'(\tau),$$

$$\vartheta_2(-1/\tau) = (-i\tau)^{\frac{1}{2}}\,\vartheta_4(\tau),$$

$$\vartheta_3(-1/\tau) = (-i\tau)^{\frac{1}{2}}\,\vartheta_3(\tau),$$

$$\vartheta_4(-1/\tau) = (-i\tau)^{\frac{1}{2}}\,\vartheta_2(\tau),$$

$$\eta(-1/\tau) = (-i\tau)^{\frac{1}{2}}\,\eta(\tau), \tag{9.17}$$

which corresponds to Eq. (9.5). Equation (9.17) can be proven with the help of the Poisson resummation formula which will be derived in the appendix.

Let us now determine the phases η. We first demand that the spin structure sum is separately modular invariant in the left- and right-moving sectors. This corresponds to a non-diagonal modular invariant. Since only relative phases are relevant we will arbitrarily set $\eta_{(--)} = +1$, i.e. $\chi_{\text{ferm}}^{(--)}(\tau) = \frac{\vartheta_3^4(\tau)}{\eta^4(\tau)}$. Using the transformation rules of the ϑ- and η-functions we easily find

$$\chi_{\text{ferm}}^{(--)}(\tau+1) = \chi_{\text{ferm}}^{(-+)}(\tau) = \frac{\vartheta_4^4(\tau)}{\eta^4(\tau)} e^{-i\pi/3} . \tag{9.18}$$

The contribution from the eight transverse bosonic degrees of freedom is $\sim \frac{1}{\eta^8(\tau)}$ which contributes an extra factor of $e^{-2i\pi/3}$ so that we get for the phase $\eta_{(-+)} = -1$. Similarly, by verifying that $\chi_{\text{ferm}}^{(-+)}(-1/\tau) = \chi_{\text{ferm}}^{(+-)}(\tau)$ we show that $\eta_{(+-)} = -1$. Clearly, $\eta_{(++)}$ cannot be determined from modular invariance; we will show below that it has to be $\eta_{(++)} = \pm 1$. With these phases the contribution of the right-moving world-sheet fermions to the superstring partition function is

$$\chi_{\text{ferm}}(\tau) = \operatorname{tr}\left(e^{2\pi i \tau H_{\text{NS}}} \frac{1}{2}\left(1 - (-1)^F\right)\right) - \operatorname{tr}\left(e^{2\pi i \tau H_{\text{R}}} \frac{1}{2}\left(1 - \eta_{++}(-1)^F\right)\right)$$

$$= \frac{1}{2} \frac{1}{\eta^4(\tau)} \{\vartheta_3^4(\tau) - \vartheta_4^4(\tau) - \vartheta_2^4(\tau) + \eta_{(++)}\vartheta_1^4(\tau)\} \tag{9.19}$$

with a similar expression for the left-movers. The relative sign between the two sectors reflects the fact that states in the NS sector are bosons whereas states in the R sector are fermions. Note that the $\frac{1}{2}(1 - (-1)^F)$ in the NS sector is just the GSO projection. In the R sector it is $(-1)^F = \pm 1$ according to $\eta_{(++)} = \pm 1$ which agrees with Chap. 7. Due to the identity Eq. (9.14a) and the vanishing of ϑ_1, the partition function vanishes. This is the hallmark of a supersymmetric spectrum: the contributions from space-time bosons and fermions cancel. Including the contribution from the bosons, the partition functions of the type II theories are

$$Z(\tau,\bar{\tau}) = \frac{1}{4} \frac{1}{(\operatorname{Im}\tau)^4} \frac{1}{|\eta|^{24}} \left|\vartheta_3^4 - \vartheta_4^4 - \vartheta_2^4\right|^2 . \tag{9.20}$$

A factor $1/|\eta|^{16}$ originates from the $8+8$ transverse bosonic oscillators and the factor $(\operatorname{Im}\tau)^{-4}$ from the Gaussian integration over the transverse momenta; cf. the discussion in Chap. 6. Even though this expression vanishes, it is useful to determine the number of physical states at each mass level in the various sectors (NS-NS, NS-R, etc.) separately.

It is worthwhile mentioning that there is one other modular invariant combination of boundary conditions: it consists of summing over the same boundary conditions for the left- and right-movers. It follows that the left- and right-moving sectors are not separately modular invariant due to the non-trivial connection between their boundary conditions. This leads to the following partition function:

9.1 Spin Structures and Superstring Partition Function

$$Z_{\text{ferm}}(\tau,\bar\tau) = \text{tr}\left(e^{2\pi i \tau H_{\text{NS}} - 2\pi \bar\tau i \bar H_{\text{NS}}} \frac{1+(-1)^{F+\bar F}}{2}\right)$$

$$+\text{tr}\left(e^{2\pi \tau i H_R - 2\pi \bar\tau i \bar H_R} \frac{1-\eta_{++}(-1)^{F+\bar F}}{2}\right). \tag{9.21}$$

Including the contribution from the bosons, we find

$$Z(\tau,\bar\tau) = \frac{1}{2}\frac{1}{(\text{Im}\tau)^4}\frac{|\vartheta_3(\tau)|^8 + |\vartheta_4(\tau)|^8 + |\vartheta_2(\tau)|^8}{|\eta(\tau)|^{24}}, \tag{9.22}$$

which defines the so-called type 0A/B string theories, depending on the sign η_{++}. Modular invariance of this expression is easily checked. This theory has only space-time bosons and contains a tachyon. Indeed, the GSO projection in the NS sector is $(-1)^{F+\bar F} = 1$ allowing the tachyon in Table 8.2.

Let us now argue why the phase $\eta_{(++)}$ can only be ± 1. For the partition function to have the interpretation as a sum over states we can only allow $\eta_{(++)} = 0$ or ± 1. If we look at the partition function at two loops it will be expressible in terms of the appropriate Riemann theta-functions, ten of which correspond to even and six to odd spin structures. In the limit where the genus two surface degenerates to two tori, the genus two theta-functions become simply products of Jacobi theta functions. Especially $\vartheta_1(\tau_1)\vartheta_1(\tau_2)$, where $\tau_{1,2}$ are the Teichmüller parameters of the two resulting tori, is the degeneration limit of an even theta-function at genus two. This has to be part of the partition function as the even theta-functions transform irreducibly under global diffeomorphisms (the Dehn twists) of the genus two surface. Thus the choice $\eta_{(++)} = 0$ is excluded.

Let us close this section with the extension of some of the results of Chap. 6 to the case of fermions. We know from Chap. 7 that the covariant derivative acts on a spin 1/2 world-sheet fermion as $\nabla_\alpha \psi = \partial_\alpha \psi - \frac{i}{2}\omega_\alpha \bar\rho \psi$ where ω_α is the spin connection (we are working in Euclidean signature, hence the i and $\bar\rho = \sigma^3$). In conformal gauge with zwei-bein $e_\alpha{}^a = e^\varphi \delta_\alpha^a$ the spin connection is $\omega_\alpha = \epsilon_\alpha{}^\beta \partial_\beta \varphi$. In local complex coordinates the covariant derivatives act on the two helicity components as $\nabla_z \psi_\pm = (\partial_z \mp \frac{1}{2}\partial_z\varphi)\psi_\pm$.

Recall now our discussion in Chap. 7, where we had to rescale the fermions to arrive at the action in conformal gauge. Under conformal transformations they transform with weight 1/2 (this follows from the invariance of the action or from Eq. (7.53)) and the factor by which we had to rescale them was exactly the square root of the zweibein. Multiplying the fermions with the square root of the zwei-bein converts their tangent space spinor index to an Einstein index (and they become half-differentials). Let us denote the transformed spinors by $e^{\varphi/2}\psi_\pm = \tilde\psi_\pm$. Since the zwei-bein is covariantly constant, the covariant derivative acts on the $\tilde\psi_\pm$ as $\nabla_z \tilde\psi_+ = (\partial_z - \partial_z\varphi)\tilde\psi_+$ and $\nabla_z \tilde\psi_- = \partial_z \tilde\psi_-$. Generalizing the notation of Chap. 6 to tensors of half-integer rank, we find that $\tilde\psi_+ \in \mathcal{T}^{(1/2)}$ and $(h^{z\bar z})^{1/2}\tilde\psi_- \in \mathcal{T}^{(-1/2)}$. We can then extend the definition of covariant derivatives Eq. (6.24) to half-integer

Table 9.1 Number of zero modes of $\nabla_z^{(n)}$ and $\nabla_{(n+1)}^z$

g	dim ker $\nabla_z^{(n)}$	dim ker $\nabla_{(n+1)}^z$	
0	$2n+1$	0	
1	1	1	Odd spin structure
1	0	0	Even spin structure
> 1	1 for $n=0$	g	
	0 for $n>0$	$(2n+1)(g-1)$	

n. The scalar product Eq. (6.19) also generalizes. Of particular interest is the Riemann-Roch theorem which holds without modification.

We can then complete Table 6.1 for half-integer n in the same way as we did for $n \in \mathbb{Z}$ in Chap. 6. The only subtlety is at genus one. For integer $n > 0$ there is always a constant zero mode of $\nabla_z^{(n)}$. For $n \in \mathbb{Z} + \frac{1}{2}$ this is only true for the odd spin structure. The results are collected in Table 9.1 We find that there are two zero modes of $\nabla_z^{(1/2)}$ at $g = 0$ corresponding to conformal Killing spinors or zero modes of the superconformal ghost γ. The zero modes of $\nabla_{(3/2)}^z$ indicate the presence of super-moduli. We will however not discuss them here. The ghost zero modes for arbitrary (integer and half-integer) n will be discussed in Chap. 13.

We end this section with a few comments on spinors on more general worldsheets with boundaries and possibly non-orientable. One can show that spin structures can always be defined.[5] This is indeed a necessary physical requirement of unitarity of the open and non-orientable type I string theories (which will be defined and discussed below). Again, for each non-contractible loop and each boundary component, one has to specify the periodicity condition of the fermions. For the explicit construction of spinors of such world-sheets one may use the fact they can all be obtained from compact Riemann surfaces without boundaries by identification of points under an involution whose fixed points are the boundaries of the quotient. The different spin structures then arise from the different ways of realizing the involution on the spinor fields combined with the spin structures on the covering space.

9.2 Boundary States for Fermions

In Sect. 4.4 we have constructed boundary states for bosons. They are a way of representing D-branes within closed string theory. We will now repeat this discussion for fermions. The main ideas are the same as those for the bosonic string, but there are differences due to the possibility of having NS and R sectors and subtleties with the zero mode structure in the latter.

---

[5]There is a subtlety here related to the fact that e.g. the second Stiefel-Whitney class w_2 of $\mathbb{RP}_2$ does not vanish. For non-orientable manifolds there is a second construction of spinors for which the obstruction is $w_2 + w_1 \cup w_1$ which vanishes on any two-dimensional manifold.

9.2 Boundary States for Fermions

Let us first find the gluing conditions. We again make the transformation from the closed to the open string channel via the exchange $\tau \leftrightarrow \sigma$, or $\sigma^{\pm} \to \sigma'^{\pm} = \pm \sigma^{\pm}$. Under this diffeomorphism the fermions transform as $\psi'_+(\sigma'^+) = \left(\frac{\partial \sigma'^+}{\partial \sigma^+}\right)^{-1/2} \psi_+(\sigma^+)$ and likewise for ψ_-. We now impose open string boundary conditions as gluing conditions on a boundary state:

$$\left(\psi'^\mu_+(\sigma'^+) + \eta \, D^\mu{}_\nu \psi'^\nu_-(\sigma'^-)\right)|B, \eta\rangle = 0 \quad \text{at} \quad \tau = 0, \tag{9.23}$$

where $\eta = \pm 1$. We will see in a moment the significance of this extra sign choice. $D^\mu{}_\nu$ is a diagonal matrix with entries $+1$ along directions with Neumann boundary conditions and entries -1 along directions with Dirichlet boundary conditions; cf. (4.169). Expressing ψ' though ψ, (9.23) becomes

$$\left(\psi^\mu_-(\sigma^-) + i\eta \, D^\mu{}_\nu \psi^\nu_+(\sigma^+)\right)|B, \eta\rangle = 0 \quad \text{at} \quad \tau = 0. \tag{9.24}$$

Using the mode expansions (7.63), we obtain the gluing conditions for the modes

$$\left(b^\mu_r + i\eta \, D^\mu{}_\nu \bar{b}^\nu_{-r}\right)|B, \eta\rangle = 0. \tag{9.25}$$

These gluing conditions imply $(L_n - \bar{L}_{-n})|B, \eta\rangle = 0$ as well as $(G_r - i\eta \bar{G}_{-r})|B, \eta\rangle = 0$. We will now determine the boundary states for the NS and R sectors in turn. We will do this for one complex fermion $\psi = \psi^1 + i\psi^2$ with the same kind of boundary conditions for both real fields. These are the building blocks for the boundary states in type II superstring theories in ten-dimensional Minkowski space.

In the NS sector one readily verifies that

$$|B_{\text{NS}}, \eta\rangle = \exp\left(\mp i \eta \sum_{i=1}^{2} \sum_{r>0} b^i_{-r} \bar{b}^i_{-r}\right)|0\rangle_{\text{NS}} \tag{9.26}$$

solves the gluing condition where the $-$ sign is for N and the $+$ sign for D boundary condition. Thus they are fermionic Ishibashi states. Here $|0\rangle_{\text{NS}}$ denotes the non-degenerate vacuum in the NS-sector. The overlap of two N or of two D boundary states is also easily computed. The contributions of the half-integer moded fermionic oscillators is determined in complete analogy to the bosonic ones and one finds

$$\langle B_{\text{NS}}, \eta | e^{-2\pi l \, H_{\text{cl}}} | B_{\text{NS}}, \eta\rangle = \frac{\vartheta_3(2il)}{\eta(2il)}, \tag{9.27a}$$

$$\langle B_{\text{NS}}, \eta | e^{-2\pi l \, H_{\text{cl}}} | B_{\text{NS}}, -\eta\rangle = \frac{\vartheta_4(2il)}{\eta(2il)}, \tag{9.27b}$$

i.e. different relative signs of η correspond to different spin structures. It is instructive to study the transformation of the overlaps under modular S-transformation, as we did for the bosonic string. This will allow us to relate the one-loop open string amplitude to the tree level closed string exchange. We find,

$$\frac{\vartheta_4(2il)}{\eta(2il)} \xrightarrow{l=\frac{1}{2t}} \frac{\vartheta_2(it)}{\eta(it)}, \quad \frac{\vartheta_3(2il)}{\eta(2il)} \xrightarrow{l=\frac{1}{2t}} \frac{\vartheta_3(it)}{\eta(it)}. \tag{9.28}$$

We see that the $(+, +)$ and $(-, -)$ overlaps yield the $\text{tr}_{NS}\, q^{L_0-c/24}$ sector, while the $(+, -)$ and $(-, +)$ overlaps yield the $\text{tr}_R\, q^{L_0-c/24}$ sector. For the overlap of an N with a D boundary state the right hand sides of (9.27a) and (9.27b) are exchanged.

The remaining two spin structures arise form the R-sector. Here the situation is complicated by the degeneracy of the ground state. We recall from Eq. (8.14) that we can pairwise combine the zero mode oscillators to fermionic creation and annihilation operators and that the $2^{d/2}$ dimensional $SO(1, d)$ spinor representation is built by acting with the creation oscillators on the highest weight state $|0\rangle_R$, which is defined to be the state annihilated by all the annihilation operators. Let us now take one such pair of operators for the left and for the right moving sector, $b_0^\pm = \frac{1}{\sqrt{2}}(b_0^1 \pm ib_0^2)$ and $\bar{b}_0^\pm$. The most general ground state is a linear combination of the states $|0\rangle_R$, $b_0^+|0\rangle_R$, $\bar{b}_0^+|0\rangle_R$ and $b_0^+\bar{b}_0^+|0\rangle_R$ and it is straightforward to construct the state which satisfies the gluing conditions:

$$|R\rangle = (b_0^+ + i\eta\bar{b}_0^+)|0_R\rangle. \tag{9.29}$$

The complete Ramond sector boundary states, including the oscillators are

$$|B_R, \eta\rangle = \exp\left(\mp i\eta \sum_{i=1}^{2} \sum_{n=1}^{\infty} b_{-n}^i \bar{b}_{-n}^i\right)|R\rangle \tag{9.30}$$

with NN and DD overlaps

$$\langle B_R, \eta|e^{-2\pi l\, H_{cl}}|B_R, \eta\rangle = \frac{\vartheta_2(2il)}{\eta(2il)}, \tag{9.31a}$$

$$\langle B_R, \eta|e^{-2\pi l\, H_{cl}}|B_R, -\eta\rangle = \frac{\vartheta_1(2il)}{\eta(2il)} = 0, \tag{9.31b}$$

with

$$\frac{\vartheta_2(2il)}{\eta(2il)} \xrightarrow{l=\frac{1}{2t}} \frac{\vartheta_4(it)}{\eta(it)}, \quad \frac{\vartheta_1'(2il)}{\eta(2il)} \xrightarrow{l=\frac{1}{2t}} t\frac{\vartheta_1'(it)}{\eta(it)}. \tag{9.32}$$

We should mention that the boundary states only involve the NS-NS and R-R sectors, but not the mixed ones. This can be seen as follows: in the open string one-loop (cylinder) diagram, we can impose either anti-periodic or periodic boundary conditions along the compactified time direction, i.e. $\psi_\pm^\mu(\sigma, 0) = \eta_\pm \psi_\pm^\mu(\sigma, t)$. Consistency with (7.62) requires $\eta_+ = \eta_-$. When translated to the closed string picture the claim follows.

9.3 D-branes

We are now ready to consider the full type IIA/B superstring theories in ten-dimensional Minkowski space. Recall that in light-cone gauge we are dealing with a superconformal field theory of eight free bosons X^i and eight free fermions together with a GSO projection. The difference between the type IIA and the type IIB superstring is the definition of the GSO projection in the Ramond sector which we can schematically write as

$$\text{IIA}: \quad (\text{NS}+,\text{NS}+) \oplus (\text{R}+,\text{R}-) \oplus (\text{NS}+,\text{R}-) \oplus (\text{R}+,\text{NS}+),$$
$$\text{IIB}: \quad (\text{NS}+,\text{NS}+) \oplus (\text{R}+,\text{R}+) \oplus (\text{NS}+,\text{R}+) \oplus (\text{R}+,\text{NS}+). \quad (9.33)$$

A Dp-brane is defined to have $(p+1)$ Neumann boundary conditions (including the time direction) and $(9-p)$ Dirichlet boundary conditions. In light cone gauge this means that one has the following boundary conditions for the eight free bosonic and fermionic fields[6]

$$(\alpha_n^i + D^i{}_j \bar{\alpha}_{-n}^j)|\text{Dp}, \eta\rangle = 0,$$
$$(b_r^i + i\eta D^i{}_j \bar{b}_{-r}^j)|\text{Dp}, \eta\rangle = 0. \quad (9.34)$$

Moreover, in the Neumann directions the center of mass momentum vanishes $k^i = 0$, $i = 0, \ldots p$, whereas in the Dirichlet directions the boundary states can carry unconstrained center of mass momentum $k_\perp = \{k^i, i = p+1, \ldots, 9\}$.

The boundary states which satisfy these gluing conditions are simply

$$|\text{Dp}, \eta\rangle = \int dk_\perp \exp\left[-\sum_{n>0} \frac{1}{n} \alpha_{-n}^i D_{ij} \bar{\alpha}_{-n}^j - i\eta \sum_{r>0} b_{-r}^i D_{ij} \bar{b}_{-r}^j\right] e^{ik_\perp \cdot x_0^\perp} |k_\perp\rangle, \quad (9.35)$$

where $x_0^\perp$ denotes the position of the D-brane in the Dirichlet directions, i.e. transverse to the brane's world-volume. In the NS-NS sector the ground state $|k_\perp\rangle$ is simply the unique ground state with transverse momentum $k_\perp$. In the R-R sector the degenerate ground state also has to satisfy the gluing condition for the zero modes

$$\left(b_0^i + i\eta D^i{}_j \bar{b}_0^j\right)|k_\perp\rangle = 0. \quad (9.36)$$

Even though the states (9.35) satisfy the gluing conditions, they cannot be the complete boundary states in type II superstring theory for the following reason.

---

[6]We choose the (0, 1)-directions as the light-cone directions.

They are not eigenstates of the GSO projections $(-1)^F$ and $(-1)^{\overline{F}}$. Let us analyze this issue in the NS-NS and R-R sectors separately.

In the NS-NS sector, acting with the GSO projections on the $|Dp, \eta\rangle_{\text{NS}}$ boundary states, we get

$$(-1)^F |Dp, \eta\rangle_{\text{NS}} = -|Dp, -\eta\rangle_{\text{NS}}, \qquad (-1)^{\overline{F}} |Dp, \eta\rangle_{\text{NS}} = -|Dp, -\eta\rangle_{\text{NS}}, \qquad (9.37)$$

where the minus sign is due to the fact that the NS ground state carries odd fermion number. Therefore, in the NS-NS sector the GSO invariant boundary state has the form

$$|Dp\rangle_{\text{NS}} = \frac{1}{\mathcal{N}_{\text{NS}}} \left(|Dp, \eta = 1\rangle_{\text{NS}} - |Dp, \eta = -1\rangle_{\text{NS}} \right). \qquad (9.38)$$

The computation of the overlap of two such boundary states is now very easy. For the fermionic part we use (9.27) and for the bosonic part the results of Chap. 6. We obtain

$$\widetilde{\mathcal{A}} = \int_0^\infty dl \; {}_{\text{NS}}\langle Dp | e^{-2\pi l \, H_{\text{cl}}} | Dp \rangle_{\text{NS}}$$

$$= \frac{1}{\mathcal{N}_{\text{NS}}^2} \int_0^\infty dl \, e^{-\frac{1}{4\pi\alpha' l} Y^2} \frac{2 V_{p+1}}{(4\pi^2 \alpha' l)^{\frac{9-p}{2}}} \left(\frac{\vartheta_3^4 - \vartheta_4^4}{\eta^{12}} \right) (2il). \qquad (9.39)$$

Transforming this expression to the loop-channel by the modular S-transformation $l = \frac{1}{2t}$ yields

$$\mathcal{A} = \frac{2^5}{\mathcal{N}_{\text{NS}}^2} (4\pi^2 \alpha')^{p-4} \int_0^\infty \frac{dt}{t} \, e^{-\frac{t}{2\pi\alpha'} Y^2} \frac{V_{p+1}}{(8\pi^2 \alpha' t)^{\frac{p+1}{2}}} \left(\frac{\vartheta_3^4 - \vartheta_2^4}{\eta^{12}} \right) (it). \qquad (9.40)$$

Up to the prefactor, which we can normalize to one by choosing $\mathcal{N}_{\text{NS}}$ accordingly, this has precisely the form of a trace of all open string excitations between two Dp-branes. However, the sum over the ϑ-functions does not include all four possible spin structures. As a consequence, this boundary state still carries a tachyonic mode and is therefore unstable. What is missing is the contribution from the R-R sector. Due to the ground state degeneracy, here the invariance under the GSO projections is subtle though interesting.

First, let us define the following left-right combinations of the Ramond zero modes[7]

---

[7] Note that this linear combination is different from the one introduced in Eq. (8.14), which did not mix left- and right moving zero modes.

9.3 D-branes

$$b_{\pm}^i = \frac{1}{\sqrt{2}}\left(b_0^i \pm i\,\bar{b}_0^i\right), \tag{9.41}$$

which satisfy the anti-commutation relations

$$\{b_{\pm}^i, b_{\pm}^j\} = 0, \qquad \{b_+^i, b_-^j\} = \delta^{ij}. \tag{9.42}$$

Then, the zero mode gluing conditions can be written as

$$b_\eta^i |Dp, \eta\rangle_R = 0, \qquad i = 2, \ldots, p,$$
$$b_{-\eta}^j |Dp, \eta\rangle_R = 0, \qquad j = p+1, \ldots, 9. \tag{9.43}$$

Due to the anti-commutation relations we can define

$$|Dp, \eta = +1\rangle_R = \prod_{i=2}^{p} b_+^i \prod_{j=p+1}^{9} b_-^j |Dp, \eta = -1\rangle_R,$$

$$|Dp, \eta = -1\rangle_R = \prod_{i=2}^{p} b_-^i \prod_{j=p+1}^{9} b_+^j |Dp, \eta = +1\rangle_R. \tag{9.44}$$

Recall from (8.34) that the GSO projection on the R-R ground state is defined as

$$(-1)^F = 16 \prod_{i=2}^{9} b_0^i = \prod_{i=2}^{9} (b_+^i + b_-^i)$$

$$(-1)^{\bar{F}} = 16 \prod_{i=2}^{9} \bar{b}_0^i = \prod_{i=2}^{9} (b_+^i - b_-^i). \tag{9.45}$$

Now, it immediately follows that

$$(-1)^F |Dp, \eta\rangle_R = |Dp, -\eta\rangle_R$$
$$(-1)^{\bar{F}} |Dp, \eta\rangle_R = (-1)^{p+1} |Dp, -\eta\rangle_R, \tag{9.46}$$

which implies that the only $(-1)^F$ invariant boundary state is

$$|Dp\rangle_R = |Dp, \eta = 1\rangle_R + |Dp, \eta = -1\rangle_R. \tag{9.47}$$

Since the right moving GSO projection acts on this boundary state as

$$(-1)^{\bar{F}} |Dp\rangle_R = (-1)^{p+1} |Dp\rangle_R, \tag{9.48}$$

it is only $(-1)^{\overline{F}}$ invariant if

$$p = \text{odd} \quad \text{for} \quad \text{type IIB} \quad \text{with} \quad (-1)^{\overline{F}} = +1,$$
$$p = \text{even} \quad \text{for} \quad \text{type IIA} \quad \text{with} \quad (-1)^{\overline{F}} = -1. \tag{9.49}$$

Thus, from the invariance of the boundary states under the GSO projection, we have derived that the type IIB/type IIA superstring contains Dp-branes only for p odd/even.

The total boundary state is now the combination of the NS-NS and R-R parts

$$|Dp\rangle = \frac{1}{\mathcal{N}_{Dp}} \left(|Dp, \eta = 1\rangle_{NS} - |Dp, \eta = -1\rangle_{NS} \right.$$
$$\left. + i |Dp, \eta = 1\rangle_R + i |Dp, \eta = -1\rangle_R \right), \tag{9.50}$$

where the factor of $+i$ in front of the R-sector boundary state is fixed by closed-open string duality and guarantees the correct GSO projection in the loop-channel, which projects out the tachyon. Note that each term in $|Dp\rangle$ is an Ishibashi state, i.e. it satisfies the gluing condition, but only the combination (9.50) satisfies both the Cardy condition and supersymmetry. Using (9.27), (9.31) and the results from Chap. 6 for the bosonic part, the overlap of two such boundary states gives

$$\widetilde{\mathcal{A}}_{Dp,Dp} = \int_0^\infty dl \, \langle Dp | e^{-2\pi l \, H_{cl}} | Dp \rangle$$
$$= \frac{2 V_{p+1}}{\mathcal{N}_{Dp}^2} \int_0^\infty dl \, e^{-\frac{Y^2}{4\pi\alpha' l}} \frac{1}{(4\pi^2 \alpha' l)^{\frac{9-p}{2}}} \left(\frac{\vartheta_3^4 - \vartheta_4^4 - \vartheta_2^4}{\eta^{12}} \right) (2il). \tag{9.51}$$

The normalisation factor will be determined via loop-channel–tree-level equivalence. To this end we perform a modular S-transformation of this amplitude. This must coincide with the loop-channel amplitude for open strings ending on a Dp-brane

$$\mathcal{A}_{Dp,Dp} = \int_0^\infty \frac{dt}{t} \, \text{tr}_{NS-R} \left[\left(\frac{1+(-1)^F}{2} \right) q^{L_0 - \frac{c}{24}} \right]$$
$$= V_{p+1} \int_0^\infty \frac{dt}{t} \, e^{-\frac{1}{2\pi\alpha'} Y^2} \frac{1}{(8\pi^2 \alpha' t)^{\frac{p+1}{2}}} \frac{1}{2} \left(\frac{\vartheta_3^4 - \vartheta_2^4 - \vartheta_4^4}{\eta^{12}} \right) (it). \tag{9.52}$$

This fixes the normalization to

$$\mathcal{N}_{Dp}^{-1} = \frac{1}{8} (4\pi^2 \alpha')^{\frac{4-p}{2}}. \tag{9.53}$$

9.3 D-branes

The boundary state (9.50) yields the complete GSO projected loop-channel amplitude, which means, in particular, that there is no tachyonic mode and the state $|\mathrm{Dp}\rangle$ is stable. These states are also called Dp-brane states, emphasizing the target space-time point of view.

Recall that the one loop cylinder amplitude has the tree-level interpretation of closed strings being emitted on one boundary (D-brane), freely propagating to the other boundary and being absorbed there. For the case of a Dp-brane, the formula (9.51) therefore encodes that the brane couples to closed string states in the NS-NS sector and R-R sector such that these two contributions precisely cancel. Mathematically this is due to the Jacobi triple identity and physically due to supersymmetry. In particular, the Dp-brane couples to the massless modes. Since the matrix D_{ij} is symmetric, from the boundary state (9.35) we can directly read off that in the NS-NS sector it only couples to left-right symmetric modes, i.e. the graviton and the dilaton. However, it does not couple to the left-right antisymmetric NS-NS two-form $B_{\mu\nu}$. Due to the extra sign in the R-R sector, the brane couples to the appropriate antisymmetric R-R form C_{p+1}.

If we expand the ϑ- and η-functions in powers of $q = e^{-2\pi l}$ we find

$$\begin{aligned}\widetilde{\mathscr{A}}_{\mathrm{Dp,Dp}} &= \frac{2 V_{p+1}}{\mathscr{N}_{\mathrm{Dp}}^2}(16_{\mathrm{NS}} - 16_{\mathrm{R}}) \int_0^\infty e^{-\frac{1}{4\pi\alpha'l}Y^2} \frac{1}{(4\pi^2\alpha'l)^{\frac{9-p}{2}}} \left(1 + \mathcal{O}(e^{-2\pi l})\right) \\ &= \frac{\pi}{8}(4\pi^2\alpha')^{3-p} V_{p+1} (16_{\mathrm{NS}} - 16_{\mathrm{R}}) \frac{1}{\pi\alpha'} G_{9-p}(Y) + \ldots .\end{aligned} \quad (9.54)$$

The notation was defined in Eqs. (6.133) and (6.134). As in the bosonic case we can reproduce this part of the amplitude, which is due to exchange of massless closed string modes, from the low-energy effective action. For the NS-NS sector the world-volume and space-time actions are the same as in the bosonic case (see also Chap. 16) and the calculation is identical to the one in Chap. 6 with $d = 10$. This leads to

$$T_p = \frac{\sqrt{\pi}}{\tilde{\kappa}_{10}}(4\pi^2\alpha')^{\frac{3-p}{2}} = \frac{\sqrt{\pi}}{\tilde{\kappa}_{10}}\ell_s^{3-p} . \quad (9.55)$$

We now turn to the contribution of the RR-sector to $\widetilde{\mathscr{A}}_{\mathrm{Dp,Dp}}$. From a space-time point of view a Dp-brane with $(p+1)$-dimensional world-volume naturally couples to a R-R $(p+1)$-form potential via $\mu_p \int_{\mathscr{W}} C_{p+1}$ where μ_p is the Dp-brane charge. This is the immediate generalization of the coupling of a point particle of charge q to a 1-form gauge potential $q \int A$ where the integral is over the particle's worldline. In both cases there will, of course, also be a kinetic energy term for the p-form potential.[8] The result (9.49) on the existence of Dp-branes for p even/odd in type IIA/B theory is therefore consistent with the spectrum of p-form fields in the RR sector that we found earlier.

---

[8] We will say more about the effective world-volume action of D-branes in Chap. 16.

Computing the contribution due to exchange of C_{p+1}[9] one finds

$$\mu_p = T_p. \tag{9.56}$$

Two comments are in order:

1. In string theory the dimensionless ten-dimensional gravitational coupling $\tilde{\kappa}_{10}/\alpha'^4$ has no absolute meaning. A change can always be compensated by a constant shift of the dilaton, $\Phi \to \Phi + \Phi_0$. This freedom is conventionally fixed by requiring that the tension of the fundamental string $T_{F1} \equiv T = 1/(2\pi\alpha')$ is equal to the tension of the D1-string $T_{D1} = T_{F1}$. The relation $T_{F1}/T_{D1} = 1$ then implies

$$\tilde{\kappa}_{10}^2 = \frac{1}{4\pi}(4\pi^2\alpha')^4 = \frac{1}{4\pi}\ell_s^8. \tag{9.57}$$

Therefore, the tension and R-R charge of a Dp-brane is given in terms of the string scale as

$$T_p = \mu_p = 2\pi (4\pi^2\alpha')^{-\frac{p+1}{2}} = 2\pi \ell_s^{-(p+1)}. \tag{9.58}$$

2. The physically measured tension τ_p and the parameter T_p which appears in front of the world-volume action of the brane in (16.160) differ by the background value Φ_0 of the dilaton or, with $g_s = e^{\Phi_0}$

$$\tau_p = \frac{T_p}{g_s} = \frac{2\pi}{g_s \ell_s^{p+1}}. \tag{9.59}$$

The world-sheet analysis of this section has led us to the conclusion that a D-brane boundary state corresponds to a dynamical target-space object with tension T_p and R-R $(p+1)$-form charge μ_p. There are competing forces between two D-branes, namely attraction due to exchange of gravitons and dilatons and repulsion due to the exchange of the R-R form. These two forces precisely cancel, independent of the distance between the two branes. In the space-time picture this is due to the equality of charge and tension, i.e. $T_p = \mu_p$.[10] This is reminiscent of the magnetic monopole in the Georgi-Glashow model. There the magnetostatic repulsion of two monopoles is precisely canceled by the attraction due to

---

[9]The calculation is similar to the one in the appendix of Chap. 6; we have to use the propagator for C_{p+1} as derived from the gauge fixed bulk action $\frac{1}{2\kappa_{10}^2} \int d^{10}x \frac{1}{2(p+1)!} C_{\mu_1...\mu_{p+1}} \Box C^{\mu_1...\mu_{p+1}}$.

[10]The general rule is that if the exchange particle has even spin, equal sign charges attract and opposite charges repel. If the spin is odd, the situation is reversed. Exchange of an antisymmetric tensor particle leads to repulsion between equal charges.

Higgs exchange. These monopoles are also called BPS (Bogomol'nyi-Prasad-Sommerfield) monopoles and for the same reason branes with the described properties are called BPS branes. As in the monopole case, supersymmetry is the mechanism responsible for the BPS property. In the annulus partition function space-time supersymmetry is evident by the vanishing of these amplitudes due to the Jacobi triple identity. Here it depends very much on the fact that we consider two branes with the same orientation and same charge, rather than a brane and an anti-brane or branes at angles. We will discuss the condition under which different kinds of branes preserve some supersymmetry in Chap. 10.

In Chap. 18 we will encounter branes as classical solutions to the supergravity equations of motion. There it will become obvious that they are sources of the gravitational field and of p-form fields. This will play a crucial role for the AdS-CFT correspondence. Another context in which D-branes play a central role is in the realization of standard-like models in compactifications of type II string theories to four space-time dimensions. Consistency of the latter constructions needs one more ingredient, which are so-called orientifold planes. The prototype example of such an orientifold construction is the type I string, which we are proceeding to discuss in some detail in the following section. Generalizations and compactifications will be considered in Chap. 12.

9.4 The Type I String

So far we have discussed the ten-dimensional type II superstring theories. These theories contain oriented strings. However, there exists one consistent superstring theory in ten-dimensions which contains non-oriented strings.[11] This is the type I superstring, which we are going to present in this section. We will see that the definition of this theory is only the simplest example of a very general construction.

The starting point is the type IIB superstring theory. Since one is taking the same GSO projection in the left- and right moving Ramond sector, in contrast to the type IIA superstring, it enjoys the symmetry of exchanging the left-moving and the right-moving sectors of the theory. We have encountered this symmetry in previous chapters: the world-sheet parity which acts on the world-sheet coordinates as

$$\Omega : (\sigma, \tau) \to (\ell - \sigma, \tau) . \qquad (9.60)$$

This transformation reverses the orientation of the string. Being a symmetry of type IIB, we can consider its quotient by Ω. This means that we identify the two

---

[11] We have already discussed aspects of non-oriented strings in Chaps. 4, 6 and 8, but so far did not construct a complete model.

Fig. 9.1 The Klein bottle

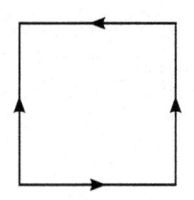

orientations of a string so that in the quotient theory we are dealing with non-oriented strings.

Since Ω is a symmetry, the whole Hilbert space of the type IIB superstring allows for an action Ω. The operator $\frac{1}{2}(1+\Omega)$ projects to Ω-even states. In the one-loop partition function we insert this projection operator into the trace (cf. Eq. (6.155))

$$Z^{\Omega}(\tau,\bar{\tau}) = \text{tr}_{\mathcal{H}}\left(\frac{1+\Omega}{2} P_{\text{GSO}} q^{L_0 - \frac{c}{24}} \bar{q}^{\bar{L}_0 - \frac{c}{24}}\right)$$

$$= \frac{1}{2} Z^{\text{torus}}(\tau,\bar{\tau}) + \frac{1}{2} \text{tr}_{\mathcal{H}}\left(\Omega\, P_{\text{GSO}}\, q^{L_0 - \frac{c}{24}} \bar{q}^{\bar{L}_0 - \frac{c}{24}}\right)$$

$$= \frac{1}{2} Z^{\text{torus}}(\tau,\bar{\tau}) + \frac{1}{2} \text{tr}_{\mathcal{H}}\left(\Omega\, P_{\text{GSO}}\, e^{-4\pi t(L_0 - \frac{c}{24})}\right), \quad (9.61)$$

where in the last line we have used that all states which contribute to the trace with the Ω insertion have $L_0 = \bar{L}_0$ and we have set $\tau_2 = t$. We have denoted the projector onto GSO-even states by P_{GSO}. In the Hamiltonian formulation of the partition function the second term has the interpretation that we have a twist Ω in the time direction, i.e. geometrically the world-sheet is a Klein bottle as shown in Fig. 9.1.

Including the measure factor the Klein bottle vacuum diagram is defined as

$$\mathcal{K} = \int_0^\infty \frac{dt}{2t} \text{tr}_{\mathcal{H}_{\text{cl}}}\left(\Omega\, P_{\text{GSO}}\, e^{-4\pi t(L_0 - \frac{c}{24})}\right). \quad (9.62)$$

The evaluation of this amplitude for the type IIB superstring is straightforward and gives

$$\mathcal{K} = \frac{V_{10}}{(4\pi^2\alpha')^5} \int_0^\infty \frac{dt}{2t} \frac{1}{t^5} \left(\frac{\vartheta_3^4 - \vartheta_4^4 - \vartheta_2^4}{2\,\eta^{12}}\right)(2it), \quad (9.63)$$

where the t^{-5} factor comes from the Gaussian integration over the ten-dimensional closed string center of mass momentum. In the following we will use the dimensionless volume $v_{10} = V_{10}/(4\pi^2\alpha')^5 = V_{10}/\ell_s^{10}$. As discussed in Chap. 6, applying the cut and shift transformations shown in Fig. 6.16, one can transform the Klein bottle into a tube stretched between two so-called crosscaps at the two ends. Therefore, very similar to the loop-channel–tree-channel equivalence for the cylinder diagram, there also exist two descriptions of the Klein bottle amplitude which are related via a

9.4 The Type I String

modular S-transformation. For the Klein bottle amplitude in tree-channel we obtain from (9.63), with $t = 1/(4l)$,

$$\widetilde{\mathcal{K}} = v_{10} \int_0^\infty dl\, 2^4 \left(\frac{\vartheta_3^4 - \vartheta_4^4 - \vartheta_2^4}{\eta^{12}} \right) (2il) . \tag{9.64}$$

In analogy to the tree-channel annulus diagram for two D-branes, we interpret this result as coupling of closed string modes to the crosscaps at the end of the tube. Their locus in the target space-time is called an orientifold plane. In the case considered here, the objects are space-time filling so that we are dealing with O9-planes. Similar to D9-branes these objects have tension and carry R-R 10-form charge. This means that there exist non-vanishing one-point functions (tadpoles) of the NS-NS graviton/dilaton and of the R-R 10-form on the brane. The exchange of massless closed string modes leads to divergences in the tree-channel amplitude which can be seen by expanding the integrand of (9.64), where the assignment R vs. NS can be read off from the ϑ-functions

$$\widetilde{\mathcal{K}} = v_{10} \int_0^\infty dl\, 2^4\, (1_{NS} - 1_R) \left[16\, q^0 + \mathcal{O}(e^{-2\pi l}) \right] . \tag{9.65}$$

Even though the divergences due to exchange of massless particles in the two sectors cancel, we will now argue that this is not sufficient for consistency. Indeed, we learn from the above discussion that an O9-plane is charged under a R-R-field which must be a ten-form C_{10}. Its coupling to an O9-plane with charge μ_{O9} can be included in the effective action through the term

$$S_{\text{eff}} = \mu_{O9} \int_{\mathbb{R}^{1,9}} C_{10} . \tag{9.66}$$

There is no kinetic term for C_{10} because in ten dimensions its field strength vanishes identically, $dC_{10} \equiv 0$ in $d = 10$. Therefore, the equation of motion for C_{10} is $\mu_{O9} = 0$, i.e. the presence of the charged O9-plane violates the equation of motion. In fact, this is just a special case of the general feature that one cannot put a non-vanishing net charge on a compact space. If the space transverse to the charged object is compact, the field lines have nowhere to end on.[12]

The source for this inconsistency is already at the level of the disc diagrams, where there exist non-vanishing one-point functions (tadpoles) of the C_{10} gauge field and the graviton. This C_{10} tadpole leads to a divergence in the tree-channel Klein bottle amplitude in the R-R sector and therefore, as it stands, the Ω quotient of the type IIB superstring is not consistent. However, the analysis above already contains the seed for the resolution of this divergence. If the O9-plane has tension

---

[12] While in the case considered here only D9 branes and O9 planes exhibit this problem, in the compactified theory also lower dimensional D-branes and orientifold planes will lead to potential inconsistencies whenever their transverse space is compact.

and carries R-R charge, then we need to introduce additional objects into the theory which also have tension and R-R C_{10}-charge so that the total charge adds up to zero.

The objects at our disposal are of course the D9-branes of which we now take N. If we put a certain number of D-branes on top of each other, we call it a stack of D-branes. However, as already explained in Chap. 3, multiple branes are accompanied by non-trivial Chan-Paton factors.

$$|\mathbf{m}; i, j\rangle = |\mathbf{m}\rangle \otimes |i, j\rangle = \lambda_{ij}^a |\mathbf{m}; a\rangle, \tag{9.67}$$

where $|\mathbf{m}\rangle$ denotes the states for a single string and $i, j = 1, \ldots, N$ label the branes on which the string starts and ends, respectively. There are N^2 such open strings. The N^2 matrices λ^a can be chosen Hermitian and to form a basis of $N \times N$ matrices which we normalize as $\text{tr}(\lambda^a \lambda^b) = \delta^{ab}$. Hermitian conjugation is $(|i, j\rangle)^\dagger \equiv \langle i, j| = \langle a|\lambda_{ji}^a$. The normalization

$$\langle i, j \mid k, l \rangle = \delta_{ik}\delta_{jl} \tag{9.68}$$

follows from $\langle a|b\rangle = \delta^{ab}$ and $\sum_a \lambda_{ji}^a \lambda_{kl}^a = \delta_{ik}\delta_{jl}$. We now formalize the action of Ω on the Chan-Paton factors. Since Ω changes the orientation of the world-sheet, it clearly interchanges start and end points of open strings. But we can allow for a more general action on the CP indices of the form

$$\Omega |i, j\rangle = \sum_{i', j'=1}^{N} \gamma_{jj'} |j', i'\rangle (\gamma^{-1})_{i'i}. \tag{9.69}$$

This form is dictated by the requirement that an open (ij) string can join with a (jk) string to form a (ik) string. Here (ij) denotes an oriented open string which starts at the i-th brane and ends at the j-th brane.[13] To preserve the normalization (9.68), γ must be unitary.

We now require that the action of Ω on the Chan-Paton labels squares to the identity. This is necessary if we want $\frac{1}{2}(1 + \Omega)$ to be a projection operator. For this we calculate

---

[13] While for this interaction the two endpoints of the string which join have to meet on a brane, there is another interaction between open strings: $(ij) + (kl) \rightarrow (il) + (kj)$. Here the two strings touch at interior points where they split and join while respecting the orientation. This can happen anywhere in the bulk. This local interaction allows for the emission of closed strings from D-branes: two interior points of an (ij) string touch, split and joint, leaving an (ij) open string plus a closed string. The reverse process is the absorption of a closed string. In the CFT description these processes are described by various correlation functions of open and closed string vertex operators.

9.4 The Type I String

$$\Omega^2 |i, j\rangle = \gamma_{jj'} \Omega |j', i'\rangle (\gamma^{-1})_{i'i}$$
$$= \gamma_{jj'} \gamma_{i'i''} |i'', j''\rangle (\gamma^{-1})_{j''j'} (\gamma^{-1})_{i'i}$$
$$= [(\gamma^{-1})^T \gamma]_{ii'} |i', j'\rangle [(\gamma)^{-1} \gamma^T]_{j'j} \stackrel{!}{=} |i, j\rangle, \quad (9.70)$$

from which we infer the constraint on the matrices

$$\gamma^T = \pm \gamma. \quad (9.71)$$

After these preliminaries, we now compute the loop-channel annulus amplitude for open strings stretched between a stack of N D9-branes:

$$\mathscr{A} = N^2 v_{10} \int_0^\infty \frac{dt}{2t} \frac{1}{(2t)^5} \left(\frac{\vartheta_3^4 - \vartheta_4^4 - \vartheta_2^4}{2 \eta^{12}} \right)(it). \quad (9.72)$$

The factor N^2 is from the trace over the Chan-Paton factors. The transformation to tree-channel via $t = \frac{1}{2l}$ gives

$$\widetilde{\mathscr{A}} = N^2 v_{10} \int_0^\infty dl \, 2^{-6} \left(\frac{\vartheta_3^4 - \vartheta_4^4 - \vartheta_2^4}{\eta^{12}} \right)(it)$$
$$= N^2 v_{10} \int_0^\infty dl \, 2^{-6} (1_{\text{NS}} - 1_{\text{R}}) \left[16 q^0 + O(e^{-2\pi l}) \right]. \quad (9.73)$$

Since in the closed string sector we took the Ω quotient, we also have to do so in the open string sector. Its action on the world-sheet coordinates and on the bosonic fields X^μ was given on page 50. On the world-sheet fermions it is

$$\Omega \, \psi_+^\mu(\sigma, \tau) \, \Omega^{-1} = \psi_-^\mu(\ell - \sigma, \tau), \quad (9.74)$$

for DD and NN boundary conditions. In terms of the modes this is

$$\Omega \, \alpha_n^\mu \, \Omega^{-1} = \pm(-1)^n \, \alpha_n^\mu, \qquad \Omega \, b_r^\mu \, \Omega^{-1} = \pm e^{\pi i r} \, b_r^\mu, \quad (9.75)$$

where the upper (lower) sign is for NN (DD) boundary conditions. In the following we will need the former. We also have to specify the action of Ω on the NS and R open string ground states. Since we are interested in supersymmetric string theories, we choose this action such that the total Möbius strip amplitude vanishes.[14] With hindsight we choose

---

[14] Different relative signs for the massless NS and R states correspond to O9-planes whose tension has the opposite sign as the R-R charge.

Fig. 9.2 The Möbius strip

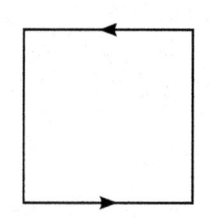

$$\Omega|0\rangle_{NS} = -i|0\rangle_{NS}, \qquad \Omega|a\rangle_R = -|a\rangle_R. \tag{9.76}$$

We will see in a moment that the overall sign is arbitrary and can be compensated by the sign choice in (9.71).

As it stands, Ω does not square to one on all open string states but it does square to one on all states which survive the GSO projection. Also, Ω commutes with P_{GSO}. Therefore, $\frac{1}{2}(1+\Omega)$ is a projection operator in GSO projected amplitudes. If we introduce it into the open string trace, we get the additional contribution

$$\mathcal{M} = \int_0^\infty \frac{dt}{2t}\, \mathrm{tr}_{\mathcal{H}_{op}}\left(\Omega\, P_{GSO}\, e^{-2\pi t(L_0-\frac{c}{24})}\right), \tag{9.77}$$

which is the one-loop partition function on the Möbius strip, as shown in Fig. 9.2. Next we have to analyze the contribution from the Chan-Paton factors. We straightforwardly compute

$$\sum_{i,j=1}^N \langle i,j|\Omega|i,j\rangle = \sum_{i,j,i',j'=1}^N \langle i,j|\gamma_{jj'}|j',i'\rangle(\gamma^{-1})_{i'i}$$

$$= \sum_{i,j,i',j'=1}^N \delta_{ij'}\delta_{ji'}\gamma_{jj'}(\gamma^{-1})_{i'i}$$

$$= \mathrm{Tr}\left(\gamma^T \gamma^{-1}\right) = \pm N, \tag{9.78}$$

where in the final step we used (9.71). In summary, by including a factor of $\pm N$ in the Möbius strip partition function, we can account for a stack of N D-branes. We can now compute the complete Möbius partition function; we obtain[15]

$$\mathcal{M} = \mp N\, v_{10} \int_0^\infty \frac{dt}{2t}\frac{1}{(2t)^5}\left(\frac{\vartheta_3^4 - \vartheta_4^4 - \vartheta_2^4}{2\eta^{12}}\right)\left(it+\frac{1}{2}\right). \tag{9.79}$$

---

[15] This is most easily seen by representing $\Omega = e^{i\pi(N_\alpha+N_b)}$ where N_α and N_b are the number operators for the bosonic and fermionic oscillators, respectively.

9.4 The Type I String

We have seen in Chap. 6 that also the Möbius strip amplitude has a tree-channel interpretation. By applying the cut and shift operations shown in Fig. 6.17, one transforms the Möbius strip into a tube with a boundary on one and a crosscap on the other end. For the transformation to tree-channel one can either use the $P = T^{\frac{1}{2}} S T^2 S T^{\frac{1}{2}}$ transformation introduced in Eq. (6.164) or alternatively one can use the relations

$$\frac{\vartheta_2}{\eta}\left(it + \frac{1}{2}\right) = \frac{\vartheta_2}{\eta}(2it) \frac{\vartheta_4}{\eta}(2it)$$

$$\eta^2\left(it + \frac{1}{2}\right) = \eta^2(2it) \frac{\vartheta_3}{\eta}(2it) \tag{9.80}$$

to avoid dealing with the half-shifted arguments in the ϑ-functions. The tree-channel Möbius strip amplitude can then be written as ($t = 1/8l$)

$$\begin{aligned}\tilde{\mathcal{M}} &= \mp N\, v_{10} \int_0^\infty dl\, 2^8\, l^4\, (1_{\text{NS}} - 1_{\text{R}})\, \frac{\vartheta_2^4\, \vartheta_4^4}{\vartheta_3^4\, \eta^{12}}\left(\frac{i}{4l}\right)\\ &= \mp N\, v_{10} \int_0^\infty dl\, (1_{\text{NS}} - 1_{\text{R}})\, \frac{\vartheta_2^4\, \vartheta_4^4}{\vartheta_3^4\, \eta^{12}}(4il)\\ &= \mp N\, v_{10} \int_0^\infty dl\, (1_{\text{NS}} - 1_{\text{R}})\, \frac{\vartheta_2^4}{\eta^{12}}\left(2il + \frac{1}{2}\right)\\ &= \mp N\, v_{10} \int_0^\infty dl\, (1_{\text{NS}} - 1_{\text{R}})\left[16\, q^0 + O(e^{-2\pi l})\right]. \end{aligned} \tag{9.81}$$

If we combine the divergences from the Klein bottle, the annulus and the Möbius strip amplitude, we find that they are proportional to a perfect square

$$2^{-6}\left(N^2 \mp 2^6 N + 2^{10}\right) = 2^{-6}\left(N \mp 2^5\right)^2. \tag{9.82}$$

Therefore, if we choose the plus sign in Eq. (9.71) and correspondingly the minus sign in the Möbius amplitude, we can cancel the tadpole for $N = 32$. This means that in this case the tension/R-R charge of the O9-plane is negative and that it is canceled by the positive tension/R-R charge of the $N = 32$ D9-branes, i.e.

$$\mu_{O9} = -32\, \mu_{D9}. \tag{9.83}$$

Tadpole cancellation is schematically indicated in Fig. 9.3. The sum of the cylinder, the two Möbius and the Klein bottle diagrams must be a perfect square such that the sum of the disc and the crosscap tadpoles cancel.

We want to stress once more the importance of the cancellation of R-R tadpoles. Quite generally, a tadpole means that the chosen background does not satisfy the

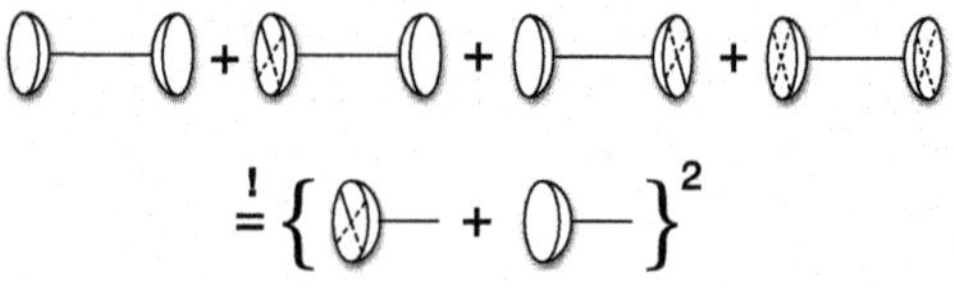

Fig. 9.3 Tadpole cancellation

equations of motion. For NS-NS tadpoles we can shift the background into a consistent one which does satisfy the equations of motion. A R-R tadpole, on the other hand, violates Gauss' law, i.e. charge conservation. It can only be eliminated by a discrete modification of the background, e.g. by adding 32 D9 branes on top of the orientifold plane. To summarize: the Ω quotient of the type IIB superstring can be made one-loop consistent by the addition of $N = 32$ D9-branes. Therefore, this ten-dimensional superstring theory contains non-oriented closed and open strings.

Let us mention with reference to Sects. 4.5 and 6.7 that one can also consider orientifold projections ΩR, where $R : X^i \to -X^i$ for $i = p+1, \ldots, d-1$. Performing an analogous computation as above, one realizes that the tree-channel Klein-bottle and Möbius strip amplitudes can be expressed in terms of crosscap states, which take the form

$$|\text{Op}, \eta\rangle_{\text{NS,R}} = i \frac{2^{p-4}}{\mathcal{N}_{\text{Dp}}} \int dk_\perp \exp\left[-\sum_{n>0} \frac{e^{i\pi n}}{n} \alpha^i_{-n} D_{ij} \overline{\alpha}^j_{-n}\right.$$

$$\left. -i\eta \sum_{r>0} e^{i\pi r} b^i_{-r} D_{ij} \overline{b}^j_{-r}\right] e^{ik_\perp \cdot x_0^\perp} |k_\perp\rangle,$$

(9.84)

where the normalization is such as to lead to tadpole cancellation. The crosscap states satisfy the gluing conditions

$$\left(\alpha^i_n + (-1)^n D^i{}_j \overline{\alpha}^j_{-n}\right) |\text{Op}, \eta\rangle = 0,$$
$$\left(b^i_r + i\eta e^{i\pi r} D^i{}_j \overline{b}^j_{-r}\right) |\text{Op}, \eta\rangle = 0,$$

(9.85)

equipped with the momentum conditions $k^i = 0$, $i = 0, \ldots p$ and unconstrained center of mass momentum in the directions $i = p+1, \ldots, 9$. They can be written as

$$\left(X^i(\sigma) - D^i{}_j X^j(\sigma + \pi)\right) |\text{Op}, \eta\rangle = 0,$$
$$\left(\partial_\tau X^i(\sigma) + D^i{}_j \partial_\tau X^j(\sigma + \pi)\right) |\text{Op}, \eta\rangle = 0,$$
$$\left(\psi^i_-(\sigma) + i\eta D^i{}_j \psi^j_+(\pi + \sigma)\right) |\text{Op}, \eta\rangle = 0.$$

(9.86)

These are to be evaluated at $\tau = 0$ and should be compared to (9.24).

9.4 The Type I String

Combining the NS-NS and R-R contribution as in Eq. (9.50), the overlap of the crosscap state (9.84) with itself gives the tree-channel Klein bottle amplitude

$$\widetilde{\mathscr{K}} = \int_0^\infty dl \, \langle O9|e^{-2\pi l \, H_{\text{cl}}}|O9\rangle \,. \tag{9.87}$$

Similarly, the overlap of a crosscap state with a boundary state gives the tree-channel Möbius strip amplitude

$$\widetilde{\mathscr{M}} = \int_0^\infty dl \, \left(\langle O9|e^{-2\pi l \, H_{\text{cl}}}|D9\rangle + \langle D9|e^{-2\pi l \, H_{\text{cl}}}|O9\rangle\right) \,. \tag{9.88}$$

The massless spectrum of the type I theory can be read off from the Ω projected closed and open string partition functions. In the closed string sector, we get the left-right even states in the NS-NS sector and the left-right odd states in the R-R sector. Moreover, under Ω the NS-R sector is mapped to the R-NS sector and a symmetric linear combination survives. The massless closed string spectrum of the type I string is therefore

$$\text{Bosons}: [(1) + (35)] + [(28)] \,,$$
$$\text{Fermions}: [(8) + (56)] \,, \tag{9.89}$$

i.e. it contains the graviton, the dilaton and the R-R two-form, one gravitino and one dilatino. The projection has reduced the ten-dimensional $\mathscr{N} = 2$ supersymmetry to $\mathscr{N} = 1$, which has 16 supercharges. It is also worth noting that the NS-NS B-field is projected out.

The spectrum is enlarged by extra massless states from the open string sector. Prior to the orientifold projection a stack of $N = 32$ D9-branes supports ten-dimensional $U(32)$ gauge bosons $A_\mu^{(ij)}$, each carrying eight on-shell degrees of freedom. In addition there are gauginos, Majorana-Weyl fermions with eight on-shell degrees of freedom. The gauge fields, together with the gauginos, comprise a vector-multiplet of ten-dimensional $\mathscr{N} = 1$ supersymmetry. However, the orientifold projection acts non-trivially on these gauge fields. Using the notation of the Chan-Paton factors in Eq. (3.55), the corresponding massless open string states transforms under Ω as

$$\Omega : \lambda \, b^\mu_{-\frac{1}{2}}|0\rangle_{\text{NS}} \to -\gamma \, \lambda^T \gamma^{-1} \, b^\mu_{-\frac{1}{2}}|0\rangle_{\text{NS}} \,. \tag{9.90}$$

Therefore, for the invariant gauge bosons, the 32×32 Hermitian matrix of Chan-Paton factors λ has to satisfy $\lambda = -\gamma \, \lambda^T \gamma^{-1}$. A possible representation for γ satisfying Eq. (9.71) with plus sign is $\gamma = \mathbb{1}$. The Chan-Paton factors then have to satisfy $\lambda = -\lambda^T$. Together with $\lambda^\dagger = \lambda$ this gives $\frac{1}{2}N(N-1) = 496$ real degrees of freedom of an $SO(32)$ gauge field.[16]

Since both the annulus and the Möbius strip vanish, the R-sector gives the massless on-shell degrees of freedom of the corresponding gaugino. The massless open string spectrum of the type I superstring therefore reads

$$\text{Bosons}: \quad 496 \times [(8_v)] ,$$
$$\text{Fermions}: \quad 496 \times [(8_s)] . \tag{9.91}$$

We emphasize that, in contrast to D-branes, O-planes do not specify loci of open string end-points and they therefore do not support gauge fields on their world-volume. But due to the projection with which they go along, the gauge group on the D-branes which are required by tadpole cancellation is projected from $U(2N)$ to $SO(2N)$. The intuitive understanding is as follows. Before projection, there are $2N$ branes on top of an O9-plane supporting $2N \times 2N$ oriented open strings. Under Ω an open string from brane a to brane b is identified with the open string (ba) with the opposite orientation. It is clear that depending on the sign Ω acts on the (aa) open strings, the latter are either projected out or are left invariant. In the first case, $N(2N-1) = \dim(SO(2N))$ massless open string states survive while in the second the other $N(2N+1) = \dim(SP(2N))$ states survive. We have just shown, that supersymmetry and tadpole cancellation enforces the first kind of projection for the type I superstring. But if we give up supersymmetry, symplectic groups are also possible.

Let us discuss the tension of the type I D9-brane. Due to the factor $\frac{1}{2}$ in the projection operator, for the orientifold the normalization of the one-loop amplitudes differs by a factor of $1/2$ from those of the type II theories.[17] Thus, the tension μ_p^I of the D-branes in type I is

$$T_p^I = \mu_p^I = \sqrt{\frac{\pi}{2}} \frac{1}{\tilde{\kappa}_{10}} (4\pi^2 \alpha')^{\frac{3-p}{2}} = \frac{1}{\sqrt{2}} T_p^{II} , \quad p = 1, 5, 9 . \tag{9.92}$$

Finally, let us mention that already at the field theory level, the $\mathcal{N} = 1$ supergravity theory with the field content of (9.89) is inconsistent due to the existence of gravitational anomalies in ten dimensions. These anomalies can be canceled by amending the supergravity theory with a super Yang-Mills theory with gauge group $SO(32)$ or $E_8 \times E_8$. The $SO(32)$ gauge group is present in the type I superstring discussed in this section and the $SO(32)$ heterotic superstring theory. The $E_8 \times E_8$ gauge group is perturbatively realized only in the $E_8 \times E_8$ heterotic superstring theory. We will discuss heterotic strings in Chap. 10.

---

[16] It is straightforward to show that if we change the overall sign in (9.76) we need the lower sign in (9.71) and $\lambda = \gamma \lambda^T \gamma^{-1}$. In this case we can choose $\gamma = \begin{pmatrix} 0 & 1 \\ -1 & 0 \end{pmatrix}$ and we also find 496 states in the adjoint of $SO(32)$.

[17] In other words, we do not sum over the two orientations as we did in the orientable case.

This closes our discussion of the type I string which, as was shown, is a string theory which combines closed and open strings in an intricate way. In fact the Ω projection makes the introduction of D-branes a necessity. Such so-called orientifold constructions can be generalized in many ways and have been under intense investigation, also for realizing four-dimensional compactifications of string theory with quasi-realistic gauge groups and matter content. These compact orientifold models will be further discussed in Chap. 15. They contain D-branes and O-planes for various dimensions.

9.5 Stable Non-BPS Branes

From our discussion in the previous section it is clear that the only type IIB BPS Dp-branes which survive the orientifold projection are the D9, D5 and D1 branes. The D1 brane couples 'electrically' to the R-R two-form C_2, D5 branes couples 'magnetically' to C_2 (i.e. it couples to the dual six-form); the D9 branes couple to the non-dynamical C_{10}. These D-branes and their respective boundary states are supersymmetric, i.e. they are BPS-branes.

Using the boundary state formalism we will now argue that besides these supersymmetric BPS D-branes, there also exist non-supersymmetric though perturbatively stable, i.e. tachyon free D-branes. These stable non-BPS branes are a peculiar feature of orientifold and orbifold models and show that geometrically D-branes on compact spaces are not classified by homology classes, as one might naively have guessed, but by K-theory classes. These correctly take into account that there exist always D-branes and their oppositely charged anti-D-branes and that, in addition, both branes and anti-branes support vector bundles on their world-volume. Mathematically K-theory is the framework to make sense of the notion of adding and subtracting vector bundles. D-branes are a very concrete and intuitive physical realization of these notions because D-branes carry gauge fields and (sometimes tachyonic) charged matter fields on their world-volume and the process of putting D-branes on top of each other or annihilating D-branes and anti-D-branes corresponds to adding and subtracting vector bundles.

As we have seen before, the pure NS-NS boundary state (9.38), though GSO invariant, is not stable in type II string theory. This can be detected by computing from the overlap of two boundary states the tree-channel annulus diagram, and then transform it to loop-channel, where a tachyonic mode appears. Recall that due to the lack of a R-R part in this boundary state, there is no GSO projection in the loop-channel.

However, in the type I string one performs not only a GSO but also an orientifold projection on the spectrum. It is therefore conceivable that the tachyonic mode, though not projected out by the GSO projection, is eventually removed by the Ω projection. We will show in this section that for certain Dp branes this is indeed the case.

Let us reconsider the pure NS boundary state

$$|\widetilde{\mathrm{D}p}\rangle = \frac{\lambda}{\mathcal{N}_{\mathrm{D}p}} \left(|\mathrm{D}p, \eta = 1\rangle_{\mathrm{NS}} - |\mathrm{D}p, \eta = -1\rangle_{\mathrm{NS}}\right), \tag{9.93}$$

where $\mathcal{N}_{\mathrm{D}p}$ is the normalization found for BPS Dp-branes in (9.53) and λ sets the overall normalization which, as usual, we will later constrain by open-closed string duality. Since we are looking for a dynamical D-brane with positive tension, we need to require $\lambda > 0$.

One can straightforwardly compute the tree-channel and loop-channel annulus diagrams for such a single brane

$$\mathcal{A} = \frac{\lambda^2}{2} \int_0^\infty \frac{dt}{t} \frac{V_{p+1}}{(8\pi^2 \alpha' t)^{\frac{p+1}{2}}} \frac{(\vartheta_3^4 - \vartheta_2^4)}{\eta^{12}}(it). \tag{9.94}$$

Comparison with (9.52) teaches us the following two properties of the $\widetilde{\mathrm{D}p}$ branes. They do not couple to R-R fields and their tension is λT_p.

The next step is to compute the Möbius strip amplitude from the overlap of boundary state (9.93) and the O9-plane from Eq. (9.84). This overlap can be written as

$$\widetilde{\mathcal{M}} = \int_0^\infty dl \left(\langle \mathrm{O}9|e^{-2\pi l H_{\mathrm{cl}}}|\widetilde{\mathrm{D}p}\rangle + \langle \widetilde{\mathrm{D}p}|e^{-2\pi l H_{\mathrm{cl}}}|\mathrm{O}9\rangle\right)$$

$$= -\frac{32\lambda V_{p+1}}{\mathcal{N}_{\mathrm{D}p}\mathcal{N}_{\mathrm{D}9}} \int_0^\infty dl\, 2^{\frac{13-p}{2}} \left[\frac{\left(\frac{\vartheta_3}{\eta}\right)^{\frac{p-1}{2}} \left(\frac{\vartheta_4}{\eta}\right)^{\frac{9-p}{2}}}{\eta^{p-1}\left(\frac{\vartheta_2}{\eta}\right)^{\frac{9-p}{2}}} \left(2il + \frac{1}{2}\right) \right.$$

$$\left. - \frac{\left(\frac{\vartheta_4}{\eta}\right)^{\frac{p-1}{2}} \left(\frac{\vartheta_3}{\eta}\right)^{\frac{9-p}{2}}}{\eta^{p-1}\left(\frac{\vartheta_2}{\eta}\right)^{\frac{9-p}{2}}} \left(2il + \frac{1}{2}\right) \right]. \tag{9.95}$$

The transformation to loop-channel via $l = \frac{1}{8t}$ proceeds via the P-transformation

$$\left(\frac{\vartheta_3}{\eta}\right)\left(\frac{i}{4t} + \frac{1}{2}\right) = e^{i\frac{\pi}{4}} \left(\frac{\vartheta_4}{\eta}\right)\left(it + \frac{1}{2}\right),$$

$$\left(\frac{\vartheta_4}{\eta}\right)\left(\frac{i}{4t} + \frac{1}{2}\right) = e^{-i\frac{\pi}{4}} \left(\frac{\vartheta_3}{\eta}\right)\left(it + \frac{1}{2}\right),$$

9.5 Stable Non-BPS Branes

$$\left(\frac{\vartheta_2}{\eta}\right)\left(\frac{i}{4t}+\frac{1}{2}\right) = \left(\frac{\vartheta_2}{\eta}\right)\left(it+\frac{1}{2}\right),$$

$$\eta\left(\frac{i}{4t}+\frac{1}{2}\right) = \frac{1}{\sqrt{2t}}\,\eta\left(it+\frac{1}{2}\right) \tag{9.96}$$

and yields, after a few elementary steps,

$$\mathcal{M} = \frac{\lambda}{2}\int_0^\infty \frac{dt}{t}\,\frac{V_{p+1}}{(8\pi^2\alpha' t)^{\frac{p+1}{2}}}\, 2^{\frac{9-p}{2}} \left[e^{i\frac{\pi}{4}(5-p)}\,\frac{\left(\frac{\vartheta_3}{\eta}\right)^{\frac{p-1}{2}}\left(\frac{\vartheta_4}{\eta}\right)^{\frac{9-p}{2}}}{\eta^{p-1}\left(\frac{\vartheta_2}{\eta}\right)^{\frac{9-p}{2}}}\left(it+\frac{1}{2}\right) \right.$$

$$\left. - e^{-i\frac{\pi}{4}(5-p)}\,\frac{\left(\frac{\vartheta_4}{\eta}\right)^{\frac{p-1}{2}}\left(\frac{\vartheta_3}{\eta}\right)^{\frac{9-p}{2}}}{\eta^{p-1}\left(\frac{\vartheta_2}{\eta}\right)^{\frac{9-p}{2}}}\left(it+\frac{1}{2}\right) \right]. \tag{9.97}$$

Extracting the tachyonic mode from the annulus (9.94) and Möbius strip (9.97) partition functions, one finds

$$\mathcal{A} + \mathcal{M} \simeq \int_0^\infty \frac{dt}{t}\,\frac{V_{p+1}}{(8\pi^2\alpha' t)^{\frac{p+1}{2}}}\,\frac{q^{-1/2}}{2}\left(\lambda^2 + 2\lambda\sin\left[\frac{\pi}{4}(5-p)\right]\right). \tag{9.98}$$

With the brane tension λ restricted to be positive, this implies that only for the cases $p \in \{-1, 0, 6, 7, 8\}$ the tachyon has a chance to be removed by the orientifold projection. For p even a more detailed analysis shows that the world-sheet parity operation acts on the open string ground states with eigenvalue $e^{-i\frac{\pi}{4}p}$. For $p = 6$, $\Omega^2 = -1$ and $(1 + \Omega)/2$ is not a projection operator and the above analysis is meaningless. There is no such obstruction in the other cases so that we are left with the stable non-BPS branes shown in Table 9.2.

Recall from (9.93) that $\lambda = 1$ corresponds to the normalization of the BPS Dp-branes of the type I superstring. The non-BPS D(−1) instanton and the non-BPS D7 brane both have $\lambda = 2$ and can be considered as a brane–anti-brane pair, whose tachyonic mode is projected out by Ω.

If one computes the loop-channel annulus diagram for an open string stretched between such a non-BPS brane and the 32 D9-branes present in the type I string theory, one finds that here only the D(−1) and D0-branes are tachyon free.

Table 9.2 Stable non-BPS branes

p	−1	0	7	8
λ	2	$\sqrt{2}$	2	$\sqrt{2}$

Table 9.3 K-theory groups for Type I superstring

p	-1	0	1	5	7	8	9
$KO(S^{(9-p)})$	$\mathbb{Z}_2$	$\mathbb{Z}_2$	$\mathbb{Z}$	$\mathbb{Z}$	$\mathbb{Z}_2$	$\mathbb{Z}_2$	$\mathbb{Z}$

Of particular interest is the non-BPS particle, i.e. the non-BPS D0-brane, whose existence has been inferred as a consequence of the proposed $SO(32)$ heterotic-type I strong-weak duality.[18] Indeed the massless Ramond open string states between the non-BPS D0-brane and the 32 D9-branes indicate that the D0 brane transforms in the spinor representation of $SO(32)$. Such spinorial states are present in the perturbative heterotic string, but are not among the perturbative states of the type I string.

With a stack of at least two non-BPS branes, one can show that the Ω projection cannot remove all tachyonic modes, so that a pair of such branes is unstable and decays into something else. For completeness, but without further explanation, we want to mention that this is related to K-theory classes, $KO(S^{(9-p)})$, which are all $\mathbb{Z}_2$ valued. In Table 9.3 we list the relevant K-theory classes for stable D-brane in type I string theory. They reflect the fact that BPS branes carry $\mathbb{Z}$-valued charge and stable multiple brane configurations exist (only branes and anti-branes can annihilate) while non-BPS branes carry $\mathbb{Z}_2$ charge and annihilate in pairs. The K-theory classes for other values of p are trivial.

9.6 Appendix: Theta-Functions and Twisted Fermionic Partition Functions

In this appendix we collect results on ϑ-functions which will prove useful and we prove the Poisson resummation formula which is used to derive the modular transformation properties of partition functions. Finally, we generalize the computation of the fermionic partition functions to general boundary conditions.

Theta-Functions

Theta-functions are actually functions of two variables $\vartheta\begin{bmatrix}\alpha\\\beta\end{bmatrix}(z|\tau)$. They have a series expansion ($\operatorname{Im}\tau > 0$)[19]

---

[18] Duality symmetries will be discussed in Chap. 18.

[19] This representation of the theta-function has an immediate generalization to higher genus Riemann surfaces Σ_g:

$$\vartheta\begin{bmatrix}a\\b\end{bmatrix}(z|\Omega) = \sum_{n\in\mathbb{Z}} \exp\left[i\pi(n+a)^T \Omega(n+a) + 2\pi i(n+a)^T(z+b)\right]. \quad (9.99)$$

9.6 Appendix: Theta-Functions and Twisted Fermionic Partition Functions

$$\vartheta\begin{bmatrix}\alpha\\\beta\end{bmatrix}(z|\tau) = \sum_{n\in\mathbb{Z}} \exp\left[i\pi(n+\alpha)^2\tau + 2\pi i(n+\alpha)(z+\beta)\right] . \tag{9.100}$$

From the discussion in Sect. 9.1 it is clear that the parameters (α,β) encode the periodicity conditions in the (ξ^1,ξ^2) directions of the torus. From (9.100) one finds

$$\vartheta\begin{bmatrix}\alpha\\\beta\end{bmatrix}(z|\tau) = \vartheta\begin{bmatrix}\alpha\\\beta+z\end{bmatrix}(0|\tau) , \tag{9.101a}$$

$$\vartheta\begin{bmatrix}\alpha\\\beta\end{bmatrix}(z+m+n\tau|\tau) = e^{2\pi i m\alpha - i\pi n^2\tau - 2\pi i n(z+\beta)}\vartheta\begin{bmatrix}\alpha\\\beta\end{bmatrix}(z|\tau) , \tag{9.101b}$$

$$\vartheta\begin{bmatrix}\alpha\\\beta\end{bmatrix}(z|\tau) = e^{i\pi\alpha^2\tau + 2\pi i\alpha(z+\beta)}\vartheta\begin{bmatrix}0\\0\end{bmatrix}(z+\alpha\tau+\beta|\tau) , \tag{9.101c}$$

$$\vartheta\begin{bmatrix}\alpha+m\\\beta+n\end{bmatrix}(z|\tau) = e^{2\pi i n\alpha}\vartheta\begin{bmatrix}\alpha\\\beta\end{bmatrix}(z|\tau) \quad (m,n\in\mathbb{Z}) , \tag{9.101d}$$

$$\vartheta\begin{bmatrix}-\alpha\\-\beta\end{bmatrix}(z|\tau) = \vartheta\begin{bmatrix}\alpha\\\beta\end{bmatrix}(-z|\tau) . \tag{9.101e}$$

The first argument, or the shift in the ξ^2 boundary condition, is important if one couples the fermions to an external field. We will use the notation $\vartheta(\tau) \equiv \vartheta(0|\tau)$. An alternative representation of the theta-functions is as an infinite product ($-1/2 \leq \alpha,\beta \leq 1/2$)

$$\frac{\vartheta\begin{bmatrix}\alpha\\\beta\end{bmatrix}(\tau)}{\eta(\tau)} = e^{2\pi i\alpha\beta}q^{\frac{\alpha^2}{2}-\frac{1}{24}}\prod_{n=1}^{\infty}\left(1+q^{n+\alpha-\frac{1}{2}}e^{2\pi i\beta}\right)\left(1+q^{n-\alpha-\frac{1}{2}}e^{-2\pi i\beta}\right) \tag{9.102}$$

with $q = e^{2\pi i\tau}$. For completeness we repeat the definition of the Dedekind η-function

$$\eta(\tau) = q^{1/24}\prod_{n=1}^{\infty}(1-q^n) . \tag{9.103}$$

For $\alpha,\beta \in \{0,\tfrac{1}{2}\}$ the ϑ-functions have special names,

$$\vartheta\begin{bmatrix}1/2\\1/2\end{bmatrix} = \vartheta_1, \quad \vartheta\begin{bmatrix}1/2\\0\end{bmatrix} = \vartheta_2, \quad \vartheta\begin{bmatrix}0\\0\end{bmatrix} = \vartheta_3, \quad \vartheta\begin{bmatrix}0\\1/2\end{bmatrix} = \vartheta_4 . \tag{9.104}$$

$a,b,z \in \mathbb{R}^g$, $n \in \mathbb{Z}^g$ and Ω is the period matrix. a and b denote the spin-structure or, more generally, encode the periodicity conditions around the a and b cycles of the Riemann surface.

It is not hard to show the transformation properties under S and T transformations, the generators of the modular group:

$$\vartheta\begin{bmatrix}\alpha\\\beta\end{bmatrix}(\tau+1) = e^{-\pi i(\alpha^2-\alpha)}\,\vartheta\begin{bmatrix}\alpha\\\alpha+\beta-\frac{1}{2}\end{bmatrix}(\tau), \tag{9.105a}$$

$$\vartheta\begin{bmatrix}\alpha\\\beta\end{bmatrix}\left(-\frac{1}{\tau}\right) = \sqrt{-i\tau}\,e^{2\pi i\alpha\beta}\,\vartheta\begin{bmatrix}-\beta\\\alpha\end{bmatrix}(\tau), \qquad |\arg\sqrt{-i\tau}| < \frac{\pi}{2}. \tag{9.105b}$$

The first equation follows directly from the sum representation. To show the second one, one uses the Poisson resummation formula, which we will prove below. For completeness we also repeat the modular transformations of the Dedekind η-function:

$$\eta(\tau+1) = e^{i\pi/12}\eta(\tau), \tag{9.106a}$$

$$\eta(-1/\tau) = \sqrt{-i\tau}\,\eta(\tau). \tag{9.106b}$$

The transformation rules for a general $SL(2,\mathbb{Z})$ transformation are more difficult to derive. We simply state the result in a form which will be useful later. The combination

$$Z\begin{bmatrix}\alpha\\\beta\end{bmatrix}(\tau) = e^{i\pi\alpha(\beta+1)}\,\frac{\vartheta\begin{bmatrix}\alpha+\frac{1}{2}\\-\beta-\frac{1}{2}\end{bmatrix}(\tau)}{\eta(\tau)} \tag{9.107}$$

transforms in a simple way:

$$Z\begin{bmatrix}\alpha\\\beta\end{bmatrix}\left(\frac{a\tau+b}{c\tau+d}\right) = \varepsilon_g\,Z\begin{bmatrix}a\alpha-c\beta\\d\beta-b\alpha\end{bmatrix}(\tau), \tag{9.108}$$

where ε_g depends only on the $SL(2,\mathbb{Z})$ element $g = \begin{pmatrix}a&b\\c&d\end{pmatrix}$. It satisfies $\varepsilon_g^{12} = 1$; e.g. $\varepsilon_T = e^{i\pi/6}$ and $\varepsilon_S = e^{i\pi/2}$ for the two generators $T = \begin{pmatrix}1&1\\0&1\end{pmatrix}$ and $S = \begin{pmatrix}0&1\\-1&0\end{pmatrix}$ of $SL(2,\mathbb{Z})$.

The theta-functions satisfy the following useful identities:

$$\vartheta_2^4(\tau) - \vartheta_3^4(\tau) + \vartheta_4^4(\tau) = 0 \qquad \text{(Riemann identity)}, \tag{9.109a}$$

$$\vartheta_2(\tau)\,\vartheta_3(\tau)\,\vartheta_4(\tau) = 2\,\eta^3(\tau) \qquad \text{(Jacobi triple product identity)}, \tag{9.109b}$$

$$\vartheta_1'(\tau) = 2\pi\,\eta^3(\tau). \tag{9.109c}$$

9.6 Appendix: Theta-Functions and Twisted Fermionic Partition Functions

The physical content of the Riemann identity is the vanishing of the one-loop partition function of the type II theories due to a space-time supersymmetric spectrum.

A generalization of this supersymmetry identity, which will be needed in Chap. 10, is the generalized Riemann identity

$$\sum_{\alpha,\beta\in\{0,\frac{1}{2}\}} e^{2\pi i(a+b)} \prod_{i=1}^{4} e^{2\pi i\alpha h_i} \vartheta\begin{bmatrix}\alpha+g_i\\\beta+h_i\end{bmatrix}(\tau) = 0, \quad \text{if} \quad \begin{array}{l}\sum_i g_i = 0 \bmod 1\\ \sum_i h_i = 0 \bmod 1\end{array}. \tag{9.110}$$

Poisson Resummation Formula

The transformation properties of partition functions under the modular S-transformation can be derived via the Poisson resummation formula. In its general form, this formula needs some notions from the theory of lattices, which we will discuss in more detail in Chaps. 10 and 11. Consider the function

$$F(x) = \sum_{p\in\Lambda} e^{-\pi(p+x)^T A(p+x) + 2\pi i y\cdot(p+x)}, \tag{9.111}$$

where the sum runs over the points of an n-dimensional lattice Λ and A is a symmetric positive definite $n\times n$ matrix. x and y are arbitrary vectors. Since

$$F(x+p) = F(x) \tag{9.112}$$

for $p\in\Lambda$, we can expand it in a Fourier series

$$F(x) = \sum_{q\in\Lambda^*} e^{2\pi i x\cdot q} F^*(q), \tag{9.113}$$

where

$$F^*(q) = \frac{1}{\text{vol}(\Lambda)} \int_{\text{unit cell}} d^n x\, e^{-2\pi i q\cdot x} F(x). \tag{9.114}$$

Here Λ^* is the dual lattice. Inserting Eq. (9.111), combining the sum over Λ and the integral over the unit cell to an integral over all of $\mathbb{R}^n$, we find, after doing the Gaussian integral, the Poisson resummation formula

$$\sum_{p\in\Lambda} e^{-\pi(p+x)^T A(p+x) + 2\pi i y\cdot(p+x)}$$
$$= \frac{1}{\text{vol}(\Lambda)\sqrt{\det A}} \sum_{q\in\Lambda^*} e^{-2\pi i q\cdot x - \pi(y+q)A^{-1}(y+q)}. \tag{9.115}$$

Twisted Fermions

In the first part of this chapter we have seen that the four theta-functions ϑ_i, $i = 1, \ldots, 4$ are in one-to-one correspondence with the four possible spin-structures on a torus. The spin-structures are defined in terms of the periodicities of a fermion along the two homology cycles of the torus. For real fermions the only allowed phases are ± 1. For complex fermions more general phases are allowed. We will now compute the one-loop partition function of a complex fermion with so-called twisted boundary conditions. A similar discussion for twisted bosons will be given in Chap. 10.

We define $\psi = \frac{1}{\sqrt{2}}(\psi^1 + i\psi^2)$ and $\bar\psi = \frac{1}{\sqrt{2}}(\psi^1 - i\psi^2)$.[20] The action for one chirality (say the right-movers) is

$$S = \frac{i}{\pi}\int d^2\sigma\, \bar\psi \partial_+ \psi \tag{9.116}$$

with energy-momentum tensor

$$T = \frac{i}{2}\left(\bar\psi \partial_- \psi + \psi \partial_- \bar\psi\right). \tag{9.117}$$

The only non-zero anti-commutator is

$$\{\bar\psi(\sigma^0,\sigma^1), \psi(\sigma^0,\sigma^{1\prime})\} = 2\pi\delta(\sigma^1 - \sigma^{1\prime}). \tag{9.118}$$

Impose the twisted boundary conditions[21]

$$\psi(\sigma^0, \sigma^1 + 2\pi) = -e^{+2\pi i\alpha}\psi(\sigma^0, \sigma^1),$$
$$\bar\psi(\sigma^0, \sigma^1 + 2\pi) = -e^{-2\pi i\alpha}\bar\psi(\sigma^0, \sigma^1). \tag{9.119}$$

The mode expansions compatible with (9.119) are

$$\psi(\sigma^-) = \sum_{n\in\mathbb{Z}} b_{n+\alpha+\frac{1}{2}}\, e^{-i(n+\alpha+\frac{1}{2})(\sigma^0-\sigma^1)},$$
$$\bar\psi(\sigma^-) = \sum_{n\in\mathbb{Z}} \bar b_{n-\alpha-\frac{1}{2}}\, e^{-i(n-\alpha-\frac{1}{2})(\sigma^0-\sigma^1)} \tag{9.120}$$

---

[20] Here a bar means complex conjugate, not left-movers.
[21] In terms of the coordinates ξ^1, ξ^2 the boundary conditions (9.119) and (9.125) were given in (9.12). $\xi^1 \to \xi^1 + 1$ corresponds to $(\sigma^0, \sigma^1) \to (\sigma^0, \sigma^1 + 2\pi)$ and $\xi^2 \to \xi^2 + 1$ to $(\sigma^0, \sigma^1)) \to (\sigma^0 + 2\pi\tau_2, \sigma^1 + 2\pi\tau_1)$. The complex coordinates on the torus are $z = \sigma^1 + i\sigma^0 = 2\pi(\xi^1 + \tau\xi^2)$.

9.6 Appendix: Theta-Functions and Twisted Fermionic Partition Functions

where

$$\bar{b}_{n-\alpha-\frac{1}{2}} = \left(b_{-n+\alpha+\frac{1}{2}}\right)^\dagger. \tag{9.121}$$

The non-trivial anti-commutator of the modes is

$$\left\{\bar{b}_{n-\alpha-\frac{1}{2}}, b_{m+\alpha+\frac{1}{2}}\right\} = \delta_{m+n,0}. \tag{9.122}$$

Here, b_η and $\bar{b}_\eta$ are annihilation operators for $\eta > 0$ and creation operators for $\eta < 0$. The Hamiltonian is

$$H = \sum_{m \in \mathbb{Z}} \left(m + \alpha + \frac{1}{2}\right) :b^\dagger_{m+\alpha+\frac{1}{2}} b_{m+\alpha+\frac{1}{2}}: + \frac{\alpha^2}{2} - \frac{1}{24} = L_0 - \frac{c}{24} \tag{9.123}$$

with

$$L_0 = \sum_{m=1}^\infty \left\{ \left(m + \alpha - \frac{1}{2}\right) \bar{b}_{-m-\alpha+\frac{1}{2}} b_{m+\alpha-\frac{1}{2}} \right.$$

$$\left. + \left(m - \alpha - \frac{1}{2}\right) b_{-m+\alpha+\frac{1}{2}} \bar{b}_{m-\alpha-\frac{1}{2}} \right\} + \frac{\alpha^2}{2} \tag{9.124}$$

and $c = 1$ for one complex fermion. Using zeta-function regularization (3.44), the normal ordering constant in (9.123) is $-\sum_{m \geq 0}(m - \alpha + \frac{1}{2}) = \frac{\alpha^2}{2} - \frac{1}{24}$. The constant in L_0 can be computed directly as the one-point function of the energy-momentum tensor on the plane in the vacuum annihilated by the b_η and $\bar{b}_\eta$ with $\eta > 0$ via $\langle T(z) \rangle = \frac{\alpha^2}{2z^2} = \frac{1}{z^2}\langle L_0 \rangle$. These expressions are valid for $-\frac{1}{2} < \alpha < \frac{1}{2}$. The expressions for $\frac{1}{2} \leq \alpha < 1$ follow via the replacement $\alpha \to \alpha - 1$.

We now compute the partition function for one complex fermion on a sector where, in addition to (9.119), the fermions satisfy

$$\psi(\sigma^0 + 2\pi\tau_2, \sigma^1 + 2\pi\tau_1) = -e^{+2\pi i \beta} \psi(\sigma^0, \sigma^1),$$

$$\bar{\psi}(\sigma^0 + 2\pi\tau_2, \sigma^1 + 2\pi\tau_1) = -e^{-2\pi i \beta} \bar{\psi}(\sigma^0, \sigma^1), \tag{9.125}$$

where $\tau = \tau_1 + i\tau_2$ is the modular parameter of the world-sheet torus. The minus signs in (9.125) are automatic in the trace as it corresponds to a path-integral with anti-periodic boundary conditions. If we want periodic boundary conditions we have to insert $(-1)^F$. But we still need to implement the β-twist, i.e. we look for an operator U which satisfies

$$U b_{n+\alpha+\frac{1}{2}} U^{-1} = e^{+2\pi i \beta} b_{n+\alpha+\frac{1}{2}},$$

$$U \bar{b}_{n+\alpha+\frac{1}{2}} U^{-1} = e^{-2\pi i \beta} \bar{b}_{n+\alpha+\frac{1}{2}}. \tag{9.126}$$

This operator is easily found to be

$$U = e^{2\pi i \beta (N-\overline{N})}, \qquad (9.127)$$

where $N, \overline{N}$ are the number operators

$$N = \sum_{\eta>0} b_{-\eta}\bar{b}_{\eta}, \qquad \overline{N} = \sum_{\eta>0} \bar{b}_{-\eta}b_{\eta}. \qquad (9.128)$$

The partition function is ($q = e^{2\pi i \tau}$)

$$Z\begin{bmatrix}\alpha\\\beta\end{bmatrix}(\tau) = \mathrm{Tr}\left(e^{2\pi i \beta(N-\overline{N})} q^{L_0 - \frac{1}{24}}\right)$$

$$= q^{\frac{\alpha^2}{2}-\frac{1}{24}} \prod_{n=1}^{\infty} (1 + q^{n+\alpha-\frac{1}{2}} e^{-2\pi i \beta})(1 + q^{n-\alpha-\frac{1}{2}} e^{+2\pi i \beta})$$

$$= e^{2\pi i \alpha \beta} \frac{\vartheta\begin{bmatrix}\alpha\\-\beta\end{bmatrix}(\tau)}{\eta(\tau)} = \det(\partial_{\alpha,\beta}). \qquad (9.129)$$

Up to a phase this result is valid for all α, β. The last equality expresses the relation between the Hamiltonian formulation and the functional integral of the vacuum transition amplitude.

Comparison with (9.108) shows that the transformation of the boundary conditions under modular transformations, Eq. (9.3), are encoded in the partition function (9.129).

Further Reading

Spin structures in string theory and their relation to modular invariance were first discussed in

- N. Seiberg, E. Witten, Spin structures in string theory. Nucl. Phys. **B276**, 272 (1986)

Further insight into the subject is given in

- L. Alvarez-Gaumé, G.W. Moore, C. Vafa, Theta functions, modular invariance, and strings. Comm. Math. Phys. **106**, 1 (1986)

The need for the $(-1)^F$ insertion into the trace with periodic boundary conditions for fermions is explained in

- R. Dashen, B. Hasslacher, A. Neveu, Phys. Rev. **D12**, 2443 (1975)

9.6 Appendix: Theta-Functions and Twisted Fermionic Partition Functions

A discussion on spin structures for open and non-orientable world sheets and all one-loop vacuum amplitudes of the superstring can be found in

- C.P. Burgess, T.R. Morris, Open superstrings à la Polyakov. Nucl. Phys. **B291**, 285 (1987)

For crosscap states for type II theories see

- J. Polchinski, Y. Cai, Consistency of open superstring theories. Nucl. Phys. **B296**, 91 (1988)
- C.G. Callan Jr., C. Lovelace, C.R. Nappi, S.A. Yost, Adding holes and crosscaps to the superstring. Nucl. Phys. **B293**, 83 (1987)

A good review on D-branes and boundary states is

- P. Di Vecchia, A. Liccardo, *D-branes in string theory I*, NATO Adv.Study Inst.Ser.C.Math.Phys.Sci. 556 (2000) 1 [arXiv:hep-th/9912161]; *D-branes in string theory II*, YITP Proceedings Series No. 4, p. 7, Kyoto 1999 [arXiv:hep-th/9912275]

The action of symmetries on CP-factors has been developed in

- E. Gimon, J. Polchinski, Consistency conditions for orientifolds and D-manifolds. Phys. Rev. **D54**, 1667 (1996) [arXiv:hep-th/9601038]

A review on general orientifolds is

- C. Angelantonj, A. Sagnotti, Open strings. Phys. Rep. **371**, 1 (2002) [arXiv:hep-th/0204089]

Non-BPS branes have been discussed in

- A. Sen, Stable non BPS states in string theory. JHEP **9806**, 007 (1998) [arXiv:hep-th/9803194]
- A. Sen, Stable non BPS bound states of BPS D-branes. JHEP **9808**, 010 (1998) [arXiv:hep-th/9805019]
- M. Frau, L. Gallot, A. Lerda, P. Strigazzi, Stable non BPS D-branes in type I string theory. Nucl. Phys. **B564**, 60 (2000) [arXiv:hep-th/9903123]

Chapter 10
Toroidal Compactifications: 10-Dimensional Heterotic String

Abstract So far we discussed the 26-dimensional bosonic string and three kinds of 10-dimensional superstring theories, the type IIA/B theories and the type I theory. One option to make contact with the four-dimensional world is to compactify the closed string theories on compact spaces. We first study the simplest examples, toroidal compactifications of the bosonic string and the type II superstring theories. These feature a new symmetry, called T-duality. To break supersymmetry, however, one has to compactify on non-flat spaces. The simplest such class are toroidal orbifolds. Moreover, we introduce two additional superstring theories in ten-dimensions, which are hybrid theories of a right-moving superstring and a left-moving bosonic string, whose additional sixteen dimensions are compactified on the weight-lattice of $\mathrm{Spin}(32)/\mathbb{Z}_2$ or $E_8 \times E_8$. We then study D-branes on toroidal spaces and how they transform under T-duality. We introduce intersecting D-branes, their T-dual images and simple orientifolds on such toroidal spaces.

10.1 Motivation

In previous chapters we have described two kinds of closed oriented string theories. First, the closed bosonic string theory was formulated consistently in 26 dimensional Minkowski space-time. The spectrum of physical states contains a negative $(\mathrm{mass})^2$ scalar tachyon and, at the next level, with $(\mathrm{mass})^2 = 0$, a symmetric traceless tensor (the graviton), an antisymmetric tensor field and a scalar dilaton. These states are accompanied by an infinite tower of massive excitations. As such, the closed bosonic string has several serious drawbacks: flat 26-dimensional space-time, a tachyon and no space-time fermions.

Some of these difficulties are overcome with the 10-dimensional fermionic string theories. Modular invariance requires a GSO projection of the spectrum. This projects out the tachyon and leads to a space-time supersymmetric spectrum. The lightest states are again a massless graviton, a dilaton and several anti-symmetric tensor fields. These bosonic fields are accompanied by their superpartners, namely two gravitini and two dilatini. The theory has $\mathcal{N} = 2$ space-time supersymmetry in ten dimensions. A further projection, which involves the world-sheet parity transformation, leads to a ten-dimensional theory with non-Abelian $SO(32)$ gauge group and $\mathcal{N} = 1$ supersymmetry.

To relate these higher dimensional fundamental string theories to the physics in our four-dimensional world, one has to explain why we do not see the extra space-time dimensions. This can be achieved by compactifying some of the string coordinates on an internal compact space of sufficiently small size so that the masses of the so-called Kaluza-Klein modes are beyond the energies reached by particle physics experiments.

The notion of compactification should not be taken too literally. With the most general ansatz of constructing four-dimensional string theories, the concept of critical dimension is replaced by the requirement that the total central charge of the Virasoro algebra be zero. Four-dimensional space-time can be realized by introducing only four flat string coordinates (i.e. free CFT bosons/fermions) corresponding to four-dimensional Minkowski space-time and in addition a two-dimensional (super) conformal field theory (including a ghost sector which reflects the consistent coupling to world-sheet gravity) which has to satisfy the consistency constraints of unitarity, locality, modular invariance etc. Simple realizations of these internal conformal field theories are two-dimensional bosons whose target space is a torus.

Due to the decoupling of the left- and right-moving world-sheet sectors, one can define a hybrid superstring theory whose right-moving sector is that of the 10-dimension fermionic string and whose left-moving sector that of the 26-dimensional bosonic string compactified on special 16-dimensional toroidal spaces. Absence of tachyons and modular invariance imposes further constraints and eventually leads to the two heterotic superstring theories. These aspects of toroidal compactifications will be the main subject of this chapter. An interesting generalization are strings on orbifolds which we also introduce in this chapter. In later chapters we will expand on this and also discuss more general compactifications both as conformal field theories and by geometric means.

Let us mention that many constructions may turn out to be quantum mechanically equivalent, e.g. via the two-dimensional equivalence between bosons and fermions which we discuss in Sect. 11.4. String dualities lead to further redundancies in the description of physically inequivalent compactifications. They will be discussed in some detail in Chap. 18.

In this chapter, for economy of notation, we use $\ell = 2\pi$ for the closed string and $\ell = \pi$ for the open string; σ and τ are dimensionless.

10.2 Toroidal Compactification of the Closed Bosonic String

For illustrative purposes we first consider the simplest case of one coordinate compactified on a circle of radius R. It means that for one spatial coordinate, e.g. X^{25}, we require periodicity such that points on the real axis are identified

10.2 Toroidal Compactification of the Closed Bosonic String

according to
$$X^{25} \sim X^{25} + 2\pi R L, \quad L \in \mathbb{Z}. \tag{10.1}$$

X^{25} parametrizes a one-dimensional circle S^1 of radius R. Here S^1 is the quotient of the real line by the integers times $2\pi R$, i.e.
$$S^1 = \mathbb{R}/(2\pi R \, \mathbb{Z}) \tag{10.2}$$

which also defines the equivalence relation Eq. (10.1). The coordinate $X^{25}(\sigma, \tau)$, $0 \leq \sigma \leq 2\pi$, maps the closed string onto the spatial circle $0 \leq X^{25} \leq 2\pi R$. The identification (10.1) requires to modify the periodicity condition of a closed string to
$$X^{25}(\sigma + 2\pi, \tau) = X^{25}(\sigma, \tau) + 2\pi R L. \tag{10.3}$$

The second term gives rise to strings which are closed on the circle S^1 but not on its covering space $\mathbb{R}$. After quantization this leads to new states, called winding states; they are characterized by the winding number L that counts how many times the string wraps around the circle. This phenomenon has no counterpart in the theory of point particles on a circle. The winding states are topologically stable solitons; the winding number cannot be changed without breaking the string; it is a conserved charge. Such solitons always exist if the internal manifold contains non-contractible loops. States with non-trivial winding constitute the so-called twisted sector: twisted sector states are closed on the quotient manifold but not on the original manifold. As we will see in detail below, the partition function is only modular invariant, if we include the twisted sector states.

The following mode expansion for $X^{25}(\sigma, \tau)$ respects Eq. (10.3)
$$X^{25}(\sigma, \tau) = x^{25} + \alpha' p^{25} \tau + L R \sigma$$
$$+ i\sqrt{\frac{\alpha'}{2}} \sum_{n \neq 0} \frac{1}{n} \left(\alpha_n^{25} e^{-in(\tau - \sigma)} + \bar{\alpha}_n^{25} e^{-in(\tau + \sigma)} \right). \tag{10.4}$$

x^{25} and p^{25} obey the usual commutation relation
$$[x^{25}, p^{25}] = i. \tag{10.5}$$

p^{25} generates translations of x^{25}. Single-valuedness of the wave function $\exp(ip^{25} x^{25})$ restricts the allowed internal momenta to discrete values:
$$p^{25} = \frac{M}{R}, \quad M \in \mathbb{Z}. \tag{10.6}$$

The quantized momentum states are called Kaluza-Klein modes. We split $X^{25}(\sigma, \tau)$ into left and right-movers

$$X_R^{25}(\tau - \sigma) = \frac{1}{2}(x^{25} - c) + \frac{\alpha'}{2}\left(\frac{M}{R} - \frac{LR}{\alpha'}\right)(\tau - \sigma) + i\sqrt{\frac{\alpha'}{2}}\sum_{n \neq 0}\frac{1}{n}\alpha_n^{25}e^{-in(\tau-\sigma)},$$

$$= x_R^{25} + \alpha' p_R^{25}(\tau - \sigma) + \text{oscillators},$$

$$X_L^{25}(\tau + \sigma) = \frac{1}{2}(x^{25} + c) + \frac{\alpha'}{2}\left(\frac{M}{R} + \frac{LR}{\alpha'}\right)(\tau+\sigma) + i\sqrt{\frac{\alpha'}{2}}\sum_{n \neq 0}\frac{1}{n}\bar{\alpha}_n^{25}e^{-in(\tau+\sigma)},$$

$$= x_L^{25} + \alpha' p_L^{25}(\tau + \sigma) + \text{oscillators}. \tag{10.7}$$

In the quantized theory x^{25}, M and c, L are operators whose eigenvalues we denote by the same letters. The mass operator receives contributions from the winding states:

$$\alpha' m_L^2 \equiv \frac{\alpha'}{2}\left(\frac{M}{R} + \frac{LR}{\alpha'}\right)^2 + 2(N_L - 1) = 2(\bar{L}_0 - 1),$$

$$\alpha' m_R^2 \equiv \frac{\alpha'}{2}\left(\frac{M}{R} - \frac{LR}{\alpha'}\right)^2 + 2(N_R - 1) = 2(L_0 - 1),$$

$$\alpha' m^2 = \alpha'(m_L^2 + m_R^2) = \alpha'\frac{M^2}{R^2} + \frac{1}{\alpha'}L^2R^2 + 2(N_L + N_R - 2), \tag{10.8}$$

where $m^2 = -\sum_{\mu=0}^{24} p_\mu p^\mu$. Moreover, $\frac{M^2}{R^2}$ is the contribution to the mass from the momentum in the compact dimension and the term $\frac{1}{\alpha'^2}L^2R^2$ is the energy required to wrap the string around the circle L times. Physical states have to satisfy the reparametrization constraint Eq. (3.22)

$$m_L^2 = m_R^2 \quad \leftrightarrow \quad N_R - N_L = ML. \tag{10.9}$$

Let us examine the spectrum of this theory which is effectively 25-dimensional. First consider states with zero winding and zero internal momentum.

1. The lowest energy state is again the scalar tachyon with $\alpha' m^2 = -4$ (we suppress the space-time momentum)

$$|\text{tachyon}\rangle = |0\rangle. \tag{10.10}$$

2. At the massless level with $N_L = N_R = 1$ there are now the 25-dimensional graviton, antisymmetric tensor and dilaton[1]

$$|G^{\mu\nu}\rangle = \alpha_{-1}^\mu \bar{\alpha}_{-1}^\nu |0\rangle \quad \mu, \nu = 0, \ldots 24 \tag{10.11}$$

with the oscillators in the 25 uncompactified space-time directions.

---

[1] Previously we had considered the spectrum in light-cone gauge. Here we give the Lorentz-covariant form of the states. The unphysical degrees of freedom decouple in the critical dimension as a result of the physical state conditions $L_n = \bar{L}_n = 0$ for $n > 0$.

10.2 Toroidal Compactification of the Closed Bosonic String

3. In addition to these states, which were already present in the uncompactified theory, there are also new states which arise from the compactification. We can replace one space-time oscillator by an internal oscillator to get two vector states:

$$|V_1^\mu\rangle = \alpha^\mu_{-1}\bar\alpha^{25}_{-1}|0\rangle,$$
$$|V_2^\mu\rangle = \alpha^{25}_{-1}\bar\alpha^\mu_{-1}|0\rangle. \qquad (10.12)$$

These massless vectors originate from the Kaluza-Klein compactification of the bosonic string on the circle—they are the $(\mu, 25)$ components of the graviton $(G_{\mu,25})$ and antisymmetric tensor field $(B_{\mu,25})$. They give rise to a $U(1)_L \times U(1)_R$ gauge symmetry. Since these gauge bosons arise by dimensional reduction of the higher dimensional graviton and antisymmetric tensor field, they already appear in any field theoretical compactification. The identification of massless vector states in the excitation spectrum of the string with gauge bosons will be explained in Chap. 16 when we discuss the low-energy effective field theory.

4. Finally, acting with two internal oscillators on the vacuum we obtain a massless scalar field which is an internal degree of freedom of the 26-dimensional metric; its vacuum expectation value corresponds to the radius R of the circle:

$$|\phi\rangle = \alpha^{25}_{-1}\bar\alpha^{25}_{-1}|0\rangle. \qquad (10.13)$$

More interesting are the states with non-trivial internal momentum and winding number, obtained by acting with the oscillators on the soliton vacua $|M, L\rangle$. We will concentrate on the first winding sector, $M = \pm L = \pm 1$. Choosing $M = L = \pm 1$ we derive from Eq. (10.8) that

$$\alpha' m_L^2 = \frac{\alpha'}{2R^2} + \frac{R^2}{2\alpha'} + 2N_L - 1,$$

$$\alpha' m_R^2 = \frac{\alpha'}{2R^2} + \frac{R^2}{2\alpha'} + 2N_R - 3,$$

$$\alpha' m^2 = \frac{\alpha'}{R^2} + \frac{R^2}{\alpha'} + 2(N_L + N_R - 2). \qquad (10.14)$$

The level matching constraint Eq. (10.9) is satisfied if $N_L = 0$, $N_R = 1$. We therefore find two vector states of the form

$$|V_a^\mu\rangle = \alpha^\mu_{-1}|\pm 1, \pm 1\rangle, \quad a = 1, 2, \quad \mu = 0, \ldots, 24 \qquad (10.15)$$

and two scalars

$$|\phi_a\rangle = \alpha^{25}_{-1}|\pm 1, \pm 1\rangle, \quad a = 1, 2, \qquad (10.16)$$

whose mass depends on the radius of the circle

$$\alpha' m^2(R) = \frac{\alpha'}{R^2} + \frac{R^2}{\alpha'} - 2. \tag{10.17}$$

Analogously we can set $M = -L = \pm 1$. Then Eq. (10.9) is satisfied if $N_L = 1$, $N_R = 0$ and we obtain again two vectors and two scalars

$$|V'^{\mu}_a\rangle = \bar{\alpha}^{\mu}_{-1}|\pm 1, \mp 1\rangle, \quad a = 1, 2, \quad \mu = 0, \ldots, 24, \tag{10.18}$$

$$|\phi'_a\rangle = \bar{\alpha}^{25}_{-1}|\pm 1, \mp 1\rangle, \tag{10.19}$$

whose mass is also given by Eq. (10.17). It is easy to see that $m^2(R) \geq 0$ with equality for $R = \sqrt{\alpha'}$. This means that for this particular radius, which is determined by the string tension, we get extra massless states, of which the massless vectors are of particular interest. This phenomenon is of utmost importance in the theory of the compactified bosonic string. Together with the massless vectors (10.12), these four additional massless vectors complete the adjoint representation of $SU(2)_L \times SU(2)_R$. The oscillator excitations (10.12) with zero winding number and zero Kaluza-Klein momentum correspond to the $U(1)_L \times U(1)_R$ Cartan subalgebra generators of $SU(2)_L \times SU(2)_R$, the soliton states of Eqs. (10.15) and (10.18) to the roots (this will be put on more rigorous grounds in the next chapter where we also review some background on Lie algebras). $(M + L, M - L)$ are the $U(1)_L \times U(1)_R$ quantum numbers. It is easy to convince oneself that these additional massless vectors are the only ones possible for any choice of M, L and R. We also have extra tachyons $M = 0$, $L = \pm 1$ and $M = \pm 1$, $L = 0$. It is important to stress once more that while the gauge bosons corresponding to the $U(1)_L \times U(1)_R$ Cartan subalgebra generators are Kaluza-Klein states familiar from field theoretical compactifications on a circle, the non-Abelian gauge bosons are a characteristic string effect and cannot appear in any point-particle compactification.

At the special radius $R = \sqrt{\alpha'}$, where the gauge symmetry is enhanced, there are eight additional massless scalars

$$\alpha^{25}_{-1}|\pm 1, \pm 1\rangle, \quad \bar{\alpha}^{25}_{-1}|\pm 1, \mp 1\rangle, \quad |\pm 2, 0\rangle, \quad |0, \pm 2\rangle \tag{10.20}$$

which, together with $\alpha^{25}_{-1}\bar{\alpha}^{25}_{-1}|0, 0\rangle$ (cf. Eq. (10.13)) form the $(\mathbf{3}, \mathbf{3})$ representation of $SU(2)_L \times SU(2)_R$. However, for arbitrary values of the radius, both the four non-commuting gauge bosons of $SU(2)_L \times SU(2)_R$ and the four scalars with internal oscillator excitations are massive—the gauge symmetry is broken to $U(1)_L \times U(1)_R$. Therefore this phenomenon can be interpreted as a stringy Higgs effect. For arbitrary radii these four massive scalars are the longitudinal components of the four massive vector particles. Two of the remaining scalars in Eq. (10.20) become massive and the other two become tachyons.

The $U(1)_L \times U(1)_R$ gauge bosons on the other hand stay massless for all values of R. This is also the case for the single scalar of Eq. (10.13). In a low energy effective field theory this neutral scalar will have a completely flat potential which corresponds to the freedom of choosing the radius of the circle as a free

10.2 Toroidal Compactification of the Closed Bosonic String

parameter. Said differently, the spectrum of the bosonic string compactified on S^1 is characterized by a single parameter, also called a modulus, which is the radius of the circle. In a low-energy effective field theory description (cf. Chap. 16) it is the vacuum expectation value of a scalar field which is associated with the state (10.13).

One might be tempted to conclude that the space of physically inequivalent S^1-compactifications is parametrized by $R/\alpha' \in (0, \infty)$. However, inspection of the mass formula Eq. (10.8) shows that the spectrum is invariant under the transformation $R \to \frac{\alpha'}{R}$. Simultaneous interchange of winding and momentum numbers $L \leftrightarrow M$ maps winding states to momentum states and vice versa. The transformation, which is a symmetry of the S^1 compactified bosonic string,

$$R \to \frac{\alpha'}{R}, \qquad L \leftrightarrow M \tag{10.21}$$

is an example of a duality transformation, called T-duality. The radius $R = \sqrt{\alpha'}$ is a fixed point of this transformation. It means, at least as far as the excitation spectrum is concerned, the compactified string theory looks the same regardless whether we consider it at large or small radius of the internal circle. Therefore, the spectrum of the compactified bosonic string is completely characterized by the values $R \geq \sqrt{\alpha'}$, or, equivalently, $0 < R \leq \sqrt{\alpha'}$. In other words, the moduli space of this theory is not the whole positive real axis but only one of the above intervals.

If we consider not only the spectrum but also the mode expansion (10.7), we realize that under T-duality, Eq. (10.21), the momentum modes transform as $(p_L, p_R) \to (p_L, -p_R)$. For T-duality to be a symmetry of the entire string theory and not only its zero mode sector, we need to extend this action to the oscillator modes such that

$$(X_L, X_R) \to (X_L, -X_R) \tag{10.22}$$

or

$$X(\sigma, \tau) = X_L(\sigma^+) + X_R(\sigma^-) \to X'(\sigma, \tau) = X_L(\sigma^+) - X_R(\sigma^-). \tag{10.23}$$

This also requires $c \leftrightarrow x$. Therefore, T-duality acts as an asymmetric $\mathbb{Z}_2$ reflection of the right moving world-sheet boson leaving its left-moving partner invariant. This again reflects the fact that such a symmetry is of truly string origin. Since X and X' have the same energy-momentum tensor and the same operator product expansions, the conformal field theories of X and X' are identical. This demonstrates that T-duality is a symmetry of the bosonic string theory and not just its excitation spectrum.

It is now also clear how this transformation generalizes to the superstring. Let us introduce the notation $\psi_L = \psi_+$ and $\psi_R = \psi_-$. Due to world-sheet supersymmetry T-duality must also act on the world-sheet fermions as an asymmetric reflection

$$(\psi_L, \psi_R) \to (\psi_L, -\psi_R). \tag{10.24}$$

This simple transformation has an important consequence. Recall from Eq. (8.34) that the action of the right-moving GSO projection on the Ramond sector ground state can be expressed in light cone gauge as $(-1)^F = 16 \prod_{i=2}^{9} b_0^i$. The b_0^i being the zero-modes of ψ_R^i we conclude that T-duality along a single S^1 changes the sign of the right-moving GSO projection in the Ramond sector. Therefore it exchanges the type IIB and the type IIA superstring. Summarizing, T-duality acts on the type II superstrings as

$$\text{type IIB on } S^1 \text{ with radius } R \leftrightarrow \text{ type IIA on } S^1 \text{ with radius } (\alpha'/R). \quad (10.25)$$

For compactifications on higher dimensional tori, this generalizes straightforwardly. For an odd number of T-dualities, i.e. asymmetric reflections, the type IIB and type IIA superstring theories are exchanged, whereas for a even number they are mapped to themselves.

An important consequence of the T-duality symmetry is that we cannot distinguish a string theory compactified on a large circle with radius R from one compactified on a small circle with radius α'/R. To understand this one has to recall that in ordinary quantum mechanics position eigenstates are defined as the Fourier transformed momentum eigenstates. Quantization of momentum on the circle leads to $|x + 2\pi R\rangle = |x\rangle$. In string theory we could, however, have defined position eigenstates as Fourier transformed of winding eigenstates which would lead to $|\tilde{x} + \frac{2\pi\alpha'}{R}\rangle = |\tilde{x}\rangle$ with radius α'/R. The first definition is useful for large R where there are low mass momentum states but only high mass winding states. For small R (or large α'/R) the situation is reversed. This leads to an effective minimum size of $R = \sqrt{\alpha'}$, which is the fixed point of the duality symmetry.

We now want to generalize this discussion to the compactification of D bosonic coordinates on a D-dimensional torus T^D. The resulting theory is effectively $(26-D)$-dimensional. The torus is defined by identifying points in the D-dimensional internal space as follows (compact dimensions are labeled with capital letters):

$$X^I \sim X^I + 2\pi \sum_{i=1}^{D} n^i e_i^I = X^I + 2\pi L^I, \quad n_i \in \mathbb{Z} \quad (10.26)$$

with

$$L^I = \sum_{i=1}^{D} n^i e_i^I, \quad n_i \in \mathbb{Z}. \quad (10.27)$$

The $e_i = \{e_i^I\}$ ($i = 1, \ldots, D$) are D linear independent vectors which generate a D-dimensional lattice Λ_D. The $\boldsymbol{L} = \{L^I\}$ are lattice vectors,[2] i.e. $\boldsymbol{L} \in \Lambda_D$. The torus on which we compactify is obtained by dividing $\mathbb{R}^D$ by $2\pi\Lambda_D$:

$$T^D = \mathbb{R}^D/2\pi\Lambda_D. \quad (10.28)$$

---

[2] Some basis facts about lattices are collected at the beginning of Sect. 11.2.

10.2 Toroidal Compactification of the Closed Bosonic String

Center of mass position and momentum satisfy canonical commutation relations

$$[x^I, p_J] = i\delta^I_J, \tag{10.29}$$

i.e. p_I generates translations of x^I and single-valuedness of $e^{ix^I p_I}$ requires that $L^I p_I \in \mathbb{Z}$, i.e. the allowed momenta have to lie on the lattice which is dual to Λ_D, denoted by Λ_D^*, with basis vectors e^{*i}:

$$p_I = m_i\, e^{*i}_I, \qquad m_i \in \mathbb{Z}. \tag{10.30}$$

The $e^{*i} = \{e^{*i}_I\}$ are dual to the e_i, i.e.

$$e_i \cdot e^{*j} = e^I_i\, e^{*j}_I = \delta^j_i. \tag{10.31}$$

From this

$$e^I_i\, e^{*i}_J = \delta^I_J. \tag{10.32}$$

follows. For later use we define the metric on Λ_D

$$g_{ij} = e_i \cdot e_j = e^I_i\, e^J_j\, \delta_{IJ} \tag{10.33}$$

and on Λ_D^*:

$$g^*_{ij} \equiv g^{ij} = e^{*i} \cdot e^{*j} = e^{*i}_I\, e^{*j}_J\, \delta^{IJ} = (g^{-1})_{ij}. \tag{10.34}$$

Note that $e^{*i} = g^{ij} e_j$. The volumes of the unit cells are $\text{vol}(\Lambda_D) = \sqrt{\det g}$ and $\text{vol}(\Lambda_D^*) = \sqrt{\det g^*} = \frac{1}{\sqrt{\det g}}$.

The condition which a closed string in the compact directions has to satisfy is

$$X^I(\sigma + 2\pi, \tau) = X^I(\sigma, \tau) + 2\pi L^I. \tag{10.35}$$

Nontrivial L^I means nontrivial winding. If we define dimensionless fields $X_{L,R}$ via $X^I = \sqrt{\frac{\alpha'}{2}}(X^I_L + X^I_R)$, then their mode expansions are

$$X^I_L = x^I_L + p^I_L(\tau + \sigma) + i\sum_{n\neq 0}\frac{1}{n}\tilde{\alpha}^I_n e^{-in(\tau+\sigma)},$$

$$X^I_R = x^I_R + p^I_R(\tau - \sigma) + i\sum_{n\neq 0}\frac{1}{n}\alpha^I_n e^{-in(\tau-\sigma)}, \tag{10.36}$$

where we have also defined dimensionless left and right momenta

$$p^I_{L,R} = \frac{1}{\sqrt{2}}(\sqrt{\alpha'} p^I \pm \frac{1}{\sqrt{\alpha'}} L^I) \qquad (10.37)$$

and dimensionless left and right center-of-mass coordinates

$$x^I_{L,R} = \frac{1}{\sqrt{2\alpha'}} x^I \pm \sqrt{\frac{\alpha'}{2}} Q^I \quad \text{with} \quad Q^I = \frac{1}{\alpha'} c^I. \qquad (10.38)$$

The mass formula is[3] $(m^2 = -\sum_{\mu=0}^{25-D} p_\mu p^\mu)$

$$\alpha' m_L^2 = \frac{\alpha'}{2} \sum_{I=1}^{D} \left(p^I + \frac{1}{\alpha'} L^I \right)^2 + 2(N_L - 1) = \boldsymbol{p}_L^2 + 2(N_L - 1) = 2(\bar{L}_0 - 1),$$

$$\alpha' m_R^2 = \frac{\alpha'}{2} \sum_{I=1}^{D} \left(p^I - \frac{1}{\alpha'} L^I \right)^2 + 2(N_R - 1) = \boldsymbol{p}_R^2 + 2(N_R - 1) = 2(L_0 - 1)$$

$$(10.39)$$

so that

$$\alpha' m^2 = \alpha'(m_L^2 + m_R^2)$$

$$= 2(N_R + N_L - 2) + \sum_{I=1}^{D} \left(\alpha' p^I p^J + \frac{1}{\alpha'} L^I L^J \right) \delta_{IJ}$$

$$= 2(N_L + N_R - 2) + \sum_{i,j=1}^{D} \left(\alpha' m_i g^{ij} m_j + \frac{1}{\alpha'} n^i g_{ij} n^j \right). \quad (10.40)$$

The constraint Eq. (10.9) generalizes to

$$N_R - N_L = \boldsymbol{p} \cdot \boldsymbol{L} = \sum_{i=1}^{D} m_i n^i. \qquad (10.41)$$

Using this information one shows that the $2D$-dimensional vectors $\boldsymbol{P} = (\boldsymbol{p}_L, \boldsymbol{p}_R)$ build an even self-dual lattice[4] $\Gamma_{D,D}$ if we choose the signature of the metric of this lattice to be $((+1)^D, (-1)^D)$; i.e. $\boldsymbol{P} \cdot \boldsymbol{P'} = \sum_I (p_L^I p_L^{\prime J} - p_R^I p_R^{\prime J}) \delta_{IJ}$. Therefore $\Gamma_{D,D}$ is called a Lorentzian lattice. With (10.27) and (10.30) the scalar product between

---

[3] These and other expressions given below are easily generalized to the compactified fermionic string. $N_{L,R}$ then include the fermionic oscillators and the zero-mode part (-2) has to be modified (0 in R and -1 in NS sectors).

[4] Our notation is such that we denote the left- and right-moving momentum lattices by $\Gamma_{L,R}$ and the winding vector lattice by Λ.

10.2 Toroidal Compactification of the Closed Bosonic String

any two lattice vectors is

$$\boldsymbol{P} \cdot \boldsymbol{P}' = \sum_{i=1}^{D}(m_i\, n'^i + m'_i\, n^i) \in \mathbb{Z} \tag{10.42}$$

and, clearly, $\boldsymbol{P} \cdot \boldsymbol{P} \in 2\mathbb{Z}$. A lattice with these properties is called even. Below we will show that the lattice is self-dual. The momentum lattices which occur in torus compactifications of the string are also called Narain lattices. Note however that, given the vectors $(\boldsymbol{p}_L, \boldsymbol{p}_R) \in \Gamma_{D,D}$, the set of vectors $\boldsymbol{p}_L$, $\boldsymbol{p}_R$ do not, for a general metric on the torus, form separate lattices Γ_L, Γ_R in spite of closing under addition. For example, consider the two-dimensional even Lorentzian self-dual lattice $\Gamma_{1,1}$ consisting of points

$$\left(\sqrt{\frac{\alpha'}{2}}\frac{M}{R} + \frac{1}{\sqrt{2\alpha'}}LR,\; \sqrt{\frac{\alpha'}{2}}\frac{M}{R} - \frac{1}{\sqrt{2\alpha'}}LR \right) \in \Gamma_{1,1}. \tag{10.43}$$

The left (or right) components alone, even though they close under addition, do not form a one-dimensional lattice. For general real values of R we cannot write all possible $\boldsymbol{p}_L$ as the integer multiple of one basis vector. A torus compactification where the left and right momenta $\boldsymbol{p}_L$, $\boldsymbol{p}_R$ build separately Euclidean lattices, Γ_L and Γ_R, is called rational. The notation 'rational' is used since Γ_L, Γ_R can be decomposed into a finite number of cosets (see Chap. 11). In this case the possible $[U(1)]_L^D \times [U(1)]_R^D$ charges (see (10.44)) of the Kaluza-Klein and winding states are rational numbers and the corresponding conformal field theory is rational. For example, the lattice $\Gamma_{1,1}$ is only rational if R^2/α' is a rational number.

As before, the sector without any winding and internal momentum contains a $(26 - D)$-dimensional tachyon, a massless graviton, antisymmetric tensor and dilaton. Furthermore there exist $2D$ massless vectors

$$|V_1^{\mu I}\rangle = \alpha_{-1}^{\mu}\overline{\alpha}_{-1}^{I}|0\rangle,$$
$$|V_2^{\mu I}\rangle = \overline{\alpha}_{-1}^{\mu}\alpha_{-1}^{I}|0\rangle \tag{10.44}$$

which are the gauge bosons of $[U(1)]_L^D \times [U(1)]_R^D$. Finally, there are D^2 massless scalars

$$|\phi^{IJ}\rangle = \alpha_{-1}^{I}\overline{\alpha}_{-1}^{J}|0\rangle. \tag{10.45}$$

These scalar fields are associated to the moduli of D-dimensional torus compactifications of the bosonic string. Of the D^2 scalars, $\frac{1}{2}D(D+1)$ are the internal graviton components; their vacuum expectation values give the constant background parameters G_{IJ} which specify the shape of the D-dimensional torus T^D. The remaining $\frac{1}{2}D(D-1)$ scalars are the internal components of the antisymmetric tensor field B_{IJ} which may also acquire constant vacuum expectation values. This kind of background fields will also influence the string spectrum and therefore enter the mass formula Eq. (10.39).

The anti-symmetric tensor background field $B_{\mu\nu}$ couples to the bosonic fields X^μ via an additional term in the string action

$$S = -\frac{1}{4\pi\alpha'} \int d^2\sigma \left(\eta^{\alpha\beta} G_{\mu\nu} \partial_\alpha X^\mu \partial_\beta X^\nu + \epsilon^{\alpha\beta} B_{\mu\nu} \partial_\alpha X^\mu \partial_\beta X^\nu \right). \quad (10.46)$$

The B-field does not affect the energy-momentum tensor,[5] and if B is constant it does not contribute to the equations of motion for X^μ and neither does it change the commutation relations of the oscillators. But the momentum canonically conjugate to X^μ is modified (we use $\epsilon^{01} = -1$):

$$\Pi_\mu = \frac{1}{2\pi\alpha'}(G_{\mu\nu}\dot{X}^\nu + B_{\mu\nu}X'^\nu) \quad (10.47)$$

which leads to the center of mass momentum

$$\pi_\mu = \int_0^{2\pi} \Pi_\mu d\sigma = G_{\mu\nu}P^\nu + \frac{1}{\alpha'}B_{\mu\nu}L^\nu. \quad (10.48)$$

This shows that winding states ($L^I \neq 0$) are charged under $B_{\mu I}$ while KK momentum states ($P^I \neq 0$) are charged under $G_{\mu I}$.

In the following we take a constant antisymmetric tensor with non-vanishing components only in the compactified directions and $G_{IJ} = \delta_{IJ}$. The center of mass momentum along the compact directions is $\pi_I = G_{IJ}p^J + \frac{1}{\alpha'}B_{IJ}L^J$. It is π_I which generates translations and it must therefore lie on the lattice Λ_D^* which is dual to Λ_D, i.e. $\pi_I = m_i e_I^{*i}$. But it is p^I which enters the mass formula. p_L and p_R can be expressed in terms of π^I instead of p^I:

$$(p_I)_{L,R} = \sqrt{\frac{\alpha'}{2}} \left(\pi_I \pm \frac{1}{\alpha'}(G_{IJ} \mp B_{IJ}) L^J \right)$$

$$= e_I^{*i} \left(\sqrt{\frac{\alpha'}{2}} m_i \pm \frac{1}{\sqrt{2\alpha'}} g_{ij} n^j - \frac{1}{\sqrt{2\alpha'}} b_{ij} n^j \right), \quad (10.49)$$

where $b_{ij} = e_i^I B_{IJ} e_j^J$ with inverse $B_{IJ} = e_I^{*i} b_{ij} e_J^{*j}$ and, as before, $g_{ij} = e_i^I G_{IJ} e_j^J$. Employing an obvious matrix notation, the mass formula is $m^2 = m_L^2 + m_R^2$ with

$$\alpha' m_{L,R}^2 = p_{L,R}^2 + 2(N_{L,R} - 1)$$
$$= \frac{\alpha'}{2} \mathbf{m}^T \mathbf{g}^{-1} \mathbf{m} + \frac{1}{2\alpha'} \mathbf{n}^T (\mathbf{g} - \mathbf{b}\mathbf{g}^{-1}\mathbf{b}) \mathbf{n} + \mathbf{n}^T \mathbf{b} \mathbf{g}^{-1} \mathbf{m} \pm \mathbf{n}^T \mathbf{m} + 2(N_{L,R} - 1)$$
$$(10.50)$$

and level matching $(L_0 - \bar{L}_0) = \int d\sigma \, \Pi \cdot X' = 0$ is the same as (10.41).

---

[5] This can e.g. be seen by noting that the coupling to B can be written, prior to fixing conformal gauge, in the form $\int B_{\alpha\beta} d\sigma^\alpha \wedge d\sigma^\beta$ where $B_{\alpha\beta} = \partial_\alpha X^I \partial_\beta X^J B_{IJ}$ is the pull-back to the world-sheet. This expression in independent of the world-sheet metric.

10.2 Toroidal Compactification of the Closed Bosonic String

It is again instructive to calculate the inner product of two vectors $\boldsymbol{P} = (\boldsymbol{p}_L, \boldsymbol{p}_R)$, $\boldsymbol{P}' = (\boldsymbol{p}'_L, \boldsymbol{p}'_R) \in \Gamma_{D,D}$:

$$\boldsymbol{P} \cdot \boldsymbol{P}' = \sum_{i=1}^{D}(m_i n'^i + m'_i n^i). \tag{10.51}$$

It does not depend on the choice of the D^2 background parameters g_{ij} and b_{ij}. Using this background independence we can take, for example, $e_i^I = \sqrt{\alpha'}\delta_i^I$ and $b_{ij} = 0$. The lattice $(\boldsymbol{p}_L, \boldsymbol{p}_R)$ is then spanned by the vectors $\frac{1}{\sqrt{2}}(m^I + n^I, m^I - n^I)$ which is clearly even Lorentzian self-dual.[6] We conclude that $\Gamma_{D,D}$ is self-dual for any value of the background fields.

We have seen that the torus compactification of the bosonic string is described by D^2 real parameters. The D^2 dimensional parameter space of the background fields is the moduli space of the torus compactification. Different values of the D^2 parameters correspond to different choices of the Lorentzian, self-dual lattice $\Gamma_{D,D}$. This fact is very useful to obtain more information about the geometrical structure of the moduli space. It is known that all possible Lorentzian self-dual lattices $\Gamma_{D,D}$ can be obtained by a $O(D, D)$ rotations of some reference lattice Γ_0, which can always be chosen to correspond to $b_{ij} = 0$ and $g_{ij} = \alpha'\delta_{ij}$. However, not every Lorentz rotation leads to a different string theory since the string spectrum is invariant under separate rotations $O(D)_L$, $O(D)_R$ of the vectors $\boldsymbol{p}_L$ and $\boldsymbol{p}_R$ (see the mass formula (10.39)). Therefore, distinct compactified string theories, i.e. different points in the moduli space correspond to points in the coset manifold $\frac{O(D,D)}{O(D) \times O(D)}$ which is of dimension D^2. We conclude that the geometrical structure of the moduli space is given by this manifold. However the string spectrum is again invariant under generalized, discrete duality transformations involving the background fields g_{ij} and b_{ij}. The global structure of the moduli space is quite complicate as those points in the above coset, which are connected by the duality transformations, have to be identified.

It is not difficult to identify the transformations which generalize (10.21). Inspection of (10.50) reveals that $p_{L,R}^2$ are invariant under the map

$$\boldsymbol{m} \leftrightarrow \boldsymbol{n} \quad \text{and} \quad \alpha'\mathbf{g}^{-1} \leftrightarrow \frac{1}{\alpha'}(\mathbf{g} - \mathbf{b}\mathbf{g}^{-1}\mathbf{b})$$

$$\mathbf{b}\mathbf{g}^{-1} \leftrightarrow -\mathbf{g}^{-1}\mathbf{b}, \tag{10.52}$$

which is equivalent to

$$\frac{1}{\alpha'}(\mathbf{g} + \mathbf{b}) \leftrightarrow \alpha'(\mathbf{g} + \mathbf{b})^{-1}. \tag{10.53}$$

---

[6] The lattice is of the form $\otimes \Gamma_{1,1}$ where the two basis vectors of each $\Gamma_{1,1}$ are $\frac{1}{\sqrt{2}}(1, 1)$ and $\frac{1}{\sqrt{2}}(1, -1)$. They also span $\Gamma_{1,1}^*$.

For $\mathbf{b} = 0$ and a diagonal metric $\mathbf{g}$, in which case $T^D = S^1 \times \cdots \times S^1$, this is becomes (10.21) for each circle. But $p_{L,R}^2$ are also invariant under

$$\mathbf{m} \to \mathbf{m} - \mathbf{Nn} \quad \text{and} \quad \mathbf{b} \to \mathbf{b} + \alpha' \mathbf{N}, \qquad (10.54)$$

where $\mathbf{N} = (N_{ij})$ is an arbitrary antisymmetric matrix with integer elements. One can show that the transformations (10.52) and (10.54) generate the group $O(D, D; \mathbb{Z})$ and that the moduli space is $\frac{O(D,D)}{O(D) \times O(D)} / O(D, D; \mathbb{Z})$. The discrete symmetries leave $p_{L,R}^2$ invariant, but not $p_{L,R}$. Their change has again to be compensated by corresponding changes in $X_{L,R}$ to ensure that the symmetry is a true symmetry of the theory, i.e. not only of the spectrum but also of the CFT.

A non-trivial instructive example is the 2-torus $T^2 = \mathbb{R}^2 / \Lambda$ which has one Kähler modulus and one complex structure modulus.[7] e_1, e_2 are the two generators of Λ and the metric is $g_{ij} = e_i \cdot e_j$. The Kähler modulus is the area $\sqrt{\det g}$. In the presence of an antisymmetric field $b_{ij} = B \epsilon_{ij}$ it is natural to introduce the complexified Kähler modulus T via

$$T = \frac{1}{\alpha'}(B + i\sqrt{\det g}) = T_1 + i\, T_2. \qquad (10.55)$$

The complex structure modulus, denoted U, is

$$U = \frac{|e_2|}{|e_1|} e^{i\varphi(e_1, e_2)} = \frac{g_{12} + i\sqrt{\det g}}{g_{11}} = U_1 + i\, U_2. \qquad (10.56)$$

This is what we had called τ when we discussed the world-sheet torus in Chap. 6. Both U and T are defined to lie in the upper half-plane. One can express the metric in terms of the moduli as

$$g_{ij} = \alpha' \frac{T_2}{U_2} \begin{pmatrix} 1 & U_1 \\ U_1 & |U|^2 \end{pmatrix}. \qquad (10.57)$$

We compute $p_{L,R}^2$ and find

$$\frac{2}{\alpha'} p_L^2 = \frac{1}{2 T_2 U_2} \left| m_2 - U m_1 + \overline{T}(n_1 + U n_2) \right|^2,$$

$$\frac{2}{\alpha'} p_R^2 = \frac{1}{2 T_2 U_2} \left| m_2 - U m_1 + T(n_1 + U n_2) \right|^2. \qquad (10.58)$$

There are various symmetries of the spectrum, i.e. symmetries which leave $p_{L,R}^2$ invariant. One easily checks that $SL(2, \mathbb{Z})_T$, $T \to \frac{aT+b}{cT+d}$, if we also transform the

---

[7] The notions of complex and Kähler manifolds will be discussed in Chap. 14.

10.2 Toroidal Compactification of the Closed Bosonic String

momentum and winding numbers appropriately, is such a symmetry. More interestingly, there is also a $\mathbb{Z}_2$ symmetry, $T \leftrightarrow U$ (accompanied by $(m_1, m_2, n_1, n_2) \to (-n_1, m_2, -m_1, n_2)$), which is an exchange of the complex structure and the Kähler modulus. This is a simple example of mirror symmetry (which we discuss further in Chaps. 12 and 14). It also implies an $SL(2; \mathbb{Z})_U$ symmetry. There is a second $\mathbb{Z}_2$ symmetry, $(T, U) \to (-\bar{T}, -\bar{U})$ and the moduli space of the T^2 compactification is $\frac{\mathscr{F} \times \mathscr{F}}{\mathbb{Z}_2 \times \mathbb{Z}_2}$ rather than the product of two copies of the upper half-plane as one would naively expect. The discrete transformations which are responsible for this generate the T-duality group of the T^2 compactifications. The presence of the two moduli also means that compactification on T^2 will lead to two massless complex fields with completely arbitrary vacuum expectation values (vevs). Their couplings to other fields are, however restricted by invariance of the low-energy effective action under $SL(2, \mathbb{Z})_U$ and $SL(2, \mathbb{Z})_T$.

We now take a closer look at the soliton states. We start with $b_{ij} = 0$. By the same arguments as before we might expect additional massless states for special lattices. We are again particularly interested in massless vectors. Inspection of Eq. (10.39) shows that we need either $p_L^2 = 2$, $N_L = 0$ with $p_R = 0$, $N_R = 1$ or L and R interchanged. Together with Eq. (10.41) this means that the only possibilities to get massless vectors are $m_i = \pm n_i = \pm 1$, $m_j = n_j = 0$ for $i \neq j$ and $g_{ij} = \alpha' \delta_{ij}$, i.e. $e_i^I = \sqrt{\alpha'} \delta_i^I$. The gauge group is $[SU(2)]_L^D \times [SU(2)]_R^D$. In this case $p_{L,R}$ build the weight lattice of $[SU(2)]^D$. This is a trivial extension of the case considered before in the sense that the bosonic string is compactified on D orthogonal circles with radii $R = \sqrt{\alpha'}$. If we want to get more general and in particular larger gauge groups $G_{L,R}$, we need $\Gamma_{L,R}$ to contain the root lattice of $G_{L,R}$. For this we have to consider non-trivial antisymmetric tensor field backgrounds b_{ij}.

As the simplest non-trivial example with non-vanishing b_{ij} consider the toroidal compactification of two dimensions. Choose for the two-dimensional lattice Λ_2 the root lattice of $SU(3)$ with basis vectors $e_1 = R(\sqrt{2}, 0)$ and $e_2 = R(-\frac{1}{\sqrt{2}}, \sqrt{\frac{3}{2}})$ (the two simple roots of $SU(3)$). This fixes two metric background fields, namely the ratio of the two radii and the relative angle between e_1 and e_2. Only the overall scale R is left as a free parameter. The most general antisymmetric tensor field background is $b_{ij} = B\epsilon_{ij}$ which provides a second free parameter. For generic R and B the gauge group is $[U(1)]_L^2 \times [U(1)]_R^2$. However at the critical point $R = \sqrt{\frac{\alpha'}{2}}$, $B = \frac{\alpha'}{2}$ the bosonic string has an enlarged gauge symmetry $[SU(3)]_L \times [SU(3)]_R$. In this case the lattice $\Gamma_{2,2}$ contains lattice vectors $\boldsymbol{P} = (\boldsymbol{p}_L, 0)$ and $\boldsymbol{P}' = (0, \boldsymbol{p}_R)$ with $\boldsymbol{p}_L, \boldsymbol{p}_R$ being the six root vectors of $SU(3)$. These states correspond to the non-Abelian gauge bosons of $[SU(3)]_L \times [SU(3)]_R$. In fact, one can easily verify that the lattice $\Gamma_{2,2}$ is the weight lattice of $[SU(3)]_L \times [SU(3)]_R$ specified by the three allowed conjugacy classes $(0, 0)$, $(1, 1)$ and $(2, 2)$ where 0, 1 and 2 are the three conjugacy classes of $SU(3)$ (see next chapter). We are therefore dealing with a rational lattice for this choice of background fields.

Rather than discussing more examples with non-vanishing antisymmetric tensor field background, we consider the problem from a different point of view, which also provides the key for the construction of the heterotic string. Consider again the mode expansion Eq. (10.36). So far, only the oscillators α_n^I, $\bar{\alpha}_n^I$ were treated as independent variables, but the center of mass coordinate x^I and the momenta p^I were not, as we wanted to maintain the interpretation that the $X^I(\sigma,\tau)$ are coordinates in some D-dimensional manifold, e.g. Euclidean space. In this case the left- and right-moving modes must have common center of mass and common momentum. However, for general two-dimensional world-sheet bosons this is not necessary; we are free to regard X_L^I, X_R^I as completely independent (dimensionless) two-dimensional fields with mode expansions

$$X_L^I(\tau+\sigma) = x_L^I + p_L^I(\tau+\sigma) + i\sum_{n\neq 0}^{\infty}\frac{1}{n}\bar{\alpha}_n^I e^{-in(\tau+\sigma)},$$

$$X_R^I(\tau-\sigma) = x_R^I + p_R^I(\tau-\sigma) + i\sum_{n\neq 0}^{\infty}\frac{1}{n}\alpha_n^I e^{-in(\tau-\sigma)}. \quad (10.59)$$

Since for closed strings the fields have to satisfy $X_{L,R}^I(\sigma+2\pi) \simeq X_{L,R}^I(\sigma)$ where the identification is up to a vector of a lattice $\Lambda_{L,R}$, we find that treating X_L^I and X_R^I as independent necessitates that they are compactified on a torus, however not necessarily on the same torus; in other words, chiral bosons must be compactified on a torus. Considering chiral bosons means that one gives up the naive (geometric) picture of compactifying the string on an internal manifold in the sense that there is no decompactification limit. The proper way of understanding this is to regard the resulting theory as a string theory in $(26 - D)$ space-time dimensions; X_L^I, X_R^I are (chiral and anti-chiral) scalar fields which are needed as internal degrees of freedom to cancel the conformal anomaly. The periodicity requirement entails that p_L^I and p_R^I have to be interpreted as winding vectors, i.e. $\boldsymbol{p}_{L,R} \in \Lambda_{L,R}$. But $p_{L,R}^I$ also generate translations of $x_{L,R}^I$. The commutation relations are

$$[x_{L,R}^I, p_{L,R}^J] = i\delta^{IJ},$$

$$[x_{L,R}^I, p_{R,L}^J] = 0. \quad (10.60)$$

The second commutator follows from our assumption that left- and right-movers are independent. There is actually a subtlety. Equation (10.60) are not the canonical commutation relations. For a scalar field $X = \sqrt{\frac{\alpha'}{2}}(X_L + X_R)$ as in (10.59) with action $S = -\frac{1}{4\pi\alpha'}\int \partial^\alpha X \partial_\alpha X$, the canonical momentum is $\Pi^I = \frac{1}{2\pi\alpha'}\partial_\tau X^I$ and $\pi^I = \frac{1}{\sqrt{2\alpha'}}(p_L^I + p_R^I)$. The center-of-mass position is $q^I = \sqrt{\frac{\alpha'}{2}}(x_L^I + x_R^I)$ and the canonical commutator $[p_L^I + p_R^I, x_L^J + x_R^J] = 2i\delta^{IJ}$. Under the assumption that X is purely left-moving, we would simply set $x_R = p_R = 0$ and conclude $[x_L^I, p_L^J] = 2i\delta^{IJ}$. However, requiring the X^I to be purely left-moving constitutes a constraint: $\phi_L^I \equiv (\partial_\tau - \partial_\sigma)X^I = 0$. From

10.2 Toroidal Compactification of the Closed Bosonic String

$$C^{I,J}(\sigma,\sigma') = \{\phi_L^I(\sigma,\tau), \phi_L^J(\sigma',\tau)\}_{\text{P.B.}} = -8\pi\delta^{IJ}\partial_\sigma\delta(\sigma-\sigma') \tag{10.61}$$

(with inverse $C_{IJ}(\sigma,\sigma') = -\frac{1}{8\pi}\partial_\sigma^{-1}\delta(\sigma-\sigma')\delta_{IJ}$), we find that it is second class. As described in Chap. 7, we have to replace the Poisson bracket by a Dirac bracket. The relevant Dirac bracket is

$$\{X^I(\sigma,\tau), \Pi_J(\sigma',\tau)\}_{\text{D.B.}} = \frac{1}{2}\delta^i_J\delta(\sigma-\sigma') \tag{10.62}$$

which leads to Eq. (10.60).

An alternative way of arriving at the same result is to treat Q^I and L^I as conjugate variables, to promote them to operators and to impose canonical commutation relations

$$[Q^I, L^J] = i\delta^{IJ} \quad \text{and} \quad [x^I, p^J] = i\delta^{IJ}. \tag{10.63}$$

Then (10.60) follows from the definitions of $x_{L,R}^I$ and $p_{L,R}^I$, Eqs. (10.37) and (10.38).

Single-valuedness of $e^{ip_{L,R}^I x_{L,R}^I}$ requires that $\boldsymbol{p}_{L,R} \in \Lambda_{L,R}^*$; i.e. we find that $\boldsymbol{p}_{L,R} \in \Lambda_{L,R} \cap \Lambda_{L,R}^* := \Gamma_{L,R}$. The $2D$-dimensional vectors $\boldsymbol{p} = (\boldsymbol{p}_L, \boldsymbol{p}_R)$ again build a Lorentzian lattice $\Gamma_{D,D} = \Gamma_L \otimes \Gamma_R$, and modular invariance forces this lattice to be even and self-dual. We will show this below.

Let us now discuss the spectrum of this theory. The mass formula and reparametrization constraint are

$$\alpha' m_{L,R}^2 = \boldsymbol{p}_{L,R}^2 + 2(N_{L,R} - 1),$$
$$N_L - N_R = \frac{1}{2}(\boldsymbol{p}_R^2 - \boldsymbol{p}_L^2). \tag{10.64}$$

Clearly, we still have the $U(1)$ gauge bosons of Eq. (10.44). Additional massless $(26 - D)$-dimensional vectors are obtained if there exist lattice vectors $\boldsymbol{p} = (\boldsymbol{p}_L, \boldsymbol{p}_R) \in \Gamma_{D,D}$ with the property $\boldsymbol{p}_L^2 = 2$, $\boldsymbol{p}_R = 0$ or $\boldsymbol{p}_R^2 = 2$, $\boldsymbol{p}_L = 0$. The massless vector states are

$$|V_L^\mu\rangle = \alpha_{-1}^\mu |\boldsymbol{p}_L^2 = 2, \boldsymbol{p}_R = 0\rangle,$$
$$|V_R^\mu\rangle = \bar{\alpha}_{-1}^\mu |\boldsymbol{p}_L = 0, \boldsymbol{p}_R^2 = 2\rangle. \tag{10.65}$$

Therefore, if $\Gamma_{L,R}$ contains $l_{L,R}$ vectors $\boldsymbol{p}_{L,R}$ with $\boldsymbol{p}_{L,R}^2 = 2$, we find $l_{L,R}$ massless vectors $|V_{L,R}^\mu\rangle$. These vectors correspond to the non-commuting generators of a non-Abelian Lie group $G_{L,R}$. The $\boldsymbol{p}_{L,R}^2 = 2$ vectors must therefore be roots of $G_{L,R}$ and

$G_{L,R}$ must be simply laced.[8] This means that $\Gamma_{L,R}$ must contain the root lattice of a simply laced group $G_{L,R}$. Then the massless vectors of Eq. (10.65) are, together with the states Eq. (10.44), the gauge bosons of the non-Abelian gauge group $G_L \times G_R$ with $\dim(G_{L(R)}) = l_{L(R)} + D$ and $\text{rank}(G_L) = \text{rank}(G_R) = D$. The oscillator excitations Eq. (10.44) correspond to the $[U(1)]_L^D \times [U(1)]_R^D$ Cartan subalgebra of $G_L \times G_R$. Note that G_L and G_R are in general different.

In conclusion, toroidal compactification of the bosonic string may be viewed in two ways: as compactification of independent left- and right movers on different tori or as compactification on the same torus in the presence of b_{ij} and g_{ij} background fields.

An important aspect of compactification of the ten-dimensional superstring theories is the supersymmetry which is preserved by the compact geometry. Since for toroidal compactifications all the massless modes which were present in ten dimensions descend to massless modes in the non-compact lower dimensional theory, all components of the gravitino survive. Therefore, torus compactification does not break space-time supersymmetry. The vanishing of the partition function is one consequence of this which we will demonstrate below. If we want more realistic lower dimensional theories, we have to break some supersymmetries. This requires more general, in particular curved compact spaces. A first step in this direction will be discussed in Sect. 10.5. A more general account has to wait until Chaps. 14 and 15.

10.3 Toroidal Partition Functions

The partition function of the bosonic string compactified on a torus T^D is easy to write down:

$$Z(\tau,\bar\tau) = \frac{1}{\tau_2^{(24-D)/2}} \frac{1}{|\eta(\tau)|^{48}} \sum_{(p_L,p_R)\in\Gamma_{D,D}} \bar q^{\frac{1}{2}p_L^2} q^{\frac{1}{2}p_R^2}. \tag{10.66}$$

$|\eta(\tau)|^{-48}$ is the bosonic oscillator contribution and $\tau_2^{(24-D)/2}$ the contribution from the transverse non-compact momenta. Recall from Chap. 6 that the torus amplitude is $\mathscr{T} = \frac{V_{26-D}}{\ell_s^{26-D}} \int_{\mathscr{F}} \frac{d^2\tau}{4\tau_2^2} Z(\tau,\bar\tau)$. This is easy to generalize to the type II string compactified on T^D:

---

[8] A Lie group G is simply laced if all its roots α_i have the same length which can be normalized to $\alpha_i^2 = 2$, $\forall i = 1,\ldots,\dim G - \text{rank}(G)$. Dots in the Dynkin diagram (one for each simple root) are then either disconnected or connected by a single line. This leaves only $D_n \sim SO(2n)$, $A_n \sim SU(n+1)$ and E_n, $n = 6,7,8$ (of rank n) or products thereof.

10.3 Toroidal Partition Functions

$$Z(\tau, \bar{\tau}) = \frac{1}{\tau_2^{(8-D)/2}} \frac{1}{|\eta(\tau)|^{24}} \frac{1}{4} \left| \vartheta_3^4(\tau) - \vartheta_4^4(\tau) - \vartheta_2^4(\tau) \right|^2$$
$$\times \sum_{(p_L, p_R) \in \Gamma_{D,D}} \bar{q}^{\frac{1}{2} p_L^2} q^{\frac{1}{2} p_R^2}, \tag{10.67}$$

which vanishes by virtue of the Riemann identity (9.109a). With the help of the Poisson resummation formula and the known transformation properties of ϑ and η one shows that (10.66) and (10.67) are modular invariant provided $\Gamma_{D,D}$ is an even, self-dual lattice. We will explicitly perform a similar computation in the next section.

Let us now, for simplicity and for illustrative purposes consider a single free boson compactified on a circle of radius R. Using the mass formulas (10.8), its partition function can be straightforwardly computed

$$Z_{\text{circ}}(\tau, \bar{\tau}) = \frac{1}{|\eta(\tau)|^2} \sum_{M, L \in \mathbb{Z}} q^{\frac{1}{4} \left(\sqrt{\alpha'} \frac{M}{R} - \frac{RL}{\sqrt{\alpha'}} \right)^2} \bar{q}^{\frac{1}{4} \left(\sqrt{\alpha'} \frac{M}{R} + \frac{RL}{\sqrt{\alpha'}} \right)^2}. \tag{10.68}$$

Since we sum over all Kaluza-Klein and winding modes, this partition function is invariant under the transformation $R \to \alpha'/R$.

We now return to the simple case of a free boson on a circle and exhibit several interesting features of its one-loop partition function (10.68). We consider the situation where the Narain lattice is rational. More specifically, we chose $R = \sqrt{k\alpha'}$ with $k \in \mathbb{Z}_+$. We start with states with $(\bar{h}, h) = (0, h)$, i.e.

$$\bar{L}_0 |M, L\rangle = \bar{h} |M, L\rangle = \frac{1}{4} \left(\frac{M}{\sqrt{k}} + \sqrt{k} L \right)^2 |M, L\rangle = 0. \tag{10.69}$$

Such a chiral state $|M, L\rangle$ is thus specified by $M = -kL$ and the remaining sum over L in the partition function (10.68) becomes

$$\sum_{n \in \mathbb{Z}} q^{kn^2} =: \Theta_{0,k}(\tau) \tag{10.70}$$

which one defines as a Θ-function. Under a modular S-transformation the chiral part $\Theta_{0,k}(\tau)$ of the partition function transforms into a finite sum of more general $\Theta_{m,k}$-functions defined as

$$\Theta_{m,k}(\tau) := \sum_{n \in \mathbb{Z} + \frac{m}{2k}} q^{kn^2}, \qquad -k+1 \leq m \leq k. \tag{10.71}$$

This Θ-function is related to the ones which we defined in the previous chapter via $\Theta_{m,k}(\tau) = \vartheta \begin{bmatrix} \frac{m}{2k} \\ 0 \end{bmatrix} (\frac{\tau}{2k})$. With the help of the Poisson resummation formula (9.115) one finds the following modular S-transformation of the chiral characters

$$\Theta_{m,k}\left(-\frac{1}{\tau}\right) = \sqrt{-i\tau} \sum_{m'=-k+1}^{k} S_{m,m'} \Theta_{m',k}(\tau) \tag{10.72}$$

with the modular S-matrix

$$S_{m,m'} = \frac{1}{\sqrt{2k}} \exp\left(-\pi i \frac{m m'}{k}\right). \tag{10.73}$$

The transformation

$$\Theta_{m,k}(\tau + 1) = e^{\frac{i\pi m^2}{k}} \Theta_{m,k}(\tau) \tag{10.74}$$

follows immediately from the definition (10.71).

The partition function (10.68) for one free boson on a circle of radius $R = \sqrt{k\alpha'}$ can be expressed in terms of the finite set of $\Theta_{m,k}$-functions as

$$Z_{\text{circ}}(\tau, \bar{\tau}) = \frac{1}{|\eta(\tau)|^2} \sum_{m=-k+1}^{k} \left|\Theta_{m,k}(q)\right|^2, \tag{10.75}$$

which can be checked to be modular invariant. The rational conformal field theories which correspond to these partition functions are commonly denoted as $\hat{u}(1)_k$ and show up in many instances, as e.g. in the realization of $\mathcal{N} = 2$ supersymmetric conformal field theories, which will be discussed in Chap. 12.

It is instructive to rewrite the partition function (10.68) by performing a Poisson resummation on M. This leads to the expression

$$Z_{\text{circ}}(\tau, \bar{\tau}) = \frac{1}{\sqrt{\tau_2}} \frac{R}{\sqrt{\alpha'}} \frac{1}{|\eta(\tau)|^2} \sum_{m,n} e^{-\frac{\pi R^2}{\tau_2 \alpha'}|m+\tau n|^2} \tag{10.76}$$

This is the form which one obtains if one computes the path integral directly. The sum is over classical solutions which satisfy the periodicity conditions

$$X(\sigma^0, \sigma^1 + 2\pi) = X(\sigma^0, \sigma^1) + 2\pi R m,$$
$$X(\sigma^0 + 2\pi\tau_2, \sigma^1 + 2\pi\tau_1) = X(\sigma^0, \sigma^1) + 2\pi R n. \tag{10.77}$$

The solutions are

$$X = X_0 + \frac{R}{\tau_2}(n - \tau_1 m)\sigma^0 + Rm\sigma^1 \tag{10.78}$$

10.3 Toroidal Partition Functions

with classical action

$$S_{cl} = \frac{\pi R^2}{\alpha' \tau_2} \left| (n - \tau_1 m)^2 + \tau_2^2 m^2 \right|, \qquad (10.79)$$

where we have used that the area of the world-sheet torus is $4\pi^2 \tau_2$. Note that in the form (10.77) T duality is not manifest.

We can also consider boundary states for the free boson on S^1. There are two possible boundary states, $|D0\rangle$ and $|D1\rangle$. We start with the former. For a D0 brane at position x_0 it has to satisfy

$$X(0, \sigma)|D0\rangle = x_0|D0\rangle \qquad (10.80)$$

which is solved by

$$|D0\rangle = \frac{1}{\mathcal{N}} \exp\left(\sum_{n=1}^{\infty} \frac{1}{n} \alpha_{-n} \bar{\alpha}_{-n} \right) |x_0\rangle, \qquad (10.81)$$

where $x|x_0\rangle = x_0|x_0\rangle$. $|x_0\rangle$ can be expanded in momentum eigenstates and we find

$$|D0\rangle = \frac{1}{\mathcal{N}} \exp\left(\sum_{n=1}^{\infty} \frac{1}{n} \alpha_{-n} \bar{\alpha}_{-n} \right) \sum_{M=-\infty}^{\infty} e^{i \frac{M}{R} x_0} |M, 0\rangle. \qquad (10.82)$$

To fix the normalization, we compute the annulus amplitude $\mathcal{A} = \langle D0|q^H|D0\rangle$, where ($q = e^{-2\pi l}$)

$$H = L_0 + \bar{L}_0 - \frac{2}{24}$$

$$= \sum_{n=1}^{\infty} (\alpha_{-n} \alpha_n + \bar{\alpha}_{-n} \bar{\alpha}_n) + \left(\frac{\alpha' M^2}{2R^2} + \frac{L^2 R^2}{2\alpha'} \right) - \frac{1}{12}. \qquad (10.83)$$

One easily finds

$$\mathcal{A} = \frac{1}{\mathcal{N}^2} \frac{1}{\eta(2il)} \sum_{M \in \mathbb{Z}} e^{-2\pi l \left(\frac{\alpha' M^2}{2R^2} \right)}$$

$$= \frac{1}{\mathcal{N}^2} \frac{1}{\eta(it)} \left(\frac{2R^2}{\alpha'} \right)^{1/2} \sum_{L \in \mathbb{Z}} e^{-2\pi t L^2 R^2 / \alpha'}, \qquad (10.84)$$

where in going from the first to the second line we have made a modular transformation $2l = 1/t$ from the tree to the one-loop channel and performed a Poisson resummation which trades the sum over quantized momenta for a sum over winding numbers. In the loop channel, $\mathscr{A}$ has the interpretation of a partition function, i.e. as a sum over states with positive integer multiplicities. Furthermore, there should be a unique ground states. This fixes the normalization to

$$\mathscr{N} = \left(\frac{2R^2}{\alpha'}\right)^{1/4}. \tag{10.85}$$

We could now construct the D1-brane state $|D1\rangle$ by solving

$$\partial_\tau X(0,\sigma)|D1\rangle = 0. \tag{10.86}$$

Alternatively we perform T-duality on $|D0\rangle$. This acts as

$$R \to \alpha'/R, \quad x_0 \leftrightarrow c, \quad (M,L) \to (L,M), \quad \alpha_n \to -\alpha_n, \quad \bar{\alpha}_n \to \bar{\alpha}_n \tag{10.87}$$

which leads to

$$|D1\rangle = \left(\frac{R^2}{2\alpha'}\right)^{1/4} \exp\left(-\sum_{n=1}^\infty \frac{1}{n}\alpha_{-n}\bar{\alpha}_{-n}\right) \sum_{L\in\mathbb{Z}} e^{i\frac{L_R}{\alpha'}c}|0,L\rangle. \tag{10.88}$$

Note the change of sign in the exponent, which corresponds to the change from D to N boundary conditions. One easily checks that each term in the sum satisfies (10.86) and is thus an Ishibashi state. Only the whole sum is a Cardy state which, when we compute the annulus amplitude and transform to the loop channel, gives the correct spectrum. c has the interpretation as a constant background gauge field (continuous Wilson line) on the D1-brane wrapped around the circle.

10.4 The $E_8 \times E_8$ and $SO(32)$ Heterotic String Theories

After our discussion of toroidal compactifications, we are now prepared to introduce the ten-dimensional heterotic string. The important observation is that, since we have treated the left- and right-moving compactified coordinates as completely independent, we can drop either one of them. This is the starting point of the heterotic string construction.

The heterotic string is a hybrid construction of the left-moving sector of the 26-dimensional bosonic string combined with the right-moving sector of the 10-dimensional superstring. By the arguments given before it is a string theory in 10-dimensions. We deal with the following two-dimensional fields: As left moving coordinates we have 10 uncompactified bosonic fields $X_L^\mu(\tau+\sigma)$ ($\mu = 0,\ldots,9$) and 16 internal bosons $X_L^I(\tau+\sigma)$ ($I = 1,\ldots,16$) which live on a 16-dimensional torus. The right-moving degrees of freedom consist of 10 uncompactified bosons $X_R^\mu(\tau-\sigma)$ ($\mu = 0,\ldots,9$), and their two-dimensional fermionic superpartners

10.4 The $E_8 \times E_8$ and $SO(32)$ Heterotic String Theories

$\psi_R^\mu(\tau - \sigma)$. Finally, we have left- and right-moving reparametrization ghosts b, c, $\bar{b}$ and $\bar{c}$ and right-moving superconformal ghosts β, γ. $X_L^\mu(\tau + \sigma)$ and $X_R^\mu(\tau - \sigma)$ have common center of mass coordinates and common space-time momentum with continuous spectra. On the other hand, the momenta of the additional chiral bosons $X_L^I(\tau + \sigma)$ are discrete; they are vectors of a 16-dimensional lattice Γ_{16}:

$$\boldsymbol{p}_L \in \Gamma_{16}, \quad p_L^I = p_i e_i^I, \quad I = 1, \ldots, 16, \quad p_i \in \mathbb{Z}. \tag{10.89}$$

The e_i^I are the basis vectors of Γ_{16}, whose metric is

$$g_{ij} = \sum_{I=1}^{16} e_i^I e_j^I. \tag{10.90}$$

Γ_{16} cannot be a chosen arbitrarily, as modular invariance imposes severe restrictions. We will only consider the one-loop vacuum amplitude, i.e. the one-loop partition function. In the Hamiltonian formalism it is given by the by now well known expression

$$Z_{\text{het}}(\tau, \bar{\tau}) = \text{Tr}_{\mathcal{H}_{\text{het}}} \left(\bar{q}^{H_L} q^{H_R} \right), \quad q = e^{2\pi i \tau}, \tag{10.91}$$

where H_L and H_R are the left- and right-moving Hamiltonians in light cone gauge:

$$H_L = \frac{1}{2} p_\mu^2 + N_L + \frac{1}{2} p_L^2 - 1$$

$$H_R = \frac{1}{2} p_\mu^2 + N_R + H^{\text{NS(R)}} - \frac{1}{3}. \tag{10.92}$$

Note that here p_μ ($\mu = 2 \ldots 9$) are the transverse components of the space-time momentum and $\boldsymbol{p}_L$ the 16 internal left-moving momentum; N_L contains the 8 transverse space-time as well as the 16 internal left-moving bosonic oscillators, while N_R receives contribution only from the 8 right-moving space-time bosonic oscillators. Finally, H^{NS}, H^R are the Neveu-Schwarz and Ramond Hamiltonians of the fermionic string, as in (9.8). This leads to the following partition function:

$$Z_{\text{het}}(\tau, \bar{\tau}) = \frac{1}{(\text{Im}\,\tau)^4} \left(\frac{1}{[\eta(\bar{\tau})]^{24}} \sum_{\boldsymbol{p}_L \in \Gamma_{16}} \bar{q}^{\frac{1}{2} p_L^2} \right)$$

$$\times \left(\frac{1}{[\eta(\tau)]^{12}} \left(\vartheta_3^4(\tau) - \vartheta_4^4(\tau) - \vartheta_2^4(\tau) \right) \right). \tag{10.93}$$

Here $\eta(\bar{\tau})^{-24} \eta(\tau)^{-8}$ is the bosonic oscillator contribution, the $(\text{Im}\,\tau)^{-4}$ factor arises from the zero modes of the uncompactified transverse coordinates and $\eta(\tau)^{-4} \times$

(ϑ-functions) comes from the world-sheet fermions (cf. Chaps. 6 and 9). The novel and most interesting part of this partition function is the lattice (or soliton) sum[9]

$$P(\tau) = \sum_{p_L \in \Gamma_{16}} q^{\frac{1}{2}p_L^2}. \tag{10.94}$$

The summation is over all lattice vectors of Γ_{16}. From the known modular transformation properties of $\text{Im}\,\tau$ and $\eta(\tau)$ under S and T we conclude that, in order for $Z_{\text{het}}(\tau, \bar{\tau})$ to be modular invariant, $P(\tau)$ must be invariant under the T transformation,

$$P(\tau + 1) = P(\tau) \tag{10.95}$$

and must transform under S like

$$P\left(-\frac{1}{\tau}\right) = \tau^8 P(\tau). \tag{10.96}$$

We first check Eq. (10.95):

$$P(\tau + 1) = \sum_{p_L \in \Gamma_{16}} q^{\frac{1}{2}p_L^2} e^{\pi i p_L^2}. \tag{10.97}$$

Invariance under T clearly demands that $p_L^2 \in 2\mathbb{Z}$, $\forall \, p_L \in \Gamma_{16}$, which means that Γ_{16} must be an even lattice. Note that this already follows from Eq. (10.64) with $p_R = 0$. Since $p_L^2 = \sum_I p_L^I p_L^I = \sum_{ij} p^i g_{ij} p^j = \sum_i (p^i)^2 g_{ii} + 2 \sum_{i<j} p^i g_{ij} p^j$ ($p^i, p^j \in \mathbb{Z}$) we find that for an even lattice the diagonal elements of the metric g_{ij} must be even integers: $g_{ii} \in 2\mathbb{Z}$, $\forall i = 1, \ldots, 16$.

The more subtle part is the transformation of $P(\tau)$ under S. To study it, we recall the Poisson resummation formula (9.115)

$$\sum_{p \in \Lambda} e^{-\pi\alpha(p+x)^2 + 2\pi i y \cdot (p+x)} = \frac{1}{\text{vol}(\Lambda)\alpha^{n/2}} \sum_{q \in \Lambda^*} e^{-2\pi i q \cdot x} e^{-\frac{\pi}{\alpha}(y+q)^2}. \tag{10.98}$$

This is the key formula to derive the implications of modular invariance for $P(\tau)$. Applying it to $P(-1/\tau)$ we get

$$P\left(-\frac{1}{\tau}\right) = \frac{\tau^8}{\sqrt{\det g}} \sum_{p \in (\Gamma_{16})^*} q^{\frac{1}{2}p^2}. \tag{10.99}$$

---

[9] For notational simplicity we consider the complex conjugate of the lattice sum in (10.93).

10.4 The $E_8 \times E_8$ and $SO(32)$ Heterotic String Theories

Therefore, in order to satisfy Eq. (10.96), Γ_{16} must be a self-dual lattice, i.e.

$$\Gamma_{16} = (\Gamma_{16})^*. \tag{10.100}$$

Then $\det g = 1$ and $\text{vol}(\Gamma) = \text{vol}(\Gamma^*) = 1$.

In summary, modular invariance of the one-loop partition function implies that the internal 16-dimensional momentum lattice Γ_{16} must be an even self-dual Euclidean lattice. One can show that no further conditions arise from higher loops. These lattices are very rare. We will study them more carefully in the next chapter. The result is that in 16 dimensions there are only two even self-dual Euclidean lattices, namely the direct product lattice $\Gamma_{E_8} \otimes \Gamma_{E_8}$, where Γ_{E_8} is the root lattice of E_8, and $\Gamma_{D_{16}}$ which is the weight lattice of $\text{Spin}(32)/\mathbb{Z}_2$ which contains the root lattice of $SO(32)$. The metric g_{ij} of Γ_{E_8} is the Cartan matrix of E_8:

$$g_{ij}^{E_8} = \begin{pmatrix} 2 & -1 & & & & & & \\ -1 & 2 & -1 & & & & & \\ & -1 & 2 & -1 & & & & \\ & & -1 & 2 & -1 & & & \\ & & & -1 & 2 & -1 & & -1 \\ & & & & -1 & 2 & -1 & \\ & & & & & -1 & 2 & \\ & & & & -1 & & & 2 \end{pmatrix}. \tag{10.101}$$

One may check that $\det g_{ij} = 1$. The construction of the weight lattice of $\text{Spin}(32)/\mathbb{Z}_2$ will be discussed in the next chapter. Both, the root lattice of $E_8 \times E_8$ and the weight lattice of $\text{Spin}(32)/\mathbb{Z}_2$ contain 480 vectors of $(\text{length})^2 = 2$ which are the roots of $E_8 \times E_8$ and $SO(32)$, respectively. Therefore, according to our previous discussion the dimension 496 gauge group of the heterotic string is either $E_8 \times E_8$ or $SO(32)$. This is required by modular invariance.

Let us investigate the on-shell spectrum of the heterotic string more carefully. First consider left-moving excitations. As usual, there is the tachyonic vacuum of the bosonic string. At the massless level we have oscillator excitations $\bar{\alpha}_{-1}^\mu|0\rangle$, $\bar{\alpha}_{-1}^I|0\rangle$. The former transform like space-time vectors whereas the internal oscillator excitations correspond to the left-moving part of the Abelian $U(1)^{16}$ gauge boson. They build the Cartan subalgebra of $E_8 \times E_8$ or $SO(32)$. We also have the states in the soliton sector with non-trivial internal momenta p_L. The states $|p_L^2 = 2\rangle$, $N_L = 0$ are massless, p_L is a $(\text{length})^2 = 2$ root vector of $E_8 \times E_8$ or $SO(32)$ and generate the non-Abelian gauge bosons of these groups.

The right-moving excitations are those of the 10-dimensional superstring—the spectrum is space-time supersymmetric. The NS tachyon $|0\rangle_{\text{NS}}$ is projected out by the GSO projection which was enforced by modular invariance. Therefore the lowest states are the vector $b_{-1/2}^\mu|0\rangle_{\text{NS}}$ and the spinor $|S^\alpha\rangle$ (previously denoted by $|a\rangle$).

Finally we take the tensor product of the left- and right-moving sectors to obtain the spectrum of the heterotic string. It is clear that there is no tachyon since the left-moving tachyonic vacuum does not satisfy the left-right level matching constraint

$$N_L + \frac{1}{2}p_L^2 - 1 = \begin{cases} N_R & \text{R sector} \\ N_R - \frac{1}{2} & \text{NS sector.} \end{cases} \qquad (10.102)$$

Due to the right-moving supersymmetry the spectrum is $\mathcal{N} = 1$ supersymmetric in 10 dimensions. We have four kinds of massless states:

1. The states corresponding to the ten-dimensional graviton, antisymmetric tensor and dilaton

$$\bar{\alpha}_{-1}^\mu |0\rangle \otimes b_{-\frac{1}{2}}^\nu |0\rangle_{NS}, \qquad (10.103)$$

2. And their supersymmetric partners, the gravitino and dilatino

$$\bar{\alpha}_{-1}^\mu |0\rangle \otimes |S^\alpha\rangle_R. \qquad (10.104)$$

3. In addition we have the gauge bosons of $E_8 \times E_8$ or $SO(32)$

$$\bar{\alpha}_{-1}^I |0\rangle \otimes b_{-\frac{1}{2}}^\mu |0\rangle_{NS},$$
$$|p_L^2 = 2\rangle \otimes b_{-\frac{1}{2}}^\mu |0\rangle_{NS}, \qquad (10.105)$$

where in the first line we have the gauge bosons corresponding to the Cartan subalgebra and in the second line the gauge bosons corresponding to the root vectors.

4. Finally there are the 496 supersymmetric partners of the gauge bosons, the gaugini

$$\bar{\alpha}_{-1}^I |0\rangle \otimes |S^\alpha\rangle_R,$$
$$|p_L^2 = 2\rangle \otimes |S^\alpha\rangle_R. \qquad (10.106)$$

It is straightforward to work out the massive spectrum, but we will not do it here. However, it is useful to remember that the number of (massive) solitonic states is encoded in the lattice partition function of the root and weight lattices of $E_8 \times E_8$ and $\text{Spin}(32)/\mathbb{Z}_2$, respectively. To see this we calculate the partition functions

$$P = \sum_\lambda e^{i\pi\tau\lambda^2} = \sum_\lambda q^{\frac{1}{2}\lambda^2}, \qquad (10.107)$$

where the sum extends over all vectors in the E_8 and $\text{Spin}(32)/\mathbb{Z}_2$ weight lattices respectively. We first compute P for E_8. The 240 roots of E_8 are the eight-

10.4 The $E_8 \times E_8$ and $SO(32)$ Heterotic String Theories

component vectors[10]

$$\alpha = \begin{cases} (\pm 1, \pm 1, 0, \ldots, 0) & + \text{ permutations} \\ \left(\pm \frac{1}{2}, \ldots, \pm \frac{1}{2}\right) & \text{even number of "} - \text{" signs} \end{cases} \quad (10.108)$$

and the weight vectors are of the form

$$\lambda = \begin{cases} (n_1, \ldots, n_8) \\ \left(n_1 + \frac{1}{2}, \ldots, n_8 + \frac{1}{2}\right) \end{cases}, \quad \sum_{i=1}^{8} n_i = \text{even integer}. \quad (10.109)$$

To implement the condition on the n_i we insert a factor $\frac{1}{2}(1 + e^{i\pi \sum n_i})$. We then get

$$P_{E_8} = \frac{1}{2}\left[\prod_{i=1}^{8}\sum_{n_i \in \mathbb{Z}} e^{i\pi n_i^2 \tau} + \prod_{i=1}^{8}\sum_{n_i \in \mathbb{Z}} e^{i\pi n_i^2 \tau} e^{i\pi n_i}\right.$$
$$\left. + \prod_{i=1}^{8}\sum_{n_i \in \mathbb{Z}} e^{i\pi(n_i+\frac{1}{2})^2 \tau} + \prod_{i=1}^{8}\sum_{n_i \in \mathbb{Z}} e^{i\pi(n_i+\frac{1}{2})^2 \tau} e^{i\pi(n_i+\frac{1}{2})}\right]$$
$$= \frac{1}{2}\left[\vartheta_3^8(\tau) + \vartheta_4^8(\tau) + \vartheta_2^8(\tau)\right], \quad (10.110)$$

where we have used the definitions of the ϑ-functions given in Chap. 9. The contribution of the last term is $\vartheta_1^8(\tau) = 0$. Expanding P_{E_8} in powers of q we find

$$P_{E_8}(\tau) = 1 + 240q + 2160q^2 + 6720q^3 + \ldots . \quad (10.111)$$

It shows that the E_8 root lattice has 240 points of $(\text{length})^2 = 2$ corresponding to the roots, 2160 points of $(\text{length})^2 = 4$, etc.

For the $\text{Spin}(32)/\mathbb{Z}_2$ case, using results from Chap. 11, one derives in a similar way that

$$P_{\text{Spin}(32)/\mathbb{Z}_2} = \frac{1}{2}\left[\vartheta_3^{16}(\tau) + \vartheta_4^{16}(\tau) + \vartheta_2^{16}(\tau)\right]. \quad (10.112)$$

Here we used that the $\text{Spin}(32)/\mathbb{Z}_2$ weight lattice consists of vectors

$$\lambda = \begin{cases} (n_1, \ldots, n_{16}) \\ (n_1 + \frac{1}{2}, \ldots, n_{16} + \frac{1}{2}) \end{cases}, \quad \sum_{i=1}^{16} n_i = \text{even integer}, \quad (10.113)$$

---

[10] Properties of root and weight lattices will be discussed in the next chapter.

as will be further explained in Chap. 11. With the help of the Riemann identity (9.14a) $\vartheta_3^4 = \vartheta_2^4 + \vartheta_4^4$ we show that

$$P_{\text{Spin}(32)/\mathbb{Z}_2} = [P_{E_8}]^2$$
$$= 1 + 480q + 61920q^2 + 1050240q^3 + \ldots. \quad (10.114)$$

It follows that the $E_8 \times E_8$ and the $SO(32)$ heterotic string theories have the same number of states at every mass level which are however differently organized under the internal gauge symmetries. So, even though the partition functions are identical, the theories are nevertheless different. The differences show up in correlation functions.

Let us finally write down the bosonic part of the sigma-model action for the heterotic string. In addition to the terms present in (10.46), there is the action for the 16 chiral bosons which are compactified on the maximal torus of the gauge group and their coupling to a background gauge field. The new terms are

$$S_h = -\frac{1}{8\pi} \int d^2\sigma \left(\eta^{\alpha\beta} \partial_\alpha X^A \partial_\beta X^A + 2\epsilon^{\alpha\beta} A_\mu^A \partial_\alpha X^\mu \partial_\beta X^A \right), \quad (10.115)$$

where $A, B = 1, \ldots, 16$ and the (dimensionless) bosons X^A are restricted to satisfy $\phi^A \equiv \partial_+ X^A = 0$. For constant A_μ^I the coupling to the gauge field is a total derivative. This term could also be written with $\eta^{\alpha\beta}$ instead of $\epsilon^{\alpha\beta}$. The difference between these two terms vanishes for chiral bosons. The last term in (10.115) is a $j \cdot A$ type coupling where A is the pullback of the gauge field to the world-sheet and j is a current in the Cartan subalgebra. Note that we can only describe the coupling to the gauge bosons in the Cartan subalgebra.

We will soon consider the heterotic string compactified on a torus T^D in the presence of constant background fields, in particular constant gauge fields with components in the compact directions and vanishing field strength. The latter implies that the gauge fields A_I all commute, i.e. they lie in the Cartan subalgebra (the maximal torus) of the gauge group, that is in $U(1)^{16}$. There are therefore $16D$ independent components. Even though the field strength vanishes, i.e. the gauge field is locally pure gauge, this is not necessarily true globally, if the manifold is not simply connected, which is the case considered here. For example, one can show that a constant $U(1)$ gauge field on S^1 cannot be gauged away, as this would require a gauge transformation not respecting the periodicity of the circle. Gauge field backgrounds give rise to Wilson lines

$$W(\gamma) = \mathcal{P} \exp\left(i \oint_\gamma A \right), \quad (10.116)$$

10.4 The $E_8 \times E_8$ and $SO(32)$ Heterotic String Theories

where γ is a closed path in space-time and $\mathscr{P}$ is path ordering.[11] For a constant (non-zero) gauge field $A = A_I dx^I$, $W(\gamma)$ is clearly non-trivial (i.e. $W(\gamma) \neq 1$) if γ is a non-contractible loop even. if $F_{IJ} = 0$. In this case it only depends on the homotopy class of the loop. In our context there are D independent non-contractible loops on T^D and we can have D independent Wilson lines with values in $U(1)^{16}$.

In the presence of a background gauge field, charged fields, i.e. those fields which couple to the background, will receive a background dependent mass. This is most easily seen by considering the gauge covariant derivative $D_I \phi = \partial_I \phi + q A_I \phi$. If I is a compact direction, there is a contribution to the mass from the KK momentum and from the coupling to the gauge background.

While the preceding discussion was field theoretical, we will find the same features for the compactified heterotic string in the presence of a gauge background, namely contributions to the masses from the coupling to the gauge background. Here the charges are the quantized winding and momentum numbers in the internal directions.

We now consider the heterotic string compactified on a torus T^D. We denote the coordinates along the torus directions by X^I and those along of the sixteen dimensional self-dual lattice by X^A. We choose a constant background fields $G_{IJ} = \delta_{IJ}$, B_{IJ} and A_I^A. The canonical momenta derived from the action (10.115) are

$$\Pi_I = \frac{1}{2\pi\alpha'}\left(\dot{X}_I + B_{IJ} X'^J + \tfrac{\alpha'}{2} A_I^A X'^A\right),$$
$$\Pi_A = \frac{1}{2\pi}(\dot{X}_A - A_{IA} X'^I). \tag{10.117}$$

Because of the chirality constraint for X^A, the canonical quantization procedure is via Dirac brackets. They are straightforward to work out. We have normalized Π_A such that $\{X^A, \Pi_B\}_{\text{D.B}} = 1$ without the factor $1/2$ on the r.h.s. of (10.62). One also finds that X^I and Π_J, which are unconstrained, are not canonically conjugate variables; their Dirac brackets are not the canonical ones, in particular $\{\Pi_I, \Pi_J\}_{\text{D.B.}}$ does not vanish. This can be remedied by defining

$$\tilde{\Pi}_I = \Pi_i + A_I^A \Pi_A, \tag{10.118}$$

while leaving Π_A unchanged. For the center of mass momenta p_I and p_A, which appear in the expansion of the coordinates

$$X^I = q^I + \alpha' p^I \tau + L^I \sigma + \text{osc.},$$
$$X^A = q^A + p^A(\tau + \sigma) + \text{osc.}, \tag{10.119}$$

---

[11] $\gamma \to W(\gamma)$ is a homomorphism of the fundamental group of the compact manifold into the gauge group. Since $\pi_1(T^D)$ is Abelian, Wilson lines on the torus must commute, i.e. they are in the maximal torus.

we find

$$(p_I)_{L,R} = \sqrt{\frac{\alpha'}{2}}\left(\tilde{\pi}_I - \frac{1}{\alpha'}(B_{IJ} \mp G_{IJ})L^J - \pi_A A_I^A - \frac{1}{2}A_I^A A_J^A L^J\right),$$

$$p_A = \pi_A + A_{IA}L^I. \tag{10.120}$$

As in (10.37), we have defined dimensionless left- and right-moving momenta. The equations (10.120) and (10.123) below are the generalizations of (10.49) to the heterotic string.

The canonical momenta and the windings are quantized. L^I are vectors in the torus lattice and $\tilde{\pi}_I$ and π_A are vectors in the dual torus lattice and in the (self-dual) weight lattice of $E_8 \times E_8$ or $\text{Spin}(32)/\mathbb{Z}_2$, respectively. They are the components w.r.t. to a Cartesian basis of $\mathbb{R}^D$ and $\mathbb{R}^{16}$, respectively. (For the weight lattices they are vectors of the form (10.109) and (10.113).) We can expand them in a (dual) lattice basis as follows

$$\pi_A = l_a\, E_A^{*a}, \qquad \tilde{\pi}_I = m_i\, e_I^{*i}, \qquad L^I = n^i\, e_i^I, \qquad l_a, m_i, n^i \in \mathbb{Z}. \tag{10.121}$$

We also expand the background fields

$$\delta_{IJ} = e_I^{*i} e_J^{*j} g_{ij}, \quad B_{IJ} = e_I^{*i} e_J^{*j} b_{ij}, \quad \delta_{AB} = E_A^{*a} E_B^{*b} \tilde{g}_{ab}, \quad A_{IA} = e_I^{*i} E_A^{*a} a_{ia}. \tag{10.122}$$

The lattice vectors e_i^I and their duals were defined previously. They generate T^D. The E_A^{*a} are the dual basis of the root lattice of $E_8 \times E_8$ or of the $\text{Spin}(32)/\mathbb{Z}_2$ weight lattice and $\tilde{g}_{ab}$ is the lattice metric. If we define $p_I = e_I^{*i} p_i$ and $p_A = E_A^{*a} p_a$ and $p_L = (p_{Li}, p_a)$ and $p_R = (p_{Ri})$ we obtain

$$p_L = \sqrt{\frac{\alpha'}{2}}\left(\mathbf{m} + \frac{1}{\alpha'}(\mathbf{g}-\mathbf{b})\mathbf{n} - \mathbf{a}\tilde{\mathbf{g}}^{-1}\mathbf{l} + \frac{1}{2}\mathbf{a}\tilde{\mathbf{g}}^{-1}\mathbf{a}^T\mathbf{n},\ \sqrt{\frac{2}{\alpha'}}(\mathbf{l} + \mathbf{a}^T\mathbf{n})\right),$$

$$p_R = \sqrt{\frac{\alpha'}{2}}\left(\mathbf{m} - \frac{1}{\alpha'}(\mathbf{g}+\mathbf{b})\mathbf{n} - \mathbf{a}\tilde{\mathbf{g}}^{-1}\mathbf{l} + \frac{1}{2}\mathbf{a}\tilde{\mathbf{g}}^{-1}\mathbf{a}^T\mathbf{n}\right), \tag{10.123}$$

where we have employed an obvious matrix notation. Defining the Lorentzian scalar product with signature $(D+16, D)$ as

$$p \cdot p \equiv p_L^2 - p_R^2 = p_{Li}\, g^{ij}\, p_{Lj} + p_{La}\, \tilde{g}^{ab}\, p_{Lb} - p_{Ri}\, g^{ij}\, p_{Rj}, \tag{10.124}$$

we find

$$p \cdot p' = \mathbf{m}^T \mathbf{n}' + \mathbf{n}^T \mathbf{m}' + \mathbf{l}^T \tilde{\mathbf{g}}^{-1} \mathbf{l}'. \tag{10.125}$$

10.4 The $E_8 \times E_8$ and $SO(32)$ Heterotic String Theories

Thus, we conclude that the momentum lattice is an even, self-dual Lorentzian lattice, $\Gamma_{D+16,D}$, for all values of the background fields g_{ij}, b_{ij} and a_{ia}. The momentum lattice of toroidal compactifications of the heterotic string are also called Narain lattices. While the momenta p and p' depend on the background fields, their Lorentzian scalar product doesn't. Contrarily, the mass formula, which contains $p_L^2 + p_R^2$, does. The partition function of the heterotic string compactified on T^D is

$$Z(\tau,\bar{\tau}) = \frac{1}{\tau_2^{(8-D)/2}} \frac{1}{\eta(\bar{\tau})^{24}\eta(\tau)^{12}}$$
$$\times \sum_{(p_L,p_R)\in\Gamma_{16+D,D}} \bar{q}^{\frac{1}{2}p_L^2} q^{\frac{1}{2}p_R^2} \left(\vartheta_3^4(\tau) - \vartheta_4^4(\tau) - \vartheta_2^4(\tau)\right). \quad (10.126)$$

The Narain lattice being even and self-dual guarantees that the partition function is modular invariant. This follows as a straightforward generalization of our discussion in Sect. 10.4 to lattices with Lorentzian signature.

Even Lorentzian lattices with signature (p,q) only exist if $p - q = 8n$. If they exists, they are unique, up to $SO(p,q)$ Lorentz transformations. As a reference lattice one can choose $n\,\Gamma_{E_8} \otimes q\,\Gamma_{1,1}$ where $\Gamma_{1,1}$ is the lattice from Footnote 6 on page 275. Consider the $E_8 \times E_8$ heterotic string compactified on S^1. The uniqueness of the lattice then states that we can rotate $\Gamma_{E_8} \otimes \Gamma_{E_8} \otimes \Gamma_{1,1}$ to $\Gamma_{\text{Spin}(32)/\mathbb{Z}_2} \otimes \Gamma_{1,1}$ via an $SO(17,1)$ rotation.

One important consequence of this is that we can continuously interpolate between the compactified $E_8 \times E_8$ and $\text{Spin}(32)/\mathbb{Z}_2$ theories via a background dependent Lorentz transformation. In other words, the compactified $E_8 \times E_8$ and $SO(32)$ theories are on the same moduli space and the two uncompactified theories are two special points on this moduli space. Since the lattice contribution to the mass^2 of a state is proportional to $p_L^2 + p_R^2$, which is only invariant under $O(D+16) \times O(D) \subset O(D+16, D)$, the moduli space of heterotic Narain lattices is locally $\frac{O(D+16,D)}{O(D)\times O(D+16)}$. It is parametrized by the components of the internal metric, g_{ij}, the components of the internal anti-symmetric tensor b_{ij} and 16 Wilson lines a_{ia} along the maximal torus of the gauge group. There are discrete identifications of points in this space which relate physically indistinguishable toroidal compactifications. These are the invariances of the reference lattice. They are the generalizations of the T duality transformation of the compactification on S^1 and of the discrete symmetries of the torus compactifications which we discussed on page 275. Without giving the details of the derivation, we simply state that the moduli space of toroidally compactified heterotic strings can be shown to be

$$\frac{O(D, D+16)}{O(D) \times O(D+16)} \Big/ O(D, D+16; \mathbb{Z}). \quad (10.127)$$

Different points in the moduli space correspond to physically distinct compactifications, e.g. the gauge groups are different, although always of rank $16 + 2D$. At generic points it is $U(1)^{16+2D}$, but there are special subspaces of the moduli space where it is enhanced. The non-Abelian gauge bosons correspond to momentum vectors with $p_L^2 = 2$. The gauge group can be enhanced beyond the original $E_8 \times E_8$ or $SO(32)$. For instance, for particular background parameters one finds an $SO(32 + 2D)$ gauge group. Also, the torus compactification does not break supersymmetry. This is reflected in the spectra which always organize into complete SUSY multiplets.

Consider the compactification of the heterotic $E_8 \times E_8$ theory on S^1 of radius R. In the absence of a gauge background, it is invariant under T-duality: $R \to \alpha'/R$ and exchange of momentum and winding quantum numbers. The same is true for the heterotic $SO(32)$ theory. Even more interestingly, we can transform the two heterotic theories on S^1 into each other, if we choose in each theory an appropriate gauge background breaking the gauge symmetry to $SO(16) \times SO(16)$ and relate the radii of the two circles as $R_1 R_2 = \alpha'/2$. One can show that the two theories have identical spectra and symmetries. This is the T-duality between the two heterotic theories compactified on S^1. Details can be found in the references.

What we have used in this section is the bosonic formulation of the heterotic string. We have represented the gauge sector in terms of chiral bosons which are coordinates on the maximal torus of the gauge group. In Sect. 11.5 we will briefly discuss the alternative fermionic description. It uses 32 left-moving world-sheet fermions which transform as a vector of $SO(32)$ and of $SO(16) \times SO(16) \subset E_8 \times E_8$, respectively.

10.5 Toroidal Orbifolds

So far we have discussed strings compactified on tori. These compactifications correspond to simple conformal field theories and are under complete computational control, but for the superstring they have the disadvantage that all 32 supercharges of the fundamental ten-dimensional theory are preserved. With realistic particle spectra in four space-time dimensions in mind, the existence of chirality in the Standard Model of particle physics forces us to consider compactifications with at most four supercharges, i.e. $\mathcal{N} = 1$ supersymmetry in four dimensions. Therefore it is necessary to understand strings moving on non-toroidal spaces.

Various techniques, both geometric and conformal field theoretic, have been developed to describe such spaces at large radii and at string scale sizes, respectively. Probably the simplest class of spaces which (partially) break supersymmetry are descendants of the toroidal spaces and are called toroidal orbifolds. We start with a simple one-dimensional prototype example which already shows several of the relevant new features of compactification on orbifolds. We then generalize this to compactifications on higher dimensional orbifolds. In Chap. 15 we discuss supersymmetric compactifications on orbifolds and orientifolds which is a step towards string model building.

10.5 Toroidal Orbifolds

Fig. 10.1 Illustration of the $\mathbb{Z}_2$-orbifold of the circle. Effectively, the circle becomes a line with a fixed point at each end

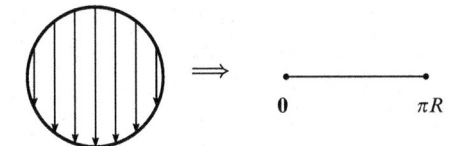

Consider once more the free boson $X(\sigma^1, \sigma^0)$[12] compactified on a circle of radius R and realize that the theory has a certain discrete $\mathbb{Z}_2$ symmetry acting as

$$I_1 : X(\sigma^1, \sigma^0) \mapsto -X(\sigma^1, \sigma^0). \tag{10.128}$$

The idea now is similar to the orientifold construction of the type I string in Chap. 9. Once we have identified a discrete symmetry we can gauge it and take the quotient theory. As illustrated in Fig. 10.1, this amounts to identifying the fields $X(z, \bar{z})$ and $-X(z, \bar{z})$ which means that effectively the circle becomes a line segment with a fixed point of the $\mathbb{Z}_2$ action at each end. This space is called the $\mathbb{Z}_2$-orbifold of the circle. Taking the quotient means that the Hilbert space contains only states which are invariant under the orbifold action. For the calculation of the partition function this means that one projects onto states invariant under I_1. Therefore, at this stage the partition function of the orbifold theory is tentatively

$$Z(\tau, \bar{\tau}) = \text{Tr}_{\mathscr{H}} \left(\frac{1 + I_1}{2} \, q^{L_0 - \frac{c}{24}} \, \bar{q}^{\bar{L}_0 - \frac{c}{24}} \right)$$

$$= \frac{1}{2} Z_{\text{circ.}}(\tau, \bar{\tau}) + \frac{1}{2} \text{Tr}_{\mathscr{H}} \left(I_1 \, q^{L_0 - \frac{c}{24}} \, \bar{q}^{\bar{L}_0 - \frac{c}{24}} \right). \tag{10.129}$$

The first term is the familiar partition function of the free boson on the circle. We now compute the second term. The action of I_1 on the oscillator modes α_n is easily found to be $I_1 \alpha_n I_1^{-1} = -\alpha_n$ and similarly for the left-movers. What remains is to determine the action of I_1 on the bosonic zero modes, i.e. on the momentum and winding modes. For the states $|M, L\rangle$ we easily find from (10.7)

$$I_1 |M, L\rangle = |-M, -L\rangle, \tag{10.130}$$

i.e. under the reflection I_1 momentum and winding quantum numbers change sign. Therefore, in the calculation of the partition function only states with $|M = 0, L = 0\rangle$ will contribute. We then obtain

$$\frac{1}{2} \text{Tr}_{\mathscr{H}} \left(I_1 \, q^{L_0 - \frac{c}{24}} \, \bar{q}^{\bar{L}_0 - \frac{c}{24}} \right) = \frac{1}{2} (q\bar{q})^{-\frac{1}{24}} \prod_{n=1}^{\infty} \frac{1}{(1 + q^n)} \frac{1}{(1 + \bar{q}^n)}$$

$$= \left| \frac{\eta(\tau)}{\vartheta_2(\tau)} \right|. \tag{10.131}$$

---

[12]In this section we use (σ^1, σ^0) rather than (σ, τ) as world-sheet coordinates and reserve τ for the modular parameter of the torus.

However this cannot be the full partition function because this term is not invariant under modular transformations. In particular, it transforms under modular S and T transformations as

$$T \left|\frac{\eta}{\vartheta_2}\right| \xleftrightarrow{S} \left|\frac{\eta}{\vartheta_4}\right| \xleftrightarrow{T} \left|\frac{\eta}{\vartheta_3}\right| S \qquad (10.132)$$

so that modular invariance forces us to introduce a twisted sector

$$Z_{\text{tw}}(\tau, \bar{\tau}) = \left|\frac{\eta(\tau)}{\vartheta_4(\tau)}\right| + \left|\frac{\eta(\tau)}{\vartheta_3(\tau)}\right|. \qquad (10.133)$$

This terminology becomes apparent if we note the following explicit form of, for instance, the right-moving character $\sqrt{\eta/\vartheta_4}$

$$\sqrt{\frac{\eta(\tau)}{\vartheta_4(\tau)}} = q^{\frac{1}{16}-\frac{1}{24}} \prod_{n=0}^{\infty} \frac{1}{\left(1-q^{n+\frac{1}{2}}\right)}. \qquad (10.134)$$

We can interpret this expression as the chiral partition function in a sector with ground state energy $L_0|0\rangle = \frac{1}{16}|0\rangle$ and half-integer modes $\alpha_{n+\frac{1}{2}}$

$$X_R(\sigma^0 - \sigma^1) = \frac{x_0}{2} + i\sqrt{\frac{\alpha'}{2}} \sum_{n=-\infty}^{\infty} \frac{1}{\left(n+\frac{1}{2}\right)} \alpha_{n+\frac{1}{2}} e^{-i\left(n+\frac{1}{2}\right)(\sigma^0-\sigma^1)}, \qquad (10.135)$$

where only $x_0 = 0$ and $x_0 = \pi R$, i.e. the two fixed points, are compatible with the $\mathbb{Z}_2$-symmetry. An analogous expansion holds for the left-moving sector. We observe that this mode expansion respects the symmetry

$$X(\sigma^1 + 2\pi, \sigma^0) = -X(\sigma^1, \sigma^0) = I_1 X(\sigma^1, \sigma^0) I_1^{-1} \qquad (10.136)$$

so that in this sector a string is periodic only up to the action of the discrete symmetry I_1. Clearly, in the orbifold, where two points related by the action of I_1 are identified, such a string is closed. Such a sector is called an I_1 twisted sector and the partition function in this sector can be defined as

$$Z_{\text{tw}}(\tau, \bar{\tau}) = \text{Tr}_{\mathcal{H}_{\text{tw}}} \left(\frac{1+I_1}{2} q^{L_0-\frac{c}{24}} \bar{q}^{\bar{L}_0-\frac{c}{24}}\right) = \left|\frac{\eta(\tau)}{\vartheta_4(\tau)}\right| + \left|\frac{\eta(\tau)}{\vartheta_3(\tau)}\right|, \qquad (10.137)$$

which also makes the meaning of the second term in (10.133) evident. States in the Hilbert space $\mathcal{H}_{\text{tw}}$ are created from the vacuum with half-integer moded oscillators. To summarize, the modular invariant partition function of a free boson on the $\mathbb{Z}_2$-orbifold of the circle is

$$Z_{\text{orb.}}(\tau, \bar{\tau}) = \frac{1}{2} Z_{\text{circ.}}(\tau, \bar{\tau}) + \left|\frac{\eta(\tau)}{\vartheta_2(\tau)}\right| + \left|\frac{\eta(\tau)}{\vartheta_4(\tau)}\right| + \left|\frac{\eta(\tau)}{\vartheta_3(\tau)}\right|. \qquad (10.138)$$

10.5 Toroidal Orbifolds

Note that the states in the twisted sector have an overall twofold degeneracy. This is due to the fact that the twisted sector states are localized at the fixed points of the orbifold action which give equal contributions to the partition function.

One important lesson is that modular invariance of the one-loop partition function requires the inclusion of the twisted sectors. This can already be seen for the compactification of the bosonic string on S^1 where by the same logic the sum over the windings is the sum over the twisted sectors. Without it, the partition function would not be modular invariant.

The orbifold construction we have just explained for the $\mathbb{Z}_2$ quotient of the circle compactification of the bosonic string can be generalized in many ways. We start with some general definitions. An orbifold $\mathcal{O}$ is the quotient of a Riemannian manifold $(\mathcal{M}, g)$ by the action of a discrete group G which preserves the metric g, i.e. G is an isometry:

$$\mathcal{O} = \mathcal{M}/G. \tag{10.139}$$

For $g \in G$ and $x \in \mathcal{M}$, the points x and gx are equivalent in the quotient, i.e. points in $\mathcal{O}$ are orbits of G in $\mathcal{M}$, hence the name orbifold. The action of G on $\mathcal{M}$ is not required to be free. Fixed points, or more generally, fixed subspaces of $\mathcal{M}$, are singular points (subspaces) on $\mathcal{O}$.

The simplest orbifolds are those with $\mathcal{M} = \mathbb{R}^D$. Strings compactified on these orbifolds lead to tractable conformal field theories; e.g. the excitation spectrum can be determined and correlation functions of physical state vertex operators can be computed.

The isometry group of $\mathbb{R}^D$ is the Euclidean group $E(D)$ which consists of rotations and translations. For the heterotic string G also acts on the gauge degrees of freedom. A general element $g \in G$ is

$$g = (\theta, v; \gamma), \tag{10.140}$$

where $\theta \in O(D)$ is a rotation, v a shift and γ the action on the gauge degrees of freedom. The space group S, whose elements are $g = (\theta, v)$, is a discrete subgroup of $E(D)$ and the orbifold is

$$\mathcal{O} = \mathbb{R}^D/S. \tag{10.141}$$

For the bosonic and the type II strings $G = S$. $g \in S$ acts on the space-time coordinates as

$$x \mapsto gx = \theta x + v \tag{10.142}$$

and the multiplication law for elements in S is

$$(\theta_1, v_1)(\theta_2, v_2) = (\theta_1 \theta_2, \theta_1 v_2 + v_1) \tag{10.143}$$

and the inverse of (θ, v) is

$$(\theta, v)^{-1} = (\theta^{-1}, -\theta^{-1}v). \tag{10.144}$$

The point group P is defined as the discrete subgroup of $O(D)$ consisting of rotations θ such that $(\theta, v) \in S$ for some v. The lattice Λ is defined to be the subgroup of S which consists of pure translations $(1, v)$ and we use it to define the torus

$$T^D = \mathbb{R}^D/\Lambda. \tag{10.145}$$

Tori are the simplest examples of orbifolds for which $S = \Lambda$ and the point group is trivial. There are no fixed points and in T^D the points x and $x + V$, $V \in \Lambda$ are identified.

More generally, since $(\theta, v)(\theta, u)^{-1} = (1, v - u) \in \Lambda$, the point group has a well defined action $\bar{P}$ on T^D, where $\bar{P} = S/\Lambda$ is a subgroup of the isometry group of the torus. The orbifolds can then be equivalently defined as

$$\mathcal{O} = T^D/\bar{P}. \tag{10.146}$$

In general P and $\bar{P}$ are different as elements of $\bar{P}$ might include translations which are not in Λ. In most explicitly worked out examples in the literature $\bar{P} = P$. From $(\theta, u)(1, v)(\theta, u)^{-1} = (1, \theta v)$ one concludes that the point group acts crystallographically on the torus lattice Λ, i.e. it is a discrete lattice automorphism: for any $L \in \Lambda$ and $g \in P$, $gL \in \Lambda$.

Consider two points x and y which are distinct in $\mathbb{R}^D$ but such that $y = \theta x + v$. Then x and y are identified on $\mathcal{O}$ and moreover the tangent vectors at x are identified with the tangent vectors at y rotated by θ. Parallel-transport some vector along a path from x to y which is closed on the orbifold. The torus is flat and hence this vector remains constant but since the tangent basis is rotated by θ, the final vector is rotated by θ with respect to the initial vector. This means that the point group P is the holonomy group of the orbifold. We will discuss holonomy groups further in Chap. 14. For the holonomy group of $\mathcal{O} = \mathbb{R}^D/S$ to be non-trivial, the action of S has to have fixed points at which the orbifold fails to be a manifold: it has an orbifold singularity.

A simple (non-compact) example is $\mathbb{R}^2/\mathbb{Z}_2$ where $P = \mathbb{Z}_2$ acts as a rotation by π around the origin O which is a fixed point. This space is a cone with a deficit angle π. A vector transported around the origin is rotated by π. This is illustrated in Fig. 10.2.

Generically orbifolds are singular limits of smooth manifolds which, conversely, are obtained from the orbifold by a process which is referred to as 'blowing up' the singularity. The smooth manifold is curved and its holonomy group $\mathcal{H}$ is a continuous subgroup of $O(D)$ such that $P \subset \mathcal{H}$. In the orbifold limit the

10.5 Toroidal Orbifolds

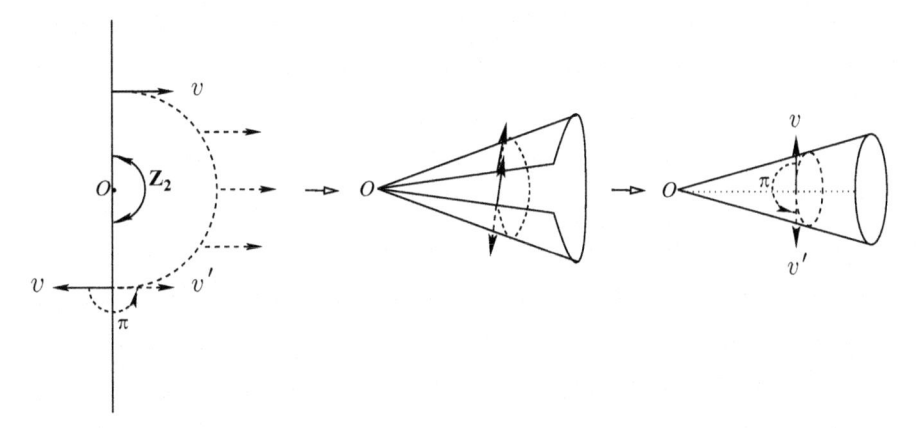

Fig. 10.2 The $\mathbb{R}^2/\mathbb{Z}_2$ orbifold with holonomy group $\mathbb{Z}_2$. The parallel transported vector v' is rotated by π w.r.t. to the image of v under the $\mathbb{Z}_2$ orbifold action

curvature gets squeezed into the fixed points. The requirement that the compactified string theory has some unbroken space-time supersymmetry imposes restrictions on the allowed holonomy groups and therefore the allowed space-groups. Furthermore, the blowing-up of the singularities must be such that the holonomy group of the resulting smooth manifold satisfies the condition imposed by supersymmetry. The conditions for space-time supersymmetry for compactifications on smooth manifolds will be discussed in Chap. 14 and on orbifolds in Chap. 15.

So far we have discussed the orbifold geometry. If one discusses point particles on orbifolds, the geometric singularities lead to singularities in the dynamics. Strings on orbifolds are, however, completely well behaved. This is ultimately due to the extended nature of strings.

We will not present an exhaustive discussion of strings on orbifolds. This would require a detailed treatment of the conformal field theory of orbifolds. But we will show how to compute the string partition function, which contains information about the spectrum of the theory and we will discuss some of its properties.

When computing the partition function, two issues have to be taken into consideration. First, since x and gx are the same point on $\mathcal{O}$, we only consider the subspace of the string Hilbert space which is invariant under G. We therefore have to project onto invariant states. In other words, we are gauging G. As G is discrete, there are no gauge fields. Here and elsewhere we use the same symbol g for an element of the point group and the operator which acts on the string Hilbert space.[13] For instance, for strings on tori $\mathbb{R}^D/\Lambda$, for each translation $\boldsymbol{L} \in \Lambda$ there is

---

[13] We could consider the situation where the action of the orbifold group is different on the left-movers than on the right-movers. This would lead to an asymmetric orbifold compactification. We will only consider symmetric orbifolds where the (geometric) action of the orbifold group is the same for left- and right-movers. For the heterotic string there is, of course, a difference but for symmetric orbifolds this only affects the gauge sector.

an operator $e^{i p \cdot L}$. Projection on the invariant subspace restricts the eigenvalues of p to lie in Λ^*. For groups of finite order $|G|$ this is implemented by the insertion of the projection operator

$$P = \frac{1}{|G|} \sum_{g \in G} g \qquad (10.147)$$

which satisfies $gP = P$. An example which will be studied in some detail below is $\mathcal{M} = T^D$ and $G = \mathbb{Z}_N$ which is an Abelian group with a single generator θ and N is the smallest integer such that $\theta^N = 1$.

The second feature is the appearance of twisted sectors in which the coordinate X closes up to a non-trivial transformation $h \in G$, i.e. the string is closed on $\mathcal{O}$ but not on $\mathcal{M}$:

$$X(\sigma^0, \sigma^1 + 2\pi) = h X(\sigma^0, \sigma^1) h^{-1}. \qquad (10.148)$$

For instance, in the example T^D, the twisted sectors have boundary conditions (10.35), i.e. the twisted sectors are simply the winding sectors.

For an Abelian orbifold with point group G, for the torus amplitude this leads to (d is the number of non-compact space-time dimensions)

$$\mathcal{T} = \frac{V_d}{\ell_s^d} \int_{\mathcal{F}} \frac{d^2\tau}{4\tau_2^2} Z(\tau, \bar{\tau}) \qquad (10.149)$$

with partition function

$$Z(\tau, \bar{\tau}) = \sum_{g, h \in G} \epsilon(h, g) Z[h, g](\tau, \bar{\tau})$$

$$= \sum_{h \in G} \left[\frac{1}{|G|} \sum_{g \in G} \epsilon(h, g) \mathrm{Tr}_h \left(g P_{\mathrm{GSO}}\, q^{L_0(h) - \frac{c}{24}}\, \bar{q}^{\bar{L}_0(h) - \frac{c}{24}} \right) \right]. \qquad (10.150)$$

The coefficients $\epsilon(h, g)$ are phases which are fixed by requiring modular invariance. They are called discrete torsion.[14] The sum over h is the sum over twisted sectors and the eigenvalues of L_0 depend on h. The insertion of g in the trace implies that we sum over states which satisfy the boundary conditions

$$X(\sigma^0 + 2\pi\tau_2, \sigma^1 + 2\pi\tau_1) = g X(\sigma^0, \sigma^1) g^{-1}. \qquad (10.151)$$

---

[14] They can be interpreted to arise from the contribution of the B field in the Polyakov action $\int_\Sigma \phi^* B$ where ϕ is the embedding of the world-sheet into space-time. If $dB = 0$ and the embedding is topologically non-trivial, this introduces topological phases into the Polyakov path integral.

10.5 Toroidal Orbifolds

For non-Abelian G the sum in (10.150) is over h and g such that $[h, g] = 0$ since otherwise (10.148) and (10.151) are incompatible. For the bosonic string P_{GSO} is, of course, absent.

The inclusion of twisted sectors is enforced by the requirement of modular invariance. Indeed, from Sect. 9.1 we know that under a modular transformation $\tau \to \tau' = \frac{a\tau+b}{c\tau+d}$ the boundary conditions change according to

$$(h, g)\left(\frac{a\tau + b}{c\tau + d}\right) = (h', g')(\tau) = (h^a g^{-c}, h^{-b} g^d)(\tau). \tag{10.152}$$

In particular $(h, g)(\tau + 1) = (h, gh^{-1})(\tau)$ and $(h, g)(-1/\tau) = (g, h^{-1})(\tau)$. Note that $[g, h] = 0 \to [g', h'] = 0$. We can thus generate contributions to Z by applying modular transformations to one of the $Z[h, g]$. At one loop it can happen that there are sectors which do not mix. This leaves some of the phases $\epsilon(h, g)$ arbitrary and different choices generically lead to different compatifications with different spectra. (This is similar to the situation in Chap. 9 where not all spin structures mixed under modular transformations, a fact that allowed for different GSO projections.) For instance, for a $\mathbb{Z}_N \times \mathbb{Z}_M$ orbifold with generators θ ($\theta^N = 1$) and ω ($\omega^M = 1$) the sector (θ, ω) never mixes with the untwisted sectors $(1, \theta^a \omega^b)$. As in Chap. 9, higher loop (in fact two-loop is sufficient) modular invariance and factorization imply restrictions on the phases. In general there can be several solutions leading to different compactifications with different spectra, etc. For the $\mathbb{Z}_N$ orbifold all sectors mix under modular transformations and all phases can be chosen $\epsilon(h, g) = 1$.

In the following we restrict to even dimensional toroidal $\mathbb{Z}_N$ orbifolds, i.e. $\mathcal{O} = T^D/\mathbb{Z}_N$ with D even. $\mathbb{Z}_N$ has one single generator θ with $\theta^N = \mathbb{1}$. θ is a $SO(D)$ rotation with eigenvalues $e^{\pm 2\pi i v_j}$ in the vector representation. $\theta^N = \mathbb{1}$ requires $v_j = k_j/N$ for some integers k_i, $i = 1, \ldots, D/2$. It is diagonal in the complex basis

$$Z^j = \frac{1}{\sqrt{2}}(X^{2j-1} + i X^{2j}), \qquad Z^{\star j} = \frac{1}{\sqrt{2}}(X^{2j-1} - i X^{2j}), \tag{10.153}$$

where it acts as

$$Z^j \to e^{2\pi i v_j} Z^j, \qquad Z^{\star j} \to e^{-2\pi i v_j} Z^{\star j}, \quad j = 1, \ldots, D/2. \tag{10.154}$$

Since P must act crystallographically on the torus lattice and since $\boldsymbol{L} = n_i \boldsymbol{e}_i$ with integer coefficients n_i, in the lattice basis θ must be a matrix of integers. Hence the quantities

$$\text{Tr}\theta = \sum_{j=1}^{D/2} 2\cos(2\pi v_j), \tag{10.155a}$$

Table 10.1 Irreducible crystallographic actions

$D = 2$	$D = 4$	$D = 6$
(v_1)	(v_1, v_2)	(v_1, v_2, v_3)
$\frac{1}{3}(1)$	$\frac{1}{5}(1, 2)$	$\frac{1}{7}(1, 2, 3)$
$\frac{1}{4}(1)$	$\frac{1}{8}(1, 3)$	$\frac{1}{9}(1, 2, 4)$
$\frac{1}{6}(1)$	$\frac{1}{10}(1, 3)$	$\frac{1}{14}(1, 3, 5)$
	$\frac{1}{12}(1, 5)$	$\frac{1}{18}(1, 5, 7)$

All entries are up to a sign

$$\chi(\theta) = \det(1 - \theta) = \prod_{j=1}^{D/2} 4\sin^2(\pi v_j) \tag{10.155b}$$

must be integers. In fact, by the Lefschetz fixed point theorem, $\chi(\theta)$ is the number of fixed points of θ. If there are fixed tori, i.e. if $v_j = 0 \mod 1$ for some j, omitting that factor in (10.155b) gives the number of fixed tori.

It turns out that the requirement of crystallographic action is very restrictive. For instance, for $D = 2$ only $N = 2, 3, 4, 6$ are allowed. In Table 10.1 we collect the irreducible possibilities for the v_j when $D = 2, 4, 6$. By irreducible we mean that the corresponding θ cannot be written in a block form in the lattice basis. Notice that the case $D = 2$, $v_1 = \frac{1}{2}$ is reducible since already in a one-dimensional lattice a $\mathbb{Z}_2$ (and only a $\mathbb{Z}_2$) is allowed, as we saw in the $S^2/\mathbb{Z}_2$ example.

We now turn to the evaluation of the partition function of the type II string compactified on a symmetric $\mathbb{Z}_N$ orbifold. For simplicity we assume that θ^ℓ for $\ell \neq 0$ rotates all directions, i.e. there are no fixed tori, only isolated fixed points. This is e.g. the case for $D = 6$ and $N = 7$ and for the $\mathbb{Z}_3$ orbifold which we will discuss in Chap. 15.

We start with the bosonic contributions. The mode expansions are

$$Z^j(\sigma^0, \sigma^1) = z_0^j + i\sqrt{\frac{\alpha'}{2}} \sum_s \frac{\alpha_s^j}{s} e^{-is(\sigma^0 - \sigma^1)} + i\sqrt{\frac{\alpha'}{2}} \sum_t \frac{\bar{\alpha}_t^j}{t} e^{-it(\sigma^0 + \sigma^1)}. \tag{10.156}$$

Imposing

$$Z^j(\sigma^0, \sigma^1 + 2\pi) = e^{2\pi i k v_j} Z^j(\sigma^0, \sigma^1), \tag{10.157}$$

which is valid for a complex boson in the k-th twisted sector, fixes the frequencies to $s = n + kv_j$ and $t = n - kv_j$ with n integer. Furthermore, z_0^j must satisfy $(1 - e^{2\pi i k v_j})z_0^j = 0 \mod 2\pi\Lambda$, i.e. it must be a fixed point of the orbifold action and, therefore, states in the twisted sectors are localized at the fixed points.

10.5 Toroidal Orbifolds

For the complex conjugate $Z^{\star j}$ there is an analogous expansion with coefficients $\alpha^{\star j}_{n-kv_j} = (\alpha^j_{-n+kv_j})^\dagger$ for the right-movers, $\overline{\alpha}^{\star j}_{n+kv_j} = (\overline{\alpha}^j_{-n-kv_j})^\dagger$ for the left-movers and $z^{\star j}_0 = (z^j_0)^\dagger$ for the center-of-mass position. Canonical quantization results in the following commutator relations for the oscillators:

$$[\alpha^i_{m+kv_i}, \alpha^{\star j}_{n-kv_j}] = (m+kv_i)\delta^{ij}\delta_{m+n,0},$$
$$[\overline{\alpha}^i_{m-kv_i}, \overline{\alpha}^{\star j}_{n+kv_j}] = (m-kv_i)\delta^{ij}\delta_{m+n,0}. \tag{10.158}$$

There are several Fock vacua $|f, 0\rangle_k$ where $f = 1, \ldots, \chi(\theta^k)$ is the fixed point label and k labels the different twisted sectors. Each vacuum is annihilated by all positive frequency modes and $\theta^\ell|f, 0\rangle = |f', 0\rangle$ where f' is also a fixed point of θ^k. The creation operators are thus $\alpha^j_{-n+kv_j}$, $n > 0$ and $\alpha^{\star j}_{-n-kv_j}$, $n \geq 0$ for the right-movers and $\overline{\alpha}^j_{-n-kv_j}$, $n \geq 0$ and $\overline{\alpha}^{\star j}_{-n+kv_j}$, $n > 0$ for the left movers. Here we consider the case where $0 < kv_j < 1$. The occupation number operators are

$$N_R = \sum_{n=-\infty}^{+\infty} :\alpha^j_{n+kv_j}\alpha^{\star j}_{-n-kv_j}:,$$
$$N_L = \sum_{n=-\infty}^{+\infty} :\overline{\alpha}^{\star j}_{n+kv_j}\overline{\alpha}^j_{-n-kv_j}:, \tag{10.159}$$

where normal ordering means that we move annihilation operators to the right. Note that the eigenvalues of N_L and N_R in the twisted sectors are multiples of $1/N$.

To compute the partition function we start with the untwisted sector ($k = 0$) where $L_0(\mathbb{1}) = \frac{1}{2}p_R^2 + N_R$ and $\overline{L}_0(\mathbb{1}) = \frac{1}{2}p_L^2 + N_L$. Then $Z[\mathbb{1}, \mathbb{1}]$ is simply

$$Z[\mathbb{1}, \mathbb{1}] = \frac{1}{|\eta(\tau)|^{2D}} \sum_{p\in\Lambda^*}\sum_{w\in\Lambda^*} q^{\frac{1}{2}(p+\frac{1}{2}w)^2}\overline{q}^{\frac{1}{2}(p-\frac{1}{2}w)^2}. \tag{10.160}$$

For $\ell \neq 0$, i.e. for bosons which satisfy the boundary conditions

$$Z^j(\sigma^0 + 2\pi\tau_2, \sigma^1 + 2\pi\tau_1) = e^{2\pi i \ell v_j} Z^j(\sigma^0, \sigma^1), \tag{10.161}$$

we need to evaluate the trace with an θ^ℓ insertion. Since we assume that θ^ℓ leaves no directions unrotated, neither quantized momenta nor windings survive the trace. We only need to consider states obtained from the Fock vacuum by acting with creation operators for which the complex coordinates are eigenvectors of θ^ℓ. The Fock vacuum is defined to be invariant under θ. Then, for instance, for the right movers in Z^j we find the contribution

$$\text{Tr}(\theta^\ell q^{L_0^j(\mathbb{1})-\frac{2}{24}}) = q^{-1/12}(1 + qe^{2\pi i \ell v_j} + qe^{-2\pi i \ell v_j} + \ldots), \tag{10.162}$$

where the first term is the contribution from the vacuum, the second and third from states obtained by acting with α^j_{-1} and α^{*j}_{-1} on the vacuum, and so on. It is not hard to see that the whole expansion can be cast into the form

$$\mathrm{Tr}(\theta^\ell q^{L_0^j(1)-\frac{2}{24}}) = q^{-1/12} \prod_{n=1}^\infty (1 - q^n e^{2\pi i \ell v_j})^{-1} (1 - q^n e^{-2\pi i \ell v_j})^{-1}$$

$$= -2\sin(\ell \pi v_j) \frac{\eta(\tau)}{\vartheta\begin{bmatrix}\frac{1}{2}\\ -\ell v_j - \frac{1}{2}\end{bmatrix}(\tau)}. \tag{10.163}$$

Notice that for $\ell = 0$ this becomes $1/\eta(\tau)^2$, as it should. Taking into account left- and right-movers for all (compact) coordinates we obtain

$$Z[\mathbb{1}, \theta^\ell] = \chi(\theta^\ell) \left| \prod_{j=1}^{D/2} \frac{\eta}{\vartheta\begin{bmatrix}\frac{1}{2}\\ -\ell v_j - \frac{1}{2}\end{bmatrix}} \right|^2, \tag{10.164}$$

where $\chi(\theta^\ell) = \prod_{j=1}^{D/2} 4\sin^2(\pi \ell v_j)$ is the number of fixed points of θ^ℓ. As is clear from (10.163), the coefficient of the first term in the expansion of (10.164) is one. This means that in the full untwisted sector, i.e. fixing $k = 0$ and summing over ℓ, the untwisted vacuum appears with the correct multiplicity one.

To obtain the other pieces $Z[\theta^k, \theta^\ell]$ we use modular transformations.[15] For example, $Z[\theta^\ell, \mathbb{1}]$ follows by applying $\tau \to -1/\tau$ to (10.164) (cf. (10.152)). Using the modular properties of the theta functions gives

$$Z[\theta^k, \mathbb{1}] = \chi(\theta^k) \left| \prod_{j=1}^{D/2} \frac{\eta}{\vartheta\begin{bmatrix}\frac{1}{2}+ kv_j\\ \frac{1}{2}\end{bmatrix}} \right|^2$$

$$= \chi(\theta^k)(q\bar{q})^{-\frac{D}{24}+E_k} \left| \prod_{j=1}^{D/2} \prod_{n=1}^\infty (1-q^{n-1+\{kv_j\}})^{-1}(1-q^{n-\{kv_j\}})^{-1} \right|^2, \tag{10.165}$$

where

$$E_k = \sum_{j=1}^{D/2} \frac{1}{2} \{kv_j\}(1-\{kv_j\}) \tag{10.166}$$

---

[15] It is not difficult to compute the traces in any $[\theta^k, \theta^\ell]$ sector directly, as we did for fermions in the Appendix to Chap. 9. The only subtlety is the degeneracy factor in the twisted sectors.

10.5 Toroidal Orbifolds

is the vacuum expectation value of L_0 in the twisted Fock vacuum which is annihilated by all positive oscillator modes. We defined $0 \leq \{x\} < 1$ as the fractional value of x: $\{x\} = x - \lfloor x \rfloor$. E_k can also be obtained independently from the vacuum expectation value of the energy-momentum tensor on the plane, $T = -\frac{2}{\alpha'} \sum_i :\partial Z^i \partial Z^{*i}::$

$$-\frac{2}{\alpha'} \sum_{i=1}^{D/2} \langle 0|\partial Z^i(w)\partial Z^{*i}(z)|0\rangle \overset{w \to z}{\sim} \frac{D/2}{(w-z)^2} + \frac{E_k}{z^2} + \mathcal{O}((w-z)). \quad (10.167)$$

Yet another way to derive $(E_k - \frac{D}{24})$ is from the oscillator expression for L_0 using ζ-function regularization. In any case, the Hamiltonian for one twisted boson is

$$H = N + E_k - \frac{c}{24} = L_0 - \frac{c}{24}, \quad (10.168)$$

where N is the oscillator contribution and $c = 24$. Here we have neglected possible center-of-mass momentum contributions. The Hamiltonian for a twisted fermion was already given in (9.123).

The lowest order term in the expansion (10.165) has coefficient $\chi(\theta^k)$, in agreement with the fact that in the θ^k twisted sector the center of mass coordinate can be any fixed point. From the higher order terms in the q-expansion we read off the eigenvalues of $L_0(\theta^k)$ and therefore the contribution of the twisted oscillators to the masses. In the sectors in which quantized momenta and windings are allowed, we must include in the partition function sums over the allowed momenta and windings. They will contribute to the eigenvalues of $L_0(\theta^k)$ and for particular shape and size of the torus their contribution might lead to extra massless states that signal enhanced symmetries as in the example of the circle compactification at the self-dual radius.

We can continue generating pieces of the partition function by employing modular transformations. The general result can be written as

$$Z[\theta^k, \theta^\ell] = \chi(\theta^k, \theta^\ell) \prod_{j=1}^{D/2} \left| \frac{\eta}{\vartheta\left[\begin{smallmatrix}\frac{1}{2}+kv_j\\ \frac{1}{2}-\ell v_j\end{smallmatrix}\right]} \right|^2, \quad (10.169)$$

where $\chi(\theta^k, \theta^\ell)$ is the number of simultaneous fixed points of θ^k and θ^ℓ. This formula is valid when θ^k leaves no fixed directions, otherwise a sum over momenta and windings could appear. This is important when determining the $\mathbb{Z}_N$-invariant states. The correct result can be found by carefully determining the untwisted sector pieces and then performing modular transformations. In addition, $\chi(\theta^k, \theta^\ell)$ should be replaced by $\tilde{\chi}(\theta^k, \theta^\ell)$, the number of fixed points in the sub-lattice effectively rotated by θ^k. χ and $\tilde{\chi}$ differ because when $kv_j = $ integer, the expansion of

$\vartheta\begin{bmatrix}\frac{1}{2}+kv_j\\\frac{1}{2}-\ell v_j\end{bmatrix}/\eta$ has a prefactor $(2\sin\pi\ell v_j)$, as follows from the product representation of the ϑ-function, just as in (10.163). Thus the actual coefficient in the expansion of (10.169) is $\tilde{\chi}(\theta^k,\theta^\ell) = \chi(\theta^k,\theta^\ell)/\prod_{j,kv_j\in\mathbb{Z}} 4\sin^2\pi\ell v_j$.

To summarize, the bosonic piece of the partition function of the type II string compactified on a symmetric $\mathbb{Z}_N$ orbifold is

$$Z_B(\theta^k,\theta^\ell) = \left(\frac{1}{\sqrt{\tau_2}\eta\bar\eta}\right)^{8-D} \chi(\theta^k,\theta^\ell) \left|\prod_{j=1}^{D/2} \frac{\eta}{\vartheta\begin{bmatrix}\frac{1}{2}+kv_j\\\frac{1}{2}-\ell v_j\end{bmatrix}}\right|^2, \tag{10.170}$$

where the first term is the contribution of the non-compact coordinates. Possible contributions from quantized momenta and windings are omitted.

We now add (complex) world-sheet fermions. In light-cone gauge there are four complex internal fermions with boundary conditions

$$\psi^j(\sigma^0,\sigma^1+2\pi) = -e^{2\pi i\alpha}e^{2\pi ikv_j}\psi^j(\sigma^0,\sigma^1),$$
$$\psi^j(\sigma^0+2\pi\tau_2,\sigma^1+2\pi\tau_1) = -e^{2\pi i\beta}e^{2\pi i\ell v_j}\psi^j(\sigma^0,\sigma^1), \tag{10.171}$$

where $\alpha,\beta \in \{0,\frac{1}{2}\}$ are the spin structures. The action on the light-cone gauge Ramond ground states $|s\rangle$, which were constructed in Chap. 8 is then defined as

$$\theta : |s_1,s_2,s_3,s_4\rangle \to e^{2\pi i\,s\cdot v}|s_1,s_2,s_3,s_4\rangle. \tag{10.172}$$

The degenerate Ramond ground states transform as a spinor of $SO(8)$, generated from $|0\rangle$ by the action of the b_i^- with $i = 1,\ldots,4$. More generally,

$$\theta = \exp\left(2\pi i \sum_{i=1}^{4} v_i J_{2i-1,2i}\right), \tag{10.173}$$

where $J_{2i-1,2i}$ are the generators of the Cartan subalgebra of $SO(8)$. On a state which is characterized by the $SO(8)$ weight[16] λ it has eigenvalue $e^{2\pi i\lambda\cdot v}$. Requiring $\theta^N = \mathbb{1}$ on spinors leads to the condition

$$N\sum v_i = 0 \mod 2. \tag{10.174}$$

The full partition function still has the form (10.150). Each contribution is explicitly evaluated as

---

[16]This will be explained in detail in Chap. 11.

10.5 Toroidal Orbifolds

$$Z[\theta^k, \theta^\ell] = \text{Tr}_{(NS\oplus R)\otimes(NS\oplus R)}\left(P_{GSO}\, \theta^\ell\, q^{L_0(\theta^k)-\frac{D}{16}}\, \bar{q}^{\bar{L}_0(\theta^k)-\frac{D}{16}}\right). \quad (10.175)$$

The trace is over the left and right NS and R sectors for the fermions. This is equivalent to summing over $\alpha = 0 \in \{0, \frac{1}{2}\}$. Similarly, the GSO projection amounts to summing over $\beta \in \{0, \frac{1}{2}\}$.

The fermionic partition function for general twisted boundary conditions was computed in the appendix of Chap. 9. This yields

$$Z_F[\theta^k, \theta^\ell] = \frac{1}{4}\left|\sum_{\alpha,\beta} s_{\alpha\beta}(k,\ell) \prod_{j=1}^{4} \frac{\vartheta\begin{bmatrix}\alpha+kv_j\\-\beta-\ell v_j\end{bmatrix}}{\eta}\right|^2. \quad (10.176)$$

The complete partition function is the product of the bosonic and the fermionic piece

$$Z_{\text{orb}} = \frac{1}{N}\sum_{k,l=0}^{N-1} Z_B[\theta^k, \theta^l]\, Z_F[\theta^k, \theta^l]. \quad (10.177)$$

Modular invariance imposes relations among the spin structure coefficients $s_{\alpha\beta}(k, \ell)$. One checks that

$$s_{00}(k, \ell) = -s_{\frac{1}{2}0}(k, \ell) = 1, \quad s_{0\frac{1}{2}}(k, \ell) = -e^{i\pi k \sum v_j} = \mp s_{\frac{1}{2}\frac{1}{2}}(k, \ell) \quad (10.178)$$

leads to a modular invariant partition function.

Note that $k = N$ should give the same solution as $k = 0$. This gives, once more, condition (10.174). Note further that the sign of $s_{\frac{1}{2}\frac{1}{2}}(k, \ell)$ is not fixed by modular invariance. Choosing opposite (equal) signs in the left and right sectors corresponds to orbifold compactifications of type IIA(B) strings (as one can see by looking at the $k = \ell = 0$ sector).

Using the representation (9.100) for the theta-functions one finds that states in the NS sector ($\alpha = 0$) are labelled by four integers $|n_1, n_2, n_3, n_4\rangle$ such $\sum n_i = $ odd and states in the R sector by four half-integers $|r_1, r_2, r_3, r_4\rangle$ such that $\sum r_i = $ even or odd, depending on the sign choice in (10.178). States which do not satisfy these conditions are projected out. These are, of course, the two possible GSO projections. We will see in the next chapter that this corresponds to states in the vector and either one of the two spinor conjugacy classes of $SO(8)$.

Two comments are in order. First, there are other choices for the spin structure coefficients, e.g. those which correspond to compactifications of the type 0 string on orbifolds. Second, with the help of the generalized Riemann identity (9.110) one can show that $Z_F[\theta^k, \theta^l] = 0$ if the twist vector satisfies

$$\pm v_1 \pm v_2 \pm v_3 \pm v_4 = 0 \bmod 2 \quad (10.179)$$

for some choice of signs.[17] This signals a space-time supersymmetric spectrum. Indeed, in Chap. 15 we will rederive this condition as the condition for space-time supersymmetry of the type II and heterotic theories compactified on a $\mathbb{Z}_N$ orbifold. We postpone further discussion of space-time supersymmetry in orbifold compactifications until then.

$\mathbb{Z}_N$ orbifolds are the simplest examples of toroidal orbifolds. Generalizations involve $\mathbb{Z}_M \times \mathbb{Z}_N$ orbifolds, where a new, discrete degree of freedom, discrete torsion, can be switched on. Moreover, one can consider non-Abelian orbifolds with non-Abelian point-group and asymmetric orbifolds, for which the discrete symmetries act differently on the left- and right-moving sectors.

As already pointed out, the attraction of orbifold compactifications of string theory stems from the fact that they have a description in terms of solvable conformal field theories, i.e. they are under complete calculational control. However, orbifold compactifications of the type II string do not lead to phenomenologically interesting string vacua. To find those without leaving the class of tractable CFT's we either have to consider orientifolds of the above or orbifold compactifications of the heterotic string. Both will be discussed in Chap. 15

10.6 D-branes on Toroidal Compactifications

We have seen that an open string can have either Neumann or Dirichlet boundary conditions at its end. In this section we will point out that in the case that space is compactified more possibilities exist.

To begin with, let us consider an open string with NN boundary conditions moving on a circle of radius R, i.e. we identify $X \simeq X + 2\pi R$. Using that the center of mass momentum is quantized in units $p = M/R$, the mode expansion reads

$$X(\sigma, \tau) = x_0 + 2\alpha' \frac{M}{R} \tau + i\sqrt{2\alpha'} \sum_{n \neq 0} \frac{1}{n} \alpha_n e^{-in\tau} \cos n\sigma. \tag{10.180}$$

What happens to this expansion under T-duality? For the closed string we found that it acts as an asymmetric $\mathbb{Z}_2$ reflection $T : X = (X_L, X_R) \to X' = (X_L, -X_R)$. Applying this transformation to (10.180) leads to

$$X'(\sigma, \tau) = c_0 + 2\alpha' \frac{M}{R} \sigma + \sqrt{2\alpha'} \sum_{n \neq 0} \frac{1}{n} \alpha_n e^{-in\tau} \sin n\sigma. \tag{10.181}$$

---

[17]This is true if one of the $v_i = 0$, in which case the $\alpha = \beta = \frac{1}{2}$ term in (10.176) vanishes. This is always the case if we compactify to four dimensions. If we compactify to two dimensions, Z_F vanishes only if the sign of the $\alpha = \beta = \frac{1}{2}$ term is $(-1)^m$ where m is the number of minus signs in (10.179).

10.6 D-branes on Toroidal Compactifications

But this is the mode expansion for an open string with Dirichlet boundary conditions in the X direction. It starts at position c_0 and winds M times around the dual circle with T-dual radius $R' = \alpha'/R$. Thus, T-duality acts on an open string by exchanging N with D boundary conditions with quantized momentum becoming integer winding around the dual circle. The exchange of the boundary conditions can also be seen directly:

$$\partial_\sigma X = \partial_+ X_L - \partial_- X_R \quad \to \quad \partial_+ X_L + \partial_- X_R = \partial_\tau X'. \tag{10.182}$$

The exchange of boundary conditions means that T-duality transforms a Dp-brane into a D(p∓1)-brane, depending on whether it acts along a direction on the brane (−) or transverse to the brane (+). (The more complicated situation where T-duality acts along a direction at an angle with the brane will be discussed below.)[18]

In the T-dual picture (10.181), both end points of the open string end on a single brane located at c_0. We can consider the more general situation where there are two branes, located at c_0 and c_1 and the open string stretches between them. In this case (10.181) will be modified to

$$X'(\sigma, \tau) = c_0 + \frac{1}{\pi}(c_1 - c_0)\sigma + 2MR'\sigma + \sqrt{2\alpha'} \sum_{n \neq 0} \frac{1}{n} \alpha_n e^{-in\tau} \sin n\sigma. \tag{10.183}$$

Since the c_i are positions on a circle, it is clear that they are only defined mod $2\pi R'$. Let us therefore parametrize the position as $\theta_i = c_i/(2\pi R')$ with $\theta_i \in [0, 1]$. We now apply T-duality to (10.183), i.e. to the open string stretched between two D-branes along a DD direction:

$$X(\sigma, \tau) = x_0 + 2\alpha' \frac{\theta_1 - \theta_0}{R}\tau + 2\alpha' \frac{M}{R}\tau + i\sqrt{2\alpha'} \sum_{n \neq 0} \frac{1}{n} \alpha_n e^{-in\tau} \cos n\sigma. \tag{10.184}$$

It appears that the distance between the two D-branes T-dualizes to a finite shift of the center of mass momentum of the open NN string.

We want to understand the origin of this new term from first principles, i.e. from canonically quantizing an open string with NN boundary conditions. For this purpose we recall the discussion of Sect. 10.2 where we coupled the closed string to a constant background field, the antisymmetric tensor B_{IJ}, which led to a modification of the center of mass momentum. For the open string there is another option, namely to couple it to a background gauge field on the D-branes on which it ends.

---

[18] T-duality acts on the world-sheet fermions of the superstring as $(\bar\psi, \psi) \to (\bar\psi, -\psi)$. As is clear from our discussion in Chap. 7, this also corresponds to the exchange of D and N boundary conditions.

For the present discussion, which concerns the center of mass momentum, only the zero-mode sector, in which the string reduces to a point particle, is relevant and we will restrict to it. A more complete discussion will follow later in this section.

The action, including the coupling to a constant background gauge field is

$$S = -\frac{1}{4\alpha'} \int d\tau \left(x'^2 - \dot{x}^2\right) - \int d\tau \left(A_1 - A_0\right) \dot{x}, \tag{10.185}$$

where $A_{0,1}$ are the components of the $U(1) \times U(1) \subset U(2)$ gauge fields on the two D-branes along the compact dimension. The relative minus sign between A_0 and A_1 reflects the fact that the string is oriented and that the two ends carry opposite charge. The canonical center-of-mass momentum of the string, which must be quantized in units $\widetilde{M}/R$, $\widetilde{M} \in \mathbb{Z}$ on the circle with radius R is

$$\pi = \frac{1}{2\alpha'} \dot{x} - (A_1 - A_0) = \frac{\theta_1 - \theta_0}{R} + \frac{M}{R} - (A_1 - A_0) = \frac{\widetilde{M}}{R}. \tag{10.186}$$

From this we conclude $A_i = \theta_i/R$ mod $\mathbb{Z}/R$, which allows us to identify the θ_i with non-trivial Wilson lines $\exp(i\theta_i)$, with

$$\theta_i = \frac{1}{2\pi} \oint_{S^1} A_i, \quad i = 1, 2. \tag{10.187}$$

To summarize, the T-dual description of the positions of the D-branes is in terms of Wilson lines in the Cartan subalgebra of the gauge-group on the world-volume of the branes along the compact direction.

This simple analysis, whose main message is that T-duality along a compact circle exchanges Neumann and Dirichlet boundary conditions, straightforwardly extends to the complete supersymmetric Dp-branes in type II string theory. Since T-duality along a circle S^1 therefore exchanges p = odd branes with p = even branes, it must better exchange the type IIB and the type IIA string. This is what we already found for the closed string sector in (10.25). There we have also discussed the action of T-duality on the world-sheet fermions. Recalling our discussion of boundary conditions for fermions in Chap. 7, we find that it is consistent with T-duality exchanging N↔D.

Let us now generalize the above discussion and compactify two directions X and Y on a two-dimensional torus T^2 and let us consider D-branes which wrap a 1-cycle on this T^2. Such a brane is specified by Neumann boundary conditions along one direction and Dirichlet boundary conditions along the orthogonal direction. For two such D-branes we encounter the possibility that the branes intersect at an angle $\Delta\Phi$. Let us first discuss the case of two intersecting D-branes on non-compact $\mathbb{C}$, which is shown in Fig. 10.3. Later we will extend this result to the case of branes wrapping 1-cycles of a compact T^2.

Denote by X and Y the two real coordinates of $\mathbb{R}^2$. Placing for convenience one D-brane along the X-axis, an open string stretched between two intersecting D-branes

10.6 D-branes on Toroidal Compactifications

Fig. 10.3 Intersecting D-branes

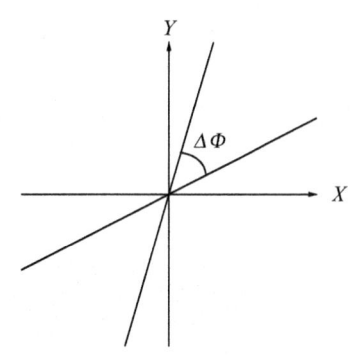

satisfies the following boundary conditions

$$\sigma = 0 : \partial_\sigma X = \partial_\tau Y = 0,$$
$$\sigma = \pi : \cos(\Delta\Phi)\,\partial_\sigma X + \sin(\Delta\Phi)\,\partial_\sigma Y = 0,$$
$$-\sin(\Delta\Phi)\,\partial_\tau X + \cos(\Delta\Phi)\,\partial_\tau Y = 0. \tag{10.188}$$

In order to implement these boundary conditions for the mode expansions of $X(\sigma, \tau)$ and $Y(\sigma, \tau)$ it is convenient to introduce complex target space coordinates $Z = \frac{1}{\sqrt{2}}(X + iY)$ and its complex conjugate Z^*. Now the boundary conditions at the two ends of the open string read

$$\sigma = 0 : \partial_\sigma(Z + Z^*) = \partial_\tau(Z - Z^*) = 0,$$
$$\sigma = \pi : \partial_\sigma Z + e^{2i\,\Delta\Phi}\,\partial_\sigma Z^* = 0,$$
$$\partial_\tau Z - e^{2i\,\Delta\Phi}\,\partial_\tau Z^* = 0. \tag{10.189}$$

Implementing these boundary conditions in the mode expansion of the fields Z and Z^*, one first realizes that the center of mass position is fixed at the intersection point, which is the origin in the $\mathbb{C}$-plane. Concerning the oscillator modes, we make the usual ansatz

$$Z(\sigma, \tau) = i\sqrt{\frac{\alpha'}{2}} \sum_s \frac{1}{s} \alpha_s\, e^{-is(\tau-\sigma)} + i\sqrt{\frac{\alpha'}{2}} \sum_\nu \frac{1}{s} \tilde\alpha_s\, e^{-is(\tau+\sigma)},$$

$$Z^*(\sigma, \tau) = i\sqrt{\frac{\alpha'}{2}} \sum_s \frac{1}{s} \alpha_s^*\, e^{-is(\tau-\sigma)} + i\sqrt{\frac{\alpha'}{2}} \sum_s \frac{1}{s} \tilde\alpha_s^*\, e^{-is(\tau+\sigma)}, \tag{10.190}$$

where we defined the modes as $\alpha_s = \frac{1}{\sqrt{2}}(\alpha_s^X + i\alpha_s^Y)$, $\alpha_s^* = \frac{1}{\sqrt{2}}(\alpha_s^X - i\alpha_s^Y)$ and similarly for $\tilde\alpha_s$ and $\tilde\alpha_s^*$. Implementing the boundary conditions at $\sigma = 0$ leads to the relations $\alpha_s^* = \tilde\alpha_s$ and $\tilde\alpha_n^* = \alpha_n$. The boundary conditions at $\sigma = \pi$ can be solved

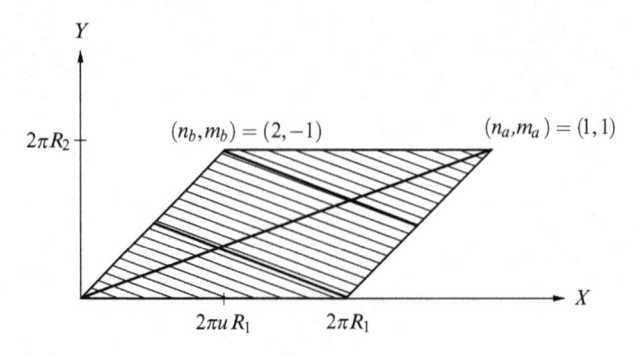

Fig. 10.4 Intersecting D-branes on T^2

for modes $\bar{\alpha}_{n-\epsilon}$ and $\alpha_{n+\epsilon}$ with $\epsilon = \Delta\Phi/\pi$ and $n \in \mathbb{Z}$. Therefore, the solution to the boundary conditions for two D-branes at angle $\Delta\Phi$ reads

$$Z(\sigma,\tau) = i\sqrt{\frac{\alpha'}{2}} \sum_{n \in \mathbb{Z}} \left\{ \frac{1}{(n+\epsilon)} \alpha_{n+\epsilon}\, e^{-i(n+\epsilon)(\tau-\sigma)} \right. $$
$$\left. + \frac{1}{(n-\epsilon)} \bar{\alpha}_{n-\epsilon}\, e^{-i(n-\epsilon)(\tau+\sigma)} \right\} \tag{10.191}$$

and similarly for $Z^\star(\sigma,\tau)$. Imposing $Z^\star = Z^\dagger$ requires $(\alpha_{n+\epsilon})^\dagger = \bar{\alpha}_{-n-\epsilon}$ and canonical quantization gives $[\alpha_{n+\epsilon}, \bar{\alpha}_{m-\epsilon}] = (n+\epsilon)\delta_{n+m}$. Analogously, for the complexified world-sheet fermions the modes are $\psi_{r+\epsilon}$ and $\overline{\psi}_{r-\epsilon}$ with $r \in \mathbb{Z}$ in the R sector and $r \in \mathbb{Z} + \frac{1}{2}$ in the NS sector. Canonical quantization gives $\{\psi_{r+\epsilon}, \overline{\psi}_{s-\epsilon}\} = \delta_{r+s}$.

The shift of the modes by ϵ is very reminiscent of what we found for the mode expansion of a closed string with twisted boundary conditions, e.g. in a twisted sector of an orbifold. For this reason, an open string between two intersecting D-branes can be considered as a twisted open string.

Consider now two intersecting D-branes on a compact T^2. This is shown in Fig. 10.4. The complex structure of this target space torus is defined as $U = U_1 + iU_2 = u + i\frac{R_2}{R_1}$. Each brane is specified by two wrapping numbers (n_a, m_a), $a = 1, 2$ of the brane around the two basic 1-cycles. Using the holonomy basis introduced in Chap. 6 the brane wraps the cycle $n_a\, a + m_a\, b$. The angle at which it intersects the x-axis is fixed by the wrapping numbers and the complex structure of the torus

$$\cot(\Phi_a) = \frac{n_a + m_a U_1}{m_a U_2} \tag{10.192}$$

and the relative angle between the two branes is $\Delta\Phi_{ab} = \Phi_b - \Phi_a$. Quantizing an open string stretched between two such branes, one finds the same fractional oscillator modes as in (10.191). In addition, the center of mass position is still fixed

10.6 D-branes on Toroidal Compactifications

at an intersection point of the two branes. The only difference to the non-compact case is that on a compact space the two branes can intersect more than once. In fact, in the present case the intersection number between the two branes is (cf. (6.44) on page 139)

$$I_{ab} = n_a\, m_b - m_a\, n_b. \tag{10.193}$$

Thus, we have established that D-branes intersecting at angles contain new open string states, which are localized at their, in general multiple, intersections points.

We now generalize the above discussion by considering the type IIA string compactified on a six-dimensional torus which is the direct product of three two-tori, i.e.

$$T^6 = T^2 \times T^2 \times T^2. \tag{10.194}$$

This corresponds to a partial fixing of the complex structure[19] of T^6. Consider D6-branes whose world-volume completely fills the remaining four uncompactified dimensions and which wrap a 1-cycle on each of the three T^2-factors. Clearly, each such brane is specified by three pairs of wrapping numbers (n_a^I, m_a^I) with $I = 1, 2, 3$. The total intersection number of two D6-branes is

$$I_{ab} = \prod_{I=1}^{3} \left(n_a^I\, m_b^I - m_a^I\, n_b^I\right). \tag{10.195}$$

If the two branes intersect on the I-th torus at an angle Φ_a^I, then the complex fields $Z^I(\sigma, \tau)$ carry modes $\alpha_{n+\epsilon^I}^I$ and $\bar\alpha_{n-\epsilon^I}^I$ with $\epsilon^I = \Delta\Phi^I/\pi$. Due to the similarity to twisted closed strings, it is now straightforward to compute the annulus diagram for an open string stretched between two intersecting D6-branes

$$\mathcal{A}_{D6_a, D6_b} = \int_0^\infty \frac{dt}{t}\, I_{ab}\, \frac{V_4}{(8\pi^2\alpha' t)^2}\, \frac{\sum_{\alpha,\beta}(-1)^{2(\alpha+\beta)}\, \vartheta\begin{bmatrix}\alpha\\\beta\end{bmatrix}\prod_I \vartheta\begin{bmatrix}\epsilon^I+\alpha\\\beta\end{bmatrix}}{2\eta^3\, \prod_I \vartheta\begin{bmatrix}\epsilon^I+1/2\\1/2\end{bmatrix}}$$
$$\tag{10.196}$$

with argument $q = e^{-2\pi t}$. As usual, $\alpha, \beta \in \{0, \tfrac{1}{2}\}$ is the sum over the four spin-structures. From this partition function we conclude two important aspects of intersecting branes. First, due to the generalized Riemann identity (9.110), the sum over the spin-structures vanishes for $\sum_I \Delta\Phi^I = 0 \bmod \pi$. However, this is merely a necessary condition for supersymmetry. In Sect. 13.3, a thorough analysis

---

[19] We will discuss the notion of complex structures of higher dimensional manifolds in Chap. 14.

Table 10.2 It is shown which 4D supersymmetry is preserved by intersecting D6-branes

Condition	$d = 4$ susy	Supermultiplet
$\Phi^1 = \Phi^2 = \Phi^3 = 0$	$\mathcal{N} = 4$	Vector
$\Phi^I + \Phi^J = 0, \quad \Phi^K = 0 \ I \neq J \neq K \neq I$	$\mathcal{N} = 2$	Hyper
$\Phi^1 \neq \Phi^2 \neq \Phi^3 \neq \Phi^1 \ \sum_I \Phi^I = 0$	$\mathcal{N} = 1$	Chiral
$\sum_I \Phi^I \neq 0$	Non-susy	Weyl fermion

of supersymmetry for intersecting D-branes will be carried out. The result is that supersymmetry is preserved under the stronger condition

$$\sum_I \Delta \Phi^I = 0 \mod 2\pi. \tag{10.197}$$

Second, for any choice of angles one gets at least one fermionic zero mode from the Ramond sector (in the uncompactified part of space-time, reflected by the factor $\vartheta[{\alpha \atop \beta}]$ in (10.196)). Taking the two orientations of the open string into account this gives rise to one chiral Weyl-fermion in four dimensions. In case of supersymmetry this is accompanied by additional zero modes from the Neveu-Schwarz sector and, depending on the amount of supersymmetry, by additional zero modes from the Ramond sector. The various possibilities are collected in Table 10.2. Supermultiplets in $d = 4$ will be summarized in Chap. 15. Therefore, with at most $\mathcal{N} = 1$ supersymmetry, intersecting D-branes support chiral fermions on each intersection point on T^6. If one now places stacks of D6-branes at angles supporting a $U(N_a) \times U(N_b)$ gauge group on their world-volume, it is clear that the I_{ab} chiral fermions transform in the bi-fundamental representation $(\overline{N}_a, N_b)$. This fact can be used to engineer intersecting D-brane configurations with standard model like world-volume actions, both on simple toroidal compactifications and more complicated backgrounds such as toroidal orbifolds. This is described in more detail in Chaps. 15 and 17.

Returning to the general discussion of D-branes on toroidal spaces, one might wonder how intersecting branes on a T^2 behave under T-duality, say in the X-direction. Since T-duality exchanges Neumann and Dirichlet boundary conditions, the boundary conditions for two intersecting D-branes (10.188) become

$$\sigma = 0: \quad \partial_\tau X = \partial_\tau Y = 0,$$
$$\sigma = \pi: \quad \cot(\Delta\Phi) \, \partial_\tau X + \partial_\sigma Y = 0,$$
$$\partial_\sigma X - \cot(\Delta\Phi) \, \partial_\tau Y = 0. \tag{10.198}$$

10.6 D-branes on Toroidal Compactifications

Therefore at $\sigma = 0$ we now have Dirichlet boundary conditions describing a D-brane localized in the (X, Y) plane. However, at $\sigma = \pi$ we have obtained a new boundary condition which is a mixture of Neumann and Dirichlet boundary conditions. In order for this T-duality induced result to make sense, this mixed ND boundary condition has to arise from a boundary term in the variation of the string world-sheet action.

We already know that background gauge fields on D-branes modify the quantization of the open string. Let us therefore extend the non-linear world-sheet sigma-model action (10.46) to include the gauge fields on the D-branes at the end of the open string world-sheet

$$S = -\frac{1}{4\pi\alpha'} \int_\Sigma d^2\sigma \left(\eta^{\alpha\beta} G_{IJ}(X) \partial_\alpha X^I \partial_\beta X^J + \epsilon^{\alpha\beta} B_{IJ}(X) \partial_\alpha X^I \partial_\beta X^J \right)$$
$$- \int_{\partial\Sigma} d\tau \, A_I(X) \, \partial_\tau X^I. \tag{10.199}$$

Here we have introduced a coupling of the boundary $\partial\Sigma$ of the open string to the gauge field on the D-brane. The boundary is at fixed $\sigma = 0, \pi$. The action has the Abelian gauge invariance of the vector potential at the boundary $\delta A_I = \partial_I \lambda$ and the combined two-form gauge invariance of the antisymmetric tensor B_{IJ}, which also involves a boundary term,

$$\delta B_{IJ} = \partial_I \zeta_J - \partial_J \zeta_I, \qquad \delta A_I = -\frac{1}{2\pi\alpha'} \zeta_I. \tag{10.200}$$

For constant metric G_{IJ}, antisymmetric tensor B_{IJ} and gauge field strength F_{IJ} the action (10.199) simplifies. First we notice that, using

$$\int_\Sigma d^2\sigma \, \epsilon^{\alpha\beta} B_{IJ} \, \partial_\alpha X^I \partial_\beta X^J = \int_{\partial\Sigma} d\tau \, B_{IJ} \, X^I \partial_\tau X^J, \tag{10.201}$$

the term involving B_{IJ} can be written as a boundary term. For constant F_{IJ} one can also write

$$\int_{\partial\Sigma} d\tau \, A_I \, \partial_\tau X^I = \frac{1}{2} \int_{\partial\Sigma} d\tau \, F_{IJ} \, X^I \, \partial_\tau X^J \tag{10.202}$$

so that the total world-sheet action becomes

$$S = -\frac{1}{4\pi\alpha'} \int_\Sigma d^2\sigma \, \eta^{\alpha\beta} G_{IJ} \, \partial_\alpha X^I \partial_\beta X^J$$
$$- \frac{1}{4\pi\alpha'} \int_{\partial\Sigma} d\tau \left(B_{IJ} + (2\pi\alpha') F_{IJ} \right) X^I \partial_\tau X^J. \tag{10.203}$$

Therefore, the end points of the open string couple to the gauge invariant field strength

$$2\pi\alpha' \mathscr{F}_{IJ} = B_{IJ} + 2\pi\alpha' F_{IJ}. \tag{10.204}$$

The boundary conditions that follow from the variation of the world-sheet action are

$$G_{IJ} \partial_\sigma X^J + (2\pi\alpha' \mathscr{F}_{IJ}) \partial_\tau X^J = 0. \tag{10.205}$$

For the target space T^2 with flat metric $G_{IJ} = \delta_{IJ}$ and the constant field strength $\mathscr{F} = \mathscr{F}_{YX}$, we obtain the boundary conditions

$$\partial_\sigma Y + (2\pi\alpha' \mathscr{F}) \partial_\tau X = 0,$$
$$\partial_\sigma X - (2\pi\alpha' \mathscr{F}) \partial_\tau Y = 0. \tag{10.206}$$

This looks precisely like the mixed boundary conditions (10.198) we derived via T-duality from the intersecting D-brane configuration. The intersection angle is related to the gauge invariant field strength via

$$\cot \Delta\Phi = 2\pi\alpha' \mathscr{F}. \tag{10.207}$$

Indeed, applying T-duality $R_1 \to \alpha'/R_1$ to relation (10.192), we obtain

$$\cot \Delta\Phi \to \frac{u\alpha'}{R_1 R_2} + \frac{n\alpha'}{mR_1 R_2} = 2\pi\alpha' \mathscr{F} = B + 2\pi\alpha' F. \tag{10.208}$$

Defining $B = 2\pi\alpha' \tilde{B}$, this relation very much suggests to identify

$$F = \frac{n}{2\pi m R_1 R_2}, \quad \tilde{B} = \frac{u}{2\pi R_1 R_2}. \tag{10.209}$$

This means that the closed string background B-field is related to the tilt of the T-dual torus and the gauge field F on the brane is related to the wrapping numbers of the T-dual brane. This makes sense: first, for $m = 1$ the first Chern class of this gauge field is[20]

$$c_1(F) = \frac{1}{2\pi} \int_{T^2} F = n \tag{10.210}$$

which shows that, as expected, the gauge field is quantized and the wrapping number n is related to the first Chern class of the gauge field. Second, the tilt of the torus is

---

[20] We will introduce Chern-classes in Chap. 14.

10.6 D-branes on Toroidal Compactifications

Table 10.3 Supersymmetric orientifolds of type II strings on T^6 with local tadpole cancellation

Superstring	Orientifold	O-planes	CP-gauge group
Type IIB	Ω	O9	$SO(32)$
	$\Omega I_2(-1)^{F_L}$	O7	$SO(8)^4$
	ΩI_4	O5	$SO(2)^{16}$
Type IIA	ΩI_1	O8	$SO(16)^2$
	$\Omega I_3(-1)^{F_L}$	O6	$SO(4)^8$

The reflection of n coordinates is denoted as I_n

not constrained further, which is also expected for the B-field. It can be shown that for $m > 1$ the T-dual description is given by a certain class of $U(m)$ bundles on T^2. Thus, we have seen that the T-dual picture of a D-brane at angle is a T^2-wrapping D-brane supporting non-trivial gauge flux on its world-volume. Such D-branes are also called magnetized D-branes.

The final topic of this section concerns the fate of the world-sheet parity transformation $\Omega : (\sigma, \tau) \rightarrow (\ell - \sigma, \tau)$ under T-duality. As we emphasized in Chap. 9, Ω is a symmetry of the ten-dimensional type IIB superstring but not of the type IIA superstring. Since T-duality acts as an asymmetric reflection $T : (X_L, X_R) \rightarrow (X_L, -X_R)$, world-sheet parity reversal transforms as

$$T \Omega T^{-1} = \Omega I_1, \quad \text{with} \quad I_1 : X \rightarrow -X. \tag{10.211}$$

For this to be consistent with (10.25), the combined $\mathbb{Z}_2$ transformation ΩI_1 must be a symmetry of the type IIA string compactified on a circle. This is indeed the case. Analogous to the type I string, one can now take the quotient of the type IIA string by ΩI_1 and compute the Klein-bottle amplitude. Instead of one O9-plane we now have two O8 planes, one at each fixed point of the reflection I_1. Again there will be divergences from graviton/dilaton exchange and RR 9-form exchange. The first is associated to an NS-NS tadpole and the second to an R-R tadpole. These can be canceled by introducing 32 D8-branes. Local tadpole cancellation is guaranteed, if we place 16 D8-branes on each O8-plane. This will result in a Chan-Paton gauge group $SO(16) \times SO(16)$. This defines the so-called type I' superstring theory.

By successively applying T-duality one can generate more types of orientifolds. The models on T^6, for which local cancellation of tadpoles is possible, are listed in Table 10.3.[21]

F_L is the left-moving space-time fermion number, i.e. $(-1)^{F_L} = -1$ on states in the left-moving R-sector and $(-1)^{F_L} = +1$ on states in the left-moving NS-sector.

---

[21] As we discuss below, further T-dualities will result in more O-planes than D-branes in which case local tadpole cancellation, i.e. local charge neutrality, in not possible.

The appearance of this factor can be understood as follows, e.g. in the second line of the table. I_2, the reflection of two directions, is a rotation by π in the plane spanned by these two directions. On a spinor, I_2^2 should act as -1 while it acts as $+1$ on a vector. This follows indeed from (10.172): $I_2|s_1, s_2, s_3, s_4\rangle \to e^{i\pi s_4}|s_1, s_2, s_3, s_4\rangle$. This can be summarized as $I_2^2 = (-1)^{F_L+F_R}$. From this, together with $\Omega^2 = 1$ on all physical states, it follows that $(\Omega I_2)^2 = (-1)^{F_L+F_R}$. This means that ΩI_2 cannot have eigenvalue $+1$ on space-time fermions, i.e. the orientifold IIB/ΩI_2 is not space-time supersymmetric. On the other hand, $(\Omega I_2(-1)^{F_L})^2 = 1$ (use $\Omega(-1)^{F_L} = (-1)^{F_R}\Omega$) which allows for eigenvalues $+1$, i.e. invariant spinors. In fact, half of the eigenvalues are $+1$ and half are -1. In Sect. 15.3 we discuss the action of T-duality on spin fields. We can use the results obtained there to show that on all states the various orientifold actions in Table 10.3 are indeed related by T-duality; e.g. $T_{89}\Omega T_{89}^{-1} = \Omega I_{89}(-1)^{F_L}$ where T_{89} and I_{89} are T-duality and reflection of the (89)-directions, respectively. All theories obtained in this fashion have the same number of space-time supercharges (16) as the type I= type IIB/Ω theory.

The models in Table 10.3 are only special solutions to the tadpole cancellation conditions, which are global constraints. Therefore, one does not need to place the D-branes on top of the O-planes. A simple example of global rather than local tadpole cancellation is if the 32 D8 branes of the ΩI_1 type IIA orientifold theory are distributed differently. In the generic case the gauge group is broken to $U(1)^{16}$. One might at first have thought that the gauge group is $U(1)^{32}$, but this is not correct. The number of branes plus their images under the I_1 action is 32, but the number of independent and freely adjustable brane positions is 16 or, in other words, the $U(1)$ gauge bosons which one would naively associate with each of the 32 branes, are pairwise identified by the action of I_1.

We will discuss four-dimensional orientifolds of type $T^6/\Omega I_3(-1)^{F_L}$ with intersecting D6-brane configurations in Chap. 17 which, as explained, are T-dual to models with magnetized D9-branes.

We now compute the tension and charge of Op-planes. In the previous chapter we have computed the tension and charge of type I branes and of the O9 plane; cf. Eq. (9.83). As we have just argued, by applying T-dualities to the type I theory one obtains Op-planes and Dp-branes for any p. The ratios between the O-plane and D-brane tensions can be easily determined. O-planes are located at the fixed-points of the $\mathbb{Z}_2$ action. In type I theory this is simply the world-sheet parity Ω but after T-duality in $(9-p)$ directions, it also contains a geometric piece, namely refection along p direction. To do T-duality we first have to compactify on a torus. On the torus the reflection has 2^p fixed points which are the locations of the Op-planes. The number of Dp-branes, on the other hand, does not change under T-duality. This leads to

$$\mu_{\text{O}p} = -2^{p-4}\mu_{\text{D}p}, \qquad T_{\text{O}p} = -2^{p-4}T_{\text{D}p}. \qquad (10.212)$$

Note that the counting is such that we count a brane and its image under $\mathbb{Z}_2$ separately. This is also referred to as counting in the 'upstairs geometry', i.e. on

10.6 D-branes on Toroidal Compactifications

T^p rather than on T^p/I_p where we would count only half as many branes, each carrying twice the charge.

Finally we note that the orientifold planes discussed so far have negative tension and charge. For a space-time filling plane this is denoted as $O9^{(-,-)}$. However, by considering the dressed type IIB orientifold projection $\Omega(-1)^F$ one gets an $O9^{(-,+)}$-plane with negative tension and positive charge. Here $(-1)^F$ is the space-time fermion number F. The flip of charge can be seen as follows. In the Klein-bottle amplitude the insertion of $\Omega(-1)^F$ has no effect relative to just Ω, as only NS-NS and R-R states contribute. However, in the Möbius amplitude the R-sector receives an extra overall minus sign. After transformation to tree-channel this leads to an extra minus sign for the R-sector part in the total crosscap state (9.84). Therefore, the R-R-charge is flipped.

On the level of the crosscap states, it is clear that we can also flip the overall sign in front of its NS-NS-part leading to $O9^{(+,+)}$ and $O9^{(+,-)}$-planes. The type IIB orientifold with an $O9^{(+,+)}$-plane instead of the $O9^{(-,-)}$-plane is called the Sugimoto model. R-R tadpole cancellation can be achieved by the introduction of 32 anti D9-branes. However, these also have positive tension so that one is left with an uncancelled NS-NS tadpole. The positive vacuum energy also signals that supersymmetry is (explicitly) broken by the introduction of the anti D9-branes. In contrast to the R-R tadpole, an uncancelled NS-NS tadpole is not yet a disaster for the theory, but just an indication that the configuration does not solve the string equation of motion for the NS-NS fields, which however can in principle adjust themselves. This goes under the name Fischler-Susskind mechanism.

Further Reading

Torus compactifications of the closed string were constructed in

- E. Cremmer, J. Scherk, Dual models in four dimensions with internal symmetries. Nucl. Phys. **B103**, 399 (1976)
- M.B. Green, J.H. Schwarz, L. Brink, N = 4 Yang-Mills and N = 8 supergravity as the limit of string theories. Nucl. Phys. **B198**, 474 (1982)

The heterotic string was invented in

- D.J. Gross, J. Harvey, E. Martinec, R. Rohm, Heterotic string. Phys. Rev. Lett. **54**, 502 (1985)
- D.J. Gross, J. Harvey, E. Martinec, R. Rohm, Heterotic string theory (I). The free heterotic string. Nucl. Phys. **B256**, 253 (1985)
- D.J. Gross, J. Harvey, E. Martinec, R. Rohm, Heterotic string theory (II). The interacting heterotic string. Nucl. Phys. **B267**, 75 (1986)

Torus compactifications of the heterotic string were studied in

- K.S. Narain, New heterotic string theories in uncompactified dimensions $d < 10$. Phys. Lett. **169B**, 41 (1986)

- K.S. Narain, M.H. Sarmadi, E. Witten, A note on toroidal compactification of heterotic string theory. Nucl. Phys. **B278**, 369 (1987)
- P. Ginsparg, Comment on toroidal compactification of heterotic superstrings. Phys. Rev. **D35**, 648 (1987)

T-duality of the closed string is reviewed in

- A. Giveon, M. Porrati, E. Rabinovici, Target space duality in string theory. Phys. Rep. **244**, 77–202 (1994) [arXiv:hep-th/9401139]

The observation that T-duality maps between type IIA and type IIB was made in

- M. Dine, P. Huet, N. Seiberg, Large and small radius in string theory. Nucl. Phys. **B322**, 301 (1989)
- J. Dai, R.G. Leigh, J. Polchinski, New connections between string theories. Mod. Phys. Lett. **A4**, 2073–2083 (1989)

Orbifold compactifications were first discussed in

- L. Dixon, J. Harvey, C. Vafa, E. Witten, Strings on orbifolds. Nucl. Phys. **B261**, 678 (1985)
- L. Dixon, J. Harvey, C. Vafa, E. Witten, Strings on orbifolds. 2. Nucl. Phys. B **274**, 285 (1986)

Discrete torsion was introduced in

- C. Vafa, Modular invariance and discrete torsion on orbifolds. Nucl. Phys. **B273**, 592 (1986)

D-branes at angles were discussed in the non-compacts case in

- M. Berkooz, M.R. Douglas, R.G. Leigh, Branes intersecting at angles. Nucl. Phys. **B480**, 265 (1996) [arXiv:hep-th/9606139]

On tori they were analyzed in

- R. Blumenhagen, L. Görlich, B. Körs, D. Lüst, Noncommutative compactifications of type I strings on tori with magnetic flux. JHEP **0010**, 006 (2000) [arXiv:hep-th/0007024]

Chapter 11
Conformal Field Theory II: Lattices and Kač-Moody Algebras

Abstract In the previous chapter we have learned that in toroidal compactifications of the bosonic string there are, in addition to the Kaluza-Klein gauge bosons familiar from field theory, further massless vectors of purely stringy origin. However, we did not show that these massless vectors are gauge bosons of a non-Abelian gauge group G, transforming in the adjoint representation. The necessary mathematical tool to do this is the theory of infinite dimensional (current) algebras, the so-called affine Kač-Moody algebras. They are the subject of this chapter for which we assume some familiarity with the structure of finite dimensional Lie algebras.

11.1 Kač-Moody Algebras

A Kač-Moody algebra is the infinitesimal version of a certain infinite dimensional Lie group $\mathscr{G}$, namely the group of mappings from the circle S^1 into a finite dimensional Lie group G. We will only consider compact connected G. $\mathscr{G}$ is the so-called loop group of G.

We represent S^1 as the unit circle in the complex plane

$$S^1 = \{z \in \mathbb{C} : |z| = 1\} \tag{11.1}$$

and denote a map from S^1 into G by $z \mapsto \gamma(z) \in G$. The group operation on $\mathscr{G}$ is defined by point-wise multiplication; i.e. given two maps $\gamma_1, \gamma_2 \in \mathscr{G}$, the product of γ_1 and γ_2 is $\gamma_1 \cdot \gamma_2 \in \mathscr{G}$, where

$$\gamma_1 \cdot \gamma_2(z) = \gamma_1(z)\,\gamma_2(z). \tag{11.2}$$

The infinite dimensional algebra $\hat{g}_0$ of $\mathscr{G}$ can be obtained from the finite dimensional Lie algebra g of G,[1]

$$[T^a, T^b] = if^{abc} T^c, \tag{11.3}$$

where f^{abc} are the structure constants of g, by writing

$$\gamma(z) = \exp\left[-i \sum_{a=1}^{\dim g} T^a \Theta_a(z)\right]. \tag{11.4}$$

$\Theta_a(z)$ are dim g functions defined on the unit circle. Expanding these functions into modes

$$\Theta_a(z) = \sum_{n=-\infty}^{\infty} \Theta_a^{-n} z^n, \tag{11.5}$$

we can introduce generators T_n^a

$$T_n^a = T^a z^n \tag{11.6}$$

such that

$$\gamma(z) = \exp\left[-i \sum_{n,a} T_{-n}^a \Theta_a^n\right]. \tag{11.7}$$

We see that the Θ_a^n's are an infinite set of parameters for $\mathscr{G}$ and the T_n^a an infinite set of generators of $\mathscr{G}$ satisfying the following algebra

$$[T_m^a, T_n^b] = if^{abc} T_{m+n}^c \tag{11.8}$$

which follows from Eqs. (11.3) and (11.6). This is the (untwisted) affine Kač-Moody algebra $\hat{g}_0$, also called loop algebra. (We will not consider twisted Kač-Moody algebras.) Note that the T_0^a generate a subalgebra isomorphic to g. It corresponds to $\Theta_a(z) = $ const.

If the T^a are Hermitian generators of g with

$$T^{a\dagger} = T^a, \tag{11.9}$$

then the Kač-Moody generators satisfy ($z^* = z^{-1}$ for $|z| = 1$)

$$T_n^{a\dagger} = T_{-n}^a. \tag{11.10}$$

A representation of $\hat{g}_0$ satisfying this Hermiticity condition is called unitary.

---

[1] Since we consider only compact G's, we can take the Cartan-Killing metric to be δ^{ab} in which case $f^{abc} = f^{ab}{}_c$ is antisymmetric in all indices. The generators T^a are Hermitian and the structure constants are real.

11.1 Kač-Moody Algebras

In Chap. 2 we have seen that for the closed bosonic string the (say) right movers are functions of $(\tau - \sigma)$ only and periodicity allows an expansion in Fourier modes $e^{in(\tau-\sigma)}$. This means that the fields are defined on S^1. The Virasoro algebra $\hat{v}_0$ generates reparametrizations of S^1. The generators can be represented as

$$L_n = -z^{n+1}\partial_z, \tag{11.11}$$

where $z = e^{i(\tau-\sigma)} \in S^1$. With the $\hat{g}_0$ being the algebra of the group of maps $S^1 \to G$ and $\hat{v}_0$ the algebra of $\mathrm{Diff}(S^1)$, it is clear that they are not unrelated and, as we will show below, to every Kač-Moody algebra there is an associated Virasoro algebra. Using the explicit form for the generators Eqs. (11.6) and (11.11), we easily find

$$[L_m, T_n^a] = -n\, T_{n+m}^a, \tag{11.12}$$

i.e. $\mathcal{A}_0$ is the semi-direct sum $\mathcal{A}_0 = \hat{v}_0 \Subset \hat{g}_0$.

So far we have constructed the Kač-Moody algebra from the classical Lie algebra Eq. (11.3). When going to the quantum theory, we have to be careful. We saw in Chap. 3 that the Virasoro algebra receives a central extension parametrized by the central charge c. This possibility also arises for Kač-Moody algebras.[2] Allowing a general central extension, it has the form

$$[T_m^a, T_n^a] = if^{abc}T_{m+n}^c + d^{ab}_{mnj}\, k^j,$$
$$[T_m^a, k^j] = [k^i, k^j] = 0, \tag{11.13}$$

where k^i, $(i = 1, \ldots, M)$ are central elements. For G compact and simple one can show that up to redefinitions of the generators by terms linear in the k^i the only possible choice for $d^{ab}_{mni}k^i$ consistent with the Jacobi identities is $km\delta^{ab}\delta_{m+n}$ and the central extension is one-dimensional. We can then define the untwisted affine Kač-Moody algebra $\hat{g}$ by the following commutation relations:

$$[T_m^a, T_n^b] = if^{abc}T_{m+n}^c + km\,\delta^{ab}\delta_{m+n}. \tag{11.14}$$

By Schur's lemma, the central element k is a constant in every irreducible representation. One calls

$$x = \frac{2k}{\psi^2} \tag{11.15}$$

the level of the Kač-Moody algebra. Here ψ^2 is the length of the highest root of g (cf. below). The level is independent of any normalizations (e.g. of the structure constants). When considering irreducible unitary representations of the Kač-Moody algebra, the allowed values of x are not arbitrary but constrained to $x \geq 0$; for $G \neq U(1)$, $x \in \mathbb{Z}_+$. This will be shown in Sect. 11.5. From Eq. (11.14) we see that the

---

[2] Semi-simple finite dimensional Lie algebras do not possess non-trivial central extensions.

finite dimensional Lie algebra g does not allow a non-trivial central extension. If G is compact but not simple we get a different k for each simple factor. The level of any $U(1)$ factor can always be set to one by suitable redefinition of the generators.

The Kač-Moody algebra Eq. (11.14) is closely related to a two-dimensional current algebra. Consider the conserved chiral currents $J^a(z)$ which carry the adjoint representation index of some Lie group. Since $\partial_{\bar{z}} J^a(z) = 0$ we have, as in the case of the Virasoro algebra, an infinite number of conserved charges

$$J_n^a = \oint \frac{dz}{2\pi i} z^n J^a(z) \tag{11.16}$$

which satisfy the same algebra as the corresponding generators, namely an affine Kač-Moody algebra:

$$[J_m^a, J_n^b] = i f^{abc} J_{m+n}^c + mk \, \delta^{ab} \delta_{m+n} \tag{11.17}$$

and also

$$[L_m, J_n^a] = -n J_{m+n}^a , \tag{11.18}$$

reflecting the semi-direct sum structure $\mathscr{A} = \hat{v} \oplus \hat{g}$. Inverting Eq. (11.16) we get

$$J^a(z) = \sum_n z^{-n-1} J_n^a . \tag{11.19}$$

Using the techniques of Chap. 4, we easily find that Eq. (11.17) is equivalent to the following current operator algebra:

$$J^a(z) J^b(w) = \frac{k \delta^{ab}}{(z-w)^2} + i f^{abc} \frac{J^c(w)}{(z-w)} + \ldots \tag{11.20}$$

In this equation, the central charge appears as a so-called Schwinger term which is proportional to $\hbar$, which we set to one. In conformal field theory language Eq. (11.18) means that the currents $J^a(z)$ are primary fields of weight $h = 1$ of the Virasoro algebra (cf. Eq. (4.58b) on page 75). Indeed, Eq. (11.18) is equivalent to the operator product

$$T(z) J^a(w) = \frac{J^a(w)}{(z-w)^2} + \frac{\partial_w J^a(w)}{z-w} + \ldots \tag{11.21}$$

We can now define primary fields of $\mathscr{A}$ by

$$T(z) \phi_i(w) = \frac{h \phi_i(w)}{(z-w)^2} + \frac{\partial \phi_i(w)}{z-w} + \ldots ,$$

$$J^a(z) \phi_i(w) = \frac{(T^a)_i{}^j \phi_j(w)}{z-w} + \ldots , \tag{11.22}$$

11.1 Kač-Moody Algebras

where $(T^a)_i{}^j$ are representation matrices of g. Comparison with Eq. (11.20) gives that $J^a(z)$ is not primary with respect to the combined chiral algebra $\mathscr{A}$. Indeed, in the notation of Chap. 4 we have $J^a(z) = \hat{J}^a_{-1} I(z)$.

Following Chap. 4, we define primary or highest weight states of $\mathscr{A}$ by

$$|\phi_i\rangle = \phi_i(0)|0\rangle, \tag{11.23}$$

where i is a representation index of G and $\phi_i(z)$ is primary according to Eq. (11.22). The vacuum is $SL(2,\mathbb{C})$ and G invariant. The highest weight states satisfy

$$L_n|\phi_i\rangle = 0, \quad n > 0,$$
$$J^a_n|\phi_i\rangle = 0, \quad n > 0,$$
$$L_0|\phi_i\rangle = h_i|\phi_i\rangle,$$
$$J^a_0|\phi_i\rangle = (T^a)_i{}^j|\phi_j\rangle. \tag{11.24}$$

To each primary field there exists an infinite number of descendant fields of the form

$$L_{-k_1}\ldots L_{-k_m} J^{a_1}_{-l_1}\ldots J^{a_n}_{-l_n}|\phi_i\rangle, \quad k_i, l_i > 0. \tag{11.25}$$

The conformal dimensions of the descendant states are $h_i + \sum_{i=1}^m k_i + \sum_{i=1}^n l_i$. We denote the totality of fields in a given highest weight representation by $[\phi_i]_{\mathscr{A}}$. The $[\phi_i]_{\mathscr{A}}$ are also called current algebra families.

There is an explicit construction of the Virasoro algebra in terms of the Kač-Moody generators. This is the Sugawara construction. For simplicity we will only consider the case when g is simple. Define the energy momentum tensor as

$$T(z) = \frac{1}{2k + C_2} \sum_a :J^a(z)J^a(z): \tag{11.26a}$$

or

$$L_n = \frac{1}{2k + C_2} \sum_a \sum_m :J^a_m J^a_{n-m}: . \tag{11.26b}$$

Up to the normalization this is the only candidate with the correct conformal weight. C_2 is the quadratic Casimir of the adjoint representation defined by

$$f^{acd} f^{bcd} = C_2 \delta^{ab}. \tag{11.27}$$

It clearly depends on the normalization of the structure constants. A normalization independent quantity is the dual Coxeter number

$$C_g = \frac{C_2}{\psi^2} \tag{11.28}$$

which can be shown to be an integer.

In Eq. (11.26a) normal ordering is defined as

$$: J^a(z) J^b(z) := \lim_{w \to z} \left\{ J^a(w) J^b(z) - \frac{k \delta^{ab}}{(w-z)^2} - i \frac{f^{abc} J^c(z)}{(w-z)} \right\}. \quad (11.29)$$

For the modes this implies (see also (4.111))

$$: J_n^a J_m^b := \begin{cases} J_n^a J_m^b & n \leq 1 \\ J_m^b J_n^a & n \geq 0 \end{cases} \quad (11.30)$$

which can be written in the form

$$J_n^a J_m^b =: J_n^a J_m^b : +[J_n^a, J_m^b] \theta(n), \quad \text{where} \quad \theta(n) = \begin{cases} 1 & n \geq 0 \\ 0 & n < 0. \end{cases} \quad (11.31)$$

It can then be shown that $T(z)$ satisfies (11.22) and the Virasoro algebra with central charge

$$c = \frac{2k \dim G}{2k + C_2} = \frac{x \dim G}{x + C_g}. \quad (11.32)$$

For these the prefactor in (11.26a) is crucial. Note that the classical value C_2 is modified to $C_2 + 2k$ by quantum effects. We will give details of the computation in the appendix of this chapter.

The conformal dimension of the primary field defined in Eq. (11.24) is

$$h = \frac{C_2^R}{2k + C_2}, \quad (11.33)$$

where C_2^R is the quadratic Casimir of the representation R under which $|\phi\rangle$ transforms:

$$(T^a)_i{}^j (T^a)_j{}^k = C_2^R \delta_i^k. \quad (11.34)$$

If we recall that the generators in the adjoint representation are $(T^a)_b{}^c = i f^{abc}$, we see that (11.27) is a special case of (11.34). The explicit expression for C_2^R is

$$C_2^R = \frac{\dim R}{\dim G} \frac{\sum \lambda^2}{\operatorname{rank} G}, \quad (11.35)$$

where the sum is over all weights in the representation R. Equation (11.33) is easily derived using the explicit expression for L_0 given in Eq. (11.26b).

We now specialize to simply laced groups G of rank n, cf. Footnote 8 on page 280. We normalize the roots to have (length)$^2 = 2$. In particular $\psi^2 = 2$ and $x = k$. This corresponds to a particular normalization of the structure constants (e.g. for $SU(2)$, $f^{abc} = \sqrt{2} \epsilon^{abc}$). Then C_g is given by the formula

11.2 Lattices and Lie Algebras

$$C_g = \left(\frac{\dim G}{n} - 1\right). \tag{11.36}$$

It implies that for level one Kač-Moody algebras, i.e. $k = 1$, the central charge of the corresponding Virasoro algebra is an integer, namely the rank n of G. This suggests that the level one Kač-Moody currents of the simply laced group G can be constructed from n free bosonic fields. This construction, known as the Frenkel-Kač-Segal construction, uses n bosonic fields "compactified" on the root lattice of the finite dimensional Lie algebra g and establishes the appearance of non-Abelian gauge symmetries in the heterotic string theory introduced in Chap. 10. The final step in the identification will be made in Chap. 16, when we compare string scattering amplitudes with field theory amplitudes.

First, let us recall some basic facts about lattices and Lie algebras. Some of these have already been introduced and used in Chap. 10.

11.2 Lattices and Lie Algebras

A lattice Λ is defined as a set of points in an n-dimensional real vector space V:

$$\Lambda = \left\{\sum_{i=1}^{n} n_i \, e_i \, | \, n_i \in \mathbb{Z}\right\}. \tag{11.37}$$

The e_i ($i = 1 \ldots n$) are n basis vectors of V. We will only be interested in the cases where V is $\mathbb{R}^n$ with Euclidean inner product or $\mathbb{R}^{p,q}$ ($p + q = n$) with Lorentzian inner product; i.e. for v, w being two lattice vectors we have $v \cdot w = \sum_{I=1}^{n} v^I w^I$ for the Euclidean case and $v \cdot w = \sum_{I=1}^{p} v^I w^I - \sum_{I=p+1}^{n} v^I w^I$ for the Lorentzian case. An element of the lattice Λ has an expansion $V^I = n_i e_i^I$. The matrix $g_{ij} = e_i \cdot e_j$ is the metric on Λ; it contains all information about the angles between the basis vectors and about their lengths. The volume of the unit cell, which contains exactly one lattice point, vol(Λ), is also determined by g_{ij}:

$$\text{vol}(\Lambda) = \sqrt{|\det g|}. \tag{11.38}$$

The dual lattice Λ^* is defined as

$$\Lambda^* = \{w \in V, w \cdot v \in \mathbb{Z}, \forall v \in \Lambda\}. \tag{11.39}$$

The basis vectors e_i^* of Λ^* satisfy

$$e_i^* \cdot e_j = \delta_{ij} \tag{11.40}$$

and the metric on Λ^* is $g_{ij}^* = e_i^* \cdot e_j^*$ which is the inverse of g_{ij}. The volume of the unit cell of Λ^* is then

$$\text{vol}(\Lambda^*) = (\text{vol}(\Lambda))^{-1}. \tag{11.41}$$

A lattice is called unimodular if it has one point per unit volume, i.e. if $\text{vol}(\Lambda) = 1$; then also $\text{vol}(\Lambda^*) = 1$. It is integral if $v \cdot w \in \mathbb{Z}$, $\forall v, w \in \Lambda$. Clearly Λ is integral if and only if $\Lambda \subset \Lambda^*$. Furthermore an integral lattice is called even if all lattice vectors have even (length)^2; it is called odd otherwise. Finally, Λ is self-dual if $\Lambda = \Lambda^*$. A necessary and sufficient condition for Λ being self-dual is to be unimodular and integral.

If Λ_s is a sublattice of equal dimension as Λ, we can decompose Λ into cosets with respect to Λ_s. To do so, choose a set of vectors m_i ($i = 2, \ldots, N_s$), such that

$$m_i \in \Lambda, \quad m_i \notin \Lambda_s,$$
$$m_i - m_j \notin \Lambda_s \text{ if } i \neq j. \tag{11.42}$$

Then the lattice Λ can be written as sum over cosets

$$\Lambda = \Lambda_s \oplus (m_2 + \Lambda_s) \oplus \ldots \oplus (m_{N_s} + \Lambda_s). \tag{11.43}$$

This notation means that every vector in Λ can be written as $m_i + v_s$, $v_s \in \Lambda_s$, $i = 1, \ldots, N_s$, if we define $m_1 = 0$, the zero vector. The vectors m_i are called coset representatives. The volume of Λ can be expressed as

$$\text{vol}(\Lambda) = \frac{1}{N_s} \text{vol}(\Lambda_s). \tag{11.44}$$

The lattices we are most interested in are the so-called Lie algebra lattices. To discuss them we need some basic facts about Lie algebras, which we will now review. We will especially concentrate on the properties of their root and weight lattices. This is most conveniently done in the so-called Cartan-Weyl basis. Choose a maximal set of Hermitian and commuting generators H^I ($I = 1, \ldots, n$)

$$[H^I, H^J] = 0, \tag{11.45}$$

where the dimension n of this subalgebra is called the rank of G. The H^I generate the Cartan subalgebra. Given a choice of a Cartan subalgebra we can diagonalize the remaining generators in the sense that they have definite eigenvalues with respect to the H^I:

$$[H^I, E^\alpha] = \alpha^I E^\alpha. \tag{11.46}$$

The real non-zero n-dimensional vector α is called a root and E^α a step operator corresponding to α. Note that from Eq. (11.46) it follows that the E^α are necessarily non-Hermitian. Indeed, we find that

$$E^{-\alpha} = (E^\alpha)^\dagger, \tag{11.47}$$

11.2 Lattices and Lie Algebras

i.e. if α is a root then so is $-\alpha$. A root is called positive if its first non-zero component is positive. The E^α with α positive are called raising operators and lowering operators otherwise. If all roots have the same length, the group is called simply laced. We can then normalize the roots to $\alpha^2 = \alpha^I \alpha^I = 2$. In the following we will only consider simply laced groups. Some of the expressions given below will have to be modified for the general case.

To complete the Lie algebra we need to determine the commutation relation between the step operators E^α, E^β. The commutators are constrained by the Jacobi identities and the result can be summarized as follows:

$$[E^\alpha, E^\beta] = \begin{cases} \epsilon(\alpha, \beta) E^{\alpha+\beta} & \text{if } \alpha + \beta \text{ is a root} \\ \alpha \cdot H & \text{if } \alpha = -\beta \\ 0 & \text{otherwise.} \end{cases} \quad (11.48)$$

The constants $\epsilon(\alpha, \beta)$, antisymmetric in α and β, can be arranged to be ± 1. With each root we can associate an $su(2)$ subalgebra generated by E^α, $E^{-\alpha}$ and $\alpha \cdot H$. If we identify them with J_+, J_- and $2J_3$, we recognize the angular momentum algebra. It is well known that for unitary representations the eigenvalues of $2J_3$ or $\alpha \cdot H$ have to be integer.

Taking arbitrary integer linear combinations of root vectors one generates an n-dimensional Euclidean lattice, called root lattice Λ_R. Since the number of root pairs $\pm \alpha$ in general exceeds the rank n of G, it is convenient to select a set of roots α_i ($i = 1, \ldots, n$) which serve as a basis for Λ_R. These are the so-called simple roots. They are those positive roots which cannot be written as a sum of two positive roots. The Cartan matrix, defined by[3]

$$g_{ij} = \alpha_i \cdot \alpha_j \quad (11.49)$$

is an integer $n \times n$ matrix; its diagonal elements are 2 and its off diagonal elements are -1 or 0; i.e. the root lattice of any simply laced Lie algebra is an integral, even lattice. Therefore it is contained in its dual lattice Λ_R^*. g_{ij} is the metric on Λ_R. From a given Cartan matrix one can construct a basis of simple roots and from that all roots.

Let us look at the classification of simply laced Lie algebras. The first class is the D_n ($n \geq 1$) series[4] corresponding to the orthogonal groups $SO(2n)$ with rank n and dimension $n(2n-1)$. The n-component root vectors are

$$(\ldots, \pm 1, \ldots, \pm 1, \ldots) \qquad \text{all other entries zero.} \quad (11.50)$$

---

[3]The Cartan matrix is often denoted $A_{ij} = 2\alpha_i \cdot \alpha_j / \alpha_j^2$. For simply laced Lie algebras it reduces to our g_{ij}.
[4]We also include the case $n = 1$ with $D_1 \sim SO(2) \sim U(1)$.

Counting all combinations of distributing two "±1" entries, one easily verifies that there are $2(n^2 - n)$ root vectors. The E^{α_i}, together with the n Cartan subalgebra generators H^I, are the $2n^2 - n$ generators of D_n in the Cartan-Weyl basis. The n simple roots of D_n are

$$(1, -1, 0^{n-2}), (0, 1, -1, 0^{n-3}), \ldots, (0^{n-2}, 1, -1), (0^{n-2}, 1, 1), \tag{11.51}$$

where 0^i denotes the i-dimensional null vector.

The next class of simply laced Lie algebras is the A_n series ($n \geq 1$) corresponding to $SU(n+1)$ with rank n and dimension $n^2 + 2n$. Let us take as an example A_2. The six roots of $SU(3)$ are

$$\alpha_1 = \left(\sqrt{\tfrac{1}{2}}, \sqrt{\tfrac{3}{2}}\right), \quad \alpha_2 = \left(\sqrt{\tfrac{1}{2}}, -\sqrt{\tfrac{3}{2}}\right), \quad \alpha_3 = \left(\sqrt{2}, 0\right),$$

$$\alpha_4 = -\left(\sqrt{\tfrac{1}{2}}, \sqrt{\tfrac{3}{2}}\right), \quad \alpha_5 = -\left(\sqrt{\tfrac{1}{2}}, -\sqrt{\tfrac{3}{2}}\right), \quad \alpha_6 = -\left(\sqrt{2}, 0\right). \tag{11.52}$$

The two simple roots are α_1 and α_2.

Besides the D_n and A_n series there are the exceptional simply laced Lie algebras E_6, E_7 and E_8 of dimensions 78, 133 and 248 and rank 6, 7 and 8, respectively. The 72 roots of E_6 (in a suitably chosen basis) are

$$(\pm 1, \pm 1, 0^3; 0) + 36 \text{ permutations}$$

$$\left(\pm\tfrac{1}{2}, \pm\tfrac{1}{2}, \pm\tfrac{1}{2}, \pm\tfrac{1}{2}, \pm\tfrac{1}{2}; \tfrac{\sqrt{3}}{2}\right) \quad \text{even number of "$-$" signs,}$$

$$\left(\pm\tfrac{1}{2}, \pm\tfrac{1}{2}, \pm\tfrac{1}{2}, \pm\tfrac{1}{2}, \pm\tfrac{1}{2}; -\tfrac{\sqrt{3}}{2}\right) \quad \text{odd number of "$-$" signs,} \tag{11.53}$$

and the root vectors of E_7 are

$$(\pm 1, \pm 1, 0^4; 0) + 56 \text{ permutations}$$

$$\left(0^6; \pm\sqrt{2}\right)$$

$$\left(\pm\tfrac{1}{2}, \pm\tfrac{1}{2}, \pm\tfrac{1}{2}, \pm\tfrac{1}{2}, \pm\tfrac{1}{2}, \pm\tfrac{1}{2}; \pm\tfrac{\sqrt{2}}{2}\right) \quad \begin{array}{l}\text{even number of "$-$" signs}\\ \text{in first six components.}\end{array} \tag{11.54}$$

The roots of E_8 have a particularly simple form; they are given by the 112 root vectors of D_8 (see Eq. (11.50)) and in addition to these the following 128 eight-dimensional vectors:

$$\left(\pm\tfrac{1}{2}, \pm\tfrac{1}{2}, \ldots, \pm\tfrac{1}{2}\right) \quad \text{even number of "$-$" signs .} \tag{11.55}$$

11.2 Lattices and Lie Algebras

So far we have only considered root lattices Λ_R. They are constructed from the adjoint representation of G whose non-zero weights are the roots. However, any Lie group G has infinitely many irreducible representations which are characterized by their weight vectors. Consider a finite dimensional irreducible representation of G. The states which transform in a specific representation are denoted by $|m_l, D\rangle$ where D is the dimension of the representation and k runs from 1 to D. These states are eigenstates of the Cartan subalgebra generators:

$$H^I |m_l, D\rangle = m_l^I |m_l, D\rangle. \tag{11.56}$$

The n-dimensional vector m_l is called weight vector of $|m_l, D\rangle$. The weight vectors characterize the representation. We can reach all states in a given representation by acting with lowering operators on the highest weight state.[5] Thus the weights in a given representation differ by vectors in the root lattice. For simply laced groups $\alpha_i \cdot m \in \mathbb{Z}$ for all roots and weights. Also, if $\beta \cdot m \in \mathbb{Z}$, $\forall m$, then $\beta \in \Lambda_R$.

Irreducible representations fall into different conjugacy classes. Two different representations are said to be in the same conjugacy class if the difference between their weight vectors is a vector of the root lattice. Of course, all weights of a given representation belong to the same conjugacy class. Therefore, a particular conjugacy class k is specified by choosing a representative weight vector m_k, e.g. the highest weight of the lowest dimensional representation contained in this conjugacy class. Then all vectors within the same conjugacy class are constructed by adding the whole root lattice to this representative. Since the root lattice is an integral lattice and also the scalar product between the roots and m_k are integer, it follows that the mutual inner product modulo integers of any two vectors of two conjugacy classes depends only on the scalar product of their representative vectors, i.e. on the conjugacy classes.

All weights of all conjugacy classes including the (0) conjugacy class (the root lattice itself) form the weight lattice Λ_w. Our results above clearly imply

$$\Lambda_R \subset \Lambda_w, \quad \Lambda_R = \Lambda_w^*, \tag{11.57}$$

which entails that

$$\text{vol}(\Lambda_R) = (\text{vol}(\Lambda_w))^{-1}. \tag{11.58}$$

The decomposition of the weight lattice into conjugacy classes is then simply a coset decomposition of Λ_w with respect to Λ_R and we can write

$$\Lambda_w = \Lambda_R \oplus (\Lambda_R + m_2) \oplus \ldots \oplus (\Lambda_R + m_{N_c}), \tag{11.59}$$

---

[5]A highest weight state $|m_0, D\rangle$ satisfies $E^\alpha |m_0, D\rangle = 0$ $\forall$ positive α. It means that $\alpha + m_0$ is not a weight vector for any positive root α. The other states in the same representation are obtained by acting with lowering operators on the highest weight state. Any irreducible representation of g has a unique highest weight state—the other weights m have the property that $m_0 - m$ is a sum of positive roots. The highest weight of the adjoint representation is called highest root ψ with $\psi^2 = 2$.

where the m_k ($k = 2, \ldots, N_c$) are the representative vectors of each non-trivial conjugacy class and N_c is the number of conjugacy classes which is finite for finite dimensional Lie algebras. The cosets form an Abelian discrete group under addition, isomorphic to the center of G. The direct sum decomposition Eq. (11.59) attributes to each root N_c weights (including the root itself); it follows that (cf. Eq. (11.44))

$$\text{vol}(\Lambda_w) = \frac{1}{N_c} \text{vol}(\Lambda_R) \tag{11.60}$$

and with Eq. (11.58)

$$\text{vol}(\Lambda_w) = \frac{1}{\sqrt{N_c}},$$
$$\text{vol}(\Lambda_R) = \sqrt{N_c}. \tag{11.61}$$

One can also consider so-called Lie algebra lattices, which are the direct sum of only a subset of all possible conjugacy classes. The choice of possible conjugacy classes is restricted; it must be closed under addition of all lattice vectors which in particular means that the root lattice is always present. The possible subsets of conjugacy classes are in one-to-one correspondence with the subgroups of the center of G.

Since the volume of the unit cell of Λ_R is $\sqrt{N_c}$, the Lie algebra lattice is unimodular if it contains $\sqrt{N_c}$ conjugacy classes (cf. Eq. (11.44) with $\Lambda_s = \Lambda_R$). This means that N_c must be the square of an integer. Furthermore, the Lie algebra lattice will be self-dual if all mutual scalar products between the different conjugacy classes are integer.

Let us illustrate this by considering specific simply laced Lie algebras. The D_n algebras have four inequivalent conjugacy classes. The (0) conjugacy class, which we have already discussed, i.e. the root lattice, contains vectors of the form

$$(k_1 \ldots k_n), \quad k_i \in \mathbb{Z}, \quad \sum_{i=1}^{n} k_i = 0 \mod 2. \tag{11.62}$$

Next, the vector conjugacy class, denoted by (V), contains as smallest representation the vector representation of dimension $2n$. Its weight vectors are

$$\boldsymbol{m} = (\pm 1, 0, \ldots, 0) + \text{all permutations}. \tag{11.63}$$

A representative vector of the (V) conjugacy class can be chosen to be $(1, 0^{n-1})$ which implies that all vectors of the (V) conjugacy class have the form

$$(k_1 \ldots k_n), \quad k_i \in \mathbb{Z}, \quad \sum_{i=1}^{n} k_i = 1 \mod 2. \tag{11.64}$$

It also follows that the (length)^2 of any vector in the (V) conjugacy class is 1 mod 2. The spinor conjugacy class, denoted by (S), has as smallest representation

11.2 Lattices and Lie Algebras

Table 11.1 Addition rules for D_n conjugacy classes

n even	(0)	(V)	(S)	(C)	n odd	(0)	(V)	(S)	(C)
(0)	(0)	(V)	(S)	(C)	(0)	(0)	(V)	(S)	(C)
(V)	(V)	(0)	(C)	(S)	(V)	(V)	(0)	(C)	(S)
(S)	(S)	(C)	(0)	(V)	(S)	(S)	(C)	(V)	(0)
(C)	(C)	(S)	(V)	(0)	(C)	(C)	(S)	(0)	(V)

Table 11.2 Multiplication rules (mod1) for D_n conjugacy classes

	(0)	(V)	(S)	(C)
(0)	0	0	0	0
(V)	0	0	$\frac{1}{2}$	$\frac{1}{2}$
(S)	0	$\frac{1}{2}$	$n/4$	$(n-2)/4$
(C)	0	$\frac{1}{2}$	$(n-2)/4$	$n/4$

the (Weyl) spinor representation of dimension 2^{n-1}. The corresponding weight vectors are

$$m = \left(\pm\frac{1}{2}, \pm\frac{1}{2}, \ldots, \pm\frac{1}{2}\right), \quad \text{even number of "$-$" signs.} \tag{11.65}$$

Thus, a representative vector of the (S) conjugacy class can be chosen to be $((\frac{1}{2})^n)$. Finally, the (C) conjugacy class possesses as lowest dimensional representation the conjugate-spinor representation with weights

$$m = \left(\pm\frac{1}{2}, \pm\frac{1}{2}, \ldots, \pm\frac{1}{2}\right), \quad \text{odd number of "$-$" signs.} \tag{11.66}$$

Its representative vector is $(-\frac{1}{2}, (\frac{1}{2})^{n-1})$. General weight vectors of the two spinor conjugacy classes are

$$\left(k_1 + \frac{1}{2}, \ldots, k_n + \frac{1}{2}\right) \quad \text{with} \quad \begin{cases} \sum_{i=1}^n k_i = 0 \mod 2 & \text{(S)} \\ \sum_{i=1}^n k_i = 1 \mod 2 & \text{(C)}. \end{cases} \tag{11.67}$$

The (length)^2 of all vectors in the (S) and (C) conjugacy classes is $\frac{n}{4}$ mod 2.

The center of D_n is $\mathbb{Z}_4$ for n odd and $\mathbb{Z}_2 \times \mathbb{Z}_2$ for n even. It has the same number of elements as there are conjugacy classes, namely four. The addition rules of the conjugacy classes (which correspond to the tensor products of the representations) are determined by the addition rules of the different representative vectors and are summarized in Table 11.1. The mutual scalar products (defined modulo 1) between the four conjugacy classes are shown in Table 11.2.

From the above discussion it is clear that the D_n Lie algebra lattices are unimodular, if they contain in addition to the root lattice one further conjugacy class. Inspection of Tables 11.1 and 11.2 shows that the (0) and (V) conjugacy classes of D_n together form an odd self-dual lattice for any value of n. This lattice is identical to the n-dimensional cubic lattice $\mathbb{Z}^n$. For even n, the lattice with (0) and either (S) or (C) conjugacy class is also a unimodular Lie algebra lattice. It is the weight lattice of $\text{Spin}(2n)/\mathbb{Z}_2$.[6] For $n = 8$ this is also the root lattice of E_8. Furthermore, we obtain an odd self-dual lattice of this type for $n = 4 \mod 8$ and an even self-dual lattice if $n = 0 \mod 8$.

The A_n weight lattice consists of $n + 1$ conjugacy classes denoted by (p) ($p = 0, \ldots, n$) where the (0) conjugacy class corresponds again to the root lattice. The center of A_n is $\mathbb{Z}_{n+1}$. The smallest representations in each conjugacy class are the symmetric rank p tensors. The addition rules of the conjugacy classes are very simple:

$$(p) + (q) = (p + q), \tag{11.68}$$

where $(p + q)$ is defined modulo $n + 1$. The mutual scalar products are

$$(p) \cdot (q) = \frac{p(n + 1 - q)}{n + 1} \mod 1, \quad p \leq q. \tag{11.69}$$

Using this one can verify that the Lie algebra lattices A_{k^2-1} with conjugacy classes $(0), (k), (2k), \ldots, ((k-1)k)$ are odd self-dual for k even and even self-dual for k odd.

The E_6 weight lattice contains three conjugacy classes (0), (1) and $(\bar{1})$ corresponding to the singlet, the **27** and the $\overline{\mathbf{27}}$ representations of E_6. The addition rules of these conjugacy classes are the same as for A_2, the (length)^2 of the weights of the **27**, $\overline{\mathbf{27}}$ is $4/3$ and the mutual scalar product between (1) and $(\bar{1})$ is $2/3 \mod 1$.

E_7 has two conjugacy classes, (0) and (1) where the minimal representation of the (1) is the **56** with weights of (length)$^2 = \frac{3}{2}$.

Finally E_8 has only one, namely the (0) conjugacy class. Therefore, the weight lattice of E_8 is identical to its root lattice which implies that it is even self-dual. Recall that the root vectors of E_8 are of the form

$$\alpha = \begin{cases} (\pm 1, \pm 1, 0^6) & + \text{ permutations} \\ \left(\pm\frac{1}{2}, \pm\frac{1}{2}, \ldots, \pm\frac{1}{2}\right) & \text{even number of ``--'' signs.} \end{cases} \tag{11.70}$$

We recognize that the E_8 root lattice is identical to the D_8 lattice with (0) and (S) conjugacy classes and one can also show that (up to a rotation) it is the A_8 Lie

---

[6]$\text{Spin}(2n)$, n even, is simply connected and has center $\mathbb{Z}_2 \times \mathbb{Z}_2$. If we divide by the diagonal $\mathbb{Z}_2$ we get $SO(2n)$ with only (0) and (V) conjugacy classes. If we divide by one of the $\mathbb{Z}_2$ we are left with the (0) and one of the spinor conjugacy classes.

11.2 Lattices and Lie Algebras

algebra lattice with (0), (3) and (6) conjugacy classes. On the other hand, if a Lie algebra lattice contains besides the roots additional weight vectors of (length)$^2 = 2$, then these lattice vectors are roots of a larger Lie algebra with fewer conjugacy classes. E.g. the adjoint of E_8 decomposes into $SO(16)$ as

$$\mathbf{248} = \mathbf{120} + \mathbf{128}, \tag{11.71}$$

where the **128** is the spinor representation and the **120** the adjoint representation of $SO(16)$.

Now consider direct products of Lie algebra lattices like $D_n \otimes D_m$, $A_n \otimes A_m$, $D_n \otimes A_m$ etc. These contain of course the root system of the corresponding semi-simple Lie algebras. Furthermore, the Lie algebra lattice is specified by the so-called glue vectors which generate upon addition all other conjugacy classes. Take e.g. $D_2 \otimes D_3$ with glue vector (V,S). Then we obtain, according to Table 11.1, a Lie algebra lattice with the following conjugacy classes: (0,0), (V,S), (0,V), (V,C). Since $D_2 \otimes D_3$ contains 16 conjugacy classes, the above specified Lie algebra lattice is unimodular. It is however not self-dual. A different example is $E_6 \otimes A_2$ with glue vector $(1, 1)$. The conjugacy classes are now $(0, 0)$, $(1, 1)$ and $(\bar{1}, 2)$. We now get additional (length)$^2 = 2$ vectors, and in fact this lattice is again the even self-dual E_8 root lattice which can also be seen by decomposition of E_8 to $E_6 \times SU(3)$:

$$\mathbf{248} = (\mathbf{78}, \mathbf{1}) + (\mathbf{1}, \mathbf{8}) + (\mathbf{27}, \mathbf{3}) + (\overline{\mathbf{27}}, \overline{\mathbf{3}}). \tag{11.72}$$

We now state (without proof) the classification of Euclidean even self-dual lattices. They only exist in dimensions which are a multiple of 8. In eight dimensions the only Euclidean even self-dual lattice is the root lattice of E_8. In 16 dimensions there are two even self-dual lattices, namely the root lattice of $E_8 \otimes E_8$ and the Lie algebra lattice of D_{16} with (0) and (S) conjugacy classes, called Spin(32)/$\mathbb{Z}_2$. These two lattices arise in the construction of the 10-dimensional heterotic string theory. They are the only ones satisfying the constraint of one-loop modular invariance, as discussed in Chap. 10.

For the covariant lattice construction of heterotic theories in Sect. 13.5 we will need self-dual lattices in 24 dimensions. There are 24, called Niemeier lattices. Twenty-three of them are Lie algebra lattices of semi-simple Lie groups. Table 11.3 summarizes them together with the relevant glue vectors. The 24th even self-dual lattice is the so-called Leech lattice. It contains no vectors of (length)$^2 = 2$. Its shortest vectors have (length)$^2 = 4$. Above 24 dimensions the number of even self-dual lattices increases rapidly and most of them are not known explicitly.

So far we have considered only Euclidean Lie algebra lattices. However there exist two types of Lorentzian "Lie algebra" lattices.[7] First consider $\Gamma_{n,m} = \Gamma_{n_L} \otimes \Gamma_{m_R}$ in $\mathbb{R}^{n_L, m_R}$ where Γ_{n_L} and Γ_{m_R} are semi-simple Lie algebra lattices

---

[7] We already encountered Lorentzian lattices when we discussed the torus compactification of the bosonic and the heterotic strings.

Table 11.3 The 23 Euclidean self-dual semi-simple Lie algebra lattices in 24 dimensions (conjugacy classes in square brackets should be cyclically permuted)

Lie algebra	Glue vector
D_{24}	(S)
$D_{16}E_8$	(S,0)
E_8^3	(0,0,0)
A_{24}	(5)
D_{12}^2	(S,V), (V,S)
$A_{17}E_7$	(3, 1)
$D_{10}E_7^2$	(S,1,0), (C,0,1)
$A_{15}D_9$	(2, S)
D_8^3	(S,V,V), (V,S,V), (V,V,S)
A_{12}^2	(1, 5)
$A_{11}D_7E_6$	(1,S,1)
E_6^4	$(1, 0, 1, \bar{1}), (1, \bar{1}, 0, 1), (1, 1, \bar{1}, 0)$
$A_9^2 D_6$	(2, 4, 0), (5,0,S), (0,5,C)
D_6^4	Even permutations of (0,S,V,C)
A_8^3	(1, 1, 4), (4, 1, 1), (1, 4, 1)
$A_7^2 D_5^2$	(1,1,S,V), (1,7,V,S)
A_6^4	(1, 2, 1, 6), (1, 6, 2, 1), (1, 1, 6, 2)
$A_5^4 D_4$	(2, [0, 2, 4], 0), (3,3,0,0,S), (3,0,3,0,V), (3,0,0,3,C)
D_4^6	(S,S,S,S,S,S), (0,[0,V,C,C,V])
A_4^6	(1, [0, 1, 4, 4, 1])
A_3^8	(2, [2, 0, 0, 1, 0, 1, 1])
A_2^{12}	(2, [1, 1, 2, 1, 1, 1, 2, 2, 2, 1, 2])
A_1^{24}	(1, [0, 0, 0, 0, 0, 1, 0, 1, 0, 0, 1, 1, 0, 0, 1, 1, 0, 1, 0, 1, 1, 1, 1])

of dimensions n_L and m_R respectively. Again $\Gamma_{n,m}$ is completely specified by the knowledge of the relevant glue vectors. The second type of Lorentzian Lie algebra lattice is obtained, if the sign of the signature of the metric changes within a given Lie algebra lattice. $D_{n,m}$ with (0) and (S) conjugacy classes is a suitable example. Actually, Lorentzian lattices of the second type are always also of the first type. For example, $D_{n,m}$ with (0) and (S) conjugacy classes is identical to $D_{n_L} \otimes D_{m_R}$ with (0,0), (V,V), (S,S) and (C,C) conjugacy classes.

Lorentzian even self-dual lattices $\Gamma_{n,m}$ exist for $n - m = 0 \mod 8$. They are unique up to Lorentz transformations. For $n = m + 8p$ they are Lorentz transformations of $(E_8)^p \otimes D_{m,m}$ where $D_{m,m}$ is defined by the (0) and (S) conjugacy classes. Alternatively, we can choose as reference lattice $(E_8)^p \otimes (D_{1,1})^m$ where $D_{1,1}$ is the $D_1 \otimes D_1$ lattice with conjugacy classes (0,0)+(V,V)+(S,S)+(C,C). It is not hard to find the explicit Lorentz transformation which transforms this $D_{1,1}$ lattice to $\Gamma_{1,1}$ of Footnote 6 on page 275.

11.3 Frenkel-Kač-Segal Construction

Let us now return to our original problem namely to provide an operator construction of the level one Kač-Moody algebra from free chiral boson fields moving on an n-dimensional torus. This construction is due to Frenkel and Kač and to Segal.

The level k Kač-Moody algebra in the Cartan-Weyl basis for a simply laced group G reads:

$$[H_m^I, H_n^J] = m\, \delta^{IJ}\, \delta_{m+n},$$

$$[H_m^I, E_n^\alpha] = \alpha^I\, E_{m+n}^\alpha,$$

$$[E_m^\alpha, E_n^\beta] = \begin{cases} \epsilon(\alpha, \beta)\, E_{m+n}^{\alpha+\beta} & \text{if } \alpha \cdot \beta = -1 \\ \alpha \cdot H_{m+n} + km\, \delta_{m+n} & \text{if } \alpha \cdot \beta = -2 \\ 0 & \text{if } \alpha \cdot \beta \geq 0 \end{cases} \quad (11.73)$$

with Hermiticity properties

$$H_n^{I\,\dagger} = H_{-n}^I, \quad E_n^{\alpha\,\dagger} = E_{-n}^{-\alpha}, \quad k^\dagger = k. \quad (11.74)$$

Note that $\alpha \cdot \beta = -1$ implies that $\alpha + \beta$ is a root and $\alpha \cdot \beta = -2$ that $\alpha + \beta = 0$. Let us now try to construct conformal fields $H^I(z)$ and $E^\alpha(z)$ from free chiral bosons $X^I(z)$, $(I = 1, \ldots, \text{rank } G)$ with mode expansion[8]

$$X^I(z) = q^I - i\, p^I\, \ln z + i \sum_{n \neq 0} \frac{1}{n} \alpha_n^I\, z^{-n}, \quad (11.75)$$

two-point function

$$\langle X^I(z) X^J(w) \rangle = -\delta^{IJ}\, \ln(z - w) \quad (11.76)$$

and energy momentum tensor

$$T(z) = -\frac{1}{2} \sum_i :\partial X^I(z)\, \partial X^I(z): \quad (11.77)$$

(cf. Chap. 4). The moments of $H^I(z)$ and $E^\alpha(z)$ are the Kač-Moody generators H_m^I, E_m^α (cf. Eq. (11.16)):

$$E_m^\alpha = \oint_{C_0} \frac{dz}{2\pi i}\, z^m\, E^\alpha(z),$$

$$H_m^I = \oint_{C_0} \frac{dz}{2\pi i}\, z^m\, H^I(z). \quad (11.78)$$

---

[8] This is the dimensionless chiral field X_R, introduced in (10.59), mapped to the complex z-plane.

Consider the conformal field

$$\tilde{E}^\alpha(z) =: e^{i\alpha \cdot X(z)} :, \qquad (11.79)$$

where α is a root vector of G with (length)$^2 = 2$. This implies that $\tilde{E}^\alpha(z)$ has conformal dimension $h = 1$. We will now show that $\tilde{E}^\alpha(z)$ is almost the desired field $E^\alpha(z)$. The operator product expansion of $\tilde{E}^\alpha(z)$ and $\tilde{E}^\alpha(w)$ has the form (see Eq. (4.120a)):

$$\tilde{E}^\alpha(z) \tilde{E}^\beta(w) = (z-w)^{\alpha\cdot\beta} \tilde{E}^{\alpha+\beta}(w) +$$
$$+ (z-w)^{\alpha\cdot\beta+1} \alpha \cdot i\partial X(w) \tilde{E}^{\alpha+\beta}(w) + \ldots \qquad (11.80)$$

Since $\alpha \cdot \beta$ is an integer one derives that

$$\tilde{E}^\alpha(z) \tilde{E}^\beta(w) = (-1)^{\alpha\cdot\beta} \tilde{E}^\beta(w) \tilde{E}^\alpha(z). \qquad (11.81)$$

This implies that with the contour integration trick of Chap. 4 we can only calculate

$$\tilde{E}^\alpha_m \tilde{E}^\beta_n - (-1)^{\alpha\cdot\beta} \tilde{E}^\beta_n \tilde{E}^\alpha_m$$

$$= \left\{ \oint_{C_0} \frac{dz}{2\pi i} \oint_{C_0} \frac{dw}{2\pi i} - \oint_{C_0} \frac{dz}{2\pi i} \oint_{C_0} \frac{dw}{2\pi i} \right\} z^m \tilde{E}^\alpha(z) w^n \tilde{E}^\beta(w)$$
$$\phantom{=\{} |z|>|w| |w|>|z|$$

$$= \oint_{C_0} \frac{dw}{2\pi i} \oint_{C_w} \frac{dz}{2\pi i} (z-w)^{\alpha\cdot\beta} z^m w^n \tilde{E}^{\alpha+\beta}(w)$$
$$\times [1 + i(z-w)\alpha \cdot \partial X(w) + \ldots]$$

$$= \begin{cases} \tilde{E}^{\alpha+\beta}_{m+n} & \alpha\cdot\beta = -1 \\ m\delta_{m+n} + i\alpha \cdot \oint_{C_0} \frac{dw}{2\pi i} w^{m+n} \partial X(w) & \alpha\cdot\beta = -2 \\ 0 & \text{otherwise.} \end{cases} \qquad (11.82)$$

Comparing this with Eq. (11.73) we recognize that, the unwanted factor $(-1)^{\alpha\cdot\beta}$ and the structure constants $\epsilon(\alpha,\beta)$ aside, we have reached our goal, if we identify $H^I(z)$ with the derivative of the free bosons $X^I(z)$:

$$H^I(z) = i\, \partial X^I(z). \qquad (11.83)$$

This is in fact the correct identification; $i\partial X^I(z)$ are conformal fields of dimension one and are the $[U(1)]^n$ Cartan-subalgebra currents. They satisfy the operator algebra

11.3 Frenkel-Kač-Segal Construction

$$H^I(z) H^J(w) = \frac{\delta^{IJ}}{(z-w)^2} + \text{finite} \tag{11.84}$$

which immediately leads to the correct commutator Eq. (11.73) between H_m^I and H_n^J. We can also check that, using

$$H^I(z) \tilde{E}^\alpha(w) = \frac{\alpha^I}{(z-w)} \tilde{E}^\alpha(w) + \ldots, \tag{11.85}$$

the correct algebra between the H_m^I and $\tilde{E}_n^\alpha$ is obtained. Finally, we have to compensate the factor $(-1)^{\alpha\cdot\beta}$ in Eq. (11.3) and to produce the structure constants $\epsilon(\alpha,\beta)$. This is done by introducing so-called cocycle factors (also called Klein factors) c_α. We define $E^\alpha(z)$ as:

$$E^\alpha(z) = \tilde{E}^\alpha(z)\, c_\alpha. \tag{11.86}$$

There are various explicit constructions of c_α. One is in terms of the zero modes $p^I = (2\pi i)^{-1} \oint i\,\partial X^I(z)$ in which case the $c_\alpha(p)$ have to satisfy[9]

$$c_\alpha(p+\beta)\, c_\beta(p) = (-1)^{\alpha\cdot\beta} c_\beta(p+\alpha)\, c_\alpha(p)$$
$$c_\alpha(p+\beta)\, c_\beta(p) = \epsilon(\alpha,\beta)\, c_{\alpha+\beta}. \tag{11.87}$$

We will not present this or any other construction.

In conclusion the level one Kač-Moody algebra of a simply laced group G with rank n possesses an explicit operator construction with n free bosonic fields $X^I(z)$:

$$H_m^I = \oint_{C_0} \frac{dz}{2\pi i}\, z^m\, H^I(z), \quad H^I(z) = i\,\partial X^I(z),$$
$$E_m^\alpha = \oint_{C_0} \frac{dz}{2\pi i}\, z^m\, E^\alpha, \quad E^\alpha(z) = :e^{i\alpha\cdot X(z)}:\, c_\alpha. \tag{11.88}$$

H_0^I, E_0^α obviously generate a finite dimensional subalgebra isomorphic to the Lie algebra g of Eq. (11.48). In this case the Cartan sub-algebra generators H_0^I are simply given by the momentum operators p^I in Eq. (11.75). The non-commuting E_0^α's are characterized by the momentum eigenvalues which are the roots α of the Lie algebra. It means that the allowed momentum eigenvalues of the bosons $X^I(z)$ are quantized, or, equivalently, the bosons $X^I(z)$ live on an n-dimensional torus. This is also required for the field (11.75) to be well defined (because of the logarithm).

We can now explicitly construct the vertex operators of the gauge boson states which results from the compactification of the bosonic coordinates. The vertex

---
[9] Use the relation $e^{iq\cdot\beta} f(p) e^{-iq\cdot\beta} = f(p-\beta)$ for any function $f(p)$.

operators of the Cartan subalgebra $[U(1)]^n$ gauge bosons are the currents $i\partial X^I(z)$. The asymptotic states are

$$|I\rangle = \oint_{C_0} \frac{dz}{2\pi i} \frac{i}{z} \partial X^I(z)|0\rangle = \alpha_{-1}^I |0\rangle \qquad (11.89)$$

which is nothing but an internal oscillator excitation in agreement with our considerations in Chap. 10.

The vertex operators of the non-Abelian gauge bosons are again given by the currents $E^\alpha =: e^{i\alpha \cdot X(z)} : c_\alpha$. Since $\alpha^2 = 2$, this vertex operator has conformal dimension $h = \frac{\alpha^2}{2} = 1$ as required for a physical state. The gauge boson state corresponding to the soliton vacuum has the form

$$|\alpha\rangle = \oint_{C_0} \frac{dz}{2\pi i} \frac{1}{z} :e^{i\alpha \cdot X(z)}: c_\alpha |0\rangle. \qquad (11.90)$$

This state carries quantized internal momentum corresponding to the roots of G. It is a winding state on the maximal torus of G defined by $\mathbb{R}/\Lambda_R$. The allowed quantized momenta of the $X^I(z)$ are however not restricted to lie in the root lattice of G but can a priori be any weight vector. So we consider vertex operators

$$V_\lambda(z) =: e^{i\lambda \cdot X(z)}: c_\lambda, \qquad (11.91)$$

where the λ's are weight vectors of some (irreducible) representation R of G and c_λ is again a cocycle factor. From now on we will not write the cocycle factors explicitly. The operators (11.91) create states which transform under the representation R. The operator product algebra is

$$V_\lambda(z) V_{\lambda'}(w) = (z-w)^{\lambda \cdot \lambda'} V_{\lambda+\lambda'}(w) + \ldots . \qquad (11.92)$$

The operator algebra is completely determined by the addition rules of the different conjugacy classes. The requirement for a closed operator algebra implies that the quantized momenta λ are vectors of a Lie algebra lattice of G. Additional requirements from string theory, like locality and modular invariance, imply that the lattice is integral and self-dual.

11.4 Fermionic Construction of the Current Algebra: Bosonization

In the previous section we presented the bosonic construction of the level one Kač-Moody algebra of a simply laced group G of rank n. It was motivated by the fact that the CFT of n free bosons and of the level one Kač-Moody algebra, via the

11.4 Fermionic Construction of the Current Algebra: Bosonization

Sugawara construction, have central charge $c = n$. On the other hand, the CFT of a "real" chiral (Majorana-Weyl) fermion has $c = \frac{1}{2}$. We thus expect that one can realize the level one Kač-Moody algebra with $2n$ fermions. Furthermore, we want to establish that the conformal field theory of $2n$ fermions with specific boundary conditions is equivalent to a conformal field theory of n bosons compactified on a torus.

Consider a system of $2n$ two-dimensional real fermions $\psi^i(z)$ ($i = 1, \ldots, 2n$), all either in the R or the NS sector, transforming as a vector of $SO(2n)$. They have the mode expansions

$$\psi^i(z) = \sum b_r^i \, z^{-r-\frac{1}{2}}, \tag{11.93}$$

where r is integer (R) or half-integer (NS).[10] Their operator products are

$$\psi^i(z)\psi^j(w) = \frac{\delta^{ij}}{z-w} + \ldots. \tag{11.94}$$

We build the fermion bilinears

$$J^a(z) = \frac{1}{2} : \psi^i(z) \, T_{ij}^a \, \psi^j(z) : \qquad a = 1, \ldots, 2n^2 - n, \tag{11.95}$$

where the antisymmetric $2n \times 2n$ matrices T_{ij}^a are the generators of $SO(2n)$ in the vector representation. An explicit representation of the $SO(2n)$ generators in the vector representation is $(T^{kl})_{ij} = i \, (\delta_i^k \delta_j^l - \delta_j^k \delta_i^l)$ and the currents become $J^{kl}(z) = i : \psi^k \psi^l(z) :$. Using Eq. (11.94), it is not difficult to show that the fermionic currents $J^a(z)$ in Eq. (11.95) obey the level one Kač-Moody algebra of $D_n \sim SO(2n)$ as given in Eq. (11.20).

This suggests that the bilinears of $2n$ free fermions are identical to the currents Eq. (11.88) constructed from n free bosons with momentum vectors being root vectors of D_n. This is basically the statement of bosonization. It means that the conformal field theories of n free bosons and $2n$ free fermions are equivalent in the sense that the conformal properties of their operators and all correlation functions are identical. However, this is only true if the bosons are compactified on a special torus. For instance a single boson is only equivalent to two free fermions if the radius of the circle the boson is compactified on takes special values.

Let us work out the bosonization prescription in more detail. For this purpose it is useful to convert the real basis for the fermions ψ^i to a complex basis. Define

$$\psi^{\pm i} = \frac{1}{\sqrt{2}}(\psi^{2i-1} \pm i\psi^{2i}), \qquad i = 1, \ldots, n. \tag{11.96}$$

---

[10] The chiral world-sheet fermions mapped to the complex plane are of this kind. We will discuss the conformal field theory of fermions in detail in Chap. 12.

The Cartan subalgebra currents of D_n are then

$$J^{+i,-i}(z) = :\Psi^{+i}(z)\Psi^{-i}(z): \qquad (11.97)$$

and the non-commuting D_n currents are

$$J^{\pm i,\pm j}(z) = :\Psi^{\pm i}(z)\Psi^{\pm j}(z): \quad (i < j). \qquad (11.98)$$

Bosonization then consists of identifying $J^{i,-i}(z)$ with the derivative of the bosons $X^i(z)$

$$:\Psi^{+j}(z)\Psi^{-j}(z): := \lim_{w \to z}\left[e^{iX^j(w)}e^{-iX^j(z)} - \frac{1}{w-z}\right] = i\partial X^j(z) \qquad (11.99)$$

and the non-commuting currents with the operators $E^\alpha(z)$

$$:\Psi^{\pm i}(z)\Psi^{\pm j}(z): =: e^{i(\pm X^i \pm X^j(z))}: c_{\pm i,\pm j} \quad (i < j). \qquad (11.100)$$

We recognize that the corresponding root vector of D_n is $\boldsymbol{\alpha} = \pm \boldsymbol{e}_i \pm \boldsymbol{e}_j$ ($\boldsymbol{e}_i$ are the Euclidean unit vectors, $\boldsymbol{e}_i \cdot \boldsymbol{e}_j = \delta_{ij}$).

One can also give a bosonic representation of the fermions $\Psi^i(z)$ themselves. Since they transform as a vector of D_n they are bosonized according to

$$\Psi^{\pm j}(z) =: e^{\pm i X^j(z)}: c_{\pm j}, \qquad (11.101)$$

where the quantized bosonic momentum is now a vector weight of D_n: $\boldsymbol{\lambda} = \pm \boldsymbol{e}_i$. This gives the correct conformal dimension $h = \frac{\lambda^2}{2} = \frac{1}{2}$ for the fermions $\Psi^{\pm i}(z)$. The cocycle factors $c_{\pm i}$ are necessary to implement that also fermions with $i \neq j$ anti-commute.

As already discussed in Chap. 8, the Hilbert space of the fermionic theory splits into two sectors, namely the NS and R sector. In the NS sector, the fermions have anti-periodic boundary conditions on the cylinder which means that they are periodic on the complex plane. For R states the situation is reversed. This change of boundary conditions is due to the Jacobian factor in Eq. (4.10) for $h = \frac{1}{2}$.

All states in the NS sector are (even or odd rank) tensors of $SO(2n)$. Therefore, in bosonic language, their vertex operators are

$$V_\lambda(z) =: e^{i \boldsymbol{\lambda} \cdot X(z)}: c_\lambda, \qquad (11.102)$$

where $\boldsymbol{\lambda}$ is either in the (0) or (V) conjugacy class of D_n. The currents Eq. (11.100) or the fermions Eq. (11.101) are of this type. Since the mutual scalar products between the (0) and (V) conjugacy classes are integer (see Table 11.2), the operator algebra Eq. (11.92) in the NS sector is local, i.e. contains no branch cuts.

On the other hand, states in the R sector are built on the vacua $|S^\alpha\rangle$ ($|S^{\dot\alpha}\rangle$) which transform as spinors (conjugate-spinors) of D_n. Thus the vertex operators in the R sector are of the form Eq. (11.102) but now with λ a weight either in the (S) or (C) conjugacy class of D_n. We can also give an explicit construction for the spinorial vacuum $|S^\alpha\rangle$ ($|S^{\dot\alpha}\rangle$). It is created from the $SL(2,\mathbb{Z})$ invariant vacuum by an operator Eq. (11.102) with λ_α ($\lambda_{\dot\alpha}$) a (conjugate) spinor weight of D_n as displayed in Eqs. (11.65), (11.66). This operator has conformal dimension $h = \frac{\lambda^2}{2} = \frac{n}{8}$.

Inspection of Table 11.2 shows that the operator algebra Eq. (11.92) in the R sector is non-local. Furthermore, states in the NS sector corresponding to the (V) conjugacy class of D_n are non-local with respect to the R sector. Their operator algebra contains branch cuts.

In conclusion, it is evident that using the techniques of bosonization the vertex operators of the CFT of free fermions become extremely simple. This leads to many simplifications, as we will discuss in more detail later. The draw-back is that covariance is not manifest in the bosonic formulation.

11.5 Unitary Representations and Characters of Kač-Moody Algebras

We now return to the discussion at the beginning of this chapter about the representations of the chiral algebra $\mathscr{A} = \hat{v} \in \hat{g}$ and determine the restrictions on the level k imposed by unitarity. As before, we restrict to simply laced g and normalize the roots to have (length)$^2 = 2$. Consider the $SU(2)$ subalgebra of $\hat{g}$ generated by $E_1^{-\alpha}$, E_{-1}^α and $(k - \alpha \cdot H_0)$ where α is any root. Again, if we identify these generators with J_+, J_- and $2J_3$ we have the angular momentum algebra and know that unitarity requires that the eigenvalues of $(k - \alpha \cdot H_0)$ have to be integer. Acting on a primary state $|m\rangle$ with weight m we find $k - \alpha \cdot m \in \mathbb{Z}$ or, since $\alpha \cdot m \in \mathbb{Z}$

$$k \in \mathbb{Z}. \tag{11.103}$$

Since $|m\rangle$ is primary, $E_{+1}^{-\alpha}|m\rangle = 0$ and

$$||E_{-1}^\alpha|m\rangle||^2 = \langle m|E_1^{-\alpha} E_{-1}^\alpha|m\rangle = (k - \alpha \cdot m)|| |m\rangle||^2. \tag{11.104}$$

Positivity of the Hilbert space then gives $k \geq \alpha \cdot m$. The right hand side of this inequality is maximized if we chose α to be the highest root ψ and m to be the highest weight m_0 of the given representation. We then have

$$k \geq \psi \cdot m_0. \tag{11.105}$$

Clearly, for a given k only a finite number of highest weights satisfies this criterion.[11] For $k = 1$ they are those belonging to the lowest dimensional representation in each conjugacy class. So, in spite of the fact that we are considering conformal field theories with c larger than one, the combined Kač-Moody and Virasoro algebra has only a finite number of primary fields or current algebra families. Therefore, we call them also rational conformal field theories. Of course, the irreducible representations of the combined algebra are highly reducible under the Virasoro algebra. A current algebra family contains an infinite number of conformal fields which all transform under some representation of G. On the other hand, any current algebra family is generated from a specific lowest dimensional representation; states in all other representations are obtained by acting with the E^α_{-n}.

As usual, the operator product algebra between the primary fields is determined by the conformal dimensions and the fusion rules among the fields (cf. (4.71) on page 78)

$$[\phi_i] \times [\phi_j] = \sum_l N_{ij}{}^l [\phi_l]. \tag{11.106}$$

For arbitrary level k the fusion coefficients $N_{ij}{}^l$ are quite difficult to determine. Of course, the fusion rules obey the decomposition rules of the tensor products between two irreducible representations. Therefore, the $N_{ij}{}^l$ are necessarily zero if the corresponding Clebsch-Gordon coefficients vanish. However, the converse is not true which makes a systematic discussion highly non-trivial.

For this reason we will concentrate on the simplest case $k = 1$. We have already seen that one can explicitly construct the level one currents and also the states which transform under a specific representation by n free bosons compactified on the weight lattice of G. Equation (11.105), which determines the number of current algebra families, is satisfied only for the lowest dimensional representation in each conjugacy class of G. Therefore, the number of primary fields is identical to the number of conjugacy classes. The highest weight state has the form:

$$|\phi_i\rangle = \oint_{C_0} \frac{dz}{2\pi i} \frac{1}{z} : e^{i\lambda_i \cdot X(z)} : |0\rangle. \tag{11.107}$$

λ_i is the highest weight vector of the lowest representation in the i-th conjugacy class. The fusion rules Eq. (11.106) are simply determined by the addition rules for the different conjugacy classes resp. highest weight vectors. All non-vanishing coefficients $N_{ij}{}^l$ are one.

Let us finally discuss the generalized characters of $\mathscr{A}$. The characters are defined as

$$\chi_i(\tau) = \mathrm{Tr}_{\phi_i} q^{L_0 - c/24}, \tag{11.108}$$

where the trace is over all states of the current algebra family $[\phi_i]$. This definition holds for all k. Again, for $k = 1$ these characters have a rather simple form.

---

[11] The corresponding highest weight representations are also called integrable representations.

11.5 Unitary Representations and Characters of Kač-Moody Algebras

Since the L_0 eigenvalue of the field $\phi(z) =: e^{i\lambda \cdot X(z)} :$ is $h = \frac{\lambda^2}{2}$, the level one characters are

$$\chi_i(\tau) = \frac{1}{[\eta(\tau)]^n} \sum_{\lambda \in (i)} e^{i\pi\tau\lambda^2} = \frac{1}{[\eta(\tau)]^n} P_i(\tau). \tag{11.109}$$

The sum $P_i(\tau)$ is over all lattice vectors within the conjugacy class (i). The factor $[\eta(\tau)]^{-n}$, where n is the rank of G, takes into account the contribution of the L_{-k}'s and $q^{-c/24}$ ($c = n$).

For the D_n algebras at level one we can give a closed expression for the lattice sums. From the explicit expressions for the simple roots and the representative weights, one can construct all weights in each of the four conjugacy classes. One then finds that the lattice sums corresponding to the (0), (V), (S), (C) conjugacy classes of D_n are

$$P_0(\tau) = \frac{1}{2}\{[\vartheta_3(\tau)]^n + [\vartheta_4(\tau)]^n\},$$

$$P_V(\tau) = \frac{1}{2}\{[\vartheta_3(\tau)]^n - [\vartheta_4(\tau)]^n\},$$

$$P_S(\tau) = \frac{1}{2}\{[\vartheta_2(\tau)]^n + i^n [\vartheta_1(\tau)]^n\},$$

$$P_C(\tau) = \frac{1}{2}\{[\vartheta_2(\tau)]^n - i^n [\vartheta_1(\tau)]^n\}. \tag{11.110}$$

The ϑ-functions were defined in Chap. 9. It is illustrative to compare these expressions with the partition functions of the fermionic string theory (cf. Chap. 9). Again, the equivalence between n bosons compactified on a D_n lattice and $2n$ periodic resp. anti-periodic fermions becomes manifest. Explicitly we obtain the following identities between the bosonic and fermionic partition functions[12]

$$\chi_{PP}(\tau) = (i)^n(\chi_S(\tau) - \chi_C(\tau)),$$
$$\chi_{PA}(\tau) = \chi_S(\tau) + \chi_C(\tau),$$
$$\chi_{AA}(\tau) = \chi_0(\tau) + \chi_V(\tau),$$
$$\chi_{AP}(\tau) = \chi_0(\tau) - \chi_V(\tau). \tag{11.111}$$

With the help of Eq. (11.110) the partition function for the $SO(32)$ heterotic string, given in Chap. 10, follows immediately.

---

[12] These identities remain also true for higher genus partition functions.

We now consider the contribution of the world-sheet fermions to the partition function of the superstring in the light of bosonization. In the two sectors we found $(r^2 = \sum r_a^2)$

NS sector

$$\frac{1}{2}\left\{\frac{\vartheta_3^4}{\eta^4} - \frac{\vartheta_4^4}{\eta^4}\right\} = \frac{1}{\eta^4}\sum_{r_a \in \mathbb{Z}} q^{\frac{1}{2}r^2}\frac{1}{2}\left(1 - e^{i\pi(r_1+r_2+r_3+r_4)}\right), \quad (11.112)$$

R sector

$$\frac{1}{2}\left\{\frac{\vartheta_2^4}{\eta^4} \mp \frac{\vartheta_1^4}{\eta^4}\right\} = \frac{1}{\eta^4}\sum_{r_a \in \mathbb{Z}+\frac{1}{2}} q^{\frac{1}{2}r^2}\frac{1}{2}\left(1 \mp e^{i\pi(r_1+r_2+r_3+r_4)}\right). \quad (11.113)$$

We see that the fermionic degrees of freedom of a given NS sector state depend on a vector r with four integer entries. This is an $SO(8)$ weight in the (0) and (V) conjugacy class, i.e. space-time bosons. But only the (V) conjugacy class ($\sum r_a$ odd) contributes. This is, of course, the GSO projection. Similarly, a given R sector state depends on a vector with half-integer entries, i.e. to an $SO(8)$ spinor weight. Depending on the relative sign only states with $\sum r_a$ odd or even appear, which corresponds to the (C) and (S) conjugacy class, respectively. This is once more the GSO projection. The η^{-4} factors are now interpreted as the oscillator contributions of the four bosons which arise in the bosonization of the eight fermions in the light-cone directions. The sums are the zero mode contributions of the chiral bosons, i.e. over momenta and windings.

Similarly, one can also express the level one characters of E_6, E_7 and E_8 by theta functions; this requires again the explicit forms of the simple roots and the representative weights. For E_8 they have been given before. For E_6 and E_7 we simply state the results for the lattice sums. One finds

$$P_{(E_6)0}^{c=6} = \frac{1}{2}\left(\vartheta_3(3\tau)\vartheta_3^5(\tau) + \vartheta_4(3\tau)\vartheta_4^5(\tau) + \vartheta_2(3\tau)\vartheta_2^5(\tau)\right),$$

$$P_{(E_6)1}^{c=6} = \frac{1}{2}\left(\vartheta\begin{bmatrix}\frac{1}{6}\\0\end{bmatrix}(3\tau)\vartheta_2^5(\tau) + \vartheta\begin{bmatrix}\frac{2}{3}\\0\end{bmatrix}(3\tau)\vartheta_3^5(\tau) + e^{-2\pi i/3}\vartheta\begin{bmatrix}\frac{2}{3}\\\frac{1}{2}\end{bmatrix}(3\tau)\vartheta_4^5(\tau)\right),$$

$$P_{(E_6)\bar{1}}^{c=6} = \frac{1}{2}\left(\vartheta\begin{bmatrix}\frac{5}{6}\\0\end{bmatrix}(3\tau)\vartheta_2^5(\tau) + \vartheta\begin{bmatrix}\frac{1}{3}\\0\end{bmatrix}(3\tau)\vartheta_3^5(\tau) - e^{-i\pi/3}\vartheta\begin{bmatrix}\frac{1}{3}\\\frac{1}{2}\end{bmatrix}(3\tau)\vartheta_4^5(\tau)\right),$$

$$P_{(E_7)0}^{c=7} = \frac{1}{2}\left(\vartheta_2(2\tau)\vartheta_2^6(\tau) + \vartheta_3(2\tau)\left(\vartheta_3^6(\tau) + \vartheta_4^6(\tau)\right)\right),$$

11.5 Unitary Representations and Characters of Kač-Moody Algebras

$$P^{c=7}_{(E_7)1} = \frac{1}{2}\left(\vartheta_3(2\tau)\vartheta_2^6(\tau) + \vartheta_2(2\tau)\left(\vartheta_3^6(\tau) - \vartheta_4^6(\tau)\right)\right),$$

$$P^{c=8}_{(E_8)0} = \frac{1}{2}\left(\vartheta_2^8(\tau) + \vartheta_3^8(\tau) + \vartheta_4^8(\tau)\right). \tag{11.114}$$

We conclude this section with the discussion of modular properties of the level one Kač-Moody characters. This is important because they appear in string theory as one-loop partition functions of n bosonic coordinates compactified on a specific Lie algebra lattice. The Hilbert space then contains, in general, several current algebra families (or conjugacy classes) including the identity family which of course corresponds to the (0) conjugacy class.

The characters of the Kač-Moody algebra $\hat{g}$ at level one form a finite dimensional representation of the modular group, i.e. the characters transform under the generators of the modular group $T : (\tau \to \tau + 1)$ and $S : (\tau \to -\frac{1}{\tau})$ like ($i = 1, \ldots, N_C =$ number of conjugacy classes):

$$\chi_i(\tau + 1) = T_{ij}\,\chi_j(\tau), \qquad \chi_i(-1/\tau) = S_{ij}\,\chi_j(\tau). \tag{11.115}$$

The matrix T is easy to determine. From Eq. (11.108) we immediately derive

$$T_{ij} = e^{-\pi i\left(\frac{c}{12} - \lambda_i^2\right)}\delta_{ij}, \tag{11.116}$$

where λ_i is some vector in the conjugacy class (i). The S transformation, on the other hand, is more involved. We obtain for the level one characters the following result:

$$S: \quad \chi_i\left(-\frac{1}{\tau}\right) = \frac{1}{\sqrt{N_C}} \sum_{j=1}^{N_C} e^{2\pi i\,\lambda_i\cdot\lambda_j}\,\chi_j(\tau). \tag{11.117}$$

This is actually not difficult to derive. The weights in the i-th conjugacy class can be written as

$$\{\lambda^{(i)}\} = \overline{\lambda}^{(i)} + \Lambda_R, \tag{11.118}$$

where $\overline{\lambda}^{(i)}$ is some representative weight and Λ_R the root lattice. We then write

$$\chi_i\left(-\frac{1}{\tau}\right) = \frac{1}{\eta^n(-1/\tau)} \sum_{\lambda \in \Lambda_R} e^{-\frac{i\pi}{\tau}(\overline{\lambda}^{(i)} + \lambda)^2}$$

$$= \frac{1}{\eta^n(\tau)}\frac{1}{\sqrt{N_C}} \sum_{\lambda \in \Lambda_w} e^{2\pi i\,\lambda\cdot\overline{\lambda}^{(i)}} e^{i\pi\tau\lambda^2}, \tag{11.119}$$

where we have used Eq. (9.115) and the fact that $\Lambda_w = (\Lambda_R)^*$. With $\lambda^{(j)} \cdot \overline{\lambda}^{(i)} = \lambda^{(j)} \cdot \lambda^{(i)}$ mod 1 we get Eq. (11.117). It shows that the S transformation acts as a Fourier transformation on the characters $\chi_i(\tau)$: For the D_n algebra we obtain, using Table 11.2, the following matrix:

$$S_{ij} = \frac{1}{2}\begin{pmatrix} 1 & 1 & 1 & 1 \\ 1 & 1 & -1 & -1 \\ 1 & -1 & e^{i\pi n/2} & -e^{i\pi n/2} \\ 1 & -1 & -e^{i\pi n/2} & e^{i\pi n/2} \end{pmatrix} \qquad (11.120)$$

($i = 1$ corresponds to (0), $i = 2$ to (V), $i = 3$ to (S) and $i = 4$ to (C)). One can verify that $S^2 = 1$ for n even. For n odd, acting with S^2 interchanges the S and C conjugacy classes. Recall from Chap. 4 that this is a general feature: if complex representations are present, $S^2 = C$ with $C^2 = 1$ and C is called the charge conjugation matrix. The difference between n even and odd is that for n even, the vertex operator $e^{i\lambda \cdot X}$ and its conjugate $e^{-i\lambda \cdot X}$ belong to the same conjugacy class while for n odd, $\lambda \in (S) \leftrightarrow -\lambda \in (C)$.

For the A_n series S_{kl} is ($k, l = 1, \ldots, n+1$)

$$S_{kl} = \frac{1}{\sqrt{n+1}} e^{-\frac{2\pi i}{n+1}(k-1)(l-1)}. \qquad (11.121)$$

Finally the E_6 characters transform like A_2 (E_6 and A_2 have the same fusion rules), the E_7 characters like the A_1 characters and for E_8, S and T are the identity. The E_8 character is a modular invariant in agreement with our result of Chap. 10 that the partition function of an even self-dual lattice is modular invariant.

The agreement between the partition function for chiral bosons on a lattice and of free fermions, which we pointed out below Eq. (11.110), hints at an alternative formulation of the heterotic string. In fact, one can 'fermionize' the 16 chiral bosons which leads to the fermionic description of the heterotic strings in terms of 32 left-moving Weyl fermions. The world-sheet action for the 32 fermions is

$$S = \frac{1}{4\pi} \int d^2z \, \bar{\lambda}^A \bar{\partial} \bar{\lambda}^A. \qquad (11.122)$$

For the $SO(32)$ heterotic string A is an $SO(32)$ vector index, for the $E_8 \times E_8$ string we replace $A \to (A, A')$ where $A, A' = 1, \ldots, 16$ are vector indices of the $SO(16) \times SO(16) \subset E_8 \times E_8$ subgroup. The full $E_8 \times E_8$ symmetry cannot be made manifest. The partition functions of these free fermions, summed over all four spin structures, are the same as those of 16 chiral bosons on the $\mathrm{Spin}(32)/\mathbb{Z}_2$ and $E_8 \times E_8$ lattices.

It is now straightforward to couple the chiral fermions to a background gauge field: one simply covariantizes $\partial \to \partial + A_\mu \partial X^\mu$ leading to $A_\mu^a \partial X^\mu j^a$ with $j^a = \bar{\lambda}^A (T^a)^{AB} \bar{\lambda}^B$. For the $E_8 \times E_8$ string we can only couple to the gauge fields of the $SO(16) \times SO(16)$ subgroup. If we restrict to gauge bosons in the Cartan subalgebra this becomes, via the bosonization formula (11.99), $\bar{\partial} X^I \partial X^\mu A_\mu^I$. This is the coupling of the heterotic string in the bosonic formulation to a gauge field background, cf. (10.115).

11.6 Highest Weight Representations of $\widehat{su}(2)_k$

For later use (see last part of Sect. 12.2) we discuss as an example of a higher level Kač-Moody algebra and its highest weight representations the case $\widehat{su}(2)_k$. Recall the algebra

$$[j_m^a, j_n^b] = i\sqrt{2} \sum_c \epsilon^{abc} j_{m+n}^c + k\, m\, \delta_{m+n,0}\, \delta^{ab} \qquad (11.123)$$

with structure constants $f^{abc} = \sqrt{2}\epsilon^{abc}$ and $C_2 = 4$, i.e. the two roots $\alpha = \pm\sqrt{2}$ with $\alpha^2 = 2$. The central charge is according to (11.32)

$$c = \frac{3k}{k+2}. \qquad (11.124)$$

Define the raising and lowering operators

$$\hat{j}_m^3 = \frac{1}{\sqrt{2}} j_m^3, \qquad \hat{j}_m^\pm = \frac{1}{\sqrt{2}} \left(j_m^1 \pm i\, j_m^2\right). \qquad (11.125)$$

The commutation relations (11.123) become

$$[\hat{j}_m^3, \hat{j}_n^3] = \frac{m\,k}{2} \delta_{m+n,0}, \qquad [\hat{j}_m^3, \hat{j}_n^\pm] = \pm \hat{j}_{m+n}^\pm,$$
$$[\hat{j}_m^+, \hat{j}_n^-] = k\, m\, \delta_{m+n,0} + 2\, \hat{j}_{m+n}^3. \qquad (11.126)$$

A highest weight state $|h, q\rangle$ is defined via the requirements

$$\hat{j}_n^3 |h, q\rangle = \hat{j}_n^\pm |h, q\rangle = 0 \qquad \text{for} \qquad n > 0,$$
$$\hat{j}_0^3 |h, q\rangle = \frac{q}{2} |h, q\rangle,$$
$$\hat{j}_0^+ |h, q\rangle = 0. \qquad (11.127)$$

This definition also implies that it is a highest weight state w.r.t. the zero mode algebra which is the angular momentum algebra $su(2)$. From the $su(2)$ representation theory we know that the unitary highest weight representations have spin $\frac{l}{2}$, $l = 0, 1, 2, \ldots$ and dimension $l + 1$. The states $|h, q\rangle$ in the spin-$\frac{l}{2}$ representation are $|h, l - 2m\rangle = (\hat{j}_0^-)^m |h, l\rangle$, $m = 0, \ldots, l$ with $q = l - 2m$. Note that since $[L_0, j_0^a] = 0$ all states $|h, l - 2m\rangle$ have the same L_0 eigenvalue. From Eq. (11.33) we know that

$$L_0 |h, l - 2m\rangle = \frac{l(l+2)}{4(k+2)} |h, l - 2m\rangle. \qquad (11.128)$$

To obtain this, we used that in the normalization of (11.123), which defines k, the weights of the spin-$\frac{1}{2}$ representation are $\frac{1}{\sqrt{2}}(l-2m)$ with $m=0$ being the highest weight. We then know from (11.105) that unitary representations of $\widehat{su}(2)_k$ only exist for $k \in \mathbb{Z}^+$ and $0 \leq l \leq k$. Here this follows directly from requiring that the norm of $\hat{j}^+_{-1}|h,l\rangle$

$$\langle h,l \,|\, \hat{j}^-_1 \,\hat{j}^+_{-1} \,|\, h,l\rangle = k - l \tag{11.129}$$

is non-negative.

In order to construct the characters of the $\widehat{su}(2)_k$ unitary highest weight representations, we first define

$$\chi_l^{(k)}(\tau, z) = \operatorname{tr} e^{2\pi i \tau L_0 - \frac{c}{24}} e^{-2\pi i z \hat{j}^3}, \tag{11.130}$$

where z is a 'chemical potential' for the $U(1)$-charge. These can be computed with the help of the Weyl-Kač character formula. The result, which we state without proof, is:

$$\chi_l^{(k)}(\tau, z) = \frac{\Theta_{l+1,k+2}(\tau,z) - \Theta_{-l-1,k+2}(\tau,z)}{\Theta_{1,2}(\tau,z) - \Theta_{-1,2}(\tau,z)} \tag{11.131}$$

with $0 \leq l \leq k$ and the generalized $SU(2)$ Θ-functions defined as ($q = e^{2\pi i \tau}$)

$$\Theta_{l,k}(\tau, z) = \sum_{n \in \mathbb{Z} + \frac{l}{2k}} q^{k n^2} e^{-2\pi i n k z}. \tag{11.132}$$

They are generalizations of the Θ-functions (10.71) which appeared in the discussion of the $\widehat{u}(1)_k$ rational CFT's. At $z = 0$ the ratio of Θ-functions in (11.131) is indeterminate but the limit $\chi_l^{(k)}(\tau) = \lim_{z \to 0} \chi_l^{(k)}(\tau, z)$ has a well-defined q-expansion.

One can expand the $\widehat{su}(2)_k$ characters in terms of the Θ-functions in the following way

$$\chi_l^{(k)}(\tau) = \sum_{\substack{m=-k+1 \\ l+m=0 \bmod 2}}^{k} C_{l,m}^{(k)}(\tau) \, \Theta_{m,k}(\tau). \tag{11.133}$$

This expansion, which is the character decomposition of $\widehat{su}(2)_k$ representations in terms on $\widehat{u}(1)_k$ characters, provides an implicit definition of the so-called string-functions $C_{l,m}^{(k)}(\tau)$.

With the help of (10.73) we determine the modular S-matrix for the characters $\chi_l^{(k)}(\tau)$ as

$$S_{l\,l'}^{(k)} = \sqrt{\frac{2}{k+2}} \sin\left(\frac{\pi}{k+2}(l+1)(l'+1)\right) \quad \text{with} \quad l, l' = 0, \ldots, k. \tag{11.134}$$

11.6 Highest Weight Representations of $\widehat{su}(2)_k$

Table 11.4 All $\widehat{su}(2)_k$ modular invariant partition functions

Level	Partition function	Name
$k = n$	$\mathscr{L} = \sum_{l=0}^{n} |\chi_l|^2$	$A_{n+1}, \; n \geq 1$
$k = 4n$	$\mathscr{L} = \sum_{l=0}^{n-1} |\chi_{2l} + \chi_{k-2l}|^2 + 2|\chi_{\frac{k}{2}}|^2$	$D_{2n+2}, \; n \geq 1$
$k = 4n - 2$	$\mathscr{L} = \sum_{l=0}^{\frac{k}{2}} |\chi_{2l}|^2 + \sum_{l=0}^{2n-2} \chi_{2l+1} \overline{\chi}_{k-2l-1}$	$D_{2n+1}, \; n \geq 2$
$k = 10$	$\mathscr{L} = |\chi_0 + \chi_6|^2 + |\chi_3 + \chi_7|^2 + |\chi_4 + \chi_{10}|^2$	E_6
$k = 16$	$\mathscr{L} = |\chi_0 + \chi_{16}|^2 + |\chi_4 + \chi_{12}|^2 + |\chi_6 + \chi_{10}|^2$ $+ (\chi_2 + \chi_{14}) \overline{\chi}_8 + \chi_8 (\overline{\chi}_2 + \overline{\chi}_{14}) + |\chi_8|^2$	E_7
$k = 28$	$\mathscr{L} = |\chi_0 + \chi_{10} + \chi_{18} + \chi_{28}|^2$ $+ |\chi_6 + \chi_{12} + \chi_{16} + \chi_{22}|^2$	E_8

For ease of notation, the (k)-labels on the characters and partition functions have been omitted

Plugging this into the Verlinde formula, one obtains for the fusion rules

$$N_{l_1 l_2}{}^{l_3} = \begin{cases} 1 \text{ if } \; |l_1 - l_2| \leq l_3 \leq \min(l_1 + l_2, 2k - l_1 - l_2) \\ \quad \text{and } l_1 + l_2 + l_3 = 0 \mod 2, \\ 0 \text{ otherwise}. \end{cases} \quad (11.135)$$

For the case of $\widehat{su}(2)_k$ a complete classification of modular invariant partition functions satisfying Eq. (6.99) has been achieved. The corresponding partition functions are listed in Table 11.4 which is known as the A-D-E classification.

The subscripts of the characters label again the highest weight representation. Furthermore, the name of each class corresponds to a Lie group to which the partition function can be associated, and the dual Coxeter number of each algebra is $k + 2$.

Appendix

In this appendix we give some details of the Sugawara construction. We start with determining the normalization of the energy momentum tensor by requiring (11.18). We compute the commutator (summation over repeated group indices is implied)

$$\left[\sum_p :J_p^b J_{n-p}^b:, J_m^a\right] = \sum_p [J_p^b J_{n-p}^b, J_m^a] = \sum_p \left\{[J_p^b, J_m^a]J_{n-p}^b + J_p^b[J_{n-p}^b, J_m^a]\right\}$$

$$= \sum_p \left\{if^{bac}J_p^b J_{m+n-p}^c + if^{bac}J_{p+m}^c J_{n-p}^b\right.$$

$$\left. + k(n-p)J_p^a \delta_{n+m-p} + kp J_{n-p}^a \delta_{p+m}\right\}. \quad (11.136)$$

In the first line normal ordering is irrelevant as it produces a c-number which commutes with J_m^a. The last line gives $-2km J_{m+n}^a$. In the second line one might be tempted to shift the summation variable in one of the two terms and to use the antisymmetry of the structure constants to obtain zero. But one has to be careful. This is analogous to the situation one meets in the evaluation of anomalies in QFT where one encounters linearly divergent integrals. Schematically, $\Delta(a) = \int_{-\infty}^{\infty} dx[f(x+a) - f(x)]$ is naively zero. However, expanding $f(x+a)$ around x one obtains $\Delta(a) = a[f(\infty) - f(-\infty)] + \frac{a^2}{2}[f'(\infty) - f'(-\infty)] + \ldots$ which is not necessarily zero. To avoid such pitfalls we use (11.31) to express all bilinears in the currents in terms of well-defined normal ordered products for which the shift in the summation variable is allowed. This gives for the second line

$$-if^{abc}\sum_p \left\{[J_p^b, J_{n+m-p}^c]\theta(p) + [J_{p+m}^c, J_{n-p}^b]\theta(p+m)\right\}$$

$$= C_2 J_{n+m}^a \sum_p [\theta(p) - \theta(p+m)] = -mC_2 J_{n+m}^a \quad (11.137)$$

which reproduces (11.18), if we fix the normalization as in (11.26).

Next we compute ($\gamma = (C_2 + 2k)^{-1}$)

$$[L_n, L_m] = \gamma \sum_p [L_n, :J_p^a J_{m-p}^a:] = \gamma \sum_p [L_n, J_p^a J_{m-p}^a]$$

$$= \gamma \sum_p \left\{-pJ_{n+p}^a J_{m-p}^a - (m-p)J_p^a J_{n+m-p}^a\right\}$$

$$= \gamma \sum_p \left\{-p :J_{n+p}^a J_{m-p}^a: -(m-p):J_p^a J_{n+m-p}^a:\right.$$

$$\left. -p[J_{n+p}^a, J_{m-p}^a]\theta(n+p) - (m-p)[J_p^a, J_{n+m-p}^a]\theta(p)\right\}. \quad (11.138)$$

In the first line, once again the normal ordering symbol is irrelevant. In the third line we are allowed to shift the summation variable and obtain $(n-m)L_{n+m}$. The last line gives

11.6 Highest Weight Representations of $\widehat{su}(2)_k$

$$\gamma \sum_p \left\{ -p\delta^{aa}k(n+p)\delta_{n+m}\theta(n+p) - (m-p)\delta^{aa}kp\delta_{n+m}\theta(p) \right.$$

$$= \gamma\, k \dim(G)\, \delta_{n+m} \sum_p p(n+p)[\theta(p) - \theta(n+p)]$$

$$= \frac{1}{6}\gamma\, k \dim(G)\, n(n^2 - 1)\, \delta_{n+m}, \tag{11.139}$$

where we used (3.102). This verifies (11.32).

An alternative way to determine the correct normalization of the current bilinear and to compute the central charge of the Virasoro algebra would be to compute operator products where the basic product is (11.20). However, because $J^a(z)$ are not free fields, the use of Wick's theorem is not straightforward.

Further Reading

Comprehensive references for the material covered in this chapter are

- P. Di Francesco, P. Mathieu, D, Sénéchal, *Conformal Field Theory* (Springer, Berlin, 1996)
- J. Fuchs, *Affine Lie Algebras and Quantum Groups* (Cambridge University Press, Cambridge, 1992)
- P. Goddard, D. Olive, Kač-Moody and Virasoro algebras in relation to quantum physics. Int. J. Mod. Phys. **A1**, 303–414 (1986)

The last reference and many of the original articles are reprinted in

- P. Goddard, D. Olive (eds.), *Kač-Moody and Virasoro Algebras*. Advanced Series in Mathematical Physics, vol. 3 (World Scientific, Singapore, 1988)

Properties of lattices and their role for string compactification are reviewed in

- P. Goddard, D. Olive, *Algebras, Lattices and Strings*, reprinted in previous reference pp. 210–255
- W. Lerche, A.N. Schellekens, N. Warner, Lattices and strings. Phys. Rep. **177**, 1 (1989)

The partition function for strings compactified on simply laced groups associated to level one Kač-Mood algebras were computed in:

- A. Redlich, H. Schnitzer, K. Tsokos, Bose-Fermi equivalence on the two-dimensional torus for simply laced groups. Nucl. Phys. **B289**, 397 (1987)

The generalization of the Frenkel-Kač construction to nonsimply laced groups was worked out in

- P. Goddard, W. Nahm, D. Olive, A. Schwimmer, Vertex operators for nonsimply laced algebras. Comm. Math. Phys. **107**, 179 (1986)

Chapter 12
Conformal Field Theory III: Superconformal Field Theory

Abstract In Chap. 4 we have demonstrated the usefulness of conformal field theory as a tool for the bosonic string. In the same way as conformal symmetry was a remnant of the reparametrization invariance of the bosonic string in conformal gauge, superconformal invariance is a remnant of local supersymmetry of the fermionic string in super-conformal gauge. This leads us to consider superconformal field theory. In many aspects our discussion of superconformal field theory parallels that of conformal field theory, and we will treat those rather briefly. Of special interest are $N = 2$ superconformal field theories, as they are related to space-time supersymmetry. These theories show some new features which we will present in more detail.

12.1 $N = 1$ Superconformal Symmetry

In Chaps. 7 and 8 we have met the superconformal algebra when we studied the fermionic string in superconformal gauge. As in the bosonic string, this is a particular realization of a much more general structure, that of superconformal field theories (SCFT). Recall that the generators of superconformal transformations are the conserved energy-momentum tensor $T(z)$ and the conserved world-sheet supercurrent $T_F(z)$.[1] The basic objects of superconformal field theory are chiral conformal (or primary) superfields. Their transformation properties under superconformal transformations are most convincingly motivated in superspace.[2] The

---

[1] As in Chap. 4 we will mainly only consider the holomorphic part of the theory. Note that whereas both sectors of the theory are conformally invariant, it is possible that only one of them, say the holomorphic one, exhibits superconformal invariance. This is for instance the case in the heterotic string theory. We should also mention that superconformal invariance can also appear in the internal sector of the bosonic string.

[2] We are considering $N = 1$ superspace. If one introduces several Grassmann odd coordinates $\theta_i, \bar\theta_i$ one arrives at extended superspaces.

coordinates of superspace are $\mathbf{z} = (z, \theta)$ and $\bar{\mathbf{z}} = (\bar{z}, \bar{\theta})$. The Grassmann parity of $z, \bar{z}$ is even and that of θ and $\bar{\theta}$ is odd and $\theta^2 = \bar{\theta}^2 = 0$. Due to this property any superfield, i.e. function on superspace, has a finite Taylor expansion in θ and $\bar{\theta}$:

$$\Phi(\mathbf{z}, \bar{\mathbf{z}}) = \phi_0(z, \bar{z}) + \theta\phi_1(z, \bar{z}) + \bar{\theta}\bar{\phi}_1(z, \bar{z}) + \theta\bar{\theta}\phi_2(z, \bar{z}), \qquad (12.1)$$

where the Grassmann parity of ϕ_0 and ϕ_2 is opposite to that of ϕ_1 and $\bar{\phi}_1$. The Grassmann parity of a superfield Φ is that of its lowest component ϕ_0. The (holomorphic) super-interval between two points $\mathbf{z}_1$ and $\mathbf{z}_2$ in superspace is defined as

$$z_{12} = z_1 - z_2 - \theta_1\theta_2 \qquad (12.2)$$

and similarly for the anti-holomorphic super interval. It is invariant under translations in superspace whose generators are the super derivatives

$$D = \partial_\theta + \theta\partial_z, \qquad \bar{D} = \partial_{\bar{\theta}} + \bar{\theta}\partial_{\bar{z}}. \qquad (12.3)$$

These indeed satisfy $(D_1 + D_2)z_{12} = (\bar{D}_1 + \bar{D}_2)\bar{z}_{12} = 0$ and have the additional properties

$$D^2 = \partial_z, \qquad \bar{D}^2 = \partial_{\bar{z}}, \qquad \{D, \bar{D}\} = 0. \qquad (12.4)$$

Thus, D can be considered as the square-root of $\partial = \partial_z$. There exists one more translationally invariant combination

$$\theta_{12} = \theta_1 - \theta_2. \qquad (12.5)$$

A chiral superfield is defined by the property $\bar{D}\Phi = 0$, which imposes restrictions on its components (12.1):

$$\Phi(\mathbf{z}) = \phi_0(z) + \theta\phi_1(z), \qquad (12.6)$$

i.e. a chiral superfield has only two instead of four independent component fields.

Recall that we could define conformal transformations by the relations $z \to z'(z)$ with $\partial_{\bar{z}} z' = 0$. In this case, the derivative transforms as $\partial = \frac{\partial z'}{\partial z}\partial'$. Analogously, superconformal transformations are holomorphic transformations on superspace such that the super-derivative transforms accordingly:

$$\mathbf{z} = (z, \theta) \to \mathbf{z}' = (z'(z, \theta), \theta'(z, \theta)), \qquad \text{with} \quad D = (D\theta')D'. \qquad (12.7)$$

The second relation is now non-trivial, as generally the super-derivative transforms as $D = (D\theta')D' + (Dz' - \theta'D\theta')(D')^2$. This leads to the condition $Dz' - \theta'D\theta' = 0$ whose solution can be written as

12.1 N = 1 Superconformal Symmetry

$$z' = f(z) + \frac{1}{2}\theta g(z)\epsilon(z),$$

$$\theta' = \frac{1}{2}\epsilon(z) + \theta g(z), \qquad g^2 = \partial f + \frac{1}{4}\epsilon\partial\epsilon, \qquad (12.8)$$

where f, g are Grassmann even and ϵ is Grassmann odd. For infinitesimal transformations $f(z) = z + \xi$ with ξ and ϵ small, (12.8) becomes

$$\delta z = \xi + \frac{1}{2}\theta\epsilon, \qquad \delta\theta = \frac{1}{2}\epsilon + \frac{1}{2}\theta\partial\xi. \qquad (12.9)$$

The definition of superconformal primary fields is the straightforward generalization of (4.6). A holomorphic superconformal primary field transforms under superconformal transformations as

$$\Phi(z) = (D\theta')^{2h}\,\Phi'(z'). \qquad (12.10)$$

The generalization to the non-holomorphic case is obvious. Under infinitesimal conformal transformations with parameter $\xi(z)$ the component fields transform as

$$\delta_\xi \phi_0(z) = -[h\partial\xi + \xi\partial]\,\phi_0(z),$$

$$\delta_\xi \phi_1(z) = -\left[\left(h + \frac{1}{2}\right)\partial\xi + \xi\partial\right]\phi_1(z), \qquad (12.11)$$

i.e. the fields ϕ_0 and ϕ_1 have conformal weights h and $(h + \frac{1}{2})$, respectively. Under two-dimensional supersymmetry transformations with parameter $\epsilon(z)$ the two components of the superfield transform into each other. For infinitesimal transformations,

$$\delta_\epsilon \phi_0(z) = -\frac{1}{2}\epsilon(z)\,\phi_1(z),$$

$$\delta_\epsilon \phi_1(z) = -\frac{1}{2}\epsilon(z)\,\partial\phi_0(z) - h\,\partial\epsilon(z)\,\phi_0(z). \qquad (12.12)$$

Similarly to the conformal transformations (4.17) and (4.20), one can introduce an infinite set of generators T_{F_ϵ} for these infinitesimal superconformal transformations $\delta_\epsilon \phi(z) = -[T_{F_\epsilon}, \phi(z)]$, where

$$T_{F_\epsilon} = \oint \frac{dz}{2\pi i}\,\epsilon(z)\,T_F(z) \qquad (12.13)$$

with T_F denoting the anticommuting generator of the superconformal algebra. Note that the supersymmetry transformation Eq. (12.12) is the "square root" of conformal transformations Eq. (12.11) in the sense that

$$[\delta_{\epsilon_1}, \delta_{\epsilon_2}] \phi_0(z) = [h \, \partial \xi + \xi \, \partial] \phi_0(z) \, ,$$

$$[\delta_{\epsilon_1}, \delta_{\epsilon_2}] \phi_1(z) = \left[\left(h + \frac{1}{2} \right) \partial \xi + \xi \, \partial \right] \phi_1(z) \, , \quad (12.14)$$

where $\xi = \frac{1}{2} \epsilon_1 \epsilon_2$. We can again translate the transformation rules for conformal superfields to operator product expansions. Using the techniques of Chap. 4 we easily find

$$T(z) \phi_0(w) = \frac{h \, \phi_0(w)}{(z-w)^2} + \frac{\partial \phi_0(w)}{z-w} + \ldots \, ,$$

$$T(z) \phi_1(w) = \frac{(h + \frac{1}{2}) \phi_1(w)}{(z-w)^2} + \frac{\partial \phi_1(w)}{z-w} + \ldots \, ,$$

$$T_F(z) \phi_0(w) = \frac{\frac{1}{2} \phi_1(w)}{z-w} + \ldots \, ,$$

$$T_F(z) \phi_1(w) = \frac{h \, \phi_0(w)}{(z-w)^2} + \frac{\frac{1}{2} \partial \phi_0(w)}{z-w} + \ldots \, . \quad (12.15)$$

The superconformal algebra (SCA) is specified by the operator products of the generators of superconformal transformations:

$$T(z) T(w) = \frac{\frac{3}{4} \hat{c}}{(z-w)^4} + \frac{2 T(w)}{(z-w)^2} + \frac{\partial T(w)}{z-w} + \ldots \, , \quad (12.16a)$$

$$T(z) T_F(w) = \frac{\frac{3}{2} T_F(w)}{(z-w)^2} + \frac{\partial T_F(w)}{z-w} + \ldots \, , \quad (12.16b)$$

$$T_F(z) T_F(w) = \frac{\frac{\hat{c}}{4}}{(z-w)^3} + \frac{\frac{1}{2} T(w)}{(z-w)} + \ldots \, . \quad (12.16c)$$

This is the $N = 1$ superconformal or super-Virasoro algebra. It is $N = 1$ because we have only one supercurrent. We will encounter extended superconformal algebras in the next section.

The operator products can be verified by commuting various combinations of conformal and superconformal transformations. Equation (12.16b) simply states that T_F is a primary field of the Virasoro algebra of weight $h = 3/2$. Equation (12.16c) reflects Eq. (12.14). Note that the central charges in (12.16a) and (12.16c) are related. The reason for this will be given below. A central extension in (12.16b) is forbidden by scale invariance and the Grassmann properties of T (even) and T_F (odd). We have normalized the central charge such that a free superfield $\mathscr{X}(z, \theta) = X(z) + \theta \psi(z)$, where $X(z)$ and $\psi(z)$ are free world-sheet bosons and fermions respectively, has $\hat{c} = 1$. (The central charge c in Eq. (4.29) and $\hat{c}$ are related by $c = \frac{3}{2} \hat{c}$.) We will explicitly verify the algebra for this case below.

12.1 $N = 1$ Superconformal Symmetry

Comparing Eqs. (12.15) and (12.16) we find that, apart from the central charge terms, T_F and T are the two components of an $h = 3/2$ Grassmann odd superfield

$$\mathcal{T}(\mathbf{z}) = T_F(z) + \theta T(z) \,. \tag{12.17}$$

Using this superfield the OPEs (12.16) can be compactly written in the form of a super operator product expansion

$$\mathcal{T}(\mathbf{z}_1)\,\mathcal{T}(\mathbf{z}_2) = \frac{\frac{1}{4}\hat{c}}{z_{12}^3} + \frac{\frac{3}{2}\theta_{12}\,\mathcal{T}(\mathbf{z}_2)}{z_{12}^2} + \frac{\frac{1}{2}D\mathcal{T}(\mathbf{z}_2)}{z_{12}} + \frac{\theta_{12}\,D^2\mathcal{T}(\mathbf{z}_2)}{z_{12}} + \dots, \tag{12.18}$$

where we used the super-intervals defined in (12.2) and (12.5). Using also the definition of a chiral superfield (12.6), the OPEs (12.15) for the definition of a superconformal primary field can be written as

$$\mathcal{T}(\mathbf{z}_1)\,\Phi(\mathbf{z}_2) = \frac{h\,\theta_{12}\,\Phi(\mathbf{z}_2)}{z_{12}^2} + \frac{\frac{1}{2}D\Phi(\mathbf{z}_2)}{z_{12}} + \frac{\theta_{12}\,D^2\Phi(\mathbf{z}_2)}{z_{12}} + \dots\,. \tag{12.19}$$

As in the bosonic case, we now expand T and T_F in modes and derive their algebra from the operator products. This will of course give the super-Virasoro algebra. The modes of $T(z)$ are defined as in Chap. 4. We expand T_F as

$$T_F(z) = \frac{1}{2} \sum_{r \in \mathbb{Z}+a} z^{-\frac{3}{2}-r}\, G_r \tag{12.20}$$

or

$$G_r = 2 \oint \frac{dz}{2\pi i}\, T_F(z)\, z^{r+\frac{1}{2}}. \tag{12.21}$$

We have introduced the parameter a to distinguish NS and R sectors. Integer modings ($a = 0$) correspond to the R sector and half-integer modings ($a = \frac{1}{2}$) to the NS sector.[3] From the reality of T_F we get the Hermiticity condition

$$G_r^\dagger = G_{-r}. \tag{12.22}$$

Notice that $T_F(z)$ is single-valued on the complex plane in the NS sector and double-valued in the R sector. In general, the fermionic (anti-commuting) components of NS superfields are single-valued on the plane whereas the fermionic components of R superfields are double-valued. That is $\phi_f^{NS}(e^{2\pi i}z) = +\phi_f^{NS}$ and

---

[3] In later chapters we will repeatedly encounter the situation where T_F is the sum of several terms, for instance a space-time part and an internal part. In all cases all parts of T_F must satisfy the same boundary conditions. This is because it is the total T_F which couples to the world-sheet gravitino.

$\phi_f^R(e^{2\pi i}z) = -\phi_f^R$. This is the reversed situation we had on the cylinder. The reason is that, when we map a field of dimension h from the cylinder to the complex plane, we have the Jacobian factor $(1/z)^h$ which changes the analyticity properties of world-sheet fields with non-integer conformal weight. This discussion especially applies to the world-sheet fermions $\psi(z)$.

Using the contour deformation trick of Chap. 4, we easily show that the operator algebra Eq. (12.16) is equivalent to (cf. (8.10) on page 197)

$$[L_m, L_n] = (m-n)L_{m+n} + \frac{\hat{c}}{8}(m^3 - m)\delta_{m+n},$$

$$[L_m, G_r] = \left(\frac{1}{2}m - r\right)G_{m+r},$$

$$\{G_r, G_s\} = 2L_{r+s} + \frac{\hat{c}}{2}\left(r^2 - \frac{1}{4}\right)\delta_{r+s}. \quad (12.23)$$

Note that, since T_F is an anti-commuting field, the operator product $T_F(z)T_F(w)$ leads to an anti-commutator of the modes. Only for $a = \frac{1}{2}$ (i.e. in the NS sector) exists a finite dimensional subalgebra, generated by $L_0, L_{\pm 1}$ and $G_{\pm \frac{1}{2}}$. This is the super-algebra $osp(1|2)$. As in the $N=0$ case discussed in Chap. 4 it leads to global Ward identities.[4] The Jacobi identity $[\{G_r, G_s\}, L_n] + \{[L_n, G_r], G_s\} - \{[G_s, L_n], G_r\} = 0$ implies that the central charge parameters $\hat{c}$ in Eqs. (12.16a) and (12.16c) have to be the same.

We can now also work out the commutation relations of the L_n and G_r with the modes of the primary fields. Define

$$\phi_0(z) = \sum z^{-n-h}\phi_{0,n},$$

$$\phi_1(z) = \sum z^{-n-h-\frac{1}{2}}\phi_{1,n}. \quad (12.24)$$

Fields with integer conformal weight always have integer mode numbers while fields with half-integer weights have integer mode numbers, if they are in the

---

[4]The classical orthosymplectic super-Lie-algebras $osp(m|n)$ with $n = 2p$ are matrix algebras defined by $MG + GM^{st} = 0$ where $G = \begin{pmatrix} 1_m & 0 \\ 0 & J_n \end{pmatrix}$ with $J_n = \begin{pmatrix} 0 & 1 \\ -1 & 0 \end{pmatrix}$ is the invariant metric. A general element X has the form $M = \begin{pmatrix} A & B \\ C & D \end{pmatrix}$ of which the $m \times m$ matrix A and the $n \times n$ matrix D are associated with the bosonic generators while B and C are associated with the fermionic generators. The supertranspose of X is defined as $X^{st} = \begin{pmatrix} A^T & C^T \\ -B^T & D^T \end{pmatrix}$. The maximal bosonic subalgebra is $o(m) \times sp(n)$. $osp(m|n)$ has $\frac{1}{2}[(m+n)^2 + n - m]$ generators of which mn are fermionic. An explicit matrix representation of $osp(1|2)$ is $L_0 = \frac{1}{2}\begin{pmatrix} 0 & 0 & 0 \\ 0 & -1 & 0 \\ 0 & 0 & 1 \end{pmatrix}$, $G_{-1/2} = \begin{pmatrix} 0 & 1 & 0 \\ 0 & 0 & 0 \\ 1 & 0 & 0 \end{pmatrix}$, $G_{+1/2} = G^{st}_{-1/2}$, $L_{\pm 1} = (G_{\pm 1/2})^2$.

12.1 $N = 1$ Superconformal Symmetry

R sector, and half-integer mode numbers, if they are in the NS sector. We find from (12.16)

$$[L_m, \phi_{0,n}] = (m(h-1) - n) \phi_{0,m+n},$$
$$[L_m, \phi_{1,n}] = \left(m\left(h - \frac{1}{2}\right) - n\right) \phi_{1,m+n},$$
$$[\epsilon G_r, \phi_{0,n}] = \epsilon \phi_{1,r+n},$$
$$[\epsilon G_r, \phi_{1,n}] = \epsilon ((2h-1)r - n) \phi_{0,r+n}, \qquad (12.25)$$

where we have introduced a constant anti-commuting parameter ϵ to make ϵG_r a commuting quantity. If a superfield satisfies these relations only for the generators of the $osp(1|2)$ subalgebra, it is called a superconformal quasi-primary field. Analogous to the bosonic CFT case, for such fields one can derive the general form of the holomorphic two- and three-point functions. For the two-point function, in addition to invariance under global conformal transformations, to be also super-translational invariant, it has to be of the form

$$\langle \Phi_i(z_1) \Phi_j(z_2) \rangle = \frac{d_{ij+}}{z_{12}^{2h_i}} \delta_{h_i, h_j}, \qquad (12.26)$$

where d_{ij+} is defined as in (4.85). For three points in superspace, one finds that the combination

$$\theta_{123} = \theta_1 z_{23} + \theta_2 z_{31} + \theta_3 z_{12} + \theta_1 \theta_2 \theta_3 \qquad (12.27)$$

is super-translational invariant, i.e. it satisfies $(D_1 + D_2 + D_3)\theta_{123} = 0$. For the three-point function one has to distinguish the two cases

$$\langle \Phi_i(z_1) \Phi_j(z_2) \Phi_k(z_3) \rangle = \begin{cases} \frac{C_{ijk}}{z_{12}^{h_i+h_j-h_k} z_{13}^{h_i+h_k-h_j} z_{23}^{h_j+h_k-h_i}} & \sum_i h_i \in \mathbb{Z} \\ \frac{C_{ijk} \theta_{123}}{z_{12}^{h_i+h_j-h_k+\frac{1}{2}} z_{13}^{h_i+h_k-h_j+\frac{1}{2}} z_{23}^{h_j+h_k-h_i+\frac{1}{2}}} & \sum_i h_i \in \mathbb{Z} + \frac{1}{2}. \end{cases}$$
(12.28)

The superspace formalism is a powerful tool to keep track of the relations among the component fields imposed by supersymmetry.

We now turn to the Hilbert space of the superconformal field theory. From the commutation relations

$$[L_0, L_m] = -m L_m,$$
$$[L_0, G_r] = -r G_r, \qquad (12.29)$$

we conclude that $L_m, G_m, m > 0$ are lowering operators. Ground states are the highest weight states $|h\rangle$ of the superconformal algebra; they are annihilated by all the lowering operators and have conformal weight h:

$$L_0|h\rangle = h|h\rangle,$$
$$L_m|h\rangle = 0 \quad \text{for } m > 0,$$
$$G_r|h\rangle = 0 \quad \text{for } r > 0. \qquad (12.30)$$

We will treat the action of G_0 in the R sector below.

We have already seen in Chap. 4 that unitarity requires $\hat{c} \geq 0$ and $h \geq 0$. This can be refined for the superconformal case by considering $(r > 0)$

$$\langle h|G_r G_{-r}|h\rangle = \langle h|\{G_r, G_{-r}\}|h\rangle$$
$$= 2\langle h|L_0|h\rangle + \frac{\hat{c}}{2}\left(r^2 - \frac{1}{4}\right)\langle h|h\rangle$$
$$= \left[2h + \frac{\hat{c}}{2}\left(r^2 - \frac{1}{4}\right)\right]\langle h|h\rangle \geq 0, \qquad (12.31)$$

from which we find

$$h \geq 0 \quad \text{(NS)}, \qquad (12.32a)$$
$$h \geq \frac{\hat{c}}{16} \quad \text{(R)}. \qquad (12.32b)$$

The result for the R sector follows from $G_0^2 = L_0 + \frac{\hat{c}}{4}(r^2 - \frac{1}{4})$. For $\hat{c} \geq 1$ these are indeed the only restrictions imposed by unitarity (cf. $c \geq 1$ in the conformal case). For $\hat{c} < 1$ there is again a discrete set of allowed $\hat{c}$ and h values, namely

$$\hat{c} = 1 - \frac{8}{m(m+2)},$$
$$h_{p,q} = \frac{[(m+2)p - mq]^2 - 4}{8m(m+2)} + \frac{1}{32}(1 - (-1)^{p-q})$$
$$m = 3, 4, \ldots, \quad 1 \leq p < m,\ 1 \leq q < m+2, \qquad (12.33)$$

where $p - q \in 2\mathbb{Z}$ in the NS sector and $p - q \in 2\mathbb{Z} + 1$ in the R sector.

We will now look at the two sectors separately starting with the NS sector. In analogy with the conformal case, we define the $OSp(1|2)$ invariant in- and out-vacua $|0\rangle$ and $\langle 0|$ to be the states annihilated by the generators of $OSp(1|2)$. Clearly this vacuum is in the NS sector as it satisfies $L_0|0\rangle = 0$. Regularity of T and T_F at the origin and infinity requires

12.1 $N = 1$ Superconformal Symmetry

$$L_n|0\rangle = 0 \quad n \geq -1, \qquad \langle 0|L_n = 0 \quad n \leq 1,$$
$$G_r|0\rangle = 0 \quad r \geq -\frac{1}{2}, \qquad \langle 0|G_r = 0 \quad r \leq \frac{1}{2}. \qquad (12.34)$$

The correspondence between highest weight states and conformal superfields $\Phi(z,\theta)$ of conformal weight h is made as follows:

$$\phi_0(0)|0\rangle = |h\rangle,$$
$$\phi_1(0)|0\rangle = (\hat{G}_{-\frac{1}{2}}\phi_0)(0)|h\rangle = G_{-\frac{1}{2}}|h\rangle, \qquad (12.35)$$

where the second equation follows from $[\epsilon G_{-\frac{1}{2}}, \phi_0(z)] = \epsilon\phi_1(z)$ and we have defined (cf. Chap. 4)

$$\hat{G}_r\phi_0(z) = 2\oint \frac{dw}{2\pi i} T_F(w)\phi_0(z)(w-z)^{r+\frac{1}{2}}. \qquad (12.36)$$

It is straightforward to show that $|h\rangle$ satisfies the highest weight conditions (12.30) and for $\phi_1(0)|0\rangle$ we find

$$L_0\left(G_{-\frac{1}{2}}|h\rangle\right) = \left(h + \frac{1}{2}\right)\left(G_{-\frac{1}{2}}|h\rangle\right), \qquad L_m\left(G_{-\frac{1}{2}}|h\rangle\right) = 0 \quad m > 0,$$
$$G_{\frac{1}{2}}\left(G_{-\frac{1}{2}}|h\rangle\right) = 2h|h\rangle, \qquad G_m\left(G_{-\frac{1}{2}}|h\rangle\right) = 0 \quad m > \frac{1}{2}. \qquad (12.37)$$

Note that the relation

$$G_{-\frac{1}{2}}^2 = L_{-1} \qquad (12.38)$$

is the global supersymmetry algebra on the complex plane. It is the global algebra since $G_{-\frac{1}{2}} = 2\oint \frac{dz}{2\pi i} T_F(z)$ and $L_{-1} = \oint \frac{dz}{2\pi i} T(z)$ and it is the supersymmetry algebra since L_{-1} is the translation operator on the plane (cf. Chap. 4). Since the NS vacuum satisfies $G_{-\frac{1}{2}}|0\rangle = 0$, NS supersymmetry is unbroken.

In the R sector we can define a global supersymmetry charge on the cylinder; it is simply G_0 as can be most easily seen by referring to Eq. (7.66). The global supersymmetry algebra on the cylinder is

$$G_0^2 = L_0 - \frac{\hat{c}}{16}. \qquad (12.39)$$

Recall the transformation law of the energy-momentum tensor from the cylinder to the plane (cf. Eq. (4.47))

$$(L_0)_{\text{cyl.}} = (L_0)_{\text{plane}} - \frac{\hat{c}}{16}, \qquad (12.40)$$

where $(L_0)_{\text{cyl.}}$ is the translation operator on the cylinder. Note that this shift in L_0 in the R sector is the one described in Chap. 8, which was necessary to bring the NS and R super-Virasoro algebras into identical form. This is automatic here as the operator products in Eq. (12.16) are the same for the two sectors.

Clearly, supersymmetry in the R-sector can only be unbroken if there exists a ground state which satisfies $G_0|h\rangle = 0$, implying $L_0|h\rangle = \frac{\hat{c}}{16}|h\rangle$.[5] This is in agreement with Eq. (12.32b). Since $[G_0, L_0] = 0$, highest weight states come in pairs

$$|h^+\rangle \quad \text{and} \quad |h^-\rangle \equiv G_0|h^+\rangle. \qquad (12.41)$$

It is easy to see that if $|h^+\rangle$ is a highest weight state then so is $|h^-\rangle$. If $|h^+\rangle$ is the R ground state with $h = \frac{\hat{c}}{16}$, then $|h^-\rangle$ is a null state: $\langle h^-|h^-\rangle = \langle h^+|G_0^2|h^+\rangle = 0$. It decouples as it is perpendicular to all descendants of $|h^+\rangle$. So we can formally set $|h^-\rangle = 0$ in the case of unbroken supersymmetry.

Since the Virasoro algebra is a subalgebra of the R-algebra, the above highest weight states must be created from the vacuum by ordinary conformal fields, the so-called spin fields $S^\pm(z)$; i.e.

$$|h^+\rangle = S^+(0)|0\rangle \quad \text{and} \quad |h^-\rangle = S^-(0)|0\rangle. \qquad (12.42)$$

This, together with Eq. (12.41) implies that $\hat{G}_0 S^+(z) = S^-(z)$ from which we derive

$$T_F(w) S^+(z) = \frac{\frac{1}{2}S^-(z)}{(w-z)^{\frac{3}{2}}} + \text{less singular}. \qquad (12.43)$$

Likewise $\hat{G}_0 S^-(z) = (h - \frac{\hat{c}}{16})S^+(z)$ leads to

$$T_F(w) S^-(z) = \frac{1}{2}\left(h - \frac{\hat{c}}{16}\right)\frac{1}{(w-z)^{\frac{3}{2}}}S^+(z) + \text{less singular}. \qquad (12.44)$$

We see that the spin fields introduce branch cuts into the operator algebra. In addition, since they transform the NS ground state into the R ground state, they interpolate between the two sectors. This means that if we take a NS fermion and carry it around the spin field it feels the branch cut and changes sign; i.e.

$$\phi_f^{\text{NS}}(e^{2\pi i} z) S^\pm(0) = -\phi_f^{\text{NS}}(z) S^\pm(0) \qquad (12.45)$$

---

[5] Note that in the $N = 1$ discrete series R supersymmetry is unbroken for even m only.

12.1 $N = 1$ Superconformal Symmetry

which means that $\phi_f^{NS}(z) \, S^\pm(0)$ has an expansion in half-integer powers of z so that $z^{\frac{1}{2}}\phi_f^{NS}(z) \, S^\pm(0)$ is single valued on the plane. Now we define the operators $\phi_{f,n}^R$ via

$$\phi_{f,n}^R \, S^\pm(z) = \oint \frac{dw}{2\pi i} \phi_f^{NS}(w) \, (w-z)^{n+h-1} \, S^\pm(z) \,, \quad n+h \in \mathbb{Z} + \frac{1}{2}, \quad (12.46)$$

where h is the conformal weight of ϕ_f^{NS}. The $\phi_{f,n}^R$ are the modes of the fermionic component of an R superfield $\phi_f^R(w)$. Thus one should not think of the NS and R superfields as two separate sets of superfields, but rather as one superfield whose fermionic component gets modified in the presence of a spin field. The superfields themselves act diagonally on states in the two sectors of the Hilbert space, i.e.

$$\begin{pmatrix} |NS\rangle' \\ |R\rangle' \end{pmatrix} = \begin{pmatrix} \phi & 0 \\ 0 & \phi \end{pmatrix} \begin{pmatrix} |NS\rangle \\ |R\rangle \end{pmatrix}, \quad (12.47)$$

whereas spin fields act off-diagonally

$$\begin{pmatrix} |NS\rangle' \\ |R\rangle' \end{pmatrix} = \begin{pmatrix} 0 & S \\ S & 0 \end{pmatrix} \begin{pmatrix} |NS\rangle \\ |R\rangle \end{pmatrix}. \quad (12.48)$$

In both sectors the descendants of a given ground state are obtained by acting with the lowering operators of the superconformal algebra. The n-th level of the superconformal Verma module is spanned by the following vectors with L_0 eigenvalue $h + n$

$$G_{-r_1} G_{-r_2} \ldots L_{-m_1} L_{-m_2} \ldots |h\rangle, \quad (12.49)$$

where

$$0 < r_1 < r_2 \ldots \quad \text{and} \quad 0 < m_1 \leq m_2 \ldots,$$

$$n = \sum r_i + \sum m_i, \quad r_i \in \mathbb{Z} + \frac{1}{2} \quad (\text{NS}), \quad r_i \in \mathbb{Z} \quad (\text{R}).$$

The condition on the fermionic oscillators takes into account that $G_m^2 = L_{2m}$.

Let us now return to string theory. The world-sheet bosons and fermions form two-dimensional superfields of the type (12.1). Explicitly

$$\mathscr{X}^\mu(z,\bar{z}) = X^\mu + i\theta\psi^\mu + i\bar{\theta}\bar{\psi}^\mu + \theta\bar{\theta}F^\mu \quad (12.50)$$

with action[6]

---

[6]This is the action of Chap. 7, after going to superconformal gauge and performing a Wick rotation to an Euclidean world-sheet, i.e. replacing $\tau \to -i\tau$ and mapping from the Euclidean cylinder

$$S = \frac{1}{4\pi} \int d^2z\, d^2\theta\, \bar{D}\mathcal{X}^\mu D\mathcal{X}_\mu$$
$$= \frac{1}{4\pi} \int d^2z \left(\partial X^\mu \bar{\partial} X_\mu + \psi^\mu \bar{\partial}\psi_\mu + \bar{\psi}^\mu \partial\bar{\psi}_\mu + F^\mu F_\mu \right). \quad (12.51)$$

The equation of motion for the auxiliary scalar is $F^\mu = 0$ and we will set it to zero.

The mode expansion of $\psi^\mu(z)$ follows from the expansion on the (Euclidean) cylinder given in Chap. 7. The Jacobian from the map to the complex plane contributes a factor $z^{-1/2}$ and one obtains

$$\psi^\mu(z) = \sum_{r \in \mathbb{Z}+\phi} b_r^\mu\, z^{-r-\frac{1}{2}} \qquad \phi = \begin{cases} 0 & \text{(R)} \\ \frac{1}{2} & \text{(NS)} \end{cases} \quad (12.52)$$

with $(b_r^\mu)^\dagger = b_{-r}^\mu$.

In the NS sector the modes of ψ are half-integer and $\psi^\mu(e^{2\pi i}z) = \psi^\mu(z)$. The basic operator product is

$$\psi^\mu(z)\,\psi^\nu(w) = \frac{\eta^{\mu\nu}}{z-w} + \text{regular} \quad (12.53)$$

which is equivalent to

$$\{b_r^\mu, b_s^\nu\} = \delta_{r+s}\, \eta^{\mu\nu}. \quad (12.54)$$

We can also calculate the propagator. Using $b_r^\mu|0\rangle = 0$ for $r \geq 1/2$, we get

$$\langle \psi^\mu(z)\,\psi^\nu(w)\rangle_{\text{NS}} = \sum_{m,n \in \mathbb{Z}} z^{-n} w^{-m} \langle 0|b_{n-\frac{1}{2}}^\mu b_{m-\frac{1}{2}}^\nu|0\rangle$$
$$= \sum_{n \geq 0} \frac{1}{z}\left(\frac{w}{z}\right)^n \eta^{\mu\nu}$$
$$= \frac{\eta^{\mu\nu}}{z-w} \qquad \text{for } |z| > |w|. \quad (12.55)$$

The propagator can also be derived from the free field action Eq. (12.51) using standard field theory arguments. We should note that this is the propagator on

---

to the plane. For the fermions we replace $\rho^0 \to -i\rho^0$ and $\bar{\psi} \to i\psi^T \rho^0$. In contrast to previous chapters we use dimensionless scalar fields X^μ. This amounts to setting $\alpha' = 2$. To restore α' one replaces $X^\mu \to \sqrt{\frac{2}{\alpha'}} X^\mu$ and $k^\mu \to \sqrt{\frac{\alpha'}{2}} k^\mu$. The integration over the Grassmann coordinates is $\int d^2\theta\, \bar{\theta}\theta = 1$.

12.1 $N = 1$ Superconformal Symmetry

the Riemann sphere. On higher genus Riemann surfaces the propagator is more complicated due to the global structure of the surfaces. The short distance limit will however be the same as in Eq. (12.55). Using the relation (cf. (4.99) on page 85)

$$\frac{1}{2\pi} \partial_{\bar{z}} \left(\frac{1}{z-w} \right) = \delta^2(z-w), \qquad (12.56)$$

we find that the propagator satisfies

$$\partial_{\bar{z}} \langle \psi^\mu(z) \psi^\nu(w) \rangle_{\text{NS}} = 2\pi \, \delta^2(z-w) \, \eta^{\mu\nu} . \qquad (12.57)$$

In the R sector the fields $\psi^\mu(z)$ have integer modings (cf. also Eq. (12.46)) and $\psi^\mu(e^{2\pi i}z) = -\psi^\mu(z)$. One readily shows that they also satisfy the anti-commutation relations Eq. (12.54). This already follows from the fact that the short distance expansion for NS and R fermions should be the same, as the branch cut in the R case cannot be felt locally. Of particular interest is the relation

$$\{b_0^\mu, b_0^\nu\} = \eta^{\mu\nu} , \qquad (12.58)$$

which is just the Dirac algebra and allows us to identify the ψ^μ zero modes with Dirac matrices. Since $[L_0, b_0^\mu] = 0$, the b_0^μ act on the R ground state which is therefore degenerate. The spin fields $S_\alpha(z)$ that create them are spinors of $SO(9, 1)$ with conformal weight $h = \frac{\hat{c}}{16} = \frac{5}{8}$ ($\hat{c} = d = 10$, see below). The fact that spinors of $SO(9, 1)$ have conformal weight $\frac{5}{8}$ can be seen by going to the Wick-rotated Lorentz group $SO(10)$ and bosonize them (cf. Chap. 11, Eq. (11.102) on page 342 and the discussion following it). The R ground states are then labeled by a spinor index of $SO(9, 1)$:

$$|\alpha\rangle = S_\alpha(0)|0\rangle \qquad (12.59)$$

and the b_0^μ act as

$$b_0^\mu |\alpha\rangle = \frac{1}{\sqrt{2}} (\Gamma^\mu)_\alpha{}^\beta |\beta\rangle , \qquad (12.60)$$

where Γ^μ are the $SO(9, 1)$ Dirac matrices which satisfy $\{\Gamma^\mu, \Gamma^\nu\} = 2\eta^{\mu\nu}$.

We can now also calculate the propagator of ψ^μ in the R sector. Using $b_n^\mu |\alpha\rangle = 0$ for $n > 0$ and Eq. (12.58), we find

$$\langle \psi^\mu(z) \psi^\nu(w) \rangle_{\text{R}} = \sum_{m,n \in \mathbb{Z}} z^{-n-\frac{1}{2}} w^{-m-\frac{1}{2}} \langle b_n^\mu b_m^\nu \rangle_{\text{R}}$$

$$= \frac{1}{\sqrt{zw}} \left\{ \sum_{n>0} z^{-n} w^n \langle b_n^\mu b_{-n}^\nu \rangle_{\text{R}} + \langle b_0^\mu b_0^\nu \rangle_{\text{R}} \right\}$$

$$= \frac{1}{\sqrt{zw}} \left\{ \sum_{n=1}^{\infty} \left(\frac{w}{z}\right)^n + \frac{1}{2} \right\} \eta^{\mu\nu}$$

$$= \frac{1}{2} \frac{1}{z-w} \left(\sqrt{\frac{z}{w}} + \sqrt{\frac{w}{z}} \right) \eta^{\mu\nu}, \qquad |z| > |w|. \tag{12.61}$$

Note that this is just the four-point function $\langle 0|S_\alpha(\infty) \psi^\mu(z) \psi^\nu(w) S_\alpha(0)|0\rangle$ (no sum over α). The propagators in the NS and the R sectors have the same short distance behavior, as they should.

We still have to discuss the question of locality. As we have seen, the spin fields introduce branch cuts into the theory. A suitable string theory however has to have local correlation functions to give well defined S-matrix elements. To arrive at a local string theory one performs a generalized GSO projection. This is a consistent truncation of the spectrum such that the resulting superconformal field theory is local. In fact, not only is the truncation consistent but also required by modular invariance.

The energy-momentum tensor and supercurrent for the fermionic string are

$$T(z) = -\frac{1}{2} \partial X^\mu \, \partial X_\mu(z) - \frac{1}{2} \psi^\mu \, \partial \psi_\mu(z),$$

$$T_F(z) = \frac{i}{2} \psi^\mu \, \partial X_\mu(z). \tag{12.62}$$

They are the Noether currents associated with the transformations Eq. (12.11) and (12.12), i.e.

$$\delta_\xi S = \frac{1}{2\pi} \int d^2z \, T \, \bar{\partial}\xi,$$

$$\delta_\epsilon S = \frac{1}{2\pi} \int d^2z \, T_F \, \bar{\partial}\epsilon. \tag{12.63}$$

X and ψ are free fields and we can use Wick's theorem to evaluate operator products of composite operators such as T and T_F. For notational convenience, we still use dimensionless fields X^μ, or, equivalently, we set $\alpha' = 2$ and we have dropped normal ordering symbols. One straightforwardly verifies the algebra Eq. (12.16) with $\hat{c} = \frac{2}{3}c = d$, where d is the range of the index μ. Moreover, one can confirm the transformation rules for X^μ and ψ^μ under holomorphic transformations, i.e. (12.15) with $h = 0$.

The ghost system of the fermionic string is another example of an $N = 1$ superconformal field theory. It consists of the (anti-commuting) conformal ghosts b and c and the (commuting) superconformal ghosts β and γ. Together they form two conformal superfields $B = \beta + \theta b$ and $C = c + \theta\gamma$ with conformal weights $h = 3/2$ and $h = -1$ respectively. The action is (cf. Chaps. 3 and 8)

12.1 $N = 1$ Superconformal Symmetry

$$S = \frac{1}{2\pi} \int d^2z\, d^2\theta\; B\overline{D}C = \frac{1}{2\pi}\int d^2z\,(b\bar{\partial}c + \beta\bar{\partial}\gamma)\,. \tag{12.64}$$

We can generalize the ghost system by considering superfields B and C with conformal weights $\lambda - \frac{1}{2}$ and $(1 - \lambda)$ $(\lambda \in \mathbb{Z})$ respectively. The action is still given by Eq. (12.64) and the superconformal transformations of the components fields follow from (12.11) and (12.12). (We will consider these generalized b, c systems in more detail in Chap. 13.) The energy momentum tensor and supercurrent are calculated as in (12.63); one finds

$$T^\lambda = -\lambda b \partial c + (1-\lambda)(\partial b)c - \left(\lambda - \frac{1}{2}\right)\beta \partial \gamma + \left(\frac{3}{2} - \lambda\right)(\partial \beta)\gamma\,,$$

$$T_F^\lambda = \frac{1}{2}b\gamma + (1-\lambda)(\partial\beta)c - \left(\lambda - \frac{1}{2}\right)\beta \partial c\,, \tag{12.65}$$

which can be written as a superfield

$$\mathcal{T}^\lambda = T_F^\lambda + \theta T^\lambda = -\left(\lambda - \frac{1}{2}\right) B\partial_z C + (1-\lambda)(\partial_z B)C + \frac{1}{2}(DB)(DC)\,. \tag{12.66}$$

Using the operator products

$$c(z)\,b(w) = \gamma(z)\,\beta(w) = \frac{1}{z-w}\,, \tag{12.67}$$

it is straightforward to show that the above system generates the $N = 1$ superconformal algebra (12.16) with central charge $c = -c^\lambda + c^{\lambda - 1/2}$. Here $c^\lambda = 12\lambda^2 - 12\lambda + 2$. The first contribution, $-c^\lambda$, is that of the anti-commuting b, c system and the second, $c^{\lambda-1/2}$, that of the commuting β, γ system. For the fermionic string ($\lambda = 2$) we find the ghost contribution to the anomaly $c^{\text{ghost}} = -26 + 11 = -15$. The matter fields come in chiral multiplets of the $N = 1$ superconformal symmetry. Each of them contributes $c = 3/2$ to the conformal anomaly. This means that we need ten of them for anomaly cancellation. Hence $d = 10$ as the critical dimension of the fermionic string. It is however by no means necessary to represent the $\hat{c} = 10$ algebra in terms of free superfields (12.50). For a d-dimensional space-time interpretation it suffices to represent d of them as free superfields and have a $\hat{c} = 10 - d$ internal $N = 1$ superconformal theory (cf. Chap. 15).

One may wonder whether extended superconformal algebras play a role in string theory. Let us begin with the case $N = 2$. The $N = 2$ gravity multiplet is $(h_{\alpha\beta}, \chi_\alpha, A_\alpha)$ where besides the graviton we have a gravitino which is, in contrast to the $N = 1$ case, a complex Dirac fermion, and a $U(1)$ gauge current. Since we have now gauge fields of spin 2,3/2 and 1, Faddeev-Popov quantization will give b, c systems with $\lambda = 2$ and $\lambda = 1$ and two β, γ systems with $\lambda = 3/2$. We get two of the latter since a Dirac spinor is equivalent to two Majorana spinors. Their combined contribution to the conformal anomaly is $c^{\text{ghost}} = -26 + 2 \times 11 - 2 = -6$. This has to be canceled by the matter fields. The $N = 2$ matter multiplets are (X, ψ) where X is a complex scalar and ψ a spin $1/2$ Dirac-Weyl fermion, which is equivalent to two real scalars and two Majorana-Weyl fermions, thus contributing $c = 3$ to the conformal anomaly. We then conclude that the critical (real) dimension of the theory with local $N = 2$ superconformal invariance is $d = 4$. Since we have two complex bosons the signature is either Euclidean or $(++--)$. String theory with local $N = 2$ supersymmetry has been studied in detail. It has only a finite number of excitations and all (perturbative) scattering amplitudes beyond three point functions vanish. It describes self-dual gravity and is not rich enough to be of interest as a model for particle physics and quantum gravity.

The situation becomes even worse when we consider the $N = 4$ case. The $N = 4$ gravity multiplet is $(h_{\alpha\beta}, \chi_\alpha^i, A_{\alpha j}^i)$ where i is an $SU(2)$ index. The four gravitini form a complex doublet of $SU(2)$ and the $SU(2)$ currents transform in the adjoint. The ghost contribution to the anomaly is therefore $c^{\text{ghost}} = -26 + 4 \times 11 - 3 \times 2 = 12$. $N = 4$ matter multiplets $(4X, \psi)$ contain four real scalars and a complex doublet of spinors which gives $c = 6$ contribution to the conformal anomaly. This leads to a (real) critical dimension $d = -8$. This is clearly unacceptable for our purposes.

Nevertheless, extended superconformal algebras do play an important role in string theory, however only as global symmetries, i.e. not as residual symmetries of world-sheet theories with extended local supersymmetry (supergravity). In particular there are no ghost systems which result from gauge fixing these extended local symmetries which would affect the value of the critical dimension. The maximal gauged supersymmetry is $N = 1$. There exists a deep connection between space-time symmetries and world-sheet symmetries. In particular, as will be shown in Chap. 15, $\mathcal{N} = 1$ space-time supersymmetry in four dimensions requires (global) $N = 2$ superconformal invariance on the world-sheet.[7]

12.2 $N = 2$ Superconformal Symmetry

As just mentioned, among the extended superconformal algebras, $N = 2$ plays an important role in string theory due to its relation to minimal space-time supersymmetry in the compactified theory. As we are going to discuss in this section, this is intimately related to the existence of the so-called spectral flow for $N = 2$ superconformal field theories. Further extensions exist, e.g. the $N = 4$ algebra is relevant for extended space-time supersymmetry of the heterotic string. There are, in fact, several versions of the $N = 4$ algebra, so-called long and short versions, but we will not discuss them. We concentrate on the most relevant case, $N = 2$. As we will

---

[7]More accurately, for the heterotic string one needs $N = (0, 2)$ two-dimensional world-sheet supersymmetry. Similarly, for orientifolds of type II superstring theory, one needs $N = (2, 2)$ supersymmetry on the world-sheet.

12.2 $N=2$ Superconformal Symmetry

see in detail, the $N=2$ algebra has two bosonic and two fermionic generators: on the bosonic side, besides the energy-momentum tensor there is a dimension 1 $U(1)$ current. The symmetry generated by it is called $\mathscr{R}$-symmetry. On the fermionic side there are two dimension 3/2 supercurrents with charges ± 1 under the $\mathscr{R}$-symmetry.

Similar to the $N=1$ case, there exists a superfield formulation. We present it only for one sector, say the holomorphic one. There are now two Grassmann variables, i.e. the coordinates of a point in (holomorphic) superspace are the triple $Z = (z, \theta, \bar{\theta})$. On this space, one defines the two super derivatives

$$D = \partial_\theta + \frac{1}{2}\bar{\theta}\partial_z, \qquad \bar{D} = \partial_{\bar{\theta}} + \frac{1}{2}\theta\partial_z \qquad (12.68)$$

which are nilpotent and satisfy $\{D, \bar{D}\} = \partial_z$. One finds the super-translationally invariant interval

$$Z_{12} = z_1 - z_2 - \frac{1}{2}(\theta_1\bar{\theta}_2 + \bar{\theta}_1\theta_2) \qquad (12.69)$$

of two points in superspace, i.e. $(D_1 + D_2)Z_{12} = (\bar{D}_1 + \bar{D}_2)Z_{12} = 0$. Two other translationally invariant intervals can be defined:

$$\theta_{12} = \theta_1 - \theta_2 \quad \text{and} \quad \bar{\theta}_{12} = \bar{\theta}_1 - \bar{\theta}_2 . \qquad (12.70)$$

$N=2$ superconformal transformations can be defined as generalizations of the $N=1$ case:

$$Z \to Z'(Z) \quad \text{with} \quad D = (D\theta')D' \quad \text{and} \quad \overline{D} = (\overline{D\bar{\theta}'})\overline{D}' . \qquad (12.71)$$

The most general solution of these conditions is

$$z' = f + \frac{1}{2}\theta\bar{\epsilon}g + \frac{1}{2}\bar{\theta}\epsilon\bar{g} + \frac{1}{4}\theta\bar{\theta}\partial(\epsilon\bar{\epsilon}),$$

$$\theta' = \epsilon + \theta g + \frac{1}{2}\theta\bar{\theta}\partial\epsilon, \qquad \partial f - g\bar{g} + \frac{1}{2}(\epsilon\partial\bar{\epsilon} + \bar{\epsilon}\partial\epsilon) = 0,$$

$$\bar{\theta}' = \bar{\epsilon} + \bar{\theta}\bar{g} - \frac{1}{2}\theta\bar{\theta}\partial\bar{\epsilon}. \qquad (12.72)$$

Here f, g are Grassmann even and $\epsilon, \bar{\epsilon}$ are Grassmann odd functions of z. For infinitesimal transformations $f = z + \xi$, $g = 1 + \frac{1}{2}\partial\xi - \alpha$, $\bar{g} = 1 + \frac{1}{2}\partial\xi + \alpha$ with $\xi, \alpha, \epsilon, \bar{\epsilon}$ small, (12.72) becomes

$$z' = z + \xi + \frac{1}{2}\theta\bar{\epsilon} + \frac{1}{2}\bar{\theta}\epsilon ,$$

$$\theta' = \theta + \epsilon + \frac{1}{2}\theta\partial\xi - \theta\alpha + \frac{1}{2}\theta\bar{\theta}\partial\epsilon ,$$

$$\bar{\theta}' = \bar{\theta} + \bar{\epsilon} + \frac{1}{2}\bar{\theta}\partial\xi + \bar{\theta}\alpha - \frac{1}{2}\theta\bar{\theta}\partial\bar{\epsilon}. \tag{12.73}$$

One of the bosonic functions, $\xi(z)$, parametrizes infinitesimal conformal transformations which are generated by the energy-momentum tensor $T(z)$, while the second, $\alpha(z)$, parametrizes infinitesimal $U(1)$ transformations under which θ and $\bar{\theta}$ carry charge $+1$ and -1. Its generator $J(z)$ has conformal weight $h = 1$. The two Grassmann odd functions $\epsilon(z)$ and $\bar{\epsilon}(z)$ with $U(1)$ charges $+1$ and -1 parametrize two supersymmetry transformations whose generators $T_F^-(z)$ and $T_F^+(z)$ have conformal dimension $h = 3/2$ and $U(1)$ charges -1 and $+1$ such that the combinations $\oint \epsilon T_F^-$ and $\oint \bar{\epsilon} T_F^+$ are neutral.

A holomorphic $N = 2$ superfield can be expanded into four component fields

$$\Phi(Z) = \phi_0(z) + \frac{1}{\sqrt{2}}(\theta\bar{\phi}_1(z) - \bar{\theta}\phi_1(z)) + \theta\bar{\theta}\phi_2(z). \tag{12.74}$$

The normalizations of the terms linear in $\theta, \bar{\theta}$ are chosen such that the operator product expansions of the superfields (cf. (12.79) and (12.81) below) have the familiar form when expressed in component fields. A superconformal primary field of conformal weight h and $U(1)$ charge q is defined to transform as (cf. (12.10))

$$\Phi(Z) = \left(D\theta'\right)^{h-\frac{q}{2}} \left(\overline{D\theta}'\right)^{h+\frac{q}{2}} \Phi'(Z') \tag{12.75}$$

under superconformal transformations. The weight and charge are those of its lowest component. It is straightforward, though somewhat tedious, to work out the infinitesimal transformation of its component fields. We will not present all the details. Considerable simplification occurs if its conformal weight and $U(1)$ charge are related as $|q| = 2h$. This leads to a shortening of the general supermultiplet to chiral ($q = 2h$) and anti-chiral ($q = -2h$) superfields.

A chiral superfield is a function of $y^+ = z + \frac{1}{2}\theta\bar{\theta}$ and θ (note that $\overline{D}y^+ = \overline{D}\theta = 0$):

$$\Phi^+(y^+, \theta) = \phi_0^+(z) + \theta\,\phi_1^+(z) + \frac{1}{2}\theta\bar{\theta}\partial\phi_0^+(z). \tag{12.76}$$

For an anti-chiral multiplet with $y^- = z - \frac{1}{2}\theta\bar{\theta}$,

$$\Phi^-(y^-, \bar{\theta}) = \phi_0^-(z) + \bar{\theta}\,\phi_1^-(z) - \frac{1}{2}\theta\bar{\theta}\partial\phi_0^-(z). \tag{12.77}$$

Chiral and anti-chiral primary superfields will play a prominent role in the discussion below.

12.2 $N = 2$ Superconformal Symmetry

A simplification also occurs for uncharged superfields. An example is the superfield whose components are the generators of the conformal, superconformal and $U(1)$ transformations:

$$\mathscr{T}(Z) = J(z) + \frac{1}{\sqrt{2}}(\theta \, T_F^-(z) - \bar{\theta} \, T_F^+(z)) + \theta\bar{\theta} \, T(z) \,. \tag{12.78}$$

Of course, this is not a primary field because of the superconformal anomaly. The generalization of the superfield OPE (12.18) for the $N = 1$ superconformal algebra to the $N = 2$ case is

$$\mathscr{T}(Z_1) \, \mathscr{T}(Z_2) = \frac{c/3}{Z_{12}^2} + \frac{\theta_{12}\bar{\theta}_{12}}{Z_{12}^2} \mathscr{T}(Z_2) + \frac{\theta_{12} D - \bar{\theta}_{12} \bar{D}}{Z_{12}} \mathscr{T}(Z_2)$$

$$+ \frac{\theta_{12}\bar{\theta}_{12}}{Z_{12}} \{D, \bar{D}\} \mathscr{T}(Z_2) + \ldots \,. \tag{12.79}$$

In terms of the superfield OPE, we can now reformulate the condition for a general superfield Φ of conformal dimension h and $U(1)$ charge q to be primary as

$$\mathscr{T}(Z_1) \, \Phi(Z_2) = \frac{h \, \theta_{12}\bar{\theta}_{12}}{Z_{12}^2} \Phi(Z_2) + \frac{\theta_{12} D - \bar{\theta}_{12} \bar{D}}{Z_{12}} \Phi(Z_2)$$

$$+ \frac{\theta_{12}\bar{\theta}_{12}}{Z_{12}} \{D, \bar{D}\}\Phi(Z_2) + \frac{q}{Z_{12}}\Phi(Z_2) + \ldots \,. \tag{12.80}$$

Expanding (12.79) in components we find the following operator algebra

$$T(z) T(w) = \frac{\frac{c}{2}}{(z-w)^4} + \frac{2\, T(w)}{(z-w)^2} + \frac{\partial T(w)}{z-w} + \ldots,$$

$$T(z) T_F^\pm(w) = \frac{\frac{3}{2} T_F^\pm(w)}{(z-w)^2} + \frac{\partial T_F^\pm(w)}{z-w} + \ldots,$$

$$T(z) J(w) = \frac{J(w)}{(z-w)^2} + \frac{\partial J(w)}{z-w} + \ldots,$$

$$J(z) J(w) = \frac{\frac{c}{3}}{(z-w)^2} + \ldots,$$

$$J(z) T_F^\pm(w) = \pm \frac{T_F^\pm(w)}{(z-w)} + \ldots,$$

$$T_F^+(z) T_F^-(w) = \frac{\frac{2c}{3}}{(z-w)^3} + \frac{2\, J(w)}{(z-w)^2} + \frac{2\, T(w) + \partial J(w)}{z-w} + \ldots,$$

$$T_F^\pm(z) T_F^\pm(w) = \text{finite} \,. \tag{12.81}$$

It is easy to see that T and $T_F = \frac{1}{2\sqrt{2}}(T_F^+ + T_F^-)$ form an $N = 1$ superconformal sub-algebra.

The superconformal transformations of (anti)chiral primary fields can be expressed as the following operator products with the generators:

$$T(z)\phi_0^\pm(w) = \frac{h\phi_0^\pm(w)}{(z-w)^2} + \frac{\partial \phi_0^\pm(w)}{z-w} + \dots,$$

$$T(z)\phi_1^\pm(w) = \frac{(h+\frac{1}{2})\phi_1^\pm(w)}{(z-w)^2} + \frac{\partial \phi_1^\pm(w)}{z-w} + \dots,$$

$$J(z)\phi_0^\pm(w) = \pm \frac{2h\phi_0^\pm(w)}{z-w} + \dots,$$

$$J(z)\phi_1^\pm(w) = \pm \frac{(2h-1)\phi_1^\pm(w)}{z-w} + \dots,$$

$$T_F^\pm(z)\phi_0^\pm(w) = 0 + \dots,$$

$$T_F^\pm(z)\phi_0^\mp(w) = \frac{\phi_1^\mp(w)}{z-w} + \dots,$$

$$T_F^\pm(z)\phi_1^\pm(w) = \frac{2h\phi_0^\pm(w)}{(z-w)^2} + \frac{\partial \phi_0^\pm(w)}{z-w} + \dots,$$

$$T_F^\pm(z)\phi_1^\mp(w) = 0 + \dots. \tag{12.82}$$

Chiral primary fields will be further discussed later in this section and in Sect. 12.3. Further simplification occurs for $h = \frac{1}{2}$. The special role played by (anti)chiral primary fields with $h = \frac{1}{2}$ will be discussed in Chap. 15. By expanding (12.80) in components one can also write the component OPEs for a primary superfield of conformal dimension h and $U(1)$ charge q.

The operator algebra (12.81) can be converted to (anti)commutators of the modes of the generators defined by

$$T(z) = \sum_{n \in \mathbb{Z}} z^{-n-2} L_n, \qquad L_n = \oint \frac{dz}{2\pi i} z^{n+1} T(z),$$

$$T_F^\pm(z) = \sum_{n \in \mathbb{Z}} z^{-n-\frac{3}{2} \mp a} G_{n \pm a}^\pm, \longleftrightarrow G_{n \pm a}^\pm = \oint \frac{dz}{2\pi i} z^{n+\frac{1}{2} \pm a} T_F^\pm(z),$$

$$J(z) = \sum_{n \in \mathbb{Z}} z^{-n-1} J_n, \qquad J_n = \oint \frac{dz}{2\pi i} z^n J(z) \tag{12.83}$$

with Hermiticity conditions

12.2 $N = 2$ Superconformal Symmetry

$$L_n^\dagger = L_{-n}, \quad (G_{n+a}^+)^\dagger = G_{-n-a}^-, \quad J_n^\dagger = J_{-n}. \tag{12.84}$$

The operator algebra is then equivalent to the $N = 2$ super-Virasoro algebra

$$[L_m, L_n] = (m-n) L_{m+n} + \frac{c}{12}(m^3 - m)\delta_{m+n},$$

$$[L_m, G_{n\pm a}^\pm] = \left(\frac{1}{2}m - n \mp a\right) G_{m+n\pm a}^\pm,$$

$$[L_m, J_n] = -n J_{m+n},$$

$$[J_m, J_n] = \frac{c}{3} m\, \delta_{m+n},$$

$$[J_m, G_{n\pm a}^\pm] = \pm G_{m+n\pm a}^\pm,$$

$$\{G_{m+a}^+, G_{n-a}^-\} = 2L_{m+n} + (m-n+2a) J_{m+n} + \frac{c}{3}\left[(m+a)^2 - \frac{1}{4}\right]\delta_{m+n},$$

$$\{G_{m+a}^+, G_{n+a}^+\} = \{G_{m-a}^-, G_{n-a}^-\} = 0. \tag{12.85}$$

The relation between the central charges in the $T(z)\,T(w)$, $T_F^+(z)\,T_F^-(w)$ and $J(z)\,J(w)$ operator products are again fixed by Jacobi identities.

We have introduced the real parameter a. From the mode expansion of $T_F^\pm$ we find

$$T_F^\pm(e^{2\pi i}\, z) = -e^{\mp 2\pi i a}\, T_F^\pm(z), \tag{12.86}$$

i.e. a labels the boundary conditions of the fermionic operators $T_F^\pm$. For $a \in \mathbb{Z}$ we are in the R sector whereas $a \in \mathbb{Z} + \frac{1}{2}$ corresponds to the NS sector. The algebras for a and $a+1$ are clearly isomorphic and we can restrict a to $a \in [0, 1)$. Notice that only for $a = 0$ and $a = \frac{1}{2}$ we can form a real $T_F = \frac{1}{2\sqrt{2}}(T_F^+ + T_F^-)$ with definite boundary conditions. In the $N = 1$ case we had only two sectors, as there the supercurrent was real allowing only for periodic or anti-periodic boundary conditions. In the $N = 2$ case we can interpolate between the two sectors by varying a from 0 to $\frac{1}{2}$. For $a = \frac{1}{2}$, i.e. in the NS sector, there exists again a finite dimensional subalgebra generated by $L_{0,\pm 1}$, $G_{1/2}^\pm$, $G_{-1/2}^\pm$ and J_0. This is the algebra $osp(2|2)$. The maximally commuting subalgebra is generated by L_0 and J_0 and each state $|\phi\rangle$ in the Hilbert space of an $N = 2$ superconformal field theory carries two quantum numbers (h, q): $L_0|\phi\rangle = h|\phi\rangle$ and $J_0|\phi\rangle = q|\phi\rangle$.

Let us present three simple realizations of the $N = 2$ algebra in terms of free fields. The first one consists of two free real $N = 1$ superfields $\mathscr{X}^{1,2}(z) = X^{1,2}(z) + i\theta \psi^{1,2}(z)$. Define the complex fields

$$X^\pm = \frac{1}{\sqrt{2}}\left(X^1 \pm i X^2\right) \quad \text{and} \quad \psi^\pm = \frac{1}{\sqrt{2}}\left(\psi^1 \pm i \psi^2\right). \tag{12.87}$$

Energy momentum tensor and supercurrent are then

$$T(z) = -\partial X^+ \, \partial X^-(z) - \frac{1}{2}\left(\psi^+ \, \partial \psi^-(z) + \psi^- \, \partial \psi^+(z)\right),$$

$$T_F(z) = \frac{1}{2\sqrt{2}}\left(T_F^+(z) + T_F^-(z)\right), \tag{12.88}$$

where

$$T_F^+ = i\sqrt{2}\,\psi^+ \, \partial X^- \quad \text{and} \quad T_F^- = i\sqrt{2}\,\psi^- \, \partial X^+. \tag{12.89}$$

We also define a $U(1)$ current

$$J(z) = \psi^+ \psi^-(z). \tag{12.90}$$

It is straightforward to verify that T, T_F and J generate a $c = 3$, $N = 2$ superconformal algebra (12.81). The partition function of this theory can be expressed in terms of the characters (11.109) and (11.110) for $\widehat{so}(2)_1$ as

$$Z_{N=2}(\tau,\bar{\tau}) = \left(|\chi_0^{(0,0)}|^2 + |\chi_V^{(\frac{1}{2},1)}|^2 + |\chi_S^{(\frac{1}{8},+\frac{1}{2})}|^2 + |\chi_C^{(\frac{1}{8},-\frac{1}{2})}|^2\right), \tag{12.91}$$

where the upper index is (h, q), i.e. it shows the conformal dimension and $U(1)$ charge of the ground state in the respective character. With the modular S-matrix in (11.120) it is straightforward to compute the fusion rules via the Verlinde formula (6.100). The result is Table 11.1 on page 333 for $n = 1$. This example is easily generalized to the case of $2k$ free superfields from which we can build k complex ones. This leads to $c = 3k$, $N = 2$ algebra. In the string context this example arises in $2k$-dimensional torus compactifications.

The second example consists of only one real free boson:

$$T(z) = -\frac{1}{2}\partial\phi\,\partial\phi,$$

$$T_F^\pm(z) = \sqrt{\frac{2}{3}}\,e^{\pm i\sqrt{3}\phi(z)},$$

$$J(z) = \frac{i}{\sqrt{3}}\,\partial\phi(z) \tag{12.92}$$

which generate a $c = 1$, $N = 2$ superconformal algebra. Yet another realization can be found in the superconformal ghost system if we define

$$T_F^+ = \sqrt{2}\,b\gamma, \quad T_F^- = -2\sqrt{2}\,(\partial\beta)c - 3\sqrt{2}\,\beta\,\partial c, \quad J = -2bc - 3\beta\gamma. \tag{12.93}$$

This shows that the (local) $N = 1$ superconformal ghost system possesses an $N = 2$ extended global supersymmetry.

12.2 $N = 2$ Superconformal Symmetry

Let us now discuss highest weight representations of the $N = 2$ super Virasoro algebra which show some new features compared to the $N = 1$ and $N = 0$ cases. As usual, we define highest weight states by the conditions

$$G_r^\pm |\phi\rangle = 0 \qquad\qquad r > 0,$$
$$L_n |\phi\rangle = J_n |\phi\rangle = 0 \qquad n > 0,$$
$$L_0 |\phi\rangle = h |\phi\rangle, \qquad J_0 |\phi\rangle = q |\phi\rangle. \tag{12.94}$$

Here q is the $U(1)$ charge of the state. Actually, the conditions for L_n and J_n follow from $G_r^\pm |\phi\rangle = 0$ and the use of the algebra. Requiring unitarity of the representation of the algebra restricts the allowed values of (h, q). For states in the NS sector we have

$$0 \leq \left|G_{-1/2}^\mp |\phi\rangle\right|^2 + \left|G_{1/2}^\pm |\phi\rangle\right|^2 = \langle\phi|\{G_{1/2}^\pm, G_{-1/2}^\mp\}|\phi\rangle$$
$$= 2\langle\phi|L_0 \pm \frac{1}{2} J_0|\phi\rangle = 2\left(h \pm \frac{q}{2}\right)\langle\phi|\phi\rangle, \tag{12.95}$$

i.e. $h \geq \frac{1}{2}|q|$ in the NS sector. When the bound is satisfied, there are two states which satisfy

$$G_{-1/2}^\pm |h = \pm\tfrac{q}{2}\rangle = 0. \tag{12.96}$$

To each highest weight state there corresponds a primary field $\phi^{(h,q)}$ of the $N = 2$ algebra. For the fields with $h = \frac{1}{2}|q|$, Eq. (12.96) is equivalent to

$$T_F^\pm(z)\, \phi^{(\pm\frac{1}{2}q,q)}(w) \sim \text{regular}. \tag{12.97}$$

These are the chiral (anti-chiral) primary fields which were introduce earlier in this section. For a (anti-)chiral primary field the action of only one of the two supercharges $G_{-1/2}^\pm$ results in a new field, cf. Eq. (12.82).

We can simply characterize chiral primary fields by the requirement

$$G_{-1/2}^+ |\phi\rangle = G_{+1/2}^- |\phi\rangle = 0. \tag{12.98}$$

Indeed, from $\langle\phi|\{G_{1/2}^-, G_{-1/2}^+\}|\phi\rangle = 0 = 2\langle\phi|(h - \frac{1}{2}q)|\phi\rangle$, $h = \frac{1}{2}q$ follows. To show that such a state is primary, we observe that any operator which lowers the L_0 eigenvalue but does not change the $U(1)$ charge of a state, must annihilate the state, since otherwise the bound $h \geq \frac{1}{2}|q|$ is violated. Since $L_0 J_n|\phi\rangle = (h - n)J_n|\phi\rangle$, we find $J_n|\phi\rangle = 0$ for $n > 0$. Finally, the anti-commutator $\{J_n, G_r^\pm\} = \pm G_{r+n}^\pm$ applied to the state $|\phi\rangle$ leads to the remaining relations in (12.94). Conversely, any state which satisfies $h = \frac{1}{2}q$ is chiral primary and satisfies $h \leq \frac{c}{6}$. The latter property follows from $0 \leq |G^+_{-3/2}|\phi\rangle|^2$ and the algebra. The notion of (anti) chiral fields and states of the $N = 2$ superconformal field theory is more general and does not only apply to primary fields and highest weight states. Any field whose operator product with T^+_F (T^-_F) is regular is called (anti)chiral. Equivalently, any NS state with is annihilated by $G^+_{-1/2}$ ($G^-_{-1/2}$) is called (anti)chiral.

Even though the representation spaces of the $N = 2$ super Virasoro algebra for different values of a are different, the algebras are isomorphic for all values of a. In fact, there is a one-parameter isomorphism U_η of the $N = 2$ superconformal algebra. To see this let us write the $\{G^+, G^-\}$ anti-commutator in the following form

$$\{G^+_{m+a+\eta}, G^-_{n-a-\eta}\} = 2\left[L_{m+n} + \eta J_{m+n} + \frac{c}{6}\eta^2 \delta_{m+n}\right]$$
$$+ (m - n + 2a)\left[J_{m+n} + \frac{c}{3}\eta \delta_{m+n}\right] + \frac{c}{3}\left[(m+a)^2 - \frac{1}{4}\right]\delta_{m+n}. \quad (12.99)$$

One now verifies that the generators ($L^\eta_n = U_\eta L_n U^{-1}_\eta$, etc.)

$$L^\eta_n = L_n + \eta J_n + \frac{c}{6}\eta^2 \delta_n,$$
$$G^{\eta\pm}_{n\pm a} = G^\pm_{n\pm(a+\eta)},$$
$$J^\eta_n = J_n + \frac{c}{3}\eta \delta_n \quad (12.100)$$

satisfy the $N = 2$ superconformal algebra. Therefore the algebras are equivalent for all a. We will now investigate the behavior of the representations of the $N = 2$ superconformal algebra as we change η. Consider a state $|\phi\rangle$ with weight h and $U(1)$ charge q. We are interested in the weight and charge of the transformed state $|\phi_\eta\rangle = U_\eta|\phi\rangle$ with respect to the original theory, i.e. we are interested in h_η and q_η defined by

$$L_0|\phi_\eta\rangle = h_\eta|\phi_\eta\rangle, \qquad J_0|\phi_\eta\rangle = q_\eta|\phi_\eta\rangle. \quad (12.101)$$

Using

$$U_\eta L_0|\phi\rangle = U_\eta L_0 U^{-1}_\eta|\phi_\eta\rangle = L^\eta_0|\phi_\eta\rangle = h|\phi_\eta\rangle \quad (12.102)$$

and

$$U_\eta J_0|\phi\rangle = U_\eta J_0 U^{-1}_\eta|\phi_\eta\rangle = J^\eta_0|\phi_\eta\rangle = q|\phi_\eta\rangle \quad (12.103)$$

together with (12.100), we can express the conformal weight and J_0 charge of the transformed state with respect to the original operators in the following way:

12.2 N = 2 Superconformal Symmetry

$$h_\eta = h - \eta q + \frac{\eta^2}{6}c, \qquad q_\eta = q - \frac{c}{3}\eta. \tag{12.104}$$

This change of the spectrum is called spectral flow. and U_η the spectral flow operator. It has conformal weight $h = \frac{c}{6}\eta^2$ and charge $q = -\frac{c}{3}\eta$. This can be seen by considering the weight and charge of $U_\eta |0\rangle_{NS}$.

Let us now investigate the spectral flow for chiral primaries. In particular, using equation (12.104) with $\eta = \frac{1}{2}$ and employing $h = \frac{q}{2}$ for a chiral primary, we obtain

$$\left|h_0 = \frac{q_0}{2}, q_0\right\rangle_{NS} \xrightarrow{\eta=\frac{1}{2}} \left|h_{\frac{1}{2}} = \frac{c}{24}, q_{\frac{1}{2}} = q_0 - \frac{c}{6}\right\rangle_R. \tag{12.105}$$

In particular, the NS vacuum $|0\rangle_{NS}$ with $(h,q) = (0,0)$ is mapped to a state with $(h,q) = (\frac{c}{24}, -\frac{c}{6})$. The conformal weight of the state in the Ramond sector in (12.105), $h = \frac{c}{24}$, is independent of its J_0 charge q. Therefore, this state is generally degenerate. We now show that the state in the Ramond sector is actually a ground state. Indeed, in a unitary theory,

$$0 \leq \left|G_0^+|h,q\rangle_R\right|^2 + \left|G_0^-|h,q\rangle_R\right|^2 = {}_R\langle h,q|\{G_0^+, G_0^-\}|h,q\rangle_R$$
$$= 2\left(h - \frac{c}{24}\right)\||h,q\rangle_R\|^2, \tag{12.106}$$

i.e. the lowest possible value of any R sector state is $h = \frac{c}{24}$ which is attained by the states (12.105). There is thus a one-to-one correspondence between chiral primary states and Ramond ground states and, by a similar argument, between anti-chiral primaries and Ramond ground states.

A further transformation with $\eta = \frac{1}{2}$ maps the Ramond sector back to the Neveu-Schwarz sector. Employing Eq. (12.104) with $\eta = 1$, we find

$$\left|h_0 = \frac{q_0}{2}, q_0\right\rangle_{NS} \xrightarrow{\eta=1} \left|h_1 = -\frac{q_1}{2}, q_1 = q_0 - \frac{c}{3}\right\rangle_{NS}, \tag{12.107}$$

which is an anti-chiral primary state in the NS sector. Therefore, spectral flow with $\eta = 1$ maps chiral states in the NS sector to anti-chiral states in the same sector.

Since the spectral flow connects the NS with the R sector, and states in these sectors of the fermionic string are space-time bosons and fermions respectively, we have here an indication of the connection between $N = 2$ world-sheet and space-time supersymmetry. We will elaborate on this in more detail in Chap. 15.

Besides the spectral flow, there also exists an outer automorphism of the $N = 2$ super Virasoro algebra, which acts as

$$L_n \to L_n, \quad J_n \to -J_n, \quad G_r^+ \to G_r^-, \quad G_r^- \to G_r^+. \tag{12.108}$$

This allows us to impose in any $(2,2)$ superconformal field theory two distinctive boundary conditions. Recall from Sect. 6.6 that a boundary state $|B\rangle$ is specified by gluing conditions (6.138) for the left- and right-moving generators of the chiral algebra $\mathscr{A} = \overline{\mathscr{A}}$. We can therefore define gluing conditions for the (L_n, G_r^+, G_r^-, J_n) of the superconformal algebra, which read

$$(L_n - \overline{L}_{-n})|B\rangle = 0, \qquad (J_n + \overline{J}_{-n})|B\rangle = 0,$$
$$(G_r^+ - i\eta \overline{G}_{-r}^+)|B\rangle = 0, \qquad (G_r^- - i\eta \overline{G}_{-r}^-)|B\rangle = 0 \qquad (12.109)$$

with $\eta = \pm 1$. These are called B-type boundary conditions. However, if there exist an automorphism of the chiral algebra, one can define a second set of gluing conditions (6.139). Thus, the automorphism (12.108) leads to

$$(L_n - \overline{L}_{-n})|B\rangle = 0, \qquad (J_n - \overline{J}_{-n})|B\rangle = 0,$$
$$(G_r^+ - i\eta \overline{G}_{-r}^-)|B\rangle = 0, \qquad (G_r^- - i\eta \overline{G}_{-r}^+)|B\rangle = 0 \qquad (12.110)$$

which are called A-type boundary conditions. In the geometric setting the boundary states are interpreted as D-branes on CY manifolds which preserve $\mathcal{N} = 1 \subset \mathcal{N} = 2$ space-time supersymmetry. They are called A-branes and B-branes, respectively.

Let us finally comment on the representation theory of the $N = 2$ super Virasoro algebra. The fact that the Cartan subalgebra (L_0, J_0) is two-dimensional is responsible for significant differences compared to the $N = 0$ and $N = 1$ algebras. In particular, while for these cases the unitarity constraints following from the algebra are sufficient to guarantee that representations with $c \geq 1$ and $c \geq \frac{3}{2}$ are unitarity, this is no longer true for $N = 2$ and $c \geq 3$ which is the central charge for a free $N = 2$ superfield. However, as in the $N = 0, 1$ cases there exists a discrete series of rational unitary models in the regime $0 < c < 3$ with

$$c = \frac{3k}{k+2}, \qquad k \in \mathbb{Z}, \quad k \geq 1. \qquad (12.111)$$

For each value of k in the unitary series (12.111), there exists a finite number $((k + 1)(k + 2))$ of highest weight representations of the $N = 2$ algebra. They are labeled by two integers. However one is often interested in the highest weight representations of the bosonic subalgebra. These representations are generated from a highest weight state by the action of an arbitrary number of bosonic generators and an even number of fermionic generators. They are labeled by three integers $\phi_{m,s}^l \equiv (l, m, s)$, where $0 \leq l \leq k$, m is defined modulo $2(k + 2)$ and s is defined modulo four. The labels have to satisfy $l + m + s \in 2\mathbb{Z}$. Furthermore there is a $\mathbb{Z}_2$ identification

$$\phi_{m,s}^l \sim \phi_{m+k+2,s+2}^{k-l}. \qquad (12.112)$$

The conformal weights and $U(1)$ charges of these states are

12.2 N = 2 Superconformal Symmetry

$$h^l_{m,s} = \frac{l(l+2) - m^2}{4(k+2)} + \frac{s^2}{8} + \text{integer},$$

$$Q_{m,s} = -\frac{m}{k+2} + \frac{s}{2} + \text{even integer}. \quad (12.113)$$

Representations with $s \in 2\mathbb{Z}$ are in the NS sector while those with $s \in 2\mathbb{Z}+1$ are in the R sector. The integers in (12.113) are zero if, with the help of the identifications (12.112) the triplet of parameters (l, m, s) can be brought to the range $0 \le |m - s| \le l$ and $s \in \{-1, 0, 1, 2\}$. This is not always the case. In the NS sector the only exceptions are representations with labels $(l, -l, 2)$. For $l > 0$ (12.113) still apply with vanishing integers if we use $s = -2$ for these representations. The highest weight state $(0, 0, 2)$ has, in fact, $(h, Q) = (\tfrac{3}{2}, 1)$. It is obtained by acting with $G^+_{-3/2}$ on the $OSp(2|2)$ invariant vacuum $(0, 0, 0)$. The chiral primaries of the $N = 2$ minimal models are $(l, -l, 0)$, the anti-chiral primaries are $(l, l, 0)$. Here we have identified a representation with the primary operator which creates the highest weight state from the vacuum.

To compute characters and modular properties of these highest weight representations, one utilizes a realization of the unitary $N = 2$ models in terms of a certain quotient of Kač-Moody algebras.[8] This allows to determine the characters as

$$\chi^l_{m,s}(\tau) = \sum_{j=1}^{k+2} C^{(k)}_{l, m-4j-s}(\tau)\, \Theta_{-2m+(4j+s)(k+2),\, 2k(k+2)}(\tau). \quad (12.115)$$

The $SU(2)$ string-functions $C^{(k)}_{l,m}$ have been implicitly defined in (11.133) and the Θ-functions in (11.132). The modular S-matrix is

$$S^{N=2}_{(lms)(l'm's')} = \frac{1}{2k+4} \sin\left(\frac{\pi}{k+2}(l+1)(l'+1)\right) e^{-\pi i \left(\frac{ss'}{2} - \frac{mm'}{k+2}\right)} \quad (12.116)$$

which implies the fusion rules

$$[\phi^{l_1}_{m_1, s_1}] \times [\phi^{l_2}_{m_2, s_2}] = \sum_{l_3 = |l_1 - l_2|}^{\min(l_1 + l_2, 2k - l_1 - l_2)} [\phi^{l_3}_{m_1 + m_2, s_1 + s_2}], \quad (12.117)$$

where we used the $\widehat{su}(2)_k$ fusion coefficients (11.135).

---

[8] Concretely, this realization is given by the coset

$$\frac{\widehat{su}(2)_k \times \widehat{u}(1)_2}{\widehat{u}(1)_{k+2}}, \quad (12.114)$$

where $\widehat{u}(1)_k$ is the CFT of a free boson compactified on a circle with radius $\sqrt{\alpha' k}$ introduced in Sect. 10.3.

In our discussion on space-time supersymmetric compactifications of the superstring in Chap. 15, two special operators of the $N=2$ minimal models will play a prominent role. The first is $(0, 1, 1)$ with (use (12.113)) $(h, Q) = (\frac{c}{24}, \frac{c}{6})$, which apparently coincide with the values for the spectral flow operator $U_{-1/2}$. This identification is further supported by the fusion rules

$$[\phi_{1,1}^0] \times [\phi_{m,s}^l] = [\phi_{l+1,s+1}^l] \qquad (12.118)$$

which shows that $(0, 1, 1)$ interpolates between NS and R sectors and that it is a simple current. Likewise $(0, -1, -1)$ is identified with $U_{1/2}$. The second special operator is $(0, 0, 2)$ with $(h, Q) = (\frac{3}{2}, 1)$. Its fusion rules are

$$[\phi_{0,2}^0] \times [\phi_{m,s}^l] = [\phi_{m,s+2}^l] \qquad (12.119)$$

which indicates that it is a simple current which acts diagonally on the NS and R sectors. It increases the conformal weight of a NS state by a half-integer and that of an R state by an integer. This is the supercurrent $T_F^+ \sim G_{-3/2}^+|0\rangle$. Note that $T_F^- \sim G_{-3/2}^- G_{+3/2}^- G_{-3/2}^+|0\rangle$. From the fusion rules (12.117) it follows that all fields $\phi_{m,s}^0$ with $m + s \in 2\mathbb{Z}$ are simple currents.[9]

These results on $N=2$ minimal models will be employed when we discuss Gepner models in Chap. 15. These special superconformal field theories have a correspondence to geometric compactifications on Calabi-Yau manifolds. An important role for establishing this relationship is played by the chiral ring, which we discuss next.

12.3 Chiral Ring and Topological Conformal Field Theory

Let us consider the OPE of two chiral primary fields $\phi_i(z)$ and $\phi_j(w)$. On general grounds (cf. Chap. 4) it has the form

$$\phi_i(z)\phi_j(w) = \sum_k C_{ij}{}^k (z-w)^{h_k - h_i - h_j} \psi_k(w), \qquad (12.120)$$

where ψ_k are not necessarily (chiral) primaries. $U(1)$ charge conservation requires $q_k = q_i + q_j$. In addition, we need to satisfy $h_k \geq \frac{q_k}{2}$ with equality for chiral primaries. This implies that

$$h_i + h_j - h_k = \frac{q_i}{2} + \frac{q_j}{2} - h_k \leq \frac{q_k}{2} - \frac{|q_k|}{2} \leq 0. \qquad (12.121)$$

---

[9] The spectral flow operators $U_{\pm\frac{1}{2}}$ and the supercurrents $T_F^\pm$ are simple currents of any $N=2$ SCFT.

12.3 Chiral Ring and Topological Conformal Field Theory

Therefore, the OPE between two chiral primaries contains only regular terms. This allows us to define a product between two such fields as

$$(\phi_i \cdot \phi_j)(w) \equiv \lim_{z \to w} \phi_i(z)\phi_j(w) = \sum_k C_{ij}{}^k \phi_k(w), \qquad (12.122)$$

where all fields on the right hand side are chiral primaries. In other words, chiral primaries can be multiplied point-wise. They generate a ring, the so-called chiral ring.

In complete analogy there is a ring of anti-chiral primary fields. If the anti-holomorphic sector of the theory also possesses $N=2$ superconformal symmetry, i.e. if we are dealing with a $(2,2)$ superconformal theory, then there are four chiral rings, denoted as (c,c), (a,c), (c,a) and (a,a). We have already hinted at the relation between world-sheet and space-time supersymmetry. In Chap. 14 we will discuss string compactification on Calabi-Yau manifolds which lead to a space-time supersymmetric theory in the four uncompactified dimensions. The following discussion is the CFT analogue of (geometric) space-time properties which we will encounter when we study Calabi-Yau manifolds.

In a unitary theory, every state in the NS sector has a unique orthogonal decomposition of the form

$$|\phi\rangle = |\phi_0\rangle + G^+_{-1/2}|\phi_1\rangle + G^-_{+1/2}|\phi_2\rangle, \qquad (12.123)$$

where $|\phi_0\rangle$ is chiral primary. If $|\phi\rangle$ has conformal weight h and $U(1)$ charge q,[10] $|\phi_1\rangle$ and $|\phi_2\rangle$ have $(h-\frac{1}{2}, q-1)$ and $(h+\frac{1}{2}, q+1)$, respectively. There is an alternative decomposition of the same state as

$$|\phi\rangle = |\phi'_0\rangle + G^-_{-1/2}|\phi'_1\rangle + G^+_{+1/2}|\phi'_2\rangle \qquad (12.124)$$

with $|\phi'_0\rangle$ anti-chiral primary. Assuming for the moment that the decomposition (12.123) exists, orthogonality is easy to demonstrate. Uniqueness is also straightforward to prove: If there were two different decompositions, their difference is of the form $0 = |\phi_0\rangle + G^+_{-1/2}|\phi_1\rangle + G^-_{+1/2}|\phi_2\rangle$. Multiplying with $\langle\phi_2|G^+_{-1/2}$ we find $|G^-_{+1/2}|\phi_2\rangle|^2 = 0$ and likewise $G^+_{-1/2}|\phi_1\rangle = 0$ and hence also $|\phi_0\rangle = 0$. Thus the decomposition is unique.

We now show the existence of the decomposition (12.123). The statement of the decomposition, after its orthogonality and uniqueness have been established, is that any state which is not chiral primary, i.e. for which $h \neq \frac{q}{2}$, can be decomposed as $|\phi\rangle = G^+_{-1/2}|\phi_1\rangle + G^-_{+1/2}|\phi_2\rangle$. To show this, we observe that on such a state $\{G^+_{-1/2}, G^-_{+1/2}\}|\phi\rangle = (2L_0 - J_0)|\phi\rangle = (2h-q)|\phi\rangle \neq 0$. We can

---

[10]Since L_0 and J_0 commute, they can be simultaneously diagonalized and we can restrict the discussion to eigenstates of L_0 and J_0.

therefore write $|\phi\rangle = \{G^+_{-1/2}, G^-_{+1/2}\}|\eta\rangle = G^+_{-1/2}(G^-_{+1/2}|\eta\rangle) + G^-_{+1/2}(G^+_{-1/2}|\eta\rangle)$ with $|\eta\rangle = \frac{1}{2h-q}|\phi\rangle$.

If $|\phi\rangle$ is chiral (i.e. $G^+_{-1/2}|\phi\rangle = 0$) but not necessarily primary, $|\phi_2\rangle = 0$ in the decomposition (12.123), i.e. for chiral fields we have

$$|\phi\rangle_{\text{chiral}} = |\phi_0\rangle + G^+_{-1/2}|\phi_1\rangle. \tag{12.125}$$

The decomposition (12.123) is reminiscent of the Hodge decomposition of differential forms on compact Riemannian manifolds. If we combine the chiral and the anti-chiral sectors and consider $(2, 2)$ SCFTs, the analogy to complex manifolds, their (p, q) forms and Dolbeault cohomology becomes even more striking (the relevant mathematical concepts and notation will be introduced in Chap. 14). Here we have the correspondence of nilpotent operators

$$(G^+_{-1/2}, G^-_{+1/2}) \Leftrightarrow (\partial, \partial^*), \{G^+_{-1/2}, G^-_{1/2}\} = 2(L_0 - \tfrac{1}{2}J_0) \Leftrightarrow \Delta_\partial, \tag{12.126a}$$

$$(\bar{G}^+_{-1/2}, \bar{G}^-_{+1/2}) \Leftrightarrow (\bar{\partial}, \bar{\partial}^*), \{G^+_{1/2}, G^-_{-1/2}\} = 2(L_0 + \tfrac{1}{2}J_0) \Leftrightarrow \Delta_{\bar{\partial}} \tag{12.126b}$$

and

$$\text{chiral fields} \iff \text{closed forms},$$
$$\text{chiral primary fields} \iff \text{harmonic forms}. \tag{12.127}$$

For the compactification to four dimensions, (2,2) SCFTs with $(c, \bar{c}) = (9, 9)$ are of interest. Chiral primary fields then satisfy $h \leq \tfrac{3}{2}$ and $\bar{h} \leq \tfrac{3}{2}$ and $|q| = 2h$ and $|\bar{q}| = 2\bar{h}$. Fields with integer $U(1)$ charge form a subring of the chiral ring. It contains only fields with $q, \bar{q} \in \mathbb{Z}$, $q, \bar{q} \in [-3, +3]$ where $q, \bar{q} \geq 0$ for the (c, c) ring and $-q, \bar{q} \geq 0$ for the (a, c) ring. We thus find that the range of integer charges in the (c, c) ring is precisely the range (p, q) in the Dolbeault cohomology groups $H^{p,q}(M)$ of a complex manifold M of (complex) dimension three. The dimensions of these cohomology groups are denoted as $h^{p,q}(M)$. Furthermore, since the field with $q = 2h = \tfrac{c}{3} = 3$ is unique—it flows to the unique NS ground state under spectral flow with $\eta = 1$—the correspondence is with manifolds M with $h^{0,0}(M) = 1$ and $h^{3,0}(M) = h^{0,3}(M) = h^{3,3}(M) = 1$. These are Calabi-Yau manifolds which we will discuss in some detail in Chap. 14.

If in the superconformal field theory we define $h^{p,q}$ as the number of chiral primaries with $(U(1)_L, U(1)_R)$ charges (p, q), we can construct the Poincaré polynomial of the chiral ring as

$$P_{(c,c)} = \text{Tr}_{\text{NS}}[t^{J_0} \bar{t}^{\bar{J}_0}]_{(c,c)} = \sum_{p,q=0}^{3} h^{p,q} t^p \bar{t}^q. \tag{12.128}$$

12.3 Chiral Ring and Topological Conformal Field Theory

As we have seen, under the action of the spectral flow with parameters $(\eta, \bar{\eta}) = (-1, 0)$, elements in the (c, c) ring flow to elements in the (a, c) ring. The $U(1)$ charges change as $(q, \bar{q}) \to (q - 3, \bar{q})$. The Poincaré polynomial of the (a, c) ring is

$$P_{(a,c)} = \text{Tr}_{\text{NS}}[t^{-J_0}\bar{t}^{\bar{J}_0}]_{(a,c)} = \sum_{p,q=0}^{3} h^{3-p,q} t^p \bar{t}^q. \tag{12.129}$$

Two crucial observations follow:

(1) we can interpret the Poincaré polynomial $P_{(c,c)}$ as that of a Calabi-Yau manifold M where $h^{p,q}$ are the Hodge numbers. The Polyakov action coupled to the Calabi-Yau metric leads, in superconformal gauge, to a $(2, 2)$ SCFT with $c = 9$. (2) the distinction between the (c, c) and (a, c) rings is merely conventional. Applying the outer automorphism (12.108) to, say, only the left-moving sector, changes the relative sign of the two $U(1)$ charges and therefore interchanges the two rings. This leads to the far reaching conclusion that we should be able to associate the Poincaré polynomial $P_{(a,c)}$ of the (a, c) ring also with a manifold M^*. M^* is called the mirror manifold of M, whose existence is suggested (predicted) by superconformal field theory. Moreover, the relation between the Hodge numbers of M and M^* must be

$$h^{3-p,q}(M^*) = h^{p,q}(M). \tag{12.130}$$

Once the relation between $(2, 2)$ superconformal symmetry and Calabi-Yau manifolds, in particular between chiral primary fields and Dolbeault cohomology, has been established, the existence of mirror pairs of Calabi-Yau manifolds, while trivial from the CFT point of view, is a highly non-trivial fact from the point of view of geometry. Moreover, the $U(1)$ flip interchanges the B-type (12.109) and A-type (12.110) boundary conditions so that an $A(B)$-brane on the Calabi-Yau M is mapped to a $B(A)$-brane on the mirror manifold M^*.

The subsector of chiral primary fields in an $N = 2$ superconformal field theory gives a realization of so-called topological field theories (TFT) which are a wide subject with ramifications in physics and mathematics. One of their defining properties is that they possess a nil-potent fermionic charge, called BRST-charge and that the energy momentum tensor is BRST-exact. We will not enter a general discussion of TFTs, but will limit ourselves to two-dimensional topological conformal field theories (TCFT) which are obtained by twisting $N = 2$ SCFTs. They share many properties with more general TFTs. (Topological) CFTs can be viewed as fixed points of non-conformal theories where the coupling constants are tuned such that all β-functions vanish.

Consider the $N = 2$ algebra (12.81) and define the topological energy momentum tensor

$$T^{\text{top}} = T + \frac{1}{2}\partial J. \tag{12.131}$$

This new energy-momentum tensor is obtained from the old one by twisting with the $U(1)$ current. Notice that this is not possible for $N < 2$ but there are several possible twistings for $N = 4$. Defining in addition

$$Q(z) = \frac{1}{\sqrt{2}} T_F^+(z), \qquad G(z) = \frac{1}{\sqrt{2}} T_F^-(z), \qquad (12.132)$$

one derives the following OPEs from (12.81)[11]

$$T^{\text{top}}(z) T^{\text{top}}(w) = \frac{2 T^{\text{top}}(w)}{(z-w)^2} + \frac{\partial T^{\text{top}}(w)}{z-w} + \ldots, \qquad (12.133\text{a})$$

$$T^{\text{top}}(z) Q(w) = \frac{Q(w)}{(z-w)^2} + \frac{\partial Q(w)}{z-w} + \ldots, \qquad (12.133\text{b})$$

$$T^{\text{top}}(z) G(w) = \frac{2 G(w)}{(z-w)^2} + \frac{\partial G(w)}{z-w} + \ldots, \qquad (12.133\text{c})$$

$$T^{\text{top}}(z) J(w) = \frac{-\frac{c}{3}}{(z-w)^3} + \frac{J(w)}{(z-w)^2} + \frac{\partial J(w)}{z-w} + \ldots, \qquad (12.133\text{d})$$

$$J(z) J(w) = \frac{\frac{c}{3}}{(z-w)^2} + \ldots, \qquad (12.133\text{e})$$

$$J(z) Q(w) = +\frac{Q(w)}{(z-w)} + \ldots, \qquad (12.133\text{f})$$

$$J(z) G(w) = -\frac{G(w)}{(z-w)} + \ldots, \qquad (12.133\text{g})$$

$$Q(z) G(w) = \frac{\frac{c}{3}}{(z-w)^3} + \frac{J(w)}{(z-w)^2} + \frac{T^{\text{top}}(w)}{z-w} + \ldots, \qquad (12.133\text{h})$$

$$Q(z) Q(w) = \text{regular}, \qquad G(z) G(w) = \text{regular}. \qquad (12.133\text{i})$$

The twisted algebra (12.133) is the algebraic structure of a TCFT. One important consequence is that the conformal weight of any field is shifted by minus half its $U(1)$ charge ($L_0^{\text{top}} = L_0 - \frac{1}{2} J_0$). Equations (12.133b) and (12.133c) exemplify this for the supercurrents, whose conformal weights w.r.t. T^{top} are 1 (T_F^+) and 2 (T_F^-). We further learn from (12.133a) that T^{top} satisfies a Virasoro algebra without central charge, i.e. it is a primary field. However, J is no longer primary due to the background charge $-\frac{c}{3}$. This interpretation of (12.133d) will become clear in the next chapter. In Table 12.1 we collect the properties of the generators of the twisted algebra.

---

[11] It is straightforward to work out the algebra in terms of the modes. All mode numbers are integers and there is a finite-dimensional subalgebra generated by $L_{0,\pm 1}$ and $G_{0,\pm 1}$.

12.3 Chiral Ring and Topological Conformal Field Theory

Table 12.1 Data of topologically twisted $N=2$ superconformal algebra

Current	$U(1)$ charge	$h^{N=2}$	h^{top}	Statistics
$T(z)$	0	2	2	Bosonic
$G(z)$	-1	$\frac{3}{2}$	2	Fermionic
$Q(z)$	$+1$	$\frac{3}{2}$	1	Fermionic
$J(z)$	0	1	1	Bosonic

There are two fermionic fields with integer conformal weights so that their expansions are in terms of integer modes. In particular the modes of Q and G are related to those of $T_F^\pm$ in the NS sector as $Q_n = \frac{1}{\sqrt{2}} G^+_{n-\frac{1}{2}}$ and $G_n = \frac{1}{\sqrt{2}} G^-_{n+\frac{1}{2}}$.

Now one can define a nilpotent charge

$$Q = \oint \frac{dz}{2\pi i}\, Q(z) = Q_0, \tag{12.134}$$

i.e. $Q^2 = 0$, as a consequence of (12.133i). From (12.133h) we conclude

$$\{Q, G(z)\} = T^{\text{top}}(z) \tag{12.135}$$

and from (12.133f) we derive

$$\{Q, J(z)\} = -Q(z), \tag{12.136}$$

i.e. $T(z)$ and $Q(z)$ are both Q-exact. This is the same algebraic structure which we met in Chap. 5 and therefore we interpret Q as a BRST operator. Then $Q(z)$ is the BRST current and $J(z)$ the ghost current so that the $U(1)$ charge is the ghost charge. A nilpotent charge Q and a Q-exact energy-momentum tensor are hallmarks of topological field theories.

Before we proceed let us give one specific example of a TCFT of the type (12.133), which can be constructed from the familiar bosonic string by simple modifications of the BRST current and the ghost current introduced in Chap. 5. The BRST current is constrained by the requirements that it has ghost number one and that it is Hermitian. There is one more term we can add to Eq. (5.32) leading to

$$Q(z) = c\left(T^{(X)}(z) + \frac{1}{2}T^{(b,c)}(z)\right) + \frac{3}{2}\partial^2 c(z) + i\lambda_\mu \partial(c\partial X^\mu)(z). \tag{12.137}$$

The third term was added to make $Q(z)$ a primary field. Without the last term the $Q(z)Q(w)$ operator product has third and second order poles. They are removed by the addition of the last term provided $\lambda^2 = 2$. If we further modify the ghost current (5.16) and define

$$J(z) = -bc(z) - i\lambda_\mu \partial X^\mu(z) \tag{12.138}$$

and

$$G(z) = b(z), \qquad (12.139)$$

we can verify that the four fields $T^{\text{top}}(z) \equiv T^{(X)}(z) + T^{(b,c)}(z)$, $Q(z)$, $J(z)$ and $G(z)$ generate the algebra (12.133) with $c = 9$.

To continue our general discussion, we note that physical states are defined as the cohomology classes of Q, i.e. they have to be Q-closed

$$Q|\phi\rangle = 0, \qquad (12.140)$$

where states differing by a Q exact field are identified (cf. (12.125))

$$|\phi\rangle \sim |\phi\rangle + Q|\lambda\rangle. \qquad (12.141)$$

The physical Hilbert space is then given by the Q-cohomology $\mathcal{H}_{\text{phys}} = \ker Q / \operatorname{im} Q$. We can select a unique representative in each equivalence class by requiring $G_0|\phi_i\rangle = 0$. Together with (12.140) these are the same conditions as (12.98), i.e. there is a one-to-one correspondence between chiral primary states of the untwisted theory and equivalence classes of physical states in the twisted theory and the chiral primary operators are the BRST invariant observables in the topological version of $N = 2$ theories. Notice that $h^{\text{top}} = 0$ for chiral primaries.

Let us close this brief discussion of topological field theories by pointing out important properties of correlation functions of physical operators. An obvious requirement is that the correlation functions vanish unless they respect the background charge of the $U(1)$-current. Furthermore, recall that correlation functions of the energy-momentum tensor are defined as

$$\langle T_{\alpha\beta} \mathcal{O}_1 \ldots \mathcal{O}_n \rangle \sim \frac{\delta}{\delta h^{\alpha\beta}} \langle \mathcal{O}_1 \ldots \mathcal{O}_n \rangle. \qquad (12.142)$$

This vanishes if $T_{\alpha\beta} = \{Q, G_{\alpha\beta}\}$ and $[Q, \mathcal{O}_i] = 0$. In other words, correlation functions of physical observables do not depend on the world-sheet metric. This is why these theories are topological. In fact, the above property is generic for all TFT, in particular topological CFTs which are required to have traceless energy-momentum tensor.

Another distinctive property is that correlation functions of physical states are constants on the world-sheet. To show this, we use

$$\partial_z \phi(z) = [L_{-1}, \phi(z)] = [\{Q, G_{-1}\}, \phi(z)] = [Q, [G_{-1}, \phi(z)]], \qquad (12.143)$$

where we have used the Jacobi-identity and $[Q, \phi(z)] = 0$. Together with the fact that Q annihilates the vacuum this leads to the conclusion that correlation functions of physical operators are independent of their insertion points. Note that the critical bosonic string has a Q-exact energy-momentum tensor. And indeed, correlation functions of vertex operators of physical states are independent of the insertion points. For instance on a world-sheet with the topology of the sphere: all but three vertex operators are integrated and the positions of the remaining three are arbitrary due to $SL(2,\mathbb{C})$ invariance. The dependence on their position is canceled by the correlation function of the three c-ghost zero modes. Similar remarks hold for the fermionic string whose BRST quantization will be discussed in the next chapter.

Of course, application of the outer automorphism (12.108) leads to a second topological twist, where one replaces $J \to -J$ and interchanges G and Q. In $(2,2)$ SCFTs there are thus two inequivalent twistings: we can perform the same twist or opposite twists on both sides. These two possibilities are referred to as A and B twists, respectively. Mirror symmetry interchanges them.

To summarize, we have demonstrated how one can construct from every $N = 2$ SCFT, via the twisting procedure, a TCFT. Its physical states are precisely the chiral primary states of the original (i.e. untwisted) $N = 2$ SCFT.

Further Reading

The $N = 1$ superconformal algebra first appeared in

- P. Ramond, Dual theory for free fermions. Phys. Rev. **D3**, 2415 (1971)
- A. Neveu, J.H. Schwarz, Factorizable dual model of pions. Nucl. Phys. **B31**, 86 (1971)

Extended superconformal algebras were constructed in

- M. Ademollo et al., Dual string with U(1) colour symmetry. Nucl. Phys. **B111**, 77 (1976)
- M. Ademollo et al., Dual string models with non-Abelian colour and flavour symmetries. Nucl. Phys. **B114**, 297 (1976)
- M. Ademollo et al., Supersymmetric strings and color confinement. Phys. Lett. **62B**, 105 (1976)

The superspace formalism of superconformal theories in two dimensions can be found in

- P.C. West, *Introduction to Supersymmetry and Supergravity* (World Scientific, Singapore, 1990)

Superconformal $N = 1$ and $N = 2$ minimal models were constructed in

- D. Friedan, Z. Qiu, S. Shenker, Superconformal invariance in two dimensions and the Tricritical Ising model. Phys. Lett. **15IB**, 37 (1985)
- P. Di Vecchia, J.L. Petersen, M. Yu, On the unitary representations of N = 2 superconformal theory. Phys. Lett. **B172**, 211 (1986)

- W. Boucher, D. Friedan, A. Kent, Determinant formulae and unitarity for the N = 2 superconformal algebras in two dimensions or exact results on string compactification. Phys. Lett. **B172**, 316 (1986)
- S. Nam, The Kac formula for the N = 1 and the N = 2 superconformal algebras. Phys. Lett. **B172**, 323 (1986)

The spectral flow of $N = 2$ SCFT was discovered in

- A. Schwimmer, N. Seiberg, Comments on the N = 2, 3, 4 superconformal algebras in two dimensions. Phys. Lett. **B184**, 191 (1987)

Our discussion of chiral primary operators, their ring structure and the conjectured mirror symmetry follows the original paper

- W. Lerche, C. Vafa, N.P. Warner, Chiral rings in N = 2 superconformal theories. Nucl. Phys. **B324**, 427 (1989)

Topological CFTs and twisted $N = 2$ theories were first discussed in

- T. Eguchi, S.K. Yang, N = 2 superconformal models as topological field theories. Mod. Phys. Lett. **A5**, 1693 (1990)

A review of TCFTs in a more general context including the coupling to gravity, i.e. topological string theory, is

- R. Dijkgraaf, H. Verlinde, E. Verlinde, *Notes on topological string theory and 2-D quantum gravity*, in *String Theory and Quantum Gravity*, ed. by M. Green et al., World Scientific 1991
- M. Marino, *Chern-Simons Theory, Matrix Models, and Topological Strings* (Oxford University Press, Oxford, 2005)

Chapter 13
Covariant Vertex Operators, BRST and Covariant Lattices

Abstract We reexamine the 10-dimensional type II and heterotic superstring theories using the bosonic language. The aim of this bosonic formulation is the construction of the covariant fermion vertex operators, which involves a proper treatment of the (β, γ) ghost system, This will in turn lead to the introduction of the so-called covariant lattices.

13.1 Bosonization and First Order Systems

Recall[1] the action of the fermionic string in superconformal gauge:

$$S = \frac{1}{4\pi} \int d^2z \left(\partial X^\mu \bar\partial X_\mu + \psi^\mu \bar\partial \psi_\mu + \bar\psi^\mu \partial \bar\psi_\mu \right), \tag{13.1}$$

where we have only written the matter (X^μ, ψ^μ) part. We will turn to the ghost part below. These fields generate a superconformal field theory with $\hat c = 10$ ($c = 15$) where from now on we discuss only the right-moving, holomorphic part of the theory. The two-dimensional supercharge, also called supercurrent, is

$$T_F(z) = \frac{i}{2} \partial X_\mu(z) \psi^\mu(z). \tag{13.2}$$

Applying the techniques of bosonization as introduced in Chap. 11, we replace the ten real fermions $\psi^\mu(z)$ ($\mu = 1, \ldots, 10$) by five chiral bosons $\phi^i(z)$ ($i = 1, \ldots, 5$) with momentum eigenvalues being lattice vectors of the D_5 weight lattice. The bosonization is performed by converting the ten real fermions $\psi^\mu(z)$ to the complex Cartan-Weyl basis:

$$\psi^{\pm i}(z) = \frac{1}{\sqrt{2}} (\psi^{2i-1} \pm i \psi^{2i})(z), \quad i = 1, \ldots, 5. \tag{13.3}$$

---

[1] The first part of this section is a review of material from Sect. 11.4 with a slight change of notation: $X^i \to \phi^i$.

The action for the complex fermions is

$$S = \frac{1}{4\pi} \int d^2z \, (\Psi^{+i} \bar{\partial} \Psi^{-i} + \Psi^{-i} \bar{\partial} \Psi^{+i}). \tag{13.4}$$

The part of the generators of the Wick rotated Lorentz group $SO(10)$ which are constructed from the world-sheet fermions are bosonized according to

$$J^{+i,-i}(z) = :\Psi^{+i} \Psi^{-i}(z): = i\partial\phi^i(z)$$
$$J^{\pm i, \pm j}(z) = :\Psi^{\pm i} \Psi^{\pm j}: = :e^{\pm i\phi^i \pm i\phi^j}(z): \quad (i < j), \tag{13.5}$$

where the complex fermions themselves are expressed as

$$\Psi^{\pm i}(z) = :e^{\pm i\phi^i(z)}:. \tag{13.6}$$

Here and in the following we drop cocycle factors. They are necessary to get anti-commuting fermions $\Psi^{\pm i}$ for $i \neq j$ and to produce manifestly covariant results, as e.g. in Eq. (13.8) below.

The states of the spinning string theory are created from the $SL(2, \mathbb{C})$ invariant vacuum state by vertex operators which contain the five bosons $\phi^i(z)$. Let us concentrate on expressions which are exponentials of these bosons; possible derivative terms $\partial \phi^i(z)$ play only a trivial role in the following. We will not discuss the X^μ-dependent part of the vertex operators either.

In the NS sector the states are space-time bosons. The ground state is the NS vacuum $|0\rangle$ which is a tachyon as discussed in Chap. 8. The first excited state is the massless ten-dimensional vector $|\psi^\mu\rangle = b^\mu_{-1/2}|0\rangle$ with the corresponding vertex operator $\psi^\mu(z)$. Thus, the vector vertex operator in the bosonized version of the theory is simply given by Eq. (13.6). In general, states in the NS sector are created by vertex operators

$$V_\lambda(z) =: e^{i\lambda \cdot \phi(z)} :, \tag{13.7}$$

where λ are D_5 lattice vectors in the (0) or (V) conjugacy class. For instance, for the tachyon $\lambda = 0$ and for the $SO(10)$ vector $\lambda = (0, \ldots, \pm 1, \ldots, 0)$.

On the other hand, states in the R sector of the theory have Fermi statistics. They are created again by vertex operators of the form Eq. (13.7) but now with λ a lattice vector of the (S) or (C) conjugacy classes of D_5. The ground states are the two massless spinors of opposite chirality denoted by $|S^\alpha\rangle$ and $|S^{\dot\alpha}\rangle$ with D_5 weights $\lambda = (\pm\frac{1}{2}, \pm\frac{1}{2}, \pm\frac{1}{2}, \pm\frac{1}{2}, \pm\frac{1}{2})$ with an even and odd number of minus signs, respectively.

The vertex operator of the massless vector has conformal dimension $h = \frac{\lambda^2}{2} = \frac{1}{2}$. Remember that vertex operators of physical massless states must have $h = 1$. Similarly, the massless spinors belong to vertex operators with conformal dimension $h = \frac{5}{8}$, also in contradiction to the physical state condition. (Note that for massless

13.1 Bosonization and First Order Systems

states $e^{ik \cdot X}$ has vanishing conformal weight.) This discrepancy clearly indicates that the vertex operators of Eq. (13.7) with $\lambda \in (0), (V), (S), (C)$ of D_5 are not the full vertex operators of the fermionic string theory and have to be complemented by an additional piece.

The mismatch of the conformal dimension is however not the only problem. Consider the operator algebra between the vector ψ^μ and the spinor fields S_α, $S_{\dot\alpha}$ in a $SO(10)$ covariant chiral basis:

$$\psi^\mu(z) S_\alpha(w) = \frac{1}{\sqrt{2}} \frac{(\Gamma^\mu)_\alpha{}^{\dot\beta}}{(z-w)^{1/2}} S_{\dot\beta}(w) + \ldots,$$

$$S_\alpha(z) S_\beta(w) = \frac{1}{\sqrt{2}} \frac{(\Gamma_\mu)_{\alpha\beta}}{(z-w)^{3/4}} \psi^\mu(w) + \ldots,$$

$$S_\alpha(z) S_{\dot\beta}(w) = \frac{C_{\alpha\dot\beta}}{(z-w)^{5/4}} + \frac{1}{2} \frac{(\Gamma_\mu \Gamma_\nu)_{\alpha\dot\beta}}{(z-w)^{1/4}} :\psi^\mu \psi^\nu(w): + \ldots. \quad (13.8)$$

Here $C_{\alpha\dot\beta}$ is the $SO(10)$ charge conjugation matrix and $(\Gamma_\mu)_\alpha{}^{\dot\beta}$ are Dirac matrices in ten dimensions. Consult the appendix of Chap. 8 for details. There are several ways to derive these operator products. One uses bosonization and an explicit form for the cocycle factors. The normalization factors are easy to get this way. Another is to use $SO(10)$ covariance.

These equations show that $S_\alpha(z)$ $(S_{\dot\alpha}(z))$ creates a branch cut which renders the theory non-local. Locality of the operator product algebra is, however, necessary e.g. to get well-defined scattering amplitudes, which require an integration over the positions of the vertex operators. The appearance of branch cuts can be traced back to the D_5 inner-product rule $(V) \cdot (S) = \frac{1}{2} + n$, $n \in \mathbb{Z}$ (cf. Table 11.2 on page 333). Furthermore, S_α does not anticommute with itself, as $(S) \cdot (S) = 5/4 + n$. Also note the branch cut in the operator product between S_α and $S_{\dot\beta}$ due to $(S) \cdot (C) = 3/4 + n$.

In summary, without further modifications the fermionic string theory is non-local and therefore ill-defined. This is simply a reflection of the properties of the D_5 weight lattice. Therefore one expects that the complete vertex operators of the fermionic string require a modification of the D_5 lattice. The missing piece will be provided by the superconformal ghost system. Before demonstrating this, let us first present a general discussion of first order systems such as the b, c and β, γ ghost systems. Their bosonization will be of particular importance for the construction of the vertex operators of the superstring.

We introduce a common notation and consider b and c as conjugate fields with first order action

$$S = \frac{1}{2\pi} \int d^2z \, b \, \bar\partial c. \quad (13.9)$$

The field b has conformal weight λ and c has weight $1 - \lambda$ and the action is conformally invariant. The statistics of b and c is parametrized by ϵ: $\epsilon = 1$ for Fermi statistics and $\epsilon = -1$ for Bose statistics. For $\lambda = 2$, $\epsilon = 1$, this is the conformal ghost system, discussed in detail in Chap. 5. For $\lambda = 3/2$, $\epsilon = -1$ this is the superconformal ghost system, called (β, γ) in Chap. 8. The system of one complex fermion is an example of a fermionic first order system with $\lambda = 1/2$. Here we identify $b = \Psi^+$ and $c = \Psi^-$.

The equations of motion and their solutions are

$$\bar{\partial} b = 0, \quad \text{i.e. } b = b(z),$$
$$\bar{\partial} c = 0, \quad \text{i.e. } c = c(z). \tag{13.10}$$

The propagator is

$$\langle c(z) b(w) \rangle = \frac{1}{z - w} \tag{13.11}$$

and the basic operator products are

$$c(z) b(w) = \frac{1}{z - w} + \text{regular}, \quad b(z) c(w) = \frac{\epsilon}{z - w} + \text{regular}. \tag{13.12}$$

The $b(z) b(w)$ and $c(z) c(w)$ products are non-singular. We have the mode expansions

$$b(z) = \sum_{n \in a - \lambda + \mathbb{Z}} z^{-n-\lambda} b_n,$$

$$c(z) = \sum_{n \in a + \lambda + \mathbb{Z}} z^{-n-(1-\lambda)} c_n \tag{13.13}$$

with the following Hermiticity conditions

$$b_n^\dagger = \epsilon b_{-n}, \quad c_n^\dagger = c_{-n} \tag{13.14}$$

and (anti) commutator

$$[c_m, b_n]_\epsilon = \delta_{m+n}. \tag{13.15}$$

For the case of half-integer λ, there are two sectors, the R sector specified by $a = \frac{1}{2}$ and the NS sector with $a = 0$.[2] The action of the modes on the $SL(2, \mathbb{C})$ invariant vacuum is

$$b_n |0\rangle = 0 \quad \text{for } n \geq 1 - \lambda,$$
$$c_n |0\rangle = 0 \quad \text{for } n \geq \lambda. \tag{13.16}$$

---

[2] We can also have $a = \frac{1}{2}$ for integer λ corresponding to a twisted sector.

13.1 Bosonization and First Order Systems

The energy-momentum tensor can be derived either by first coupling the theory to a world-sheet metric or as the Noether current for the conformal symmetry. Either way we find

$$T = -\lambda :b\partial c: + (1-\lambda) :(\partial b)c:$$
$$= \frac{1}{2}(:(\partial b)c: - :b\partial c:) + \frac{1}{2}\epsilon Q\, \partial(:bc:), \qquad (13.17)$$

where we have introduced

$$Q = \epsilon(1 - 2\lambda). \qquad (13.18)$$

Normal ordering is defined in the usual way, cf. e.g. Eq. (5.10) on page 108. The significance of Q as a background charge will become clear below. The Virasoro generators are[3]

$$L_n = \sum_m (m - (1-\lambda)n) :b_{n-m} c_m: . \qquad (13.19)$$

The central charge is found by computing the operator product between $T(z)$ and $T(w)$:

$$c = -\epsilon(12\lambda^2 - 12\lambda + 2) = \epsilon(1 - 3Q^2) \qquad (13.20)$$

and it is straightforward to verify that

$$T(z)\, b(w) = \frac{\lambda b(w)}{(z-w)^2} + \frac{\partial b(w)}{z-w} + \text{regular} \qquad (13.21)$$

and

$$T(z)\, c(w) = \frac{(1-\lambda)c(w)}{(z-w)^2} + \frac{\partial c(w)}{z-w} + \text{regular}, \qquad (13.22)$$

as expected. In Table 13.1 we collect the values of the various parameters for the conformal and super-conformal ghost systems and complex NSR fermions.

The action Eq. (13.9) is invariant under a chiral $U(1)$ acting as $c \to e^{i\alpha}c$, $b \to e^{-i\alpha}b$ with Noether current $\delta S = -\frac{i}{2\pi}\int d^2z\, j\, \bar\partial\alpha$:

$$j(z) = - :b(z)\, c(z): = \sum_n z^{-n-1} j_n, \qquad (13.23)$$

---

[3]The generalization of (5.13) is $L_0 = \sum_{n=1}^{\infty} n(b_{-n}c_n + c_{-n}b_n) - \frac{1}{2}\lambda(\lambda - 1)$.

Table 13.1 Familiar first order systems

	ϵ	λ	Q	c
b, c	1	2	-3	-26
β, γ	-1	$\frac{3}{2}$	2	11
$\psi^{\pm i}$	1	$\frac{1}{2}$	0	1

where

$$j_n = \sum_m \epsilon :c_{n-m} b_m: . \tag{13.24}$$

The operator products of b and c with j reflect the fact that they have $U(1)$ charges -1 and $+1$ respectively:

$$j(z) b(w) = \frac{-1}{z-w} b(w) + \text{regular},$$

$$j(z) c(w) = \frac{1}{z-w} c(w) + \text{regular}. \tag{13.25}$$

Furthermore one finds

$$j(z) j(w) = \frac{\epsilon}{(z-w)^2} + \text{regular} \tag{13.26}$$

or, equivalently,

$$[j_m, j_n] = \epsilon m \, \delta_{n+m}. \tag{13.27}$$

The operator product of the chiral current and the energy-momentum tensor is anomalous:

$$T(z) j(w) = \frac{Q}{(z-w)^3} + \frac{j(w)}{(z-w)^2} + \frac{\partial j(w)}{z-w} + \ldots \tag{13.28}$$

and only for $Q = 0$, $(\lambda = \frac{1}{2})$ is j a true primary field with $h = 1$. In fact, (13.28) is equivalent to the following transformation of the ghost current under infinitesimal conformal transformations:

$$\delta j(z) = -\partial \xi(z) j(z) - \xi(z) \partial j(z) - \frac{Q}{2} \partial^2 \xi(z), \tag{13.29}$$

which can be integrated to (cf. (4.40) on page 72)

$$j'(w) = \left(\frac{\partial w}{\partial z}\right)^{-1} \left\{ j(z) - \frac{Q}{2} \frac{\partial^2 w}{\partial w} \right\} . \tag{13.30}$$

13.1 Bosonization and First Order Systems

Equation (13.28) also implies that the ghost number current is not conserved

$$\nabla^z j_z(z) = \frac{1}{4} Q\, R(h). \tag{13.31}$$

This can be seen as follows. From (13.28) we derive $\bar{\partial}_{\bar{w}} \langle T(z) j(w) \rangle = -\pi Q \partial_w^2 \delta^2(z-w)$. On the other hand, if we couple the theory to a metric of the form $ds^2 = 2dz d\bar{z} + h_{\bar{z}\bar{z}} d\bar{z}^2$ we find $\langle j(w) \rangle = -\frac{1}{4\pi} \int d^2z\, h^{zz}(z) \langle j(w) T(z) \rangle_0 + \mathcal{O}(h^2)$ where the subscript 0 indicates that the correlation function is computed at $h_{\bar{z}\bar{z}} = 0$ and we have also used that $\langle j(z) \rangle_0 = 0$. From this we derive $\bar{\partial} \langle j(w) \rangle = \frac{Q}{4} \partial_w^2 h_{\bar{w}\bar{w}}(w)$ whose covariant form is (13.31).

The anomaly in the $U(1)$ current is related to the existence of b, c zero modes. Their number is determined by the Riemann-Roch theorem, cf. (6.32) on page 132,

$$N_c - N_b = \epsilon Q(g-1) = (1-2\lambda)(g-1), \tag{13.32}$$

where g is the genus of the Riemann surface. The situation is very similar to the $U(1)$ anomaly in gauge theory. There the anomalous divergence of the chiral $U(1)$ current is also a topological density whose integral measures the difference between the number of massless left- and right-handed fermions, i.e. the zero modes of the chiral Dirac operator. A heuristic way to fix the normalization of the r.h.s. of Eq. (13.32) is to explicitly construct the normalizable c-zero modes on the sphere as we did in Chap. 6. They are the monomials $1, z, \ldots, z^{2\lambda-2}$. There are none for b.[4]

The operator product Eq. (13.28) is equivalent to the anomalous commutators

$$[L_n, j(z)] = \frac{1}{2} Q n(n+1) z^{n-1} + (n+1) z^n\, j(z) + z^{n+1} \partial j(z) \tag{13.33}$$

and

$$[L_n, j_m] = \frac{1}{2} Q n(n+1)\, \delta_{n+m} - m\, j_{m+n}. \tag{13.34}$$

From Eq. (13.33) it is apparent that $j(z)$ transforms covariantly under translations (L_{-1}) and dilations (L_0) but not under the transformations generated by L_{+1}; i.e. $j(z)$ is not quasi-primary. It is easy to show using the Hermiticity conditions Eq. (13.14) that $j_n^\dagger = -j_{-n}, \forall n \neq 0$. The case $n=0$ is delicate because of normal ordering ambiguities. We can use Eq. (13.34) to find $j_0^\dagger$:

$$j_0^\dagger = -[L_{-1}, j_1]^\dagger = -[L_1, j_{-1}] = -j_0 - Q. \tag{13.35}$$

---

[4] Alternatively, we require that $\delta_\alpha \int Db Dc\, b^{N_b} c^{N_c} e^{-S[b,c]} = 0$ where the b and c insertions are necessary to absorb the zero modes. This emphasizes the meaning of Q as a background charge, cf. below.

Then, if O_p is an operator with $U(1)$ charge p, i.e. $[j_0, O_p] = pO_p$ and $|q\rangle$ a state with $U(1)$ charge q, we find $p\langle q'|O_p|q\rangle = \langle q'|[j_0, O_p]|q\rangle = -(q' + q + Q)\langle q'|O_p|q\rangle$; i.e. we have to insert an operator with $U(1)$ charge $p = -(q+q'+Q)$ in order to get a non-vanishing result. We then normalize the states such that

$$\langle -q - Q|q\rangle = 1 \tag{13.36}$$

for the non-vanishing inner products. This makes the meaning of Q as a background charge manifest.[5]

We now bosonize the b, c system by a chiral scalar field

$$j(z) = \epsilon \partial \phi(z) \tag{13.37}$$

with

$$\phi(z)\phi(w) \sim \epsilon \ln(z - w). \tag{13.38}$$

In terms of ϕ the action is

$$S = -\frac{1}{8\pi} \int d^2z \sqrt{h} \left(\epsilon h^{\alpha\beta} \partial_\alpha \phi \partial_\beta \phi + QR\phi\right), \tag{13.39}$$

from which the anomalous current conservation law Eq. (13.31) follows as the equation of motion for ϕ. The energy momentum tensor derived from above action is (after going to conformal gauge)

$$T^{(j)} = \epsilon \left(\frac{1}{2} :jj: -\frac{1}{2}Q \, \partial j\right). \tag{13.40}$$

As a check we can reproduce Eq. (13.28).

Conformal (primary) fields $V(z)$ are exponentials of $\phi(z)$:

$$V(z) =: e^{q\phi(z)}:, \tag{13.41}$$

where q is (half-) integer for the NS (R) sector. This will become clear below. (We again suppress possible derivative terms and will drop normal ordering symbols from now on.) Therefore, the allowed bosonic momenta are points in a D_1

---

[5]A completely equivalent way is to consider $\langle \oint_C \frac{dz}{2\pi i} j(z)...\rangle$ where ... stands for some operator insertions on the sphere which the contour C encloses. In the absence of a background charge this would be zero as we can contract the contour on the back of the sphere. This requires a change of variables $z = 1/u$ and the contour now encircles $u = 0$. However, due to the inhomogeneous term in (13.30) we get a non-zero contribution. This means that only correlation functions of operators with a total charge Q can be non-zero.

13.1 Bosonization and First Order Systems

lattice where integer lattice points belong to the (0) or (V) conjugacy classes and half-integer elements to the (S) and (C) conjugacy classes of D_1. The $U(1)$ charge of $e^{q\phi(z)}$ can be determined as

$$j(z) e^{q\phi(w)} = \frac{q}{z-w} e^{q\phi(w)} + \ldots, \qquad (13.42)$$

and its conformal dimension follows directly from

$$T(z) e^{q\phi(w)} = \left[\frac{\frac{1}{2}\epsilon q(q+Q)}{(z-w)^2} + \frac{\partial_w}{z-w} \right] e^{q\phi(w)} + \ldots. \qquad (13.43)$$

The contribution to the conformal weight linear in q has its origin in the ghost number anomaly. The operator $e^{q\phi(z)}$ shifts the ghost charge of the vacuum by q units. It is the vertex operator for a state $|q\rangle$:

$$|q\rangle = \oint_{C_0} \frac{dz}{2\pi i} \frac{1}{z} e^{q\phi(z)} |0\rangle = e^{q\phi(0)} |0\rangle, \qquad (13.44)$$

which satisfies $j_0|q\rangle = q|q\rangle$. According to Eq. (13.36) one therefore obtains

$$\langle 0 | e^{-Q\phi(z)} | 0 \rangle = 1. \qquad (13.45)$$

So far we have bosonized only the $U(1)$ current $j(z)$ which is a bilinear in the b, c fields. In the case of Fermi statistics ($\epsilon = +1$), the fields b, c themselves can be bosonized in a straightforward way—they are given by the exponentials of $\phi(z)$ (compare also with Eq. (13.6)):

$$b(z) = e^{-\phi(z)}, \quad c(z) = e^{\phi(z)}$$

$$\epsilon = +1:$$

$$\phi(z)\phi(w) \sim \ln(z-w). \qquad (13.46)$$

One readily verifies that for $\epsilon = 1$ Eq. (13.40) follows from Eq. (13.17).

On the other hand, in case of Bose statistics the "bosonization" of $b(z)$, $c(z)$ is more complicated. In fact, the energy-momentum tensor $T^{(j)}$ is not complete for $\epsilon = -1$. If we calculate the anomaly in the $T^{(j)}T^{(j)}$ operator product we find

$$c^{(j)} = (1 - 3\epsilon Q^2) = \begin{cases} c & \epsilon = +1 \\ c+2 & \epsilon = -1, \end{cases} \qquad (13.47)$$

where c refers to the value of the anomaly given in Eq. (13.20). This means that in the bosonic case the field ϕ does not give a complete description of the system. This can already be deduced from the fact that the solitons $e^{\pm\phi}$ are always fermions and cannot accommodate Bose statistics. One requires extra fields for the "bosonization" of b, c. What is needed is a fermionic system with $c = -2$. Let us therefore define two conjugate free fermions $\eta(z)$, $\xi(z)$ of conformal weight 1 and 0 respectively. They constitute a first order system with $\epsilon = 1$, $\lambda^{\eta\xi} = 1$, $Q^{\eta\xi} = -1$, $c^{\eta\xi} = -2$. Their operator product is

$$\eta(z)\xi(w) \sim \xi(z)\eta(w) \sim \frac{1}{z-w}. \tag{13.48}$$

Then the Bose fields b, c can be "bosonized" as

$$b(z) = e^{-\phi(z)}\partial\xi(z), \quad c(z) = \eta(z)e^{\phi(z)},$$

$$\epsilon = -1:$$

$$\phi(z)\phi(w) \sim -\ln(z-w). \tag{13.49}$$

The correct operator products Eq. (13.12) between b and c are easily verified using Eq. (13.49).[6] One may also show that the 'bosonized' version of the energy-momentum tensor (13.17) is a sum of two terms of the form (13.40), one for a first order system with λ and $\epsilon = -1$ and one for a system with $\lambda = 1$ and $\epsilon = +1$.

The η, ξ system contains its own chiral $U(1)$ current which provides a second scalar field $\chi(z)$:

$$:\xi(z)\eta(z): = \partial\chi(z), \quad \chi(z)\chi(w) \sim \ln(z-w),$$

$$\eta(z) = e^{-\chi(z)}, \quad \xi(z) = e^{\chi(z)}. \tag{13.50}$$

Thus in terms of ϕ and χ, b and c for $\epsilon = -1$ are expressed as:

$$b(z) = e^{-\phi(z)}e^{\chi(z)}\partial\chi(z), \quad c(z) = e^{-\chi(z)}e^{\phi(z)}. \tag{13.51}$$

It is important to note that the irreducible representations of the b, c algebra are built only from ϕ, η and $\partial\xi$; the zero mode field ξ_0 never appears in the b, c algebra. On the sphere we have $N_\xi - N_\eta = 1$ implying that as long as we do not include ξ_0 we do not have to neutralize the background charge of the η, ξ system. One distinguishes between the 'small' and the 'large' algebra, where the former only includes $\partial\xi$ while the latter also includes the ξ zero-mode. The (β, γ) system belongs to the small

---

[6]We adopt the convention that different types of fermions anti-commute. This is implemented by multiplying each bosonized fermion by a cocycle factor, which we do not write explicitly. For instance, the order of the factors in (13.49) matters in order to reproduce (13.12).

13.1 Bosonization and First Order Systems

algebra and any operator which belongs to the small algebra can be constructed in terms of (β, γ).[7]

Let us now address the question of vacuum states. A hint that the $SL(2,\mathbb{C})$ invariant vacuum $|0\rangle$ might not be the only possible one came already from our discussion of the conformal ghost system. We know that regularity requires that $c_n|0\rangle = 0$ for $n \geq \lambda$. From $[L_0, c_n] = -nc_n$ we find that $L_0 c_n|0\rangle = -n c_n|0\rangle \neq 0$ for $n < \lambda$. In particular, for $0 < n < \lambda$, c_n lowers the energy of the vacuum $|0\rangle$. If c_n is a bosonic operator, we can apply it to $|0\rangle$ an arbitrary number of times thus lowering the vacuum energy by an arbitrary amount. We find that in the bosonic case the spectrum is unbounded from below. It is clearly also unbounded from above. Consequently, the question of the vacuum is ambiguous. In the Fermi case we can also build states with negative energy, but not without lower bound. Here the situation is familiar from Dirac theory; we can define different vacua depending on to what level the states are filled, and these vacua are stable due to the exclusion principle. In the Bose case the situation is unfamiliar and would be a disaster were it not for the fact that we are dealing with a free, i.e. non-interacting theory, which does not allow transitions from one vacuum to another.

We define an infinite number of vacua $|q\rangle$ which can be viewed as Bose/Fermi seas, by requiring

$$b_n|q\rangle = 0, \quad n > \epsilon q - \lambda,$$
$$c_n|q\rangle = 0, \quad n \geq -\epsilon q + \lambda \tag{13.52}$$

where $q \in \mathbb{Z}$ for the NS sector and $q \in \mathbb{Z} + \frac{1}{2}$ for the R sector. Since in the Fermi case the different vacua are distinguished by the occupation of a finite number of states, we can go from one vacuum to another by application of a finite number of creation or annihilation operators. This is not so in the Bose case. Here a finite number of operators will never bring us from one vacuum to another. The q-vacua are in fact identical to the states $|q\rangle = e^{q\phi(0)}|0\rangle$ we have encountered before. It is straightforward to show that they satisfy Eq. (13.52). For instance, for Bose statistics ($\epsilon = -1$) we have $b_n|q\rangle = \oint \frac{dz}{2\pi i} z^{n+\lambda-1} e^{-\phi(z)} \partial \xi(z) e^{q\phi(0)}|0\rangle = \oint \frac{dz}{2\pi i} z^{n+\lambda+q-1} :$ $e^{-\phi(z)} \partial \xi(z) e^{q\phi(0)} : |0\rangle$. Equation (13.52) then follows from regularity of the normal ordered product at $z = 0$. We also see that when acting with an operator in the NS sector ($n + \lambda =$ integer) on a state with half-integer q we get a branch cut. Hence these states belong to the R sector. From the expressions of j_n and L_n in terms of modes of b and c it follows that

$$j_n|q\rangle = L_n|q\rangle = 0, \quad n > 0. \tag{13.53}$$

---

[7]Using the integral representation of the δ-function and formal manipulations, one can show, by verifying the OPEs, that $e^{\phi(z)} = \delta(\beta(z))$, $e^{-\phi(z)} = \delta(\gamma(z))$, $\eta(z) = \partial \gamma(z) \delta(\gamma(z))$ and $\partial \xi(z) = \partial H(\beta(z))$ (H is the Heaviside step function). Another way to verify this is to show that the states $\delta(\beta(0))|0\rangle$ and $\delta(\gamma(0))|0\rangle$ have the properties (13.52) for $\epsilon = -1$ and $\lambda = 3/2$ of the $|+1\rangle$ and $|-1\rangle$ vacua, respectively. Note that the zero-mode of ξ cannot be expressed in terms of β, γ.

The propagator receives a finite correction from the vacuum charge:

$$\langle -q - Q | c(z) b(w) | q \rangle \equiv \langle c(z) b(w) \rangle_q$$
$$= \sum_{n \leq \epsilon q - \lambda} \langle -q - Q | [c_{-n}, b_n]_\epsilon | q \rangle z^{-n-(1-\lambda)} w^{-n-\lambda}$$
$$= \left(\frac{z}{w}\right)^{\epsilon q} \frac{1}{z - w}. \tag{13.54}$$

The conformal properties of the current $j(z)$ and the energy momentum tensor are also modified. Using the above propagator we find

$$\langle j(z) j(w) \rangle_q = \frac{\epsilon}{(z-w)^2},$$

$$\langle T(z) j(w) \rangle_q = \frac{Q}{(z-w)^3} + \frac{q}{z(z-w)^2},$$

$$\langle T(z) T(w) \rangle_q = \frac{c}{2(z-w)^4} + \frac{\epsilon q(Q+q)}{zw(z-w)^2}, \tag{13.55}$$

where c is the central charge (13.20) of the bc-system. Comparing this with Eq. (13.28) and the $T(z) T(w)$ operator product gives

$$\langle j(z) \rangle_q = \frac{q}{z} = \frac{1}{z} \langle j_0 \rangle_q,$$

$$\langle T(z) \rangle_q = \frac{1}{2} \epsilon q (q + Q) \frac{1}{z^2} = \frac{1}{z^2} \langle L_0 \rangle_q, \tag{13.56}$$

i.e.

$$j_0 |q\rangle = q |q\rangle,$$
$$L_0 |q\rangle = \frac{1}{2} \epsilon q(q + Q) |q\rangle. \tag{13.57}$$

Equation (13.57) also follows directly from Eqs. (13.43) and (13.44) or via the point-splitting method

$$\lim_{w \to z} \left(\langle -\lambda b(w) \partial c(z) + (1-\lambda) \partial b(w) c(z) \rangle_q + \frac{\epsilon}{(w-z)^2} \right) = \frac{\langle L_0 \rangle_q}{z^2}. \tag{13.58}$$

We see that the L_0 eigenvalue is bounded from below for fermions ($\epsilon = +1$) and unbounded for bosons ($\epsilon = -1$), in agreement with our discussion above. The $SL(2, \mathbb{C})$ invariant vacuum has $h = 0$. There are two states which satisfy this condition, namely $|0\rangle$ and $|-Q\rangle$. However it is easy to show that $L_{-1} |q\rangle \neq 0$ for $q \neq 0$; i.e. $|0\rangle$ is the unique $SL(2, \mathbb{C})$ invariant state.

13.2 Covariant Vertex Operators, BRST and Picture Changing

Let us now apply the bosonization of the super-conformal ghosts β, γ to the construction of the vertex operators of the fermionic string theory. As already mentioned, the conformal fields of the NSR (ψ^μ) part of the fermionic string have to be completed by conformal fields of the β, γ system. This is in analogy to the bosonic string where physical states contain also ghost excitations. The fundamental reason for including the superconformal ghosts is again the requirement of BRST invariance of the physical states. We will see this in the following.

The complete ghost sector of the fermionic string consists of two first order systems: the conformal ghosts (b, c) for a fermionic $\lambda = 2$ system and the superconformal ghosts (β, γ) form a bosonic $\lambda = \frac{3}{2}$ system, whose 'bosonization' we have discussed above and which we repeat here for completeness:

$$b = e^{-\sigma}, \quad c = e^{\sigma}, \quad \sigma(z)\sigma(w) \sim \ln(z-w) \tag{13.59}$$

and

$$\beta = e^{-\phi}\partial\xi = e^{-\phi}e^\chi \partial\chi, \quad \gamma = \eta e^\phi = e^{-\chi}e^\phi,$$

$$\phi(z)\phi(w) \sim -\ln(z-w), \quad \chi(z)\chi(w) \sim \ln(z-w), \quad \xi(z)\eta(w) \sim \frac{1}{z-w}, \tag{13.60}$$

Energy-momentum tensor and supercurrent of the ghost sector are

$$T^{gh} = -2b\partial c - (\partial b)c - \frac{3}{2}\beta\partial\gamma - \frac{1}{2}(\partial\beta)\gamma,$$

$$T^{gh}_F = -(\partial\beta)c - \frac{3}{2}\beta\partial c + \frac{1}{2}b\gamma, \tag{13.61}$$

and the total ghost number operator reads

$$N_g = -\oint \frac{dz}{2\pi i}(bc + \beta\gamma) = \oint \frac{dz}{2\pi i}(cb + \partial\phi). \tag{13.62}$$

For the superconformal ghost system there exist a new quantum number, called the (ghost) picture charge, which is defined as

$$N_p = \oint \frac{dz}{2\pi i}(\xi\eta - \partial\phi) = \oint \frac{dz}{2\pi i}(\partial\chi - \partial\phi) \tag{13.63}$$

such that the original β, γ ghosts have picture number zero.

Consider states of the form $|\lambda\rangle_\psi \otimes |q\rangle_{\beta,\gamma}$ with corresponding vertex operators in bosonized form:

$$V_{\lambda,q}(z) = e^{i\lambda\cdot\phi(z)} e^{q\phi(z)}, \tag{13.64}$$

where λ is a weight vector of D_5 and the picture number q is a D_1 "lattice vector". The conformal dimension of (13.64) is (cf. Eq. (13.43) with $\epsilon = -1$ and $Q = 2$)

$$h = \frac{1}{2}\lambda^2 - \frac{1}{2}q^2 - q + N \tag{13.65}$$

(N counts possible oscillator excitations which we neglected in Eq. (13.64).). The superghosts β, γ satisfy the same periodicity conditions as the world-sheet gravitino and therefore, due to the coupling of the gravitino to the supercurrent, the same boundary conditions as the NSR fermions ψ^μ. This implies that we must couple the R (NS) sector of the ψ^μ system to the R (NS) sector of the superconformal ghost system. This is to say that for $\lambda \in$ (S), (C) of D_5, q must be half-integer, and for $\lambda \in$ (0), (V) of D_5, q is integer.

Let us first look at the NS sector of the theory. Here, massless vectors are characterized by D_5 lattice vectors $\lambda = (0,\ldots,\pm 1,\ldots,0)$ and the corresponding vertex operators have conformal dimension $\frac{1}{2}$. Using formula Eq. (13.65) we see that the full vertex operator Eq. (13.64) describes a massless vector with $h = 1$ if

$$q = -1. \tag{13.66}$$

We will call this the canonical choice for the picture number and NS vertex operators with $q = -1$ are said to be in the canonical ghost picture. Thus, a general NS state in the canonical ghost picture is created by a vertex operator

$$V_{\lambda,-1}(z)\, e^{ik\cdot X(z)}, \qquad \lambda \in (0), (V) \text{ of } D_5 \tag{13.67}$$

and has (right-moving) mass

$$\frac{\alpha'}{2} m_R^2 = \frac{1}{2}\lambda^2 - \frac{1}{2}q^2 - q - 1 = \frac{1}{2}\lambda^2 - \frac{1}{2}. \tag{13.68}$$

Here we have again neglected oscillator contributions. The -1 contribution to the zero point energy is due to the requirement that physical vertex operators have conformal dimension $h = +1$ or, from the point of view of Chap. 5, due to the reparametrization ghosts. Together with the $\frac{1}{2}$ mass unit from the superconformal ghosts β, γ one obtains the correct normal ordering constant, the tachyon mass in the NS sector. The ground state in the NS sector is thus $e^{-\phi}c(0)|0\rangle$.

Let us now consider the R sector, where massless spinors correspond to the weight vectors $\lambda = (\pm\frac{1}{2}, \pm\frac{1}{2}, \pm\frac{1}{2}, \pm\frac{1}{2}, \pm\frac{1}{2})$ with an even and odd number of minus

13.2 Covariant Vertex Operators, BRST and Picture Changing

signs for S_α and $S_{\dot\alpha}$ respectively. We now find that

$$q = -\frac{1}{2} \tag{13.69}$$

for the vertex operator Eq. (13.64) of the massless spinors to have conformal dimension $h = 1$. Again, we call this the canonical ghost picture in the R sector. All vertex operators in this ghost picture are of the form

$$V_{\lambda,-\frac{1}{2}}(z)\, e^{ik\cdot X(z)}, \quad \lambda \in S, C \text{ of } D_5 \tag{13.70}$$

with mass

$$\frac{\alpha'}{2} m_R^2 = \frac{1}{2}\lambda^2 - \frac{1}{2}q^2 - q - 1 = \frac{1}{2}\lambda^2 - \frac{5}{8}. \tag{13.71}$$

Now consider operator products between two different vertex operators as given in Eq. (13.64):

$$V_{\lambda,q}(z) V_{\lambda',q'}(w) = (z-w)^{\lambda\cdot\lambda' - qq'}\, V_{\lambda+\lambda', q+q'}(w) + \ldots. \tag{13.72}$$

Since the ghost charges add, it is apparent that vertex operators in non-canonical ghost pictures appear in the operator product expansion. What is their meaning? Recall that in the bosonic string we found two vertex operators for each physical state, the integrated $\int V$ and the unintegrated, cV which satisfy $[Q, V] = \partial(cV)$. One can use either one in correlation functions as long as the ghost zero modes are saturated. A similar redundancy is found for the fermionic string. However, since the superconformal ghosts are Grassmann even, we find an infinite degeneracy on top of the twofold degeneracy due to the conformal ghost system. For each physical state we have an infinite number of integrated and unintegrated vertex operators with different picture numbers. They are related by the so-called picture changing operation.

For better understanding of the meaning of states with arbitrary, non-canonical ghost charge q, which are required for closure of the operator algebra Eq. (13.72), we need to use the BRST formalism. For this we need to construct the BRST charge Q of the fermionic string theory. It receives contributions from the fields X, b, c as well as from their superpartners ψ, β, γ. One can straightforwardly generalize the discussion of Chap. 5 to graded symmetry algebras, such as the super-Virasoro algebra (12.23) and construct a nilpotent BRST charge. One finds

$$Q = \oint \frac{dz}{2\pi i} \left\{ c\left(T^{X,\psi} + \frac{1}{2}T^{b,c,\beta,\gamma}\right) - \gamma\left(T_F^{X,\psi} + \frac{1}{2}T_F^{b,c,\beta,\gamma}\right) \right\}$$
$$= Q_0 + Q_1 + Q_2, \tag{13.73}$$

where

$$Q_0 = \oint \frac{dz}{2\pi i} \left(c\, T^{X,\psi,\beta,\gamma} + c(\partial c)b\right),$$

$$Q_1 = -\oint \frac{dz}{2\pi i} \gamma\, T_F^{X,\psi} = -\frac{i}{2}\sqrt{\frac{2}{\alpha'}} \oint \frac{dz}{2\pi i} e^{-\chi} e^{\phi} \psi_\mu \partial X^\mu,$$

$$Q_2 = -\oint \frac{dz}{2\pi i} \frac{1}{4} b\gamma^2 = +\oint \frac{dz}{2\pi i} \frac{1}{4} b\, e^{2\phi} e^{-2\chi}. \tag{13.74}$$

The subscript on Q denotes the superconformal ghost charge. Q_0 is the BRST operator of the bosonic theory, if we treat β, γ as extra matter fields. Q_1 generates world-sheet supersymmetry transformations with parameter given by the supersymmetry ghost γ. Finally, Q_2 is needed for the closure of the BRST algebra. Note that Q has picture number zero. For the BRST transformations of the various fields we find

$$[Q, X^\mu(z)] = c\partial X^\mu(z) + \frac{i}{2}\sqrt{\frac{\alpha'}{2}} \gamma \psi^\mu(z),$$

$$\{Q, \psi^\mu(z)\} = \left(\frac{1}{2}\partial c \psi^\mu(z) + c\partial \psi^\mu(z)\right) - \frac{i}{2}\sqrt{\frac{2}{\alpha'}} \gamma \partial X^\mu(z),$$

$$\{Q, c(z)\} = c\partial c(z) - \frac{1}{4}\gamma^2(z),$$

$$[Q, \gamma(z)] = -\frac{1}{2}\partial c\gamma(z) + c\partial \gamma(z),$$

$$\{Q, b(z)\} = T^{\text{tot}}(z),$$

$$[Q, \beta(z)] = -T_F^{\text{tot}}(z),$$

$$[Q, T^{\text{tot}}(z)] = \frac{1}{8}(d-10)\partial^3 c(z),$$

$$\{Q, T_F^{\text{tot}}(z)\} = -\frac{1}{8}(d-10)\partial^2 \gamma. \tag{13.75}$$

Here T^{tot} and T_F^{tot} are the energy-momentum tensor and the supercurrent for all the fields involved (matter and ghosts). On the matter fields Q acts as a combined conformal transformation with parameter $-c$ and a superconformal transformation with parameter γ. One can verify that $Q^2 = 0$ for $d = 10$ and that the action is BRST invariant.

Consider the holomorphic part of the vertex operator for a massless spinor in the canonical $q = -\frac{1}{2}$ ghost picture:

$$V_{-\frac{1}{2}}(z) = u^\alpha S_\alpha(z) e^{-\frac{1}{2}\phi(z)} e^{ik\cdot X(z)}. \tag{13.76}$$

13.2 Covariant Vertex Operators, BRST and Picture Changing

u_α is the spinor wave function. Let us first show that it is BRST invariant i.e. $[Q, V_{-\frac{1}{2}}]$ vanishes up to a total derivative which is irrelevant upon integration over z. First we have (cf. Eq. (5.44) on page 115)

$$[Q_0, V_{-\frac{1}{2}}] = \partial(c V_{-\frac{1}{2}}). \tag{13.77}$$

Then, using Eqs. (13.8) and (13.49), we obtain

$$e^{\phi-\chi} T_F(z) V_{-\frac{1}{2}}(w) \sim \frac{1}{2\sqrt{2}} \sqrt{\frac{\alpha'}{2}} (z-w)^{-1} (\not{k} u)^{\dot\alpha} e^{\frac{1}{2}\phi} e^{-\chi} S_{\dot\alpha}(w) e^{ik \cdot X(w)}. \tag{13.78}$$

Thus $[Q_1, V_{-\frac{1}{2}}] = 0$ if we demand that u_α satisfies the on-shell condition $\not{k} u = 0$. This is the massless Dirac equation in momentum space for the polarization spinor $u_\alpha(k)$. Finally

$$e^{2\phi} e^{-2\chi} b(z) V_{-\frac{1}{2}}(w) \sim (z-w) e^{\frac{3}{2}\phi} e^{-2\chi} b \, u^\alpha S_\alpha e^{ik \cdot X(w)} \tag{13.79}$$

shows that $[Q_2, V_{-\frac{1}{2}}] = 0$. Therefore $V_{-\frac{1}{2}}$ is a BRST invariant vertex operator. In general, BRST invariance of physical state vertex operators requires that they satisfy on-shell conditions.

However we can create a second version of the fermion vertex operator with different ghost number which is also BRST invariant. This operator is defined as

$$V_{+\frac{1}{2}} = 2[Q, \xi V_{-\frac{1}{2}}]. \tag{13.80}$$

The subscript $\frac{1}{2}$ denotes that this vertex operator now has picture number $q = +\frac{1}{2}$. $V_{+\frac{1}{2}}$ is obviously BRST closed. One might think that it is also BRST exact, in which case it would decouple in correlation functions of BRST invariant operators. This is however not the case since the β, γ algebra and consequently also Q only contain $\partial \xi$, but not the zero mode of ξ.[8] We can then compute[9]

---

[8] The BRST cohomology in the large algebra is trivial because Q is invertible there. Indeed, $\{Q, c\xi \partial \xi e^{-2\phi}\} = \text{constant}$.

[9] In the derivation of Eq. (13.82) we need the subleading term of the first of the operator products in Eq. (13.8). One can show that in d dimensions

$$k_\mu u^\alpha \psi^\mu(z) S_\alpha(w) \sim \frac{1}{\sqrt{2}} (z-w)^{1/2} \frac{1}{\frac{d}{2}-1} u^\alpha k_\mu \psi^\mu \psi^\nu (\Gamma_\nu)_\alpha{}^{\dot\beta} S_{\dot\beta} \tag{13.81}$$

for on-shell u^α.

$$V_{+\frac{1}{2}} = [Q_0 + Q_1 + Q_2, 2\xi V_{-\frac{1}{2}}]$$

$$= -2\partial(\xi c V_{-\frac{1}{2}}) + \frac{1}{\sqrt{\alpha'}} u^\alpha \{e^{\frac{1}{2}\phi}(i\partial X^\mu + \tfrac{\alpha'}{8}(k\cdot\psi)\psi^\mu)(\Gamma_\mu)_\alpha{}^{\dot\beta} S_{\dot\beta}\} e^{ik\cdot X}$$

$$+ \frac{1}{2} b\eta e^{\frac{3}{2}\phi} u^\alpha S_\alpha e^{ik\cdot X}. \tag{13.82}$$

The derivative term is the contribution from Q_0 and the term with ghost charge $\tfrac{3}{2}$ is generated by Q_2. Both of these terms will never contribute to correlation functions (at least at tree level); the former vanishes upon integration over z and the latter contains one b field which will not be absorbed by a c field. Also note that the zero mode of ξ does not contribute to $V_{+\frac{1}{2}}$, so its ghost part is also in the β, γ algebra. The physical fermion vertex operator in the $+\tfrac{1}{2}$ ghost picture which is relevant for tree-level computations is therefore

$$V_{+\frac{1}{2}} = [Q_1, 2\xi V_{-\frac{1}{2}}]$$

$$= -2 \oint \frac{dw}{2\pi i} e^{-\chi(w)} e^\phi T_F(w) e^{\chi(z)} V_{-\frac{1}{2}}(z)$$

$$= +2 \lim_{w\to z} e^{\phi(w)} T_F(w) V_{-\frac{1}{2}}(z)$$

$$= +\frac{1}{\sqrt{\alpha'}} u^\alpha \{e^{\frac{1}{2}\phi}\left(i\partial X^\mu + \tfrac{\alpha'}{8}(k\cdot\psi)\psi^\mu\right)(\Gamma_\mu)_\alpha{}^{\dot\beta} S_{\dot\beta}\} e^{ik\cdot X}, \tag{13.83}$$

where we have restored factors of α'.

This discussion generalizes in the following way. As we already mentioned, we have both integrated and unintegrated vertex operators. The above was an example of an integrated vertex operator, i.e. $\int V_{-1/2}$ had to be BRST invariant and $V_{-1/2}$ had to have conformal weight one.[10] There are also unintegrated vertex operators V' which are BRST invariant, i.e. they satisfy $[Q, V'] = 0$ and are related to the integrated ones as $[Q, V] = \partial V'$. Using that $[Q, L_{-1}] = 0$ together with the requirement that the BRST cohomology be non-trivial shows that Q-closure of V' is a consequence of $[Q, V] = \partial V'$.

In the fermionic string, V and V' come in infinitely many version of different picture number, which are related as follows. Since Q has zero picture number q, at fixed picture number we still have the above relations between V_q and V'_q. One defines the picture changing operator via

$$V'_{q+1} = P_{+1} V'_q = [Q, 2\xi V'_q]. \tag{13.84}$$

From $[Q, V'_q] = 0$ it follows that

$$P_{+1} V'_q(z) \equiv \lim_{w\to z} Z(w) V'_q(z) \tag{13.85}$$

---

[10] In Chap. 5 we wrote ϕ for integrated and $\psi = c\phi$ for unintegrated vertex operators.

13.2 Covariant Vertex Operators, BRST and Picture Changing

with

$$Z = 2\{Q, \xi\} = 2c\partial\xi + 2T_F^{(X,\psi)}e^\phi + \frac{1}{2}e^{2\phi}b\partial\eta + \frac{1}{2}\partial(e^{2\phi}b\eta)$$

$$= 2T_F^{\text{tot}}e^\phi - \frac{1}{2}(\partial b)\eta e^{2\phi}$$

$$= 2T_F^{(X,\psi)}e^\phi + \ldots \tag{13.86}$$

is the picture raising operator. It carries one unit of picture charge. In the last line we have explicitly written only the part of Z which are relevant for all vertex operators which we will encounter in this and, in particular, in Chap. 16. We will often simply call this the picture changing operator. For integrated vertex operators one defines

$$V_{q+1} = [Q, 2\xi V_q] + \partial(2\xi V_q') = P_{+1}V_q + 2\partial\xi V_q'. \tag{13.87}$$

We have seen in the example above that $[Q_0, 2\xi V_q]$ is a total derivative. It has to be subtracted since we want V_{q+1} to be in the small algebra. The subtraction of the zero-mode part of ξ is such that it does not affect BRST invariance. Indeed, $\{Q, V_{q+1}\} = \partial[Q, 2\xi V_q'] = \partial V_{q+1}'$, which vanishes upon integration.

We have seen in the example above that the relevant part of the picture raising operator acts like a world-sheet supersymmetry transformation. In fact, apart from the ghost contribution, it essentially acts like $\hat{G}_{-\frac{1}{2}}$ (cf. (12.35) on 363). This also explains the factor of two in the definition of the picture raising operation, which is the same as the one in the defining relation between G_r and T_F.

The inverse of picture raising is picture lowering. The picture lowering operator P_{-1} is the inverse of P_{+1}. Defining a field $Y(w)$ by $P_{-1}V_q' = \lim_{w \to z} Y(w)V_q'(z) = V_{q-1}'(z)$ one gets $\lim_{z \to w} Z(z)Y(w) = 1$. One verifies that $Y = 2c\partial\xi e^{-2\phi}$. We will not go into any further details. In summary, the picture changing operations provide infinitely many BRST invariant vertex operators for every physical state with arbitrary picture number q where $q \in \mathbb{Z}$ ($\mathbb{Z} + \frac{1}{2}$) for space-time bosons (fermions).

As a second example, picture changing of the vertex operator of the massless vector in the -1 picture,

$$V_{-1} = e^{-\phi} \epsilon_\mu \psi^\mu e^{ik \cdot X} \tag{13.88}$$

leads to its copy in the 0 picture:

$$V_0 = \sqrt{\frac{2}{\alpha'}} \epsilon_\mu \left(i \partial X^\mu + \frac{\alpha'}{2}(k \cdot \psi)\psi^\mu \right) e^{ik \cdot X}. \tag{13.89}$$

There is no contribution from Q_2. For the tachyon we derive

$$V_0 = \sqrt{\frac{\alpha'}{2}} k \cdot \psi \, e^{ik \cdot X}. \tag{13.90}$$

We will discuss other vertex operators and also their normalization in Chap. 16 when we use them to compute scattering amplitudes.

One can immediately check that the mass formula Eq. (13.65) gives the same answer in any ghost picture. Z, which has conformal weight 0, always acts in a way that the effect of changing the ghost charge q is compensated e.g. by oscillators ∂X^μ or extra ψ^μ factors. Thus the mass spectrum of the fermionic string is in fact bounded from below unlike the spectrum of the β, γ theory. The reason is that only BRST invariant combinations are accepted as physical states. Also, the physical vertex operators, no matter in what ghost picture, never contain the zero mode of ξ; i.e. they are always in the β, γ algebra.

Note that it follows from Eq. (12.15) that a NS sector vertex operator in the zero ghost picture, which depends only on the fields from the matter sector, is BRST invariant if it is the upper component of a $h = \frac{1}{2}$ superfield. The lower component is provided by the vertex operator in the canonical ghost picture (apart from the factor $e^{-\phi}$).

The picture changing operation is important for the evaluation of correlation functions. We have seen that correlation functions in the $SL(2, \mathbb{C})$ invariant vacuum vanish if we do not insert an operator with ghost charge $q = -Q = -2$. This means that to get non-vanishing scattering amplitudes we have to choose the ghost pictures for the vertex operators such that the total superconformal ghost charge adds up to -2. This is analogous to the situation we encountered when discussing the conformal ghosts in Chap. 6. There the vacuum carried three units of ghost charge and they had to be absorbed by ghost zero mode insertions. If we denote by σ the boson that arises from bosonizing the conformal ghosts b, c and by ϕ the boson from the β, γ system, we have[11]

$$\langle -Q^{b,c} - Q^{\beta,\gamma}|0\rangle = \langle 0|e^{3\sigma(0)-2\phi(0)}|0\rangle = 1. \tag{13.91}$$

BRST invariance of the states $|0\rangle$ and $e^{3\sigma(0)-2\phi(0)}|0\rangle$ is easy to verify. For instance, since in the critical dimension Q commutes with the L_n's (cf. Eq. (13.75)), we find that $L_{0,\pm 1} Q|0\rangle = 0$. Since the $SL(2, \mathbb{C})$ invariant vacuum state is unique, we conclude $Q|0\rangle = 0$. Invariance of $e^{3\sigma(0)-2\phi(0)}|0\rangle$ is shown by computing $[Q, e^{3\sigma(0)-2\phi(0)}]$. Hence the vacuum expectation values of BRST invariant operators will be BRST invariant.

It is now important that correlation functions of physical vertex operators are independent of how we distribute the ghost charges among them as long as $\sum_i q_i = -2$, at tree level, say. The reason is the following. Since none of the vertex operators depends on the zero mode ξ_0, we can insert it in the functional integral and integrate over it; i.e. we insert $1 = \int d\xi_0 \xi_0$. Since ξ_0 is a Grassmann variable, we can replace it by $\xi(z)$ for arbitrary z. (This follows from $\int \mathcal{D}\xi' d\xi_0 \xi_0 F(\xi') = \int \mathcal{D}\xi' d\xi_0 \xi(z) F(\xi')$ where ξ' denotes the non-zero mode

---

[11]If we include the zero mode of the η, ξ system, we also have to neutralize its background charge and get $\langle 0|e^{3\sigma-2\phi+\chi}|0\rangle = 1$. $e^{3\sigma-2\phi+\chi}|0\rangle$ is however not BRST invariant.

part of ξ.) So we can attach $\xi(z)$ to any of the vertex operators in the correlation function, say to $V_{q_1}(z_1)$. Now rewrite any of the other vertex operators, say $V_{q_2}(z_2)$, as $V_{q_2}(z_2) = 2\oint \frac{dw}{2\pi i}\xi(w)j_{\text{BRST}}(w)V_{q_2-1}(z_2)$. We deform the integration contour by pulling it off the back of the sphere. Due to BRST invariance it passes through all vertex operators except for $\xi(z_1)V_{q_1}(z_1)$ which becomes $V_{q_1+1}(z_1)$. Then $\int d\xi_0 \xi(w) = 1$. In summary, we have traded one unit of picture charge between two vertex operators within a BRST invariant correlation function without changing its value.

Most of our discussion in this section included only one chiral sector of the fermionic string. For the type II theories we have to combine the two chiral sectors, in particular when constructing vertex operators, which are now doubly infinitely degenerate. For the heterotic string we have to combine the chiral vertex operators of the fermionic string with anti-chiral vertex operators of the bosonic string. As we will discuss in more detail in Chap. 15, one can construct covariant vertex operators also for states of compactified string theories. This does not affect the ghost sector but only the structure of the $c = 15$ ($c = 26$) conformal field theory.

For open strings, i.e. for world-sheets with boundary, there are some changes, e.g. for the BRST transformation we employ the doubling trick, etc. Rather than presenting a general discussion, we will point out those aspects which will be important for the computation of open string tree level amplitudes in Chap. 16. On the disk, the superconformal ghost charges of the vertex operators have to add up to -2. However, due to the presence of the boundary, the holomorphic and anti-holomorphic parts are no longer independent, i.e. there is no independent left- and right-moving background charge. In this respect, the situation is completely analogous to that of the b, c ghosts. In the presence of D-branes of various dimensions, the $c = 15$ ($c = 26$) conformal field theory changes, while the ghost sector is the one discussed here.

As we have discussed in Chap. 6, one-point functions of the disk do not necessarily vanish. An example are one-point functions (tadpoles) of RR fields which will play an important role later on. Even though we will not need to compute one-point functions explicitly, it is worth pointing out it would require the RR vertex operator in an asymmetric ghost picture, e.g. $(-\frac{1}{2}, -\frac{3}{2})$. The construction of the appropriate vertex operator is rather complicated and will not be pursued here. We will discuss other open string vertex operators in Chap. 16.

13.3 D-branes and Space-Time Supersymmetry

The vertex operator (13.76) is the holomorphic part of the gravitino vertex operator of the type II and heterotic strings (cf. Chap. 16 for the complete vertex operator). The right-moving space-time supercharge is the integral of this vertex operator at zero momentum

$$Q_{R\alpha} \sim \oint \frac{dz}{2\pi i} S_\alpha e^{-\phi/2}(z). \qquad (13.92)$$

$Q_{R\alpha}$ is a Majorana spinor. In this section the spin fields S_α, $\alpha = 1,\ldots,32$ are Majorana, not Majorana-Weyl. The latter will be denoted as $S^\pm = \frac{1}{2}(1 \pm \Gamma)S \equiv \Gamma_\pm S$, rather than S_α and $S_{\dot\alpha}$ with $\alpha, \dot\alpha = 1,\ldots, 16$, as we did previously. Here $\Gamma \equiv \Gamma_{11} = \Gamma^0 \ldots \Gamma^9$.

A sensible algebra for the charges results only, if the operator products have no branch cuts. This is implemented by the GSO projection which only allows, say, $Q_R^{(+)}$. Using the operator product of two spin fields one verifies the chiral supersymmetry algebra

$$\{Q_{R\alpha}^{(+)}, Q_{R\beta}^{(+)}\} \sim \left(\Gamma_+ \Gamma_\mu C^{-1}\right)_{\alpha\beta} p_{(-1)}^\mu. \tag{13.93}$$

Here

$$p_{(-1)}^\mu \sim \oint \frac{dz}{2\pi i} \psi^\mu e^{-\phi}(z) \tag{13.94}$$

is the zero-momentum part of the holomorphic piece of the graviton vertex operator in the canonical ghost picture. (In this discussion we drop factors of α' and numerical factors.) It is related to the space-time momentum

$$p^\mu \sim \oint \frac{dz}{2\pi i} i\partial X^\mu(z) \tag{13.95}$$

via picture changing. The supercharges (13.92) are in the canonical ghost picture. It is not difficult to work out their version in the $(+1/2)$ picture:

$$Q_{R\alpha} \sim \oint \frac{dz}{2\pi i} i\partial X^\mu (\Gamma_\mu)_\alpha{}^\beta S_\beta \, e^{\phi/2}(z). \tag{13.96}$$

The supercharge transforms a space-time fermion into a space-time boson and vice versa:

$$\{Q_{R\alpha}^{(+)}, u^\alpha S_\alpha^{(+)} e^{-\phi/2} e^{ik\cdot X}\} \sim \left(\Gamma_+ \Gamma_\mu\right)_{\alpha\beta} u^\beta \psi^\mu e^{-\phi} e^{ik\cdot X} \tag{13.97}$$

and

$$[Q_{R\alpha}^{(+)}, \epsilon_\mu \psi^\mu e^{-\phi} e^{ik\cdot X}] \sim k^\mu \epsilon^\nu \left(\Gamma_+ \Gamma_{\mu\nu}\right)_\alpha{}^\beta S_\beta^{(-)} e^{-\phi/2} e^{ik\cdot X}. \tag{13.98}$$

For type II strings a second gravitino comes from the left-movers: $Q_L^{(+)}$ for type IIB and $Q_L^{(-)}$ for type IIA.

How is this modified in type I theory or, more generally, in the presence of Dp-branes? To analyze this, we need to understand spin fields in the presence of a boundary. Recall the discussion of boundary conditions of world-sheet fermions in Sect. 7.3. If we map them to the upper half plane and take into account that the

13.3 D-branes and Space-Time Supersymmetry

fermions are half-differentials which leads to a non-trivial contribution from the Jacobian, we find

$$\psi^\mu(z) = \eta\, D^\mu{}_\nu\, \overline{\psi}^\nu(\bar z)\Big|_{\mathrm{Im}\, z=0}, \qquad \eta = \begin{cases} +1 & (\mathrm{NS}), \\ \mathrm{sign}(\mathrm{Re}\, z) & (\mathrm{R}). \end{cases} \tag{13.99}$$

The matrix $D^\mu{}_\nu$ was defined previously: it has eigenvalues $+1$ along $(p+1)$ NN and -1 along $(9-p)$ DD directions, i.e. these are the boundary conditions relevant in the presence of a Dp-brane. For the spin fields we make the ansatz

$$S_\alpha(z) = P_\alpha{}^\beta\, \overline{S}_\beta(\bar z)\Big|_{\mathrm{Im}\, z=0}. \tag{13.100}$$

Our task is to determine P. This can be done by requiring the consistency of the OPE

$$\psi^\mu(z)\, S_\alpha(w) \sim \frac{1}{\sqrt{2}(z-w)}(\Gamma^\mu)_\alpha{}^\beta\, S_\beta(w) \tag{13.101}$$

and its analogue for the left-movers with the boundary conditions (13.100). This leads to the condition

$$P(D^\mu{}_\nu\, \Gamma^\nu) = \Gamma^\mu P \tag{13.102}$$

which is solved by

$$P = \prod_i (\Gamma_i\, \Gamma). \tag{13.103}$$

The product is over all D-directions. The sign in (13.103) is arbitrary but otherwise the phase of P is fixed by the requirement that it transforms Majorana spinors to Majorana spinors.[12] One can check that (13.103) is also consistent with the $S_\alpha(z) S_\beta(w)$ OPE.

One important observation is that P reverses the chirality, if the number of Dirichlet directions is odd. Recall from Chap. 4 the discussion of conserved charges in the presence of a boundary. It follows from this discussion that Q_L and PQ_R must have the same chirality and that the conserved supercharges are

$$Q_\alpha = Q_{L\alpha} + (PQ_R)_\alpha, \tag{13.104}$$

---

[12] We use $C = C_+$ in which case Majorana spinors and the Dirac matrices are real in the Majorana representation.

where Q_L and Q_R are represented by integrals over semi-circles, as in (4.147).[13] This reproduces the familiar result that the spectrum of SUSY preserving Dp-branes is p even for IIA and p odd for IIB. For the type I theory we need symmetry under $S \leftrightarrow \bar{S}$. With (13.100) this requires $P^2 = 1$ which leaves only $p = 1, 5, 9$, another familiar result.

We can parametrize SUSY transformations by two Majorana fermions $\epsilon_{L,R}$ and define the supercharges

$$\bar{\epsilon}_L Q_L + \bar{\epsilon}_R Q_R. \tag{13.105}$$

For type IIB $\epsilon_{L,R}$ have the same chirality and for type IIA they have opposite chiralities. In either case there are 16+16 independent real components and hence as many independent supercharges. In the presence of D-branes $\epsilon_{L,R}$ are not independent. They have to satisfy

$$\epsilon_L = \pm P \, \epsilon_R \tag{13.106}$$

and solutions to this equation parametrize unbroken supercharges. The two signs distinguish branes from anti-branes.

Since $\epsilon_{L,R}$ are chiral spinors, the condition for the preserved supercharge of a Dp-brane along the directions $(x^0, x^1, \ldots, x^p)$ can be conveniently written in the form

$$\epsilon_L = \pm \Gamma^0 \Gamma^1 \cdots \Gamma^p \, \epsilon_R. \tag{13.107}$$

If there are several branes, either of different dimensions or spanning different directions in space, one obtains several conditions on $\epsilon_{L,R}$. These might or might not have non-trivial solutions. The number of parameters in the solution counts the number of supercharges left unbroken by the brane configuration.

Consider two Dp-branes, rotated against each other. If R is the $SO(9)$ matrix which rotates the world-volume of one into the other—the time-direction is common to both—the supersymmetries preserved by the two branes are

$$\epsilon_L = \Gamma^0 \cdots \Gamma^p \, \epsilon_R \quad \text{and} \quad \epsilon_L = R^{-1} \Gamma^0 \cdots \Gamma^p R \, \epsilon_R. \tag{13.108}$$

While generically they have no non-trivial common solution,[14] there are special cases for which they do. To see this, we consider the case of two D6-branes which intersect along four-dimensional Minkowski space-time, already studied in Sect. 10.6. Here $R \in SO(6)$ and the relations to analyze are ($\Gamma_{11} \epsilon_R = \epsilon_R$)

---

[13]Using the boundary condition (13.100) on the real axis, we can analytically continue the spin field to the lower half plane and use the doubling trick to represent the supercharge by a closed-contour integral. The superconformal ghost is continuous across the real axis.

[14]The condition is $\det(\Gamma^0 \cdots \Gamma^p - R^{-1} \Gamma^0 \cdots \Gamma^p R) = 0$.

13.4 The Covariant Lattice

$$\epsilon_L = \Gamma^7 \Gamma^8 \Gamma^9 \epsilon_R \quad \text{and} \quad \epsilon_L = R^{-1} \Gamma^7 \Gamma^8 \Gamma^9 R \epsilon_R. \tag{13.109}$$

Furthermore is suffices to consider the case where $\epsilon_{L,R}$ are chiral and anti-chiral spinors of $SO(6)$, respectively.[15] The two conditions (13.109) have a non-trivial solution, if $R \in SU(3) \subset SO(6)$. To see this, recall our discussion in Appendix 8.5 and choose for ϵ_R the highest weight state $|\Omega\rangle$ which is annihilated by $b^k = \frac{1}{2}(\Gamma^k + i\Gamma^{k+3})$, $k = 4, 5, 6$. Since this state is a singlet of $SU(3)$, it is invariant under $R \in SU(3)$. $\Gamma^7 \Gamma^8 \Gamma^9$ transforms $|\Omega\rangle$ into $|\overline{\Omega}\rangle$, the lowest weight state which is annihilated by $b_k = \frac{1}{2}(\Gamma^k - i\Gamma^{k+3})$. This state is also a singlet and is therefore also invariant under $SU(3)$. In other words, the two conditions in (13.109) become identical and are simultaneously solved by $\epsilon_R \sim |\Omega\rangle$ and $\epsilon_L \sim |\overline{\Omega}\rangle$. We conclude that for $R \in SU(3)$ the two branes preserve one quarter of the supersymmetries which are preserved by either one of them, i.e. four supercharges. This corresponds to $\mathcal{N} = 1$ space-time supersymmetry in the four dimensions which are common to both branes. The same is true, if we add additional D6 branes as long as they are all related by $SU(3)$ rotations.

To make contact with our discussion in Chap. 10, we note that a general $SO(6)$ rotation matrix can be diagonalized to

$$\text{diag}(e^{+i\phi_1}, e^{+i\phi_2}, e^{+i\phi_3}, e^{-i\phi_1}, e^{-i\phi_2}, e^{-i\phi_3}). \tag{13.110}$$

For $\phi_1 + \phi_2 + \phi_3 = 0 \bmod 2\pi$ this is an element of $SU(3)$ in the $\mathbf{3} + \overline{\mathbf{3}}$, which is the decomposition of the $\mathbf{6}$ of $SO(6)$ under $SU(3) \subset SO(6)$. This is the condition (10.197). Replacing in the above discussion $SU(3)$ by $SU(2)$, we find that two D7-branes intersect along $\mathbb{R}^{1,5}$ and that they preserve eight supercharges. The results obtained can also be generalized to intersecting Dp-branes at angles for other values of p: one finds that as long as they are related by $SU(n)$ rotations, some supersymmetry is preserved.

The discussion of unbroken supercharges in the presence of branes given here relies on CFT arguments. We will reproduce these results using space-time arguments in Sect. 18.5.

13.4 The Covariant Lattice

Let us now return to the operator product expansion Eq. (13.72). It strongly suggests to combine the D_5 weight vectors λ with the D_1 weights q to a six-dimensional vector $w = (\lambda, q)$. Doing this we can write the operator algebra as

$$V_{w_1}(z) V_{w_2}(w) = (z - w)^{w_1 \cdot w_2} V_{w_1 + w_2}(w) + \ldots. \tag{13.111}$$

---

[15] To see this, refer to the decomposition (14.40).

Table 13.2 Decomposition of $D_{5,1}$ conjugacy classes

$D_{5,1}$	D_5	$\otimes$	D_1
(0)	(V,V)	$\oplus$	(0,0)
(V)	(V,0)	$\oplus$	(0,V)
(S)	(S,S)	$\oplus$	(C,C)
(C)	(C,S)	$\oplus$	(S,C)

Closure of the algebra implies that w_1, w_2 are vectors of a six-dimensional Lorentzian lattice $D_{5,1}$ with metric signature $(+ + + + +, -)$. The minus sign is due to the ghost sector. This enlarged Lorentzian lattice is usually called covariant lattice, as it describes the covariant vertex operators of the fermionic string. Since the only allowed vectors of $D_{5,1}$ decompose under $D_5 \otimes D_1$ as $w = (\lambda; q)$, where λ and q belong both to either the NS sector or both to the R sector, the lattice $D_{5,1}$ contains four conjugacy classes: (0), (V), (S) and (C). Now consider the weights of $D_{5,1}$ which correspond to the states of lowest mass in the canonical ghost picture:

$$w_{\text{tachyon}} = (0, 0, 0, 0, 0; -1) \in (V) \text{ of } D_{5,1}$$

$$w_{\text{vector}} = (0, \ldots, \pm 1, 0, \ldots, 0; -1) \in (0) \text{ of } D_{5,1}$$

$$w_{\text{spinor}} = (\pm \frac{1}{2}, \pm \frac{1}{2}, \pm \frac{1}{2}, \pm \frac{1}{2}, \pm \frac{1}{2}; -\frac{1}{2}) \in (C) \text{ of } D_{5,1} \quad \text{odd number of minus signs}$$

$$w_{\text{antispinor}} = (\pm \frac{1}{2}, \pm \frac{1}{2}, \pm \frac{1}{2}, \pm \frac{1}{2}, \pm \frac{1}{2}; -\frac{1}{2}) \in (S) \text{ of } D_{5,1} \quad \text{even number of minus signs}.$$

We recognize that for states in the canonical ghost picture, all vectors in the (V) [(S)] conjugacy class of D_5 belong to the (0) [(C)] conjugacy class of $D_{5,1}$ and vice versa. The decomposition of the conjugacy classes of $D_{5,1}$ into those of $D_5 \otimes D_1$ is given in Table 13.2. The mass of a state $|w\rangle$ can be written in terms of lattice vectors of $D_{5,1}$ as (remember the negative metric of this lattice)

$$\frac{\alpha'}{2} m_R^2 = \frac{1}{2} w^2 + w \cdot e_6 - 1, \qquad (13.112)$$

where e_6 is the basis vector $(0, 0, 0, 0, 0; 1)$. From the discussion above we understand the meaning of lattice vectors with different ghost charge q. They all correspond to copies of the same physical state but in different ghost pictures. However there is no one-to-one relation between $D_{5,1}$ lattice vectors and physical states. Only states in the canonical ghost picture are directly related to lattice vectors of $D_{5,1}$. In other ghost pictures there is no clear relation between lattice vectors and physical vertex operators. These are in general given by linear combinations of vertex operators of the form Eq. (13.64). For instance, the relevant $D_{5,1}$ lattice vectors for the massless vector in the 0 picture are the null vector and the vectors $(\alpha; 0)$ where α is a root of D_5. The picture changing operation does not change the $D_{5,1}$ conjugacy class of a state. This is so because the picture raising operator P_{+1}

13.4 The Covariant Lattice

Table 13.3 Mutual scalar products of $D_{5,1}$ conjugacy classes

	(0)	(V)	(S)	(C)
(0)	$\mathbb{Z}$	$\mathbb{Z}$	$\mathbb{Z}$	$\mathbb{Z}$
(V)		$\mathbb{Z}$	$\mathbb{Z}+\frac{1}{2}$	$\mathbb{Z}+\frac{1}{2}$
(S)			$\mathbb{Z}$	$\mathbb{Z}+\frac{1}{2}$
(C)				$\mathbb{Z}$

corresponds to the $D_{5,1}$ lattice vector[16]

$$w_{PC} = (0,\ldots,\pm 1,\ldots,0;+1), \tag{13.113}$$

which is a root of $D_{5,1}$. Picture changing acts on the lattice $D_{5,1}$ by simply adding w_{PC}.

Let us return to the question of locality, i.e. absence of branch cuts in the operator product algebra of the fermionic string. The exponents of $(z - w)$ are determined by the inner product rules of conjugacy classes of the Lorentzian lattice $D_{5,1}$ which are summarized in Table 13.3. We realize that we have almost reached our aim to obtain a local theory by extending D_5 to $D_{5,1}$. The massless spinor ($\in$ (C) of $D_{5,1}$) is now local with respect to the massless vector ($\in$ (0) of $D_{5,1}$). The branch cut in the operator product $\psi^\mu(z) S^\alpha(w)$ is canceled by the branch cut in $e^{-\phi(z)} e^{-\frac{1}{2}\phi(w)}$. However there are still some sources of non-locality. The tachyon ($\in$ (V) of $D_{5,1}$) is non-local with respect to the spinor (and also to the antispinor) and also the spinor is non-local with respect to the antispinor ($\in$ S of $D_{5,1}$). Thus a projection is needed which eliminates half of the conjugacy classes of $D_{5,1}$ and makes the theory local. We see that the NS sector with (0) and (V) of $D_{5,1}$ leads to a local, closed operator algebra as well as the projection onto the (0) and (C) (or equivalently (S)) conjugacy classes of $D_{5,1}$. The latter projection is identical to the GSO projection introduced within the fermionic formulation of the spinning string. It leads to a space-time supersymmetric spectrum; the (0) and (C) conjugacy classes contain each others supersymmetric partners.

We now show that this projection on the lattice $D_{5,1}$ with (0) and (C) (or (S)) conjugacy classes is enforced by modular invariance of the one loop partition function of the fermionic string. This is in complete analogy to the fact that in the fermionic language the GSO projection was necessary when summing over a modular invariant combination of different spin structures. The holomorphic part of the one loop partition function of the fermionic string in Hamiltonian description has the form

$$\chi(\tau) \sim \text{Tr}\, e^{2\pi i \tau (L_0 - 1)} (-1)^{F_R}. \tag{13.114}$$

---

[16] This is for the relevant part of P_{+1}. The other pieces also correspond to vectors in the (0) conjugacy class of $D_{5,1}$.

F_R is the right-moving space-time fermion number and the phase factor $(-1)^{F_R}$ takes into account the correct space-time statistics, i.e. $(-1)^{F_R}$ is 1 for space-time bosons and -1 for space-time fermions. Let us first discuss the non-trivial part of Eq. (13.114), namely the contribution of the zero modes of the bosons ϕ^i ($i = 1, \ldots, 5$) and ϕ. They just give the sum over the lattice vectors of the Lorentzian lattice $D_{5,1}$ (recall Eq. (13.65)):

$$\tilde{\chi}(\tau) = \sum_{w=(\lambda,q) \in D_{5,1}} e^{2\pi i \tau (\frac{1}{2}\lambda^2 - \frac{1}{2}q^2 - q)} e^{-2\pi i q}. \qquad (13.115)$$

The factor $e^{-2\pi i q}$ ensures the correct space-time statistics using the fact that q is (half) integer for space-time bosons (fermions).

However this expression for $\tilde{\chi}(\tau)$ cannot represent the physical partition function, as the sum extends over arbitrarily high ghost numbers q. In other words, Eq. (13.115) sums over all possible equivalent ghost pictures. In addition we know also that the physical (light-cone) partition function should only count the transverse degrees of freedom. In the covariant lattices language these physical light-cone states in the canonical ghost picture can be characterized by decomposing $D_{5,1}$ to a part which describes the transverse Lorentz group $SO(8)$ and a two-dimensional part which describes the longitudinal, time-like and ghost degrees of freedom of any state:

$$D_{5,1} = D_4 \otimes D_{1,1}, \quad w = (u, x),$$
$$w \in D_{5,1}, \quad u \in D_4, \quad x \in D_{1,1}. \qquad (13.116)$$

Then the physical state condition is to consider lattice vectors $w = (u, x)$ with fixed vector x_0 in the following way:

$$x_0 = (0, -1) \quad \text{NS sector},$$
$$x_0 = \left(\frac{1}{2}, -\frac{1}{2}\right) \quad \text{R sector}. \qquad (13.117)$$

The trace in the partition function should then run only over those states satisfying Eq. (13.117). To realize this constraint let us write a general vector $w \in D_{5,1}$ in a convenient form

$$w = (u, x_0) + \Delta = w_0 + \Delta, \qquad (13.118)$$

where Δ is the sum of two light-like (picture changing) vectors:

$$\Delta = m\Delta_1 + n\Delta_2, \quad n, m \in \mathbb{Z},$$
$$\Delta_1 = (0, 0, 0, 0, 1, 1),$$
$$\Delta_2 = (0, 0, 0, 1, 0, 1). \qquad (13.119)$$

13.4 The Covariant Lattice

Substituting Eq. (13.118) into Eq. (13.115) we are left with the following expression:

$$\tilde{\chi}(\tau) = \sum_{w_0} e^{2\pi i \tau (\frac{1}{2} w_0^2 + w_0 \cdot e_6)} e^{2\pi i w_0 \cdot e_6} \sum_{\Delta} e^{\pi i \tau \Delta^2} \quad (13.120)$$

where we have shifted Δ without affecting the infinite sum. The first part is simply the trace over all physical states. Therefore, to obtain the physical light-cone partition function, one has to divide $\tilde{\chi}(\tau)$ by

$$\Theta_{1,1}(\tau) = \sum_{\Delta} e^{\pi i \tau \Delta^2}. \quad (13.121)$$

$\Theta_{1,1}$ is the partition function of the two-dimensional Lorentzian even self-dual lattice $D_{1,1}$ which is, up to a $SO(1,1)$ Lorentz transformation the same as $\Gamma_{1,1}$ in Footnote 6 on page 275. Although the two functions $\Theta_{1,1}(\tau)$ and $\tilde{\chi}(\tau)$ are separately ill defined because of the Lorentzian metric, their ratio is nevertheless well defined and describes the physical partition function. Finally we also have to take into account the contribution of the bosonic X and ϕ oscillators. Then the complete (purely holomorphic) result is:

$$\chi(\tau) \sim \frac{1}{\eta(\tau)^{12} \Theta_{1,1}(\tau)} \sum_{w \in D_{5,1}} e^{2\pi i \tau \frac{1}{2}(w+e_6)^2} e^{2\pi i w \cdot e_6}. \quad (13.122)$$

Let us now check the modular invariance of $\chi(\tau)$. First consider the transformation $\tau \to \tau + 1$. Since $\eta(\tau)^{12}$ changes sign under this transformation while $\Theta_{1,1}(\tau)$ remains invariant, we require that the lattice sum also changes sign. This requires

$$\frac{1}{2} w^2 - q \in \mathbb{Z}, \quad (13.123)$$

which is satisfied, if states with (half) integer q corresponding to (R) NS states are associated with (odd) even points on the lattice. This is also what the spin statistics theorem demands. The operator product between vertex operators in the NS (R) sector ($\boldsymbol{\Phi} = (\boldsymbol{\phi}; -i\phi)$) is

$$e^{iw \cdot \boldsymbol{\Phi}(z)} e^{-iw \cdot \boldsymbol{\Phi}(w)} \sim (z-w)^{-w^2} + \ldots. \quad (13.124)$$

NS (R) states have to be (anti) commuting, i.e. $w^2 = 2n$ ($w^2 = 2n+1$), $n \in \mathbb{Z}$. It is easily checked that all vectors in the (S), (C) and (V) conjugacy classes of $D_{5,1}$ have odd (length)^2 whereas the vectors of the (0) conjugacy class have even (length)^2. Thus, requiring invariance under $\tau \to \tau + 1$ discards the (V) conjugacy class.

For the second transformation, $\tau \to -\frac{1}{\tau}$ we use that

$$\eta^{-1}(\tau) = (-i\tau)^{1/2} \eta^{-1}\left(-\frac{1}{\tau}\right),$$

$$\Theta_{1,1}^{-1}(\tau) = -i\tau \Theta_{1,1}^{-1}\left(-\frac{1}{\tau}\right). \qquad (13.125)$$

The second relation follows since $\Theta_{1,1}$ is the lattice sum of the even self-dual lattice $D_{1,1}$. Using now the Poisson resummation formula (9.115) we find

$$\chi(\tau) = \frac{\tau^4}{\text{vol}(D_{5,1})} \frac{1}{\eta^{12}\left(-\frac{1}{\tau}\right) \Theta_{1,1}\left(-\frac{1}{\tau}\right)} \sum_{w \in D_{5,1}^*} e^{-\frac{2\pi i}{\tau}\frac{1}{2}(w+e_6)^2} e^{-2\pi i w \cdot e_6}. \qquad (13.126)$$

Thus, apart from a factor $e^{4\pi i q}$, which is irrelevant for integer or half-integer q, $\chi(\tau)$ is invariant under $\tau \to -\frac{1}{\tau}$, if the covariant lattice $D_{5,1}$ is self-dual. (The factor τ^4 will be compensated by the transformation of $(\text{Im}\,\tau)^{-4}$.

In summary, modular invariance implies that $D_{5,1}$ must be an odd self-dual Lorentzian lattice which contains only the (0) and (S) (or (0) and (C)) conjugacy classes. Then the spinning string is automatically local as discussed above. These two conjugacy classes contain as lowest states a massless vector and massless spinor. They are the on-shell degrees of freedom of a ten-dimensional supermultiplet and one can show that also the massive states can be arranged into supermultiplets. Thus the requirement of self-duality of the covariant lattice $D_{5,1}$ is equivalent to the GSO projection.

The covariant lattice description also allows for a straightforward derivation of the light-cone partition function of the fermionic string in terms of Jacobi theta functions. Since the lattice $D_{5,1}$ contains the (0) and (S) conjugacy classes, the light-cone partition function is given by the difference between the lattice sums of the (V) and (C) conjugacy classes of the D_4 weight lattice. The (V) $\in D_4$ conjugacy class is obtained from the (0) $\in D_{5,1}$ by truncation to light-cone states according to the physical state selection rule Eq. (13.117). Likewise this gives (C) $\in D_4$ from (S) $\in D_{5,1}$. The relative minus sign takes into account spin-statistics. Thus, using the expressions Eq. (11.110) on page 345 for the sums over the lattice vectors of the D_n-weight lattice, the contribution of the world-sheet fermions to the light-cone partition function becomes:

$$\chi(\tau) \sim \frac{1}{2} \frac{1}{\eta(\tau)^4} [\vartheta_3^4(0|\tau) - \vartheta_4^4(0|\tau) - \vartheta_2^4(0|\tau)]. \qquad (13.127)$$

This expression is identical to the one obtained from the sum over all spin structures of the world-sheet fermions in the fermionic description (see Chap. 9) and vanishes due to the triality relation (9.14a) among the (V), (C) and (S) conjugacy classes of D_4. This reflects the underlying space-time supersymmetry: the contributions of space-time bosons and space-time fermions cancel.

13.4 The Covariant Lattice

The covariant Lorentzian lattice $D_{5,1}$ with (0) and (S) conjugacy classes is very similar to the root lattice of E_8 which can also be thought of being the weight lattice of D_8, again with (0) and (S) conjugacy classes. In both cases the restriction to these two conjugacy classes implies self-duality. Therefore, due to this analogy, we may call the self-dual lattice $D_{5,1}$ also $E_{5,1}$. Both, E_8 and $E_{5,1}$, contain spinorial generators which, for E_8, correspond to length$^2 = 2$ vectors and commute; for $E_{5,1}$ however they correspond to length$^2 = 1$ vectors and therefore anticommute. They generate the space-time supersymmetry algebra (cf. the discussion in Chap. 15). In fact, it turns out to be useful to consider instead of the Lorentzian lattice $E_{5,1}$ the Euclidean covariant lattice E_8. This is equivalent to replacing the superconformal ghost lattice D_1 by a D_3 lattice while simultaneously changing the signature of the metric of the lattice.[17] In other words, we replace the Lorentzian covariant lattice $D_{5,1}$ by the Euclidean covariant lattice D_8. The requirement of modular invariance implies in both cases that only the (0) and (S) conjugacy classes must be present such that we are dealing with the lattices $E_{5,1}$ resp. E_8. To describe the states in the canonical ghost picture we are forced to decompose E_8 (D_8) to $D_5^{\text{Lorentz}} \otimes D_3^{\text{ghost}}$ and consider only vectors of D_3^{ghost} with fixed entries. Conventionally we choose them as

$$w = (u, v) \in D_8, \quad u \in D_5^{\text{Lorentz}}, \quad v \in D_3^{\text{ghost}},$$

$$v_0 = \left(\frac{1}{2}, \frac{1}{2}, -\frac{1}{2}\right), \quad \text{R sector,}$$

$$v_0 = (0, 0, -1), \quad \text{NS sector.} \tag{13.128}$$

Note that $\frac{1}{2}v_0^2$ is exactly the superconformal ghost contribution to the conformal weight in the canonical ghost picture. Furthermore, if we are only interested in the physical light-cone states we have to decompose D_8 to $D_4 \otimes D_4$ and consider states which have fixed entries in second D_4:

$$x_0 = \left(\frac{1}{2}, \frac{1}{2}, \frac{1}{2}, -\frac{1}{2}\right) \quad \text{R sector,}$$

$$x_0 = (0, 0, 0, -1) \quad \text{NS sector.} \tag{13.129}$$

This means that we have replaced the Lorentzian lattice $D_{1,1}$, which describes the longitudinal, time-like and superconformal ghost degrees of freedom, by the Euclidean lattice D_4. The physical light-cone partition function can now be written as a lattice sum over the D_8 (E_8) lattice counting only those vectors which satisfy Eq. (13.129) and taking into account the correct spin statistics assignment. Since under $D_8 \to D_4$ the conjugacy classes are interchanged according to (V) $\leftrightarrow$ (0) and (S) $\leftrightarrow$ (C), the truncation to physical light cone states acts on the theta functions

---

[17] These lattice maps will be discussed in more detail below.

as $\vartheta_3^8 \to \vartheta_3^4$, $\vartheta_4^8 \to -\vartheta_4^4$ and $\vartheta_2^8 \to -\vartheta_2^4$. Given the known expression for the E_8 (chiral) partition function, $\chi_{E_8}(\tau) \sim \frac{1}{2}\frac{1}{\eta(\tau)^8}[\vartheta_2^8(0|\tau) + \vartheta_3^8(0|\tau) + \vartheta_4^8(0|\tau)]$ one immediately derives Eq. (13.127) as the physical light-cone partition function. (Note that one also has to drop the contribution of four oscillators.) At this point it is important to realize that the replacement of the Lorentzian lattice $D_{5,1}$ by the Euclidean lattice D_8 does not mean that the non-unitary ghost Hilbert space is contained in the positive definite Hilbert space of the D_8 Kač-Moody algebra. So far this procedure is just a convenient technical trick since the Euclidean lattices are much nicer to handle. Both descriptions lead, with the conditions Eq. (13.117) and Eq. (13.129) respectively, to the correct light-cone degrees of freedom.

13.5 Heterotic Strings in Covariant Lattice Description

In this section we also formulate the ten-dimensional heterotic strings in the covariant lattice description. In Chap. 9 we have discussed the original version of the heterotic string which has a space-time supersymmetric spectrum and gauge groups $E_8 \times E_8$ or $SO(32)$. Subsequently, additional (non-supersymmetric) heterotic string theories in ten-dimensions were discovered. They can also be formulated in the covariant lattice approach.

The holomorphic (right-moving) fermionic string in its bosonized version is characterized by the lattice $(D_{5,1})_R$ corresponding to the world-sheet fermions $\psi^\mu(z)$ and superconformal ghosts β, γ. The lattice vectors $w_R = (\lambda_R, q) \in D_{5,1}$ describe the Lorentz transformation properties and superconformal ghost charge of the right-moving part of any string state. To obtain the heterotic string theory we have to combine the right-moving fermionic string with the left-moving bosonic string. As discussed in Chap. 9, it consists of ten bosonic space-time coordinates $X^\mu(\bar{z})$ and of 16 'compactified' bosonic variables $X^A(\bar{z})$ ($A = 1, \ldots, 16$). The corresponding quantized momenta build a 16-dimensional Euclidean lattice $(\Gamma_{16})_L$ whose lattice vectors we denote by w_L. Thus the (soliton) vertex operators of the heterotic string theory have the general form (neglecting contributions from bosonic oscillators):

$$V_{w_L;\lambda_R,q}(z,\bar{z}) = e^{iw_L \cdot X(\bar{z})} e^{i\lambda_R \cdot \phi(z)} e^{q\phi(z)}. \tag{13.130}$$

Here we have once more dropped normal ordering symbols and cocycle factors. The operator product expansion of two such vertex operators

$$V_{w_{L_1};\lambda_{R_1},q_1}(z,\bar{z}) V_{w_{L_2};\lambda_{R_2},q_2}(w,\bar{w})$$
$$= (\bar{z}-\bar{w})^{w_{L_1} \cdot w_{L_2}} (z-w)^{\lambda_{R_1} \cdot \lambda_{R_2} - q_1 q_2} V_{w_{L_1}+w_{L_2};\lambda_{R_1}+\lambda_{R_2},q_1+q_2}(w,\bar{w}) + \ldots \tag{13.131}$$

shows that the condition for locality, i.e. the absence of branch cuts, reads

$$-w_{L_1} \cdot w_{L_2} + \lambda_{R_1} \cdot \lambda_{R_2} - q_1 q_2 \in \mathbb{Z}. \tag{13.132}$$

13.5 Heterotic Strings in Covariant Lattice Description

This suggests to combine the 16-dimensional left-moving lattice Γ_{16} and the six-dimensional right-moving lattice $D_{5,1}$ into a Lorentzian lattice

$$\Gamma_{16;5,1} = (\Gamma_{16})_L \otimes (D_{5,1})_R \tag{13.133}$$

(the semicolon separates left- from right-movers) with lattice vectors $w = (w_L; \lambda_R, q)$, where the inner product $w_1 \cdot w_2$ is defined with the metric diag $[(-1)^{16}, (+1)^5, (-1)]$. Locality demands that this lattice be integral with respect to this Lorentzian metric.

Combining the left- and right-moving sectors, the partition function of ten-dimensional heterotic string theories is essentially given by the sum over all lattice vectors of $\Gamma_{16;5,1}$:

$$\chi(\tau, \bar{\tau}) = \mathrm{Tr}\, e^{2\pi i \tau (L_0 - 1)} e^{-2\pi i \bar{\tau}(\bar{L}_0 - 1)} (-1)^{F_R}$$

$$= \frac{(\mathrm{Im}\tau)^{-4}}{\eta^{24}(\bar{\tau})\eta^{12}(\tau)\Theta_{1,1}(\tau)} \sum_{w \in \Gamma_{16;5,1}} e^{-2\pi i \bar{\tau} \frac{1}{2} w_L^2} e^{2\pi i \tau (\frac{1}{2}\lambda_R^2 - \frac{1}{2}q^2 - q - \frac{1}{2})} e^{-2\pi i q}. \tag{13.134}$$

Again, as in the fermionic string theory, modular invariance forces $\Gamma_{16;5,1}$ to be an odd self-dual Lorentzian lattice. This can be proven by Poisson resummation.

The requirement of self-duality can be trivially satisfied if both $(\Gamma_{16})_L$ and $(D_{5,1})_R$ are self-dual separately, i.e. $\Gamma_{16;5,1}$ is a direct product of two self-dual lattices of which $(D_{5,1})_R$ must be odd. Then (Γ_{16}) must be either the root lattice of $E_8 \times E_8$ or the weight lattice of $\mathrm{Spin}(32)/\mathbb{Z}_2$ implying $E_8 \times E_8$ or $SO(32)$ as the two possible gauge groups. On the other hand, self-duality of $(D_{5,1})_R$ implies that the spectrum is space-time supersymmetric in ten dimensions.

However this is by far not the most general case—it is possible to obtain a self-dual lattice $\Gamma_{16;5,1}$ without self-dual sublattices $(\Gamma_{16})_L$ and $(D_{5,1})_R$. Then Eq. (13.133) does not represent a direct product decomposition. Instead, $\Gamma_{16;5,1}$ is specified by non-trivial correlations between the various conjugacy classes of $(\Gamma_{16})_L$ and $(D_{5,1})_R$ given by the glue vectors as explained in Chap. 10. Thus, in this case there is a non-trivial interplay between the left- and right-moving degrees of freedom. This implies that in the fermionic description of the right-moving string the GSO projections have to be modified. This destroys space-time supersymmetry. Analogously, the left-moving gauge group will be different from $E_8 \times E_8$ or $SO(32)$.

Let us classify all possible odd self-dual lattices $\Gamma_{16;5,1}$ which lead to a sensible heterotic string theory in ten dimensions. In general, the classification of Lorentzian self-dual lattices $\Gamma_{p,q}$ of a given dimension and metric diag $[(+1)^p, (-1)^q]$ is meaningless, as they are unique up to Lorentz rotations in $\mathbb{R}^{p,q}$. However, if we add the requirement of having a sensible space-time interpretation, a classification becomes possible. Because of Lorentz invariance all states are classified (off-shell) according to $SO(10)$ representations, and we have to demand that $(D_{5,1})_R$ builds the right-moving part of $\Gamma_{16;5,1}$. The non-trivial question is now how the four conjugacy classes (0), (V), (S) and (C) $\in (D_{5,1})_R$ are coupled to the conjugacy classes of Γ_{16}.

We analyze this problem by converting the Lorentzian lattices into Euclidean ones. Any even self-dual lattice consisting of one or several D_n factors can be replaced by another even self-dual lattice by changing the dimension of any D_n factor by multiples of eight and keeping all conjugacy classes the same, i.e. $D_n \to D_{n+8p}$ ($p \in \mathbb{Z}$). Such a transformation changes the (length)^2 of all vectors only modulo 2 and all mutual scalar products modulo 1, so that it does not affect self-duality. For example, the self-dual lattice D_8 with (0) and (S) conjugacy classes (i.e. the E_8 root lattice) can be mapped to the D_{16} weight lattice with the same conjugacy classes. As we know from Chap. 10, this is the self-dual weight lattice of Spin(32)/$\mathbb{Z}_2$. On the other hand, changing the dimension by multiples of four, maps an even self-dual lattice to an odd self-dual lattice and vice versa. Finally, one may even subtract multiples of eight or four to make the dimension of a D_n factor negative. This can be interpreted as a change of signature, i.e. as a map of an Euclidean self-dual lattice to a Lorentzian self-dual lattice or vice versa. This is what happened when we replaced $D_{5,1}$ ($E_{5,1}$) by D_8 (E_8)—we changed the dimension by minus four units, thus converting an odd self-dual Lorentzian lattice to an even self-dual Euclidean lattice. The reason for doing so is the possibility of classifying Euclidean lattices as discussed in Chap. 10.

We are now ready to apply these techniques to the lattice $\Gamma_{16;5,1}$ which describes the heterotic string theories in ten dimensions. First we map $(D_{5,1})_R$ to $(D_8)_R$ obtaining the Lorentzian even self-dual lattice $\Gamma_{16;8} = (\Gamma_{16})_L \otimes (D_8)_R$. This lattice can in turn be mapped to those 24-dimensional Euclidean even self-dual lattices Γ_{24} which can be decomposed as

$$\Gamma_{24} = \Gamma_{16} \otimes D_8. \tag{13.135}$$

Thus, our aim is to find all even self-dual Euclidean 24-dimensional lattices which allow for this decomposition. Each solution will be completely specified by the Lie algebra lattice Γ_{16} together with the glue vectors to D_8.

In Table 11.3 we have listed all possible even self-dual lattices of dimension 24, the so-called Niemeier lattices. Seven of them contain D_8 as regular subalgebra and therefore lead to a heterotic string theory in ten dimensions. These are displayed in the first column of Table 13.4. They lead to eight different heterotic string theories, as D_8 can be embedded in two different ways in $E_8 \otimes D_{16}$. In all the other cases there is only one possible regular embedding of D_8. The algebras which commute maximally with D_8 build the left lattice Γ_{16} and are displayed in the third column of the table. The appearing conjugacy classes of Γ_{16} as well as their coupling to the D_8 conjugacy classes are shown in the last four columns.

The root vectors of Γ_{16} give rise to massless gauge bosons of the heterotic string theory and the corresponding gauge group. They are of course always of rank 16 and can be read off from the third column of the table. We recognize that the first two models are the two original supersymmetric heterotic string theories. Here Γ_{16} is self-dual and so is D_8 which is thus the root lattice of E_8. All other theories are not supersymmetric. Only one other model, the last one in the table, is tachyon free and has gauge group $SO(16) \times SO(16)$. In all models, the tachyon, if present, comes

13.5 Heterotic Strings in Covariant Lattice Description

Table 13.4 Ten dimensional heterotic strings; $k = 1, 2$ and 1_{E_7} is the E_7 conjugacy class with the $\underline{56}$

Niemeier lattice		Heterotic string				
Roots	Weights	Algebra			D_8-sector	
			(0)	(V)	(S)	(C)
E_8^3	(0,0,0)	$E_8 \times E_8$	(0+S,0+S)	–	(0+S,0+S)	–
$E_8 \times D_{16}$	(0,S)	D_{16}	(0+S)	–	(0+S)	–
$E_8 \times D_{16}$	(0,S)	$E_8 \times D_8$	(0+S,0)	(0+S,V)	(0+S,S)	(0+S,C)
D_{24}	(S)	D_{16}	(0)	(V)	(S)	(C)
D_{12}^2	[V,S]	$D_4 \times D_{12}$	(0,0)	(V,0)	(S,V)	(S,C)
	(C,C)		(V,S)	(0,S)	(C,C)	(C,V)
$D_{10} \times E_7^2$	(S,1,0)	$D_2 \times E_7^2$	(0,0,0)	(V,0,0)	(S,1,0)	(C,1,0)
	(C,0,1)		(V,1,1)	(0,1,1)	(C,0,1)	(S,0,1)
	(V,1,1)					
$D_9 \times A_{15}$	(0, 8k)	$D_1 \times A_{15}$	(0, 8k)	(V, 8k)	(0, 8k + 2)	(C, 8k + 2)
	(S, 8k + 2)		(V, 8k + 4)	(0, 8k + 4)	(C, 8k + 6)	(S, 8k + 2)
	(V, 8k + 4)					
	(C, 8k + 6)					
D_8^3	[S,V,V]	$D_8 \times D_8$	(0,0)	(S,V)	(V,V)	(C,0)
	[C,0,C]		(C,C)	(V,S)	(S,S)	(0,C)
	(S,S,S)					

from the (V) conjugacy class of $(D_8)_R$. For the $SO(16) \times SO(16)$ model, however, the (V) conjugacy class is coupled to the (V,S) and (S,V) conjugacy classes of $(D_8 \otimes D_8)_L$. The lowest states within these two conjugacy classes have $\frac{\alpha'}{2}m_L^2 = \frac{1}{2}$ such that the right-moving tachyonic state with $\frac{\alpha'}{2}m_R^2 = -\frac{1}{2}$ does not satisfy the left-right level matching constraint. In the five remaining non-supersymmetric models it does.

In summary, via bosonization one obtains eight different modular invariant heterotic string theories in ten dimensions. There exists one additional tachyonic model with rank eight gauge group E_8 which cannot be described within the covariant lattice formalism. The reason is that this model involves real fermions with different spin structures which however cannot be bosonized with the methods described in Chap. 10 and therefore do not lead to a covariant lattice. For lattice models, the Kač-Moody algebra corresponding to the gauge group is always at level one. The theory with gauge group E_8 has a E_8 Kač-Moody algebra at level two and can thus not be represented by free bosons in the way described in Chap. 10.

Further Reading

Covariant quantization of the superstring was developed in

- D. Friedan, E. Martinec, S. Shenker, Covariant quantization of superstrings. Phys. Lett. **160B**, 55 (1985); Conformal invariance, super symmetry and string theory. Nucl. Phys. **B271**, 93 (1986)

- V.G. Knizhnik, Covariant fermionic vertex in superstrings. Phys. Lett. **160B**, 403 (1985)
- W. Lerche, D. Lüst, Covariant heterotic strings and odd selfdual lattices. Phys. Lett. **187B**, 45 (1987)

The first reference contains a general discussion of first order systems, which is also reviewed in

- D. Friedan, in *Notes on String Theory and Two-Dimensional Conformal Field Theory*, ed. by M. Green, D. Gross. Proceedings of the Workshop on Unified String Theories, Santa Barbara, 1985 (World scientific, Singapore, 1986)

The first and the third references are the main sources for the first three sections of this chapter.

The content of Footnote 7 is elaborated in

- E. Verlinde, H. Verlinde, *Lectures on String Perturbation Theory*. Presented at Spring School and Workshop on Superstrings, Trieste, 1988

RR vertex operators in the non-symmetric ghost picture, which are needed for the one-point function on the disc, were constructed in

- M. Billo, P. Di Vecchia, M. Frau, A. Lerda, I. Pesando, R. Russo, S. Sciuto, Microscopic string analysis of the D0-D8 brane system and dual R-R states. Nucl. Phys. **B526**, 199 (1998) [arXiv:hep-th/9802088]

The covariant lattice construction is reviewed in

- W. Lerche, A.N. Schellekens, N. Warner, Lattices and strings. Phys. Rep. **177**, 1 (1989)

Chapter 14
String Compactifications

Abstract An alternative to describing compactifications via a solvable conformal field theory is the perturbative approach around a geometric supergravity background at large radius. For this purpose one analyzes the string equations of motion at leading order in a typical length scale $L/\sqrt{\alpha'}$. We describe this approach in detail for a class of backgrounds which preserve some amount of space-time supersymmetry in four-dimensions: compactification on Calabi-Yau manifolds. But we start with a brief discussion of the string equations of motion as the requirement of vanishing beta-functions of the non-linear sigma model for a string moving in a curved background. We then derive a generalization of T-duality to manifolds with isometries. This leads to the so-called Buscher rules. We then introduce some of the mathematical tools which are required for an adequate treatment of Calabi-Yau compactifications. With them at hand we consider compactifications of the type II and heterotic superstring on Calabi-Yau manifolds and discuss the structure of their moduli spaces. In an appendix we fix our notation and review some concepts of Riemannian geometry. The derivations of some results which are used in the main text are also relegated to the appendix.

14.1 Conformal Invariance and Space-Time Geometry

In the first part of this book we have constructed string theories in Minkowski space-time. Quantum consistency required the number of space-time dimensions to be 10 for the fermionic and 26 for the bosonic string. Both are higher than the four large dimensions which we observe around us.

The main consistency condition, the absence of a conformal anomaly of the combined matter and ghost system, has other solutions besides the one which leads to Minkowski space in the critical dimension. Other solutions are more complicated CFTs with or without a geometric interpretation as a compactification from d_{crit} to four dimensions. We have already discussed simple examples in Chap. 10: tori and toroidal orbifolds. In the present chapter we study more general compactifications and some of their properties, using geometrical tools. In the next chapter we will discuss compactifications using CFT methods.

We start with a discussion of the implications which follow from the requirement of conformal invariance of the world sheet theory. This can be illustrated by considering the bosonic string propagating through a space-time M with metric $G_{\mu\nu}(X)$, rather than $\eta_{\mu\nu}$. In this case the world-sheet action is

$$S_{\text{P}} = -\frac{1}{4\pi\alpha'} \int_\Sigma d^2\sigma \sqrt{-h} \, h^{\alpha\beta} \, G_{\mu\nu}(X) \, \partial_\alpha X^\mu \partial_\beta X^\nu. \tag{14.1}$$

The σ-model metric is also called the string frame metric. In conformal gauge (14.1) no longer reduces to a free scalar field theory. One has to deal with an interacting theory, a non-linear sigma model with target space metric $G_{\mu\nu}(X)$:

$$S_{\text{P}} = -\frac{1}{4\pi\alpha'} \int_\Sigma d^2\sigma \, \eta^{\alpha\beta} \, G_{\mu\nu}(X) \, \partial_\alpha X^\mu \partial_\beta X^\nu. \tag{14.2}$$

Note that if we expand the background metric around Minkowski space as $G_{\mu\nu}(X) = \eta_{\mu\nu} + i\epsilon_{\mu\nu}(k)e^{ik\cdot X}$, we are led to interpret the non-trivial background as arising from a coherent insertion of graviton vertex operators.

The classical action is invariant under Weyl rescaling of the world-sheet metric but this classical symmetry is anomalous in the quantum theory. The absence of a Weyl anomaly, which is necessary for the decoupling of the degrees of freedom of the world-sheet metric, lead, for $G_{\mu\nu} = \eta_{\mu\nu}$, to the condition $d = d_{\text{crit}}$. For a general background metric, requiring Weyl invariance imposes constraints on $G_{\mu\nu}(X)$, which can be expressed as the requirement that the β-function for the coupling function $G_{\mu\nu}(X)$ vanishes.

In two dimensions scalar fields are dimensionless and the non-linear sigma model is renormalizable. Here the dimensionless scalar fields are the rescaled fields $\frac{1}{L}X$, where L is a typical length scale of the background geometry, e.g. its size or the radius of curvature. The metric $G_{\mu\nu}$ is a function of the rescaled fields. The dimensionless ratio $(l_s/L)^2$, which multiplies the action, plays the role of a loop counting parameter of the two-dimensional field theory.[1]

For dimensional reasons, all counter-terms must be of second order in world-sheet derivatives. The fact that an n-loop counter-term contains a factor $(l_s/L)^{2n}$ means that it must be a symmetric second rank tensor composed of the curvature tensor of $G_{\mu\nu}$ with altogether $2n$ (space-time) derivatives.

The coefficients of the allowed counter-terms are computed in perturbation theory and from them the β-function, which has a covariant expansion in orders of derivatives of the background metric. In this way one finds, at one-loop order, that the background metric must be Ricci-flat, i.e. it must be a solution of the vacuum

---

[1] There are also contributions from string loops and, of course, from both world-sheet and space-time non-perturbative effects.

14.1 Conformal Invariance and Space-Time Geometry

Einstein equations. There are, however, higher order (in $(l_s/L)^2$) corrections to this condition. The order of the first correction and the precise coefficients depend on the kind of string theory in question. For instance, for the heterotic string there are two-loop, i.e. (curvature)^2 corrections while for the type II strings the first corrections appear at four-loop in the σ-model perturbation theory, i.e. (curvature)^4.

We will not present the calculations which lead to these results. They involve loop calculations in two-dimensional non-linear sigma models. Instead we will add a few more comments. In addition to the metric, there are other background fields to which one can couple the string. We limit ourselves to fields corresponding to massless bosonic string excitations.[2] Within the RNS formulation for the superstring, which is the one we have discussed, there is a problem with fields from the R-R sector of the closed string. Their vertex operators contain spin fields which introduce branch cuts on the world-sheet which leads to a violation of locality.[3] Backgrounds for the massless NS-NS fields can, however, be straightforwardly incorporated in the world-sheet action. For the closed string the bosonic part of the most general action with at most two world-sheet derivatives is[4]

$$S_P = -\frac{1}{4\pi\alpha'} \int_\Sigma d^2\sigma \sqrt{-h} \left(h^{\alpha\beta} \partial_\alpha X^\mu \partial_\beta X^\nu G_{\mu\nu}(X) \right.$$
$$\left. + \epsilon^{\alpha\beta} \partial_\alpha X^\mu \partial_\beta X^\nu B_{\mu\nu}(X) + \alpha' \Phi R(h) \right). \tag{14.3}$$

Our conventions here are such that $\varepsilon^{\alpha\beta} = \pm 1/\sqrt{-h}$. The coupling of the dilaton needs explanation. It breaks classical Weyl invariance. For dimensional reasons it must be proportional to α' and should be considered together with the one-loop corrections to the other coupling functions. Its classical Weyl non-invariance is cancelled by one-loop Weyl anomalies from the other terms.

For such a general background, the trace of the world-sheet energy momentum tensor no longer vanishes and the Weyl anomaly can be parametrized as

$$2\alpha' T^\alpha_\alpha = \alpha' \beta^\Phi R(h) + \beta^G_{\mu\nu} h^{\alpha\beta} \partial_\alpha X^\mu \partial_\beta X^\nu + \beta^B_{\mu\nu} \epsilon^{\alpha\beta} \partial_\alpha X^\mu \partial_\beta X^\nu. \tag{14.4}$$

This expression is valid for the bosonic string and for the fermionic string, if we set the world-sheet fermions to zero. Consistency of the quantization demands that

---

[2] Massive ones contain more than two world-sheet derivatives and therefore lead to irrelevant operators in the two-dimensional CFT.

[3] In the pure spinor formulation of the superstring one uses, in addition to X^μ, fermionic world-sheet scalars which transform as spinors under the space-time Lorentz group. In this formulation there is no problem to couple the R-R fields of the RNS formulation directly to the world sheet.

[4] We will not write down the σ-models for the superstring and the heterotic string. This is most succinctly done in (1, 1) and (0, 1) superspace, respectively. In the heterotic σ-model a background gauge field appears. For the bosonic string one should include the tachyon which couples without world-sheet derivatives.

this vanishes on a flat world-sheet, i.e. we have to require $\beta^G = \beta^B = 0$.[5] One can show (to all orders in α') that on solutions of $\beta^{G,B} = 0$ the dilaton β^Φ is constant and $T^\alpha_\alpha = -\frac{c}{12}R(h)$, i.e. the anomaly reduces to the Weyl anomaly of a CFT coupled to a curved world-sheet metric and it is characterized by the central charge of the Virasoro algebra.

Without giving any details of the derivation, we quote the lowest order results of the β-functions of the type II superstring theories

$$\beta^G_{\mu\nu} = \alpha'\left(R_{\mu\nu} - \frac{1}{4}H_\mu{}^{\rho\sigma}H_{\nu\rho\sigma} + 2\nabla_\mu\nabla_\nu\Phi\right) + \mathcal{O}(\alpha'^2),$$

$$\beta^B_{\mu\nu} = \alpha'\left(-\frac{1}{2}\nabla_\lambda H^\lambda{}_{\mu\nu} + H^\lambda{}_{\mu\nu}\nabla_\lambda\Phi\right) + \mathcal{O}(\alpha'^2),$$

$$\beta^\Phi = \frac{1}{4}(d - d_{\text{crit}}) + \alpha'\left((\nabla\Phi)^2 - \frac{1}{2}\nabla^2\Phi - \frac{1}{24}H^2\right) + \mathcal{O}(\alpha'^2). \quad (14.5)$$

Here $R_{\mu\nu}$ is the Ricci tensor of the space-time metric, ∇_μ its covariant derivative and $H = dB$ the field strength of the anti-symmetric tensor B. The term in β^Φ with d_{crit} is the contribution from the superconformal ghost system. For the heterotic string (in the fermionic formulation) there is a coupling of the gauge field to the world-sheet $A^a_\mu \bar\partial X^\mu j^a$ with $j^a = \lambda^i T^{ij}_a \lambda^j$ a Kač-Moody current and correspondingly a β-function

$$\beta^A_\mu = \alpha'\left(\frac{1}{2}D^\nu F_{\nu\mu} + \frac{1}{4}H_\mu{}^{\lambda\rho}F_{\lambda\rho} + F_\mu{}^\lambda\partial_\lambda\Phi\right) + \mathcal{O}(\alpha'^2). \quad (14.6)$$

Now D_ν denotes a gauge and diffeomorphism covariant derivative and there is also a corresponding term in (14.4), $\beta^A_{\mu a}\bar\partial X^\mu j^a$. Note that no one-loop correction to (14.5) from the gauge field arises, i.e. to the order displayed they are unchanged.

For the open string a background gauge field couples to the boundary of the world-sheet and, at $\mathcal{O}(\alpha')$, the dilaton couples to the extrinsic curvature of the boundary. This leads to the boundary terms which should be added to (14.3)

$$S_P = -\oint_{\partial\Sigma} ds \left(A_\mu(X)\partial_s X^\mu + \frac{1}{2\pi}k(s)\Phi(X)\right). \quad (14.7)$$

Here k is the trace of the extrinsic curvature of the boundary: $k = t^\alpha t^\beta \nabla_\alpha n_\beta$, where t^α and n^α are unit tangent and normal vectors to the boundary, respectively.

The first term in (14.7) is simply $\oint A$. In the path integral it appears as an Abelian Wilson loop of a background gauge field. This is nothing else than the exponentiation of the photon vertex operator. For non-Abelian gauge fields we have to incorporate Chan-Paton factors. This is done by inserting a non-Abelian Wilson loop in the

---

[5]The Weyl anomaly functions β are not quite the renormalization group β-functions. Their relation is explained in the references at the end of this chapter.

14.1 Conformal Invariance and Space-Time Geometry

path integral. For constant dilaton the curvature terms in (14.3) and (14.7) combine to compute the Euler number of an (Euclidean) world-sheet; cf. (6.3) on page 125.

We want to close this qualitative discussion with a few remarks:

1. An important observation is that the conditions $\beta^{G,B,\Phi} = 0$ allow for non-trivial solutions. These conditions are also called the string equations of motion. In fact these conditions on the background fields can be derived from an effective action of the massless string excitations. One easily verifies that, to the order displayed in (14.5), the effective action is, up to an overall normalization,

$$S_{\text{eff}} = \int d^d x \, \sqrt{-G} \, e^{-2\Phi}$$
$$\times \left(-\frac{(d - d_{\text{crit}})}{\alpha'} + R - \frac{1}{12} H_{\mu\nu\rho} H^{\mu\nu\rho} + 4\nabla^\mu \Phi \nabla_\mu \Phi \right). \quad (14.8)$$

This is the effective action in string frame. An alternative way to arrive at the effective action, which allows to incorporate RR background fields also in the RNS formulation of the superstring, is to compute scattering amplitudes of on-shell string excitations and to find a space-time action which reproduces them. This approach will be pursued in Chap. 16.

2. The perturbative computation of the β-functions is an expansion in derivatives of the background fields. It is valid when the curvatures are small. The requirement $\beta^\Phi = 0$ allows, besides the critical string solution $d = d_{\text{crit}}$, also so-called non-critical strings with $d \neq d_{\text{crit}}$. For these solutions the curvature of space-time must be of the order of the string scale and the expansion on which the above computations rely breaks down. The often repeated statement that string theory requires 10 or 26 space-time dimensions is only correct if one requires that the corresponding conformal field theory has a geometric interpretation with a large volume limit, i.e. that it possesses an exactly marginal deformation which can be interpreted as a change of volume. A simple example is the compactification on a circle, where the vertex operator of the internal component of the graviton is such a deformation. However, string theory neither requires this nor is it necessary in order to eventually find four macroscopically large space-time dimensions. There are CFTs leading to non-critical strings which have no higher dimensional interpretation and others with no interpretation in terms of space-time dimensions at all. These CFTs are usually not weakly coupled and are thus harder to study. In fact, the right question to ask seems to be why there are four large dimensions rather than none.

We should keep in mind that, when we discuss a particular compactification, this is merely a specific ansatz for the background configurations motivated by the properties of the resulting lower dimensional theory, which we want to examine in a string theory context. There is as yet no dynamical principle known which would favor a particular background configuration.

3. Different background configurations do not correspond to different string theories. They correspond to different string vacua. By this one means string

backgrounds which are compatible with Weyl invariance of the world-sheet theory. In the α' expansion this condition is known explicitly only to the first few orders. Nevertheless some non-trivial string backgrounds are known to be exact. Examples are Calabi-Yau manifolds, to be discussed later in this chapter, or the Gepner models from Chap. 15. Another class of models are strings on group manifolds, whose CFT is given by WZW models with fixed level k. They do not have a large volume limit. All these examples have only non-trivial NS-NS background fields. A more general compactification is given by $AdS_5 \times S^5$, which contains in addition to a metric a self-dual five-form R-R-flux. This is a notable example of an exact perturbative type IIB superstring background with RR-fields and provides the basis of the AdS/CFT correspondence to be discussed in Chap. 18.

One might wonder about the analogous computation in the open string sector. Here the vanishing of β^A_μ leads to an equation of motion for the gauge field on the world-volume of the D-branes on which the boundary of the open string world-sheet ends. This is the DBI action which will be discussed in Chap. 16.

14.2 T-Duality and Buscher Rules

Recall that when we discussed string compactification on a circle in Chap. 10, we observed a striking discrete symmetry, T-duality. Given the coupling of the world-sheet degrees of freedom to the background fields, we can now give a more systematic discussion of T-duality, which is valid whenever the background has isometries. In general the isometry group of the background can be non-Abelian, but we will only discuss the case with a single Killing vector k^μ.

We start with the closed string and will later generalize to the open string. For simplicity we choose a flat metric on the world-sheet. For this choice we loose control over the dilaton whose transformation under T-duality will be discussed later.

Assume that the action (14.3) is invariant under a diffeomorphism $X^\mu \to X^\mu + k^\mu$. This is a condition on the background fields which have to satisfy

$$\mathscr{L}_k G = \mathscr{L}_k \Phi = 0 \quad \text{and} \quad \mathscr{L}_k H = 0, \quad \text{i.e.} \quad \mathscr{L}_k B = d\omega, \qquad (14.9)$$

where $\mathscr{L}_k$ is the Lie derivative in the direction k. We choose coordinates, called adapted coordinates, $X^\mu = (\theta, X^i)$ such that $k^\mu = (1, 0)$ and $\mathscr{L}_k = \partial_\theta$, i.e. the isometry acts as a shift in the θ-direction.[6] Using the gauge invariance of the action under $B \to B + d\omega$, we can choose all background fields to be independent of the

---

[6] Note that θ is not necessarily meant to denote an angle-variable. Our units are such that the metric is dimensionless.

14.2 T-Duality and Buscher Rules

coordinate θ. In these coordinates the action becomes

$$S = -\frac{1}{4\pi\alpha'} \int_\Sigma d^2\sigma \Big[\big(G_{\theta\theta} \partial_\alpha\theta\partial_\beta\theta + 2G_{\theta i} \partial_\alpha\theta\partial_\beta X^i + G_{ij} \partial_\alpha X^i \partial_\beta X^j \big)\eta^{\alpha\beta}$$
$$+ \big(2B_{\theta i} \partial_\alpha\theta\partial_\beta X^i + B_{ij} \partial_\alpha X^i \partial_\beta X^j \big) \epsilon^{\alpha\beta} \Big]. \tag{14.10}$$

We now rewrite this action in first order form[7]

$$S' = -\frac{1}{4\pi\alpha'} \int_\Sigma d^2\sigma \Big[\big(G_{\theta\theta} V_\alpha V_\beta + 2G_{\theta i} V_\alpha \partial_\beta X^i + G_{ij} \partial_\alpha X^i \partial_\beta X^j \big)\eta^{\alpha\beta}$$
$$+ \big(2B_{\theta i} V_\alpha \partial_\beta X^i + B_{ij} \partial_\alpha X^i \partial_\beta X^j \big)\epsilon^{\alpha\beta} + 2\tilde{\theta}\epsilon^{\alpha\beta}\partial_\alpha V_\beta \Big], \tag{14.11}$$

where in the functional integral we now integrate over X^i, θ and $\tilde{\theta}$. Integrating over the Lagrange multiplier field $\tilde{\theta}$ enforces $dV = 0$ with solution $V_\alpha = \partial_\alpha \theta$. Inserting this into the action leads back to (14.10). Here we ignore global issues which arise if the world-sheet has non-contractible cycles.

Alternatively we can integrate out V_α, i.e. eliminate it via its equation of motion. Indeed, if we define

$$L^\alpha = \frac{1}{G_{\theta\theta}} \Big(G_{\theta i} \partial_\beta X^i \eta^{\alpha\beta} + (B_{\theta i} \partial_\beta X^i + \partial_\beta \tilde{\theta})\epsilon^{\alpha\beta} \Big), \tag{14.12}$$

we can rewrite (14.11) in the form

$$S' = -\frac{1}{4\pi\alpha'} \int_\Sigma d^2\sigma \Big[\big(G_{\theta\theta} \tilde{V}_\alpha \tilde{V}_\beta - G_{\theta\theta} L_\alpha L_\beta + G_{ij} \partial_\alpha X^i \partial_\beta X^j \big)\eta^{\alpha\beta}$$
$$+ B_{ij} \partial_\alpha X^i \partial_\beta X^j \epsilon^{\alpha\beta} \Big], \tag{14.13}$$

where $\tilde{V}_\alpha = V_\alpha + L_\alpha$ and $\tilde{V}_\alpha = 0$ is the equation of motion for V_α. Making a change of variables and integrating over $\tilde{V}$ one obtains the dual action

$$\tilde{S} = -\frac{1}{4\pi\alpha'} \int_\Sigma d^2\sigma \Big[\big(\tilde{G}_{\theta\theta} \partial_\alpha\tilde{\theta}\partial_\beta\tilde{\theta} + 2\tilde{G}_{\theta i} \partial_\alpha\tilde{\theta}\partial_\beta X^i + \tilde{G}_{ij} \partial_\alpha X^i \partial_\beta X^j \big)\eta^{\alpha\beta}$$
$$+ \big(2\tilde{B}_{\theta i} \partial_\alpha\tilde{\theta}\partial_\beta X^i + \tilde{B}_{ij} \partial_\alpha X^i \partial_\beta X^j \big) \epsilon^{\alpha\beta} \Big] \tag{14.14}$$

---

[7] We can also view this action as a gauged version of (14.10) where the isometry has been gauged, i.e. we have replaced $\partial_\alpha \theta \to D_\alpha \theta = \partial_\alpha \theta + V_\alpha$ where $V \to V - d\epsilon$ under $\theta \to \theta + \epsilon$. The Lagrange multiplier $\tilde{\theta}$ enforces that the gauge field V_α has vanishing field strength. We obtain (14.11) after fixing the gauge $\theta = 0$. The dual theory is obtained by integrating over the gauge field and fixing $\theta = 0$.

with

$$\tilde{G}_{\theta\theta} = \frac{1}{G_{\theta\theta}} \qquad \tilde{G}_{\theta i} = \frac{1}{G_{\theta\theta}} B_{\theta i}, \qquad \tilde{B}_{\theta i} = \frac{1}{G_{\theta\theta}} G_{\theta i}$$
$$\tilde{G}_{ij} = G_{ij} - \frac{1}{G_{\theta\theta}} (G_{\theta i} G_{\theta j} - B_{\theta i} B_{\theta j}) \qquad \tilde{B}_{ij} = B_{ij} - \frac{1}{G_{\theta\theta}} (G_{\theta i} B_{\theta j} - B_{\theta i} G_{\theta j}). \tag{14.15}$$

These are the so-called Buscher rules for the string frame metric and anti-symmetric tensor field. The Jacobian of the change of variables in the Polyakov path integral, which is necessary in the dualization procedure, is non-trivial and, after appropriate regularization, can be shown to lead to a shift of the dilaton:

$$\tilde{\Phi} = \Phi - \frac{1}{2} \log |G_{\theta\theta}| = \Phi - \frac{1}{4} \log \left| \frac{G_{\theta\theta}}{\tilde{G}_{\theta\theta}} \right|. \tag{14.16}$$

Combining this with

$$\det \tilde{G} = \frac{1}{G_{\theta\theta}^2} \det G, \tag{14.17}$$

one finds the T-duality invariant combination

$$e^{-2\tilde{\Phi}} \sqrt{\det \tilde{G}} = e^{-2\Phi} \sqrt{\det G}. \tag{14.18}$$

Equation (14.16) implies that the string coupling transforms as

$$g_s \to \tilde{g}_s = g_s / \sqrt{|G_{\theta\theta}|}. \tag{14.19}$$

For instance, compactifying on a product of n circles with radii R_i (14.18) becomes

$$e^{-2\tilde{\Phi}} = e^{-2\Phi} \prod_{i=1}^{n} \left(\frac{R_i^2}{\alpha'} \right), \qquad \tilde{g}_s = g_s \prod_{i=1}^{n} \left(\frac{\sqrt{\alpha'}}{R_i} \right). \tag{14.20}$$

The dilaton shift can be understood as follows: in Chap. 16 we will find that Newton's constant depends as $1/g_s^2$ on the string coupling. Then, Eq. (14.18) means that Newton's constant of the compactified theory is invariant under T-duality. This could be imposed as a physical requirement to 'derive' (14.16).

Another way to confirm (14.16) is to check consistency with the D-brane tension. Consider a Dp-brane wrapped on a circle with radius R. For the observer in the uncompactified space-time this appears as a D(p−1)-brane with tension $\frac{1}{g_s} T_p 2\pi R$. In the T-dual situation we have an unwrapped D(p−1)-brane with tension

$$\frac{1}{\tilde{g}_s} T_p (2\pi \tilde{R}) = \frac{1}{g_s} T_p (2\pi \sqrt{\alpha'}) = \frac{1}{g_s} T_{p-1}, \tag{14.21}$$

14.2 T-Duality and Buscher Rules

which is the expected result $T_{p-1}/T_p = \ell_s$.

An alternative derivation of the Buscher rules is based on the requirement that starting from a background which leads to vanishing beta functions, the T-dual background should have the same property. The explicit calculation is rather cumbersome. We will instead demonstrate that the effective action (14.8), after reduction along the dualized direction, is T-duality invariant. This gives us also the opportunity to introduce a few basic ideas of dimensional reduction of higher-dimensional theories.

The Kaluza-Klein ansatz for the d-dimensional metric is

$$G_{\mu\nu} = \begin{pmatrix} k^2 & k^2 A_j \\ k^2 A_i & g_{ij} + k^2 A_i A_j \end{pmatrix}, \tag{14.22}$$

where A_i is the Kaluza-Klein gauge boson and k a KK scalar. We use $k^2 = G_{\theta\theta}$ to simplify expressions. All fields are independent of the coordinate θ. The inverse of (14.22) is

$$G^{\mu\nu} = \begin{pmatrix} A^k A_k + k^{-2} & -A^j \\ -A^i & g^{ij} \end{pmatrix}, \tag{14.23}$$

where $g^{ij} g_{jk} = \delta^i_k$ and $A^i = g^{ij} A_j$. The metric (14.22) satisfies $\det(G) = k^2 \det(g)$ and under diffeomorphism in the θ-direction, i.e. under $\theta \to \theta + \xi^\theta$ with $\partial_\theta \xi^\theta = 0$, A_i transforms as a gauge field $A_i \to A_i - \partial_i \xi^\theta$ while k and g_{ij} are invariant. Under the diffeomorphisms generated by ξ^i all fields transform according to their tensor structure. Furthermore, a straightforward but somewhat tedious calculation gives

$$R(G) = R(g) - \frac{1}{4} k^2 F(A)^{ij} F(A)_{ij} - \frac{2}{k} \nabla^i \nabla_i k. \tag{14.24}$$

A convenient ansatz for the B-field is

$$B_{i\theta} = B_i \qquad B_{ij} = b_{ij} - \frac{1}{2}(A_i B_j - A_j B_i). \tag{14.25}$$

The vector field B_i transforms like a gauge field under $B \to B + d\omega$ with gauge parameter $\omega = \omega_\theta d\theta$. The transformation of b_{ij} under the two gauge transformations generated by ω_θ and ξ^θ is $\delta b_{ij} = \frac{1}{2}\left(B_i \partial_j \xi^\theta + A_i \partial_j \omega_\theta - (i \leftrightarrow j)\right)$.

With the help of the Buscher rules (14.15) one shows that T-duality leaves g_{ij} and b_{ij} invariant, exchanges A_i and B_i and inverts k. If we define the T-duality and gauge invariant field strength

$$H_{ijk} = 2\big(\partial_{[i} b_{jk]} - \frac{1}{2}(A_{[i} F(B)_{jk]} + B_{[i} F(A)_{jk]})\big), \tag{14.26}$$

where e.g. $F(A)_{ij} = \partial_i A_j - \partial_j A_i$, we can write the effective action

$$S = L \int d^{d-1}x \sqrt{-g} e^{-2\varphi} \left(R(g) + 4(\partial\varphi)^2 - (\partial \ln k)^2 - \frac{1}{12} H_{ijk} H^{ijk} \right.$$
$$\left. - \frac{k^2}{4} F(A)^{ij} F(A)_{ij} - \frac{1}{4k^2} F(B)^{ij} F(B)_{ij} \right), \tag{14.27}$$

where L is the size of the dualized dimension and

$$\varphi = \Phi - \frac{1}{2} \ln k \tag{14.28}$$

is invariant under (14.16). In fact the complete dimensionally reduced action is manifestly invariant under T-duality.

We now turn to the open string whose T-duality we have already discussed in Chap. 10. There we found that dualization transforms the gauge field background on the world-volume of a Dp-brane to the position of the D(p-1)-brane along the dualized coordinate. Here we will derive the same result from the σ-model, which now consists of the two terms (14.3) and (14.7). For simplicity we only consider a $U(1)$ gauge field and, to avoid global issues, choose the world-sheet to be a disc (or the upper half plane) with a flat metric. In addition to (14.9) we require

$$\mathscr{L}_k F = 0, \quad \text{i.e.} \quad \mathscr{L}_k A = da. \tag{14.29}$$

In adapted coordinates we can use $U(1)$ gauge invariance to set $\partial_\theta A_\mu = 0$. This leaves the residual gauge invariance $A_\mu \to A_\mu + \partial_\mu \alpha$ with $\alpha = \tilde{\alpha}(X^i) + c\theta$.

The dualization now proceeds as in the closed string case except that the field $\tilde{V}_\alpha$ no longer decouples. There is a boundary term $\int_{\partial \Sigma} \left(A_i \partial_\alpha X^i + (A_\theta + \frac{1}{2\pi\alpha'}\tilde{\theta}) V_\alpha \right) d\sigma^\alpha$. Integrating over the boundary values of $\tilde{V}$ leads to the constraint

$$2\pi\alpha' A_\theta + \tilde{\theta} = 0 \quad \text{on} \quad \partial \Sigma \tag{14.30}$$

and the dual action has the form of the original action, but with the dual backgrounds $(\tilde{G}_{\mu\nu}, \tilde{B}_{\mu\nu}, \tilde{\Phi})$ as in (14.15) and (14.16) and, in addition,

$$\tilde{A}_\mu = (\tilde{A}_\theta, \tilde{A}_i) = (0, A_i). \tag{14.31}$$

If we started from a $(p+1)$-dimensional gauge background, the dual gauge field is in p dimensions. The component along the dualized direction becomes the position of the boundary of the world-sheet along the dual direction, (14.30). The residual gauge invariance is mapped to translational invariance of the position of the D-brane in the dual theory, $\tilde{\theta} \to \tilde{\theta} + c$. In other words, T-duality along (transverse to) the world-volume of a Dp-brane results in a D(p-1) (D(p+1)) brane.

We now examine how the effective world-volume action for a D-brane, the Dirac-Born-Infeld action, transforms under T-duality. The DBI action for a space-time

14.2 T-Duality and Buscher Rules

filling brane with tension T is (cf. (16.160) on page 630)

$$S = T \int d^d X e^{-\Phi} \sqrt{\det(G + B + 2\pi\alpha' F)_{\mu\nu}}. \tag{14.32}$$

Using the formula

$$\det \begin{pmatrix} A & B \\ C & D \end{pmatrix} = \det A \, \det(D - CA^{-1}B), \tag{14.33}$$

one finds

$$\det[(G + B + 2\pi\alpha' F)_{\mu\nu}] = G_{\theta\theta} \det[(\hat{G} + \hat{B} + 2\pi\alpha' \hat{F})_{ij}], \tag{14.34}$$

where

$$\hat{G}_{ij} = \partial_i \tilde{X}^\mu \partial_j \tilde{X}^\nu \tilde{G}_{\mu\nu} \tag{14.35}$$

and likewise for the other fields. In particular, $\hat{F}_{ij} = F_{ij}$ are the fields induced on the world-volume of the brane and $\tilde{X}^\mu = (\tilde{\theta}, X^i)$ are the embedding functions of the $(d-1)$-dimensional world-volume of the brane into d dimensional space-time. The T-dual world-volume action is thus

$$\tilde{S} = LT \int d^{d-1} X \, e^{-\tilde{\Phi}} \sqrt{\det(\hat{G} + \hat{B} + 2\pi\alpha' \hat{F})_{ij}}, \tag{14.36}$$

where L is the size of the dualized direction.

We close this section with a few comments:

1. In our discussion of the dualization of the σ-model action we have neglected global issues which arise on world-sheets Σ with non-trivial topology, e.g. the torus. In this case, the solution to $dV = 0$ allows other than pure gauge solutions, namely non-trivial gauge holonomies $\oint V \neq 0$ around the non-contractible loops of Σ. In other words, $V = d\theta$ holds only locally, i.e. θ is not globally defined. If the isometry is a compact $U(1)$, i.e. θ parametrizes a circle of radius R, the holonomies $\frac{1}{2\pi R} \oint V = \frac{1}{2\pi R} \oint d\theta$ must be integers. One can then show that the functional integral is independent of the holonomies, if $\frac{1}{2\pi \tilde{R}} \oint d\tilde{\theta} \in \mathbb{Z}$ where $\tilde{R} = \alpha'/R$, i.e. the dual theory is compactified on the dual circle.

2. We have only considered T-duality along a single isometry direction. Generalization to several directions, in particular to a higher-dimensional torus, is straightforward. The duality group is again a discrete symmetry which relates physically indistinguishable string backgrounds. In Chap. 9 we have mentioned that the coset manifold $\frac{O(D,D)}{O(D)\times O(D)}$ is the moduli space of compactifications of the bosonic string on a D-dimensional torus, which is parametrized by the $D \times D$ components of the constant background metric and B-field. This manifold has to

be further divided by the T-duality group in order to arrive at the parameter-space of physically distinct torus compactifications.

3. In general, T-duality is a symmetry which relates geometrically and sometimes even topologically distinct backgrounds. Nevertheless, string theory does not distinguish between them. A simple but non-trivial example is the background ($\mathbb{R}^2 \times \mathbb{R}^{1,d-3}$ in polar coordinates)

$$ds^2 = dr^2 + r^2 d\varphi^2 + d\mathbf{y}^2, \qquad \Phi = 0. \tag{14.37}$$

Dualizing along the isometry direction ∂_φ we arrive at the dual background

$$d\tilde{s}^2 = dr^2 + \frac{\alpha'^2}{r^2} d\tilde{\varphi}^2 + d\mathbf{y}^2, \qquad \Phi = -\ln(r/\sqrt{\alpha'}) \tag{14.38}$$

with curvature scalar $R = -4/r^2$ which is singular at the origin. One checks that the equations of motion derived from the action (14.8) are indeed satisfied.

4. Our derivation of the Buscher rules a priori applies to any sigma model with a target space with an Abelian isometry. However, in general the two dual models are not equivalent as quantum field theories. For conformal field theories it has been shown that the two dual descriptions are equivalent functional integral representations of the same CFT. If one starts with a σ-model where the target space metric is the solution of the lowest order (in α') equations of motion but the α'-corrected β-function does not vanish, the metric and Buscher rules will receive α'-corrections.

5. In superstring theories R-R fields and space-time spinors (R-NS and NS-R) fields also transform under T-duality. In the RNS formulation, they cannot be directly coupled to the world-sheet, and their transformation rules cannot be obtained along the lines we have presented here. One can use space-time arguments and the fact that T-duality is a map between type IIA and type IIB theories to find the generalization of the Buscher rules to R-R fields and the T-duality transformation of space-time spinors; for details we refer to the literature. The relevant features can, however, be inferred from the following simple argument: we know that T-duality along a world-volume direction of a Dp brane transforms it into a D(p-1) brane and if the T-duality acts in a direction orthogonal to the world-volume it becomes a D(p+1) brane. Since D-branes couple to R-R fields this means that T-duality transforms a p-form into a combination of $p \pm 1$-forms. This follows also from the form of the R-R vertex operators and the action of T-duality on spin fields. We will comment on this in Chap. 15.

   For the generalization of the Buscher rules to heterotic string backgrounds we also refer to the literature.

6. T-duality of type II backgrounds relates solutions of the supergravity equations of motion to solutions of the same equations and is thus a solution generating technique. However, while the original solution might correspond to a weakly coupled string theory in a large background geometry for which supergravity is a good effective low-energy description, the dual background, which still solves the equations of motion, might correspond to a strongly coupled string and/or

a small background geometry where the SUGRA approximation is no longer valid. Similarly, if the dual background has strong curvature, the α' expansion breaks down. The simple example in (14.37) and (14.38) demonstrates both of these points.
7. Non-Abelian T-duality (in the presence of non-Abelian isometries of the background) has also been studied in detail but this and many other interesting aspects will not be covered here.

14.3 Compactification

We now come to the main theme of this chapter and discuss strings in a non-trivial metric background. All other possible background fields are, for time being, turned off. We will see later that this is inconsistent for the heterotic string, for which we will need a non-trivial gauge bundle. We work in the critical dimension of the superstring. The notion of dimension makes sense, as these compactifications possess a large volume, i.e. a decompactification limit.

We will use the following notation: d_c is the dimension of space-time prior to compactification. For string theory it is the critical dimension $d_c = 10$. We compactifiy on a D-dimensional manifold which leaves a $d = d_c - D$ dimensional non-compact space-time. For the real world this is $d = 4$.

As we have just seen, the requirement of conformal invariance of the world-sheet theory imposes restrictions on the allowed background metrics. These conditions can be further tightened by imposing physically motivated conditions on the resulting string vacuum. For instance, one requires that the vacuum leads to realistic low-energy physics, i.e. to a realistic effective field theory of the light string excitations. We should stress that at the present stage these conditions have to be imposed by hand, possibly with reference to some kind of anthropic reasoning or with the expectation that some day we will find a dynamical principle leading to these specific choices.

One obvious requirement is to resolve the discrepancy between the critical dimension $d_c = 10$ and the number of observed dimensions $d = 4$. The second requirement is that the four-dimensional theory has minimal supersymmetry. This is based on the well-motivated believe that the standard model of particle physics is incomplete. This is obvious because it does not contain the gravitational interaction. But even without gravity a manifestation of its incompleteness is the instability of the weak scale against quantum corrections. Supersymmetry, even if softly broken, can avoid this so-called gauge hierarchy problem through the absence of quadratic divergences in loop corrections to the Higgs mass.

We will now discuss the consequences of these two requirements where we are implicitly assuming that a large volume limit exists where methods of classical geometry are applicable. Whether this is justified for geometries of the size of the string scale is certainly questionable. In fact, various string dualities, such as T-duality, suggest a notion of 'string geometry' which is very different from the geometry probed by point particles. To deal with string size 'geometries' one has to treat the internal degrees of freedom with the methods of conformal field theory, e.g. in the form of Gepner models. The geometric (space-time) picture is, nevertheless, very useful. Geometric notions are rather intuitive and powerful mathematical tools are available, though one has to keep in mind that there are perturbative and non-perturbative corrections, both σ-model (i.e. α') and string loop (i.e. g_s) unless the classical results are protected by some symmetry.

The first requirement naturally leads to the idea of compactification. Since the space-time geometry $\mathcal{M}_{d_c}$ is dynamical, there can exist solutions, consistent with the requirements imposed by local scale invariance on the world-sheet, which make the world appear four-dimensional. The simplest possibility is to have only a background metric and to require that the manifold has the product form $\mathcal{M}_{d_c} = \mathcal{M}_d \times K_D$ where e.g. $\mathcal{M}_d$ is Minkowski space and K_D is a compact space which admits a Ricci-flat metric and $d + D = d_c$. Moreover, to escape detection, the size of K_D must be smaller than the length scales already probed by particle accelerators. The physics observed in $\mathcal{M}_d$ will depend on properties of the compact space K_D. Of course, we are mostly interested in $d_c = 10$ and $d = 4$. If the string equations of motion have such a solution we say that the system admits spontaneous compactification. The background metric G_{MN} then satisfies

$$\bar{G}_{MN}(x, y) = \begin{pmatrix} \bar{g}_{\mu\nu}(x) & 0 \\ 0 & \bar{g}_{mn}(y) \end{pmatrix}, \qquad (14.39)$$

where $X^M = (x^\mu, y^m)$ with x^μ and y^m denoting the coordinates of $\mathcal{M}_d$ and K_D, respectively. The bars indicate background values of the metric. Note that with this ansatz there are no non-zero components of the Christoffel symbols and the Riemann tensor which carry both Latin and Greek indices. An interesting generalization of (14.39) is to keep the product form but with the metric components on $\mathcal{M}_d$ replaced by $e^{2A(y)} \bar{g}_{\mu\nu}(x)$, where $A(y)$ is a so-called 'warp factor'. This does not affect the d-dimensional space-time isometries, e.g. $\bar{G}_{\mu\nu}(x, y) = e^{2A(y)} \eta_{\mu\nu}$ is compatible with d-dimensional Poincaré symmetry. Such warped metrics appear, if additional objects like D-branes or fluxes are present in the background. Such solutions will be discussed in Chaps. 17 and 18.

Consider the low-energy effective field theories for the massless string modes. These are d_c-dimensional supergravity theories and we will search for compactifications that preserve some degree of supersymmetry. As long as there is only a background metric, we must assume that the manifolds are spin manifolds, so that spinor fields can be defined (otherwise $Spin_{\mathbb{C}}$ might be an option). Instead of analyzing whether the equations of motion, which are highly nonlinear, admit solutions of the form (14.39), it is more convenient to demand (14.39) and require unbroken supersymmetries in $\mathcal{M}_d$. A posteriori it can be checked that the background or vacuum expectation values (vevs) obtained for all fields are compatible with the equations of motion and Bianchi identities.

One way to obtain four-dimensional supersymmetric theories is to start in $d_c = 11$ or $d_c = 10$ and compactify on a D-dimensional torus, keeping only the constant modes on the torus, i.e. those which do not depend on the torus coordinates

14.3 Compactification

y^m. This is referred to as dimensional reduction. The reason why we consider also $d_c = 11$ is the existence of a unique supergravity theory in this dimension. Its role in string theory will be discussed in Chap. 18. A Majorana spinor ψ in $d_c = 11$ decomposes under $SO(1,3) \times SO(7)$ as **32** = (**4**, **8**), where **4** and **8** are Majorana spinors of $SO(1,3)$ and $SO(7)$, respectively. Hence, ψ gives rise to eight Majorana spinors in $d = 4$. This means that starting with $d_c = 11$, $\mathcal{N} = 1$, in which the space-time supercharge Q is Majorana, gives a $d = 4$, $\mathcal{N} = 8$ theory upon dimensional reduction. Here and below we suppress spinor indices. As another interesting example, consider $d_c = 10$, $\mathcal{N} = 1$ in which Q is a Majorana-Weyl spinor. The Weyl representation of $SO(1,9)$ decomposes under $SO(1,3) \times SO(6)$ as

$$\mathbf{16} = (\mathbf{2_L}, \bar{\mathbf{4}}) + (\mathbf{2_R}, \mathbf{4}) \,, \tag{14.40}$$

where **4**, $\bar{\mathbf{4}}$ are Weyl spinors of $SO(6)$ and $\mathbf{2_{L,R}}$ are Weyl spinors of $SO(1,3)$. If we further impose the Majorana condition in $d_c = 10$, then dimensional reduction of Q gives rise to four Majorana spinors in $d = 4$. Thus, $\mathcal{N} = 1, 2$ supersymmetric theories in $d_c = 10$ yield $\mathcal{N} = 4, 8$ supersymmetric theories in $d = 4$ upon dimensional reduction.

The large amount of supersymmetry renders toroidal compactifications of superstrings unrealistic because they are non-chiral, i.e. they cannot have the chiral matter content and gauge interactions observed in nature. Supersymmetric extensions of the Standard Model require $d = 4$, $\mathcal{N} = 1$. To obtain such theories we must go beyond toroidal compactification. As a guiding principle, we demand that some supersymmetry is preserved. To this end we must understand the relation between the properties of K_D and d-dimensional space-time supersymmetry. As we will see, this leads to a more precise characterization of the internal manifold.

Unbroken supersymmetry requires that the vacuum satisfies $\bar{\epsilon} Q |0\rangle = 0$ where $\epsilon(x^M)$ parametrizes the supersymmetry transformation which is generated by Q, both Q and ϵ being spinors of $SO(1, d_c - 1)$. This, together with $\delta_\epsilon \Phi = [\bar{\epsilon} Q, \Phi]$, means that $\langle \delta_\epsilon \Phi \rangle \equiv \langle 0| [\bar{\epsilon} Q, \Phi] |0\rangle = 0$ for every field generically denoted by Φ. If we want $\mathcal{M}_d$ to be Minkowski space, we can allow a vev for the metric $\bar{g}_{\mu\nu} = \eta_{\mu\nu}$ and a d-form $\bar{F}_{\mu_1 \ldots \mu_D} \propto \epsilon_{\mu_1 \ldots \mu_D}$. A non-zero background value of any field which is not a $SO(1, d-1)$ scalar would reduce the symmetries of Minkowski space. In particular, since fermionic fields are spinors that transform non-trivially under $SO(1, d-1)$, $\langle \Phi_{\text{Fermi}} \rangle = 0$. Hence, $\langle \delta_\epsilon \Phi_{\text{Bose}} \rangle \sim \langle \Phi_{\text{Fermi}} \rangle = 0$ and we only need to worry about $\langle \delta_\epsilon \Phi_{\text{Fermi}} \rangle$. Now, among the Φ_{Fermi} in supergravity there is always the gravitino ψ_M (or $\mathcal{N}$ gravitini if there are $\mathcal{N}$ supersymmetries in higher dimensions) that transforms as

$$\delta_\epsilon \psi_M = \nabla_M \epsilon + \cdots, \tag{14.41}$$

where ∇_M is the covariant derivative on spinors defined in the appendix of this chapter. The $\ldots$ stand for terms which contain other bosonic fields (dilaton, B_{MN} and p-form fields) whose vevs are taken to be zero. Then, $\langle \delta_\epsilon \psi_M \rangle = 0$ gives

$$\langle \nabla_M \epsilon \rangle \equiv \bar{\nabla}_M \epsilon = 0 \quad \Rightarrow \quad \bar{\nabla}_m \epsilon = 0 \quad \text{and} \quad \bar{\nabla}_\mu \epsilon = 0. \tag{14.42}$$

Notice that in $\bar{\nabla}_M$ there appears the vev of the spin connection $\bar{\omega}$. Spinor fields ϵ, which satisfy (14.42) are covariantly constant (in the background metric). They are also called Killing spinors or parallel spinors. The existence of Killing spinors is thus a necessary requirement for supersymmetric compactifications.

So far we have studied the supersymmetry transformation of the gravitini. In type II supergravities there are also two dilatini. Their supersymmetry transformations vanish in the case that there are no other non-trivial background fields besides the metric. The full supersymmetry variations are given in Eq. (17.64). In heterotic and type I SUGRA there are also gaugini. Their transformation is also zero, if we choose a trivial background configuration for the gauge field. This is, however, inconsistent with the Bianchi identities. We will come back to this issue later. The transformation of the dilatino is not affected and still vanishes.

The existence of covariantly constant spinors restricts the class of manifolds on which we may compactify. To see this explicitly, we iterate (14.42) to obtain the integrability condition (since the following manipulations are completely general, we drop the bar)

$$[\nabla_M, \nabla_N]\epsilon = \frac{1}{4} R_{MN}{}^{AB} \Gamma_{AB}\epsilon = \frac{1}{4} R_{MNPQ} \Gamma^{PQ}\epsilon = 0, \quad (14.43)$$

where $\Gamma_{AB} = \frac{1}{2}[\Gamma_A, \Gamma_B]$ and R_{MNPQ} is the Riemann tensor in the d-dimensional background.

Next we multiply by Γ^N and use the property

$$\Gamma^N \Gamma^{PQ} = \Gamma^{NPQ} + G^{NP}\Gamma^Q - G^{NQ}\Gamma^P, \quad (14.44)$$

where $\Gamma^{NPQ} = \frac{1}{3!}(\Gamma^N \Gamma^P \Gamma^Q \pm \cdots)$. The Bianchi identity

$$R_{MNPQ} + R_{MQNP} + R_{MPQN} = 0 \quad (14.45)$$

implies that $\Gamma^{NPQ} R_{MNPQ} = 0$. In this way, we arrive at the necessary (but not sufficient) requirement for the existence of covariantly constant spinors $R_{MQ}\Gamma^Q \epsilon = 0$, or for a metric of the form (14.39)

$$R_{\mu\nu}\Gamma^\nu \epsilon = 0, \qquad R_{mn}\Gamma^n \epsilon = 0. \quad (14.46)$$

If we require that the non-compact space-time is a maximally symmetric space-time, i.e. that $R_{\mu\nu\rho\sigma} = k(g_{\mu\rho}g_{\nu\sigma} - g_{\mu\sigma}g_{\nu\rho})$, then the first of the above conditions requires $k = 0$ and $g_{\mu\nu} = \eta_{\mu\nu}$. Taking the Hermitian conjugate of the second condition, multiplying by $\Gamma^p \epsilon$ from the right and using (14.46) one obtains $\epsilon^\dagger \epsilon R_m{}^p = 0$, from where we conclude that a necessary condition for the existence of a covariantly constant spinor on a Riemannian manifold is the vanishing of its Ricci tensor[8]:

---

[8]Another way to see this is to consider $\Gamma \equiv a_n \Gamma^n$. Then $\det(\Gamma^2) = g^{mn}a_n a_m$ which, on a Riemannian manifold with Euclidean signature, vanishes iff $a_n = 0$, $\forall n$. Therefore, $a_n \Gamma^n \epsilon =$

14.3 Compactification

$$R_{mn} = 0. \tag{14.47}$$

Thus, the internal K_D has to be a compact Ricci-flat manifold.

Recall that the condition of Ricci flatness also follows from the requirement of Weyl invariance of the world-sheet theory and it is also the equation of motion derived from the supergravity action, if all fields except the metric are set to zero.[9] One allowed solution is $K_D = T^D$, i.e. a D-dimensional torus which is compact and flat. Because in this case $\bar{\nabla}_m \epsilon = \partial_m \epsilon = 0$, i.e. ϵ is constant, it gives the maximum number of supersymmetries in the lower dimensions.

We will now analyze sufficient conditions for the existence of covariantly constant spinors. There are two aspects of this issue. The first obvious condition to have such a spinor is that there exists a globally defined nowhere vanishing spinor field ϵ. This imposes topological restrictions on the manifold. For the case $D = 6$, which we will be mostly interested in, this condition is that the manifold has $SU(3)$ structure. More aspects of G-structures are shown in Sect. 17.4. On a manifold with $SU(3)$ structure the bundle of orthonormal frames can be patched together by elements of $SU(3) \subset SO(6)$. The decomposition of the spinor representation **4** of $SO(6) \simeq SU(4)$ contains an $SU(3)$ singlet which implies the existence of a globally defined nowhere vanishing spinor. Requiring the existence of a covariantly constant spinor is a condition on the connection or, more precisely, on the holonomy group of the manifold, i.e. it is a differential condition.[10]

Upon parallel transport along a closed curve on an m-dimensional manifold, a vector v is rotated into Uv. The set of matrices U obtained in this way forms the holonomy group $\mathcal{H}$. The U's are necessarily matrices in $O(m)$ which is the structure group of a generic Riemannian manifold K_m. Hence $\mathcal{H} \subseteq O(m)$. For manifolds with an orientation the stronger condition $\mathcal{H} \subseteq SO(m)$ holds. Now, from (14.99) it follows that for a simply-connected manifold to have non-trivial holonomy it has to have curvature. Indeed, the Riemann tensor (and its covariant derivatives), when viewed as a Lie-algebra valued two-form, generate $\mathcal{H}$. If the manifold is not simply connected, the Riemann tensor and its covariant derivatives only generate the identity component of the holonomy group, called the restricted holonomy group $\mathcal{H}_0$ for which $\mathcal{H}_0 \subseteq SO(m)$. $\mathcal{H}_0$ is trivial if and only if the Riemann tensor vanishes. Non-simply connected manifolds can have non-trivial $\mathcal{H}$ without curvature (gravitational Wilson lines).

Under parallel transport along a loop in K_6, spinors are also rotated by elements of $\mathcal{H}$. But a covariantly constant spinor such as ϵ remains unchanged. This means

---

0 requires, for non-zero ϵ, that $a_n = 0$, $\forall n$. On Lorentzian manifolds Ricci-flatness is not a necessary condition for the existence of covariantly constant spinors.

[9] These conditions and also (14.41) receive α' corrections. We will comment on them later.

[10] The discussion of the condition of having Killing spinors is much richer than the corresponding question for Killing vectors. In the latter case it is known that Ricci-flat compact manifolds do not admit Killing vectors other than those associated with tori. Equivalently, the first Betti number b_1 receives contributions only from non-trivial cycles associated to torus factors in K_D.

that ϵ is a singlet under $\mathcal{H}$. But ϵ is an $SO(6)$ spinor and hence it has right- and left-chirality pieces that transform respectively as **4** and $\bar{\mathbf{4}}$ of $SO(6) \simeq SU(4)$. How can ϵ be an $\mathcal{H}$-singlet? Suppose that $\mathcal{H} = SU(3)$. Under $SU(3)$ the **4** decomposes into a triplet and a singlet: $\mathbf{4}_{SU(4)} = (\mathbf{3} + \mathbf{1})_{SU(3)}$. Thus, if $\mathcal{H} = SU(3)$ there is one covariantly constant spinor of positive and one of negative chirality, which we denote $\epsilon_\pm$. If $\mathcal{H}$ were $SU(2)$, there would exist two right-handed and two left-handed covariantly constant spinors, as under $SU(2)$ the **4** decomposes into a doublet and two singlets. There could be as many as four covariantly constant spinors of each chirality, as occurs when $K_6 = T^6$ and $\mathcal{H}_0$ is trivial.

If the manifold K_6 has $SU(3)$ holonomy, the resulting theory in $d = 4$ has precisely $\mathcal{N} = 1$ supersymmetry, if it had $\mathcal{N} = 1$ in $d_c = 10$. Indeed, taking into account the decomposition (14.40) and the discussion of the previous paragraph, we see that the allowed supersymmetry parameter takes the form

$$\epsilon = \epsilon_R \otimes \epsilon_+ + \epsilon_L \otimes \epsilon_-. \tag{14.48}$$

Since ϵ is also Majorana, $\epsilon_R = \epsilon_L^*$ and $\epsilon_+ = (\epsilon_-)^*$ and hence (ϵ_R, ϵ_L) form a single $SO(1,3)$ Majorana spinor, associated to a single supersymmetry generator. Similarly, if K_6 has $SU(2)$ holonomy, the resulting $d = 4$ theory will have $\mathcal{N} = 2$ supersymmetry. Obviously, the number of supersymmetries in $d = 4$ is doubled in case we start with $\mathcal{N} = 2$ in $d_c = 10$. The supersymmetry parameter ϵ is an anti-commuting object. In the decomposition (14.48) this is achieved by treating $\epsilon_{R,L}$ as anti-commuting and $\epsilon_\pm$ as commuting spinors.

We define Calabi-Yau manifolds CY_n as $2n$-dimensional compact Riemannian manifolds with $SU(n) \subset SO(2n)$ holonomy. We will give a slightly refined definition later. We have seen that Calabi-Yau manifolds admit covariantly constant spinors and that they are Ricci-flat. For $n = 1$ there is only one[11] CY_1, namely the torus T^2. The only CY_2 is the K3 manifold. For $n \geq 3$ there is a large number of topologically distinct CY_3's. We will give simple examples of CY_3 in Sect. 14.5. We want to remark that except for the trivial case $n = 1$ no metric with $SU(n)$ holonomy on any compact CY_n is known explicitly. But their existence and uniqueness have been proved.

Calabi-Yau manifolds are a class of manifolds with special holonomy. Generically on an oriented manifold $\mathcal{H} \simeq SO(m)$. This raises the following question: which subgroups $G \subset SO(m)$ do occur as holonomy groups of Riemannian manifolds? For the case of simply connected manifolds which are neither symmetric nor locally a product of lower dimensional manifolds, there exists a complete classification (known as Berger's list).

Returning to the important case, $\mathcal{N} = 1$, $d_c = 10$, $d = 4$, and the requirement of unbroken supersymmetry we find the following possibilities. The internal K_6 can be a torus T^6 with trivial holonomy and hence ϵ leads to $D = 4$, $\mathcal{N} = 4$

---

[11]More precisely, there is only one family which is parametrized by a moduli space. For CY_1 this was discussed in Chap. 6 and will be further discussed later in this section and for CY_3 in Sect. 14.5.

14.3 Compactification

supersymmetry. K_6 can be a product K3 $\times$ T^2 with $SU(2)$ holonomy and ϵ leads to $\mathcal{N} = 2$ in $d = 4$. Finally, K_6 can be CY$_3$ with $SU(3)$ holonomy so that ϵ gives $d = 4$, $\mathcal{N} = 1$ supersymmetry. These are the results for heterotic and type I superstrings. For type II superstrings the number of supersymmetries in $d = 4$ is doubled, as we start from $\mathcal{N} = 2$ in $d_c = 10$.[12] In the case of the heterotic string, the supersymmetry conditions and the Bianchi identity for the (NS,NS) two-form also requires a non-trivial background for the gauge field. We will discuss this in Sect. 14.7. Type II compactifications with $\mathcal{N} = 1$ space-time supersymmetry also require additional background fields which break $\mathcal{N} = 2 \to \mathcal{N} = 1$. They will be discussed in Chap. 17 and in Sect. 18.8 in the context of F-theory.

For completeness, and because it is of interest for the discussion of string dualities, let us also consider compactifications from $\mathcal{N} = 1$, $d_c = 10$ to $d = 6$. In this case unbroken supersymmetry requires K_4 to be the flat torus T^4 or the K3 surface with $SU(2)$ holonomy. Toroidal compactification does not reduce the number of real supercharges (16 in $\mathcal{N} = 1$, $d_c = 10$) and for T^4 the theory in $d = 6$ has $\mathcal{N} = 2$, or rather (1,1) supersymmetry. Here the notation indicates that one supercharge is a left-handed and the other a right-handed Weyl spinor. The $SO(1, 5)$ Weyl spinors are complex, as a Majorana-Weyl condition cannot be imposed in $d = 6$. Compactification on K3 gives $d = 6$, $\mathcal{N} = 1$, or rather (1,0), supersymmetry. This can be understood from the decomposition of the **16** Weyl representation of $SO(1, 9)$ under $SO(1, 5) \times SO(4)$,

$$\mathbf{16} = (\mathbf{4_L}, \mathbf{2}) + (\mathbf{4_R}, \mathbf{2'}) , \tag{14.49}$$

where $\mathbf{4_{L,R}}$ and $\mathbf{2}$, $\mathbf{2'}$ are Weyl spinors of $SO(1, 5)$ and $SO(4)$. In both groups each Weyl representation is its own conjugate. Since the supersymmetry parameter ϵ in $d_c = 10$ is Majorana-Weyl, its $(\mathbf{4_L}, \mathbf{2})$ piece has only eight real components which form one complex $\mathbf{4_L}$ and likewise for $(\mathbf{4_R}, \mathbf{2'})$. Then, if the holonomy is trivial, ϵ gives one $\mathbf{4_L}$ plus one $\mathbf{4_R}$ supersymmetry in $d = 6$. Instead, if the holonomy is $SU(2) \subset SO(4) \simeq SU(2) \times SU(2)$, only one $SO(4)$ spinor, say $\mathbf{2}$, is covariantly constant and then ϵ gives only one $\mathbf{4_L}$ supersymmetry. Starting from $\mathcal{N} = 2$ in $d_c = 10$ there are the following possibilities. Compactification on T^4 gives (2,2) supersymmetry for both the non-chiral IIA and the chiral IIB superstrings. However, compactification on K3 gives (1,1) supersymmetry for IIA but (2,0) supersymmetry for IIB.

Let us finally mention that compactification of 11-dimensional supergravity on a manifold with G_2 holonomy leads to $\mathcal{N} = 1$ supersymmetry in $d = 4$. This follows from the decompositions $\mathbf{32}_{SO(1,10)} = (\mathbf{4}, \mathbf{8})_{SO(1,3)\times SO(7)}$ and $\mathbf{8}_{SO(7)} = (\mathbf{7} + \mathbf{1})_{G_2}$. All spinors involved are Majorana spinors.

When we compactify a higher-dimensional theory, we would like to know to what extent the choice of K_D determines the $d = d_c - D$ dimensional world and we would like to find the resulting theory in d dimensions. As a simple example, which

---

[12] We have $\epsilon^{1,2} = \epsilon_R^{1,2} \otimes \epsilon_+ + \epsilon_L^{1,2} \otimes \epsilon_-$ for type IIB. For type IIA we interchange $\epsilon_R^2 \leftrightarrow \epsilon_L^2$.

exhibits most relevant features, consider a free massless scalar in five dimensions with action

$$S = -\frac{1}{2} \int_{\mathcal{M}_5} d^5x \, \eta^{MN} \partial_M \phi \partial_N \phi. \qquad (14.50)$$

The flat metric is consistent with the five dimensional space being the product $\mathcal{M}_5 = \mathcal{M}_4 \times S^1$ of four-dimensional Minkowski space and a circle of radius R, $x^M = (x^\mu, y)$ with $y \in [0, 2\pi R]$. The equation of motion for ϕ is

$$\Box \phi = 0 \quad \Rightarrow \quad \partial_\mu \partial^\mu \phi + \partial_y^2 \phi = 0. \qquad (14.51)$$

Since $\phi(x, y) = \phi(x, y + 2\pi R)$ we can Fourier expand

$$\phi(x, y) = \frac{1}{\sqrt{2\pi R}} \sum_{n=-\infty}^{\infty} \phi_n(x) \, e^{iny/R}. \qquad (14.52)$$

Note that $Y_n(y) \equiv \frac{1}{\sqrt{2\pi R}} e^{iny/R}$ are orthonormalized eigenfunctions of ∂_y^2 on S^1. Substituting (14.52) into (14.51) gives

$$\partial_\mu \partial^\mu \phi_n - \frac{n^2}{R^2} \phi_n = 0, \qquad (14.53)$$

and we find that $\phi_n(x)$ are four-dimensional scalar fields with masses n/R. This can also be seen at the level of the action. Substituting (14.52) into (14.50) and integrating over y (using orthonormality of the Y_n) gives

$$S = -\sum_{n=-\infty}^{\infty} \frac{1}{2} \int_{\mathcal{M}_4} d^4x \left[\partial_\mu \phi_n \, \partial^\mu \phi_n^* + \frac{n^2}{R^2} \phi_n^* \phi_n \right]. \qquad (14.54)$$

This again shows that in four dimensions there is one massless scalar ϕ_0 plus an infinite tower or massive scalars ϕ_n with masses n/R. We are usually interested in the limit $R \to 0$ in which only ϕ_0 remains light while the ϕ_n, $n \neq 0$ become very heavy and can be discarded. We refer to this limit, in which only the zero mode ϕ_0 is kept, as dimensional reduction. This is for the reason that we could obtain the same result by demanding that $\phi(x^M)$ is independent of y. More generally, dimensional reduction in this restricted sense means a compactification on a torus T^D, discarding all massive modes, i.e. all states which carry momentum along the directions of the torus.

The important concept of zero modes generalizes to the case of curved internal compact spaces. However, it is only in the case of the torus compactification that all zero modes are independent of the internal coordinates. This guarantees the consistency of the procedure of discarding the heavy modes in the sense that a

14.3 Compactification

solution of the lower-dimensional equations of motion is also a solution of the full higher-dimensional ones.

We now consider Kaluza-Klein reduction when compactifying on a curved internal space with the aim of determining the theory in d dimensions. To this end, we expand all d_c-dimensional tensor fields, generically denoted $\Phi^{mn\cdots}_{\mu\nu\cdots}(x, y)$, around their vacuum expectation values

$$\Phi^{mn\cdots}_{\mu\nu\cdots}(x, y) = \langle \Phi^{mn\cdots}_{\mu\nu\cdots}(x, y) \rangle + \varphi^{mn\cdots}_{\mu\nu\cdots}(x, y). \tag{14.55}$$

Next we substitute this into the d_c-dimensional equations of motion and use the splitting (14.39) of the metric. Keeping only linear terms, and possibly fixing gauge, gives generic equations

$$\mathcal{O}_d \varphi^{mn\cdots}_{\mu\nu\cdots} + \mathcal{O}_{\text{int}} \varphi^{mn\cdots}_{\mu\nu\cdots} = 0, \tag{14.56}$$

where $\mathcal{O}_d$, $\mathcal{O}_{\text{int}}$ are differential operators of order p ($p = 2$ for bosons and $p = 1$ for fermions) that depend on the specific field.

We next expand $\varphi^{mn\cdots}_{\mu\nu\cdots}$ in terms of eigenfunctions $Y^{mn\cdots}_a(y)$ of $\mathcal{O}_{\text{int}}$ in K_D. That is

$$\varphi^{mn\cdots}_{\mu\nu\cdots}(x, y) = \sum_a \varphi_{a\mu\nu\cdots}(x) Y^{mn\cdots}_a(y). \tag{14.57}$$

Since $\mathcal{O}_{\text{int}} Y^{mn\cdots}_a(y) = \lambda_a Y^{mn\cdots}_a(y)$, from (14.56) we see that the eigenvalues λ_a determine the masses of the d-dimensional fields $\varphi_{a\mu\nu\cdots}(x)$. With R being a typical length scale of K_D, $\lambda_a \sim 1/R^p$. We again find that for finite R only the zero modes of $\mathcal{O}_{\text{int}}$ correspond to massless fields $\varphi_{0\mu\nu\cdots}(x)$ and for small R the masses of all other modes are large. When compactifying string theories, there appear two hierarchies of masses, one set by the string scale and the other by the inverse size of the compact manifold. If we start with the supergravity approximation we have already discarded the massive string excitatons.

To obtain the effective d-dimensional action for the massless fields φ_0, in general it is not consistent to simply set the massive fields, i.e. the coefficients of the higher harmonics, to zero. The problem with such a truncation is that the heavy fields, denoted φ_h, might induce interactions of the φ_0 that are not suppressed by inverse powers of the heavy mass. This occurs for instance when there are cubic couplings $\varphi_0 \varphi_0 \varphi_h$. Setting φ_h to zero would then not be a solution to the φ_h equations of motion. When the zero modes $Y_0(y)$ are constant or covariantly constant a product of them is also a zero mode and then by orthogonality of the $Y_a(y)$ terms linear in φ_h cannot appear after integrating over the extra dimensions, but otherwise they might be present and generate corrections to quartic and higher order couplings of the φ_0. Even if the heavy fields cannot be discarded, it might be possible to consistently determine the effective action for the massless fields.

We have already seen that for massless scalar fields in $d_c = 10$, $\mathcal{O}_{\text{int}}$ is the Laplacian Δ. On a compact manifold Δ has only one scalar zero mode, namely a constant and hence a scalar in d_c dimensions produces just one massless scalar in

d dimensions. An important and interesting case is that of massless Dirac fields Ψ in $d_c = 10$ for which both $\mathcal{O}_d$ and $\mathcal{O}_{\text{int}}$ are Dirac operators. Let us be more specific and consider the case $d_c = 10$ and $d = 4$. The Dirac operator is $\slashed{D} = \Gamma^M D_M$ where the covariant derivative depends on the background metric and possibly on background gauge fields, etc. Using the representation of the Dirac matrices given in the appendix and expanding Ψ in terms of eigenfunctions $\eta_a(y)$ of $\slashed{D}_{\text{int}} = \gamma^m D_m$ with eigenvalues λ_a (we suppress spinor indices)

$$\Psi(x, y) = \sum_a \psi_a(x) \otimes \eta_a(y), \tag{14.58}$$

one finds

$$\Gamma^M D_M \Psi = \sum_a \left(\gamma^\mu D_\mu \psi_a(x) \otimes \eta_a(y) + \gamma^5 \lambda_a \psi_a(x) \otimes \eta_a(y) \right) = 0. \tag{14.59}$$

Zero modes η_0 of $\slashed{D}_{\text{int}}$ lead to massless spinors in $d = 4$.

The number of zero modes of $\slashed{D}_{\text{int}}$ depends only on topological properties of the internal manifold K_D and can be determined using index theorems. When the internal manifold is Calabi-Yau we can also exploit the existence of covariantly constant spinors. For instance, from the formula $\slashed{\nabla}^2 = \nabla^m \nabla_m$, which is valid on a Ricci-flat manifold, it follows that when K_6 is a CY$_3$, the Dirac operator has only two zero modes, namely the covariantly constant ϵ_+ and ϵ_-.

Among the massless higher dimensional fields there are usually p-form gauge fields $A^{(p)}$ with field strength $F^{(p+1)} = dA^{(p)}$ and action

$$S_p = -\frac{1}{2(p+1)!} \int_{M_d} F^{(p+1)} \wedge *F^{(p+1)}. \tag{14.60}$$

After fixing the gauge freedom $A^{(p)} \to A^{(p)} + d\Lambda^{(p-1)}$ by imposing $d^* A^{(p)} = 0$, the equations of motion are

$$\Delta_d A^{(p)} = 0, \qquad \Delta_d = dd^* + d^*d. \tag{14.61}$$

If the metric splits into a d-dimensional and a D-dimensional part, as in (14.39), the Laplacian Δ_{d_c} also splits $\Delta_{d_c} = \Delta_d + \Delta_D$. Then, $\mathcal{O}_{\text{int}}$ is the Laplacian Δ_D. The number of massless d-dimensional fields is thus given by the number of zero modes of the internal Laplacian. This is a cohomology problem, as we will see in detail in Sect. 14.4. In particular, the numbers of zero modes are the Betti numbers b_r. For example, there is a 2-form which decomposes $B_{MN} \to B_{\mu\nu} \oplus B_{\mu m} \oplus B_{mn}$. Each term is an n-form with respect to the internal manifold, where n is easily read from the decomposition. Thus, from $B_{\mu\nu}$ we obtain only one zero mode since $b_0 = 1$, from $B_{\mu m}$ we obtain b_1 modes that are vectors in d dimensions and from B_{mn} we obtain b_2 modes that are scalars in d dimensions. In general, from a p-form in d

14.3 Compactification

dimensions we obtain b_n massless fields, $n = 0, \cdots, p$, that correspond to $(p-n)$-forms in d dimensions.

Let us now consider zero modes of the metric which decomposes as $g_{MN} \to g_{\mu\nu} \oplus g_{\mu m} \oplus g_{mn}$. From $g_{\mu\nu}$ there is only one zero mode, namely the lower dimensional graviton. Massless modes coming from $g_{\mu m}$, that would behave as gauge bosons in d dimensions, can appear only when $b_1 \neq 0$. Massless modes arising from g_{mn} correspond to scalars in D dimensions. To analyze these modes we write $g_{mn} = \bar{g}_{mn} + h_{mn}$. We know that a necessary condition for the fluctuations h_{mn} not to break supersymmetry is $R_{mn}(\bar{g} + h) = 0$ just as $R_{mn}(\bar{g}) = 0$. Thus, the h_{mn} are degeneracies of the vacuum, they preserve the Ricci-flatness.

The h_{mn} are called (metric) moduli. They are free parameters in the compactification which change the size and shape of the manifold but not its topology. For instance, a circle S^1 has one modulus, namely its radius R. The fact that any value of R is allowed manifests itself in the space-time theory as a massless scalar field with vanishing potential. Another instructive example is the the 2-torus T^2 which has one Kähler modulus and one complex structure modulus. This is explicit in Eq. (10.57). The split into complex structure and Kähler moduli generalizes to compactifications on more complicated manifolds, such as Calabi-Yau manifolds. This will be explained in Sect. 14.5.

Up to now our discussion of compactification has been almost entirely in terms of field theory, rather than string theory which has an infinite tower of massive modes in $d_c = 10$. Of course, what we have learned about compactification is also relevant for string theory, since at low energies, where the excitation of massive string modes can be ignored, the dynamics of the massless modes is described by a supergravity theory in ten dimensions (for type II superstrings) coupled to supersymmetric Yang-Mills theory (for type I and heterotic superstrings).

But there are striking differences between compactifications of field theories and string theories. When dealing with strings, it is not the classical geometry (or even topology) of the space-time manifold $\mathcal{M}$ which is relevant. One-dimensional objects, such as strings, probe $\mathcal{M}$ differently than point particles. Much of the attraction of string theory relies on the hope that the modification of the concept of classical geometry to 'string geometry' at distances smaller than the string scale $\sqrt{\alpha'}$ (which in most scenarios is of the order of the Planck length, i.e. $\sim 10^{-33}$ cm) will lead to interesting effects and eventually to an understanding of physics in this distance range. At distances large compared to ℓ_s a description in terms of point particles should be valid and one should recover classical geometry.

One particular property of string compactification as compared to point particles is that there might be more than one manifold K which lead to identical theories. No experiment can be performed to distinguish between the manifolds. T-duality of the torus compactification is one simple example. A particularly interesting example which arose from the study of Calabi-Yau compactifications and which involves topologically different manifolds, is mirror symmetry. It states that for any Calabi-Yau manifold X there exists a mirror manifold $\hat{X}$ such that $\text{IIA}(X) = \text{IIB}(\hat{X})$. Here the notation $\text{IIA}(X)$ means the full type IIA string theory, including all perturbative and non-perturbative effects, compactified on X. For the heterotic string with the so-called standard embedding of the spin connection in the gauge connection mirror symmetry means $\text{het}(X) = \text{het}(\hat{X})$. The manifolds comprising a mirror pair are very different, e.g. their Euler numbers satisfy $\chi(X) = -\chi(\hat{X})$. In particular, the mirror map $X \leftrightarrow \hat{X}$ exchanges complex structure and Kähler moduli between X and $\hat{X}$. We have encountered a simple example of mirror symmetry when we discussed the discrete symmetries of the compactification on T^2 in Chap. 10: it was the $\mathbb{Z}_2$ symmetry which exchanged the two moduli $T \leftrightarrow U$. But it mapped a torus to a torus, which is therefore its own mirror manifold.

We will come back to mirror symmetry, which is a perturbative duality, i.e. it holds order by order in the g_s expansion, later in this chapter. Non-perturbative so called S-dualities will be discussed in Chap. 18.

The notion of geometric dualities also exists for (classical) point particle theories through the existence of isospectral manifolds. Paraphrasing Mark Kac who asked 'Can you hear the shape of a drum?' we can say that string theory does not even hear the topology of a manifold. This is due to the fact that extended objects probe a space quite differently than point particles.

14.4 Mathematical Preliminaries

The discussion of manifolds of $SU(n)$ holonomy requires mathematical prerequisites which we discuss in this section. In the appendix we summarize our notion and summarize some results of differential geometry on real manifolds.

Complex Manifolds

Manifolds with $SU(n)$ holonomy have very special properties. In particular they are complex manifolds. We have discussed two-dimensional complex manifolds, Riemann surfaces, in Chap. 6. For Riemann surfaces the conformal structures and complex structures are one and the same. This is no longer true in higher dimensions. Neither it is true that higher dimensional manifolds always admit a complex structure.

A real manifold looks locally like $\mathbb{R}^n$ and transition functions between overlapping cooordinate patches are smooth functions. A complex manifold looks locally like $\mathbb{C}^n$. Transition functions from one coordinate patch to another are holomorphic functions. More precisely, a complex manifold M is a differentiable manifold which admits an open cover $\{U_a\}_{a \in A}$ and coordinate maps $z_a : U_a \to \mathbb{C}^n$ such that $z_a \circ z_b^{-1}$ is holomorphic on $z_b(U_a \cap U_b) \subset \mathbb{C}^n$ for all a, b. $z_a = (z_a^1, \ldots, z_a^n)$ are local holomorphic coordinates and on overlaps $U_a \cap U_b$, $z_a^i = f_{ab}^i(z_b)$ are holomorphic (complex analytic) functions, i.e. they do not depend on $\bar{z}_b^i$. (When considering local coordinates we will usually drop the subscript which refers to a particular patch.) An atlas $\{U_a, z_a\}_{a \in A}$ with the above properties defines a complex structure on M. If the union of two such atlases has again the same properties, they are said to define the same complex structure; cf. differential structure in the real case, which

14.4 Mathematical Preliminaries

is defined by equivalence classes of C^∞ atlases. n is called the complex dimension of M: $n = \dim_\mathbb{C}(M)$. Clearly, a complex manifold can be viewed as a real manifold with even (real) dimension $m = \dim_\mathbb{R}(M) = 2n$. We write M^n or M_n to indicate the complex dimension of the manifold M.

Two manifolds might be the same as real manifolds but not as complex manifolds. For instance, all two-dimensional manifolds ($\dim_\mathbb{R}(M) = 2$) have a unique differentiable structure, but we know already from Chap. 6 that T^2 allows for different complex structures which are in $1-1$ correspondence with points in the fundamental region. $S^2 \simeq \mathbb{R}^2 \cup \{\infty\}$ is the only compact Riemann surface with a unique complex structure.

As already said, all complex manifolds are locally like $\mathbb{C}^n$, which is the simplest complex manifold of dimension n. It requires only one single coordinate patch. We can consider $\mathbb{C}^n$ as a real manifold if we identify it with $\mathbb{R}^{2n}$ in the usual way by decomposing the complex coordinates into their real and imaginary parts ($i = \sqrt{-1}$):

$$z^j = x^{2j-1} + i\, x^{2j}, \quad \bar{z}^j = x^{2j-1} - i\, x^{2j}, \quad j = 1\ldots, n. \tag{14.62}$$

For later use we give the decomposition of the partial derivatives

$$\partial_j \equiv \frac{\partial}{\partial z^j} = \frac{1}{2}\left(\frac{\partial}{\partial x^{2j-1}} - i \frac{\partial}{\partial x^{2j}}\right),$$

$$\bar{\partial}_j \equiv \frac{\partial}{\partial \bar{z}^j} = \frac{1}{2}\left(\frac{\partial}{\partial x^{2j-1}} + i \frac{\partial}{\partial x^{2j}}\right) \tag{14.63}$$

and the differentials

$$dz^j = dx^{2j-1} + i\, dx^{2j}, \quad d\bar{z}^j = dx^{2j-1} - i\, dx^{2j}. \tag{14.64}$$

Locally, on any complex manifold, we can always choose real coordinates as the real and imaginary parts of the holomorphic coordinates. A complex manifold is thus also a real manifold. Moreover, since $\det \frac{\partial(x_a^1,\ldots,x_a^{2n})}{\partial(x_b^1,\ldots,x_b^{2n})} = \left|\det \frac{\partial(z_a^i,\ldots,z_a^n)}{\partial(z_b^1,\ldots,\partial z_b^n)}\right|^2 > 0$ on $U_a \cap U_b$, any complex manifold is orientable.

A very important example is the n-dimensional complex projective space $\mathbb{CP}^n$, or simply $\mathbb{P}^n$. $\mathbb{P}^n$ is defined as the set of (complex) lines through the origin of $\mathbb{C}^{n+1}$. A line through the origin can be specified by a single point and two points z and w define the same line, iff there exists $\lambda \in \mathbb{C}^* \equiv \mathbb{C}-\{0\}$ such that $z = (z^0, z^1 \ldots, z^n) = (\lambda w^0, \lambda w^1, \ldots, \lambda w^n) \equiv \lambda \cdot w$. We thus have

$$\mathbb{P}^n = \frac{\mathbb{C}^{n+1} - \{0\}}{\mathbb{C}^*}. \tag{14.65}$$

The coordinates $z^0, \ldots, z^n$ are called homogeneous coordinates on $\mathbb{P}^n$. One writes $[z] = [z^0 : z^1 : \cdots : z^n]$. $\mathbb{P}^n$ can be covered by $n+1$ coordinate patches $U_i = \{[z] : z^i \neq 0\}$, i.e. U_i consists of those lines through the origin which do not lie in the hyperplane $z^i = 0$.[13] In U_i we can choose local coordinates as $\xi^k_{(i)} = \frac{z^k}{z^i}$. They are well defined on U_i and satisfy

$$\xi^k_{(i)} = \frac{z^k}{z^i} = \frac{z^k}{z^j} \bigg/ \frac{z^i}{z^j} = \frac{\xi^k_{(j)}}{\xi^i_{(j)}}, \tag{14.66}$$

which is holomorphic on $U_i \cap U_j$, where $\xi^i_{(j)} \neq 0$. $\mathbb{P}^n$ is thus a complex manifold. The coordinates $\xi_{(i)} = (\xi^1_{(i)}, \ldots, \xi^n_{(i)})$ are called inhomogeneous coordinates. Alternatively to (14.65) we can also define $\mathbb{P}^n$ as $\mathbb{P}^n = S^{2n+1}/U(1)$, where $U(1)$ acts as $z \to e^{i\phi} z$. This shows that $\mathbb{P}^n$ is compact. One shows that $\mathbb{P}^1 \simeq S^2$ by examining transition functions between the two coordinate patches that one obtains after stereographically projecting the sphere onto $\mathbb{C} \cup \{\infty\}$.

A complex submanifold X of a complex manifold M^n is a set $X \subset M^n$ which is given locally as the zeroes of a collection $f_1, \ldots, f_k$ of holomorphic functions such that $\text{rank}(J) \equiv \text{rank}\left(\frac{\partial(f_1, \ldots, f_k)}{\partial(z^1, \ldots, z^n)}\right) = k$. X is a complex manifold of dimension $n-k$, or, equivalently, X has co-dimension k in M^n. Submanifolds of co-dimension one are called hypersurfaces. The easiest way to show that X is indeed a complex manifold is to choose local coordinates on M such that X is given by $z^1 = z^2 = \cdots = z^k = 0$. It is then clear that, if M is a complex manifold, so is X. A point $p \in X$ is a smooth point, if $\text{rank}(J(p)) = k$. Otherwise p is called a singular point. For instance, for $k = 1$, at a smooth point there is no simultaneous solution of $f = 0$ and $df = 0$. Submanifolds which satisfy this condition everywhere are called transverse.

The importance of projective space, besides being a concrete example of a compact complex manifold, lies in the fact that the best studied constructions of Calabi-Yau manifolds are submanifolds of projective space and generalizations thereof. We will come back to this later when we discuss specific examples.[14]

We have seen that any complex manifold M can be viewed as a real manifold. The tangent space at a point p is denoted by $T_p(M)$ and the tangent bundle by $T(M)$. The complexified tangent bundle $T_{\mathbb{C}}(M) = T(M) \otimes \mathbb{C}$ consists of all tangent vectors of M with complex coefficients, i.e. $v = \sum_{\mu=1}^{2n} v^\mu \frac{\partial}{\partial x^\mu}$ with $v^\mu \in \mathbb{C}$. With the help of (14.63) we can write this as

$$v = \sum_{\mu=1}^{2n} v^\mu \frac{\partial}{\partial x^\mu} = \sum_{j=1}^{n} (v^{2j-1} + i\, v^{2j}) \partial_j + \sum_{j=1}^{n} (v^{2j-1} - i\, v^{2j}) \bar{\partial}_j$$

$$\equiv v^{1,0} + v^{0,1}. \tag{14.67}$$

---

[13] $n-1$-dimensional submanifolds of $\mathbb{P}^n$ or, more generally, co-dimension-one submanifolds of $\mathbb{P}^n$ are called hyperplanes.

[14] In this context the following result is of interest: there are no compact complex submanifolds of $\mathbb{C}^n$. This is an immediate consequence of the fact that any global holomorphic function on a compact complex manifold is constant, applied to the coordinate functions.

14.4 Mathematical Preliminaries

We have thus the decomposition

$$T_{\mathbb{C}}(M) = T^{1,0}(M) \oplus T^{0,1}(M) \qquad (14.68)$$

into vectors of type $(1,0)$ and of type $(0,1)$: $T^{1,0}(M)$ is spanned by $\{\partial_i\}$ and $T^{0,1}(M)$ by $\{\bar{\partial}_i\}$. Note that $T^{0,1}_p(M) = \overline{T^{1,0}_p(M)}$ and that the splitting into the two subspaces is preserved under holomorphic coordinate changes. The transition functions of $T^{1,0}(M)$ are holomorphic, and we therefore call it the holomorphic tangent bundle. A holomorphic section of $T^{1,0}(M)$ is called a holomorphic vector field; its component functions are holomorphic.

$T^{1,0}(M)$ is a particular example of a holomorphic vector bundle $E \xrightarrow{\pi} M$. Holomorphic vector bundles of rank k are characterized by their holomorphic transition functions which are elements of $Gl(k, \mathbb{C})$ (rather than $Gl(n, \mathbb{R})$ as in the real case) with holomorphic matrix elements.

In the same way as in (14.68) we decompose the dual space, the space of one-forms:

$$T^*_{\mathbb{C}}(M) = T^{*1,0}(M) \oplus T^{*0,1}(M). \qquad (14.69)$$

$T^{*1,0}(M)$ and $T^{*0,1}(M)$ are spanned by $\{dz^i\}$ and $\{d\bar{z}^i\}$, respectively. The analogue of (14.67) is

$$\omega = \sum_{\mu=1}^{2n} \omega_\mu dx^\mu = \sum_{j=1}^n \frac{1}{2}(\omega_{2j} - i\,\omega_{2j-1}) dz^j + \sum_{j=1}^n \frac{1}{2}(\omega_{2j} + i\,\omega_{2j-1}) d\bar{z}^j$$

$$\equiv \omega^{(1,0)} + \omega^{(0,1)}. \qquad (14.70)$$

By taking tensor products, we define differential forms of type (p,q) as sections of $\wedge^p T^{*1,0}(M) \wedge^q qT^{*0,1}(M)$. The space of (p,q)-forms will be denoted by $A^{p,q}$. Clearly $\overline{A^{p,q}} = A^{q,p}$. If we denote the space of sections of $\wedge^r T^*_{\mathbb{C}}(M)$ by A^r, we have the decomposition

$$A^r = \bigoplus_{p+q=r} \Pi^{p,q} A^r = \bigoplus_{p+q=r} A^{p,q}, \qquad (14.71)$$

where $\Pi^{p,q} : A^r \to A^{p,q}$ is a projection operator. This decomposition is independent of the choice of local coordinate system.

Using the underlying real structure, we can define the exterior derivative d. If $\omega \in A^{p,q}$, i.e.

$$\omega = \frac{1}{p!q!} \omega_{i_1 \ldots i_p, \bar{j}_1 \ldots \bar{j}_q} dz^{i_1} \wedge \cdots \wedge dz^{i_p} \wedge d\bar{z}^{j_1} \wedge \cdots \wedge d\bar{z}^{j_q}, \qquad (14.72)$$

then
$$d\omega \in A^{p+1,q} \oplus A^{p,q+1}. \qquad (14.73)$$

We write $d\omega = \partial\omega + \bar\partial\omega$ with $\partial\omega \in A^{p+1,q}$ and $\bar\partial\omega \in A^{p,q+1}$. This defines the two operators

$$\partial: A^{p,q} \to A^{p+1,q}, \qquad \bar\partial: A^{p,q} \to A^{p,q+1} \qquad (14.74)$$

and

$$d = \partial + \bar\partial. \qquad (14.75)$$

The explicit expressions for the action of ∂ and $\bar\partial$ on a (p,q)-form ω are

$$\partial\omega = \frac{1}{p!q!}\partial_i \omega_{i_1\cdots i_p \bar\jmath_1\cdots \bar\jmath_q} dz^i \wedge dz^{i_1} \wedge \cdots \wedge d\bar z^{j_q},$$

$$\bar\partial\omega = \frac{1}{p!q!}\bar\partial_j \omega_{i_1\ldots i_p \bar\jmath_1\ldots \bar\jmath_q} d\bar z^j \wedge dz^{i_1} \wedge \cdots \wedge d\bar z^{j_q}$$

$$= \frac{(-1)^p}{p!q!}\bar\partial_j \omega_{i_1\ldots i_p \bar\jmath_1\ldots \bar\jmath_q} \wedge dz^{i_1} \cdots \wedge dz^{i_p} \wedge d\bar z^j \wedge d\bar z^{j_1} \wedge \cdots \wedge d\bar z^{j_q}. \qquad (14.76)$$

The following results are easy to verify:

$$d^2 = (\partial + \bar\partial)^2 \equiv 0 \quad \Rightarrow \quad \partial^2 = 0, \quad \bar\partial^2 = 0, \quad \partial\bar\partial + \bar\partial\partial = 0. \qquad (14.77)$$

Here we used that $\partial^2 : A^{p,q} \to A^{p+2,q}$, $\bar\partial^2 : A^{p,q} \to A^{p,q+2}$, $(\partial\bar\partial + \bar\partial\partial) : A^{p,q} \to A^{p+1,q+1}$, i.e. that the three operators map to three different spaces. They must thus vanish separately.

ω is called a holomorphic p-form, if it is of type $(p,0)$ and $\bar\partial\omega = 0$, i.e. if it has holomorphic coefficient functions. Likewise, $\bar\omega$ of type $(0,q)$ with $\partial\bar\omega = 0$ is called anti-holomorphic. $\Omega^p(M)$ denotes the vector-space of holomorphic p-forms.

Alternative to the way we have defined complex structures, we could have started with an almost complex structure, which is define as a globally defined differentiable endomorphism (i.e. a linear map),

$$J: T(M) \to T(M), \qquad v^\mu \mapsto J^\mu_\nu v^\nu \quad \text{with} \quad J^2 = -\mathbb{1} \qquad (14.78)$$

such that the splitting (14.68) of $T(M)$ is into eigenspaces of J with eigenvalues $+i$ and $-i$, respectively. In local holomorphic coordinates, $J^j_i = i$, $J^{\bar\jmath}_{\bar\imath} = -i$, $J^{\bar\jmath}_i = J^j_{\bar\imath} = 0$, i.e.

$$J = \mathrm{diag}(i, \ldots, i, -i, \ldots, -i), \qquad (14.79)$$

while in real coordinates

14.4 Mathematical Preliminaries

$$J = \begin{pmatrix} 0 & 1 & & & \\ -1 & 0 & & & \\ & & \ddots & & \\ & & & 0 & 1 \\ & & & -1 & 0 \end{pmatrix}. \tag{14.80}$$

The projection operators

$$P_\mu^\nu = \frac{1}{2}(\delta_\mu^\nu - iJ_\mu^\nu), \qquad Q_\mu^\nu = \frac{1}{2}(\delta_\mu^\nu + iJ_\mu^\nu),$$
$$P^2 = P, \quad Q^2 = Q, \quad PQ = 0, \quad P + Q = 1 \tag{14.81}$$

project on the eigenspaces of J with eigenvalues $\pm i$.

One defines the Nijenhuis tensor, also called the torsion of J,

$$N(J): T(M) \times T(M) \to T(M),$$

$$N(v, w) = [v, w] + J[Jv, w] + J[v, Jw] - [Jv, Jw], \tag{14.82}$$

whose component expression is

$$N_{\mu\nu}^\rho = J_\sigma^\rho \partial_\mu J_\nu^\sigma - J_\mu^\sigma \partial_\sigma J_\nu^\rho - (\mu \leftrightarrow \nu). \tag{14.83}$$

The almost complex structure J is called integrable if $N(J) = 0$. From

$$[Pv, Pw] = \frac{1}{4} N(v, w) - \frac{i}{2} P([Jv, w] + [v, Jw]) \tag{14.84}$$

and likewise for $[Qv, Qw]$ we conclude that if $N(J) = 0$ the Lie-bracket of two vector fields of, say, type $(1, 0)$ is again of type $(1, 0)$.

Equation (14.75) is not always true on an almost complex manifold. Given $\omega \in T^*(M)$, we define[15] $\omega_\mu^{(1,0)} = P_\mu^\nu \omega_\nu \in T^{*1,0}$ and $\omega_\mu^{(0,1)} = Q_\mu^\nu \omega_\nu \in T^{*0,1}$. It is straightforward to show that the $(0, 2)$ part of $d\omega^{(1,0)}$ (with components $Q_\mu^\rho Q_\nu^\sigma (d\omega^{(1,0)})_{\rho\sigma}$) vanishes provided $N_{\mu\nu}^\rho \omega_\rho^{(1,0)} = 0$. Likewise, the vanishing of the $(2, 0)$ part of $d\omega^{(0,1)} = d(Q\omega)$ requires $N_{\mu\nu}^\rho \omega_\rho^{(0,1)} = 0$. More generally, on an almost complex manifold, for $\omega \in A^{p,q}$, $d\omega \in A^{p+2,q-1} \oplus A^{p+1,q} \oplus A^{p,q+1} \oplus A^{p-1,q+2}$, but $N(J) = 0$ implies (14.73) and (14.75). One checks that $P_\rho^\mu Q_\sigma^\nu N_{\mu\nu}^\kappa = 0$, i.e. the $(1, 1)$ part of N, when viewed as a two-form with values in the tangent bundle, vanishes. Therefore in two dimensions any almost integrable structure is integrable.

---

[15] $P = \Pi^{1,0}$, $Q = \Pi^{0,1}$ in terms of our previous definition.

A theorem of Newlander and Nierenberg states that integrable almost complex structures and complex structures are essentially identical. That is, if M is endowed with an integrable almost complex structure J, we can construct on M an atlas of holomorphic charts and M is a complex manifold in the sense of the definition that we have given above with J the canonical almost integrable structure (14.79). If the almost complex structure is real-analytic, the proof is elementary. But, if it is C^∞ or satisfies even weaker smoothness conditions, the proof is quite difficult.

Not all real manifolds of even dimension can be endowed with an (almost) complex structure. For instance, among the even-dimensional spheres S^{2n}, only S^2 and S^6 admit an almost complex structure. While S^2 has a unique complex structure, it is not known whether S^6 is a complex manifold. However, direct products of odd-dimensional spheres always admit a complex structure.

Kähler Manifolds

The next step is to introduce additional structures on a complex manifold: a Hermitian metric and a Hermitian connection.

A Hermitian metric is a covariant tensor field of the form $ds^2 = 2\sum_{i,j=1}^{n} g_{i\bar{j}} dz^i \otimes d\bar{z}^j$, where $g_{i\bar{j}} = g_{i\bar{j}}(z)$ (here the notation is not to indicate that the components are holomorphic functions; they are not!) such that $g_{j\bar{i}}(z) = \overline{g_{i\bar{j}}(z)}$ and $g_{i\bar{j}}(z)$ is a positive definite matrix, that is, for any $\{v^i\} \in \mathbb{C}^n$, $v^i g_{i\bar{j}} \bar{v}^j \geq 0$ with equality only if all $v^i = 0$.[16]

To any Hermitian metric we associate a (1, 1)-form

$$\omega = i \sum_{i,j=1}^{n} g_{i\bar{j}} \, dz^i \wedge d\bar{z}^j. \qquad (14.85)$$

ω is called the fundamental form associated with the Hermitian metric g. ω is a real (1, 1)-form, i.e. $\omega = \bar{\omega}$. From

$$\frac{\omega^n}{n!} = (i)^n g(z) \, dz^1 \wedge d\bar{z}^1 \wedge \cdots \wedge dz^n \wedge d\bar{z}^n = 2^n g(z) \, dx^1 \wedge \cdots \wedge dx^{2n}, \qquad (14.86)$$

where $\omega^n = \overbrace{\omega \wedge \cdots \wedge \omega}^{n \text{ factors}}$, $g(z) = \det(g_{i\bar{j}}) > 0$ and $2^n \det(g_{i\bar{j}}) = \sqrt{\det(g_{\mu\nu})}$. It follows that ω^n is a good volume form on M, just reflecting that complex manifolds always possess an orientation.

An alternative way of defining the Hermitian metric, which does not use local holomorphic coordinates, is via the requirement $g(u, w) = g(Ju, Jv)$ for any two vector fields; in components $g_{\mu\nu} = J_\mu^\rho J_\nu^\sigma g_{\rho\sigma}$. Likewise, ω is defined

---

[16]Given a Riemannian metric $g_{\mu\nu}$, its components in local complex coordinates are g_{ij}, $g_{\bar{i}\bar{j}} = (g_{ij})^*$, $g_{i\bar{j}}$, $g_{\bar{i}j} = (g_{i\bar{j}})^*$. It is Hermitian if $g_{ij} = 0$.

14.4 Mathematical Preliminaries

via $\omega(v,w) = g(Jv,w)$ or, in components, $\omega_{\mu\nu} = J_\mu^\rho g_{\rho\nu} = -\omega_{\nu\mu}$. Given any Riemannian metric g on a complex manifold, $\tilde{g}(v,w) = \frac{1}{2}(g(v,w) + g(Ju,Jv))$ is a Hermitian metric which therefore always exists. An (almost) complex manifold with a Hermitian metric is called an (almost) Hermitian manifold.

The inverse of the Hermitian metric is $g^{i\bar{j}}$ which satisfies $g^{j\bar{i}}g_{j\bar{k}} = \delta^{\bar{i}}_{\bar{k}}$ and $g_{i\bar{j}}g^{k\bar{j}} = \delta^k_i$ (summation convention used). We use the metric to raise and lower indices, whereby they change their type. As for any other tensor field, under holomorphic coordinate changes, the index structure of the metric is preserved.

A Hermitian metric g whose associated fundamental form ω is closed, i.e. $d\omega = 0$, is called a Kähler metric. The set of all closed, real, positive definite (1, 1) forms on M is the set of all Kähler forms on M. It is straightforward to show that $d\omega = 0$ is equivalent to $\nabla J = 0$, where ∇ is the covariant derivative with the Riemannian connection.

A complex manifold endowed with a Kähler metric is called a Kähler manifold, where ω is called the Kähler form. An immediate consequence of $d\omega = 0 \Rightarrow \partial\omega = \bar{\partial}\omega = 0$ is

$$\partial_i g_{j\bar{k}} = \partial_j g_{i\bar{k}}, \qquad \bar{\partial}_{\bar{i}} g_{j\bar{k}} = \bar{\partial}_{\bar{k}} g_{j\bar{i}}. \qquad (14.87)$$

From this one finds that the only non-zero components of the Riemannian connection are

$$\Gamma^k_{ij} = g^{k\bar{l}}\partial_i g_{j\bar{l}}, \qquad \Gamma^{\bar{k}}_{\bar{i}\bar{j}} = g^{l\bar{k}}\bar{\partial}_{\bar{i}} g_{l\bar{j}}. \qquad (14.88)$$

A Hermitian connection is a connection which is compatible with the Hermitian metric, i.e. $\nabla_i g_{j\bar{k}} = 0$. Among the Hermitian connections there is precisely one with the property that its mixed components vanish. This connection is called Chern connection. The requirement that the mixed components vanish immediately gives (14.88). If g is a Kähler metric the Chern connection is symmetric and coincides with the Riemannian (Levi-Civita) connection.

The vanishing of the connection coefficients with mixed indices is a necessary and sufficient condition that under parallel transport the holomorphic and the anti-holomorphic tangent spaces do not mix (see below).

While all complex manifolds admit a Hermitian metric, this is not true for Kähler metrics. Counter-examples are quaternionic manifolds which appear as moduli spaces of type II compactifications on Calabi-Yau manifolds, or $S^{2p+1} \otimes S^{2q+1}$ with $q > 1$. A complex submanifold of a Kähler manifold M is again a Kähler manifold, with the induced Kähler metric. This follows easily if one goes to local coordinates on M where X is given by $z^1 = \cdots = z^k = 0$.

From (14.87) we also infer the local existence of a real Kähler potential K in terms of which the Kähler metric can be written as

$$g_{i\bar{j}} = \partial_i \bar{\partial}_j K \qquad (14.89)$$

or, equivalently, $\omega = i\partial\bar{\partial}K$. The Kähler potential is not uniquely defined: $K(z,\bar{z})$ and $K(z,\bar{z}) + f(z) + \bar{f}(\bar{z})$ lead to the same metric, where f and $\bar{f}$ are holomorphic and anti-holomorphic functions (on the patch on which K is defined), respectively.

From now on, unless stated otherwise, we will restrict ourselves to Kähler manifolds; some of the results are, however, true for arbitrary complex manifolds. Also, if in doubt, assume that the manifold is compact.

The components of the Riemann tensor of a Kähler manifold are easily computed from (14.251) and (14.88). One finds that the only non-vanishing components are those with index structure $R_{i\bar{j}k\bar{l}}$ and those related by symmetries. In particular the components of the type R_{ij**} are zero. Explicitly

$$R_{i\bar{j}k\bar{l}} = \partial_i \bar{\partial}_{\bar{j}} g_{k\bar{l}} - g^{m\bar{n}}(\partial_i g_{k\bar{n}})(\bar{\partial}_{\bar{j}} g_{m\bar{l}}). \tag{14.90}$$

For the Ricci tensor one obtains

$$R_{i\bar{j}} = -\partial_i \bar{\partial}_{\bar{j}} (\log \det g). \tag{14.91}$$

Note that $R_{i\bar{j}} = R_{i\mu\bar{j}}{}^{\mu} = R_{i\bar{k}\bar{j}}{}^{\bar{k}} = -R_{i\bar{j}}{}^{\bar{k}}{}_{\bar{k}} = -R_{i\bar{j}k}{}^{k}$. One also defines the Ricci-form (of type $(1, 1)$) as

$$\mathscr{R} = i\, R_{j\bar{k}}\, dz^j \wedge d\bar{z}^k = -i\partial\bar{\partial} \log(\det g), \tag{14.92}$$

which satisfies $d\mathscr{R} = 0$. Note that $\log(\det g)$ is not a globally defined function, as $\det g$ transforms as a density under change of coordinates. $\mathscr{R}$, however, is globally defined.

We learn from (14.92) that the Ricci form depends only on the volume form of the Kähler metric and on the complex structure (through ∂ and $\bar{\partial}$). Under a change of metric, $g \to g'$, the Ricci form changes as

$$\mathscr{R}(g') = \mathscr{R}(g) - i\partial\bar{\partial} \log\left(\frac{\det(g'_{k\bar{l}})}{\det(g_{k\bar{l}})}\right), \tag{14.93}$$

where the ratio of the two determinants is a globally defined non-vanishing function on M.

An explicit example of a Kähler manifold is the complex projective space. To demonstrate this we give an explicit metric, called Fubini-Study metric. Recall that $\mathbb{P}^n = \{[z^0 : \cdots : z^n]; 0 \neq (z^0 : \cdots : z^n) \in \mathbb{C}^{n+1}\}$ and $U_0 = \{[1, z^1 : \cdots : z^n]\} \simeq \mathbb{C}^n$ is an open subset of $\mathbb{P}^n$. Set

$$g_{i\bar{j}} = \partial_i \bar{\partial}_{\bar{j}} \log(1 + |z^1|^2 + \cdots + |z^n|^2) \equiv \partial_i \bar{\partial}_{\bar{j}} \ln(1 + |z|^2)$$
$$= \frac{1}{(1 + |z|^2)^2}\left((1 + |z|^2)\delta_{ij} - \bar{z}_i z_j\right) \tag{14.94}$$

or, equivalently,

$$\omega = i\partial\bar{\partial} \log(1 + |z|^2) = i\left(\frac{dz^i \wedge d\bar{z}^i}{1 + |z|^2} - \frac{\bar{z}^i dz^i \wedge z^j d\bar{z}^j}{(1 + |z|^2)^2}\right). \tag{14.95}$$

14.4 Mathematical Preliminaries

With the help of the Schwarz inequality one shows that the metric (14.94) is positive definite, i.e. for any non-zero vector u, $g_{i\bar{j}} u^i \bar{u}^j > 0$. Closure of ω is obvious if one uses (14.77). From (14.94) we also immediately read off the Kähler potential of the Fubini-Study metric (cf. (14.89)) on U_0. Clearly this is only defined locally.

On the other hand, ω is globally defined on $\mathbb{P}^n$. To see this, consider the open set $U_1 = \{(w^0, 1, w^2, \ldots, w^n)\} \subset \mathbb{P}^n$ and check what happens to ω on the overlap $U_0 \cap U_1 = \{[1 : z^1 : \cdots : z^n] = [w^0 : 1 : w^2 : \cdots : w^n]\}$, where $z^i = \frac{w^i}{w^0}$, for all $i \neq 1$ and $z^1 = \frac{1}{w^0}$:

$$\omega = i\partial\bar{\partial}\log(1 + |z^1|^2 + \cdots + |z^n|^2) = i\partial\bar{\partial}\log\left(1 + \frac{1}{|w^0|^2} + \sum_{i=2}^{n} \frac{|w^i|^2}{|w^0|^2}\right)$$

$$= i\left(\partial\bar{\partial}\log(1 + |w|^2) - \partial\bar{\partial}\log(|w^0|^2)\right) = i\partial\bar{\partial}\log(1 + |w|^2), \quad (14.96)$$

as w^0 is holomorphic on $U_0 \cap U_1$. Thus, ω and the corresponding Kähler metric are globally defined. The complex projective space is therefore a Kähler manifold and so is every complex submanifold. With

$$\det(g_{i\bar{j}}) = \frac{1}{(1 + |z|^2)^{n+1}} \quad (14.97)$$

one finds

$$R_{i\bar{j}} = -\partial_i \bar{\partial}_{\bar{j}} \log\left(\frac{1}{(1 + |z|^2)^{n+1}}\right) = (n+1)g_{i\bar{j}}, \quad (14.98)$$

which shows that the Fubini-Study metric is a Kähler-Einstein metric and $\mathbb{P}^n$ a Kähler-Einstein manifold.

Complex submanifolds of $\mathbb{P}^n$ are called algebraic manifolds. Algebraic manifolds are Kähler.

Holonomy Group of Kähler Manifolds

Consider a connected Riemannian manifold M of real dimension m endowed with the Levi-Civita connection. A vector V is said to be parallel transported along a curve $x^\mu(t)$, if the directional derivative vanishes: $\gamma^\mu \nabla_\mu V^\nu = 0$ where $\gamma^\mu = \dot{x}^\mu$ is the tangent to the curve. Under infinitesimal parallel transport V changes by $\Delta V^\mu = -\Gamma^\mu_{\nu\rho} V^\nu dx^\rho$. When V is transported around an infinitesimal contractible loop in the (μ, ν)-plane with area $a^{\mu\nu} = -a^{\nu\mu} = \oint x^\mu dx^\nu$ it changes by the amount

$$\Delta V^\rho = \frac{1}{2} a^{\mu\nu} R_{\mu\nu}{}^\rho{}_\sigma V^\sigma. \quad (14.99)$$

Under parallel transport the length $|V|$ remains constant since $|V|^2 = V^\mu V^\nu g_{\mu\nu}$ and $\nabla_\rho g_{\mu\nu} = 0$. The generalization to the parallel transport of tensors and spinors is obvious.

Fix a point p in M and parallel transport a vector $V \in T_p(M)$ along all closed paths starting and ending at p. This induces a map $T_p(M) \to T_p(M)$. The set of these transformations forms a group $\mathcal{H}(p)$, called the holonomy group at p. Since parallel transport with the Levi-Civita connection does not change the length of the vector, the holonomy group is a subgroup of $O(m)$. For manifolds with an orientation it is a subgroup of $SO(m)$. The holonomy group at two different points p and p' are related by conjugation by the map $T_p(M) \to T_{p'}(M)$ induced by the parallel transport from p to p'. $\mathcal{H}_p$ and $\mathcal{H}_{p'}$ are thus isomorphic and we simply speak of the holonomy group $\mathcal{H}$ of the manifolds M. The restricted holonomy group $\mathcal{H}_0$ is generated by transport around contractible paths.

For Kähler manifolds it follows immediately from the index structure of the connection coefficients that under parallel transport elements of $T^{1,0}(M)$ and $T^{0,1}(M)$ do not mix. The holonomy group of a Kähler manifold is therefore a subgroup of $U(n) \subset SO(2n)$.[17] In particular, elements of $T^{1,0}(M)$ transform as **n** and elements of $T^{0,1}(M)$ as $\bar{\mathbf{n}}$ of $U(n)$. More specifically,

$$\Delta V^i = a^{k\bar{l}} R_{k\bar{l}}{}^i{}_j V^j \qquad (14.100)$$

which is (14.99) expressed in a complex basis. From what we said above it follows that on a Kähler manifold the matrix $a^{k\bar{l}} R_{k\bar{l}}{}^i{}_j$ must be an element of the Lie algebra $u(n)$. The trace of this matrix, which is proportional to the Ricci tensor, generates the $u(1)$ part in the decomposition $u(n) \simeq su(n) \oplus u(1)$. The holonomy group of a Ricci-flat Kähler manifold is therefore a subgroup of $SU(n)$. Conversely, one can show that any $2n$-dimensional manifold with $U(n)$ holonomy admits a Kähler metric and that, if it has $SU(n)$ holonomy, it admits a Ricci-flat Kähler metric. This uses the fact that holomorphic and anti-holomorphic indices do not mix, which implies that all connection coefficients with mixed indices must vanish. One then proceeds with the explicitly construction of an almost complex structure with vanishing Nijenhuis tensor.

The arguments which rely on (14.99) are a priori only valid for the restricted holonomy group $\mathcal{H}_0$. The holonomy around non-contractible loops can be non-trivial even if the connection has vanishing curvature. Furthermore, only the holonomy around infinitesimal loops is generated by the Riemann tensor. For finite (but still contractible) loops, derivatives of the Riemann tensor of arbitrary order will appear. But for Kähler manifolds we have the $U(n)$ invariant split of the indices $\mu = (i, \bar{i})$ and $U(n)$ is a maximal compact subgroup of $SO(2n)$. The restricted

---

[17] The unitary group $U(n)$ is the set of all complex $n \times n$ matrices which leave invariant a Hermitian metric $\overline{g_{i\bar{j}}} = g_{j\bar{i}}$, i.e. $UgU^\dagger = g$. For the choice $g_{i\bar{j}} = \delta_{ij}$ one obtains the familiar condition $UU^\dagger = \mathbf{1}$.

14.4 Mathematical Preliminaries

holonomy group is therefore not bigger than $U(n)$. For Ricci-flat Kähler manifolds it follows from the Calabi-Yau theorem, which we will state below, that the full holonomy group is contained in $SU(n)$ even if the manifold is multiply connected.

(Co)homology of Kähler Manifolds

Before treating the (co)homology of complex manifolds, we review the real case. On a smooth, connected manifold M one defines p-chains a_p as formal sums $a_p = \sum_i c_i N_i$ of p-dimensional oriented submanifolds N_i of M. If the coefficients c_i are real (complex, integer), one speaks of real (complex, integral) chains. Define ∂ as the operation of taking the boundary with the induced orientation. $\partial a_p \equiv \sum c_i \partial N_i$ is then a $p-1$-chain. Let $Z_p = \{a_p | \partial a_p = \emptyset\}$ be the set of cycles, i.e. the set of chains without boundary and let $B_p = \{\partial a_{p+1}\}$ be the set of boundaries. Since $\partial \partial a_p = \emptyset$, $B_p \subset Z_p$. The p-th homology group of M is defined as

$$H_p = Z_p / B_p. \tag{14.101}$$

Depending on the coefficient group, one gets $H_p(M, \mathbb{R})$, $H_p(M, \mathbb{C})$, $H_p(M, \mathbb{Z})$. Elements of H_p are equivalence classes of cycles $z_p \simeq z_p + \partial a_{p+1}$ and are called homology classes denoted by $[z_p]$.

One version of Poincaré duality is the following isomorphism between homology groups, valid on orientable connected smooth manifolds of real dimension m:

$$H_p(M, \mathbb{R}) \simeq H_{m-p}(M, \mathbb{R}). \tag{14.102}$$

One defines the p-th Betti number b_p as

$$b_p = \dim(H_p(M, \mathbb{R})). \tag{14.103}$$

They are topological invariants of M. As a consequence of Poincaré duality,

$$b_p(M) = b_{m-p}(M). \tag{14.104}$$

We now turn to de Rham cohomology, which is defined with the exterior derivative operator $d : A^p \to A^{p+1}$. Let Z^p be the set of closed p-forms, i.e. $Z^p = \{\omega_p | d\omega_p = 0\}$ and let B^p be the set of exact p-forms $B^p = \{d\omega_{p-1}\}$. The de Rham cohomology groups H^p are defined as the quotients

$$H^p_{\text{D.R.}} = Z^p / B^p. \tag{14.105}$$

Elements of H^p are equivalence classes of closed forms $\omega_p \simeq \omega_p + d\alpha_{p-1}$, called cohomology classes and denoted by $[\omega_p]$.

The Hodge decomposition theorem states that on a compact orientable Riemannian manifold any p-form ω_p has a unique decomposition

$$\omega_p = h_p + d\alpha_{p-1} + d^*\beta_{p+1}, \tag{14.106}$$

where h_p denotes a harmonic p-form. For closed forms $d^*\beta_{r+1} = 0$. In particular, each equivalence class possesses a unique harmonic representative, i.e. a zero mode of the Laplacian $\Delta = dd^* + d^*d$. The action of d^* and Δ on p-forms is given in Eqs. (14.287) and (14.289) in the appendix. Among the closed forms in a given equivalence class $[\alpha]$, α is harmonic if and only if it minimizes (α, α) (cf. (14.284)). The number of (normalizable) harmonic p-forms on a compact manifold is finite.

The property of a form being closed does not depend on the definition of a metric. The harmonic representative varies as we vary the metric. Given two metrics we have two different harmonic representatives, h and h'. Their difference is always an exact form. Even though it might seem that the de Rham cohomology groups depend on the differentiable structure of the manifold, they don't. They encode topological information of the manifold.

Denote by $\mathcal{H}^p(M)$ the space of harmonic p-form on M. Then, by Hodge's theorem

$$H^p(M) \simeq \mathcal{H}^p(M). \tag{14.107}$$

Given both the homology and the cohomology classes, we can define an inner product

$$\pi(z_p, \omega_p) = \int_{z_p} \omega_p, \tag{14.108}$$

where $\pi(z_p, \omega_p)$ is called a period (of ω_p). We speak of an integral cohomology class $[\omega_p] \in H_{\text{D.R.}}(M, \mathbb{Z})$, if the period over any integral cycle is integer. With the help of Stoke's theorem one shows that the integral does not depend on which representatives of the two classes are chosen.

A theorem of de Rham ensures that the above inner product between homology and cohomology classes is bilinear and non-degenerate, thus establishing an isomorphism between homology and cohomology. The following facts are consequences of de Rham's theorem:

1. Given a homology basis $\{[z_i]\}$ of H_p there exists a basis $\{[\omega_i]\}$ of H^p such that

$$\int_{z_i} \omega_j = \delta_{ij}. \tag{14.109}$$

In other words, given $\{[z_i]\}$ and b_p real numbers v_i, we can always find a closed p-form ω such that $\pi(z_i, \omega) = v_i$. Indeed, $\omega = \sum v_i \omega_i$.
2. If all periods of a p-form vanish, ω is exact.

14.4 Mathematical Preliminaries

3. Given any p-cycle a, there exists a closed $(m - p)$-form α, called the Poincaré dual of a, such that for any closed p-form ω

$$\int_a \omega = \int_M \alpha \wedge \omega. \tag{14.110}$$

Since ω is closed, α is only defined up to an exact form.

In terms of their Poincaré duals α and β we can define the intersection number $a \cdot b$ between a p-cycle a and an $(m - p)$-cycle b as

$$a \cdot b = \int_M \alpha \wedge \beta. \tag{14.111}$$

This notion is familiar from Riemann surfaces.

So much for the collection of some facts about homology and cohomology on real manifolds. They are also valid on complex manifolds, when viewed as real manifolds. However one can use the complex structure to define (among several others) the so-called Dolbeault cohomology or $\bar\partial$-cohomology. As the name indicates, it is defined with respect to the operator $\bar\partial : A^{p,q}(M) \to A^{p,q+1}(M)$. A (p,q)-form α is $\bar\partial$-closed, if $\bar\partial\alpha = 0$. The space of $\bar\partial$-closed (p,q)-forms is denoted by $Z^{p,q}_{\bar\partial}$. A (p,q)-form β is $\bar\partial$-exact, if it is of the form $\beta = \bar\partial\gamma$ for $\gamma \in A^{p,q-1}$. Since $\bar\partial^2 = 0$, $\bar\partial(A^{p,q}(M)) \subset Z^{p,q+1}_{\bar\partial}(M)$. The Dolbeault cohomology groups are then defined as

$$H^{p,q}_{\bar\partial}(M) = \frac{Z^{p,q}_{\bar\partial}(M)}{\bar\partial(A^{p,q-1}(M))}. \tag{14.112}$$

On a compact Kähler manifold the decomposition (14.71) extends to the cohomology, i.e.

$$H^k(M,\mathbb{C}) = \bigoplus_{p+q=k} H^{p,q}_{\bar\partial}(M) \tag{14.113}$$

and the decomposition is independent of the choice of Kähler metric. This is the Hodge decomposition in cohomology.

There is a lemma (by Dolbeault) analogous to the Poincaré-lemma, which ensures that the Dolbeault cohomology groups (for $q \geq 1$) are locally[18] trivial. This is also referred to as the $\bar\partial$-Poincaré lemma.

The dimensions of the (p,q) cohomology groups are called Hodge numbers

$$h^{p,q}(M) = \dim_{\mathbb{C}}(H^{p,q}_{\bar\partial}(M)). \tag{14.114}$$

---

[18] More precisely, on polydiscs $P_r = \{z \in \mathbb{C}^n : |z^i| < r,\text{ for all } i = 1, \ldots, n\}$.

They are finite for compact, complex manifolds. The Hodge numbers of a complex manifold are often arranged in the Hodge diamond

$$
\begin{array}{ccccccc}
& & & h^{0,0} & & & \\
& & h^{1,0} & & h^{0,1} & & \\
& h^{2,0} & & h^{1,1} & & h^{0,2} & \\
h^{3,0} & & h^{2,1} & & h^{1,2} & & h^{0,3} \\
& h^{3,1} & & h^{2,2} & & h^{1,3} & \\
& & h^{3,2} & & h^{2,3} & & \\
& & & h^{3,3} & & &
\end{array}
\qquad (14.115)
$$

which we have displayed for $\dim_{\mathbb{C}}(M) = 3$. We will later show that for a Calabi-Yau manifold of the same dimension the only independent Hodge numbers are $h^{1,1}$ and $h^{2,1}$.

We now introduce a scalar product between two forms, φ and ψ, of type (p,q):

$$\psi = \frac{1}{p!q!} \psi_{i_1...i_p \bar{j}_1...\bar{j}_q}(z)\, dz^{i_1} \wedge \cdots \wedge dz^{i_p} \wedge d\bar{z}^{j_1} \wedge \cdots \wedge d\bar{z}^{j_q} \qquad (14.116)$$

and likewise for φ. Define

$$\langle \varphi, \psi \rangle = \frac{1}{p!q!} \varphi_{i_1...i_p \bar{j}_1...\bar{j}_q}(z)\, \overline{\psi}^{i_1...i_p \bar{j}_1...\bar{j}_q}(z) = \overline{\langle \psi, \varphi \rangle(z)}, \qquad (14.117)$$

where

$$\overline{\psi}^{i_1...i_p \bar{j}_1...\bar{j}_q}(z) = g^{i_1 \bar{k}_1} \cdots g^{i_p \bar{k}_p} g^{l_1 \bar{j}_1} \cdots g^{l_q \bar{j}_q} \, \overline{\psi_{k_1...k_p \bar{l}_1...\bar{l}_q}(z)}. \qquad (14.118)$$

We will also need the definition

$$\bar{\psi} = \frac{1}{p!q!} \overline{\psi_{i_1...i_p \bar{j}_1...\bar{j}_q}}\, dz^{i_1} \wedge \cdots \wedge d\bar{z}^{j_q} = \frac{1}{p!q!} \overline{\psi}_{j_i...j_q \bar{i}_1...\bar{i}_p}\, dz^{j_1} \wedge \cdots \wedge d\bar{z}^{i_p}, \qquad (14.119)$$

where

$$\overline{\psi_{i_1...i_p \bar{j}_1...\bar{j}_q}} = (-1)^{pq} \overline{\psi}_{j_1...j_q \bar{i}_1...\bar{i}_p}. \qquad (14.120)$$

The inner product $(\ ,\ ): A^{p,q} \times A^{p,q} \to \mathbb{C}$ is then

$$(\varphi, \psi) = \int_M \langle \varphi, \psi \rangle \frac{\omega^n}{n!}. \qquad (14.121)$$

The following two properties are easy to verify:

$$(\psi, \varphi) = \overline{(\varphi, \psi)}, \qquad (14.122a)$$

$$(\varphi, \varphi) \geq 0 \qquad (14.122b)$$

14.4 Mathematical Preliminaries

with equality only for $\varphi = 0$.

Define the Hodge-$*$ operator $* : A^{p,q} \to A^{n-q,n-p}$, $\psi \mapsto *\psi$ by requiring[19]

$$\langle \varphi, \psi \rangle \frac{\omega^n}{n!} = \varphi \wedge \overline{*\psi}. \qquad (14.123)$$

The $*$-operator is simply the $\mathbb{C}$-linear extension of the Hodge-$\star$ on Riemannian manifolds. It enjoys the following properties:

$$* \Pi^{p,q} = \Pi^{n-q,n-p} *, \qquad (14.124a)$$

$$*\overline{\psi} = \overline{(*\psi)}, \qquad (14.124b)$$

$$**\psi = (-1)^{p+q}\psi \quad , \quad \psi \in A^{p,q} \qquad (14.124c)$$

and[20]

$$\psi \wedge *\overline{\varphi} = \overline{\varphi} \wedge *\psi. \qquad (14.125)$$

One can work out the explicit expression for $*\psi \in A^{p,q}$:

$$*\psi = \frac{(i)^n (-1)^{n(n-1)/2 + np}}{p! q! (n-p)! (n-q)!} g \, \epsilon^{m_1 \ldots m_p}{}_{\bar{j}_1 \ldots \bar{j}_{n-p}} \epsilon^{\bar{n}_1 \ldots \bar{n}_q}{}_{i_1 \ldots i_{n-q}}$$

$$\cdot \psi_{m_1 \ldots m_p \bar{n}_1 \ldots \bar{n}_q} dz^{i_1} \wedge \cdots \wedge dz^{i_{n-q}} \wedge d\bar{z}^{j_1} \wedge \cdots \wedge d\bar{z}^{j_{n-p}}. \qquad (14.126)$$

Here we defined $\epsilon_{i_1 \ldots i_n} = \pm 1$ and its indices are raised with the metric, as usual; i.e. $\epsilon^{\bar{i}_1 \ldots \bar{i}_n} = \pm 1/g$. For ω the fundamental form and α an arbitrary real $(1,1)$-form on a three-dimensional Kähler manifold one finds

$$*\alpha = \frac{1}{2} \langle \omega, \alpha \rangle \omega \wedge \omega - \alpha \wedge \omega$$

$$= \frac{3}{2} \frac{\int \alpha \wedge \omega \wedge \omega}{\int \omega \wedge \omega \wedge \omega} \omega \wedge \omega - \alpha \wedge \omega, \quad \text{if } \alpha \text{ harmonic}, \qquad (14.127a)$$

$$*\omega = \frac{1}{2} \omega \wedge \omega, \qquad (14.127b)$$

where the second relation is a special case of the first. Note that if α is harmonic (to be defined below), $\langle \omega, \alpha \rangle$ is constant. Furthermore, on a three-dimensional complex manifold for $\Omega \in A^{3,0}$ and $\alpha \in A^{2,1}$,

---

[19]There exist other definitions in the literature; e.g. Griffiths and Harris define an operator $*_{\text{GH}}$: $A^{p,q} \to A^{n-p,n-q}$. What they call $*_{\text{GH}} \psi$ we have called $*\overline{\psi}$.

[20]Proof: $\overline{\varphi} \wedge *\psi = \overline{\varphi \wedge *\overline{\psi}} = \overline{\langle \varphi, \psi \rangle \frac{\omega^n}{n!}} = \langle \psi, \varphi \rangle \frac{\omega^n}{n!} = \psi \wedge *\overline{\varphi}$.

$$*\Omega = -i\,\Omega, \tag{14.128a}$$

$$*\alpha = i\,\alpha - i\omega \wedge *(\omega \wedge \alpha). \tag{14.128b}$$

On a complex n-dimensional Kähler manifold a k-form α with $k \leq n$ is called primitive if $\omega^{n-k+1} \wedge \alpha = 0$. An equivalent characterization is that its contraction with the Kähler form vanishes, e.g. for α a $(2,1)$-form $g^{j\bar{k}}\alpha_{ij\bar{k}} = 0$. For $n = 3$ a $(3,0)$ form is thus imaginary anti-self-dual while a primitive $(2,1)$ form is imaginary self-dual. Conversely, an imaginary self-dual $(2,1)$ form is primitive and for a $(1,2)$ form to be imaginary self-dual, it cannot not be primitive. Using (14.124b) we find that the $(0,3)$ and $(1,2)$ forms have the opposite properties. From (14.127) we also learn that $*\alpha = -\alpha \wedge \omega$ for α a primitive $(1,1)$-form.[21]

We collect a few facts which are true on compact Kähler manifolds: (1) There are no primitive k-forms for $k > n$. (2) All (p,q)-forms with $p = 0$ or $q = 0$ are primitive. (3) Any k-form α has the unique Lefschetz decomposition $\alpha = \sum_{i \geq 0} \omega^i \wedge \alpha_i$ where α_i is a primitive $(k - 2i)$-form and $\omega^i = \omega \wedge \cdots \wedge \omega$ with i factors. The Lefschetz decomposition is compatible with (14.71). There is also a Lefschetz decomposition of the cohomology. The notion of primitivity and the Lefschetz decomposition of forms also hold on (almost) Hermitian manifolds, while the Lefschetz decomposition of cohomology requires $d\omega = 0$, i.e. the Kähler condition.

Given the scalar product (14.121), we can define the adjoint of the $\bar{\partial}$ operator, $\bar{\partial}^* : A^{p,q}(M) \to A^{p,q-1}(M)$ via

$$(\bar{\partial}\varphi, \psi) = (\varphi, \bar{\partial}^*\psi), \qquad \varphi \in A^{p,q-1}, \psi \in A^{p,q}. \tag{14.129}$$

On a compact manifold[22]

$$\bar{\partial}^* = -*\partial*, \tag{14.130}$$

which satisfies $\bar{\partial}^{*2} = 0$. The explicit expression on a (p,q)-form ψ is

$$(\bar{\partial}^*\psi)_{i_1\ldots i_p\bar{j}_2\ldots\bar{j}_q} = (-1)^{p+1}\nabla^{\bar{j}_1}\psi_{i_1\ldots i_p\bar{j}_1\ldots\bar{j}_q}. \tag{14.131}$$

We now define the $\bar{\partial}$-Laplacian $\Delta_{\bar{\partial}} : A^{p,q}(M) \to A^{p,q}(M)$

$$\Delta_{\bar{\partial}} = (\bar{\partial} + \bar{\partial}^*)^2 = \bar{\partial}\bar{\partial}^* + \bar{\partial}^*\bar{\partial}, \tag{14.132}$$

and call ψ a $(\bar{\partial}-)$ harmonic form if it satisfies

$$\Delta_{\bar{\partial}}\psi = 0. \tag{14.133}$$

---

[21] The generalizations of (14.127b) to n dimensions is $*\omega = \frac{1}{(n-1)!}\omega^{n-1}$; $*\Omega = (i)^n(-1)^{\frac{1}{2}n(n+1)}$ generalizes (14.128a) to $(n,0)$ forms. Finally, for α a primitive $p + q = k$-form, $*\alpha = (-1)^{\frac{1}{2}k(k+1)}(i)^{p-q}\frac{1}{(n-k)!}\omega^{n-k} \wedge \alpha$.

[22] Proof: Since $\varphi \wedge *\bar{\psi} \in A^{n,n-1}$, $d(\varphi \wedge *\bar{\psi}) = \bar{\partial}(\varphi \wedge *\bar{\psi})$. Integrating this over M leads to (14.130).

14.4 Mathematical Preliminaries

The space of $\Delta_{\bar\partial}$ harmonic (p,q)-forms on M is denoted by $\mathcal{H}^{p,q}(M)$.[23]

On a compact manifold, ψ is harmonic iff $\bar\partial\psi = \bar\partial^*\psi = 0$, i.e. a harmonic form has zero curl and zero divergence with respect to its anti-holomorphic indices.[24] Furthermore, a harmonic form is orthogonal to every exact form and is therefore never exact.

In analogy to de Rham cohomology, one has the (complex version of the) Hodge Theorem: $A^{p,q}$ has a unique orthogonal decomposition (Hodge decomposition on forms)

$$A^{p,q} = \mathcal{H}^{p,q} \oplus \bar\partial A^{p,q-1} \oplus \bar\partial^* A^{p,q+1}. \tag{14.134}$$

In other words, every $\varphi \in A^{p,q}$ has a unique decomposition

$$\varphi = h + \bar\partial\psi + \bar\partial^*\eta, \tag{14.135}$$

where $h \in \mathcal{H}^{p,q}$, $\psi \in A^{p,q-1}$ and $\eta \in A^{p,q+1}$. If $\bar\partial\varphi = 0$, then $\bar\partial^*\eta = 0$,[25] i.e. we have the unique decomposition of $\bar\partial$-closed forms

$$Z_{\bar\partial}^{p,q} = \mathcal{H}^{p,q} \oplus \bar\partial A^{p,q-1}. \tag{14.136}$$

With reference to (14.112) we thus have

$$H_{\bar\partial}^{p,q}(M) \simeq \mathcal{H}^{p,q}(M) \tag{14.137}$$

or, in words, every $\bar\partial$-cohomology class of (p,q)-forms has a unique harmonic representative $\in \mathcal{H}^{p,q}$. Conversely, every harmonic form defines a cohomology class.

The Kähler class of a Kähler form ω is the set of Kähler forms belonging to the cohomology class $[\omega]$ of ω. The Kähler form is its harmonic representative. That ω is harmonic follows from $d\omega = 0$ and $d^*\omega = 0$, the latter e.g. being a consequence of $*\omega \propto \omega^{n-1}$.[26] Furthermore, for any harmonic form α, $\omega \wedge \alpha$ is also harmonic. Of course, it could be zero.

In addition to the $\bar\partial$-Laplacian $\Delta_{\bar\partial}$, one defines two further Laplacians on a complex manifold: $\Delta_\partial = \partial\partial^* + \partial^*\partial$ and the familiar $\Delta_d = dd^* + d^*d$. The importance of the Kähler condition is manifest in the following result which is valid on Kähler manifolds but not generally on complex manifolds:

$$\Delta_{\bar\partial} = \Delta_\partial = \frac{1}{2}\Delta_d, \tag{14.138}$$

---

[23] One can similarly define $\partial^* = -*\bar\partial *$ and Δ_∂. On a Hermitian manifold, $d^* = \partial^* + \bar\partial^*$.
[24] Proof: $(\psi, \Delta_{\bar\partial}\psi) = \|\bar\partial\psi\|^2 + \|\bar\partial^*\psi\|^2$ which vanishes iff $\bar\partial\psi = \bar\partial^*\psi = 0$.
[25] From $\bar\partial\varphi = \bar\partial\bar\partial^*\eta$ it follows that $(\bar\partial\varphi, \eta) = (\bar\partial\bar\partial^*\eta, \eta) = (\bar\partial^*\eta, \bar\partial^*\eta)$.
[26] On a Kähler manifold the three conditions $d\omega = 0$, $\nabla J = 0$ and ω is harmonic, are equivalent.

i.e. the $\bar\partial$-, ∂- and d-harmonic forms coincide. An elementary proof of (14.138) proceeds by working out the three Laplacians in terms of covariant derivatives and Riemann tensors on a Kähler manifold.

One immediate consequence of (14.138) is that Δ_d does not change the index type of a form. Another consequence is that on Kähler manifolds every holomorphic p-form is harmonic and vice-versa, every harmonic $(p,0)$ form is holomorphic. Indeed, if $\alpha \in \Omega^p \subset A^{p,0}$, $\bar\partial \alpha = 0$ and $\bar\partial^* \alpha = 0$. The latter is true since $\bar\partial^* : A^{p,q} \to A^{p,q-1}$ and can also be seen directly from (14.131). Conversely, $\Delta \alpha = 0$ implies $\bar\partial \alpha = 0$ which, for $\alpha \in \mathcal{H}^{p,0}$, means $\alpha \in \Omega^p$. The exact piece in the decomposition (14.136) of holomorphic $(p,0)$-forms therefore always vanishes and we conclude $H_{\bar\partial}^{p,0}(M) \simeq \Omega^p(M)$.[27]

It follows from (14.138) that on Kähler manifolds

$$\sum_{p+q=r} h^{p,q} = b_r, \qquad (14.139a)$$

$$\sum_{p,q}(-1)^{p+q} h^{p,q} = \sum_r (-1)^r b_r = \chi(M), \qquad (14.139b)$$

where $\chi(M)$ is the Euler number of M. The decomposition of the Betti numbers into Hodge numbers corresponds to the $U(n)$ invariant decomposition $\mu = (i, \bar i)$. Equation (14.139b) also holds in the non-Kähler case where (14.139a) is replaced by an inequality ($\geq$); i.e. the decomposition of forms (14.71) does not carry over to cohomology. Note that (14.139a) relates real and complex dimensions.

While Hodge numbers generally depend on the complex structure, on compact manifolds which admit a Kähler metric they do not change under continuous deformations of the complex structure. They also do not depend on the metric. What does depend on the metric is the harmonic representative of each class, but the difference between such harmonic representatives is always an exact form.

In Sect. 14.5 we will be led to consider the cohomology of forms with values in the holomorphic tangent bundle $T^{1,0}(M)$. To keep the discussion more general, and because it will be needed in Sect. 14.7, when we consider compactifications of the heterotic string, we consider forms with values in a holomorphic rank r vector bundle V. For $V = T^{1,0}(M)$ we will reproduce familiar results and the generalization to general V is straightforward.

A holomorphic vector bundle V of rank r over a complex manifold M is a complex vector bundle (i.e. the fibers are isomorphic to $\mathbb{C}^r$) with additional structure: if we denote the fiber coordinates of V in the patch U by $\xi = (\xi^1, \ldots, \xi^r) \in \mathbb{C}^r$, holomorphicity of V means that the transition or gluing functions

---

[27] One can proof that a non-zero $(p,0)$ form, not necessarily holomorphic, can never be exact. Indeed, considering $\alpha = d\beta \in A^{p,0}$, it is straightforward to show that $\alpha \wedge \bar\alpha \wedge \omega^{n-p} = f\omega^n$, where ω is the Kähler form and f a positive function. The r.h.s., when integrated over the manifold, is non-zero, while the l.h.s. integrates to zero, because it can be written as $d(\beta \wedge \bar\alpha \wedge \omega^{n-p})$. It then follows that any $\alpha \in \Omega^p$ satisfies $d\alpha = 0$: clearly, $d\alpha = \partial\alpha$ is an exact (holomorphic) $(p+1)$ form which vanishes by the above argument.

14.4 Mathematical Preliminaries

$f(z)$ on $U \cap U' \neq \emptyset$, $\xi^\alpha = f^\alpha_\beta \xi'^\beta$ are $r \times r$ matrices of holomorphic functions. For instance, for $V = T^{1,0}$, $f^i_j = \partial z^i / \partial z'^j$. A holomorphic line bundle is a holomorphic vector bundle of rank one.

Denote by $A^{p,q}(V)$ the space of (p,q)-forms with values in V. An element $\varphi \in A^{p,q}(V)$ is represented by a vector

$$\varphi = (\varphi^\alpha) = (\varphi^1, \ldots, \varphi^r) \in A^{p,q}(V), \tag{14.140}$$

where each component is a (p,q)-form and where the transition between overlapping patches is via $(f_{ba})^\alpha_\beta$. For $\varphi \in A^{p,q-1}(V)$ one defines $\bar{\partial}\varphi = (\bar{\partial}\varphi^\alpha) \in A^{p,q}(V)$. This allows us to define the bundle valued cohomology groups as

$$H^{p,q}_{\bar{\partial}}(M, V) = \frac{Z^{p,q}_{\bar{\partial}}(M, V)}{\bar{\partial}(A^{p,q-1}(M, V))}, \tag{14.141}$$

where $\psi \in Z^{p,q}_{\bar{\partial}}(M, V)$ satisfies $\bar{\partial}\psi = 0$.

A Hermitian metric on the fibers of V is a positive definite quadratic form $h(z)$ on U

$$h_{\alpha\bar{\beta}} \xi^\alpha \bar{\xi}^{\bar{\beta}} \tag{14.142}$$

with

$$h_{\alpha\bar{\beta}}(z) = (f^{-1}(z))^\gamma_\alpha \overline{(f^{-1}(z))^\delta_\beta} h_{\gamma\bar{\delta}}(z) \tag{14.143}$$

for $z \in U \cap U'$. This identifies h as a section of $V^* \otimes \bar{V}^*$ where V^* is the dual vector bundle with transition functions f^{-1}. E.g. for $V = T^{1,0}$, $V^* = T^{*1,0}$, $\bar{V} = T^{0,1}$ and $\bar{V}^* = T^{*0,1}$. Alternatively we can view $h_{\alpha\bar{\beta}}$ as a linear isomorphism $\bar{V} \to V^*$, $\bar{\xi}^{\bar{\beta}} \mapsto h_{\alpha\bar{\beta}} \bar{\xi}^{\bar{\beta}}$ and likewise $V \to \bar{V}^*$, $\xi^\alpha \mapsto h_{\alpha\bar{\beta}} \xi^\alpha$; A Hermitian metric always exists on V. For $V = T^{1,0}(M)$ it is $h_{i\bar{j}} = g_{i\bar{j}}$. V equipped with a Hermitian metric is called a Hermitian vector bundle (V, h).

Given h we define the inner product

$$(\varphi, \psi) = \int h_{\alpha\bar{\beta}} \varphi^\alpha \wedge *\bar{\psi}^{\bar{\beta}} \qquad \text{for} \quad \varphi, \psi \in A^{p,q}(V), \tag{14.144}$$

which satisfies the same properties as (14.122). Note that $\bar{\psi} \in A^{q,p}(V^*)$. Using the inner product one can define the adjoint of $\bar{\partial}$ via

$$(\bar{\partial}\varphi, \psi) = (\varphi, \bar{\partial}^* \psi), \qquad \varphi \in A^{p,q-1}(V), \psi \in A^{p,q}(V), \tag{14.145}$$

for which one finds

$$\bar{\partial}^* \psi^\alpha = -h^{\alpha\bar{\gamma}} * \partial(h_{\beta\bar{\gamma}} * \psi^\beta), \tag{14.146}$$

where $h^{\alpha\bar\beta}$ is the inverse metric. One also defines the Laplace operator (14.132) and harmonic forms and one can show that the decompositions (14.134) and (14.136) as well as (14.137) also hold for the V-valued forms and that dim $H^{p,q}_{\bar\partial}(M, V) < \infty$.

As for the holomorphic tangent bundle, on an arbitrary holomorphic vector bundle a Hermitian connection D is defined by the requirement that it is compatible with the Hermitian metric, i.e. $D_i h_{\alpha\bar\beta} = 0$. Among the Hermitian connections there is a unique connection, called the Chern connection, with the additional property that it is compatible with the holomorphic structure, $D = D^{1,0} + D^{0,1} = D^{1,0} + \bar\partial$. If we write $D = d + A$ its only non-vanishing components are $A_{i\alpha}^{\ \beta}$ ($-A_{i\alpha}^{\ \beta}$ for V^* and $A_{\bar i\alpha}^{\ \bar\beta}$ for $\bar V$). This leads to $D = d + A = d + h^{-1}\partial h$. The Hermitian curvature two-form

$$F = dA + A \wedge A = h^{-1}\bar\partial\partial h - (h^{-1}\bar\partial h) \wedge (h^{-1}\partial h) = \bar\partial A \qquad (14.147)$$

is of type $(1, 1)$ (cf. (14.88) and (14.90)).[28] With the Chern connection the Laplace operator on (p,q)-forms with values in V, $\Delta_{\bar\partial} = \bar\partial\bar\partial^* + \bar\partial^*\bar\partial$, is

$$\Delta_{\bar\partial}\omega^\alpha_{i_1\ldots i_p \bar j_1\ldots \bar j_q} = -g^{i\bar j}D_i\bar D_{\bar j}\omega^\alpha_{i_1\ldots i_p \bar j\ldots \bar j_q} + \sum_{s=1}^q F^{\bar j}{}_{\bar j_s}{}^\alpha{}_\beta\, \omega^\beta_{i_1\ldots i_p \bar j_1\ldots \bar j\ldots \bar j_q}$$

$$+ \sum_{s=1}^q R^{\bar j}{}_{\bar j_s}\omega^\alpha_{i_1\ldots i_p \bar j_1\ldots \bar j\ldots \bar j_q} - \sum_{r=1}^p\sum_{s=1}^q R_{i_r\bar j_s}{}^{i\bar j}\, \omega^\alpha_{i_1\ldots i\ldots i_p \bar j_1\ldots \bar j\ldots \bar j_q}, \qquad (14.148)$$

where $D_i = \nabla_i + A_i$ and $\bar D_{\bar i} = \nabla_{\bar i}$ and ∇_i is the Levi-Civita connection. The zero modes of $\Delta_{\bar\partial}$ are the harmonic forms and each bundle valued cohomology group $H^{p,q}_{\bar\partial}(M, V)$ has a unique harmonic representative.

A holomorphic vector bundle of rank r over a compact Kähler manifold is called Hermitian-Yang-Mills, if there exists a Hermitian metric h for which the Hermitian curvature F satisfies

$$F_{ij} = F_{\bar i\bar j} = 0, \quad g^{i\bar j}F_{i\bar j\alpha}{}^\beta = \mu\, \delta^\beta_\alpha, \qquad (14.149)$$

where μ is a constant. For the holomorphic tangent bundle the l.h.s. is the Ricci curvature and (14.149) becomes the Kähler-Einstein condition. One can show that a complex vector bundle admits a holomorphic structure, if there exists a $U(r)$ connection whose curvature is of type $(1, 1)$.

The Hodge numbers (14.114) of Kähler manifolds are not all independent. From $\overline{A^{p,q}} = A^{q,p}$ we learn

$$h^{p,q} = h^{q,p}. \qquad (14.150)$$

---

[28] The Chern connection is obviously not the only connection with the property that the $(0, 2)$ and $(2, 0)$ parts of its curvature vanish, since this is a gauge invariant statement: for $A \to UAU^{-1} + U\, dU^{-1}$ the curvature transforms as $F \to UFU^{-1}$.

14.4 Mathematical Preliminaries

This symmetry ensures that all odd Betti numbers of Kähler manifolds are even (possibly zero). Furthermore, since $[\Delta_d, *] = 0$ and since $* : A^{p,q} \to A^{n-q,n-p}$ we conclude

$$h^{p,q} = h^{n-q,n-p} \stackrel{(14.150)}{=} h^{n-p,n-q}. \qquad (14.151)$$

The existence of a closed $(1, 1)$-form, the Kähler form ω (which is in fact harmonic), ensures that

$$h^{p,p} > 0 \quad \text{for} \quad p = 0, \ldots, n. \qquad (14.152)$$

Indeed, ω^p is obviously closed. If it were exact for some p, then ω^n were also exact. But this is impossible since ω^n is a volume form. Moreover one has $h^{0,0} = 1$, in case the manifold is connected. The elements of $H^{0,0}_{\bar{\partial}}(M, \mathbb{C})$ are the complex constants. One can show that on $\mathbb{P}^n$ the Kähler form generates the whole cohomology, i.e. $h^{p,p}(\mathbb{P}^n) = 1$ for $p = 0, \ldots n$, with all other Hodge numbers vanishing.

On a connected complex three-dimensional Kähler manifold the symmetries (14.150), (14.151) leave only five independent Hodge numbers, e.g. $h^{1,0}, h^{2,0}, h^{1,1}, h^{2,1}$ and $h^{3,0}$. For Ricci-flat Kähler manifolds, which we will consider in detail below, we will establish three additional restrictions on its Hodge numbers.

We have already encountered one important cohomology class on Kähler manifolds: from (14.92) we learn that $\mathscr{R} \in H^{1,1}_{\bar{\partial}}(M, \mathbb{C})$ and from (14.93) that under a change of metric $\mathscr{R}$ varies within a given cohomology class. In fact, one can show that, if properly normalized, the Ricci form defines an element on $H^{1,1}_{\bar{\partial}}(M, \mathbb{Z})$. This leads us to a discussion of Chern classes.

Given a Kähler metric, we can define a matrix valued 2-form Θ of type $(1, 1)$ by

$$\Theta^j_i = g^{j\bar{p}} R_{i\bar{p}k\bar{l}} dz^k \wedge d\bar{z}^l. \qquad (14.153)$$

One defines the Chern form or total Chern class

$$c(M) = 1 + \sum_i c_i(M) = \det\left(\mathbb{1} + \frac{it}{2\pi}\Theta\right)\Big|_{t=1}$$
$$= (1 + t\phi_1(g) + t^2\phi_2(g) + \ldots)\Big|_{t=1} \qquad (14.154)$$

which has the following properties:

- $d\phi_i(g) = 0$ and $[\phi_i] \in H^{i,i}_{\bar{\partial}}(M, \mathbb{C}) \cap H^{2i}_{DR}(M, \mathbb{R})$,
- $[\phi_i(g)]$ is independent of g,
- $c_i(M)$ is represented by $\phi_i(g)$.

$c_i(M)$ is the ith Chern class of the manifold M. We will often also refer to one of its representatives as the ith Chern-class.

Below we will need $c_1(M)$ which is expressed in terms of the Ricci form:

$$\phi_1(g) = \frac{i}{2\pi}\Theta_i^i = \frac{i}{2\pi}R_{k\bar{l}}dz^k \wedge d\bar{z}^l = \frac{1}{2\pi}\mathscr{R} = -\frac{i}{2\pi}\partial\bar{\partial}\log\det(g_{k\bar{l}}). \quad (14.155)$$

For $c_1(M)$ the first two properties have been proven in (14.92) and (14.93). Moreover, if

$$dv = v\,dz^1 \wedge d\bar{z}^1 \wedge \cdots \wedge dz^n \wedge d\bar{z}^n \quad (14.156)$$

is any volume form on M, we can represent $[c_1(M)]$ by

$$c_1(M) = -\left[\frac{i}{2\pi}\partial\bar{\partial}\log(v)\right]. \quad (14.157)$$

This is so, as $v = f \det(g)$ for a non-vanishing positive function f on M. As an example, consider $M = \mathbb{P}^n$, endowed with the Fubini-Study metric. We then have (cf. (14.98)) $\mathscr{R} = (n+1)\omega$, i.e. $c_1(\mathbb{P}^n) = \frac{1}{2\pi}(n+1)[\omega]$.

We say that $c_1(M) > 0 (< 0)$ if $c_1(M)$ can be represented by a positive (negative) form ϕ_1. In local coordinates this means

$$\phi_1 = i\phi_{k\bar{l}}dz^k \wedge d\bar{z}^l, \quad (14.158)$$

where $\phi_{k\bar{l}}$ is a positive (negative) definite matrix. We say that $c_1(M) = 0$, if the first Chern class is cohomologous to zero. Clearly $c_1(\mathbb{P}^n) > 0$. Note that, e.g. $c_1(M) > 0$ means that $\int_\mathscr{C} c_1 > 0$ for any curve $\mathscr{C}$ in M.

The definition for the Chern classes can be generalized to any complex vector bundle V over M with fiber $\mathbb{C}^r$ and structure group $G \subseteq GL(r, \mathbb{C})$. When we speak of the Chern classes of a manifold, we mean the Chern classes of the holomorphic tangent bundle T_M with the Hermitian connection. (Here and below we often use the notation $T_M \equiv T^{1,0}(M)$ for the holomorphic tangent bundle and $T_M^* \equiv T^{*1,0}(M)$ for its dual bundle, the holomorphic cotangent bundle.) But for any vector bundle with a connection A we can define the Lie-algebra valued curvature two-form $F(A)$ and simply replace Θ in (14.154) by F. One uses the notation $c_i(V)$. For instance we could take $V = T_M^*$ with $c_1(T_M^*) = -c_1(T_M)$.[29] For a holomorphic vector bundle V, $c_1(V) = \frac{i}{2\pi}h^{\alpha\bar{\beta}}F_{i\bar{j}\alpha\bar{\beta}}dz^i \wedge d\bar{z}^j$.

The Chern classes do not depend on the choice of connection. They encode topological information about the bundle and help to measure how non-trivial a bundle is. For trivial bundles they are trivial. We always have $c_0(V) = 1$ and $c_i(V) = 0$ for $i > r = \text{rank}(V)$ and $c_i(V) = 0$ for $i > \dim_\mathbb{C}(M)$.

---

[29]This follows from $[\nabla_i, \nabla_{\bar{j}}]V_k = R_{i\bar{j}k}{}^l V_l$ and $[\nabla_i, \nabla_{\bar{j}}]V^k = R_{i\bar{j}}{}^k{}_l V^l$ and $R_{i\bar{j}k}{}^k = -R_{i\bar{j}}{}^k{}_k$.

One also defines the Chern characters of a holomorphic vector bundle V as

$$\text{ch}(V) = \text{tr} \exp\left(\frac{it}{2\pi}F\right)\Big|_{t=1} = \left(\sum_i t^i \text{ch}_i(V)\right)\Big|_{t=1}. \quad (14.159)$$

One easily establishes the relation with the Chern classes,

$$\text{ch}_0(V) = r,$$
$$\text{ch}_1(V) = \left[\frac{i}{2\pi}\text{tr}\, F\right] = c_1(V),$$
$$\text{ch}_2(V) = \left[-\frac{1}{8\pi^2}\text{tr}\, F \wedge F\right] = \frac{1}{2}(c_1^2(V) - 2c_2(V)), \quad \text{etc.} \quad (14.160)$$

Chern classes have the property that their integrals over cycles are integers, i.e. they are elements of $H_{\text{DR}}^{2i}(M, \mathbb{Z})$ while Chern characters are in $H_{\text{DR}}^{2i}(M, \mathbb{Q})$.

14.5 Calabi-Yau Manifolds

We have now collected all the prerequisites to define: a Calabi-Yau manifold is a compact Kähler manifold with vanishing first Chern class.

While it is obvious that any Ricci-flat Kähler manifold has vanishing first Chern class, the opposite is far from trivial. This problem was first considered by Calabi in a more general context. He asked the question, whether any representative of $c_1(M)$ is the Ricci-form of some Kähler metric. One can show that any two such representatives differ by a term of the form $i\partial\bar{\partial}f$ where $f \in C^\infty(M, \mathbb{R})$. This is the content of the $\partial\bar{\partial}$-lemma which states that on a compact Kähler manifold any d-closed (p, q)-form η, which is either d-, ∂- or $\bar{\partial}$-exact, is of the form $\eta = \partial\bar{\partial}\gamma$. If $p = q$ and η is real, $i\gamma$ is also real. Calabi also showed that, if such a Kähler metric exists, then it must be unique. Yau proved that such a metric always exists, if M is compact.

The precise statement of Yau's theorem is: let M be a compact Kähler manifold, ω its Kähler form, $c_1(M)$ its first Chern class. Any closed real two-form of type $(1,1)$ belonging to $2\pi c_1(M)$ is the Ricci form of one and only one Kähler metric in the class of ω.

For vanishing first Chern class, which is the case we are interested in, this means that given any Kähler metric g with associated Kähler form ω, one can always find a unique Ricci-flat Kähler metric g' with Kähler form ω' such that $[\omega] = [\omega']$, i.e. a Kähler manifold with $c_1(M) = 0$ admits a unique Ricci-flat Kähler metric in each Kähler class.

Since the first Chern class is represented by the Ricci form, and since the latter changes under change of metric by an exact form, i.e. $\mathcal{R}(g') = \mathcal{R}(g) + d\alpha$ (cf. (14.93)), vanishing of the first Chern class is necessary for having a Ricci-flat metric.

This is the easy part of the theorem. To prove that this is also sufficient is the hard part. Yau's proof is an existence proof. In fact no exact Calabi-Yau metric has ever been constructed explicitly on a compact manifold.[30] Let us mention that for non-compact manifolds the situation is better.[31]

The compact Kähler manifolds with zero first Chern class are thus precisely those which admit a Kähler metric with zero Ricci curvature, or equivalently, with restricted holonomy group contained in $SU(n)$. Following common practice, we will speak of Calabi-Yau manifolds, if the holonomy group is precisely $SU(n)$. This excludes tori and direct product spaces. We want to mention in passing that any compact Kähler manifold with $c_1(M) = c_2(M) = 0$ is flat, i.e. $M = \mathbb{C}^n/\Gamma$. This shows that while Ricci-flatness is characterized by the first Chern class, flatness is characterized by the second Chern class.

We should also mention that the analysis, which led to considering Ricci-flat manifolds, was based on a perturbative string theory analysis which was further restricted to lowest order in α'. If one includes α'-corrections, both the beta-function equations and the supersymmetry transformations will be corrected and the Ricci-flatness condition is modified. One finds the requirement $R_{i\bar{j}} + \alpha'^3 (R^4)_{i\bar{j}} + \cdots = 0$, where (R^4) is a certain tensor composed of four powers of the curvature. It has been shown that the α'-corrections to the Ricci-flat metric, which are necessary to solve the corrected equations, do not change the cohomology class. They are always of the form $\partial\bar{\partial}(\ldots)$ and are thus cohomologically trivial. The proof uses extended world-sheet supersymmetry which is necessary for space-time supersymmetry.

One often defines Calabi-Yau manifolds as compact complex Kähler manifolds with trivial canonical bundle. Let us explain this. An important class of vector bundles over a complex manifold are those with fibers of (complex) dimension one, i.e. line bundles with fiber $\mathbb{C}$ (complex vector bundles of rank one). Holomorphic line bundles have holomorphic transition functions and a holomorphic section is given in terms of local holomorphic functions. Each holomorphic section defines a local holomorphic frame (which is one-dimensional for a line-bundle). One important and canonically defined line bundle is the canonical line bundle $K(M) = \wedge^n T^{*1,0}(M)$ whose sections are forms of type $(n, 0)$, where $n = \dim_\mathbb{C}(M)$. It is straightforward to verify that $[\nabla_i, \nabla_{\bar{j}}]\omega_{i_1 \ldots i_n} = R_{i\bar{j}}\omega_{i_1 \ldots i_n}$, which shows that $c_1(M) = -c_1(K(M))$, i.e. if $c_1(M) = 0$, the first Chern class of the canonical line bundle also vanishes. For a line bundle this means that it is topologically trivial. Consequently there must exist a globally defined nowhere vanishing section, i.e. a globally defined nowhere vanishing n-form on M.

We will now show that on a manifold with $\mathcal{H} \subseteq SU(n)$ the canonical bundle has a unique nowhere vanishing holomorphic section, i.e. that there exists a nowhere

---

[30]There exists an algorithm, due to Donaldson, for an approximate construction which converges to the exact CY metric.

[31]Examples of explicit non-compact Ricci-flat Kähler metrics are the Eguchi-Hanson metrics and the metric on the deformed and the resolved conifold. They play a role in the resolution of singularities (orbifold and conifold singularities, respectively) which can occur in compact CY manifolds at special points in their moduli space.

14.5 Calabi-Yau Manifolds

vanishing holomorphic n-form Ω. This can be seen by explicit construction: pick any point p on M and define $\Omega_p = dz^1 \wedge \cdots \wedge dz^n$ where $\{z^i\}$ are local coordinates. Parallel transport Ω to every other point on M. This is independent of the path, since, when transported around a closed path (starting and ending at p) Ω, being a singlet of $SU(n)$, is unchanged. This defines Ω everywhere on M and furthermore shows that Ω is covariantly constant. Since M is Ricci-flat and Kähler, it follows from (14.289) that Ω is harmonic. It is also clear that it vanishes nowhere on M. A covariantly constant $(n,0)$ form is called a complex volume form.

Being covariantly constant implies that Ω is holomorphic because $\nabla_{\bar{i}} \Omega_{i_1\ldots i_n} = \bar{\partial}_{\bar{i}} \Omega_{i_1\ldots i_n} = 0$. This means that locally we can write

$$\Omega_{i_1\ldots i_n} = f(z)\, \epsilon_{i_1\ldots i_n} \tag{14.161}$$

with f a non-vanishing holomorphic function in a given coordinate patch and $\epsilon_{i_1\ldots i_n} = \pm 1$. Another important property of Ω is that it is essentially unique. Assume that given Ω there is a Ω' with the same properties. As Ω is a form of the top degree, we must have $\Omega' = f\Omega$ where f is a non-singular function. Since we require $\bar{\partial}\Omega' = 0$, f must be holomorphic. On a compact manifold this implies that f is constant.

To summarize, on a compact Ricci-flat Kähler manifold of dimension n the following statements are equivalent:

- There exists a unique holomorphic $(n,0)$ form Ω;
- There exists a unique covariantly constant $(n,0)$ form Ω;
- K_M is trivial, i.e. it has a holomorphic section Ω;
- $\mathcal{H} \subseteq SU(n)$.

Conversely, the existence of Ω implies $c_1 = 0$. Indeed, with (14.161), we can write the Ricci form as

$$\mathcal{R} = -i\partial\bar{\partial} \log \det(g_{k\bar{l}}) = i\partial\bar{\partial} \log \left(\Omega_{i_1\ldots i_n} \overline{\Omega}_{\bar{j}_1\ldots \bar{j}_n} g^{i_1 \bar{j}_1} \cdots g^{i_n \bar{j}_n} \right). \tag{14.162}$$

The argument of the logarithm is a globally defined function and the Ricci form is thus trivial in cohomology, implying $c_1 = 0$.

The same argument which showed the existence of Ω can be used construct other covariantly constant tensors: if α_p is an $\mathcal{H}$-invariant tensor of $T_p(M)$, it can be extended to a covariantly constant tensor field. We apply this to the metric and the Kähler form: If we normalize

$$\omega_p = \frac{i}{2} \sum_i dz^i \wedge d\bar{z}^i = \sum_i dx^{2i-1} \wedge dx^{2i}, \quad g_p = \sum_i |dz^i|^2 = \sum_\mu (dx^\mu)^2,$$

$$\tag{14.163}$$

then
$$\omega^n = \frac{n!}{2^n}(i)^n(-1)^{\frac{1}{2}n(n-1)}\Omega \wedge \overline{\Omega} \tag{14.164}$$

is a local statement which holds at point p and therefore everywhere on M. In terms of
$$\|\Omega\|^2 = \frac{1}{n!}\Omega_{i_1...i_n}\overline{\Omega}^{i_1...i_n} \tag{14.165}$$

which is easily shown to be a constant,
$$\int_M \Omega \wedge \overline{\Omega} = (i)^n(-1)^{\frac{1}{2}n(n-1)}\|\Omega\|^2 \text{Vol}(M). \tag{14.166}$$

We now complete the discussion of Hodge numbers of Calabi-Yau manifolds. We have just established the existence of a unique harmonic $(n, 0)$-form, Ω, and thus
$$h^{n,0} = h^{0,n} = 1. \tag{14.167}$$

With the help of Ω we can establish one further relation between the Hodge numbers. Given a holomorphic and hence harmonic $(p, 0)$-form, we can, via contraction with Ω, construct a $(0, n - p)$-form, which can be shown to be again harmonic. This works as follows. Given
$$\alpha = \alpha_{i_1...i_p}\, dz^{i_1} \wedge \cdots \wedge dz^{i_p}, \qquad \bar{\partial}\alpha = 0, \tag{14.168}$$

α being $(\Delta_{\bar{\partial}})$-harmonic means
$$\partial \alpha = 0 \quad \Leftrightarrow \quad \nabla_{[j_i}\alpha_{j_2...j_{p+1}]} = 0, \tag{14.169}$$

$$\partial^*\alpha = 0 \quad \Leftrightarrow \quad \nabla^{i_1}\alpha_{i_1...i_p} = 0. \tag{14.170}$$

We then define the $(0, n - p)$-form
$$\beta_{\bar{j}_{p+1}...\bar{j}_n} = \frac{1}{p!}\overline{\Omega}_{\bar{j}_1...\bar{j}_n}\alpha^{\bar{j}_1...\bar{j}_p}. \tag{14.171}$$

With the help of
$$\overline{\Omega}^{\bar{j}_1...\bar{j}_p \bar{i}_{p+1}...\bar{i}_n}\Omega_{\bar{j}_1...\bar{j}_p \bar{j}_{p+1}...\bar{j}_n} = p!\,\|\Omega\|^2\,\delta^{\bar{i}_{p+1}...\bar{i}_n}_{\bar{j}_{p+1}...\bar{j}_n} \tag{14.172}$$

14.5 Calabi-Yau Manifolds

we can be invert (14.171):

$$\alpha^{\bar{j}_1...\bar{j}_p} = \frac{1}{(n-p)!}\frac{1}{\|\Omega\|^2}\Omega^{\bar{j}_1...\bar{j}_p\bar{j}_{p+1}...\bar{j}_n}\beta_{\bar{j}_{p+1}...\bar{j}_n}. \tag{14.173}$$

From this we derive

$$\nabla^{\bar{j}_{p+1}}\beta_{\bar{j}_{p+1}...\bar{j}_n} = \frac{1}{p!}\overline{\Omega}_{\bar{j}_1...\bar{j}_n}\nabla^{\bar{j}_{p+1}}\alpha^{\bar{j}_1...\bar{j}_p} = 0, \tag{14.174}$$

where we used (14.169). Similarly

$$\nabla_{\bar{j}_1}\alpha^{\bar{j}_1...\bar{j}_p} = \frac{1}{(n-p)!}\frac{1}{\|\Omega\|^2}\Omega^{\bar{j}_1...\bar{j}_p\bar{j}_{p+1}...\bar{j}_n}\nabla_{\bar{j}_1}\beta_{\bar{j}_{p+1}...\bar{j}_n} = 0 \tag{14.175}$$

by virtue of (14.170). It follows that β is also harmonic.

We have thus shown the following relation between Hodge numbers

$$h^{p,0} = h^{0,p} = h^{0,n-p} = h^{n-p,0}. \tag{14.176}$$

Let us now look at $h^{p,0}$. Elements of $H_{\bar{\partial}}^{p,0}(M)$ are holomorphic p-forms. On page 467 we have shown that they are harmonic. If the manifold is Ricci-flat, they are covariantly constant. Indeed, from (14.289) we see that on a Ricci-flat Kähler manifold, where $R_{i\bar{j}} = R_{ij\bar{k}\bar{l}} \equiv 0$, we find $\nabla^\mu \nabla_\mu \alpha_{i_1...i_p} = 0$. On a compact manifold this means that α is parallel, $\nabla_\mu \alpha_{i_1...i_p} = 0$, i.e. α transforms as a singlet under the holonomy group. We now assume that the holonomy group is exactly $SU(n)$, i.e. not a proper subgroup of it.[32] Since $\alpha_{i_1...i_p}$ transforms in the $\wedge^p \mathbf{n}$ of $SU(n)$, the singlet only appears in the decomposition if $p = 0$ or $p = n$. We thus learn that on Calabi-Yau manifolds with holonomy group $SU(n)$

$$h^{p,0} = 0 \quad \text{for} \quad 0 < p < n. \tag{14.177}$$

The fact that $h^{1,0}(M) = 0$ implies that there are no continuous isometries on M. Indeed, an isometry would mean that the Killing equation $\nabla_\mu \xi_\nu + \nabla_\nu \xi_\mu = 0$ has a non-trivial solution. An immediate consequence is $\nabla^\mu \nabla_\mu \xi_\nu + \nabla^\mu \nabla_\nu \xi_\mu = 0$. However, on a Ricci-flat manifold $\nabla^\mu \nabla_\nu \xi_\mu = \nabla_\nu \nabla^\mu \xi_\mu = 0$ where the second equality follows from the trace of the Killing equation. With reference to (14.289) we conclude that the existence of a Killing vector implies the existence of a harmonic one-form.

We will be mainly interested in Calabi-Yau threefolds, i.e. $n = 3$, which we will denote by X. In this case, the only independent Hodge numbers are $h^{1,1} \geq 1$ and $h^{2,1} \geq 0$ so that the Hodge diamond for Calabi-Yau threefolds is

---

[32]Calabi-Yau orbifolds have discrete holonomy groups. There the condition is that it is not contained in any continuous subgroup of $SU(n)$.

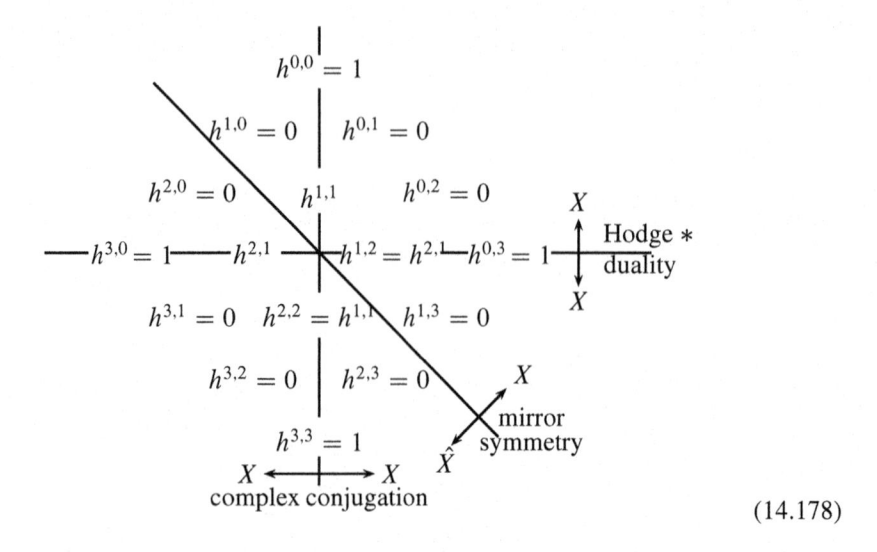

$$
\begin{array}{c}
h^{0,0} = 1 \\
h^{1,0} = 0 \quad h^{0,1} = 0 \\
h^{2,0} = 0 \quad h^{1,1} \quad h^{0,2} = 0 \\
-h^{3,0} = 1 - h^{2,1} - h^{1,2} = h^{2,1} - h^{0,3} = 1 \\
h^{3,1} = 0 \quad h^{2,2} = h^{1,1} \quad h^{1,3} = 0 \\
h^{3,2} = 0 \quad h^{2,3} = 0 \\
h^{3,3} = 1
\end{array}
$$

with arrows indicating: Hodge $*$ duality ($X \leftrightarrow X$), mirror symmetry ($X \leftrightarrow \hat{X}$), complex conjugation ($X \leftrightarrow X$). (14.178)

The Euler number of a Calaci-Yau threefold X is (cf. (14.139b))

$$\chi(X) = 2(h^{1,1}(X) - h^{1,2}(X)). \tag{14.179}$$

The significance of the elements of $H^{1,1}$ and $H^{1,2}$ on Calabi-Yau threefolds will be explained at the end of this section. The fact that $h^{2,3} = 0$ means that all elements of $\mathcal{H}^{1,2}$ are primitive because given a harmonic $(1,2)$-form α, $\omega \wedge \alpha$ is also harmonic. Furthermore, they are imaginary anti-self-dual.

The number of non-trivial Hodge numbers increases with dimension. For instance in $n = 4$, we have $h^{1,1}$, $h^{2,1}$, $h^{3,1}$ and $h^{2,2}$, but they can be shown to satisfy $h^{2,2} = 2(22 + 2h^{1,1} + 2h^{3,1} - h^{2,1})$ such that there are three independent Hodge numbers.

In (14.5) we have indicated operations which relate Hodge numbers to each other. In addition to complex conjugation (14.150) and the Hodge $*$-operation (14.151), which act on the Hodge numbers of a given CY manifold, we have also shown the action of mirror symmetry: given a CY manifold X, there exists a mirror manifold $\hat{X}$ such that

$$h^{p,q}(X) = h^{3-p,q}(\hat{X}). \tag{14.180}$$

This in particular means that the two non-trivial Hodge numbers $h^{1,1}$ and $h^{2,1}$ are interchanged between X and $\hat{X}$ so that $\chi(X) = -\chi(\hat{X})$. Within the class of Calabi-Yau manifolds constructed as hypersurfaces in toric varieties (which we have not yet discussed) they manifestly come in mirror pairs.

We close this section with a discussion of spinors on Calabi-Yau manifolds. This will be important for the identification of the massless spectrum of the compactified string theory.

14.5 Calabi-Yau Manifolds

From the Dirac algebra $\{\gamma^i, \gamma_j\} = 2\delta^i_j$, $\{\gamma^i, \gamma^j\} = \{\gamma_i, \gamma_j\} = 0$ it follows that we can view γ^i and $\gamma^{\bar{i}} = \gamma_j g^{j\bar{i}}$ as fermionic annihilation and creation operators. We can then define a vacuum state $|\Omega\rangle$ such that $\gamma^i |\Omega\rangle = 0$, $\forall i$. A general state has the decomposition

$$\eta|\Omega\rangle + \eta_{\bar{i}}\gamma^{\bar{i}}|\Omega\rangle + \eta_{\bar{i}\bar{j}}\gamma^{\bar{i}\bar{j}}|\Omega\rangle + \eta_{\bar{i}\bar{j}\bar{k}}\gamma^{\bar{i}\bar{j}\bar{k}}|\Omega\rangle, \tag{14.181}$$

where the different parts transform as $\mathbf{1}_R^{-3} + \bar{\mathbf{3}}_L^{-1} + \mathbf{3}_R^{1} + \mathbf{1}_L^{+3}$ of $SU(3) \times U(1) \subset SO(6)$; see the discussion in Sect. 8.5. The superscripts are the $U(1)$ charges and the subscript indicates the chirality of the states. For manifolds with $SU(3)$ holonomy the $U(1)$ charge decouples and we have thus found that spinors on Calabi-Yau manifolds can be associated with $(0, q)$-forms, i.e. elements of $A^{0,q}(X)$. The two singlets correspond to the Killing spinors of opposite chirality, i.e. we identify $|\Omega\rangle \sim \epsilon_+$ and $\gamma^{\bar{1}\bar{2}\bar{3}}|\Omega\rangle \sim \epsilon_-$.

We can construct Ω and ω in terms of the Killing spinors $\epsilon_\pm$ in the following way:

$$\omega_{i\bar{j}} = \mp i\epsilon_\pm^\dagger \gamma_{i\bar{j}} \epsilon_\pm, \qquad \Omega_{ijk} = -i\epsilon_-^\dagger \gamma_{ijk} \epsilon_+. \tag{14.182}$$

If we normalize the Killing spinors such that $\epsilon_\pm^\dagger \epsilon_\pm = 1$, they satisfy

$$\omega \wedge \omega \wedge \omega = \frac{3i}{4} \Omega \wedge \bar{\Omega}. \tag{14.183}$$

It can be verified that these expressions have the required properties, e.g. $d\omega = 0$. An easy way to check the normalization is to work at a point p where the metric is as in (14.163).

Let us now look at the Dirac-operator on a Kähler manifold, $\gamma^\mu D_\mu = \gamma^i D_i + \gamma^{\bar{i}} D_{\bar{i}}$ where the covariant derivatives contain the Hermitian connection. Consider the two operators $\gamma^i D_i$ and $\gamma^{\bar{i}} D_{\bar{i}}$ separately. Both square to zero because the $(2, 0)$ and $(0, 2)$ components of the curvature vanish on a Kähler manifold. Acting with $\gamma^i D_i$ and $\gamma^{\bar{i}} D_{\bar{i}}$ on the state (14.181), it is straightforward to show that these operators act on the coefficients as

$$\gamma^{\bar{i}} D_{\bar{i}} \sim \bar{\partial} \qquad \gamma^i D_i \sim \bar{\partial}^* \qquad \gamma^\mu D_\mu \sim \bar{\partial} + \bar{\partial}^*. \tag{14.184}$$

Furthermore, zero modes of the Dirac operator are harmonic forms, they are elements of $H^{0,q}_{\bar{\partial}}(X)$ and their number is $h^{0,q}$.

Later we will be interested in spinors carrying additional indices, i.e. spinors with values in a holomorphic vector bundle. If we construct the Dirac operator with the Hermitian connection, then it is not difficult to show that it satisfies (14.184) where $\bar{\partial}^*$ is as in (14.146). In this case the zero modes are the harmonic representatives of $H^{p,q}_{\bar{\partial}}(X, V)$.

Constructions of Calabi-Yau Manifolds

The construction of explicit examples of Calabi-Yau manifolds touches upon many contemporary developments in algebraic geometry. We will only scratch the surface of this extensive subject with a few examples and comments.

We have already mentioned that compact complex manifolds cannot be written as complex submanifolds of $\mathbb{C}^n$. One therefore constructs compact CY manifolds as submanifolds in projective space or generalizations thereof. The simplest examples are hypersurfaces X_d of projective space. They are defined as the zero locus of a homogeneous polynomial $p(z)$ of degree d: $p(\lambda \cdot z) = \lambda^d p(z)$:

$$X_d = \left\{ [z^0 : \cdots : z^d] \in \mathbb{P}^n \;\middle|\; p(z) = 0 \right\}. \tag{14.185}$$

The polynomial must be chosen such that the hypersurface is non-singular. Its first Chern class is

$$2\pi c_1(X_d) = (n + 1 - d)[\omega], \tag{14.186}$$

which is positive, zero or negative according to $d < n+1$, $d = n+1$ and $d > n+1$, respectively. For instance, for $n = 2, d = 3$ we obtain a torus, for $n = 3, d = 4$ a K3 surface and for $n = 4, d = 5$ a CY threefold. The quintic in $\mathbb{P}^4$, defined by

$$z_1^5 + z_2^5 + z_3^5 + z_4^5 + z_5^5 = 0 \tag{14.187}$$

is the simplest and most studied CY threefold.

There are various ways to prove (14.186). Besides an elementary proof which uses (14.157), where the volume form is the pull-back of the $(n-1)$-st power of the Kähler form on $\mathbb{P}^n$, there exist more elegant proofs which employ more sophisticated methods of algebraic geometry. We will give an indirect argument, valid for the case $c_1(X) = 0$, by explicitly constructing the holomorphic three-form Ω. This argument also works for a more general class of CY constructions, namely that of hypersurfaces in weighted projective spaces.

A weighted projective space is defined much in the same way as a projective space, but with the generalized $\mathbb{C}^*$ action on the homogeneous coordinates

$$\lambda \cdot z = \lambda \cdot (z^0, \ldots, z^n) = (\lambda^{w_0} z^0, \ldots, \lambda^{w_n} z^n), \tag{14.188}$$

where, as before, $\lambda \in \mathbb{C}^*$ and the non-zero integer w_i is called the weight of the homogeneous coordinate z^i. We will consider cases where all weights are positive. When one is interested in non-compact spaces, one also allows some of the weights to be negative. We write $\mathbb{P}^n[w_0, \ldots, w_n] \equiv \mathbb{P}^n[\mathbf{w}]$.

A hypersurface $X_d[\mathbf{w}]$ in weighted projective space is defined as the vanishing locus of a quasi-homogeneous polynomial, $p(\lambda \cdot z) = \lambda^d p(z)$, where d is the degree of $p(z)$, i.e.

14.5 Calabi-Yau Manifolds

$$X_d[\mathbf{w}] = \{[z^0 : \cdots : z^n] \in \mathbb{P}^n[\mathbf{w}] \mid p(z) = 0\}. \tag{14.189}$$

The generalization of (14.186) is that the first Chern class of $X_d[\mathbf{w}]$ vanishes if $d = \sum_{i=1}^{n} w_i$. One can show that there are precisely 7,555 weighted projective spaces $\mathbb{P}^4[\mathbf{w}]$ which admit transverse CY-type hypersurfaces.

For hypersurfaces in weighted projective spaces one can explicitly construct Ω by extending the construction of holomorphic differentials on a Riemann surface. Once constructed, we know that Ω is essentially unique (up to a multiplicative constant on the hypersurface).

Consider first the torus defined as a hypersurface in $\mathbb{P}^2$, specified by the vanishing locus of a cubic polynomial, $p(x, y, z) = 0$. This satisfies (14.186).[33] The unique (up to normalization) holomorphic differential (written in a patch with $z = 1$) is $\omega = -dy/(\partial p/\partial x) = dx/(\partial p/\partial y) = dx/(2y)$. The first equality follows from $dp = 0$ along the hypersurface and the second equality holds, if the hypersurface is defined by an equation of the form $p = y^2 - f(x) = 0$, e.g. the Weierstrass and Legendre normal forms. An interesting observation is that ω can be represented as a residue: $\omega = \frac{1}{2\pi i} \int_\gamma \frac{dx \wedge dy}{p(x,y)}$. The integrand is a two-form in the embedding space with a first order pole on the hypersurface $p = 0$ and the contour γ surrounds the hypersurface. Changing coordinates $(x, y) \to (x, p)$ and using $\frac{1}{2\pi i} \int_\gamma \frac{dp}{p} = 1$ we arrive at ω as given above.

The above construction of the holomorphic differential for a cubic hypersurface in $\mathbb{P}^2$ can be generalized to obtain the holomorphic three-form on a Calabi-Yau manifold realized as a hypersurface $p = 0$ in weighted $\mathbb{P}^4[\mathbf{w}]$. Concretely,

$$\Omega = \int_\gamma \frac{\mu}{p} \tag{14.190}$$

with

$$\mu = \sum_{i=0}^{4} (-1)^i w_i z^i \, dz^0 \wedge \cdots \wedge \widehat{dz^i} \wedge \cdots \wedge dz^4, \tag{14.191}$$

where the term under the $\widehat{}$ is omitted. The contour γ now surrounds the hypersurface $p = 0$ inside the weighted projective space. Note that the numerator and the denominator in μ/p scale in the same way under (14.188). In the patch U_i where $z^i = $ const, only one term in the sum survives. One can perform the integration by replacing one of the coordinates, say z^j, by p and using $\int_\gamma \frac{dp}{p} = 2\pi i$. In this way one finds an expression for Ω directly on the embedded hypersurface. For instance in the patch U_0 one finds (no sum on (i, j, k) implied)

$$\Omega = \frac{w_0 z^0 \, dz^i \wedge dz^j \wedge dz^k}{\Delta_0^{ijk}}, \tag{14.192}$$

---

[33] Alternatively we could represent the torus as a homogeneous polynomial of degrees 4 and 6 in $\mathbb{P}^2[1, 1, 2]$ and $\mathbb{P}^2[1, 2, 3]$, respectively and the following discussion would also be valid.

where $\Delta_0^{ijk} = \frac{\partial(z^i,z^j,z^k,p)}{\partial(z^1,z^2,z^3,z^4)}$. From our derivation it is clear that this representation of Ω is independent of the choice of $\{i, j, k\} \subset \{1, 2, 3, 4\}$ and of the choice of coordinate patch. Furthermore, it is everywhere non-vanishing and well defined at every non-singular point of the hypersurface.

Further generalizations are to hypersurfaces in products of weighted projective spaces and to submanifolds of co-dimension higher than one which are specified by the vanishing of more than one quasi-homogeneous polynomial.

The largest known class of CY constructions is as submanifolds in toric varieties. They are generalizations of projective space. One way to define them is by generalizing (14.65):

$$\frac{\mathbb{C}^{n+k+1} - \mathscr{P}}{(\mathbb{C}^*)^k}, \qquad (14.193)$$

where the $\mathbb{C}^*$'s act on the coordinates of $\mathbb{C}^{n+k+1}$ and $\mathscr{P}$ is called the excluded set.

Weighted projective spaces and more general toric varieties are generically singular[34] and, if the hypersurface intersects the singular locus, then it is also singular. In many cases the singularities can be resolved leading to smooth Calabi-Yau manifolds. We will not enter this very interesting but mostly mathematical subject.

While compact CY manifolds can never be toric varieties themselves, this is not the case in the non-compact case. An illustrative example is the so-called resolved conifold

$$\frac{\mathbb{C}^4 - \{z_1 = z_2 = 0\}}{(z_1, z_2, z_3, z_4) \sim (\lambda z_1, \lambda z_2, \lambda^{-1} z_3, \lambda^{-1} z_4)}. \qquad (14.194)$$

We observe that the $\mathbb{C}^*$-invariant coordinates $x_1 = z_1 z_3$, $x_2 = z_1 z_4$, $x_3 = z_2 z_3$ and $x_4 = z_2 z_4$, satisfy

$$x_1 x_4 - x_2 x_3 = 0, \qquad (14.195)$$

which describes a hypersurface in $\mathbb{C}^4$ and gives an isomorphic description of the conifold.[35]

The conifold plays an important role in the local description of compact CY manifolds at points in their moduli space where they develop singularities. Zooming into the vicinity of the singularity leads to considering non-compact CY spaces and

---

[34] A simple example is $\mathbb{P}^2[1, 1, 2]$, i.e. (z^0, z^1, z^2) and $(\lambda z^0, \lambda z^1, \lambda^2 z^2)$ denote the same point. For $\lambda = -1$ the point $[0:0:z^2] \equiv [0:0:1]$ is fixed but λ acts non-trivially on its neighbourhood. There is a $\mathbb{Z}_2$ orbifold singularity at this point.

[35] The toric variety (14.194) can be considered as the $\mathcal{O}(-1) \oplus \mathcal{O}(-1)$ bundle over a $\mathbb{P}^1$ defined by the two homogeneous coordinates z_1 and z_2. Here $\mathcal{O}(n)$ denotes the holomorphic line bundle over $\mathbb{P}^1$ with first Chern number $\int_{\mathbb{P}^1} c_1(\mathcal{O}(n)) = n$. For the size of this $\mathbb{P}^1$ going to zero, one gets the conifold singularity.

to a decoupling of gravity. Much interesting physics resides at these points, e.g. D-branes at singularities.

Other examples of non-compact CY manifolds are the cotangent bundle of projective space or of spheres. For instance, the cotangent bundle of S^3 is the deformed conifold.

There exists a realization of toric varieties as vacuum manifolds of two-dimensional supersymmetric linear σ-models but their discussion is beyond the scope of this book. The same applies to more sophisticated algebraic-geometric tools which are used in the construction of CY manifolds and the study of their properties.

Calabi-Yau Moduli Space

We specifically discuss three dimensional Calabi-Yau manifolds, but some results of this section straightforwardly generalize to higher dimensions.

In view of Yau's theorem, the parameter space of CY manifolds is that of Ricci-flat Kähler metrics. We thus ask the following question: given a Ricci-flat Riemannian metric $g_{\mu\nu}$ on a manifold M, what are the allowed infinitesimal variations $g_{\mu\nu} + \delta g_{\mu\nu}$ such that

$$R_{\mu\nu}(g) = 0 \quad \Rightarrow \quad R_{\mu\nu}(g + \delta g) = 0 \quad ? \tag{14.196}$$

Clearly, if g is a Ricci-flat metric, then so is any metric which is related to g by a diffeomorphism (coordinate transformation). We are not interested in those δg which are generated by a change of coordinates. To eliminate them we have to fix the diffeomorphism invariance and impose a coordinate condition. This is analogous to fixing a gauge in electromagnetism. An appropriate choice is to demand that $\nabla^\mu \delta g_{\mu\nu} = 0$. Any $\delta g_{\mu\nu}$ which satisfies this condition also satisfies $\int_M d^d x \sqrt{g}\, \delta g^{\mu\nu} (\nabla_\mu \xi_\nu + \nabla_\nu \xi_\mu) = 0$, and is thus orthogonal to any change of the metric induced by a diffeomorphism generated by the vector field ξ^μ. We now expand (14.196) to first order in δg (cf. Eq. (18.250)) and use $R_{\mu\nu}(g) = 0$ and the coordinate condition. Taking the trace of the resulting equation gives $\nabla^\rho \nabla_\rho (g^{\mu\nu} \delta g_{\mu\nu}) = 0$. On a compact manifold this means $g^{\mu\nu} \delta g_{\mu\nu} = $ const. We then arrive at

$$\Delta_L \delta g_{\mu\nu} \equiv \nabla^\rho \nabla_\rho \delta g_{\mu\nu} + 2 R_\mu{}^\rho{}_\nu{}^\sigma \delta g_{\rho\sigma} = 0. \tag{14.197}$$

Δ_L is called the Lichnerowicz operator.

We now analyze (14.197) for the case that (M, g) is a Kähler manifold. Given the index structure of the metric and the Riemann tensor on Kähler manifolds, one immediately finds that the conditions imposed on the components $\delta g_{i\bar{j}}$ and δg_{ij} decouple and can thus be studied separately. This is what we now do in turn.

1. $\delta g_{i\bar{j}}$: With the help of (14.289), it is easy to see that the condition (14.197), which now reads $\nabla^\mu \nabla_\mu \delta g_{i\bar{j}} + 2 R_i{}^k{}_{\bar{j}}{}^{\bar{l}} \delta g_{k\bar{l}} = 0$, $\mu = (k, \bar{k})$, is equivalent to $(\Delta \delta g)_{i\bar{j}} = 0$. Here we view $\delta g_{i\bar{j}}$ as the components of a (1, 1)-form. We see that harmonic (1, 1)-forms correspond to the metric variations of the form $\delta g_{i\bar{j}}$ and to cohomologically non-trivial changes of the Kähler form. Of course, we already knew from Yau's theorem that for any $[\omega + \delta\omega]$ there is again a Ricci-flat Kähler metric. Expanding $\delta g_{i\bar{j}}$ in a basis of real (1, 1)-forms, which we will denote by b^α, $\alpha = 1, \ldots, h^{1,1}$, we obtain the following general form of the deformations of the Kähler structure of the Ricci flat metric:

$$\delta g_{i\bar{j}} = \sum_{\alpha=1}^{h^{1,1}} \tilde{t}^\alpha b^\alpha_{i\bar{j}}, \quad \tilde{t}^\alpha \in \mathbb{R}. \tag{14.198}$$

Using (14.131) one may check that these δg satisfy the coordinate condition.

For $g + \delta g$ to be a Kähler metric, the Kähler moduli $\tilde{t}^\alpha$ have to be chosen such that the deformed metric is still positive definite. Positive definiteness of a metric g with associated Kähler form ω is equivalent to the conditions

$$\int_C \omega > 0, \quad \int_S \omega^2 > 0, \quad \int_M \omega^3 > 0 \tag{14.199}$$

for all curves C and surfaces S on the Calabi-Yau threefold X. The subset in $\mathbb{R}^{h^{1,1}}$ which is spanned by the parameters $\tilde{t}^\alpha$ such that (14.199) is satisfied, is called the Kähler cone. It is a cone because, if ω satisfies (14.199), then so does $\lambda\omega$ for $\lambda \in \mathbb{R}_+$.

2. $\delta g_{\bar{i}\bar{j}}$: Now (14.197) reads $\nabla^\mu \nabla_\mu \delta g_{\bar{i}\bar{j}} + 2 R_{\bar{i}}{}^k{}_{\bar{j}}{}^{\bar{l}} \delta g_{k\bar{l}} = 0$. With (14.148) this is seen to be equivalent to

$$\Delta_{\bar{\partial}} \delta g^i = 0, \tag{14.200}$$

where

$$\delta g^i = \delta g^i_{\bar{j}} d\bar{z}^{\bar{j}}, \quad \delta g^i_{\bar{j}} = g^{i\bar{k}} \delta g_{\bar{k}\bar{j}} \tag{14.201}$$

is a (0, 1)-form with values in $T^{1,0}(M)$, the holomorphic tangent bundle, which we will simply denote as T_M. The corresponding cohomology group is $H^{0,1}_{\bar{\partial}}(M, T_M) \simeq H^1(M, T_M)$.[36] We conclude that (14.200) implies that $\delta g^i \in H^1(M, T_M)$. Again one may verify that these deformations of the metric satisfy the coordinate condition.

---

[36]More generally, by a theorem of Dolbeault, $H^q(M, \wedge^p T^*_M \otimes V) \simeq H^{p,q}_{\bar{\partial}}(M, V)$ where T^*_M is the holomorphic cotangent bundle and V any holomorphic vector bundle. In the case above, we have $p = 0$, $q = 1$ and $V = T_M$. $H^q(M, \wedge^p T^*_M \otimes V)$ is a sheaf cohomology group and V is, more accurately, any sheaf of germs of a holomorphic vector bundle. For details we refer to the cited mathematical literature.

14.5 Calabi-Yau Manifolds

What is the significance of these metric deformations? For the new metric to be again Kähler, there must be a coordinate system in which it has only mixed components. Since holomorphic coordinate transformations do not change the type of index, it is clear that δg_{ij} can only be removed by a non-holomorphic transformation. But this means that the new metric is Kähler with respect to a different complex structure compared to the original metric. Therefore, the elements of $H^1(M, T_M)$ correspond to deformations of the complex structure.[37]

With the help of the unique holomorphic $(3,0)$ form, we can define an isomorphism between $H^1(X, T_X)$ and $H^{2,1}_{\bar{\partial}}(X)$ by defining the complex $(2,1)$-forms[38]

$$\Omega_{ijk}\delta g^k_{\bar{l}}\, dz^i \wedge dz^j \wedge d\bar{z}^{\bar{l}} \in \mathcal{H}^{3,0}. \tag{14.202}$$

Harmonicity follows from (14.200) and the properties of Ω. These complex structure deformations can be expanded in a basis $b^a_{ij\bar{k}}$, $a = 1, \ldots, h^{2,1}$, of harmonic $(2,1)$-forms:

$$\Omega_{ijk}\delta g^k_{\bar{l}} = \sum_{a=1}^{h^{2,1}} t^\alpha\, b^\alpha_{ij\bar{l}}, \tag{14.203}$$

where the complex parameters t^α are called complex structure moduli.[39]

If we were geometers, we would only be interested in the deformations of the metric and the number of real deformation parameters (moduli) would be $h^{1,1} + 2h^{1,2}$. However, in string theory we have additional massless scalar degrees of freedom from the internal components of the antisymmetric tensor field in the (NS,NS) sector of the type II string. Its equations of motion in the gauge $d^*B = 0$ are $\Delta B = 0$, i.e. excitations of the B-field, above the background where it vanishes, are harmonic two-forms on the Calabi-Yau manifold. We can now combine these with the Kähler deformations of the metric and form

$$(i\delta g_{i\bar{j}} + \delta B_{i\bar{j}})\, dz^i \wedge d\bar{z}^{\bar{j}} = \sum_{\alpha=1}^{h^{1,1}} t^\alpha\, b^\alpha, \tag{14.204}$$

---

[37] Another way to reach at this conclusion is to deform the almost complex structure $\tilde{J} = J + \delta J$ and to require, to first order in δJ, $\tilde{J}^2 = -\mathbb{1}$ and $N(\tilde{J}) = 0$ where N is the Nijenhuis tensor. This leads to the conditions $\delta J^j_i = 0$ and $\partial_{[\bar{i}}\delta J^k_{\bar{j}]} = 0$. However, deformations δJ are trivial, if they can be generated by a coordinate transformation. It is easy to check that under a coordinate transformation the components $J^j_{\bar{i}}$ change as $\delta J^j_{\bar{i}} = -2i\,\bar{\partial}_{\bar{i}}\xi^j$. This leaves the closed modulo the exact forms, i.e. elements of the cohomology.

[38] More generally, there is an isomorphism $H^q(X, T_M \otimes V) \simeq H^q(M, \wedge^{n-1}T^*_M \otimes V) \simeq H^{n-1,q}_{\bar{\partial}}(M, V)$.

[39] Our discussion of complex structure moduli is not complete. We have only considered the linearized deformation equation. It still needs to be shown that they can be integrated to finite deformations. It can be shown that this is indeed the case for Calabi-Yau manifolds. For a general complex manifold the number of complex structure deformations is less than $h^{2,1}$.

where the parameters t^α are now complex with their imaginary part still restricted by the condition discussed before. This is referred to as the complexification of the Kähler cone.

To summarize, there is a moduli space associated with the different Kähler and complex structures, compatible with the Calabi-Yau condition. The former are parametrized by $H^{1,1}_{\bar\partial}(M)$ and the latter by of $H^1(M,T) \simeq H^{2,1}_{\bar\partial}(M)$. The moduli space of Ricci-flat Kähler metrics is parametrized by the harmonic representatives of these cohomology groups.[40]

Let us now exemplify this for the quintic in $\mathbb{P}^4$. Here we have $h^{1,1} = 1$: this is the Kähler form induced from the ambient space $\mathbb{P}^4$. (The metric induced from the Fubini-study metric is, however, not the Ricci-flat one.) To determine $h^{1,2}$ we consider the hypersurface constraint. Equation (14.187) is a special case of the most general quintic polynomial $p = \sum a_{ijklm} z^i z^j z^k z^l z^m$ which has 126 coefficients.[41] But this parametrization is redundant: polynomials which are related by a homogeneous change of coordinates of $\mathbb{P}^4$ should not be counted as different. These are parametrized by $\dim_{\mathbb{C}}(GL(5,\mathbb{C})) = 25$ coefficients. We therefore conclude that there are 101 independent monomial deformations of the quintic hypersurface constraint. These are precisely the complex structure moduli of the quintic hypersurface in $\mathbb{P}^4$.[42] One way to understand this is from the construction of Ω in terms of the defining polynomial p and the fact that Ω depends on the complex structure moduli. This latter aspect will be discussed in more detail in Sect. 14.6.

The situation for hypersurfaces in weighted projective spaces is more complicated. If the hypersurface passes through the singular loci of the embedding space, it is itself singular and the singularities have to be repaired in such a way that the Calabi-Yau condition $c_1(M) = 0$ is maintained. This introduces additional elements in the cohomology, so that in general $h^{1,1} > 1$. Furthermore, the simple counting method of hypersurface deformations does not give the right number of complex structure moduli. For instance, $X_8[2,2,2,1,1]$ has $h^{1,1} = 2$ and $h^{2,1} = 86$ while the most general degree eight polynomial has $105 - 22 = 83$ monomial deformations (22 is the number of parameters of the most general homogeneous coordinate transformations). To arrive at the correct number requires more sophisticated mathematical tools. They use exact sequences and are also valid

---

[40]For n-dimensional manifolds with vanishing first Chern class, the number of geometric moduli, i.e. deformations of the Ricci-flat Kähler metric, is $h^{1,1} + 2h^{1,n-1} - 2h^{2,0}$. The need to subtract $2h^{2,0}$ can be understood as follows: a harmonic $(0,2)$ form $b_{\bar i \bar j}$ gives, as $g^{i\bar j} b_{\bar j k}$, rise to an element of $H^1(M,T)$. It is instructive to verify the counting by comparing the number of metric components of T^{2n}, $n(2n+1)$, with $h^{1,1} + h^{1,n-1} = n^2 + 2n^2$ and $2h^{2,0} = n(n-1)$.

[41]For special values of these coefficients the hypersurface is singular, i.e. there are solutions of $p = dp = 0$.

[42]This might be familiar from the torus which can be realized as a degree three hypersurface in $\mathbb{P}^3$. The most general hypersurface constraint can be brought to the form $p = z_1^3 + z_2^3 + z_3^3 - 3az_1 z_2 z_3 = 0$. The parameter a can be shown to be related to the modular parameter τ via $j(\tau) = \frac{216a^3(8+a^3)^3}{(a^3-1)^3}$.

for more general holomorphic bundles V over the Calabi-Yau manifold, i.e. for computing $H^q(M, \wedge^p T^* \otimes V) \simeq H^{p,q}_{\bar\partial}(M, V)$. For hypersurfaces in toric varieties the counting of both types of moduli can be reduced to simple combinatorics.

We will not address questions of global properties of the moduli space of string compactifications on Calabi-Yau manifolds, except for mentioning a few aspects. Mirror symmetry, which connects topologically distinct manifolds, is certainly relevant. Another issue is that of transitions among topologically different manifolds, the prime example being the conifold transition. While one encounters singular geometries in the process, string theory is well behaved and the transition is smooth. Indeed, it has been speculated that the moduli space of all Calabi-Yau compactifications is smoothly connected.

Orbifolds can be viewed as singular limits of smooth Calabi-Yau manifolds which are attained at special points in the moduli space. Some of the moduli, namely those from the twisted sector, are related to smoothing the orbifold singularities. The holonomy of Calabi-Yau orbifolds is a discrete subgroup of $SU(3)$ such that it is not contained in any continuous subgroup of $SU(3)$, e.g. $\mathcal{H} = \mathbb{Z}_3$ for the $\mathbb{Z}_3$-orbifold. They are Ricci-flat and, furthermore, flat away from the fixed loci of the orbifold group action.

14.6 Compactification of the Type II String on CY Threefolds

Hodge numbers count harmonic forms. From Sect. 14.3 we know that harmonic forms lead to massless fields in the compactified theory. We now consider compactification of type II supergravity on Calabi-Yau manifolds. This discussion is also relevant for string compactification, as long as the restriction to the massless string modes is justified, i.e. for energies $E^2\alpha' \ll 1$. Furthermore we take the large volume limit, $L^2/\alpha' \gg 1$ where L is the length scale which characterizes the size of the CY manifold. With this assumption, we can neglect string induced α'-corrections of the low-energy effective supergravity action. While information, which is obtained from the cohomology of the internal manifold, e.g. the spectrum of massless particles, is unaffected by α-corrections, other details such as interactions are modified. In the $\alpha' \to 0$ limit string effects which are absent in field theory compactifications, such as topological non-trivial embeddings of the string world-sheet into the CY manifold, are also suppressed. These world-sheet instantons are stringy effects without analogy in theories based on point-like constituents. Their classical world-sheet action scales as L^2/α' and their contribution to scattering amplitudes is suppressed as $e^{-S_{\text{inst}}} \sim e^{-L^2/\alpha'}$. They are non-perturbative in the σ-model coupling $\sqrt{\alpha'}/L$. World-sheet instanton contributions are small for a large internal manifold, but they become relevant for $L \sim \sqrt{\alpha'}$. This is the string regime, which cannot be fully treated with the methods of classical geometry, but is captured for instance by exact conformal field theories like Gepner models. In the supergravity approximation string compactification reduces to Kaluza-Klein compactification of supergravity with the restriction to the massless modes of the wave-operator in the internal manifold. This was discussed in Sect. 14.3. However, the size of the manifold must not be too big, as then the Kaluza-Klein excitation, which are neglected, become light. Thus, the use of the supergravity approximation always has to be justified.

The four dimensional observer can obtain information about the internal manifold via the zero-mode spectrum of various operators and also via the dimensionally reduced (Yukawa) couplings, whose texture and strength reveals details about the compact space. More detailed information requires access to the massive Kaluza-Klein excitations and stringy effects such as world-sheet instantons. We now discuss compactification of type IIA and IIB supergravities.

Type IIA supergravity is a non-chiral $\mathcal{N}=2$ theory (i.e. 32 supercharges). The only supersymmetry multiplet is the gravity multiplet with field content ($M, N = 0, \ldots, 9$, spinor indices suppressed)

$$\mathcal{G}_{\text{IIA}}(10) = \{G_{MN}, \psi_M^{(+)}, \psi_M^{(-)}, \lambda^{(+)}, \lambda^{(-)}, B_{MN}, (C_3)_{MNP}, (C_1)_M, \Phi\}. \quad (14.205)$$

The superscripts on the fermions are their chiralities. These fields correspond to the massless states of the type IIA string. The fermionic fields arise in the two Neveu-Schwarz-Ramond sectors, i.e. (NS,R) and (R,NS), they are the two Majorana-Weyl gravitini of opposite chirality $\psi_M^{(\pm)}$, and the two Majorana-Weyl dilatini $\lambda^{(\pm)}$. The metric G_{MN}, the antisymmetric tensor B_{MN} and the dilaton Φ are in the (NS,NS) sector. The remaining bosonic fields, the vector C_1 and the 3-index antisymmetric tensor C_3, appear in the (R,R) sector.

Type IIB supergravity also has $\mathcal{N}=2$ supersymmetry but it is chiral, i.e. the two gravitini have the same chirality. The gravity multiplet consists of

$$\mathcal{G}_{\text{IIB}}(10) = \{G_{MN}, \psi_M^{(-)}, \tilde{\psi}_M^{(-)}, \lambda^{(+)}, \tilde{\lambda}^{(+)}, B_{MN}, (C_2)_{MN}, (C_4^+)_{MNPQ}, C_0, \Phi\}. \quad (14.206)$$

The bosonic fields from the (R,R) sector are the axion C_0, the two-form C_2 and C_4^+ which is completely antisymmetric and has self-dual field strength (with respect to the little group $SO(8)$, C_4^+ itself is self-dual, i.e. it carries $\frac{1}{2} \cdot \frac{8 \cdot 7 \cdot 6 \cdot 5}{1 \cdot 2 \cdot 3 \cdot 4} = 35$ on-shell physical degrees of freedom). The gravitini and dilatini have opposite chirality.

We know from Chap. 10 that type IIA and type IIB strings compactified on a circle are related by T-duality. Therefore, whenever the internal manifold contains a circle, type IIA and type IIB give T-dual theories, that clearly must have the same supersymmetric structure. In particular, compactification on T^4 gives maximal (2,2) supersymmetry in $D = 6$, compactification on T^6 gives maximal $\mathcal{N} = 8$ supersymmetry in $D = 4$ and compactification on K3 $\times$ T^2 gives $D = 4$, $\mathcal{N} = 4$ supersymmetry with 22 $U(1)$ vectormultiplets. We will not consider these compactifications here in any detail, but rather turn to the CY case.

Compactification of the type IIA theory on CY_3 leads to a theory in $D = 4$ with $\mathcal{N} = 2$ supersymmetry (eight supercharges). We can determine the resulting

14.6 Compactification of the Type II String on CY Threefolds

massless fields by looking at the zero modes of the ten-dimensional multiplets given above. The four-dimensional fields must organize into appropriate massless $\mathcal{N}=2$ supersymmetry multiplets. For massless fields of spin less or equal to two the relevant irreducible multiplets are the gravity multiplet, hypermultiplets and vectormultiplets. Their propagating degrees of freedom are

$$\left(2, 2 \times \tfrac{3}{2}, 1\right)_\pm \qquad \text{gravity multiplet}$$
$$\left(1, 2 \times \tfrac{1}{2}, 0\right)_\pm \qquad \text{vectormultiplet}$$
$$\left(\tfrac{1}{2}, 2 \times 0, -\tfrac{1}{2}\right) + \text{h.c.} \qquad \text{hypermultiplet.} \qquad (14.207)$$

We have listed the helicities; the subscript $\pm$ means that we have to add the states with reversed helicity. All multiplets are CPT self-conjugate.

To show how the massless fields arise, we split the ten-dimensional indices in a $SU(3)$ covariant way, $M = (\mu, i, \bar{\imath})$[43] and then use the known results for the number of harmonic (p,q) forms on the Calabi-Yau manifold. After gauge fixing the massless fields are zero modes of the appropriate Laplace or Dirac operator, i.e. elements of the appropriate cohomology group.

We discuss the type IIA superstring first. The zero modes of $g_{\mu\nu}$, $\psi_\mu^{(+)}$, $\psi_\mu^{(-)}$ and the graviphoton $(C_1)_\mu$ form the gravity multiplet. Here we denoted the fluctuations of the metric around its background value by $g_{\mu\nu}$. Both $\psi_\mu^{(\pm)}$ have an expansion of the form (14.48) so that we obtain two Majorana gravitini in four dimensions. For the remaining fields and components it is simpler to analyze the bosonic states and to infer the fermions via $\mathcal{N}=2$ supersymmetry and the known field content of the various multiplets. Of course they can also be obtained by a zero mode analysis. We will present this analysis below. Altogether one finds for the bosons, in addition to those in the gravity multiplet,

(IIA) $\qquad\qquad A_\mu^\alpha,\ t^a,\ \tilde{t}^\alpha,\ C^a,\ S,\ C\ ,\qquad\qquad (14.208)$

where A_μ^α arises from $(C_3)_{\mu i \bar{\jmath}}$. All other fields are complex scalars. The $\tilde{t}^\alpha$ correspond to $g_{i\bar{\jmath}}$ and $B_{i\bar{\jmath}}$; the t^a to g_{ij}; C^a to the $(C_3)_{ij\bar{k}}$ modes; S to Φ and $B_{\mu\nu}$; C to $(C_3)_{ijk}$. We now group these fields into (the bosonic components of) supermultiplets: A_μ^α and $\tilde{t}^\alpha$ combine to $h^{1,1}$ vectormultiplets, whereas t^a and C^a to $h^{2,1}$ hypermultiplets. The two complex scalars S and C form an additional hypermultiplet, the compactification independent universal hypermultiplet. There are $(h^{2,1}+1)$ hypermultiplets altogether. Note that supersymmetry requires that the real scalars which arise from $g_{i\bar{\jmath}}$ and $b_{i\bar{\jmath}}$ combine into complex scalars and therefore requires the complexification of the Kähler cone.

---

[43]From now on we use indices $(i, j, \ldots, \bar{\imath}, \bar{\jmath}, \ldots)$ for the internal space and μ for the four uncompactified space-time dimensions.

A comment is in order. First, we have dualized the massless antisymmetric tensor $B_{\mu\nu}$ to a real massless scalar χ and defined the complex scalar $S = \chi + ie^\Phi$. B and χ carry one (real) propagating degree of freedom. The dualization is possible in $D = 4$, if the B-field appears in the action only via its field strength $H = dB$. In this case it has the gauge symmetry $B \to B + d\lambda$, which reduces the number of physical degrees of freedom to one. The dualization proceeds as in our discussion of T-duality in Sect. 14.2. First, we write the action in first order form

$$S = \int -\frac{1}{2} H \wedge *H + \chi \, dH, \tag{14.209}$$

where χ is a real pseudo-scalar (Lagrange multiplier) field. The equation of motion for χ, $dH = 0$ is solved by $H = dB$ which, when inserted into the action, gives the action for B: $S = \int -\frac{1}{2} H \wedge *H$. The equation of motion for H is $H = *d\chi$ which is invariant under $\chi \to \chi + c$ where c is a real constant. Inserted into the action gives the action for a real scalar field: $S = \int -\frac{1}{2} d\chi \wedge *d\chi$.[44] The physical degrees of freedom of the universal hypermultiplet coincide with those of a tensor multiplet. Moreover, the reverse dualization $\chi \to B$ is not possible, if the field χ appears in the action without derivatives. This is the case if we include non-perturbative effects. They break the perturbative (Peccei-Quinn) symmetry $\chi \to \chi + c$ to a discrete shift symmetry.

In the type IIB compactification the gravity multiplet is formed by the zero modes of $g_{\mu\nu}$, $\psi_\mu^{(+)}$, $\tilde{\psi}_\mu^{(+)}$ and $(C_4^+)_{\mu ijk}$. From the rest of the fields we obtain

(IIB) $\quad\quad\quad\quad A_\mu^a, \; t^a, \; \tilde{t}^\alpha, \; C^\alpha, \; S, \; C.$ \hfill (14.210)

Here the fields A_μ^a arise from $(C_4^+)_{\mu i j \bar{k}}$ and t^a from $g_{i\bar{j}}$; $(\tilde{t}^\alpha, C^\alpha)$ from g_{ij}, $B_{i\bar{j}}$, $(C_2)_{i\bar{j}}$ and $(C_4^+)_{\mu\nu i \bar{j}}$ and (S, C) from $\Phi, a, B_{\mu\nu}$ and $(C_2)_{\mu\nu}$. The fields which arise from the four-form are real, due to the self-duality constraint of its field-strength. These bosonic fields combine with their fermionic superpartners to $(h^{1,1} + 1)$ hypermultiplets and $h^{2,1}$ vectormultiplets. The universal hypermultiplet (it contains S and C) is equivalent, via dualization, to a double tensor multiplet. However, non-perturbative corrections will modify the action such that the continuous shift symmetries in S and C are destroyed. In this case we have to use the formulation in terms of the hypermultiplet.

We have collected the massless spectra of the compactified type II theories in Table 14.1. The fields actually denote the fluctuations around the background which is non-trivial only for the metric. To obtain the fermions, we use the following decompositions of the right-handed dilatino and gravitino (cf. (14.181)); their left handed counterparts are obtained by flipping the four-dimensional chiral spinor

---

[44] A massive B-field does not have this invariance and carries three on-shell degrees of freedom (the antisymmetric second-rank tensor of its little group $SO(3)$). It can be dualized to a massive vector with three degrees of freedom.

14.6 Compactification of the Type II String on CY Threefolds

Table 14.1 Massless spectrum of the type II theories on CY_3

Multiplet	Component fields	Multiplicity
\multicolumn{3}{c}{Type IIA}	

Multiplet	Component fields	Multiplicity
Gravity	$g_{\mu\nu}, \psi_{\mu\alpha}\eta, \bar{\tilde{\psi}}_{\mu\dot\alpha}\eta, \tilde{\psi}_{\mu\dot\alpha}\eta_{\bar{i}\bar{j}\bar{k}}, \tilde{\psi}_{\mu\alpha}\eta_{\bar{i}\bar{j}\bar{k}}, (C_1)_\mu$	1
Hyper	$\lambda_\alpha\eta, \bar{\tilde\lambda}_{\dot\alpha}\eta, \tilde{\lambda}_\alpha\eta_{\bar{i}\bar{j}\bar{k}}, \bar{\lambda}_\alpha\eta_{\bar{i}\bar{j}\bar{k}}, \Phi, B_{\mu\nu}, (C_3)_{ijk}, (C_3)_{\bar{i}\bar{j}\bar{k}}$	1
Hyper	$\psi_\alpha\eta_{i,\bar{j}\bar{k}}, \bar{\psi}_{\dot\alpha}\eta_{\bar{i},jk}, \bar{\tilde{\psi}}_{\dot\alpha}\eta_{i,\bar{j}\bar{k}}, \tilde{\psi}_\alpha\eta_{\bar{i},j}, g_{ij}, g_{\bar{i}\bar{j}}, (C_3)_{i\bar{j}\bar{k}}, (C_3)_{\bar{i}jk}$	$h^{2,1}$
Vector	$(C_3)_{\mu i\bar{j}}, \bar{\psi}_{\dot\alpha}\eta_{i,\bar{j}}, \psi_\alpha\eta_{\bar{i},j}, \psi_\alpha\eta_{i,\bar{j}}, \bar{\tilde\psi}_{\dot\alpha}\eta_{\bar{i},jk}, g_{i\bar{j}}, B_{i\bar{j}}$	$h^{1,1}$

	Type IIB	
Gravity	$g_{\mu\nu}, \bar{\tilde\psi}_{\mu\dot\alpha}\eta, \psi_{\mu\alpha}\eta_{\bar{i}\bar{j}\bar{k}}, \bar{\tilde\tilde\psi}_{\mu\dot\alpha}\eta, \tilde{\psi}_{\mu\alpha}\eta_{\bar{i}\bar{j}\bar{k}}, (C_4^+)_{\mu ijk}$	1
Hyper	$\lambda_\alpha\eta, \bar{\lambda}_{\dot\alpha}\eta_{\bar{i}\bar{j}\bar{k}}, \tilde{\lambda}_\alpha\eta, \bar{\tilde\lambda}_{\dot\alpha}\eta_{\bar{i}\bar{j}\bar{k}}, \Phi, a, B_{\mu\nu}, (C_2)_{\mu\nu}$	1
Hyper	$\psi_\alpha\eta_{i,\bar{j}}, \bar\psi_{\dot\alpha}\eta_{\bar{i},\bar{j}\bar{k}}, \tilde\psi_\alpha\eta_{i,\bar{j}}, \bar{\tilde\psi}_{\dot\alpha}\eta_{\bar{i},\bar{j}\bar{k}}, g_{i\bar{j}}, B_{i\bar{j}}, (C_2)_{i\bar{j}}, (C_4^+)_{\mu\nu i\bar{j}}$	$h^{1,1}$
Vector	$(C_4^+)_{\mu i\bar{j}\bar{k}}, \bar\psi_{\dot\alpha}\eta_{i,\bar{j}\bar{k}}, \psi_\alpha\eta_{\bar{i},j}, \bar{\tilde\psi}_{\dot\alpha}\eta_{i,\bar{j}\bar{k}}, \tilde\psi_\alpha\eta_{\bar{i},j}, g_{ij}, g_{\bar{i}\bar{j}}$	$h^{2,1}$

indices $\alpha \leftrightarrow \dot\alpha$.

$$\lambda^{(+)} \sim \lambda_\alpha\eta \oplus \bar\lambda_{\dot\alpha}\eta_{\bar{i}} \oplus \lambda_\alpha\eta_{\bar{i}\bar{j}} \oplus \bar\lambda_{\dot\alpha}\eta_{\bar{i}\bar{j}\bar{k}},$$

$$\psi_M^{(+)} \sim \psi_{\mu\alpha}\eta \oplus \bar\psi_{\mu\dot\alpha}\eta_{\bar{i}} \oplus \psi_{\mu\alpha}\eta_{\bar{i}\bar{j}} \oplus \bar\psi_{\mu\dot\alpha}\eta_{\bar{i}\bar{j}\bar{k}}$$

$$\oplus\ \psi_\alpha\eta_m \oplus \bar\psi_{\dot\alpha}\eta_{m,\bar{i}} \oplus \psi_\alpha\eta_{m,\bar{i}\bar{j}} \oplus \bar\psi_{\dot\alpha}\eta_{m,\bar{i}\bar{j}\bar{k}}$$

$$\oplus\ \psi_\alpha\eta_{\bar m} \oplus \bar\psi_{\dot\alpha}\eta_{\bar m,\bar{i}} \oplus \psi_\alpha\eta_{\bar m,\bar{i}\bar{j}} \oplus \bar\psi_{\dot\alpha}\eta_{\bar m,\bar{i}\bar{j}\bar{k}}. \quad (14.211)$$

Notice that the massless spectra for type IIA and IIB are related via the exchange $h^{1,1}$ and $h^{1,2}$. This observation, in fact, extends to the full compactified string theory: compactification of type IIB strings on a CY threefold X gives the same four-dimensional theory that appears upon compactification of type IIA strings on the mirror $\hat{X}$ whose Hodge numbers are flipped compared to those of X, i.e. $h^{1,1}(\hat{X}) = h^{2,1}(X)$ and $h^{2,1}(\hat{X}) = h^{1,1}(X)$. This is the statement of mirror symmetry, whose proof is trivial from the point of view of conformal field theory, but its implications are highly non-trivial, when expressed and analyzed in geometric terms. Of course this simply says that classical geometry is a redundant language in the context of string compactification. But the geometric language is well developed and useful. On the other hand, the insight provided by string theory has led to surprising developments in algebraic geometry. The stringy input is the relevance of worldsheet instantons, i.e. topologically non-trivial embeddings of the world-sheet into the Calabi-Yau manifold. The geometric version of mirror symmetry only works, if they are taken into account, i.e. the two string compactifications are only identical, if non-perturbative world-sheet effects are taken into account. In fact, one of the great achievements of mirror symmetry is that one can use it to compute the effects of world-sheet instantons on various quantities which are relevant for the low-energy effective action of the massless string excitations.

Mirror symmetry is a particular case of T-duality, where T-duality in the wider sense used here is a duality which is perturbative in the string coupling constant g_s but not in the α'-expansion. In other words, it is a duality which holds at every order in string perturbation theory, but not in σ-model perturbation theory, where one has to take into account the effect of world-sheet instantons. But there is a formulation of mirror symmetry, the SYZ (Strominger-Yau-Zaslow) conjecture, where mirror symmetry is T-duality along a certain three-torus which is a homology cycle of CY manifolds which can be constructed as hypersurfaces in toric varieties. Under T-duality odd-dimensional branes are mapped to even-dimensional branes and vice versa.

One would like to know the low-energy effective action of string theory on Calabi-Yau manifolds. The procedure which we will describe in Chap. 16 and which is based on the calculation of string scattering amplitudes, is generally not available here because the CFT is not a free theory and the vertex operators are not known. An exception are toroidal orbifolds. But the structure of the low-energy effective action of type II theories on Calabi-Yau manifolds can be worked out in the supergravity limit by making a Kaluza-Klein ansatz for the fields in the ten-dimensional SUGRA action, expanding the zero modes into a basis of the appropriate cohomology and integrating over the Calabi-Yau manifold. We will not pursue this strategy here but instead infer some generic features from space-time supersymmetry.

For instance, the moduli of the Calabi-Yau manifold give rise to neutral massless fields in the low-energy effective action. $\mathcal{N}=2$ supersymmetry in $D=4$ imposes stringent restrictions on the action and consequently on the geometry of the moduli space. In particular, the moduli fields have no potential and hence their vevs are free parameters. Moreover, in the kinetic terms the scalars from vectormultiplets do not mix with the scalars from hypermultiplets, and the scalar manifold is a direct product of the form

$$\mathcal{M}_{SK} \times \mathcal{Q}. \tag{14.212}$$

Here $\mathcal{M}_{SK}$ is a (special) Kähler manifold (to be defined later) for the vectormultiplets and $\mathcal{Q}$ a quaternionic manifold for the hypermultiplets.[45] While $\mathcal{M}_{SK}$ contains only moduli scalars, $\mathcal{Q}$ contains moduli scalars and non-moduli scalars which, in string theory, come from the (R,R) sector of the left-right superconformal algebra. For the type IIA and IIB theories we thus have

$$\mathcal{M}^A = \mathcal{M}^A_{h^{1,1}} \times \mathcal{Q}^A_{h^{2,1}+1},$$
$$\mathcal{M}^B = \mathcal{M}^B_{h^{2,1}} \times \mathcal{Q}^B_{h^{1,1}+1}, \tag{14.213}$$

where in both cases the first factor is a special Kähler manifold. The indices give the complex and quaternionic dimensions, respectively.

---

[45] A quaternionic manifold is a complex manifold of real dimension $4m$ and holonomy group $Sp(2) \times Sp(2m)$.

14.6 Compactification of the Type II String on CY Threefolds

In string perturbation theory the gauge group is always a product of $U(1)$ factors and there are no charged fields. Non-Abelian gauge symmetries and charged fields appear, if we take non-perturbative effects into account, e.g. by wrapping branes around appropriate cycles and/or switching on fluxes of various p-form fields. In the field theory this corresponds to gauging isometries of the metric on moduli space, leading to gauged supergravities. In this case the low-energy effective field theory has a potential for the scalars which leads to interactions between scalars from the two types of multiplets but the kinetic terms still exhibit the product structure (14.213). The potential also gives vevs to some of the scalars which are therefore no longer moduli and leads to (partial) supersymmetry breaking.

The quaternionic dimension of the hypermultiplet moduli spaces is always ≥ 1, as in both type II theories there is at least the universal hypermultiplet with scalars (S, C). Its component fields are not related to the cohomology of a Calabi-Yau manifold. Most importantly, it contains the dilaton Φ which organizes the string perturbation theory via the relation $g_s = e^{\langle\Phi\rangle}$. This means that the hypermultiplet moduli space receives (perturbative and non-perturbative) stringy corrections in type IIA and IIB. In contrast to this, the vectormultiplet moduli space is exact at string tree level. In type IIB and IIA this concerns the complex structure moduli and Kähler moduli, respectively. The metric of the Kähler moduli space of type IIA receives a perturbative correction starting at order $(\alpha'/L^2)^3$ and non-perturbative corrections, powers of $e^{-L^2/\alpha'}$, from world-sheet instantons. In contrast, the metric of the complex structure moduli space of type IIB is exact at both, string and world-sheet σ-model, tree level. It is thus determined by classical geometry. The vectormultiplet moduli space of the type IIA theory, on the other hand, is not determined by classical geometry, but rather by 'string geometry'. The string effects are suppressed at large distances, i.e. when the Calabi-Yau manifold, on which we compactify, becomes large. At small distances, of the order of the string scale $\sqrt{\alpha'}$, the intuition derived from classical geometry fails.

It thus looks hopeless to compute the vectormultiplet moduli space of the type IIA theory. Here mirror symmetry comes to rescue. It relates, via the mirror map, the vectormultiplet moduli space of the type IIA theory on X to the vector multiplet moduli space of the type IIB theory on the mirror CY manifold $\hat{X}$. Roughly speaking, mirror symmetry are the local isomorphisms

$$\mathcal{M}^X_{\text{compl.str.}} \simeq \mathcal{M}^{\hat{X}}_{\text{Kähler}}, \quad \mathcal{M}^X_{\text{Kähler}} \simeq \mathcal{M}^{\hat{X}}_{\text{compl.str.}}. \tag{14.214}$$

They are given by the mirror map, which map the complex structure moduli space of X ($\hat{X}$) to the Kähler moduli space of $\hat{X}$ (X). Thus, with the help of mirror symmetry the vectormultiplet moduli spaces of both type II theories on CY can be constructed. We refer to the literature for a detailed discussion of mirror symmetry and the explicit construction of the mirror map.

We will now briefly explain the notion of special Kähler geometry and special Kähler manifolds which arises in the construction of $\mathcal{N} = 2$ supersymmetric couplings of vectormultiplets to supergravity. One can show that the entire Lagrangian can be locally (in field space) encoded in a holomorphic function $F(t)$, where t^a are (so-called special) coordinates on the space spanned by the scalar fields inside the vectormultiplets. For instance, in type IIB compactification on a CY_3, this is the complex structure moduli space and $a = 1, \ldots, h^{2,1}$. Supersymmetry requires that this space is Kähler and furthermore, that its Kähler potential can be expressed through F via

$$K = -\ln Y,$$
$$Y = 2(F - \bar{F}) - (t^a - \bar{t}^a)(F_a + \bar{F}_a), \qquad (14.215)$$

where $F_a = \partial_a F$. For this reason F is called the (holomorphic) prepotential. If we introduce projective coordinates $Z^I = (Z^0, Z^a)$ via $t^a = Z^a/Z^0$ and define $\mathscr{F}(Z) = (Z^0)^2 F(t)$, we find that the Kähler potential (14.215) can be written, up to a Kähler transformation, as

$$K = \ln i \left(\bar{Z}^I \mathscr{F}_I - Z^I \bar{\mathscr{F}}_I \right), \qquad (14.216)$$

where $I = 0, \ldots, h^{2,1}$, and $\mathscr{F}_I = \frac{\partial \mathscr{F}}{\partial Z^I}$. Supersymmetry requires furthermore that $\mathscr{F}$ is a homogeneous functions of degree two.

The notion of special Kähler geometry is more general then discussed here and a prepotential for the Kähler potential does not always exist. However, the low-energy effective $\mathscr{N} = 2$ supersymmetric field theories which arise in string compactifications are always such that $\mathscr{F}$ exists.

We will now show how special geometry arises in CY compactifications of type II superstrings. We begin by introducing a basis of $H^3(X, \mathbb{Z})$ with generators α_I and β^J ($I, J = 0, \ldots, h^{2,1}(X)$) which are (Poincaré) dual to a canonical homology basis (A^I, B_I) of $H_3(X, \mathbb{Z})$ with intersection numbers $A^I \cdot A^J = B_I \cdot B_J = 0$, $A^I \cdot B_J = \delta^I_J$. Then

$$\int_{A^J} \alpha_I = \int_X \alpha_I \wedge \beta^J = -\int_{B_I} \beta^J = \delta^J_I. \qquad (14.217)$$

All other pairings vanish. This basis is unique up to $Sp(2h^{2,1}+2, \mathbb{Z})$ transformations and is called a symplectic basis.

Mathematicians have shown that the A-periods of the holomorphic $(3,0)$-form Ω, i.e. $Z^I = \int_{A^I} \Omega$ are local projective coordinates on the complex structure moduli space of Calabi-Yau threefolds. We then have for the B-periods $\mathscr{F}_I = \int_{B_I} \Omega = \mathscr{F}_I(Z)$ and

$$\Omega = Z^I \alpha_I - \mathscr{F}_I \beta^I. \qquad (14.218)$$

Furthermore, under an infinitesimal change of complex structure Ω, which was pure $(3,0)$ to start with, becomes a mixture of $(3,0)$ and $(2,1)$ (because dz in the old complex structure is a linear combination of $d\tilde{z}$ and $d\bar{\tilde{z}}$ in the new complex structure). An immediate consequence is $\int \Omega \wedge \partial_I \Omega = 0$. Inserting (14.218) one finds $\mathscr{F}_I = \frac{1}{2}\partial_I(Z^J \mathscr{F}_J)$ or $\mathscr{F}_I = \partial_I \mathscr{F}$ with $\mathscr{F} = \frac{1}{2} Z^I \mathscr{F}_I$, $\mathscr{F}(\lambda Z) = \lambda^2 \mathscr{F}(Z)$.

14.6 Compactification of the Type II String on CY Threefolds

We thus identify Z^I with the special coordinates of supergravity and $\mathscr{F}$ with the prepotential and we can express the Kähler potential (14.216) in terms of Ω as

$$K = -\ln\left(i \int \Omega \wedge \overline{\Omega}\right). \tag{14.219}$$

We will show in the appendix that

$$\frac{\partial \Omega}{\partial t^a} = k_a \Omega + \chi_a, \tag{14.220}$$

where $\chi_a \in H^{(2,1)}_{\bar\partial}(X)$ and k_a is a function of the moduli but independent of the coordinates on X. Moreover

$$\chi_{aij\bar k} = -\frac{1}{2}\Omega_{ij}{}^{\bar l}\frac{\partial g_{\bar l k}}{\partial t^a} \quad\Longleftrightarrow\quad \delta g_{\bar k \bar l} = -\frac{1}{\|\Omega\|^2}\overline{\Omega}^{ij}{}_{\bar k}\chi_{aij\bar l}\delta t^a, \tag{14.221}$$

which, up to a constant factor, is the same as (14.203). It is an immediate consequence of its definition that χ_a is a primitive $(2,1)$-form, i.e. that $g^{j\bar k}\chi_{aij\bar k} = 0$. One defines the covariant derivative

$$D_a \Omega \equiv (\partial_a + (\partial_a K))\Omega = \chi_a. \tag{14.222}$$

It transforms homogeneously under holomorphic (in the moduli) rescalings $\Omega \to e^f \Omega$ under which the Kähler potential transforms as $K \to K - f - \bar f$. The Kähler metric is invariant.

With the help of the above results we find the following expression for the Kähler metric on complex structure moduli space

$$G_{a\bar b} = \frac{\partial^2}{\partial t^a \partial \bar t^b} K = -\frac{\int \chi_a \wedge \bar\chi_{\bar b}}{\int \Omega \wedge \overline{\Omega}}, \tag{14.223}$$

which we can rewrite as

$$2G_{a\bar b}\delta t^a \delta \bar t^b = \frac{1}{2V}\int_X g^{i\bar j}g^{k\bar l}\delta g_{ik}\delta g_{\bar j \bar l}\, dV \tag{14.224}$$

with $V = \frac{1}{3!}\int \omega^3$. This should be compared to (6.11).

In complete analogy to (14.224) we define the metric on the space of Kähler deformations of type IIA theory as

$$2G_{\alpha\bar{\beta}} = \frac{1}{2V}\int_X g^{i\bar{j}}g^{k\bar{l}}b^\alpha_{i\bar{l}}b^\beta_{k\bar{j}}dV = \frac{1}{2V}\int_X b^\alpha \wedge *b^\beta, \qquad (14.225)$$

where b^α are the basis of real $(1,1)$ forms in (14.198). Using (14.127) and the definition of the triple intersection numbers[46]

$$\kappa(\alpha,\beta,\gamma) = \int_X \alpha \wedge \beta \wedge \gamma \qquad (14.226)$$

with $\frac{1}{3!}\kappa(\omega,\omega,\omega) = V$, we find

$$\begin{aligned}G_{\alpha\bar{\beta}} &= -\frac{3}{2}\left\{\frac{\kappa(b^\alpha,b^\beta,\omega)}{\kappa(\omega,\omega,\omega)} - \frac{3}{2}\frac{\kappa(b^\alpha,\omega,\omega)\kappa(b^\beta,\omega,\omega)}{\kappa(\omega,\omega,\omega)^2}\right\} \\ &= -\frac{\partial^2}{\partial t^\alpha \partial \bar{t}^\beta}\ln\kappa(\omega,\omega,\omega),\end{aligned} \qquad (14.227)$$

where t^α are the complexified Kähler moduli and $\omega = \sum_\alpha \tilde{t}^\alpha b^\alpha$ (recall $\tilde{t}^\alpha = \mathrm{Im}\,t^\alpha$). This shows that $G_{\alpha\bar{\beta}}$ is a Kähler metric. If we define furthermore

$$f = \frac{1}{3!}\kappa_{\alpha\beta\gamma}t^\alpha t^\beta t^\gamma \qquad (\kappa_{\alpha\beta\gamma} = \kappa(b^\alpha,b^\beta,b^\gamma)), \qquad (14.228)$$

we can rewrite the Kähler potential in terms of a prepotential

$$\kappa(\omega,\omega,\omega) = \kappa_{\alpha\beta\gamma}\tilde{t}^\alpha\tilde{t}^\beta\tilde{t}^\gamma = -\frac{3i}{4}\left\{2(f-\bar{f}) - (t^\alpha - \bar{t}^\alpha)(f_\alpha + \bar{f}_\alpha)\right\}. \qquad (14.229)$$

The prepotential $\mathscr{F}$ for the complex structure moduli space of type IIB theory can be explicitly computed in terms of the periods of the holomorphic three-form. This is a calculation in classical geometry and it does not receive any corrections, neither in the σ-model expansion parameter nor in the string coupling constant. The Kähler moduli space of type IIA theory is also characterized by a prepotential. However, the classical piece (14.228), which is determined by the triple intersection product in H_4, receives corrections due to world-sheet instantons, i.e. due to worldsheets which wrap two-cycles in the CY manifold. The direct calculation of these corrections is difficult. Mirror symmetry relates $\mathscr{F}^{\mathrm{Kähler}}(X)$ to the prepotential of the complex structure moduli space on the mirror manifold $\mathscr{F}^{\mathrm{complex}}(\hat{X})$. The latter can be computed classically and mapped, via the mirror map, to $\mathscr{F}^{\mathrm{Kähler}}(X)$. In other words, the metric on the Kähler part of the moduli space of type II Calabi-Yau

---

[46]Three four-cycles $a,b,c \in H^4$ in a six-dimensional manifold generically intersect in finitely many points. If $\alpha,\beta,\gamma \in H^2$ are their Poincaré dual two-forms, the triple intersection number is $\kappa(\alpha,\beta,\gamma)$.

compactifications can be computed explicitly, in type IIB via classical geometry and in type IIA via mirror symmetry.

While, as we have seen, a great deal is known about the (local) geometry of the vectormultiplet moduli space, the question about the structure of the hypermultiplet moduli space, which supersymmetry dictates to be a quaternionic manifold, is still not well understood. The difficulty comes, of course, from the fact that it receives perturbative and non-perturbative quantum corrections. They are due to branes wrapping compact cycles inside the CY space. The wrapped cycles are not arbitrary representatives of the appropriate homology groups. They must be so-called supersymmetric cycles. In this case the wrapped brane do preserve some supersymmetry. Even dimensional holomorphic cycles, i.e. cycles which are holomorphically embedded in the CY are supersymmetric. The only odd-dimensional cycles are three-cycles (recall: $b_1 = b_5 = 0$). Supersymmetric three-cycles are special Lagrange submanifolds. They satisfy that the restriction of the Kähler form to these cycles vanishes and that $\text{Im}(e^{i\varphi}\Omega)$ restricts to zero as well. Here φ is an arbitrary phase. In type IIA theory the possibilities are Euclidean NS and D2 brane instantons. Their world-volumes are purely Euclidean and they wrap the whole CY (NS5) or supersymmetric three-cycles. In type IIB they are Euclidean D(-1), D1, D3, D5 or NS5 brane instantons wrapping the corresponding holomorphic cycles. These two types of non-perturbative configurations are mapped to each other under mirror symmetry.

14.7 Compactification of the Heterotic String on CY Threefolds

So far we have considered compactification of type II superstring theories. This led to $\mathcal{N} = 2$ supersymmetry in $D = 4$ which is unacceptable for phenomenological reasons. Breaking to $\mathcal{N} = 1$ can be achieved by turning on additional background fields besides the metric. This leads to flux-compactifications which we will discuss in Chap. 17.

Another possibility is to start from a theory with $\mathcal{N} = 1$ in $d = 10$ and to compactify it on a Calabi-Yau manifold. This leads to $\mathcal{N} = 1$ in $D = 4$, as the theory in $d = 10$ has only one gravitino. This is the case of the heterotic string. The massless fields in $d = 10$ consist of the $\mathcal{N} = 1$ supergravity multiplet and the vectormultiplet

$$\mathcal{G}(10) + \mathcal{V}(10) = \{G_{MN}, B_{MN}, \Phi, \psi_M^{(-)}, \lambda^{(+)}\} + \{A_M, \chi^{(-)}\}, \quad (14.230)$$

where gauge fields A_M and gaugini χ transform in the adjoint representation of either $E_8 \times E_8$ or $SO(32)$. The fermions are Majorana-Weyl spinors. The chiralities of Ψ_M and χ are those of the Ramond vacuum of the supersymmetric side of the heterotic string and, as usual, the dilatino has the opposite chirality from the gravitino. Of course, we can choose the GSO projection such that all chiralities are reversed.

Our discussion of supersymmetry preserving backgrounds for the type II string, which led to Ricci-flat manifolds with Killing spinors, was based on the assumption that the only field with a non-trivial vacuum configuration is the metric. In the case of the heterotic string this condition is too strong due to the Bianchi identity for the three-form field strength (cf. also the discussion in Sect. 16.4)

$$dH = \frac{\alpha'}{4}\bigl(\text{tr}(R \wedge R) - \text{tr}(F \wedge F)\bigr). \qquad (14.231)$$

Here

$$\text{tr}(F \wedge F) = \frac{1}{30}\text{Tr}(F \wedge F), \qquad (14.232)$$

where Tr is the trace in the adjoint representation and, for gauge group $SO(32)$, tr is the trace in the fundamental. For the $E_8 \times E_8$ heterotic string (14.232) is simply a definition of tr. The trace in tr$(R \wedge R)$ is in the **6** of $O(6)$.

The Bianchi identity (14.231) implies that $H \neq dB$ but rather $H = dB + \frac{\alpha'}{4}(\Omega_L - \Omega_{YM})$ where Ω_L and Ω_{YM} are the Lorentz and Yang-Mills Chern-Simons terms which, by definition, satisfy $d\Omega_L = \text{tr}(R \wedge R)$ and $d\Omega_{YM} = \text{tr}(F \wedge F)$. Chern-Simons forms play a central role in the analysis of anomalies (here gravitational and gauge anomalies) and (14.231) can be obtained from such an analysis for the heterotic string or rather of ten-dimensional super-Yang-Mills theory coupled to supergravity. In this field theory limit it turns out that the only anomaly free gauge groups are $SO(32)$ and $E_8 \times E_8$, just what we found from the requirement of modular invariance of the heterotic string. It can be shown that the above combination for H appears in the low-energy effective action. Clearly, the term involving ω_L can only be seen, if one includes higher derivative corrections.

Since tr$(R \wedge R)$ is non-trivial in cohomology—it is the second Chern character of the tangent bundle of the compactification manifold—we must choose a nontrivial background field strength F in order to satisfy tr$(F \wedge F) = \text{tr}(R \wedge R)$. This renders the supersymmetry transformations for the gaugini nontrivial:

$$\delta\chi^a = -\frac{1}{4}F^a_{MN}\Gamma^{MN}\epsilon \stackrel{\text{SUSY}}{=} 0, \qquad (14.233)$$

where a is a gauge index and the last equality has to be satisfied if we want to preserve supersymmetry in the four-dimensional space-time.

In the presence of an H-field, the supersymmetry transformation laws for the gravitino and the dilatino are modified by H-dependent terms and have to be reconsidered. We will discuss such flux compactifications in Chap. 17. The simplest way to proceed is therefore to set $H = 0$. From $R = d\omega + \omega \wedge \omega$ and $F = dA + A \wedge A$ it follows immediately that $A = \omega$ is a solution which also solves (14.149) with $\mu = 0$. This solution means that we identify the gauge connection

14.7 Compactification of the Heterotic String on CY Threefolds

A with the spin-connection ω which, for a CY manifold, is an element of the $su(3)$ Lie-algebra which is clearly a sub-algebra of the gauge algebra of either heterotic theory. There are several embeddings of $SU(3)$ in $E_8 \times E_8$ or $SO(32)$, but $\text{tr}(F \wedge F) = \text{tr}(R \wedge R)$ is only satisfied for the minimal embedding which leads to the maximal subgroups $SU(3) \times E_6 \times E_8$ and $SU(3) \times SO(26) \times U(1)$.

We will check this for the $E_8 \times E_8$ heterotic string. For this we need a little bit of group theory. Under $SU(3) \subset O(6) : \mathbf{6} = \mathbf{3} + \bar{\mathbf{3}}$ and under $SU(3) \times E_6 \subset E_8$:

$$\mathbf{248} = (\mathbf{3}, \mathbf{27}) + (\bar{\mathbf{3}}, \overline{\mathbf{27}}) + (\mathbf{1}, \mathbf{78}) + (\mathbf{8}, \mathbf{1}). \tag{14.234}$$

The index of the representation R is defined as $\text{tr}_R(T^a T^b) = l(R)\delta^{ab}$ and it satisfies $l(R_1 + R_2) = l(R_1) + l(R_2)$ and $l(R_1 \times R_2) = \dim(R_1)l(R_2) + \dim(R_2)l(R_1)$. The index of the $\mathbf{1}$ is zero and $l(\mathbf{n}) = l(\bar{\mathbf{n}})$. This suffices to show that $l(27 \cdot 3 + 27 \cdot \bar{3} + 8) = 30\, l(3 + \bar{3})$. Note that one of the E_8 factors is left untouched.

The background gauge field breaks the gauge symmetry. The unbroken gauge group is the commutant of $SU(3)$ in $E_8 \times E_8$ or $SO(32)$ which is $E_8 \times E_6$ and $SO(26) \times U(1)$, respectively. In the latter case the $U(1)$ gauge symmetry is generally anomalous. As we will show Chap. 17 in a slightly different context, such Abelian anomalies are cured in string theory by gauging axionic shift symmetries. In this process the gauge boson becomes massive, where the longitudinal polarization mode is provided by the axion: the gauge boson acquires a Stückelberg mass. In the $SO(32)$ heterotic case, the axion in question is the Hodge dual of the anti-symmetric $D = 4$ tensor $B_{\mu\nu}$. This way of cancelling anomalies is called Green-Schwarz mechanism. For the compactified heterotic string, the unbroken gauge groups in $D = 4$ are thus $E_6 \times E_8$ and $SO(26)$, respectively. The choice $A = \omega$ is referred to as the standard embedding.

Before considering more general possibilities of satisfying (14.231), let us study the standard embedding further. Since it is phenomenologically more interesting, we will consider the massless spectrum of the $E_8 \times E_8$ theory. Again, massless fields correspond to harmonic forms on the internal manifold and their multiplicities are the appropriate Hodge numbers.

The spectrum must arrange itself into massless supersymmetry multiplets. The relevant ones are gravity, vector and chiral multiplets with physical field contents

$$\left(2, \tfrac{3}{2}\right)_{\pm} \qquad \text{gravity},$$
$$\left(1, \tfrac{1}{2}\right)_{\pm} \qquad \text{vector},$$
$$\left(\tfrac{1}{2}, 0\right) + \text{h.c.} \qquad \text{chiral}. \tag{14.235}$$

We start with the bosons. The gauge bosons are clear; they are simply the space-time components A_μ of the ten-dimensional $E_6 \times E_8$ gauge bosons. Their internal parts are constants. Scalars arise from the internal components of the gauge fields, $A_{\bar{j}}^{(i,A)}$, $A_{\bar{j}}^{(\bar{i},\bar{A})}$ (and their complex conjugates $A_j^{(i,A)}$, $A_j^{(\bar{i},\bar{A})}$). Here i is an index in the **3** of $SU(3)$ and A an index in the **27** of E_6. The barred indices belong to the complex conjugate representations. They are $(0, 1)$-forms with values in the holomorphic tangent bundle T_X and cotangent bundle T_X^*, respectively.[47] Via dimensional reduction of the ten-dimensional SUGRA action on a Calabi-Yau manifold X, one can show that massless scalar fields in the **27** and $\overline{\mathbf{27}}$ of E_6 are in 1-1 correspondence with the harmonic representatives of $H^1(X, T_X) \simeq H_{\bar{\partial}}^{2,1}(X)$ and $H^1(X, T_X^*) \simeq H_{\bar{\partial}}^{1,1}(X)$. We will not present this analysis here. A simpler way to relate the number of massless fields to the dimensions of appropriate cohomology groups, is to analyse the zero mode spectrum of the Dirac operator and then invoke space-time supersymmetry. We will sketch this analysis when we discuss the non-standard embedding later in this section.

We still need to discuss the components $A_{\bar{i}}^{(8,1)} +$ c.c. The **8** of $SU(3)$ are traceless 3×3 matrices $u^i{}_j$. They act as traceless linear maps $T_X \to T_X$, i.e. they are endomorphisms of the holomorphic tangent bundle. The bundle of traceless endomorphisms of T_X is denoted by $\text{End}(T_X)$. $\text{End}(T_X)$ is a holomorphic subbundle of $T_X \otimes T_X^*$. The generalization to $\text{End}(V)$, where V is any holomorphic vector bundle, is immediate. We will now argue that the massless modes arsing from $A_{\bar{i}}^{(8,1)}$ mathematically correspond to holomorphic deformation of the gauge bundle. Since on a Calabi-Yau manifold the curvature satisfies $R_{i\bar{j}} = 0$, for the standard embedding $A = \omega$, the gauge connection $A_{(0,1)} = A_{\bar{j}} d\bar{z}_{\bar{j}}$ is holomorphic, i.e. $F_{(0,2)} = \bar{\partial} A_{(0,1)} + A_{(0,1)} \wedge A_{(0,1)} = 0$. In the heterotic string the gauge field A is a priori an independent degree of freedom and one can try to deform away from this special choice $A = \omega$ without spoiling the supersymmetry conditions. This means in particular that the deformed bundle still has to be holomorphic so that at linear order, under a deformation $A_{(0,1)} \to A_{(0,1)} + \delta A_{(0,1)}$

$$\delta F_{(0,2)} = \bar{\partial}(\delta A_{(0,1)}) + A_{(0,1)} \wedge \delta A_{(0,1)} + \delta A_{(0,1)} \wedge A_{(0,1)}$$
$$= \bar{D}(\delta A_{(0,1)}) = 0. \qquad (14.236)$$

We find that the allowed linear order deformations of the holomorphic vector bundle away from T_X are $\bar{D}$ closed. One notices that $(\bar{D})^2 = 0$ and that a gauge transformation can be written as $\delta A_{(0,1)} \to \delta A_{(0,1)} + \bar{D}\chi_{(0,0)}$. We can use the Chern-connection in which case $\bar{D} = \bar{\partial}$ and non-trivial solutions to (14.236) are in one-to-one correspondence to elements in the cohomology group $H^1(X, \text{End}(T_X))$. They are of the form $\alpha = d\bar{z}^{\bar{i}} \alpha_{\bar{i}}{}^j{}_k dz^k \partial_j$ with $\alpha_{\bar{i}}{}^j{}_j = 0$ and $\bar{\partial}\alpha = 0$. Here we have used $H^1(X) = 0$ and the isomorphism (cf. Footnote 36 on page 484) $H^1(X, \text{End}(T_X)) \simeq H_{\bar{\partial}}^{1,1}(X, T_X)$. The harmonic representatives of

---

[47] Here we use the isomorphisms between the anti-holomorphic tangent bundle and the holomorphic cotangent bundle, i.e. raising and lowering the index with the Hermitian metric.

14.7 Compactification of the Heterotic String on CY Threefolds

$H^1(X, \text{End}(T_X))$ can be shown to be in 1-1 correspondence with massless scalar fields in four dimensions. Giving them non-zero vevs breaks $(2, 2)$ world-sheet supersymmetry to $(0, 2)$ but does not break space-time supersymmetry. We will discuss the relation between space-time and world-sheet symmetries in Chap. 13.

For a given CY manifold X, $\dim(H^1(X, \text{End}(T_X)))$ is generally difficult to determine. For instance for the quintic in $\mathbb{P}^4$ one finds $h^1(\text{End}\, T_X) = 224$. We will not discuss the issue of computing $h^1(\text{End}\, T_X)$. It requires more sophisticated mathematical tools than the ones introduced here.

To summarize, the massless bosonic fields are

$$g_{\mu\nu},\ A_\mu^{E_6\times E_8},\ t^a,\ \tilde{t}^\alpha,\ \phi_a^A,\ \phi_\alpha^{\bar{A}},\ S \tag{14.237}$$

with $(a = 1,\ldots, h^{2,1},\ \alpha = 1,\ldots, h^{1,1},\ A \in \mathbf{27}, \bar{A} \in \overline{\mathbf{27}})$ and where t^a correspond to $g_{i\bar{j}}, B_{i\bar{j}}$; $\tilde{t}^\alpha$ to g_{ij}, S to Φ and $B_{\mu\nu}$. All scalars are complex. In addition there are the bundle moduli which we will not discuss any further.

We now determine the E_6 quantum numbers and chiralities of the fermions which combine with the charged scalars into chiral multiplets. To obtain this information, we need the zero modes of the Dirac operator in an external gauge background which, for the standard embedding, is the $SU(3)$ spin connection. Consider the decomposition of the gaugini χ. It is similar to the decomposition of the dilatino (14.211), except that it carries an $SU(3) \times E_6$ index according to the decomposition of the $\mathbf{248}$ of E_8. By our choice of gauge background the E_6 index is a dummy index. The zero modes of the Dirac operator again correspond to harmonic forms. Let us consider the possibilities in turn. The results of this analysis are summarized in Table 14.2.

Let us explain some of the entries. The modes $\bar{\chi}_{\dot{\alpha}}^{(1,78)}\eta$ and $\chi_\alpha^{(1,78)}\eta_{\bar{i}\bar{j}\bar{k}}$ are the positive and negative chirality gaugini of the E_6 gauge factor. They correspond to the harmonic representatives of $H_{\bar{\partial}}^{3,0}$ and $H_{\bar{\partial}}^{0,3}$. Of course there are also the gaugini of the unbroken E_8 factor. Consider $\chi_\alpha^{(i,A)}\eta_{\bar{i}}$ where i, A enumerate the $(\mathbf{3}, \mathbf{27})$ representation of $SU(3) \times E_6$. As the standard embedding identifies i with a holomorphic tangent space index, the zero modes are elements of $H^1(X, T_X)$, the number of which is already known to be equal to $h^{2,1}$. We have thus found that there are $h^{2,1}$ massless right-handed fermions which transform in the $\mathbf{27}$ of E_6. Likewise, $\chi_\alpha^{(\bar{i},\bar{A})}\eta_j$ lead to $h^{1,1}$ right-handed fermions in the $\overline{\mathbf{27}}$ of E_6. The fermions combine with scalars into $h^{2,1}$ chiral multiplets transforming in the $\mathbf{27}$ and into $h^{1,1}$ chiral multiplets in the $\overline{\mathbf{27}}$ of E_6.

For $h^{2,1} \neq h^{1,1}$, the matter spectrum is chiral and $|h^{2,1} - h^{1,1}|$ chiral multiplets will be protected from becoming massive. There are thus $N_{\text{gen}} = |N_{\mathbf{27}} - N_{\overline{\mathbf{27}}}| = |h^{2,1} - h^{1,1}| = \frac{1}{2}\chi$ massless charged matter multiplets or, in other words, the number of massless generations is half the Euler number of the Calabi-Yau manifold.

The fields (B, Φ) together with a Majorana fermion, are the components of a linear multiplet which can be dualized to a chiral multiplet. The gauge symmetry $B \to B + d\lambda$ translates to a shift symmetry of the complex scalar $S = \chi + ie^\Phi \to S + c$ where χ is the pseudo-scalar dual to $B_{\mu\nu}$. The auxiliary field structure of the linear and the chiral multiplets are different and at the level of the interacting

Table 14.2 Massless spectrum of the $E_8 \times E_8$ heterotic string on CY_3 with standard embedding

Multiplets	Component fields	Multiplicity
Gravity	$g_{\mu\nu}, \bar{\psi}_{\mu\dot\alpha}\eta, \psi_{\mu\alpha}\eta_{\bar{i}\bar{j}\bar{k}}$	1
Chiral	$\Phi, B_{\mu\nu}, \lambda_\alpha\eta, \bar{\lambda}_{\dot\alpha}\eta_{\bar{i}\bar{j}\bar{k}}$	1
Chiral	$g_{ij}, g_{\bar{i}\bar{j}}, \psi_\alpha\eta_{\bar{i},\bar{j}}, \bar{\psi}_{\dot\alpha}\eta_{i,\bar{j}\bar{k}}$	$h^{2,1}$
Chiral	$g_{i\bar{j}}, B_{i\bar{j}}, \psi_\alpha\eta_{i,\bar{j}}, \bar{\psi}_{\dot\alpha}\eta_{\bar{i},j\bar{k}}$	$h^{1,1}$
Vector	$A_\mu^{(248)}, \bar{\chi}_{\dot\alpha}^{(248)}\eta, \chi_\alpha^{(248)}\eta_{\bar{i}\bar{j}\bar{k}}$	1
Vector	$A_\mu^{(1,78)}, \bar{\chi}_{\dot\alpha}^{(1,78)}\eta, \chi_\alpha^{(1,78)}\eta_{\bar{i}\bar{j}\bar{k}}$	1
Chiral	$(A_i^{(8,1)}, \chi_\alpha^{(8,1)}\eta_{\bar{i}\bar{j}}) + (A_{\bar{i}}^{(8,1)}, \bar{\chi}_{\dot\alpha}^{(8,1)}\eta_{\bar{i}})$	$h^1(\mathrm{End}\,T)$
Chiral	$(A_{\bar{i}}^{(\overline{3,27})}, \chi_\alpha^{(\overline{3,27})}\eta_{\bar{i}}) + (A_i^{(3,27)}, \bar{\chi}_{\dot\alpha}^{(3,27)}\eta_{\bar{i}\bar{j}})$	$h^{1,1}$
Chiral	$(A_{\bar{i}}^{(3,27)}, \chi_\alpha^{(3,27)}\eta_{\bar{i}}) + (A_i^{(\overline{3,27})}, \bar{\chi}_{\dot\alpha}^{(\overline{3,27})}\eta_{\bar{i}\bar{j}})$	$h^{2,1}$

theory the two formulations are not equivalent. For instance, it is believed that non-perturbative g_s effects break the axionic shift symmetry and generate a potential for the dilaton which, due to supersymmetry, must be derived from a holomorphic (in S) superpotential. In this case S cannot be dualized to a linear multiplet.

The structure of the moduli space of heterotic CY compactifications with standard embedding can be obtained from the type II results. Notice that the Kähler factors of the type II moduli spaces contain the same moduli fields which appear in the heterotic string, whereas the quaternionic factors are obtained by gluing together moduli scalars with non-moduli scalars which, in string theory, come from the RR-sector of the left-right superconformal field theory. Therefore, one obtains the heterotic moduli space by setting the (R,R) fields to zero. This gives

$$\mathcal{M}^{\mathrm{het}} = \frac{SU(1,1)}{U(1)} \times \mathcal{M}_{h^{1,1}} \times \mathcal{M}_{h^{2,1}}, \tag{14.238}$$

where the second and third factors are special-Kähler manifolds and the first factor is parametrized by the complex scalar S with positive imaginary part. Due to the shift symmetry the Kähler potential must be of the form $K(\mathrm{Im}S)$. But $\mathrm{Im}S$ is the dilaton coupling whose power is fixed at tree level to give the kinetic term

$$\frac{1}{(\mathrm{Im}S)^2}\partial^\mu S\,\partial_\mu \bar{S} \tag{14.239}$$

which is a sigma model with target space $SU(1,1)/U(1) \simeq H_+$. Note that the structure (14.238) is valid at string tree level. Perturbative loop corrections destroy the product structure.

Our discussion of the moduli space relied on space-time supersymmetry arguments, but its product structure and the special geometry can also be derived from superconformal Ward identities of the world-sheet theory with CY target space. World-sheet aspects of supersymmetric compactifications are discussed in Chap. 15.

14.7 Compactification of the Heterotic String on CY Threefolds

So far we have ignored the SUSY condition (14.233). For our choice of gauge field background this equation reads $(\gamma^{ij} F_{ij} + \gamma^{\bar{i}\bar{j}} F_{\bar{i}\bar{j}} + 2\gamma^{i\bar{j}} F_{i\bar{j}})\epsilon = 0$ where ϵ is a Killing spinor. From (14.181) and the comments given there, it follows that (14.233) splits into the three conditions

$$F_{ij} = F_{\bar{i}\bar{j}} = 0, \qquad g^{i\bar{j}} F_{i\bar{j}} = 0 \qquad (14.240)$$

on the gauge bundle. This set of equations, which follow from requiring unbroken space-time supersymmetry, are the Hermitian-Yang-Mills equations for the $H \subset E_8 \times E_8$ valued connection with $\mu = 0$, see Eq. (14.149). For the standard embedding, Eq. (14.240) are clearly satisfied but for more general solutions of the Bianchi identity they impose restrictions. By the Uhlenbeck-Yau theorem a stable holomorphic vector bundle over compact Kähler manifolds M admits a unique Hermitian connection which solves (14.240). Stability is a technical criterion which is formulated in terms of the slope $\mu(V)$

$$\mu(V) = \frac{1}{\text{vol } M} \frac{\int_M c_1(V) \wedge *\omega}{\text{rank}(V)}, \qquad (14.241)$$

where ω is the Kähler form on M. It is not difficult to show that this is the same as the constant on the r.h.s. of (14.149). A holomorphic vector bundle is called stable, if $\mu(V') < \mu(V)$ for any subbundle (subsheaf) $V' \subset V$ of lower rank. All solutions of (14.240) are furnished by stable vector bundles or by direct sums of stable vector bundles of equal slope. We are interested in stable bundles with $\mu(V) = 0$. While it is easy to construct holomorphic vector bundles on M, stability is generally hard to show. For the holomorphic tangent bundle, equation (14.240) is the condition for Ricci flatness, for which Yau proved that there exists a unique solution. By the Uhlenbeck-Yau theorem this guarantees that the holomorphic tangent bundle is stable.

A gauge field background which solves (14.240) or, more generally, (14.149), is also a solution to the Yang-Mills equations $D^\mu F_{\mu\nu} = 0$. For a Hermitian vector bundle on a Kähler manifold they are $D^i F_{i\bar{j}} = g^{i\bar{k}} D_{\bar{k}} F_{i\bar{j}} = 0$; D is the gauge and and general coordinate covariant derivative. If we use the Bianchi identity $D_{\bar{k}} F_{i\bar{j}} + D_{\bar{j}} F_{\bar{k}i} + D_i F_{\bar{j}\bar{k}} = 0$ and (14.240) we see that the YM equations are satisfied.

It remains to look at the conditions imposed by (14.231).[48] For $V \neq T$, the Bianchi identity (14.231) is no longer solved locally. However, since the left hand side is exact, only those vector bundles are admissible which satisfy

$$-\text{ch}_2(V) = c_2(T), \qquad (14.242)$$

---

[48]Note that the H field which solves (14.231) is $\mathcal{O}(\alpha')$ and therefore does not contribute to the lowest order YM equations $\beta^A = 0$, cf. (14.6).

where $c_1(T) = 0$ has been used. This is the so-called tadpole cancellation condition for the two-form B. For V a vector bundle with structure group $SU(n)$, $c_1(V) = 0$ and (14.242) simplifies to $c_2(V) = c_2(T)$. Note that if $h^{1,1}(M) > 1$, $\mu(V) = 0$ does not imply $c_1(V) = 0$. Another condition on V comes from requiring the existence of spinors. This can be shown to lead to $c_1(V) \in H^2(M, 2\mathbb{Z})$.

As we have seen, the choice $V = T$ leads to the gauge group $E_6 \times E_8$ in four dimensions. Phenomenologically, E_6 might be an interesting GUT gauge group. Within string theory we can envision a unification of all forces at the string scale without any unification of just the Yang-Mills gauge couplings. But even if we adhere to the GUT scenario, there are more compelling choices than E_6, namely $SU(5)$ and $SO(10)$, in which the Standard Model matter can be embedded without any extra exotic matter particles. In order to get these gauge groups, one needs vector bundles with structure groups $SU(5)$ and $SU(4)$, whose commutants in E_8 are $SU(5)$ and $SO(10)$, respectively. Since on a CY one always has $b^1(X) = 0$, one cannot get matter fields in the adjoint representation of the GUT group. Thus, e.g. breaking an $SU(5)$ GUT group to the Standard Model with a Higgs field in the adjoint representation of $SU(5)$ is not possible. However, the extra dimensions allow for other ways of gauge symmetry breaking. One possibility is that the CY has a discrete (torsion) piece in the fundamental group, e.g. $\pi_1(X) = \mathbb{Z}_n$, so that a nontrivial $U(1) \subset SU(5)$ valued discrete Wilson line $W = \exp(i \oint_C A_Y)$ around this cycle C allows for the breaking of $SU(5) \to SU(3) \times SU(2) \times U(1)_Y$. A way to get a Calabi-Yau manifold with non-trivial fundamental group is to take a quotient of a smooth Calabi-Yau space Y with trivial $\pi_1(Y)$ by a freely acting discrete symmetry group G, i.e. $X = Y/G$. In this case $\pi_1(X) = G$.

The massless GUT spectrum is most easily determined by looking at the zero modes of the Dirac operator and invoking space-time supersymmetry. For the gravitino and dilatino, which do not carry gauge indices, the analysis is the same as for the standard embedding. Differences occur for the gaugini. Denote by H the structure group of the vector bundle and by G the commutant in E_8. Here we consider the case where the second E_8 is unbroken. It is the gauge group of the hidden sector which communicates with the visible sector with gauge group G only gravitationally. We then have the decomposition under $E_8 \supset H \times G$

$$\mathbf{248} = \sum_i (R_i(H), R_i(G)). \tag{14.243}$$

The gaugini are zero modes of the Dirac operator which contains the Hermitian H-connection and the coefficients in the decomposition (14.181) carry an $H \times G$ index according to (14.243), where the G part just goes along for the ride. It follows that massless fermions in four dimensions correspond to elements of $H^q(X, V_{R_i(H)}) \otimes R_i(G)$, $q = 0, \ldots, 3$. The notation $V_{R(H)}$ indicates that the transition functions of V are matrices in the representation $R(H)$ (with holomorphic matrix elements). In order to find the number of generations and anti-generations we need to compute the dimensions of these cohomology groups. There exits efficient methods for doing so but we will not present them here. However, there are a few general results which we

14.7 Compactification of the Heterotic String on CY Threefolds

state without proof: (1) For stable vector bundles with $\mu = 0$, $H^0(X, V_R) = 0$ for any non-trivial representation and, of course, $H^1(X, V_1) = H^2(X, V_1) = 0$ for the trivial representation; (2) $h^q(X, V) = h^{3-q}(X, V^*)$ where V^* is the dual bundle.

To be specific, consider $H = SU(4)$, $G = SO(10)$ and

$$\mathbf{248} = (\mathbf{15}, \mathbf{1}) + (\mathbf{1}, \mathbf{45}) + (\mathbf{6}, \mathbf{10}) + (\mathbf{4}, \mathbf{16}) + (\overline{\mathbf{4}}, \overline{\mathbf{16}}). \tag{14.244}$$

In this case there is matter in the **10** and in the **16** of $SO(10)$; the net number of generations is

$$\begin{aligned} N_{10} &= h^1(X, V_6), \\ N_{16} &= |h^1(X, V) - h^1(X, V^*)| = |h^1(X, V) - h^2(X, V)|, \\ N_1 &= h^1(X, \text{End } V), \end{aligned} \tag{14.245}$$

where we used $V \equiv V_4$, $V_{\bar 4} = V^*$; note also that $V_6 = \wedge^2 V$ and $H^1(X, \text{End } V) \simeq H^1(X, V^* \otimes V)$; End V corresponds to the adjoint representation.[49] An explicit example of a stable $SU(4)$ bundle satisfying the tadpole condition can be constructed e.g. over the Calabi-Yau threefold $X_{4,4}[1, 1, 1, 1, 2, 2]$, i.e. the complete intersection of two quartics in $\mathbb{P}^5[1, 1, 1, 1, 2, 2]$. The relevant Hodge numbers can be computed: $h^{2,1}(X) = 73$, $h^{1,1}(X) = 1$, $h^1(X, V) = 80$, $h^1(X, \wedge^2 V) = 72$, $h^2(X, V) = 0$ and $h^1(X, \text{End } V) = 255$.

Note that the Hermitian Yang-Mills equations are w.r.t. a fixed complex structure. It might happen that for some changes of the complex structure the bundle cannot adjust to satisfy the Hermitian Yang-Mills equations w.r.t. the new complex structure. If this is the case, these deformations break space-time supersymmetry. With the help of the projection operators introduced in (14.81) we can write the first of the conditions in (14.240) as

$$Q_\mu^\rho Q_\nu^\sigma F_{\rho\sigma} = 0. \tag{14.246}$$

Now vary the complex structure infinitesimally: $J \to J + \delta J$. We want to find the condition for δA such that (14.246) is also satisfied w.r.t. the new complex structure. Recall Footnote 37 and expand $i\delta J_{\bar i}^{j} = \sum_a \delta t^a (b^a)_{\bar i}^{j}$ where the b^a are a basis of $H^1(M, T)$ and δt^a parametrize infinitesimal changes of the complex structure. If we also use $\delta F_{\bar i \bar j} = D_{\bar i} \delta A_{\bar j} - D_{\bar j} \delta A_{\bar i}$, we can write the $(\bar i, \bar j)$ component of the change of (14.246) as

$$\sum \delta t^a (b^a)_{[\bar i}^{k} F_{|k|\bar j]} + 2 D_{[\bar i} \delta A_{\bar j]} = 0. \tag{14.247}$$

The $(i, \bar j)$ component is identically satisfied. If for a given variation δt^a Eq. (14.247) has no solution $\delta A_{\bar j}$, then this variation will break supersymmetry. In the low energy

---

[49] More precisely, $H^1(X, V^* \otimes V) \simeq H^1(X, \text{End} V) \oplus H^1(X)$; but $H^1(X) \simeq H^{0,1}_{\bar\partial}(X)$ is trivial on a CY manifold.

effective field theory the corresponding flat direction will be lifted. One can show that if δA exists, then the bundle is still stable and $g^{i\bar{j}}\delta F_{i\bar{j}} = 0$ can also be satisfied. For further details we refer to the literature, were one also finds a discussion from an effective field theory point of view.

The lifting of flat directions which correspond to moduli of the Calabi-Yau manifold also happens in type II compactifications if one switches on RR fluxes. This will be discussed in Chap. 17.

We want to remark that for non-standard embedding there is no relation between the number of matter fields and the number of moduli fields. This can also be understood from the CFT description of these compactifications. While the standard embedding corresponds to a symmetric internal $(2, 2)$ SCFT with $(\bar{c}, c) = (9, 9)$, the more general solutions to the SUSY conditions are $(0, 2)$ SCFTs. The fermionic sector of the heterotic string must be an $N = 2$ SCFT to have $\mathcal{N} = 1$ space-time supersymmetry,[50] but the superconformal symmetry in the other sector is not required and turns out to be a luxury which allows for many explicit constructions and calculations. CY compactifications of the type II strings always have $(2, 2)$ SCFTs. Here the extended space-time SUSY leads to many restrictions on the structure of the low-energy effective action, such as the appearance of special geometry. As a consequence $(2, 2)$ heterotic compactifications inherit much of this structure, such as special geometry of the moduli space, but there are perturbative and non-perturbative quantum corrections.

We will see in Chap. 15 that a special class of such $(2, 2)$ SCFTs are given by Gepner models. There the strong relationship between $(2, 2)$ type II and heterotic SCFT is reflected by the application of the bosonic string map. For $V \neq T$ we get genuine left-right asymmetric $(0, 2)$ superconformal field theories. The challenge from the world-sheet perspective is to construct modular invariant partition functions. Using the simple current construction it is however possible to generalize the Gepner construction to also include $(0, 2)$ models with $SO(10)$ and $SU(5)$ gauge groups. For instance, the geometric (X, V) compactification on the Calabi-Yau threefold $X_{4,4}[1, 1, 1, 1, 2, 2]$ discussed above, can be realized by a $(0, 2)$ Gepner like model based upon the $(k = 3)^5$ tensor product of unitary $N = 2$ models.

For compactifications of the heterotic string on Calabi-Yau manifolds with standard embedding, mirror symmetry is a very powerful tool for constructing the low-energy effective action. The statement is that $\text{het}(X) = \text{het}(\hat{X})$ with the role of **27**s and $\overline{\mathbf{27}}$s exchanged. In these $\mathcal{N} = 1$ compactifications, the cubic superpotentials and hence the (unnormalized) $\mathbf{27}^3$ and $\overline{\mathbf{27}}^3$ Yukawa couplings are determined by the prepotentials of the complex structure and Kähler moduli spaces, respectively. One can show that neither type of Yukawa coupling receives corrections from string and sigma-model loops. The **27** Yukawa couplings are, furthermore, uncorrected by world-sheet instanton effects and can be computed exactly in classical geometry. The precise form can be obtained by dimensional

---

[50]This will be shown in Chap. 15.

14.7 Compactification of the Heterotic String on CY Threefolds

reduction from the ten-dimensional action. The $\overline{\mathbf{27}}^3(X)$ couplings can also be obtained by dimensional reduction, but there are corrections from world-sheet instantons. These are difficult to compute directly, but they can be very efficiently determined with the help of mirror symmetry via the mirror map from the classically exact $\mathbf{27}^3(\hat{X})$ couplings. In fact both couplings can be written as the third derivative with respect to the moduli (with a suitable choice of coordinates) of the respective prepotential. One of them is exactly computable via classical geometry, while the other receives instanton corrections which can be obtained via the mirror map. Furthermore, the prepotential also determines the kinetic energy of the moduli and matter fields and one can therefore compute the normalized Yukawa couplings of the generations and anti-generations. However, there are two other types of Yukawa couplings, $\mathbf{1}^3$ and $\mathbf{1} \cdot \mathbf{27} \cdot \overline{\mathbf{27}}$ of which much less is known.

The low-energy effective action of heterotic CY compactifications with non-standard embeddings is much less constrained. The only known general constraints are those which follow from $\mathcal{N} = 1$ space-time supersymmetry. Special geometry of the type which we have discussed, is not relevant and does not govern the low-energy effective action. If there is any special geometry, it is less special. A precise formulation of mirror symmetry, which is a powerful tool in the heterotic compactifications with standard embedding, is not known. The same comments apply for breaking the space-time supersymmetry in type II compactifications by switching on NS-NS and R-R background fluxes. This is, of course, a consequence of the reduced world-sheet and space-time supersymmetry.

One of the main lessons of this chapter is that supersymmetric compactifications of type II and heterotic strings on Calabi-Yau manifolds have generally a large number of moduli which make their appearance in the four-dimensional low-energy world as neutral massless scalar fields without potential. For particle physics phenomenology and also for cosmology this poses two serious problems. On the one hand, massless fields mean long range forces which are not observed. On the other hand, the vacuum expectation values of the moduli are arbitrary free parameters, on which e.g. the Yukawa couplings depend. This means that at this level, even if we forget for the moment the huge degeneracy imposed by the large number of CY manifolds, no definite predictions can be derived from string theory. One has to find vacua with few or even no moduli. Loosely speaking, starting from a background with moduli, one has to find a way to generate a potential for the moduli which fixes their vacuum expectation values. This can indeed be done, if one allows for further background fields. We have seen that in heterotic compactifications non-trivial gauge bundles can lift some of the flat directions associated with complex structure moduli. But, of course, one then also to deal with the bundle moduli which themselves give rise to new flat directions in the low-energy effective action. In type II compactifications flat directions can be lifted by switching on R-R background fields which also break the $\mathcal{N} = 2$ supersymmetry to $\mathcal{N} = 1$. But these additional background fields modify the SUSY transformation rules of the gravitino and generate an energy-momentum tensor which modifies the Einstein equations away from $R_{MN} = 0$, and they have a back-reaction on the geometry already at zeroth order in α'. We will discuss some aspects of these more general flux compactifications in Chap. 17. In heterotic compactifications with non-standard embedding the Bianchi identity (14.231) implies that the NS B-field must have a non-trivial profile which also leads to a back-reaction on the geometry, but at $\mathcal{O}(\alpha')$ and in such a way that the CY condition ($c_1 = 0$) can be maintained. This is generally not true for type IIB flux compactifications.

There are also world-sheet versions of the various restrictions on the background fields. For instance, $(0, 2)$ global world-sheet supersymmetry of the non-linear sigma model in a metric, antisymmetric tensor and gauge background, can be shown to require the target space to be a complex manifold with a Hermitian metric and the vector bundle to be holomorphic. The world-sheet version of (14.242) in the fermionic formulation of the heterotic string, is the absence of an anomaly of the total fermionic measure $\mathcal{D}\bar{\psi}\,\mathcal{D}\lambda$. In the bosonic formulation the anomaly of the 16 left-moving chiral bosons cancels that of the right-moving fermions. The condition $c_1(V) \in H^2(M, 2\mathbb{Z})$ follows from the requirement to have an anomaly free right-moving $\mathbb{Z}_2$ symmetry which is needed for the GSO projection. Finally, the condition that the R-symmetry of the $(0, 2)$ SCFT is anomaly free requires $c_1(M) = 0$. Further conditions which follow from the existence of a space-time supersymmetry generator will be derived in the Chap. 15.

14.8 Appendix: Some Riemannian Geometry, Hodge Duals, etc.

In this appendix we collect some results on Riemannian geometry of real manifolds. We also present details of the derivation of (14.220) and (14.221) and compute the Hodge dual of the Poincaré dual of the symplectic homology basis in terms of the prepotential. This will be useful in Chap. 17.

Riemannian Geometry

A d-dimensional Riemannian (pseudo-Riemannian or Lorentzian) manifold (M, g) is a manifold M endowed with a metric $g_{\mu\nu}$ with signature $(+, \ldots, +)$ $((-, +, \ldots, +))$. Curved indices $\mu, \nu, \ldots$ are raised and lowered with $g^{\mu\nu}$ and $g_{\mu\nu}$ ($g^{\mu\nu} g_{\nu\rho} = \delta^\mu_\rho$). Flat tangent space indices $a, b, \ldots$ are raised and lowered with δ^{ab} and δ_{ab} (η^{ab} and η_{ab} in the Lorentzian case). The d-bein e^a_μ and its inverse e^μ_a are used to convert curved to flat indices, e.g. $V^a = e^a_\mu V^\mu$ and $V^\mu = e^\mu_a V^a$. They satisfy $g_{\mu\nu} = e^a_\mu e^b_\nu \delta_{ab}$ (or $g_{\mu\nu} = e^a_\mu e^b_\nu \eta_{ab}$), $e^a_\mu e^\mu_b = \delta^a_b$, $e^a_\mu e^\nu_a = \delta^\nu_\mu$, $\eta^{ab}\eta_{bc} = \delta^a_c$. The metric is invariant under local $SO(d)$ ($SO(1, d-1)$) transformations, under which the viel-bein transforms as a vector.

A Riemannian connection $\Gamma^\rho_{\mu\nu}$ is defined by imposing

$$\nabla_\rho g_{\mu\nu} \equiv \partial_\rho g_{\mu\nu} - \Gamma^\sigma_{\rho\mu} g_{\sigma\nu} - \Gamma^\sigma_{\rho\nu} g_{\mu\sigma} = 0 \qquad \text{(metricity)},$$

14.8 Appendix: Some Riemannian Geometry, Hodge Duals, etc.

$$\Gamma^\rho_{\mu\nu} = \Gamma^\rho_{\nu\mu} \quad \text{(no torsion).} \tag{14.248}$$

One finds for the Christoffel symbols

$$\Gamma^\rho_{\mu\nu} = \frac{1}{2} g^{\rho\sigma} \left(\partial_\mu g_{\sigma\nu} + \partial_\nu g_{\mu\sigma} - \partial_\sigma g_{\mu\nu} \right). \tag{14.249}$$

The Riemann tensor is defined as

$$[\nabla_\mu, \nabla_\nu] V^\rho = R_{\mu\nu}{}^\rho{}_\sigma V^\sigma, \tag{14.250}$$

which gives

$$R_{\mu\nu}{}^\rho{}_\sigma = \partial_\mu \Gamma^\rho_{\nu\sigma} - \partial_\nu \Gamma^\rho_{\mu\sigma} + \Gamma^\rho_{\mu\tau} \Gamma^\tau_{\nu\sigma} - \Gamma^\rho_{\nu\tau} \Gamma^\tau_{\mu\sigma}. \tag{14.251}$$

Defining the connection one-form $\Gamma^\mu{}_\nu = \Gamma^\mu_{\rho\nu} dx^\rho$, we can write this as

$$R^\mu{}_\nu = \frac{1}{2} R_{\rho\sigma}{}^\mu{}_\nu \, dx^\rho \wedge dx^\sigma = d\Gamma^\mu{}_\nu + \Gamma^\mu{}_\rho \wedge \Gamma^\rho{}_\nu. \tag{14.252}$$

The Ricci tensor and the Ricci scalar are $R_{\mu\nu} = g^{\rho\sigma} R_{\mu\rho\nu\sigma}$ and $R = g^{\mu\nu} R_{\mu\nu}$. The curvature tensors enjoy the symmetries

$$R_{\mu\nu\rho\sigma} = R_{\rho\sigma\mu\nu} = -R_{\nu\mu\rho\sigma}, \quad R_{\mu\nu} = R_{\nu\mu}, \quad R_{(\mu\nu\rho)\sigma} = 0, \tag{14.253}$$

and they satisfy the identities

$$\nabla_{(\mu} R_{\nu\rho)\sigma\tau} = 0 \implies \nabla^\mu R_{\mu\nu\rho\sigma} = \nabla_\rho R_{\nu\sigma} - \nabla_\sigma R_{\nu\rho}$$

$$\implies \nabla^\mu R_{\mu\nu} = \frac{1}{2} \nabla_\nu R. \tag{14.254}$$

Indices are symmetrized and anti-symmetrized with unit weight, e.g.

$$A_{[\mu\nu]} = \frac{1}{2}(A_{\mu\nu} - A_{\nu\mu}), \quad B_{(\mu\nu\rho)} = \frac{1}{3!}(B_{\mu\nu\rho} + 5 \text{ permutations}), \tag{14.255}$$

i.e. we divide by the number of terms.

The Lie algebra ($so(d)$ or $so(1, d-1)$) valued spin connection ω_μ is defined via the (no-torsion) condition

$$\nabla_\mu e^a_\nu = \partial_\mu e^a_\nu - \Gamma^\rho_{\mu\nu} e^a_\rho + \omega^a_{\mu b} e^b_\nu = 0 \tag{14.256}$$

which leads to the following explicit expression for its components

$$\omega^{ab}_\mu = \frac{1}{2} \left(\Omega_{\mu\nu\rho} - \Omega_{\nu\rho\mu} + \Omega_{\rho\mu\nu} \right) e^{\nu a} e^{\rho b}, \tag{14.257}$$

where

$$\Omega_{\mu\nu\rho} = \left(\partial_\mu e_\nu^a - \partial_\nu e_\mu^a\right) e_{a\rho}. \tag{14.258}$$

In terms of ω_μ^{ab}, the components of the Lie-algebra valued curvature 2-form are

$$R_{\mu\nu}{}^{ab} = e^{a\rho}e^{b\sigma}R_{\mu\nu\rho\sigma} = \partial_\mu \omega_\nu^{ab} - \partial_\nu \omega_\mu^{ab} + \omega_\mu^{ac}\omega_{\nu c}{}^b - \omega_\nu^{ac}\omega_{\mu c}{}^b, \tag{14.259}$$

or, more succinctly, in differential form notation ($\omega = \omega_\mu dx^\mu$) and using a matrix notation for the Lorentz indices

$$\mathscr{R} = d\omega + \omega \wedge \omega. \tag{14.260}$$

Here, the connection one-form is a solution of the no-torsion condition ($e^a = e_\mu^a dx^\mu$)

$$T^a = de^a + \omega^a{}_b\, e^b = 0. \tag{14.261}$$

Under local Lorentz-transformations, which act on the viel-bein as

$$\delta e^a = \Lambda^a{}_b e^e, \tag{14.262}$$

ω transforms like a connection:

$$\delta \omega = -d\Lambda + [\Lambda, \omega]. \tag{14.263}$$

The covariant derivative, acting on an object with only tangent-space indices, is generically

$$\nabla_\mu = \partial_\mu + \frac{i}{2}\omega_\mu^{ab}\, T_{ab}, \tag{14.264}$$

where T_{ab} is a generator of the tangent space group $SO(1, D-1)$. They satisfy, e.g. in the Lorentzian case,

$$[T_{ab}, T_{cd}] = i\left(\eta_{ac}T_{bd} - \eta_{ad}T_{bc} - \eta_{bc}T_{ad} + \eta_{bd}T_{ac}\right). \tag{14.265}$$

For example

$$(T_{ab})_c{}^d = -i(\eta_{ac}\delta_b^d - \eta_{bc}\delta_a^d) \quad \text{(vector rep.)},$$
$$T_{ab} = -\frac{i}{2}\Gamma_{ab} \quad \text{(spinor rep.)}. \tag{14.266}$$

14.8 Appendix: Some Riemannian Geometry, Hodge Duals, etc.

Defining

$$[\nabla_\mu, \nabla_\nu] = \frac{i}{2} R_{\mu\nu}{}^{ab} T_{ab}, \qquad (14.267)$$

one recovers (14.259).

The Dirac operator is

$$\nabla\!\!\!\!/ = \Gamma^\mu \nabla_\mu = \Gamma^\mu \left(\partial_\mu + \frac{1}{4} \omega_\mu^{ab} \Gamma_{ab} \right) \qquad (14.268)$$

which squares to

$$\nabla\!\!\!\!/^2 = \nabla^\mu \nabla_\mu - \frac{1}{4} R. \qquad (14.269)$$

Dirac matrices with flat and curved indices are related via the d-bein, e.g. $\Gamma^\mu = e_a^\mu \Gamma^a$ and they satisfy the algebra $\{\Gamma^a, \Gamma^b\} = 2\eta^{ab}$ and $\{\Gamma^\mu, \Gamma^\nu\} = 2g^{\mu\nu}$. A particular representation of the Dirac-matrices in ten dimensions, which is appropriate for the compactification to four dimensions, is $\Gamma^\alpha = \gamma^\alpha \otimes \mathbb{1}$, $\Gamma^a = \gamma^5 \otimes \gamma^a$ where the 4×4 matrices γ^α satisfy $\{\gamma^\alpha, \gamma^\beta\} = 2\eta^{\alpha\beta}$ and the 8×8 matrices γ^a satisfy $\{\gamma^a, \gamma^b\} = 2\delta^{ab}$ where $\gamma^5 = i\gamma^0\gamma^1\gamma^2\gamma^3$ with $\{\gamma^5, \gamma^\alpha\} = 0$ and $\alpha = 0, \ldots, 3$ and $a = 4, \ldots 9$. If the metric decomposes as in (14.39), with a corresponding factorized form for the zehn-bein, one also has $\Gamma^\mu = \gamma^\mu \times \mathbb{1}$ and $\Gamma^m = \gamma^5 \otimes \gamma^m$.

Differential Forms, Hodge-Star, etc.

A p-form $\alpha \in A^p$ can be expanded in components as

$$\alpha = \frac{1}{p!} \alpha_{\mu_1 \ldots \mu_p} dx^{\mu_1} \wedge \cdots \wedge dx^{\mu_p}. \qquad (14.270)$$

Its exterior derivative is the $(p+1)$-form $d\alpha$ with components

$$d\alpha = \frac{1}{p!} \partial_{\mu_1} \alpha_{\mu_2 \ldots \mu_{p+1}} dx^{\mu_1} \wedge \cdots \wedge dx^{\mu_{p+1}}, \qquad (14.271)$$

where d satisfies the Leibniz rule

$$d(\alpha \wedge \beta) = d\alpha \wedge \beta + (-1)^p \alpha \wedge d\beta \qquad (14.272)$$

for $\alpha \in A^p$ and $\beta \in A^q$ and

$$d^2 = 0. \qquad (14.273)$$

We define the totally antisymmetric tangent space ε-tensor as

$$\varepsilon_{a_1\ldots a_d} = \pm 1, \qquad \varepsilon_{12\ldots d} = 1,$$
$$\varepsilon^{a_1\ldots a_d} = \pm 1, \qquad \varepsilon^{12\ldots d} = \xi, \qquad (14.274)$$

where

$$\xi = \begin{cases} -1 & \text{(Minkowskian)} \\ +1 & \text{(Euclidean)}. \end{cases} \qquad (14.275)$$

The curved space epsilon tensor is defined as

$$\varepsilon_{\mu_1\ldots\mu_d} = e^{a_1}_{\mu_1}\cdots e^{a_d}_{\mu_d}\varepsilon_{a_1\ldots a_d} = \pm\sqrt{|g|}, \qquad \varepsilon_{12\ldots d} = \sqrt{|g|},$$

$$\varepsilon^{\mu_1\ldots\mu_d} = g^{\mu_1\nu_1}\cdots g^{\mu_d\nu_d}\varepsilon_{\nu_1\ldots\nu_d} = \pm\frac{1}{\sqrt{|g|}}, \qquad \varepsilon^{12\ldots d} = \frac{\xi}{\sqrt{|g|}}. \qquad (14.276)$$

One also defines the ϵ-symbol

$$\epsilon_{\mu_1\ldots\mu_d} = \pm 1, \qquad \epsilon_{12\ldots d} = +1,$$

$$\epsilon^{\mu_1\ldots\mu_d} = g^{\mu_1\nu_1}\cdots g^{\mu_d\nu_d}\epsilon_{\nu_1\ldots\nu_d} = \pm\frac{1}{|g|}, \qquad \epsilon^{12\ldots d} = \frac{\xi}{|g|}. \qquad (14.277)$$

The ϵ-symbol satisfies

$$\epsilon_{\mu_1\ldots\mu_p\mu_{p+1}\ldots\mu_d}\,\epsilon^{\mu_1\ldots\mu_p\nu_{p+1}\ldots\nu_d} = \frac{\xi}{|g|}\,p!\,\delta^{\nu_{p+1}\ldots\nu_d}_{\mu_{p+1}\ldots\mu_d}, \qquad (14.278)$$

where $\delta^{\nu_1\ldots\nu_p}_{\mu_1\ldots\mu_p} = p!\,\delta^{[\nu_1}_{\mu_1}\cdots\delta^{\nu_p]}_{\mu_p} = \pm 1$.

The Hodge-$*$ is a linear map $A^p \to A^{d-p}$ with the defining property

$$\alpha \wedge \star\beta = \frac{1}{p!}\alpha_{\mu_1\ldots\mu_p}\beta^{\mu_1\ldots\mu_p}\,v \equiv \langle\alpha,\beta\rangle\,v, \qquad \alpha,\beta \in A^p. \qquad (14.279)$$

v is the volume form

$$v = \frac{1}{n!}\sqrt{|g|}\,\epsilon_{\mu_1\ldots\mu_d}\,dx^{\mu_1}\wedge\cdots\wedge dx^{\mu_d} = \sqrt{|g|}\,dx^1\wedge\cdots\wedge dx^d. \qquad (14.280)$$

The explicit expression of $\star\alpha$ is

14.8 Appendix: Some Riemannian Geometry, Hodge Duals, etc.

$$*\alpha = \frac{\sqrt{|g|}}{p!(d-p)!} \alpha_{\mu_1\ldots\mu_p} \epsilon^{\mu_1\ldots\mu_p}{}_{\nu_{p+1}\ldots\nu_d} dx^{\nu_{p+1}} \wedge \cdots \wedge dx^{\nu_d}$$

$$= \frac{1}{p!(d-p)!} \alpha_{\mu_1\ldots\mu_p} \varepsilon^{\mu_1\ldots\mu_p}{}_{\nu_{p+1}\ldots\nu_d} dx^{\nu_{p+1}} \wedge \cdots \wedge dx^{\nu_d}. \quad (14.281)$$

In particular, one has

$$*1 = v \quad (14.282)$$

and on p-forms

$$** = \xi(-1)^{p(d-p)}. \quad (14.283)$$

The Hodge-$*$ satisfies $\alpha \wedge *\beta = \beta \wedge *\alpha$ and $\langle \alpha, \beta \rangle = \xi \langle *\alpha, *\beta \rangle$.

Define the symmetric inner product of two p-forms

$$(\alpha, \beta) = (\beta, \alpha) = \int_M \alpha \wedge *\beta = \frac{1}{p!} \int_M \alpha_{\mu_1\ldots\mu_p} \beta^{\mu_1\ldots\mu_p} \sqrt{|g|} d^d x. \quad (14.284)$$

Denote by $d^* : A^p \to A^{p-1}$ the adjoint of d, i.e.

$$(d\beta, \alpha) = (\beta, d^*\alpha) \quad (14.285)$$

for $\alpha \in A^p$ and $\beta \in A^{p-1}$. Then, one finds

$$d^* = \xi(-1)^{d(p+1)+1} * d *. \quad (14.286)$$

It obviously satisfies $(d^*)^2 = 0$. Its action on a p-form is

$$d^*\alpha = -\frac{1}{(p-1)!} \nabla^{\mu_1} \alpha_{\mu_1\mu_2\ldots\mu_p} dx^{\mu_2} \wedge \cdots \wedge dx^{\mu_p}. \quad (14.287)$$

The Laplace-Beltrami operator is defined as

$$\Delta = d\,d^* + d^*d, \quad (14.288)$$

which maps p-forms to p-forms and commutes with $*$, d and d^*. Its action on the components of a p-form α is

$$\Delta \alpha_{\mu_1\ldots\mu_p} = -\nabla^\mu \nabla_\mu \alpha_{\mu_1\ldots\mu_p} + p\, R_{\nu[\mu_1} \alpha^\nu{}_{\mu_2\ldots\mu_p]}$$

$$- \frac{1}{2} p(p-1) R_{\mu\nu[\mu_1\mu_2} \alpha^{\mu\nu}{}_{\mu_3\ldots\mu_p]}. \quad (14.289)$$

A simple way to derive this is to use $(d\beta, d\alpha) + (d^*\beta, d^*\alpha) = (\beta, \Delta\alpha)$.

When working with (p, q) forms on Hermitian manifolds, the following relation is useful:

$$\varepsilon_{i_1...i_n \bar{j}_1...\bar{j}_n} = g\, \epsilon_{i_1...i_n \bar{j}_1...\bar{j}_n} = (i)^n (-1)^{\frac{1}{2}n(n-1)} g\, \epsilon_{i_1...i_n} \epsilon_{\bar{j}_1...\bar{j}_n}$$

$$= (i)^n (-1)^{\frac{1}{2}n(n-1)} (g_{i_1 \bar{j}_1} \cdots g_{i_n \bar{j}_n} \pm \ldots), \qquad (14.290)$$

where

$$\varepsilon_{i_1...i_n \bar{j}_1...\bar{j}_n} = \left(\frac{\partial x^{\mu_1}}{\partial z^{i_1}}\right) \cdots \left(\frac{\partial x^{\mu_{2n}}}{\partial \bar{z}^{\bar{j}_n}}\right) \varepsilon_{\mu_1...\mu_{2n}},$$

$$\epsilon_{i_1...i_n} = \pm 1, \qquad g = \det(g_{i\bar{j}}). \qquad (14.291)$$

Note that (14.290) depends on the choice of complex coordinates (14.62). We will always use conventions such that (14.128) holds.

Often the Hodge-$\star$ of a p-form is defined via

$$(\star\alpha)_{\mu_1...\mu_{d-p}} = \frac{1}{p!} \varepsilon_{\mu_1...\mu_{d-p}}{}^{\nu_1...\nu_p} \alpha_{\nu_1...\nu_p}, \qquad (14.292)$$

which differs from (14.281) by a factor $(-1)^{p(d-p)}$. This is the definition used in much of the string literature. The important property (14.283) holds without change.

Variation of Ω Under Change of Complex Structure

We derive Eqs. (14.220) and (14.221). X_J is the Calabi-Yau threefold with complex structure J and Ω the unique holomorphic $(3,0)$ three-form on X_J. The family of complex manifolds is parametrized by the complex structure moduli t^a. If $z^i \equiv z^i(t^a)$ are the local holomorphic coordinates on X_J, the holomorphic coordinates on $X_{J+\delta J}$ are $z^i(t^a + \delta t^a) \equiv z^i + m_a^i \delta t^a$ and therefore

$$\partial_a(dz^k) = dm_a^k = \partial_l m_a^k\, dz^l + \bar{\partial}_l m_a^k\, d\bar{z}^l. \qquad (14.293)$$

This is true in the patch on which the local holomorphic coordinates z^i are defined. We have suppressed the label which distinguishes between different patches. In particular the m_a^k are not globally defined. If they were, we would be considering a diffeomorphism.

From (14.293) it follows that the holomorphic three-from on X_J changes as

$$\partial_a \Omega = \frac{\partial}{\partial t^a}\left(\frac{1}{3!}\Omega_{ijk} dz^i \wedge dz^j \wedge dz^k\right)$$

14.8 Appendix: Some Riemannian Geometry, Hodge Duals, etc.

$$= \frac{1}{3!}(\partial_a \Omega_{ijk}) \, dz^i \wedge dz^j \wedge dz^k + \frac{1}{2!} \Omega_{ijk} \, dz^i \wedge dz^j \wedge (\partial_a dz^k)$$

$$= \left(\frac{1}{3!} \partial_a \Omega_{ijk} + \frac{1}{2} \Omega_{ijl} \partial_k m_a^l \right) dz^i \wedge dz^j \wedge dz^k + \frac{1}{2} \Omega_{ijl} \bar{\partial}_{\bar{k}} m_a^l dz^i \wedge dz^j \wedge d\bar{z}^{\bar{k}}.$$
(14.294)

We write this as

$$\partial_a \Omega = \Omega_a' + \chi_a \in H^{3,0}(X_J) \oplus H^{2,1}(X_J). \tag{14.295}$$

The fact that $\partial_a \Omega$ is an element of the cohomology follows from $[\partial_a, d] = 0$. We will now determine Ω_a and χ_a.

On X_J the Kähler metric and Ω have index structure $g_{i\bar{j}}$ and Ω_{ijk} w.r.t. to the coordinates $z^i(t_a)$ and on $X_{J+\delta J}$ the same is true w.r.t. to the coordinates $z^i(t^a + \delta t^a)$. If we express this w.r.t. to $z^i(t^a)$, we find

$$\delta g_{i\bar{j}} + (g_{k\bar{i}} \partial_{\bar{j}} m_a^k + g_{k\bar{j}} \partial_i m_a^k) \delta t^a = 0 \tag{14.296}$$

and

$$\delta \Omega_{ij\bar{k}} + \Omega_{ijl} \partial_{\bar{k}} m_a^l \delta t^a = 0. \tag{14.297}$$

Consider the (0, 2) form $\alpha \equiv g_{k\bar{j}} \bar{\partial}_{\bar{i}} m_a^k \, d\bar{z}^{\bar{i}} \wedge d\bar{z}^{\bar{j}}$. Recalling (14.89), we find that $\bar{\partial} \alpha = 0$ and that $\alpha = \bar{\partial} \beta$ with $\beta = m_a^k g_{k\bar{j}} d\bar{z}^{\bar{j}}$. But this is only true locally because m_a^k are not globally defined. We conclude that $\alpha \in H^{2,0}$ and therefore it vanishes on CY_3 where $h^{2,0} = 0$. This allows us to rewrite (14.296) as

$$\bar{\partial}_{\bar{j}} m_a^k = -\frac{1}{2} g^{k\bar{i}} \partial_a g_{\bar{i}j}. \tag{14.298}$$

This determines χ_a.

From (14.202) we know that χ_a are harmonic forms. On a compact Kähler manifold this means that $\partial \chi_a = \bar{\partial} \chi_a = 0$. It then follows from (14.295) that $\partial \Omega_a' = 0$, i.e. that Ω_a' is a holomorphic (3, 0) from, i.e. $\Omega_a' = k_a \Omega$ where k_a is a constant on X.

The Hodge Dual of the Poincaré Dual of the Symplectic Homology Basis

In Chap. 17 we will consider RR-forms and their Hodge duals on the world-volume of D-branes which are wrapped on homology cycles of Calabi-Yau manifolds. In addition of providing the necessary results needed there, the following analysis also reveals some of the structure which is implied by special geometry.

The symplectic homology basis on a CY threefold was introduced in (14.217):

$$\int_X \alpha_I \wedge \beta^J = \delta_I^J, \quad \int_X \alpha_I \wedge \alpha_J = 0 = \int_X \beta^I \wedge \beta^J, \tag{14.299}$$

where $I, J = 0, \ldots, h^{1,2}(X)$. α_I and β^I are real three-forms. We want to compute $*\alpha_I$ and $*\beta^I$. If α_I and β^I are harmonic representatives of H^3, i.e. they satisfy $d\alpha_I = d*\alpha_I = 0$ and likewise for β^I, then so are $*\alpha_I$ and $*\beta^I$. They can therefore be expanded

$$*\alpha_I = A_I{}^J \alpha_J + B_{IJ}\, \beta^J,$$
$$*\beta^I = C^{IJ} \alpha_J + D^I{}_J\, \beta^J. \tag{14.300}$$

Note that

$$\int \alpha_J \wedge *\alpha_I = B_{IJ} = B_{JI},$$

$$\int \beta^J \wedge *\alpha_I = -A_I{}^J,$$

$$\int \alpha_J \wedge *\beta^I = D^I{}_J = -A_J{}^I,$$

$$\int \beta^J \wedge *\beta^I = -C^{IJ} = -C^{JI}, \tag{14.301}$$

where we have used that $\int \alpha \wedge *\beta = \int \beta \wedge *\alpha$, which is valid for real 3-forms. The condition $** = -1$, valid on three-forms, imposes additional conditions on A, B, C, D which one can use to show that $\begin{pmatrix} A & B \\ C & D \end{pmatrix} \in Sp(2(h^{1,2}+1))$.

We want to determine A, B, C, D. Expand the holomorphic three-form Ω in the symplectic basis

$$\Omega = Z^I \alpha_I - \mathcal{F}_I \beta^I, \qquad \bar\Omega = \bar Z^I \alpha_I - \bar{\mathcal{F}}_I \beta^I, \tag{14.302}$$

where Z^I are homogeneous coordinates on the complex structure moduli space, $\mathcal{F}(\lambda z) = \lambda^2 \mathcal{F}(Z)$ is the homogeneous prepotential and $\mathcal{F}_I = \frac{\partial}{\partial Z^I} \mathcal{F}$, i.e.

$$\frac{\partial}{\partial Z^I} \Omega = \alpha_I - \mathcal{F}_{IJ}\, \beta^J. \tag{14.303}$$

With (14.303) and (14.302) we find

$$\int \partial_I \Omega \wedge \bar\Omega = 2i\, \mathrm{Im}(\mathcal{F}_{IJ})\, \bar Z^J. \tag{14.304}$$

14.8 Appendix: Some Riemannian Geometry, Hodge Duals, etc.

On the other hand, we know that[51]

$$\partial_I \Omega = k_I \Omega + \chi_I, \tag{14.305}$$

where k_I is a function of Z^I independent of the coordinates of X, and χ_I are primitive $(2, 1)$-forms. Therefore (use $z^I \mathscr{F}_{IJ} = \mathscr{F}_J$)

$$\int \partial_I \Omega \wedge \bar{\Omega} = k_I \int \Omega \wedge \bar{\Omega} = 2i \operatorname{Im}(\mathscr{F}_{KL}) \bar{Z}^K Z^L k_I, \tag{14.306}$$

and comparison with (14.304) gives

$$k_I = \frac{\operatorname{Im}(\mathscr{F}_{IJ}) \bar{Z}^J}{\operatorname{Im}(\mathscr{F}_{KL}) \bar{Z}^K Z^L}. \tag{14.307}$$

From (14.128) we know that $*\Omega = -i\Omega$ and $*\chi_I = i\chi_I$. Combining this with (14.303) and (14.305) we find

$$*(\partial_I \Omega) = -ik_I \Omega + i\chi_I = i\partial_I \Omega - 2ik_I \Omega$$
$$= i(\alpha_I - \mathscr{F}_{IJ} \beta^J) - 2ik_I(Z^J \alpha_J - \mathscr{F}_I \beta^J)$$
$$= *\alpha_I - \mathscr{F}_{IJ} * \beta^J. \tag{14.308}$$

Inserting the ansatz (14.300), this becomes

$$(A_I{}^J - \mathscr{F}_{IK} C^{KJ})\alpha_J + (B_{IJ} - \mathscr{F}_{IK} D^K{}_J)\beta^J$$
$$= i(\delta_I^J - 2k_I Z^J)\alpha_J + i(-\mathscr{F}_{IJ} + 2K_I \mathscr{F}_J)\beta^J. \tag{14.309}$$

Separating the real and imaginary parts of the coefficients of α_I and β^J, we obtain a system of four coupled equations for A, B, C, D

$$A_I{}^J - \operatorname{Re}(\mathscr{F}_{IK}) C^{KJ} = 2\operatorname{Im}(k_I Z^J)$$
$$-\operatorname{Im}\mathscr{F}_{IK} C^{KJ} = \delta_I^J - 2\operatorname{Re}(k_I Z^J),$$
$$B_{IJ} - \operatorname{Re}(\mathscr{F}_{IK}) D^K{}_J = \operatorname{Im}(\mathscr{F}_{IJ}) - 2\operatorname{Im}(k_I \mathscr{F}_J)$$
$$-(\operatorname{Im}\mathscr{F}_{IK}) D^K{}_J = -\operatorname{Re}\mathscr{F}_{IJ} + 2\operatorname{Re}(k_I \mathscr{F}_J), \tag{14.310}$$

whose solution reads

---

[51] Here we use homogeneous coordinates on moduli space, $Z^I = (Z^0, Z^0 t^a)$. The relation between the quantities used in (14.305) and (14.220) is $\Omega|_{\text{here}} = Z^0 \Omega|_{\text{there}}$, $k_I = \left(\frac{1}{Z^0}(1 - t^a k_a), \frac{1}{Z^0} k_a\right)$ and $\chi_I = (-t^a \chi_a, \chi_a)$.

$$A_I{}^J = -\left[(\text{Re}\mathscr{F})(\text{Im}\mathscr{F})^{-1}\right]_I{}^J + \frac{1}{\langle Z|\bar{Z}\rangle}(\mathscr{F}_I \bar{Z}^J + \bar{\mathscr{F}}_I Z^J),$$

$$B_{IJ} = \left[\text{Re}\mathscr{F}(\text{Im}\mathscr{F})^{-1}\text{Re}\mathscr{F} + \text{Im}\mathscr{F}\right]_{IJ} - \frac{1}{\langle Z|\bar{Z}\rangle}(\mathscr{F}_I \bar{\mathscr{F}}_J + \bar{\mathscr{F}}_I \mathscr{F}_J),$$

$$C^{IJ} = -(\text{Im}\mathscr{F}^{-1})^{IJ} + \frac{1}{\langle Z|\bar{Z}\rangle}(Z^I \bar{Z}^J + \bar{Z}^I Z^J),$$

$$D = -A^\text{T}, \tag{14.311}$$

where

$$\langle Z|\bar{Z}\rangle = \mathscr{F}_{IJ} \bar{Z}^I Z^J. \tag{14.312}$$

If we define

$$N_{IJ} = \bar{\mathscr{F}}_{IJ} + \frac{2i}{\langle Z|\bar{Z}\rangle}(\text{Im}\mathscr{F}_{IK}) Z^k (\text{Im}\mathscr{F}_{JL}) Z^L, \tag{14.313}$$

A, B, C, D can be rewritten as

$$A = \text{Re}N \, (\text{Im}N)^{-1},$$
$$B = -\text{Im}N - \text{Re}N \, (\text{Im}N)^{-1} \, \text{Re}N = B^\text{T},$$
$$C = (\text{Im}N)^{-1} = C^\text{T},$$
$$D = -A^\text{T}. \tag{14.314}$$

Further Reading

The computation of beta-functions is reviewed in

- A.A. Tseytlin, Sigma model approach to string theory. Int. J. Mod. Phys. A **4**, 1257 (1989)
- C. Callan, L. Thorlacius, *Sigma Models and String Theory*, in *Particles, Strings and Supernovae*; eds. A. Jevicki, C.-I. Tan, World Scientific 1989, p. 795

T-duality of RR-fields are discussed in the book by Ortin. Further references are

- A. Giveon, M. Porrati, E. Rabinovici, Target space duality in string theory. Phys. Rep. **244**, 77 (1994) [arXiv:hep-th/9401139]
- H. Dorn, H.J. Otto, On T-duality for open strings in general abelian and nonabelian gauge field backgrounds. Phys. Lett. B **381**, 81 (1996) [arXiv:hep-th/9603186]
- E. Alvarez, L. Alvarez-Gaumé, Y. Lozano, An introduction to T duality in string theory. Nucl. Phys. Proc. Suppl. **41**, 1 (1995) [arXiv:hep-th/9410237]

14.8 Appendix: Some Riemannian Geometry, Hodge Duals, etc.

- S.F. Hassan, Supersymmetry and the systematics of T duality rotations in type II superstring theories. Nucl. Phys. Proc. Suppl. **102**, 77–82 (2001) [arXiv: hep-th/0103149]

Good references on Calabi-Yau manifolds which are written physicists, besides the relevant chapters in the second volume of Green, Schwarz and Witten, are

- P. Candelas, *Lectures on Complex Manifolds*, in *Superstrings '87*, eds. L. Alvarez-Gaume et al, World Scientific, Singapore, 1988, pp. 1–88
- T. Hübsch, *Calabi-Yau Manifolds: A Bestiary for Physicists* (World Scientific, Singapore, 1992)
- K. Hori et al., *Mirror Symmetry*. Clay Mathematics Monographs, vol. 1 (AMS, Providence, 2003)

Of the original papers on CY compactifications we particularly recommend

- P. Candelas, G.T. Horowitz, A. Strominger, E. Witten, Vacuum configurations for superstrings. Nucl. Phys. B **258**, 46 (1985)
- A. Strominger, E. Witten, New manifolds for superstring compactification. Comm. Math. Phys. **101**, 341 (1985)

Mathematical references for most of the concepts used here (and much more) are

- K. Kodaira, *Complex Manifolds and Deformation of Complex Structures*, Ergebnisse der Mathematik und Ihrer Grenzgebiete, Vol. 10, Springer 1987, reprinted as Classics in Mathematics, Springer 2005
- D. Huybrechts, *Complex Geometry* (Springer, Berlin, 2005)
- P. Griffiths, J. Harris, *Principles of Algebraic Geomeyty* (Wiley, NY, 1978)
- A.L. Besse, *Einstein Manifolds*, Grundlehren der mathematischen Wissenschaften, Vol. 283, Springer 1986, reprinted as Classics in Mathematics, Springer 2008

The first of these books, whose notation and conventions we mostly follow, is very explicit. We also recommend Huybrechts for its readability.

That α'-corrections of the β-function modify the Ricci-flatness condition without changing the first Chern-class was shown in

- D. Nemeschansky, A. Sen, Conformal invariance of supersymmetric σ-models on Calabi-Yau manifolds. Phys. Lett. **B178**, 365 (1986)
- P.S. Howe, G. Papadopoulos, K.S. Stelle, Quantizing the N=2 super sigma model in two-dimensions. Phys. Lett. **B176**, 405–410 (1986)

The $\mathcal{N} = 2$ SUSY restrictions on the effective action were found in

- B. de Wit, A. Van Proeyen, Potentials and symmetries of general gauged N=2 supergravity – Yang-Mil ls models. Nucl. Phys. **B245**, 89–117 (1984)
- B. de Wit, P.G. Lauwers, A. Van Proeyen, Lagrangians of N=2 supergravity – matter systems. Nucl. Phys. **B255**, 569–608 (1985)

- E. Cremmer, C. Kounnas, A. Van Proeyen, J.P. Derendinger, S. Ferrara, B. de Wit, L. Girardello, Vector multiplets coupled to N=2 supergravity: Superhiggs effect, flat potentials and geometric structure. Nucl. Phys. **B250**, 385–426 (1985)

Special geometry was further developed in

- A. Strominger, Special geometry. Comm. Math. Phys. **133**, 163–180 (1990)
- P. Candelas, X. de la Ossa, Moduli space of Calabi-Yau manifolds. Nucl. Phys. **B355**, 455–481 (1991)

Our discussion of the moduli space of CY compactifications relies heavily on the last reference.

Special geometry was derived from the structure of CFT correlation functions in

- L.J. Dixon, V. Kaplunovsky, J. Louis, On effective field theories describing (2,2) vacua of the heterotic string. Nucl. Phys. **B329**, 27–82 (1990)

The discussion of the stabilization of the complex structure moduli of the heterotic string on Calabi-Yau manifolds follows

- L.B. Anderson, J. Gray, A. Lukas, B. Ovrut, Stabilizing the complex structure in heterotic Calabi-Yau vacua. JHEP **1102**, 088 (2011) [arXiv:1010.0255 [hep-th]]

Chapter 15
CFTs for Type II and Heterotic String Vacua

Abstract In this chapter we present various conformal field theory constructions which describe string theories in six and four space-time dimensions. We start with some general comments about strings moving in compactified spaces and then continue our investigation from Sect. 10.5 on strings in orbifold spaces. We then generalize the construction of non-oriented string theories to compact dimensions and discuss the prototype example of an orientifold on the compact space $T^4/\mathbb{Z}_2$. In this model we introduce fractional D-branes to cancel the tadpoles. Next, on a more abstract level, we outline the general structure a CFT must at least have in order to lead to a space-time supersymmetric compactification. Finally, we provide two concrete four dimensional realizations in terms of certain classes of $N = 2$ superconformal field theories. The first are the so-called Gepner models and the second are heterotic generalizations of the covariant lattice approach from Chap. 13.

15.1 Motivation

In the previous chapter we discussed supersymmetric compactifications of type II and heterotic strings. Our emphasis there was on target space, i.e. geometric, aspects rather than on world-sheet, i.e. CFT, methods. In this chapter we will discuss supersymmetric compactifications from the CFT point of view. While the geometric approach allows the use of powerful mathematical tools to compute spectra of massless particles and, to some extend, also couplings in the low-energy effective action, it largely ignores typical stringy aspects, as it assumes that the internal geometry is large compared to the string scale. This is justified as far as topological properties of the manifold are concerned. For instance, the massless spectrum of CY compactifications could be reliably computed but their interactions in the low-energy effective action depends on the size of the manifold and typical string effects are suppressed in the supergravity approximation. Furthermore, we already know from the construction of the heterotic string that strings can be consistently described in non-geometric backgrounds. This includes some of the covariant lattices, asymmetric orbifolds and also spaces with more general background fields. Such non-geometric compactifications are possible due to the decoupling of left- and right-moving degrees of freedom which may therefore live on different background spaces. The appropriate tool for studying these backgrounds is their formulation as CFTs. CFT methods are also appropriate to probe string scale geometries where geometric notions and intuition are likely to fail. The drawback of the CFT description of string vacua is that one has a manageable description, e.g. in terms of free fields or in terms of tensor products of minimal models, only at special points in the moduli space. Orbifolds and Gepner models are examples. At these points explicit calculations of correlation functions are possible.

Phenomenologically interesting compactifications are those with $\mathcal{N} = 1$ space-time supersymmetry in four dimensions. This can be achieved by considering toroidal orbifolds of the heterotic string or, for type II strings, by performing in addition an orientifold projection, under which only one linear combination of the space-time supercharges, which arise from the left- and the right-moving sectors, is invariant.

15.2 Supersymmetric Orbifold Compactifications

In this section we resume the discussion of orbifolds which we started in Chap. 10. We will first determine the condition under which some space-time supercharges survive the orbifold projection and then turn to heterotic orbifolds. Once again we will mostly consider symmetric toroidal $\mathbb{Z}_N$ orbifolds.

From our discussion of Calabi-Yau manifolds we know that supersymmetric compactifications require restricted holonomy groups. For instance, compactification on manifolds with $D = 4$ and $D = 6$ with minimal supersymmetry requires the holonomy groups to be $SU(2)$ and $SU(3)$, respectively. For the six-dimensional orbifold this means that the point group P must be a discrete subgroup of $SU(3)$ such that $P \not\subset SU(2)$.

We can show this explicitly as follows. We will specify to $D = 6$, the results for $D = 2, 4$ then come as by-products. As in Chap. 10, Eq. (10.153), we introduce complex coordinates. The action of $\mathbb{Z}_N$ on the complex coordinates is specified by the twist vector v. Supersymmetry demands the existence of a Killing spinor ϵ. On the orbifold, which is flat away from the fixed loci of G, the condition for ϵ to be a Killing spinor is $\theta \epsilon = \epsilon$. Here θ is an $SO(6)$ rotation with eigenvalues $e^{\pm 2\pi i v_i}$ in the vector representation of $SO(6)$, whose six weights are $(\underline{\pm 1, 0, 0})$ where the underlining means to include permutations. We can write θ as (cf. 10.173)

$$\theta = \exp(2\pi i (v_1 J_{12} + v_2 J_{34} + v_3 J_{56})), \qquad (15.1)$$

where $J_{2i-1,2i}$ are the generators of the Cartan subalgebra. Since the spinor weights of $SO(6)$ are $(\pm\frac{1}{2}, \pm\frac{1}{2}, \pm\frac{1}{2})$, in the spinor representation the group action θ has eigenvalues $\exp(i\pi(\pm v_1 \pm v_2 \pm v_3))$. Hence to have invariant spinors we need

15.2 Supersymmetric Orbifold Compactifications

Table 15.1 Supersymmetric $\mathbb{Z}_N$ actions in $D = 6$ with minimal supersymmetry, represented by the shift vector $\boldsymbol{v} = (v_i)$

$\mathbb{Z}_3$	$\frac{1}{3}(1, 1, -2)$	$\mathbb{Z}_6'$	$\frac{1}{6}(1, -3, 2)$	$\mathbb{Z}_8'$	$\frac{1}{8}(1, 3, -2)$
$\mathbb{Z}_4$	$\frac{1}{4}(1, 1, -2)$	$\mathbb{Z}_7$	$\frac{1}{7}(1, 2, -3)$	$\mathbb{Z}_{12}$	$\frac{1}{12}(1, -5, 4)$
$\mathbb{Z}_6$	$\frac{1}{6}(1, 1, -2)$	$\mathbb{Z}_8$	$\frac{1}{8}(1, 3, -4)$	$\mathbb{Z}_{12}'$	$\frac{1}{12}(1, 5, -6)$

$$\pm v_1 \pm v_2 \pm v_3 = 0 \bmod 2 \tag{15.2}$$

for some choice of signs. This condition guarantees that the holonomy group is contained in $SU(3)$. In Chap. 10 we showed that this condition also guarantees that the one-loop partition function vanishes. In fact, orbifolds which admit Killing spinors are singular limits of smooth Calabi-Yau manifolds. Conversely, one can 'repair' the orbifold singularities to produce smooth manifolds of $SU(3)$ holonomy. To do this one excises the singular points or surfaces and replaces them by plugs that patch the holes smoothly. This can be done in such a way that the result is a smooth Calabi-Yau manifold, i.e. a smooth manifold with vanishing first Chern class. This procedure is also called a resolution of the (orbifold) singularity.

Recall from Chap. 10 the condition (10.174) on the shift vector. Imposing all the conditions on $\boldsymbol{v}$ (see also Table 10.1) one finds a small number of possible supersymmetric $\mathbb{Z}_N$ actions. If $v_3 = 0$ the only solution is $v_1 = -v_2 = 1/N$, $N = 2, 3, 4, 6$. The associated four-dimensional orbifolds are singular limits of the K3 surface. The inequivalent possibilities with all three $v_i \neq 0$, which lead to minimal supersymmetry in $d = 4$, i.e. $\mathcal{N} = 2$ for the type II and $\mathcal{N} = 1$ for the heterotic string, are summarized in Table 15.1.

Note that in all cases the holomorphic three form of the torus

$$\Omega = dz^1 \wedge dz^2 \wedge dz^3 \tag{15.3}$$

is left invariant under $z^j \to e^{2\pi i v_j} z^j$. The holomorphic three-from of the torus thus descends to the holomorphic three-form of the Calabi-Yau orbifold. In fact, in all cases it is straightforward to work out which elements of the Dolbeault cohomology of the torus survive the orbifold projection. This is referred to as the untwisted cohomology of the orbifold. There will be additional contributions from the twisted sectors which are related to moduli controlling the repair of the orbifold singularities. They can be determined from a careful analysis of the massless spectrum, which will be done in two examples below.

Once the allowed twist vectors have been found, one still needs to specify the lattice and the $\mathbb{Z}_N$ automorphism, i.e. one has to find lattices on which θ can be represented as an integer transformation which leaves the lattice metric invariant. It turns out that the lattices Λ such that $T^6 = \mathbb{R}^6/\Lambda$, can all be represented as Lie algebra root lattices. A detailed analysis, which we will not present, shows that there are 18 inequivalent possibilities. They are listed in Table 15.2. $h_0^{p,q}$ are the contributions to the Hodge numbers from the untwisted sector. In the last column

Table 15.2 The 18 supersymmetric, crystallographic $\mathbb{Z}_N$ orbifolds of T^6

Case	$\mathbf{v}=(v^1,v^2,v^3)$	Lie algebra root lattice	$h^{1,1}$	$(h_0^{1,1})$	$h^{1,2}$	$(h_0^{1,2})$	Orbifold action Θ
1	$\mathbb{Z}_3$ $(\frac{1}{3},\frac{1}{3},-\frac{2}{3})$	$A_2 \times A_2 \times A_2$	36	(9)	0	(0)	ω
2	$\mathbb{Z}_4$ $(\frac{1}{4},\frac{1}{4},-\frac{1}{2})$	$A_1 \times A_1 \times B_2 \times B_2$	31	(5)	7	(1)	ω
3		$A_1 \times A_3 \times B_2$	27	(5)	3	(1)	ω
4		$A_3 \times A_3$	25	(5)	1	(1)	ω
5	$\mathbb{Z}_6$ $(\frac{1}{6},\frac{1}{6},-\frac{1}{3})$	$A_2 \times G_2 \times G_2$	29	(5)	5	(0)	ω
6		$G_2 \times A_2 \times A_2$	25	(5)	1	(0)	$\Gamma_1\Gamma_2\Gamma_3\Gamma_4 P_{36} P_{45}$
7	$\mathbb{Z}_6'$ $(\frac{1}{6},\frac{1}{3},-\frac{1}{2})$	$A_1 \times A_1 \times A_2 \times G_2$	35	(3)	11	(1)	ω
8		$A_2 \times D_4$	29	(3)	5	(1)	ω
9		$A_1 \times A_1 \times A_2 \times A_2$	31	(3)	7	(1)	$\Gamma_1\Gamma_2\Gamma_3\Gamma_4 P_{36} P_{45}$
10		$A_1 \times A_5$	25	(3)	1	(1)	ω
11	$\mathbb{Z}_7$ $(\frac{1}{7},\frac{2}{7},-\frac{3}{7})$	A_6	24	(3)	0	(0)	ω
12	$\mathbb{Z}_8$ $(\frac{1}{8},\frac{1}{4},-\frac{3}{8})$	$B_2 \times B_4$	27	(3)	3	(0)	ω
13		$A_3 \times A_3$	24	(3)	0	(0)	$\Gamma_1\Gamma_2\Gamma_3 P_{16} P_{25} P_{34}$
14	$\mathbb{Z}_8'$ $(\frac{1}{8},\frac{3}{8},-\frac{1}{2})$	$B_4 \times D_2$	31	(3)	7	(1)	ω
15		$A_1 \times D_5$	27	(3)	3	(1)	ω
16	$\mathbb{Z}_{12}$ $(\frac{1}{12},\frac{1}{3},-\frac{5}{12})$	$A_2 \times F_4$	29	(3)	5	(0)	ω
17		E_6	25	(3)	1	(0)	ω
18	$\mathbb{Z}_{12}'$ $(\frac{1}{12},\frac{5}{12},-\frac{1}{2})$	$D_2 \times F_4$	31	(3)	7	(1)	ω

we have specified the orbifold action. Γ_i is a Weyl reflection on the simple root $\boldsymbol{\alpha}_i$ and P_{ij} the exchange of the two simple roots $\boldsymbol{\alpha}_i$ and $\boldsymbol{\alpha}_j$. For 15 cases the orbifold action can be realized as the Coxeter element $\omega = \Gamma_1\Gamma_2\Gamma_3\Gamma_4\Gamma_5\Gamma_6$ acting on the root lattice of an appropriate Lie algebra. For the remaining three, the orbifold action is realized as a combination of Weyl reflections and outer automorphisms acting on the Lie algebra root lattice.

The mass spectrum can be read off from the partition function whose computation was outlined in Chap. 10. In the general case the situation is complicated by the presence of fixed tori in some twisted sectors, which leads to quantized momenta and winding modes. If they are absent one finds that the masses of states in the k-th twisted sector are of the general form

$$\frac{\alpha'}{2} m_R^2(\theta^k) = \langle L_0 \rangle - \frac{1}{2} = N_R + \frac{1}{2}(\boldsymbol{r}+k\boldsymbol{v})^2 + E_k - \frac{1}{2} \tag{15.4}$$

with a similar expression for the left-moving sector of the type II string. Here E_k is as in (10.166) and $\boldsymbol{r}$ is an $SO(8)$ weight vector. This result has a very natural interpretation in the light of the discussion in Chap. 13. If we extend the D_4 weight vector $\boldsymbol{r}$ to a $D_{5,1}$ weight vector $\boldsymbol{w} = (\boldsymbol{r},\boldsymbol{x}_o)$, the shift vector $\boldsymbol{v}$ to $\boldsymbol{V} = (\boldsymbol{v},0,0)$ and $\boldsymbol{\phi}$ to $\boldsymbol{\Phi} = (\boldsymbol{\phi},\phi_0,-i\phi)$, this is simply the condition that the conformal weight, i.e. L_0 eigenvalue, of a vertex operator of the form

$$e^{i(\boldsymbol{w}+k\boldsymbol{V})\cdot\boldsymbol{\Phi}(z)} e^{ik\cdot X(z)} \sigma_k \tag{15.5}$$

15.2 Supersymmetric Orbifold Compactifications

be one. Possible oscillator contributions are not shown and the space-time momentum k^μ along the uncompactified directions should not be confused with the label k of the twisted sector. σ_k is a twist field for the bosons which creates the twisted sector vacuum $|0\rangle_k = \sigma_k(0)|0\rangle$ out of the $SL(2,\mathbb{C})$ invariant vacuum $|0\rangle$. Its conformal dimension is E_k. Indeed, the computation of E_k as the vacuum expectation value of L_0 in the twisted vacuum can be interpreted as the computation of the three point function $\langle 0|\sigma_k^\dagger(\infty)T(z)\sigma_k(0)|0\rangle$. The twist field for the RNS fermions is $e^{ik\boldsymbol{v}\cdot\boldsymbol{\phi}}$ which is also part of (15.5). The θ-eigenvalue of the state created by the vertex operator (15.5) is $\exp(2\pi i(\boldsymbol{r}+k\boldsymbol{v})\cdot\boldsymbol{v})$, again neglecting the contribution from oscillators. The oscillators in the expansion of ϕ^i have integer frequencies. The twisting is realized as a shift of the $SO(8)$ weight lattice; see also the discussion of the heterotic string below. N_R in (15.4) is the oscillator occupation number which receives integer contributions from the fermionic degrees of freedom (i.e. the $\boldsymbol{\phi}$ oscillators) and fractional contributions (multiples of $1/N$) from the Z oscillators.

To find physical states we have to impose the level matching constraint $m_L^2 = m_R^2$ and project onto invariant states. In type IIB theory the states which survive the GSO projection satisfy $\sum r_a = $ odd for all states, while in type IIA they satisfy $\sum r_a = $ even for the state in the left-moving Ramond sector.

We now turn to the discussion of heterotic orbifolds. This requires that we specify the action of the orbifold group on the gauge degrees of freedom. They are represented as level one Kač-Moody algebras. Just as the space-group S with elements (θ, v) was a symmetry of the geometric background, its realization $\gamma(\theta, v)$ on the gauge part should be an automorphism of the corresponding Lie algebra. If we discard the possibility of an outer automorphism which exchanges the two E_8 factors, γ must be an element of the heterotic gauge group, $E_8 \times E_8$ or $\text{Spin}(32)/\mathbb{Z}_2$ and the unbroken gauge group is the commutant of $\gamma(S)$ in $E_8 \times E_8$ or $SO(32)$.

Non-trivial realizations of the lattice Λ are Wilson lines on $T^D = \mathbb{R}^D/\Lambda$. Except for a few comments later on, we will not discuss Wilson lines any further and set them to zero. The realization of the rotations θ can be diagonalized such that $\gamma(\theta): J^a \mapsto e^{2\pi i \eta_a} J^a$ with $N\eta_a \in \mathbb{Z}$ and $\eta_a + \eta_b = \eta_c$, if $f^{abc} \neq 0$. In the untwisted sector only the invariant states survive the orbifold projection. This means in the orbifold theory only the gauge bosons corresponding to the J^a with $\eta_a = 0$ survive. They form a closed subalgebra which generates the gauge group of the orbifold theory. In the θ-twisted sector the boundary conditions of the left-moving Kač-Moody currents are

$$J^a(e^{-2\pi i}\bar{z}) = e^{2\pi i \eta_a} J^a(\bar{z}) \tag{15.6}$$

which leads to mode expansions with mode numbers $J^a_{m+\eta_a}$.

We have encountered two formulations of the heterotic string. In the bosonic formulation the currents J^a were presented in terms of 16 left-moving bosons X^I which span the $E_8 \times E_8$ or $\text{Spin}(32)/\mathbb{Z}_2$ weight lattices while in the fermionic formulation they were constructed with 32 left-moving fermions $\bar{\lambda}^A$ which transform as vectors of $SO(16) \times SO(16) \subset E_8 \times E_8$ and $SO(32)$, respectively. In the fermionic formulation we can easily present $\gamma(\theta)$ by embedding $\theta \in O(D)$ in $SO(16) \times SO(16)$ or $SO(32)$. We will not discuss this construction any further but concentrate on the bosonic construction.

For $\mathbb{Z}_N$ orbifolds where the space-group is generated by a single element θ with $\theta^N = 1$, we can always choose $\gamma(\theta)$ to lie in the maximal torus which is generated by the Cartan subalgebra. This means that γ commutes with the Cartan currents $i\bar{\partial} X^I$ and restricts the action of γ to a constant shift of the chiral bosons X^I, i.e.

$$X^I \mapsto \gamma X^I = X^I + V^I . \tag{15.7}$$

Therefore, in the sector twisted by $\gamma(\theta^k)$ the bosons satisfy

$$X^I(\sigma + 2\pi) = X^I(\sigma) + kV^I \bmod \Lambda , \tag{15.8}$$

where $\theta^N = 1$ requires that the shift vector V^I satisfies

$$N V^I \in \Lambda . \tag{15.9}$$

Λ is the weight lattice of $E_8 \times E_8$ or $\mathrm{Spin}(32)/\mathbb{Z}_2$. These weight lattices were characterized in Chap. 10.

The condition (15.8) leads to integer moded oscillators, which implies that in the partition function the internal oscillators contribute as before, whereas the momentum lattice sum in the k-th twisted sector is over shifted vectors kV^I. Therefore, the partition function is as in the type II string except that the left-moving part is replaced by

$$\frac{1}{\bar{\eta}^2} \frac{\bar{\eta}^3}{\prod\limits_{i=1}^{3} \bar{\vartheta}\left[\frac{\frac{1}{2}+kv_i}{\frac{1}{2}+\ell v_i}\right]} \left(\frac{1}{\eta^8} \sum_{\alpha,\beta} \prod_{I=1}^{8} \vartheta\left[\begin{smallmatrix}\alpha+kV^I\\ \beta+\ell V^I\end{smallmatrix}\right]\right) \left(\frac{1}{\eta^8} \sum_{\alpha,\beta} \prod_{I=1}^{8} \vartheta\left[\begin{smallmatrix}\alpha+kV'^I\\ \beta+\ell V'^I\end{smallmatrix}\right]\right), \tag{15.10}$$

where we have specified to the $E_8 \times E_8$ heterotic string where (V^I, V'^I) are the shift vectors in the two E_8 factors. The first factor is the contribution from the two non-compact transverse space-time bosons, the second factor from the six internal right-moving bosons and the last two factors from the sums of the E_8 lattices shifted by V^I and V'^I, respectively.

The mass formula for the right-movers is (15.4), while for the left movers one finds

$$\frac{\alpha'}{2} m_L^2(\theta^k) = \langle \bar{L}_0 \rangle - 1 = N_L + \frac{1}{2}(\boldsymbol{P} + k\boldsymbol{V})^2 + E_k - 1 , \tag{15.11}$$

15.2 Supersymmetric Orbifold Compactifications

where P is a lattice vector and E_k is as in (15.4). N_L is the total oscillator number. The contribution from the bosons X^I is integer while that of the twisted bosons Z^i is a multiple of $1/N$. Note that Eq. (15.11) is correct in the absence of Wilson lines and fixed tori. We again have to impose level matching $m_L^2 = m_R^2$. This can be written in the form

$$N_R - N_L - \frac{1}{2}P^2 + \frac{1}{2}r^2 - kP \cdot V + kr \cdot v + \frac{1}{2} = \frac{1}{2}k^2(V^2 - v^2) \,. \quad (15.12)$$

If one uses that the gauge lattice is even and self-dual, condition (10.174) and that states which survive the GSO projection satisfy $r^2 = \text{odd}$, one can show that the l.h.s. of this equations is an integer mod $1/N$. Then level matching can only be achieved if, in addition to (10.174) and (15.9), the v_i and V^I satisfy the conditions

$$N \sum v_i = N \sum V^I = N \sum V'^I = 0 \quad \text{mod } 2 \quad (15.13)$$

and

$$N(v^2 - V^2 - V'^2) = 0 \quad \text{mod } 2 \,. \quad (15.14)$$

(For the $SO(32)$ heterotic string $V' = 0$.) These conditions are also be obtained from the requirement of modular invariance. For instance, the boundary conditions $(h, g) = (\theta, \mathbb{1})$ transform into themselves under $\tau \to \tau + N$. The partition function in this sector, $Z[\theta, \mathbb{1}] = \text{Tr}\,(q^{L_0(\theta) - \frac{c}{24}} \bar{q}^{\bar{L}_0(\theta) - \frac{\bar{c}}{24}})$, is only invariant if $(\langle L_0 \rangle - \frac{1}{2}) - (\langle \bar{L}_0 \rangle - 1) = 0 \mod 1/N$, which is the condition we have just analyzed. Note that $V = 0$ is not a solution (unless $v = 0$), i.e. the gauge symmetry is necessarily broken. An easy way to satisfy (15.14) is to choose $V = (v_1, v_2, v_3, 0, \ldots, 0)$ and $V' = 0$. It is called the standard embedding of the point group into the gauge lattice. This is orbifold version of the embedding of the spin connection in the gauge connection which we discussed in Chap. 14. While for smooth CY compactifications this leads to the gauge group $E_6 \times E_8$, in the orbifold limit it is enhanced to at least $E_6 \times G \times E_8$ where G can be $SU(3)$, $SU(2) \times U(1)$ or $U(1) \times U(1)$, depending on the shift vector. The blow-up modes, which are the elements of the cohomology in the twisted sectors, are charged under G. In the effective field theory description, the blowing up procedure means that these scalar fields acquire a non-vanishing vev and G is broken.

So far we have neglected the possibility of non-trivial Wilson lines. They arise, if we choose non-trivial embeddings of the space group into the gauge group and not just of the point group. One distinguishes between Abelian and non-Abelian embeddings. In the Abelian embedding $g = (\theta, v_i) \mapsto \gamma = (V^I, a^I)$ the full space-group is realized as shifts of the gauge lattice, i.e. $\gamma X^I = X^I + V^I + a^I$. It preserves the rank of the gauge group. The group law and modular invariance restricts the allowed values of the Wilson lines. Consider for instance for the two-dimensional $\mathbb{Z}_3$ orbifold $\mathscr{O} = \mathbb{R}^2/S = T^2/\mathbb{Z}_3$ where the space group is generated by the $SU(3)$ root lattice with generators (e_1, e_2) and $2\pi/3$ rotations θ with $\theta^3 = \mathbb{1}$ and $\theta e_1 = e_2$. This requires that the two Wilson lines along the two cycles a_1 and a_2 of the torus,

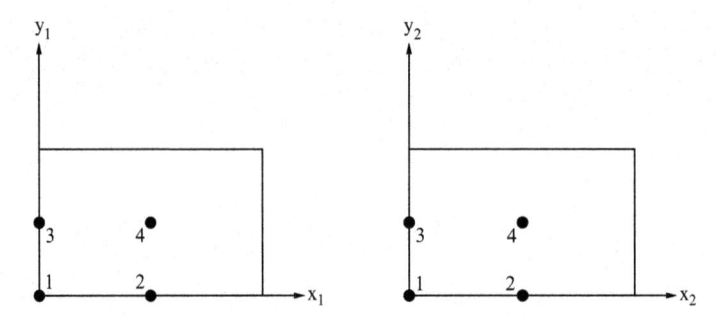

Fig. 15.1 Geometry of the torus T^4

$\int_{a_i} A^I = 2\pi a_i^I$ are the same. Furthermore, any element of the space group satisfies $g^3 = (\theta, e_i)^3 = (\theta^3, (\theta^2 + \theta + \mathbb{1})e_i) = (\mathbb{1}, 0)$ which demands that $3a^I$ is a vector in the gauge lattice, i.e. the Wilson lines are quantized. The non-Abelian embedding $g = (\theta, v_i) \mapsto \gamma = (\Theta, a^I)$ generalizes (15.7) to $\gamma : X^I \to (\Theta X)^I + a^I$. E.g. for the $\mathbb{Z}_3$-orbifold, there is again only one independent Wilson line but because of $(\Theta, a^I)^3 = (\Theta^3, \Theta^2 a^I + \Theta a^I + a^I) = (\mathbb{1}, 0)$ for any a^I it is not quantized. This embedding allows for the reduction of the rank of the gauge group and orbifold compactifications with the standard model gauge group have been constructed.

At the beginning of this section we have determined the condition for spacetime supersymmetry: the existence of invariant spinors which, e.g. for the heterotic string, is the right-moving part of the gravitino vertex operator. One might wonder whether additional gravitini can appear in twisted sectors, thus enhancing spacetime supersymmetry. That this is not possible can be seen by considering the left-moving part of the gravitino vertex operator which carries the space-time vector index, i.e. ∂X^μ which already has conformal dimension one, without any room for a contribution from the twist fields. The same argument forbids gauge bosons arising from the twisted sectors.

We now turn to examples. We begin with the compactification of type II string theory to six dimensions on $T^4/\mathbb{Z}_2$ and will then consider the type II and the heterotic strings to four dimensions on $T^6/\mathbb{Z}_3$. No Wilson lines will be turned on.

The $T^4/\mathbb{Z}_2$ Orbifold

We consider the type IIB superstring compactified on a four-dimensional torus which, for simplicity, we assume to be factorized, i.e. $T^4 = T^2 \times T^2$. Furthermore, we take each T^2 to be generated by a rectangular lattice. We introduce complex coordinates $z_1 = x_1 + iy_1$ and $z_2 = x_2 + iy_2$ and define the orbifold action as

$$I_4 : z_i \to -z_i, \qquad \text{for } i = 1, 2, \qquad (15.15)$$

i.e. $v = (+\frac{1}{2}, -\frac{1}{2})$. The geometry of this torus is shown in Fig. 15.1. As indicated in the figure, there exist 16 fixed points of the $\mathbb{Z}_2$ action. To simplify notation, we

15.2 Supersymmetric Orbifold Compactifications

Table 15.3 GSO projected chiral closed string sector ($s_i = \pm\frac{1}{2}$)

State	$SU(2) \times SU(2)$	I_4
Untwisted NS-sector		
$\|\pm 1, 0; 0, 0\rangle$	$(2, 2)$	$+1$
$\|0, 0; \pm 1, 0\rangle$	$4 \times (1, 1)$	-1
Untwisted R-sector		
$\|s_1, s_2; s_3, s_4\rangle$ $s_1 = +s_2; s_3 = -s_4$	$2 \times (2, 1)$	-1
$\|s_1, s_2; s_3, s_4\rangle$ $s_1 = -s_2; s_3 = +s_4$	$2 \times (1, 2)$	$+1$
Twisted NS-sector		
$\|0, 0; s_3, s_4\rangle$ $s_3 = s_4$	$2 \times (1, 1)$	$+1$
Twisted R-sector		
$\|s_1, s_2; 0, 0\rangle$ $s_1 = s_2$	$(2, 1)$	$+1$

Table 15.4 $D = 6$ massless multiplets with 16 supercharges

	(2,0) SUSY
Gravity multiplet	(3,3) + 4 (2,3) + 5 (1,3)
Tensormultiplet	(3,1) + 4 (2,1) + 5 (1,1)
	(1,1) SUSY
Gravity multiplet	(3,3) + (3,1) + (1,3) + 4 (2,2) + (1,1)
	+ 2 (3,2) + 2 (2,3) + 2 (2,1) + 2 (1,2)
Vectormultiplet	(2,2) + 4 (1,1) + 2 (2,1) + 2 (1,2)

choose the two lattices to be equal, i.e. we take the two x (y)-cycles to have radii R_x (R_y). We will also use dimensionless radii $\rho_x = R_x/\sqrt{\alpha'}$ and $\rho_y = R_y/\sqrt{\alpha'}$. The $\mathbb{Z}_2$ quotient breaks the supersymmetry by half, i.e. 16 of the 32 supercharges of the type II theory on T^4 are invariant under the orbifold projection. Using the methods from Chap. 9 it is straightforward to compute the torus partition function

$$Z(\tau, \bar{\tau}) = \sum_{h=1, I_4} \text{Tr}_h \left(\frac{1 + I_4}{2} P_{\text{GSO}} \, q^{L_0(h) - \frac{c}{24}} \, \bar{q}^{\bar{L}_0(h) - \frac{c}{24}} \right) \qquad (15.16)$$

of this model from which we can extract information on the massless spectrum. Let us construct the massless states explicitly. They are classified with respect to the little group $SO(4) \simeq SU(2) \times SU(2)$. The states with $m_R^2 = 0$ are summarized in Table 15.3 where we also include their I_4 charges.

Before we present the complete massless spectrum, we list in Table 15.4 the possible massless multiplets in $D = 6$ with 16 supercharges. We have to distinguish between chiral (2, 0) and non-chiral (1, 1) supersymmetry. Their physical degrees of freedom are given as $SU(2) \times SU(2)$ representations (m,n) with $m+n$ even (odd) for space-time bosons (fermions). We note that the tensor in the (2,0) tensormultiplet is self-dual while the tensor in the (2,0) gravity multiplet is anti-self-dual.

The massless spectrum of the type IIB string compactified on the $T^4/\mathbb{Z}_2$ orbifold is found by taking the I_4 invariant states in the tensor product of the states in Table 15.3. One finds

$$\begin{aligned}&\text{untwisted sector:} &&\text{gravity} + 5 \text{ tensormultiplets} \\ &\text{twisted sector:} &&16 \text{ tensormultiplets}\end{aligned} \qquad (15.17)$$

where the multiplicity in the twisted sector reflects the 16 fixed points.

For the type IIA string the GSO projection for the left movers is different. In terms of the $SU(2) \times SU(2)$ representation content the only difference is the exchange $(m,n) \to (n,m)$. This leads to the following type IIA spectrum

$$\begin{aligned}&\text{untwisted sector:} &&\text{gravity} + 4 \text{ vectormultiplets} \\ &\text{twisted sector:} &&16 \text{ vectormultiplets}\,.\end{aligned} \qquad (15.18)$$

The $T^4/\mathbb{Z}_2$ orbifold, as well as all other four-dimensional toroidal orbifolds which preserve half of the supercharges, is a singular limit of the (topologically) unique two-dimensional Calabi-Yau manifold, the K3 surface. Its non-trivial Hodge numbers are $h^{0,0} = h^{2,0} = 1$ and $h^{1,1} = 20$. The K3 surface thus has 22 two-cycles, where in the $T^4/\mathbb{Z}_2$ orbifold limit six descent from the T^4 and the remaining 16 arise from the smoothing of the I_4 fixed points. In other words, the $T^4/\mathbb{Z}_2$ orbifold limit is reached from the smooth surface by shrinking 16 two-cycles to zero size. For this orbifold $h_0^{1,1} = 4 = h^{1,1}(T^4)$, i.e. all harmonic $(1,1)$-form on T^4 are invariant under T^4 and so are the unique $(2,0)$ and $(0,2)$ forms. For other orbifold limits of K3, a different number of two-cycles appears in the untwisted sector. E.g. for the $T^4/\mathbb{Z}_3$ orbifold with $v = (+\frac{1}{3}, -\frac{1}{3})$, there are only four.

It is straightforward, at least for the bosonic part, to re-derive the massless spectra (15.17) and (15.18) from dimensional reduction of the ten-dimensional IIB/A supergravity multiplets on the K3 surface.

The $T^6/\mathbb{Z}_3$ Orbifold

We now consider a type II compactification to four dimensions. As torus lattice we choose the direct product of three $SU(3)$ root lattices which has a $\mathbb{Z}_3$ automorphism shown in Fig. 15.2. This corresponds to a partial fixing of the Kähler moduli and to a complete fixing of the complex structure moduli of a generic six-torus which has $h^{1,1}(T^6) = 9$, $h^{2,1}(T^6) = 9$, $h^{1,0} = 3$, $h^{2,0} = 3$ and $h^{3,0} = 1$. The $\mathbb{Z}_3$ action on the complex coordinates is

$$\Theta: \quad z_i \to e^{2\pi i v_i} z_i, \qquad \text{for } i = 1, 2, 3 \qquad (15.19)$$

15.2 Supersymmetric Orbifold Compactifications

Fig. 15.2 The $SU(3)$ root-lattice. The *dots* indicate the three fixed points of the $\mathbb{Z}_3$ automorphism

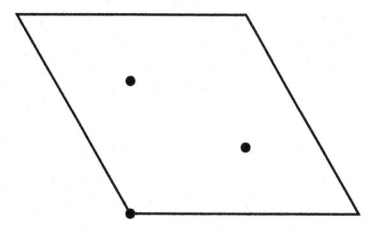

Table 15.5 GSO projected chiral closed string sector of $T^6/\mathbb{Z}_3$ orbifold; $\alpha = e^{2\pi i/3}$

State	Θ	State	Θ
Untwisted NS		Untwisted R	
$\lvert \pm 1; 0,0,0 \rangle$	1	$\lvert \pm (\frac{1}{2}; -\frac{1}{2}, -\frac{1}{2}, -\frac{1}{2}) \rangle$	1
$\lvert 0; \underline{+1, 0, 0} \rangle$	α	$\lvert \frac{1}{2}; \frac{1}{2}, \frac{1}{2}, -\frac{1}{2} \rangle$	α^2
$\lvert 0; \underline{-1, 0, 0} \rangle$	α^2	$\lvert -\frac{1}{2}; \frac{1}{2}, -\frac{1}{2}, -\frac{1}{2} \rangle$	α
θ-twisted NS		θ-twisted R	
$\lvert 0; \frac{1}{3}, \frac{1}{3}, \frac{1}{3} \rangle$	1	$\lvert -(\frac{1}{2}; \frac{1}{6}, \frac{1}{6}, \frac{1}{6}) \rangle$	1
θ^2-twisted NS		θ^2-twisted R	
$\lvert 0; -(\frac{1}{3}, \frac{1}{3}, \frac{1}{3}) \rangle$	1	$\lvert \frac{1}{2}; \frac{1}{6}, \frac{1}{6}, \frac{1}{6} \rangle$	1

Table 15.6 $D = 4$ massless multiplets with eight supercharges

Gravity multiplet	$(\pm 2) + 2(\pm \frac{3}{2}) + (\pm 1)$
Vectormultiplet	$(\pm 1) + 2(\pm \frac{1}{2}) + 2(0)$
Hypermultiplet	$[(\pm \frac{1}{2}) + 2(0)] + \text{h.c.}$

with twist vector $\boldsymbol{v} = (\frac{1}{3}, \frac{1}{3}, -\frac{2}{3})$. This orbifold has 27 fixed points, which are all equivalent, and the orbifold CFT has two isomorphic twisted sectors. There are no fixed tori and therefore the evaluation of the partition function of this orbifold compactification

$$Z(\tau, \bar{\tau}) = \sum_{h=1, \Theta, \Theta^2} \text{Tr}_h \left(\frac{1}{3} (1 + \Theta + \Theta^2) P_{\text{GSO}} \, q^{L_0(h) - \frac{c}{24}} \, \bar{q}^{\bar{L}_0(h) - \frac{c}{24}} \right) \quad (15.20)$$

proceeds along the lines outlined in Chap. 10.

It is again straightforward to determine the massless spectrum explicitly. In Table 15.5 we list the GSO projected sector of one chirality. For the states in the untwisted sector we list $\boldsymbol{r}$ and for the twisted sector states we give $(\boldsymbol{r} + k\boldsymbol{v})$, where $\boldsymbol{r}$ is an $SO(8)$ weight vector with $\sum r_i = $ odd (this is the GSO projection) and $\boldsymbol{v}$ is the twist vector. The underlining means to include all states obtained by permutation. We also use $E_1 = E_2 = \frac{1}{3}$.

The massless spectrum is obtained from the tensor product of left- and right-movers. They can be organized into massless $\mathcal{N} = 2$ supermultiplets in $D = 4$. In Table 15.6 we display their propagating helicity states.

For the type IIB string, with the same GSO projections for left- and right-movers and taking into account the fixed point multiplicity, we find the following invariant spectrum

untwisted sector: gravity + 10 hypermultiplets

twisted sectors: 27 hypermultiplets .   (15.21)

One of the hypermultiplets in the untwisted sectors contains as bosonic degrees of freedom the (dualized) anti-symmetric tensor field and the dilaton from the (NS,NS) sector and two real scalars from the (R,R) sector. This is the universal hypermultiplet which is present in every supersymmetric type II Calabi-Yau compactification.

For type IIA we have to change the GSO projection for the left-moving Ramond sector states to $\sum r_i =$ even. This merely reverses the first entry in the R-states in Table 15.5. Combining left- with right-movers leads to the massless spectrum

untwisted sector: gravity + 1 hypermultiplet + 9 vectormultiplets

twisted sectors: 27 vectormultiplets   (15.22)

There is again a universal hypermultiplet in the untwisted sector.

Let us now comment on the massless spectra which we just found. From the discussion of the previous chapter we know that they are those of the type II string compactified on a CY threefold with Hodge numbers $h^{1,1} = 36$ and $h^{2,1} = 0$. Indeed, of the 15 two-forms on the torus, nine are invariant under (15.19). These are the elements of $h_0^{1,1}$. Likewise, two of the 20 three-forms are invariant. They are the $(3, 0)$ and $(0, 3)$ forms. These forms descent to the orbifold where the $(1, 1)$ forms makes their appearance in the massless spectrum as the (NS,NS) scalars of the untwisted (non-universal) hypermultiplets (IIB) and vector-multiplets (IIA). The (NS,NS) scalars of the 27 twisted sector multiplets are in one-to-one correspondence with the Kähler moduli of a smooth Calabi-Yau manifold, which has the $\mathbb{Z}_3$ orbifold as a particular singular submanifold of co-dimension 27 in its 36 dimensional Kähler moduli space. This particular CY manifold has no complex structure moduli. Such Calabi-Yau manifolds are called rigid.

All type II orbifold models preserve at least $\mathcal{N} = 2$ supersymmetry in four-dimensions. To obtain a chiral spectrum, as dictated by particle physics phenomenology, we can at most allow $\mathcal{N} = 1$ supersymmetry. This can be achieved in two different ways: either consider orbifolds of the heterotic string or type II orbifolds with an additional orientifold quotient which leaves invariant only one combination of the left- and right-moving space-time supercharges. In either case, the possible massless multiplets are shown in Table 15.7. Here we consider the heterotic $T^6/\mathbb{Z}_3$ orbifold with standard embedding and postpone the discussion of type II orientifolds to the next section.

The invariant spectrum of the heterotic $\mathbb{Z}_3$ orbifold with standard embedding is also easy to construct explicitly. For the right-movers, the discussion is unchanged as compared to the type II string. For the left-movers we have to look for solutions of

15.2 Supersymmetric Orbifold Compactifications

Table 15.7 $D = 4$ massless multiplets with four supercharges

Gravity multiplet	$(\pm 2) + (\pm \frac{3}{2})$
Vector (gauge) multiplet	$(\pm 1) + (\pm \frac{1}{2})$
Chiral (matter) multiplet	$[(\frac{1}{2}) + (0)] +$ h.c.

Table 15.8 Untwisted states with m_L^2 of the heterotic $T^6/\mathbb{Z}_3$ orbifold; $\boldsymbol{P}$ are E_8 weights

State	N_L	Θ		
		1	α	α^2
Untwisted sector				
$\tilde{\alpha}_{-1}^{\mu}\|0\rangle$, $\mu = 2, 3$	1	2		
$\tilde{\alpha}_{-1}^{i}\|0\rangle$, $i = 1, 2, 3$	1		3	
$\overline{\tilde{\alpha}}_{-1}^{i}\|0\rangle$, $i = 1, 2, 3$	1			3
$\tilde{\alpha}^I$, $I = 1, \ldots, 8$	1	6+2		
$\|P\rangle$, $P^2 = 2$	0	72+6	3×27	3×27
θ-twisted sector				
$\tilde{\alpha}_{-\frac{1}{3}}^{i}\|P+V\rangle$, $(P+V)^2 = \frac{2}{3}$	$\frac{1}{3}$	3×3		
$\|P+V\rangle$, $(P+V)^2 = \frac{4}{3}$	0	27		
θ^2-twisted sector				
$\tilde{\alpha}_{-\frac{1}{3}}^{i}\|P+2V\rangle$, $(P+V)^2 = \frac{2}{3}$	$\frac{1}{3}$	3×3		
$\|P+V\rangle$, $(P+2V)^2 = \frac{4}{3}$	0	27		

$$N_L + \frac{1}{2}(\boldsymbol{P} + k\boldsymbol{V})^2 - \frac{2}{3} = 0 \quad \text{for} \quad k = 0, 1, 2, \tag{15.23}$$

where $\boldsymbol{P}$ is an E_8 weight vector. It is clear that the second E_8 factor is unaffected and simply leads to E_8 vectormultiplets. In the following we consider only the first E_8 factor in which the shift $\boldsymbol{V}$ acts. The states in the untwisted sector are collected in Table 15.8 where the numbers denote the multiplicities. These states combine into the following representations of $E_6 \times SU(3)$: $(\mathbf{78}, \mathbf{1}) + (\mathbf{27}, \mathbf{3}) + (\overline{\mathbf{27}}, \overline{\mathbf{3}})$.

To construct the massless states in the twisted sectors, note that $(\boldsymbol{P} + k\boldsymbol{V})^2 \geq \frac{2}{3}$ for $k = 1, 2$. Furthermore, the non-zero eigenvalues of N_L are $\frac{1}{3} + n$ and $\frac{2}{3} + m$ where n, m are non-negative integers. This leaves the possibilities (i) $N_L = \frac{1}{3}$ and $(\boldsymbol{P} + k\boldsymbol{V})^2 = \frac{2}{3}$ and (ii) $N_L = 0$ and $(\boldsymbol{P} + k\boldsymbol{V})^2 = \frac{4}{3}$. For (i) one finds the states $3 \times [(\mathbf{1}, \overline{\mathbf{3}})_1 + (\mathbf{1}, \mathbf{3})_2]$ where the multiplicity is due to the three possible oscillator directions and for (ii) the states $(\mathbf{27}, \mathbf{1})_1 + (\overline{\mathbf{27}}, \mathbf{1})_2$. The subscripts indicate in which twisted sector the states arise.

We now combine right- and left-movers into invariant states and find the following massless supermultiplets[1]:

untwisted sector: gravity

gauge multiplet of $E_6 \times SU(3) \times E_8$

chiral multiplets: $9 \times (\mathbf{1,1}) + 3 \times (\mathbf{27,3})$ of

twisted sectors: chiral multiplets: $27 \times [3 \times (\mathbf{1,\bar{3}}) + (\mathbf{27,1})]$.   (15.24)

For the matter multiplets we have given the $E_6 \times SU(3)$ representations. The multiplicity in the twisted sector is the number of fixed points. The states in the two twisted sectors are complex conjugates of each other and combine into chiral multiplets. Note that the spectrum is chiral, which is a prerequisite for successful model building.

Altogether there are 36 families, i.e. **27**'s of E_6, nine in the untwisted and 27 in the twisted sectors. Furthermore, the gauge singlet scalars in the untwisted sector are the nine moduli of the torus which survive the orbifold projection. In the twisted sector there are nine E_6 singlets per fixed point. One of these nine scalars corresponds to a Kähler modulus, whose vev is related to the blowing-up of the orbifold singularity. Thus, among the twisted E_6 singlets, 27 are Kähler moduli. When we move away (in moduli space) from the orbifold point, the $SU(3)$ gauge group gets completely broken and eight of the chiral multiplets combine with the $SU(3)$ vector-multiplets into eight massive vector multiplets. This is a string theory realization of the Higgs effect. The remaining 208 massless scalars can be shown to correspond to bundle moduli, i.e. to elements of $H^1(\text{End }T_X)$. Their fate is more complicated. The number of bundle moduli is not constant across the moduli space of X. Some, but not all, can acquire masses due to world-sheet instanton effects. For the $\mathbb{Z}_3$ orbifold it has been shown that, as we move away from the orbifold point, 108 of the bundle moduli become massive. The analysis of bundle moduli is rather technical, even in the case of the standard embedding where the relevant bundle is a deformation of the holomorphic tangent bundle.

15.3 Orientifold Compactifications

A way to reduce the amount of space-time supersymmetry in type II compactifications are orientifolds. Since all salient features can be demonstrated much clearer in an example with six non-compact space-time dimensions, we will discuss an orientifold of the $T^4/\mathbb{Z}_2$ orbifold of the previous section. We will then comment on the generalization to four-dimensional models.

---

[1] The grouping of the states into multiplets of $E_6 \times SU(3)$ can be checked if one uses the explicit expressions for the roots and weights as given e.g. in N. Bourbaki, *Elements of Mathematics, Lie Groups and Lie Algebras*, Chaps. 4–6.

Type IIB Orientifold of the $T^4/\mathbb{Z}_2$ Orbifold

Recall that the $\mathbb{Z}_2$ quotient breaks half of the supersymmetry, i.e. the orbifold preserves 16 of the 32 supercharges. As the uncompactified theory, the type IIB theory on the $T^4/\mathbb{Z}_2$ orbifold also possesses a non-geometric $\mathbb{Z}_2$ symmetry. Gauging this symmetry amounts to performing an orientifold projection under which eight of the 16 supercharges are invariant. We therefore arrive at a chiral $\mathcal{N} = (1,0)$ theory in six dimensions

Define the orientifold action[2] $\Omega R(-1)^{F_L}$ with

$$R : y_i \to -y_i \quad \text{for} \quad i = 1,2. \quad (15.25)$$

Note that this is $z_i \to \bar{z}_i$, i.e. it is different from the orbifold action (15.15).

We need the action of R on the fields $X^\mu, \psi^\mu, \overline{\psi}^\mu$ and also on the spin fields S_α. The action on the states then follows if we have specified the action on the vacuum. Define the reflection matrix $R^\mu{}_\nu = (\delta^\alpha{}_\beta, -\delta^i{}_j)$ where i, j are the reflected directions whose number we keep unspecified for this discussion. Then clearly $X^\mu \to R^\mu{}_\nu X^\nu$ and $\psi^\mu \to R^\mu{}_\nu \psi^\nu$ and likewise for $\overline{\psi}^\mu$. To find the representation of R on the spin fields we proceed as in Sect. 13.3. We make the ansatz $R : S_\alpha \to P_\alpha{}^\beta S_\beta$ and require that, given the transformation of ψ, this is consistent with (13.101). One then obtains $P(R^\mu{}_\nu \Gamma^\nu) = \Gamma^\mu P$ with solution[3]

$$P = \prod \Gamma^i \Gamma, \quad (15.26)$$

where the product is over the reflected directions and $\Gamma = \Gamma^0 \cdots \Gamma^9$ is the chirality operator which commutes with all Γ^μ. Up to a sign, the phase of P is fixed by the requirement that it maps Majorana spinors to Majorana spinors. Of course, for the right-moving spin fields $\overline{S}_\alpha$ we find the same result.

A few comments are in order:

- This is the same P that we found when we specified the boundary conditions of a spin field in Sect. 13.3. There the product was over the Dirichlet directions.
- An odd (even) number of reflections reverses (preserves) the chirality of S.
- Reflection along two directions is a rotation by π in the plane spanned by these directions. This operation squares to -1, as expected for spinors.
- In Euclidean signature the Dirac matrices in a Majorana representation are imaginary. In even dimensions, the condition that P maps Majorana spinors to Majorana spinors then leads to $P = \prod(i\Gamma^i \Gamma)$. This is relevant if we work in light cone gauge and if we reflect an odd number of directions. In light cone gauge $\Gamma = \Gamma_2 \ldots \Gamma_9$.

---

[2] This defines an orientifold which is related by T-duality to the Bianchi-Pradisi-Sagnotti model (which is sometimes also called the Gimon-Polchinski model).

[3] The same condition can be found by requiring that condition (8.19) is invariant under $p_i \to -p_i$ and $|\alpha\rangle = P_\alpha{}^\beta|\beta\rangle$.

Before returning to the discussion of the orientifold we observe that the transformation of spin fields under reflections immediately provides their transformation rule under T-duality. Indeed, as we have repeatedly seen, T duality is a reflection of the right-moving fields X_R and ψ in the dualized directions while the left-moving X_L and $\bar\psi$ are invariant. This leads to

$$S_\alpha(z) \to P_\alpha{}^\beta S_\beta(z), \qquad \bar S_\alpha(\bar z) \to \bar S_\alpha(\bar z), \tag{15.27}$$

for T duality along the directions X^i with P as in (15.26). This has interesting implications. For an odd number of dualized directions, it changes the chirality of the right-moving Ramond ground state and therefore transforms the type IIA to the type IIB string and vice versa. Furthermore, it also tells us that under T-duality in, say one direction, a R-R p-form transforms into a linear combination of $(p \pm 1)$-forms. This follows if we apply (15.27) to the R-R vertex operators (16.21) and use the identity

$$\Gamma^{\mu_1 \cdots \mu_p} \Gamma^{\mu_q} = \Gamma^{\mu_1 \cdots \mu_p \mu_q} + \left(\Gamma^{\mu_1 \cdots \mu_{p-1}} g^{\mu_p \mu_q} \mp \cdots \right). \tag{15.28}$$

With the help of (15.27) one can also check that the orientifolds in Table 10.3 on page 317 are related by T-dualities. For instance, one verifies that $T_2 \Omega T_2^{-1} = \Omega I_2 (-1)^{F_L}$, e.g. by considering its action on massless sates. Here I_2 is the reflection along two directions and T_2 is T-duality along the same directions. For the type IIA entries in this table we have written the action of $T \Omega T^{-1}$ on light-cone states. For instance, one verifies that with $I_1 = i\Gamma^9 \Gamma$ on spinors, $(\Omega I_1)^2 = 1$ on all physical states. In particular, there are invariant fermions, as required for the T-dual of the type I theory.[4] One can also directly determine the action $T \Omega T^{-1}$ on states of the type I theory and verify that the orientifold actions in Table 10.3 are related by T duality.

We now return to the orientifold. With the explicit representation of the Dirac matrices (8.51), it is straightforward to show that the action of $R = (\Gamma \Gamma^7)(\Gamma \Gamma^9)$ on the Ramond ground states is

$$R : |s_1, s_2, s_3, s_4\rangle \to (-1)^{\frac{1}{2}+s_4} |s_1, s_2, -s_3, -s_4\rangle \tag{15.29}$$

The eigenvalues of R on the Ramond ground states are $\pm i$.

We still have to specify the action of $\Omega R(-1)^{F_L}$ on the twisted sectors. We choose it such that there is an additional minus sign on states in the twisted sector. This can be viewed as assigning the twist field $\sigma(z, \bar z)$ in the bosonic sector charge -1.

Table 15.3 gives the action of I_4 on the (chiral) massless states. We now have to combine the left- and right moving sectors and project onto I_4 and $\Omega R(-1)^{F_L}$

---

[4] In the covariant formulation we would use $I_1 = \Gamma^9 \Gamma$ (with $\Gamma = \Gamma^0 \cdots \Gamma^9$) on spin fields. In this case, $\Omega I_1 (-1)^{F_L}$ squares to one.

15.3 Orientifold Compactifications

Table 15.9 $D = 6$ massless multiplets with eight supercharges

(1,0) SUSY	
Gravity multiplet	(3,3) + 2 (2,3) + (1,3)
Tensormultiplet	(3,1) + 2 (2,1) + (1,1)
Vectormultiplet	(2,2) + 2 (1,2)
Hypermultiplet	[(2,1) + 2 (1,1)] + h.c.

even states. As for the type I superstring in Chap. 9, Ω symmetrizes the NS-NS sector and anti-symmetrizes the R-R sector. We find the following $SU(2) \times SU(2)$ representations for the massless closed string states:

Untwisted	(NS,NS)	$(3,3) + 11 \times (1,1)$
	(R,R)	$(3,1) + (1,3) + 6 \times (1,1)$
	(NS,R)+(R,NS)	$2 \times (2,3) + 10 \times (2,1)$
Twisted	(NS,NS)	$48 \times (1,1)$
	(R,R)	$16 \times (1,1)$
	(NS,R)+(R,NS)	$32 \times (2,1)$

These can be combined into the following supermultiplets:

$$\text{untwisted sector}: \quad \text{gravity} + \text{tensor} + 4 \text{ hypermultiplets}$$
$$\text{twisted sector}: \quad 16 \text{ hypermultiplets}. \tag{15.30}$$

We have collected the multiplets of $\mathcal{N} = (1,0)$ supersymmetry in $D = 6$ in Table 15.9.

As mentioned before, the $T^4/\mathbb{Z}_2$ orbifold is a singular K3-surface which inherits all two-forms from the T^4, $h_0^{1,1} = 4$ and $h_0^{2,0} = h_0^{0,2} = 1$ and it has 16 harmonic $(1,1)$-forms in the twisted sector. Under $\Omega R(-1)^{F_L}$ only two of the six untwisted two-forms or, equivalently, two-cycles, are invariant ($dx^1 \wedge dx^2$ and $dy^1 \wedge dy^2$). Inspection of the spectrum leads to the conclusion that the invariant ones give rise to one self-dual and one anti-self-dual tensor, where the tensor fields are given by the dimensional reduction of the R-R four form on the invariant two-cycles. Contrarily, the anti-invariant two-cycles lead to scalars in hypermultiplets residing in the four (16) hypermultiplets in the untwisted (twisted) sector. If we had chosen $\Omega R(-1)^{F_L}$ to act with the opposite sign on the states in the twisted sectors, we had found tensormultiplets instead. This particular sign will be important later, when we need to determine how the orientifold projection acts on the D7-branes.

Just as the orbifold projection requires the inclusion of twisted sectors, the orientifold projection requires the inclusion of D-branes to cancel Ramond-Ramond tadpoles. In order to detect these tadpoles we carry out the same steps that we did in Chap. 9 for the type I superstring. First, we compute the loop-channel Klein-bottle amplitude

$$\mathcal{K} = \int_0^\infty \frac{dt}{2t} \, \text{tr} \left(\Omega R(-1)^{F_L} \left(\frac{1+I_4}{2} \right) P_{\text{GSO}} \, e^{-4\pi t (L_0 - \frac{c}{24})} \right). \quad (15.31)$$

Only left-right symmetric states contribute to the Klein-bottle. As the reflections R and I_4 act trivially on the oscillator parts of such states, their contribution to $\mathcal{K}$ in the untwisted sector is therefore identical to the one for the type I string. The only difference lies in the zero mode contributions along the compact directions. On a state with momentum m and winding n, orientation reversal Ω and reflection I_4 act as

$$\Omega|m, n\rangle = |m, -n\rangle, \quad I_4|m, n\rangle = |-m, -n\rangle$$
$$\Rightarrow \quad I_4\Omega|m, n\rangle = |-m, n\rangle. \quad (15.32)$$

Invariance under Ω therefore requires zero winding and invariance under $I\Omega$ requires zero momentum. It is now straightforward to evaluate the zero mode contribution to $\mathcal{K}$. We find for the loop-channel Klein-bottle amplitude in the untwisted sector

$$\mathcal{K} = v_6 \int_0^\infty \frac{dt}{t^4} \frac{1}{8} (1-1) \left(\frac{\vartheta_4^4}{\eta^{12}} \right) (2it) \left[\left(\sum_{m \in \mathbb{Z}} e^{-\pi t \left(\frac{m}{\rho_x} \right)^2} \right)^2 \left(\sum_{n \in \mathbb{Z}} e^{-\pi t (\rho_y n)^2} \right)^2 \right.$$

$$\left. + \left(\sum_{m \in \mathbb{Z}} e^{-\pi t \left(\frac{m}{\rho_y} \right)^2} \right)^2 \left(\sum_{n \in \mathbb{Z}} e^{-\pi t (\rho_x n)^2} \right)^2 \right] \quad (15.33)$$

with $v_6 = V_6/(4\pi^2 \alpha')^3$ and the short-hand notation $(1-1) = (1_{\text{NS}} - 1_{\text{R}})$ introduced in Eq. (9.65). In the twisted sector there are no zero-modes and for the contribution of the oscillators we find

$$\mathcal{K}_{\text{tw}} \sim v_6 \int_0^\infty \frac{dt}{t^4} (1-1) \left(\frac{\vartheta_4^2 \vartheta_1^2}{\eta^6 \vartheta_4^2} \right) (2it). \quad (15.34)$$

Since $\vartheta_1 = 0$ this term clearly does not give rise to any tree-channel tadpoles. Transforming (15.33) to the tree-channel via $t = 1/(4l)$ yields

$$\widetilde{\mathcal{K}} = v_6 \int_0^\infty dl \, 8 \, (1-1) \left(\frac{\vartheta_2^4}{\eta^{12}} \right) (2il)$$

$$\times \left[\left(\frac{\rho_x}{\rho_y} \right)^2 \left(\sum_{n \in \mathbb{Z}} e^{-4\pi l (\rho_x n)^2} \right)^2 \left(\sum_{m \in \mathbb{Z}} e^{-4\pi l \left(\frac{m}{\rho_y} \right)^2} \right)^2 + (x \leftrightarrow y) \right]. \quad (15.35)$$

Comparison with the computation of the type I string, we realize that the prefactor $v_6(\rho_x)^2$ in front of the massless tadpole indicates that it corresponds to an O7-plane

15.3 Orientifold Compactifications

which occupies the six non-compact dimensions and the two internal directions along the two x-axes. Similarly, the term multiplied by $v_6(\rho_y)^2$ signals the presence of an O7-plane along the two y-axes.

These dangerous tadpoles need to be canceled by introducing appropriate D7-branes. The simplest possibility is to place a number of D7-branes on top of the two orientifold planes. We will discuss this possibility in more detail using conformal field theory models. Let us mention that more general tadpole canceling D-brane configurations are possible.

Fractional D7-Branes

First we need to discuss D7-branes on the $T^4/\mathbb{Z}_2$ background. We have seen in Chap. 9 how to describe D-branes on flat backgrounds in terms of boundary states and computed the loop-channel and tree-channel annulus amplitudes for open strings with their two end points localized on parallel D-branes. In order to see what happens under the action of I_4, consider a D7-brane which is parallel to the two x-axes. If the brane is placed at (y_1, y_2), I_4 maps it to a D7-brane localized at $(-y_1, -y_2)$. Therefore for performing the I_4 orbifold, a second brane needs to be introduced at $(-y_1, -y_2)$, unless the brane is placed at one of the four fixed positions $(y_1, y_2) \in \{(0,0), (\pi R_y, 0), (0, \pi R_y), (\pi R_y, \pi R_y)\}$. For the loop-channel annulus diagram for an open string starting and ending on the same D7-brane, which is defined as

$$\mathscr{A} = \int_0^\infty \frac{dt}{2t} \, \text{tr}_{(D7,D7)} \left(\left(\frac{1+I_4}{2} \right) P_{\text{GSO}} \, e^{-2\pi t (L_0 - \frac{c}{24})} \right), \qquad (15.36)$$

this means that the part with the I_4 insertion vanishes except for D7-branes invariant under I_4.

Consider now such an invariant D7-brane localized at, say, $(y_1, y_2) = (0,0)$. Taking into account that along the x-directions the open strings have Kaluza-Klein momentum and in the orthogonal y-directions winding modes, it is straightforward to compute the annulus amplitude as

$$\mathscr{A}_{xx} = v_6 \int_0^\infty \frac{dt}{t^4} \frac{1}{2^6} (1-1) \left[\left(\frac{\vartheta_4^4}{\eta^{12}} \right) (2it) \left(\sum_{m \in \mathbb{Z}} e^{-2\pi t \left(\frac{m}{\rho_x}\right)^2} \right)^2 \right.$$

$$\left. \times \left(\sum_{n \in \mathbb{Z}} e^{-2\pi t (\rho_y n)^2} \right)^2 + 4 \left(\frac{\vartheta_4^2 \vartheta_3^2}{\eta^6 \vartheta_2^2} \right) (2it) \right]. \qquad (15.37)$$

The last term in brackets is the contribution from the I_4 insertion (the prefactor of four in the second line cancels the same factor contained in ϑ_2^2). Since ϑ_4 and ϑ_3 differ by the sign for the first excited level, this term implies that in the massless sector only the excitations $\psi^\mu_{-1/2}$ in the six non-compact directions survive the I_4 projection, whereas the excitations $\psi^i_{-1/2}$ in the internal directions $i = 5, \ldots, 8$ are projected out. The D7-brane therefore supports a six-dimensional gauge field and cannot be moved in the internal directions, i.e. it is frozen at $(y_1, y_2) = (0, 0)$ and has no continuous Wilson-line in the x-directions.

Transforming (15.37) to the tree-channel via $t = 1/(2l)$ gives

$$\widetilde{\mathscr{A}}_{xx} = v_6 \int_0^\infty dl \, \frac{1}{2^7} (1 - 1) \left[\left(\frac{\rho_x}{\rho_y}\right)^2 \left(\frac{\vartheta_2^4}{\eta^{12}}\right) (2il) \left(\sum_{n \in \mathbb{Z}} e^{-\pi l (\rho_x n)^2}\right)^2 \right.$$

$$\left. \times \left(\sum_{m \in \mathbb{Z}} e^{-\pi l \left(\frac{m}{\rho_y}\right)^2}\right)^2 + 4^2 \left(\frac{\vartheta_2^2 \vartheta_3^2}{\eta^6 \vartheta_4^2}\right) (2il) \right]. \qquad (15.38)$$

Let us interpret this result. The first term has a very similar form as the tree-channel Klein-bottle amplitude and again results from exchange of I_4-untwisted states between the two D7-branes. Since it is multiplied by the same volume factor, an appropriate number of such branes have a chance to cancel the $v_6(\rho_x/\rho_y)^2$ tadpole from the Klein-bottle. The second term in (15.38) is new. Inspection of the ϑ-functions reveals that it arises from the exchange of closed string modes from the I_4-twisted sector. In other words, this kind of D7-branes, without any transverse moduli, do not only carry charge under the untwisted RR-forms C_8 but also under the twisted ones, which are given by the reduction of C_8 along the collapsed S^2 at the fixed point, $\int_{S^2} C_8$. This makes sense, as the D7-brane runs precisely through the four fixed points on the x-axes. Geometrically this D7-brane not only wraps the two-cycle spanned by (x^1, x^2) but also the four two-spheres which have collapsed to zero size at the orbifold point. Since these S^2's are localized at the fixed points, the brane cannot move away. Such branes are also called fractional D-branes.

In the same way as one defines boundary states in the untwisted sector (9.35), one can define them in the twisted sector. Noting that such states can carry positive and negative twisted sector charge, the entire boundary state for a fractional D7-brane can be schematically written as

$$|D7_x^\pm\rangle_{\text{frac}} = \frac{1}{2\mathcal{N}_{D7}} \left(\left(\frac{\rho_x}{\rho_y}\right) |D7\rangle_{\text{ut}} \pm 2 \sum_{i=1}^{4} |D7, i\rangle_{\text{tw}} \right), \qquad (15.39)$$

where the sum runs over the four fixed points through which the I_4 invariant D7-brane runs; in Chap. 9 we have fixed $\mathcal{N}_{D7} = 8 \, (4\pi^2 \alpha')^{\frac{3}{2}}$. Similarly, we can introduce branes $|D7_y^\pm\rangle_{\text{frac}}$ wrapping the y-directions. The numerical prefactors in (15.39) are a result of loop-channel–tree-channel equivalence, which means that in loop-channel one has the correct interpretation as a trace with $(1 + I_4)/2$ insertion.

The twisted sector boundary states in for instance the NS sector are

15.3 Orientifold Compactifications

$|D7_{x,NS}, \eta, i\rangle_{tw}$

$$= \exp\left[-\sum_{n=1}^{\infty}\sum_{\mu=1}^{4}\frac{1}{n}\alpha_{-n}^{\mu}\overline{\alpha}_{-n}^{\mu} - \sum_{n=1/2}^{\infty}\sum_{m=5}^{6}\frac{1}{n}\alpha_{-n}^{m}\overline{\alpha}_{-n}^{m} + \sum_{n=1/2}^{\infty}\sum_{m=7}^{8}\frac{1}{n}\alpha_{-n}^{m}\overline{\alpha}_{-n}^{m}\right.$$

$$\left. +i\eta\sum_{r=1/2}^{\infty}\left(-\sum_{\mu=1}^{4}b_{-r}^{\mu}\overline{b}_{-r}^{\mu} - \sum_{m=5}^{6}b_{-r+\frac{1}{2}}^{m}\overline{b}_{-r+\frac{1}{2}}^{m} + \sum_{m=7}^{8}b_{-r+\frac{1}{2}}^{m}\overline{b}_{-r+\frac{1}{2}}^{m}\right)\right]|t_i\rangle.$$

(15.40)

Here $|t_i\rangle$ denote the closed string twisted sector ground states in the NS-NS sector, which are localized at the 16 I_4 fixed points.

One needs two fractional branes of opposite twisted sector charge ± 1 to make one ordinary bulk D7-brane. Said differently, moving a bulk D7-brane in an I_4 invariant position, it can split into two fractional D7-branes with opposite twisted sector charge. Computing the overlap of two parallel fractional D7-branes with opposite twisted sector charges and transforming the result to loop-channel, one finds (15.37) with the sign in front of the last term reversed. Therefore, this time the six-dimensional gauge field is projected out and what remains are eight massless scalars making two hypermultiplets transforming in the bi-fundamental $(1, -1)$ representation of the $U(1) \times U(1)$ gauge symmetry on the pair of fractional D7-branes.

Tadpole Cancellation

We would like to introduce fractional branes $|D7_x^{\pm}\rangle_{\text{frac}}$ and $|D7_y^{\pm}\rangle_{\text{frac}}$ to cancel the Klein-bottle divergences. First we note that with our choice of the action of $\Omega R(-1)^{F_L}$ in the I_4 twisted sectors, a fractional brane is mapped to another parallel fractional brane with opposite twisted sector charges

$$\Omega R : |D7^{\pm}\rangle_{\text{frac}} \to |D7^{\mp}\rangle_{\text{frac}}.$$

(15.41)

Therefore, we have to introduce these branes in pairs, which automatically guarantees that the twisted sector tadpoles cancel. Let us mention that in the model where one gets 16 tensormultiplets from the closed string twisted sector, i.e. if we choose the opposite action of $\Omega R(-1)^{F_L}$ in the twisted sectors, a fractional brane is invariant under the orientifold projection.

We introduce two stacks of M pairs of fractional D7-branes $|D7_x^{\pm}\rangle_{\text{frac}}$ and N pairs of fractional D7-branes $|D7_y^{\pm}\rangle_{\text{frac}}$. Summing over all annulus diagrams

$$\mathcal{A}_{\text{tot}} = \sum_{i,j \in \{+,-\}} \left(\mathcal{A}_{D7_x^i, D7_x^j} + \mathcal{A}_{D7_y^i, D7_y^j}\right)$$

(15.42)

and transforming the amplitude to tree-channel yields

$$\widetilde{\mathscr{A}}_{\text{tot}} = v_6 \int_0^\infty dl \, \frac{1}{32} (1-1) \left(\frac{\vartheta_2^4}{\eta^{12}} \right) (2il)$$

$$\times \left[M^2 \left(\frac{\rho_x}{\rho_y} \right)^2 \left(\sum_{n\in\mathbb{Z}} e^{-\pi l(\rho_x n)^2} \right)^2 \left(\sum_{m\in\mathbb{Z}} e^{-\pi l\left(\frac{m}{\rho_y}\right)^2} \right)^2 \right.$$

$$\left. + N^2 \left(\frac{\rho_y}{\rho_x} \right)^2 \left(\sum_{m\in\mathbb{Z}} e^{-\pi l\left(\frac{m}{\rho_x}\right)^2} \right)^2 \left(\sum_{n\in\mathbb{Z}} e^{-\pi l(\rho_y n)^2} \right)^2 \right]. \quad (15.43)$$

Note that the structure of this expression is very similar to the tree-channel Klein-bottle amplitude (15.35).

Since the orientifold projection exchanges two types of fractional branes, it is clear that the Chan-Paton gauge group of this model is $U(M) \times U(N)$. Moreover, the open string sectors which are invariant under the orientifold action are

$$|D7_x^+\rangle_{\text{frac}} \longleftrightarrow |D7_x^-\rangle_{\text{frac}} \qquad |D7_y^+\rangle_{\text{frac}} \longleftrightarrow |D7_y^-\rangle_{\text{frac}}. \quad (15.44)$$

These states are either symmetrized or anti-symmetrized depending on the sign in the Möbius strip amplitude, which we now compute. Let us consider the $D7_x$ branes, for which the loop-channel Möbius strip amplitude is defined as

$$\mathscr{M}_{D7_x^+, D7_x^-} = \pm \int_0^\infty \frac{dt}{2t} \, \text{tr}\left(\Omega R(-1)^{F_L} \left(\frac{1+I_4}{2} \right) P_{\text{GSO}} \, e^{-2\pi t \left(L_0 - \frac{c}{24}\right)} \right), \quad (15.45)$$

where the trace is over all excitations of the open string stretched between the brane $D7_x^+$ and $D7_x^-$. Since R reflects winding and momentum in the y-directions, it is clear that under the $\Omega R(-1)^{F_L}$ all x-KK-momentum and y-winding modes are invariant and contribute to the trace. Contrarily, for the $\Omega R \, I_4(-1)^{F_L}$ insertion all non-vanishing KK-momenta and windings are projected out. Moreover, in the latter case, similar to the twisted sector contributions to the Klein-bottle tadpole, no tree-channel tadpole appears, so that for $D7_x$ branes only the $\Omega R(-1)^{F_L}$ contributes to the tadpole and similarly for $D7_y$ branes only the $\Omega R I_4(-1)^{F_L}$ gives a non-vanishing tadpole. With the methods from Chap. 9 it is now straightforward to evaluate the loop-channel Möbius strip amplitudes and transform them via $t = 1/(8l)$ into tree-channel. The final result is

$$\widetilde{\mathscr{M}}_{D7_x^+, D7_x^-} + \widetilde{\mathscr{M}}_{D7_y^+, D7_y^-}$$

$$= \pm v_6 \int_0^\infty dl \, (1-1) \left(\frac{\vartheta_2^4}{\eta^{12}} \right) \left(\tfrac{1}{2} + 2il \right)$$

15.3 Orientifold Compactifications

$$\times \left[M \left(\frac{\rho_x}{\rho_y}\right)^2 \left(\sum_{n\in\mathbb{Z}} e^{-4\pi l(\rho_x n)^2}\right)^2 \left(\sum_{m\in\mathbb{Z}} e^{-4\pi l\left(\frac{m}{\rho_y}\right)^2}\right)^2 \right.$$

$$\left. + N \left(\frac{\rho_y}{\rho_x}\right)^2 \left(\sum_{m\in\mathbb{Z}} e^{-4\pi l\left(\frac{m}{\rho_x}\right)^2}\right)^2 \left(\sum_{n\in\mathbb{Z}} e^{-4\pi l(\rho_y n)^2}\right)^2 \right]. \quad (15.46)$$

What we have neglected so far are the open strings between fractional D7$_x$ and D7$_y$ branes. Using the boundary states (15.39) and taking into account that $|D7_x^\pm\rangle_{\text{frac}}$ and $|D7_y^\pm\rangle_{\text{frac}}$ always share precisely one fixed point, one can straightforwardly compute the overall loop-channel annulus amplitude

$$\mathscr{A}_{xy} = MN\, v_6 \int_0^\infty \frac{dt}{2(2t)^4}\, 4\left[\left(\frac{\vartheta_3^2 \vartheta_2^2}{\eta^6 \vartheta_4^2}\right)(2it) - \left(\frac{\vartheta_2^2 \vartheta_3^2}{\eta^6 \vartheta_4^2}\right)(2it)\right], \quad (15.47)$$

where one factor of two is due to the two different orientations of the open string, which are identified under the orientifold projection. First we note that transformation to tree-channel reveals that this amplitude does not induce further massless tadpoles. Up to numerical factors, this annulus amplitude looks precisely like (the square root of) the closed string partition function in the $\mathbb{Z}_2$ twisted sector. From Eq. (10.166) we know that in such a sector we have $E_1 = 1/2$ so that only the (degenerate) ground state contributes to the massless spectrum, i.e. without any oscillator excitations. Clearly, these are scalars so that, taking also the numerical factors into account, one finds one hypermultiplet in the bi-fundamental (N, M) representation.

We are now ready to collect all pieces and extract the divergences from the tree-channel Klein-bottle, annulus and Möbius strip amplitudes. These are

$$\text{Tad} = \frac{1}{32}\int_0^\infty dl \left[\left(\frac{\rho_x}{\rho_y}\right)^2 (M \pm 16)^2 + \left(\frac{\rho_y}{\rho_x}\right)^2 (N \pm 16)^2\right] \quad (15.48)$$

which means that, for the choice of the minus sign in the Möbius strip amplitude, we need $M = N = 16$ pairs of fractional D7-branes to cancel all tadpoles. Therefore, the gauge group is $U(16) \times U(16)$ and the states in the D7$^+$ − D7$^-$ sectors are anti-symmetrized. In Table 15.10 we list the massless open string matter spectrum. Combining this with the massless spectrum from the closed string sector and denoting the number of tensor, vector and hyper multiplets by n_T, n_V and n_H, we have

$$n_T = 1, \quad n_V = 512, \quad n_H = 756. \quad (15.49)$$

The spectrum satisfies

$$n_H - n_V + 29 n_T = 273. \quad (15.50)$$

This is a consistency condition which every six-dimensional string compactification with $\mathcal{N} = (1,0)$ supersymmetry has to satisfy. It is the condition that the irreducible

Table 15.10 Massless hypermultiplets from open string sectors

Sector	Reps. of $U(16) \times U(16)$
$D7_x^+ - D7_x^-$	$2 \times (\mathbf{120}, \mathbf{1})$
$D7_y^+ - D7_y^-$	$2 \times (\mathbf{1}, \mathbf{120})$
$D7_x^i - D7_y^j$	$1 \times (\mathbf{16}, \mathbf{16})$

gravitational anomaly cancels. The remaining anomalies (gauge and mixed) then cancel via the Green-Schwarz mechanism, which we discuss in more detail for the four-dimensional case in Chap. 17.

Note that we have considered a very simple D7-brane configuration, namely a number of fractional D7-branes placed right on top of the orientifold planes. We can also move N_a pairs of such branes away from the orientifold locus, so that they become bulk branes. This corresponds to giving a vacuum expectation value to the scalars in the $(\mathbf{120}, \mathbf{1})$ and $(\mathbf{1}, \mathbf{120})$ representations. The resulting bulk branes carry $Sp(2N_a)$ Chan-Paton factors with still one hypermultiplet in the antisymmetric representation of dimension $N_a(2N_a - 1)$. However, one can also try to cancel the tadpoles with D7-branes, which are not parallel to the orientifold planes, but instead intersect them at non-trivial angles. This is indeed possible in a supersymmetric fashion and gives rise to so-called intersecting D7-brane models.

All these constructions can be generalized in many ways. One can consider more general orbifolds both in four and six dimensions, where each case needs a separate treatment of the twisted sectors and fractional D-branes. Moreover, one can also cancel the tadpoles by D7-branes which are not parallel to the O7-planes, but intersect them. Constructions of this kind are called intersecting D-brane models and will be discussed in more generality for intersecting D6-branes in type IIA in Sect. 17.2.

15.4 World-Sheet Aspects of Supersymmetric Compactifications

In the previous two sections we have presented relatively simple examples of CFTs which describe strings on four-dimensional and six-dimensional compact spaces with various amounts of space-time supersymmetry. As already repeatedly emphasised, four-dimensional theories with $\mathcal{N} > 1$ supersymmetry cannot have a chiral spectrum, i.e their low-energy limits cannot be good descriptions of the standard model. With possible phenomenological applications in mind, we are therefore interested in $\mathcal{N} = 1$ space-time supersymmetric compactifications to four dimensions, i.e. compactifications with four unbroken supercharges.[5] We have encountered two ways to achieve this in perturbative string theory.

---

[5] Of course, for phenomenological reasons they also have to be broken, but we assume here that this is achieved at a different level, e.g. by fluxes or non-perturbative effects (see Chap. 17).

15.4 World-Sheet Aspects of Supersymmetric Compactifications

- One starts with the type II superstring compactified to four-dimensions such that it preserves $\mathcal{N} = 2$ supersymmetry and performs an orientifold quotient. This breaks half of the eight supercharges. D-branes are necessary for tadpole cancellation and gauge degrees of freedom and chiral matter arise from open strings connecting them.
- One considers the ten-dimensional heterotic string theories and compactifies them on the same geometric backgrounds, consistently including the $E_8 \times E_8$ or $SO(32)$ gauge sector. Since the heterotic theories start with half the supersymmetry in ten-dimensions, one directly gets $\mathcal{N} = 1$ space-time supersymmetry.

Both classes of models (prior to the orientifold quotient) have to satisfy a number of general conditions, which we want to elaborate on in this section. Some of the relevant discussion is restricted to one chiral sector and applies to the heterotic and the type II string. When combining left- and right-movers we have to treat them separately.

For a specific CFT to qualify as a perturbative string vacuum, it has to pass certain consistency requirements such as the correct value of the central charge, modular invariance, etc. One may then impose additional four-dimensional requirements, like space-time supersymmetry or a chiral fermionic spectrum, which constrain the internal CFT further. We will, in particular, clarify the intricate relation between space-time supersymmetry and the required structure of the world-sheet superconformal field theory.

Any four-dimensional heterotic string theory possesses four space-time string coordinates $X^\mu(z,\bar{z})$. In addition there are right-moving world-sheet fermions $\psi^\mu(z)$. Finally, there are the conformal ghosts $\bar{b}(\bar{z})$, $\bar{c}(\bar{z})$ and $b(z)$, $c(z)$ and the superconformal ghosts $\beta(z)$ and $\gamma(z)$. These fields constitute the external CFT. The left-moving part is a CFT with central charge $\bar{c}^{\text{ext}} = -22$ and the right-moving part an $N=1$ SCFT with $c^{\text{ext}} = -9$. The requirements $c^{\text{tot}} = \bar{c}^{\text{tot}} = 0$ mean that we have a left-moving internal CFT with $\bar{c}^{\text{int}} = 22$ and a right-moving $N=1$ SCFT with $c^{\text{int}} = 9$. For type II theories the two sectors have the structure of the holomorphic sector of the heterotic string.

It is crucial to realize that the left-moving and right-moving internal CFTs can be chosen largely independently of each other; only locality and modular invariance link them in a definite way. Moreover, it is not required that the internal CFT admits an interpretation as a compactification on some manifold.

It is once again convenient to replace the four external world-sheet fermions $\psi^\mu(z)$ by bosonic fields $\phi^i(z)$ ($i = 1, 2$) and the superconformal ghosts by a scalar $\phi(z)$. Then all states are characterized by vectors $(\boldsymbol{\lambda}_R, q)$ of the covariant lattice $(D_{2,1})_R$ and heterotic vertex operators are sums of terms of the form

$$V(z,\bar{z}) = (\bar{\partial}^{n_L} X^\mu(\bar{z}))^{m_L} \, (\partial^{n_{1R}} X^\nu(z))^{m_{1R}} \, (\partial^{n_{2R}} \phi^i(z))^{m_{2R}}$$
$$\times e^{i\boldsymbol{\lambda}_R \cdot \boldsymbol{\phi}(z)} \, e^{q\phi(z)} V_{\text{int}}(z,\bar{z}) \, e^{ik \cdot X(z,\bar{z})}. \tag{15.51}$$

$V_{\text{int}}(z,\bar{z})$ are conformal fields of the internal CFT with weight $(h_{\text{int}}, \bar{h}_{\text{int}})$. V creates a state with mass

$$\frac{\alpha'}{2} m_L^2 = N_L + \overline{h}_{\text{int}} - 1,$$

$$\frac{\alpha'}{2} m_R^2 = \frac{1}{2}\lambda_R^2 - \frac{1}{2}q^2 - q + N_{1R} + N_{2R} + h_{\text{int}} - 1,$$

$$m^2 = m_L^2 + m_R^2. \tag{15.52}$$

The $N_{L,R}$ are oscillator excitation numbers. We require that the internal CFT is unitary which implies $h_{\text{int}}, \overline{h}_{\text{int}} \geq 0$ with equality only for the vacuum state $|0\rangle$. Furthermore, we have to insist that the theory is local. This requirement provides a link between the external and the internal parts of the theory. Consider the operator product

$$V_1(z,\overline{z}) V_2(w,\overline{w}) \sim (\overline{z}-\overline{w})^{-\overline{h}_{\text{int}}^1 - \overline{h}_{\text{int}}^2 + \overline{h}_{\text{int}}^3} (z-w)^{-h_{\text{int}}^1 - h_{\text{int}}^2 + h_{\text{int}}^3 + \lambda_{1R}\cdot\lambda_{2R} - q_1 q_2}$$
$$\times e^{i(\lambda_{1R}+\lambda_{2R})\cdot\phi(w)} e^{(q_1+q_2)\phi(w)} V_{\text{int}}^3(w,\overline{w}) + \ldots \tag{15.53}$$

where we have neglected the bosonic oscillators which always give rise to integer powers of $(z-w)$. Also, the $e^{ik\cdot X}$ factors give $|z-w|^{k_1\cdot k_2}$ which does not have branch cuts. Locality demands that

$$\overline{h}_{\text{int}}^1 + \overline{h}_{\text{int}}^2 - \overline{h}_{\text{int}}^3 - h_{\text{int}}^1 - h_{\text{int}}^2 + h_{\text{int}}^3 + \lambda_{1R} \cdot \lambda_{2R} - q_1 q_2 \in \mathbb{Z}. \tag{15.54}$$

The complete Hilbert space, containing the left-moving, right-moving, the internal and external states, must obey the above condition. Modular invariance of the one-loop partition function constrains the possible combinations of internal and external highest weight states further. But to analyze the constraints of modular invariance for a general model is not an easy task. The generalization of (15.54) to the type II string is

$$\overline{h}_{\text{int}}^1 + \overline{h}_{\text{int}}^2 - \overline{h}_{\text{int}}^3 - h_{\text{int}}^1 - h_{\text{int}}^2 + h_{\text{int}}^3 + \lambda_{1R} \cdot \lambda_{2R} - \lambda_{1L} \cdot \lambda_{2L} - q_1 q_2 + \overline{q}_1 \overline{q}_2 \in \mathbb{Z}. \tag{15.55}$$

Space-Time Versus World-Sheet Supersymmetry

The requirement of space-time supersymmetry has interesting implications for the two-dimensional world-sheet theory. Space-time supercharges arise in the heterotic string only from the right moving sector, whereas for type II theories from both sectors.

The relevant sector for the discussion of space-time supersymmetries is the right-moving one. We start the discussion with general $\mathcal{N}$, but will soon specify to $\mathcal{N} = 1$. The relevant $D_{2,1}$ conjugacy classes are the two spinor classes, but, as we will see momentarily, more input is needed. We denote the supercharges by Q_α^A

15.4 World-Sheet Aspects of Supersymmetric Compactifications

($A = 1, \ldots, \mathcal{N}$). They are the contour integrals of the holomorphic part of the gravitino vertex operators at zero space-time momentum, i.e. the zero-mode of the space-time supersymmetry currents:

$$Q^A_{\alpha(q)} = \oint \frac{dz}{2\pi i} V^A_{\alpha(q)}(z), \quad \overline{Q}_{\dot\alpha A(q)} = \oint \frac{dz}{2\pi i} \overline{V}_{\dot\alpha A(q)}(z), \tag{15.56}$$

where q is the ghost picture charge. In the canonical ghost picture they are

$$V^A_{\alpha(-\frac{1}{2})}(z) = 2(\alpha')^{-1/4} S_\alpha\, e^{-\phi/2}\, \Sigma^A(z)$$

$$\overline{V}_{\dot\alpha A(-\frac{1}{2})}(z) = 2(\alpha')^{-1/4} S_{\dot\alpha}\, e^{-\phi/2}\, \Sigma_A(z). \tag{15.57}$$

$S_\alpha(z)$ and $S_{\dot\alpha}(z)$ are spin fields of the Lorentz group $SO(1,3)$ characterized by the spinor weights of D_2:

$$S_\alpha(z) = e^{i\lambda_\alpha \cdot \phi(z)}, \quad \lambda_\alpha = \pm\left(\frac{1}{2}, \frac{1}{2}\right),$$

$$S_{\dot\alpha}(z) = e^{i\lambda_{\dot\alpha} \cdot \phi(z)}, \quad \lambda_{\dot\alpha} = \pm\left(\frac{1}{2}, -\frac{1}{2}\right). \tag{15.58}$$

The fields $\Sigma^A(z)$ and $\Sigma_A = (\Sigma^A)^\dagger$ are the (degenerate) Ramond ground states of dimension $h_{int} = \frac{3}{8}$ of the internal SCFT with $c = 9$. They are the Killing spinors in CFT language.

The normalization of the supercharges are chosen for later convenience and to make it independent of the number of non-compact space-time dimensions. In four dimensions they satisfy the space-time supersymmetry algebra[6]

$$\{Q^A_\alpha, \overline{Q}_{\dot\beta B}\} = 2\sigma^\mu_{\alpha\dot\beta} P_\mu \delta^A_B,$$

$$\{Q^A_\alpha, Q^B_\beta\} = \epsilon_{\alpha\beta} Z^{AB}, \tag{15.59}$$

where $Z^{AB} = -Z^{BA}$ are central charges and we have used that the momentum operator in the canonical ghost picture is $\sqrt{\frac{2}{\alpha'}} \oint e^{-\phi} \psi^\mu$. Alternatively we could

---

[6] We use the two-component notation for spinors and Dirac matrices in four dimensions. The Dirac matrices are $\sigma^\mu_{\alpha\dot\alpha} = (\mathbb{1}, \sigma^i)$ and $\overline{\sigma}^{\mu\dot\alpha\alpha} = (\mathbb{1}, -\sigma^i)$, where σ^i are the three Pauli matrices. Two-component spinor indices are raised and lowered with $\epsilon^{\alpha\beta} = \begin{pmatrix} 0 & 1 \\ -1 & 0 \end{pmatrix}$, $\epsilon_{\alpha\beta}\epsilon^{\beta\gamma} = \delta^\gamma_\alpha$ and likewise for the dotted indices. We also define $\sigma^{\mu\nu}{}_\alpha{}^\beta = \frac{1}{4}(\sigma^\mu_{\alpha\dot\gamma}\overline{\sigma}^{\nu\dot\gamma\beta} - \sigma^\nu_{\alpha\dot\gamma}\overline{\sigma}^{\mu\dot\gamma\beta})$ and $\overline{\sigma}^{\mu\nu\dot\alpha}{}_{\dot\beta} = \frac{1}{4}(\overline{\sigma}^{\mu\dot\alpha\gamma}\sigma^\nu_{\gamma\dot\beta} - \overline{\sigma}^{\nu\dot\alpha\gamma}\sigma^\mu_{\gamma\dot\beta})$. These are the Wess-Bagger conventions.

have built one of the supercharges from the gravitino vertex in the $q = +\frac{1}{2}$ picture at zero momentum. We would then find $p^\mu = \frac{2}{\alpha'} \oint i \partial X^\mu(z)$ on the r.h.s. of (15.59).

Using the expression for the supercharges, this translates to the following operator product expansions:

$$\Sigma^A(z) \Sigma_B(w) \sim (z-w)^{-\frac{3}{4}} \delta^A_B + (z-w)^{\frac{1}{4}} J^A_B(w) + \ldots$$

$$\Sigma^A(z) \Sigma^B(w) \sim (z-w)^{-\frac{1}{4}} \psi^{AB} + \mathcal{O}\left((z-w)^{\frac{3}{4}}\right). \quad (15.60)$$

The dimension one fields J^A_B can be shown to be currents of internal level one Kač-Moody algebras $\hat{g}$ of rank k ($k = 1, 2, 3$ for $\mathcal{N} = 1, 2, 4$). The dimension $1/2$ fields $\psi^{AB} = -\psi^{BA}$ are related to the central charges in the $q = -1$ picture by $Z^{AB}_{-1} = \oint \frac{dz}{2\pi i} e^{-\phi} \psi^{AB}(z)$. Using the Frenkel-Kač construction, the currents J^A_B can be expressed by k free internal bosons $H^{\text{int}}_i(z)$ ($i = 1, \ldots, k$). The internal vertex operators $V_{\text{int}}(z)$ can then always be written as (neglecting derivatives of H^{int}_i):

$$V_{\text{int}}(z) = e^{i w_{\text{int}} \cdot H_{\text{int}}(z)} \tilde{V}_{\text{int}}(z). \quad (15.61)$$

The vectors w_{int} are the weights of the internal algebra g spanning the weight lattice of g. The $\tilde{V}_{\text{int}}(z)$ belong to a CFT with $\tilde{c} = 9 - k$ and commute with H^{int}_i.

Let us fill in some of the details for the (phenomenologically) most interesting case $\mathcal{N} = 1$. Equation (15.60) simplifies to

$$\Sigma(z) \Sigma^\dagger(w) \sim (z-w)^{-3/4} + \frac{1}{2}(z-w)^{1/4} J(w) + \ldots,$$

$$\Sigma(z) \Sigma(w) \sim \mathcal{O}\left((z-w)^{3/4}\right). \quad (15.62)$$

Consider the four point correlation function

$$f(z_1, \ldots, z_4) = \langle \Sigma(z_1) \Sigma^\dagger(z_2) \Sigma(z_3) \Sigma^\dagger(z_4) \rangle$$

$$= \left(\frac{z_{13} z_{24}}{z_{12} z_{34} z_{14} z_{23}}\right)^{3/4} \tilde{f}(x), \quad (15.63)$$

where $x = \frac{z_{12} z_{34}}{z_{13} z_{24}}$ (cf. Eq. (4.91) on page 82). The leading orders of the operator product expansions Eq. (15.62) determine the behavior of $f(z_1, \ldots, z_4)$ as $z_{ij} \to 0$ for any pair i, j. One finds that $\tilde{f}(x)$ is an analytic function, bounded at $x = 0, 1, \infty$. It is thus a constant. If we normalize $\langle 1 \rangle = 1$ we find $\tilde{f}(x) = 1$. In the limit $z_1 \to z_2$, Eq. (15.63) becomes

$$f(z_1, \ldots, z_4) \to \langle \Sigma(z_3) \Sigma^\dagger(z_4) \rangle (z_{12})^{-3/4} + \frac{1}{2} \langle J(z_2) \Sigma(z_3) \Sigma^\dagger(z_4) \rangle (z_{12})^{1/4} + \ldots$$

$$= (z_{12} z_{34})^{-3/4} + \frac{3}{4} (z_{12} z_{34})^{1/4} (z_{23} z_{24})^{-1} + \ldots, \quad (15.64)$$

15.4 World-Sheet Aspects of Supersymmetric Compactifications

where the second line is the expansion of (15.63). If we now take the limits $z_3 \to z_4$, $z_2 \to z_3$ and $z_2 \to z_4$ we find

$$J(z) J(w) \sim \frac{3}{(z-w)^2} + \text{finite},$$

$$J(z) \Sigma(w) \sim \frac{\frac{3}{2} \Sigma(w)}{z-w} + \text{finite},$$

$$J(z) \Sigma^\dagger(w) \sim -\frac{\frac{3}{2} \Sigma^\dagger(w)}{z-w} + \text{finite}. \tag{15.65}$$

Therefore, $J(z)$ is a $U(1)$ Kač-Moody current which can be written in terms of a free boson as

$$J(z) = i\sqrt{3}\partial H(z). \tag{15.66}$$

The operators Σ and $\Sigma^\dagger$ can be expressed as exponentials of H:

$$\Sigma(z) = e^{i\frac{\sqrt{3}}{2}H(z)}, \quad \Sigma^\dagger(z) = e^{-i\frac{\sqrt{3}}{2}H(z)}. \tag{15.67}$$

For supersymmetric orbifold compactifications one can express $H(z)$ explicitly through the bosonized internal world-sheet fermions as $\sqrt{3}H = \sum v_i \phi^i$. $\Sigma(z)$ corresponds to the Killing spinor.

Any operator with definite $U(1)$ charge Q, i.e.

$$J(z) V_Q(w, \bar{w}) = \frac{Q V_Q(w, \bar{w})}{z-w} + \ldots \tag{15.68}$$

can be written as

$$V_Q(z, \bar{z}) = e^{i\frac{Q}{\sqrt{3}}H(z)} P(J(z)) \tilde{V}(z, \bar{z}), \tag{15.69}$$

where $P(J)$ is a polynomial in J and its derivatives and $\tilde{V}$ is independent of the boson H. In particular, physical states are characterized by vertex operators

$$V(z, \bar{z}) = e^{i\lambda_R \cdot \phi(z)} e^{q\phi(z)} e^{i\frac{Q}{\sqrt{3}}H(z)} \tilde{V}_{\text{int}}(z, \bar{z}), \tag{15.70}$$

where we suppress possible oscillator contributions. $\tilde{V}_{\text{int}}$ belongs to the remaining CFT with $\tilde{c} = 8$ and $\bar{c} = 22$ (heterotic) or $\bar{c} = 9$ (type II). (The boson H contributes one unit to the central charge.) The numbers $\frac{Q}{\sqrt{3}}$ form the internal weight lattice Γ_1 of the $U(1)$ Kač-Moody algebra. BRST invariance of the gravitino vertex operator demands that the operator product between $T_F^{\text{int}}(z)$ and $\Sigma(w)$, $\Sigma^\dagger(w)$ contains a branch cut of order $\frac{1}{2}$:

$$T_F^{\text{int}}(z)\,\Sigma(w) = \frac{\hat{\Sigma}(w)}{(z-w)^{\frac{1}{2}}} + \ldots ,$$

$$T_F^{\text{int}}(z)\,\Sigma^\dagger(w) = \frac{\hat{\Sigma}^\dagger(w)}{(z-w)^{\frac{1}{2}}} + \ldots , \qquad (15.71)$$

where $\hat{\Sigma}(w)$, $\hat{\Sigma}^\dagger(w)$ are operators of dimension 11/8. These equations, together with Eq. (15.67), imply that the internal supercurrent does not have a definite $U(1)$ charge but splits into two parts with $U(1)$ charges $Q = \pm 1$ respectively:

$$T_F^{\text{int}}(z) = e^{\frac{i}{\sqrt{3}}H(z)}\tilde{T}_F^+(z) + e^{-\frac{i}{\sqrt{3}}H(z)}\tilde{T}_F^-(z) = \frac{1}{2\sqrt{2}}\left(T_F^+(z) + T_F^-(z)\right) \quad (15.72)$$

with

$$J(z)\,T_F^\pm(w) \sim \pm\frac{T_F^\pm(w)}{z-w}. \qquad (15.73)$$

$\tilde{T}_F^\pm$ are fields of conformal dimension 4/3 of an internal CFT with $\tilde{c} = 8$. We also define

$$T_F'^{\text{int}} = \frac{1}{2\sqrt{2}}(T_F^+ - T_F^-) \qquad (15.74)$$

and immediately show

$$J(z)\,T_F^{\text{int}}(w) = \frac{1}{(z-w)}T_F'^{\text{int}}(w) + \ldots \qquad (15.75\text{a})$$

$$J(z)\,T_F'^{\text{int}}(w) = \frac{1}{(z-w)}T_F^{\text{int}}(w) + \ldots \qquad (15.75\text{b})$$

With reference to Eq. (12.15), from (15.75a) we conclude that J and $-2T_F'^{\text{int}}$ are the lower and upper components of an $N=1$ superfield with $h=1$. This also fixes

$$T_F^{\text{int}}(z)\,T_F'^{\text{int}}(w) = -\frac{\frac{1}{2}J(w)}{(z-w)^2} - \frac{\frac{1}{4}\partial J(w)}{4(z-w)}. \qquad (15.76)$$

Furthermore,

$$T_F^{\text{int}}(z)\,T_F^{\text{int}}(w) \sim -T_F'^{\text{int}}(z)T_F'^{\text{int}}(w) + \mathcal{O}(1) \qquad (15.77)$$

which is equivalent to the following relation for the modes of T_F^{int} and $T_F'^{\text{int}}$, which are defined in the usual way:

15.4 World-Sheet Aspects of Supersymmetric Compactifications

$$\{G'^{\text{int}}_r, G'^{\text{int}}_s\} = \{[J_0, G^{\text{int}}_r], G'^{\text{int}}_s\} = -[\{G^{\text{int}}_r, G'^{\text{int}}_s\}, J_0] + \{[G'^{\text{int}}_s, J_0], G^{\text{int}}_r\}$$
$$= -\{G^{\text{int}}_s, G^{\text{int}}_r\}. \tag{15.78}$$

We have used (15.75a) and the (graded) Jacobi-Identity and (15.76), which translates to $\{G^{\text{int}}_r, G'^{\text{int}}_s\} = -(r-s)J_{r+s}$. From (15.77) and (15.76) it follows that the $T_F^+ T_F^+$ and $T_F^- T_F^-$ operator products are regular. Using these results and the fact that T^{int}, T_F^{int} satisfy an $N = 1$ SCA with $c = 9$, one shows that T^{int}, $T_F^{\pm}$, J satisfy the $N = 2$ SCA (12.81) with $c = 9$. Note that

$$T^{\text{int}}(z) = -\frac{1}{2}(\partial H(z))^2 + \tilde{T}^{\text{int}}(z), \tag{15.79}$$

where $\tilde{T}_{\text{int}}$ satisfies a Virasoro algebra with $\tilde{c} = 8$. We have thus shown that $\mathcal{N} = 1$ space-time supersymmetry requires an $N = 2$ extended superconformal algebra in the supersymmetric sector of the world-sheet theory.

The fact that the $U(1)$ current J does not commute with the $N = 1$ supercharge T_F means that it is not the upper component of an $N = 1$ superfield (in fact, it is the lower component of a superfield) and does not commute with Q_{BRST}. It therefore does not correspond to a physical state.

This is a consequence of the fact that the $U(1)$ symmetry generated by J is a global symmetry on the world-sheet which does not descend from a local symmetry. (Local symmetries would give rise to additional ghosts which lower the critical dimension. This was discussed in Chap. 12.). Furthermore, while $T_F \sim T_F^+ + T_F^-$ is a remnant of a local gauge symmetry on the world-sheet—local $N = 1$ supersymmetry—the enhancement to a global $N = 2$ symmetry is a property of a particular class of string vacua.

In addition to having extended $N = 2$ superconformal symmetry, space-time supersymmetry demands that all states have quantized (integer or half-integer) $U(1)$ charge Q. This can be derived from the requirement that the operator product expansion of an arbitrary state Eq. (15.70) with the gravitino vertex be local:

$$\left(\frac{1}{2}, \frac{1}{2}\right) \cdot \lambda_R + \frac{q}{2} + \frac{Q}{2} \in \mathbb{Z}. \tag{15.80}$$

The following combinations of $D_{2,1}$ conjugacy classes and $U(1)$ charges are then allowed:

$$(0, 2\mathbb{Z}), \quad (V, 2\mathbb{Z} + 1), \quad \left(S, 2\mathbb{Z} + \frac{1}{2}\right), \quad \left(C, 2\mathbb{Z} - \frac{1}{2}\right). \tag{15.81}$$

This selection rule, which is a consequence of locality, is nothing but the GSO projection. If we had started with supercharges (15.57) with S_α and $S_{\dot{\alpha}}$ interchanged, the quantization condition would be (15.81) with the two spinor conjugacy classes interchanged. For type IIB we take the same condition for both sectors and different ones for type IIA. Note that locality forbids to keep both types of supercharges in one sector.

The internal part $\Sigma(z)$ of the space-time supercharge Q_α is the spectral flow operator of the $N=2$ superconformal algebra with $\eta = -\frac{1}{2}$. In general the spectral flow operator has conformal weight $h = \frac{c}{24}$ and $U(1)$ charge $Q = \frac{c}{6}$, which indeed agrees with the conformal dimension $h = \frac{3}{8}$ and $Q = \frac{3}{2}$ of Σ. Also, one easily verifies (12.104) on a general operator of the form $\exp(i \frac{Q}{\sqrt{3}})\mathcal{O}$. Hence, space-time supersymmetry transformations are nothing else than applying the spectral flow operator in the SCFT. For general parameter η, the spectral flow operator is $U_\eta = \exp(-i\sqrt{3}\eta H)$. For $\eta = -1$ this is essentially the square of Σ. Geometrically speaking, U_{-1} can be associated with the holomorphic three form Ω_3 of a Calabi-Yau manifold, which we know can be written as the square of the Killing spinor.

Our analysis so far mainly dealt with supercharges emanating from the right-moving sector. In type II theories also the left-moving sector can provide supercharges. It also possesses local $N=1$ world sheet supersymmetry and for left-right symmetric models one gets at least $\mathcal{N}=2$ space-time supersymmetry, which arises from a CFT with global $N=(2,2)$ supersymmetry with quantized $U(1)_L \times U(1)_R$ charges. Here we assume that there is no open string sector as e.g. in orientifold compactifications which we have discussed in the previous section. Using the same left-right symmetric $N=(2,2)$ SCFT for the $(c^{\text{int}}, \overline{c}^{\text{int}}) = (9,9)$ internal part of a heterotic string, the enhanced left-moving superconformal symmetry leads to the heterotic gauge group E_6. The connection between exceptional groups and space-time supersymmetry will be discussed later in this section.

We now discuss the possible massless spectra of four-dimensional string theories with minimal supersymmetry; $\mathcal{N}=1$ for the heterotic string and $\mathcal{N}=2$ for the type II string. For the latter case we discuss symmetric $(2,2)$ CFTs. The similarity to the geometric discussion in Chap. 14 will be apparent. In fact, the discussion here is the simply the translation into the CFT language.

Universal Sector

This sector is present in any four-dimensional string theory. Therefore, it gives a general prediction for the physics consequences of any string theory in four dimensions.

Any unitary internal CFT contains the identity operator with $h_{\text{int}} = \overline{h}_{\text{int}} = 0$. Also, the 0 conjugacy class of $(D_{2,1})_R$ must always be present and it leads in the NS sector with canonical ghost charge $q = -1$ to a space-time vector with $\lambda = (\pm 1, 0), (0, \pm 1)$. Furthermore, $\partial X^\mu(\overline{z})$ is always possible. This combination of conformal fields yields the graviton $G_{\mu\nu}$, the anti-symmetric tensor $B_{\mu\nu}$ and the dilaton D. In addition there is the space-time supercharge which is the zero-momentum part of one chiral half of the gravitino vertex operator. It leads in the canonical ghost picture to a space-time spinor with $\lambda = \pm(\frac{1}{2}, \frac{1}{2}), \pm(\frac{1}{2}, -\frac{1}{2})$. We

15.4 World-Sheet Aspects of Supersymmetric Compactifications

then get the following states in the canonical ghost picture for the heterotic and the type II superstring theories.[7]

Heterotic

In the heterotic string we obtain

$$V^{G,B,D} = \epsilon_{\mu\nu} \bar{\partial}X^\mu \, \psi^\nu \, e^{-\phi} \, e^{ik\cdot X}$$
$$V^{\psi,\lambda} = \bar{\partial}X^\mu \left[u_\mu^\alpha S_\alpha \Sigma + \bar{u}_{\mu\dot{\alpha}} S^{\dot{\alpha}} \Sigma^\dagger \right] e^{-\frac{1}{2}\phi} e^{ik\cdot X}, \qquad (15.82)$$

where $\epsilon_{\mu\nu}$ is the polarization tensor. It is symmetric and traceless for the graviton $G_{\mu\nu}$, antisymmetric for $B_{\mu\nu}$ (which is a pseudo-scalar χ in four dimensions after a Hodge-duality transformation $\epsilon_\mu{}^{\nu\rho\sigma} \partial_\nu B_{\rho\sigma} = \partial_\mu \chi$) and the trace part for the dilaton D. u_μ^α, $\bar{u}_\mu^{\dot{\alpha}}$ are right- and left-handed polarization vector-spinors. They are not irreducible representations of the Lorentz groups. The two irreducible components correspond to the gravitino ψ with $u_{\mu\alpha}$ such that $\bar{\sigma}^{\mu\dot{\alpha}\alpha} u_{\mu\alpha} = 0$ and the dilatino λ with $\bar{u}_\mu^{\dot{\alpha}} = \bar{\sigma}_\mu^{\dot{\alpha}\alpha} u_\alpha$. BRST invariance imposes the on-shell conditions $k^2 = 0$, $k^\mu \epsilon_{\mu\nu} = k^\nu \epsilon_{\mu\nu} = 0$, $k^\mu u_{\mu\alpha} = 0$ and $\not{k} u_\mu = 0$.

These states are the on-shell components of the $\mathcal{N} = 1$ supergravity multiplet (G, Ψ) and of a linear multiplet (B, D, λ) which can be dualized to a chiral multiplet (χ, D, λ). One can also explicitly construct the vertex operators for the auxiliary fields in the various multiplets. They do not correspond to physical states and are not BRST invariant.

Type II

In type II theories the universal sector is richer, due to the extended supersymmetry whose supermultiplets contain more states. One has to combine the conjugacy classes of $(D_{2,1})_L \times (D_{2,1})_R$. Various combinations lead to states in different sectors. Analogous to (15.82) there are

$$V^{G,B,D} = \epsilon_{\mu\nu} \bar{\psi}^\mu e^{-\bar{\phi}} \psi^\nu e^{-\phi} e^{ik\cdot X},$$
$$V^{\psi_1,\lambda_1} = \bar{\psi}^\mu \left[u_{1\mu}^\alpha S_\alpha \Sigma + \bar{u}_{1\mu}^{\dot{\alpha}} S_{\dot{\alpha}} \Sigma^\dagger \right] e^{-\frac{1}{2}(\bar{\phi}+\phi)} e^{ik\cdot X},$$
$$V^{\psi_2,\lambda_2} = \left[u_{2\mu}^\alpha \bar{S}_\alpha \bar{\Sigma} + \bar{u}_{2\mu}^{\dot{\alpha}} \bar{S}_{\dot{\alpha}} \bar{\Sigma}^\dagger \right] \psi^\mu e^{-\frac{1}{2}(\bar{\phi}+\phi)} e^{ik\cdot X}. \qquad (15.83)$$

There are two gravitini and two dilatini. We have given their vertex operators for the type IIB theory. For type IIA, $\overline{\Sigma}$ and $\overline{\Sigma}^\dagger$ are exchanged. In addition there is a vector field, the graviphoton A, and a complex scalar field $C, \overline{C}$ with vertex operators

---

[7]In this section we use unnormalized vertex operators. We will discuss their normalization in Chap. 16.

$$V^A = \epsilon_\mu k_\nu \left[\bar{S}^\alpha \,\overline{\Sigma}\, \sigma^{\mu\nu}{}_\alpha{}^\beta \, S_\beta \, \Sigma + \bar{S}_{\dot\alpha} \,\overline{\Sigma}^\dagger\, \bar{\sigma}^{\mu\nu\dot\alpha}{}_{\dot\beta} \, S^{\dot\beta} \, \Sigma^\dagger \right] e^{-\frac{1}{2}(\phi+\bar\phi)} e^{ik\cdot X} \,,$$

$$V^C = k_\mu \, \bar{S}_{\dot\alpha} \,\overline{\Sigma}^\dagger\, \bar{\sigma}^{\mu\dot\alpha\beta} \, S_\beta \, \Sigma \, e^{-\frac{1}{2}(\phi+\bar\phi)} \, e^{ik\cdot X} \,,$$

$$V^{\bar C} = k_\mu \, \bar{S}^\alpha \,\overline{\Sigma}\, \sigma^\mu{}_{\alpha\dot\beta} \, S^{\dot\beta} \, \Sigma^\dagger \, e^{-\frac{1}{2}(\phi+\bar\phi)} \, e^{ik\cdot X} \,. \tag{15.84}$$

Again, these expressions are valid for the type IIB theory, the ones for type IIA being obtained by the interchange $\overline{\Sigma} \leftrightarrow \overline{\Sigma}^\dagger$.

A naive guess for the vertex operator, say for the scalar C, would contain $\bar{S}^\alpha S_\alpha$ without any explicit dependence on the space-time momentum. This is, however, not BRST invariant. The explicit k-dependence of the (R,R) vertex operators indicates that the corresponding massless fields have only derivative couplings in the low-energy effective action. In other words they only appear through their field strength which implies that there are e.g. no fields charged with respect to the graviphoton.[8] Also note that the two contributions in V^A correspond to the self-dual and anti-selfdual parts of the graviphoton's field-strength.

The fields in the universal sector combine into the gravity multiplet (G, ψ_1, ψ_2, A) and, after dualizing the antisymmetric tensor, into the universal hypermultiplet $(D, \chi, C, \bar C, \lambda_1, \lambda_2)$.

Non-universal Sector

In addition, there are gauge fields, gaugini and matter fields (scalars and spin-1/2 fermions). We will discuss all of them in turn. There are again differences between heterotic and type II theories.

Heterotic

Massless space-time scalars $\Phi_i \, (i = 1, \ldots, N_S)$ have $\boldsymbol{\lambda}_R = \mathbf{0}$ and $q = -1$. Therefore their existence implies the presence of internal fields $\Lambda^i(z, \bar z)$ with conformal dimension $(\bar h, h) = (1, \frac{1}{2})$. Their vertex operators in the -1 ghost picture are

$$V^{\Phi_i} = \Lambda^i \, e^{-\phi} \, e^{ik\cdot X}. \tag{15.85}$$

The fields Λ^i transform under some representation of the left-moving gauge symmetry of the heterotic string.

BRST invariance implies that operator products between $T_F^{\text{int}}(z)$ and $\Lambda^i(w, \bar w)$ have the structure

---

[8] The same comment applies to the ten-dimensional effective action of the uncompactified theories. The fields which correspond to vertex operators in the (R,R) sector appear only through their field strengths.

15.4 World-Sheet Aspects of Supersymmetric Compactifications

$$T_F^{\text{int}}(z)\, \Lambda^i(w,\overline{w}) \sim \frac{\hat{\Lambda}^i(w,\overline{w})}{z-w} + \text{finite} , \qquad (15.86)$$

where $\hat{\Lambda}^i(w,\overline{w})$ are internal conformal fields of dimension $(h, \overline{h})_{\text{int}} = (1, 1)$. This means that $\Lambda^i(z, \overline{z})$ and $\hat{\Lambda}^i(z, \overline{z})$ are the two components of a two-dimensional superfields with respect to the right-moving $N = 1$ world-sheet supersymmetry. Then the scalar vertex operator in the 0 ghost picture is

$$V^{\Phi_i} = [\Lambda^i\, k \cdot \psi + 2\hat{\Lambda}^i]\, e^{ik \cdot X} , \qquad (15.87)$$

where the first term comes from picture changing with the space-time supercurrent $T_F^{\text{s.t.}} = \frac{i}{2} \psi_\mu \partial X^\mu$.

We now show that Λ^i are chiral or anti-chiral primaries of the right-moving $N = 2$ algebra. To do this we use the fact that a primary field of the $N = 1 \subset N = 2$ algebra with $h = \frac{1}{2}$ is automatically a primary field of the $N = 2$ algebra. This claim can be proven as follows: $[L_0, G_r^\pm] = -r G_r^\pm$ applied to a state $|\phi\rangle$ with $L_n|\phi\rangle = 0$ for $n > 0$ and $L_0|\phi\rangle = \frac{1}{2}|\phi\rangle$, gives $L_0(G_r^\pm|\phi\rangle) = (\frac{1}{2} - r)(G_r^\pm|\phi\rangle)$. For a unitary CFT the spectrum of L_0 is positive, hence $G_r^\pm|\phi\rangle = 0$ for $r > \frac{1}{2}$. Furthermore, $L_0 G_{1/2}^\pm|\phi\rangle = 0$. This either implies $G_{1/2}^\pm|\phi\rangle = 0$ or that $G_{1/2}^\pm|\phi\rangle$ is the state created by the identity operator. The only candidate for $|\phi\rangle$ is then $|\phi\rangle = G_{-1/2}^\mp|0\rangle$, but this state can be shown (use the $N = 2$ algebra) to be annihilated by $G_{1/2}^\pm$. Using $[L_0, J_n] = -n J_n$ and once again positivity of L_0 allows us to show $J_n|\phi\rangle = 0$ for $n > 0$. This concludes the proof of our claim. Combined with the integrality of the $U(1)$ charge (15.81) this leads to $|Q| = 1$, i.e. Λ^i is chiral or anti-chiral primary with respect to the right-moving $N = 2$ algebra. Below, when we discuss special features of heterotic $(2, 2)$ compactifications, we will see how (anti-) chiral primaries also appear on the left-moving side.

Massless space-time spin-1/2 fermions have to be spinors with $\lambda_R = (\pm\frac{1}{2}, \pm\frac{1}{2})$ and $Q = -\frac{1}{2}$. Their internal conformal fields $\Sigma^i(z, \overline{z})$ must be operators of dimension $(\overline{h}, h)_{\text{int}} = (1, \frac{3}{8})$. These fermions are obtained from the massless scalar fields Φ_i ($i = 1, \ldots, N_S$, $N_S = N_F$) by applying a space-time supersymmetry transformation. In the language of $N = 2$ SCFT the $\Sigma^i(z, \overline{z})$ are obtained from the chiral primaries $\Lambda^i(z, \overline{z})$ with $(h, Q) = (\frac{1}{2}, \pm 1)$ by applying the holomorphic spectral flow operator $U_{\pm 1/2}$. This leads to operators Σ^i with $(h, Q) = (\frac{3}{8}, \mp\frac{1}{2})$, the Ramond ground states of the right-moving SCFT.

The complete fermion vertex operator in the canonical ghost picture is

$$V^{\Psi_i} = \Sigma^i\, S_\alpha\, e^{-\frac{1}{2}\phi}\, e^{ik \cdot X} . \qquad (15.88)$$

BRST invariance requires

$$T_F^{\text{int}}(z)\, \Sigma^i(w, \overline{w}) \sim \frac{\hat{\Sigma}^i(w, \overline{w})}{(z-w)^{1/2}} + \text{finite} , \qquad (15.89)$$

where $\hat{\Sigma}^i$ are conformal fields of dimension $(\bar{h}, h) = (1, \frac{11}{8})$. The presence of the branch cut confirms that Σ^i is in the Ramond sector of the internal SCFT. For the fermions to be chiral the fields $\Sigma^i(z, \bar{z})$ must transform under a complex representation of the gauge group G, and it must be ensured that the spinors of negative chirality are not coupled to the same fields $\Sigma^i(z, \bar{z})$ as the spinors of positive chirality.

Constraints from SCFT also exclude the presence of fermionic tachyons. The contribution to the conformal dimension of an R state in the canonical ghost picture is $\geq \frac{5}{8}$ and the internal CFT contributes $h_{\text{int}} \geq \frac{c_{\text{int}}}{24} = \frac{3}{8}$ (cf. Eq. (12.32b) on page 362) so that $h \geq 1$. The contribution from $e^{ik \cdot X}$ must then be $\frac{1}{2}k^2 \leq 0$. This especially excludes the presence of tachyons in space-time supersymmetric theories.

We finally turn to vectors. Massless vectors are gauge fields and we assume that the space-time gauge symmetry has its world-sheet origin in an internal Kač-Moody algebra of level k. The affine Kač-Moody algebra Eq. (11.17) can be realized either on the left-moving side by currents $\bar{J}^a(\bar{z})$ ($a = 1, \ldots, \dim G$) or on the right-moving side by currents $J^a(z)$. Recall that the currents have conformal weight one. In the former case the vertex operators of the gauge bosons are

$$V^a = \epsilon_\mu \bar{J}^a \psi^\mu e^{-\phi} e^{ik \cdot X} \tag{15.90}$$

with $k^\mu \epsilon_\mu = 0$. In the latter case, BRST invariance demands that $J^a(z)$ is the upper component of a two-dimensional superfield, i.e. it can be obtained via picture changing from a dimension $\frac{1}{2}$ (fermionic) field $\hat{J}^a(z)$. Then the right-moving heterotic gauge boson vertex operator in the canonical ghost picture is

$$V^a = \epsilon_\mu \bar{\partial} X^\mu \hat{J}^a e^{-\phi} e^{ik \cdot X} , \tag{15.91}$$

whereas in the 0-ghost picture we have

$$V^a = \epsilon_\mu \bar{\partial} X^\mu J^a e^{ik \cdot X} . \tag{15.92}$$

Vertex operators of the form Eq. (15.92) with $J^a(z)$ not being the upper component of a two-dimensional superfield are not BRST invariant. They are "auxiliary" non-propagating fields which are needed for the (off-shell) description of space-time superfields.

If J^a is a Kač-Moody current and if $\hat{J}^a$ and J^a are the lower and upper components of a fermionic $N = 1$ superfield, they can be shown to generate a super-Kač-Moody algebra. Indeed, using (12.25) with $h = 1/2$ and $(\phi_0, \phi_1) \to (\hat{J}^a, J^a)$ and (11.17), one finds with the help of the graded Jacobi-identity that the modes of $\hat{J}^a$ and J^a satisfy

$$[J_m^a, J_n^b] = if^{abc} J_{m+n}^c + m k \delta^{ab} \delta_{m+n} ,$$

15.4 World-Sheet Aspects of Supersymmetric Compactifications

$$[J_m^a, \hat{J}_r^b] = if^{abc} \hat{J}_{m+r}^c, \qquad r \in \begin{cases} \mathbb{Z} & \text{R} \\ \mathbb{Z} + \frac{1}{2} & \text{NS} \end{cases}$$

$$\{\hat{J}_r^a, \hat{J}_s^b\} = k\delta^{ab}\delta_{r+s}. \tag{15.93}$$

In terms of the current superfield $\mathscr{J}^a = \hat{J}^a + \theta J^a$, this is equivalent to the OPE

$$\mathscr{J}^a(z_1)\mathscr{J}^b(z_2) = \frac{k\delta^{ab}}{z_{12}} + i\frac{\theta_{12}}{z_{12}} f^{abc} \mathscr{J}^c(z_2). \tag{15.94}$$

One feature worth pointing out is that the fermionic currents $\hat{J}^a$ are free fermions.

Consider now the constraint (12.32b) for the internal SCFT with $\hat{c} = 6$. As massless space-time fermions have $h_{\text{int}} = \frac{3}{8}$, the inequality must be saturated. If the internal SCFT is a direct sum of several SCFTs, this equality must hold for each of them separately, i.e.

$$h_{\text{int}}^i = \frac{\hat{c}_{\text{int}}^i}{16}. \tag{15.95}$$

This, in particular, applies to the case that one of the internal SCFTs is a super-Kač-Moody algebra. Using properties of super-Kač-Moody algebras, it is possible to show that the equality is always violated for non-Abelian currents, i.e. there cannot be massless fermions which are charged under non-Abelian gauge symmetries from the right-moving sector. For Abelian currents the equality can be saturated, but only for fermions which are neutral with respect to the corresponding charges. Finally, gauge symmetries based on super-Kač-Moody currents exclude a chiral spectrum. This is easy to prove: if there are n fermionic $U(1)$ currents $\hat{J}^i$, they combine with the free space-time fermions ψ^μ into an $SO(1, 3 + n)$ vector. The Ramond ground state is therefore a spinor of $SO(1, 3 + n)$ whose decomposition with respect to $SO(1, 3)$ includes space-time spinors of both chiralities.

Let us now return to the left gauge bosons, Eq. (15.90). The internal Kač-Moody algebra $\hat{g}$ contributes to the central charge of the internal CFT with (cf. (11.32) on page 326)

$$\bar{c} = \frac{k \dim G}{k + C_g}. \tag{15.96}$$

Clearly this value must not exceed the total internal central charge $\bar{c} = 22$. This sets a limit on the dimension of the gauge group, depending on the level k and the dual Coxeter number C_g. If k is one, the rank of the gauge group must be less than or equal to 22. For simply laced groups at level one the internal CFT can be realized by free bosons or fermions. For these cases the internal CFT has a very simple structure.

Type II

For type II theories with $N = (2,2)$ superconformal symmetry the massless scalar fields $\Lambda^i(z,\bar{z})$ have conformal dimensions $\bar{h} = h = \frac{1}{2}$ and $U(1)$ charges $\bar{Q} = \pm Q = \pm 1$, i.e. they are elements of one of the four chiral rings which we denoted in Chap. 12 by (c,c), (a,a), (a,c) and (c,a). The rings (c,c) and (a,a) as well as (a,c) and (c,a) are complex conjugate and isomorphic pairs. As a result we get two non-equivalent chiral rings, e.g. (c,c) and (a,c). We denote their elements by Λ^α and Λ^a. We will comment on the ranges of the indices α and a momentarily. With respect to either $N = 2$ SCA, Λ is the lowest component of an (anti)chiral superfield (cf. (12.76) on page 372), whose upper component (either $[G^+_{-1/2}, \Lambda^i]$ or $[G^-_{-1/2}, \Lambda^i]$) has charge $Q = 0$ and $h = 1$.

Denote by $U_{\bar{\eta},\eta}$ the spectral flow with parameter η ($\bar{\eta}$) for the (anti)holomorphic flow. For $\eta, \bar{\eta} \in \{0, \pm\frac{1}{2}\}$, Λ flows to fields in the (NS,R), (R,NS) and (R,R) sectors of the internal CFT. The flow patterns and the weights of the resulting fields are as follows:

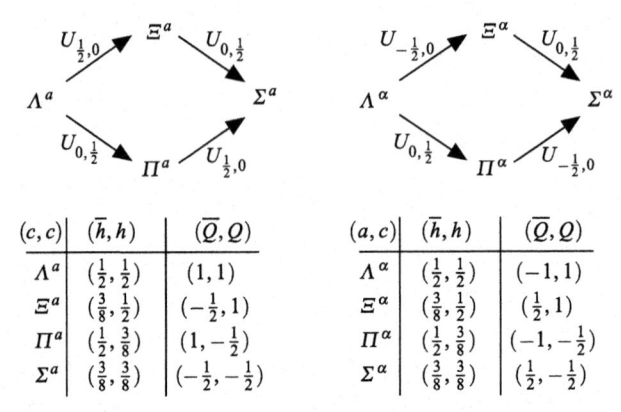

(c,c)	$(\bar{h},h)$	$(\bar{Q},Q)$
Λ^a	$(\frac{1}{2},\frac{1}{2})$	$(1,1)$
Ξ^a	$(\frac{3}{8},\frac{1}{2})$	$(-\frac{1}{2},1)$
Π^a	$(\frac{1}{2},\frac{3}{8})$	$(1,-\frac{1}{2})$
Σ^a	$(\frac{3}{8},\frac{3}{8})$	$(-\frac{1}{2},-\frac{1}{2})$

(a,c)	$(\bar{h},h)$	$(\bar{Q},Q)$
Λ^α	$(\frac{1}{2},\frac{1}{2})$	$(-1,1)$
Ξ^α	$(\frac{3}{8},\frac{1}{2})$	$(\frac{1}{2},1)$
Π^α	$(\frac{1}{2},\frac{3}{8})$	$(-1,-\frac{1}{2})$
Σ^α	$(\frac{3}{8},\frac{3}{8})$	$(\frac{1}{2},-\frac{1}{2})$

These fields make up the internal part of the vertex operators of massless bosons and fermions. For instance, in type IIA and IIB we have the (NS,NS) scalars[9]

$$V^{t^a} = (\Lambda^a + \bar{\Lambda}^a) e^{-(\bar{\phi}+\phi)} e^{ik\cdot X}$$
$$V^{t^\alpha} = (\Lambda^\alpha + \bar{\Lambda}^\alpha) e^{-(\bar{\phi}+\phi)} e^{ik\cdot X} \ . \tag{15.97}$$

Applying space-time supersymmetry transformations to these fields, we obtain spin-1/2 fermions in the (R,NS) and (NS,R) sectors:

Type IIB

$$V^{\psi^a} = k_\mu (\bar{S}^\beta \sigma^\mu_{\beta\dot{\beta}} u^{\dot{\beta}} \bar{\Xi}^a + \bar{S}_{\dot{\beta}} \bar{\sigma}^{\mu\dot{\beta}\beta} u_\beta \Xi^a) e^{-(\frac{1}{2}\bar{\phi}+\phi)} e^{ik\cdot X} ,$$

---

[9]In these and the following expressions α is not a spinor index.

15.4 World-Sheet Aspects of Supersymmetric Compactifications

$$V^{\lambda^a} = k_\mu(u_{\dot\beta}\,\overline\sigma^{\mu\dot\beta\beta}\,S_\beta\,\overline\Pi^a + u^\beta\,\sigma^\mu_{\beta\dot\beta}\,S^{\dot\beta}\,\Pi^a)\,e^{-(\overline\phi+\frac{1}{2}\phi)}\,e^{ik\cdot X}\,,$$

$$V^{\psi^\alpha} = k_\mu(\overline S^\beta\,\sigma^\mu_{\beta\dot\beta}\,u^{\dot\beta}\,\Xi^\alpha + \overline S_{\dot\beta}\,\overline\sigma^{\mu\dot\beta\beta}\,u_\beta\,\overline\Xi^\alpha)\,e^{-(\frac{1}{2}\overline\phi+\phi)}\,e^{ik\cdot X}\,,$$

$$V^{\lambda^\alpha} = k_\mu(u^\beta\,\sigma^\mu_{\beta\dot\beta}\,S^{\dot\beta}\,\Pi^\alpha + u_{\dot\beta}\,\overline\sigma^{\mu\dot\beta\beta}\,S_\beta\,\overline\Pi^\alpha)\,e^{-(\overline\phi+\frac{1}{2}\phi)}\,e^{ik\cdot X}\,. \tag{15.98}$$

Type IIA

$$V^{\psi^a} = k_\mu(\overline S^\beta\,\sigma^\mu_{\beta\dot\beta}\,u^{\dot\beta}\,\Xi^a + \overline S_{\dot\beta}\,\overline\sigma^{\mu\dot\beta\beta}\,u_\beta\,\Xi^a)\,e^{-(\frac{1}{2}\overline\phi+\phi)}\,e^{ik\cdot X}\,,$$

$$V^{\lambda^a} = k_\mu(u_{\dot\beta}\,\overline\sigma^{\mu\dot\beta\beta}\,S_\beta\,\overline\Pi^a + u^\beta\,\sigma^\mu_{\beta\dot\beta}\,S^{\dot\beta}\,\Pi^a)\,e^{-(\overline\phi+\frac{1}{2}\phi)}\,e^{ik\cdot X}\,,$$

$$V^{\psi^\alpha} = k_\mu(\overline S^\beta\,\sigma^\mu_{\beta\dot\beta}\,u^{\dot\beta}\,\Xi^\alpha + \overline S_{\dot\beta}\,\overline\sigma^{\mu\dot\beta\beta}\,u_\beta\,\overline\Xi^\alpha)\,e^{-(\frac{1}{2}\overline\phi+\phi)}\,e^{ik\cdot X}\,,$$

$$V^{\lambda^\alpha} = k_\mu(u^\beta\,\sigma^\mu_{\beta\dot\beta}\,S^{\dot\beta}\,\Pi^\alpha + u_{\dot\beta}\,\overline\sigma^{\mu\dot\beta\beta}\,S_\beta\,\overline\Pi^\alpha)\,e^{-(\overline\phi+\frac{1}{2}\phi)}\,e^{ik\cdot X}\,. \tag{15.99}$$

Applying a second supersymmetry transformation yields vectors and scalars in the (R,R) sectors (scalars and vectors):

Type IIB

$$V^{A^a} = k_\mu\epsilon_\nu(\overline S_{\dot\beta}\,\overline\sigma^{\mu\nu\dot\beta}{}_{\dot\gamma}\,S^{\dot\gamma}\,\Sigma^a + \overline S^\beta\,\sigma^{\mu\nu}{}_\beta{}^\gamma\,S_\gamma\,\overline\Sigma^a)\,e^{-\frac{1}{2}(\overline\phi+\phi)}\,e^{ik\cdot X}\,,$$

$$V^{C^\alpha} = k_\mu\,(\overline S^\beta\,\sigma^\mu_{\beta\dot\beta}\,S^{\dot\beta}\,\Sigma^\alpha + \overline S_{\dot\beta}\,\overline\sigma^{\mu\dot\beta\beta}\,S_\beta\,\overline\Sigma^\alpha)\,e^{-\frac{1}{2}(\overline\phi+\phi)}\,e^{ik\cdot X} \tag{15.100}$$

Type IIA

$$V^{A^a} = k_\mu\epsilon_\nu(\overline S^\beta\,\sigma^{\mu\nu}{}_\beta{}^\gamma\,S_\gamma\,\Sigma^\alpha + \overline S_{\dot\beta}\,\overline\sigma^{\mu\nu\dot\beta}{}_{\dot\gamma}\,S^{\dot\gamma}\,\overline\Sigma^\alpha)\,e^{-\frac{1}{2}(\overline\phi+\phi)}\,e^{ik\cdot X}\,,$$

$$V^{C^a} = k_\mu(\overline S_{\dot\beta}\,\overline\sigma^{\mu\dot\beta\beta}\,S_\beta\,\Sigma^a + \overline S^\beta\,\sigma_{\mu\beta\dot\beta}\,S^{\dot\beta}\,\overline\Sigma^a)\,e^{-\frac{1}{2}(\overline\phi+\phi)}\,e^{ik\cdot X}\,. \tag{15.101}$$

For the bosons we have used the same notation as we did in Sect. 14.6. (Do not confuse the index on Σ^α with a spinor index.) The states organize into multiplets of $\mathcal{N} = 2$ space-time supersymmetry. We also indicate from which of the four rings of chiral primaries they originate:

Type IIB   hyper: $(\tilde t^\alpha, C^\alpha, \psi^\alpha, \lambda^\alpha)$ $(a,c)+(c,a)$ (Kähler)

vector: $(t^a, \psi^a, \lambda^a, A^a)$ $(c,c)+(a,a)$ (complex structure)

Type IIA   hyper: $(t^a, C^a, \psi^a, \lambda^a)$ $(c,c)+(a,a)$ (complex structure)

vector: $(\tilde t^\alpha, \psi^\alpha, \lambda^\alpha, A^\alpha)$ $(a,c)+(c,a)$ (Kähler)

Comparing this with Sect. 14.6 and asserting that the choice of internal CFT has an alternative description in terms of a geometric compactification on a Calabi-Yau manifold X, we find that for type IIB the range of the indices is $a = 1, \ldots, h^{2,1}(X)$ and $\alpha = 1, \ldots, h^{1,1}(X)$, while it is reversed in type IIA. This also suggests (as already alluded to in Chap. 14) the isomorphism between the various elements of chiral rings and the cohomology of CY manifolds.

The scalars from the (NS,NS) sector are moduli of the low energy effective field theory, i.e. they are massless and there is no potential which would fix their vacuum expectation value. They parametrize flat directions. In CFT language this means that there is a one-to-one correspondence with exactly marginal operators of conformal dimension $(\bar{h}, h) = (1, 1)$ which can be added to the world-sheet action without destroying conformal invariance. Indeed, the upper components of the chiral superfields are exactly marginal. They are the vertex operators of the massless scalars in the zero ghost picture at zero momentum. They are protected from receiving an anomalous dimension because of the relation $|Q| = 2h$; i.e. they satisfy a so-called BPS property (see Chap. 18). Equivalently, they are in short multiplets of the extended world-sheet superconformal algebra, while a generic multiplet with $h > \frac{1}{2}|Q|$ contains (at least) twice as many component fields.

The CFT description of CY compactifications led to a very interesting proposal: mirror symmetry. Going from type IIA compactifications to type IIB compactifications merely exchanges the roles of the two chiral rings. The difference between type IIA versus type IIB in four dimensions is merely a change of the arbitrary relative sign between the right- and left-moving $U(1)$ charges Q and $\bar{Q}$. The four-dimensional spectrum of massless scalar fields is the same in $N = (2, 2)$ type IIA/B string compactifications. Nevertheless, the two distinct chiral rings have very different geometric interpretations when viewing the $N = (2, 2)$ SCFT with $c = \bar{c} = 9$ as a six-dimensional Calabi-Yau manifold. In this case, the (a, c) chiral primary fields correspond in type IIB to the Kähler deformations of the Calabi-Yau manifold, and the (c, c) chiral primary fields correspond to the complex structure deformations. The associated four-dimensional massless scalar fields are the Kähler and complex structure moduli fields. In type IIA compactifications their geometric role is precisely reversed.

Switching from type IIB to type IIA and vice versa by reversing the sign of one of the $U(1)$ charges now has a striking consequence: the replacement of one particular Calabi-Yau space X by another one, denoted by $\hat{X}$, which exchanges complex structure and Kähler deformations. This mirror transformation between X and $\hat{X}$, which we already discussed in Chap. 14 in the geometric setting is, in terms of the SCFT, a trivial relabeling of two-dimensional $U(1)$ charges. The existence of mirror pairs of Calabi-Yau manifolds has meanwhile been proven mathematically, at least for a large class of Calabi-Yau hypersurfaces in toric varieties.

The two different chiral rings have also an interpretation in topological CFT (see Sect. 12.3). There they correspond to the finite number of BRST invariant operators. The two topological sectors are also called the topological A and B model, respectively, which correspond to the two possible independent topological twists of $\bar{T}(\bar{z})$ relative to $T(z)$.

15.4 World-Sheet Aspects of Supersymmetric Compactifications

The vectors we have discussed so far are $U(1)$ gauge fields. In perturbative string theory there are no states which are charged w.r.t. them. In particular, the gauge group cannot be enhanced to a non-Abelian group. This is no longer the case if we go beyond perturbation theory. There can be additional gauge bosons arising from branes wrapping non-trivial cycles of the internal manifold. These generate non-Abelian gauge groups where the $U(1)$ gauge fields in the (R,R) sector generate the Cartan subalgebra. For the heterotic string additional non-perturbative gauge symmetries arise, if instantons in the $E_8 \times E_8$ and $SO(32)$ gauge group have zero size. For these non-perturbative gauge fields and charged matter fields a CFT description is not known.

Gauge fields of the type (15.91) can also exist for type II superstrings, both from the left- and right-moving sectors, with $c = \bar{c} = 9$. Their vertex operators are

$$V^a = \epsilon_\mu \overline{\psi}^\mu \hat{J}^a e^{-(\bar{\phi}+\phi)} e^{ik \cdot X} \tag{15.102}$$

and a similar expression with left-and right-movers interchanged. As in the heterotic case, cf. (15.91,15.92), the fermionic currents $\hat{J}^a$ are the lower components of $h = \frac{1}{2}$ superfields, whose higher components are the bosonic Kač-Moody currents J^a. Together they generate a super-Kač-Moody algebra. The discussion of right-moving gauge symmetries for the heterotic string can be applied to the type II string and leads to the following conclusions. First, if there are any non-Abelian gauge groups based on a super-Kač-Moody algebra, there cannot be any massless R-R scalars. Second, there cannot be any charged massless fermions in left-right symmetric compactifications with super-Kač-Moody algebras in the holomorphic sector and in the anti-holomorphic sector. Even if one considers asymmetric compactifications with a super-Kač-Moody symmetry in only one of the two sectors, one can prove that it is impossible to obtain the standard model with gauge group $SU(3) \times SU(2) \times U(1)$ and fermions in the required representations.

Note, however, that this no-go theorem is only valid in perturbative type II theory. Beyond perturbation theory D-branes and orientifold planes provide a different mechanism for generating gauge symmetries and one can indeed construct compactifications with the standard-model gauge group and fermions in the correct representations. The principles of these constructions will be outlined in later chapters.

Space-Time Supersymmetry and Exceptional Groups

There exists an intriguing connection between superstrings with extended space-time supersymmetry and bosonic strings with extended exceptional gauge symmetries. In fact one can map one to the other via the so-called bosonic string map. We explain the basic group theoretical origin of this relation in the covariant formulation. The light-cone gauge version of this map will be further explored in the Gepner model construction in the next section.

The key point is that the conjugacy classes Eq. (15.81) generate the Lorentzian lattice $E_{3,1}$. Recall from Chap. 13 that we can map the Lorentzian $D_{2,1} \subset E_{3,1}$ lattice to the Euclidean D_5 lattice. Recall further that D_5 can be decomposed to $D_2^{\text{Lorentz}} \otimes D_3^{\text{ghost}}$ where states in the canonical ghost picture correspond to fixed D_3^{ghost} lattice vectors as shown in Eq. (13.128). Thus, all states Eq. (15.70) are characterized by vectors w in the D_5 lattice. (In the canonical ghost picture the space-time and superconformal ghost contribution to the conformal weight of any state is $\frac{1}{2}w^2$.) In this Euclidean version we are now dealing with a level one $SO(10)$ Kač-Moody algebra. $\mathcal{N}=1$ space-time supersymmetry, i.e. the existence of the free internal boson $H(z)$, further enlarges $SO(10)$ to a $SO(10) \times U(1)$ Kač-Moody algebra with lattice vectors $(w, \frac{Q}{\sqrt{3}})$. The supercharges are represented by the lattice vectors $(\pm(\frac{1}{2},\frac{1}{2}), \frac{1}{2}, \frac{1}{2}, -\frac{1}{2}, \frac{\sqrt{3}}{2})$ and $(\pm(\frac{1}{2},-\frac{1}{2}), \frac{1}{2}, \frac{1}{2}, -\frac{1}{2}, -\frac{\sqrt{3}}{2})$. However these lattice vectors are among the "spinorial" roots of E_6 (c.f. Eq. (11.53)). Their existence enlarges the $SO(10) \times U(1)$ Kač-Moody algebra to the level one E_6 Kač-Moody algebra. Locality of the operator algebra demands that all vectors $(w, \frac{Q}{\sqrt{3}})$ have integer scalar product with these roots, i.e. they span the weight lattice of E_6.

E_6 possesses three conjugacy classes, denoted by 0, 1 and $\bar{1}$, with lowest representations **1**, **27** and $\overline{\mathbf{27}}$, respectively. Under $D_5 \otimes U(1)$ the conjugacy classes decompose as

$$0 = (0,0) \oplus \left(V, \sqrt{3}\right) \oplus \left(S, -\frac{\sqrt{3}}{2}\right) \oplus \left(C, \frac{\sqrt{3}}{2}\right),$$

$$1 = \left(0, \frac{2}{3}\sqrt{3}\right) \oplus \left(V, -\frac{\sqrt{3}}{3}\right) \oplus \left(S, \frac{\sqrt{3}}{6}\right) \oplus \left(C, -\frac{5}{6}\sqrt{3}\right),$$

$$\bar{1} = \left(0, -\frac{2}{3}\sqrt{3}\right) \oplus \left(V, \frac{\sqrt{3}}{3}\right) \oplus \left(C, -\frac{\sqrt{3}}{6}\right) \oplus \left(S, \frac{5}{6}\sqrt{3}\right). \quad (15.103)$$

In this notation the 12 $U(1)$ conjugacy classes

$$q = -\sqrt{3}, -\frac{5}{6}\sqrt{3}, \ldots, \frac{5}{6}\sqrt{3} \quad (15.104)$$

define the elements α_q of the one-dimensional $U(1)$ weight lattice by

$$\alpha_q = q + 2\sqrt{3}k \quad (k \in \mathbb{Z}). \quad (15.105)$$

The conjugacy classes in Eq. (15.103) are those allowed by locality (cf. Eq. (15.81)), if we replace $D_{2,1}$ by D_5 and rescale the $U(1)$ charge by $\sqrt{3}$.

Space-time supersymmetry transformations act on a particular state (vertex operator) V_w ($w \in E_{3,1}$ resp. E_6) as

15.4 World-Sheet Aspects of Supersymmetric Compactifications

$$V'_{\mathbf{w}'}(w) = \oint_{C_w} \frac{dz}{2\pi i} Q_\alpha(z) V_{\mathbf{w}}(w). \tag{15.106}$$

Since the supercharge corresponds to spinorial root vectors α of E_6, $V'_{\mathbf{w}'}$ is characterized by the E_6 vector $\mathbf{w} + \alpha$. Thus the supersymmetric partners correspond to vectors within the same E_6 conjugacy class. It follows that the supermultiplet structure is encoded in the representations of the exceptional group E_6.

Reversing arguments, it is due to the appearance of E_6 that $\mathcal{N}=1$ space-time supersymmetry implies the $N=2$ superconformal algebra, together with the $U(1)$ quantization condition which is dictated by locality with the gravitino vertex. The quantization condition, the correlation between $U(1)$ charges and space-time transformation properties, are contained in the $E_{3,1}$ or, equivalently, in the E_6 weight lattice. The supercharge, being one of the roots, connects different states. It is also the operator which generates the spectral flow between the NS and R sectors as described in Chap. 12.

Let us exemplify the above by looking at the massless states of an arbitrary $\mathcal{N}=1$ supersymmetric (heterotic) string theory. Here only the weights of the fundamental and adjoint representations of E_6 are relevant. We start with the latter. It decomposes under $SO(10) \times U(1)$ as

$$\mathbf{78} = (\mathbf{45}, 0) + \left(\mathbf{16}, -\frac{\sqrt{3}}{2}\right) + \left(\overline{\mathbf{16}}, \frac{\sqrt{3}}{2}\right) + (\mathbf{1}, 0). \tag{15.107}$$

The $(\overline{\mathbf{16}}, \frac{\sqrt{3}}{2})$, $(\mathbf{16}, -\frac{\sqrt{3}}{2})$ representations correspond to a holomorphic spinor and antispinor, respectively, the supercharges Q_α, $\overline{Q}_{\dot\alpha}$. Group theoretically it is clear that the roots which correspond to the supercharges interpolate between the different $SO(10) \times U(1)$ conjugacy classes. But this can also be verified at the level of vertex operators if we make the truncation $D_5 \to D_{2,1}$. Indeed, acting with the supercharge on the holomorphic spinor we obtain the vector $\psi^\mu(z) e^{-\phi(z)}$. This state corresponds to the $(\mathbf{45}, 0)$ representation of $SO(10) \times U(1)$. But note that the spinor itself can be reached from a vertex operator $\sim i\sqrt{3}\partial H(z)$, i.e. from the $U(1)$ current. This state is not BRST invariant; it does not have a copy in the canonical ghost picture. It corresponds to an auxiliary field. The three states, the vector, the spinor and the auxiliary scalar are the off-shell degrees of freedom of an $\mathcal{N}=1$ vector supermultiplet. Multiplying these with a left-moving Kač-Moody current $j^a(\bar{z})$ leads to an $\mathcal{N}=1$ gauge vectormultiplet. Multiplying instead with $\bar\partial X^\mu(\bar{z})$ gives the $\mathcal{N}=1$ supergravity multiplet whose auxiliary vector is again $\bar\partial X^\mu(\bar{z})$ times the auxiliary $U(1)$ current.

The massless matter sector of the $\mathcal{N}=1$ supersymmetric heterotic string theory is obtained from the $\mathbf{27}$ ($\overline{\mathbf{27}}$) representation of E_6. The weights of this representation lead to conformal fields $e^{i\mathbf{w}\cdot\boldsymbol{\phi}(z)}$ ($\mathbf{w} \in E_{3,1}$) of dimension $\frac{2}{3}$. (The weights in the $\mathbf{27}$ and $\overline{\mathbf{27}}$ of E_6 have (length)$^2 = \frac{4}{3}$.) Therefore, to obtain massless fields one has to multiply this operator by a conformal field $\tilde{G}^i(z,\bar{z})$ ($i=1,\ldots,N$, N being the number of "families") of conformal dimension $\bar{h} = 1$, $\tilde{h} = \frac{1}{3}$ of the internal CFT with $\bar{c} = 22$ and $\tilde{c} = 8$. The fundamental representation of E_6 decomposes under $SO(10) \times U(1)$ as

$$27 = \left(16, \frac{\sqrt{3}}{6}\right) + \left(10, -\frac{\sqrt{3}}{3}\right) + \left(1, \frac{2\sqrt{3}}{3}\right). \tag{15.108}$$

The first term corresponds to a space-time spinor. Acting with the supercharge on this state gives the second term in Eq. (15.108) which corresponds to a physical massless scalar in the -1 ghost picture. Let us stress again that when determining the space-time properties of a state given by a particular $SO(10) \times U(1)$ representation, one always has to make the truncation $D_5 \to D_2$. The scalar vertex operators $V_i(z, \bar{z}) \sim e^{-i\frac{\sqrt{3}}{3}H_{\text{int}}(z)} \tilde{G}^i(z, \bar{z})$ of conformal dimension $(h, \bar{h}) = (\frac{1}{2}, 1)$ and $U(1)$ charge $Q = -1$ are the chiral primary fields of the internal $N = 2$ superconformal algebra. Finally, the last term in Eq. (15.108) corresponds to an unphysical, i.e. not BRST invariant, scalar in the 0 ghost picture. Together, these three fields are the off-shell degrees of freedom of an $\mathcal{N} = 1$ chiral multiplet (if we add also the CPT conjugate states in the $\overline{27}$ representation). The $U(1)$ charge is the R-charge of supersymmetric field theory.

The appearance of the exceptional group E_6 can also be used to show explicitly that the one-loop partition function of any four-dimensional $\mathcal{N} = 1$ supersymmetric heterotic string theory vanishes. Since every state is characterized by an E_6 weight vector it is clear that the partition function contains the sum over the E_6 weight lattice, i.e. over the three level one characters of E_6 which are in one-to-one correspondence with the three different conjugacy classes (the level one Kač-Moody characters were introduced in Chap. 11). The only subtlety which arises is the fact that one must sum only over physical transverse states, namely those states which have fixed D_4 lattice vectors x_0 (see Eq. (13.129)) when decomposing E_6 to $D_1 \otimes D_4 \otimes U(1)$. We will call these restricted E_6 characters $\chi_i(\tau)$ where i denotes the three conjugacy classes 0, 1 and $\bar{1}$. Then the partition function has the following general form:

$$Z(\tau, \bar{\tau}) \sim \frac{1}{\text{Im}\tau} \frac{1}{|\eta(\tau)|^4} \sum_{ij} a_{ij} \chi_i(\tau) \chi_j^{\tilde{c}=8,\bar{c}=22}(\tau, \bar{\tau}) \tag{15.109}$$

($i = 0, 1, \bar{1}$), where $\chi_j^{\tilde{c}=8,\bar{c}=22}$ are the characters of the internal CFT without the free boson H. The choice of the coefficients a_{ij} depends on the particular model and has to satisfy the constraints of modular invariance and spin statistics. The "true" (unrestricted) E_6 characters were given in Eq. (11.114). From them it is easy to derive the restricted, physical E_6 characters, if we make the truncation as described at the end of Sect. 13.4. We find

15.4 World-Sheet Aspects of Supersymmetric Compactifications

$$\chi_0(\tau) = \frac{1}{\eta^2(\tau)} \{\vartheta_3(0|3\tau)\,\vartheta_3(0|\tau) - \vartheta_4(0|3\tau)\,\vartheta_4(0|\tau) - \vartheta_2(0|3\tau)\,\vartheta_2(0|\tau)\},$$

$$\chi_1(\tau) = \frac{1}{\eta^2(\tau)} \left\{ -\vartheta\begin{bmatrix}1/6\\0\end{bmatrix}(0|3\tau)\,\vartheta_2(0|\tau) + \vartheta\begin{bmatrix}2/3\\0\end{bmatrix}(0|3\tau)\,\vartheta_3(0|\tau) \right.$$

$$\left. -e^{-2\pi i/3}\vartheta\begin{bmatrix}2/3\\1/2\end{bmatrix}(0|3\tau)\,\vartheta_4(0|\tau) \right\}$$

$$\chi_{\bar{1}}(\tau) = \frac{1}{\eta^2(\tau)} \left\{ -\vartheta\begin{bmatrix}5/6\\0\end{bmatrix}(0|3\tau)\,\vartheta_2(0|\tau) + \vartheta\begin{bmatrix}1/3\\0\end{bmatrix}(0|3\tau)\,\vartheta_3(0|\tau) \right.$$

$$\left. +e^{-\pi i/3}\vartheta\begin{bmatrix}1/3\\1/2\end{bmatrix}(0|3\tau)\,\vartheta_4(0|\tau) \right\}. \qquad (15.110)$$

Because of space-time supersymmetry, these characters (not however the original characters Eq. (11.114)) are supposed to vanish identically. This can indeed be proven using the theory of modular forms.

The appearance of the exceptional group E_6 in the holomorphic sector of four-dimensional heterotic vacua was dictated by space-time supersymmetry. However, there are no requirements on the left-moving, i.e. the bosonic sector, as far as global symmetries on the world-sheet are concerned. Except for conformal invariance, of course. But one can also impose conditions such as extended superconformal symmetry in the internal left-moving sector. For instance, one may consider left-right *symmetric* $\mathcal{N} = 1$ supersymmetric 'compactifications' of the heterotic string which have $(2,2)$ world-sheet superconformal symmetry. The holomorphic sector consists of a CFT with world-sheet fields $(b, c, \beta, \gamma, X^\mu, \psi^\mu)$ with $c = -9$ plus an internal $N = 2$ SCFT with $c^{\text{int}} = 9$. The anti-holomorphic sector consists of a CFT with $(\bar{b}, \bar{c}, \overline{X}^\mu)$ of central charge $\bar{c} = -22$ and the *same* internal SCFT with $\bar{c}^{\text{int}} = 9$. In particular, it has the quantization condition (15.81) but with $D_{2,1}$ replaced by D_5 (or D_{13}) conjugacy classes. The simplest possibility is that the remaining 13 units of the central charge to be provided by an $E_8 \times SO(10)$ or an $SO(26)$ level one current algebra.

We first show that for the first case the $U(1)$ current in the $N = 2$ SCA extends the $SO(10)$ to the E_6 gauge group.[10] This and the following discussion is very close to the one given above for the holomorphic sector. The left-moving part of the vertex operators for the missing E_6 roots (cf. (15.107)) are $\exp(i\lambda_s \cdot \bar{\phi})(U_{+\frac{1}{2}}\mathbb{1})$ and $\exp(i\lambda_c \cdot \bar{\phi})(U_{-\frac{1}{2}}\mathbb{1})$. $\lambda_{s,c}$ are weight vectors in the **16** and $\overline{\mathbf{16}}$ of $SO(10)$, U_η the spectral operator and $\mathbb{1}$ the unit operator. The singlet in (15.107) is of course the current $i\sqrt{3}\,\partial H$.

---

[10] Similarly, $SO(26)$ is simply enhanced to $SO(26) \times U(1)$, where the $U(1)$ appears to be anomalous and becomes massive via the Green-Schwarz mechanism (see Chap. 17).

So much for the gauge bosons. For type II compactifications we have clarified the role of the elements of the (c,c) and (a,c) rings with conformal weights $(\bar{h}, h) = (\frac{1}{2}, \frac{1}{2})$. Each such (2,2) superfield consists of four component fields which are essentially the vertex operators of the moduli fields in the canonical and the zero ghost pictures. This interpretation still applies to the holomorphic sector of the heterotic string, but what is the meaning of the two components with $\bar{h} = \frac{1}{2}$ and $\bar{h} = 1$ on the anti-holomorphic side? The latter is still the vertex operator of the moduli fields. Consider the component with $(\bar{h}, \overline{Q}) = (\frac{1}{2}, 1)$, ϕ_0^+ (cf. Sect. 12.2; we suppress the superscripts relating to the holomorphic sector and the index which enumerates the different moduli fields). Then $\exp(i\lambda_v \cdot \bar{\phi})\phi_0^+$ is the holomorphic part of the vertex operator for a scalar which transforms in **10** of $SO(10)$. We can also construct the vertex operators $\exp(i\lambda_c \cdot \bar{\phi})(U_{+\frac{1}{2}}\phi_0^+)$ and $(U_{+1}\phi_0^+)$. Together they fill out the $\overline{\mathbf{27}}$ of E_6. We thus find that the elements of the (c,c) ring are in one-to-one correspondence with massless scalar fields in the $\overline{\mathbf{27}}$ of E_6. Likewise one shows that the elements of the (a,c) ring are in one-to-one correspondence with scalar fields in the **27** of E_6. The fermionic partners of the scalar are obtained by applying the right-moving space-time supersymmetry generator which, in the internal sector is the holomorphic spectral flow operator. Together the physical states combine into chiral multiplets. The (a,a) and (c,a) rings correspond to anti-chiral multiplets. They contain the anti-particles. To summarize, symmetric $(2,2)$ compactifications of the $E_8 \times E_8$ heterotic string lead to gauge group $E_6 \times E_8$ with chiral multiplets in the **27** and $\overline{\mathbf{27}}$ of E_6.

The one-to-one pairings $(c,c) \leftrightarrow \overline{\mathbf{27}}$ and $(a,c) \leftrightarrow \mathbf{27}$ depends on the arbitrary choice of the relative sign between the two $U(1)$ charges of the holomorphic and anti-holomorphic SCA's. In the geometric interpretation of these compactifications this leads to the following statement of heterotic mirror symmetry: compactifications of the heterotic string on a Calabi-Yau manifold X with $h^{1,1}(X)$ chiral multiplets in the **27** of E_6 and $h^{2,1}(X)$ chiral multiplets in the $\overline{\mathbf{27}}$ is indistinguishable from compactification on the mirror CY $\hat{X}$ with $h^{2,1}(\hat{X}) = h^{1,1}(X)$ chiral multiplets in the **27** of E_6 and $h^{1,1}(\hat{X}) = h^{2,1}(X)$ in the $\overline{\mathbf{27}}$ of E_6.

There are other massless scalar fields (and their fermionic partners) in the theory which are not related to elements of the chiral rings. They are worldsheet superfields with conformal weights $(\bar{h}, h) = (1, \frac{1}{2})$ and charge $(\overline{Q}, Q) = (0, \pm 1)$. The left-moving charge follows from the conditions $\bar{h} > \frac{|Q|}{2}$ and the quantization condition (15.81). With respect to the holomorphic SCA they are (anti)chiral primaries. With respect to the anti-holomophic SCA they are general (long) superfields. In the geometric interpretation these scalar fields are the bundle moduli. They are singlets under the E_6 gauge group.

In the next section we will present an explicit construction of symmetric $(2,2)$ heterotic 'compactifications'. We should stress that they are very special classes of space-time supersymmetric compactifications which, generically, only require $(2,0)$. Requiring $(2,2)$ leads to significant simplifications, from the geometric and the CFT points of view.

By similar arguments as for the case of $\mathcal{N} = 1$ supersymmetry one can show that the presence of two right-moving holomorphic supercharges ($\mathcal{N} = 2$ heterotic space-time supersymmetry) in four dimensions implies the existence of an internal right-moving $SU(2) \times U(1)$ Kač-Moody algebra which extends $D_{2,1}$ to $E_{4,1}$ or D_5 to E_7. The supersymmetry on the world-sheet is an $N = 4$, $c = 6$ superconformal algebra plus a superconformal system with $N = 2$, $c = 3$. Finally $\mathcal{N} = 4$ space-time supersymmetry implies an internal $SO(6)$ Kač-Moody algebra which extends $D_{2,1}$ to $E_{5,1}$ respectively D_5 to E_8. For heterotic compactifications with left-right symmetric extended superconformal invariance this implies the gauge groups $E_7 \times E_8$ and unbroken $E_8 \times E_8$. The former are $(4, 4)$ compactifications on the K3 surface with standard embedding of the spin connection into the gauge connection. The latter are compactifications on T^6 which preserve all supercharges. For the $SO(32)$ heterotic string the resulting gauge groups are $SO(26) \times U(1)$, $SO(28) \times U(1)$ and $SO(32)$ for $\mathcal{N} = 1, 2, 4$ supersymmetry. The $U(1)$ factors become massive via the Green-Schwarz mechanism (cf. footnote 10).

15.5 Gepner Models

We now give a concrete realization of the connection between the $N = 2$ worldsheet supersymmetry and $\mathcal{N} = 1$ space-time supersymmetry. We will continue to consider superstring compactifications from ten to four dimensions preserving $\mathcal{N} = 2$ supersymmetry for the type II superstring and $\mathcal{N} = 1$ supersymmetry for the heterotic superstring. The SCFTs we consider correspond to solutions of the two-dimensional non-linear sigma model governing the motion of a string moving on strongly curved Calabi-Yau manifolds (with curvature radius of the order of the string scale $\sqrt{\alpha'}$). They were first constructed by Gepner.

The starting point of this construction is the bosonic string in light-cone gauge with central charges $(c, \bar{c}) = (24, 24)$ with a particular choice for the CFT. For this CFT we use the simple current method developed in Sect. 6.3 to generate modular invariant partition functions and, at the same time, to implement the spectral flow between NS and R sectors. We then apply the light-cone gauge version of the bosonic string map to generate space-time supersymmetric four-dimensional string vacua. If we apply the map to both the right- and left-moving sectors we get an $\mathcal{N} = 2$ supersymmetric type II compactification. Applying it only to the holomorphic sector leads to an $\mathcal{N} = 1$ supersymmetric heterotic compactification with a spectral flow enhancing the gauge group to E_6.

Instead of taking 24 free bosons, which would be appropriate for the light-cone bosonic string in 26-dimensional Minkowski space, we choose the $(c, \bar{c}) = (24, 24)$ CFT to consist of the following building blocks.

- Four-dimensional Minkowski space with coordinates X^μ where $\mu = 0, \ldots, 3$. Two of these, say X^0 and X^1, are arranged into light-cone coordinates $X^\pm = \frac{1}{\sqrt{2}}(X^0 \pm X^1)$ which are gauged away. We are thus left with X^2 and X^3 to which

we associate two copies of the free boson CFT with total central charge $(c, \bar{c}) = (2, 2)$.
- An $N = (2, 2)$ SCFT in the left- and right-moving sector with central charges $(c, \bar{c}) = (9, 9)$.
- An $(\widehat{e}_8)_1 \times \widehat{so}(10)_1$ Kač-Moody algebra with $(c, \bar{c}) = (13, 13)$. Via the Frenkel-Kač construction this can be realized by 13 free bosons compactified on the root-lattice of $E_8 \times SO(10)$. The $SO(10)$ current algebra can alternatively be realized in terms of five complex free fermions. The $\widehat{so}(10)_1$ characters for the four conjugacy classes were determined in Eqs. (11.109) and (11.110).

The idea is now to construct a modular invariant partition function for this bosonic string and then map it, via the bosonic string map, to a superstring theory. Let us explain this map for the current situation.

The modular S-matrix for $\widehat{so}(10)_1$, (11.120), is the same as for $\widehat{so}(2)_1$ and, as a consequence of the Verlinde formula, so are the fusion rules, cf. Table 11.1 on page 333. Furthermore, the difference between the central charges of $(\widehat{e}_8)_1 \times \widehat{so}(10)_1$ and $\widehat{so}(2)_1$ is

$$c_{(\widehat{e}_8)_1 \times \widehat{so}(10)_1} - c_{\widehat{so}(2)_1} = 13 - 1 = 12 \,. \tag{15.111}$$

Therefore, one might hope that replacing $(\widehat{e}_8)_1 \times \widehat{so}(10)_1$ by $\widehat{so}(2)_1$ maps the partition function of the bosonic string with $(c, \bar{c}) = (24, 24)$ to a partition function of the $E_8 \times E_8$ heterotic string with $(c, \bar{c}) = (12, 24)$ or to a partition function of the superstring with $(c, \bar{c}) = (12, 12)$ Indeed, $\widehat{so}(2)_1$ is the Kač-Moody algebra formed by two free fermions, which under such a map could become the superpartners of the two free bosons X^3 and X^4 in four-dimensions. However, there is a problem with this idea, related to the modular T-matrix. For representations in the spinor conjugacy classes (S) and (C) the conformal weights in the $(\widehat{e}_8)_1 \times \widehat{so}(10)_1$ theory and in the $\widehat{so}(2)_1$ theory are not equal, for instance

$$h\left(S^{\widehat{so}(10)_1}\right) - h\left(S^{\widehat{so}(2)_1}\right) = \frac{1}{2} \text{ mod } 1 \,. \tag{15.112}$$

Therefore, in a modular invariant partition function Eq. (4.78), which for $M_{i\bar{j}} \neq 0$ requires $h_i - \bar{h}_j - \frac{c-\bar{c}}{24} \in \mathbb{Z}$ in order to be invariant under $\tau \to \tau + 1$, we cannot simply replace the $\widehat{so}(10)_1$ characters $(0, V, S, C)_{\widehat{so}(10)_1}$ by $(0, V, S, C)_{\widehat{so}(2)_1}$. For instance, if a left moving 0 or V character is combined with a right moving S or C character, or vice versa, invariance under T would be spoiled by the map. The following observation saves the day. The modular S-matrix on the vector of $\widehat{so}(10)_1$ characters $(0, V, S, C)^T$ is identical to the one on the vectors of $\widehat{so}(2)_1$ characters $(V, 0, -C, -S)^T$ and $(V, 0, -S, -C)^T$ where e.g. $-C$ denotes $-\chi_C$. Noting that $(\widehat{e}_8)_1$ has only the singlet representation (1) with conformal weight $h = 0$, which is invariant under modular S-transformations, we can finally formulate two versions of the bosonic string map:

15.5 Gepner Models

$$(1)_{\widehat{(e_8)}_1} \otimes (0, V, S, C)_{\widehat{so(10)}_1} \longrightarrow \begin{cases} (V, 0, -C, -S)_{\widehat{so(2)}_1} & (a) \\ (V, 0, -S, -C)_{\widehat{so(2)}_1} & (b) \end{cases}$$
(15.113)

It is now easy to check that the problem with the modular T-transformation has been solved by switching the 0 and V conjugacy classes. The bosonic string map maps a modular invariant partition function constructed from the characters of $(\widehat{e_8})_1$ and $\widehat{so}(10)_1$ to a modular invariant partition function with the characters of $\widehat{so}(2)_1$. This transforms a partition function of the bosonic string to a supersymmetric one of either the heterotic string, when applying the mapping only to the right-moving sector, or to the type II string when applying it both to the left- and right-moving sector. If we use the same map on both sides we arrive at the type IIB theory, if we use different ones at the type IIA theory. Note that the spinor conjugacy classes come with the correct minus sign.

Let us now be more specific about the $N = (2,2)$ SCFT with central charge $(c, \bar{c}) = 9$. A possible, though surely not the most general, choice for this SCFT is the tensor product of $N = 2$ minimal models with $0 < c < 3$ (c.f. Eq. (12.111) ff) in the following way

$$(N = 2)_{c=9} = \bigotimes_{i=1}^{r} (N = 2)_{c_i} \quad \text{with} \quad \sum_{i=1}^{r} c_i = \sum_{i=1}^{r} \frac{3k_i}{k_i + 2} = 9 \, .$$
(15.114)

Because the k_i are integers, it turns out that there are only 168 combinations which have total central charge $c = 9$. For example, one can choose $r = 5$ factors with $k_i = 3$ giving $c_i = \frac{9}{5}$ and thus in total $c = 9$. An explicit classification shows that most cases have $r = 4$ or $r = 5$ with a few exceptions with $r = 6, 9$.

Ignoring the two free bosons X^2 and X^3, the CFT in the holomorphic sector of the bosonic string (before the bosonic string map) therefore has the following tensor product structure

$$\bigotimes_{i=1}^{r} (N = 2)_{c_i} \otimes \widehat{so}(10)_1 \otimes (\widehat{e_8})_1 \, .$$
(15.115)

The highest weight representations of this product CFT are again tensor products of the individual HWRs which we will denote as

$$\bigotimes_{i=1}^{r} (l_i, m_i, s_i) \otimes (s_0) \otimes (1) \, ,$$
(15.116)

where the s_0 label for the (0,V,C,S) characters of $\widehat{so}(10)_1$ is $(0, 2, 1, -1)$ mod 4. The label (1) for the singlet representation of $(\widehat{e_8})_1$ is often omitted. The energy momentum tensor and $U(1)$ current of this CFT are

$$T(z) = \sum_{i=1}^{r} T_i(z) + T_{\widehat{so(10)}_1 \otimes \widehat{(e_8)}_1}(z),$$

$$j(z) = \sum_{i=1}^{r} j_i(z) + j_{\widehat{so(10)}_1}(z) + j_{\widehat{(e_8)}_1}(z), \qquad (15.117)$$

where T_i and j_i are the energy-momentum tensor and the $U(1)$ current for each $N = 2$ tensor factor in (15.115). $j_{\widehat{so(10)}_1}(z)$ is the diagonal $U(1)$ current in the Cartan subalgebra of $so(10)$. In can be expressed either in terms of five real bosons or five complex fermions (c.f. Eqs. (11.83) and (11.97))

$$j_{\widehat{so(10)}_1}(z) = \sum_{i=1}^{5} :\Psi^{+i}\Psi^{-i}(z) := i\sum_{i=1}^{5} \partial\phi^i(z). \qquad (15.118)$$

Likewise $j_{\widehat{(e_8)}_1}$ is the diagonal $U(1)$ current in the Cartan subalgebra of e_8. For the theory (15.115) one can construct the trivial diagonal modular invariant partition function. However, after applying the bosonic string map (15.113) this partition function will not correspond to a space-time supersymmetric string compactification. In particular there is not the same number of states in the NS as in the R sector, which leads to an unequal number of bosons and fermions in the four-dimensional space-time. What is needed to ensure space-time supersymmetry is a GSO projection which can be implemented by using the spectral flow for the $N = 2$ SCFT. What is specific about the Gepner construction is that this flow can be implemented by the method of simple currents for rational CFTs introduced in Sect. 6.3. As we will see below, the relevant simple currents are orbit simple currents, i.e. they have positive integer conformal dimension before applying the bosonic string map. We can then use the results of Sect. 6.3 to construct modular invariant partition functions. This is the reason why we started with a bosonic string theory in the first place.

Prior to even implementing the flow between the NS and R sectors, we need to implement the requirement that only R and only NS sectors are coupled between the various factors in the tensor product theory (15.115). After all, the distinction between the two sectors lies in the moding of the world-sheet supercurrents which have to be either integer (R) or half-integer (NS) for the total supercurrents, which are simply the sum of the supercurrents of the constituent theories. Therefore, we need to perform a projection such that only NS-states of the minimal models couple to the (0) and (V) conjugacy classes of $SO(10)$ and only R-states to the two spinor classes. This projection can be achieved in the CFT via a simple current construction. The set of simple currents to consider is

$$J_i = (0,0,0) \otimes \ldots \otimes \underbrace{(0,0,2)}_{i^{\text{th}} \text{ pos.}} \otimes \ldots \otimes (0,0,0) \otimes (V), \qquad i = 1, \ldots, r, \quad (15.119)$$

where the notation is as in (15.116). Their conformal weight and $U(1)$ charge is $(h, Q) = (2, 2)$ so that these are indeed orbit simple currents which we discussed

15.5 Gepner Models

in Sect. 6.3. Using the fusion rules (12.117) and Table 11.1, one finds the fusion of the simple current J_i with an arbitrary state in the NS sector of the CFT:

$$[J_i] \times \left[(NS)_{mm} \otimes (0, V)_{\widehat{so}(10)_1}\right] = \left[(NS')_{mm} \otimes (V, 0)_{\widehat{so}(10)_1}\right], \quad (15.120)$$

where $(NS)_{mm}$ denotes a state in the NS sector of the minimal model theory and $(0,V)$ stands for a state in either the 0 or the V conjugacy class of the $\widehat{so}(10)_1$ current algebra. The monodromy charge of an NS-state w.r.t. J_i is then

$$Q_i(NS) = h(J_i) + h(NS) - h(J_i \times NS) \mod 1$$

$$= 2 + \left(h_{mm} + \left(0, \frac{1}{2}\right)_{\widehat{so}(10)_1}\right) - \left(h_{mm} + \frac{1}{2} + \left(\frac{1}{2}, 0\right)_{\widehat{so}(10)_1}\right) \mod 1$$

$$= 0. \quad (15.121)$$

Since it is integer, these states will survive the simple current projection. For states in the R sector

$$[J_i] \times \left[(R)_{mm} \otimes (S, C)_{\widehat{so}(10)_1}\right] = \left[(R')_{mm} \otimes (C, S)_{\widehat{so}(10)_1}\right] \quad (15.122)$$

we find likewise

$$Q_i(R) = 2 + \left(h_{mm} + \frac{5}{8}\right)_{\widehat{so}(10)_1} - \left(h_{mm} + \left(\frac{5}{8}\right)_{\widehat{so}(10)_1}\right) \mod 1 = 0, \quad (15.123)$$

i.e. these states also survive. On the other hand,

$$[J_i] \times \left[(R)_{mm} \otimes (0, V)_{\widehat{so}(10)_1}\right] = \left[(R')_{mm} \otimes (V, 0)_{\widehat{so}(10)_1}\right],$$

$$[J_i] \times \left[(NS)_{mm} \otimes (S, C)_{\widehat{so}(10)_1}\right] = \left[(NS')_{mm} \otimes (C, S)_{\widehat{so}(10)_1}\right], \quad (15.124)$$

and it is straightforward to show that these states have half-integer monodromy charge and are therefore projected out.

In order to guarantee degeneracy of space-time bosons and space-time fermions, we have to project on orbits of the spectral flow. In the tensor product CFT the spectral flow operator is also a simple current:

$$J_{sf} = (0, 1, 1)^r \otimes (S) \quad (15.125)$$

with conformal dimension and $U(1)$ charge $(h, Q) = (1, 2)$. Its fusion with a general state in the CFT is

$$[(0, 1, 1)^r \otimes (S)] \times \left[\bigotimes_i (l_i, m_i, s_i) \otimes (s_0)\right] = \bigotimes_i (l_i, m_i + 1, s_i + 1) \otimes (s_0 + 1). \quad (15.126)$$

With the help of (12.113) it is straightforward to compute the monodromy charge Q_{sf} and the $U(1)$ charge Q of a state (15.116) with the result

$$Q_{\text{sf}} = -\frac{Q}{2} \mod 1 . \qquad (15.127)$$

Therefore, the simple current J_{sf} projects onto states with charge $Q \in 2\mathbb{Z}$.

After having studied the simple currents and corresponding monodromy charges of the states, we are ready to construct the modular invariant partition function of the bosonic string. According to Sect. 6.3 it reads schematically

$$Z^{\text{bos}}_{(24,24)}(\tau, \bar{\tau}) = \frac{1}{\kappa} \boldsymbol{\chi}^T(\tau) \, M(J_{\text{sf}}) \prod_{i=1}^{r} M(J_i) \, \overline{\boldsymbol{\chi}}(\bar{\tau}) , \qquad (15.128)$$

where κ is an overall normalization constant fixed by the requirement that the vacuum appears precisely once. The information about which holomorphic characters couple to which antiholomorphic characters due to the extension by the simple current J is encoded in the matrix $M(J)$. From this bosonic string partition function we generate a supersymmetric partition function by applying the bosonic string map (15.113). As a consequence, instead of dealing with an $\widehat{so}(10)_1 \times (\widehat{e_8})_1$ Kač-Moody algebra, we work with the algebra $\widehat{so}(2)_1$ of the world-sheet fermions corresponding to the uncompactified dimensions. Moreover, since the $SO(2)$ spinors (S,C) receive a minus sign in (15.113), we are guaranteed that the space-time fermions contribute with a negative sign to the supersymmetric partition function.

In order to give an explicit expression for the supersymmetric modular invariant partition function, the following notation is useful. We define the characters and the so-called charge vector $\boldsymbol{\lambda}$ as

$$\chi^l_{\lambda}(\tau) := \prod_{i=1}^{r} \chi^{l_i}_{m_i, s_i}(\tau) \cdot \chi^{\widehat{so}(10)_1 \times (\widehat{e_8})_1}_{s_0}(\tau) \quad \text{with} \quad \begin{cases} \boldsymbol{\lambda} = (s_0; m_1, \ldots, m_r; s_1, \ldots, s_r), \\ \boldsymbol{l} = (l_1, \ldots, l_r). \end{cases}$$
$$(15.129)$$

The scalar product between two charge vectors is defined as

$$\boldsymbol{\lambda} \cdot \boldsymbol{\lambda}' = \frac{s_0 s'_0}{4} + \frac{1}{2} \sum_{i=1}^{r} \left(-\frac{m_i m'_i}{k_i + 2} + \frac{s_i s'_i}{2} \right) . \qquad (15.130)$$

The charge vector for the simple currents J_i is

$$\boldsymbol{\beta}_i = (2; 0, \ldots, 0; 0, \ldots, 2, \ldots, 0) .$$
$$\uparrow$$
$$i^{\text{th}} \text{ position} \qquad (15.131)$$

We have seen above that only states either purely in the NS or purely in the R sector survive the simple current projection with respect to J_i. In terms of the charge vector

15.5 Gepner Models

this is expressed as
$$\beta_i \cdot \lambda \in \mathbb{Z}. \tag{15.132}$$

The charge vector for the spectral flow simple current J_{sf} is

$$\beta_0 = (1; 1, \ldots, 1; 1, \ldots, 1), \tag{15.133}$$

and the condition $Q \in 2\mathbb{Z}$ can be expressed as

$$2\beta_0 \cdot \lambda \in 2\mathbb{Z}. \tag{15.134}$$

This projection generalizes the GSO projection to the $N = 2$ SCFT context. We can express the simple current extended, modular invariant partition function from above more explicitly as

$$Z^{\text{bos}}_{(24,24)}(\tau, \bar{\tau}) = \frac{1}{\kappa} \sum_{\substack{l, \lambda \\ \lambda \cdot \beta_i \in \mathbb{Z} \\ \lambda \cdot \beta_0 \in \mathbb{Z}}} \sum_{v_i = 0,1} \sum_{v_0 = 0}^{L-1} \chi^l_{\lambda + \sum_{i=1}^r v_i \beta_i + v_0 \beta_0}(\tau) \, \overline{\chi}^l_\lambda(\bar{\tau}), \tag{15.135}$$

where L is the length of the simple current J_{sf} and κ is again an overall normalization constant. The supersymmetric type IIA/IIB and heterotic Gepner model partition functions $Z^{\text{IIA/IIB}}_{(12,12)}(\tau, \bar{\tau})$ and $Z^{\text{het}}_{(24,12)}(\tau, \bar{\tau})$ simply follow from $Z^{\text{bos}}_{(24,24)}(\tau, \bar{\tau})$ by applying the bosonic string map (15.113) to both left- and right moving sectors or only to the right moving sector, respectively. This introduces extra minus signs for states in the R-sectors, i.e. for space-time fermions.

The charge quantization condition (15.127) changes under the bosonic string map (15.113). For version (a) it is $Q \in 2\mathbb{Z} + 1$ for all states while for version (b) it is $Q \in 2\mathbb{Z} + 1$ for states in the (0) and (V) conjugacy classes and $Q \in 2\mathbb{Z}$ for states in the (S) and (C) classes. This agrees with Eq. (15.81) and the discussion following it if we translate it to light-cone gauge and include in Q also the contribution from the space-time $SO(2)$.

As a concrete example we now discuss the massless spectra for these two classes of models with $\mathcal{N} = 2$ and $\mathcal{N} = 1$ space-time supersymmetry in more detail.

Type II Gepner Models

Returning to our general discussion from Sect. 15.4, one is interested in the massless excitations of the string which correspond to states in the partition function with conformal weight $(h, \bar{h}) = (\frac{1}{2}, \frac{1}{2})$. Since the simple current construction has arranged all states into orbits, we can distinguish for instance the massless vacuum orbit in the partition function $Z^{\text{IIA/IIB}}_{(12,12)}(\tau, \bar{\tau})$ from other massless orbits via the $U(1)$ charge (mod 2). We start with type IIB, whose partition function is obtained from (15.135) by applying the same bosonic string map, say (15.113a) on both sides.

Consider the vacuum orbit. The vacuum with $(h, Q) = (0, 0)$ in the bosonic theory is $(0, 0, 0)^r(0) \otimes \overline{(0, 0, 0)^r(0)}$. It is mapped to

$$(0, 0, 0)^r(V) \otimes \overline{(0, 0, 0)^r(V)} \qquad (15.136)$$

under the bosonic string map. Taking into account that the ground state (V) of $\widehat{so}(2)_1$ is two-dimensional, these are four states corresponding in four dimensions to the graviton $G_{\mu\nu}$, the anti-symmetric two-form $B_{\mu\nu}$ and the dilaton D. However, due to the simple current construction, there are further states in that orbit. In particular, we find four additional massless states in the (R,R) sector

$$(0, +1, +1)^r(C) \otimes \overline{(0, +1, +1)^r(C)}, \quad (0, -1, -1)^r(S) \otimes \overline{(0, -1, -1)^r(S)},$$
$$(0, +1, +1)^r(C) \otimes \overline{(0, -1, -1)^r(S)}, \quad (0, +1, +1)^r(C) \otimes \overline{(0, -1, -1)^r(S)},$$
$$(15.137)$$

which are space-time bosons. Since $[C] \times [C] = [S] \times [S] = [V]$ and $[S] \times [C] = [C] \times [S] = [0]$, these four-states are identified in four dimensions as one massless vector and one massless complex scalar. The fermionic superpartners of these bosonic fields arise in the (NS,R) and (R,NS) sector

$$(0, 0, 0)^r(V) \otimes \overline{(0, +1, +1)^r(C)}, \quad (0, +1, +1)^r(C) \otimes \overline{(0, 0, 0)^r(V)},$$
$$(0, 0, 0)^r(V) \otimes \overline{(0, -1, -1)^r(S)}, \quad (0, -1, -1)^r(S) \otimes \overline{(0, 0, 0)^r(V)}. \qquad (15.138)$$

These eight space-time bosons and eight space-time fermions constitute the contents of the $\mathcal{N} = 2$ gravity supermultiplet and of one $\mathcal{N} = 2$ hypermultiplet. Note that the gravity supermultiplet contains in addition to the spin-2 graviton one spin-1 vectorfield. The hypermultiplet contains the dilaton and for this reason it is called the universal hypermultiplet.

The vacuum orbit is present in all Gepner models whereas the structure of the massless charged orbits depends on the specific details of the $c = 9$ tensor product theory. However, since massless states have $h = \bar{h} = 1/2$ and due to the odd $U(1)$ charge, all these states must be (anti-)chiral primary states in the $N = 2$ SCFT. Concretely, in the charged orbits we find massless states of the form

$$\left(\left(\tfrac{1}{2}, -1\right)(0) + \left(\tfrac{3}{8}, +\tfrac{1}{2}\right)(S) \right) \otimes \overline{\left(\left(\tfrac{1}{2}, -1\right)(0) + \left(\tfrac{3}{8}, +\tfrac{1}{2}\right)(S) \right)},$$
$$\left(\left(\tfrac{1}{2}, +1\right)(0) + \left(\tfrac{3}{8}, -\tfrac{1}{2}\right)(C) \right) \otimes \overline{\left(\left(\tfrac{1}{2}, +1\right)(0) + \left(\tfrac{3}{8}, -\tfrac{1}{2}\right)(C) \right)}, \qquad (15.139)$$

where we used (h, Q) to denote the conformal weight and charge of the corresponding state. For each such orbit, including all combinations of NS and R sectors, we obtain one vector, one complex boson and four fermionic states forming one $\mathcal{N} = 2$ vectormultiplet $(\phi^c, \lambda^\alpha, A_\mu)$. However, it can also happen that in an orbit

15.5 Gepner Models

of $(\frac{1}{2}, -1)(0)$, there appears a state with $(\frac{1}{2}, +1)(0)$ so that by including all NS and R sectors one gets the additional massless states

$$\left((\tfrac{1}{2}, -1)\,(0) + (\tfrac{3}{8}, +\tfrac{1}{2})\,(S) \right) \otimes \overline{\left((\tfrac{1}{2}, 1)\,(0) + (\tfrac{3}{8}, -\tfrac{1}{2})\,(C) \right)},$$

$$\left((\tfrac{1}{2}, +1)\,(0) + (\tfrac{3}{8}, -\tfrac{1}{2})\,(C) \right) \otimes \overline{\left((\tfrac{1}{2}, -1)\,(0) + (\tfrac{3}{8}, +\tfrac{1}{2})\,(S) \right)}. \tag{15.140}$$

These four bosonic and four fermionic states fill out one $\mathcal{N} = 2$ hypermultiplet. Whether this happens or not, depends on the concrete model.

Let us finally discuss the massless spectrum for the $(k = 3)^5$ Gepner model in more detail. Each $(k = 3)$ tensor factor has chiral states $(0, 0, 0)_0$, $(1, -1, 0)_{\frac{1}{5}}$, $(2, -2, 0)_{\frac{2}{5}}$ and $(3, -3, 0)_{\frac{3}{5}}$, where the subscript denotes the $U(1)$ charge. We can now make a list of all the combinatorial possibilities to form chiral states with $(h, Q) = (\frac{1}{2}, 1)$ in the tensor product $(k = 3)^5$ theory. This list reads

$$\begin{array}{l|l}
(3, -3, 0)\ (2, -2, 0)\ (0, 0, 0)^3 & 20 \\
(3, -3, 0)\ (1, -1, 0)^2 (0, 0, 0)^2 & 30 \\
(2, -2, 0)^2 (1, -1, 0)\ (0, 0, 0)^2 & 30 \\
(2, -2, 0)\ (1, -1, 0)^3 (0, 0, 0) & 20 \\
(1, -1, 0)^5 & 1 \\
\hline
 & 101
\end{array} \tag{15.141}$$

and by counting states, we see that there are 101 vectormultiplets. A more detailed investigation reveals that only the orbit of $(1, -1, 0)^5(0)$ contains a state with $(h, Q) = (\frac{1}{2}, -1)$, namely $(1, 1, 0)^5(0)$, giving one hypermultiplet. To summarize, the massless spectrum of the type IIB superstring 'compactified' on the $(k = 3)^5$ Gepner model consists of the gravity supermultiplet, the universal hypermultiplet, 101 vectormultiplets and more hypermultiplet.

For the type IIA superstring one has opposite GSO projections in the left- and right-moving R-sectors, say (15.113a) for the right-movers and (15.113b) for the left-movers. The massless spectrum for the gravity sector stays the same whereas in the charged orbits the massless modes have the form

$$\left((\tfrac{1}{2}, -1)\,(0) + (\tfrac{3}{8}, +\tfrac{1}{2})\,(S) \right) \otimes \overline{\left((\tfrac{1}{2}, -1)\,(0) + (\tfrac{3}{8}, +\tfrac{1}{2})\,(C) \right)},$$

$$\left((\tfrac{1}{2}, +1)\,(0) + (\tfrac{3}{8}, -\tfrac{1}{2})\,(C) \right) \otimes \overline{\left((\tfrac{1}{2}, +1)\,(0) + (\tfrac{3}{8}, -\tfrac{1}{2})\,(S) \right)}. \tag{15.142}$$

Therefore, compared to the type IIB case, hypermultiplets and vectormultiplets are exchanged. The massless type IIA spectrum for the $(k = 3)^5$ Gepner model consists of the gravity sector (supergravity + universal hypermultiplet), 1 vectormultiplet and 101 hypermultiplets.

Geometrically type IIB (IIA) string theory compactified on a Calabi-Yau manifold with Hodge numbers (h^{21}, h^{11}) gives rise to $h^{21}(h^{11})$ vectormultiplets and $h^{11}(h^{21})$ hypermultiplets. A Calabi-Yau manifold with Hodge numbers $(h^{21}, h^{11}) = (101, 1)$ is the quintic hypersurface in $\mathbb{P}^4$. Apart from the counting of multiplets illustrated above, more evidence has been collected that the Gepner model $(k = 3)^5$ exactly solves the non-linear sigma model on the quintic (at a fixed value of the moduli and in particular at a fixed string scale size). While the geometric construction allows to explore the moduli space, its validity is limited to large manifolds (compared to the string scale) where geometric notions are valid. The CFT Gepner construction, on the other hand, is rigid but explores the truly stringy regime.

Heterotic Gepner Models

Applying the bosonic string map (15.113a) only to the right-moving sector of (15.135) one arrives at the partition function of the heterotic Gepner models. The massless states are now given by states in this partition function with $(h, \bar{h}) = (\frac{1}{2}, 1)$. The vacuum orbit contains again the universal massless graviton, two-form and dilaton

$$(0, 0, 0)^r(V) \otimes (\overline{\partial X}^\mu)_{-1} . \tag{15.143}$$

However, in addition it contains the massless states

$$(0, 0, 0)^r(V) \otimes \begin{cases} \overline{(0, 0, 0)^r(0)} & \textbf{Adj} \text{ of } E_8 \times SO(10) \times U(1)^r \\ \overline{(0, 1, 1)^r(S)} & \textbf{16} \\ \overline{(0, -1, -1)^r(C)} & \overline{\textbf{16}} \end{cases} \tag{15.144}$$

which are vectors under the four-dimensional Lorentz group and, as expected from the general analysis of Sect. 15.4, indeed extends the naive gauge symmetry $SO(10) \times U(1)^r$ to $E_6 \times U(1)^{r-1}$. Under the spectral flow in the right-moving sector, these bosonic states are mapped to the massless fermionic states corresponding to the gravitino, dilatino and gauginos. Together these massless states fill out one $\mathcal{N} = 1$ gravity multiplet and one chiral supermultiplet.

Similar to the type II string, the charged orbits also contain further massless states

$$\left((\tfrac{1}{2}, 1)(0) + (\tfrac{3}{8}, -\tfrac{1}{2})(C) \right) \otimes \begin{cases} \overline{(\tfrac{1}{2}, 1)(V)} & \textbf{10} \text{ of } SO(10) \\ \overline{(\tfrac{3}{8}, -\tfrac{1}{2})(S)} & \textbf{16} \\ \overline{(1, -2)(0)} & \textbf{1} \end{cases} \tag{15.145}$$

15.5 Gepner Models

which, together with their CPT conjugates, combine into chiral supermultiplets transforming in the $\overline{\mathbf{27}}$ representation of E_6. Therefore, the (c,c) chiral ring is related to the number $n_{\overline{27}}$ of matter multiplets in $\overline{\mathbf{27}}$ and similarly the (a,c) chiral ring to the number n_{27} of matter multiplets in $\mathbf{27}$ representation. The net number of generations is $n_{27} - n_{\overline{27}}$. For the example of the $(3)^5$ Gepner model we therefore obtain $n_{\overline{27}} = 101$ and $n_{27} = 1$ and the gauge group $E_8 \times E_6 \times U(1)^4$. The Yukawa couplings $\mathbf{27}^3$ and $\overline{\mathbf{27}}^3$ are governed by the structure of the (a,c) and (c,c) chiral ring. The second bosonic string map (15.113b) simply exchanges $\mathbf{27}$ and $\overline{\mathbf{27}}$.

In the left-moving sector, unitarity implies that there can be contributions from states with $(\overline{h},\overline{q}) = (1,\overline{q})$ with $|q| \leq 2$. Thus, in addition to the states mentioned so far, there arise additional massless states of the type

$$\left(\left(\tfrac{1}{2},1\right)(0) + \left(\tfrac{3}{8},-\tfrac{1}{2}\right)(C)\right) \otimes \overline{(1,0)(0)} \quad \mathbf{1} \text{ of } SO(10) \tag{15.146}$$

which, together with their CPT conjugate states, form chiral supermultiplets in the singlet representation of $E_8 \times E_6$. These correspond to the moduli of the background. Let us compute their number for the $(3)^5$ Gepner model. In the following table we list all the left-moving $(\overline{h},\overline{q}) = (1,0)$ states coupling to the corresponding right-moving $(h,q) = (1/2,1)$ states.

$$\begin{array}{l|l}
(3,-3,2)(2,-2,0)(0,0,0)^3 & 2 \cdot 20 \\
\hline
(3,-3,2)(1,-1,0)^2(0,0,0)^2 & 3 \cdot 30 \\
\hline
(2,-2,2)^2(1,-1,0)(0,0,0)^2 & 3 \cdot 30 \\
\hline
(2,-2,2)(1,-1,0)^3(0,0,0) & 4 \cdot 20 \\
\hline
\underline{(1,-1,2)(1,-1,0)^4} & 3 \cdot 5 \cdot 1 \\
\hline
\underline{(1,1,2)(1,1,0)^4} & 3 \cdot 5 \cdot 1 \\
\hline
& 330
\end{array} \tag{15.147}$$

The entry $s = 2$ is permuted among the underlined nontrivial triplets. This accounts for the multiplicities $2,\ldots,5$. The extra multiplicity of $\mathbf{3}$ for the last two contributions arises from the simple current orbits and cannot be seen easily. The practical way of finding the spectrum is to use a computer. In total we get 330 gauge singlets in the matter sector. From the Hodge numbers for the quintic we compute $h^{21}(X) + h^{11}(X) + h^1(\text{End}(T_X)) = 101 + 1 + 224 = 326$ moduli. This precisely coincides with the Gepner model result, after Higgsing of the additional $U(1)^4$ gauge symmetry. This is broken away from the Gepner point, which is a special point in the complex structure, bundle and Kähler moduli space of the Calabi-Yau manifold.

Finally, let us mention that the Gepner construction of modular invariant partition functions for Calabi-Yau compactifications can be generalized to genuine $(0,2)$ models. The breaking of the left-right symmetric $N = (2,2)$ world-sheet supersymmetry can also be implemented by the simple current construction.

15.6 Four-Dimensional Heterotic Strings via Covariant Lattices

As a second concrete realization of the general structure discussed in Sect. 15.4, we present the the basic steps in the generalization of the covariant lattice approach of Chap. 13 to $\mathcal{N} = 1$ heterotic superstring compactifications to four dimensions. Like the Gepner models, these models are in general special points in families of string vacua which are parametrized by a higher dimensional moduli space At these special points we have an exact CFT description and can explore stringy effects.

Going from ten to d dimensions only d bosonic fields $X^\mu(z,\bar{z})$ ($\mu = 1,\ldots,d$) play the role of space-time coordinates. The d-dimensional center-of-mass momenta k^μ, which are canonically conjugate to the center-of-mass-positions, have continuous eigenvalues. We are left with $26 - d$ left-moving bosonic fields $X^I(\bar{z})$ ($I = 1,\ldots,26-d$) and $10-d$ right-moving fields $X^J(z)$ ($J = 1,\ldots,10-d$). For these variables we assume that the holomorphic and antiholomorphic fields move independently on a $(26-d)$-dimensional left and a $(10-d)$-dimensional right torus respectively. This implies that the corresponding momentum eigenvalues w_L, w_R are quantized and generate a $26-d$ dimensional lattice $(\Gamma_{26-d})_L$ and a $10-d$ dimensional lattice $(\Gamma_{10-d})_R$. (For general background values the left- and right-moving momenta do not always separately generate lattices. Only when combined they form a Lorentzian lattice. We will, however, discard this possibility for simplicity.)

Now we replace the ten right-moving world-sheet fermions $\psi^I(z)$ ($I = 1,\ldots,10$) by five free bosonic fields. However only d (d even) of the ten fermions are the two-dimensional world-sheet superpartners of the bosonic coordinates $X^\mu(z)$. These d fermionic fields generate a level one $SO(d)$ Kač-Moody algebra which corresponds to the Wick rotated Lorentz group $SO(d-1,1)$. Via bosonization these d fermions $\psi^\mu(z)$ are replaced by $\frac{d}{2}$ bosons $\phi^i(z)$ ($i = 1,\ldots,\frac{d}{2}$). To obtain a sensible space-time interpretation, we have to insist that their zero mode momentum vectors are elements of the weight lattice of $SO(d)$: $\lambda_R \in D_{\frac{d}{2}}$. Then, as in ten dimensions, states in the R sector are characterized by $\lambda_R \in$ S,C of $D_{\frac{d}{2}}$. These states are d-dimensional fermions. The NS sector with d-dimensional space-time bosons is obtained from $\lambda_R \in$ 0,V of $D_{\frac{d}{2}}$. As before, we also bosonize the superconformal ghosts by introducing a free boson $\phi(z)$. The ghost charge is either half-integer or integer. The canonical choice is $q = -\frac{1}{2}$ for space-time fermions and $q = -1$ for space-time bosons. As before, we combine the D_1 ghost lattice with the $SO(d)$ weight lattice $D_{\frac{d}{2}}$ to form the covariant lattice $D_{\frac{d}{2},1}$ with lattice vectors (λ_R, q) and signature $((+1)^{\frac{d}{2}}, -1)$.

We are left with $10-d$ additional, internal world-sheet fermions $\tilde{\psi}^J(z)$ ($J = 1,\ldots,10-d$) which we replace by $\frac{1}{2}(10-d)$ bosons $\tilde{\phi}^i(z)$ ($i = 1,\ldots,\frac{1}{2}(10-d)$). For these bosons it is not necessary that the momenta $\tilde{\lambda}$ are vectors of a $D_{5-\frac{d}{2}}$ lattice, as we have given up the notion of ten-dimensional space-time. We treat these fields, as the bosons $X^J(z)$, as internal degrees of freedom which are needed

15.6 Four-Dimensional Heterotic Strings via Covariant Lattices

to ensure conformal invariance, i.e. the vanishing of the central charge of the Virasoro algebra. So let us denote the lattice with vectors $\tilde{\lambda}$ by $\Gamma_{5-\frac{d}{2}}$. In fact, there is no reason to treat $X^J(z)$ and $\tilde{\phi}^i(z)$ differently. From a CFT point of view there is no difference between compactified bosons and bosonized internal fermions. We therefore combine the internal right-moving bosonic degrees of freedom to $\boldsymbol{X}_R(z) = (X^1(z), \ldots, X^{10-d}(z), \tilde{\phi}^1(z), \ldots, \tilde{\phi}^{5-\frac{d}{2}}(z))$ with momentum vectors $\boldsymbol{w}_R = (w_R^1, \ldots, w_R^{10-d}, \tilde{\lambda}^1, \ldots, \tilde{\lambda}^{5-\frac{d}{2}})$; they build a $(15 - \frac{3}{2}d)$-dimensional lattice $(\Gamma_{15-\frac{3}{2}d})_R = \Gamma_{10-d} \otimes \Gamma_{5-\frac{d}{2}}$.

We would like to emphasize that the d-dimensional heterotic string constructed in this way cannot, in general, be regarded as a geometric compactification of the 10-dimensional heterotic string in the sense that the string is moving on a $(10 - d)$-dimensional internal compact manifold, even when taking into account possible background fields. This is due to the asymmetric treatment of the left- and right-moving fields and also due to the fact that we treat the internal bosonized world-sheet fermions and compactified bosons on equal footing. On the contrary, these theories provide, in general, truly four-dimensional string vacua where only d bosons and world-sheet fermions play the role of (flat) space-time coordinates and their two-dimensional superpartners.

The covariant vertex operators of the d-dimensional string states can be written as (neglecting bosonic oscillators and space-time momentum dependence):

$$V_{\boldsymbol{w}_L;\boldsymbol{w}_R,\lambda_R,q}(z,\bar{z}) = e^{i\boldsymbol{w}_L \cdot \boldsymbol{X}_L(\bar{z})} \, e^{i\boldsymbol{w}_R \cdot \boldsymbol{X}_R(z)} \, e^{i\lambda_R \cdot \boldsymbol{\phi}(z)} \, e^{q\phi(z)} \,,$$

$$\boldsymbol{w}_L \in (\Gamma_{26-d})_L, \quad \boldsymbol{w}_R \in (\Gamma_{15-\frac{3}{2}d})_R, \quad (\lambda_R, q) \in D_{\frac{d}{2},1}. \quad (15.148)$$

The conformal dimension of this operator is

$$\bar{h} = h_L = \frac{1}{2}\boldsymbol{w}_L^2,$$
$$h = h_R = \frac{1}{2}\boldsymbol{w}_R^2 + \frac{1}{2}\lambda_R^2 - \frac{1}{2}q^2 - q\,. \quad (15.149)$$

It follows that the mass of a d-dimensional string state created by this vertex operator is given by the following expressions, where N_L and N_R count the number of left- and right-moving space-time as well as internal oscillators:

$$\frac{\alpha'}{2}m_L^2 = \frac{1}{2}\boldsymbol{w}_L^2 + N_L - 1,$$
$$\frac{\alpha'}{2}m_R^2 = \frac{1}{2}\boldsymbol{w}_R^2 + \frac{1}{2}\lambda_R^2 - \frac{1}{2}q^2 - q + N_R - 1,$$
$$m^2 = m_L^2 + m_R^2\,. \quad (15.150)$$

Physical states have to satisfy $m_L^2 = m_R^2$. The operator product expansion between two vertex operators Eq. (15.148) reads:

$$V_{\mathbf{w}_{L_1};\mathbf{w}_{R_1},\lambda_{R_1},q_1}(z,\bar{z})\, V_{\mathbf{w}_{L_2};\mathbf{w}_{R_2},\lambda_{R_2},q_2}(w,\bar{w})$$
$$= (\bar{z}-\bar{w})^{\mathbf{w}_{L_1}\cdot\mathbf{w}_{L_2}}(z-w)^{\mathbf{w}_{R_1}\cdot\mathbf{w}_{R_2}+\lambda_{R_1}\cdot\lambda_{R_2}-q_1q_2}$$
$$\times V_{\mathbf{w}_{L_1}+\mathbf{w}_{L_2};\mathbf{w}_{R_1}+\mathbf{w}_{R_2},\lambda_{R_1}+\lambda_{R_2},q_1+q_2}(w,\bar{w}) + \ldots. \quad (15.151)$$

In complete analogy with the heterotic string theories in ten dimensions, it is convenient to combine $(\mathbf{w}_L;\mathbf{w}_R,\lambda_R,q)$ to form vectors of a $(42-2d)$-dimensional Lorentzian lattice $\Gamma_{26-d;15-d,1} = (\Gamma_{26-d})_L \otimes (\Gamma_{15-\frac{3}{2}d} \otimes D_{\frac{d}{2},1})_R$. As in Sect. 13.5, modular invariance of the partition function forces $\Gamma_{26-d;15-d,1}$ to be an odd self-dual Lorentzian lattice.

For an odd self-dual lattice $\Gamma_{26-d;15-d,1}$ to represent a physically sensible heterotic string theory additional constraints have to be imposed. One requirement, namely that all states are classified according to representations of the Lorentz group $SO(d)$, is already satisfied, if we demand that $(D_{\frac{d}{2},1})_R$ is part of $\Gamma_{26-d;15-d,1}$. This requirement was essentially sufficient to classify all heterotic string theories in ten dimensions; see Sect. 13.5. However in lower dimensions this is not the end of the story. Remember that for decoupling of ghosts the right-moving fermionic string must possess a two-dimensional (local) world-sheet supersymmetry. However this is lost in lower dimensions when we treat the internal bosons and the internal fermions on the same footing. In other words, having bosonized all right-moving internal degrees of freedom, which leads to the lattice $(\Gamma_{15-\frac{3}{2}d})_R$, we have to find a way to realize the right-moving world-sheet supersymmetry entirely in terms of $15-\frac{3}{2}d$ internal bosonic fields. In contrast to the ten-dimensional string theory, the world-sheet supersymmetry will now be manifest only in a more complicated, in general non-linearly realized way. Since the realization of the two-dimensional supersymmetry is one of the key points in the construction of the lower dimensional string theories, let us be more precise.

The supercurrent of the fermionic string theory has to satisfy the following operator algebra:

$$T(z)T_F(w) = \frac{\frac{3}{2}T_F(w)}{(z-w)^2} + \frac{\partial T_F(w)}{z-w} + \ldots,$$
$$T_F(z)T_F(w) = \frac{\frac{5}{2}}{(z-w)^3} + \frac{\frac{1}{2}T(w)}{z-w} + \ldots. \quad (15.152)$$

This shows that $T_F(z)$ belongs to a SCFT with central charge $c = 15$ which can be realized in the simplest case by ten free bosons plus ten free fermions. This kind of linear realization of world-sheet supersymmetry appears in the fermionic string theory in 10 dimensions. However in lower dimensional theories there are only d free bosons $X^\mu(z)$ together with their superpartners $\psi^\mu(z)$. These fields contribute $\frac{3}{2}d$ units to the central charge of the superconformal algebra and the space-time part of the supercurrent has its standard form:

15.6 Four-Dimensional Heterotic Strings via Covariant Lattices

$$T_F^{\text{space-time}}(z) = -\frac{1}{2}\psi^\mu(z)\partial X_\mu(z) . \tag{15.153}$$

The missing $15 - \frac{3}{2}d$ units to the central charge are provided by the internal fields $\boldsymbol{X}_R(z)$. The internal supercurrent $T_F^{\text{int}}(z)$ must be built entirely from the bosons $\boldsymbol{X}_R(z)$ and it must satisfy

$$T_F^{\text{int}}(z) T_F^{\text{int}}(w) = \frac{\frac{1}{6}(15 - \frac{3}{2}d)}{(z-w)^3} + \frac{\frac{1}{2}T^{\text{int}}(w)}{z-w} + \ldots , \tag{15.154}$$

where $T^{\text{int}}(w)$ is the internal energy momentum tensor. The most general ansatz for T_F^{int}, which is a conformal field of dimension $3/2$ and built entirely from the $15 - \frac{3}{2}d$ free bosons $\boldsymbol{X}_R(z)$, is

$$T_F^{\text{int}}(z) = \sum_t A(t)\, e^{i t \cdot \boldsymbol{X}_R(z)} + i \sum_l \boldsymbol{B}(l) \cdot \partial \boldsymbol{X}_R(z)\, e^{i l \cdot \boldsymbol{X}_R(z)} \tag{15.155}$$

with

$$t^2 = 3, \quad l^2 = 1, \quad \boldsymbol{B} \cdot \boldsymbol{l} = 0 .$$

The coefficients A and $\boldsymbol{B}$ have to be determined such that Eq. (15.154) is satisfied. One can show that a necessary condition to arrive at theories with chiral fermions is $\boldsymbol{B} = 0$. Since these are the most interesting theories from a phenomenological point of view, we will limit our discussion to this case.

The (length)$^2 = 3$ vectors $\boldsymbol{t}$ play an important role in the construction of lower dimensional heterotic string theories. This becomes clear, if we consider the picture changing operator (13.85,13.86) which is itself a sum of a space-time and an internal part:

$$Z = Z^{\text{s.t.}} + Z^{\text{int}} = 2\left(T_F^{\text{s.t.}} + T_F^{\text{int}}\right) e^\phi . \tag{15.156}$$

Because of BRST invariance, $Z^{\text{int}}(w)$ must act properly on all states, i.e. it maps a physical state to its picture changed image. Thus consider a state of the form Eq. (15.148) characterized by a lattice vector $(\boldsymbol{w}_L; \boldsymbol{w}_R, \lambda_R, q) \in (\Gamma_{26-d})_L \otimes (\Gamma_{15-\frac{3}{2}d} \otimes D_{\frac{d}{2},1})_R$. The operator product with $Z^{\text{int}}(w)$ then reads:

$$Z^{\text{int}}(w)\, V_{\boldsymbol{w}_L;\boldsymbol{w}_R,\lambda_R,q}(z,\bar{z})$$
$$= 2 \sum_t A(t)\, (w-z)^{t \cdot \boldsymbol{w}_R - q}\, V_{\boldsymbol{w}_L;\boldsymbol{w}_R + t, \lambda_R, q+1}(z,\bar{z}) + \ldots . \tag{15.157}$$

We find that the picture changed state is characterized by lattice vectors $(\boldsymbol{w}_L; \boldsymbol{w}_R + t, \lambda_R, q+1)$ (if $\boldsymbol{w}_R \cdot t = -1\, (-\frac{1}{2})$ in the NS (R) sector for states in the canonical ghost picture). Furthermore, locality demands that every lattice vector $\boldsymbol{w}_R \in (\Gamma_{15-\frac{3}{2}d})_R$ satisfies $\boldsymbol{w}_R \cdot t \in \mathbb{Z}$ (NS) and $\boldsymbol{w}_R \cdot t \in \mathbb{Z} + \frac{1}{2}$ (R). On the other hand we know that $\Gamma_{26-d;15-d,1}$ is a self-dual lattice. It then follows immediately that $(\mathbf{0};\mathbf{t},\mathbf{0},1) \in (\Gamma_{26-d})_L \otimes (\Gamma_{15-\frac{3}{2}d})_R \otimes (D_{\frac{d}{2},1})_R$.

The upshot of this discussion is that the vectors $\mathbf{t}$ must themselves be lattice vectors of the right-moving internal lattice $(\Gamma_{15-\frac{3}{2}d})_R$. They always appear in connection with the V conjugacy class of $(D_{\frac{d}{2},1})_R$, since Z^{int} has superconformal ghost charge 1. The vectors $\mathbf{t}$ are called constraint vectors. Any sensible heterotic string theory, whose right-moving part possesses world-sheet supersymmetry, requires the existence of these constraint vectors. In summary, the classification of d-dimensional heterotic strings within the covariant lattice construction amounts to determine all Lorentzian odd self-dual lattices $\Gamma_{22-d;15-d,1}$ which contain $(D_{\frac{d}{2},1})_R$ and allow for a proper realization of the internal supercurrent. For theories with chiral fermions the internal supercurrent must have the form of Eq. (15.155) with constraint vectors $\mathbf{t}$, $\mathbf{t}^2 = 3$. Only a finite but very large number of Lorentzian lattices satisfies this constraint and these lattices can be shown to possess only a finite number of conjugacy classes, i.e. they correspond to rational CFTs. One could now proceed and discuss the massless spectrum of such construction but we do not do it here.

Further Reading

In writing large parts of this chapter we have greatly profited from L. Dixon's Trieste lecture notes

- L.J. Dixon, *Some World-Sheet Properties of Superstring Compactifications, on Orbifolds and Otherwise*. Lectures given at the 1987 ICTP Summer Workshop in High Energy Physics and Cosmology, Trieste 1987. Published in Trieste HEP Workshop, 1987, pp. 67–126

There one also finds the discussion of right-moving perturbative gauge symmetries of four-dimensional heterotic strings and a proof of the no-go theorem of the standard model for the perturbative type II string.

Orbifold models were introduced in

- L. Dixon, J. Harvey, C. Vafa, E. Witten, Strings on orbifolds. Nucl. Phys. **B261**, 651 (1985); Strings on orbifolds 2. Nucl. Phys. **B274**, 285 (1986)

A careful study of the spectrum of toroidal $\mathbb{Z}_N$-orbifolds was performed in

- J. Erler, A. Klemm, Comment on the generation number in orbifold compactifications. Comm. Math. Phys. **153**, 579 (1993) [arXiv:hep-th/9207111]

The discussion of the massless E_6 singlets of the $\mathbb{Z}_3$ orbifold is from

- P.S. Aspinwall, M.R. Plesser, Elusive worldsheet instantons in heterotic string compactifications [arXiv:1106.2998 [hep-th]]

Some basic references on orientifold models are

- C. Angelantonj, A. Sagnotti, Open strings. Phys. Rep. **371**, 1–150 (2002) [arXiv:hep-th/0204089]
- E.G. Gimon, J. Polchinski, Consistency conditions for orientifolds and d manifolds. Phys. Rev. **D54**, 1667–1676 (1996) [arXiv:hep-th/9601038]

The general structure of the SCFT of supersymmetric string models is explored in

- T. Banks, L. Dixon, D. Friedan, E. Martinec, Phenomenology and conformal field theory or can string theory predict the weak mixing angle? Nucl. Phys. **B299**, 613 (1988)
- T. Banks, L.J. Dixon, Constraints on string vacua with space-time supersymmetry. Nucl. Phys. **B307**, 93–108 (1988)
- J. Lauer, D. Lüst, S. Theisen, Supersymmetric string theories, superconformal algebras and exceptional groups. Nucl. Phys. **B309**, 771 (1988)
- S. Cecotti, S. Ferrara, L. Girardello, Geometry of type II superstrings and the moduli of superconformal field theories. Int. J. Mod. Phys. **A4**, 2475 (1989)

Their realizations as tensor products of $N = 2$ minimal models was developed in

- D. Gepner, Space-time supersymmetry in compactified string theory and superconformal models. Nucl. Phys. **B296**, 757 (1988); Exactly solvable string compactifications on manifolds of SU(N) holonomy. Phys. Lett. **B199**, 380–388 (1987)

Generalization of the Gepner construction to $(0, 2)$ heterotic models were presented in

- R. Blumenhagen, A. Wißkirchen, Exactly solvable (0,2) supersymmetric string vacua with GUT gauge groups. Nucl. Phys. **B454**, 561–586 (1995) [arXiv:hep-th/9506104]

The covariant lattice construction of four-dimensional heterotic strings was discussed in

- W. Lerche, D. Lüst, A.N. Schellekens, Chiral four-dimensional heterotic strings from self-dual lattices. Nucl. Phys. **B287**, 477 (1987)
- W. Lerche, A.N. Schellekens, N. Warner, Lattices and strings. Phys. Rep. **177**, 1 (1989)

Chapter 16
String Scattering Amplitudes and Low Energy Effective Field Theory

Abstract To relate string theory to the usual description of particles and their interactions in terms of quantum field theories, it is important to have tools at hand to derive the effective point particle interactions for the massless excitation modes of the string. Such effective actions can be deduced from on-shell string scattering amplitudes which are computed as correlation functions of physical state vertex operators. We construct the vertex operators and compute various three-point functions which are needed to extract e.g. the interactions of graviton, two-form, dilaton and of gauge fields at leading order. We also compute the four-point functions of open and closed string tachyons and discuss some of their properties. Often the leading order (in α') effective actions are already uniquely determined by symmetries, such as gauge symmetries or supersymmetry. We present the bosonic sectors of the ten-dimensional supergravity theories which are related to the ten-dimensional superstring theories. We also include a discussion of eleven-dimensional supergravity. The Dirac-Born-Infeld action, which governs the dynamics of the gauge field on a D-brane, will also be discussed.

16.1 Generalities

String theory is claimed to be a unifying framework for the description of all particles and their interactions, including gravity. So far our exposition of the subject has been rather formal, and it is not at all apparent why it might be relevant for particle physics. The only hint we got so far was from looking at the spectrum. The occurrence of a massless spin two symmetric tensor excitation indicated that gravity might be contained in string theory. To confirm this at the level of interactions will be one of the goals of this chapter.

The theoretical framework for particle physics at energies, which are accessible in experiments, is quantum field theory. At these energies, gravity is many orders of magnitude weaker than the other gauge interactions mediated by spin one gauge bosons. The dimensionless gravitational coupling constant is $G_N^{(4)} E^2 \sim E^2/M_{\text{Pl}}^2$

where $G_N^{(4)}$ is the four-dimensional Newton constant, M_{Pl} the Planck mass and E the relevant energy. At high energy, gravity is thus expected to become stronger and must be included in the theoretical description of particle physics. However, a quantum field theory of gravity, which is constructed along the lines of quantum field theories of Yang-Mills gauge interactions, is not renormalizable and does not make sense at very high energies. (The same is true for higher dimensional Yang-Mills theories where the gauge coupling has negative mass dimension.) New concepts and ideas are required, and this is what string theory is believed to achieve. At low energies, though, string theory should have a description in terms of an *effective field theory*. For such a theory string theory also provides a natural cutoff, the string scale, above which the field theory description is no longer adequate. From the low energy point of view, the ultraviolet completion is provided by the full string theory. Put differently, in string theory the effective field theory is an emergent concept. To compute corrections to it, we have to resort to the microscopic theory, i.e. to string theory.

Field theory imposes severe constraints e.g. on the interaction of massless spin one, spin 3/2 or spin two fields. At lowest order in a derivative expansion, their dynamics is that dictated by supersymmetric Yang-Mills theory coupled to supergravity. Thus, to make the just described picture consistent, string theory must precisely generate these actions. As long as we work in the critical dimension, these considerations lead to essentially unique low-energy effective actions. However, once we compactify and consider effectively four-dimensional theories, the uniqueness is lost. Nevertheless, once a particular string vacuum has been chosen, the low energy effective action can, at least in principle, be computed. For several classes of vacua this has been done, e.g. for toroidal orbifolds and covariant lattices using CFT methods and for CY compactifications using geometric methods.

Four-dimensional string vacua generically have massless vectors that transform in the adjoint representation of a (simply-laced) gauge group and plenty of massless scalars and fermions in non-trivial representations of the gauge group. Beyond knowing the spectrum, we would also like to deduce the dynamics of these fields. In this chapter we learn how to extract a low energy effective field theory directly from string theory. This will be done by deriving a point particle field theory Lagrangian which represents the low energy description of string theory in the sense that it reproduces string scattering amplitudes. The latter will be computed as correlation functions of vertex operators. Here low energy means energies far below the string scale, in which case on-shell massive string excitations cannot be produced in the collision of massless modes and their contribution from internal lines in Feynman diagrams is suppressed by powers on $\alpha' k^2$ where k is a typical momentum of the external (on-shell) states. In the limit $\alpha' \to 0$ the effect of massive modes is eliminated completely and we obtain a description of the massless modes only. This action can then be used as an effective low-energy description of string compactifications to four space-time dimensions, e.g. heterotic string compactifications on Calabi-Yau manifolds or type II intersecting D-brane models and type II flux compactifications, which will be discussed in subsequent chapters.

16.1 Generalities

We can then relate various parameters in the effective action to the string scale and the string coupling constant. For instance, in heterotic string compactifications, the string scale will turn out to be of the order of the Planck scale, whereas in D-brane models the string scale is essentially a free parameter, experimentally only constrained to be higher than the known scale of the Standard Model, i.e. $\simeq 1$ TeV. Therefore in both schemes all known particles like quarks, leptons and all the know gauge bosons have to be contained among the massless string states.

Of course we are still far away from extracting the Standard Model Lagrangian with all its particular masses and coupling constants but at least we can see whether we find something semi-realistic, i.e. a field theory with the correct generic features.

It is worth mentioning that there exist several ways to derive the low energy effective action. They all rely of the validity of string perturbation theory or, better, that there is a duality frame in which the theory is weakly coupled. We will discuss (non-)perturbative dualities in Chap. 18. The most direct approach is the one just sketched, namely to compute string theory scattering amplitudes of the massless modes and to construct a field theory action which reproduces them in the limit $\alpha' \to 0$. Starting with an exact conformal field theory description, this procedure is the most precise way to derive the effective action. However, taken by itself, this method is of limited use, since in principle one has to compute an infinite number of string scattering amplitudes if one wants to derive the full effective action, e.g. if one wants to derive the non-linear effective action for the graviton, or the moduli fields or the dilaton couplings to the fields of the Standard Model. One must also keep in mind that the matching between string amplitudes and the field theory action is not unique in the sense that one can perform field redefinitions which change the form of the effective action but not the physics. Furthermore, the scattered string excitations are represented by vertex operators. The condition of BRST invariance requires the state created by a vertex operators to be on-shell. This means that we can only compute on-shell scattering amplitudes, a fact which also leads to ambiguities in the construction of the low energy effective action.

A second method is to exploit space-time symmetries such as general coordinate invariance and supersymmetry. One can infer the space-time symmetries from the structure of the world-sheet theory, e.g. via existence of conserved currents. The space-time symmetries then restrict the action of the massless string modes. Additional arguments like vanishing of anomalies also play an important role. In this way the ten-dimensional effective supergravity theories of the various string theories with maximal supersymmetry are completely fixed, at least up to higher derivative corrections which are also constrained by supersymmetry.

It is interesting to note that many of the additional field theory conditions reflect themselves in certain string theory consistency requirements. For instance, absence of space-time anomalies in the context of the ten-dimensional supergravity action is related to world-sheet modular invariance in string theory. The fact that both are one-loop effects in the respective description is, of course, no accident. More generally, any effective field theory derived from string theory is automatically consistent (e.g. anomaly free) because its ultraviolet completion is a consistent string theory.

Having found the effective supergravity action in ten dimensions, one can often use dimensional reduction over a compact six-dimensional space, in order to derive the effective action in four space-time dimensions. In case of unbroken supersymmetry, the effective supergravity formalisms again acts as a very useful guiding principle. However, with fewer symmetries there is more freedom in building effective actions. Given a particular four-dimensional string vacuum, this freedom can be fixed by comparison with string scattering amplitudes, provided we have a CFT description. Large classes of vacua with rather generic features can be treated in this way.

Another method of arriving at the low energy effective action is based on the requirement of world-sheet conformal invariance, namely the condition that the β-functions of the two-dimensional string σ-model coupled to target space background fields have to vanish. This was explained in Sect. 14.1.

The most common way at arriving at the effective action is a combination of the first two methods: computing string scattering amplitudes and exploiting space-time symmetries. This is the route we will take.

16.2 String Scattering Amplitudes

Our discussion of scattering amplitudes will be restricted to the string tree level. Furthermore, we will scatter massless modes as they are the most interesting ones from the point of view of the low-energy effective action. In the large extra dimension scenario, where the string scale can be much lower than the Planck scale, some of the lightest massive states might also be of interest, as they could be directly detected in high energy experiments.

There are many amplitude calculations in the literature and the explicit computations are generally quite tedious. The CFT correlation functions can become quite involved and for more than four external states the integration over the insertion points of the vertex operators is a technical challenge. Fortunately, for the extraction of the two-derivative low-energy effective action, four point amplitudes is all one needs to compute. One particular subtlety is the correct normalization. It can either be computed by a careful evaluation of the Polyakov path integral or inferred from the requirement of unitarity. More concretely, a four-point tree amplitude has simple poles which are due to the exchange of string modes. At these poles the amplitude must factorize into a product of two three-point amplitudes. This fixes the normalization of both the three-point and the four-point amplitudes. We will demonstrate this in two simple examples. The precise normalizations are important e.g. for relating the string coupling constant and the string scale to the Yang-Mills and the gravitational coupling constant which appear in the low-energy effective field theory.

We now discuss closed and open string tree amplitudes in turn. Our treatment will not be exhaustive but the few examples which we present illustrate essential points and techniques.

Closed String Tree Amplitudes

For closed oriented string tree amplitudes the only relevant world-sheet topology is the sphere S^2 with various vertex operator insertions representing external physical on-shell states. We will make some remarks on amplitude calculations on the projective plane, after we have discussed open string amplitudes.

In the language of conformal field theory, an N-particle amplitude is essentially given by the correlation function

$$A \sim g_s^{N-2} \int \frac{d^2z_1 \ldots d^2z_N}{V_{\text{CKG}}} \langle V_1 \ldots V_N \rangle_{S^2}, \tag{16.1}$$

where g_s is the string coupling constant and V_{CKG} is the volume of the conformal Killing group, which is $SL(2,\mathbb{C})$ for the sphere. There is one power of the string coupling constant for each splitting of one string into two or two strings merging into one. The total power of g_s is the Euler number of the N-punctured sphere. Alternatively we can attach one power of g_s to each closed string vertex operator and an overall power of g_s^{-2} to the correlation function on the sphere.

We thus have to compute correlation functions of vertex operators on the sphere. We will only consider string vacua where the vertex operators can be expressed in terms of free fields. This is the case for strings in Minkowski space in the critical dimension, where the fields are X^μ, ψ^μ and the ghosts, but also e.g. for covariant lattice constructions. In orbifold compactifications one has to introduce twist fields. They can be handled but we will not discuss this. Correlation functions including fields in the Ramond sector can be a challenge. In simple cases covariance and general analytic properties are sufficient to determine them. Another way is to bosonize world-sheet fermions and spin fields. Correlation functions are then easy to compute, but to restore covariance, which is hidden in the cocycle factors, can be quite cumbersome. The superconformal ghosts are discussed in their bosonized form.

For free fields it suffices to know the two-point functions. Higher correlation functions follow via Wick's theorem. The basic two-point functions on the sphere were given before and are repeated here:

$$\langle X^\mu(z) X^\nu(w) \rangle = -\frac{\alpha'}{2} \eta^{\mu\nu} \ln(z-w),$$

$$\langle \psi^\mu(z) \psi^\nu(w) \rangle = \frac{\eta^{\mu\nu}}{(z-w)},$$

$$\langle \phi^i(z), \phi^j(w) \rangle = -\delta^{ij} \ln(z-w),$$

$$\langle \phi(z) \phi(w) \rangle = -\ln(z-w). \tag{16.2}$$

ϕ arises from the bosonization of the superconformal ghosts and ϕ^i from the bosonization of the world sheet fermions ψ^μ. There is a similar set for the right movers. Note that we have separated X into left and right-movers (cf.

Chap. 2): $X^\mu(z,\bar{z}) = X^\mu(z) + \overline{X}^\mu(\bar{z})$. In tree level amplitudes we can treat them as independent and factorize correlation functions into a holomorphic and an antiholomorphic part. For more general internal CFTs, this is not always the case.

As in ordinary quantum field theory, where in the calculation of matrix elements we have to integrate over the positions of the field operators, we have to integrate over the insertion points of the vertex operators. The integrated vertex operators are the momentum space representatives of vertex operators in position space. For instance, a general vertex operator has the form

$$V(k) = \int d^2z\, \mathcal{V}(z,\bar{z})\, e^{ik_\mu X^\mu(z,\bar{z})}, \tag{16.3}$$

where $\mathcal{V}(z,\bar{z})$ carries all the quantum numbers of the state, such as Lorentz indices and charges with respect to internal symmetries (gauge symmetries) and k is its space-time momentum. Fourier transforming it we obtain

$$V(x) = \int d^d k \int d^2z\, \mathcal{V}(z,\bar{z})\, e^{ik\cdot X(z,\bar{z})} e^{ik\cdot x} = \int d^2z\, \mathcal{V}(z,\bar{z})\, \delta^{(d)}(X(z,\bar{z}) - x). \tag{16.4}$$

In ordinary field theory translation invariance results in an infinite volume factor. Here we have to deal with the invariance under $SL(2,\mathbb{C})$. We have learned in Chap. 6 how to factor out the $SL(2,\mathbb{C})$ volume. That led to the following prescription. We formally insert a factor

$$\frac{|\langle c(z_i)\, c(z_j)\, c(z_k) \rangle|^2}{\int dz_i^2\, dz_j^2\, dz_k^2} \tag{16.5}$$

with

$$|\langle c(z_1)\, c(z_2)\, c(z_3) \rangle|^2 = |z_{12}\, z_{23}\, z_{13}|^2, \tag{16.6}$$

where we have defined $z_{ij} = z_i - z_j$. This means that we arbitrarily fix the positions of three of the vertex operators, drop the corresponding integrations and insert a reparametrization ghost at these positions. This takes care of the Jacobian of the transformation from z_1, z_2, z_3 to the three (complex) parameters of $SL(2,\mathbb{C})$. For the remaining vertex operators one uses their integrated version. For three-point functions no integral is left to do. One usually chooses for the fixed positions $z = 0, 1$ and ∞. For three-point amplitudes, one may keep them at arbitrary (but fixed) values; the fact that the final amplitude has to be independent of them then serves as a check. The ghost insertions do not change the conformal weights of the vertex operators, as both dz and $c(z)$ have weight -1. Moreover, we have shown in Chap. 5 that, if $\int d^2z\, V(z,\bar{z})$, for V independent of the b, c ghosts, is BRST invariant, then so is $c(z)\bar{c}(\bar{z}) V(z,\bar{z})$.

16.2 String Scattering Amplitudes

For fermionic strings we also have to deal with conformal Killing spinors. As discussed in Chap. 13, they are dealt with by choosing the ghost pictures for the vertex operators such that their superconformal ghost charges satisfy $\sum_i q_i = -2$. As explained there, the amplitudes are independent of the way how the superconformal ghost charges are distributed among the various vertex operators. The particular choice is a matter of convenience.

If we bosonize the world-sheet fermions, then all states are characterized by a weight vector of $SO(1, 9)$. These weight vectors must add up to zero. The covariant meaning of this is that the tensor product of the various $SO(1, 9)$ representations must contain the trivial representation. Finally the space-time momenta k_i must add to zero.

The construction of vertex operators for string excitations uses the operator-state correspondence. For instance, a tachyon of the bosonic string corresponds to the state $|k\rangle$ with $k^2 = 4/\alpha'$. The vertex operator which creates this state from the $SL(2, \mathbb{Z})$ invariant vacuum is[1]

$$V(k) = e^{ik \cdot X(z,\bar{z})}, \qquad k^2 = -m^2 = \frac{4}{\alpha'}, \qquad (16.7)$$

cf. Chap. 4. The graviton of the bosonic string is the state $\epsilon_{\mu\nu}(k)\alpha^\mu_{-1}\bar{\alpha}^\nu_{-1}|k\rangle$ with associated vertex operator

$$V(k, \epsilon) = -\frac{2}{\alpha'} \epsilon_{\mu\nu}(k) \, \bar{\partial} X^\mu(\bar{z}) \, \partial X^\nu(z) \, e^{ik \cdot X(z,\bar{z})}, \qquad (16.8)$$

where we used $\lim_{z \to 0} i \partial X(z)|k\rangle = \sqrt{\frac{\alpha'}{2}} \alpha_{-1}|k\rangle$. Here, $\epsilon_{\mu\nu}$ denotes a polarization tensor, on which BRST invariance imposes the on-shell conditions $k^\mu \epsilon_{\mu\nu} = \epsilon_{\mu\nu} k^\nu = 0$ and $k^2 = 0$. The polarization tensor is the wave function of the string excitation in momentum space. The above vertex operators represent either a graviton ($h_{\mu\nu}$), an anti-symmetric tensor ($B_{\mu\nu}$) or a dilaton (D), depending on whether the polarization tensor is symmetric and traceless, anti-symmetric or transverse diagonal. We decompose a general transverse $\epsilon_{\mu\nu}$ accordingly and call the three irreducible, pieces $\epsilon^{(h)}_{\mu\nu}$, $\epsilon^{(B)}_{\mu\nu}$ and $\epsilon^{(D)}_{\mu\nu}$:

graviton: $\qquad \epsilon^{(h)}_{\mu\nu} = \epsilon^{(h)}_{\nu\mu}, \qquad \epsilon^{(h)}_{\mu\nu} \eta^{\mu\nu} = k^\mu \epsilon^{(h)}_{\mu\nu} = 0,$

antisym. tensor: $\quad \epsilon^{(B)}_{\mu\nu} = -\epsilon^{(B)}_{\nu\mu}, \qquad k^\mu \epsilon^{(B)}_{\mu\nu} = 0,$

Dilaton: $\qquad \epsilon^{(D)}_{\mu\nu} = \frac{1}{\sqrt{d-2}} (\eta_{\mu\nu} - k_\mu \bar{k}_\nu - \bar{k}_\nu k_\mu), \qquad k^\mu \epsilon^{(D)}_{\mu\nu} = 0.$

$$(16.9)$$

---

[1] Here and below we often drop normal ordering symbols.

$\bar{k}_\mu$ is an auxiliary vector which satisfies $\bar{k}^2 = 0$ and $\bar{k} \cdot k = 1$.[2] Sometimes we will refer to these polarizations collectively as $\epsilon_{\mu\nu}^{(G)}$. If we normalize the vertex operators such that for each of the three irreducible components the state $|\epsilon, k\rangle = \lim_{z,\bar{z}\to 0} V(k, \epsilon, z, \bar{z})|0\rangle = \epsilon_{\mu\nu}^{(G)} \alpha_{-1}^\mu \bar{\alpha}_{-1}^\nu |k\rangle$ is normalized, i.e. $\langle k', \epsilon|k, \epsilon\rangle = (2\pi)^d \delta^{(d)}(k+k')$, we need to require $\epsilon_{\mu\nu}^{(G)} \epsilon^{(G)\mu\nu} = 1$. The three polarization tensors are mutually orthogonal, e.g. $\epsilon_{\mu\nu}^{(D)} \epsilon^{(G)\mu\nu} = 0$.

Consider the holomorphic operator

$$V(k, \epsilon, z) = \sqrt{\frac{2}{\alpha'}} \epsilon_\mu(k)\, i\, \partial X^\mu(z)\, e^{ik \cdot X(z)}, \qquad (16.10)$$

which is essentially the holomorphic part of (16.8). Under $\epsilon_\mu \to \epsilon_\mu + \lambda k_\mu$ it changes as[3]

$$\delta V(k, \epsilon, z) = \lambda \sqrt{\frac{2}{\alpha'}} i k_\mu \partial X^\mu(z) e^{ik \cdot X(z)} = \hat{L}_{-1}\left(\lambda \sqrt{\frac{2}{\alpha'}} e^{ik \cdot X(z)}\right), \qquad (16.11)$$

i.e. by a descendant field. In Chap. 4 we have shown that Virasoro descendants decouple from string scattering amplitudes. In other words, correlation functions containing vertex operators (16.8) should be invariant under $\epsilon_{\mu\nu} \to \epsilon_{\mu\nu} + k_\mu \xi_\nu$ (with $k \cdot \xi = 0$ to preserve transversality). But these are just the changes of the graviton under infinitesimal diffeomorphisms, $\delta h_{\mu\nu} = \partial_\mu \xi_\nu + \partial_\nu \xi_\mu$ and of the antisymmetric tensor under gauge transformations $\delta B = d\underline{\xi}$. This also shows the independence of the amplitudes of the choice of the vector $\bar{k}$. Changing it by $\bar{k} \to \bar{k} + \bar{l}$, with $k \cdot \bar{l} = 0$, leads to a change of the polarization tensors of the form just discussed. Taking into account this gauge freedom, one verifies that the number of physical degrees of freedom for the graviton, anti-symmetric tensor and dilaton is the same as what we derived in Chap. 3 when we analyzed the massless spectrum of the closed string in light-cone gauge.[4]

For the fermionic string and the holomorphic sector of the heterotic string the construction of vertex operators is more involved. This is due to the appearance of spin fields and the issue of ghost pictures. For instance, the superstring graviton vertex operator in the canonical ghost picture is

---

[2] Given $\epsilon_{\mu\nu}$ the symmetric-traceless piece is $\frac{1}{2}(\epsilon_{\mu\nu} + \epsilon_{\nu\mu}) - \frac{\epsilon^\rho_\rho}{d-2}(\eta_{\mu\nu} - k_\mu \bar{k}_\nu - \bar{k}_\nu k_\mu)$.

[3] λ is a constant which makes λk dimensionless.

[4] The anti-symmetric tensor is slightly tricky, so let us give some details. An unconstrained antisymmetric tensor $\epsilon_{\mu\nu}$ in d dimensions has $\frac{1}{2}d(d-1)$ independent components. The transversality condition $k^\mu \epsilon_{\mu\nu} = 0$ imposes $(d-1)$ constraints (because of the identity $k^\mu \epsilon_{\mu\nu} k^\nu \equiv 0$ not all components of $k^\mu \epsilon_{\mu\nu}$ are linearly independent). Furthermore there is the gauge freedom $\epsilon_{\mu\nu} \to \epsilon_{\mu\nu} + k_\mu \zeta_\nu - k_\nu \zeta_\mu$ where transversality imposes $k \cdot \zeta = 0$. However there is a 'gauge invariance for the gauge invariance', i.e. the choice $\zeta_\mu = a k_\mu$ does not change $\epsilon_{\mu\nu}$. Taking all of this into account leaves $\frac{1}{2}(d-2)(d-3)$ components.

16.2 String Scattering Amplitudes

$$V_{(-1,-1)}(k,\epsilon) = \epsilon_{\mu\nu}(k)\, e^{-\bar{\phi}(\bar{z})}\, \overline{\psi}^{\mu}(\bar{z})\, e^{-\phi(z)}\, \psi^{\nu}(z)\, e^{ik\cdot X(z,\bar{z})}$$

$$\equiv \epsilon_{\mu\nu}(k)\, \overline{V}^{\mu}_{(-1)}(k,\bar{z})\, V^{\nu}_{(-1)}(k,z) \tag{16.12}$$

with

$$V^{\mu}_{(-1)}(k,z) = e^{-\phi}\psi^{\mu}(z)\, e^{ik\cdot X(z)}. \tag{16.13}$$

Its space-time part creates the state $\epsilon_{\mu\nu}(k)\, b^{\mu}_{-1/2}\bar{b}^{\nu}_{-1/2}|k\rangle$. For this state to be normalized to one, we again require $\epsilon^{(G)}_{\mu\nu}\epsilon^{(G)\mu\nu} = 1$.

For non-vanishing correlation functions we also need the graviton vertex operator in the zero ghost picture which is obtained via the picture changing operation from the one in the canonical ghost picture. This was done in Chap. 13 with the result

$$V_{(0,0)}(k,\epsilon) = \frac{2}{\alpha'}\epsilon_{\mu\nu}(k)\, \left(i\bar{\partial}X^{\mu} + \frac{\alpha'}{2}(k\cdot\overline{\psi})\,\overline{\psi}^{\mu}\right)(\bar{z})$$

$$\left(i\partial X^{\nu} + \frac{\alpha'}{2}(k\cdot\psi)\,\psi^{\nu}\right)(z)\, e^{ik\cdot X(z,\bar{z})}$$

$$\equiv \epsilon_{\mu\nu}(k)\overline{V}^{\mu}_{(0)}(k,\bar{z})\, V^{\nu}_{(0)}(k,z) \tag{16.14}$$

with

$$V^{\mu}_{(0)}(k,z) = \sqrt{\frac{2}{\alpha'}}\left(i\partial X^{\mu} + \frac{\alpha'}{2}(k\cdot\psi)\,\psi^{\mu}\right) e^{ik\cdot X}(z). \tag{16.15}$$

For the heterotic string, the graviton vertex operators in the two ghost pictures are

$$V_{(-1)}(k,\epsilon) = \sqrt{\frac{2}{\alpha'}}\epsilon_{\mu\nu}(k)\, i\bar{\partial}X^{\mu}(\bar{z})\, e^{-\phi}\psi^{\nu}(z)\, e^{ik\cdot X(z,\bar{z})},$$

$$V_{(0)}(k,\epsilon) = \frac{2}{\alpha'}\epsilon_{\mu\nu}(k)\, i\bar{\partial}X^{\mu}(\bar{z})\left[i\partial X^{\nu} + \frac{\alpha'}{2}(k\cdot\psi)\,\psi^{\nu}\right](z)\, e^{ik\cdot X(z,\bar{z})}, \tag{16.16}$$

and the heterotic gauge bosons vertex operators are

$$V^{a}_{(-1)}(k,\epsilon) = \epsilon_{\mu}(k)\, J^{a}(\bar{z})\, e^{-\phi}\psi^{\mu}(z)\, e^{ik\cdot X(z,\bar{z})},$$

$$V^{a}_{(0)}(k,\epsilon) = \sqrt{\frac{2}{\alpha'}}\epsilon_{\mu}(k)\, J^{a}(\bar{z})\left[i\partial X^{\mu} + \frac{\alpha'}{2}(k\cdot\psi)\psi^{\mu}\right](z)\, e^{ik\cdot X(z,\bar{z})}, \tag{16.17}$$

where ϵ_{μ} is a polarization vector which satisfies the on-shell condition $k^{\mu}\epsilon_{\mu} = 0$ and the normalization $\epsilon_{\mu}\epsilon^{\mu} = 1$. Here, J^{a} are Kač-Moody currents normalized as $\langle J^{a}(1)\, J^{b}(0)\rangle = \delta^{ab}$ (see Eqs. (11.17) and (11.20)). For differently normalized currents J^{a} we have to rescale the above vertex operators in order to be able to compare the string amplitude with field theory amplitudes with canonically normalized gauge fields.

So far all vertex operators were in the NS sector. For the R sector we need spin fields, where from Eq. (13.76) we already know the vertex operator of the gravitino (at zero momentum). We therefore have the vertex operator for the heterotic gravitino and dilatino in the canonical ghost picture

$$V_{(-\frac{1}{2})}(k,u) = (\alpha')^{-\frac{1}{4}} u^\alpha_\mu(k) \, i\bar\partial X^\mu(\bar z) \, e^{-\frac{1}{2}\phi} S_\alpha(z) \, e^{ik \cdot X(z,\bar z)} \qquad (16.18)$$

with $k^2 = 0$, $k^\mu u_{\mu\alpha} = 0$ and $\slashed{k} u_\mu = 0$ as required by BRST invariance. Here S_α is a chiral spinor, say of the (S) conjugacy class of $SO(1,9)$. Spinors in the (C) conjugacy class are $S_{\dot\alpha}$. u^α_μ is the spinor wave function in momentum space. It splits into two irreducible pieces of opposite chiralities, $u_{\mu\alpha} = \left(u_{\mu\alpha} - \frac{1}{10}(\gamma_\mu)_\alpha{}^{\dot\beta}(\gamma \cdot u)_{\dot\beta}\right) + \frac{1}{10}(\gamma_\mu)_\alpha{}^{\dot\beta}(\gamma \cdot u)_{\dot\beta} \equiv \tilde u_{\mu\alpha} + (\gamma_\mu)_\alpha{}^{\dot\beta} u_{\dot\beta}$, the gravitino $\tilde u_{\mu\alpha}$ with $\gamma^\mu \tilde u_\mu = 0$ and the dilatino $u_{\dot\alpha}$. The vertex operator (16.18) follows from the graviton vertex operator by the action of the space-time supercharge. We have not fixed the absolute normalization of the fermion vertex operators. This could be done by comparing scattering amplitudes, e.g. the two gravitino-one graviton amplitude, with the field theory amplitude, but we will not do this.

The gravitini and dilatini vertex operators for the type IIA/B superstrings easily follow from this expression by the replacement $i\bar\partial X^\mu \to \sqrt{\frac{\alpha'}{2}} e^{-\bar\phi} \bar\psi^\mu$. This yields the vertex operators $V_{(-1,-\frac{1}{2})}$. In addition, there is a second gravitino and dilatino with vertex operators

$$V_{(-\frac{1}{2},-1)}(k,u) = \begin{cases} (\alpha')^{-\frac{1}{4}} u^\alpha_\mu(k) \, \overline{S}_\alpha^{(-1/2)}(\bar z) \, \psi^{\mu(-1)}(z) \, e^{ik \cdot X(z,\bar z)} & \text{(IIB)} \\ (\alpha')^{-\frac{1}{4}} u^{\dot\alpha}_\mu(k) \, \overline{S}_{\dot\alpha}^{(-1/2)}(\bar z) \, \psi^{\mu(-1)}(z) \, e^{ik \cdot X(z,\bar z)} & \text{(IIA).} \end{cases}$$

$$(16.19)$$

For the gravitino in the $+\frac{1}{2}$ picture we replace the holomorphic part according to Eq. (13.83).

Often it is convenient to consider the spin fields and the ψ^μ together with the superconformal ghost factor and to define e.g. $\psi_\mu^{(-1)} = e^{-\phi}\psi_\mu$ and $S_\alpha^{(-\frac{1}{2})} = e^{-\frac{1}{2}\phi} S_\alpha$. As we have discussed in Chap. 15, these combinations are characterized by weight vectors of $D_{5,1}$, where states in the NS sector are in the (0) conjugacy class while states in the R sector are either in the (S) or the (C) conjugacy class, depending on the GSO projection. Vertex operators in the (0) conjugacy class are space-time bosons and vertex operators in a spinor conjugacy class are space-time fermions.

The vertex operators for heterotic gaugini are

$$V^a_{(-\frac{1}{2})}(k,u) = \frac{1}{\sqrt{2}} (\alpha')^{\frac{1}{4}} u^\alpha(k) \, \overline{J}^a(\bar z) \, S_\alpha^{(-\frac{1}{2})}(z) \, e^{ik \cdot X(z,\bar z)}. \qquad (16.20)$$

16.2 String Scattering Amplitudes

Vertex operators for RR fields can also be constructed. For a p-form potential it is, in the canonical ghost picture,

$$V_{(-\frac{1}{2},-\frac{1}{2})}(k,\epsilon) = \frac{\sqrt{2\alpha'}}{8\sqrt{p!}} F_{\mu_1...\mu_{p+1}} \overline{S}_\alpha^{(-\frac{1}{2})} (\Gamma^{\mu_1...\mu_{p+1}})^{\alpha\beta} S_\beta^{(-\frac{1}{2})} e^{ik\cdot X} \quad p \text{ even,}$$

$$V_{(-\frac{1}{2},-\frac{1}{2})}(k,\epsilon) = \frac{\sqrt{2\alpha'}}{8\sqrt{p!}} F_{\mu_1...\mu_{p+1}} \overline{S}_\alpha^{(-\frac{1}{2})} (\Gamma^{\mu_1...\mu_{p+1}})^{\alpha\dot\beta} S_{\dot\beta}^{(-\frac{1}{2})} e^{ik\cdot X} \quad p \text{ odd,}$$

(16.21)

where the p-even/odd forms are for the type IIB/IIA superstring. BRST invariance requires

$$k^{\mu_1} F_{\mu_1...\mu_{p+1}} = 0 \quad \text{and} \quad k_{[\mu_1} F_{\mu_2...\mu_{p+2}]} = 0. \quad (16.22)$$

These physical state conditions imply that the vertex operator represents the field strength of the p-form rather than the field itself; they are the field equation and the Bianchi identity, respectively. They imply $F_{\mu_1...\mu_{p+1}} = (p+1)k_{[\mu_1}\epsilon_{\mu_2...m_{p+1}]}$ with $k^2 = 0$ and $k^{\mu_1}\epsilon_{\mu_1...\mu_p} = 0$.[5] The normalization of the vertex operators will be checked when we compare to field theory in the next section.

The vertex operators we have constructed so far are for the string in its critical dimension. For the graviton, anti-symmetric tensor and dilaton they are also valid for the compactified theory, with the only change that the range of the space-time vector index is restricted to be along the non-compact directions and the amplitudes which contain only fields in the NS sectors are therefore also correct for the compactified theory. Other vertex operators receive explicit dependence of the internal CFT. For further details we refer to our discussion in Sect. 15.4.

GSO projection guarantees that correlation functions have no branch cuts, i.e. are single valued. Three point functions are independent of the three arbitrarily chosen positions of the vertex operators. Four-point functions, after the choice $z_1 = 0$, $z_2 = 1$, $z_4 = \infty$ lead to an integral of the general form

---

[5]This is shown as follows: $k^\mu = (k^0, \mathbf{k})$, together with $\overline{k}^\mu = (-k^0, \mathbf{k})$ and $(d-2)$ transverse polarization vectors e^i_μ allow us to decompose

$$F_{\mu_1...\mu_{p+1}} = a_{i_1...i_{p+1}} e^{i_1}_{[\mu_1} \cdots e^{i_{p+1}}_{\mu_{p+1}]} + (b_{i_1...i_p} k_{[\mu_1} + c_{i_1...i_p} \overline{k}_{[\mu_1}) e^{i_1}_{\mu_2} \cdots e^{i_p}_{\mu_{p+1}]}$$

$$+ d_{i_1...i_{p-1}} k_{[\mu_1} \overline{k}_{\mu_2} e^{i_1}_{\mu_3} \cdots e^{i_{p-1}}_{\mu_{p+1}]}.$$

If we impose (16.22) we find $a = c = d = 0$ and $k^2 b = 0$. This also shows that the number of propagating degrees of freedom equals the number of components of $b_{i_1...i_p}$, i.e. $\binom{d-2}{p}$.

$$\int d^2z\, |z|^\alpha |1-z|^\beta z^n (1-z)^m$$

$$= 2\pi(-1)^{m+n} \frac{\Gamma\left(1+n+\tfrac{1}{2}\alpha\right) \Gamma\left(1+m+\tfrac{1}{2}\beta\right) \Gamma\left(-1-n-m-\tfrac{1}{2}(\alpha+\beta)\right)}{\Gamma\left(-\tfrac{1}{2}\alpha\right) \Gamma\left(-\tfrac{1}{2}\beta\right) \Gamma\left(2+\tfrac{1}{2}(\alpha+\beta)\right)}.$$
(16.23)

The integral will be performed in the appendix. It is a meromorphic function of the parameters m, n, α, β, which in case of string amplitudes are functions of the momenta. The integral is potentially divergent at $z = 0, 1, \infty$. It exists for a certain region in complex momentum space and outside this region it is defined by analytic continuation. This is a general feature of higher point string amplitudes, where it is not a priori clear that there always exists a region in momentum space where the integral converges.

We are now almost ready to compute a few scattering amplitudes. But we still have to comment on the normalization of the vertex operators and the scattering amplitudes. First, we multiply each closed string vertex operator by a constant $g_c \sim g_s$. Normalization factors related to the positions of the vertex operators, (e.g. d^2z vs. $\frac{1}{2\pi}d^2z$) are absorbed in g_c. Comparison of the string scattering amplitudes with the amplitudes computed from the low-energy effective action will establish the relation between g_c and the gauge and gravitational coupling constants appearing in the low-energy effective action. Note that all closed string vertex operators are multiplied by the same g_c, because they all correspond to excitations of the closed string. Second, we know from the general discussion of Chap. 6 that all tree-level amplitudes are multiplied by C_{S^2}. Normalization constants from the measure on the CKG are absorbed into C_{S^2} as well as various other measure factors from the functional integrals. C_{S^2}, which should be proportional to $g_c^{-2} \sim g_s^{-2}$, will be determined by requiring unitarity of the amplitudes. Finally, we assume space-time momentum conservation in all correlation functions and drop the momentum conserving delta-function which arises from the integration over the zero modes of X^μ in the functional integral.

The simplest three-point amplitude is the three-tachyon amplitude of the bosonic string

$$A = C_{S^2} g_c^3 \langle c\bar{c} V(1)\, c\bar{c} V(2)\, c\bar{c} V(3) \rangle = g_c^3\, C_{S^2},$$
(16.24)

where we used the vertex operators (16.7) and the on-shell constraint $\frac{\alpha'}{2} k_i \cdot k_j = -2$ for $i \neq j$.

A less trivial example is the three gauge boson amplitude in the heterotic string

$$A_{\text{het}}^{abc}(k_1, \epsilon_1; k_2, \epsilon_2; k_3, \epsilon_3) = C_{S^2} g_c^3\, \langle c\bar{c} V_{(-1)}^a(1)\, c\bar{c} V_{(-1)}^b(2)\, c\bar{c} V_{(0)}^c(3) \rangle \quad (16.25)$$

with the vertex operators (16.17). For gauge group $SO(2n)$ the level-one Kač-Moody currents can be expressed in terms of free fermions (cf. Eq. (11.95) on

16.2 String Scattering Amplitudes

page 341)

$$J^a = \frac{1}{2} :\psi^i \psi^j: T^a_{ij},\qquad(16.26)$$

where T^a_{ij} are representations matrices in the vector representation normalized such that $\text{tr}(T^a T^b) = 2\delta^{ab}$, i.e. $\langle J^a(1) J^b(0)\rangle = \delta^{ab}$. This corresponds to root vectors with (length)$^2 = 2$ which implies a normalization of the weight vectors in the vector representation to (length)$^2 = 1$.[6]

The evaluation of (16.25) is straightforward. It factors into several independent pieces. They are easily computed using Wick's theorem (some details are given in the next example). The three point function of the level-one Kač-Moody currents is

$$\langle J^a(\bar{z}_1) J^b(\bar{z}_2) J^c(\bar{z}_3)\rangle = \frac{1}{\bar{z}_{12}\bar{z}_{13}\bar{z}_{23}}\frac{1}{2}\text{tr}_v([T^a, T^b]T^c) = i\frac{f^{abc}}{\bar{z}_{12}\bar{z}_{13}\bar{z}_{23}}\qquad(16.27)$$

with $\bar{z}_{ij} = \bar{z}_i - \bar{z}_j$. The final answer for this amplitude is[7]

$$A^{abc}_{\text{het}}(k_1, \epsilon_1; k_2, \epsilon_2; k_3, \epsilon_3) = -i C_{S^2} g_c^3 \sqrt{\frac{\alpha'}{2}} f^{abc} t^{\mu_1\mu_2\mu_3} \epsilon^{(1)}_{\mu_1}\epsilon^{(2)}_{\mu_2}\epsilon^{(3)}_{\mu_3},\qquad(16.28)$$

where we have abbreviated

$$t^{\mu_1\mu_2\mu_3} = \eta^{\mu_1\mu_2} k_2^{\mu_3} + \eta^{\mu_2\mu_3} k_3^{\mu_1} + \eta^{\mu_3\mu_1} k_1^{\mu_2}.\qquad(16.29)$$

In the derivation of this result we have used momentum conservation $k_1 + k_2 + k_3 = 0$ from which, together with $k_i^2 = 0$ for massless particles, it follows that $k_i \cdot k_j = 0$. We have also used the fact that the momentum vectors will be contracted with transverse polarization vectors, i.e. we can replace e.g. $k_1^{\mu_2} \to -k_3^{\mu_2}$ since $k_2^{\mu_2}\epsilon_{\mu_2} = 0$.[8] One easily verifies that the amplitude is invariant under $\epsilon^{(i)}_\mu \to \epsilon^{(i)}_\mu + \lambda k^{(i)}_\mu$, which is nothing but gauge invariance.

The generalization to gauge groups G which arise from the left-moving sector of the heterotic string with a level k Kač-Moody algebra is straightforward. The currents $J^a(z)$ satisfy

$$J^a(z) J^b(w) = \frac{k\delta^{ab}}{(z-w)^2} + \frac{if^{abc} J^c(w)}{(z-w)} + \ldots\qquad(16.30)$$

---

[6] For a general compact simple Lie algebra we can choose Hermitian generators such that $\text{tr}_R(T^a T^b) = C_R \delta^{ab}$ where $C_R = \frac{1}{\text{rank}(G)}\sum(\lambda,\lambda)$ and the sum is over all weights of the irreducible representation R. The structure constants are defined as $[T^a, T^b] = if^{abc} T^c$. They depend on the normalization of the roots.

[7] We should caution the reader that we are not careful about overall phases of the amplitudes.

[8] Strictly speaking this amplitude vanishes, because for three massless particles the momenta must be co-linear. This can be avoided by allowing complex momenta.

so that the normalization of the Kač-Moody currents is

$$\langle J^a(z) J^b(w) \rangle = \frac{k \delta^{ab}}{(z-w)^2}. \tag{16.31}$$

Therefore, to relate the three gauge boson amplitude for level k to the one computed before, which was at level one, we need to define rescaled currents $\tilde{J}^a(\bar{z}) = J^a(\bar{z})/\sqrt{k}$ whose two point function is normalized to δ^{ab} and we have to replace J^a in the vertex operators by $\tilde{J}^a$.

This leads to

$$\langle \tilde{J}^a(\bar{z}_1) \tilde{J}^b(\bar{z}_2) \tilde{J}^c(\bar{z}_3) \rangle = \frac{i f^{abc}/\sqrt{k}}{\bar{z}_{12} \bar{z}_{13} \bar{z}_{23}} \tag{16.32}$$

and the three gauge boson amplitude becomes

$$A_{\text{het}}^{abc}(k_1, \epsilon_1; k_2, \epsilon_2; k_3, \epsilon_3)) = -i C_{S^2} g_c^3 \sqrt{\frac{\alpha'}{2}} \frac{1}{\sqrt{k}} f^{abc} t^{\mu_1 \mu_2 \mu_3} \epsilon^{(1)}_{\mu_1} \epsilon^{(2)}_{\mu_2} \epsilon^{(3)}_{\mu_3}. \tag{16.33}$$

Next, consider the scattering of two gaugini and one gauge boson of the heterotic string

$$A_{\text{het}}^{abc}(k_1, u_1; k_2, u_2; k_3, \epsilon_3) = C_{S^2} g_c^3 \langle c\bar{c} V^a_{(-\frac{1}{2})}(1) \, c\bar{c} V^b_{(-\frac{1}{2})}(2) \, c\bar{c} V^c_{(-1)}(3) \rangle. \tag{16.34}$$

The gaugino vertex operators are those of (16.20). The only correlator which deserves attention is

$$\langle S_\alpha(z_1) S_\beta(z_2) \psi^\mu(z_3) \rangle = \frac{1}{\sqrt{2}} \gamma^\mu_{\alpha\beta} \frac{1}{z_{12}^{3/4} z_{13}^{1/2} z_{23}^{1/2}}. \tag{16.35}$$

The z_i dependence follows from (4.86) and the overall normalization from the OPEs (13.8). The Lorentz structure of these relations is fixed by covariance and the normalization e.g. by using the explicit bosonization formulas for S_α and ψ^μ. The amplitude (16.34) is then

$$A_{\text{het}}^{abc}(k_1, u_1; k_2, u_2; k_3, \epsilon_3) = i g_c^3 C_{S^2} \sqrt{\frac{\alpha'}{8}} \epsilon_\mu(k_3) \, u^\alpha(k_1) \gamma^\mu_{\alpha\beta} u^\beta(k_2) \, f^{abc}. \tag{16.36}$$

Our next example is the heterotic three graviton amplitude. We need to compute

$$A_{\text{het}}(k_1, \epsilon_1; k_2, \epsilon_2; k_3, \epsilon_3) = g_c^3 C_{S^2} \langle c\bar{c} V_{(-1)}(1) \, c\bar{c} V_{(-1)}(2) \, c\bar{c} V_{(0)}(3) \rangle \tag{16.37}$$

with the vertex operators of Eq. (16.16). This requires the evaluation of the correlator $\langle \prod_{i=1}^{3} i \bar{\partial} X^{\nu_i}(\bar{z}_i) e^{i k_i \cdot X(\bar{z}_i)} \rangle$. An efficient way to evaluate it and similar

16.2 String Scattering Amplitudes

correlators is to use the representation

$$i\rho_j \cdot \partial X e^{ik_j \cdot X(z_j)} = \exp\left(i[k_j \cdot X(z_j) + \rho_j \cdot \partial X(z_j)]\right)\Big|_{\text{linear in } \rho_j} \quad (16.38)$$

and

$$\left\langle \prod_i \exp\left(i[k_{i\mu} X^\mu(z_i) + \rho_{i\mu} \partial X^\mu(z_i)]\right)\right\rangle$$

$$= \prod_{i<j}(z_i - z_j)^{\frac{\alpha'}{2} k_i \cdot k_j} \exp\left\{\frac{\alpha'}{2} \sum_{i<j} \frac{\rho_i \cdot \rho_j}{(z_i - z_j)^2} + \frac{\alpha'}{2} \sum_{i \neq j} \frac{k_j \cdot \rho_i}{(z_i - z_j)}\right\}. \quad (16.39)$$

Upon expanding, using momentum conservation and the fact that the indices will be contracted with transverse polarization tensors, one finds

$$\left\langle \prod_{i=1}^{3} i\partial X^{\nu_i}(z_i) e^{ik_i \cdot X(z_i)}\right\rangle = \left(\frac{\alpha'}{2}\right)^2 \left(t^{\nu_1\nu_2\nu_3} + \frac{\alpha'}{2} k_3^{\nu_1} k_1^{\nu_2} k_2^{\nu_3}\right)(z_{12}z_{13}z_{23})^{-1}. \quad (16.40)$$

The final expression for the amplitude is

$$A_{\text{het}}(1;2;3) = \frac{1}{2}\alpha' g_c^3 C_{S^2} t^{\mu_1\mu_2\mu_3} t^{\nu_1\nu_2\nu_3} \epsilon^{(1)}_{\mu_1\nu_1}(k_1) \epsilon^{(2)}_{\mu_2\nu_2}(k_2) \epsilon^{(3)}_{\mu_3\nu_3}(k_3) + \mathcal{O}(k^4)$$

$$= \frac{1}{2}\alpha' g_c^3 C_{S^2} \Big\{(\epsilon_1^T \epsilon_2)(k_2\epsilon_3 k_2) + (k_2\epsilon_3\epsilon_1^T\epsilon_1 k_3) + (k_2\epsilon_3\epsilon_1^T\epsilon_2 k_1)$$

$$+\text{two permutations}\Big\} + \mathcal{O}(k^4), \quad (16.41)$$

where the notation implies index contractions on neighboring objects inside each bracket, e.g. $(k_2\epsilon_3 k_2) = k_2^\mu \epsilon^{(3)}_{\mu\nu} k_2^\nu$ and $(\epsilon_2\epsilon_3) = \epsilon^{(2)}_{\mu\nu}\epsilon^{(3)\nu\mu}$. This amplitude is also correct for the compactified heterotic string. Dropping the $\mathcal{O}(k^4)$ terms, it is also valid for type II strings. Since we only used transversality of the polarization tensors, (16.41) is also valid for antisymmetric tensors and dilatons. This amplitude is invariant under $\epsilon^{(1)}_{\mu\nu} \to \epsilon^{(1)}_{\mu\nu} + k_\mu^{(1)}\xi_\nu$ (with $k^{(1)} \cdot \xi = 0$), etc.

The last closed string three-point amplitude which we will compute is the scattering of two RR p-form fields and one graviton in type IIB theory. The first part of the computation is straightforward. It gives

$$A(k_1, \epsilon^{(1)}; k_2, \epsilon^{(2)}; k_3, \epsilon^{(G)})$$

$$= g_c^3 C_{S^2} \frac{\alpha'}{64p!} \epsilon^{(G)}_{\mu\nu} k_{1\mu_1} \epsilon^{(1)}_{\mu_2...\mu_{p+1}} k_{2\nu_1} \epsilon^{(2)}_{\nu_2...\nu_{p+1}} \, \text{tr}\left(\Gamma^\mu \Gamma^{\mu_1...\mu_{p+1}} \Gamma^\nu \Gamma^{\nu_1...\nu_{p+1}}\right), \quad (16.42)$$

where $\epsilon^{(1)}$ and $\epsilon^{(2)}$ are the polarization tensors of the two p-forms. What remains is mainly Dirac algebra. In the case where $G = h$, we find

$$A = \frac{\alpha'}{2} g_c^3 C_{S^2} h_{\mu\nu}(k_3) F^{(1)\mu\mu_2...\mu_{p+1}} F^{(2)\nu}{}_{\mu_2...\mu_{p+1}} \tag{16.43}$$

with $F_{\mu_1...\mu_{p+1}} = k_{\mu_1}\epsilon_{\mu_2...\mu_{p+1}} + k_{\mu_2}\epsilon_{\mu_3...\mu_{p+1}\mu_1} + \text{cycl.}$. For the dilaton, i.e. $G = D$, (16.42) becomes

$$A = \frac{\alpha'}{4\sqrt{2}} g_c^3 C_{S^2} \frac{(p-4)}{(p+1)} D F^{(1)}_{\mu_1...\mu_{p+1}} F^{(2)\mu_1...\mu_{p+1}}, \tag{16.44}$$

where the factor D is inserted to keep track of the dilaton polarization.

Four point amplitudes are cumbersome to compute, either because of the presence of spin fields and/or the large number of contractions one has to perform which lead to a rather complicated kinematic structure. The simplest four-point amplitude is the four-tachyon amplitude of the closed bosonic string which is

$$A_{\text{bos}}^{(4)}(k_1; k_2; k_3; k_4) = g_c^4 C_{S^2} \int d^2 z_3 \langle c\bar{c}V(1) c\bar{c}V(2) V(3) c\bar{c}V(4) \rangle \tag{16.45}$$

with the tachyon vertex operator as in (16.7). The correlation function is easy to compute and results in

$$|z_{12}z_{14}z_{24}|^2 \prod_{i<j} |z_i - z_j|^{\alpha' k_i \cdot k_j}. \tag{16.46}$$

If we now fix the positions of the three unintegrated vertex operators to $z_1 = 0$, $z_2 = 1$, $z_4 = \infty$ and use momentum conservation, i.e. $k_1 + k_2 + k_3 + k_4 = 0$, we obtain, with the help of (16.23)

$$A_{\text{bos}}^4(k_1; k_2; k_3; k_4) = g_c^4 C_{S^2} \int d^2 z |z|^{\alpha' k_1 \cdot k_3} |1-z|^{\alpha' k_2 \cdot k_3}$$

$$= 2\pi g_c^4 C_{S^2} \frac{\Gamma(\alpha(s)) \Gamma(\alpha(t)) \Gamma(\alpha(u))}{\Gamma(\alpha(t) + \alpha(u)) \Gamma(\alpha(s) + \alpha(u)) \Gamma(\alpha(s) + \alpha(t))}, \tag{16.47}$$

where

$$\alpha(s) = -1 - \frac{\alpha'}{4} s \tag{16.48}$$

and s, t, u are the Mandelstam variables, the three kinematic invariants of a four-point scattering amplitude

16.2 String Scattering Amplitudes

$$s = -(k_1 + k_2)^2 = -2k_1 \cdot k_2 + m_1^2 + m_2^2 = -2k_3 \cdot k_4 + m_3^2 + m_4^2,$$
$$t = -(k_1 + k_3)^2 = -2k_1 \cdot k_3 + m_1^2 + m_3^2 = -2k_2 \cdot k_4 + m_2^2 + m_4^2,$$
$$u = -(k_1 + k_4)^2 = -2k_1 \cdot k_4 + m_1^2 + m_4^2 = -2k_2 \cdot k_3 + m_2^2 + m_3^2. \quad (16.49)$$

They satisfy

$$s + t + u = m_1^2 + m_2^2 + m_3^2 + m_4^2. \quad (16.50)$$

For the closed bosonic string tachyon $m^2 = -\frac{4}{\alpha'}$.

Recall the properties of the Gamma-function $\Gamma(x)$. It has no zeros but single poles at $x = -n = 0, -1, -2, \ldots$ with residues $(-1)^n/n!$. We thus see that in each of the three channels the amplitude has an infinite number of simple poles but there are no higher order poles. For instance, in the s-channel, the poles are at $s = -\frac{4}{\alpha'}, 0, \frac{4}{\alpha'}, \frac{8}{\alpha'}, \ldots$. But these are precisely the masses of the excitations of the closed bosonic string. Their vertex operators appear in the operator product of two closed string tachyon operators. In other words, if we consider tachyons 1 and 2 as incoming and choose their momenta such that $(k_1 + k_2)^2$ is $-m^2$ of a closed string excitation, the amplitude has a pole. The same discussion applies to the other two channels. The fact that the amplitude is completely symmetric in the three channels and that it exhibits an infinite number of poles in each channel is called (channel) duality and this is the origin for the name dual resonance models. String theory started from amplitudes of this type which were postulated in order to reproduce scattering results of hadronic resonances. Shortly afterward, it was realized that they can be derived from a theory of relativistic strings. The amplitude (16.47) is called Virasoro-Shapiro amplitude.

We can factorize the amplitude (16.47) on the tachyon pole in the s-channel

$$A(k_1; k_2; k_3; k_4) \overset{s \to -\frac{4}{\alpha'}}{\sim} g_c^4 C_{S^2} \frac{8\pi}{\alpha'} \frac{1}{(k_1 + k_2)^2 - \frac{4}{\alpha'}}$$

$$\overset{!}{=} A(k_1, k_2, -(k_1 + k_2)) \frac{1}{(k_1 + k_2)^2 - \frac{4}{\alpha'}} A(k_3, k_4, -k_3 - k_4)$$

$$= \frac{g_c^6 \, C_{S^2}^2}{(k_1 + k_2)^2 - \frac{4}{\alpha'}}. \quad (16.51)$$

By unitarity we can conclude that

$$C_{S^2} = \frac{8\pi}{\alpha' g_c^2}. \quad (16.52)$$

The same result is obtained if we factorize a four-point amplitude of the heterotic or the type II string, but the computations are more involved.

Open String Scattering Amplitudes

So far we have considered scattering amplitudes of closed oriented strings. This is all that is needed for the heterotic string. In theories with D-branes, however, there are open strings and there are massless open string excitations. We therefore need to compute string scattering amplitudes on world-sheets with boundaries. As for the closed string, we compute tree amplitudes of massless excitations.

The oriented open string tree level world-sheet has one boundary. It may be mapped to the disk $D_2 = \{z \in \mathbb{C} : |z| \leq 1\}$ or to the upper half plane $\mathbb{H}_+ = \{z \in \mathbb{C} \mid \text{Im}(z) > 0\}$. The map between them was given in (4.140). The disk and the upper half plane are quotients of the sphere by $\mathbb{Z}_2$, where $\mathbb{Z}_2$ acts as $z \to 1/\bar{z}$ and $z \to \bar{z}$ with fixed points $|z| = 1$ and $z = \bar{z}$, respectively. These representations of D_2 and $\mathbb{H}_+$ are useful to determine the propagators via the method of images. The conformal Killing group is the subgroup of $PSL(2, \mathbb{C})$ which commutes with the $\mathbb{Z}_2$ action. For D_2 it is $SU(1, 1)$ (the matrices (4.39) with $c = \bar{b}, d = \bar{a}$) and for $\mathbb{H}_+$ it is $PSL(2, \mathbb{R})$ (the matrices (4.39) with $a, b, c, d \in \mathbb{R}$). Of course, these two groups are isomorphic. $PSL(2, \mathbb{R})$ ($SU(1, 1)$) maps the upper half plane (D_2) 1−1 to itself and the real axis (unit circle) to the real axis (unit circle). In the following we work on the upper half plane, but we will often refer to it as the disk.

In theories with D-branes massless fields like gauge fields or matter fields are open string excitations which are localized on the D-brane world-volume. Hence the boundary of the disk diagram—the end-points of an open string—is attached to the D-brane world-volume and open string vertex operators are inserted at boundary points of the disk. On the other hand, graviton, dilaton and anti-symmetric tensor fields, etc. are closed string excitations and are not confined to D-branes. They can propagate through the entire space-time and their vertex-operators are inserted in the interior of the disk. To summarize, vertex operators for open string states are inserted at the boundary of the disk while the closed string vertex operators are inserted in the interior of the disk. A general amplitude has N_c closed and N_o open string insertions and has the general form

$$A \sim g_s^{N_c + \frac{1}{2}N_o - 1} \int \frac{\prod_1^{N_o} dx_i \prod_i^{N_c} d^2 z_i}{V_{\text{CKG}}} \langle V_1(x_1) \cdots V_{N_o}(x_{N_o}) V_1(z_1, \bar{z}_1)$$
$$\cdots V_{N_c}(z_{N_c}, \bar{z}_{N_c}) \rangle_{D_2}. \tag{16.53}$$

The closed string vertex operators are integrated over the upper half plane whereas the open string vertex operators are integrated over the real axis.

The three parameters of $PSL(2, \mathbb{R})$ can be used in various ways, e.g. to fix the positions of three open string vertex operators, or the position of one closed string and one open string vertex operator or to fix the position of one closed string vertex operator and to restrict a second closed string vertex to the positive imaginary axis. The Jacobian of the transformation from the fixed coordinates to the parameters of $PSL(2, \mathbb{R})$ can be represented as the correlation function of three ghost insertions. The fixed positions are, of course, not integrated over. Note that we have to make

16.2 String Scattering Amplitudes

sure that the fixed positions can be obtained from generic positions by means of a $PSL(2,\mathbb{R})$ transformation. For instance, $PSL(2,\mathbb{R})$ transformations preserve the cyclic order of points on the compactified real axis.[9] We therefore have to sum over the different cyclic orders of the unintegrated vertex operators where the different terms come with relative minus signs if the vertex operators are Grassmann odd which is the case for states in the R sector.

Due to the presence of the boundary along the real axis, there are non-trivial two-point functions between left- and right-moving fields; see the discussion in Chaps. 3 and 4. We can summarize the boundary conditions on the world-sheet fields along the real axis in the presence of a Dp-brane as $(X(z,\bar{z}) = X(z) + \overline{X}(\bar{z}))$

$$X^\mu(z) = D^\mu{}_\nu \overline{X}^\nu(\bar{z}), \qquad \psi^\mu(z) = D^\mu{}_\nu \overline{\psi}^\nu(\bar{z}) \qquad \text{at } z = \bar{z} = x, \qquad (16.54)$$

where we have used the matrix defined in (4.169)

$$D^\mu{}_\nu = (\delta^\alpha{}_\beta, -\delta^i{}_j), \qquad \begin{array}{l} \alpha, \beta = 0, \ldots, p-1 \text{ (NN directions)} \\ i, j = p, \ldots, 9 \text{ (DD directions)} \end{array}. \qquad (16.55)$$

One should not confuse the indices $\alpha, \beta, \ldots$ along the N directions with spinor indices for which we also used small Greek letters. The two-point functions are (16.2) and

$$\langle X^\mu(z) \overline{X}^\nu(\bar{w}) \rangle = -\frac{\alpha'}{2} D^{\mu\nu} \ln(z - \bar{w}),$$

$$\langle \psi^\mu(z) \overline{\psi}^\nu(\bar{w}) \rangle = \frac{D^{\mu\nu}}{(z - \bar{w})},$$

$$\langle \phi^i(z) \overline{\phi}^j(\bar{w}) \rangle = -\delta^{ij} \ln(z - \bar{w}),$$

$$\langle \phi(z) \overline{\phi}(\bar{w}) \rangle = -\ln(z - \bar{w}). \qquad (16.56)$$

A simple way to get these is to use the doubling trick. For instance, define $\psi(w') = \overline{\psi}(\bar{w})$, $w' \in \mathbb{H}_-$ and use (16.2).[10] The additional sign in the fermionic two point functions for Dirichlet conditions is due to the relative sign between left- and right-movers in (7.70).

---

[9] On the disk the conformal Killing group preserves the cyclic order of points on the boundary of the disk.

[10] If we work on D_2 rather than $\mathbb{H}_+$, there is a nontrivial Jacobian, i.e. $\overline{\psi}(\bar{z}) = \psi(z') \left(\frac{\partial z'}{\partial \bar{z}} \right)^{1/2}$. With $z' = 1/\bar{z}$ one finds $\langle \psi(z) \overline{\psi}(\bar{w}) \rangle_{D_2} = \frac{i}{1 - z\bar{w}}$. One verifies that this maps to the result on $\mathbb{H}_+$ under (4.140). An alternative way to get the disc propagator is to use the boundary state formalism and to compute e.g. $\langle Dp | \psi(z) \overline{\psi}(\bar{w}) | 0 \rangle$.

We also need the correlation functions for the ghosts:

$$\langle c(z_1) c(z_2) c(z_3) \rangle = (z_1 - z_2)(z_1 - z_3)(z_2 - z_3),$$
$$\langle c(z_1) c(z_2) \bar{c}(\bar{z}_3) \rangle = (z_1 - z_2)(z_1 - \bar{z}_3)(z_2 - \bar{z}_3). \quad (16.57)$$

For a ghost field inserted on the boundary $c(x) = \bar{c}(x)$

The correlation functions (16.56) are also valid, if we move insertion points to the boundary. Specifically, we obtain

$$\langle X^\alpha(x) X^\beta(x') \rangle = -2\alpha' \eta^{\alpha\beta} \log|x - x'| \quad (N),$$

$$\langle \partial_n X^i(x) \partial_n X^j(x') \rangle = 2\alpha' \frac{\delta^{ij}}{(x-x')^2} \quad (D),$$

$$\langle \psi^\mu(x) \psi^\nu(x') \rangle = \frac{\eta^{\mu\nu}}{x - x'} \quad (N \text{ and } D), \quad (16.58)$$

where $x, x' \in \mathbb{R}$. The factor four in the first two lines relative to the two-point functions in (16.2) and (16.56) is due to $X(x) \equiv X(z,\bar{z})|_{z=\bar{z}=x}$.

With the help of Wick's theorem one then obtains the following basic correlator

$$\left\langle \prod_{i=1}^{N_o} e^{il_i \cdot X(x_i)} \prod_{j=1}^{N_c} e^{ik_j \cdot X(z_j, \bar{z}_j)} \right\rangle$$

$$= \prod_{i<j}^{N_c} |z_i - z_j|^{\alpha' k_i \cdot k_j} |z_i - \bar{z}_j|^{\alpha' k_i \cdot D \cdot k_j} \prod_i^{N_c} |z_i - \bar{z}_i|^{\frac{\alpha'}{2} k_i \cdot D \cdot k_i}$$

$$\times \prod_{i<j}^{N_o} |x_i - x_j|^{2\alpha' l_i \cdot l_j} \prod_{i=1}^{N_o} \prod_{j=1}^{N_c} |x_i - z_j|^{\alpha' l_i \cdot k_j}. \quad (16.59)$$

Note the contributions from self-contractions between left- and right-moving parts of closed string vertex operators.

Let us now discuss open string vertex operators. The simplest is the tachyon of the bosonic string:

$$V(k, x) = e^{ik \cdot X(x)}, \quad k^2 = -m^2 = \frac{1}{\alpha'}. \quad (16.60)$$

More interesting are the massless excitations of D-branes. The bosonic fields on a space-time filling D-brane, where the boundary conditions in all directions are Neumann, are the gauge bosons. For a Dp-brane, which one obtains e.g. by T-dualities along $9 - p$ directions, the directions perpendicular to the brane have D-boundary conditions and the components of the vector fields in these directions become scalar fields. They also transform in the adjoint representation of the gauge group. In the canonical ghost picture we have the following vertex operators:

16.2 String Scattering Amplitudes

$$V^a_{(-1)}(k, \epsilon, x) = \lambda^a \, \epsilon_\alpha(k) \, \psi^\alpha \, e^{-\phi} \, e^{ik \cdot X}(x) \qquad \text{(gauge boson)},$$

$$V^{i,a}_{(-1)}(k, x) = \lambda^a \, \psi^i \, e^{-\phi} \, e^{ik \cdot X}(x) \qquad \text{(transverse scalar)}, \qquad (16.61)$$

where λ^a is a Chan-Paton factor. In contrast to closed string vertex operators, the momentum k of open string operators has only components along the world-volume of the D-brane, i.e. $k \cdot X = k_\alpha X^\alpha$. (Recall that there is no center-of-mass momentum in D-directions.) The Chan-Paton factors are normalized to tr $(\lambda^a \lambda^b) = \delta^{ab}$.

We also need the vertex operators in the zero-ghost picture. We use the doubling trick to write the picture changing operator as a holomorphic operator in the complex plane (rather than a holomorphic and an antiholomorphic operator in the upper half plane) and define $V_{(0)}(x) = \lim_{z \to x} Z(z) V_{(-1)}(x)$ where Z is the picture changing operator. In this way we obtain

$$V^a_{(0)}(k, \epsilon, x) = \frac{1}{\sqrt{2\alpha'}} \lambda^a \, \epsilon_\alpha(k) \left(i \partial_t X^\alpha + 2\alpha'(k \cdot \psi) \psi^\alpha \right) e^{ik \cdot X}(x),$$

$$V^{i,a}_{(0)}(k, x) = \frac{1}{\sqrt{2\alpha'}} \lambda^a \left(\partial_n X^i + 2\alpha'(k \cdot \psi) \psi^i \right) e^{ik \cdot X}(x). \qquad (16.62)$$

Here we have used that on the real axis, $\partial X^\alpha = \frac{1}{2} \partial_t X^\alpha$ and $\partial X^i = -\frac{i}{2} \partial_n X^i$. If we parametrize the complex plane as $z = x + iy$, then $\partial_t = \partial_x$ and $\partial_n = \partial_y$. Furthermore, $k \cdot \psi = k_\alpha \psi^\alpha$.

The vertex operator for a gaugino in the canonical ghost picture is

$$V^a_{(-\frac{1}{2})}(k, u, x) = \alpha'^{1/4} \, \lambda^a \, u^\alpha(k) \, e^{-\frac{1}{2}\phi} \, S_\alpha \, e^{ik \cdot X}(x). \qquad (16.63)$$

The normalization will be justified later. This is valid for gaugini on D9 branes. For lower dimensional Dp-branes, e.g. for $p = 3$, one decomposes the $SO(1, 9)$ positive chirality spin fields S_α as $(S_a \Sigma^I, S_{\dot{a}} \overline{\Sigma}^I)$ where S_a and $S_{\dot{a}}$ are positive and negative chirality $SO(1, 3)$ spinors and Σ^I, $I = 1, 2, 3, 4$ are fields with conformal weight $\frac{3}{8}$. Recalling the bosonized form of the spin fields, the Σ^I can be represented in terms of three free bosons, $(\Sigma^I, \overline{\Sigma}^I) = \exp(i(\pm \phi^1 \pm \phi^2 \pm \phi^3)/2)$ with an (even,odd) number of minus signs (c.f. the discussion in Chap. 15).

Since D-branes are treated as infinitely heavy objects, they can absorb momentum in the transverse directions. Consequently only momentum parallel to the branes is conserved. If we define

$$k = \frac{1}{2}(1 + D)k + \frac{1}{2}(1 - D)k = k_\parallel + k_\perp, \qquad (16.64)$$

then

$$\sum k^{(i)}_\parallel = 0 \qquad (16.65)$$

for non-vanishing correlation functions. Of course, open string states have $k_\perp = 0$.

We now turn to examples. In analogy to the closed string, we multiply each vertex operator by g_o and each amplitude by C_{D_2}. The coupling g_o will be determined by comparison with the effective field theory and C_{D_2} by unitarity.

The simplest example is once again the three tachyon amplitude of the open bosonic string: via an $PSL(2, \mathbb{R})$ transformation, we can map the positions of the three vertex operators to some fixed positions on the real axis as long as we do not change their cyclic order. There are two different orders for three points and we have to sum over them. The correlator of the three ghost insertions reproduces the Jacobian $|x_{12} x_{13} x_{23}|$ of the change of variables from (x_1, x_2, x_3) to the parameters of $PSL(2, \mathbb{R})$. The normalization which has to do with the change of variables is absorbed into the factor C_{D_2}. But we have to keep track of relative normalizations which arise from different choices of fixing $PSL(2, \mathbb{R})$; cf. Footnote 11 on page 611. The normalization of the integration over the vertex operator positions is absorbed in g_o. This leads to the expression

$$A\ (k_1; k_2; k_3)$$
$$= g_o^3 C_{D_2} \{\langle cV(k_1, x_1)\, cV(k_2, x_2)\, cV(k_3, x_3)\rangle + (k_1 \leftrightarrow k_2)\}. \quad (16.66)$$

Using $k_i \cdot k_j = -\frac{1}{2\alpha'}$ for $i \neq j$, which is implied by momentum conservation and the on-shell conditions, one verifies that the amplitude is independent of the x_i (as it must) and one finds

$$A(k_1, k_2, k_3) = 2 g_o^3\, C_{D_2}. \quad (16.67)$$

Our next example is the three gluon amplitude in superstring theory:

$$A_{\text{open}}^{abc}(k_1, \epsilon_1; k_2, \epsilon_2; k_3, \epsilon_3)$$
$$= C_{D_2}\, g_o^3\, \text{tr}\, \Big\{\langle cV_{(-1)}^a(k_1, \epsilon_1, x_1)\, cV_{(-1)}^b(k_2, \epsilon_2, x_2)\, cV_{(0)}^c(k_3, \epsilon_3, x_3)\rangle$$
$$+ (a, k_1, \epsilon_1) \leftrightarrow (b, k_2, \epsilon_2)\Big\}$$
$$= \sqrt{2\alpha'}\, g_o^3\, C_{D_2}\, \text{tr}\,([\lambda^a, \lambda^b]\lambda^c)\, \epsilon_{\mu_1}^{(1)} \epsilon_{\mu_2}^{(2)} \epsilon_{\mu_3}^{(3)}\, t^{\mu_1 \mu_2 \mu_3} \quad (16.68)$$

with $t^{\mu_1 \mu_2 \mu_3}$ as in (16.29). Note that Eq. (16.68) holds for a D-brane of any dimension as long as we restrict the polarization vectors and the momenta to be parallel to the brane.

The amplitude for the scattering of two gaugini and one gauge boson is

$$A^{abc}(k_1, u_1; k_2, u_2; k_3, e_3)$$
$$= C_{D_2} g_o^3\, \text{tr}\, \Big\{\langle cV_{(-\frac{1}{2})}^a(k_1, u_1, x_1)\, cV_{(-\frac{1}{2})}^b(k_2, u_2, x_2)\, cV_{(-1)}^c(k_3, u_3, x_3)\rangle$$
$$- (a, k_1, u_1) \leftrightarrow (b, k_2, u_2)\Big\}. \quad (16.69)$$

16.2 String Scattering Amplitudes

The origin of the relative sign is that when we go to the second cyclic ordering we exchange the positions of the two gluini whose vertex operators are Grassmann odd. The final result for this amplitude is

$$A^{abc}(k_1, u_1; k_2, u_2; k_3, e_3) = \sqrt{\frac{\alpha'}{2}} g_0^3 \, C_{D_2} \, \epsilon_\mu(k_3) \, u^\alpha(k_1) \, \gamma^\mu_{\alpha\dot\beta} \, u^{\dot\beta}(k_2) \, \text{tr}\,([\lambda^a, \lambda^b]\lambda^c). \tag{16.70}$$

As it stands, the result is valid for a D9 brane and all indices are $SO(1,9)$ indices. But it can be generalized to, say, a D3-brane by simply replacing the kinematic factor by $u^\alpha(k_1)\gamma^\mu_{\alpha\dot\beta}u^{\dot\beta}(k_2)$ where all indices are $SO(1,3)$ indices.

Four and higher point amplitudes of open strings are more involved and in general rather tedious to compute. The general form of open string 4-point amplitudes is

$$A^{(4)} = \sum_{\pi \in S_4/\mathbb{Z}_4} V_{\text{CKG}}^{-1} \int_{\mathscr{I}_\pi} \prod_{k=1}^{4} dx_k \, \langle V_{\Phi^1}(x_1) \, V_{\Phi^2}(x_2) \, V_{\Phi^3}(x_3) \, V_{\Phi^4}(x_4) \rangle. \tag{16.71}$$

One has to sum over all six in-equivalent cyclic orderings of the four vertex operators along the boundary of the disk, or, equivalently, along the real axis which is compactified by adding the point at infinity. Each permutation π gives rise to an integration region $\mathscr{I}_\pi = \{x_{\pi(1)} < x_{\pi(2)} < x_{\pi(3)} < x_{\pi(4)}\}$. The factor V_{CKG} will be canceled by fixing three vertex positions and introducing the respective c-ghost correlator. Depending on the ordering of the vertex operator positions, we obtain six partial amplitudes. The first set of three partial amplitudes may be obtained by the choice

$$x_1 = 0, \quad x_3 = 1, \quad x_4 = \infty, \tag{16.72}$$

while for the second set we choose:

$$x_1 = 1, \quad x_3 = 0, \quad x_4 = \infty. \tag{16.73}$$

The two choices imply the ghost factor $|\langle c(0)c(1)c(x_\infty)\rangle| = x_\infty^2$. The remaining vertex position x_2 has to be integrated along the real axis. The resulting six arrangement of four vertex operators are indicated in Fig. 16.1.

After computing the correlation function, the integral in Eq. (16.71) can be reduced to the Euler Beta-function:

$$\int_0^1 dx \, x^{\alpha-1} (1-x)^{\beta-1} \equiv B(\alpha, \beta) = \frac{\Gamma(\alpha)\,\Gamma(\beta)}{\Gamma(\alpha+\beta)}. \tag{16.74}$$

This integral is also computed in the appendix.

If the vertex operators carry Chan-Paton factors, the amplitude's group theoretical factor is determined by the trace of the product of individual Chan-Paton factors,

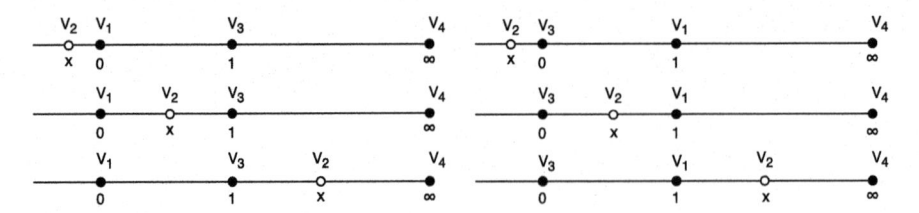

Fig. 16.1 The six orderings of four open string vertex operators on the real axis

ordered in the same way as the vertex positions. If the vertex operators are fermionic, we have to take into account signs when changing their order.

The simplest four-point amplitude is the four-tachyon amplitude of the open bosonic string

$$A(k_1; k_2; k_3; k_4) = g_0^4 C_{D_2} \int_{-\infty}^{\infty} dx \, \langle cV(k_1, 0) cV(k_2, x) V(k_3, 1) cV(k_4, \infty) \rangle$$
$$+ (k_3 \leftrightarrow k_1)$$
$$= 2 g_0^4 C_{D_2} \int_{-\infty}^{\infty} dx \, |x|^{2\alpha' k_1 \cdot k_2} |1 - x|^{2\alpha' k_2 \cdot k_3}. \tag{16.75}$$

We split the integral into three regions: $-\infty < x < 0$, $0 < x < 1$ and $1 < x < \infty$. The integral over the second interval can be done directly with (16.74). The integrals over the other regions are done via change of variables: $x = \frac{y-1}{y}$ maps $[-\infty, 0]$ to $[0, 1]$ and $x = 1/y$ maps $[1, \infty]$ to $[0, 1]$. Performing the integrals, we obtain for the four-tachyon amplitude

$$A(k_1; k_2; k_3; k_4) = 2 g_0^4 \, C_{D_2} \Big\{ B(\alpha(s), \alpha(t)) + B(\alpha(t), \alpha(u))$$
$$+ B(\alpha(u), \alpha(s)) \Big\}, \tag{16.76}$$

where $\alpha(s) = -1 - \alpha' s$ and $s + t + u = -\frac{4}{\alpha'}$ and s, t, u were defined in (16.49). This is the Veneziano amplitude. It has an infinite number of simple poles in each of the three channels. In particular there is a tachyonic pole in the s-channel. Repeating the same unitarity argument as for the closed string tachyon, we find

$$C_{D_2} = \frac{1}{\alpha' g_0^2}. \tag{16.77}$$

This is also the correct value for the type I string.

It is not difficult to check that the three-point amplitude of two open and one closed string tachyon on the disk does not vanish. One might therefore wonder why the Veneziano amplitude has no pole corresponding to the exchange of the closed string tachyon. The answer is simple: the closed string tachyon exchange

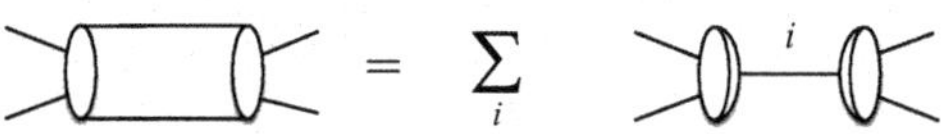

16.2 String Scattering Amplitudes

Fig. 16.2 Closed string exchange

occurs in the cylinder or, equivalently, annulus diagram where two open string vertex operators are inserted on each boundary. This is clear from Fig. 16.2

The Veneziano amplitude played a central role in the early period of string theory. It actually predated string theory and was constructed in order to reproduce properties of measured hadronic S-matrix elements. The crucial property of the Euler Beta-function is that it is completely determined by its poles in either of its variables. Generally, by the Mittag-Leffler theorem, a meromorphic function can be represented as a sum of its poles plus an entire function. For instance, the Gamma function has the representation

$$\Gamma(z) = \int_0^\infty dt\, e^{-t}\, t^{z-1} = \int_0^1 dt\, e^{-t}\, t^{z-1} + \int_1^\infty dt\, e^{-t}\, t^{z-1}$$

$$= \sum_{n=0}^\infty \frac{(-1)^n}{n!(z+n)} + \Gamma(z,1), \qquad (16.78)$$

where the incomplete Gamma-function $\Gamma(z,1)$ is analytic on the complex plane. For the Beta-function the entire function piece is zero and one finds, using the integral representation of the Gamma function twice

$$B(\alpha,\beta) = \frac{1}{\Gamma(1-\beta)} \int_0^\infty ds \int_0^1 dx\, s^{-\beta}\, e^{-(1-x)s}\, x^{\alpha-1}$$

$$= \frac{1}{\Gamma(1-\beta)} \int_0^\infty ds\, s^{-\beta}\, e^{-s} \int_0^1 dx \sum_{n=0}^\infty \frac{s^n}{n!} x^{n+\alpha-1}$$

$$= \frac{1}{\Gamma(1-\beta)} \sum_{n=0}^\infty \frac{1}{n!} \frac{1}{n+\alpha} \int_0^\infty ds\, s^{n-\beta}\, e^{-s}$$

$$= \sum_{n=0}^\infty \frac{1}{n!(n+\alpha)} \frac{\Gamma(n-\beta+1)}{\Gamma(1-\beta)} = \sum_{n=0}^\infty \frac{1}{n!(n+\beta)} \frac{\Gamma(n-\alpha+1)}{\Gamma(1-\alpha)}. \qquad (16.79)$$

The residue at the pole at $\alpha = -n$ is an n-th order polynomial in β. This expansion gives e.g. $B(\alpha(s),\alpha(t))$ a very direct interpretation in terms of exchange processes of an infinite number of massive string excitations. It can be decomposed either as a sum over an infinite number of s-channel or t channel poles.

A less trivial four-point amplitude is the scattering of four gauge-bosons. It is straightforward to compute, but the fact that two of the vertex operators must be in the zero ghost picture leads to a large number of possible Wick contractions and therefore to a rather complicated kinematic expression:

$$A(k_1, \epsilon_1, \lambda^{a_1}; \ldots ; k_4, \epsilon_4, \lambda^{a_4})$$
$$= g_o^4 C_{D_2} T^{a_1 a_2 a_3 a_4} \alpha' \frac{B(-\alpha' s, -\alpha' u)}{t} K_4(k_1, \epsilon_1; \ldots; k_4, \epsilon_4)$$
$$+ (1) \leftrightarrow (2) + (2) \leftrightarrow (3), \qquad (16.80)$$

where

$$T^{abcd} = \text{tr}\,(\lambda^a \lambda^b \lambda^c \lambda^d) + \text{tr}\,(\lambda^d \lambda^c \lambda^b \lambda^a) \qquad (16.81)$$

is the color ordering factor and K_4 is the kinematic factor

$$K_4(k_1, \epsilon_1; \ldots; k_4, \epsilon_4) = -us(\epsilon_1 \epsilon_3)(\epsilon_2 \epsilon_4) + 2t[(\epsilon_1 \epsilon_2)(\epsilon_3 k_2)(\epsilon_4 k_1)$$
$$+ (\epsilon_1 \epsilon_4)(\epsilon_2 k_1)(\epsilon_3 k_4) + (\epsilon_2 \epsilon_3)(\epsilon_1 k_2)(\epsilon_4 k_3) + (\epsilon_3 \epsilon_4)(\epsilon_1 k_4)(\epsilon_2 k_3)]$$
$$+ (2 \leftrightarrow 3) + (3 \leftrightarrow 4). \qquad (16.82)$$

One can show that the kinematic factor has complete symmetry in the four external lines, even though, a priori, only cyclic symmetry was expected. One checks that the amplitude satisfies the Ward identity, i.e. it vanishes, if any of the polarizations is replaced by the respective momentum.

We now examine the pole structure a little closer. We expand (use $s + u + t = 0$)

$$\frac{1}{t} B(-\alpha' s, -\alpha' u) = -\frac{1}{u} B(-\alpha' s, 1 - \alpha' u) = \sum_{n=0}^{\infty} \frac{\gamma(u, n)}{s - M_n^2}$$
$$= -\frac{1}{s} B(1 - \alpha' s, -\alpha' u) = \sum_{n=0}^{\infty} \frac{\gamma(s, n)}{u - M_n^2}, \qquad (16.83)$$

where the residues

$$\gamma(u, n) = \frac{1}{n!} \frac{\Gamma(\alpha' u + n)}{\Gamma(\alpha' u + 1)} \overset{u \to \infty}{\sim} (\alpha' u)^{n-1} \qquad (16.84)$$

are determined by the three-point couplings of two gluons to the intermediate states (with spin $j \leq n$). We find that the partial amplitude with color structure $T^{a_1 a_2 a_3 a_4}$ has poles is the s and u-channels at masses

$$M_n^2 = M_s^2 n = \frac{n}{\alpha'}, \qquad n = 0, 1, \ldots \qquad (16.85)$$

of the open string excitations. There are no tachyonic poles.

There are a few differences between open string and heterotic string gauge boson amplitudes which we want to point out. Since the open string tree level

16.2 String Scattering Amplitudes

world-sheet has only one boundary component, the amplitudes are always single traces over the Chan-Paton factors. In order to get double traces, we need two boundary components, i.e. a cylinder, but this is open string one loop. It is not difficult to see that the heterotic tree-level four-gluon amplitude has double trace terms. This difference in the structure of the amplitude will lead to differences in the gauge sector of the low-energy effective actions at higher orders in α'.

The disk amplitudes considered so far only involve open string vertex operators which are inserted on the boundary of the disk. But we can also insert closed string vertex operators in the interior of the disc. If we insert one closed string vertex operator, e.g. a graviton, this does not fix the $SL(2,\mathbb{R})$ symmetry completely. But since the remaining symmetry is $U(1)$ with finite volume the amplitude does not vanish and describes e.g. the gravitational coupling of a D-brane. Mixed open/closed string amplitudes can be considered as excited D-branes emitting and absorbing bulk closed string modes. Our only example of this type of amplitude will be the scattering of two open strings with one massless closed string state. The open string excitations can be either gauge bosons or scalars and the closed string state a graviton, $B_{\mu\nu}$ or dilaton.

There are various ways to fix the $PSL(2,\mathbb{R})$ invariance. One convenient way is to insert the closed string vertex at $z = i$ and to constrain the positions of the two open vertices to obey $x_2 = -x_1 = v$. This leads to a Jacobian $4(1+v^2)$.[11] The three-point amplitude to compute is therefore

$$A^{ab}(p_1,\xi_1;p_2,\xi_2;k,\epsilon)$$
$$= 4g_c g_o^2 C_{D_2} \int_{-\infty}^{+\infty} dv\, (1+v^2) \langle V_{(0)}^a(p_1,\xi_1,-v) V_{(0)}^b(p_2,\xi_2,+v) V_{(-1,-1)}(k,\epsilon,i) \rangle, \quad (16.86)$$

where we have chosen the canonical ghost picture for the graviton with polarization ϵ and the zero ghost picture for the two open string bosons with polarizations ξ. For a gauge boson $\xi_\perp = 0$ while for a transverse scalar excitation $\xi_\parallel = 0$. The momenta of the open string excitations have only components parallel to the brane and momentum conservation is $p_1 + p_2 + k_\parallel = 0$ with $p_1^2 = p_2^2 = 0$ and $k_\parallel^2 + k_\perp^2 = 0$. There is only one kinematic invariant,

$$t = -2p_1 \cdot p_2 = 2p_1 \cdot k = 2p_2 \cdot k = -k_\parallel^2 = k_\perp^2. \quad (16.87)$$

The computation of this type of correlation functions is simplified by the following observations: (1) open string vertex operators with momentum k have the same form as the holomorphic part of closed string vertex operators with momentum $2k$.

---

[11] Define $u = \frac{1}{2}(x_1 + x_2)$ and $v = \frac{1}{2}(x_1 - x_2)$. With $z = x + iy$ we have $d^2z\, dx_1\, dx_2 = 4\, dx\, dy\, du\, dv$. An infinitesimal $SL(2,\mathbb{R})$ transformation acts as $\delta z = \alpha + \beta z + \gamma z^2$ with α, β, γ real. Then $\left|\frac{\partial(x,y,u)}{\partial(\alpha,\beta,\gamma)}\right| = (1+v^2)$ at $x = 0, y = 1, u = 0$. Note that if we fix the positions of three open strings to x_1, x_2, x_3, the Jacobian is $|x_{12} x_{13} x_{23}|$.

To see this, replace $X(x) \to 2X(z)|_{z=x}$, where $X(z)$ is a holomorphic field with $\langle X(z) X(w) \rangle = -\frac{\alpha'}{2} \log(z - w)$ and where on holomorphic fields one has $\partial_t = \partial$, $\partial_n = i\partial$.

(2) Use the doubling trick to rewrite closed string vertex operators. Replace

$$\overline{X}^\mu(\bar{z}) \to D^\mu{}_\nu X^\nu(z'), \quad \overline{\psi}^\mu(\bar{z}) \to D^\mu{}_\nu \psi^\nu(z'), \quad \overline{\phi}(\bar{z}) \to \phi(z') \tag{16.88}$$

with $z' = \bar{z} \in \mathbb{H}_-$.

All correlators are now as in (16.2). Closed string vertex operators then factorize into two holomorphic vertex operators, e.g. for the graviton

$$V_{(-1,-1)}(k, \epsilon, z, \bar{z}) = D_\mu{}^\rho \epsilon_{\rho\nu} V^\mu_{(-1)}(Dk, z') V^\nu_{(-1)}(k, z), \qquad (z' = \bar{z}) \tag{16.89}$$

with $V^\nu_{(-1)}(k, z)$ as in (16.13). Correlation functions on the upper half plane are thus reduced to correlation functions of the holomorphic part of closed string vertex operators on the plane.[12]

The explicit evaluation of the amplitude (16.86) is straightforward but tedious. One finds[13]

$$A^{ab}(p_1, \xi_1; p_2, \xi_2; k, \epsilon) = \pi g_o^2 g_c C_{D_2} \alpha'^2 t \, \mathrm{tr}\,(\lambda^a \lambda^b) \times$$

$$K(p_1, \xi_1; p_2, \xi_2; k, \epsilon) \frac{\Gamma(-\alpha' t)}{\Gamma\left(1 - \frac{\alpha' t}{2}\right)^2} \tag{16.90}$$

with the kinematic factor

$$K(p_1, \xi_1; p_2, \xi_2; k, \epsilon) = -\big[(\xi_2 k)(\xi_1 Dk) + (\xi_1 k)(\xi_2 Dk) - t(\xi_1 \xi_2)\big] \mathrm{tr}\,(\epsilon D)$$

$$-4(\xi_1 \xi_2)\big[(p_1 \epsilon D p_2) + (p_2 \epsilon D p_1)\big] + 2t\big[(\xi_1 \epsilon D \xi_2) + (\xi_2 \epsilon D \xi_1)\big]$$

$$-2(p_1 \xi_2)(k D \epsilon D \xi_1) - 2(p_1 \xi_2)(\xi_1 \epsilon D k) + 4(k \xi_2)(\xi_1 \epsilon D p_1) + 4(\xi_2 D k)(p_1 \epsilon k \xi_1)$$

$$-2(p_2 \xi_1)(k D \epsilon D \xi_2) - 2(p_2 \xi_1)(\xi_2 \epsilon D k) + 4(k \xi_1)(\xi_2 \epsilon D p_2) + 4(\xi_1 D k)(p_2 \epsilon k \xi_2). \tag{16.91}$$

If we specify to two gluons and one dilaton, this expression simplifies to

$$K(p_1, \xi_1; p_2, \xi_2; k, \epsilon^{(D)}) = \frac{2(d + 4 - 2p)}{\sqrt{d-2}} \big[(p_1 \xi_2)(p_2 \xi_1) - (p_1 p_2)(\xi_1 \xi_2)\big] \tag{16.92}$$

which is independent of $\bar{k}$, as it should.

---

[12] This can be extended to spin fields but we will not do that.

[13] The representation $B(\alpha, \beta) = \int_{-\infty}^{\infty} dx \, \frac{(x^2)^{\alpha - 1/2}}{(1 + x^2)^{\alpha + \beta}}$ of the Beta function is useful.

16.2 String Scattering Amplitudes

The amplitude (16.90) has poles at $\alpha' t = 0, 1, \ldots$. These values of t correspond to massive open string states with

$$m^2 = \frac{n}{\alpha'}, \qquad n = 0, 1, 2, \ldots. \tag{16.93}$$

They appear in the operator product expansion of the two open string vertex operators. But for even n the residues vanish and hence these states do not propagate in internal lines.

So far we have considered only oriented strings. In the presence of orientifold planes we also need to consider non-orientable world-sheets. At tree-level this is only the real projective plane $\mathbb{PR}^2 = S^2/\mathbb{Z}_2$ where $\mathbb{Z}_2 : z \mapsto -1/\bar{z}$ (which acts without fixed points, i.e. there is no boundary). The conformal Killing group of $\mathbb{PR}^2$ is $SU(2)$ with three real parameters. It is the subgroup of $PSL(2, \mathbb{C})$ which commutes with the $\mathbb{Z}_2$ action. The volume of $SU(2)$ is finite in contrast to the volumes of the CKG's of the orientable tree-level world sheets.

There are various ways to obtain the propagators of $\mathbb{RP}^2$. One way is to use the boundary state formalism and to compute e.g. $\langle C|\psi^\mu(z)\overline{\psi}^\nu(\overline{w})|0\rangle$. Another way is to use the doubling trick and to define $\overline{\psi}(\bar{z})(d\bar{z})^{1/2} = \psi(z')(dz')^{1/2}$ where $z' = -1/\bar{z}$. A third way is to use the method of images. Either way one finds (16.2) and (16.56) is replaced by

$$\langle X^\mu(z)\overline{X}^\nu(\overline{w})\rangle = -\frac{\alpha'}{2} D^{\mu\nu} \ln(1 + z\overline{w}),$$

$$\langle \psi^\mu(z)\overline{\psi}^\nu(\overline{w})\rangle = \frac{D^{\mu\nu}}{(1 + z\overline{w})},$$

$$\langle \phi^i(z)\overline{\phi}^j(\overline{w})\rangle = -\delta^{ij} \ln(1 + z\overline{w}),$$

$$\langle \phi(z)\overline{\phi}(\overline{w})\rangle = -\ln(1 + z\overline{w}), \tag{16.94}$$

which are valid for an Op-plane. They are very similar to the two-point functions on D_2 (as compared to $\mathbb{H}_+$).

The new ghost correlator is

$$\langle c(z_1) c(z_2) \overline{c}(\bar{z}_3)\rangle = (z_1 - z_2)(1 + z_1\bar{z}_3)(1 + z_2\bar{z}_3). \tag{16.95}$$

When considering scattering amplitudes in the presence of Op-planes, the vertex operators have to be modified. Recall that Op-planes are the result of $(9 - p)$ T-dualities of type I theory that itself contains only states which are invariant under world-sheet parity Ω. We have seen in Chap. 10 that in the NS-NS sector (where $(-1)^{F_L} = 1$) T-duality transforms Ω to ΩI. Here I denotes the reflection in the directions perpendicular to the Op-plane. The orientifold projection consists of gauging ΩI, i.e. the total wave-function of a state of the orientifold theory must be invariant under ΩI. If the internal (oscillator) part is even (odd), the center of mass part must be even (odd) as well. For the graviton and antisymmetric tensors this means

$$G_{\mu\nu}(x^\alpha, -x^i) = +D_\mu{}^\rho D_\nu{}^\sigma G_{\rho\sigma}(x^\alpha, x^i),$$
$$B_{\mu\nu}(x^\alpha, -x^i) = -D_\mu{}^\rho D_\nu{}^\sigma B_{\rho\sigma}(x^\alpha, x^i), \quad (16.96)$$

and the invariant wave-functions in momentum space are

$$\frac{1}{2}\left(\epsilon_{\mu\nu} e^{ik\cdot x} + (D\epsilon^T D)_{\mu\nu} e^{ik\cdot Dx}\right), \quad (16.97)$$

where the polarization tensor $\epsilon_{\mu\nu}$ is symmetric for the graviton and anti-symmetric for the B-field. The vertex operators for these states are

$$V_{(q,\bar{q})}(k, \epsilon, z, \bar{z}) = \frac{1}{2}\Big(\epsilon_{\mu\nu}(k) V_{(q)}^\mu(k, z) \overline{V}_{(\bar{q})}^\nu(k, \bar{z})$$
$$+ (D\epsilon^T D)_{\mu\nu} V_{(q)}^\mu(Dk, z) \overline{V}_{(\bar{q})}^\nu(Dk, \bar{z})\Big) \quad (16.98)$$

with e.g. $V_{(-1)}^\mu(k, z)$ and $V_{(0)}^\mu(k, z)$ as in (16.13) and (16.15) and similarly for the anti-holomorphic pieces. We will not present any amplitude computations on $\mathbb{RP}_2$ and close this section with a few general remarks.

While in field theory, for any given amplitude, a large number of diagrams has to be computed, there is only one string diagram. By taking the field theory limit of the string amplitude, one can circumvent the computation of Feynman diagrams. Of course, not every quantum field theory is the low-energy limit of a string theory. Nevertheless, the above observation led to the development of the so-called worldline method for perturbative quantum field theory.

String amplitudes contain terms of higher order in momenta, accompanied by higher powers of α'. In fact, amplitudes where the CKG does not suffice to fix the positions of all vertex operators, i.e. where some integrations over the positions remain, have an infinite power series expansion in α', leading to an infinity of higher derivative terms in the effective action. In this way, string induced corrections to the Einstein-Hilbert and Yang-Mills actions have been computed. These calculations are quite tedious, though.

The observation that closed string vertex operators can be split into a holomorphic and an anti-holomorphic piece, where each piece looks like an open string vertex operator, suggests that closed string tree amplitudes can be factorized into (sums of) open string tree amplitudes (with various cyclic orderings). This is trivially the case for three point amplitudes and can be shown to hold for higher point amplitudes. This requires a careful analysis of the integrations involved. The relations between open and closed amplitudes are called KLT (Kawai-Lewellen-Tye) relations. They survive the field theory limit as relations between graviton and gauge boson amplitudes. They have been used, in combination which unitarity arguments which allow to cut loop diagrams into trees, to demonstrate finiteness of $\mathcal{N} = 8$ supergravity in four dimensions at the first few loop orders.

Higher loop amplitudes are much harder to evaluate. For the superstring, gauge fixing introduces Grassmann odd super-moduli and this makes the construction of the measure of the path integral, which must be independent of the gauge slice, complicated beyond one loop. This seems to be merely a technical problem which has been solved at two loops, higher loop orders require further work. The computation of CFT correlation functions is also more difficult; e.g. at one loop the propagators are elliptic functions. Furthermore, sums over spin structures have to be performed and one has to integrate over the moduli of the world-sheet. The alternative pure spinor formulation avoids some of these difficulties but there are also open issues at higher loops. In any case, except for the vacuum amplitudes computed in earlier chapters, we will not open this Pandora's Box.

16.3 From Amplitudes to the Low Energy Effective Field Theory

The general procedure to extract the low energy field theory from string theory is as follows. First calculate various string scattering amplitudes of massless string states, represented by their vertex operators. Then write down a field theory Lagrangian which reproduces these amplitudes. This is done in a perturbative fashion. Start by writing down the effective Lagrangian for the massless free particles, $\mathscr{L}_{2pt}$. Then add $\mathscr{L}_{3pt}$ to reproduce the three point string amplitudes. $\mathscr{L}_{3pt}$ already allows to relate various coupling constants of the effective action to the normalizations of the vertex operators g_c and g_o and the string tension α'. As α' is the only dimensionful constant, the expansion of the effective action in numbers of derivatives, or equivalently, in powers of momenta, is an expansion in powers of $\sqrt{\alpha'}$. At the next step consider the four-point amplitudes. Unitarity guarantees that the massless poles will be those generated by the tree graphs of $\mathscr{L}_{3pt}$. This allows to check again the relation between the coupling constants. The remainder is in general due to massive particle exchanges and will be reproduced by $\mathscr{L}_{4pt}$. The contribution to the string four point amplitudes which are due to massive particle exchange can be expanded in powers of the external momenta. Each term in this expansion generates a local four point vertex in $\mathscr{L}_{4pt}$. This procedure can now be carried on to arbitrary order.

In field theory, the presence of massless vectors and of a massless spin-two particle imply gauge and general coordinate invariance; otherwise no consistent theory is possible beyond the free theory. At each order one therefore writes $\mathscr{L}$ in a gauge and diffeomorphism invariant way so that, for instance, the kinetic energy term of any charged field, which is part of $\mathscr{L}_{2pt}$, already contains the three and four point couplings to gauge fields (via gauge invariance) and couplings to arbitrary order to gravitons (general coordinate invariance). Of course, these couplings must be reproduced by the corresponding string amplitudes.

Equipped with the amplitudes we have computed in the previous section, we will now demonstrate, on a few simple examples, how this procedure works. We start with the heterotic three gauge boson amplitude (16.28). As we just said, massless vectors imply gauge invariance and we therefore expect the effective action of the heterotic string to contain the pure gauge term (we will include coupling to the metric and the dilaton later)

$$S = -\frac{1}{4} \int d^d x \, F^a_{\mu\nu} F^{a\mu\nu}, \qquad (16.99)$$

where

$$F^a_{\mu\nu} = \partial_\mu A^a_\nu - \partial_\nu A^a_\mu + g_d \, f^{abc} A^b_\mu A^c_\nu, \qquad (16.100)$$

and g_d is the Yang-Mills coupling constant in d-dimensions. From this one derives the following cubic action

$$-\frac{1}{4} F^a_{\mu\nu} F^{a\mu\nu}|_{A^3} = -g_d \, \partial_\mu A^a_\nu A^{b\mu} A^{c\nu} f^{abc}. \qquad (16.101)$$

If we compute the tree level three gauge boson amplitude with this action, we reproduce (16.28) provided we identify

$$g_c = \frac{\sqrt{2\alpha'}}{4\pi} g_d \qquad \text{(heterotic string)}. \qquad (16.102)$$

This is also valid for the $E_8 \times E_8$ heterotic string. This can be seen by considering the $SO(16) \times SO(16)$ subgroup.

Let us now turn to the amplitude (16.41). Depending of the choice for the polarization tensor ϵ it describes the scattering of graviton, anti-symmetric tensor B and dilaton D. Their action, up to second order in derivatives, is expected to be of the form

$$S = \int d^d x \sqrt{-g} \left\{ \frac{1}{2\kappa_d^2} R - \frac{1}{6} e^{-2cD} H_{\mu\nu\rho} H^{\mu\nu\rho} - \frac{1}{2} g^{\mu\nu} \partial_\mu D \, \partial_\nu D \right\}, \qquad (16.103)$$

where, as in Chap. 6, κ_d denotes the physical gravitational coupling constant (see Eq. (6.136)). R is the curvature scalar and $H_{\mu\nu\rho} = \partial_\mu B_{\nu\rho} + \partial_\rho B_{\mu\nu} + \partial_\nu B_{\rho\mu}$, the totally antisymmetric field strength of the anti-symmetric tensor. Invariance of (16.41) under $\epsilon^{(h)}_{\mu\nu} \to \epsilon^{(h)}_{\mu\nu} + k_{(\mu} \xi_{\nu)}$ is built into (16.103) as diffeomorphism invariance and invariance under $\epsilon^{(B)}_{\mu\nu} \to \epsilon^{(B)}_{\mu\nu} + k_{[\mu} \zeta_{\nu]}$, because B only appears through its field strength $H = dB$ which is invariant under the Abelian symmetry $B \to B + d\zeta$. As we will see below, the coupling of the dilaton to the anti-symmetric tensor is necessary. The form of the coupling will also be explained. Note that in (16.103) we have not specified the dimension d of space-time. The amplitude (16.41) has

16.3 From Amplitudes to the Low Energy Effective Field Theory

the same form for any number of uncompactified space-time dimensions, the only dimension dependence being the range of the space-time indices.

We have to check whether the action (16.103) reproduces the string theory amplitudes. Only if it does, can we identify the massless spin two string mode with the graviton and justify the claim that string theory automatically incorporates gravity. For this purpose, we expand the metric around Minkowski space, $g_{\mu\nu} = \eta_{\mu\nu} + 2\kappa_d h_{\mu\nu}$. $h_{\mu\nu}$ is the graviton field, whose quadratic action, after gauge fixing, is (6.177). The normalization is such as to correspond to the normalization $\epsilon^{\mu\nu}\epsilon_{\mu\nu} = 1$ of the polarization tensors which appear in the graviton vertex operator.

Expanding the Einstein-Hilbert action to third order in the graviton field, we find

$$\frac{1}{2\kappa_d^2}\sqrt{-g}R|_{3\text{pt}} = -\kappa_d(h^{\mu\nu}h^{\rho\sigma}\partial_\mu\partial_\nu h_{\rho\sigma} + 2\partial^\sigma h_{\mu\nu}\partial^\mu h^{\nu\rho}h_{\rho\sigma}). \tag{16.104}$$

We have used the on-shell conditions for the gravitons, i.e. $k^\mu h_{\mu\nu} = 0$, $h^\mu_\mu = 0$. For a graviton ($\epsilon^{(h)}_{\mu\nu} \equiv h_{\mu\nu}$) the string amplitude (16.41) becomes

$$A^{hhh} = 4\pi g_c\left((k_2 h_1 k_2)(h_2 h_3) + 2(k_2 h_1 h_2 h_3 k_1) + \text{two cyclic perms.}\right). \tag{16.105}$$

The same amplitude follows from the Einstein-Hilbert action provided we identify

$$g_c = \frac{\kappa_d}{2\pi} \quad \text{(heterotic, type I/II)}, \tag{16.106}$$

where one has to notice that the three-point interaction vertex gives rise to six terms in the three-graviton scattering amplitude, which are pairwise equal. This expression is valid for all string theories because the $\mathcal{O}(k^2)$ part of the three-graviton amplitude from where it was derived is the same in all cases.

Comparison with (16.102) gives the relation between the gauge and the gravitational coupling constant of the heterotic string[14]

$$g_d = \sqrt{\frac{2}{\alpha'}}\kappa_d \quad \text{(heterotic string)}. \tag{16.107}$$

For the Kač-Moody algebra at level k the relations (16.102) and (16.107) are replaced by

$$g_c = \frac{\sqrt{2\alpha'k}}{4\pi}g_d, \quad g_d = \sqrt{\frac{2}{\alpha'k}}\kappa_d \quad \text{(heterotic at KM level } k\text{)}. \tag{16.108}$$

---

[14]This uses the normalization $\text{tr}(T^a T^b) = 2\delta^{ab}$ for the generators in the fundamental representation. If we use group generators normalized to $\text{tr}(T^a T^b) = \delta^{ab}$, we obtain $g_d = \frac{2}{\sqrt{\alpha'}}\kappa_d$.

For the type I theory we will determine the relation between the gauge coupling and the gravitational coupling in Sect. 16.5 and we list it here for completeness:

$$g_{10}^I = \frac{2}{\sqrt{\alpha' g_s}} \kappa_{10} \quad \text{(type I)}. \tag{16.109}$$

We now expand the $H_{\mu\nu\rho}$ term in (16.103) to third order. The quadratic term, after adding the gauge fixing term $-(\partial_\nu B^\nu{}_\mu)^2$, is $\mathscr{L} = \frac{1}{2} B^{\mu\nu} \Box B_{\mu\nu}$. For the cubic term one finds

$$-\frac{1}{6}\sqrt{-g}\, e^{-2cD} H_{\mu\nu\rho} H^{\mu\nu\rho}\Big|_{3\text{pt}}$$
$$= cD \left(\partial_\mu B_{\nu\rho} \partial^\mu B^{\nu\rho} + 2\partial_\mu B_{\nu\rho} \partial^\nu B^{\rho\mu} \right)$$
$$+ \kappa_d h^{\mu\nu} \left(\partial_\mu B_{\sigma\rho} \partial_\nu B^{\sigma\rho} + 4\partial_\rho B_{\sigma\mu} \partial_\nu B^{\rho\sigma} + 2\partial_\rho B_{\sigma\mu} \partial^\rho B^\sigma{}_\nu - 2\partial_\rho B_{\sigma\mu} \partial^\sigma B^\rho{}_\nu \right). \tag{16.110}$$

The first part reproduces (on-shell, $\epsilon_{\mu\nu}^{(B)} = B_{\mu\nu}$)

$$A^{BBD} = -\frac{16\pi}{\sqrt{d-2}} g_c (k_2 B_1 B_2 k_1) D \tag{16.111}$$

which follows from Eq. (16.41) with appropriate choice for the polarization tensors. Comparison determines the constant c as

$$c = \frac{2}{\sqrt{d-2}} \kappa_d. \tag{16.112}$$

The second part of Eq. (16.110) is to be compared with

$$A^{hBB} = 4\pi g_c \big[-(k_3 h_1 k_3)(B_3 B_2) - 2(k_2 B_3 B_2 h_1 k_3) - 2(k_3 h_1 B_3 B_2 k_1) + 2(k_1 B_2 h_1 B_3 k_2) \big]. \tag{16.113}$$

We again find the relation (16.106). Finally,

$$-\frac{1}{2}\sqrt{-g}\, g^{\mu\nu} \partial_\mu D\, \partial_\nu D \big|_{3\text{pt}} = \kappa_d\, h^{\mu\nu} \partial_\mu D\, \partial_\nu D \tag{16.114}$$

reproduces

$$A^{hDD} = -4\pi g_c (k_3 \epsilon_1 k_2) \tag{16.115}$$

yielding (16.106) once more. For all other choices of the external fields the string amplitude vanishes to $\mathcal{O}(k^2)$, and it is easy to see that (16.41) does not lead to any other three-point on-shell amplitudes either.

16.3 From Amplitudes to the Low Energy Effective Field Theory

In order to reproduce the $\mathcal{O}(k^4)$ terms in the string amplitude (16.41), higher derivative terms such as $\alpha' R_{\mu\nu\rho\sigma} R^{\mu\nu\rho\sigma}$ must be included in the effective action. They are absent in type II theories where the first higher derivative corrections involving the curvature are of order $\mathcal{O}(\alpha'^3 R^4)$, which contribute only to amplitudes with at least four gravitons.

One of the noteworthy properties of the action (16.103) is the fact that at the quadratic level the dilaton and the graviton decouple. This property is, of course, not invariant under field redefinitions. The frame (in field space) in which it is true is called Einstein frame. It has a purely gravitational part which is the Einstein-Hilbert action. The fact that the string scattering amplitudes are reproduced by the Einstein-frame action is a consequence of our choice of vertex operators. They are such that the states generated by them are orthogonal to each other.

Performing a Weyl rescaling of the metric

$$g_{\mu\nu} \to e^{-\frac{2\kappa_d}{\sqrt{d-2}}D} g_{\mu\nu} \qquad (16.116)$$

accompanied by a rescaling of the anti-symmetric tensor and a redefinition of the dilaton (cf. (6.171) on page 172)

$$B \to \frac{1}{2\kappa_d} B,$$

$$D = \frac{2}{\kappa_d \sqrt{d-2}} (\Phi - \Phi_0), \qquad (16.117)$$

the action transforms into

$$S = \frac{1}{2\tilde{\kappa}_d^2} \int d^d x \sqrt{-g}\, e^{-2\Phi} \left(R - \frac{1}{12} H_{\mu\nu\rho} H^{\mu\nu\rho} - \frac{\tilde{\kappa}_d^2}{2\tilde{g}_d^2} F^a_{\mu\nu} F^{a\mu\nu} + 4\partial^\mu \Phi \partial_\mu \Phi \right). \qquad (16.118)$$

Φ_0 is an arbitrary additive constant part of Φ and D the fluctuation around it. $\tilde{\kappa}_d$ is the gravitational coupling constant already introduced in in Chap. 6. Its relation to the physical coupling constant κ_d is

$$\kappa_d = e^{\Phi_0} \tilde{\kappa}_d \equiv g_s \tilde{\kappa}_d, \qquad (16.119)$$

where g_s is called the string coupling constant.

In (16.118) we have also included the gauge term for which we have rescaled the gauge fields $A^a \to A^a / g_d$ and we have defined

$$g_d = e^{\Phi_0} \tilde{g}_d. \qquad (16.120)$$

The coupling of the gauge boson to the dilaton and the graviton, which is implied by (16.118), can be verified by computing the two gauge boson—one graviton amplitude, which is easily done. There is no gauge invariant coupling to $B_{\mu\nu}$ at this order. If we drop the gauge term, (16.118) is also correct for the NS-NS sector of the type II theories. Of course, since the type II and heterotic dilatons are a priori unrelated, so are coupling constants g_s^{het} and g_s^{II}. The frame in which the NS-NS part of the tree-level effective action has an overall factor of $e^{-2\Phi} = e^{-\chi(S^2)\Phi}$ is called string frame. The metric which appears in the Polyakov action (14.1) is the string frame metric. In Chap. 18 we will say more about the issue of frames.

We now turn to the amplitudes (16.43) and (16.44). A good guess for the effective action with coupling of gravity and dilatons to the RR-fields is (here we work in the uncompactified theory in the critical dimension and $F^{(p+1)} = dC^{(p)}$)

$$S = -\int d^{10}x \sqrt{-g} \frac{1}{2(p+1)} e^{\tilde{c}\Phi} F_{\mu_1\ldots\mu_{p+1}} F^{\mu_1\ldots\mu_{p+1}}. \tag{16.121}$$

The normalization is such that after gauge fixing $d * C = 0$ the kinetic term for C is $\int \frac{1}{2} C_{\mu_1\ldots\mu_p} \Box C^{\mu_1\ldots\mu_p}$. S is invariant under diffeomorphisms and under $\delta C^{(p)} = d\alpha^{(p-1)}$. Expansion around flat space and vanishing vevs for the dilaton and the RR-fields gives, up to third order in fluctuations around the background values

$$\kappa_{10} h^{\mu\nu} F_{\mu\mu_1\ldots\mu_p} F_{\nu}{}^{\mu_1\ldots\mu_p} - \frac{\tilde{c}}{2(p+1)} D F_{\mu_1\ldots\mu_{p+1}} F^{\mu_1\ldots\mu_{p+1}}. \tag{16.122}$$

Comparing with (16.43) shows that the vertex operators (16.21) were correctly normalized and comparison with (16.44) determines

$$\tilde{c} = -\frac{\kappa_{10}}{\sqrt{2}}(p-4). \tag{16.123}$$

If we now perform, in addition to (16.116) and (16.117) (with $d = 10$), the field rescaling

$$C \to \frac{1}{\tilde{\kappa}_{10}\sqrt{2p!}} C, \tag{16.124}$$

the action (16.121) becomes

$$S = -\frac{1}{2\tilde{\kappa}_{10}^2} \int d^{10}\sqrt{-g} \frac{1}{2(p+1)!} F_{\mu_1\ldots\mu_{p+1}} F^{\mu_1\ldots\mu_{p+1}}. \tag{16.125}$$

There is no exponential dilaton prefactor. The tree-level exponential factor $e^{-2\Phi}$ arises after a further rescaling $C^{(p)} \to e^{-\Phi} C^{(p)}$ which leads to $F^{(p+1)} \to F^{(p+1)} - d\Phi \wedge C^{(p)}$. Note that this redefinition of the potential changes the Bianchi identity and the gauge symmetry of $F^{(p+1)}$. Both now involve the gradient of the dilaton.[15]

---

[15]One can show that in the presence of a non-trivial dilaton background these do, in fact, correspond to the physical state conditions of the vertex operator.

16.3 From Amplitudes to the Low Energy Effective Field Theory

Often the form of the action (16.125) without the exponential dilaton factor is preferred because of the simpler equations of motion and Bianchi identities.

Let us now comment on the string coupling constant. In Chap. 9 we have defined the type II coupling constant as the ratio of the tensions of the fundamental string and the D-string, i.e.

$$g_s = \frac{\tau_{F1}}{\tau_{D1}} = \frac{2\sqrt{\pi}\kappa_{10}}{(4\pi^2\alpha')^2} \qquad \text{(type II)}. \tag{16.126}$$

For the heterotic string we define it as

$$g_s = \frac{2\kappa_{10}}{(2\alpha')^2} \qquad \text{(heterotic)}. \tag{16.127}$$

Note that the arbitrariness of Φ_0 or, equivalently, of g_s, reflects the freedom to simultaneously rescale the amplitudes and the vertex operators.

We now turn to the open string three-gluon amplitude. We expect the two-derivative part of the effective field theory on the world-volume of a Dp-brane to contain (16.99) with $d = p+1$. For the vertex operators to be correctly normalized we choose tr$(\lambda^a \lambda^b) = \delta^{ab}$, which corresponds to $C_{\text{fund.}} = 1$. Comparison of the string amplitude (16.68) with the field theory amplitude derived from (16.99) we find the relation

$$g_o = \sqrt{2\alpha'} g_{\text{YM}}, \tag{16.128}$$

where g_{YM} is the gauge coupling constant on the D-brane.

The two gauge boson-one dilaton disk amplitude (16.90) and (16.92) can be used to fix the dilaton prefactor in front of the gauge kinetic term as $e^{-\Phi}$. For the comparison one must rescale the dilaton according to (16.117). This is correct for any p.

We now justify the normalization of the gaugino vertex operator (16.63). We expect the world-volume theory on the Dp-brane to be supersymmetric. E.g. for $p = 9$, i.e. for the type I string, this is the $\mathcal{N} = 1$ supersymmetric Yang-Mills theory in $d = 10$ with action

$$S = \int d^{10}x \left\{ -\frac{1}{4} F^a_{\mu\nu} F^{a\mu\nu} - \frac{1}{2}\bar{\lambda}^a \gamma^\mu D_\mu \lambda^a \right\}, \tag{16.129}$$

where λ^a is the gaugino field (a Majorana fermion) and $D_\mu \lambda^a = \partial_\mu \lambda^a + g_{\text{YM}} f^{abc} A^b_\mu \lambda^c$ is the gauge covariant derivative. From this we once again obtain (16.128), thus showing that the gaugino vertex operator was properly normalized.

The pieces in the effective action we have constructed so far, have been written in a covariant form. Expanding the actions in the fluctuating fields, i.e. graviton, gauge field etc., leads to higher than just cubic interactions. One can check that they agree with the higher point string amplitudes. The comparison is, however, not as direct as for the three-point couplings because e.g. the field theoretic four-point amplitudes consist of two parts: exchange diagrams and point-interactions which appear in the Lagrangian. The sum of all contributions must be reproduced by the string amplitude. Conversely, if we want to deduce new interaction terms in the effective action from a string amplitude, we first have to subtract the field theoretic part computed via exchange diagrams. What remains is a new interaction in the effective action.

We now indicate on one example some of the features which we have just alluded to. For this we return to the amplitude (16.80) and expand it in powers of α':

$$\frac{\alpha'}{t} B(-\alpha' s, -\alpha' u) \sim \frac{1}{us} - \frac{\pi^2}{6}(\alpha')^2 + \ldots, \qquad (16.130)$$

where the second term can be obtained by summing the contribution from all massive states either in the s or the u channel. If we multiply (16.130) with the kinematic factor (16.82) we see that in addition to massless poles there are new contact interactions which arise from terms of the structure $D^n F^m$ in the low energy effective action which represent corrections to the YM action. The terms with $m > 4$ can, of course, not be seen in the four gluon amplitude.

16.4 Effective Supergravities in Ten and Eleven Dimensions

As we have seen in the previous sections, the dynamics of the massless modes of string theory is governed by an effective field theory. After integrating out the massive string modes, which assumes energies $E \ll M_s$, the Lagrangian of the effective field theory has infinitely many terms. The higher order operators are suppressed by inverse powers of $M_s = 1/\sqrt{\alpha'}$, but the leading terms in the α' expansion are expected to be familiar field theories. So far we have derived these effective theories from tree level string scattering amplitudes. Higher loop contributions are suppressed by additional powers of g_s and can be neglected at weak coupling $g_s \ll 1$. Additional structure such as gauge symmetry or supersymmetry then allows one in some cases to fix the effective actions (up to the two-derivative level) uniquely.

In this section we discuss the ten-dimensional effective supergravity actions of the massless string modes. They had been constructed previously and it gives further credence to string theory that it reproduces these actions at leading order in α'. There are five different supersymmetric string theories in ten space-time dimensions. Their low-energy effective supergravity theories in $d = 10$ are:

- Non-chiral type IIA supergravity theory with $\mathcal{N} = (1, 1)$ space-time supersymmetry (32 supercharges)
- Chiral type IIB supergravity theory with $\mathcal{N} = (2, 0)$ space-time supersymmetry (32 supercharges)
- Chiral type I supergravity theory with $\mathcal{N} = (1, 0)$ space-time supersymmetry coupled to $G = SO(32)$ super-Yang-Mills theory (16 supercharges)

16.4 Effective Supergravities in Ten and Eleven Dimensions

- Two chiral heterotic supergravity theories with $\mathcal{N} = (1, 0)$ space-time supersymmetry coupled to $G = SO(32)$ or $G = E_8 \times E_8$ super-Yang-Mills theory (16 supercharges).

The first two are low-energy limits of type IIA and IIB string theories, respectively, while the latter are the limits of the type I and the heterotic theories. Incidentally, the two gauge groups $SO(32)$ and $E_8 \times E_8$ are the only ones which are anomaly free.[16] In addition to these ten-dimensional theories there exists a unique $\mathcal{N} = 1$ (32 supercharges) supergravity theory in eleven dimensions; its dimensional reduction to ten dimensions gives type IIA supergravity. This will be explained more thoroughly in Chap. 18. Note that these theories do not exist as ten-dimensional quantum-field theories. To get such a consistent quantum theory one must resort to the full string theory as its ultraviolet completion.

In the next subsection we list the parts of the supergravity actions which contain only bosonic fields. Some pieces were explicitly derived in the previous section from tree level string scattering amplitudes. The remaining part, which contains fermionic fields, is determined by supersymmetry. These effective supergravity actions are the starting point for many studies of string theory in the so-called supergravity limit. For instance, one has found solutions corresponding to p-brane solitons, black-holes, black rings etc. These consideration also led to the formulation of the AdS/CFT correspondence (see Chap. 18). Moreover, the only known way so far to study the important issue of moduli stabilization is in the supergravity approximation. Except in very special cases, all these solutions receive higher α' corrections and they are only reliable if these can be ignored. For instance, as these corrections are of the general form $\alpha'^n R^{n+1}$, where R is the curvature, for the moduli stabilization problem and for other aspects of string compactification, this requires working in the large volume limit of the compactification manifold. And, of course, it is assumed that the string is weakly coupled.

11-Dimensional Supergravity

Eleven-dimension supergravity contains two massless bosonic fields, the space-time metric G_{MN} and a three-form potential A_3 with field strength $F_4 = dA_3$. The bosonic part of the action is

$$S_{\text{eff}}^{(d=11)} = \frac{1}{2\kappa_{11}^2} \left[\int d^{11}x \sqrt{-G} \left(R - \frac{1}{2 \cdot 4!} F_{M_1...M_4} F^{M_1...M_4} \right) \right.$$
$$\left. - \frac{1}{6} \int A_3 \wedge F_4 \wedge F_4 \right]. \tag{16.131}$$

---

[16]The structure of anomalies is much richer in $d = 10$ than in $d = 4$, mainly because of possible gravitational anomalies. We will not present a general discussion of anomalies in string theory, but a few aspects will be mentioned as we go along.

Besides the kinetic terms for the metric and the three-form, there is a Chern-Simons term for A_3. κ_{11} denotes the eleven-dimensional gravitational coupling. The only fermionic field is the Majorana gravitino. Matching of the numbers of bosonic and fermionic on-shell degrees of freedom is easily verified with the help of (7.5).

The classification of supermultiplets in various dimensions reveals that supergravities theories in more than $10 + 1$ dimensions always contain fields fields of spin > 2, i.e. symmetric tensors of rank higher than two. (For supersymmetric gauge theories the limit is $d = 9 + 1$.) An easy way to understand this is to consider the theory compactified on a torus to $D = 4$. Starting in $d > 11$ the number of supersymmetry generators is larger than eight and, if we apply them successively to a state with helicity $s = +2$, they will transform it to a state with helicity $s < -2$.

Perturbative superstring theory, as it has been developed in this book so far, does exist in ten dimensions and so it seems that the unique maximal supergravity theory in $d = 11$ does not play a role. That this is not so is one of the most remarkable results of string theory. As we will discuss in more detail in Chap. 18, it arises when studying the strong coupling limit $g_s \to \infty$ of type IIA superstring theory.

Type IIA/B Supergravities

In general, the spectrum of any closed string theory in ten dimensions is obtained by tensoring the left- and right-moving massless states and applying the GSO projection. In the NS-NS sector of type II theories this leads to the universal result

$$\mathbf{8}_V \otimes \mathbf{8}_V \longrightarrow \{G_{MN}, B_{MN}, \Phi\}. \tag{16.132}$$

The massless spectrum consists of the ten-dimensional metric G_{MN}, the antisymmetric NS-NS two-form B_{MN} and the dilaton Φ. The massless states in the R-R sector are obtained by tensoring the two R ground states

$$\mathbf{8}_s \otimes \begin{cases} \mathbf{8}_c & \text{IIA} \\ \mathbf{8}_s & \text{IIB} \end{cases} \longrightarrow \begin{cases} \{(C_1)_M, (C_3)_{MNR}\} & \text{IIA} \\ \{C_0, (C_2)_{MN}, (C_4)_{MNRS}\} & \text{IIB} \end{cases}. \tag{16.133}$$

These are the antisymmetric R-R p-form potentials. The four-form of type IIB obeys a self-duality constraint so that only half of the degrees of freedom survive. All the fields in (16.132) and (16.133) are chosen to have engineering dimension zero. The fermionic spectrum in the R-NS and NS-R sectors consists of the two gravitini of the $\mathcal{N} = 2$ supergravities, with equal (IIB) or opposite (IIA) chiralities and two dilatini whose chiralities are opposite to that of the gravitini. All fermions are Majorana-Weyl spinors.

The string-frame ten-dimensional actions for the bosonic fields of the type IIA and IIB supergravity actions are

16.4 Effective Supergravities in Ten and Eleven Dimensions

$$S_{\text{IIA}} = \frac{1}{2\tilde{\kappa}_{10}^2} \int d^{10}x \sqrt{-G} \left[e^{-2\Phi} \left(R + 4(\nabla\Phi)^2 - \frac{1}{2}|H_3|^2 \right) - \frac{1}{2}|F_2|^2 - \frac{1}{2}|F_4|^2 \right]$$

$$- \frac{1}{4\tilde{\kappa}_{10}^2} \int B_2 \wedge dC_3 \wedge dC_3, \tag{16.134a}$$

$$S_{\text{IIB}} = \frac{1}{2\tilde{\kappa}_{10}^2} \int d^{10}x \sqrt{-G} \left[e^{-2\Phi} \left(R + 4(\nabla\Phi)^2 - \frac{1}{2}|H_3|^2 \right) - \frac{1}{2}|F_1|^2 - \frac{1}{2}|F_3|^2 \right.$$

$$\left. - \frac{1}{4}|F_5|^2 \right] - \frac{1}{4\tilde{\kappa}_{10}^2} \int C_4 \wedge H_3 \wedge F_3 \tag{16.134b}$$

with gravitational coupling, cf. Eq. (9.57),

$$\tilde{\kappa}_{10}^2 = \frac{1}{4\pi}(4\pi^2\alpha')^4. \tag{16.135}$$

We use the notation

$$|F_p|^2 = \frac{1}{p!} F_{M_1\ldots M_p} F^{M_1\ldots M_p}. \tag{16.136}$$

The last term in each of the two actions is a Chern-Simons terms.

The field strength of the NS-NS two-form is defined as[17]

$$H_3 = dB_2, \quad H_{MNR} = 3\partial_{[M} B_{NR]}. \tag{16.137}$$

The field strengths F_p in the R-R sector also involve the R-R-potentials C_q of lower degree and also the NS-NS two-form B_2:

(IIA) $\quad F_2 = dC_1, \quad F_4 = dC_3 - dB_2 \wedge C_1,$

(IIB) $\quad F_1 = dC_0, \quad F_3 = dC_2 - C_0 \, dB_2,$

$$F_5 = dC_4 - \frac{1}{2}C_2 \wedge dB_2 + \frac{1}{2}B_2 \wedge dC_2. \tag{16.138}$$

This leads to correction terms for the usual Bianchi identities $dF_p = 0$ with $F_p = dC_{p-1}$. For instance one gets $dF_3 = H_3 \wedge F_1$.

The field strengths are invariant under the gauge transformations $\delta B_2 = d\zeta$ and

(IIB) $\quad \delta C_0 = 0, \quad \delta C_2 = d\Lambda_1,$

$$\delta C_4 = d\Lambda_3 - \frac{1}{2}dB \wedge \Lambda_1 + \frac{1}{2}dC_2 \wedge \zeta,$$

---

[17] We define the anti-symmetrization symbol [...] with unit weight, e.g. $\partial_{[M} B_{NP]} = \frac{1}{3!}(\partial_M B_{NP} \pm 5 \text{ permutations})$.

(IIA) $$\delta C_1 = d\Lambda_0, \quad \delta C_3 = d\Lambda_2 - dB\,\Lambda_0, \tag{16.139}$$

while the CS terms transform into total derivatives.

One important subtlety is the self-duality constraint of F_5 in type IIB,

$$F_5 = *F_5. \tag{16.140}$$

There is no way to write a covariant action which takes it into account. Since $\int F_5 \wedge *F_5 = 0$ one has to impose the self-duality constraint separately on top of the equations of motion which follow from (16.134). The additional factor of $\frac{1}{2}$ in front of F_5^2 compensates for the fact that an unconstrained F_5 has twice as many degrees of freedom as a self-dual one.[18]

The type IIB action has a hidden $SL(2,\mathbb{R})$ invariance which becomes manifest if we express it in Einstein frame. For this purpose we rescale the metric

$$G_{MN}^{\text{E}} = e^{-\Phi/2} G_{MN}. \tag{16.141}$$

We also define a complex scalar and three-form

$$\tau = C_0 + i e^{-\Phi}, \quad G_3 = F_3 - i e^{-\Phi} H_3 = dC_2 - \tau dB_2. \tag{16.142}$$

In terms of these fields the type IIB action is[19]

$$S_{\text{IIB}} = \frac{1}{2\tilde{\kappa}_{10}^2} \int d^{10}x \sqrt{-G} \left[R - \frac{\partial_M \tau \partial^M \bar{\tau}}{2(\text{Im}\,\tau)^2} - \frac{1}{2}\frac{|G_3|^2}{\text{Im}\,\tau} - \frac{1}{4}|F_5|^2 \right]$$
$$+ \frac{1}{8i\tilde{\kappa}_{10}^2} \int \frac{1}{\text{Im}\,\tau} C_4 \wedge G_3 \wedge \bar{G}_3, \tag{16.143}$$

where we have dropped the superscript E. Which frame is being used is always clear from the form of the Einstein-Hilbert term. Self-duality of F_5 again needs to be imposed by hand. This form of the action is invariant under an $SL(2,\mathbb{R})$ symmetry which leaves the metric and four-form invariant and acts on the remaining fields as

$$\tau \to \frac{a\tau + b}{c\tau + d}, \quad \begin{pmatrix} C_2 \\ B_2 \end{pmatrix} \to \begin{pmatrix} a & b \\ c & d \end{pmatrix} \begin{pmatrix} C_2 \\ B_2 \end{pmatrix}, \quad (ad - bc = 1). \tag{16.144}$$

Generically (for $c \neq 0$) this inverts the string coupling constant $g_s = e^{\langle\Phi\rangle}$, i.e. it transforms a weakly coupled theory to a theory at strong coupling. Of course, this

---

[18] There exists a formulation which takes into account the self-duality at the level of the action, but at the expense of manifest covariance or of introducing auxiliary degrees of freedom.
[19] For a complex p-form we define $|F_p|^2 = \frac{1}{p!} F_{M_1...M_p} \overline{F}^{M_1...M_p}$.

16.4 Effective Supergravities in Ten and Eleven Dimensions

would take us out of the framework in which the effective action has been derived. We will discuss such strong-weak coupling dualities further in Chap. 18.

There exist another, more symmetric formulation of the type II SUGRA actions, called democratic or self-dual formulation. For each C_p it also includes a C_{8-p} and therefore for F_p a F_{10-p}. To cut down the number of degrees of freedom one imposes the constraints $*F_{2p} = (-1)^p F_{10-2p}$ and $*F_{2p+1} = (-1)^p F_{9-2p}$, consistent with $*^2 = +1$ on forms of odd and $*^2 = -1$ on forms of even degree. The actions, which together with the constraints leads to the same equations as (16.134), contains only kinetic terms $|F_p|^2$ (with an additional factor of 1/2) and no CS terms. This formulation is useful for the discussion of flux compactifications and of SUGRA brane solutions, because it treats electric and magnetic branes on equal footing.

The Effective Type I Supergravity Action

We now turn to the ten-dimensional type I superstring, whose massless modes include an $SO(32)$ gauge field and 32 space-time filling D9-branes. Recall that the type I superstring is the $\mathbb{Z}_2$ orientifold of the type IIB superstring. The action for the bulk fields is obtained by projecting out all degrees of freedom of the type IIB theory that are odd under the world-sheet parity. The bosonic type I open plus closed massless spectrum is thus

$$\text{type I:} \quad \{A_M^a\} \;+\; \{G_{MN}, \Phi, (C_2)_{MN}\}. \tag{16.145}$$

The fermions are a single Majorana-Weyl gravitino and dilatino in the closed string sector and Majorana-Weyl gaugini in the open sector.

The leading order action is identical to the relevant pieces of the type IIB theory plus the super Yang-Mills action for the gauge fields, i.e. in string frame

$$S_{\rm I} = \frac{1}{2\tilde{\kappa}_{10}^2} \int d^{10}x \sqrt{-G} \left\{ e^{-2\Phi} \left(R + 4(\nabla\Phi)^2 \right) - \frac{1}{2}|F_3|^2 \right\}$$
$$- \frac{1}{2\tilde{g}_{10}^2} \int d^{10}x \sqrt{-G} \, e^{-\Phi} \, {\rm tr}_v (|F_{\rm YM}|^2). \tag{16.146}$$

Here $F_{\rm YM} = F_{\rm YM}^a T^a$ is the Yang-Mills field strength and the trace is over the vector representation of $SO(32)$ with ${\rm tr}_v(T^a T^b) = \delta^{ab}$. In our conventions A_M^a carries engineering dimension one while all other bosonic fields are dimensionless. The relation between the type I gravitational and gauge coupling will be determined in Sect. 16.5 and reads

$$\frac{\tilde{\kappa}_{10}^2}{\tilde{g}_{10}^2} = \frac{\alpha'}{4}. \tag{16.147}$$

Let us emphasize that the dilaton dependence of the Einstein-Hilbert and the Yang-Mills action are different. This is because the former arises from a closed string sphere diagram and the latter from an open string disk diagram. Compared to the type IIB theory, the type I R-R three-form field strength receives additional contributions

$$F_3 = dC_2 - \frac{\alpha'}{4}(\Omega_{\text{YM}} - \Omega_{\text{L}}), \qquad (16.148)$$

where Ω_{YM} and Ω_{L} are the Yang-Mills and Lorentz the Chern-Simons three-forms

$$\Omega_{\text{YM}} = \text{tr}\left(A \wedge dA - \frac{2i}{3}A \wedge A \wedge A\right),$$

$$\Omega_{\text{L}} = \text{tr}\left(\omega \wedge d\omega + \frac{2}{3}\omega \wedge \omega \wedge \omega\right), \qquad (16.149)$$

where ω is the spin connection. The relative factors in the two CS terms are different because the YM gauge fields are Hermitian while the spin connection is real and anti-symmetric, i.e. anti-Hermitian.[20]

The Chern-Simons three-forms satisfy

$$d\Omega_{\text{YM}} = \text{tr}(F \wedge F) \quad \text{and} \quad d\Omega_{\text{L}} = \text{tr}(R \wedge R), \qquad (16.150)$$

where R denotes the Ricci two-form (14.260) and $F = dA - i\,A^2$ is the Yang-Mills field strength. Therefore

$$dF_3 = \frac{\alpha'}{4}(\text{tr } R \wedge R - \text{tr } F \wedge F). \qquad (16.151)$$

Since B_2 is projected out, there is no bulk Chern-Simons term in type I.

The Chern-Simons forms are not invariant under gauge and Lorentz transformations, $\delta A = d\Lambda + i[\Lambda, A]$ and $\delta\omega = d\Lambda - [\Lambda, \omega]$:

$$\delta\Omega_{\text{YM}} = d\,\text{tr}\,(\Lambda_{\text{YM}} dA) \quad \text{and} \quad \delta\Omega_{\text{L}} = d\,\text{tr}\,(\Lambda_{\text{L}} d\omega). \qquad (16.152)$$

Therefore, the combination (16.148) is not invariant, unless C_2 transforms as

$$C_2 \to C_2 + \frac{\alpha'}{4}\big(\text{tr}(\Lambda_{\text{YM}}\,dA) - \text{tr}\,(\Lambda_{\text{L}} d\omega)\big). \qquad (16.153)$$

---

[20] Note that Ω_{YM} is first order in a derivative expansion, while Ω_{L} is third order. For this reason, the Lorentz term is sometimes considered subleading. However, both terms are equally important for anomaly cancellation.

16.4 Effective Supergravities in Ten and Eleven Dimensions

Note that perturbatively the action for p-forms is invariant under constant shifts $C_p \to C_p + a_p$.[21] We have just seen that in type I superstring theory, a gauge transformation induces a coordinate dependent shift of C_2. Thus, one also speaks of a gauging of the axionic shift symmetry, which plays an important role for the cancellation of the ten-dimensional anomalies for the type I and also the heterotic string. This is the so-called Green-Schwarz mechanism. A four-dimensional version will be discussed in Chap. 17.

One can also write the R-R part of the type I action in a democratic version which has, in addition to C_2 a six-form C_6 with the constraint is $dC_2 = - *dC_6$. The two-form and six-form couple to D-strings and D5-branes. Beyond that, there is also a non-dynamical ten-form C_{10} to which D9-branes couple.

The Effective Heterotic Supergravity Action

The effective $\mathcal{N} = 1$ supergravity actions of the two heterotic string theories with gauge groups $G = E_8 \times E_8, SO(32)$ are quite similar to type I. Since there are only closed strings, the gauge kinetic term has the same dilaton dependence as the gravitational part of the action:

$$S_I = \frac{1}{2\tilde{\kappa}_{10}^2} \int d^{10}x \sqrt{-G}\, e^{-2\Phi} \left(R + 4(\nabla\Phi)^2 - \frac{1}{2}|H_3|^2 \right) \tag{16.154}$$

$$- \frac{1}{2\tilde{g}_{10}^2} \int d^{10}x \sqrt{-G}\, e^{-2\Phi} \, \text{Tr}\,(|F_{\text{YM}}|^2), \tag{16.155}$$

where for $\text{tr}\,_v(T^a T^b) = \delta^{ab}$ one also has

$$\frac{\tilde{\kappa}_{10}^2}{\tilde{g}_{10}^2} = \frac{\alpha'}{4}. \tag{16.156}$$

Similarly to the type I superstring one has

$$H_3 = dB_2 - \frac{\alpha'}{4}(\Omega_{\text{YM}} - \Omega_{\text{L}}) \tag{16.157}$$

with the Chern-Simons three-forms and the gauge coupling defined as for the type I supergravity theory. The Bianchi identity is

$$dH_3 = \frac{\alpha'}{4}(\text{tr}\, R \wedge R - \text{tr}\, F \wedge F) \tag{16.158}$$

---

[21] C_0 is also called an axion and the symmetry $C_0 \to C_0 + \text{const}$ is called axionic shift symmetry or Peccei-Quinn symmetry. We will use this terminology also for $p > 0$.

which we have analyzed in some detail in Sect. 14.7. The discussion below (16.151) also applies here, with the obvious replacement $C_2 \to B_2$.

The two $d = 10$ $\mathcal{N} = 1$ SUGRA+SYM theories at first glance appear to be different, e.g. the dilaton couplings are not the same. However, for any given gauge group there actually exist a unique such theory in $d = 10$. In Chap. 18 we will see how the type I and heterotic $SO(32)$ theories can be mapped to each other by a field redefinition and what this implies for the relation between them.

16.5 Effective Action for D-branes

From our discussion of string scattering amplitudes it is apparent that both the closed and the open string, i.e. the gravitational and the gauge parts of the effective action receive corrections at higher orders in α'. While it is very tedious to compute these corrections in the closed string sector, some exact results are known in the open string sector.

The dynamics of the massless open string modes, the ten-dimensional gauge field multiplet in the case of D9-branes or its lower-dimensional descendants for Dp-branes with $p < 9$, is governed by the Dirac-Born-Infeld (DBI) plus the Chern-Simons (CS) action. Together they constitute the relevant Lagrangian at leading order in the string coupling (disk level). The two pieces, DBI and CS, involve different closed string background fields to which the open string modes couple,

$$S_{\text{eff}} = S_{\text{DBI}}[G, \Phi, B] + S_{\text{CS}}[C_p]. \qquad (16.159)$$

The DBI action contains the coupling of the open string degrees of freedom to the bulk NS-NS fields, the dilaton, metric and two-form, while the Chern-Simons action involves the R-R p-forms C_p. In particular, the DBI action is only well understood for a single brane, i.e. for Abelian gauge symmetry. In this case it is a non-linear generalization of Maxwell theory. Its bosonic part in string-frame is

$$S_{\text{DBI}} = -T_p \int_{\mathcal{W}} d^{p+1}\xi \, e^{-\Phi(X)} \sqrt{-\det\bigl(g_{\alpha\beta}(X) + 2\pi\alpha' \mathcal{F}_{\alpha\beta}(X)\bigr)}, \qquad (16.160)$$

where $2\pi\alpha' \mathcal{F} = B + 2\pi\alpha' F$ was defined in Eq. (10.204). We split the space-time indices $M, N, \cdots \in \{0, \ldots, 9\}$ into the brane world-volume directions with $\alpha, \beta, \cdots \in \{0, \ldots, p\}$ and the transverse space directions with $i, j, \cdots \in \{p+1, \ldots, 9\}$. ξ^α are the intrinsic world-volume coordinates and the functions $X^M(\xi)$ describe the embedding of the world-volume $\mathcal{W}$ into space-time M_{10}. $g_{\alpha\beta}$ is the pull-back of the ten-dimensional metric G_{MN} to the world-volume and likewise for $B_{\alpha\beta}$:

$$g_{\alpha\beta} = \partial_\alpha X^M \partial_\beta X^N G_{MN}, \quad B_{\alpha\beta} = \partial_\alpha X^M \partial_\beta X^N B_{MN}. \qquad (16.161)$$

16.5 Effective Action for D-branes

The field strength $F_{\alpha\beta}$ is restricted to $\mathcal{W}$, it is not the pull-back of a bulk field. The parameter T_p has been determined in Chap. 9, Eqs. (9.58) and (9.92):

$$T_p = 2\pi \ell_s^{-(p+1)} \times \begin{cases} 1 & \text{type II} \\ \dfrac{1}{\sqrt{2}} & \text{type I} \end{cases}. \tag{16.162}$$

The massless bosonic degrees of freedom on the D-brane are the massless bosonic open string modes: the $(p+1)$ components of the gauge field $A_\alpha(\xi)$ and the $(9-p)$ fluctuations of the transverse coordinates $X^i(\xi)$. They describe the motion and deformation of the brane inside M_{10}. The dilaton prefactor identifies (16.160) as the open string tree-level effective action, i.e. resulting from disk diagrams.

In static or coordinate gauge, $p+1$ of the embedding functions are identified with the intrinsic world-volume coordinates. The remaining coordinates are split into a constant piece x^i (the position of the brane) and a fluctuating piece:

$$X^\alpha = \xi^\alpha, \quad X^i = x^i + 2\pi\alpha' \phi^i(\xi) + \cdots. \tag{16.163}$$

There is one scalar field ϕ^i for each transverse direction of the Dp-brane. The $p-1$ propagating degrees of freedom of A_α and the ϕ^i comprise the eight bosonic degrees of freedom found in the open string spectrum.

To extract the leading, two-derivative order Lagrangian from (16.160) one expands in powers of the field strength by use of

$$\det(1 + M) = 1 + \operatorname{tr}(M) + \frac{1}{2}(\operatorname{tr} M)^2 - \frac{1}{2}\operatorname{tr}(M^2) + \cdots. \tag{16.164}$$

For simplicity we set the scalars and the B-field to zero, keeping only the gauge fields and the metric. The expansion of the DBI action leads to

$$S_{\text{DBI}} = -T_p \int d^{p+1}\xi \sqrt{-g}\, e^{-\Phi} \left[1 + \frac{1}{4}(2\pi\alpha')^2 F_{\alpha\beta} F^{\alpha\beta} + \ldots\right]. \tag{16.165}$$

This is the sum of a vacuum energy and the kinetic term for the gauge fields. From this expression we can read off the gauge coupling constant of the world-volume theory

$$g_{\text{D}p}^2 = \frac{g_s}{T_p}(2\pi\alpha')^{-2} = g_s\,(2\pi)^{p-2}(\alpha')^{\frac{p-3}{2}}. \tag{16.166}$$

Note that the DBI action contains terms at all orders in α', but, in addition, there will be higher derivative corrections if the field strength $\mathcal{F}_{\alpha\beta}$ and the space-time metric are not constant. There are various ways to derive it as the effective action on the world-volume of a Dp-brane: from scattering amplitudes, from the boundary state formalism, from conformal invariance of the world-sheet theory.

To determine the gauge coupling of the non-oriented type I superstring theory, which we gave without derivation in (16.146), we could compute the relevant scattering amplitudes in the orientifold theory. However, a shorter argument uses that the type I theory is a $\mathbb{Z}_2$ quotient of the type IIB theory. From the point of view of the latter, the type I $U(N)$ gauge group appears on a stack of N D9-branes, which are mapped to N image D9 branes under the orientifold projection Ω. As discussed at the end of Sect. 10.6, $U(N)$ gauge groups arise by compactifying the type I theory on a torus and by turning on appropriate Wilson lines along the non-trivial one-cycles. The $U(N)$ gauge coupling was determined in Eq. (16.166). If we now move the two stacks of D9 branes on top of each other, we get an $SO(2N)$ gauge group and the $U(N)$ generator T is embedded into $SO(2N)$ as[22]

$$\tilde{T} = \begin{pmatrix} T & 0 \\ 0 & -T^{\mathrm{T}} \end{pmatrix}. \tag{16.167}$$

Therefore,

$$\frac{1}{2g_{SO(2N)}^2}\mathrm{tr}_{2N}(\tilde{T}^2) = \frac{1}{2g_{SO(2N)}^2} 2\,\mathrm{tr}_N(T^2) = \frac{1}{2g_{U(N)}^2}\mathrm{tr}_N(T^2) \tag{16.168}$$

implies the relation $g_{SO(2N)}^2 = 2\,g_{U(N)}^2$, which is independent of N. Thus, we get

$$(g_{10}^I)^2 = g_s\,(4\pi)\,(4\pi^2\alpha')^3 \tag{16.169}$$

so that

$$\frac{\tilde{\kappa}_{10}}{\tilde{g}_{10}^I} = \frac{\sqrt{\alpha'}}{2}, \tag{16.170}$$

and Eq. (16.109) for the relation between the two physical couplings.

Formally, (16.160) is also the form of the (disk-level) action expected for the non-Abelian case but it is not clear how to define the trace over gauge group indices (Chan-Paton labels) in that case. Various approaches have been proposed but the all-order answer is unknown. One of the difficulties lies in the fact that due to the identity $[D_\alpha, D_\beta] F_{\gamma\delta} = -i[F_{\alpha\beta}, F_{\gamma\delta}]$ there is no clear distinction between derivative and non-derivative terms. This can be eliminated by a symmetric trace

---

[22] The relation between the generators associated with the two stacks of branes, i.e. T vs. $-T^{\mathrm{T}}$, is due to the fact that the gauge field is odd under Ω and that Ω reverses the orientation of the open strings. (16.167) can be expressed in the more familiar basis where $SO(2N)$ generators are real anti-symmetric matrices. If we write $iT = A + iB$, A, B real and $A^{\mathrm{T}} = -A$ and $B^{\mathrm{T}} = B$ as a consequence of $T^\dagger = T$, (16.167) corresponds to $\begin{pmatrix} A & B \\ -B & A \end{pmatrix}$.

16.5 Effective Action for D-branes

prescription. Corrections to the DBI action from higher order string diagrams have been determined from open string scattering amplitudes. They consist of even and odd powers of the field strength and its derivative and rule out the symmetric trace prescription already at $\mathcal{O}(\alpha'^3)$. The presence of higher $D^n F^m$ terms was already inferred from the structure of the four open string disc amplitude on page 622.

The second piece of the open string effective action is the CS action on the Dp-brane world-volume $\mathscr{W}$, sometimes also called Wess-Zumino action. It is essential for supersymmetry and for anomaly cancellation via the Green-Schwarz mechanism. This CS action is

$$S_{\text{CS}} = \mu_p \int_{\mathscr{W}} \text{tr}\left(e^{2\pi\alpha'\mathscr{F}}\right) \wedge \sqrt{\frac{\hat{A}(4\pi^2\alpha' R_T)}{\hat{A}(4\pi^2\alpha' R_N)}} \wedge \bigoplus_q C_q \bigg|_{p+1} \qquad (16.171)$$

with $\mu_p = \pm T_p$ for brane and anti-brane. Here we mean by C_q the pull-back of the bulk potential to the world-volume. The subscripts T, N on R stand for the curvature form of the tangent and normal bundle of $\mathscr{W} \hookrightarrow M_{10}$, respectively.[23] $\hat{A}$ is called the the A-roof genus. It can be expressed in terms of Pontryagin classes p_n

$$\hat{A}(R) = 1 - \frac{1}{24} p_1 + \frac{1}{5670}\left(7p_1^2 - 4p_2\right) + \cdots, \qquad (16.172)$$

where

$$p_0 = 1,$$
$$p_1 = -\frac{1}{2}\frac{1}{(2\pi)^2} \text{tr } R^2,$$
$$p_2 = \frac{1}{8}\frac{1}{(2\pi)^4}\left[(\text{tr } R^2)^2 - 2 \text{tr } R^4\right] \qquad (16.173)$$

and we use the notation $R^2 = R \wedge R$, etc. Finally, $2\pi\alpha'\mathscr{F} = 2\pi\alpha' F + B\mathbb{1}$, where $\mathbb{1}$ is the $n \times n$ unit matrix with n denoting the number of Dp-branes and F taking values in the n-dimensional fundamental representation of $U(n)$.

The sum over the R-R q-forms is over all the potentials that appear in either type IIA or IIB theory. The ten-form and eight-form of type IIB and the nine-form and seven-form of type IIA are meant to be included. The integral over the world-volume picks out the $(p+1)$-form of the formal sum. To write down the CS action does not require a metric, only connections. But it is independent of the choice of connections (given the gauge groups) and is thus of topological nature. The CS action measures

---

[23]The normal bundle on the submanifold $\mathscr{W}$ of a Riemannian manifold M consists of the pairs (y, v), where y ranges over the points of $\mathscr{W}$ and v is an element of the tangent space of $T_y(M)$ which is orthogonal, w.r.t. the Riemann metric on M, to the tangent space $T_y(\mathscr{W})$.

D-brane charges. Note that in non-trivial backgrounds a Dp-brane also couples to q-forms with $q < p$. This can already be seen by applying T-duality to a D-brane at an angle (see Chap. 10) and can be made very explicit by expanding (16.171):

$$S_{CS} = \mu_p \int_{\mathcal{W}} \left(n\, C_{p+1} + 2\pi\alpha' C_{p-1} \wedge \operatorname{tr} F \right.$$
$$\left. + \frac{1}{2}(2\pi\alpha')^2 C_{p-3} \wedge \left[\operatorname{tr} F \wedge F + \frac{n}{48}(\operatorname{tr} R_T \wedge R_T - \operatorname{tr} R_N \wedge R_N) \right] + \ldots \right), \quad (16.174)$$

where n is the number of Dp-branes. We have set $B = 0$ for simplicity. If, for instance, in the compactified theory the world-volume of a D5 brane extends over four-dimensional Minkowski space and wraps a two-sphere of the internal manifold which carries a $U(1)$ bundle with first Chern class m, i.e. $\frac{1}{2\pi}\int_{S^2} F = m$, this leads to a term in the effective action in the four uncompactified dimensions of the form $4\pi^2 \alpha' m\, \mu_5 \int C_4$.

As we have seen in Chap. 9, not only Dp-branes but also orientifold planes carry charge and tension and thus couple to the bulk closed string fields, including the R-R forms. The action which describes this coupling is similar to the DBI and CS actions after setting world volume fields to zero. One can show that for an $Op^{(\epsilon_1, \epsilon_2)}$-plane with the four choices of signs $\epsilon_1, \epsilon_2 \in \{-1, +1\}$ they are

$$S_{DBI}^{Op^{(\epsilon_1,\epsilon_2)}} = -\epsilon_1\, 2^{p-4}\, T_p \int_{\mathcal{W}} d^{p+1}\xi\, e^{-\Phi} \sqrt{-\det(g_{\alpha\beta})},$$

$$S_{CS}^{Op^{(\epsilon_1,\epsilon_2)}} = \epsilon_2\, 2^{p-4}\, \mu_p \int_{\mathcal{W}} \sqrt{\frac{L(\pi^2\alpha' R_T)}{L(\pi^2\alpha' R_N)}} \wedge \bigoplus_q C_q \bigg|_{p+1}, \quad (16.175)$$

where the Hirzebruch L-polynomial is defined as

$$L(R) = 1 + \frac{1}{3}p_1 + \frac{1}{45}(-p_1^2 + 7p_2) + \cdots. \quad (16.176)$$

In Eq. (16.175) the factor 2^{p-4} reflects the tension and charge of an Op-plane relative to a Dp-brane. Note that (10.212) was derived in the 'upstairs' type II geometry.

The expansion of the Chern-Simons term for orientifold planes is

$$S_{CS}^{Op} = \epsilon_2\, 2^{p-4} \mu_p \int_{\mathcal{W}} \left(C_{p+1} - \frac{1}{192}(2\pi\alpha')^2 \right.$$
$$\left. \times C_{p-3} \wedge (\operatorname{tr} R_T \wedge R_T - \operatorname{tr} R_N \wedge R_N) + \ldots \right). \quad (16.177)$$

16.5 Effective Action for D-branes

The CS actions have also been obtained in a number of different ways, e.g. abstracted from the appropriate disc amplitudes or using the boundary state formalism.

The presence of S_{CS} has an important consequence: it leads to the tadpole cancellation condition. Consider the simplified situation with action

$$\begin{aligned} S &= -\frac{1}{2} \int_M F_{p+2} \wedge \star F_{p+2} + q \int_{\mathcal{W}} C_{p+1} \\ &= -\frac{1}{2} \int_M F_{p+2} \wedge \star F_{p+2} + q \int_M C_{p+1} \wedge \delta_{\mathcal{W}} \end{aligned} \qquad (16.178)$$

with $F_{p+2} = dC_{p+1}$. $\delta_{\mathcal{W}}$ is the Poincaré dual of $\mathcal{W}$ and is sometimes called the brane current. It is a closed $(9-p)$-form with support on $\mathcal{W} \hookrightarrow M$ and integrates to one over any normal fiber over $\mathcal{W}$. For a brane in Minkowski space with world-volume extended along $(x^0, \ldots, x^p)$, $\delta_{\mathcal{W}} = \delta^{(9-p)}(y) dy^1 \wedge \cdots \wedge dy^{9-p}$ where $(y^1, \ldots, y^{9-p})$ span the space transverse to $\mathcal{W}$. The equation of motion for C_{p+1} derived from (16.178) is

$$d \star F_{p+2} = (-1)^p q \, \delta_{\mathcal{W}}. \qquad (16.179)$$

If we integrate this over the transverse space we obtain a generalization of Gauss' law, valid for extended charged objects. Let us now assume that the transverse space is compact without boundary. Integrating over it leads to the tadpole condition $q = 0$. This is a global constraint which does not have to be satisfied locally. Intuitively this result is clear: on a compact space all field lines have to end on another charged object.

In the context of brane physics the tadpole condition can be satisfied either by having as many branes as anti-branes, but this would break supersymmetry completely, or by having branes and orientifold planes as in the type I theory. But more general situations are possible where also the bulk CS-term contributes to the tadpole equation. Such a situation will be discussed in Sect. 17.3. In the presence of non-trivial bundles there can also be induced brane charges, as in the example discussed below Eq. (16.174). They generate tadpoles which, of course, also have to be cancelled.

Note that the tadpoles we have discussed here arise from the equations of motion of R-R fields. They are therefore called R-R tadpoles. Because R-R charges are quantized, the tadpole conditions are 'integer conditions' and uncancelled R-R tadpoles render the theory inconsistent. In Sect. 17.2 we will see in a particular example that they lead to gauge anomalies in the low-energy effective field theory and it can be shown that this is a general feature. This is sharp contrast to possible NS-NS tadpoles. They signify that the equations of motion are not satisfied but this can be remedied by shifting the NS-NS background. They are not related to anomalies.

16.6 Appendix: Integrals in Veneziano and Virasoro-Shapiro Amplitudes

We derive the integrals (16.74)

$$B(\alpha, \beta) = \int_0^1 dx \, x^{\alpha-1} (1-x)^{\beta-1} \tag{16.180}$$

and (16.23)

$$I(\alpha, \beta, n, m) = \int d^2z \, |z|^\alpha \, |1-z|^\beta \, z^n \, (1-z)^m. \tag{16.181}$$

Both integrals converge for certain ranges of the parameters, and they are defined elsewhere by analytic continuation.

We start with the Euler Beta-function. Using the integral representation of the Γ-function, we have

$$\Gamma(\alpha)\,\Gamma(\beta) = \int_0^\infty ds \, e^{-s} s^{\alpha-1} \int_0^\infty dt \, e^{-t} t^{\beta-1}$$

$$= 4 \int_0^\infty ds \, e^{-s^2} s^{2\alpha-1} \int_0^\infty dt \, e^{-t^2} t^{2\beta-1}$$

$$= 4 \int_0^\infty dr \, e^{-r^2} r^{2(\alpha+\beta)-1} \int_0^{\pi/2} d\varphi \, \cos^{2\alpha-1}\varphi \, \sin^{2\beta-1}\varphi$$

$$= \Gamma(\alpha+\beta) \int_0^1 dx \, x^{\alpha-1} (1-x)^{\beta-1}, \tag{16.182}$$

where in going from the second to the third line we have changed variables $(s, t) = (r\cos\varphi, r\sin\varphi)$ and in the last line $\cos^2\varphi = x$. This is, of course, the standard textbook derivation. Alternative representations which one obtains by transforming the range $x \in [0, 1]$ to $x \in [-\infty, 0]$ and $x \in [1, \infty]$ via the change of variables $x = 1 - 1/y$ and $x = 1/y$, are

$$B(\alpha, \beta) = \int_1^\infty dx \, x^{-\alpha-\beta} (x-1)^{\beta-1} = \int_{-\infty}^0 dx \, (-x)^{\alpha-1} (1-x)^{-\alpha-\beta}. \tag{16.183}$$

The second integral (16.181) is less standard. Using the integral representation of the Γ-function, we can write

$$|z|^\alpha = \frac{1}{\Gamma\left(-\frac{\alpha}{2}\right)} \int_0^\infty ds \, s^{-\frac{1}{2}\alpha-1} e^{-s|z|^2} \tag{16.184}$$

16.6 Appendix: Integrals in Veneziano and Virasoro-Shapiro Amplitudes

and likewise for $|1-z|^\beta$. Then (16.181) becomes

$$\frac{1}{\Gamma\left(-\frac{\alpha}{2}\right)\Gamma\left(-\frac{\beta}{2}\right)}\int_0^\infty ds\, s^{-\frac{1}{2}\alpha-1}\int_0^\infty dt\, t^{-\frac{1}{2}\beta-1}\int d^2z\, z^n(1-z)^m e^{-s|z|^2-t|1-z|^2}. \tag{16.185}$$

We do the integral over z first. In fact, we first evaluate

$$J(j_1, j_2) = \int d^2z\, e^{-s|z|^2-t|1-z|^2+j_1 z+j_2(1-z)} \tag{16.186}$$

and then take n derivatives w.r.t. j_1 and m derivatives w.r.t. j_2 and set $j_1 = j_2 = 0$ at the end. J is a Gaussian integral and its evaluation is straightforward. We obtain (use $z = x + iy$)

$$J(j_1, j_2) = \frac{2\pi}{(s+t)} e^{-\frac{st}{s+t}} e^{j_1 \frac{t}{s+t} + j_2 \frac{s}{s+t}} \tag{16.187}$$

and finally arrive at

$$I(\alpha, \beta, n, m) = \frac{2\pi}{\Gamma\left(-\frac{\alpha}{2}\right)\Gamma\left(-\frac{\beta}{2}\right)}\int_0^\infty ds\, s^{-\frac{1}{2}\alpha-1} \times$$

$$\int_0^\infty dt\, t^{-\frac{1}{2}\beta-1}\left(\frac{t}{s+t}\right)^n\left(\frac{s}{s+t}\right)^m\left(\frac{1}{s+t}\right) e^{-\frac{st}{s+t}}. \tag{16.188}$$

We now make the following change of variables

$$s = \frac{u}{x}, \quad t = \frac{u}{1-x}, \quad 0 \leq x \leq 1, \quad 0 \leq u \leq \infty, \tag{16.189}$$

for which the integrals factorize and we find

$$I(\alpha, \beta, n, m)$$
$$= \frac{2\pi}{\Gamma\left(-\frac{\alpha}{2}\right)\Gamma\left(-\frac{\beta}{2}\right)}\int_0^1 dx\, x^{n+\frac{1}{2}\alpha}(1-x)^{m+\frac{1}{2}\beta}\int_0^\infty du\, u^{-2-\frac{1}{2}(\alpha+\beta)} e^{-u}$$

$$= \frac{2\pi}{\Gamma\left(-\frac{\alpha}{2}\right)\Gamma\left(-\frac{\beta}{2}\right)} \frac{\Gamma\left(1+n+\frac{\alpha}{2}\right)\Gamma\left(1+m+\frac{\beta}{2}\right)\Gamma\left(-1-\frac{1}{2}(\alpha+\beta)\right)}{\Gamma\left(2+n+m+\frac{1}{2}(\alpha+\beta)\right)}$$

$$= (-)^{n+m} 2\pi \frac{\Gamma\left(1+n+\frac{\alpha}{2}\right)\Gamma\left(1+m+\frac{\beta}{2}\right)\Gamma\left(-1-n-m-\frac{1}{2}(\alpha+\beta)\right)}{\Gamma\left(-\frac{\alpha}{2}\right)\Gamma\left(-\frac{\beta}{2}\right)\Gamma\left(2+\frac{1}{2}(\alpha+\beta)\right)},$$

(16.190)

where in the last step we have assumed integer $m+n$.

There is an alternative derivation of this integral which can be generalized to higher point functions. It relates an integral over the complex plane to a pair of integrals over the real axis and demonstrates an intriguing relation between closed and open string scattering amplitudes, the KLT relations.

Further Reading

The computation of string scattering amplitudes and the extraction of the low-energy effective action has been developed in many papers. A few references which were useful in the preparation of this chapter are

- J. Scherk, Zero slope limit of the dual resonance model. Nucl. Phys. B **31**, 222 (1971)
- A. Neveu, J. Scherk, Connection between Yang-Mills fields and dual models. Nucl. Phys. B **36**, 155 (1972)
- T. Yoneya, Connection of dual models to electrodynamics and gravidynamics. Prog. Theor. Phys. **51**, 1907 (1974)
- J. Scherk, J.H. Schwarz, Dual models for non-hadrons. Nucl. Phys. B **81**, 118 (1974)
- N. Cai, C.A. Nunez, Heterotic string covariant amplitudes and low energy effective action. Nucl. Phys. B **287**, 279 (1987)
- D. Gross, J. Sloan, The quartic effective action for the heterotic string. Nucl. Phys. B **291**, 41 (1987)
- D. Lüst, S. Theisen, G. Zoupanos, Four-dimensional heterotic strings and conformal field theory. Nucl. Phys. B **296**, 800 (1988);
  J. Lauer, D. Lüst, S. Theisen, Four-dimensional supergravity from four-dimensional strings. Nucl. Phys. B **304**, 236 (1988)
- M. Garousi, R. Myers, Superstring scattering from D-branes. Nucl. Phys. B **475**, 193 (1996) [arXiv:hep-th/9603194]
- A. Hashimoto, I. Klebanov, Decay of excited D-branes. Phys. Lett. B **381**, 437 (1996) [arXiv:hep-th/9604065]

16.6 Appendix: Integrals in Veneziano and Virasoro-Shapiro Amplitudes

Gross et al., as cited in Chap. 10, compute scattering amplitudes of the heterotic string. Our appendix is taken from there. The relation between open and closed string amplitudes, mentioned in the appendix, is due to KLT and was elaborated on and extended in the second reference below:

- H. Kawai, D.C. Lewellen, S.H.H. Tye, A relation between tree amplitudes of closed and open strings. Nucl. Phys. **B269**, 1 (1986)
- S. Stieberger, Open & Closed vs. Pure Open String Disk Amplitudes. [arXiv: 0907.2211 [hep-th]]

The low-energy effective supergravity actions are extensively discussed in the book by Ortin and in the preprint collection edited by Salam and Sezgin; cf. Chap. 1.
The DBI action was derived in

- E. Fradkin, A. Tseytlin, Nonlinear electrodynamics from quantized strings. Phys. Lett. B **163**, 123 (1985)
- A. Abouelsaood, C. Callan, C. Nappi, S. Yost, Open strings in background gauge fields. Nucl. Phys. B **280**, 599 (1987)
- R. Leigh, Dirac-born-infeld action from Dirichlet sigma model. Mod. Phys. Lett. A **4**, 2767 (1989)

and the couplings of RR fields to the brane in

- B. Stefanski, Gravitational couplings of D-branes and O-planes. Nucl. Phys. B **548**, 275 (1999) [arXiv:hep-th/9812088]
- J.F. Morales, C.A. Scrucca, M. Serone, Anomalous couplings for D-branes and O-planes. Nucl. Phys. B **552**, 291 (1999) [arXiv:hep-th/9812071]

The democratic formulation of ten-dimensional supergravity was developed in

- E. Bergshoeff, R. Kallosh, T. Ortin, D. Roest, A. Van Proeyen, New formulations of D = 10 supersymmetry and D8 – O8 domain walls. Class. Quant. Grav. **18**, 3359–3382 (2001) [arXiv:hep-th/0103233]
- M. Fukuma, T. Oota, H. Tanaka, Comments on T dualities of Ramond-Ramond potentials on tori. Prog. Theor. Phys. **103**, 425–446 (2000) [arXiv:hep-th/9907132]

Chapter 17
Compactifications of the Type II Superstring with D-branes and Fluxes

Abstract Type II compactifications with D-branes and background fluxes are viable candidates to relate string theory to the physics we observe in four dimensions. For simple toroidal orbifold backgrounds the D-brane and orientifold sector can be described by an exact CFT, but issues such as tadpole cancellation, the Green-Schwarz mechanism, determining the massless spectrum etc. arise in a broader context and can be discussed from the low-energy-effective action perspective. String compactifications with non-vanishing NS-NS and R-R p-form field strengths provide solutions to the moduli problem, as these background fluxes modify the string equations of motion at leading order so that its solutions generically generate a potential for the would-be moduli fields. Thus they receive a vacuum expectation value and a mass. Basic knowledge of $\mathcal{N} = 1$ supersymmetry in four dimensions is assumed.

17.1 Brane Worlds and Fluxes

In Chap. 15 we have discussed orientifold compactification on toroidal orbifold backgrounds. In order to cancel the tadpoles introduced by the orientifold planes, one had to introduce D-branes, generically intersecting D-branes. The localization of gauge fields on D-branes provides a concrete stringy realization of the so-called brane world scenario, in which the Standard Model fields are confined on lower dimensional branes whereas gravity propagates in all ten space-time directions (the so-called bulk). It is an immediate consequence of the Dirac-Born-Infeld action that, if the branes on which the gauge degrees of freedom propagate wrap compact cycles of the internal manifold and if they also fill out the four non-compact space-time directions, then the four-dimensional gauge couplings are determined by the volume of the wrapped cycles. The gravitational coupling, on the other hand, depends on the total volume of the six dimensional compact manifold. This allows for the possibility of having the string scale M_s parametrically smaller than the Planck scale M_P.

To see this, we compactify the ten-dimensional type II superstring theory to four dimensions on a compact six dimensional manifold X with volume V_X and consider a Dp-brane filling the flat four-dimensional Minkowski space-time and wrapping a $(p-3)$-dimensional cycle C of X of volume V_C. By dimensional reduction of the Einstein-Hilbert and Dirac-Born-Infeld actions to four dimensions, one can compute the effective four-dimensional Planck mass and gauge coupling

$$M_P^2 \equiv \frac{8\pi}{\kappa_4^2} = \frac{8\pi V_X}{\kappa_{10}^2} = \frac{8M_s^2 v_X}{g_s^2},$$

$$\frac{1}{g_{YM}^2} \equiv \frac{V_C}{g_{Dp}^2} = \frac{v_C}{2\pi g_s}. \tag{17.1}$$

$v_X = V_X/\ell_s^6$ and $v_C = V_C/\ell_s^{p-3}$ are the dimensionless volumes of X and of the wrapped cycle C in string units (recall from (2.12), $\ell_s = 2\pi\sqrt{\alpha'}$). Note that the two couplings scale differently with the string coupling g_s, but the following combination is independent of g_s:

$$M_P\, g_{YM}^2 = \sqrt{32\pi^2}\, M_s\, \frac{\sqrt{v_X}}{v_C}. \tag{17.2}$$

This is in contrast to the perturbative heterotic string where gravitational and gauge degrees of freedom propagate in ten dimensions and where the gravitational and gauge couplings scale in the same way with g_s and with the volume v_X. The combination $(M_s/M_P)^2 \sim g_{YM}^2$ is therefore volume independent (cf. (16.107)). The quantities M_P and g_{YM} are experimentally accessible coupling constants. The scale of quantum gravity is $M_P \equiv G_N^{-1/2} \simeq 1.2 \times 10^{19}$ GeV and evolving g_{YM} via the renormalization group to large scales, assuming e.g. a supersymmetric completion of the standard model, one finds $M_s/M_P \simeq 10^{-2}$.

In view of this heterotic result, the volume dependence of the type II relation (17.2) offers the interesting possibility of a large hierarchy between the string and the Planck scales, $M_s \ll M_P$, if $\sqrt{v_X}/v_C \gg 1$. It is easy to achieve this: for instance in CY manifolds there are regions in (complex structure and Kähler) moduli space where some cycles become small, e.g. in the vicinity of orbifold points, of conifold points, etc. where some cycles shrink to zero size. If we want g_s and g_{YM} in the perturbative regime at the string scale and no large hierarchy between these two coupling constants, we find from the second relation in (17.1) that we need $v_C \sim 1$ and therefore $v_X \gg 1$. This is why this kind of brane world models is also called the large extra dimension scenario. In principle one can decrease the string scale down to $M_s = 1$ TeV, but one has to worry about phenomenological implications such as light Kaluza-Klein states or deviations from Newton's law. Contrarily, Coulomb's law is not affected, because the gauge fields do not propagate along the large transverse dimensions.

To summarize, the brane-world idea can be realized with type II orientifolds with D-branes. In particular the string scale is not tied to the Planck scale, where the possible hierarchy between these two scales is generated by the existence of a large internal geometry.

The second aspect we will address in this chapter has to do with the scalar fields (moduli) notoriously present in all the supersymmetric compactifications discussed so far. Since the vevs of these fields determine the precise values of important quantities (like gauge couplings, Yukawa couplings etc.) in the four-dimensional low energy effective action, for a predictive framework it is mandatory to dynamically freeze these moduli. This is problem is known as moduli stabilization.

One expects that in the process of breaking supersymmetry a scalar potential is generated. Breaking supersymmetry explicitly at the string scale, e.g. by having branes and anti-branes, is problematic. Due to the appearance of tachyons and uncanceled NS-NS tadpoles, we loose technical control. To avoid these difficulties, we start with a supersymmetric theory and look for ways to generate a scalar potential which has vacua with spontaneously broken supersymmetry.[1] For four-dimensional $\mathcal{N} = 1$ supergravity the scalar potential arises from a holomorphic superpotential W, which is an F-term, and from D-terms. Concerning the superpotential, in supersymmetric field theory there exist powerful non-renormalization theorems: it can only be induced at tree level and non-perturbatively. There are no higher loop corrections both in the string coupling g_s and the sigma model coupling $\sqrt{\alpha'}/R$.

Strictly speaking, moduli stabilization is a misnomer. What one is really looking for are compactifications with metric and other background field configurations which have no moduli. Often this is discussed as a two step process where one first chooses a Calabi-Yau manifold and then adds further background fields such as gauge bundles and a B-field (in heterotic conpactificatons) or NS-NS and R-R fluxes (in type II) which generate potentials for the metric moduli of the manifold. In doing this, one first ignores the back-reaction of the fluxes on the geometry, i.e. one ignores the energy-momentum tensor of the non-metric fields. This can be justified if they are suppressed by powers of α', though this is often not the case (see e.g. Eq. (17.4) below).

To see how a tree-level scalar potential might arise, recall that superstring theory contains p-form gauge fields in its massless spectrum. In Chap. 14 we have considered backgrounds with trivial p-form fields. We now consider solutions of the ten-dimensional string (or rather supergravity) equations of motion with non-trivial fluxes $F_{p+1} \neq 0$. As we will see, switching on fluxes has the following consequences:

- It can generate new tadpoles for some of the other p-form fields via Chern-Simons terms in the 10D action.
- The kinetic term

$$S_{\text{kin}} \sim \int_M F \wedge \star F \qquad (17.3)$$

---

[1] These vacua can be meta-stable as long as they are long living.

induces a scalar potential in the four dimensional effective action, which in general depends on the moduli of the embedded cycle $\Sigma_{p+1} \mapsto M$ with $\int_{\Sigma_{p+1}} F \neq 0$. Moreover, this scalar potential can be derived from a tree-level superpotential W_flux and can lead to moduli stabilization.

- F_{p+1} generates a source term in the Einstein equations

$$R_{\mu\nu} - \frac{1}{2} g_{\mu\nu} R = 8\pi G T_{\mu\nu}(F_{p+1}) \qquad (17.4)$$

which implies that the metric is no longer Ricci-flat.

Hence flux compactifications are a possibility to stabilize moduli at tree level. If some moduli are not stabilized at tree level, non-perturbative corrections, which are generically present, become important. Computational control over these corrections is therefore important but will not be covered in this book.[2]

17.2 Supersymmetry, Tadpoles and Massless Spectra

One can achieve $\mathcal{N} = 1$ space-time supersymmetry in four-dimensions by performing an orientifold projection of a type II Calabi-Yau compactification. These can be orientifolds of the type IIB superstring, which require the introduction of (D9,D5) or (D7,D3) branes to cancel tadpoles. Type IIB orientifolds of toroidal orbifolds were discussed in Chap. 15. The T-dual class of models are type IIA orientifolds with O6-planes and tadpole canceling D6 branes.[3] Both are space-time filling and also wrap an internal three-cycle. In this section, we will generalize this construction to orientifolds with intersecting D6-branes on smooth Calabi-Yau manifolds.

In Chap. 15, we derived the consistency conditions and the massless spectrum for a type IIB orientifold on $T^4/\mathbb{Z}_2$ where the tadpoles were canceled by stacks of intersecting fractional D7-branes. This model was analyzed via conformal field theory methods. For smooth geometries a conformal field theory description is not available. An alternative is to use effective field theory. We will see that it allows us to derive many of the relevant features of these models.

We start with a brief general discussion of type II orientifolds. The main idea is the same as for orbifolds. There one divides a string theory by a group G of target space symmetries. For orientifolds one divides by

$$G \cup \Omega_p S, \qquad (17.5)$$

---

[2] Since the r.h.s. of (17.4) is of the same order (in α' and in g_s) as the l.h.s, the flux is as much part of the background as the metric is and it cannot be viewed as a correction controlled by a small parameter.

[3] Additional magnetized D8-branes are only possible, if there exist non-trivial 5-cycles on the six-dimensional manifold.

17.2 Supersymmetry, Tadpoles and Massless Spectra

where G is again a group of target space symmetries, Ω_p the world-sheet parity operator[4] and S is such that $\Omega_p S$ is a symmetry of the string theory on $\mathcal{M}/G$. If S includes a target space symmetry σ, the fixed-point sets of σ are the orientifold planes. The twisted sectors which have to be added include D-branes and open strings. Simple examples with $M_{10} = \mathbb{R}^{1,3} \times T^6$ are those listed in Table 10.3 where G is trivial and S can be read off from the second column.

Calabi-Yau orientifolds are obtained, if we choose type II theory on a CY manifold X (which has no continuous isometries), i.e. $M_{10} = \mathbb{R}^{1,3} \times X$. For simplicity we choose $G = \mathbb{1}$ and $\sigma : M_{10} \to M_{10}$ to be an involution ($\sigma^2 = 1$) which acts non-trivially on the CY manifold but trivially on $\mathbb{R}^{1,3}$. This, of course, restricts the allowed class of CY manifolds for this construction. To proceed, we have to distinguish between type IIA and IIB orientifolds.

Type IIB Orientifolds

Choose σ to be a discrete holomorphic isometry of X. It leaves the metric and the complex structure of M invariant and hence

$$\sigma^*(\omega) = \omega, \qquad \sigma^*(\Omega) = \pm\Omega, \tag{17.6}$$

where σ^* is the pullback of σ and the action on Ω is consistent with σ being holomorphic and involutive. In local coordinates, where $\Omega = dz^1 \wedge dz^2 \wedge dz^3$ and $\omega = \frac{i}{2} \sum_{i=1}^3 dz^i \wedge d\bar{z}^i$, σ acts by inverting an even (+) and odd (−) number of the z^i, respectively. It is now easy to determine the dimensionality of the fixed point sets and we find the following cases (cf. Table 10.3)

$$\begin{aligned} \text{O5/O9}: &\quad \sigma^*(\Omega) = +\Omega, \quad S = \sigma, \\ \text{O3/O7}: &\quad \sigma^*(\Omega) = -\Omega, \quad S = \sigma(-1)^{F_L}. \end{aligned} \tag{17.7}$$

Since σ acts trivially on $\mathbb{R}^{1,3}$, the O-planes are space-time filling. F_L, the left-moving space-time fermion number, is included in order to guarantee that $S\Omega_p$ acts as an involution on all states, i.e. that $(S\Omega_p)^2 = 1$. This is necessary in order to have invariant space-time fermions, in particular for one of the gravitini to survive the projection.

We can now check explicitly that there is always one linear combination of the two gravitini which is invariant under the action of the orientifold group. In light cone gauge the most general linear combination is

$$|a,\eta\rangle \otimes \psi^i_{-\frac{1}{2}}|0\rangle + \alpha \, \overline{\psi}^i_{-\frac{1}{2}}|0\rangle \otimes |a,\eta\rangle, \tag{17.8}$$

---

[4] In this chapter we use Ω_p for the world-sheet parity in order to avoid confusion with the three-form Ω.

where a is an $SO(2)$ spinor index, η the Killing spinor (of definite chirality) and $i = 1, 2$ enumerates the two transverse space-time directions. Recall that the transformation $z^j \to -z^j$ is a π-rotation in the z^j plane, which acts as multiplication by i on the spinors. It is then straightforward to show that in all cases considered in (17.7) there is one choice for α such that the linear combination (17.8) is invariant under S. One also verifies that without the $(-1)^{F_L}$ insertion there would not be an invariant combination.

Type IIA Orientifolds

Consider now the compactification of type IIA superstring theory on $\mathbb{R}^{1,3} \times X$. We now choose σ to be an isometric anti-holomorphic involution of the Calabi-Yau manifold, i.e.

$$\sigma^*(\omega) = -\omega, \qquad \sigma^*(\Omega) = e^{2i\theta_{O6}} \overline{\Omega}. \qquad (17.9)$$

θ_{O6} is a constant phase. The fixed point locus of this involution supports an orientifold O6-plane. By redefining the phase of Ω we can set $\theta_{O6} = 0$. In this case the action of σ in local coordinates is $\sigma : z^i \to \bar{z}^i$ and we have

$$\text{O6}: \qquad \sigma^*(\Omega) = \overline{\Omega}, \qquad S = \sigma(-1)^{F_L}. \qquad (17.10)$$

It is again straightforward to demonstrate that this preserves $\mathcal{N} = 1$ supersymmetry, i.e. that there is a linear combination

$$|a, \eta\rangle \otimes \psi^i_{-\frac{1}{2}}|0\rangle + \alpha \overline{\psi}^{\bar{i}}_{-\frac{1}{2}}|0\rangle \otimes |a, \bar{\eta}\rangle \qquad (17.11)$$

which is invariant under $\Omega_p S$. The action of σ on the Killing spinor is[5] $\sigma|\eta\rangle = |\bar{\eta}\rangle$ and $\sigma|\bar{\eta}\rangle = -|\eta\rangle$.

On Calabi-Yau threefolds, (17.7) and (17.10) exhaust all possibilities: since there are no homology one- and five-cycles, we cannot have space-time filling O4 or O8-planes.

Type IIA CY Orientifolds

We now turn to a more detailed discussion of type IIA Calabi-Yau orientifolds of the type just discussed. Unless stated otherwise, in this section O6-planes always mean $O6^{(-,-)}$-planes. They carry negative charge with respect to D6-branes and

---

[5] σ is a reflection in the (5-7-9) directions and it acts on positive and negative chirality spinors as $\pm i \gamma^5 \gamma^7 \gamma^9$. If we express the Dirac-matrices in terms of creating and annihilation operators and use that $|\eta\rangle$ and $|\bar{\eta}\rangle$ are highest and lowest weight states, respectively, the stated results follow.

17.2 Supersymmetry, Tadpoles and Massless Spectra

have negative tension. We will denote the homology class of the O6-plane by Π_{O6} and its dual Poincaré dual 3-form by π_{O6}.

Tadpole Cancellation

The orientifold plane induces a tadpole for the R-R 7-form potential C_7 with four legs on the flat uncompactified part $\mathbb{R}^{1,3}$. For the determination of this tadpole it is not necessary to compute the whole Klein-bottle amplitude of this orientifold projection. Since we are only interested in the couplings of massless states, the tadpole can already be derived from the CS-terms (16.175) on the O6-plane

$$S^{O6}_{CS} = -4\mu_6 \int_{M_{10}} C_7 \wedge \pi_{O6}. \tag{17.12}$$

In order to cancel this tadpole one adds stacks of N_a D6-branes wrapping three-cycles Π_a in the internal manifold X and with flat gauge connections on their world volume. For σ to be a symmetry, one has to introduce "mirror branes" which wrap the three-cycles Π'_a which are images under the action induced by σ on the homological classes. The CS action (16.171) on a stack of D6-branes simplifies to

$$S^{D6_a}_{CS} = \mu_6 \int_{M_{10}} C_7 \wedge \pi_a \tag{17.13}$$

with a similar contribution from the mirror cycles π'_a. Taking also the kinetic term for C_7 into account, the part of the ten-dimensional action which involves C_7 is

$$S = -\frac{1}{4\tilde{\kappa}^2_{10}} \int_{M_{10}} F_8 \wedge \star F_8 + \mu_6 \int_{M_{10}} C_7 \wedge \left(\sum_a N_a(\pi_a + \pi'_a) - 4\pi_{O6} \right). \tag{17.14}$$

The equation of motion for C_7 is

$$d \star F_8 = -\ell_s \left(\sum_a N_a (\pi_a + \pi'_a) - 4\pi_{O6} \right). \tag{17.15}$$

Since the left hand side in Eq. (17.15) is exact, the R-R tadpole cancellation condition boils down to a condition on the cohomology classes. Written in terms of the Poincaré dual homology three-cycles this constraint is

$$\sum_a N_a (\Pi_a + \Pi'_a) - 4\Pi_{O6} = 0. \tag{17.16}$$

It implies that the overall three-cycle which all the D6-branes and orientifold planes wrap is trivial in homology. This is a moderate restriction which admits non-trivial solutions with branes that are not simply placed on top of the orientifold plane.

We have seen in Chap. 9 that besides D-branes there also exist stable non-BPS branes. They are not simply classified by cohomology classes but by torsion-like K-theory classes. Thus the topological classification of D-branes via cohomology actually has to be refined by using K-theory instead. This means that besides the usual R-R tadpole cancellation conditions (which are conditions on the homology of the cycles wrapped by the branes) additional constraints arise due to torsion factors in the K-groups of the Chan-Paton bundles of the D-branes. For a compact Calabi-Yau it is in general quite difficult to explicitly compute the K-theory groups.

Supersymmetric Cycles

In Chap. 10 we have discussed the conditions under which D6-branes covering $\mathbb{R}^{1,3}$ and intersecting on a torus T^6 preserve supersymmetry; see Table 10.2. We now generalize this to intersecting D6-branes wrapping three-cycles of smooth Calabi-Yau manifolds. As explained earlier, type IIA superstring theory compactified on a Calabi-Yau threefold preserves $\mathcal{N} = 2$ supersymmetry in four dimensions, which is further broken by the orientifold projection to $\mathcal{N} = 1$ supersymmetry. The orientifold plane wrapping the fixed point loci Σ_{O6} of σ generically breaks supersymmetry completely, unless Σ_{O6} satisfies certain conditions, which we now state without derivation. For this purpose we define: a three-dimensional submanifold $\Sigma \subset M$ is called Lagrangian, if the restriction of the Kähler form of X to Σ_3 vanishes[6]

$$\omega|_{\Sigma_3} = 0. \tag{17.17}$$

A special Lagrangian submanifold is defined to satisfy the additional condition

$$\mathrm{Im}(e^{-i\theta}\Omega)|_{\Sigma_3} = 0, \tag{17.18}$$

where $\theta \in \mathbb{R}$. Writing the second equation in (17.9) as $\sigma^*(e^{-i\theta_{O6}}\Omega) = \overline{(e^{-i\theta_{O6}}\Omega)}$ it is clear that ω and $\mathrm{Im}(e^{-i\theta_{O6}}\Omega)$ restrict trivially to the O6-plane which therefore wraps a special Lagrangian submanifold of the Calabi-Yau threefold X. The parameter θ, which parametrizes a $U(1)$, determines which $\mathcal{N} = 1 \subset \mathcal{N} = 2$ supersymmetry is preserved by the O6-plane. An alternative definition of a special Lagrangian submanifold of a CY threefold is that it is calibrated w.r.t. $\mathrm{Re}(e^{-i\theta}\Omega)$. This means for the induced volume form on Σ_3,

$$\mathrm{vol}|_{\Sigma_3} = \mathrm{Re}(e^{-i\theta}\Omega)|_{\Sigma_3} \tag{17.19}$$

and implies that Σ_3 is volume minimizing in its homology class. These two definitions are equivalent if Ω is normalized such that $\omega \wedge \omega \wedge \omega = \frac{3i}{4}\Omega \wedge \overline{\Omega}$.[7]

---

[6]This is a notion from symplectic geometry where ω is the symplectic form and X the phase space, i.e. the cotangent bundle of the configuration space. The latter is a Lagrangian submanifold (by Darboux' theorem we can choose local coordinates such that $\omega = \sum dp_i \wedge dq^i$).

[7]The notion of calibrated submanifolds is more general, in particular it is defined for submanifolds of arbitrary dimensions. The supersymmetry preserving cycles of type IIB orientifolds are even cycles. They are calibrated w.r.t. $\frac{1}{p!}\omega^p$, for $p = 0, 1, 2, 3$ for O3,O5,O7 and O9 planes, respectively. This means e.g. that $\mathrm{vol}|_{\Sigma_2} = \omega|_{\Sigma_2}$. The calibrated cycles are precisely the

17.2 Supersymmetry, Tadpoles and Massless Spectra

Table 17.1 Spectrum of IIA Calabi-Yau orientifolds

Multiplicity	Multiplet
1	Gravity multiplet
$h_+^{1,1}$	Vector multiplets
$h_-^{1,1} + h^{2,1} + 1$	Chiral multiplets

For the D6-branes to also preserve supersymmetry, they should as well wrap internal special Lagrangian 3-cycles. Since 3-cycles with different values for θ_a preserve different $\mathcal{N} = 1$ supersymmetries, in order to preserve an overall $\mathcal{N} = 1$ supersymmetry, all stacks of D6-branes have to wrap special Lagrangian three-cycles with the angle defined by the orientifold plane, i.e. $\theta_a = \theta_{O6} = 0$.[8]

Massless Spectrum

In the closed string sector we have the orientifold projected massless $\mathcal{N} = 2$ supersymmetric type IIA spectrum. Recall that for type IIA the dimensional reduction on a Calabi-Yau manifold leads to $h^{1,1}$ vector multiplets, whose vectors come from the reduced R-R three-form C_3 and whose scalars come from the NS-NS two-form B_2 and the Kähler moduli. Furthermore, there are $h^{2,1} + 1$ hypermultiplets which include the internal components of C_3 and the complex structure moduli (the one extra hypermultiplet is the universal one that includes the dilaton). The operation σ acts on the cohomology groups as

$$\sigma : H^{p,q} \to H^{q,p}. \qquad (17.20)$$

This implies that $(1, 1)$-forms are mapped onto themselves, but $(3, 0)$- and $(2, 1)$-forms are swapped with their complex conjugates. Thus, $H^{1,1}$ can be split into subspaces of eigenvalues ± 1 with dimensions $h_\pm^{1,1}$. From the four bosonic on-shell degrees of freedom of an $\mathcal{N} = 2$ vectormultiplet, two survive the projection. For $H_+^{1,1}$ the vector survives whereas for $H_-^{1,1}$ the complex scalar survives. One can form linear combinations of the scalars in the hypermultiplets such that half of them are even.[9] All scalars are in $\mathcal{N} = 1$ chiral multiplets while the vectors are in $\mathcal{N} = 1$ vectormultiplets. The superfield content of the $\mathcal{N} = 1$ supergravity theory of the massless closed string modes of a type IIA Calabi-Yau orientifold with orientifold group generated by $\Omega_p \sigma (-1)^{F_L}$ is summarized in Table 17.1.

---

complex submanifolds. In the presence of other background fields, e.g. a background B field, the supersymmetry conditions are modified.

[8] On T^6, choosing $\Omega = dz_1 \wedge dz_2 \wedge dz_3$, one can easily show that the special Lagrangian condition (17.18) reduces to the angle condition $\sum_I \Phi^I = 0 \mod 2\pi$, which we have derived via CFT in Chap. 10.

[9] Half of the scalars are from the R-R sector and the other half from the NS-NS sector. The invariant combinations do not mix the two sectors.

Table 17.2 Chiral spectrum for intersecting D6-branes

Representation	Multiplicity
$\Box\!\Box_a$ (antisymmetric)	$\frac{1}{2}\left(\Pi'_a \circ \Pi_a + \Pi_{O6} \circ \Pi_a\right)$
$\Box\!\Box_a$ (symmetric)	$\frac{1}{2}\left(\Pi'_a \circ \Pi_a - \Pi_{O6} \circ \Pi_a\right)$
$(\overline{\Box}_a, \Box_b)$	$\Pi_a \circ \Pi_b$
$(\Box_a, \Box_b)$	$\Pi'_a \circ \Pi_b$

We now turn to the open string spectrum where we will find various non-Abelian gauge fields in addition to chiral charged matter. If a D6-brane wraps a submanifold $\Sigma_a \in \Pi_a$ that is invariant under σ, the gauge symmetry is $SO(2N_a)$ or $Sp(2N_a)$.[10] In general, the cycles are mapped non-trivially, $\Sigma'_a \neq \Sigma_a$, and the gauge symmetry is $U(N_a)$. The chiral massless spectrum is given by the topological intersection numbers. For a gauge group $G = \prod_a U(N_a)$, it is summarized in Table 17.2. Here, a enumerates all pairs of cycles which are mapped to each other (in homology) by σ and N_a is the number of branes wrapped around the a-th cycle.

The common situation is an open string connecting two branes which wrap two cycles which are not images (in homology) under σ. This open string is identified under $\Omega_p \sigma (-1)^{F_L}$ with the open string between the two images branes. Thus the ground state and all excitations of this open string transform in bi-fundamental representations of the two factors in the gauge group corresponding to the wrapped cycles. Only open strings stretched between a D-brane and its image under σ are invariant under the combined operation $\Omega \sigma (-1)^{F_L}$. Therefore, they transform in the antisymmetric or symmetric representation of the gauge group. More concretely, a CFT analysis of toroidal orientifolds shows that for $O6^{(-,-)}$-planes (i.e. with negative tension and R-R charge) the chiral (ground) states localized at intersection points invariant under $\Omega \sigma (-1)^{F_L}$ transform in the antisymmetric representation, while those at intersections points which are anti-invariant give rise to chiral states in the symmetric representation. This is the situation in Table 17.2.

Additional non-chiral matter transforming in the adjoint representation of $U(N_a)$ arises from open strings stretched between branes in the same stack, i.e. branes lying on top of each other. Geometrically, these correspond to deformations of the three-cycle inside the Calabi-Yau or to continuous Wilson lines along non-trivial one-cycles inside the three-cycles. They are counted by $H^0(\Sigma, N_\Sigma)$ and $H^1(\Sigma)$ respectively, where N_Σ denotes the normal bundle of Σ inside X. One can show

---

[10] Note that it is not sufficient that the cycle Π_a is mapped to itself in homology under the induced action of σ. The submanifold representative of the cycle has to be invariant. The gauge group can be determined from the Möbius strip amplitude. For tori one finds that on the world-volume of Dp-branes on top of an Op-plane, the gauge group is SO while for D(p−4) branes on an Op-plane it is Sp.

17.2 Supersymmetry, Tadpoles and Massless Spectra

that for special Lagrangian cycles the number of these two kinds of moduli are equal. They combine into complex scalars in the adjoint whose multiplicity is given by the first Betti number $b_1(\Pi_a)$ of the three-cycle.

For the spectrum of charged matter fields in Table 17.2 the R-R tadpole cancellation condition (17.16) guarantees the absence of non-Abelian gauge anomalies. Indeed, for a factor $SU(N_a)$ the $SU(N_a)^3$ triangle anomaly is proportional to[11]

$$\sum_{b \neq a} N_b \Big[-\Pi_a \circ \Pi_b + \Pi_a' \circ \Pi_b \Big]$$

$$+ \frac{(N_a - 4)}{2} \Big[\Pi_a' \circ \Pi_a + \Pi_{O6} \circ \Pi_a \Big] + \frac{(N_a + 4)}{2} \Big[\Pi_a' \circ \Pi_a - \Pi_{O6} \circ \Pi_a \Big]$$

$$= -\Pi_a \circ \Big(\sum_b N_b \big[\Pi_b + \Pi_b' \big] - 4 \Pi_{O6} \Big) = 0. \qquad (17.21)$$

Here we have used the tadpole condition (17.16). Note also the relations

$$\Pi_a \circ \Pi_a = 0, \quad \Pi_a' \circ \Pi_b = \Pi_b' \circ \Pi_a = -\Pi_a \circ \Pi_b', \quad \Pi_a \circ \Pi_b = -\Pi_a' \circ \Pi_b'. \quad (17.22)$$

There can also be non-chiral massless matter fields whose spectrum cannot be determined from topology. To find those one has to compute the number of points in which two submanifolds intersect geometrically (not just the intersection number in topology which counts intersections with orientation).

Green-Schwarz Mechanism

Given the chiral spectrum of Table 17.2, we have shown that the non-Abelian gauge anomalies of all $SU(N_a)$ factors in the gauge group cancel. However, as we will show momentarily, the Abelian and the mixed anomalies with Abelian and non-Abelian gauge fields and with Abelian gauge fields and gravitons do not cancel among the charged fields of Table 17.2. This is a very generic situation in string theory and is resolved by the so-called Green-Schwarz (GS) mechanism. This mechanism is already effective in the uncompactified theory and is at the heart of the proof of the anomaly freedom of the ten-dimensional string theories. In the ten-dimensional theory the GS mechanism relies on the fact that the anti-symmetric tensor field also transforms under gauge transformations of the gauge fields, as discussed in Chap. 16.4. In the compactified four-dimensional string models it relies on the generic appearance of axions, whose perturbative shift symmetry is gauged

---

[11] Here we use that the anomaly coefficients $\operatorname{tr}_R(\{T^a, T^b\}T^c) = A(R)\operatorname{tr}_\square(\{T^a, T^b\}T^c)$ for the relevant $SU(N)$ representations are $A_{\text{adj}} = (N-4)A_\square$, $A_{\square\square} = (n+4)A_\square$, $A_{\overline{\square}} = -A_\square$. This can be derived e.g. by looking at the decomposition w.r.t. $SU(3)$.

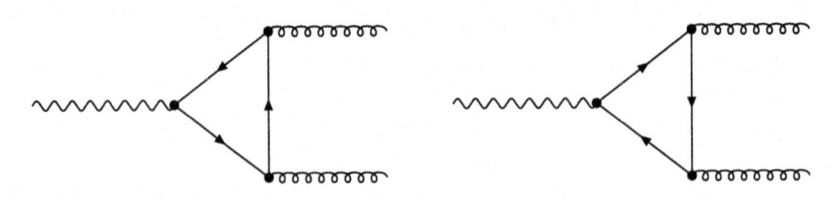

Fig. 17.1 One-loop diagrams for $U(1)_a\, SU(N_b)^2$ or for gravitational anomalies. The external *lines* on the *left* are photons; the *lines* on the *right* they are either two gluons or two gravitons

and leads to non-invariances of the action which cancel Abelian anomalies which appear as non-invariances of the measure in the functional integral.

We will convey the idea for the class of intersecting D6-brane models. A more thorough treatment would, among others, require an exposition of the formalism of anomalies in quantum field and string theories, but this is one of the topics which we have mentioned in the preface as one of the omissions.

The $U(1)_a\, SU(N_b)^2$ anomalies result from the second triangle diagram shown in Fig. 17.1. The triangle anomaly for any two stacks a and b is proportional to[12]

$$A_{abb} = N_a(-\Pi_a + \Pi'_a) \circ \Pi_b. \tag{17.23}$$

For $a \neq b$ this is immediate as the bi-fundamental fields are the only fields which run in the loop. For $a = b$ all states which couple to $U(1)_a$ can run in the loop and we have to sum over them. This gives[13]

$$A_{aaa} = (N_a - 2)\,(\Pi'_a \circ \Pi_a + \Pi_{O6} \circ \Pi_a) + (N_a + 2)\,(\Pi'_a \circ \Pi_a - \Pi_{O6} \circ \Pi_a)$$

$$+ \sum_{b \neq a} N_b\,(-\Pi_a \circ \Pi_b + \Pi'_a \circ \Pi_b)$$

$$= N_a \Pi'_a \circ \Pi_a. \tag{17.24}$$

Here we have normalized the $U(1)$ charge of the fundamental representation to $+1$. The anti-fundamental then has charge -1 and the symmetric and anti-symmetric representations have charge $+2$.

Analogously, one can compute the $U(1)_a\, U(1)_b^2$ anomalies. One finds

$$B_{abb} = \begin{cases} N_a\, N_b\,(-\Pi_a + \Pi'_a) \circ \Pi_b & a \neq b \\ \dfrac{1}{3} \cdot 3\, N_a^2\, \Pi'_a \circ \Pi_a & a = b. \end{cases} \tag{17.25}$$

---

[12] Here we define $\operatorname{tr}_R(Q^a T^b T^b) = A_{abb} \operatorname{tr}_\square(T^b T^b)$; there is no sum over b. We also define $B_{abb} = \operatorname{tr}(Q^a Q^b Q^b)$ and $A_{agg} = \operatorname{tr}(Q^a)$, where Q^a is the $U(1)_a$ charge.

[13] If $\operatorname{tr}_R(T^a T^b) = c(R)\delta^{ab}$, $c(\square\!\square) = (N+2)c(\square)$, $c(\square\!\!\!\square) = (N-2)c(\square)$, $c(\overline{\square}) = c(\square)$.

17.2 Supersymmetry, Tadpoles and Massless Spectra

For the mixed gauge-gravitational anomaly A_{agg} one obtains

$$A_{agg} = \frac{1}{2}N_a(N_a-1)(\Pi'_a \circ \Pi_a + \Pi_{O6} \circ \Pi_a) + \frac{1}{2}N_a(N_a+1)(\Pi'_a \circ \Pi_a - \Pi_{O6} \circ \Pi_a)$$

$$+ \sum_{b \neq a} N_a N_b (\Pi'_a \circ \Pi_b - \Pi_a \circ \Pi_b)$$

$$= 3 \, N_a \, \Pi_{O6} \circ \Pi_a. \tag{17.26}$$

To summarize, we find that anomalies which involve only non-Abelian gauge currents cancel as a consequence of the tadpole condition, but the triangle diagrams which involve one or three $U(1)$ currents, are anomalous.[14] In other words, the massless spectrum of the theory is not anomaly free and would, by itself, lead to a break-down of gauge invariance of the (one-loop) effective action. But there is another source of gauge variation of the (tree-level) effective action: it is related to the transformation of p-form fields (axions) under Abelian gauge transformations of the gauge field. We have encountered such transformations for the two-form in type I and heterotic theories (cf. (16.153)), as a consequence of the non-trivial Bianchi identity for their three-form field-strength.

The essence of the Green-Schwarz mechanism is that there are terms in the tree-level effective action, which involve the axions and which are not invariant under gauge transformations such as to cancel the one-loop anomalies. In the case of intersecting D6-branes, the axions are the dimensionally reduced C_3 and C_5-forms, which are Hodge-duals of each other, i.e. $dC_5 = \star dC_3$.

The three-cycles Π_a which the D6-branes wrap can be expanded in the symplectic homology basis introduced in (14.217)[15]

$$\Pi_a = m_a^I \, A_I + n_{a,I} \, B^I \tag{17.27}$$

with $m^I, n_I \in \mathbb{Z}$. Since we are working in the so-called 'upstairs geometry', i.e. prior to the orientifold quotient, we also have the orientifold mirror branes which wrap the mirror cycles $\Pi'_a = m'^I_a \, A_I + n'_{a,I} \, B^I$.[16] The topological intersection number between two three-cycles Π_a, Π_b is

$$\Pi_a \circ \Pi_b = m_a^I \, n_{b,I} - n_{a,I} \, m_b^I. \tag{17.28}$$

---

[14] We can regularize the triangle diagrams in such a way that $A^{(1)}_{abb}$ and $B^{(1)}_{abb}$ are proportional to the divergences of the $U(1)_a$ current while the other two currents in the respective triangle diagrams are conserved. Bose symmetry in the three currents accounts for the factor $\frac{1}{3}$ in $B^{(1)}_{aaa}$ (second line in (17.25)).

[15] Compared to Chap. 14 we have slightly changed the notation from A^a, B_a to A_I, B^I.

[16] The wrapping numbers m'_a, n'_a of the mirror branes are, of course, not independent of those of the branes. The precise relation depends on the action of σ on the homology basis, but it will not be needed.

On shell, C_3 and C_5 are harmonic forms and we can expand them in the Poincaré dual symplectic cohomology basis

$$C_3 = \ell_s^3 \left(C_{0,I}\, \alpha^I - D_0^I\, \beta_I\right), \qquad C_5 = \ell_s^5 \left(C_{2,I}\, \alpha^I - D_2^I\, \beta_I\right). \tag{17.29}$$

The coefficients are axionic space-time fields. From $dC_5 = \star dC_3$ one derives

$$\ell_s^2(-dD_2, dC_2) = (\star dC_0, \star dD_0)\, M \tag{17.30}$$

with

$$M = \begin{pmatrix} B & A \\ -D & -C \end{pmatrix}, \tag{17.31}$$

where A, B, C, D are the matrices defined in (14.300): they appear when we expand $(\star\alpha^I, \star\beta_J)$ in the basis (α^I, β_J). Notice that $M \in Sp(2(h^{2,1}+1))$. In (17.30), and below, we have employed an obvious matrix notation.

It is now straightforward to reduce the low-energy effective action to four dimensions. For simplicity we set the dilaton, the (R,R) 1-form and the (NS,NS) 2-form to zero. The ten-dimensional kinetic term for C_3 becomes

$$S_{\text{kin}} = -\frac{\pi}{\ell_s^2} \int_{\mathbb{R}^{1,3}} (dC_0, dD_0)\, M \wedge \star \begin{pmatrix} dC_0 \\ dD_0 \end{pmatrix}. \tag{17.32}$$

We now expand the Chern-Simons action and integrate over the cycles wrapped by the branes and the O6-plane. From the second term in (16.174) we obtain the following mixing terms between the four-dimensional axions and the $U(1)$ gauge fields $f^a = dA^a$

$$S_{\text{mix}} = \frac{\pi}{\ell_s^2} \sum_a N_a \int_{\mathbb{R}^{1,3}} \frac{1}{\pi} A^a \wedge \star(dC_0, dD_0) M \begin{pmatrix} -(n_a - n'_a) \\ (m_a - m'_a) \end{pmatrix}, \tag{17.33}$$

where we have integrated by parts and used (17.30). The fact that under world-sheet parity Ω_p the $U(1)_a$ gauge field changes sign explains the relative sign between the contributions from branes and image branes.

The second line in (16.174) reduces to

$$S_{\text{ax}}^{D6} = \frac{1}{4\pi} \sum_a \int_{\mathbb{R}^{1,3}} (C_0, D_0) \begin{pmatrix} m_a + m'_a \\ n_a + n'_a \end{pmatrix} \left(N_a f^a \wedge f^a\right.$$

$$\left. + \operatorname{tr}_{N_a} F^a \wedge F^a + \frac{1}{48} N_a \operatorname{tr} R \wedge R\right). \tag{17.34}$$

17.2 Supersymmetry, Tadpoles and Massless Spectra

F_a denotes the $SU(N_a)$ gauge field strength and f_a again the field strength of the diagonal $U(1)_a \subset U(N_a)$. Finally, the contribution from the orientifold plane is, cf. (16.177),

$$S_{ax}^{O6} = \frac{1}{48 \cdot 2\pi} \int_{\mathbb{R}^{1,3}} (C_0, D_0) \begin{pmatrix} m_{O6} \\ n_{O6} \end{pmatrix} \operatorname{tr} R \wedge R. \tag{17.35}$$

It will turn out to be convenient for the following discussion to define the $k \times (h^{1,2} + 1)$ matrices

$$Q_I^a = -N_a(n_{a,I} - n'_{a,I}), \qquad P^{a,I} = N_a(m_a^I - m'^I_a), \tag{17.36}$$

where k is the range of a, i.e. the number of wrapped branes or, equivalently, the number of pairs of homology cycles (Π_a, Π'_a). We furthermore define one-form field strengths (here and below we will not display the index I)

$$G_1 = dC_0 - \sum_a \frac{Q^a}{2\pi} A_a, \qquad H_1 = dD_0 - \sum_a \frac{P^a}{2\pi} A_a. \tag{17.37}$$

We observe that they obey the Bianchi-identities

$$dG_1 = -\sum_a \frac{Q^a}{2\pi} f^a \quad \text{and} \quad dH_1 = -\sum_a \frac{P^a}{2\pi} f^a. \tag{17.38}$$

This is reminiscent of the definitions and the Bianchi-identities of H_3 and F_3 in the ten-dimensional heterotic and type I superstring theory, respectively. There, gauge invariance of H_3 and F_3 required that B_2 and C_2 transform under gauge transformations. Likewise, gauge invariance of G_1 and H_1 can be achieved by requiring that under $A_a \to A_a + d\lambda_a$, C_0 and D_0 transform as

$$C_0 \to C_0 + \frac{Q^a}{2\pi} \lambda_a, \qquad D_0 \to D_0 + \frac{P^a}{2\pi} \lambda_a, \tag{17.39}$$

i.e. the axionic shift symmetries are gauged. Note that each zero mode $\xi^I Q_I^a + \eta_I P^{a,I} = 0$ gives rise to a gauge invariant axion $\xi^I C_{0,I} + \eta_I D_0^I$.

In terms of the gauge invariant field strengths G_1, H_1 we can rewrite the sum of (17.32) and (17.33) as

$$S_{\text{kin}} + S_{\text{mix}} = -\frac{\pi}{\ell_s^2} \int_{\mathbb{R}^{1,3}} (G_1, H_1) \, M \wedge \star \begin{pmatrix} G_1 \\ H_1 \end{pmatrix}. \tag{17.40}$$

This action includes a mass-term for the $U(1)$ gauge bosons with mass-matrix proportional to

$$A^a(Q^a, P^a) \, M \begin{pmatrix} Q^b \\ P^b \end{pmatrix} \wedge \star A^b \tag{17.41}$$

Fig. 17.2 Tree-level anomalous contribution of RR axions (*dashed line*)

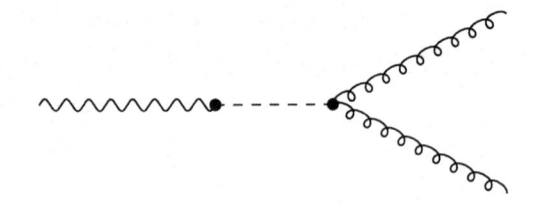

which was not present in the tree-level action which we started with. One can show that the mass terms are generated at one-loop: they arise from annulus diagrams with one gauge-boson vertex operator insertion at each boundary.

While the action is invariant (up to topological terms) under constant shifts of the axionic fields, this is no longer the case after the gauging. The Chern-Simons terms (17.34) and (17.35) lead to a tree-level non-invariance of the action which precisely cancels the non-invariance of the effective action which is due to the anomaly. This is the generalized Green-Schwarz mechanism. Indeed, we easily find

$$\delta_\lambda(S^{D6}_{ax} + S^{O6}_{ax})\Big|_{tree}$$
$$= \frac{1}{4\pi} \int_{\mathbb{R}^{1,3}} \Bigg\{ \sum_{a,b} \lambda_a \left(\frac{Q^a}{2\pi}, \frac{P^a}{2\pi}\right) \binom{m_b + m'_b}{n_b + n'_b} \left(N_b f^b \wedge f^b + \text{tr}_{N_b} F^b \wedge F^b\right.$$
$$\left. + \frac{1}{48} N_b \text{tr } R \wedge R \right) + \frac{1}{24} \sum_a \lambda_a \left(\frac{Q^a}{2\pi}, \frac{P^a}{2\pi}\right) \binom{m_{O6}}{n_{O6}} \text{tr } R \wedge R \Bigg\}$$
$$= \frac{1}{(2\pi)^2} \int_{\mathbb{R}^{1,3}} \sum_a \lambda_a \Bigg\{ N_a (\Pi_a - \Pi'_a) \circ \Pi_b \text{ tr}_{N_b} F^b \wedge F^b$$
$$+ N_b N_a (\Pi_a - \Pi'_a) \circ \Pi_b \, f^b \wedge f^b + \frac{1}{24} 3 N_a \Pi_a \circ \Pi_{O6} \text{tr } R \wedge R \Bigg\}. \quad (17.42)$$

We see that this is just the right expression to cancel all one-loop anomalies which are not already canceled as a consequence of the tadpole condition. Diagrammatically we can express the GS anomaly cancellation through the tree-level diagrams of Fig. 17.2.

In the process of canceling the anomalies, a $U(1)_a$ gauge transformation acts as a shift on the axions. Thus, using the gauge symmetry one can, for instance, gauge away one axion. Having done so, there is no gauge freedom left to gauge away one of the three polarizations of the $U(1)_a$ gauge field. Therefore, the gauge field is massive which is consistent with the generation of the mass term (17.41). Another way of saying this is that the longitudinal polarization mode of the massive anomalous Abelian gauge field is provided by the derivative of an axion. This mechanism of giving mass to a $U(1)$ gauge boson is called Stückelberg mechanism. The $U(1)$ gauge potentials and axions which remain massless have to lie in the left and right kernel, respectively, of the $k \times (2h^{2,1} + 2)$ matrix (Q, P). The $U(1)$ symmetries, which correspond to the massive gauge potentials, survive as perturbative global symmetries in that all perturbatively computed correlation functions respect them. In supersymmetry preserving configurations the massive photons and axions are, of course, members of massive vector and chiral multiplets, respectively.

17.3 Flux Compactifications

In previous chapters we have discussed the consistency conditions on possible string backgrounds with a non-constant internal metric field. The restrictions came first from conformal invariance on the string world-sheet, and second from the requirement of preserving space-time supersymmetry. The string compactifications we obtained in this way generically have many massless scalars fields, whose vevs are free parameters.

We will now generalize the compactifications in the sense that we will also allow for non-vanishing background fluxes. Since non-trivial fluxes carry energy-density, they are expected to have a back-reaction on the metric background such that the compact space is no longer Ricci-flat, i.e. it is not a Calabi-Yau manifold.[17] As we will see, such background fluxes generically introduce a tree-level scalar potential for the (used-to-be) moduli fields, thus allowing to dynamically freeze them. We distinguish three kinds of background fluxes:

- *Neveu-Schwarz background fluxes:* These are non-trivial vacuum expectation values of the NS-NS three-form field strength H. They can be included in the world-sheet σ-model action. In type II compactifications, H is the exterior derivative of the antisymmetric tensor field B,

$$H = dB \qquad (17.43)$$

and H is necessarily closed:

$$dH = 0. \qquad (17.44)$$

For heterotic string compactification, there is a non-trivial Bianchi identity (cf. (16.158))

$$dH = \frac{\alpha'}{4}(\operatorname{tr} R \wedge R - \operatorname{tr} F \wedge F), \qquad (17.45)$$

which however induces a non-trivial H-flux at subleading order in α'.

---

[17] We know that higher order α' corrections also destroy Ricci flatness but preserve the CY condition $c_1(M) = 0$. The corrections to $R_{mn} \neq 0$ generated by fluxes are not of this type. The resulting manifolds might e.g. not even be Kähler.

- *Ramond background fluxes:* In type II string compactifications, R-R field strengths F_{p+1} can have non-vanishing expectation values. In the simplest case

$$F_{p+1} = dC_p, \qquad dF_{p+1} = 0. \tag{17.46}$$

Within the RNS formulation of the fermionic string there is no known way to include them in the world-sheet action and one is therefore limited to the effective supergravity approach to find restrictions on these backgrounds and to study their effect. One must keep in mind that the results so obtained are only guaranteed to be true in the supergravity limit.[18]

- *Metric fluxes:* They are T-dual, via the Buscher rules, to a non-trivial H-flux background. We will not discuss this kind of non-trivial backgrounds.

For simplicity, but also motivated by observation, we only consider a background ansatz where the four-dimensional space-time is maximally symmetric, i.e. either Minkowski, de Sitter (dS) or anti-de Sitter (AdS). To preserve the symmetries of these space-times, fluxes must either be supported only in the internal space or on the complete four-dimensional space-time (for which the tensor field needs to be at least a four-form).

For a p-form which satisfies (17.46), the flux through a $(p+1)$-dimensional manifold Σ_{p+1} without boundary is quantized. Indeed, in a completely analogous way as one derives the (Dirac) quantization of magnetic flux in the presence of an electric charge, one finds that the coupling $\mu_{p-1} \int C_p$ leads to the condition that the flux through Σ_{p+1} satisfies $\mu_{p-1} \int_{\Sigma_{p+1}} F_{p+1} \in 2\pi \mathbb{Z}$. Using (9.58) we find

$$\frac{1}{\ell_s^p} \int_{\Sigma_{p+1}} F_{p+1} \in \mathbb{Z}. \tag{17.47}$$

Flux quantization and the resulting Dirac quantization condition for charges will be derived in Sect. 18.3. On account of the Bianchi identity the integral only depends on the homology class of Σ_{p+1} and, since Σ_{p+1} has no boundary, only on the cohomology class of F_{p+1}. Thus, one speaks of a $(p+1)$-form flux through the $(p+1)$-cycle $[\Sigma_{p+1}]$.

We will now study the effects of a non-trivial flux on an example.

Three-Form Fluxes in Type IIB

Consider the type IIB superstring compactified to four-dimensions on a manifold X with non-trivial NS-NS and R-R three-form fluxes H_3 and F_3 turned on along internal directions. The type IIB supergravity action in the presence of N_{D3} D3-branes was given in Chap. 16. In Einstein frame it is

---

[18]In the pure spinor approach it is possible to include anti-symmetric tensor backgrounds in the world-sheet action, but it has not yet been developed sufficiently far to be easily applicable to the situations we are interested in.

17.3 Flux Compactifications

$$S_{\text{IIB}} = \frac{1}{2\tilde{\kappa}_{10}^2} \int d^{10}x \sqrt{-G} \left(R - \frac{\partial_M \tau \partial^M \bar{\tau}}{2(\text{Im}\tau)^2} - \frac{|G_3|^2}{2\,\text{Im}\tau} - \frac{|F_5|^2}{4} \right)$$

$$+ \frac{1}{8i\tilde{\kappa}_{10}^2} \int \frac{C_4 \wedge G_3 \wedge \bar{G}_3}{\text{Im}\tau} - \left(N_{\text{D3}} - \frac{1}{2} N_{\text{O3}} \right) T_3 \int_{\mathbb{R}^{1,3}} d^4x \sqrt{-g}$$

$$+ \frac{1}{2} \left(N_{\text{D3}} - \frac{1}{2} N_{\text{O3}} \right) \mu_3 \int_{\mathbb{R}^{1,3}} C_4 \tag{17.48}$$

with the dilaton-axion field $\tau = C_0 + ie^{-\Phi}$ and the complex three-form flux $G_3 = F_3 - \tau H_3$ (here $F_3 = dC_2$ and $H_3 = dB_2$). In the second line we have only written the contributions of D3-branes and O3-planes which suffices for the following discussion. The Chern-Simons term in the second line provides a source term for the four-form C_4.[19] With $G_3 \wedge \bar{G}_3 = 2i\,\text{Im}(\tau)\, F_3 \wedge H_3$, the C_4 tadpole cancellation condition becomes

$$N_{\text{D3}} + N_{\text{flux}} - \frac{1}{2} N_{\text{O3}} = 0 \tag{17.49}$$

with

$$N_{\text{flux}} = \frac{1}{\ell_s^4} \int_X F_3 \wedge H_3. \tag{17.50}$$

Therefore, the G_3-form flux gives an integer valued contribution to the C_4 tadpole.

The Scalar Potential

The kinetic term for G_3 in (17.48) gives rise to a second important effect. Integrating it over the internal manifold X, it generates a positive semi-definite scalar potential in the effective four-dimensional theory

$$V_{\text{flux}} = \frac{1}{4\tilde{\kappa}_{10}^2} \int_X \frac{G_3 \wedge \star_6 \bar{G}_3}{\text{Im}\tau}. \tag{17.51}$$

Here, $\star_6$ is the Hodge-star on X. The total scalar potential is

$$V_{\text{tot}} = V_{\text{flux}} + \mu_3 \left(N_{\text{D3}} - \tfrac{1}{2} N_{\text{O3}} \right), \tag{17.52}$$

where we have used $T_3 = \mu_3$. We now rewrite V_{flux}: in six dimensions with Euclidean signature the Hodge star on a three-form satisfies $\star_6^2 = -1$ (cf. (14.283)).

---

[19] The factor of 1/2 was introduced because F_5 is self-dual, D3-branes carry both electric and magnetic charges with respect to F_5 and the action only contains the electric interaction term.

Therefore, we can split G_3 into an imaginary self-dual and an imaginary anti-self-dual part $G_3 = G_3^{\text{ISD}} + G_3^{\text{IASD}}$ with

$$G_3^{\text{ISD}} = \frac{1}{2}(1 - i \star_6) G_3, \qquad \star_6 G_3^{\text{ISD}} = +i\, G_3^{\text{ISD}},$$

$$G_3^{\text{IASD}} = \frac{1}{2}(1 + i \star_6) G_3, \qquad \star_6 G_3^{\text{IASD}} = -i\, G_3^{\text{IASD}}. \qquad (17.53)$$

Using these definitions we can rewrite the potential in two ways:

$$V_{\text{flux}} = \frac{1}{2\tilde{\kappa}_{10}^2} \int_X \frac{G_3^{\text{IASD}} \wedge \star_6 \overline{G_3^{\text{IASD}}}}{\text{Im}\,\tau} - \frac{i}{4\tilde{\kappa}_{10}^2} \int_X \frac{G_3 \wedge \overline{G}_3}{\text{Im}\,\tau}, \qquad (17.54\text{a})$$

$$V_{\text{flux}} = \frac{1}{2\tilde{\kappa}_{10}^2} \int_X \frac{G_3^{\text{ISD}} \wedge \star_6 \overline{G_3^{\text{ISD}}}}{\text{Im}\,\tau} + \frac{i}{4\tilde{\kappa}_{10}^2} \int_X \frac{G_3 \wedge \overline{G}_3}{\text{Im}\,\tau}. \qquad (17.54\text{b})$$

The first term in each expression is metric dependent and positive semi-definite and vanishes for ISD and IASD G_3, respectively. In these cases the second term, which is topological and proportional to N_{flux}

$$T_{\text{flux}} = \pm \frac{1}{2\tilde{\kappa}_{10}^2} \int_X F_3 \wedge H_3 = \pm \mu_3\, N_{\text{flux}} \qquad (17.55)$$

is also positive semi-definite (because V_{flux} is). It now follows from (17.49) and (17.52) that for ISD three-form flux the potential vanishes while for IASD flux it is positive.

In $\mathcal{N} = 1$ supersymmetric theories in four dimensions the scalar potential can be written as a sum of two terms, an F-term and a positive semi-definite D-term. The former can be derived from a holomorphic superpotential W and from the real Kähler potential K:

$$V_F = e^{\kappa_4^2 K}(G^{i\bar{j}} D_i W\, D_{\bar{j}} \overline{W} - 3\kappa_4^2 |W|^2), \qquad (17.56)$$

where

$$D_i W = \left(\partial_i + \kappa_4^2 (\partial_i K)\right) W \qquad (17.57)$$

is the Kähler covariant derivative of W, $G_{i\bar{j}} = \partial_i \partial_{\bar{j}} K$ the Kähler metric and i enumerates all chiral superfields. A non-vanishing $D_i W$ for some i signals spontaneous supersymmetry breaking. Supersymmetric vacua therefore generically have negative vacuum energy.

The Kähler potential also governs the kinetic terms of the chiral superfields. One can show, but we won't do it here, that the splitting of the potential as in (17.54a) is the splitting into an F-term (non-topological part) and a D-term (toplogical part).

17.3 Flux Compactifications

Neglecting the back-reaction of the flux and the branes on the Calabi-Yau metric,[20] one can also show that the superpotential which generates the F-term potential is

$$W = \frac{1}{\tilde{\kappa}_{10}^2} \int_X \Omega \wedge G_3. \tag{17.58}$$

The holomorphic three-form depends on the complex structure moduli, G_3 on the complex scalar τ but W is independent of the Kähler moduli. The Kähler potential, which is also needed to compute the scalar potential, depends on the complex-structure moduli, on the Kähler-moduli and on τ. The dependence on the Kähler moduli features the so-called no-scale-structure, which means that supersymmetric minima have vanishing vacuum energy.[21]

Due to the complex structure dependence of the superpotential (17.58), the condition for a supersymmetric minimum $G_3^{\text{IASD}} = 0$ imposes constraints on the complex structure moduli and the dilaton. Once the flux quanta are fixed, only for certain values of these latter moduli does G_3^{IASD} vanish.

We close with a number of remarks, some of which will be elaborated further in the next section:

- The minima of the positive semi-definite flux potential occur for imaginary self-dual three-form flux G_3. From Eq. (14.128) we know that G_3 has the decomposition $G_3 = G_3^{(2,1)} + G_3^{(1,2)} + G_3^{(0,3)}$ in a complex structure for that Ω is purely $(3, 0)$. In a supersymmetric minimum the Kähler covariant derivatives w.r.t. all chiral superfields must vanish. While the Kähler potential depends on all moduli and on τ, the superpotential W depends on the complex structure moduli only through Ω, on τ through G_3 and is independent of the Kähler moduli. The vanishing of the covariant derivative of W w.r.t. the complex structure moduli, together with (14.222) and (17.58), implies that $G_3^{(1,2)} = 0$. Independence of W on the Kähler moduli requires $W = 0$ and therefore $G_3^{(0,3)} = 0$. In summary a three-form flux which is a primitive $(2, 1)$ form breaks the bulk $\mathcal{N} = 2$ supersymmetry to $\mathcal{N} = 1$.

  As a consequence of the no-scale structure, for the Kähler moduli T the $|D_T W|^2$ contribution to the scalar potential cancels against the $-3|W|^2$ term. Therefore, not all minima with vanishing tree-level vacuum energy are supersymmetric, i.e. necessarily satisfy $D_T W = 0$. But this is not a solution to the cosmological constant problem, as higher order corrections in g_s and α' break the

---

[20]Taking the back-reaction into account one gets a warped CY. This will be explained below.

[21]For instance, if we have only one Kähler modulus T and $K = -3\ln(T + \overline{T}) + \tilde{K}(\phi_A, \overline{\phi}_A)$ and if furthermore the superpotential is independent of T, $W = W(\phi_A)$, we obtain from (17.56) $V_F = e^{\kappa_4^2 K} G^{A\bar{A}} D_A W D_{\bar{A}} \overline{W}$. Supersymmetric minima now have vanishing potential energy. Also note that the vev of T is not determined. This leads to a degenerate family of vacua and hence the name no-scale models.

no-scale structure. For general G_3 fluxes, the scalar potential allows to dynamically freeze all complex structure moduli and the complexified dilaton. However, the Kähler moduli still remain as massless fields. Imposing supersymmetry yields one more condition on the complex structure moduli and the dilaton.
- So far we have not analyzed the back-reaction of the flux on the geometry, i.e. the source term in the Einstein equations (17.4). In particular, writing the superpotential (17.58), we were implicitly assuming that a holomorphic three-form Ω exists and by talking about complex structure and Kähler moduli, we were assuming that we are on a CY. One can show that in a model with only D3 branes and G_3 fluxes the ten-dimensional geometry no longer has a direct product structure with an internal CY manifold, but it is given by a warped product of the form

$$ds_{10}^2 = e^{2A(y)} g(x)_{\mu\nu} \, dx^\mu \, dx^\nu + e^{-2A(y)} \hat{g}(y)_{mn} \, dy^m \, dy^n, \quad (17.59)$$

where $\hat{g}_{mn}$ is a CY metric and $e^{2A(y)}$ is called warp factor. Furthermore,

$$F_5 = (1 + \star)[dA(y) \wedge \text{Vol}_4], \qquad \Phi = \Phi_0, \quad (17.60)$$

where $\text{Vol}_4 = \sqrt{g(x)} dx^1 \wedge \cdots \wedge dx^4$. Note that the dilaton is constant. The equations of motion require the relation between the five-form flux and A. This will be established in the next section.

The effect of the warping is expected to be small in the so-called dilute flux limit, i.e. when one takes the large volume limit for fixed quantized fluxes. In this limit, the energy density of the flux becomes small and at leading order can be neglected on the r.h.s. of the Einstein equation.

For models with D7 branes the geometry is no longer of this simple warped form, but involves a strong back-reaction on the dilaton. This is best described via F-theory (see Chap. 18).
- Due to the warp factor the fundamental form ω and the three-form Ω are no longer closed but satisfy

$$d\omega = -2(dA) \wedge \omega, \qquad d\Omega = -3(dA) \wedge \Omega. \quad (17.61)$$

These follow from $\omega = ie^{-2A}\omega_{\text{CY}}$ and $\Omega = e^{3A}\Omega_{\text{CY}}$ (to preserve (17.69)).
- Flux vacua constructions are an appropriate starting point for a discussion of the string landscape. It was already noticed several years before flux vacua were being considered that the number of four-dimensional CFT string vacua is immensely large—e.g. the number of covariant lattices that lead to 4D heterotic strings was estimated to be of $\mathcal{O}(10^{1500})$, but at the time it was not studied whether they have marginal deformation which connect them. In flux vacua, in particular if we include non-perturbative effects, some or even all moduli of the underlying geometry are fixed. They can therefore provide a set of vacua which can be regarded as discrete points or low dimensional valleys in a huge multidimensional string landscape. In the supergravity approximation the flux vacua

are local minima of an effective field theory scalar potential, which depends on the moduli parameters of a certain background geometry. The number of supersymmetric vacua, N_{SUSY}, can in principle be computed for any given class of flux compactification. For a typical Calabi-Yau compactification the number of vacua with different 3-form fluxes through its $2(h^{2,1}+1)$ three-cycles is of the order

$$N_{\text{SUSY}} = \mathcal{O}(10^{500}). \tag{17.62}$$

One can argue that there is a good chance to find vacua (after uplift to positive vacuum energy by non-perturbative effects) with a tiny positive cosmological constant of order $\Lambda \simeq 10^{-120} M_p^4$. In fact, the vast proliferation of string vacua might suggest an explanation of the smallness of the cosmological constant via the anthropic principle. Combining flux vacua statistics with the statistics of intersecting D-brane models could also lead to an anthropic understanding of the Standard Model and its parameters. Anthropic arguments, in particular in the context of the string landscape and in combination with cosmological models of eternal inflation, are highly attractive for some people, while for others they signal the end of science. For this and other issues, an understanding of flux vacua beyond the supergravity limit is needed.

17.4 Fluxes and SU(3) Structure Group

In Chap. 14 we have seen that in a purely metric background, the condition for space-time supersymmetry is the existence of a covariantly constant spinor. We have also seen that this implies that the internal space is Ricci flat, i.e. the supersymmetry condition implies the equations of motion for the metric. In the heterotic case there was an additional condition, the Bianchi identity for the field strength of the B-field, which requires a nontrivial gauge bundle.

In the type II context, a purely metric background which does not break supersymmetry completely leaves unbroken $\mathcal{N}=2$ space-time supersymmetry, i.e. eight supercharges. However, heterotic string compactification on a CY manifold leads to $\mathcal{N}=1$. In order to achieve $\mathcal{N}=1$ in type II compactifications, one has to switch on additional background fields. In this section we mainly discuss the question under which conditions the string tree-level equations of motion and the Killing spinor equations, i.e. the supersymmetry conditions, have solutions with non-trivial background fluxes. The most direct way to answer the first question is to use the field equations of the 10-dimensional effective supergravity theory, which were derived in Chap. 16. However, as we have already seen in the case of a purely metric background, it is often not necessary to solve the full field equations, which are second order differential equations. For supersymmetric backgrounds it suffices to study the supersymmetry conditions, which are first order differential equations and therefore simpler to analyze. One can show that under rather mild assumptions, namely imposing integrability conditions and Bianchi identities, solutions of the supersymmetry conditions also solve the field equations.

In the following we present the first steps of a general discussion of the type II supersymmetry conditions, i.e. the vanishing of the supersymmetry variations of the gravitino and dilatino fields. These are local conditions in ten dimensions that guarantee the existence of Killing spinors in the given background. Even though they lead to a classification of supersymmetric geometries with fluxes, one should keep in mind that compactness of the internal space generally imposes additional global (Gauss-law) constraints. It is not known in all cases whether compact solutions of the type described actually exist. Furthermore, one has to keep in mind that it is not at all clear that the thus obtained solutions of the supergravity equations of motion lift to solutions of the full string equations of motion.

For the ten-dimensional metric one uses a warped ansatz of the form

$$ds^2 = e^{2A(y)} g_{\mu\nu}(x) dx^\mu dx^\nu + \hat{g}_{mn}(y) dy^m \otimes dy^n. \qquad (17.63)$$

The four-dimensional metric $g_{\mu\nu}$ is either that of Minkowski, de Sitter (dS_4), or anti-de Sitter (AdS_4) space, and $\hat{g}_{mn}$ is the six-dimensional, internal metric. In general other bosonic fields are also allowed to acquire non-trivial profiles (vacuum expectation values) but all fermion fields vanish.

In ten space-time dimensions the type II supersymmetry variations for the two gravitini ψ_M^A ($A = 1, 2$) and the two dilatini in the democratic formulation for the RR fields, after setting the fermions to zero, are (in string frame)[22]

$$\delta_\epsilon \psi_M = \nabla_M \epsilon + \frac{1}{4} \slashed{H}_M \mathcal{P} \epsilon + \frac{1}{16} e^\Phi \sum_n \slashed{F}_n \Gamma_M \mathcal{P}_n \epsilon,$$

$$\delta_\epsilon \lambda = \slashed{\partial} \Phi \, \epsilon + \frac{1}{2} \slashed{H} \mathcal{P} \epsilon + \frac{1}{8} e^\Phi \sum_n (-1)^n (5-n) \slashed{F}_n \mathcal{P}_n \epsilon. \qquad (17.64)$$

Setting the variations to zero are the Killing spinor equations in general bosonic backgrounds. Given a background, their solutions determine the supercharges which are unbroken in this background. In addition, the Bianchi identities on the fluxes have to be satisfied. Of course, the existence of solutions presents non-trivial constraints on the backgrounds. Generically all supercharges are broken. In (17.64) the spinors ψ_M, λ and ϵ are doublets of Majorana-Weyl spinors of equal (IIB) or opposite (IIA) chiralities. The sums are over $n \in \{0, \ldots, 10\}$ with n even for IIA[23] and n odd for IIB with the duality constraints $F_n = (-1)^{[\frac{n}{2}]} \star F_{10-n}$. One can use the duality constraints to write (17.64) with the sums restricted to $n \in \{0, \ldots, 5\}$.

---

[22] For ease of comparison with the relevant literature, in this section we are using the convention (14.292) for the Hodge-$\star$. Otherwise we would either have to modify (17.64) or the duality properties of the field strengths.

[23] $F_0 = m$ is the constant mass-parameter of Romans massive type IIA supergravity.

17.4 Fluxes and SU(3) Structure Group

This will become clear in the example discussed at the end of this section. $\mathscr{P}$ and $\mathscr{P}_n$ are 2×2 matrices which act on the doublet label (which we have suppressed). For type IIA $\mathscr{P} = \Gamma_{11}$ and $\mathscr{P}_n = (\Gamma_{11})^{n/2} \sigma^1$; for type IIB, $\mathscr{P} = -\sigma^3$, $\mathscr{P}_n = \sigma^1$ for $\frac{n+1}{2}$ even and $\mathscr{P}_n = i\sigma^2$ for $\frac{n+1}{2}$ odd. σ^i are Pauli matrices. Furthermore, $\not{F}_n = \frac{1}{n!} \Gamma^{M_1...M_n} F_{M_1...M_N}$ and $\not{H}_M = \frac{1}{2} \Gamma^{NP} H_{MNP}$. Recall that the chiralities of the dilatini is opposite to that of the gravitini and the supersymmetry parameters. The vanishing of these variations is required for supersymmetry and the number of solution ϵ determines the number of supercharges that are preserved. It is evident that without fluxes and with constant dilaton the solutions are covariantly constant spinors and we are back to Calabi-Yau compactifications. Fluxes and a non-trivial dilaton profile lead to considerably more complicated differential equations.

It is clear that for a supersymmetric compactification the internal manifold must admit a globally defined nowhere vanishing spinor η. Recall that on a CY manifold there exists a covariantly constant spinor and the existence of covariantly constant spinors implies a reduced (also called special) holonomy group. Likewise, globally defined spinors exist only on manifolds with reduced structure group. The structure group G of a d-dimensional Riemannian manifold (M, g) is the structure group of its orthonormal frame bundle whose fiber is a basis of vectors (d-beins) e_a^m such that $e_a^m e_b^n g_{mn} = \delta_{ab}$. For a generic orientable manifold $G \simeq SO(d)$, i.e. the orthonormal frames on the overlap of two coordinate patches are related by a $SO(d)$ rotation. A reduced structure group means $G \subset SO(6)$. While on a generic Riemannian manifold all tensors transform in representations of $SO(d)$, on a manifold with structure group G they transform in representations of G. Examples of manifolds with reduced structure group are flat tori $T^d = \mathbb{R}^d / \Gamma$ with trivial structure group and, less trivial examples, Kähler manifolds or, more generally, almost Hermitian manifolds, of complex dimension n, with $G \subseteq U(n)$. In the same way as not all $G \subset SO(d)$ can occur as holonomy groups, not all subgroups can be structure groups. They must be from the same list (Berger's list) as the possible holonomy groups.

On CY threefolds, i.e. manifolds with $SU(3)$ holonomy, there is one covariantly constant spinor, and two covariantly constant tensors, the Kähler form and the holomorphic three-form (both of which can be constructed in terms of the spinor, cf. (14.182)). They are singlets under $SU(3)$. Their existence can serve as an alternative definition of Calabi-Yau manifolds. Likewise, manifolds with reduced structure group G can be defined by the existence of globally defined nowhere vanishing tensors or, in the case of spin-manifolds, spinors, i.e. tensors or spinors which transform as singlets under G. For Kähler manifolds the Kähler form ω is an $U(n)$ invariant tensor.

In general the G invariant spinor/tensors ξ are not covariantly constant with respect to the Levi-Civita connection ∇, i.e. $\nabla \xi \neq 0$. It can be shown that on manifolds with G-structure there exists a metric compatible connection ∇' (i.e. $\nabla'_p g_{mn} = 0$) such that

$$\nabla' \xi = 0 \tag{17.65}$$

for all G invariant tensors/spinors. This connection differs from the Levi-Civita connection by a torsion piece. This will be discussed below.

The case of interest for us are six-manifolds with $SU(3)$ structure. They can be defined as almost complex six-manifolds with a globally defined nowhere vanishing spinor $\eta_\pm$ ($\eta_+^* = \eta_-$). In general, the spinor representation is the **4** of the frame rotation group $SO(6) \simeq SU(4)$. However, if there exists a globally defined spinor on the manifold, then there must be a singlet in the decomposition w.r.t. G. Indeed, for $G = SU(3)$ we have the decomposition

$$\mathbf{4} \to \mathbf{3} + \mathbf{1}. \tag{17.66}$$

We can decompose general tensors on the manifold with respect to $SU(3)$:

$$\begin{aligned} 1-\text{form}: & \quad \mathbf{6} \to \mathbf{3} + \bar{\mathbf{3}}, \\ 2-\text{form}: & \quad \mathbf{15} \to \mathbf{8} + \mathbf{3} + \bar{\mathbf{3}} + \mathbf{1}, \\ 3-\text{form}: & \quad \mathbf{20} \to \mathbf{6} + \bar{\mathbf{6}} + \mathbf{3} + \bar{\mathbf{3}} + \mathbf{1} + \mathbf{1}. \end{aligned} \tag{17.67}$$

We see that there are also singlets in the decomposition of 2-forms and 3-forms. This means that there is also a non-vanishing globally defined real 2-form ω, and a complex 3-form Ω. Often the pair (ω, Ω) is called the $SU(3)$ group structure of the six-dimensional manifold. With respect to the almost complex structure, Ω is a $(3,0)$-form and ω is a (real) $(1,1)$-form. Since there are no invariant vectors (or equivalently five-forms), it follows that

$$\omega \wedge \Omega = 0. \tag{17.68}$$

On the other hand, a six-form is a singlet, which implies that $\omega \wedge \omega \wedge \omega$ is proportional to $\Omega \wedge \bar{\Omega}$. So we are free to use the convention (cf. (14.183))

$$\omega \wedge \omega \wedge \omega = \frac{3i}{4} \Omega \wedge \bar{\Omega} \neq 0. \tag{17.69}$$

In summary, the $SU(3)$ structure is determined equivalently by the $SU(3)$ invariant spinor η, or by ω and Ω. Moreover, ω and Ω can be related to $\eta_\pm$ as in (14.182).[24] While on manifolds with $SU(3)$ holonomy, i.e. on CY manifolds, η, ω and Ω are covariantly w.r.t the Levi-Civita connection ∇, this is generally not the case on manifolds with $SU(3)$ structure. The non-vanishing of $\nabla \xi$ is measured by what is called intrinsic torsion.

A connection with torsion has the form

$$\nabla' = \nabla - \kappa, \tag{17.70}$$

---

[24] The spinors η_+ (η_-) correspond to the internal spin fields $\Sigma(z)$ ($\Sigma^\dagger(z)$) of Chap. 15, the $U(1)$ current $J(z)$ to the 2-form ω, and the 3-form Ω to the identity operator of the internal CFT.

17.4 Fluxes and SU(3) Structure Group

e.g. $\nabla'_m V_n = \partial_n V_m - \Gamma'^p_{mn} V_p = \partial_m V_n - \Gamma^p_{mn} V_p - \kappa_{mn}{}^p V_p$ where Γ is the Christoffel symbol and κ the contorsion tensor. Metric compatibility requires

$$\kappa_{mnp} = -\kappa_{mpn}, \qquad \kappa_{mnp} = \kappa_{mn}{}^q g_{qp}. \tag{17.71}$$

The torsion tensor is defined via

$$[\nabla'_m, \nabla'_n] V_q = R'_{mnp}{}^q V_q - 2T_{mn}{}^p \nabla'_p V_q, \tag{17.72}$$

where $R'_{mnp}{}^q$ is the curvature tensor constructed from Γ'. From (17.72) we derive the relation

$$T_{mn}{}^p = \frac{1}{2}\left(\kappa_{mn}{}^p - \kappa_{nm}{}^p\right) \quad \Leftrightarrow \quad \kappa_{mnp} = T_{mnp} + T_{pnm} + T_{pmn}, \tag{17.73}$$

and from (17.71) we learn that[25] $\kappa_{nmp} \in \Lambda^1 \otimes \Lambda^2$ or, equivalently

$$\kappa_{mn}{}^p \in \Lambda^1 \otimes so(d), \tag{17.74}$$

where we use that the elements of $so(d)$ are antisymmetric $d \times d$ matrices. Denote the Lie algebra of G by g. We split $so(d) = g \oplus so(d)/g = g \oplus g^\perp$ and likewise

$$\kappa = \kappa^0 + \kappa^g \quad \text{where} \quad \kappa^0 \in \Lambda^1 \otimes g^\perp, \; \kappa^g \in \Lambda^1 \otimes g. \tag{17.75}$$

On invariant tensors ξ

$$0 = \nabla'\xi = (\nabla - \kappa^0 - \kappa^g)\xi = (\nabla - \kappa^0)\xi. \tag{17.76}$$

where $T^0_{mn}{}^p = \kappa^0_{[mn]}{}^p$ is called the intrinsic torsion. It measures the failure of the invariant tensors to be covariantly constant w.r.t. ∇ and of the holonomy group to reduce to G.

We now decompose the intrinsic torsion into irreducible G-representations. Restricting to six-manifolds with $G = SU(3)$, we obtain with $\Lambda^1 \sim \mathbf{3} \oplus \bar{\mathbf{3}}$, $g \sim \mathbf{8}$ and $g^\perp \sim \mathbf{1} \oplus \mathbf{3} \oplus \bar{\mathbf{3}}$,

$$T^0_{mn}{}^p \in \Lambda^1 \otimes su(3)^\perp = (\mathbf{3} \oplus \bar{\mathbf{3}}) \otimes (\mathbf{1} \oplus \mathbf{3} \oplus \bar{\mathbf{3}})$$
$$= \underbrace{(\mathbf{1} \oplus \mathbf{1})}_{\mathcal{W}_1} \oplus \underbrace{(\mathbf{8} \oplus \mathbf{8})}_{\mathcal{W}_2} \oplus \underbrace{(\mathbf{6} \oplus \bar{\mathbf{6}})}_{\mathcal{W}_3} \oplus \underbrace{2\,(\mathbf{3} \oplus \bar{\mathbf{3}})}_{\mathcal{W}_4, \mathcal{W}_5}. \tag{17.77}$$

$\mathcal{W}_1, \ldots, \mathcal{W}_5$ are called the five torsion classes. They appear in the covariant derivatives of η, ω and Ω. From their $SU(3)$ representation content we can infer

---

[25] Elements of Λ^p are p-form fields on M.

their form type: $\mathcal{W}_1$ is a complex scalar, $\mathcal{W}_2$ a complex primitive $(1,1)$ form.[26] $\mathcal{W}_3$ is a real primitive $(2,1) + (1,2)$ form, $\mathcal{W}_4$ is a real $(1,0) + (0,1)$ form and $\mathcal{W}_5$ a complex $(1,0)$-form (which is equivalent to a real $(1,0) + (0,1)$ form).

Using the fact that $\nabla'\omega = \nabla'\Omega = 0$ we can express $d\omega$ and $d\Omega$ in terms of the intrinsic torsion. Recall from Chap. 14 that on an almost complex manifold $d: A^{p,q} \to A^{p+2,q-1} \oplus A^{p+1,q} \oplus A^{p,q+1} \oplus A^{p-1,q+2}$, where the first and the last term are absent on complex manifolds. Using simple $SU(3)$ representation theory we find

$$d\omega \in \underbrace{A^{3,0} \oplus A^{0,3}}_{\substack{(1\oplus 1) \\ \mathcal{W}_1}} \oplus \underbrace{A^{2,1} \oplus A^{1,2}}_{\substack{(6\oplus\bar{6})\oplus(\bar{3}\oplus 3) \\ \oplus\;\mathcal{W}_3 \oplus \mathcal{W}_4}} \tag{17.78}$$

and

$$d\Omega \oplus d\overline{\Omega} \in \underbrace{A^{3,1} \oplus A^{1,3}}_{\substack{3\oplus\bar{3} \\ \mathcal{W}_5}} \oplus \underbrace{A^{2,2} \oplus A^{2,2}}_{\substack{(1\oplus 1)\oplus(8\oplus 8) \\ \oplus\;\mathcal{W}_1 \oplus \mathcal{W}_2}}. \tag{17.79}$$

More explicitly, one decomposes $d\omega$ and $d\Omega$ as

$$d\omega = \frac{3}{2}\mathrm{Im}(\overline{\mathcal{W}}_1\Omega) + \mathcal{W}_4 \wedge \omega + \mathcal{W}_3,$$
$$d\Omega = \mathcal{W}_1\omega\wedge\omega + \mathcal{W}_2\wedge\omega + \overline{\mathcal{W}}_5\wedge\Omega. \tag{17.80}$$

These relations can be inverted and the $\mathcal{W}_i$ can be expressed though $\omega, \Omega, d\omega$ and $d\Omega$. The fact that $\mathcal{W}_1$ appears in $d\omega$ and in $d\Omega$ is due $\omega \wedge d\Omega = \Omega \wedge d\omega$, which follows from (17.68).

From Eqs. (17.78) and (17.79) we immediately conclude that a manifold with $SU(3)$ structure is complex if $\mathcal{W}_1 = \mathcal{W}_2 = 0$, i.e. $d\Omega$ is a $(3,1)$-form and $d\omega$ has no $(3,0)$ and $(0,3)$ components. It is symplectic[27] if $\mathcal{W}_1 = \mathcal{W}_3 = \mathcal{W}_4 = 0$, i.e. ω is closed. A Kähler manifold is at the same time complex and symplectic, and therefore the only non-zero torsion class can be $\mathcal{W}_5$. For Calabi-Yau manifold with $SU(3)$ holonomy all five torsion classes are zero. Various hybrid cases are also of interest and have been given separate names, but we will not discuss them, except for the example below.

---

[26] Recall from Chap. 14 that primitivity means $(\mathcal{W}_2)_{i\bar{j}}\omega^{i\bar{j}} = 0$. This corresponds to the decomposition $3\otimes\bar{3} = 8 + 1$ into traceless and trace parts with $\mathcal{W}_2 \sim 8$ and $\mathcal{W}_1 \sim 1$.

[27] A manifold is symplectic if is has a globally defined closed non-degenerate two-form, the symplectic structure. Non-degenerate means that the coefficient matrix of the two-form has an inverse.

17.4 Fluxes and SU(3) Structure Group

We can now use the supersymmetry conditions to relate the fluxes to the geometric torsion classes of the internal manifold. We will discuss this for the case considered in the previous section, i.e. type IIB flux compactification with 3-form flux $G_3 = F_3 - \tau H_3$. To be specific we will make the following ansatz

$$ds_{10}^2 = e^{2A(y)} \eta_{\mu\nu} dx^\mu dx^\nu + e^{-2A(y)} \hat{g}_{mn} dy^m dy^n. \qquad (17.81)$$

We set the dilaton to a constant (for simplicity we set it to zero) and $F_1 = 0$, i.e. $G_3 = F_3 - iH_3$. H_3 and F_3 have only components along the internal directions; otherwise we could not have space-time Poincaré invariance, as our metric ansatz requires. F_7 is fixed by the duality constraint $F_7 = - \star F_3$. F_5 is self-dual and is fixed by its components which are purely internal.

The most general ansatz for the supersymmetry parameter which leads to vacua with $\mathcal{N} = 1$ supersymmetry in four dimensions, is (cf. (14.48))

(IIB) $\quad \epsilon = \begin{pmatrix} \epsilon_1 \\ \epsilon_2 \end{pmatrix} \quad$ with $\quad \begin{aligned} \epsilon_1 &= \epsilon_R \otimes (a\eta_+) + \epsilon_L \otimes (\bar{a}\eta_-), \\ \epsilon_2 &= \epsilon_R \otimes (b\eta_+) + \epsilon_L \otimes (\bar{b}\eta_-), \end{aligned}$

$$(17.82)$$

where $\eta_\pm(y)$ is the invariant spinor on the internal $SU(3)$-structure manifold and $\epsilon_{R,L}(x)$ are space-time spinors with $\epsilon_L = \epsilon_R^*$. The chirality is indicated by the subscript. The paring of chiralities is such that $\Gamma_{11} \epsilon_{1,2} = \epsilon_{1,2}$ which is appropriate for type IIB. For type IIA the ansatz is similar, with $\epsilon_R \leftrightarrow \epsilon_L$ in ϵ_2 such that $\Gamma_{11} \epsilon_2 = -\epsilon_2$. In the absence of fluxes we have eight unbroken supercharges and two space-time spinors $\epsilon_{R,L}^{1,2}$. But since we are looking for vacua with four unbroken supercharges they must be related. In a maximally symmetric space-time this relation must be space-time independent, but it can depend on the internal coordinates. This is taken into account by the complex functions a and b. The requirement that the ansatz solves the Killing spinor equations relates them. Different relations correspond to different choices for $\mathcal{N} = 1 \subset \mathcal{N} = 2$.

The analysis is somewhat involved. In the Appendix we present it for the choice $b = -ia$. We show in some detail that in this case the Killing spinor equations have non-trivial solutions if the background satisfies (here $\eta = \eta_+$)

$$\hat{\star}_6 F_5 = 4e^{-4A} dA,$$

$$G_3 = F_3 - iH_3 \qquad \hat{\star} G_3 = iG_3 \qquad G_3^{(0,3)} = 0,$$

$$\hat{\nabla} \hat{\eta} = 0 \qquad \hat{\eta} = e^{-A/2} \eta. \qquad (17.83)$$

The hatted quantities are w.r.t. $\hat{g}_{mn}$ which, as the last equation shows, must be Calabi-Yau. Therefore, comparing with (17.61) and (17.80), this 3-form flux only allows for non-vanishing torsion classes $\mathcal{W}_4$ and $\mathcal{W}_5$. Since the 6-dimensional space is a warped Calabi-Yau manifold, it follows that $2\mathcal{W}_5 = 3\mathcal{W}_4 = -6dA$. Finally, the two non-vanishing torsion classes $\mathcal{W}_4$ and $\mathcal{W}_5$ have to be balanced by a non-trivial 5-form flux F_5, as shown in Eq. (17.60).

We want to stress that not all solutions (17.83) are solutions to the SUGRA equations of motion. We still have to impose the Bianchi identity, $dF_5 = H_3 \wedge F_3$, i.e.

$$d \hat{\star}_6 d(e^{-4A}) = H_3 \wedge F_3. \qquad (17.84)$$

A systematic analysis of the Killing spinor equations and the general relation between fluxes and torsion classes can be found in the references at the end of this chapter. One must keep in mind that most of the analysis along the lines indicated here has been done in the supergravity approximation and it is not clear how or even whether it can be lifted to the full string theory. Many of the techniques which were powerful in CY compactifications are not available or have to be refined (e.g. mirror symmetry). The structure of the moduli spaces of these more general compactifications is also largely unknown. The essential ingredient which is missing compared to CY compactifications is the (2,2) extended world-sheet supersymmetry.

17.5 Appendix

In this appendix we give some details of the analysis of the Killing spinor equations for the ansatz (17.81). We start with the variation of the gravitini. Using the duality properties of the fluxes and the identity

$$\Gamma_{M_1...M_n} = \frac{(-1)^{\frac{1}{2}n(n-1)}}{(10-n)!} \Gamma_{N_1...N_{10-n}} \varepsilon^{N_1...N_{10-n}}{}_{M_1...M_n} \Gamma_{11} \qquad (17.85)$$

which is valid in $d = 10$ with Lorentzian signature, we find for the μ and m components of the gravitino equation

$$\nabla_\mu \epsilon + \frac{1}{8}(\slashed{F}_3 \sigma^1 + \slashed{F}_5 i\sigma^2)\Gamma_\mu \epsilon = 0,$$

$$\left(\nabla_m - \frac{1}{4}\slashed{H}_m \sigma^3\right)\epsilon + \frac{1}{8}\left(\slashed{F}_3 \sigma^1 + \slashed{F}_5 i\sigma^2\right)\Gamma_m \epsilon = 0, \qquad (17.86)$$

where F_5 has only components along the internal directions and we have used $\Gamma_{11}\epsilon = \epsilon$. Given the ansatz for the metric one can work out the spin connection and the covariant derivative $\nabla_M = \partial_M + \frac{1}{4}\omega_M^{AB}\Gamma_{AB}$:

$$\nabla_\mu \epsilon = \left(\partial_\mu - \frac{1}{2}(\Gamma^n \partial_n A)\Gamma_\mu\right)\epsilon. \qquad (17.87)$$

Decompose the Dirac matrices as $\Gamma^M = (\gamma^\mu \gamma^m)$ with

$$\Gamma^\mu = \gamma^\mu \otimes \mathbb{1}, \qquad \Gamma^m = \gamma_{(4)} \otimes \gamma^m. \qquad (17.88)$$

17.5 Appendix

To avoid confusion, we have denoted the chirality matrix in four dimensions by $\gamma_{(4)}$. We work in Majorana basis where γ^μ are real and γ^m imaginary.

We now insert the ansatz (17.82) with $\epsilon_{R,L}$ constant spinors ($\eta \equiv \eta_+$, $\bar{\eta} \equiv \eta_-$)

$$\partial\!\!\!/ A(a\eta) - \frac{1}{4}(\not{F}_3 + \not{F}_5)(b\eta) = 0,$$

$$\partial\!\!\!/ A(b\eta) + \frac{1}{4}(\not{F}_3 - \not{F}_5)(a\eta) = 0,$$

$$\left(\nabla_m - \frac{1}{4}\not{H}_m\right)(a\eta) + \frac{1}{8}(\not{F}_3 + \not{F}_5)\gamma_m(b\eta) = 0,$$

$$\left(\nabla_m + \frac{1}{4}\not{H}_m\right)(b\eta) + \frac{1}{8}(\not{F}_3 - \not{F}_5)\gamma_m(a\eta) = 0. \tag{17.89}$$

Here $\partial\!\!\!/ = \gamma^m \partial_m$, $\not{F}_3 = \frac{1}{3!}\gamma^{mnp} F_{mnp}$ and likewise for F_5.

There is the a second set of equations which follows from the above with the replacements $(\eta, a, b) \to (\bar{\eta}, \bar{a}, \bar{b})$.

We now link a and b by setting $b = -ia$. If we absorb a into η, we find the equations

$$\partial\!\!\!/ A\eta + \frac{i}{4}(\not{F}_3 + \not{F}_5)\eta = 0,$$

$$\partial\!\!\!/ A\eta - \frac{i}{4}(\not{F}_3 - \not{F}_5)\eta = 0 \tag{17.90}$$

and

$$\left(\nabla_m - \frac{1}{4}\not{H}_m\right)\eta - \frac{i}{8}(\not{F}_3 + \not{F}_5)\gamma_m \eta = 0,$$

$$\left(\nabla_m + \frac{1}{4}\not{H}_m\right)\eta + \frac{i}{8}(\not{F}_3 - \not{F}_5)\gamma_m \eta = 0. \tag{17.91}$$

From (17.90) and its complex conjugates we find

$$\not{F}_3 \eta = 0 \quad \text{and} \quad \not{F}_3 \bar{\eta} = 0 \tag{17.92}$$

To extract the contents of these equations we use the fact that η is an $SU(3)$ singlet. As in Chap. 14.5 we view it as a highest state which is annihilated by the γ^i while the lowest weight state $\bar{\eta}$ is annihilated by $\gamma^{\bar{i}}$. It is then easy to see that (17.92) implies

$$F_3 = F^{(2,1)} + F^{(1,2)} \quad \text{and primitive.} \tag{17.93}$$

Recall that primitivity means $g^{i\bar{j}} F_{i\bar{j}k} = 0$. The second relation we can extract from (17.90) is

$$\partial\!\!\!/\, A\eta + \frac{i}{4} F\!\!\!\!/_5 \eta = 0 \quad \text{and} \quad \text{c.c.} \tag{17.94}$$

We now use the identity, valid in six dimensions with Euclidean signature,

$$\gamma_{m_1...m_n} = -i \frac{(-1)^{\frac{1}{2}n(n+1)}}{(6-n)!} \gamma_{n_1...n_{6-n}} \varepsilon^{n_1...n_{6-n}}{}_{m_1...m_n} \gamma_{(6)}, \tag{17.95}$$

where $\gamma_{(6)}$ is the chirality operator in six dimensions. Using also $\gamma_{(6)}\eta = \eta$ we find from (17.94)

$$4\, dA = \star_6 F_5. \tag{17.96}$$

$\star_6$ is the Hodge star in the internal six-dimensional manifold. If we denote by $\hat{\star}_6$ the Hodge star w.r.t. metric $\hat{g}_{mn}$, this can be written as

$$\hat{\star}_6 F_5 = 4 e^{-4A} dA. \tag{17.97}$$

We now analyze (17.91). The difference of these two equations is

$$H\!\!\!\!/_{m} \eta + \frac{i}{2} F\!\!\!\!/_3 \gamma_m \eta = 0. \tag{17.98}$$

Contracting with γ^m and use of the identify

$$\gamma^m \gamma^{[n]} \gamma_m = (-1)^n (D - 2n) \gamma^{[n]} \tag{17.99}$$

gives $H\!\!\!\!/\, \eta = 0$ and, together with the complex conjugate equation,

$$H_3 = H^{(2,1)} + H^{(1,2)} \quad \text{and} \quad \text{primitive.} \tag{17.100}$$

With these results we can analyze (17.98) further. Using the highest weight property of η it is straightforward to show that

$$H^{(1,2)} = -i F^{(1,2)}. \tag{17.101}$$

Define

$$G_3 = F_3 - i H_3. \tag{17.102}$$

17.5 Appendix

What we have just shown is that G_3 is primitive, purely $(2,1)$ and therefore imaginary self-dual.

It remains to analyze the sum of the two equations in (17.91):

$$\nabla_m \eta + \frac{i}{8} \slashed{F}_5 \gamma_m \eta = 0. \tag{17.103}$$

One first shows that

$$\nabla_m = \partial_m + \frac{1}{4}\omega_m^{ab}\gamma_{ab} = \hat{\nabla}_m - \frac{1}{2}\hat{\gamma}_m{}^n \partial_n A, \tag{17.104}$$

where $\hat{\nabla}$ is the covariant derivative for $\hat{g}$ and $\hat{\gamma}_m = \gamma_a \hat{e}_m^a$ where $\hat{e}_m^a \hat{e}_n^b \eta_{ab} = \hat{g}_{mn}$. With the help of (17.95) we can rewrite

$$\slashed{F}_5 = i\gamma^n (\star_6 F_5)_n \gamma_{(6)} = 4i\, \slashed{\partial} A \gamma_{(6)}, \tag{17.105}$$

where we have used (17.97). Using these results we finally find from (17.103)

$$\hat{\nabla}_m \hat{\eta} = 0 \quad \text{with} \quad \hat{\eta} = e^{-A/2}\eta. \tag{17.106}$$

We still need to show that the dilatino variation also vanishes, but this follows immediately with (17.93) and (17.100).

Further Reading

Reviews on four-dimensional intersecting D-brane models are

- R. Blumenhagen, M. Cvetič, P. Langacker, G. Shiu, Towards realistic intersecting D-brane models. Ann. Rev. Nucl. Part. Sci. **55**, 71 (2005) [arXiv:hep-th/0502005]
- R. Blumenhagen, M. Körs, D. Lüst, S. Stieberger, Four-dimensional string compactifications with D-branes, orientifolds and fluxes. Phys. Rep. **445**, 1 (2007) [arXiv:hep-th/0610327]

Reviews on intersecting brane models include

- A.M. Uranga, Chiral four-dimensional string compactifications with intersecting D-branes. Class. Quant. Grav. **20**, S373–S394 (2003) [arXiv:hep-th/0301032]
- F.G. Marchesano, Intersecting D-brane models [arXiv:hep-th/0307252]

Flux compactifications and moduli stabilization are reviewed in

- M. Graña, Flux compactifications in string theory: A comprehensive review. Phys. Rep. **423**, 91 (2006) [arXiv:hep-th/0509003]
- M.R. Douglas, S. Kachru, Flux compactification. Rev. Mod. Phys. **79**, 733 (2007) [arXiv:hep-th/0610102]

Our discussion of intrinsic torsion is based on

- S. Chiossi, S. Salamon, The intrinsic torsion of SU(3) and G(2) structures, math/0202282 [math-dg]
- S. Gurrieri, J. Louis, A. Micu, D. Waldram Mirror symmetry in generalized Calabi-Yau compactifications. Nucl. Phys. **B654**, 61 (2003) [arXiv:hep-th/0211102]
- G. Lopes Cardoso, G. Curio, G. Dall'Agata, D. Lüst, P. Manousselis, G. Zoupanos, Non-Kahler string backgrounds and their five torsion classes. Nucl. Phys. **B652**, 5 (2003) [arXiv:hep-th/0211118]

Chapter 18
String Dualities and M-Theory

Abstract The five superstring theories which we have encountered so far will be argued to be different perturbative limits of one underlying theory, connected via a web of perturbative and non-perturbative dualities. We will discuss the concept of dualities and the role of BPS states in establishing non-perturbative dualities. We construct brane solutions of supergravity, discuss their BPS properties and identify those carrying R-R charge as D-branes. We discuss the place of eleven-dimensional supergravity in this duality web and the self-consistent evidence for the existence of a so far unknown eleven-dimensional quantum gravity theory, called M-theory. The known facts include that its low-energy effective field theory is given by eleven-dimensional supergravity and that it contains membranes and five-branes, which upon compactification are related to the branes of string theory. Two examples of the power of dualities will be discussed in the final two sections: F-theory, an interesting class of non-perturbative compactifications of type IIB string theory and the AdS-CFT correspondence. They derive their appeal and usefulness from the fact that non-perturbative effects can be related to simple geometric structures.

18.1 General Remarks

A well known example of a duality is the particle-wave duality of quantum mechanics: depending on the experiment, either the particle or the wave aspect of light or matter gives the simpler description. In general, duality means an exact quantum equivalence between two theories $\mathscr{T}$ and $\tilde{\mathscr{T}}$, which thus really represent only one theory, albeit in different guises. To establish such a duality we must (1) identify the pair of theories which are proposed to be dual to each other and (2) find the duality map $\mathscr{T} \leftrightarrow \tilde{\mathscr{T}}$. Dualities in quantum field field theory and string theory are very intricate: $\mathscr{T}$ and $\tilde{\mathscr{T}}$ generally have different microscopic degrees of freedom and they have different actions. One of the attractive features of duality symmetries in string and field theory is that what is an emergent phenomenon in $\mathscr{T}$, might be fundamental in $\tilde{\mathscr{T}}$, and vice versa.

A field theoretical duality which we have already encountered is the duality between a free complex world-sheet fermion Ψ and a chiral boson ϕ. The two-dimensional world-sheet actions of these two theories indeed look very different. The map between the elementary fields appearing in the two actions is via bosonization: $\Psi(z) =: e^{i\phi(z)}:$ and $\phi(z) = -i \int^z dw : \overline{\Psi}\Psi(w) :$. These two relations demonstrate an important feature of a duality: the elementary fields of one theory might be a coherent superposition of the fields of the other theory or the relation might be a non-local one. All calculations of this $c = 1$ CFT can be done in either representation. They are quantum equivalent and the duality map is explicitly known. This example is simple in the sense that both theories are free theories. A generalization to a duality between two interacting theories, where the quantum equivalence is also established via bosonization, is between the two-dimensional massive Thirring model and the two-dimensional sine-Gordon model.

The common technique to treat interacting field theories is perturbation theory, which is also (so far) the well-established way to study string theory. This assumes that the theory $\mathcal{T}$ has a parameter, most often a coupling constant g, which is sufficiently small. One then computes correlation functions or S-matrix elements order by order in g and hopes that this is an expansion of the 'true' quantity in powers of g and that the first few terms in the expansion lead to good approximate results which are sufficient for comparison with existing experiments. The perturbation series usually does not converge and is at best an asymptotic expansion. Nevertheless, perturbation theory often works very well, for instance for QED with the fine-structure constant $\alpha \simeq \frac{1}{137}$ being the expansion parameter. As discussed in Chap. 6, the situation in string theory is that we only know the Feynman rules for computing on-shell string scattering amplitudes in a topological expansion. The expansion parameter is the string coupling constant $g = g_s$, and the expansion is certainly not meaningful for $g \simeq 1$, which is the regime where all loop effects ($\simeq g^n$) and non-perturbative effects ($\simeq \exp(-1/g^m)$) are of comparable size.

A duality between two theories $\mathcal{T}$ and $\tilde{\mathcal{T}}$ is most useful, if we can learn about the non-perturbative behavior (strong coupling) of $\mathcal{T}$ from the computable perturbative behavior (weak coupling) of $\tilde{\mathcal{T}}$. In other words, the idea behind a strong-weak duality is that for $g \gg 1$ there exists another dual theory $\tilde{\mathcal{T}}$ with a coupling constant $\tilde{g}$ so that the duality map works such that the strong coupling regime of $\mathcal{T}$ is mapped to the perturbative regime $\tilde{g} \ll 1$ of $\tilde{\mathcal{T}}$, e.g. $\tilde{g} = 1/g$. Under the duality map, the image of perturbative results of $\tilde{\mathcal{T}}$ generally includes perturbative and non-perturbative corrections of $\mathcal{T}$. Such a strong-weak duality is schematically shown in Fig. 18.1.

This is clearly an appealing idea, but obviously it is not easy to find the dual theory $\tilde{\mathcal{T}}$. Since one cannot simply solve the original theory $\mathcal{T}$ for $g \gg 1$, there seems to be nothing to compare to for testing the duality. As we will see in Sect. 18.4, supersymmetric theories provide a number of quantities which are protected against certain loop and/or non-perturbative corrections and can therefore be extrapolated from weak to strong coupling. They are good candidates to provide evidence for such dualities. Once one has a candidate for a dual pair ($\mathcal{T}, \tilde{\mathcal{T}}$), which satisfy non-trivial tests, one can move on and collect more evidence.

18.2 Simple Examples: Modular Invariance and T-Duality

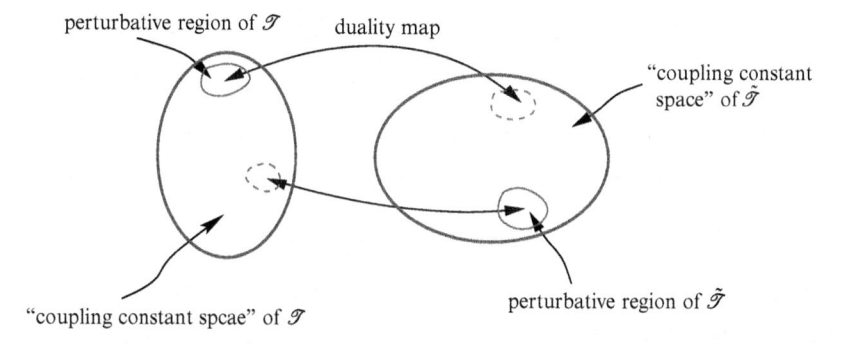

[Figure 18.1: diagram showing "perturbative region of $\mathcal{T}$", "duality map", "coupling constant space" of $\tilde{\mathcal{T}}$, "coupling constant spcae" of $\mathcal{T}$, and "perturbative region of $\tilde{\mathcal{T}}$"]

Fig. 18.1 Strong-weak duality between theories $\mathcal{T}$ and $\tilde{\mathcal{T}}$

Besides the strong-weak dualities, there also exist weak-weak dualities. Furthermore, it might be that $\mathcal{T} = \tilde{\mathcal{T}}$ and that one can map the weak coupling regime to the strong coupling regime of the same theory. A field theory example of the latter is the strong-weak self-duality of $\mathcal{N} = 4$ super Yang-Mills theory in four dimensions, which has the special feature of being conformal, i.e. the gauge coupling does not run. By this duality, electrically charged particles get exchanged with magnetically charged monopoles. We won't discuss this field theoretic example in detail, but in the course of this chapter we will present string theoretic analogues.

18.2 Simple Examples: Modular Invariance and T-Duality

To give an idea of how a strong-weak duality can work, let us discuss a very simple example. In Chap. 6 we have discussed conformal field theory on a torus with modular parameter τ. We choose for convenience $\tau = i/g$ with $g \in \mathbb{R}^+$ and consider only one chiral sector of the theory. Then the partition function of the CFT has an expansion (in powers on $q = \exp(2\pi i \tau)$)

$$Z(g) \simeq \sum_{n=0}^{\infty} a_n \, e^{-\frac{2\pi}{g} n}, \tag{18.1}$$

where a_0 can be considered a "tree-level" contribution and the exponentials for $n \geq 1$ can be interpreted as "non-perturbative" corrections in the "coupling constant" g. In this case, there are no further "perturbative" corrections, which would contribute like g^n, $n \geq 1$. Now, for $g \ll 1$, depending on the growth of a_n, there is a chance that the sum converges rapidly, whereas for $g \gg 1$ all contributions become of the same order and it seems hopeless to make any sense of the expansion in this "non-perturbative" regime. However, we know that such torus partition functions have the peculiar feature of modular invariance. Therefore, for $g \gg 1$ we simply introduce $\tilde{g} = 1/g$ and apply a modular S-transformation to the partition function $Z(g) = \tilde{Z}(\tilde{g})$ with

$$\tilde{Z}(\tilde{g}) \simeq \sum_{n=0}^{\infty} \tilde{a}_n \, e^{-\frac{2\pi}{\tilde{g}} n} = \sum_{n=0}^{\infty} \tilde{a}_n e^{-2\pi g n}, \qquad (18.2)$$

where in general $\tilde{a}_n \neq a_n$. Therefore, the behavior of (18.1) for $g \gg 1$ is given by the leading order ("tree-level") term $\tilde{a}_0$ in (18.2). This simple model illustrates the idea of a strong-weak duality. In fact, we have often made use of this fact in the course of this book, for instance for computing the tadpoles from open string diagrams.

We are also already familiar with a target-space duality, where the relevant parameter is the radius R of an S^1, on which the string theory is compactified. We found that there is no physical distinction between the bosonic string compactified on a circle of radius R or on a circle of radius $\tilde{R} = \alpha'/R$. We have also established the T-duality between the type II theories on a circle and also of the two heterotic theories. Consider, for instance, (10.25). We can view it as a strong-weak duality of the two-dimensional string sigma model with coupling constant $g_{2D} = \sqrt{\alpha'}/R$. Since it holds for any value of the string coupling constant $g_s = \exp(\langle \Phi \rangle)$, it is often called a perturbative (in g_s) duality. In other words, from the point of view of sigma-model perturbation theory it is a strong-weak duality, while it is weak-weak duality from the point of view of string perturbation theory.

The compactified theory has two types of excitations whose masses depend on the radius: the momentum or Kaluza-Klein modes, whose masses (in string units) are integer multiples of g_{2D}, and the solitonic[1] winding modes, whose masses are integer multiples of g_{2D}^{-1}. For $g_{2D} \ll 1$ the KK modes are light and the winding modes are heavy while for $g_{2D} \gg 1$ it is opposite: the solitonic winding states are light while the KK modes are heavy. The duality states that the winding/KK modes can be described as the KK/winding modes of a dual theory, which is the same world-sheet sigma model but on a circle of radius $\tilde{R} = \alpha'/R$, i.e. $\tilde{g}_{2D} = 1/g_{2D}$. A generalization of T-duality is mirror symmetry of Calabi-Yau compactifications, which was mentioned in Chap. 14.

---

[1] In field theory, solitons are non-trivial stable source-free solutions of the classical field equations with a finite action integral and non-zero topological charge. Examples are magnetic monopoles, whose energy is localized in three-dimensional space and which behave very much like point particles. In Yang-Mills theories one also has instantons, classical solutions of the equations of motion in Euclidean signature. They are localized in space and time and their non-trivial contributions to the YM path integral scale like $\exp(-S_{\text{inst}})$, where $S_{\text{inst}} \sim 1/g_{\text{YM}}^2$ is the classical action evaluated on the instanton solution. The string winding modes are solitonic in the sense that they are topologically stable. The conserved topological charge is the winding number. World-sheet instantons in string theory are topologically non-trivial embeddings of the euclideanized world-sheet into the space-time manifold. We have met them in Chap. 14.

18.2 Simple Examples: Modular Invariance and T-Duality

Beyond winding modes and world-sheet instantons, there are other solitonic objects in type II superstring theory, namely the Dp-branes which we have described as boundary states of the type II theories but which we have also defined from an open string world-sheet perspective. Computing annulus diagrams, we have shown that they are dynamical p-dimensional objects with physical tension

$$\tau_p = \frac{1}{g_s} T_p = \frac{1}{g_s} \frac{2\pi}{l_s^{p+1}} \sim 1/g_s \tag{18.3}$$

and non-vanishing R-R $(p+1)$-form charge $\mu_p = T_p$. The Dp-brane tension is non-perturbative in the string coupling constant g_s and its mass (per unit world-volume) is infinite in the free theory ($g_s = 0$). We will show in Sect. 18.5 that such objects can be found as classical solutions to the ten-dimensional type II effective supergravity equations of motion. T-duality along a direction transverse to the Dp-brane maps it to a D(p+1)-brane while T-duality along a world-volume direction maps it to a D(p-1)-brane. Thus, non-perturbative objects with tension $\sim 1/g_s$ are mapped to non-perturbative objects with tension $\sim 1/g_s$. This again indicates that T-duality does not mix different powers of the string coupling g_s constant.

Contrary to the first impression, the T-duality between type IIA and type IIB is a self-duality. Once the two theories are compactified on a circle, they are simply different points in the moduli space of one theory, namely type II string theory on S^1 where the modulus is the radius of the circle. The spectra of the two compactified theories are identical, if one theory is compactified on a circle of radius R and the other on a circle of radius $\tilde{R}$. This is easily seen for the massless modes (independent of R). Note that in nine dimensions there is no chirality, which might distinguish between IIA and IIB in the fermionic sector. The two boundary points of the moduli space are special: here the two theories are different, one being chiral the other being non-chiral. This is a consequence of an infinite number of particles becoming massless. If we follow for instance the type IIA theory from $R = \infty$ to $R = 0$ (which is the same as $\tilde{R} = \infty$), it emerges as type IIB. In the $R \to \infty$ limit the infinitely many light momentum modes are the signal of a new large dimension while in the $R \to 0$ limit the winding modes play this role.

Similarly, the two heterotic theories compactified on a torus share a common moduli space which is parametrized by the background values of the metric, the anti-symmetric tensor field and of the gauge field (Wilson lines).

As we will see below, in string theory we encounter a web of perturbative and non-perturbative dualities, which involves all five perturbatively defined superstring theories, type IIA/B, type I and the two heterotic theories, either in the critical dimension or compactified on different manifolds. The dualities are of the kind

$$\text{string theory } \mathscr{T} \text{ on } M \quad \longleftrightarrow \quad \text{string theory } \tilde{\mathscr{T}} \text{ on } \tilde{M}. \tag{18.4}$$

They are highly non-trivial and involve the non-perturbative solitonic objects in an intricate manner. There is no proof of the non-perturbative dualities, but sufficiently robust arguments have been collected to be confident that they should be true. In any case, they lead to a non-trivial web of string dualities which has passed many consistency checks.

18.3 Extended Objects: Some Generalities

Before 1995 string theory was only defined and studied via its perturbative expansion. As we discussed in detail in previous chapters, there exist five consistent ten dimensional superstring theories, which are listed in Table 18.1, together with some of their distinctive properties.

For each of the five string theories one can set up a perturbative expansion as a sum over world-sheet topologies. The expansion (loop counting) parameter is $g_s = \exp(\langle \Phi \rangle)$ and a world-sheet Σ contributes at order $g_s^{-\chi(\Sigma)}$. Of course, each of the five theories has its own dilaton, and hence its own coupling constant, and they are, a priori, unrelated. For $g_s \gtrsim 1$ all terms in the series are equally important and the expansion in powers of g_s is not meaningful. Furthermore, non-perturbative effects in g_s contribute to the scattering processes. They are loosely referred to as space-time instantons, and are quite distinct from world-sheet instantons, which are non-perturbative in the σ-model coupling constant. By analogy to field theory, the contributions of the space-time instantons are expected to be of the type

$$\mathcal{A} \sim \exp\left(-\frac{\text{const.}}{g_s^n}\right) \quad \text{with } n \in \mathbb{Z}_+. \tag{18.5}$$

In quantum field theory, we expand the path integral around field configurations solving the classical equations of motion, e.g. instantons in Yang-Mills-theories. In string theory we do not have a path integral formulation in terms of string fields and neither do we know the classical equations of motion beyond the leading order in g_s and the first orders in α'. Recall that at tree-level in g_s, the closed string equations of motion were defined as the vanishing of the beta-functional (14.5) for the non-linear sigma model. Therefore the question of classical (i.e. tree-level in g_s) solutions is mostly addressed within the framework of the low-energy effective actions. Recall that the classical world-volume action (the DBI action) of D-branes is proportional to g_s^{-1}, which reflects its origin from open string disk amplitudes. Thus, they are promising candidates for non-perturbative effects contributing to amplitudes as in (18.5) with $n = 1$. Experience from field theory, however, suggests the existence of classical solutions whose actions scale like g_s^{-2} leading to the suppression factor $\exp(-c/g_s^2)$. In Yang-Mills-theory (with $g_s \to g_{\text{YM}}$) these contributions to amplitudes arise from magnetic monopoles and instanton configurations.

We will find that string theory also contains branes, which do not carry R-R charge, but instead couple to the NS-NS two-form and its Hodge dual NS-NS six-form. Recall that the Hodge duality is between the respective field strengths so that,

18.3 Extended Objects: Some Generalities

Table 18.1 The five consistent superstring theories in d = 10

Type	Gauge group	# of supercharges	$\mathcal{N}$
Heterotic	$E_8 \times E_8$	16	1
Heterotic'	$SO(32)$	16	1
I (includes open strings)	$SO(32)$	16	1
IIA (non-chiral)	$U(1)$	$16 + \overline{16}$	2
IIB (chiral)	–	$16 + 16$	2

in this sense, the ten-dimensional Hodge dual of a p-form potential is a $(8-p)$-form potential, i.e. locally

$$dC_{8-p} = \star dC_p. \tag{18.6}$$

The object which is charged under the NS-NS two-form $B_{\mu\nu}$ is the fundamental string itself. The solitonic object which is magnetically charged with respect to $B_{\mu\nu}$ is an NS five-brane. We will find that the tension of the NS five-brane scales like g_s^{-2}.

Before we embark on the construction of explicit supergravity brane solutions which, as expected, are charged under Abelian higher form gauge fields, we discuss some of the features of these fields and their couplings. This is, to a large extent, a straightforward generalization of the well known four-dimensional Maxwell theory.

An electric point charge, e.g. an electron, with a one-dimensional world-line couples to a one-form, the $U(1)$ gauge potential. Likewise, a p-brane with a $(p+1)$-dimensional world-volume couples 'electrically' to a $(p+1)$-form gauge potential A_{p+1} with field strength $F_{p+2} = dA_{p+1}$. The action for A_{p+1} is

$$\begin{aligned} S[A] &= -\frac{1}{2} \int_M F_{p+2} \wedge \star F_{p+2} + \int_M A_{p+1} \wedge \star j^e_{p+1} \\ &= -\int_M d^d x \sqrt{|g|} \left(\frac{1}{2(p+2)!} F_{\mu_0 \ldots \mu_{p+1}} F^{\mu_0 \ldots \mu_{p+1}} - \frac{1}{(p+1)!} A_{\mu_0 \ldots \mu_p} j_e^{\mu_0 \ldots \mu_p} \right) \\ &= -\frac{1}{2} \int_M F_{p+2} \wedge \star F_{p+2} + q_e \int_{\Sigma_{p+1}} A_{p+1}, \end{aligned} \tag{18.7}$$

where the second term is the coupling to the brane. The part of the action which governs the dynamics of the brane, the DBI action, is independent of A_{p+1} and is not written in (18.7). The $(p+1)$-form 'electric' current density is the generalization of the electric current density of a charged point particle. Its contravariant components are[2]

---

[2] The δ-function satisfies $\int_M d^d x \, \delta^d(x) = \int_M d^d x \sqrt{|g|} \frac{1}{\sqrt{|g|}} \delta^d(x) = 1$. If Σ_p has no boundary, $\star j^e_{p+1}$ is essentially its Poincaré dual.

$$j_e^{\mu_0\ldots\mu_p}(x) = q_e \int_{\Sigma_{p+1}} d^{p+1}\xi \, \frac{\delta^{(d)}(x-X(\xi))}{\sqrt{|g(x)|}} \, \frac{\partial(X^{\mu_0},\ldots,X^{\mu_p})}{\partial(\xi^0,\ldots,\xi^p)}$$

$$= q_e \int_{\Sigma_{p+1}} \frac{\delta^{(d)}(x-X(\xi))}{\sqrt{|g(x)|}} dX^{\mu_0} \wedge \cdots \wedge dX^{\mu_p}. \tag{18.8}$$

q_e is the 'electric' charge of the brane, $X^\mu(\xi)$ are the embedding functions of the brane into space-time M and ξ^α are local coordinates on Σ_{p+1}. In the last term of the second line of (18.7) and elsewhere the pull-back of A_{p+1} to the brane is understood. The action is invariant under $A_{p+1} \to A_{p+1} + d\Lambda_p$ provided the current is conserved, i.e. $d \star j_{p+1}^e = 0$. This is the case if the world-volume has no boundary. Indeed,

$$\nabla_{\mu_0} j_e^{\mu_0\ldots\mu_p} = \frac{q_e}{\sqrt{|g(x)|}} \int_{\Sigma_{p+1}} d[\delta^{(d)}(x-X(\xi)) \wedge dX^{\mu_1} \wedge \cdots \wedge dX^{\mu_p}] \tag{18.9}$$

with $d = dX^\mu \partial_{X^\mu}$. The Bianchi identity, which holds by virtue of the definition the field strength, and the field equations for A_{p+1} are

$$dF_{p+2} = 0 \tag{18.10a}$$

$$d \star F_{p+2} = (-1)^p \star j_{p+1}^e \qquad (\nabla_\nu F^{\nu\mu_0\ldots\mu_p} = j_e^{\mu_0\ldots\mu_p}). \tag{18.10b}$$

These are generalizations of the homogeneous and inhomogeneous Maxwell equations, respectively. Current conservation

$$d \star j_{p+1}^e = 0 \qquad (\nabla^\nu j_{\nu\mu_1\ldots\mu_p}^e = 0) \tag{18.11}$$

follows from (18.10b). Integrating (18.11) over a $(d-p)$-dimensional manifold M with boundary ∂M one obtains $\int_{\partial M} \star j_e = 0$. Take ∂M to have topology of a cylinder with base B_{d-p-1} (and $\partial B_{d-p-1} = S^{d-p-2}$) and time-like walls. For currents which are localized in space such that the integral over the time-like walls gives zero contribution, one finds that the electric charge

$$q_e = \int_{B_{d-p-1}} \star j_{p+1}^e = (-1)^p \int_{S^{d-p-2}} \star F \tag{18.12}$$

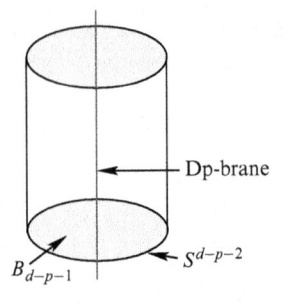

18.3 Extended Objects: Some Generalities

is independent of the choice of spatial slice through the cylinder and is therefore conserved.

The solution to the equation of motion for F_{p+2} with the p-brane as the source is generally singular at the location of the p-brane, analogous to the $1/r^2$ electric field for a point charge in four dimensions. But the source-free coupled equations of motion for the metric, the dilaton and the gauge potential also admit solitonic solutions. They have the structure of a $(d-p-4)$-brane carrying magnetic charge[3]

$$q_m = \int_{S^{p+2}} F_{p+2}. \tag{18.13}$$

For the magnetic charge to be non-zero, the potential A_{p+1} cannot be globally defined. The magnetic charge is a topological charge, i.e. it is conserved without the use of the equations of motion. This follows from $q_m = \int_{B_{p+3}} dF$ and $d^2 F \equiv 0$.

Consider the action (18.7) for a world-volume Σ with the topology of S^{p+1}. We can write the interaction term as $q_e \int_{H^+} F$ where H^+ is a hemisphere whose boundary is Σ. But we could just as well take H^- such that the interaction term becomes $-q_e \int_{H^-} F$ (the minus sign is due to the relative orientations); i.e.

$$\exp\left(iq_e \int_\Sigma A\right) = \exp\left(iq_e \int_{H^+} F\right)$$
$$= \exp\left(-iq_e \int_{H^-} F\right)$$
$$= \exp\left(iq_e \int_{H^+} F\right) \exp\left(-iq_e \int_{S^{p+2}} F\right). \tag{18.14}$$

If S encloses a magnetic charge, there is an ambiguity $\exp(-iq_e \int_{S^{p+2}} F)$ in transition amplitudes. This ambiguity is invisible provided $q_e \int_S F = 2\pi n$, or

$$q_e q_m = 2\pi n \quad \text{with} \quad n \in \mathbb{Z}. \tag{18.15}$$

This is the Dirac quantization condition.

Notice that the elementary object in this discussion is the electric p-brane. It makes an appearance in the action (18.7) through the current j_e. The dual magnetic $(d-p-4)$-brane is a composite, solitonic, object and does not appear as a degree of freedom in the action.

---

[3] The Dirac monopole satisfies $\nabla \cdot B = q_m \delta^3(x)$ and it is not a soliton. Those can only occur in non-linear systems. The solitonic brane solutions are more closely related to the 't Hooft-Polyakov monopole of the $SO(3)$ Yang-Mills-Higgs model where the non-linearity is due to the non-Abelian gauge group and the Higgs sector.

There exists a dual description of the system where the role of the elementary and the composite objects are interchanged. It is $-\frac{1}{2}\int F^{(D)} \wedge \star F^{(D)} - \int A^{(D)} \wedge \star j^m$ where $(j^m)_{d-p-3}$ is a magnetic current which couples to the dual gauge field $A^{(D)}_{d-p-3}$. For the magnetic brane solution, the relation between A and $A^{(D)}$ is $dA^{(D)}_{d-p-3} = \star dA_{p+1}$. In this dual formulation $q_m = \int \star F^{(D)}$ and therefore the $(d-p-4)$-brane is the 'electrically' charged object while the p-brane is a solitonic solution of the source-free equations of motion for $A^{(D)}$ with $q_e = \int F^{(D)}$. There is no local action which contains both the electric and the magnetic object and the respective gauge fields to which they couple.

With reference to the figure in (18.12) it is clear that an open brane cannot carry charge because if the brane ended in free space we could slide off and shrink the S^{d-p-2} to a point. Charged branes are therefore of infinite extend or wrap compact cycles (submanifolds) of space-time, in which case they are topologically stable. If they wrap submanifolds which can be contracted to a point the brane collapses under its tension unless it is stabilized e.g. by rotation or by an external field to which it couples. Another possibility is that a brane ends on another brane (e.g. string ends on a D-brane).

In Sect. 18.5 we will describe electric and magnetic branes as classical solutions of the low energy effective supergravity theories which are derived from the various superstring theories. They provide insight into the nature of the non-perturbative degrees of freedom in string theory which is a fundamental ingredient of the conjectured non-perturbative duality symmetries. Like T-duality, these dualities are supposed to establish the complete equivalence of two (seemingly different) string theories, but now the duality transformation maps the weak coupling regions $g_s \ll 1$ of one theory to the strong coupling $g_s \gg 1$ regions of the other, and vice versa. We will see that they exchange elementary string excitations and the solitonic p-branes. Several different kinds of such non-perturbative dualities have been proposed, namely the so-called S- and U-dualities, string dualities between heterotic strings and type I strings, string dualities between heterotic and type II superstrings and the duality between the ten-dimensional type IIA superstring and a new theory in eleven dimensions, called M-theory. This proposed theory, which is not a string theory, has eleven-dimensional supergravity as its low energy effective field theory. We will learn more about this still largely elusive theory later in this chapter.

18.4 Central Charges and BPS Bound

To prove a duality symmetry that relates the spectrum and scattering amplitudes of a theory at weak coupling to those of a theory at strong coupling is very difficult. Generically one expects that masses and amplitudes receive coupling constant dependent corrections. The arguments in favor of a non-perturbative duality in string theory are based on space-time supersymmetry. An important role is played by the so-called BPS (Bogomol'nyi-Prasad-Sommerfield) states. They belong to 'short' representations of the supersymmetry algebra. This implies that there is

18.4 Central Charges and BPS Bound

a precise relationship, the BPS bound, between their masses and 'charges' which is protected by supersymmetry from perturbative and non-perturbative corrections. These corrections are not expected to generate additional degrees of freedom, which would be needed to convert short into long multiplets, whose masses and charges do not saturate the BPS bound. However, the BPS property does, in general, not imply that masses and charges are not corrected. What is protected is their equality.

We first explain the idea for a simple example which is familiar from Chap. 12, where we have discussed the representation theory of two-dimensional $N = 2$ superconformal field theories. Recall that the algebra of the globally defined superconformal generators includes the anti-commutator

$$\{G^\pm_{+1/2}, G^\mp_{-1/2}\} = 2 L_0 \pm J_0. \tag{18.16}$$

In a unitary theory, for an NS sector state with $L_0|h,q\rangle = h|h,q\rangle$, $J_0|h,q\rangle = q|h,q\rangle$ and $G^\pm_{+1/2}|h,q\rangle = 0$, this implies the bound $h \geq \frac{1}{2}|q|$ for the "mass" and charge of any such state. When the bound is saturated, we found $G^\pm_{-1/2}|h = \pm\frac{q}{2}\rangle = 0$, i.e. for each choice of sign one of the supercharges acts trivially on the state. As a consequence, for such states the generic supermultiplet (i.e. $N = 2$ superprimary field)

$$|h,q\rangle \begin{array}{c} \nearrow G^+_{-\frac{1}{2}}|h,q\rangle \searrow \\ \searrow G^-_{-\frac{1}{2}}|h,q\rangle \nearrow \end{array} G^+_{-\frac{1}{2}} G^-_{-\frac{1}{2}}|h,q\rangle \tag{18.17}$$

was reduced from four to only two states, cf. Eqs. (12.76) and (12.77) on page 372.

What we just described is also a general feature of super-algebras with central charges. First, they imply a bound for the mass of a particle and second, the states saturating the bound (BPS states) come in short supermultiplets. Central charges of a (super) Lie-algebra are generators which commute with all generators. By Schur's lemma, they are constant on any irreducible representation. The familiar example is the central charge of the (super) Virasoro algebra.

Central charges appear in extended supersymmetry algebras. A simple way to construct them, and this is one way how they appear in string theory, is by starting with the supersymmetry algebra in higher dimensions and to reduce on a torus. For concreteness, we consider the $\mathcal{N} = (0,1)$ supersymmetry of the heterotic string in $d = 10$. The 16 supercharges Q_α, which are the components of a Majorana-Weyl spinor of $SO(1,9)$, satisfy

$$\{Q_\alpha, Q_\beta\} = -p_M(\Gamma^M C^{-1})_{\alpha\beta}. \tag{18.18}$$

We now work in a Majorana representation where $C = \Gamma^0$ and Majorana spinors are real. We now split the momentum into a D-dimensional space-time and a $10 - D$ dimensional internal part, i.e. $p^M = (p^\mu, p^I)$. We are interested in massive representations, where in the rest frame we simply have $p^\mu = (M, 0, \ldots, 0)$. The algebra (18.18) decomposes as[4]

$$\{Q_\alpha, Q_\beta\} = M \delta_{\alpha\beta} + p_I (\Gamma^I \Gamma^0)_{\alpha\beta}. \tag{18.19}$$

The matrix $Z = \Gamma^I \Gamma^0 p_I$ satisfies $Z^2 = p_{\text{int}}^2$ and $\operatorname{tr} Z = 0$. This means that it has eigenvalues $\pm |p_{\text{int}}|$, each eightfold degenerate. We diagonalize the r.h.s. of (18.19) with a unitary (in fact, an orthogonal) matrix U. If we define $\tilde{Q} = UQ$ and $|p_{\text{int}}| = zM$, (18.19) becomes

$$\{\tilde{Q}_\alpha, \tilde{Q}_\beta\} = M \begin{pmatrix} (1+z)\mathbb{1} & 0 \\ 0 & (1-z)\mathbb{1} \end{pmatrix}, \tag{18.20}$$

where $\mathbb{1}$ is the 8×8 unit matrix. Define ($j = 1, \ldots, 8$)

$$q_j = \frac{1}{\sqrt{2M}} (\tilde{Q}_{2j} + i\, \tilde{Q}_{2j-1}), \qquad q_j^\dagger = \frac{1}{\sqrt{2M}} (\tilde{Q}_{2j} - i\, \tilde{Q}_{2j-1}). \tag{18.21}$$

They satisfy the algebra of eight independent fermionic oscillators

$$\{q_i, q_j^\dagger\} = (1+z)\delta_{ij}, \quad i, j = 1, \ldots 4,$$
$$\{q_i, q_j^\dagger\} = (1-z)\delta_{ij}, \quad i, j = 5, \ldots 8. \tag{18.22}$$

All other anti-commutators vanish. Positive semi-definiteness of $q_i q_i^\dagger$ (no summation) leads to the BPS bound $z \leq 1$ for the ratio of the mass to the internal momentum.

The representation theory of the algebra (18.22) depends crucially on whether the bound is saturated or not. The dimension of non-BPS representations, where the bound is not satisfied, is 2^8 with 2^7 bosonic and 2^7 fermionic states (we assume that the oscillator ground state is a singlet under Lorentz transformations). They are called long multiplets. If the bound is saturated, half of the oscillators decouple and can be represented by zero and one obtains short multiplets with 2^4 states. States in short multiplets are called BPS states. They are invariant under half of the supercharges.

So far the discussion is valid for compactified field theories. If we compactify string theory, the supercharges, say of the heterotic string, are derived from the holomorphic part of the gravitino vertex operator, cf. Sect. 15.4. Therefore, what appears on the r.h.s. of the supersymmetry algebra, is the holomorphic momentum

---

[4] We could also decompose the supercharges into representations of $SO(1, D-1) \times SO(10-D)$. For instance, if we compactify on T^6, we have the decomposition (14.40) and obtain the $\mathcal{N} = 4$ SUSY algebra in $D = 4$.

18.4 Central Charges and BPS Bound

operator $p_R^M = \frac{2}{\alpha'} \oint i \partial X^M$ with eigenvalues p^μ and $p_R^I = p^I - \frac{1}{\alpha'} L^I$, the right-moving mass and momenta. With $p^2 = -m^2 = -2m_R^2$ the BPS condition is $2m_R^2 = (p_R^I)^2$. In Chap. 10 we derived the mass formula (10.39) for a toroidal compactification of the bosonic string. The generalization to the fermionic string is straightforward and gives for the holomorphic part

$$\alpha' m_R^2 = \frac{\alpha'}{2} \sum_{I=1}^{10-D} \left(p^I - \frac{1}{\alpha'} L^I \right)^2 + 2N_R - \begin{cases} 1 & \text{NS} \\ 0 & \text{R} \end{cases}. \qquad (18.23)$$

$N_R = N_{\text{tr}}^a + N_{\text{tr}}^b$ counts the transverse oscillations of the fermionic string. We find that BPS states of the toroidally compactified heterotic string are in the oscillator ground state (R) or satisfy $N_R = \frac{1}{2}$ (NS). For the left-movers there is no restriction other than level-matching. At the massless level this leaves the gauge bosons and gaugini as BPS states.

We know from Chap. 10 that p_R^I depends on the quantized momentum and winding numbers. This implies that the masses of BPS states are not quantum corrected. We also know that the winding and momentum numbers are the charges with respect to $U(1)$ Kaluza-Klein gauge fields. They are the $G_{\mu I}$ and $B_{\mu I}$ components of the metric and anti-symmetric tensor, where the former couple to momentum and the latter to winding. The winding number in the I-th direction is $L^I = \frac{1}{2\pi} \int d\sigma \, \partial_\sigma X^I = \frac{1}{2\pi} \int dX^I$. We can then write the SUSY algebra of the compactified string as

$$\{Q_\alpha, Q_\beta\} = p_M (\Gamma^M \Gamma^0)_{\alpha\beta} + (\Gamma^M \Gamma^0)_{\alpha\beta} Z_M \qquad (18.24)$$

with

$$Z^M = -\frac{1}{2\pi\alpha'} \int dX^M, \qquad (18.25)$$

where the integral is over the length of the string. We will discuss generalizations of this extension of the algebra for branes in Sect. 18.7.

We should mention that the extended SUSY algebra, which we have discussed here, is not the most general one. For instance, the $\mathcal{N}$-extended algebra in $D = 4$ allows for $\mathcal{N}/2$ and $(\mathcal{N}-1)/2$ independent central charges z_i for $\mathcal{N}$ even and odd, respectively. The number of oscillators which decouple depends on how many of the z_i satisfy the bound $M \geq |z_i|$. If n is the number of z_i which saturate the bound, the number of states in a representation is reduced to $2^{2(\mathcal{N}-n)}$. In the extremal case, e.g. for $\mathcal{N}$ even, $n = \mathcal{N}/2$, half of the supercharges are trivially represented, i.e. they leave all states invariant. One speaks of 1/2-, 1/4-, etc. BPS states if 1/2, 1/4, etc. of the supercharges act trivially. In the next section we discuss brane solutions of supergravity. We will show explicitly that they are annihilated by half of the supercharges, i.e. that they are 1/2-BPS configurations.

BPS states are of great importance for establishing dualities, in particular if one of the theories in a duality pair is strongly coupled. The reason is that most of the (quantitative) knowledge we have about a quantum field or string theory is the result of a weak coupling, i.e. perturbative analysis and an extrapolation to strong coupling is usually not controllable. In particular, results derived from the classical action can be completely spoiled by large quantum effects. But there are examples where the extrapolation to strong coupling can be reliably performed: this involves BPS states of supersymmetric theories. Their mass is determined by their charges and this relation, being a result of the representation theory of the supersymmetry algebra, must be true for all values of the coupling constant unless several short multiplets combine into longer multiplets which can violate the BPS bound. If the charge is quantized, the mass of BPS states cannot change continuously. In string theory such quantized charges can be momenta and winding numbers in compact directions. Therefore, once the BPS states are identified in the classical approximation of some superstring theory, they should exist also in the non-perturbative regime.

Due to cancellations between bosons and fermions, supersymmetric field theories often possess parameters which do not suffer from quantum corrections. These protected parameters are moduli of the theory, exactly determined by their bare values. In $\mathcal{N} = 4$ super Yang-Mills theory, the gauge coupling constant g_{YM} is such a free parameter; it does not run. If we know the exact dependence, for instance of the central charges, on these moduli at weak coupling, which we can in principle compute, we can extrapolate them safely to strong coupling. Similarly, for string theories with enough supersymmetry, the dilaton will be a modulus of the theory, which implies that the string coupling g_s is a free parameter. Thus, if we compute the dependence of a central charge, and thus of the mass of a BPS state, on g_s in the perturbative regime, this will be the dependence for all values of g_s.

After discussing BPS states in the supergravity limit of string theory in the next section, we will describe several non-perturbative string dualities. Together with T-duality and the duality to eleven-dimensional supergravity (M-theory), it is possible to connect all ten-dimensional superstring theories through a web of dualities. But they are also the seeds for more intricate duality relations between compactified string theories in lower dimensions. One is led to the conclusion that all seemingly different superstring theories are equivalent to each other and are merely different perturbative vacua of one common theory, called M-theory.

To provide an overview of what is going to be discussed in the following sections, Table 18.2 summarizes basic superstring dualities.

18.5 Brane Solutions in Supergravity

One strategy to find states of non-perturbative nature is to look for non-trivial solutions of the classical equations of motion of the low energy approximation to string theory, i.e. of the supergravity theories. These equations turn out to have brane solutions. They are non-perturbative, i.e. their masses/tensions are proportional to

18.5 Brane Solutions in Supergravity

Table 18.2 Basic superstring dualities

	Acts between	Acts on BPS states
Pert. T-duality: $R \leftrightarrow 1/R$	Self-duality of 9D heterotic string 9D type IIA $\leftrightarrow$ 9D type IIB string	KK momentum $\leftrightarrow$ winding KK momentum $\leftrightarrow$ winding
NP. S-duality: $g_{H,IIB} \leftrightarrow 1/g_{H,IIB}$	Self-duality of 4D heterotic string Self-duality of 10D type IIB string	Wrapped F1 $\leftrightarrow$ Wrapped NS5 F1 $\leftrightarrow$ D1 NS5 $\leftrightarrow$ D5 D3 $\leftrightarrow$ D3
NP. het./type IIA duality: $g_6^H \leftrightarrow 1/g_6^{IIA}$	Het on T^4 $\leftrightarrow$ type IIA on K3	F1 $\leftrightarrow$ Wrapped NS5
NP. het./type I duality: $g_H \leftrightarrow 1/g_I$	10D Het($SO(32)$) $\leftrightarrow$ 10D type I	F1 $\leftrightarrow$ D1 NS5 $\leftrightarrow$ D5
NP. type IIA/M duality: $g_{IIA} \leftrightarrow R_{11}^{3/2}$	10D type IIA $\leftrightarrow$ 11D SUGRA on S^1	F1 $\leftrightarrow$ Wrapped M2 D2 $\leftrightarrow$ M2 D4 $\leftrightarrow$ Wrapped M5 NS5 $\leftrightarrow$ M5 D6 $\leftrightarrow$ KK monopole
NP. Het/F-theory duality: $g_H \leftrightarrow \text{Vol}(\mathbb{CP}^1)$	Het. on T^2 $\leftrightarrow$ F-theory on K3	F1 $\leftrightarrow$ D1

NP means non-perturbative and F1 the fundamental string

inverse powers of the string coupling $\tau \sim 1/g_s^n$ (when measured in the string frame), while the states of the perturbative spectrum have masses independent of g_s.

We have seen in the previous section that a p-brane couples electrically to a $(p+1)$-form and magnetically to the dual $(d-p-3)$-form. Therefore, the dimensionality of the extended charged objects which we expect in the various string models can be determined by simply inspecting their bosonic field contents. The universal NS-NS sector is common to both heterotic, the IIA and the IIB model. Its 2-form potential $B_{\mu\nu}$ couples to the respective fundamental string through the familiar coupling in the world-sheet σ-model. The solitonic object which couples magnetically to the NS-NS 2-form is called NS5-brane. The solutions which are charged under R-R sector fields are expected to be the D-branes which we encountered by studying the open string with various boundary conditions. This means that quantum fluctuations around these classical solutions are excitations of open strings which end on the brane. Notice that in type IIB theory, where we have a $SL(2;\mathbb{Z})$ doublet of two-forms, we can have (p,q) strings which carry p (q) times the NS-NS (R-R) charge of F1 (D1) and likewise (p,q) five-branes. There are also (p,q) seven-branes on which (p,q) strings can end. We will discuss them later in this chapter. Also in type IIB, the D3-brane is self-dual because F_5 is. For the type I superstring only the Dp-branes with $p = 1$ mod 4 survive the orientifold projection. Since the NS-NS two-form is projected out, the theory does not contain BPS NS5-branes and (oriented) fundamental string solutions.

What remains to be shown is that the effective ten- and eleven-dimensional supergravity theories indeed admit solitonic brane solutions. A p-brane solution of the equations of motion should contain a non-trivial $A_{(p+1)}$ background configuration whose energy-momentum tensor sources a non-trivial space-time metric. We will see that for most solutions there is also a non-constant dilaton. From Chap. 16 we know the low-energy effective action for these massless string excitations:

$$S = \frac{1}{2\kappa_d^2} \int d^d x \sqrt{-g}\left(R - \frac{1}{2}g^{MN}\partial_M\phi\partial_N\phi - \frac{1}{2(p+2)!}e^{a_p\phi}\tilde{F}^2_{(p+2)}\right). \quad (18.26)$$

This action is related to the ones in Chap. 16 by first splitting off the asymptotic value Φ_0 of the dilaton via $\Phi = \Phi_0 + \phi$ with $e^{\Phi_0} \equiv g_s$ and then rescaling the metric G_{MN} via $G_{MN} = e^{\frac{1}{2}\phi}g_{MN}$. We have also rescaled the gauge fields $\tilde{F}_{RR} = g_s F_{RR}$ and $\tilde{F}_{NSNS} = F_{NSNS}$ and $\tilde{F}^2_{(p+2)} = \tilde{F}_{M_0...M_{p+2}}\tilde{F}^{M_0...M_{p+2}}$. The parameter a_p controls the dilaton coupling. For the NS-NS 2-form it is $a_{-1} = -1$ while in the R-R sector $a_p = (3-p)/2$ with $a_p + a_{p'} = 0$ for $p + p' = 6$. In (18.26) we have dropped the topological Chern-Simons term. For string theory $d = 10$, and $2\kappa_{10}^2 = (2\pi)^7\alpha'^4 g_s^2 = \frac{1}{2\pi}\ell_s^8 g_s^2$. If we drop the dilaton and set $d = 11$, $p = 2$, $a_p = 0$ the action is that of $d = 11$ supergravity, whose brane solutions will also be found in this chapter. In $d = 11$ we also define the Planck length ℓ_{11} via $2\kappa_{11}^2 = \frac{1}{2\pi}(2\pi\ell_p)^9$.

We will pause here to comment on various frames which are used in string theory. A frame is defined by the metric. The string frame metric is the one which appears in the Polyakov (sigma-model) action. Various other frames are obtained from the Einstein frame by a dilaton dependent rescaling. The Einstein frame is defined such that in the low-energy effective action the kinetic energy for dilaton and metric is diagonal or, in other words, the Ricci scalar is not multiplied by a dilaton dependent factor. The frame we have used in (18.26) is called modified Einstein frame. It is related to the Einstein frame through a rescaling of the metric by the constant (asymptotic) value of the dilaton. The solutions which we will find below are such that the metric (in the modified Einstein frame) approaches the Minkowski metric η_{MN} asymptotically. Masses, tensions, etc., which are measured by asymptotic observers, can therefore be directly compared with those measured in Minkowski space; no rescaling is necessary. In the Einstein frame the asymptotic metric of our solutions is $\eta_{MN}/\sqrt{g_s}$ and the masses measured in different frames are related via $m_S^2 = m_{mE}^2 = m_E^2/\sqrt{g_s}$. From here on, in this section, Einstein frame always means modified Einstein frame.

If only one of the $(p+1)$-form fields is non-trivial, in which case $F_{p+2} = dA_{p+1}$, the equations of motion derived from the action are

$$\Box\phi = \frac{a_p}{2(p+2)!}e^{a_p\phi}\tilde{F}^2,$$

$$\nabla_M\left(e^{a_p\phi}\tilde{F}^{MM_1...M_{p+1}}\right) = 0,$$

$$R_{MN} = \frac{1}{2}\partial_M\phi\partial_N\phi + \frac{e^{a_p\phi}}{2(p+1)!}\left(\tilde{F}^2_{MN} - \frac{(p+1)}{(d-2)(p+2)}g_{MN}\tilde{F}^2\right), \quad (18.27)$$

where $\tilde{F}_{MN}^2 = \tilde{F}_{MM_1...M_{p+1}} \tilde{F}_N{}^{M_1...M_{p+1}}$. The equation of motion for the dilaton implies that for non-vanishing $\tilde{F}^2$ it cannot be constant unless $a_p = 0$. This is the case for the $d = 11$ SUGRA action and for $d = 10$, $p = 3$. However for $p = 3$ the action (18.26) is not valid since $\tilde{F}_{(5)}^2 = 0$ for self-dual $\tilde{F}_{(5)}$. The equations of motion are nevertheless the same as those derived from (18.26) if one multiplies the $\tilde{F}_5^2$ term by $1/2$ and imposes the self-duality condition $\tilde{F}_{(5)} = \star \tilde{F}_{(5)}$ separately. This special case plays a role for the AdS/CFT correspondence which we discuss in Sect. 18.9. There is also the open string contribution to the action which acts as a source term for the electric brane solutions. More on this below.

Before we proceed to solve the equations of motion, let us comment on the validity of the effective action (18.26). It assumes that the string is weakly coupled so that only the spherical world-sheet topology contributes: otherwise the dilaton dependence is more complicated. We have furthermore assumed that all energies are small compared to the string scale: otherwise higher derivative corrections have to be included, in particular higher powers of the space-time curvature. As a consequence, the solutions which we find can only be trusted if the parameters are chosen such that the curvatures are small compared to $1/\alpha'$.

In the following n denotes the dimension of the world-volume, i.e. $n = p + 1$ for a p-brane. Given n we denote by $\tilde{n} = \tilde{p} + 1 = d - n - 2 = d - p - 3$ the dimension of the world-volume of the dual object. We look for solutions of (18.27) which are Poincaré invariant in the n dimensions along the brane and isotropic in the $(d - n)$ dimensions transverse to the brane, i.e. for solutions with (Poincaré)$_n \times SO(d - n)$ symmetry. We split the space-time coordinates as $x^M = (x^\mu, y^m)$, $\mu = 0,\ldots, n - 1$, $m = 1,\ldots, d - n$. The most general ansatz for the metric and dilaton of a brane at $y = 0$ which respects these symmetries is

$$ds^2 = e^{2A(r)} \eta_{\mu\nu} dx^\mu dx^\nu + e^{2B(r)} \delta_{mn} dy^m dy^n,$$
$$e^\phi = e^{\phi(r)}$$
(18.28)

with $r^2 = y^m y^n \delta_{mn}$ the distance from the brane. As boundary condition we impose that for $r \to \infty$ the metric approaches d-dimensional Minkowski space-time and $\phi \to 0$.

For the gauge potential, or rather its field strength, there are two possible ansätze consistent with the symmetries, one for an electric p-brane and one for a magnetic $\tilde{p} = (d - p - 4)$-brane:

$$\tilde{F}^{\text{el}}_{(p+2)} = P_e \star \left(e^{-a_p \phi} \epsilon_{S^{d-p-2}}\right), \qquad \tilde{F}^{\text{mag}}_{(\tilde{p}+2)} = P_m \, \epsilon_{S^{\tilde{p}+2}},$$
(18.29)

where

$$\epsilon_{S^n} = \frac{1}{n!} \frac{1}{r^{n+1}} \epsilon_{m_0...m_n} y^{m_0} dy^{m_1} \wedge \cdots \wedge dy^{m_n}$$
(18.30)

is the volume form on the unit n-sphere. It satisfies

$$d\epsilon_{S^n} = \Omega_n \, \delta^{(n+1)}(y) \, dy^0 \wedge \cdots \wedge dy^n \tag{18.31}$$

with

$$\Omega_{n-1} = \mathrm{Vol}(S^{n-1}) = \frac{2\pi^{n/2}}{\Gamma(n/2)} \tag{18.32}$$

denoting the volume of the unit S^{n-1}. From (18.29), which is motivated by our discussion of electric and magnetic charges, it is clear that P_e and P_m are proportional to the electric and magnetic charge of the brane. We will establish the precise relation below.

Inserting the ansatz into the equations of motion leads to a system of coupled non-linear ordinary differential equations for $A(r)$, $B(r)$ and $\phi(r)$. Imposing the condition

$$nA + \tilde{n}B = \mathrm{const} \tag{18.33}$$

leads to considerable simplifications. If we require that the metric far from the brane approaches η_{MN}, the constant must be zero. Equation (18.33) can be derived from the requirement that the solution preserves half of the supercharges, i.e. that the brane is a 1/2 BPS state.

Without going into further details of the derivation, we simply state that the brane solution can then be expressed in terms of a single harmonic function $H(r)$:

$$ds^2 = H^{-\frac{4\tilde{n}}{\Delta(d-2)}} \eta_{\mu\nu} dx^\mu dx^\nu + H^{\frac{4n}{\Delta(d-2)}} \left(dr^2 + r^2 d\Omega_{d-n-1} \right),$$

$$e^\phi = H^{\frac{2a_p}{\zeta\Delta}}, \quad \text{with} \quad \zeta = \begin{cases} +1 & \text{electric brane} \\ -1 & \text{magnetic brane,} \end{cases}$$

$$\tilde{F}^{\mathrm{el}}_{(n+1)} = (-1)^{pd+1} \frac{2N}{\sqrt{\Delta}} \alpha \tilde{n} \star \left(e^{-a_p\phi} \epsilon_{S^{\tilde{n}+1}} \right) = \frac{2}{\sqrt{\Delta}} dH^{-1} \wedge \epsilon_{\mathbb{R}^{1,n-1}},$$

$$\tilde{F}^{\mathrm{mag}}_{(n+1)} = \frac{2N}{\sqrt{\Delta}} \alpha \, n \, \epsilon_{S^{n+1}}, \quad N \in \mathbb{Z} \tag{18.34}$$

with

$$\Delta = a_p^2 + \frac{2n\tilde{n}}{d-2} \tag{18.35}$$

and

$$H(r) = 1 + \frac{N\alpha}{r^{\tilde{n}}} \quad (\tilde{n} > 0). \tag{18.36}$$

18.5 Brane Solutions in Supergravity

$H(r)$ is the Green function of the transverse Laplace operator, i.e. it satisfies

$$\Box H(r) = -N\alpha\,\tilde{n}\,\Omega_{\tilde{n}+1}\delta^{(\tilde{n}+2)}(y). \tag{18.37}$$

Two integration constants have been chosen such that $g_{MN} \to \eta_{MN}$ and $\phi \to 0$ as $r \to \infty$. There remains the third integration constant α, which, as we show below, is related to the tension of the brane via

$$\alpha = \frac{2\kappa_d^2 \tau_{n-1}}{\tilde{n}\Omega_{\tilde{n}+1}} \frac{\Delta}{4}. \tag{18.38}$$

One can check that for all cases which occur in $d = 10$ and $d = 11$, $\Delta = 4$. We will use this value in the following and also assume $\tilde{n} > 0$.[5] The meaning of the integer N as the number of branes will become clear below. The sign of the field strength in the solutions (18.34) is arbitrary and we have made an arbitrary choice. It distinguishes branes from anti-branes which carry opposite charges.

Before we proceed, let us make a few comments on the solution (18.34). For $d = 11$ the dilaton is, of course, absent. For $d = 10$, $p = 3$ self-duality has not been imposed yet. The self-dual D3 brane solution will be given below. It plays an important role in Sect. 18.9. The case $d = 10$, $p = 7$ needs special consideration and will be discussed at the end of this section. Finally, the solution for F^{el} (but not for F^{mag}) is of the form $F = dA$. We could have arrived at it with the ansatz $\tilde{A}^{\text{el}}_{\mu_0\ldots\mu_p} = -\epsilon_{\mu_0\ldots\mu_p}(e^{C(r)} - 1)$ to find $e^C = H^{-1}$.

Coming back, to establish (18.38) we start from the following definition of the tension (mass per unit world-volume) of the brane solution.

$$N\tau_{n-1} = \int d^{d-n}y\,\Theta_{00} \qquad (N \in \mathbb{Z}), \tag{18.39}$$

where τ_{n-1} is the tension of a single brane and N the number of coincident branes. The integral is over the transverse space and, as we will see below, it depends only on the asymptotic behavior of the metric far away from the brane. Θ_{MN} is the energy-momentum pseudo-tensor of the system. It is defined as follows: expand the metric which we found above around flat space $g_{MN} = \eta_{MN} + h_{MN}$ and define $R_{MN} = R^{(1)}_{MN} + \mathcal{O}(h^2)$. Explicitly this is (cf. Eq. (18.250) on page 770)

$$R^{(1)}_{MN} = \frac{1}{2}(\partial_P \partial_M h^P{}_N + \partial_P \partial_N h^P{}_M - \partial^P \partial_P h_{MN} - \partial_M \partial_N h^P{}_P),$$

$$R^{(1)} = \partial^M \partial^N h_{MN} - \partial^P \partial_P h^M{}_M, \tag{18.40}$$

where indices are raised with η^{MN}. Then Θ_{MN} is defined as

---

[5] The solutions with $\tilde{n} = 0$ and $\tilde{n} = -1$ deserve special treatment. Here the Green functions are $-\alpha \ln r$ ($\tilde{n} = 0$) and $-\alpha|y|$ ($\tilde{n} = -1$).

$$R^{(1)}_{MN} - \frac{1}{2}\eta_{MN} R^{(1)} \equiv \kappa_d^2 \Theta_{MN}. \tag{18.41}$$

It satisfies $\partial_M \Theta^{MN} = 0$. One can consider (18.41) as the wave equation of a spin-two particle (the graviton) with source given by the energy-momentum of the gravitational field and matter.[6] For the metric (18.28) one finds

$$2\kappa_d^2 \Theta_{00} = -(n-1)\Box e^{2A(r)} - (d-n-1)\Box e^{2B(r)}, \tag{18.42}$$

where $\Box = \frac{1}{r^{d-n-1}} \partial_r (r^{d-n-1} \partial_r)$ such that[7]

$$\int d^{d-n} y \, \Theta_{00}$$
$$= \frac{1}{2\kappa_d^2} \Omega_{d-n-1} \lim_{r \to \infty} r^{d-n-1} \left(-(n-1)\partial_r e^{2A(r)} - (d-n-1)\partial_r e^{2B(r)}\right)$$
$$= \frac{1}{2\kappa_d^2} \Omega_{\tilde{n}+1} N \alpha \tilde{n}$$
$$\stackrel{!}{=} N \tau_p, \tag{18.43}$$

from where (18.38) follows. What we have computed here is the ADM (Arnowitt-Deser-Misner) tension of the brane.

The conserved electric charge density of the electric p-brane is (cf. (18.12))

$$N q_e = \frac{(-1)^p}{\sqrt{2}\kappa_d} \int_{S^{\tilde{n}+1}} e^{a_p \phi} \star \tilde{F} = \sqrt{2}\kappa_d \tau_p N. \tag{18.44}$$

The normalization is with respect to a canonically normalized p-form field. The magnetic charge of the magnetic $\tilde{p} = d - p - 4$ brane is (cf. (18.13))

$$N q_m = \frac{1}{\sqrt{2}\kappa_d} \int_{S^{n+1}} \tilde{F}^{\text{mag}} = \sqrt{2}\kappa_d N \tau_{\tilde{p}} \tag{18.45}$$

and the Dirac quantization condition (18.15) becomes a relation for the tensions of an electric brane and its magnetic dual,

$$\tau_p \tau_{\tilde{p}} = \frac{\pi n}{\kappa_d^2}, \quad n \in \mathbb{Z}, \tag{18.46}$$

where for the objects of minimal charge $n = \pm 1$.

---

[6] This becomes apparent in harmonic gauge $\partial^M h_{MN} - \frac{1}{2}\partial_N h^M_M = 0$.
[7] With the appropriate modification this is also valid to $\tilde{n} = 0$. For $\tilde{n} = -1$ the integral diverges.

18.5 Brane Solutions in Supergravity

The electric charge can also be computed as the divergence of the gauge field. From (18.34) we find

$$d\left(\star e^{a_p \phi} \tilde{F}^{\text{el}}_{p+2} \right) = (-1)^p \, 2\kappa_d^2 \tau_p \, N\delta^{(d-p-1)}(y) \, dy^1 \wedge \cdots \wedge dy^{d-p-1}. \quad (18.47)$$

This is the equation of motion for the gauge field derived from (18.26), if we add the source term

$$S_{\text{brane}} = -N\tau_p \int_{\Sigma_{p+1}} d^{p+1}\xi \, e^{-a_p\phi/2} \sqrt{-\det \hat{g}_{\mu\nu}} + N\tau_p \int_{\Sigma_{p+1}} \tilde{A}_{(p+1)}. \quad (18.48)$$

$\hat{g}_{\mu\nu}$ is the induced metric on the world-volume Σ_{p+1} of the electric brane and $\tilde{A}^{(p+1)}$ the induced $(p+1)$-form potential on Σ_{p+1}. The first term, which does not follow from (18.47) is, of course, the D-brane action. This is familiar from Chap. 16, or the Nambu-Goto action for the fundamental string, both written in modified Einstein frame. Both terms have the same constant prefactor, in agreement with (18.44). A priori the brane action introduces source terms on the r.h.s of all three equations in (18.27), but the explicit calculation shows that the coefficients in front of these source terms vanish except for the equation for the gauge field. For the magnetic brane the field strength has the same form as the dual field strength of the electric brane. But this means that its dual satisfies the Bianchi identity and that therefore the equations of motion for the magnetic solution are source free, as expected for solitons.

We now discuss the evidence that the SUGRA brane solutions are related to the D-branes of string theory. D-branes with tension (9.59) satisfy (18.46) for $n = 1$. They also satisfy (18.44). Furthermore, the dimensions of the SUGRA branes which couple to RR-potentials are the same as those of the D-branes of the two type II string theories. We therefore draw the conclusion that D-branes and the R-R charged SUGRA branes are the same objects in two different descriptions: D-branes emerged from the world-sheet formulation of string theory while here we have found them from the effective space-time description. While the former involves open strings, the latter doesn't. We will give further evidence of this identification below.

For the fundamental string and its dual object, the (magnetic) NS5-brane, we can use (18.46) and the known F1 tension to determine the NS5 tension:

$$\tau_{\text{NS5}} = \frac{1}{(2\pi)^5 \alpha'^3 g_s^2} = \frac{2\pi}{\ell_s^6 g_s^2}. \quad (18.49)$$

It behaves as $1/g_s^2$, compared to $1/g_s$ for D-branes.

Let us now discuss further properties of the brane solutions in $d = 10$, postponing for the moment the discussion of $d = 11$ branes. The different g_s dependence of the D-branes and the NS5-brane has significant consequences. N coincident branes gives rise to a gravitational potential $V \sim G_N N \tau_p / r^{d-p-3}$. In string theory Newton's constant behaves as $G_N \sim \ell_s^8 g_s^2$, i.e. the strength of the potential is

$g_s N$ for Dp-branes and N for NS5-branes. This means that as long as $g_s N \ll 1$ the back-reaction of the D-branes on the space-time geometry in which they are embedded can be neglected at distances bigger than the string scale. The D-brane picture of a hypersurface in flat Minkowski-space on which open strings can end is therefore meaningful. But if we pile up many branes such that $N g_s \sim 1$ we have to take into account the curving of the space around them. For NS5-branes, on the other hand, the back-reaction is always large and cannot be neglected even for a single brane.

The brane solutions just found are only valid at weak string coupling and for slowly varying fields relative to the string scale. Since the characteristic scale L of the p-brane solutions is set by $\alpha \sim N\tau_p \kappa_d^2 \sim L^{d-p-3}$, this means that $N\tau_p \kappa_d^2$ must be large in units of the string scale. In other words, the supergravity brane solutions are valid for a large number of branes on top of each other and the metric is that of the back-reacted geometry.

One important lesson to draw from Eqs. (18.44) and (18.45) is that the tension of the branes is equal to the electric and magnetic charge. They are BPS branes. We have seen in Chap. 9 that D-branes of type II theories have two properties indicative of BPS objects: they satisfy the zero force condition and they preserve half of the space-time supersymmetry. As expected, and shown below, both conditions are met by the SUGRA brane solutions. These are so-called extremal p-brane solutions which saturate (in appropriate units) the BPS bound $mass = |charge|$ which leads to the zero-force condition. Non-extremal solutions satisfy $mass \geq |charge|$. They do not satisfy (18.33) and break all 32 supercharges. The non-extremal solutions satisfy the source-free equations of motion. They have two horizons and for $mass \geq |charge|$ there are no naked singularities. (This is the same situations as for charged, i.e. Reissner-Nordström black holes which are also characterized by two parameters, their mass and charge.) The non-extremal solutions are also called black p-branes. In addition, there are neutral solutions with trivial dilaton and gauge field. They are the generalization of Schwarzschild black holes, i.e. Schwarzschild black branes. There exist intersecting brane solutions preserving some supersymmetry, e.g. 1/4 BPS solutions, but we will not discuss them in the SUGRA context.

We first verify the zero force condition. One can generalize the brane solution to a stack of parallel branes by the replacement $H \to 1 + \sum_{i=1}^{N} \frac{\alpha}{|y - y_i|^{d-p-3}}$. The positions y_i of the branes are free parameters (moduli) of the solution; they can take arbitrary values. That these are stable solutions can also be seen as follows. Place a probe brane into the background of the SUGRA brane solution at the transverse positions $Y^m(\xi)$. In static gauge $X^\mu(\xi) = \xi^\mu$ Eq. (18.48) becomes

$$-\tau_p \int_{\Sigma_{p+1}} d^{p+1}\xi \left(e^{-a_p \Phi/2} \sqrt{-\det(e^{2A}\eta_{\mu\nu} + e^{2B}\partial_\mu Y^m \partial_\nu Y^n \delta_{mn})} - H^{-1} \right), \tag{18.50}$$

from which we derive the potential

$$V = \tau_p \left(e^{-a_p \Phi/2 + (p+1)A} - H^{-1} \right) \stackrel{(18.34)}{=} 0. \tag{18.51}$$

18.5 Brane Solutions in Supergravity

If we place a probe anti-brane in the same external field, the second term in the potential comes with the opposite sign: the probe anti-brane experiences a force in the background field of the branes.[8]

The brane solutions which we have constructed are the solutions of the bosonic sector of supergravity theories and one wonders whether they preserve some amount supersymmetry, i.e. whether they are invariant under some of the 32 supercharges. For this to be the case the brane background must admit Killing spinors, i.e. spinors ϵ which solve $\delta_\epsilon \Psi_M = 0$ and $\delta_\epsilon \lambda = 0$ in the brane background. We have already remarked that (18.33) can be derived as a consequence of perserving some supersymemtries. Therefore, we expect Killing spinors to exist.[9] We will now construct them explicitly. The supersymmetry variations of the gravitini and dilatini were given in (17.64). For a Dp-brane the (string frame) equations to solve are

$$\nabla_M \epsilon + \frac{1}{8} e^\Phi \slashed{F}_{p+2} \Gamma_M \mathcal{P}_{p+2} \epsilon = 0, \tag{18.52a}$$

$$\slashed{\partial}\Phi \epsilon + \frac{1}{4} e^\Phi (-1)^p (3-p) \slashed{F}_{p+2} \mathcal{P}_{p+2} \epsilon = 0. \tag{18.52b}$$

The notation is explained below Eq. (17.64) but here we are not using the democratic formulation. This explains the factor 1/2 difference for the contributions from the R-R fields. For the fundamental string and the NS5-brane we drop the R-R fields and keep the NS-NS three-form with the same factor as in (17.64). We must also keep in mind that these supersymmetry variations are expressed in string frame. The covariant spinor derivative $\nabla_M = \partial_m - \frac{1}{4}\omega_M^{AB}\Gamma_{AB}$ contains the spin-connection. For a metric of the form (18.28), which can be derived from the zehn-bein $E_M^A = (e_\mu^\alpha, e_m^a) = (e^A \delta_\mu^\alpha, e^B \delta_m^a)$, the non-vanishing components of the spin connection are

$$\begin{aligned}\omega_\mu^{\alpha a} &= e^{-B} \partial_m e^A \delta_\mu^\alpha \delta^{ma}, \\ \omega_m^{ab} &= e^{-B} \partial_n e^B (\delta_m^a \delta^{nb} - \delta_m^b \delta^{na}).\end{aligned} \tag{18.53}$$

From (18.52b) we find the condition

$$\left(\mathbb{1} \pm (-1)^p \Gamma^{0\ldots p} \mathcal{P}_{p+2}\right)\epsilon = 0. \tag{18.54}$$

---

[8] We can obtain the anti-brane by changing the orientation; e.g. $X^0(\xi) = -\xi^0$, $X^i(\xi) = \xi^i$. This is similar to saying that a positron is an electron going backwards in time: there the only way to change the orientation of the world-line is to change the direction of time. For higher dimensional world-volumes there are other ways.

[9] One could analyze the Killing spinor condition with the general ansatz (18.28) and (18.29). Requiring that Killing spinors exist leads to relations between the functions which appear in the ansatz, such as (18.33).

The two signs are for branes and anti-branes and the Dirac matrices inside the parenthesis are the flat ones. They are related to the curved ones as $\Gamma^\alpha = e^\alpha_\mu \Gamma^\mu$. The μ-components of (18.52a) lead once more to (18.54). The m-components give

$$\left(\partial_m - \frac{1}{8} H^{-1} \partial_m H\right)\epsilon = 0, \qquad (18.55)$$

where we used (18.54). This has the solution $\epsilon = H^{1/8}\epsilon_0$ where ϵ_0 is a constant spinor which has to satisfy (18.54). Using the explicit forms for $\mathcal{P}_p$ and $\epsilon_0 = \begin{pmatrix} \epsilon_L \\ \epsilon_R \end{pmatrix}$ we find the SUSY conditions for the type II Dp-branes and similarly for the fundamental string and the NS5 branes:

- Dp-brane along $(x^0, \ldots, x^p)$, $\epsilon = H^{1/8}\epsilon_0$:

$$\epsilon_L = +\Gamma^0 \cdots \Gamma^p \epsilon_R \quad \text{with} \quad \begin{cases} p \text{ even,} & \Gamma_{11}\epsilon_L = +\epsilon_L, \quad \Gamma_{11}\epsilon_R = -\epsilon_R \quad \text{IIA} \\ p \text{ odd,} & \Gamma_{11}\epsilon_L = +\epsilon_L, \quad \Gamma_{11}\epsilon_R = +\epsilon_R \quad \text{IIB} \end{cases}$$

- Fundamental string along (x^0, x^1), $\epsilon = H^{-1/4}\epsilon_0$:

$$\epsilon_L = +\Gamma^0 \Gamma^1 \epsilon_L, \qquad \epsilon_R = -\Gamma^0 \Gamma^1 \epsilon_R \qquad \text{IIA/B}$$

- NS5-brane along $(x^0, \ldots, x^5)$, $\epsilon = \epsilon_0$:

$$\epsilon_L = +\Gamma^0 \cdots \Gamma^5 \epsilon_L, \quad \epsilon_R = +\Gamma^0 \cdots \Gamma^5 \epsilon_R \qquad \text{IIA}$$

$$\epsilon_L = +\Gamma^0 \cdots \Gamma^5 \epsilon_L, \quad \epsilon_R = -\Gamma^0 \cdots \Gamma^5 \epsilon_R \qquad \text{IIB.}$$

We have written the SUSY conditions for one particular sign choice which we can take, by definition, as the brane rather than the anti-brane for which the signs are reversed. What is relevant are the relative signs. Notice that the chiral (non-chiral) type IIB (IIA) theory has a non-chiral (chiral) NS5-brane. For D-branes these are the same conditions which we found using boundary CFT in Sect. 13.3.

It is easy to see that the above conditions can be satisfied and that half of the 32 supercharges are preserved. Some amount of supersymmetry can be preserved even if different types of branes are present. For instance, the conditions imposed by parallel NS5-branes along (012345), parallel NS5-branes along (012389), D4-branes along (01236) can be satisfied simultaneously and this brane configuration preserves four supercharges, even if we add parallel D6-branes along (0123789).

A necessary condition for supersymmetric brane configurations can also be derived from the following light-cone gauge argument: recall that the zero-point energy is formally given by the infinite sums $\pm\frac{1}{2}\sum_{n=1}^\infty n = \mp\frac{1}{24}$ or $\pm\sum_{n=0}^\infty (n + \frac{1}{2}) = \pm\frac{1}{48}$; the upper (lower) sign is for bosonic (fermionic) oscillators. If follows that for all types of open string boundary conditions, NN, DD, DN and ND, the zero

18.5 Brane Solutions in Supergravity

point energies in the R-sector vanish whereas for the NS sector they are $-\frac{1}{16}$ (DD or NN) and $+\frac{1}{16}$ (ND and DN) for each of the eight transverse directions. To get a space-time supersymmetric theory, the zero-point energies of the two sectors can only differ by either an integer or a half-integer (because the oscillators can give either integer or half-integer contribution to the masses of any state and we have to satisfy level matching). If there are ν mixed and $(8-\nu)$ pure boundary conditions (in light-cone gauge), the total zero point energy is $\frac{\nu}{16} - \frac{8-\nu}{16} = -\frac{1}{2} + \frac{\nu}{8}$. This leads to the necessary condition $\nu = 0 \mod(4)$ for a space-time supersymmetric brane configuration.

The above argument assumed that all branes are located at fixed positions along the transverse coordinates, i.e. they intersect at right angles. Branes at arbitrary angles generically break all supersymmetries, but supersymmetric configurations are possible, as we have discussed in Chap. 10.

We have furthermore assumed that all branes have infinite extent (or are wrapped on tori). But this is not necessarily so and the rules for branes ending on branes are easily derived starting from the known fact that a fundamental string can end on a D-brane and applying T- and S-dualities. As we will discuss below, the latter exchanges F1 and D1 in type IIB and also D5 and NS5. Applying T-duality along a world-volume direction of a Dp-brane results in a D(p−1)-brane while T-duality in a transverse direction gives a D(p+1)-brane. To see this directly in the brane solution one has to find the solution for the S^1 compactified theory. This is most easily done by using the method of images, i.e. by constructing a periodic array of the solutions which we have found. We will not present the details. T-duality along the world-volume of NS5 gives back NS5. This is obvious from the string-frame NS5 solution. Furthermore, the tension of a wrapped NS5 is invariant under this T-duality. T-duality in a transverse direction produces a KK monopole.

Before we discuss further properties of the brane solutions, we have summarized them, for future convenience, in Table 18.3. For the type II $d = 10$ brane solutions we have given the metric in Einstein and in string frame. Note that for the type IIB superstring there also exist so called (p,q) strings and 5-branes, whose supergravity solutions can be generated by applying an appropriate $SL(2,\mathbb{Z}) \subset SL(2,\mathbb{R})$ duality transformation (18.76) to the solution for the F1-brane. Moreover, the solution for F1 and NS5 is also valid for the heterotic string. For the latter there exist more general BPS solutions, where also the gauge field has a non-trivial background. Supersymmetry requires this to be a gauge instanton profile. Analogously, the type IIB D1 and D5-brane solutions carry over to the type I superstring.

We now point out a few interesting features of some of the string theory brane solutions:

Fundamental string: The metric and the dilaton are

$$ds_E^2 = \left(1 + \frac{\alpha}{r^6}\right)^{-3/4} dx^2 + \left(1 + \frac{\alpha}{r^6}\right)^{1/4} dy^2,$$

$$e^{\Phi} = g_s \left(1 + \frac{\alpha}{r^6}\right)^{-1/2}. \qquad (18.56)$$

Table 18.3 1/2-BPS brane solutions of $d = 10$ and $d = 11$ SUGRA

$d = 10$	Dp	$ds_E^2 = H(r)^{\frac{p-7}{8}} dx_\parallel^2 + H(r)^{\frac{p+1}{8}} dx_\perp^2$	$(p < 7)$ $H(r) = 1 + \frac{\alpha}{r^{7-p}}$
		$ds_S^2 = H(r)^{-\frac{1}{2}} dx_\parallel^2 + H(r)^{\frac{1}{2}} dx_\perp^2$	$e^{\phi(r)} = H(r)^{\frac{3-p}{4}}$
		$\tilde{F}_{m0\ldots p} = \partial_m H(r)^{-1}$	$\alpha = (4\pi)^{\frac{1}{2}(5-p)} \Gamma(\frac{1}{2}(7-p))(\alpha')^{\frac{1}{2}(7-p)} g_s N$
	F1	$ds_E^2 = H(r)^{-\frac{3}{4}} dx_\parallel^2 + H(r)^{\frac{1}{4}} dx_\perp^2$	$H(r) = 1 + \frac{\alpha}{r^6}$
		$ds_S^2 = H(r)^{-1} dx_\parallel^2 + dx_\perp^2$	$e^{\phi(r)} = H(r)^{-\frac{1}{2}}$
		$H_{m01} = \partial_m H(r)^{-1}$	$\alpha = 2^5 \pi^2 (\alpha')^3 g_s N$
	NS5	$ds_E^2 = H(r)^{-\frac{1}{4}} dx_\parallel^2 + H(r)^{\frac{3}{4}} dx_\perp^2$	$H(r) = 1 + \frac{\alpha}{r^2}$
		$ds_S^2 = dx_\parallel^2 + H(r) dx_\perp^2$	$e^{\phi(r)} = H(r)^{\frac{1}{2}}$
		$H_{mnp} = \epsilon_{mnpq} \partial_q H(r)$	$\alpha = N \alpha'$
$d = 11$	M2	$ds^2 = H(r)^{-\frac{2}{3}} dx_\parallel^2 + H(r)^{\frac{1}{3}} dx_\perp^2$	$H(r) = 1 + \frac{\alpha}{r^6}$
		$F_{m012} = -\partial_m H(r)^{-1}$	$\alpha = 2^5 \pi^2 \ell_{11}^6 N$
	M5	$ds^2 = H(r)^{-\frac{1}{3}} dx_\parallel^2 + H(r)^{\frac{2}{3}} dx_\perp^2$	$H(r) = 1 + \frac{\alpha}{r^3}$
		$F_{mnpqr} = \epsilon_{mnpqrs} \partial_s H$,	$\alpha = \pi \ell_{11}^3 N$

We have absorbed N, the number of elementary branes, into the integration constant α. If we change the sign of the field strength of a brane solution, we obtain the anti-brane solution

Transforming to string frame, one gets $ds_S^2 = (1 + \alpha/r^6)^{-1} dx^2 + dy^2$, whose Ricci scalar is singular at the core of the string, $R \sim 1/r^2$. On the other hand, $\exp(\Phi)$ vanishes there and approaches $g_s = \exp(\Phi_0)$ far away from the string. Therefore, for small g_s the fundamental string is weakly coupled throughout and perturbation theory is valid.

NS5 brane: Metric and dilaton are

$$ds_E^2 = \left(1 + \frac{\alpha}{r^2}\right)^{-1/4} dx^2 + \left(1 + \frac{\alpha}{r^2}\right)^{3/4} dy^2,$$

$$e^\Phi = g_s \left(1 + \frac{\alpha}{r^2}\right)^{1/2}. \tag{18.57}$$

Compared to (18.56) the two regions of space-time, the brane and the transverse space are exchanged. In particular, $\exp(\Phi)$ diverges at the core of the NS5-brane, signaling that it is a non-perturbative object from the point of view of string theory. In string frame, $ds_S^2 = dx^2 + (1 + \alpha/r^2)(dr^2 + r^2 d\Omega_3)$; the transverse three-

18.5 Brane Solutions in Supergravity

sphere has a finite radius at the core of the brane and the Ricci scalar is finite,[10] $R \sim \alpha^{-1}$. Therefore, the magnetic dual to the fundamental string deserves to be called a solitonic solution.

D1 and D5 branes: The difference from F1 and NS5 is only the dependence of the dilaton,

$$\text{D5}: \quad e^\Phi = g_s \left(1 + \frac{\alpha}{r^2}\right)^{-1/2}, \qquad \text{D1}: \quad e^\Phi = g_s \left(1 + \frac{\alpha}{r^6}\right)^{1/2}. \qquad (18.58)$$

Self-dual D3 brane

$$ds_E^2 = \left(1 + \frac{\alpha}{r^4}\right)^{-1/2} dx^2 + \left(1 + \frac{\alpha}{r^4}\right)^{1/2} dy^2,$$

$$e^\Phi = g_s,$$

$$F_5 = 4\alpha(1 + \star)\,\epsilon_{S^5} = (1 + \star)\epsilon_{\mathbb{R}^{1,3}} \wedge dH^{-1}, \qquad \alpha = 4\pi g_s N(\alpha')^2, \qquad (18.59)$$

where we have included the self-dual five-form field strength. There is no distinction between the electric and magnetic solution. An important feature of this solution is that the dilaton is constant. The self-dual D3 brane will play an important role in Sect. 18.9, where the AdS/CFT correspondence will be discussed.

D7 brane: The D7-brane solution in type IIB supergravity, which is magnetically charged under the R-R scalar field C_0, deserves a more detailed discussion. Since the transverse space is two-dimensional, the harmonic function depends logarithmically on the distance from the brane, cf. Footnote 5 on page 693. A related feature is the appearance of a deficit angle: a flat seven-brane with translational invariance along the brane is essentially a $(2 + 1)$-dimensional problem. In three dimensions the Riemann tensor can be expressed in terms of the Ricci tensor.[11] Since Einstein's equations in empty space are simply $R_{\mu\nu} = 0$, this means that empty regions are flat. Curvature exactly traces the energy-density so that the exterior geometry is only affected in a global way. For instance, for a point particle at rest with mass m the transverse space space is conical with deficit angle $\Delta\varphi = m\kappa^2$ where κ is the gravitational coupling constant in three dimensions. We therefore expect a deficit angle in the asymptotic region of the brane geometry. Furthermore, the charge of a single seven brane is $\int_{S^1} F_{(1)} = \int_{S^1} dC_0 = 1$ where S^1 surrounds the seven brane. Of course, $F_{(1)} = dC_0$ can only hold locally. From this we conclude that asymptotically C_0 must behave as $C_0 \sim \frac{1}{2\pi i} \log z$ where we have combined the two transverse coordinates into a single complex coordinate $z = y^1 + iy^2$. We will now present the solution which indeed exhibits these features. To continue, as in (16.142) we also combine the type IIB dilaton and the R-R scalar field C_0 into a complex scalar

---

[10] In Einstein frame $R \sim 1/(r^{1/2}\alpha^{3/4})$.
[11] $R^{\mu\nu}{}_{\rho\sigma} = \varepsilon^{\mu\nu\alpha}\varepsilon_{\rho\sigma\beta}G_\alpha{}^\beta$ where $G_{\alpha\beta} = R_{\alpha\beta} - \frac{1}{2}R g_{\alpha\beta}$ is the Einstein-tensor.

field $\tau = C_0 + ie^{-\phi}$ which we take to be a function of the complex coordinate z. Motivated by the general discussion we expect the metric to be of the form

$$ds^2 = -dt^2 + \sum_{i=1}^{7} dx_i^2 + e^{B(z,\bar{z})} dz d\bar{z}, \qquad (18.60)$$

which we use as our ansatz. With this metric the equation of motion for τ, which we derive from the action (16.143), is

$$\partial\bar{\partial}\tau + \frac{2\partial\tau\,\bar{\partial}\tau}{\bar{\tau} - \tau} = 0. \qquad (18.61)$$

This is solved by a holomorphic function[12]

$$\bar{\partial}\tau(z,\bar{z}) = 0. \qquad (18.62)$$

Simple choices such as $\tau = z$ are unacceptable for several reasons: they violate the condition $\mathrm{Im}\,\tau > 0$; they do not lead to finite energy solutions; they are incompatible with the proposed $SL(2;\mathbb{Z})$ duality symmetry of type IIB theory. As we discussed in Chap. 16, the type IIB SUGRA action possesses an $SL(2;\mathbb{R})$ symmetry. As we will discuss further in the following section, in string theory this symmetry is expected to be broken by non-perturbative effects to $SL(2,\mathbb{Z})$ and furthermore this discrete symmetry should be a local symmetry. This implies that we must restrict τ to a fundamental domain of $PSL(2,\mathbb{Z})$ because different points on the upper half plane represent physically identical theories. This restriction to a fundamental region is also necessary in order to have finite energy solutions.

To see this, consider the contribution to the energy density arising from the kinetic term of the complex scalar τ. Integrating the respective term in (16.143) over the transverse z-plane, we get the energy density

$$\mathcal{E} = -\frac{i}{\tilde{\kappa}_{10}^2} \int d^2z\, \frac{\partial_z\tau\,\partial_{\bar{z}}\bar{\tau}}{(\tau - \bar{\tau})^2}, \qquad (18.63)$$

where we used that $\tau(z)$ is holomorphic. This integral has the remarkable feature that, independent of the specific solution for $\tau(z)$, it can be written as an integral over the complex τ-plane, which we now restrict to the fundamental domain $\mathcal{F}$ of $PSL(2,\mathbb{Z})$ shown in Fig. 6.11 on page 137

$$\mathcal{E} = -\frac{i}{\tilde{\kappa}_{10}^2} \int_{\mathcal{F}} d^2\tau\, \partial_\tau \partial_{\bar{\tau}} \log(\tau - \bar{\tau}) = \frac{i}{\tilde{\kappa}_{10}^2} \int_{\partial\mathcal{F}} d\tau\, \partial_\tau \log(\tau - \bar{\tau}). \qquad (18.64)$$

---

[12] Alternatively an anti-holomorphic function, which would lead to essentially the same analysis.

18.5 Brane Solutions in Supergravity

Here we applied Stokes' theorem to arrive at a line integral over the boundary $\partial \mathscr{F}$ of the fundamental domain. In the final integral the two contributions from the imaginary lines at $\mathrm{Re}(\tau) = \pm\frac{1}{2}$ cancel and the contribution from the segment of the unit circle is straightforward to compute. We find

$$\mathscr{E} = \frac{i}{\tilde{\kappa}_{10}^2} \int_{\partial \mathscr{F}} d\tau \, \frac{1}{(\tau - \bar{\tau})} = \frac{\pi}{6 \tilde{\kappa}_{10}^2}. \tag{18.65}$$

The contribution to the energy density from the complex scalar is indeed finite. If we had not restricted $\tau \in \mathscr{F}$ but had instead allowed $\tau \in \mathbb{H}_+$, $\mathscr{E}$ would be infinite.

We already know that $\tau(z)$ cannot be a single-valued function. As we go around the brane it changes as $\tau \to \tau + 1$. But we expect solutions with more general jumps in τ; more general $SL(2, \mathbb{Z})$ transformations should be allowed. The map between τ and the single valued z then involves a holomorphic function which maps the fundamental τ-region 1–1 to $z \in \mathbb{C} \cup \infty \simeq \mathbb{P}^1$. This is uniquely accomplished by the modular invariant j-function[13]

$$j(\tau) = \frac{\left(\vartheta_3^8(\tau) + \vartheta_4^8(\tau) + \vartheta_2^8(\tau)\right)^3}{8 \, \eta^{24}(\tau)} = \frac{1}{q} + 744 + 196884 \, q + \ldots, \tag{18.66}$$

where, as usual, $q = e^{2\pi i \tau}$. Moreover, j has a simple pole at infinity with residue 1. Modular invariance means

$$j(A\tau) = j(\tau) \quad \text{for} \quad A\tau = \frac{a\tau + b}{c\tau + d} \quad \text{with} \quad \begin{pmatrix} a & b \\ c & d \end{pmatrix} \in SL(2, \mathbb{Z}). \tag{18.67}$$

Consider the particular solution

$$j(\tau) = (z_0/z)^{-N}, \tag{18.68}$$

which, for small z, behaves as

$$\tau(z) \sim \frac{N}{2\pi i} \log(z/z_0). \tag{18.69}$$

Circling once around the origin, it gives rise to the monodromy $\tau \to \tau + N$. This is the desired behavior for N branes located at $z = 0$. Note that $z \to 0$ means $\mathrm{Im}\,\tau \to \infty$, which is the weak coupling region where the solution is reliable.

---

[13] The j-function is the third power of the partition function of eight chiral bosons compactified on the E_8 lattice or, equivalently (by bosonization) the third power of the partition function of 16 chiral fermions. It has important applications in number theory and the theory of elliptic curves. The coefficients in the power series expansion of $j(q) - 744$ are related to the dimensions of the irreducible representations of the so-called Monster group, the biggest of the sporadic finite groups.

We still have to solve Einstein's equations. Using (18.62) and (18.60) one finds only one non-trivial equation, namely the simple relation

$$\partial\bar{\partial} B = \partial\bar{\partial} \log(\mathrm{Im}\tau). \tag{18.70}$$

Solutions are $e^B = f\bar{f}\,\mathrm{Im}\tau$ where f is a holomorphic function which has to be chosen such that the metric is modular invariant (recall from Chap. 16 that the type IIB metric was $SL(2,\mathbb{R})$ invariant) and that e^B has no zeroes. A solution which satisfies these requirements is

$$e^{B(z,\bar{z})} = (\mathrm{Im}\tau)\,\frac{\eta^2(\tau)\,\bar{\eta}^2(\bar{\tau})}{|z/z_0|^{\frac{N}{6}}}, \tag{18.71}$$

where the denominator is needed because $|\eta|^4 \sim |q|^{1/6} \sim |z/z_0|^{N/6}$ for $z \to 0$. At large z, $\tau \to j^{-1}(0) = \mathrm{const}$,

$$e^{B(z,\bar{z})} \sim |z/z_0|^{-N/6}, \tag{18.72}$$

and the metric in the transverse space approaches

$$ds^2 = |(z/z_0)^{-N/12} dz|^2 = \left(\frac{r}{r_0}\right)^{-N/12}(dr^2 + r^2 d\varphi^2). \tag{18.73}$$

With a change of variables

$$\rho = \frac{1}{\alpha}\left(\frac{r}{r_0}\right)^{\alpha} r_0, \quad \theta = \alpha\varphi, \quad \alpha = 1 - N/12, \tag{18.74}$$

this can be brought to the form

$$ds^2 = d\rho^2 + \rho^2 d\theta^2 \quad \text{with} \quad \theta \in [0, 2\pi\alpha]. \tag{18.75}$$

It implies that each D7-brane produces a deficit angle $\pi/6$ in the transverse space. At first, it might be surprising that the deficit angle is independent of g_s because the D-brane tensions behave as $1/g_s$. However the deficit angle is the result of the complete energy density of the configuration which includes the energy density of the complex scalar field τ. All combined gives $\pi/6$.

For 12 parallel D7-branes the deficit angle is 2π and the transverse space far away from the branes is a cylinder. For $12 < N < 24$ D7 branes the compact space is not smooth. Finally, for 24 parallel D7-branes the deficit angle is 4π and the transverse space becomes compact: topologically it is $S^2 \simeq \mathbb{CP}^1$ whose size is determined by the integration constant r_0. One can formally combine this $\mathbb{CP}^1$ space with an auxiliary torus, whose complex structure parameter is identified with τ. The total space is a non-trivial elliptic fibration of T^2 over $\mathbb{CP}^1$ which can be shown to be a K3 surface. More details will be given in Sect. 18.8 when we discuss F-theory.

D8 brane: This brane solution also has interesting features, such as a dilaton which is discontinuous across the brane, whose transverse space is one-dimensional. We will not go into further details.

18.6 Non-perturbative Dualities

Now that we have identified non-perturbative BPS states in ten-dimensional superstring theories, we can search for dualities among these theories. We are particularly interested in dualities which act on the string coupling constant g_s and, for instance, exchange perturbative and non-perturbative states. The existence of such a strong-weak coupling duality in string theory was first conjectured in the context of the compactification of the heterotic string to four dimensions on T^6. It contains $\mathcal{N} = 4$ super Yang-Mills theory in its massless sector which was earlier conjectured to feature a (Montonon-Olive) S-duality.

There are many examples of non-perturbative dualities in string theory and we will discuss, in varying detail, a few of them, but our treatment is by no means exhaustive.

S-Duality of Type IIB

In Chap. 16 we wrote the type IIB supergravity action in Einstein-frame in a manifestly $SL(2, \mathbb{R})$ invariant form. $SL(2, \mathbb{R})$ acts on the massless ten-dimensional fields as

$$\tau' = \frac{a\tau + b}{c\tau + d}, \qquad \begin{pmatrix} C'_2 \\ B'_2 \end{pmatrix} = \begin{pmatrix} a & b \\ c & d \end{pmatrix} \begin{pmatrix} C_2 \\ B_2 \end{pmatrix} \qquad (18.76)$$

and leaves the (Einstein frame) metric and the four-form invariant. A particular $SL(2, \mathbb{R})$ transformation is the S-duality transformation with $\begin{pmatrix} a & b \\ c & d \end{pmatrix} = \begin{pmatrix} 0 & 1 \\ -1 & 0 \end{pmatrix}$. Comparing its action on the type IIB fields with the action of Ω, which is another symmetry of the type IIB theory, we observe that $S\Omega S^{-1} = (-1)^{F_L}$. For $C_0 = 0$, S-duality $\tau' = -1/\tau$ reduces to $e^\Phi \to e^{-\Phi}$ and, in addition, the three-form fluxes are exchanged.

The continuous classical symmetry group $SL(2, \mathbb{R})$ of type IIB SUGRA cannot be a symmetry of string theory. Non-perturbative effects break it to a discrete subgroup of $SL(2, \mathbb{R})$. The maximal subgroup which can be a symmetry of the full string theory is $SL(2, \mathbb{Z})$. This can be seen as follows. A fundamental string carries one unit of B_2 charge and, since there are no fractional strings, the B_2 charge must be quantized in integer units. By (18.76), under $SL(2, \mathbb{R})$ a fundamental string is transformed into a string with d units of B_2 charge. We conclude that d must be an integer. The maximal subgroup of $SL(2, \mathbb{R})$ for which d is an integer consists of matrices

$$\begin{pmatrix} a & \alpha b \\ c/\alpha & d \end{pmatrix} \tag{18.77}$$

with a, b, c, d, integers such that $ad - bc = 1$ and α denotes a real constant which can be absorbed in a rescaling of C_2. Another argument for the breakdown of $SL(2, \mathbb{R})$ is the following: D-instantons (i.e. Euclidean D(−1)-branes), whose classical action is[14] $S_{\text{inst}} = 2\pi\tau$ contribute to the functional integral as $\exp(2\pi i \tau)$ and therefore break the continuous shift symmetry $\tau \to \tau + b$, $b \in \mathbb{R}$ to discrete shifts $b \in \mathbb{Z}$. The other generator of $SL(2, \mathbb{Z})$ cannot be seen in this weak coupling analysis. In fact, the $SL(2, \mathbb{Z})$ symmetry of type IIB string theory cannot be proven and one has to rely on other arguments, such as, that assuming the symmetry leads to non-trivial consistent picture. It is a priori not clear why any symmetry survives, once all perturbative and non-perturbative modes are included. In fact, it is not known how $SL(2, \mathbb{Z})$ acts on a general state. In this sense, BPS states are rather an exception. For instance, S-duality exchanges the NS-NS two-form B_2 with the R-R two-form C_2, and one therefore expects that it maps the fundamental string to the D1-string. This is confirmed by the explicit supergravity solutions (18.56) and (18.58). Furthermore, their magnetic dual objects, the NS5-brane (18.57) and the D5-brane (18.58), are also mapped to each other and the D3-brane is self-dual. In this sense the duality is manifest in the supergravity brane solutions.

The statement of $SL(2, \mathbb{Z})$ as a duality symmetry is, however, a much more stronger one. It implies that the discrete symmetry is gauged, i.e. that background configurations which are related by an $SL(2, \mathbb{Z})$ transformation are physically indistinguishable, just as configurations related by T-duality are.

Further support for the strong-weak coupling S-duality of the type IIB superstring can be gathered by comparing the tensions of the BPS branes. Referring to the discussion in Sect. 18.3, we use the fact that the tension of these BPS objects is set by their charges and can therefore be reliably extrapolated from the weak coupling regime, where we computed them, to the strong coupling regime. We start with the tension of the fundamental string (in string frame):

$$T_{\text{F1}} = \frac{1}{2\pi \alpha'}. \tag{18.78}$$

If we want the fundamental string and the D1-brane to be exchanged under S-duality, the string length has to be transformed as well[15]:

$$g_{\text{IIB}} \to \frac{1}{g_{\text{IIB}}}, \quad \alpha' \to g_{\text{IIB}} \, \alpha'. \tag{18.79}$$

---

[14] This is obtained by evaluating the action (18.26) + (18.48) on the D(−1)-brane solution with $\tau = \langle C_0 \rangle + i e^{-\langle \Phi \rangle}$.

[15] Here we use g_{IIB} instead of g_s to emphasize that the result holds for type IIB but not for other string theories. For those we will find other relations later on.

18.6 Non-perturbative Dualities

We then obtain (again in string frame)

$$\tau_{D1} = \frac{T_{F1}}{g_{IIB}} = \frac{1}{2\pi\alpha' g_{IIB}}. \tag{18.80}$$

Notice that the rescaling of α' is also necessitated by the requirement that the physical coupling constant $\kappa_{10}^2 \sim \alpha'^4 g_{IIB}^2$ be duality invariant. In Einstein frame[16]

$$g_{IIB} \to \frac{1}{g_{IIB}} \qquad \alpha' \to \alpha', \qquad T_{F1}^{(E)} = \frac{\sqrt{g_{IIB}}}{\alpha'} \overset{S}{\longleftrightarrow} \tau_{D1}^{(E)} = \frac{1}{\sqrt{g_{IIB}}\alpha'}. \tag{18.81}$$

Similarly, the NS5-branes and the D5-branes are exchanged under S-duality. Their tensions are (again in string frame)

$$\tau_{NS5} = \frac{1}{(2\pi)^5 \alpha'^3 g_{IIB}^2} \overset{S}{\longleftrightarrow} \tau_{D5} = \frac{1}{(2\pi)^5 \alpha'^3 g_{IIB}}, \tag{18.82}$$

which are mapped to each other under (18.79). Combined with T-duality we have now related the tensions of all type II branes to the tension of the fundamental string.

The full conjectured duality group of the type IIB theory is $SL(2, \mathbb{Z})$, of which the S-duality group is a $\mathbb{Z}_2$ subgroup. As already mentioned above, it implies the existence of strings which couple electrically to both two-form potentials. We call a (p, q) string a string with p units of B_2 charge and q units of C_2 charge. Such string solutions of type IIB supergravity can be constructed and their tension can be shown to be

$$\tau_q^{(E)} = \frac{1}{\sqrt{\tau_2}} |p + \tau q| T_{F1}, \qquad T_{F1} = \frac{1}{2\pi\alpha'}. \tag{18.83}$$

Note that this expression is invariant under $SL(2, \mathbb{Z})$, if we combine the transformation (18.76) with a transformation on the charges such that $p \int B_2 + q \int C_2$ is invariant. In string frame the tension of a (p, q) string is

$$\tau_q^{(S)} = |p + \tau q| T_{F1}, \tag{18.84}$$

which has the correct limits for the F1 string and the D1 string (for the latter if $C_0 = 0$). Note that in the $g_s \ll 1$ regime the fundamental string is the lightest string whereas in the non-perturbative limit $g_s \gg 1$ it is the D1-string. Similarly, for each (p, q) string there exists a regime where it is the lightest.

---

[16] Going from string to Einstein frame means that we have to rescale all lengths: $L_S^2 = \sqrt{g_{IIB}} L_E^2$. Since L_E is S-duality invariant, L_S transforms as $L_S^2 \to L_S^2/g_{IIB}$. Alternatively we may change the length scale, i.e. $\alpha' \to g_{IIB}\alpha'$.

For $q = q' + q''$ the tension (18.84) satisfies the inequality

$$\tau_q \leq \sqrt{\tau_{q'}^2 + \tau_{q''}^2} \tag{18.85}$$

with equality if the two charge vectors are parallel. This means that a (p,q)-string with p,q relatively prime is stable against decay into two strings. In the above formulas, τ should be understood as its background value $C_0 + \frac{i}{g_s}$.

Similarly, there are two scalars (C_0, Φ) and therefore there are (p,q)-seven branes. The D7 brane is charged only under C_0 and is a $(1,0)$ 7-brane. The existence of (p,q) branes can already be seen from the solution (18.68). Recall from Chap. 6 that besides infinity the fundamental region $\mathscr{F}$ has two more orbifold points, $\tau = i$ and $\tau = \rho \equiv e^{2i\pi/3}$, the fixed points of S and ST. Around these points the function $\tau(z) = j^{-1}(z)$ has the expansions $\tau(z) = i + \alpha(z-i)^{1/2} + \ldots$ and $\tau(z) = \rho + \beta(z-\rho)^{1/3} + \ldots$ and the non-trivial monodromies $S : \tau \to -1/\tau$ (around i) and $T^{-1}S : \tau \to -(1+1/\tau)$ (around ρ). They generate $PSL(2,\mathbb{Z})$.

Since a fundamental string can end on a D7-brane, $SL(2,\mathbb{Z})$ implies that a (p,q) string can end on a (p,q) 7-brane. We know from Sect. 18.5 that the D7-brane induces an $SL(2,\mathbb{Z})$ monodromy

$$M_{D7} = \begin{pmatrix} 1 & 1 \\ 0 & 1 \end{pmatrix} \tag{18.86}$$

of $\tau = C_0 + ie^{-\Phi}$. We can find the monodromy induced by a (p,q) 7-brane by the following argument. The coupling of a (p,q) string to the two-form potentials is $p \int B_2 + q \int C_2$. Acting with $SL(2,\mathbb{Z})$ on the potentials as in (18.76), we see that the (p,q) string is transformed to a (p',q') string with

$$(q', p') = (q, p) \begin{pmatrix} a & b \\ c & d \end{pmatrix}. \tag{18.87}$$

This specifies the action of $SL(2,\mathbb{Z})$ on the charge vector. Starting with a fundamental string with $(q, p) = (0, 1)$ we obtain a general string

$$(q, p) = (0, 1) \begin{pmatrix} r & s \\ q & p \end{pmatrix}, \qquad (rp - qs = 1). \tag{18.88}$$

We then find the monodromy induced by a (p,q) 7-brane by transposing the monodromy of a D7-brane, i.e.

$$M_{(p,q)} = \begin{pmatrix} r & s \\ q & p \end{pmatrix}^{-1} \begin{pmatrix} 1 & 1 \\ 0 & 1 \end{pmatrix} \begin{pmatrix} r & s \\ q & p \end{pmatrix} = \begin{pmatrix} 1+pq & p^2 \\ -q^2 & 1-pq \end{pmatrix}, \tag{18.89}$$

which is independent of r and s (and so are all physical quantities). $M_{(p,q)}$ has the property

$$(q, p) M_{(q',p')} = (q, p), \qquad \text{if} \qquad qp' - pq' = 0. \tag{18.90}$$

18.6 Non-perturbative Dualities

Table 18.4 Monodromies around stacks of 7-branes of types A, B, C

7-branes	Number	Monodromies
A	1	$M_A = \begin{pmatrix} 1 & 1 \\ 0 & 1 \end{pmatrix}$
B	1	$M_B = \begin{pmatrix} 2 & 1 \\ -1 & 0 \end{pmatrix}$
C	1	$M_C = \begin{pmatrix} 0 & 1 \\ -1 & 2 \end{pmatrix}$
A^n	n	$M_A^n = \begin{pmatrix} 1 & n \\ 0 & 1 \end{pmatrix}$
AB	2	$M_A M_B = \begin{pmatrix} 1 & 1 \\ -1 & 0 \end{pmatrix}$
$A^2 B$	3	$M_A^2 M_B = \begin{pmatrix} 0 & 1 \\ -1 & 0 \end{pmatrix}$
$A^2 BA$	4	$M_A^2 M_B M_A = \begin{pmatrix} 0 & -1 \\ -1 & -1 \end{pmatrix}$
$A^n BC$	$n+2$	$M_A^n M_B M_C = \begin{pmatrix} -1 & -n+4 \\ 0 & -1 \end{pmatrix}$
$A^5 BCB$	8	$M_A^5 M_B M_C M_B = \begin{pmatrix} -1 & -1 \\ 1 & 0 \end{pmatrix}$
$A^6 BCB$	9	$M_A^6 M_B M_C M_B = \begin{pmatrix} 0 & -1 \\ 1 & 0 \end{pmatrix}$
$A^6 BCBA$	10	$M_A^6 M_B M_C M_B M_A = \begin{pmatrix} 0 & -1 \\ 1 & 1 \end{pmatrix}$

In this case we say that the two branes (or strings) are local w.r.t. each other. In particular, the monodromy of a (p, q) 7-brane leaves the charge of a (p, q)-string, which can end on it, invariant. We observe that (18.90) implies

$$[M_{(q,p)}, M_{(q',p')}] = 0 \quad \text{if} \quad qp' - pq' = 0 \tag{18.91}$$

and

$$A^{-1} M_{(q,p)} A = M_{(q,p)A} \quad \text{for} \quad A \in SL(2, \mathbb{Z}). \tag{18.92}$$

We will see in Sect. 18.8 that there exist inherently non-perturbative compactifications of the type IIB superstring which contain such general (p, q) 7-branes. In particular, the three types of branes $A = (1, 0)$, $B = (1, 1)$, $C = (1, -1)$ will be important. For later reference we list the monodromy matrices of certain stacks of these 7-branes in Table 18.4

S-Duality of Heterotic/Type I

The duality of the type IIB string is an example of a self-duality, i.e. it maps the theory to itself. The next example is a duality between two different perturbative string theories: the non-perturbative equivalence between the ten-dimensional heterotic and type I superstrings with gauge group $G = SO(32)$. Since the non-Abelian gauge degrees of freedom are closed strings (winding and momentum) for the heterotic string and open strings for the type I string, this duality symmetry operates between closed and open strings. Since in $d = 10$ the $N = 1$ supergravity theory coupled to Super-Yang-Mills theory is unique, up to a choice of gauge group and field redefinitions, there must be such a map. Of course, the statement here is much stronger as it implies the weak-strong duality of two string theories with a priori very different appearances.

To reveal this S-duality, we first compare the two low energy effective actions in ten dimensions, i.e. (16.146) for the type I string and (16.154) for the heterotic string. As remarked above, the two actions must be related by a field redefinition. The physical gauge couplings in the low-energy effective theories have a different dependence on the respective dilaton. This is a consequence of the fact that they are derived from string amplitudes on a sphere and a disc, for the heterotic and the type I theories, respectively. However, the two actions can be mapped into each other by

$$\Phi^H = -\Phi^I, \quad G^H_{MN} = e^{-\Phi_I} G^I_{MN} \quad H^H_3 = F^I_3, \quad F^H_{YM} = F^I_{YM}. \quad (18.93)$$

This means that the strong coupling regime of one theory is mapped to the weak coupling regime of the other and vice versa.

Further evidence for this strong-weak coupling duality can be gained by considering the BPS states. Since the duality exchanges the two-form B_2 of the heterotic string with the R-R two-form C_2 of the type I string theory, one expects that the fundamental $SO(32)$ heterotic string is mapped to the type I D-string. The same should hold for the magnetic objects, i.e. the heterotic NS5-brane is mapped to the type I D5-brane. Let us sketch the argument.

One can show that the massless excitations of the D-string are precisely the fields which appear on the world-sheet of the heterotic string. First, there are the open strings from the D-string to itself. The longitudinal bosonic modes are projected out by the orientifold projection, whereas the transverse modes remain in the spectrum. They are associated with the eight translations which are broken by the D1 string. To find the fermionic states, we need the action of Ω. For a D-string along X^1, we have NN boundary conditions in the (01) and DD boundary conditions in the (2...9) directions and all fermionic oscillators in the R sector are integer moded. The GSO projection projects to one $SO(1, 9)$ chirality, which we choose to be $\Gamma^{11} = +1$. Under $SO(1, 9) \supset SO(1, 1) \times SO(8)$ the **16** decomposes as $\mathbf{8}^+_s + \mathbf{8}^-_c$. The superscript denotes two-dimensional chirality. From (9.75) we know the action of Ω on the b^μ_0: they differ by a sign between the NN and the DD directions. Consequently (9.76), which is valid for all directions NN, has to be modified. If we

18.6 Non-perturbative Dualities

define $\Omega|0\rangle_R = -|0\rangle_R$, where $|0\rangle_R$ is the highest weight state defined below (8.15), we find that the action of Ω on a state $|s\rangle$ is

$$\Omega|s\rangle = -e^{i\pi(s_1+s_2+s_3+s_4)}|s\rangle. \tag{18.94}$$

This leaves the $\mathbf{8}_c^-$. If we impose the physical state condition (8.19), we find that they are right-moving. Altogether, the D1-D1 sector contributes the following massless states on the D1-brane world-volume: eight bosons $X^i(\sigma, \tau)$ and eight right-moving (Green-Schwarz) fermions $\psi^\alpha(\tau - \sigma)$, where α is an $SO(8)$ spinor index.

In addition, there are the D1-D9 strings stretching from the D-string to the 32 space-time filling D9-branes. Since one has eight directions with DN boundary conditions, the zero point energy in the NS-sector is positive and there are no additional bosonic zero modes. In the R-sector the zero-point energy vanishes and the massless states are $|s_0, a\rangle$. Here $s_0 = \pm\frac{1}{2}$ and $a = 1, \ldots, 32$ the $SO(32)$ Chan-Paton index. The GSO projection[17] requires $s_0 = +1/2$. Imposing the physical state condition, we find 32 left-moving fermions $\lambda^a(\tau + \sigma)$. A similar analysis can be carried out for the D9-D1 strings with DN boundary conditions replaced by ND boundary conditions. The Ω projection relates the two sectors and there are no new states. The massless fields on the D1 world-volume are thus those of the fundamental $SO(32)$ heterotic string in the Green-Schwarz formulation.

There is an interesting subtlety concerning the heterotic GSO projection for the 32 left-moving fermions. Recall that it projects out the vector and anti-spinor representation of $SO(32)$ so that the precise gauge group of the heterotic string is $Spin(32)/\mathbb{Z}_2$. In the fermionic formulation of the heterotic string this GSO projection on the spectrum results from the sum over the different spin structures for these 32 fermions on a toroidal world-sheet. How is this implemented on the D-string? If one wraps the world-sheet of the D-string on a T^2, then even though the longitudinal gauge field is projected out by Ω, one can have a discrete $\mathbb{Z}_2$ Wilson line along each of the two cycles of T^2. Summing over these four possible D1-branes in the D9-D1 sector leads to the GSO projection for the 32 fermions.

Let us comment on the fate of the heterotic/type I strong-weak duality in lower space-time dimensions. If we compactify both theories on an internal $D = 10 - d$ dimensional manifold M of normalized volume $v_D = \mathrm{Vol}(M)/(2\pi\sqrt{\alpha'})^{(D)}$, their couplings in d dimensions are determined by the d dimensional dilaton Φ_d

$$e^{-2\Phi_d} = e^{-2\Phi} v_{10-d}. \tag{18.95}$$

---

[17]This can e.g. be seen by requiring locality of the OPE of the vertex operators of the 1–1 and the 1–9 fermions.

Under the ten-dimensional S-duality transformation one obtains

$$\Phi_d^H = \left(\frac{6-d}{4}\right)\Phi_d^I + \left(\frac{2-d}{8}\right)\log v_{10-d}^I. \qquad (18.96)$$

For instance, if we compactify on a four-dimensional K3-manifold, the 6D heterotic dilaton is mapped to the volume modulus such that weak heterotic coupling, i.e. $e^{\Phi_d^H} \ll 1$, corresponds to large volume of the K3 on the type I side. Compactifying further to four-dimensions the weakly coupled type I string is mapped to a weakly coupled heterotic string for volumes not smaller than the string scale.

U-Duality

Consider the type IIA/B theories compactified on a torus T^D. As we have already discussed, for $D \geq 1$ the moduli space of the compactified type IIA and type IIB theories, which is parametrized by the constant background values of the various bosonic fields, modulo discrete symmetries, is the same. The T-duality symmetry $R \to \alpha'/R$ is the simplest example of such a discrete symmetry and compactifying the type IIA theory on a circle of radius R is the same theory as compactifying the type IIB theory on a circle of radius α'/R. For compactification on higher dimensional tori there are more general T-duality symmetries and they also act on the RR background. The defining feature of general T-dualities is that they are symmetries order by order in string perturbation theory, but it does not commute with the α'-expansion. In other words it does not commute with the sigma-model perturbation theory. In contrast to this there is the S-duality of the compactified type II theories which they inherit from type IIB in ten dimensions. S-duality is non-perturbative in the sense that it maps the weakly to the strongly coupled theory, i.e. it does not commute with the string perturbation expansion but it does commute with the α'-expansion.

It has been conjectured that type II string theory compactified on D-dimensional tori has a larger symmetry group which mixes both the g_s and the α' expansions and which contains the S- and T-duality group[18] $SL(2,\mathbb{Z}) \boxtimes O(D,D;\mathbb{Z})$ as its subgroup. These larger duality symmetries have been called U-duality. The continuous versions of these symmetry groups arise as isometries of the scalar manifolds of eleven-dimensional supergravity reduced to $d = 10 - D$-dimensions. Again, the idea is that non-perturbative effects lead to the breaking to a discrete subgroup, the U-duality group. The U-duality groups of type II string theory on a d-dimensional torus are listed in Table 18.5.

---

[18] The product $\boxtimes$ is non-commutative, as the S-duality transformation also acts non-trivially on the antisymmetric tensor.

18.6 Non-perturbative Dualities

Table 18.5 U-duality groups of type II compactified on T^D

d	9	8	7	6	5	4	3	2
U-duality group	$SL(2,\mathbb{Z}) \times \mathbb{Z}_2$	$SL(2,\mathbb{Z}) \times SL(3,\mathbb{Z})$	$SL(5,\mathbb{Z})$	$SO(5,5,\mathbb{Z})$	$E_{6(6)}(\mathbb{Z})$	$E_{7(7)}(\mathbb{Z})$	$E_{8(8)}(\mathbb{Z})$	$\hat{E}_{8(8)}(\mathbb{Z})$

$E_{n(n)}$ denotes a non-compact version of the exceptional group E_n for $n = 6, 7, 8$ and $\hat{G}$ is the loop group of G, i.e. the group of mappings from the circle S^1 into G.

Heterotic/Type IIA Duality in Six Dimensions

We now discuss a non-trivial duality in six space-time dimensions, namely between the heterotic string with gauge group $E_8 \times E_8$ compactified on T^4 and the type IIA superstring compactified on a K3 surface.[19]

The first check is that these two theories have the same supersymmetry in six dimensions. On the heterotic side, a torus compactification leads to maximal supersymmetry, namely 16 unbroken supercharges. We know from Chap. 14 that there are two possibilities. From the decomposition (14.49) of the Majorana-Weyl supercharge in $d = 10$ it follows that the heterotic string on T^4 has $\mathcal{N} = (1,1)$ supersymmetry. On the type IIA side, space-time supersymmetry is reduced by one-half due to the non-trivial $SU(2)$ holonomy of K3. This also leads to $\mathcal{N} = (1,1)$ supersymmetry in six dimensions, as we also discussed in Sect. 14.3. We have summarized the field content of massless multiplets of $\mathcal{N} = (1,1)$ supersymmetry in $D = 6$ in Table 15.4 on page 529.

The relations between the heterotic and type IIA coupling constants can be derived by comparing the effective actions in $D = 6$. On the heterotic side, for generic Wilson lines the $E_8 \times E_8$ gauge symmetry is broken to $U(1)^8 \times U(1)^8$. Including the KK gauge bosons the gauge symmetry is $U(1)^{24}$ and the effective six-dimensional supergravity action of the dimensionally reduced heterotic string in string frame is

$$S_{\text{het}} = \frac{1}{2\tilde{\kappa}_6^2} \int d^6 x \, \sqrt{|G_{\text{H}}|} \, e^{-2\Phi_6^{\text{H}}} \left(R_{\text{H}} + 4|\nabla \Phi_6^{\text{H}}|^2 - \frac{1}{2}|dB_{\text{H}}|^2 \right.$$
$$\left. - \frac{\tilde{\kappa}_6^2}{2g_6^2} \sum_{I=1}^{24} |dA_{\text{H}}^I|^2 \right) \tag{18.97}$$

---

[19]Recall from Chap. 14 that K3 surfaces are the only compact CY manifolds in complex dimension two. Their only non-trivial Hodge number is $h^{1,1} = 20$ and, of course, $h^{2,0} = 1$. The second Betti number splits according to $b^2 = b_+^2 + b_-^2$ into self-dual and anti self-dual two-forms with $b_+^2 = 3$. The Kähler form and the (2,0) and (0,2) forms are self-dual.

with $\tilde{\kappa}_6^2 = (4\pi^2\alpha')^2/4\pi$. We have not written the contribution from the moduli scalars. They will be discussed below. Notice that the entire heterotic action uniformly scales with the inverse square of the six-dimensional heterotic string coupling constant: $e^{-2\Phi_6^H} = v(T^4) e^{-2\Phi^H}$ with the normalized volume defined as $v(T^4) = \text{Vol}(T^4)/(2\pi\sqrt{\alpha'})^4$.

Type IIA superstring theory compactified on K3 also gives rise to 24 $U(1)$ gauge fields which, via KK-reduction, all arise from the R-R sector. One comes from the one-form with indices along the six non-compact directions, i.e. C_μ, and another one from the three-form with indices $C_{\mu\nu\rho}$, which is Hodge-dual to a massless vector in $D = 6$. The three-form with index structure $C_{\mu mn}$ gives 22 vectors. Since $b^1(K3) = 0$, there are no gauge fields from the KK reduction of the metric and the NS-NS anti-symmetric tensors. The effective string frame type IIA action compactified on K3 is

$$S_{\text{IIA}} = \frac{1}{2\tilde{\kappa}_6^2} \int d^6x \sqrt{|G_{\text{IIA}}|} \left(e^{-2\Phi_6^{\text{IIA}}} \left[R_{\text{IIA}} + 4|\nabla \Phi_6^{\text{IIA}}|^2 - \frac{1}{2}|dB_{\text{IIA}}|^2 \right] \right.$$

$$\left. - \frac{\tilde{\kappa}_6^2}{2g_6^2} \sum_{I=1}^{24} |dA_{\text{IIA}}^I|^2 \right). \tag{18.98}$$

Again we have not displayed contributions from the moduli scalars. We have chosen the normalization of the gauge fields for later convenience. However, since the gauge fields all arise in the R-R sector, their contribution to the action is independent of $e^{-2\Phi_6^{\text{IIA}}} = v(K3) e^{-2\Phi^{\text{IIA}}}$. It is straightforward to show that (18.97) transforms to (18.98) if we identify

$$\Phi_6^H = -\Phi_6^{\text{IIA}},$$

$$G_H = e^{2\Phi_6^H} G_{\text{IIA}} = e^{-2\Phi_6^{\text{IIA}}} G_{\text{IIA}},$$

$$dB_H = e^{-2\Phi_6^{\text{IIA}}} \star dB_{\text{IIA}},$$

$$A_H^I = A_{\text{IIA}}^I. \tag{18.99}$$

The first relation in (18.99) means that the strong coupling regime of the heterotic string in six dimensions is mapped to the weak coupling regime of the type IIA string and vice versa. The map between the two-forms proceeds in the same way as the dualization of a two-form and a scalar in $D = 4$, cf. (14.209) on page 490, except that now the Lagrange multiplier field is a two-form. Notice that the duality exchanges equations of motion and Bianchi identities for B_H and B_{IIA}. (In $d = 6$ the Hodge-$\star$ on a three-form does not change under rescaling of the metric.) Since the B-fields are Hodge-duals of each other, the duality exchanges fundamental strings on one side and wrapped NS5-branes on the other.

Further evidence for this duality comes from comparing the moduli spaces and BPS spectra of the two theories. The moduli of the heterotic string are ten components of the metric, six components of the anti-symmetric tensor plus

18.6 Non-perturbative Dualities

$4 \times 16 = 64$ Wilson line moduli in the Cartan sub-algebra of the rank 16 gauge group $E_8 \times E_8$. The 80 scalars are the scalar components of 20 $\mathcal{N} = (1, 1)$ vectormultiplets in six dimensions and parametrize the Narain moduli space

$$M = \frac{SO(4, 20)}{SO(4) \times SO(20)} \Big/ SO(4, 20; \mathbb{Z}), \qquad (18.100)$$

where $SO(4, 20, \mathbb{Z})$ is the T-duality group of the T^4 compactificaton of the heterotic string.

In the NS-NS sector of type IIA on K3 there are 58 geometrical moduli; cf. Footnote 40 on page 486. It is known from the theory of K3 surfaces that the Teichmüller space of Ricci-flat Kähler metrics on K3 is $SO(3, 19)/(SO(3) \times SO(19)) \times \mathbb{R}_+$ where the last factor is for the overall volume of the manifold. In addition there are 22 massless scalars from the dimensional reduction of the B-field. These 80 scalars are the scalars in 20 vectormultiplets and one can show that they also parametrize the coset space (18.100) which is thus the moduli space of K3 compactifications of the type IIA string on K3.[20]

To summarize, the massless spectra of the heterotic string on T^4 and the type IIA string on K3 consists of 20 $\mathcal{N} = (1, 1)$ vectormultiplets and the $\mathcal{N} = (1, 1)$ gravity multiplet in six dimensions. The latter contains one real scalar field, namely the six-dimensional heterotic and type IIA dilaton field, respectively.

The next question to address is the correspondence between BPS states on the heterotic and the type IIA side. We will concentrate on the gauge part of the BPS spectrum. As we already discussed, in heterotic toroidal compactifications, the generic gauge group at arbitrary points in the moduli space is Abelian. For T^4 it is

$$G = G_L \times G_R = U(1)_L^{20} \times U(1)_R^4. \qquad (18.101)$$

In the effective $\mathcal{N} = (1, 1)$ supergravity, the 20 left-moving vectors are part of 20 vectormultiplets, whereas the four right-moving vectors are in the supergravity multiplet. At special points in the moduli space (18.100), the left-moving gauge symmetry can be enhanced to a non-Abelian gauge group G_L. All non-Abelian gauge bosons of the heterotic string are perturbative BPS states.

As we have already discussed, in the type IIA theory the perturbative gauge symmetry is Abelian and is the same as in (18.101). Again, we have 20 vectormultiplets and a gravity multiplet. In contrast to the heterotic case, the gauge symmetry enhancement for the type IIA string on K3 cannot be perturbative. The non-Abelian gauge bosons are due to D2-branes that wrap the 2-cycles of K3, of which there are 22 homologically independent ones. In the uncompactified $D = 6$ space-time they are seen as particles which are massive for generic values of the K3

---

[20]With reference to the remarks at the end of Sect. 15.4, we mention that (18.100) is the moduli space of $(4, 4)$ superconformal field theories with $(\bar{c}, c) = (6, 6)$.

moduli, i.e. the gauge symmetry is Higgsed to the Abelian group in Eq. (18.101). But there are points in the K3 moduli space, where the K3 manifold degenerates such that a collection of intersecting 2-cycles shrinks to a point. Since the mass of a state that corresponds to a wrapped D2-brane is proportional to the volume of the wrapped 2-cycle, at points in the moduli space where 2-cycles collapse massless particles appear in $D = 6$. These are, in fact, BPS states. This will be explained in a different context is Sect. 18.7 and should be expected from the discussion of winding states in Sect. 18.4. They are charged under the 22 $U(1)$ gauge fields which descend from the R-R three-form C_3 upon dimensional reduction and they lead to a non-Abelian enhancement of the gauge symmetry. The two possible orientations for the wrapping correspond to the $W^\pm$ gauge bosons. A thorough analysis, which also shows which non-perturbative non-Abelian gauge groups can occur, uses the A-D-E classification of collapsing 2-cycles on a K3 surfaces. This refers to the feature that each cycle can be associated to the node of the Dynkin diagram and their intersection numbers are the entries of the (negative of the) Cartan matrix of the simply laced Lie algebras. For details we have to refer to the literature.

The heterotic/type IIA string-string duality can be extended to four-dimensions. First it is almost trivial to generalize it to the heterotic string compactified on a six-torus versus the type IIA string compactified on K3 $\times T^2$. This simply amounts to compactifying both sides of the previous duality on T^2. More interestingly, one can try to apply this six-dimensional duality locally for a fibration[21] of T^4 and K3 over a $\mathbb{P}^1$. Applying the previous duality fiberwise, this leads to compactifications of the heterotic string on K3 $\times T^2$ where the K3 is elliptically fibered,[22] i.e. it is the total space of a fiber bundle with base $\mathbb{P}^1$ and fiber T^2. On the type IIA side we get Calabi-Yau threefolds which are K3 fibrations over $\mathbb{P}^1$. Both compactifications have eight unbroken supercharges, i.e. four-dimensional $\mathcal{N} = 2$ supersymmetry. The heterotic side must be further specified by a choice of a gauge bundle over K3 (this corresponds to the non-trivial Wilson lines on T^4 in the six-dimensional duality). Further details on this interesting duality and its relation to dualities in $\mathcal{N} = 2$ supersymmetric field theories are beyond the scope of this book.

18.7 M-Theory

So far we have connected the two type II theories via T-duality, and also the two heterotic theories. We have also discussed the self-duality of the type IIB string and the S-duality between the type I and the heterotic $SO(32)$ theory. The latter operate directly in the ten-dimensional theories. We will now discuss a new type duality which connects the type IIA string with a new eleven-dimensional theory, called

---

[21] It cannot be a direct product with $\mathbb{P}^1$ because this would break all supersymmetries and there are no BPS states.

[22] We will discuss elliptically fibered K3 surfaces in more detail in Sect. 18.8.

18.7 M-Theory

M-theory. We will also show how the type IIB theory and the $E_8 \times E_8$ heterotic string are described within M-theory.

Type IIA/M-Theory Duality

S-duality asserts that the strongly coupled type IIB string is a again the type IIB string, i.e. it is self-dual. But what is the strong coupling limit of the type IIA string? One expects that for $g_s \gg 1$ the solitonic objects, i.e. the various branes, become light. Recall that the tension of the type IIA Dp-branes is $\tau_p = \frac{2\pi}{g_s}(4\pi^2\alpha')^{-(p+1)/2}$. The lightest D-brane is the D0-brane with mass $m = 1/(g_s\sqrt{\alpha'})$. Since this is a supersymmetric BPS object, one expects that there exists a bound state at threshold of an arbitrary number n of such D0-branes with mass

$$m = \frac{n}{g_s\sqrt{\alpha'}}. \tag{18.102}$$

This is the characteristic mass spectrum of Kaluza-Klein modes for a circle compactification of radius

$$R_{11} = g_s\sqrt{\alpha'}. \tag{18.103}$$

Thus, it looks as if for $g_s \gg 1$ a new dimension opens up, which could not be seen in perturbative ten-dimensional type IIA superstring theory.

In Chap. 16 we discussed ten-dimensional supergravity theories as low-energy effective field theories of type II string theory. There we also mentioned eleven-dimensional supergravity with the same number of 32 supercharges but its relevance for string theory was unclear. The above observation suggests the intriguing possibility that this eleven-dimensional supergravity theory is the low-energy effective action of the strong coupling limit of the type IIA superstring theory. The ten-dimensional type IIA string theory has two parameters: the string tension α' and the string coupling constant g_s. Eleven-dimensional supergravity compactified on a circle has also two parameters: the eleven-dimensional gravitational coupling κ_{11} and the radius R_{11} of the circle. Equation (18.103) establishes one relation among these parameters. A second relation, which contains κ_{11}, follows from dimensional reduction of the eleven-dimensional supergravity theory.

Its bosonic fields are the metric G_{MN} and a three-form potential A_3. Their dynamics is governed by the action

$$S_{\text{eff}}^{(d=11)} = \frac{1}{2\kappa_{11}^2} \int d^{11}x \sqrt{-G}\left(R - \frac{1}{2}|dA_3|^2\right). \tag{18.104}$$

We have suppressed the topological Chern-Simon term. We now perform a dimensional reduction to $d = 10$ dimensions. For the metric and gauge potential we make the Kaluza-Klein ansatz (cf. Chap. 14)

$$ds_{11}^2 = e^{-\frac{2}{3}\Phi} g_{\mu\nu} dx^\mu dx^\nu + e^{\frac{4}{3}\Phi} (dx^{11} + C_\mu dx^\mu)^2, \quad \mu,\nu = 0,\ldots,9,$$
$$A_3 = C_3 + B_2 \wedge dx^{11},$$
$$dA_3 = F_4 + H_3 \wedge (dx^{11} + C_1) \quad \text{where} \quad F_4 = dC_3 - H_3 \wedge C_1. \tag{18.105}$$

All fields are independent of x^{11}, i.e. shifts of x^{11} are an isometry and all forms on the r.h.s. are in $d = 10$. The ansatz is such that all fields transform as tensors under diffeomorphisms[23] generated by $\xi^\mu(x)$. Under diffeomorphisms generated by $\xi^{11}(x)$, all fields are invariant except C_1 which transforms as a $U(1)$ gauge field, $C_1 \to C_1 - d\xi^{11}$. We insert this ansatz into the action (18.104) and integrate over the eleventh dimension, for which we take a circle of radius R_{11}. After a tedious calculation one obtains precisely the type IIA supergravity action in string frame, Eq. (16.134a). One can show that this extends to the complete action, including the CS term and the fermions. As to the decomposition of the fermions: the gravitino, which is a Majorana vector-spinor in $d = 11$, decomposes into two gravitini with opposite chirality and two gravitini, also with opposite chirality: $\psi_M^a \to (\psi_\mu^a, \psi_\mu^{\dot\alpha}, \psi_{11}^\alpha, \psi_{11}^{\dot\alpha})$. This is the fermionic field content of type IIA SUGRA. In terms of representations of the little group, this is the decomposition $\mathbf{128}_{SO(9)} = (\mathbf{56}_s + \mathbf{56}_c + \mathbf{8}_s + \mathbf{8}_c)_{SO(8)}$.

The ten- and eleven-dimensional gravitational couplings are related as

$$\kappa_{11}^2 = 2\pi R_{11} \kappa_{10}^2 = \frac{1}{2}(2\pi)^8 g_s^3 (\alpha')^{\frac{9}{2}}, \tag{18.106}$$

where in the second step we have used Eqs. (18.103) and (9.57). It is convenient to define an eleven-dimensional Planck mass and Planck length $\ell_{11} = 1/M_{11}$ via

$$2\kappa_{11}^2 = \frac{1}{2\pi}\left(\frac{2\pi}{M_{11}}\right)^9 = \frac{1}{2\pi}(2\pi \ell_{11})^9. \tag{18.107}$$

Combined with (18.106) this gives

$$\ell_{11} = g_s^{1/3} \sqrt{\alpha'}. \tag{18.108}$$

This, together with (18.103), allows us to establish the following relations between the type IIA parameters g_s, α' and the (compactified) 11d supergravity parameters ℓ_{11}, R_{11}.[24]

---

[23] Recall: $\delta G_{MN} = -(\nabla_M \xi_N + \nabla_N \xi_M) = -(\xi^P \partial_P G_{MN} + \partial_M \xi^P G_{PN} + \partial_N \xi^P G_{MP})$.

[24] These relations are implied by the ansatz (18.105), if we write the length element in terms of dimensionless coordinates (for constant dilaton and zero 1-form)

$$ds^2 = \ell_{11}^2 G_{MN} dx^M dx^N = \ell_{11}^2 (g_s^{4/3} dx_{11}^2 + g_s^{-2/3} g_{\mu\nu} dx^\mu dx^\nu) \equiv R_{11}^2 dx_{11}^2 + \alpha' g_{\mu\nu} dx^\mu dx^\nu. \tag{18.109}$$

18.7 M-Theory

$$\alpha' = \ell_{11}^3/R_{11}, \qquad g_s = (R_{11}/\ell_{11})^{3/2}. \tag{18.110}$$

To summarize: The type IIA string coupling constant is determined by the radius (in units of the eleven-dimensional Planck length) of the circle on which $d = 11$ supergravity is compactified. The perturbative regime of string theory is unaware of the eleventh dimension, as the small coupling sector corresponds to the small radius region. Vice versa, the non-perturbative regime of string theory should imply a decompactification of the eleventh dimension.

We have argued that the appropriate description of type IIA string theory in the $g_s \gg 1$ limit is eleven-dimensional supergravity. However, string theory is more than just its low energy supergravity limit. Therefore, for $g_s \simeq 1$, i.e. for radius $R_{11} \simeq \ell_{11}$, there must exist a complete eleven-dimensional quantum gravity theory whose low-energy limit is eleven-dimensional supergravity. The precise structure of this conjectured theory, which is not a string theory, is not known yet. All we know about it has been derived by duality arguments of the type presented above. At least this new theory has been given a name: M-theory. There is a proposal for its quantum description as a matrix theory, but we will not pursue this here.

We have already seen that Kaluza-Klein states of the 11-dimensional theory are identified with bound states of D0-branes of type IIA superstring theory. Let us look for the M-theory origin of the other type IIA BPS branes. We start with the D2-brane. Its tension, expressed through M-theory parameters, is

$$\tau_{D2} = \frac{1}{(2\pi)^2 g_s (\alpha')^{\frac{3}{2}}} = \frac{M_{11}^3}{(2\pi)^2} \equiv T_{M2}. \tag{18.111}$$

It does not depend on the compactification scale R_{11} and should therefore correspond to an unwrapped fundamental object in M-theory, i.e. a membrane, or M2-brane, in eleven-dimensions. Indeed, the massless spectrum contains the three-form potential A_3, which naturally couples electrically to this membrane. We have constructed the M2-brane solution of $d = 11$ SUGRA in Sect. 18.5. The fundamental type IIA string can then be identified as a M2-brane which is wrapped around the circle.[25] This is supported by the fact that the NS-NS two-form B_2 arises by dimensional reduction of A_3. The tensions are related as

$$T_{F1} = \frac{1}{2\pi\alpha'} = 2\pi R_{11} T_{M2}. \tag{18.112}$$

Type IIA theory also contains the dual NS5-brane, whose tension is

$$\tau_{NS5} = \frac{1}{(2\pi)^5 g_s^2 \alpha'^3} = \frac{M_{11}^6}{(2\pi)^5} \equiv T_{M5}, \tag{18.113}$$

---

[25] The compactification reduces the dimension of space-time and the M2 world-volume. This is also called double dimensional reduction.

which is again independent of R_{11} and should therefore correspond to an unwrapped object in eleven-dimensions. We conclude that M-theory should also contain an M5-brane, which is the object which couples magnetically to the three-form A_3 and which we also constructed as a solution of $d = 11$ SUGRA. One checks the Dirac quantization $2\kappa_{11}^2 T_{M2} T_{M5} = 2\pi$. Wrapping an M5-brane around the compact circle should give the type IIA D4-brane. This is again consistent with the tensions

$$\tau_{D4} = \frac{1}{(2\pi)^4 g_s (\alpha')^{\frac{5}{2}}} = 2\pi R_{11} T_{M5}. \qquad (18.114)$$

Less obvious is the M-theory origin of the type IIA D6-brane which is the magnetic dual of the D0-brane which, in turn, is electrically charged with respect to the Kaluza-Klein gauge field $G_{\mu\,11}$. One expects, and this can indeed be shown to be true,[26] that the D6-brane is a Kaluza-Klein monopole of compactified M-theory.

A systematic and intrinsic way to see how the BPS states of M-theory arise is to analyze the $\mathcal{N} = 1$ supersymmetry algebra.[27] In $d = 11$ there are 32 supercharges Q_α which are the components of a Majorana spinor. The relevant part of the supersymmetry algebra in $d = 11$ is

$$\{Q_\alpha, Q_\beta\} = (\Gamma^\mu C^{-1})_{\alpha\beta} P_\mu. \qquad (18.116)$$

The remainder of the algebra is the Poincaré algebra and the commutators of the supercharges with the generators of the Poincaré algebra which simply state that the Q_α transform as spinors of $SO(1, 10)$ and that they are conserved, i.e. $[P_\mu, Q_\alpha] = 0$. A central extension in the strict sense, i.e. as a charge commuting with all generators of the algebra, is only possible in extended supersymmetry.

We can relax this condition and allow charges which carry Lorentz indices but we still require that they commute with the supercharges. The algebra (18.116) can be extended to the M-theory algebra[28]

$$\{Q_\alpha, Q_\beta\} = -\left(\Gamma^\mu C^{-1}\right)_{\alpha\beta} P_\mu - \frac{1}{2}\left(\Gamma^{\mu\nu} C^{-1}\right)_{\alpha\beta} Z_{\mu\nu}$$

---

[26]The metric

$$ds^2 = dx_{\parallel}^2 + V(r)(dr^2 + r^2 d\Omega_2) + V(r)^{-1}(dx^{11} + A(y) \cdot dy)^2 \qquad (18.115)$$

with $y \in \mathbb{R}^3$, $r^2 = y^2$ and $\boldsymbol{\nabla} \times \boldsymbol{A} = \boldsymbol{\nabla} V$ (which implies $\nabla^2 V = 0$) solves the vacuum Einstein equations. For $V = 1 + \alpha/r$ one finds $F = dA = \alpha \epsilon_{S^2}$. This is the KK-monopole solution. It is regular at $R = 0$, if $x^{11} \sim x^{11} + 4\pi\alpha$. If we identify $V^{1/2} = e^{-2/3\phi} = H^{1/2}$ and write it in the form (18.105), we see that $g_{\mu\nu}$ is the D6-brane solution in string frame and $dC = dA$ is its magnetic field strength.

[27]A similar analysis can be made for the IIA and IIB supersymmetry algebras.

[28]This modification of the supersymmetry algebra can be derived from the world-volume action of extended objects in the Green-Schwarz formulation.

18.7 M-Theory

$$-\frac{1}{5!} \left(\Gamma^{\mu_1 \cdots \mu_5} C^{-1} \right)_{\alpha\beta} Z_{\mu_1 \cdots \mu_5}. \tag{18.117}$$

The left hand side is a real symmetric 32×32 matrix with $32 \cdot 33/2 = 528$ independent entries. It furnishes a reducible representation of $SO(1, 10)$. The right hand side is the decomposition into irreducible components. In eleven dimensions with Lorentzian signature the $\Gamma^{[p]} C^{-1}$ with $p = 1, 2, 5, 6, 9, 10$ are symmetric, cf. appendix of Chap. 8. There are three Hodge dual pairs and the number of independent symmetric matrices is 528.

The ranks of the central charges agree with the ranks of the anti-symmetric tensor fields and with the dimensions of the supergravity M2 and M5 branes which we found earlier. The central charges are related to identically conserved topological charges associated with the branes:

$$Z^{\mu_1 \cdots \mu_p} = \frac{2\pi}{\ell_{11}^{p+1}} \int_{\mathscr{C}_p} dX^{\mu_1}(\xi) \wedge \cdots \wedge dX^{\mu_p}(\xi), \tag{18.118}$$

where the integral is over a space-like slice through the brane's world-volume. As the integrand is a closed p-form, the integral vanishes for topologically trivial $\mathscr{C}_p$. Conservation follows from $(\int_{\mathscr{C}_p} - \int_{\mathscr{C}'_p}) \omega = \int_{M_{p+1}} d\omega = 0$ where $\partial M_{p+1} = \mathscr{C}_p \cup \mathscr{C}'_p$ and ω closed. Notice that the analogue of (18.118) for the string is $\frac{1}{2\pi\alpha'} \int d\sigma \, \partial_\sigma X^\mu$ which computes its winding number in the μ-th direction.

While the M2 and M5 branes exist as solutions of the SUGRA field equations, the effective world-volume theory governing the dynamics of its fluctuations is not so straightforward to determine, in particular for several coincident branes. For M2 branes this involves a new algebraic structure called three-algebras. The world-volume theory of the M5-brane is expected to be related to the one of the NS5-brane of type IIA superstring. It should support a chiral $(0, 2)$ supersymmetric six-dimensional field theory, whose bosonic field content consists of a 6D self-dual tensor $b_{\mu\nu}$ and five scalars. The latter can be considered as the Goldstone modes of the broken transverse translations. However, the effective theory of several branes, which would involve an interacting theory of self-dual tensor fields, is still unknown and is most likely not a field theory.

From the algebra (18.117) we can derive a BPS bound. The derivation is practically identical to the one in Sect. 18.4. Without loss of generality we work in a Majorana basis where $C = \Gamma^0$ and the supercharges are real. Consider a M2 brane at rest along the (012) directions. The only non-zero components of P^μ and $Z^{\mu\nu}$ are $P^0 = M$ and $Z^{12} \equiv zM$. The algebra (18.117) becomes

$$\{Q_\alpha, Q_\beta\} = M(1 - z\Gamma^{012}). \tag{18.119}$$

The Hermitian 32×32 matrix Γ^{012} satisfies $\text{Tr}\, \Gamma^{012} = 0$ and $(\Gamma^{012})^2 = 1$. From these two properties it follows that it has eigenvalues ± 1, each with a degeneracy 16. It can be diagonalized with a unitary matrix U and if we define $\tilde{Q} = UQ$ the algebra becomes

$$\{\tilde{Q}_\alpha, \tilde{Q}_\beta\} = P_0 \begin{pmatrix} (1+z)\mathbb{1} & 0 \\ 0 & (1-z)\mathbb{1} \end{pmatrix}_{\alpha\beta}, \qquad (18.120)$$

where each block is a 16×16 matrix. Defining the combinations

$$q_j = \frac{1}{\sqrt{2P_0}}(\tilde{Q}_{2j} + i\,\tilde{Q}_{2j-1}), \qquad q_j^\dagger = \frac{1}{\sqrt{2P_0}}(\tilde{Q}_{2j} - i\,\tilde{Q}_{2j-1}), \qquad (18.121)$$

one realizes that they satisfy the anti-commutation relations of fermionic oscillator algebra

$$\{q_i, q_j^\dagger\} = (1+z)\,\delta_{ij}, \quad i,j = 1,\ldots 8$$
$$\{q_i, q_j^\dagger\} = (1-z)\,\delta_{ij}, \quad i,j = 9,\ldots 16 \qquad (18.122)$$

with all other anti-commutators vanishing. From the positive semi-definiteness of the left hand sides we obtain the BPS bound $|z| \leq 1$ for the ratio topological charge/mass. The dimension of non-BPS representations is 2^{16} with 2^{15} bosonic and 2^{15} fermionic states. If the bound is saturated, half of the oscillators decouple and one obtains short multiplets with 2^8 states. One can generalize this to several M2 branes along different directions. For judicious choices there are still unbroken supercharges, i.e. decoupled oscillators, but their number will be less than 8.

Type IIB and $E_8 \times E_8$ Heterotic String from M-Theory

If one compactifies M-theory further on a circle of radius R_9, the resulting theory is T-dual to type IIB string theory. Therefore, compactified M-theory on T^2 should be dual to type IIB on a dual circle. In order to identify the type IIB decompactification limit on the M-theory side, we need to quantify this duality.

Consider M-theory on T^2, which, for simplicity we take to be rectangular with radii R_9 and R_{11}. The complex structure and the dimensionless Kähler modulus of this torus are defined as

$$\tau = i\frac{R_9}{R_{11}}, \qquad T = M_{11}^2\,R_9\,R_{11}. \qquad (18.123)$$

This M-theory compactification corresponds to the type IIA superstring compactified on a circle of radius R_9^{IIA} with $g_s^{\text{IIA}} = (M_{11}R_{11})^{3/2}$ and $\alpha' = M_{11}^{-3}R_{11}^{-1}$. Moreover, from the metric (18.105) in the KK-reduction of eleven-dimensional supergravity and from (18.108) we we conclude $R_9^{\text{IIA}} = R_9$. Under the T-duality transformation

$$\frac{R_9^{\text{IIA}}}{(g_s^{\text{IIA}})^2} = \frac{R_9^{\text{IIB}}}{(g_s^{\text{IIB}})^2}, \qquad R_9^{\text{IIB}} = \frac{\alpha'}{R_9^{\text{IIA}}} \qquad (18.124)$$

18.7 M-Theory

this is mapped to type IIB compactified on circle of radius R_9^{IIB}. Expressed in terms of M-theory quantities, the type IIB parameters are

$$g_s^{\text{IIB}} = \frac{R_{11}}{R_9} = (\text{Im}\tau)^{-1},$$

$$R_9^{\text{IIB}} = \frac{1}{M_{11}^3 R_9 R_{11}} = \frac{1}{M_{11} T}. \quad (18.125)$$

Therefore, the ten-dimensional type IIB superstring theory can be considered as the zero size $T \to 0$ limit of M-theory compactified on a two-dimensional torus, where the type IIB string coupling is given by its complex structure. We have derived this here only for a rectangular T^2 but it holds also for a tilted one, where the real part of τ is identified with the axion C_0. The $SL(2, \mathbb{Z})$ which acts on the complex structure modulus of the torus is mapped to the type IIB self-duality group which acts on the complexified string coupling. In the zero volume limit, the M5-brane wrapped around T^2 becomes the self-dual D3-brane of type IIB.

We now briefly describe how also the $E_8 \times E_8$ heterotic string with 16 supercharges is related to M-theory. Rather than compactifying M-theory on a circle, one can compactify it on the orbifold $S^1/\mathbb{Z}_2$ where the $\mathbb{Z}_2$ acts as a reflection, $x_{11} \to -x_{11}$ such that $x^{11} \in [0, \pi R_{11}]$. The end-points of the interval are the fixed points of the $\mathbb{Z}_2$ action. This preserves ten-dimensional Poincaré symmetry and breaks one-half of the 32 supersymmetries. Similar to the type I theory, a purely ten-dimensional gravitational theory with 16 supercharges is not consistent as the gravitino and dilatino lead to chiral anomalies. For the type I superstring these were canceled by the $SO(32)$ gauge fields on the D9-branes. Therefore, for quantum consistency of this M-theory orbifold, there must exist additional degrees of freedom.

As we know from Chap. 10, string theory compactifications on orbifolds have such additional states (twisted sectors) localized at the fixed points. It is therefore natural to assume that also M-theory has such additional twisted states. Since we have two fixed points, we conjecture that one E_8 gauge factor is localized at each of the two fixed points $x_{11} = 0$ and $x_{11} = \pi R_{11}$. This eleven-dimensional orbifold compactification with the two E_8 factors 'at the end of the world' is called Hořava-Witten theory. Much evidence in its favor has been collected, but we will not to go into further details.

We close this section with a brief summary: Various dualities in various dimensions have led to the picture that all superstring theories are connected to each other and one big surprise was the appearance of an eleven-dimensional theory. All of this suggests that there is a unique underlying fundamental theory, called M-theory, and the five different superstring theories (and their compactifications) are merely different weak coupling descriptions of M-theory, which correspond to different choices of the small expansion parameter. A crucial role in establishing these dualities is played by branes of various dimensions and it is their BPS property which allowed the extrapolation from weak to strong coupling. This is certainly a fascinating scenario, but at the time of writing, a fundamental formulation of M-theory is still lacking.

18.8 F-Theory

In the previous sections we have encountered two peculiar aspects of the type IIB superstring. First, in Sect. 18.5 we have seen that the supergravity seven-brane solution is different from the lower dimensional branes, as the harmonic function depends logarithmically on the transverse distance from the brane. This had the implication that finite energy solutions invoke the $SL(2,\mathbb{Z})$ action and that the back-reaction on the geometry of the transverse space is so strong that 24 such 7-branes curve it into a compact $\mathbb{P}^1$. Furthermore, we have seen in Sect. 18.7 that the strong-weak $SL(2,\mathbb{Z})$ symmetry receives a geometric interpretation as the modular group of a T^2 compactification of M-theory in the zero volume limit.

Combining these two observations leads to F-theory. It is a way to compactify type IIB string theory where the complex scalar field τ is not constant. If we interpret τ as the complex structure modulus of an auxiliary torus whose modular group is the $SL(2,\mathbb{Z})$ symmetry of type IIB, we arrive at a twelve-dimensional formulation. However, this twelve-dimensional interpretation does not mean that the type IIB theory arises via a standard Kaluza-Klein reduction on the torus. First, there is no twelve dimensional supergravity theory with signature $(1,11)$; second, in ten dimensions there is no scalar field which corresponds to the volume modulus of the auxiliary T^2. In other words, the torus is not completely geometric. Hence the 12-dimensional interpretation only serves to provide a geometrization of the type IIB $SL(2,\mathbb{Z})$ duality symmetry rather than to imply a bona fide compactification from twelve to ten dimensions. In contrast, in the M-theory interpretation of the previous section, the torus is physical and the ten-dimensional type IIB theory arises in the zero size limit of the torus.

The true power of this F-theory picture reveals itself when one compactifies the type IIB superstring to lower dimensions. We have just recalled that there should exist a compactification of type IIB on $\mathbb{P}^1$ with 24 7-branes, preserving half the supersymmetry, i.e. 16 supercharges. M-theory compactified on a K3 surface also has 16 unbroken supercharges. Consider a K3 surface which is an elliptic fibration over $\mathbb{P}^1$. This means that over each point of the base $\mathbb{P}^1$ there is a torus whose shape is characterized by the modular parameter which varies over the base (A compact Riemann surface of genus one is also called an elliptic curve.). We can now fiber the duality between M-theory on (zero size) T^2 and type IIB theory over the $\mathbb{P}^1$ base and conjecture that F-theory on an elliptically fibered K3 is dual to the type IIB string on the base $\mathbb{P}^1$.

We still have to find the 7-branes in this description. Recall from Sect. 18.6 that type IIB theory has not only D7-branes but also (p,q) 7-branes. In Sect. 18.5 we have seen that close to a D7-brane at position $u_1 \in \mathbb{P}^1$ the complex scalar τ with $\mathrm{Im}\,\tau \geq 0$ behaves as

$$j(\tau) \simeq \frac{1}{u - u_1}. \tag{18.126}$$

18.8 F-Theory

We now elaborate on the implications of interpreting τ as the modular parameter of an elliptic curve. From Chap. 14 we know that a degree six hypersurface in $\mathbb{P}^2[1, 2, 3]$ has vanishing first Chern class. Since it is complex one-dimensional, it must be an elliptic curve. Let $[x : y : z]$ be the homogeneous coordinates with weights $(2, 3, 1)$. The most general hypersurface constraint has the form

$$y^2 + a_1 xyz + a_3 yz^3 = x^3 + a_2 x^2 z^2 + a_4 xz^4 + a_6 z^6, \tag{18.127}$$

where we have normalized the coefficients of y^2 and x^3 to one. By two successive homogeneous changes of variables,

$$y \to y - \frac{1}{2}(a_1 xz + a_3 z^3), \quad x \to x - \frac{z^2}{12}(a_1^2 + 4a_2), \tag{18.128}$$

we can bring (18.127) to the so-called Weierstrass form

$$y^2 = x^3 + f\, xz^4 + g\, z^6. \tag{18.129}$$

To express f and g in terms of the a_n, it is convenient to introduce

$$b_2 = a_1^2 + 4a_2, \quad b_4 = a_1 a_3 + 2a_4, \quad b_6 = a_3^2 + 4a_6 \tag{18.130}$$

such that

$$f = \frac{1}{48}\left(24 b_4 - b_2^2\right), \quad g = \frac{1}{864}\left(216 b_6 - 36 b_4 b_2 + b_2^3\right). \tag{18.131}$$

In Weierstrass form the freedom of making homogeneous coordinate transformations is completely fixed except for a rescaling of z.

It is a classical result in the theory of elliptic curves that its modular parameter τ is implicitly given in terms of f and g via

$$j(\tau) = \frac{4\,(24 f)^3}{4 f^3 + 27 g^2}. \tag{18.132}$$

The denominator is the discriminant Δ with

$$\begin{aligned}\Delta &= 4 f^3 + 27 g^2 \\ &= -\frac{1}{4} b_2^2 (b_2 b_6 - b_4^2) - 8 b_4^3 - 27 b_6^2 + 9 b_2 b_4 b_6. \end{aligned} \tag{18.133}$$

The curve is singular iff $\Delta = 0$. Indeed, writing (18.129) in the patch $z = 1$ in the form $y^2 = \prod_{i=1}^{3}(x - e_i)$, one finds $\Delta = -\prod_{i<j}(e_i - e_j)^2$. The curve is singular iff two of the roots e_i coincide.

To obtain an elliptically fibered K3 surface, we simply elevate the coefficients a_n to homogeneous polynomials of degree $2n$ in the two homogeneous coordinates (u_1, u_2) of $\mathbb{P}^1$. Mathematically speaking, they are sections of $K_B^{-n} \simeq \mathcal{O}(2n)$, where $K_B \simeq \mathcal{O}(-2)$ is the canonical bundle of the base $B = \mathbb{P}^1$. Then (x, y, z) are sections of K_B^{-n} with $n = (2, 3, 0)$ and f, g are homogeneous polynomials of (u_1, u_2) of degrees eight and twelve, i.e. sections of K_B^{-4} and K_B^{-6}, respectively.[29] The constraint $z = 0$ defines a section of the elliptic fibration, i.e. the divisor $z = 0$ is the base $\mathbb{P}^1$.

In the language of Sect. 14.5, the family of elliptical K3 surfaces which we are considering are hypersurfaces in the toric variety

$$\frac{\mathbb{C}^5 - \{u_1 = u_2 = 0\} \cup \{x = y = 0\}}{(\mathbb{C}^*)^2} \tag{18.134}$$

with $(\mathbb{C}^*)^2$-action

$$(u_1, u_2, x, y, z) \sim (\lambda u_1, \lambda u_2, \lambda^4 \mu^2 x, \lambda^6 \mu^3 y, \mu z). \tag{18.135}$$

The CY condition requires that the polynomial has bi-degree $(6, 12)$. Alternatively we could also describe the K3 as a hypersurface in $\mathbb{P}^4[1, 1, 4, 6]$ with coordinates $[u_1 : u_2 : x : y]$.

In the chart $z = 1, u_2 = 1$ the Weierstrass form of the fibration is ($u \equiv u_1$)

$$y^2 = x^3 + f(u) x + g(u), \tag{18.136}$$

where f, g are polynomials of orders 8 and 12, respectively.

We now have an elliptic curve over each point of $\mathbb{P}^1$ with a varying modular parameter $\tau(u)$. From Eq. (18.126) we infer that the location of the 7-branes are at the zeros of the discriminant $\Delta(u)$. These are also the points in the base over which the torus fiber degenerates. Since Δ is a polynomial of degree 24, it has 24 zeros which are generically distinct. These points on $\mathbb{P}^1$ are the positions of the 24 7-branes on $\mathbb{P}^1$. This is consistent with the fact that we need 24 branes to close off the transverse space and could have been used to establish the orders of the polynomials f and g.

---

[29] Holomorphic line bundles are completely specified by their transition functions between overlapping coordinate patches. For $\mathbb{P}^1$ there are two patches, U_i with $u_i \neq 0$, with inhomogeneous coordinates $\xi_1 = \frac{u_2}{u_1}$ and $\xi_2 = \frac{u_1}{u_2}$. A section of the canonical bundle in U_i is $\omega_i d\xi_i$ and from $d\xi_2 = -\frac{1}{\xi_1^2} d\xi_1$ we find the transformation $\omega_2 = c_{21}\omega_1 = -\xi_1^2 \omega_1$. (The minus sign is irrelevant because two line bundles are isomorphic if their transitions functions are related by $c'_{ij} = h_i c_{ij} h_j^{-1}$ where $h_i \neq 0$ on U_i.) The isomorphism class of line bundles with transition functions $c_{ij} = (\frac{u_j}{u_i})^n$ is denoted by $\mathcal{O}(n)$, i.e. $K \simeq \mathcal{O}(-2)$. Consider a homogeneous polynomial of order n, $f = \sum_{p=0}^{n} a_p u_1^p u_2^{n-p} = u_1^n f_1(\xi_1) = u_2^n f_2(\xi_2)$. They are sections of $\mathcal{O}(n) \simeq K^{-n}$. The generalization to line bundles over $\mathbb{P}^n$ is immediate. For a different but equivalent characterization see Footnote 35 on page 482.

18.8 F-Theory

Let us discuss the spectrum of gauge bosons in eight dimensions, i.e. on the non-compact world-volume of the 7-branes. We will use the fact that a vectormultiplet of eight-dimensional supersymmetry with 16 supercharges contains two real scalars and the supergravity multiplet two vectors and a single real scalar. The scalars in the vectormultiplet parametrize the position of the brane in the two transverse directions. f and g together have $22 = 9 + 13$ complex parameters. But we still have the freedom of making a coordinate transformation on the base which is $SL(2, \mathbb{C})$ acting on u. In addition, we can scale $f \to w^2 f, g \to w^3 g$ with $w \in \mathbb{C}$ without changing the torus. We therefore end up with 18 complex parameters which we can use to change the positions of the 24 7-branes. For D7 branes in generic positions this leads to an Abelian gauge group $U(1)^{18}$. In addition there is one real scalar associated to the Kähler parameter of the $\mathbb{P}^1$. This is the scalar in the supergravity multiplet which contains two more $U(1)$ vector fields (graviphotons).

At first sight it seems puzzling that not all 7-branes lead to an Abelian gauge factor. The explanation for this mismatch is that these F-theory backgrounds intrinsically contain non-trivial $SL(2, \mathbb{Z})$ actions, so that not all $U(1)$s are independent. In fact, at most 18 of the 24 7-branes can be D7-branes, while the remaining ones are more general (p, q) 7-branes. An easy way to see that not all 24 branes can be D7's is that the total monodromy must be trivial, as a cycle enclosing all 24 branes can be contracted to a point. This is nothing but the tadpole condition for the seven-branes with a compact transverse space. Since these backgrounds generically contain genuine (p, q) 7-branes, except in the so-called orientifold limit to be discussed later, they do not correspond to perturbative type IIB string compactifications but to non-perturbative ones.

So far we have discussed the generic situation where the discriminant has 24 distinct zeros, i.e. the fiber degenerates over these 24 points on $\mathbb{P}^1$. The total space is nevertheless regular, it is a smooth K3 surface.[30] Multiple zeros occur when several branes collide. This leads to a degeneration of the elliptic fibration, i.e. to a singular limit of the K3 surface. The singularity is due to 2-cycles which shrink to zero size. This can be understood as follows. Consider two points in the base with singular fibers over them. The singular fiber is a torus with one of its holonomy cycles, say the a-cycle, pinched off. Connect the two points along the base. The a-cycles over the connecting path form an S^2 which shrinks to a point, if the two points are moved towards each other. This happens at particular loci in the complex structure moduli space of the K3 surface. At the degeneration points the gauge symmetry is enhanced and non-Abelian gauge groups appear. In F-theory this is clear from the brane picture. In M-theory on K3, where the $U(1)$ bosons arise via Kaluza-Klein reduction of the three-form A_3, the off-diagonal gauge bosons arise from M2-branes wrapped around the shrunken 2-cycles.

The possible singular fibers and singularities of elliptic K3 surfaces were classified by Kodaira. His results are summarized in Table 18.6. The singularities of the K3 surface are encoded in the behavior of f, g and Δ; e.g. $f = u^{\text{ord}(f)} f_0(u)$ with $f_0(0) \neq 0$ and likewise for g and Δ. (Here we have chosen the coordinate on

---

[30]This is somewhat analogous to viewing S^2 as a S^1 fibration over an interval. At the ends of the interval the fiber degenerates but the total space is nevertheless smooth.

the base such that Δ has a zero of order ord(Δ) at $u = 0$.) In the fourth column we list the singular fibers in the notation introduced by Kodaira. They are the blow-ups of the surface singularities which have an A-D-E classification, as was already mentioned in the previous section: a collection of $\mathbb{P}^1$'s, whose intersection matrix is the negative of the Cartan matrix of the respective Lie algebra, shrink to zero size. This results in a singular K3. The singularities listed in the table are those that can be blown up into a smooth K3. If ord(f) ≥ 4 and ord(g) ≥ 6 this is no longer the case. The singular limit is attained at certain loci in complex structure moduli space where the gauge symmetry is enhanced to A_n, D_n, or E_n. At these loci $n = \mathrm{ord}(\Delta)$ branes are moved to the same position on the base. In the simplest case they are all D7 branes in which case the gauge group is $SU(n)$.[31] Other cases will be considered below. The local geometry of the singular surface is obtained from the Weierstrass form after a coordinate transformation which moves the singularity to $(x, y, u) = (0, 0, 0)$ and after dropping all higher order terms.[32] The last column is the monodromy of the a and b cycles of the torus around the points in the base above which the fiber degenerates. and hence also of the modular parameter. The monodromy of τ follows from

$$\tau = \frac{\oint_b \omega}{\oint_a \omega}, \qquad (18.137)$$

where ω is the holomorphic differential; cf. the discussion in Sect. 6.2. All the monodromies are up to conjugation by an element of $SL(2, \mathbb{Z})$ and the total monodromy must vanish. This is simply the tadpole cancellation condition.

Non-trivial monodromy only occurs along paths on the base which encircle points with a singular fiber over them. Consider e.g. a singular fiber with $\tau = i\infty$ over $u = u_0$ where $j(\tau(u))$ has a single pole (cf. (18.66) on page 703 and the discussion following it):

$$j(\tau) \sim e^{-2\pi i \tau(u)} \sim \frac{1}{u - u_0} \quad \to \quad \tau(u) \sim \frac{1}{2\pi i} \ln(u - u_0). \qquad (18.138)$$

Under $(u - u_0) \to e^{2\pi i}(u - u_0)$, the modular parameter of the auxiliary torus transforms as $\tau \to \tau + 1$, i.e. u_0 is the location of a D7-brane. From (18.137) we learn that this means that over u_0 the a-cycle of the torus is pinched of. a is called a vanishing cycle. More generally, if over $u = u_0$ there is a singular fiber with vanishing cycle $v = p\,a + q\,b$, the monodromy can be obtained from the Picard-Lefschetz theorem which says that the cycle δ undergoes the monodromy

---

[31] It is interesting to note that the gauge group is $SU(n)$ rather than $U(n)$. The explanation, which involves the Stückelberg mechanism, will not be given here.

[32] An example: the conditions which lead to the A_{n-1} singularity are satisfied if we choose $f = a$, $g = b + u^n$ with $a = -3(b^2/4)^{1/3}$ such that $\Delta = 54bu^n + \mathcal{O}(u^{2n})$. With $x \to (\frac{2}{27b})^{1/6} x + (\frac{b}{2})^{1/3}$ the Weierstrass form is transformed to $y^2 = x^2 + u^n + \mathcal{O}(x^3)$.

18.8 F-Theory

$$M_\nu : \quad \delta \to \delta - (\delta, \nu)\nu. \tag{18.139}$$

Here, (δ, ν) is the intersection number. Using $(a, b) = 1$, we find

$$\begin{pmatrix} b \\ a \end{pmatrix} \to M_{(p,q)} \begin{pmatrix} b \\ a \end{pmatrix} \tag{18.140}$$

with $M_{(p,q)}$ as in (18.89). This implies that $\nu = pa + qb$ is the vanishing cycle for a (p, q) 7-brane. The monodromy of τ around a (p, q) brane is $\tau \to M \cdot \tau$, as follows from (18.137).

One important consequence of the above discussion is that F-theory goes beyond perturbative IIB theory. As far as the torus is concerned, τ and its $SL(2, \mathbb{Z})$ transformed images describe the same torus. However, with the interpretation $\tau = C_0 + ie^{-\phi}$, an $SL(2, \mathbb{Z})$ monodromy generically transforms weak coupling to strong coupling. When we encircle the positions of all 24 7-branes on the base, the total monodromy must vanish and, as already remarked, the branes cannot all be D7-branes. At the positions of the D7 branes the string coupling goes to zero, while it becomes large at the positions of other (p, q)-branes.

Notice that also the simply laced exceptional gauge groups E_6, E_7 and E_8 are possible. Clearly they cannot be realized by fundamental open strings of the perturbative type IIB string, i.e. not just with $(1, 0)$ strings and D7-branes. These enhancements must involve more general (p, q) seven branes and suitable string-junctions between them. That string junctions exist is clear from the fact that a fundamental string can end on a D-string. But more general junctions exist. Charge conservation requires that $\sum p_i = \sum q_i = 0$ and stable configurations require either three semi-infinite strings such that the forces due to their tensions balance or more complicated string networks.[33]

To see which brane configuration leads to the various gauge groups, we compare the geometric monodromy matrices in the last column in Table 18.6 with those listed for stacks of A, B, C branes in Table 18.4. For the A-D-E series and the three fiber types II,III,IV we observe a perfectly match. As expected, the number of 7-branes is in all cases identical to the vanishing order of the discriminant. One can also show that the string junctions ending on the stacks of branes provide precisely the massless states to build the adjoint representation of A-D-E gauge groups. The D_n series is realized by $A^n BC$ 7-branes, which indicates that the BC pair can be considered as the non-perturbative description of an O7 plane. The monodromy around such an O7 is

---

[33]These rules seem to exclude the junction consisting of F1 ending on D1. But we must interpret it as a junction between $(1, 0)$ (F1), $(0, 1)$ (D1) and $(-1, -1)$ strings. The D1 string absorbs the charge carried by the F1 string and turns into a $(1,1)$ string leaving the junction (which is equivalent to a $(-1, -1)$ string entering the junction.) In the limit of zero string coupling the tension of the strings with $q \neq 0$ becomes infinite while the tension of F1 stays finite and we recover the picture of a straight D-string with a fundamental string ending on it at right angle.

Table 18.6 The Kodaira classification of singular fibers in elliptic surfaces

ord(f)	ord(g)	ord(Δ)	Fiber	Singularity	Local geometry	Monod.
≥ 0	≥ 0	0	I_0	Smooth		$\begin{pmatrix} 1 & 0 \\ 0 & 1 \end{pmatrix}$
0	0	1	I_1	Smooth	$y^2 = x^2 + u$	$\begin{pmatrix} 1 & 1 \\ 0 & 1 \end{pmatrix}$
0	0	n	I_n	A_{n-1}	$y^2 = x^2 + u^n$	$\begin{pmatrix} 1 & n \\ 0 & 1 \end{pmatrix}$
≥ 1	1	2	II	Smooth		$\begin{pmatrix} 1 & 1 \\ -1 & 0 \end{pmatrix}$
≥ 1	≥ 2	3	III	A_1	$y^2 = x^2 + u^2$	$\begin{pmatrix} 0 & 1 \\ -1 & 0 \end{pmatrix}$
≥ 2	2	4	IV	A_2	$y^2 = x^2 + u^3$	$\begin{pmatrix} 0 & 1 \\ -1 & -1 \end{pmatrix}$
2	3	6	I_0^*	D_4	$y^2 = x^2 u + u^3$	$\begin{pmatrix} -1 & 0 \\ 0 & -1 \end{pmatrix}$
2	≥ 3	$n+6$	I_n^*	D_{n+4}	$y^2 = x^2 u + u^{n+3}$	$\begin{pmatrix} -1 & -n \\ 0 & -1 \end{pmatrix}$
≥ 2	3					
≥ 3	4	8	IV*	E_6	$y^2 = x^3 + u^4$	$\begin{pmatrix} -1 & -1 \\ 1 & 0 \end{pmatrix}$
3	≥ 5	9	III*	E_7	$y^2 = x^3 + xu^3$	$\begin{pmatrix} 0 & -1 \\ 1 & 0 \end{pmatrix}$
≥ 4	5	10	II*	E_8	$y^2 = x^3 + u^5$	$\begin{pmatrix} 0 & -1 \\ 1 & 1 \end{pmatrix}$

The local geometry of the elliptic surface around such an A-D-E singularity is modeled in terms of coordinates $(x, y, u) \in \mathbb{C}^3$. In the last column the elliptic monodromy of the singular fiber is given in terms of a $SL(2; \mathbb{Z})$-matrix

$$M_{O7} = M_B M_C = \begin{pmatrix} -1 & 4 \\ 0 & -1 \end{pmatrix}. \tag{18.141}$$

All this demonstrates the power of the geometric description of non-perturbative type IIB backgrounds with 7-branes via F-theory.

One can still define a limit in which the string coupling goes to zero almost everywhere on the base. This is known as the Sen-limit. One first rescales

$$a_3 \to \epsilon \, a_3, \qquad a_4 = \epsilon \, a_4, \qquad a_6 = \epsilon^2 \, a_6 \tag{18.142}$$

and then sends $\epsilon \to 0$. In this limit one finds

$$f = \frac{1}{48} \left(24 \epsilon \, b_4 - b_2^2 \right), \qquad g = \frac{1}{864} \left(216 \epsilon^2 \, b_6 - 36 \epsilon \, b_4 b_2 + b_2^3 \right). \tag{18.143}$$

18.8 F-Theory

The discriminant becomes

$$\Delta = -\frac{\epsilon^2}{4} b_2^2 (b_2 b_6 - b_4^2) + O(\epsilon^3) \tag{18.144}$$

which leads to

$$j(\tau) \simeq \frac{b_2^4}{\epsilon^2 (b_2 b_6 - b_4^2)}. \tag{18.145}$$

We can choose an $SL(2, \mathbb{Z})$ frame where τ is in the fundamental region. Then for $\epsilon \to 0$, the type IIB string coupling constant g_s goes to zero almost everywhere except on the locus where b_2 vanishes. There the coupling diverges, signaling that this is not the location of D7-branes. Indeed, a detailed study of the monodromies reveals a D7-brane on the locus $(b_2 b_6 - b_4^2) = 0$ and an O7-plane where $b_2 = 0$. Therefore, the Sen-limit defines the region in the complex structure moduli space, where F-theory is (almost everywhere) weakly coupled and a perturbative type IIB orientifold description is justified.

It is interesting to see how perturbative type IIB orientifolds fit into the F-theory framework. Compactifying the type IIB string on a T^2 and taking the orientifold quotient $\Omega I_2 (-1)^{F_L}$ leads to four O7-planes, whose tadpoles can be canceled locally by placing four dynamical D7-branes on top of each O7-plane. This leads to the gauge group $SO(8)^4$, cf. Table 10.3 on page 317. Since we have canceled the tadpole locally, the dilaton should be constant. Thus we should look for a Weierstrass model for which $j(\tau) = \text{const}$. It is easy to see that this is the case for the choice $f = p^2(u)$ and $g = \lambda p^3(u)$ for a polynomial $p(u) = \prod_{i=1}^{4}(u - u_i)$ of order four and $\lambda \in \mathbb{C}$. Then

$$j(\tau) = \frac{55296}{4 + 27\lambda^2} \tag{18.146}$$

which indicates that by tuning $\lambda^2 \simeq -4/27$ the string coupling is small everywhere (recall that $j(\tau)$ has a simple pole at $\tau = i\infty$). The discriminant becomes

$$\Delta \simeq \prod_{i=1}^{4} (u - u_i)^6 \tag{18.147}$$

and the Kodaira classification tells us that at each of the four points u_i we have an I_0^* singular fiber which gives rise to an $SO(8)$ gauge group. In the orientifold description, the remaining four Abelian gauge bosons known to be present in F-theory, come from the dimensional reduction of the NS-NS and R-R two-form along the two internal toroidal directions, i.e. $B_{\mu a}$ and $C_{\mu a}$ with $a = 1, 2$. Therefore, we conclude that for this choice of f and g the F-theory model on K3 gives precisely the type IIB orientifold on $\mathbb{P}^1 = T^2/\mathbb{Z}_2$. The monodromy acts trivially on τ but it reverses the sign of the internal components of B_2 and C_2. This agrees with the action of the orientifold group $\Omega(-1)^{F_L}$.

This orientifold is related, via T-duality, to the type I superstring on T^2, which by itself is S-dual to the $SO(32)$ heterotic string on T^2. Another T-duality maps this to the $E_8 \times E_8$ heterotic string. Therefore, we expect a duality between F-theory on $K3$ and the $E_8 \times E_8$ heterotic string on T^2. Let us go through this chain of dualities to find the relation between the heterotic string coupling and F-theory data. Applying successively the transformation rules for T-duality and S-duality we find that the ten-dimensional dilaton and the Kähler modulus $T = R_1 R_2/\alpha'$ of the T^2 transform as shown below:

$$\Omega I_2(-1)^{F_L} \quad \overset{T}{\longleftrightarrow} \quad \text{type I} \quad \overset{S}{\longleftrightarrow} \quad SO(32) \text{ het} \quad \overset{T}{\longleftrightarrow} \quad E_8 \times E_8 \text{ het}$$

$$g_s^{\text{IIB}} \qquad\qquad g_s^I \sim \frac{g_s^{\text{IIB}}}{T^{\text{IIB}}} \qquad\qquad \tilde{g}_s^H \sim \frac{T^{\text{IIB}}}{g_s^{\text{IIB}}} \qquad\qquad g_s^H \sim T^{\text{IIB}}$$

$$T^{\text{IIB}} \qquad\qquad T^I \sim \frac{1}{T^{\text{IIB}}} \qquad\qquad \tilde{T}^H \sim \frac{1}{g_s^{\text{IIB}}} \qquad\qquad T^H \sim g_s^{\text{IIB}}$$

Thus, the heterotic string coupling is dual to the size of the base $\mathbb{CP}^1$ of the elliptically fibered $K3$. Furthermore, the gauge group for generic Wilson-lines of the heterotic string on T^2 is also $U(1)^{20}$, which can be enhanced to the A-D-E gauge groups at special point in the moduli space.

We now comment on lower dimensional compactifications of F-theory. Instead of fibering the torus over a complex one-dimensional base, one can consider fibrations over surfaces B_2 or three-dimensional bases B_3. Supersymmetry then demands that the total space must be a Calabi-Yau threefold (for B_2) or a Calabi-Yau fourfold (for B_3). One can still write down a Weierstrass model. f and g are now sections of K_B^{-4} and K_B^{-6}, respectively. The zeros of the discriminant define co-dimension one curves in B_2 and surfaces in B_3, which are the internal parts of the world-volumes of 7-branes. The singularity enhances further where these co-dimension one submanifolds, i.e. the 7-branes, intersect. Similar to intersecting D-branes, this is where additional matter fields are localized. Another feature of lower-dimensional compactifications of F-theory is that non-simply laced gauge groups are possible.

As additional information, we want to point out one interesting feature of F-theory compactifications on elliptically fibered Calabi-Yau threefolds, in particular when the base is $B_2 \simeq \mathbb{P}^2$. This is a non-perturbative compactification of the type IIB string on $\mathbb{P}^2$ with a non-trivial scalar field τ which preserves eight supercharges. Clearly there are fermions. That raises a puzzle because $\mathbb{P}^2$ is not a spin manifold. The resolution is that the coupling of the spinors to the scalar field τ is such that there is a composite $U(1)$ connection composed of the scalar τ which has just the right properties required to have a Spin$_{\mathbb{C}}$ structure on $\mathbb{P}^2$.

Of particular interest are F-theory compactifications on smooth Calabi-Yau fourfold Y. Here several new issues arise. One can show that chiral matter only arises on the intersection curve between two 7-branes, if there exists a non-trivial F_4-form background (M-theory point of view). One also finds a non-trivial D3-brane tadpole cancellation condition, which in this case reads

Table 18.7 F-theory/heterotic dualities in dimensions $D = 4, 6, 8$

Dimension	Supercharges	Heterotic	⟷	F-theory
8	16	T^2		T^2 fibered K3
6	8	T^2 fibered K3		K3 fibered CY_3
4	4	T^2 fibered CY_3		K3 fibered CY_4

$$N_{D3} + \frac{1}{2}\int_Y F_4 \wedge F_4 = \frac{\chi(Y)}{24}. \tag{18.148}$$

$\chi(Y)$ is the Euler number of the smooth fourfold Y.

A special class of lower dimensional F-theory compactifications arises when the Calabi-Yau manifold is not only elliptically fibered but also K3 fibered. This can be considered as fibering F-theory on K3 over a complex one- or two-dimensional base. The eight-dimensional duality between F-theory on K3 and the heterotic string on T^2 generalizes to lower dimensions as shown in Table 18.7.

We close this section by pointing out a relation between duality conjectures relating F-theory, M-theory and string theory. In the previous section we have discussed the duality between M-theory on T^2 and type IIB string theory on S^1 and the duality between M-theory on S^1 and the ten-dimensional type IIA string. In this section we have argued that F-theory on K3 is dual to type IIB on $\mathbb{P}^1$ and to the heterotic string on T^2. Combining these dualities leads to the conjecture that if F theory on a manifold $\mathcal{M}$ is dual to a string theory S on a manifold $\mathcal{K}$, then M theory on $\mathcal{M}$ is dual to S on $\mathcal{K} \times S^1$ and type IIA on $\mathcal{M}$ is dual to S on $\mathcal{K} \times T^2$.

18.9 AdS/CFT Correspondence

The dualities which we have discussed so far are dualities between different perturbative string theories, i.e. string-string dualities. The AdS/CFT (Anti-de-Sitter/conformal field theory) correspondence is distinct in that it is a duality between a string theory and a quantum field theory. Most fascinating is that it provides a gravitational description of theories without gravity and vice versa. And these two theories are defined in different numbers of space-time dimensions. The AdS/CFT correspondence is a wide subject with many ramifications ranging from formal developments to (possible) applications in QCD and even condensed matter physics. It has added a new aspect to string theory: to view it not only as a fundamental microscopic theory of all particles and their interactions, but as a framework which allows a detailed and quantitative study of many physical systems which is complementary (dual) to a quantum field theoretical description. Conversely, we can describe quantum gravity (strings) in asymptotically AdS geometries in terms of quantum field theories without gravity. A comprehensive treatment of the present state of this subject would require a separate book. Also, as it is still an ongoing endeavor any such treatment would soon been outdated.

We will therefore limit ourselves to presenting some of the background material which is necessary to understand the basic framework of the correspondence. We will show how it arises as a confluence of many ideas which have been essential in string theory: strong-weak duality, flux compactifications, holographic principle, large N transition, etc. We illustrate the power of the correspondence through a few examples and will close with an outlook.

We have argued that D-branes, which have a microscopic description in open string theory, are the same objects as the classical p-brane solutions of the low energy effective SUGRA theories which reproduce closed string scattering amplitudes. The two descriptions are appropriate in different regimes of the parameter space.

The D-brane picture is valid at weak string coupling, $g_s \ll 1$, where the string is perturbative and open strings end on D-branes which are heavy and themselves not in the perturbative spectrum of the theory. This picture also assumes that the back-reaction of N coincident branes on Minkowski space is small, which is the case if $g_s N \ll 1$. Otherwise their distortion of space-time cannot be neglected. In that situation we have a description as solutions of the SUGRA equations of motion with fluxes. This also requires g_s small but $g_s N$ large, because otherwise the SUGRA description (18.26) which ignores string loop effects and α' corrections due to massive string modes is not valid and the equations of motions (18.27) and therefore also their brane solutions have to be corrected.

An important difference of the two descriptions is that while D-branes couple to open strings and carry a gauge theory on their world-volume, there are no signs of open strings and gauge fields in the p-brane picture. Nevertheless, if the two descriptions are 'dual' to each other, they should describe the same physics. The AdS/CFT correspondence establishes the relation between the two pictures.

To motivate the conjecture we recall the discussion of supergravity brane solutions in Sect. 18.5. Of particular interest is the type IIB three-brane solution which we repeat here

$$ds^2 = H^{-1/2}\eta_{\mu\nu}dx^\mu dx^\nu + H^{1/2}(dr^2 + r^2 d\Omega_{S^5}), \qquad H = 1 + \frac{L^4}{r^4},$$

$$e^\Phi = g_s,$$

$$g_s F_5 = (1 + \star)\,\epsilon_{\mathbb{R}^{1,3}} \wedge dH^{-1}, \qquad L^4 = 4\pi g_s \alpha'^2 N. \qquad (18.149)$$

It is charged under the self-dual four-form which means that it is its own electric-magnetic dual. The dilaton is constant and we can choose its constant value Φ_0 such that the solution is weakly coupled throughout, i.e. $g_s = e^{\Phi_0} \ll 1$. Equation (18.149) is a fully back-reacted solution of (18.27) and it is an approximate string background if $g_s \ll 1$ and the number of coincident D3 branes is large with $L^2/\alpha' \gg 1$.

Recall that in the brane solutions one integration constant was chosen such that the asymptotic geometry at $r \to \infty$ is Minkowski space. Close to the brane, i.e. for $r/L \ll 1$, this constant can be neglected and the metric becomes

18.9 AdS/CFT Correspondence

Fig. 18.2 The two asymptotic regions of the brane geometry

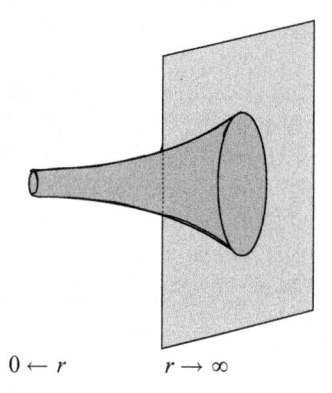

$0 \leftarrow r \qquad r \to \infty$

$$ds^2 = \frac{r^2}{L^2}d\mathbf{x}^2 + \frac{L^2}{r^2}dr^2 + d\Omega_{S^5}. \tag{18.150}$$

This is the metric on $AdS_5 \times S^5$ to which we will return below. The region $r/L \ll 1$ of the brane solution is often called the throat region.

We had already mentioned that the BPS brane solutions are extremal limits of charged black brane solutions whose two horizons (inner and outer horizon) coincide in the extremal limit. The coordinates were chosen such that $r = 0$ is the location of the degenerate horizon, hence one also calls (18.150) the near-horizon limit of the D3-brane solution. To summarize, the asymptotics of the brane geometry are

$$(ds^2)_{AdS_5 \times S^5} \stackrel{r \to 0}{\longleftarrow} (ds^2)_{\text{brane}} \stackrel{r \to \infty}{\longrightarrow} (ds^2)_{d=10 \text{ Mink.}} \tag{18.151}$$

and the D3-brane metric interpolates between them (see Fig. 18.2). The components of F_5 vanish as $1/r^5$ for $r \to \infty$.

The solutions in the asymptotic regions have higher symmetries than the brane solution. Specifically, while (18.149) preserves 16 supercharges, both Minkowski space and $AdS_5 \times S^5$ are invariant under 32 supercharges. The bosonic symmetries and therefore also the supersymmetry algebras are different, though. Minkowski space is invariant under Poincaré supersymmetry and $AdS_5 \times S^5$ under the anti-de-Sitter supergroup $PSU(2,2|4)$. Minkowski space is an exact perturbative ground state of type IIB string theory, i.e. to all orders in α' and g_s and we know how to quantize string theory in this background. It can be shown that $AdS_5 \times S^5$ is also an exact perturbative background of type IIB string theory, but the quantization of the theory in this background is a challenge. It is an example of a flux compactification of string theory which is valid beyond the SUGRA limit. The interpolating solution (18.149), however, is not an exact ground state and receives e.g. α' corrections.

Conformal Symmetry, AdS Space, etc.

Before we discuss the significance of these observations, we provide some details about the near-horizon geometry and its symmetries. In particular AdS-space and conformal symmetry play a central role in the AdS/CFT correspondence and we will discuss them in some detail.

Both factors of the near-horizon geometry are maximally symmetric spaces, or spaces of constant curvature; this means that they have as many isometries as Euclidean space of the same dimension and their curvature tensors are of the form

$$R_{ijkl} = k\,(g_{ik}\,g_{jl} - g_{il}\,g_{jk}), \quad R_{ij} = k\,(n-1)\,g_{ij}, \quad R = k\,n(n-1), \quad (18.152)$$

where k is necessarily a constant (as a consequence of the Bianchi identity) and n the dimension. Spaces which satisfy only the last two equations are called Einstein spaces.

Examples of maximally symmetric spaces with Lorentzian signature are Minkowski space ($k = 0$), Anti-de-Sitter space ($k = -\frac{1}{\ell^2}$) and de-Sitter space ($k = \frac{1}{\ell^2}$). ℓ is the curvature radius. The sphere is an example with Euclidean signature as are Wick rotations of AdS and dS. Inserting (18.152) into the explicit expression for the Weyl tensor in n dimensions[34]

$$C_{ijkl} = R_{ijkl} - \frac{1}{n-2}\left(g_{ik}R_{jl} + g_{jl}R_{ik} - g_{il}R_{jk} - g_{jk}R_{il}\right)$$
$$+ \frac{1}{(n-1)(n-2)}\left(g_{ik}g_{jl} - g_{il}g_{jk}\right)R \quad (18.153)$$

one finds $C_{ijkl} = 0$. Maximally symmetric spaces are therefore conformally flat. For S^n, defined as $\sum_{i=1}^{n+1}(y^i)^2 = \ell^2$, one sees this explicitly after defining stereographic coordinates x^i (say, on the Northern hemisphere) via $y^i = \frac{x^i}{v}$ for $i = 1, \ldots, n$ and $y^{n+1} = \ell(1 - \frac{x^2}{2v\ell^2})$, which solve the defining equation for $v = 1 + \frac{x^2}{4\ell^2}$. The metric is

$$ds^2 = \sum_{i=1}^{n+1}(dy^i)^2 = \sum_{i=1}^{n} g_{ij}\,dx^i\,dx^j \quad \text{with} \quad g_{ij} = \frac{1}{\left(1 + \frac{x^2}{4\ell^2}\right)^2}\delta_{ij}. \quad (18.154)$$

Maximally symmetric spaces are special solutions of the Einstein equations with cosmological constant which are derived from the action

$$S = \frac{1}{16\pi G_N^{(n)}} \int d^n x \sqrt{g}(R - \Lambda), \qquad \Lambda = (n-1)(n-2)\,k. \quad (18.155)$$

---

[34] The Weyl tensor has the same symmetries as the Riemann tensor, is traceless on all pairs of indices and transforms homogeneously under Weyl transformations of the metric. It vanishes identically for $n \leq 3$. A space is conformally flat iff its Weyl tensor vanishes.

18.9 AdS/CFT Correspondence

The components of the Einstein equations (18.27) along the (x^μ, r)-directions with F_5 and Φ given by the D3-brane solution in the near horizon limit follow from the action (18.155) with $n = 5$ and $\Lambda = -12/L^2$. This is the action of type IIB SUGRA on $AdS_5 \times S^5$ after integration over S^5 and all fluctuating fields except the metric set to zero. Here

$$G_N^{(5)} = G_N^{(10)}/\text{Vol}(S^5), \qquad 16\pi G_N^{(10)} = 2\kappa_{10}^2 = (2\pi)^7 g_s^2 \alpha'^4 \qquad (18.156)$$

is the Newton constant in five dimensions.

Before we discuss AdS space in some detail, we collect some background material on the conformal group. The conformal group in d dimensions is the invariance group of the light-cone in d-dimensional Minkowski space, i.e. all transformations which leave $(ds)^2 = 0$ invariant. Said differently, they are those transformations $x^\mu \to x'^\mu(x)$ for which $\eta_{\mu\nu} dx'^\mu dx'^\nu = e^{2\sigma(x)} \eta_{\mu\nu} dx^\mu dx^\nu$. Besides Poincaré transformations, constant rescalings $x^\mu \to e^\lambda x^\mu$ and inversion $x^\mu \to x^\mu/x^2$ also leave the light-cone invariant. An inversion, followed by a translation by b^μ and a second inversion leads to a special conformal transformation

$$x^\mu \to \frac{x^\mu + x^2 b^\mu}{1 + 2b \cdot x + b^2 x^2}. \qquad (18.157)$$

For infinitesimal conformal transformations $x^\mu \to x'^\mu = x^\mu + \xi^\mu(x)$ one derives the condition $\partial_\mu \xi_\nu + \partial_\nu \xi_\mu = 2\sigma(x) \eta_{\mu\nu}$. Taking the trace gives $\sigma = \frac{1}{d}(\partial \cdot \xi)$ which leads to the conformal Killing equation

$$\partial_\mu \xi_\nu + \partial_\nu \xi_\mu = \frac{2}{d}(\partial \cdot \xi)\eta_{\mu\nu}. \qquad (18.158)$$

Its solutions are the conformal Killing vectors of Minkowski space. For $d = 2$ and Euclidean signature, the conformal Killing equation are the Cauchy-Riemann equations for the two components of ξ^μ whose solutions are all holomorphic functions. For $d > 2$ the most general solution is at most quadratic in x^μ and has $\frac{1}{2}(d+1)(d+2)$ parameters[35]:

$$\xi^\mu = a^\mu + \omega^\mu{}_\nu x^\nu + \lambda x^\mu - 2(b \cdot x)x^\mu + x^2 b^\mu. \qquad (18.159)$$

λ is a constant, a^μ and b^μ are constant vectors and $\omega_{\mu\nu} = -\omega_{\nu\mu}$ a constant antisymmetric matrix. $a^\mu, \omega_{\mu\nu}, \lambda$ and b^μ parametrize infinitesimal translations, Lorentz-transformations, rescalings and special conformal transformations of x^μ, respectively. Their generators are

---

[35] One first derives $\partial_\alpha \partial_\beta \sigma = 0$ and then $\partial_\alpha \partial_\beta \partial_\gamma \xi_\mu = 0$ as consequences of the conformal Killing equation. Using the most general quadratic ansatz for ξ_μ in (18.158) leads to (18.159).

$$P_\mu = \partial_\mu, \qquad L_{\mu\nu} = x_\mu \partial_\nu - x_\nu \partial_\mu,$$
$$D = x \cdot \partial, \qquad K_\mu = -2x_\mu x \cdot \partial + x^2 \partial_\mu. \tag{18.160}$$

They satisfy the commutator algebra

$$[D, P_\mu] = -P_\mu,$$
$$[D, K_\mu] = K_\mu,$$
$$[P_\mu, K_\nu] = -2\eta_{\mu\nu} D + 2L_{\mu\nu},$$
$$[L_{\mu\nu}, P_\rho] = -\eta_{\mu\rho} P_\nu + \eta_{\nu\rho} P_\mu,$$
$$[L_{\mu\nu}, K_\rho] = -\eta_{\mu\rho} K_\nu + \eta_{\nu\rho} K_\mu,$$
$$[L_{\mu\nu}, L_{\rho\sigma}] = -\eta_{\mu\rho} L_{\nu\sigma} - \eta_{\nu\sigma} L_{\mu\rho} + \eta_{\mu\sigma} L_{\nu\rho} + \eta_{\nu\rho} L_{\mu\sigma},$$
$$[P_\mu, P_\nu] = [K_\mu, K_\nu] = [L_{\mu\nu}, D] = [D, D] = 0. \tag{18.161}$$

If one defines $L_{d,d+1} = -D$, $L_{\mu d} = \frac{1}{2}(P_\mu - K_\mu)$ and $L_{\mu,d+1} = -\frac{1}{2}(P_\mu + K_\mu)$, the above commutation relations can be combined into

$$[L_{MN}, L_{PQ}] = -\eta_{MP} L_{NQ} - \eta_{NQ} L_{MP} + \eta_{MQ} L_{NP} + \eta_{NP} L_{MQ}, \tag{18.162}$$

where $M, N, \cdots = 0, 1, \ldots, d+1$ and $\eta_{MN} = \text{diag}(-1, +1, \ldots, +1, -1)$ is the invariant metric of $SO(d, 2)$. This establishes the isomorphism of the conformal algebra of d-dimensional Minkowski space with $so(d, 2)$, the Lie algebra of $SO(d, 2)$. $SO(d, 2)$ acts linearly on $\mathbb{R}^{d,2}$ with metric $(ds)^2 = \eta_{MN} dy^M dy^N = \eta_{\mu\nu} dy^\mu dy^\nu + (dy^d)^2 - (dy^{d+1})^2$.

We now describe the non-linear action of the conformal group on d-dimensional Minkowski space $\mathbb{R}^{d-1,1}$. Since the special conformal transformation maps the point $x^\mu = -b^\mu/b^2$ to infinity, which is not part of Minkowski space, we need conformally compactified Minkowski space on which the conformal group acts properly. To this end we embed $\mathbb{R}^{d-1,1}$ with coordinates $(y^0, \ldots, y^{d-1})$ into $\mathbb{R}^{d,2}$ with coordinates $(y^0, \ldots, y^{d-1}, y^d, y^{d+1})$ where, as before, y^0, y^{d+1} are the two time-like coordinates. Consider the 'light-cone' in $\mathbb{R}^{d,2}$

$$(y^0)^2 + (y^{d+1})^2 = \sum_{i=1}^{d}(y^i)^2, \tag{18.163a}$$

or, equivalently

$$\eta_{\mu\nu} y^\mu y^\nu = uv \qquad (u = y^{d+1} + y^d, \ v = y^{d+1} - y^d). \tag{18.163b}$$

The intersection of the light-cone with the hyperplane $v = 1$ is Minkowski space. For compactification we intersect the light-cone with the sphere $(y^0)^2 + (y^{d+1})^2 +$

18.9 AdS/CFT Correspondence

$\sum_{i=1}^{d}(y^i)^2 = 2$ which leads to

$$(y^0)^2 + (y^{d+1})^2 = 1 = \sum_{i=1}^{d}(y^i)^2, \tag{18.164}$$

which is topologically $S^1 \times S^{d-1}$.

We now solve the constraint (18.163b) locally (in the patch with $u \neq 0$) by defining coordinates x^μ via

$$y^M = \begin{pmatrix} y^\mu \\ u \\ v \end{pmatrix} = \begin{pmatrix} ux^\mu \\ u \\ ux^2 \end{pmatrix}. \tag{18.165}$$

The linear $SO(d,2)$ transformations on y^M induce conformal transformations on x^μ. Specifically, we find the following relation between linearly acting $SO(d,2)$ transformations of y and conformal transformations of x:

$$\begin{pmatrix} \delta^\mu_\nu & 0 & 0 \\ 0 & \frac{1}{\Lambda} & 0 \\ 0 & 0 & \Lambda \end{pmatrix} \longleftrightarrow x^\mu \to \Lambda x^\mu, \quad \begin{pmatrix} \delta^\mu_\nu & 0 & b^\mu \\ 2b_\nu & 1 & b^2 \\ 0 & 0 & 1 \end{pmatrix} \longleftrightarrow x^\mu \to \frac{x^\mu + x^2 b^\mu}{1 + 2(b \cdot x) + b^2 x^2},$$

$$\begin{pmatrix} \Lambda^\mu{}_\nu & a^\mu & 0 \\ 0 & 1 & 0 \\ 2a_\rho \Lambda^\rho{}_\nu & a^2 & 1 \end{pmatrix} \longleftrightarrow x^\mu \to \Lambda^\mu{}_\nu x^\nu + a^\mu. \tag{18.166}$$

Each of these matrices $M_M{}^N$ satisfies $M_M{}^P M_N{}^Q \eta_{PQ} = \eta_{MN}$ with η_{MN} as given below (18.162). The infinitesimal version of these transformations are, of course, generated by the conformal Killing vectors.

Conformally invariant field theories in Minkowski space are characterized by an energy momentum tensor $T_{\mu\nu}$ which is conserved, $\partial^\mu T_{\mu\nu}$, and traceless, $T^\mu{}_\mu = 0$. In quantum field theory both equations are operational equations and lead to Ward-identities which are obeyed by correlation functions. The unavoidable violation of one of them, say of the trace condition, gives the trace anomaly. Conformal field theory was the topic of Chap. 4 where we restricted to two-dimensional Euclidean space. But $d = 2$ is very special, due to the infinite dimensional symmetry algebra. In $d > 2$ conformal symmetry still restricts correlation functions, but since the symmetry algebra is finite dimensional, to a much lesser extend.

A convenient way to study correlators of the energy-momentum tensor is to couple the energy-momentum tensor to the perturbation $h_{\mu\nu}$ around flat space of an external metric $g_{\mu\nu} \simeq \eta_{\mu\nu} + h_{\mu\nu}$. The equations satisfied by $T_{\mu\nu}$ then translate to the statement that the generating functional $W(g)$ in this background is invariant under diffeomorphisms and Weyl transformations, modulo anomalies.

The transformations corresponding to the conformal group $SO(d,2)$ are realized as transformations of the flat metric or viel-bein which leave it invariant. For instance, one easily checks that under a combined infinitesimal diffeomorphism ξ^μ, a Weyl rescaling with $\sigma = -\frac{1}{d}\partial_\nu \xi^\nu$ and a Lorentz-transformation $\Omega^a{}_b = \frac{1}{2}(\partial^a \xi_b - \partial_b \xi^a)$, where ξ^μ is a conformal Killing vector and $\partial_a = e_a^\mu \partial_\mu$, the flat viel-bein $e_\mu^a = \delta_\mu^a$ is invariant.[36] Therefore, on a flat background $e_\mu^a = \delta_\mu^a$, an action $S(e, \phi)$ which on a curved background is invariant under Lorentz-transformations, diffeomorphisms and Weyl rescaling $e_\mu^a \to e^\sigma e_\mu^a$, $\phi \to e^{-\Delta \sigma}\phi$, is invariant under[37]

$$\delta\phi = (\xi^\mu \partial_\mu - \Delta\sigma + \frac{1}{2}\Omega^{\mu\nu}\Sigma_{\mu\nu})\phi \qquad (18.167)$$

with

$$\sigma = -\lambda + 2(b \cdot x), \qquad \Omega^{\mu\nu} = -\omega^{\mu\nu} - 2(b^\mu x^\nu - b^\nu x^\mu).$$

$\Sigma_{\mu\nu}$ acts on the Lorentz-indices of ϕ which are not shown. In other words, if $S(g, \phi) = S(e^{2\sigma}g, e^{-\Delta\sigma}\phi)$ then $S(\eta, \phi)$ is a CFT. The violation of the tracelessness of $T_{\mu\nu}$ can be detected as the non-invariance of the generating functional $W(g)$ under Weyl rescaling of the metric.

In quantum field theory the transformations (18.167) are represented as unitary transformations of operators. For a group element g one has

$$U(g)\,\phi(x)\,U(g)^{-1} = D(g^{-1})\,\phi(g(x)), \qquad (18.168)$$

where D is a representation of g which acts on the Lorentz-indices of ϕ. If we define

$$U(a) = e^{ia^\mu P_\mu}, \qquad U(\omega) = e^{-\frac{i}{2}\omega^{\mu\nu} M_{\mu\nu}},$$
$$U(\lambda) = e^{i\lambda D}, \qquad U(b) = e^{ib^\mu K_\mu}, \qquad (18.169)$$

we find, by comparison with (18.167),

$$[P_\mu, \phi] = -i\,\partial_\mu \phi,$$
$$[M_{\mu\nu}, \phi] = -i\,(x_\mu \partial_\nu - x_\nu \partial_\mu + \Sigma_{\mu\nu})\phi,$$
$$[D, \phi] = -i\,(x^\mu \partial_\mu + \Delta)\phi,$$
$$[K_\mu, \phi] = -i\,(x^2 \partial_\mu - 2x_\mu x \cdot \partial - 2\Delta x_\mu - 2x^\nu \Sigma_{\mu\nu})\phi. \qquad (18.170)$$

---

[36] The general transformation is $\delta e_\mu^a = \xi^\nu \partial_\nu e_\mu^a + \partial_\mu \xi^\nu\, e_\nu^a + \sigma e_\mu^a + \Omega^a{}_b e_\mu^b$.

[37] One can construct classically Weyl-invariant actions by introducing a compensating field τ. Starting with an action $S(g, \phi)$, the action $S(e^{2\tau}g, e^{-\Delta\tau}\phi)$ is Weyl invariant if τ transforms as $\tau \to \tau - \sigma$ and $g_{\mu\nu} \to e^{2\sigma}g_{\mu\nu}$, $\phi \to e^{-\Delta\sigma}\phi$. A simple example is $S = \int \sqrt{g}R$ which leads to $S = \int \sqrt{g}e^{(d-2)\tau}(R + (d-1)(d-2)(\partial\tau)^2)$. If we define $\psi = \exp(\frac{d-2}{2}\tau)$ which has $\Delta_\psi = \frac{d-2}{2}$, we obtain the familiar action of a conformally coupled scalar $S = \int d^d x \sqrt{g}(\frac{1}{2}\nabla^i \psi \nabla_i \psi + \frac{1}{8}\frac{d-2}{d-1}R\psi^2)$. In an AdS_d background ψ has mass $m^2 = -\frac{1}{4\ell^2}d(d-2)$.

18.9 AdS/CFT Correspondence

An operator ϕ with these transformation properties is called a primary operator and Δ is its scaling dimension. As in two dimensions, one can Wick rotate to $\mathbb{R}^d$ and perform a radial quantization, where instead of equal time circles we have equal time S^{d-1}'s. D plays the role of the radial Hamiltonian (as did $L_0 + \bar{L}_0$ in $d = 2$). There is also a state-operator correspondence. Define the vacuum to be the state which is annihilated by all generators of the conformal algebra. Then we have the correspondence $\phi(x) \leftrightarrow |\phi\rangle = \phi(0)|0\rangle$. Descendant operators are obtained by repeatedly acting with P_μ on primary operators. Primaries and their descendants exhaust all local operators in the theory. The above discussion can be extended to the superconformal algebra in d dimensions, but we will not do so.

There are two subgroups of the conformal group which are relevant for the AdS/CFT correspondence:

$$SO(2)_{P_0+K_0} \otimes SO(d)_{L_{ij}}, \qquad i, j = 1, \ldots, d, \qquad (18.171a)$$

$$SO(1,1)_D \otimes SO(d-1,1)_{L_{\mu\nu}}, \qquad \mu, \nu = 0, \ldots, d-1, \qquad (18.171b)$$

where the subscripts on each factor are its generators. The first case corresponds to labeling operators by their energy, by which we mean their eigenvalue of $\mathbb{R}$, the covering space of $SO(2)$, and by their angular momentum associated to a S^{d-1}. Note that the energy is not conjugate to the time direction in Minkowski space but, as we will see below, is related to the global time in AdS whose boundary is $S^1 \times S^{d-1}$. In the second case operators are labeled by their scaling dimension Δ and their representations of the Lorentz group. It is also worth mentioning that $m^2 = -p^2$ is not a Casimir operator of the conformal group.

So much about CFT. We now discuss AdS. The $(d+1)$-dimensional Anti-de-Sitter space AdS_{d+1} is defined as the hypersurface[38] in $\mathbb{R}^{d,2}$

$$-(y^0)^2 + \sum_{i=1}^{d}(y^i)^2 - (y^{d+1})^2 = \sum_{\mu,\nu=0}^{d-1} \eta_{\mu\nu} y^\mu y^\nu - uv = -\ell^2 \qquad (18.172)$$

with the metric induced from $ds^2 = -(dy^0)^2 + \sum_i (dy^i)^2 - (dy^{d+1})^2$. This is a one-sheeted hyperboloid with a non-contractible time-like circle; cf. Fig. 18.3. To avoid the problems associated with closed time-like curves we should go to the universal covering space.

The isometries (Killing vectors) of AdS are the restrictions to the hypersurface of those isometries of the embedding space $\mathbb{R}^{d,2}$ which act tangentially to the hypersurface. These are the $SO(d,2)$ transformations and the Killing vectors of AdS_{d+1} are $\xi_m = \frac{\partial y^M}{\partial x^m} \omega_{MN} y^N$ where $y^M(x)$ are the embedding coordinates of AdS_{d+1} into $\mathbb{R}^{d,2}$, x^m are the coordinates on the hypersurface and $\omega_{MN} = -\omega_{NM}$ parametrize $SO(2,d)$ transformations. Similarly, the isometry group of S^{d+1} is $SO(d+2)$.

---

[38] de-Sitter-space dS_{d+1} is the hypersurface $\eta_{MN} y^M y^N = \ell^2$ in $\mathbb{R}^{d,2}$.

AdS space has no boundary in the topological sense. Nevertheless there is the notion of a conformal boundary which will be important in the context of the AdS/CFT correspondence. Consider solutions of (18.172) of the form $R^2 \eta_{MN} \tilde{y}^M \tilde{y}^N = -\ell^2$ for R large. In the limit $R \to \infty$ this becomes, after dividing by R^2, $\eta_{MN} \tilde{y}^M \tilde{y}^N = 0$, i.e. the 'light-cone' in $\mathbb{R}^{d,2}$. Instead of R we could have also scaled out λR and, therefore, what is meaningful are equivalence classes of points which satisfy $\eta_{MN} y^M y^N = 0$. The boundary of AdS_{d+1} is then defined as any section of the 'light-cone' in $\mathbb{R}^{d,2}$ which is pierced once by each null ray.[39] But this is the same construction which was used when we discussed the compactification of Minkowski space. The conformal boundary of AdS_{d+1} is therefore compactified d-dimensional Minkowski space and the isometries of AdS_{d+1} act as conformal transformations on its boundary, cf. below.

There are several ways to choose local coordinates on AdS space. Two frequently used coordinate systems are global coordinates and Poincaré coordinates.

- Global coordinates are introduced via

$$(y^0, y^i, y^{d+1}) = \left(\ell\sqrt{1+(r/\ell)^2}\cos(t/\ell), r\,\Omega_i, \ell\sqrt{1+(r/\ell)^2}\sin(t/\ell)\right), \tag{18.173}$$

where Ω_i are coordinates on the unit S^{d-1}, i.e. $\sum_i \Omega_i^2 = 1$, $r > 0$ and $0 \le t < 2\pi\ell$. The induced metric is

$$ds^2 = -\left(1 + \frac{r^2}{\ell^2}\right) dt^2 + \frac{dr^2}{1 + \frac{r^2}{\ell^2}} + r^2\, d\Omega_{d-1}. \tag{18.174}$$

The isometries corresponding to the subgroups (18.171a) are global time translations and the isometries of S^{d-1}, respectively. Another commonly used form of the metric in global coordinates is obtained via the transformation $r/\ell = \sinh\rho$, $t/\ell = \tau$ with metric

$$ds^2 = \ell^2\left(-\cosh^2\rho\, d\tau^2 + d\rho^2 + \sinh^2\rho\, d\Omega_{d-1}\right). \tag{18.175}$$

Define $\tilde{y}^M = (\ell/r) y^M$. In the limit $r \to \infty$, $\eta_{MN} \tilde{y}^M \tilde{y}^N = 0$ with the induced metric $ds^2 = -d\tau^2 + d\Omega_{d-1}$. We say that $r \to \infty$ is the boundary of AdS_{d+1} in global coordinates. On the covering space $t \in \mathbb{R}$ and the boundary is $\mathbb{R} \times S^{d-1}$.

- Poincaré coordinates z, x^μ are introduced via

$$(u, v, y^\mu) = \frac{1}{z}\left(\ell^2, \mathbf{x}^2 + z^2, \ell x^\mu\right) \tag{18.176}$$

---

[39] In other words, the boundary is the intersection of the hyperboloid with a plane which cuts through it in the asymptotic region. Topologically the intersection is $S^1 \times S^{d-1}$.

18.9 AdS/CFT Correspondence

for which we find the induced metric

$$ds^2 = \frac{\ell^2}{z^2}(dz^2 + \eta_{\mu\nu}dx^\mu dx^\nu). \tag{18.177}$$

Poincaré coordinates cover only part of AdS space. In these coordinates it is apparent that AdS space is conformally flat, i.e. that its Weyl tensor vanishes.

Define $(\tilde{u}, \tilde{v}, \tilde{y}^\mu) = z(u, v, y^\mu)$ which, in the limit $z \to 0$, satisfy $\eta_{\mu\nu}\tilde{y}^\mu \tilde{y}^\nu - \tilde{u}\tilde{v} = 0$ and the induced metric is $ds^2 = \eta_{\mu\nu}dx^\mu dx^\nu$. This is not all of compactified Minkowski space: 'points at infinity' have to be added. While $z = 0$ is the boundary, $z = \infty$ is a Cauchy horizon.

The Killing vectors of AdS in Poincaré coordinates are easily found either using the method outlined on page 741 of by solving the Killing equation. One finds

$$\xi^z = z(\lambda - 2(b \cdot x)),$$
$$\xi^\mu = a^\mu + \omega^\mu{}_\nu x^\nu + \lambda x^\mu - 2(b \cdot x)x^\mu + (x^2 + z^2)b^\mu. \tag{18.178}$$

Finite isometries of AdS act as

Poincaré: $(z, x^\mu) \to (z, \Lambda^\mu{}_\nu x^\nu + a^\mu)$,

dilations: $(z, x^\mu) \to (\lambda z, \lambda x^\mu)$,

sp. conf.: $(z, x^\mu) \to \frac{1}{\Omega}(z, x^\mu + (x^2 + z^2)b^\mu)$, $\Omega = 1 + 2(x \cdot b) + b^2(x^2 + z^2)$.
$$\tag{18.179}$$

They map the boundary $z = 0$ to itself and act there as conformal transformations. The isometries corresponding to (18.171b) are those parametrized by $\Lambda^\mu{}_\nu$ and by λ.

Depending on which coordinate system one uses, one obtains dual descriptions of the field theory in Minkowski space or on $\mathbb{R} \times S^{d-1}$.

The Correspondence

We now return to strings, branes, gravity and gauge theory. Take a single D3-brane. The fields which live on its world-volume arise from the excitations of open strings ending on the brane. At energies lower than the string scale $1/\sqrt{\alpha'}$, only the massless string states can be excited and their dynamics is governed by a low energy effective action on the world-volume. The massless open string excitations are the gauge field A_M and its fermionic superpartner, the gaugino, a Majorana-Weyl spinor. These are the component fields of the $\mathcal{N} = 1$, $d = 10$ Yang-Mills supermultiplet. The brane breaks ten-dimensional Lorentz invariance $SO(1, 9) \to SO(1, 3) \times SO(6)$ to Lorentz transformations along the brane and rotations in the transverse space. The gauge field decomposes as $\mathbf{10} = (\mathbf{4,1}) + (\mathbf{1,6})$: $A_M = (A_\mu, \phi^i)$. The brane also breaks translation invariance in the six transverse directions and the six scalars describe transverse fluctuations of the brane. They are the Goldstone bosons associated to the spontaneously broken translation symmetry.[40] The gaugino decomposes into four Weyl spinors which transform as $\mathbf{4}$ of $SO(6)$ and their complex conjugates. They are the Goldstinos of the 16 supercharges which are spontaneously broken in the presence of the brane. Altogether we get one $\mathcal{N} = 4$ $U(1)$ vectormultiplet on the four-dimensional world-volume of the D3-brane (16 supercharges). The generalization from one to N coincident D3 branes is straightforward leading to the gauge group $U(N)$. The $U(1)$ factor, which is a free theory, is associated to the center-of-mass motion of the branes and it decouples. We are left with an $SU(N)$ gauge theory. All fields are in the same supermultiplet and hence they all transform in the adjoint representation of $SU(N)$.

Ignoring higher derivative interactions, the dynamics of the $\mathcal{N} = 4$ $SU(N)$ vectormultiplet is uniquely governed by the action of $\mathcal{N} = 4$ Super-Yang-Mills theory (SYM). Due to the large amount of supersymmetry, this action has very few free data: the gauge group and one coupling constant for each simple factor. Here it is simply $SU(N)$, i.e. there is one coupling constant. Bosonic and fermionic contributions to divergences in Feynman diagrams cancel such that the coupling constant does not run, i.e. the β-function vanishes and $\mathcal{N} = 4$ SYM theory is a conformally invariant interacting quantum-field theory with a traceless energy-momentum operator. The $\mathcal{N} = 4$ SYM theory is actually super-conformally invariant and the traceless energy-momentum tensor is accompanied by a γ-traceless supercurrent and a conserved $\mathcal{R}$-current. All of them are anomalous and the anomalies are also members of a supermultiplet.

The fact that the coupling constant does not run means that it is a modulus of the theory, i.e. it parametrizes physically different theories. The theory is invariant under local gauge transformations and under global super-conformal transformations. The latter generate the supergroup[41]

$$PSU(2,2|4) \supset SU(2,2) \times SU(4)_\mathcal{R} \simeq SO(4,2) \times SO(6)_\mathcal{R}. \qquad (18.180)$$

On the r.h.s. we have written the maximal bosonic subgroup. The $SO(4,2)$ factor is the conformal group in $d = 4$ while $SO(6)_\mathcal{R} \simeq SU(4)_\mathcal{R}$ is the $\mathcal{R}$-symmetry group under which the supercharges transform as a $\mathbf{4}$, cf. (14.40). It can be understood in the brane picture as the rotation group of the transverse space and we have seen

---

[40]There are no additional Goldstone bosons associated with the broken rotations. The reason is that locally rotations cannot be distinguished from translations. It is a general fact that broken space-time symmetries are not accompanied 1–1 by Goldstone bosons. Another example is broken scale invariance. Even though dilations and special conformal transformations are broken, there is only one Goldstone boson, the dilaton (the field τ in Footnote 37).

[41]For N-extended superconformal theories with $N \neq 4$ in four dimensions the superconformal group is $SU(2,2|N)$ with bosonic subgroup $SU(2,2) \times U(N)$.

18.9 AdS/CFT Correspondence

how the various fields in the SYM multiplet transform under it: the gauge field (space-time vector) is a singlet, the four gaugini (Majorana fermions) transform as a **4** of $SU(4) \simeq SO(6)$ and the six real scalars as a **6** of SO(6). They can also be alternatively represented as a complex rank two self-dual tensor of $SU(4)$.

The bosonic subgroup of (18.180) is also the isometry group of $AdS_5 \times S^5$. The fermionic generators are the 16 supercharges Q of the $\mathcal{N} = 4$ supersymmetry algebra plus 16 special supersymmetries S which arise in the commutator between Q and the special conformal transformations K. The generators of the $\mathcal{R}$ symmetry appear in the $\{Q, S\}$ anti-commutators.

An obvious question to ask is: how can we recover the gauge theory dynamics, which is a central aspect of the D-brane description, in the SUGRA version of the branes? To approach this question we start with the open string picture: the low-energy effective action, which we can extract from scattering amplitudes of open and closed string massless modes, can be written schematically as

$$S = S_{\text{bulk}} + S_{\text{brane}} + S_{\text{int}}, \tag{18.181}$$

where S_{bulk} reproduces closed string scattering amplitudes, S_{brane} open string amplitudes and S_{int} mixed amplitudes. The Yang-Mills coupling constant $g_{\text{YM}}^2 \sim g_s$ is dimensionless (recall that we consider D3 branes) while the coupling between the open and closed string excitations is gravitational, i.e. the coupling constant is $\kappa_{10} \sim g_s \alpha'^2$. In the limit $\alpha' \to 0$, while keeping g_s, N and all other physical length scales, such as curvature scales, fixed, all massive string excitations decouple and higher derivative interactions can be neglected. Furthermore, open and closed string modes decouple and gravity becomes free, i.e. we arrive at a theory of free gravitons and its supersymmetry partners. Therefore, this decoupling limit, which is also referred to as the Maldacena limit, leads in the D-brane picture to two decoupled systems: free type IIB supergravity in the bulk and four-dimensional SYM theory with 16 supercharges on the world-volume of the branes.

The next step will be to find the correct decoupling limit in the SUGRA picture and to compare to the above. The following analysis gives an important clue: If a dilaton hits a D-brane, it can be absorbed, thus exciting the D-brane. The quantum excitations of the D-brane are the open string modes. Indeed, as we see from the Born-Infeld action, the dilaton couples to the gauge bosons. At lowest order this is the cubic coupling $\frac{1}{g_s}\phi F^2$, cf. (16.165). To find the strength of this coupling one should normalize the fields such that they have canonical kinetic energies. This means that we have to rescale the gauge bosons by $\sqrt{g_s}$ and the dilaton by κ_{10}. This leads to a coupling constant κ_{10} (which vanishes in the decoupling limit). The (tree level) cross-section for a dilaton is $\sigma \sim \kappa_{10}^2 E^3 N^2$ as we shall now explain. The κ_{10}^2 is clear as the cross-section involves the square of the amplitude; N^2 because there are that many gluons into which the dilaton can decay. The cross-section for the scattering of a point-particle from a three-dimensional object in nine space dimensions has dimension (length)^5; the only dimensionful quantity to fix the dimension is the dilaton's energy E, and the factor E^3 indeed arises from the kinematics of the scattering process. A careful calculation, which we will not present here, gives for the absorption cross-section of a dilaton incident at right angle (i.e. its momentum has no component parallel to the D3-brane)

$$\sigma_{D3} = 2\pi^6 g_s^2 \alpha'^4 E^3 N^2 = \frac{\pi^4}{8} E^3 L^8, \qquad L^4 = 4\pi g_s N \alpha'^2. \tag{18.182}$$

On the supergravity side one solves the wave equation for the dilaton in the s-channel in the brane geometry. This exhibits at low energies $E \ll 1/L$ a potential barrier separating the two asymptotic regions $r \ll L$ and $r \gg L$, where r is the distance from the brane. One then obtains the absorption cross section from the tunneling probability through the barrier. The explicit calculation leads to

$$\sigma_{D3} = \sigma_{SUGRA}. \tag{18.183}$$

Similar calculations have been performed for other SUGRA particles, e.g. the graviton and the RR-scalar; their absorption cross sections agree as well.

Equation (18.183) is a very interesting result: in the D-brane picture a particle incident from infinity produces excitations of the gauge theory on the brane; in the SUGRA description of the brane a particle tunnels from the region $r \gg L$ to the region $r \ll L$ and produces an excitation there. The two a priori unrelated processes occur at exactly the same rate. One is tempted to identify the $\mathcal{N} = 4$ SYM theory with gauge group $SU(N)$ with the excitations in the near horizon region, $r \ll L$, of the brane geometry, which we already know is $AdS_5 \times S^5$. This gets further support from the following identification of two types of low-energy excitations, as measured by an observer at infinity (for this observer the coordinate t appearing in (18.149) is the time coordinate as $g_{tt}(r = \infty) = -1$). Due to the energy dependence of the cross section, $\sigma \propto E^3$, low-energy SUGRA modes in the region $r \gg L$ decouple from the near-horizon region. At $r \gg L$ we thus have a free[42] SUGRA theory. On the other hand, the energy of an excitation in the near horizon region appears red-shifted for an observer at infinity, $E_\infty = [g_{tt}(r)/g_{tt}(\infty)]^{1/2} E_r = \frac{r}{L} E_r$. They cannot penetrate the energy barrier which separates the two asymptotic regions.

Therefore, in the D3 and in the SUGRA picture we get two decoupled systems. In both cases the system in the bulk is free type IIB SUGRA. But then, if the D-brane and the SUGRA brane describe the same object, we should identify the two other systems: $\mathcal{N} = 4$ SYM theory with gauge group $SU(N)$ and type IIB string theory on $AdS_5 \times S^5$.

Before pursuing this further, we need to clarify two points. Firstly, we still need to be more specific about the precise form of the near horizon limit, which zooms

---

[42] The gravitational interaction is negligible at low energies. The dimensionless coupling constant is $\kappa_{10} E^4$, where E is the typical energy of the interaction. For $E \ll 1/\sqrt{\alpha'}$ and $\kappa_{10} \propto \alpha'^2$ this is $\ll 1$.

18.9 AdS/CFT Correspondence

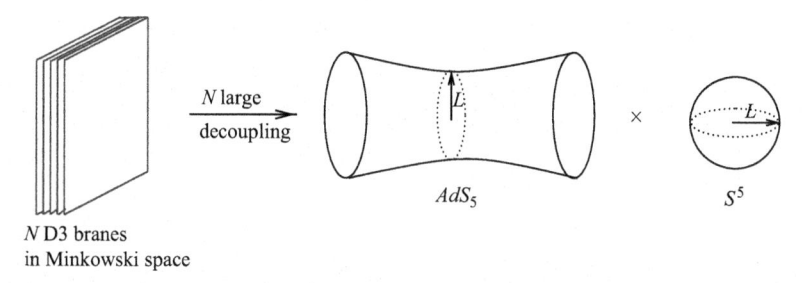

Fig. 18.3 Decoupling limit and large N transition

into the region of the three-brane SUGRA solution which we want to identify with the gauge theory of the D3 picture. This limit should involve $\alpha' \to 0$, as this was the decoupling limit for the D3-brane, and is defined as follows:

$$\alpha' \to 0, \quad r \to 0 \quad \text{such that} \quad U \equiv \frac{r}{\alpha'} \text{ fixed.} \quad (18.184)$$

In this limit α' scales out of the metric which becomes

$$ds^2/\alpha' = \frac{U^2}{\sqrt{4\pi g_s N}}(-dt^2 + d\mathbf{x}^2) + \frac{\sqrt{4\pi g_s N}}{U^2} dU^2 + \sqrt{4\pi g_s N}\, d\Omega_5. \quad (18.185)$$

Pictorially this is shown in Fig. 18.3.

The limit is taken such that for the observer at infinity the string excitations in the near horizon region, which have energies $E_0 \sim 1/\sqrt{\alpha'}$ and which are redshifted to $E_\infty \sim \frac{r}{\sqrt{\alpha'}} E_0 \sim U$ stay finite. This observer sees two decoupled systems: free SUGRA in the asymptotic region and type IIB string theory compactified on $AdS_5 \times S^5$.

If we further define $z = L^2/r = L^2/(U\alpha')$, the metric becomes

$$\frac{1}{L^2} ds^2 = \frac{1}{z^2}(dz^2 + \eta_{\mu\nu} dx^\mu dx^\nu) + d\Omega_5. \quad (18.186)$$

If we use this form of the metric in the Polyakov action, it is clear that α' cancels and its role is now played by $1/\sqrt{g_s N} \equiv \lambda^{-1/2}$, i.e. the α' expansion becomes an expansion in powers of $\lambda^{-1/2}$. We will find below that the g_s expansion will become an expansion in powers of $1/N$.

The second point we need to clarify is the range of validity of the two calculations of the absorption cross section. In both pictures we have assumed that $E \ll 1/\sqrt{\alpha'}$; otherwise massive string modes can be excited and their effect has to be taken into account. On the string side we have neglected string loop effects by assuming $g_s \ll 1$, i.e. we are in the regime of classical string theory. We also have to require that the typical length scale L of the geometry is large compared to the string scale, i.e. $L \gg \sqrt{\alpha'}$; otherwise we have to take higher derivative corrections $\sim (\alpha')^n R^{n+1}$ to the supergravity action into account, which we did not. In other words, we have also neglected σ-model loops. With $L^4 \sim g_s N \alpha'^2$ and $g_{YM}^2 \sim g_s$ we find the following correspondence between gauge theory and string theory parameters:

$$N \to \infty, \quad \lambda \equiv g_{YM}^2 N \text{ fixed} \quad \Longleftrightarrow \quad g_s \to 0,$$

$$\lambda \to \infty \quad \Longleftrightarrow \quad L^2/\alpha' \to \infty. \quad (18.187)$$

We first fix λ and send $N \to \infty$. This requires $g_s \to 0$, i.e. tree level in string theory (classical string theory) and planar limit in the YM theory.[43] The SUGRA approximation to classical string theory is attained for $L^2/\alpha' \to \infty$ and in the Yang-Mills theory this limit is controlled by λ. Therefore, the SUGRA approximation to classical string theory corresponds to strong 't Hooft coupling λ, which is the effective coupling constant in the large N limit of YM theories.

All of this suggests to identify the excitations in the near-horizon region of the three-brane geometry with the excitations of $\mathcal{N} = 4\ SU(N)$ SYM theory at large λ and $N \to \infty$. This leads to the weak form of the Maldacena conjecture, also called AdS/CFT correspondence:

$\mathcal{N} = 4$ SYM theory with $SU(N)$ gauge group at large 't Hooft coupling λ and in the limit $N \to \infty$ is equivalent (dual) to classical type IIB supergravity compactified on $AdS_5 \times S^5$ with radii $L^4 = 2\alpha'^2 \lambda$ and N units of F_5 flux.

Correlation functions of type IIB string theory compactified on $AdS_5 \times S^5$ have a double expansion in powers of g_s and $\alpha'/L^2 \sim \lambda^{-1/2}$. In the large N limit we can write this as an expansion in powers of $1/N \sim g_s(\alpha'/L^2)^2 \sim g_s \lambda^{-1}$ where each coefficient has an expansion in powers of $\lambda^{-1/2}$. Clearly, a term at some power in the g_s expansion has the same power in the $1/N$ expansion and, for fixed λ the g_s expansion of the string theory translates to the $1/N$ expansion of the gauge theory. Since the type II string theory includes only closed oriented strings, we find the same general structure as in the field theory where each correlation function has an expansion in powers of $1/N^2$, each coefficient being a function of λ. The functions of λ have different expansions from the point of view of string theory and of field theory. A stronger version of the AdS/CFT correspondence states that both expansions give rise to the same function of λ at each power of $1/N^2$. This is a correspondence between perturbative string theory and the large N expansion of the gauge theory where classical string theory corresponds to the 't Hooft limit or planar limit of the gauge theory. The fact that one expansion is in powers of λ whereas the other is in powers of $\lambda^{-1/2}$ reflects the fact that the AdS/CFT duality is a strong/weak coupling duality.

The strongest version of the conjecture is that the two theories are considered as exactly identical for all values of N and g_s with the identifications (cf. (16.166))

$$g_{YM}^2 = 2\pi g_s, \quad L^4 = 4\pi g_s N \alpha'^2. \quad (18.188)$$

---

[43]This is a notion from the large-N expansion of YM theories for which we refer to the literature.

18.9 AdS/CFT Correspondence

Here the corrections distort the space-time which is only required to be asymptotically $AdS_5 \times S^5$. The gauge theory then effectively sums over all such space-times.

Even in the weak form the conjecture has far reaching implications. It relates a classical weakly coupled supergravity theory to a strongly coupled quantum field theory. The perturbative regimes of the two theories, $g_{YM}^2 N \sim g_s N \sim L^4/\alpha'^2 \ll 1$ for the gauge theory and $L^4/\alpha'^2 \gg 1$ for the supergravity theory, do not overlap.

Support for this conjecture comes from comparing symmetries. The isometries of $AdS_5 \times S^5$ are $SO(4,2) \times SO(6)$. But these are precisely the bosonic (non-gauge) symmetries of $\mathcal{N} = 4$ SYM theory. This symmetry matching can be extended to the full Anti–de-Sitter supergroup $PSU(2,2|4)$ whose associated algebra is the maximally extended $\mathcal{N} = 8$ supersymmetry algebra on AdS_5. $PSU(2,2|4)$ is realized as global symmetry of gauged $\mathcal{N} = 8$ supergravity in $d = 5$, which can be obtained as Kaluza-Klein reduction of type IIB SUGRA on S^5: all KK modes fall into unitary representations of this group. What is gauged is the $SO(6)$ subgroup, the isometry group of S^5. On the other hand, $PSU(2,2|4)$ is also the maximally extended superconformal symmetry in four-dimensional Minkowski space, i.e. the invariance group of $\mathcal{N} = 4$ SYM theory. Furthermore, the boundary of AdS_5 is four-dimensional Minkowski space on which the isometries of AdS_5 act as conformal transformations. This leads to the statement that the field theory which is dual to the string theory 'lives' on the boundary of AdS_5 and the isometries of S^5 act as $\mathcal{R}$-symmetry transformations. Furthermore, if we simultaneously rescale $U \to \lambda U$, $(t, \mathbf{x}) \to \lambda^{-1}(t, \mathbf{x})$ the metric (18.185) does not change. This leads to the interpretation of U as the energy scale in the field theory: large U corresponds to the UV region and small U to the IR region (the boundary is at $U = \infty$ in these coordinates).

The conjectured duality has several very remarkable features. First, it is a duality between a gravity theory and a field theory. In addition, these theories live in different numbers of dimensions and have completely different degrees of freedom: gravity vs. gauge degrees of freedom. The fact that they live in different dimensions means that the duality is of holographic nature. One way of reading it is to say that the hologram of type IIB string theory compactified on asymptotic $AdS_5 \times S^5$ space-times is four-dimensional $\mathcal{N} = 4$ SYM theory.

There is the notion of the master field for Yang-Mills theory in the 't Hooft limit. One can show that fluctuations of gauge invariant observables vanish in the $N \to \infty$ limit. This is analogous to the classical limit $\hbar \to 0$ in which the functional integral is dominated by classical paths. In the same way, there should be a master field such that all Green functions are given by their value at the master field. What the Maldacena conjecture suggests is that this master field for $\mathcal{N} = 4$ SYM theory is a classical string theory in ten dimensions. Also, the gauge string of these theories is simply the fundamental type IIB string which lives, however, not in four-dimensional space-time, but in ten-dimensional $AdS_5 \times S^5$.

The duality becomes a useful tool, if we know how to map one description to the other. In particular we need to know how to relate string excitations on $AdS_5 \times S^5$ to operators in the CFT. Matters simplify in the SUGRA approximation. There we have the SUGRA fields (including the Kaluza-Klein modes) and their dynamics is governed by a classical action, the compactification of 10-dimensional type IIB SUGRA on $AdS_5 \times S^5$. Denote the fields collectively by Φ_i. We can solve their equations of motion with prescribed boundary values ϕ_i. The classical on-shell SUGRA action is then a functional of ϕ_i. These ϕ_i are interpreted as external sources for operators $\mathcal{O}_i$ of the CFT, i.e. there is a 1–1 correspondence between bulk fields and boundary operators:

$$\Phi_i \leftrightarrow \mathcal{O}_i. \qquad (18.189)$$

One then makes the following identification of partition functions:

$$Z_{\text{string}}(\phi_i) = \int_{\Phi_i|_{\text{bdy}}=\phi_i} D\Phi_i \, e^{-S_{\text{string}}(\Phi_i)} = \left\langle \exp\left(\int d^4x \, \phi_i \mathcal{O}_i\right) \right\rangle_{\text{CFT}} = e^{-W(\phi_i)} = Z_{\text{CFT}}(\phi_i). \qquad (18.190)$$

The functional integral representation of Z_{string} is, of course, formal. But in the supergravity approximation this identification can be made precise: it reduces to the classical on-shell SUGRA action[44]

$$\left. e^{-S_{\text{SUGRA}}(\Phi_i)} \right|_{\substack{\delta S_{\text{SUGRA}}=0 \\ \Phi_i|_{\text{bdy}}=\phi_i}} = Z_{\text{CFT}}(\phi_i). \qquad (18.191)$$

On the r.h.s. we have to take the limit in the CFT corresponding to the SUGRA approximation, i.e. $N \to \infty$ and $\lambda \to \infty$ in the specific example discussed so far. Once one has established the correspondence between fields in the bulk and operators in the boundary CFT, $\Phi_i \leftrightarrow \mathcal{O}_i$, this identification of partition functions is very powerful, as we will exemplify in detail on a concrete example where Φ is the bulk graviton and $\mathcal{O}$ the CFT energy-momentum tensor.

Applications

We will now consider particular applications of the AdS/CFT correspondence as a way to illustrate several general features. Recall that AdS space is a solution of Einstein's equations with negative cosmological constant Λ and that, in the present context, this arises by compactifying type IIB supergravity on S^5 in the presence of a self-dual F_5 background. This leads to gauged $\mathcal{N} = 8$ supergravity in five dimensions. The Kaluza-Klein spectrum of this theory has been worked out completely. Since we are dealing with string theory on this background, we also have the string excitations. To determine their spectrum in the background with RR-flux is still a largely open problem. A much simpler situation occurs in the limit $N \to \infty$, $\lambda \to \infty$ where all string excitations decouple, the string is free

---

[44]There can be several saddle points which lead to the possibility of phase transitions in the CFT. We will not discuss this interesting issue.

18.9 AdS/CFT Correspondence

and we have classical supergravity on $AdS_5 \times S^5$. There is a consistent truncation where all S^5 KK modes decouple and one is left with gauged $\mathcal{N} = 8$ SUGRA in five dimensions. If we set all fields but the metric to zero, what remains is five dimensional gravity with a cosmological constant. We will see below that in the dual field theory this sector of the string theory corresponds to the stress-tensor dynamics.

The following discussion is largely independent of this particular choice of gravitational action and applies to actions of the form

$$S = \int_M d^{d+1}X \sqrt{G}\, f(R), \tag{18.192}$$

where G is the bulk metric and $f(R)$ an arbitrary scalar function of the curvature and its covariant derivatives. We require that (18.192) admits AdS_{d+1} as a solution to the equations of motion. This restricts the parameters in f in such a way that, if one inserts the ansatz (18.152) into the equations of motion, the resulting equation for $k = -\frac{1}{\ell^2}$ has a real negative solution. ℓ is the AdS-radius. Below we will mention the relevance of considering more general actions of this type for obtaining more general CFTs, i.e. differing from $\mathcal{N} = 4$ SYM theory. In the simplest case we have, of course, the Einstein-Hilbert action with cosmological constant

$$S = \frac{1}{16\pi G_N^{(d+1)}} \int d^{d+1}X \sqrt{-G}\, (R(G) - \Lambda), \quad \Lambda = -\frac{d(d-1)}{L^2}, \tag{18.193}$$

which leads to

$$R_{MN} = -\frac{d}{L^2} G_{MN} \tag{18.194}$$

and $\ell = L$.

We couple the CFT to an external (non-dynamical) metric $g^{(0)}_{\mu\nu}(x)$ which acts as a source for the energy-momentum tensor of the CFT and is the metric on the d-dimensional boundary of the $(d+1)$-dimensional space-time M_{d+1} of the dual gravity theory. We choose coordinates $X^M = (\rho, x^\mu)$ such that $\rho = 0$ is the boundary of M_{d+1} and use diffeomorphisms to bring the bulk metric to the Fefferman-Graham (FG) gauge

$$ds^2 = G_{MN} dX^M dX^N = \ell^2 \left(\frac{d\rho}{2\rho}\right)^2 + \frac{\ell^2}{\rho} g_{\mu\nu}(x, \rho) dx^\mu dx^\nu \tag{18.195}$$

with $g_{\mu\nu}(x,0) = g^{(0)}_{\mu\nu}(x)$ the metric on the boundary. For $g^{(0)}_{\mu\nu} = \eta_{\mu\nu}$ this is the AdS-metric (in Poincaré coordinates (18.177) with $\rho = z^2$) with curvature radius ℓ. In the following we will set $\ell = 1$ but we will reintroduce it in the final results.

The FG-gauge is always possible locally and close to the boundary but this gauge choice does not fix the diffeomorphism invariance of the theory completely. Penrose-Brown-Henneaux (PBH) transformations are diffeomorphisms ξ^M which preserve the FG-gauge. They satisfy $\delta_\xi G_{\rho\rho} = \delta_\xi G_{\rho\mu} = 0$. The solution is parametrized by an arbitrary function $\sigma(x)$[45]:

$$\xi^\rho = 2\rho\sigma(x), \qquad \xi^\mu = a^\mu(x,\rho) = -\frac{1}{2}\int_0^\rho d\rho'\, g^{\mu\nu}(x,\rho')\partial_\nu\sigma(x). \qquad (18.196)$$

Under these diffeomorphisms $g_{\mu\nu}(x,\rho)$ changes as

$$\delta g_{\mu\nu}(x,\rho) = 2\sigma(1-\rho\partial_\rho)g_{\mu\nu}(x,\rho) - \nabla_\mu a_\nu(x,\rho) - \nabla_\nu a_\mu(x,\rho). \qquad (18.197)$$

The covariant derivatives are w.r.t. to $g_{\mu\nu}(x,\rho)$ where ρ is considered a parameter; indices are lowered with $g_{\mu\nu}(x,\rho)$. The following group property of PBH transformations can be shown:

$$\xi_1^N \partial_N \xi_2^M - \xi_2^N \partial_N \xi_1^M + \delta_2 \xi_1^M - \delta_1 \xi_2^M = 0. \qquad (18.198)$$

The last two terms are due to the dependence of the transformation parameters on $g_{\mu\nu}(x,\rho)$. At $\rho = 0$, (18.197) gives $\delta g_{\mu\nu}^{(0)} = 2\sigma g_{\mu\nu}^{(0)}$, i.e. $\sigma(x)$ is the parameter of Weyl rescalings of the boundary metric.

One can prove that for a given boundary metric $g_{\mu\nu}^{(0)}$, (18.194) has asymptotic solutions of the form (18.195), where near the boundary at $\rho = 0$, $g_{\mu\nu}(x,\rho)$ has an expansion[46]

$$g_{\mu\nu}(x,\rho) = \sum_{m\geq 0} g_{\mu\nu}^{(m)}(x)\,\rho^m + \ldots. \qquad (18.199)$$

The ellipses denote logarithmic terms of $\mathcal{O}(\rho^n \log \rho)$ which are present for even $d = 2n$. In our analysis the logarithmic terms do not play any role and we will ignore them. In $d = 2n + 1$ dimensions, starting at $\rho^{d/2}$, there are also half-integer powers. The boundary data one has to provide to obtain a unique solution are $g_{\mu\nu}^{(0)}$ and the traceless and divergenceless part (w.r.t. $g_{\mu\nu}^{(0)}$) of $g_{\mu\nu}^{(d/2)}$. All other coefficients are determined by these data via the equations of motion. In particular $g_{\mu\nu}^{(m)}(x)$ for $1 \leq m \leq [\frac{d-1}{2}]$ are completely determined by the curvature of $g_{\mu\nu}^{(0)}$ and its covariant derivatives. In even d the same holds for the coefficient of $\rho^{d/2} \log \rho$ and for the trace and divergence of $g_{\mu\nu}^{(d/2)}$, which vanishes for d being odd.

As we will see in an example below, we must make sure that the solution can be extended as a regular solution into the interior. This imposes restrictions on the allowed data, i.e. on the traceless and divergenceless part of $g_{\mu\nu}^{(d/2)}$ which, as we will show later, is related to the energy-momentum tensor of the CFT.

---

[45] The choice of a lower limit of the ρ' integral means that we do not consider diffeomorphisms of the boundary. They are of no interest here.

[46] If we include other bulk fields besides the metric, due to back-reaction the FG expansion (18.199) of the metric will, except for special masses of the fields, no longer be in integer powers of ρ only.

18.9 AdS/CFT Correspondence

We want to allow for more complicated gravitational actions and equations of motion. We will assume that the above results about asymptotic solutions is still valid (cf. also the comment in Footnote 51 on page 757).

The FG expansion of g induces an expansion of a^μ and therefore an expansion of (18.197). The first two terms are

$$\delta_\sigma g^{(0)}_{\mu\nu} = 2\sigma g^{(0)}_{\mu\nu}, \qquad \delta_\sigma g^{(1)}_{\mu\nu} = \nabla^{(0)}_\mu \nabla^{(0)}_\nu \sigma. \tag{18.200}$$

The first was already discussed above. The second relation has a unique[47] solution in terms of $g^{(0)}_{\mu\nu}$ which is local, has dimension[48] $(length)^{-2}$ and is covariant:

$$g^{(1)}_{\mu\nu} = -\frac{1}{d-2}\left(R_{\mu\nu} - \frac{1}{2(d-1)}g_{\mu\nu}R\right), \tag{18.201}$$

where R is the curvature of $g_{\mu\nu} \equiv g^{(0)}_{\mu\nu}$. The higher $g^{(m)}$ are not uniquely determined. Their PBH transformation is of the form $\delta g^{(n)}_{\mu\nu} = 2(1-n)\sigma g^{(n)}_{\mu\nu} +$ inhomogeneous terms, and expressions which transform homogeneously under Weyl transformations, e.g. $C^2 g_{\mu\nu}$ or $(C^2)_{\mu\nu}$, can appear in $g^{(2)}_{\mu\nu}$; here $C^2 = C_{\mu\nu\rho\sigma}C^{\mu\nu\rho\sigma}$ is the square of the Weyl tensor. For conformally flat boundary metrics the ambiguity disappears.

The FG-expansion (18.199) also induces an expansion of the (on-shell) integrand of (18.192) of the form

$$\sqrt{G} f(R) = \sqrt{g^{(0)}} \rho^{-\frac{d}{2}-1} b(x,\rho) = \sqrt{g^{(0)}} \rho^{-\frac{d}{2}-1} \sum_{m \geq 0} b_m(x) \rho^m. \tag{18.202}$$

One can show that b and therefore each b_n satisfies the Wess-Zumino consistency condition:

$$\int d^d x \sqrt{g^{(0)}} (\sigma_1 \delta_{\sigma_2} - \sigma_2 \delta_{\sigma_1}) b = 0. \tag{18.203}$$

A simple way to see this is as follows. For $\mathscr{O} = \sqrt{G} f(R)$ one derives $\delta_\sigma \mathscr{O} = \partial_M (\xi^M \mathscr{O})$ (cf. below) and $[\delta_{\sigma_1}, \delta_{\sigma_2}] = 0$ by virtue of the group property (18.198). Non-trivial solutions of the WZ consistency condition are candidates for anomalies. The role of b_n as the holographic Weyl anomaly in $d = 2n$ will be discussed next.

---

[47]Uniqueness is only true for the purely gravitational system. If we add other bulk fields, the situation changes; cf. also Footnote 46.

[48]The dimensions are fixed by the power of ρ which has dimension $(length)^2$ and $g_{\mu\nu}(x,\rho)$ is dimensionless.

Holographic Weyl anomaly

The essential property of the PBH transformations is that on the boundary they coincide with the action of the Weyl group. Therefore, in holography the Weyl group becomes embedded in the $d + 1$ dimensional diffeomorphisms and the study of Weyl anomalies of the boundary CFT is reduced to an analysis of how diffeomorphisms act.

Under a bulk diffeomorphism the action (18.192) is invariant up to a boundary term

$$\delta_\xi(\sqrt{G} f) = \partial_M(\sqrt{G} \xi^M f), \qquad (18.204)$$

$$\delta_\xi S = \int_{\partial M} d^d x \sqrt{G} f(R) \xi^\rho \Big|_{\rho=0} = -2 \int_{\partial M} d^d x \sqrt{G} f(R) \sigma \rho \Big|_{\rho=0}. \qquad (18.205)$$

In the second line we have restricted the diffeomorphism to a PBH transformation. The finite piece of this boundary term is the holographic Weyl anomaly. The divergent terms are covariant and can be subtracted by local counter terms. Notice that we consider passive diffeomorphisms which act on the fields rather than the coordinates. The reason for this is that we want to keep the boundary fixed.

Using the identification (18.191) the above variation of the gravitational action is interpreted as the variation of the effective action $W(g)$ of the CFT which is obtained by coupling the CFT to an external metric and integrating out the CFT. Schematically, if ϕ are the fields of the CFT with action $S_{\text{CFT}}(\phi, g)$, then

$$e^{-W(g)} = \int D\phi\, e^{-S_{\text{CFT}}(\phi, g)}. \qquad (18.206)$$

As g couples to the energy-momentum tensor, $W(g)$ is the generating functional for correlation functions of $T_{\mu\nu}$. In other words, here the map (18.189) is $G_{\mu\nu} \leftrightarrow T_{\mu\nu}$. $W(g)$ is a non-local functional of g but its Weyl variation is local and produces the anomalous trace of the energy momentum tensor

$$\delta_\sigma W = \int \frac{\delta W}{\delta g_{\mu\nu}} \delta_\sigma g_{\mu\nu} = -\frac{1}{2} \int \sqrt{g}\, \langle T^{\mu\nu}\rangle \delta_\sigma g_{\mu\nu} = -\int \sqrt{g}\, \langle T^\mu{}_\mu\rangle\, \sigma. \qquad (18.207)$$

Incidentally, using

$$\left[\frac{1}{\sqrt{g}}\frac{\delta}{\delta g^{\mu\nu}}, \delta_\sigma\right] = (d-2)\sigma \frac{1}{\sqrt{g}}\frac{\delta}{\delta g^{\mu\nu}}, \qquad (18.208)$$

one shows that

$$\delta_\sigma T_{\mu\nu} = (2-d)\sigma\, T_{\mu\nu} + \frac{2}{\sqrt{g}}\frac{\delta}{\delta g^{\mu\nu}} \delta_\sigma W(g). \qquad (18.209)$$

18.9 AdS/CFT Correspondence

For $d = 2n$, the coefficient of the boundary term at $\mathcal{O}(\rho^0)$ is b_n (cf. (18.202)) and it therefore represents the Weyl anomaly of the dual CFT. b_n depends on the $g^{(m)}$. To find $g^{(m)}$ generally requires information contained in the equations of motion. On dimensional grounds it is a priori clear that only the $g^{(m)}$ with $m \leq n$ contribute to b_n. The fact that AdS_{d+1} is a solution of the equations of motion can be used to show[49] that for b_n we only need $g^{(m)}$ with $m < n$, which are all local covariant expressions of $g^{(0)}_{\mu\nu}$. In $d = 4$ this means that all we need is $g^{(1)}_{\mu\nu}$ which, as we have seen, is universal, i.e. independent of the choice of $f(R)$ in the action. The above argument also reproduces the well-known fact that there is no Weyl anomaly in odd dimensions because we can never obtain a boundary term at $\mathcal{O}(\rho^0)$.

As a concrete example, we consider

$$f = \frac{1}{16\pi G_N^{(5)}}\left(R + \frac{12}{L^2} + \gamma L^2 R^{MNPQ} R_{MNPQ}\right), \quad d = 4, \tag{18.210}$$

where γ is a dimensionless parameter. The equations of motion have an AdS_5 solution with radius ℓ provided

$$1 - q^{-2} - \frac{2}{3}\gamma q^2 = 0, \quad q = \frac{L}{\ell}. \tag{18.211}$$

With the help of the result collected in the appendix we find that the dependence on $g^{(2)}$ cancels between different terms, if (18.211) is satisfied and that the anomaly is

$$\langle T^\mu{}_\mu \rangle = \frac{\ell^3}{128\pi G_N^{(5)}}\left((1 + 4\gamma q^2) C^2 - (1 - 4\gamma q^2) E_4\right) \equiv \frac{1}{16\pi^2}(c\, C^2 - a\, E_4). \tag{18.212}$$

Here

$$E_4 = R^{\mu\nu\rho\sigma} R_{\mu\nu\rho\sigma} - 4 R^{\mu\nu} R_{\mu\nu} + R^2,$$

$$C^2 = R^{\mu\nu\rho\sigma} R_{\mu\nu\rho\sigma} - 2 R^{\mu\nu} R_{\mu\nu} + \frac{1}{3} R^2, \tag{18.213}$$

where E_4 is the Euler density and, as before, C^2 is the square of the Weyl tensor in $d = 4$. These are the only non-trivial solutions of the WZ-consistency conditions in $d = 4$, i.e. they satisfy (18.203). $\Box R$ is also a solution but it is trivial in the sense

---

[49] On dimensional grounds b_n can at most be linear in $g^{(n)}$ as both carry length-dimension $2n$. By assumption, $f(R)$ is such that Anti-de-Sitter space is a solution of the equations of motion. Expand the action around this solution as $g_{\mu\nu}(x, \rho) = \eta_{\mu\nu} + \rho^n g^{(n)}_{\mu\nu}(x)$. In this expansion the term linear in the fluctuations around the AdS-metric can only be a total derivative (or vanish altogether). Consider the terms $\nabla^M \nabla^N \delta G_{MN}$ and $\nabla^M \nabla_M \text{tr}(\delta G)$. For fluctuations $\delta G_{\mu\nu} = \rho^{n-1} g^{(n)}_{\mu\nu}$ the possibly dangerous terms, i.e. those which might contribute to b_n, are of the type $\rho^n \text{tr}\, g^{(n)}$. Explicit calculation shows that their coefficient is zero for $d = 2n$. Higher derivative terms in the variation of the action will involve coefficients $g^{(m)}$ for $m < n$.

that $\int \sqrt{g}\Box R\sigma$ is the variation of a local expression, namely of $\int \sqrt{g}R^2$. a and c in (18.212) are the two Weyl-anomaly coefficients in $d = 4$.[50] One also checks that the two divergent terms $\int \sqrt{g}\sigma$ and $\int \sqrt{g}R\sigma$ are indeed the Weyl variations of local terms.

Consider $\gamma = 0$, i.e. $\ell = L$. In this case we know that the dual CFT is $\mathcal{N} = 4$ SYM theory with gauge group $SU(N)$ and we also know that

$$\frac{1}{16\pi G_N^{(5)}} = \frac{\text{Vol}(S^5)}{16\pi G_N^{(10)}}. \tag{18.214}$$

With $16\pi G_N^{(10)} = 2\kappa_{10} = (2\pi)^7 \alpha'^4 g_s^2$ and $\text{Vol}(S^5) = \pi^3 L^5$ we find

$$a = c = \frac{1}{4} N^2. \tag{18.215}$$

In the free theory the anomaly can be computed in field theory and gives the above result with $N^2 \to N^2 - 1$. But the dual description via classical SUGRA is valid in the $N \to \infty$ limit where the difference is suppressed by $\mathcal{O}(1/N^2)$ effects. The N^2 behavior of the anomaly, which is a measure of the number of degrees of freedom, reflects, of course, the non-Abelian symmetry of the CFT on a stack of N D3-branes. We see that the anomaly of the strongly interacting theory, as computed via AdS/CFT, is the same as that of the free theory. In $\mathcal{N} = 4$ SYM theory the gauge coupling is a modulus and the Weyl anomaly seems to be independent of it.

If one does the calculation for $AdS_7 \times S^4$, which is the near horizon limit of the M5-brane solution, one finds that the anomaly of the six-dimensional dual CFT scales as $\frac{\ell^5}{\kappa_7^2} \sim \frac{\ell^9}{\kappa_{11}^2} \sim N^3$ where we used $\ell \sim N^{1/3}\ell_{11}$ (cf. Table 18.3 on page 700). While the N-dependence is easy to determine, to obtain the precise value of the four anomaly coefficients requires more work: one needs $g_{\mu\nu}^{(2)}$ which is not universal. While the world-volume theory on a single M5-brane contains five scalars and one self-dual two-form potential (plus the $(0,2)$ SUSY partners), the N^3 behavior of the anomaly reflects the 'non-Abelian' nature of the world-volume theory of interacting self-dual tensor multiplets, but the precise mechanism is unknown. There seems to be no way of constructing a field theory of interacting tensor multiplets and the N^3 behavior of the anomaly, which is expected to reveal information of the number of degrees of freedom, is a valuable hint for this still elusive theory.

It is a special property of $\mathcal{N} = 4$ SYM theory that the two anomaly coefficients agree, but this is no longer the case for generic (super)conformal field theories. Their dual gravity descriptions therefore have to include higher derivative corrections. One can show that if we include R^2 and $R^{MN}R_{MN}$ terms in (18.210), this changes a and c but not $a - c$. To check this, the results collected in the appendix are helpful.

---

[50] In superconformal field theories they also characterize the anomalous divergence of the $\mathcal{R}$ current and the anomalous γ-trace of the supercurrent.

18.9 AdS/CFT Correspondence

In higher dimensions the situation becomes more complicated for two reasons. (1) one needs to use the equations of motion to compute higher $g^{(m)}$ as they are no longer universal and (2) the number of solutions to the WZ consistency conditions proliferates. E.g. in $d = 6$ we need to compute $g^{(2)}_{\mu\nu}$ and there are four anomaly coefficients.

In any $d = 2n$ dimension there is always one solution of WZ consistency which is distinct: it is the Euler density

$$E_{2n} = \frac{1}{2^n} R_{\mu_1\nu_1\rho_1\sigma_1} \cdots R_{\mu_n\nu_n\rho_n\sigma_n} \epsilon^{\mu_1\nu_1\ldots\mu_n\nu_n} \epsilon^{\rho_1\sigma_1\ldots\rho_n\sigma_n} = R^n + \ldots. \tag{18.216}$$

The infinitesimal Weyl variation of $\sqrt{g}E_{2n}$ is a total derivative and its integral is a topological invariant (on manifolds without boundary). Then there is a dimension dependent number of solutions which transform homogeneously. In $d = 2$ there is none, in $d = 4$ there is one, C^2, in $d = 6$ there are three, etc. (How many there are in $d = 2n$ is not known in general.). The former are also called type A and the latter type B Weyl anomalies. The coefficient a of type A in any even dimension $d = 2n$ of a CFT which has a dual gravity description in terms of an action of the type (18.192) can be determined:

$$\langle T^\mu{}_\mu \rangle = (-)^n \frac{\ell^{2n+1}}{2^{2n}(n!)^2} f(R)\Big|_{\text{AdS}} E_{2n} + \text{type B}, \tag{18.217}$$

where $f(R)|_{\text{AdS}}$ means $f(R)$ evaluated on the AdS solution. The type B anomaly always vanishes on conformally flat spaces. Using this fact is one way to derive (18.217); details can be found in the references.

Holographic Reconstruction of Space-Time

If we view the Einstein equations (18.194) as differential equations in the radial variable ρ, we should expect two branches of solutions with two sets of 'integration constants' or boundary data.[51] In fact, recall from the discussion below (18.199) that one set is the boundary metric and that the second set is a traceless and conserved tensor at $\mathcal{O}(\rho^{d/2})$. But these are the properties of the energy-momentum tensor of a CFT. We will make this relation more concrete.

One defines the holographic energy-momentum tensor as the variation of the dual gravity action under changes of the boundary metric:

$$\langle T^{\mu\nu} \rangle = -\frac{2}{\sqrt{g^{(0)}}} \frac{\delta S}{\delta g^{(0)}_{\mu\nu}}. \tag{18.218}$$

---

[51] For more general actions the equations of motion are of higher order and have more branches. The additional branches are spurious in the sense that, if the higher derivative terms are considered as small perturbations of the action (18.193), we should only consider perturbations of the two branches of solutions of the unperturbed equations of motion.

This is in line with the identification of the generating functional of the CFT with the on-shell SUGRA action. The variation of the action produces a bulk term, which vanishes on-shell, and a boundary term from which the energy-momentum tensor can be read off. Recall that on-shell $g^{(m)}_{\mu\nu}$ with $m \leq [\frac{d-1}{2}]$ and the trace of $g^{(d/2)}_{\mu\nu}$ are functions of $g^{(0)}_{\mu\nu}$. The situation simplifies, if we seek the energy-momentum tensor in Minkowski space. In this case, it suffices to consider $g_{\mu\nu} = g^{(0)}_{\mu\nu} + \rho^{d/2} g^{(d/2)}_{\mu\nu}$ and to set $g^{(0)}_{\mu\nu} = \eta_{\mu\nu}$ at the end. The lower order terms vanish in a flat background and the higher order terms never contribute to the boundary term at $\rho = 0$. Starting from the action (18.193) and using results from the appendix, one finds the boundary terms

$$\delta S = \frac{\ell^3}{16\pi G_N^{(d+1)}} \int d^d x \sqrt{g} \rho^{-\frac{d}{2}} \left(g^{\mu\nu} \delta g_{\mu\nu} - 2\rho g^{\mu\nu} \delta g'_{\mu\nu} \right.$$
$$\left. + \rho g'^{\mu\nu} \delta g_{\mu\nu} \right) \bigg|_{\rho=0}. \qquad (18.219)$$

A divergent term has to be subtracted (it is the variation of a local counter term) and one finds, after 'integrations by parts'

$$\langle T_{\mu\nu} \rangle = \frac{d\ell^3}{16\pi G_N^{(d+1)}} g^{(d/2)}_{\mu\nu} \qquad \text{for } g^{(0)}_{\mu\nu} = \eta_{\mu\nu}. \qquad (18.220)$$

For more general boundary metrics, $\langle T_{\mu\nu} \rangle$ also contains terms which are quadratic in $g^{(1)}$ and when we computed the holographic Weyl anomaly we effectively computed its trace w.r.t. $g^{(0)}$.

We now consider a simple example with $d = 2n = 4$ where we can solve the equations exactly starting from the following boundary data:

$$g^{(0)}_{\mu\nu} = \eta_{\mu\nu}, \qquad g^{(2)}_{\mu\nu} = c \begin{pmatrix} 3 & & & \\ & 1 & & \\ & & 1 & \\ & & & 1 \end{pmatrix}; \qquad (18.221)$$

c is a constant. $g^{(2)}_{\mu\nu}$ is clearly traceless and conserved (w.r.t. $\eta^{\mu\nu}$). For the metric $g_{\mu\nu}(x, \rho)$ (cf. (18.195)) we make the ansatz

$$g_{\mu\nu}(x, \rho) = \begin{pmatrix} f(\rho) & & & \\ & g(\rho) & & \\ & & g(\rho) & \\ & & & g(\rho) \end{pmatrix}. \qquad (18.222)$$

Inserting this into the Einstein equations we obtain a system of three coupled differential equations for the two functions $f(\rho)$ and $g(\rho)$. The solution which

18.9 AdS/CFT Correspondence

satisfies the boundary conditions is

$$f(\rho) = -\frac{\left(1-\left(\frac{\rho}{\rho_0}\right)^2\right)^2}{1+\left(\frac{\rho}{\rho_0}\right)^2}, \qquad g(\rho) = 1+\left(\frac{\rho}{\rho_0}\right)^2, \qquad (18.223)$$

where we have defined $c = \frac{1}{\rho_0^2}$. The metric is

$$ds^2 = \left(\frac{d\rho}{2\rho}\right)^2 - \frac{1}{\rho}\frac{\left(1-\left(\frac{\rho}{\rho_0}\right)^2\right)^2}{1+\left(\frac{\rho}{\rho_0}\right)^2}dt^2 + \frac{1}{\rho}\left(1+\left(\frac{\rho}{\rho_0}\right)^2\right)dx^2. \qquad (18.224)$$

This is a static, asymptotically AdS solution with a regular horizon at $\rho = \rho_0$ with topology $\mathbb{R}^3 \times \mathbb{R}$. The metric (18.224) is in ADM form with vanishing shift and lapse $N^2 = -\frac{1}{\rho}f(\rho)$ and the Hawking temperature is easily computed

$$T_{\text{BH}} = \frac{1}{4\pi}\frac{(N^2)'}{\sqrt{g_{\rho\rho}N^2}}\bigg|_{\text{horizon}} = \frac{1}{\pi}\sqrt{\frac{2}{\rho_0}}. \qquad (18.225)$$

With the change of variables

$$\frac{\rho}{\rho_0} = \frac{U^2}{U_0^2}\left(1+\sqrt{1-\frac{U_0^4}{U^4}}\right), \qquad \rho_0 = \frac{2}{U_0^2} \qquad (18.226)$$

the metric can be brought to the more familiar form

$$ds^2 = \frac{dU^2}{U^2\left(1-\frac{U_0^4}{U^4}\right)} - U^2\left(1-\frac{U_0^4}{U^4}\right)dt^2 + U^2 dx^2. \qquad (18.227)$$

This is the AdS part of a ten-dimensional IIB supergravity solution which also has an S^5 factor. This solution is the decoupling limit of a non-extremal black D3-brane solution (in this limit we keep U_0 constant).

The interpretation of this result is as follows. On the boundary we have $\mathcal{N} = 4$ SYM theory at finite temperature whose energy-momentum tensor is (use (18.220))

$$\langle T_{\mu\nu}\rangle = \frac{\pi^2}{8}T^4 N^2 \,\text{diag}(3,1,1,1). \qquad (18.228)$$

This is the expectation value of the energy-momentum tensor of a relativistic plasma at rest in a thermal state. Here the plasma is a hot gas of massless particles, the gauge bosons of $\mathcal{N} = 4$ SYM theory and their supersymmetric partners. The energy density is

$$\epsilon = \frac{3\pi^2}{8} N^2 T^4 = \frac{3}{4}\epsilon_{\text{free}}, \tag{18.229}$$

where ϵ_{free} is the value of the free $\mathcal{N} = 4$ SYM theory. The holographic result is a prediction for the strongly coupled theory.

Let us recapitulate: starting with the boundary data, which consist of specifying a metric and energy-momentum tensor of the four-dimensional CFT on the boundary, we have reconstructed the dual five-dimensional space-time of the gravity theory. It is an AdS black-brane whose Hawking temperature we have identified with the temperature of the field theory.

Hydrodynamics

The above procedure can be extended to other situations: given an energy-momentum tensor of the d-dimensional CFT, we can reconstruct the dual $(d + 1)$-dimensional asymptotic AdS space of the dual gravity theory. Of course, one is guaranteed to obtain a consistent set-up only in those cases which can be embedded in string or M-theory. For instance, one might wonder whether any $T_{\mu\nu} \sim g_{\mu\nu}^{(d/2)}$ which is traceless and conserved represents good boundary data. This is indeed not the case, if we require that the resulting geometry is regular, which means e.g. that singularities, if present, are shielded by a horizon. This restricts the class of CFTs which have a gravity dual. An interesting example, which we will merely sketch, is the following. In the previous example the energy-momentum tensor of the field theory was that of a relativistic plasma in $d = 4$ at rest, i.e. there was no energy or momentum flow. We can easily boost this solution to describe it in an arbitrary inertial frame. The bulk metric becomes

$$ds^2 = \left(\frac{d\rho}{2\rho}\right)^2 + \frac{1}{\rho}\left(1 + \left(\frac{\rho}{\rho_0}\right)^2\right)(d\mathbf{x}^2 - dt^2) + \frac{4\left(\frac{\rho}{\rho_0}\right)^2}{\rho\left(1 + \left(\frac{\rho}{\rho_0}\right)^2\right)}\gamma^2(dt + \boldsymbol{\beta}\cdot d\mathbf{x})^2, \tag{18.230}$$

which corresponds to

$$t_{\mu\nu} \equiv 4g_{\mu\nu}^{(2)} = (\pi T)^4(\eta_{\mu\nu} + 4u_\mu u_\nu), \tag{18.231}$$

where $(u_0, u_i) = (\gamma, \gamma\beta_i)$ is the four-velocity of the plasma and, as usual, $\gamma = 1/\sqrt{1 - \beta^2}$. As expected, this metric also solves Einstein's equations.

In this solution the temperature and the four-velocity are constants. We now want to explore the possibility of space-time dependent temperature and velocity. Clearly, if we replace in (18.230) $T \to T(x)$ and $\boldsymbol{\beta} \to \boldsymbol{\beta}(x)$, the resulting five-dimensional metric is no longer a solution of Einstein's equations. If, however, T and $\boldsymbol{\beta}$ vary

18.9 AdS/CFT Correspondence

slowly, i.e. if their space-time gradients are small, we can make a gradient expansion of the metric and satisfy Einstein's equations order by order in numbers of space-time derivatives. The dimensionless small quantities are $\frac{1}{T}\partial u$ and $\frac{1}{T^2}\partial T$.

In the Fefferman-Graham expansion of the metric, $g^{(0)}_{\mu\nu} = \eta_{\mu\nu}$ still holds; this is one of the boundary conditions which we impose, and therefore also $g^{(1)}_{\mu\nu} = 0$. However, $g^{(2)}_{\mu\nu}$ has to be modified. We make the ansatz

$$t_{\mu\nu} = t^{(0)}_{\mu\nu} + t^{(1)}_{\mu\nu} + \ldots . \tag{18.232}$$

The superscripts give the number of space-time derivatives. We already know that

$$t^{(0)}_{\mu\nu} = (\pi T)^4 \left(\eta_{\mu\nu} + 4 u_\mu u_\nu \right), \tag{18.233}$$

where we have suppressed the x-dependence of T and u_μ. $t^{(0)}_{\mu\nu}$ is still traceless but the conservation equation is modified to

$$\partial^\mu t^{(0)}_{\mu\nu} = 4\pi^4 T^3 (\partial_\nu T + 4 u_\nu u^\mu \partial_\mu T + T u_\nu \partial_\mu u^\mu + T u^\mu \partial_\mu u_\nu) = 0. \tag{18.234}$$

The condition $u^\nu \partial^\mu t^{(1)}_{\mu\nu} = 0$ leads to

$$u^\mu \partial_\mu T = -\frac{1}{3} T \partial_\mu u^\mu. \tag{18.235}$$

Inserting this into (18.234) gives

$$\partial_\mu T = \frac{1}{3} T (\partial_\nu u^\nu) u_\mu - T u^\nu \partial_\nu u_\mu, \tag{18.236}$$

which means that the derivatives of the temperature can be expressed in terms of derivatives of the velocity.

We now make the most general ansatz for $t^{(1)}_{\mu\nu}$:

$$t^{(1)}_{\mu\nu} = (\pi T)^3 \left[\alpha (\partial_\mu u_\nu + \partial_\nu u_\mu) + \beta (u_\mu u^\lambda \partial_\lambda u_\nu + u_\nu u^\lambda \partial_\lambda u_\mu) \right.$$
$$\left. + \gamma \, u_\mu u_\nu \partial_\lambda u^\lambda + \delta \, \eta_{\mu\nu} \partial_\lambda u^\lambda \right]. \tag{18.237}$$

The overall factor carries the dimension and α, β, γ and δ are real numbers. Tracelessness imposes $\gamma = 2\alpha + 4\delta$. We find further relations between the coefficients by writing $t^{(1)}_{\mu\nu}$ in a general metric background and requiring that it transforms homogeneously under Weyl transformations, after we restrict to a Minkowski background, cf. (18.209). This leads to

$$\beta = \alpha, \qquad \delta = -\frac{2}{3}\alpha, \qquad \gamma = -\frac{2}{3}\alpha, \tag{18.238}$$

which leaves one coefficient undetermined, say α. We now proceed, as we did in the case of constant T and u^μ, and make an ansatz for $g_{\mu\nu}(x, \rho)$ with the above specified boundary conditions. The ansatz contains several as yet undetermined functions of

ρ which we determine by requiring that Einstein's equations are satisfied to first non-trivial order in gradients. Remarkably the equations can be solved and α is one of the integration constants. This integration constant can be fixed by requiring that the bulk metric is regular outside and on the horizon. It is not hard to see that the FG coordinates which we have been using so far are not suitable for this step of the analysis. A suitable coordinate system are Eddington-Finkelstein coordinates which have no coordinate singularity at the horizon. This analysis indeed fixes $\alpha = -1$ and one finds for the energy-momentum tensor

$$t_{\mu\nu} = (\pi T)^4 (\eta_{\mu\nu} + 4 u_\mu u_\nu)$$
$$- (\pi T)^3 \left(\partial_\mu u_\nu + \partial_\nu u_\mu + u_\mu (u \cdot \partial) u_\nu + u_\nu (u \cdot \partial) u_\mu - \frac{2}{3} (u_\mu u_\nu + \eta_{\mu\nu}) \partial \cdot u \right). \tag{18.239}$$

To analyze this result we note that what we have considered here is a CFT plasma in the long wavelength limit. This allows for an effective description as a relativistic fluid. The conservation of the energy-momentum tensor is nothing but the relativistic Navier-Stokes equations. The zeroth order term represents an ideal fluid. The first correction contains viscosity effects. In general they are determined by two coefficients, the shear viscosity η and the bulk viscosity ξ. The energy-momentum tensor of a relativistic fluid can be shown to be (p is the pressure)[52]

$$T_{\mu\nu} = p(u_\mu u_\nu + \eta_{\mu\nu}) + \epsilon u_\mu u_\nu \tag{18.240}$$
$$- \eta \left(\partial_\mu u_\nu + \partial_\nu u_\mu + u_\mu u^\lambda \partial_\lambda u_\nu + u_\nu u^\lambda \partial_\lambda u_\mu \right) - \left(\xi - \frac{2}{3} \eta \right) (u_\mu u_\nu + \eta_{\mu\nu}) \partial_\lambda u^\lambda.$$

Conformality requires $p = \frac{1}{3}\epsilon$ and $\xi = 0$ and the requirement of having a gravity dual fixes the ratio between η and ϵ. Using the thermodynamic relation $d\epsilon = Tds$, where s is the entropy density, one derives

$$\frac{\eta}{s} = \frac{1}{4\pi}, \tag{18.241}$$

which is valid for a four-dimensional conformal field theory which has a hydrodynamical description and a gravity dual with action (18.193), e.g. strongly coupled $\mathcal{N} = 4$ SYM theory in the planar limit. Higher curvature corrections to the action modify the ratio in either direction, but there are limits from causality which seem to impose restrictions on the coefficients for these corrections in any dual gravity theory which can be embedded in string theory which we expect to be causal.

The result (18.241) is one of the most beautiful results which came out of the AdS/CFT correspondence. It is one of several examples where calculations in the

---

[52]This is the form in the Landau-frame where u^μ is the velocity of energy transport. It satisfies $u^\mu T^{(1)}_{\mu\nu} = 0$.

field theory cannot be done reliably and where AdS/CFT provides a viable and powerful alternative. The question is, of course, whether it also applies to realistic physical systems. Conventional fluids are not conformal, their value of η/s depends on the temperature and is much bigger than the AdS/CFT value. For instance, for water at room temperature it is $380 \cdot \frac{1}{4\pi}$. For a photon gas it is infinite. Surprisingly there are experimental indications that for the quark-gluon plasma it is close to $1/4\pi$.

Generalizations and Extensions

Our treatment of the AdS/CFT correspondence was by way of a few specific examples which, however, already point to various generalizations, some of which we will briefly discuss.

- We can consider other near horizon geometries, e.g. of the M2 and M5 brane solutions which lead to $AdS_4 \times S^7$ and $AdS_7 \times S^4$ space-times and dual CFTs in $d = 3$ and $d = 7$ dimensions. While quite a bit is known about the former, not much is known about the latter and it is believed that it is not a local field theory. One piece of information we have about this elusive theory is the Weyl anomaly which behaves as N^3 for N parallel branes. The problem is that the theory on the M5 brane world-volume has a self-dual anti-symmetric rank-two tensor field whose non-Abelian structure (generalizing that of interacting gauge fields) is not known.

    The near horizon geometry of other D-branes can also be studied but they do not lead to AdS geometries. For these solutions the dilaton is not constant and the string theory is no longer weakly coupled throughout.

    It is, of course, interesting to have a dual gravity description of non-conformal field theories, in particular in four dimensions. Various SUGRA solutions have been constructed, but we will not discuss them here.

- For $d = 4$ one can consider other conformal field theories and their dual geometry. One possibility is to replace S^5 by a five-dimensional Sasaki-Einstein-manifold X^5. Their defining property is that the cone over X^5 with metric $ds^2 = dr^2 + r^2 ds^2_{X_5}$ is a (non-compact) Calabi-Yau manifold. In this case $G_N^{(5)} = G_N^{(10)}/\text{Vol}(X^5)$ which changes the values of the anomaly coefficients. The simplest example is, of course, $X^5 = S^5$ where the CY manifold is simply $\mathbb{R}^6$. In general, the dual conformal field theories are $\mathcal{N} = 1$ quiver theories, i.e. with gauge groups products of $SU(N)$ factors and matter fields in bi-fundamental representations. Another way to construct SCFTs with less supersymmetry is to consider the near-horizon limit of N D3-branes on a $\mathbb{Z}_2$ orientifold seven-plane with eight D7 branes stuck on the O7 plane to ensure conformal invariance. The dual field theory is an $\mathcal{N} = 2$ superconformal theory with gauge group $USp(2N)$, four matter multiplets in the fundamental (**2N**) and one in the anti-symmetric traceless representation (**2(2N − 1) − 1**). The compact manifold is $S^5/\mathbb{Z}_2$. The $\mathbb{Z}_2$ does not act freely, its fixed locus is an S^3 which the D7's and the O7 wrap. They also fill the whole AdS_5. The fact that the compact manifold is no longer S^5 is due to the back-reaction of the D7 branes and the O7 plane. The presence of D7 branes introduces open strings and the presence of an O7 plane non-orientable world-sheets and therefore odd powers in the $1/N$ expansion, corresponding to odd powers in the g_s expansion. The dual gravity theory is also modified at this order: there are terms quadratic in the curvature which are induced from such terms in the world-volume action of the D7 branes and the O7 plane. In this case, the two Weyl anomaly coefficients no longer agree and $a - c = \mathcal{O}(N)$ which is $1/N$ suppressed compared to the leading contribution in a and c. This is also what one finds in the field theory.

- So far we have only considered purely gravitational actions on the AdS side of the correspondence. We have seen that there are two sets of boundary data: at leading order the boundary metric and, subleading, the expectation value of the boundary energy-momentum tensor. These are conjugate in the sense that the metric is a source for the energy-momentum tensor and in the CFT there is a coupling in the action of the form $\int g^{\mu\nu} T_{\mu\nu}$. The effective action, $W(g)$ is the generating functional for correlation functions of the energy-momentum tensor.

From the field theory point of view, if we are interested in correlators involving other operators, we couple them to sources and the generating function depends on the metric and these other sources. To include them in the dual description we generalize the gravity action by including other supergravity fields besides the fluctuations of the metric around the AdS background. In particular there is an infinite tower of Kaluza-Klein modes of the IIB SUGRA fields compactified on S^5 or, more generally, on X^5. Consider, for simplicity, a bosonic field $\Phi(x, \rho)$ which satisfies a second order differential equation. In the vicinity of the boundary its solution is of the form

$$\Phi(x, \rho) = \rho^{\frac{1}{2}\Delta_-}(\phi(x) + \dots) + \rho^{\frac{1}{2}\Delta_+}(\varphi(x) + \dots), \quad (18.242)$$

where we have only indicated the leading terms of the two power series. The higher order terms are fixed by the equations of motion. $\frac{1}{2}\Delta_\pm$ are the two solutions of the indicial equation with $\Delta_+ > \Delta_-$.[53] ϕ and φ are the boundary data one has to provide to obtain a unique solution. Once we specify ϕ, φ is not arbitrary: it is restricted by the requirement to obtain a regular solution in the interior.

The simplest example is a massive scalar field in AdS with equation of motion[54] $(\Box - m^2)\Phi = 0$ with $\Box$ the scalar Laplacian in AdS_{d+1}. One finds $\Delta_\pm = \frac{d}{2} \pm \sqrt{(\frac{d}{2})^2 + \ell^2 m^2}$. Here we have used the AdS metric in FG coordinates. The roots of the indicial equation are real as long as the mass is not below the

---

[53] For special values of the mass, there is only one power series solution and one solution containing $\log(\rho)$. This always happens if $\Delta_+ = \Delta_-$.

[54] In general one has to solve the coupled system of equations for the metric and the scalar field. Here we neglect the backreaction of the scalar on the metric.

18.9 AdS/CFT Correspondence

Breitenlohner-Freedman bound, i.e. as long as $m^2 \geq -\frac{(D-1)^2}{2\ell^2}$ where $D = d+1$ is the space-time dimension. One can show that tachyonic scalars, as long as they respect this bound, do not lead to instabilities. A conformally coupled scalar respects this bound.

In analogy with the metric, one identifies $\phi(x)$ as the source for the dual operator $\mathcal{O}$ whose vev is $\langle \mathcal{O} \rangle = \varphi$. On a fixed AdS background one can solve the equations of motion in terms of Bessel functions and there is one linear combination of the two solutions which leads to a regular solution in the bulk. If we insert this into the bulk action, from which the equation of motion was derived, we obtain a functional of $\phi(x)$ which is the generating functional for correlation functions of the dual operator. In this way one can compute its two-point functions from which one can read off the scaling dimension to be Δ_+. One can establish a dictionary between gauge invariant operators and their sources in $\mathcal{N} = 4$ SYM theory and KK modes in the gravity theory. Morever, there is a relation between the mass of the bulk fields and the scaling dimension of the boundary operator.

In the generic situation the leading solution, which starts with $\rho^{\Delta_-/2}$, has a boundary behavior which makes it non-normalizable while the other solution is normalizable. While the non-normalizable solution is related to the external source, the normalizable solution is associated with the vev of the dual operator and therefore encodes information about the dynamics of the field theory. However, for a certain range of masses or, equivalently, conformal dimensions Δ, both solutions are normalizable and the roles of the source and the vev of the dual operator can be interchanged. We will not discuss the implications of this fact for the CFT except for mentioning that it corresponds, for this range of masses, to two ways of quantizing the field in AdS space which lead to a well defined and conserved energy.

Recall that the holographic stress-energy tensor was defined as the functional derivative of the on-shell gravity action viewed as a functional of the boundary value of the graviton. If we include other fields besides the metric we can again evaluate the on-shell gravity action, which is now a functional of the boundary values of all the fields which we have included and is thus the generating functional for the correlation functions of the operators which are dual to them. The on-shell SUGRA action is divergent. The divergences can be subtracted with suitable counter terms, much in the same way as we described in the case of the metric.

Many examples have been worked out explicitly and correlation functions of various operators which are dual to bulk fields have been computed. We refer to the literature for further details.

What makes the identification between field theory operators and Kaluza-Klein states relatively straightforward is that the KK excitations of gauged $\mathcal{N} = 8$ SUGRA in $d = 5$ are half-BPS states and the operators in the CFT are chiral primary operators of the superconformal algebra and their descendants. This is all there is in the supergravity limit, which is what we have considered so far. Beyond that limit one also has to consider multi-particle states, bound states

- and, in particular, also string excitations in the bulk. To each of them there is a dual operator in the boundary CFT but to establish the correspondence is much harder.
- One can use the correspondence between operators and their sources in the field theory on the one hand and and bulk fields on the other, to engineer gravity duals of various (conformal) field theories. For instance, if one wants to have a field theory with a $U(1)$ current, one has to add a massless vector field in the bulk for which one can choose the gauge $A_\rho = 0$. Its leading boundary value is the source for the current and its subleading boundary value the vev of the current. Or if one wants a scalar condensate in the boundary theory, one needs to include a scalar field in the bulk. Using these ideas one can construct gravity duals of (conformal) superconductors and other condensed matter systems. In general this is quite challenging because to engineer e.g. the gravity dual of a superconductor one needs an asymptotically AdS black hole solution which looses its hair (the bulk scalar whose subleading boundary value is the condensate) above a certain temperature T_c. Whether these systems can be embedded in string theory is not clear a priori. But e.g. for the case of the superconductor, relevant bulk solutions and their embedding in string theory have been found.
- The holographic reconstruction of space-time required only field theory data in Minkowski space. Within the framework of the strong version of the AdS/CFT correspondence it can be viewed as a background independent formulation of type IIB string theory on asymptotically $AdS_5 \times S^5$ space-times. The metric in the bulk emerges from the dynamics of a field theory without gravity on the boundary. In the example given, we only considered a vev of the energy-momentum tensor but this can be extended to other operators.
- One of the fundamental observations in support of the AdS/CFT correspondence is the matching of the symmetries. One might wonder whether one can construct gravity duals for field theories with other symmetries. One example is a non-relativistic version of conformal symmetry, which is called Schrödinger symmetry, where e.g. space and time scale differently, the difference being characterized by the dynamical exponent z, i.e. $x^i \to \lambda x^i$, $t \to \lambda^z t$. One can indeed construct space-time metrics which have Schrödinger symmetry as an isometry and they are therefore, within the spirit of the AdS/CFT correspondence, dual to field theories with the same symmetry. There are some important differences though which have to do with the notion of a boundary of these metrics. The holography seems to be of co-dimension two rather than co-dimension one.
- One of the attractive features of the AdS/CFT correspondence is that it provides a simple geometric picture for a variety of physical processes and allows for an quantitative treatment. This is most convincingly demonstrated by the SUGRA computation of the gauge theory Wilson loop. To describe this, we put an additional D3 brane into the background of the near-horizon geometry of the stack of branes. On this brane open strings may end. If we place this brane at the boundary, we get the following picture (Fig. 18.4):

  a static, well separated $q\bar{q}$ pair is viewed as the endpoints of an open type IIB string at the boundary of AdS_5. In order to minimize its length (and hence its

18.9 AdS/CFT Correspondence

Fig. 18.4 Holographic Wilson loop

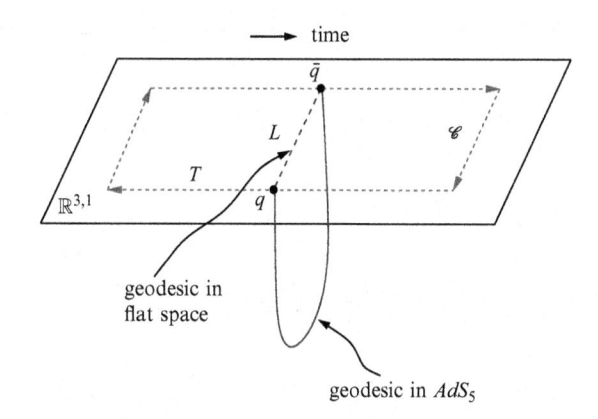

energy) the string follows a geodesic. The geodesic does not lie in the boundary, which it would if the space were flat, but extends deep into the AdS space. To describe the open string we need to add the Nambu-Goto action to the SUGRA action (as we did in Sect. 18.5) and the correspondence is now

$$\langle W(\mathscr{C})\rangle_{CFT} = e^{-V_{q\bar{q}}} = e^{-\frac{A_{\min}}{2\pi\alpha'}}, \quad (18.243)$$

where $A_{\min}$, the minimal surface of the world-sheet, is $T \times$ length of the geodesic. The $q\bar{q}$ potential is then given by the (regularized) length of this geodesic where the regularization is interpreted as the subtraction of the infinite bare mass of the static quarks. One finds

$$V_{q\bar{q}} \sim \frac{\lambda^{1/2}}{L}, \quad (18.244)$$

where the numerical factor can be easily worked out. The $1/L$ behavior is an obvious consequence of conformal invariance: there is no other scale in the theory except for L to provide the dimension for the potential. The $\lambda^{1/2}$ behavior follows from the discussion below Eq. (18.186).

It is not difficult to find gravity duals of confining theories with $V_{q\bar{q}} \sim L$. One way to get a confining theory in four dimensions is to start with the near-extremal D4-brane solution in the decoupling limit and perform a double Wick rotation of the time direction and one space direction along the world-volume. Compactification on the Euclidean time direction leads to a four-dimensional theory whose Wilson loop, again computed as the minimal area of a fundamental string, exhibits an area law, indicating confinement with a linearly rising potential.

- Recall that the comparison of the physics of coincident D-branes in Minkowski space and a particular flux-compactification of type IIB string theory led to the Maldacena conjecture. This is a particular example of a large N transition.

Similar transitions can also be found in more complicated backgrounds where e.g. N D6-branes are wrapped around S^3's which are contained in a Calabi-Yau manifold on which the string theory is compactified. An $\mathcal{N}=1$ $U(N)$ gauge theory is localized in the four large dimensions of the world-volume of the D6-brane. One can decouple gravity from the gauge degrees of freedom by zooming into the region close to the branes. In this case one ends up with the deformed conifold geometry, which is a particular non-compact Calabi-Yau manifold. If the number of D6-branes becomes large their back-reaction on the geometry must be taken into consideration. This leads to a transition to a different Calabi-Yau geometry, the resolved conifold. This is the conifold transition. After this transition the branes and thus the open strings which gave rise to the gauge bosons have disappeared. From the gauge theory perspective this is interpreted as a transition to the confining phase.

- In attempts to prove the AdS/CFT correspondence for $\mathcal{N}=4$ SYM theory, the uncovering of integrable structures on both sides of the duality has led to many interesting developments and highly non-trivial tests of the correspondence. We refer the interested reader to the vast literature on this exciting but also rather technical subject.

The big open challenge is to find gravity duals of interesting and relevant physical systems, such as QCD with gauge group $SU(3)$, where calculations can be done reliably and in a controlled way. We will not discuss any of the attempts at constructing models of holographic QCD. Despite of much effort it seems fair to say that this goal has not been reached (yet?). However, the ability to study quantitatively strong coupling phenomena which are out of reach by other methods, makes the AdS/CFT correspondence a powerful tool to explore such systems, e.g. their phase diagrams, to compute transport coefficients, etc. This might lead to new insights which can be a guide for the understanding of real physical systems.

Appendix

In this appendix we collect some expressions which were needed for the calculation of the holographic anomalies in Sect. 18.9.

For a $(d+1)$-dimensional metric of the FG form[55]

$$ds^2 = G_{MN} dX^M dX^N = \ell^2 \left(\frac{d\rho}{2\rho}\right)^2 + \frac{\ell^2}{\rho} g_{ij}(x,\rho) dx^i dx^j \qquad (18.245)$$

one finds the following expressions for various curvature components where the hat means $\hat{R} = R(G)$ while $R = R(g)$, the covariant derivatives are associated with

---

[55] For ease of notation we use small Latin letters to label the coordinates of the boundary, rather than small Greek letters, as we did previously.

18.9 AdS/CFT Correspondence

$g_{ij}(x,\rho)$, a prime stands for ∂_ρ and the curvature conventions are those of Sect. 14.8.

$$\hat{R}_{i\rho j\rho} = -\frac{\ell^2}{4\rho^3}g_{ij} - \frac{\ell^2}{2\rho}g''_{ij} + \frac{\ell^2}{4\rho}g'_{ik}g'_{jl}g^{kl},$$

$$\hat{R}_{\rho ijk} = \frac{\ell^2}{2\rho}(\nabla_k g'_{ij} - \nabla_j g'_{ik}),$$

$$\hat{R}_{ijkl} = \frac{\ell^2}{\rho}R_{ijkl} - \frac{\ell^2}{\rho^2}\left\{(g_{jl} - \rho g'_{jl})(g_{ik} - \rho g'_{ik}) - (g_{jk} - \rho g'_{jk})(g_{il} - \rho g'_{il})\right\},$$

$$\hat{R}_{ij} = R_{ij} - 2\rho g''_{ij} + 2\rho g'_{il}g^{kl}g'_{kj} - \rho g'_{ij}g^{kl}g'_{kl} + (d-2)g'_{ij} - g_{ij}g^{kl}g'_{kl} - \frac{d}{\rho}g_{ij},$$

$$\hat{R}_{\rho\rho} = -\frac{d}{4\rho^2} - \frac{1}{2}g^{ij}g''_{ij} + \frac{1}{4}g^{ij}g^{kl}g'_{ik}g'_{jl},$$

$$\hat{R}_{\rho i} = \frac{1}{2}g^{jk}(\nabla_k g'_{ij} - \nabla_i g'_{jk}),$$

$$\ell^2 \hat{R} = -d(d+1) + \rho R + 2(d-1)\rho g^{ij}g'_{ij} + 3\rho^2 g^{ij}g^{kl}g'_{ik}g'_{jl}$$
$$- 4\rho^2 g^{ij}g''_{ij} - \rho^2(g^{ij}g'_{ij})^2. \tag{18.246}$$

All other components vanish or are related by symmetries.

The FG expansion of the metric

$$g_{ij}(x,\rho) = \sum_{n\geq 0} g^{(n)}_{ij}(x,\rho)\rho^n \tag{18.247}$$

induces an expansion of the curvatures. The first few terms of some curvature invariants are (indices are raised with $g^{(0)\,ij}$, the inverse of $g^{(0)}_{ij}$, $\operatorname{tr}(g^{(1)}) = g^{(0)\,ij}g^{(1)}_{ij}$ and we also use $g_{ij} \equiv g^{(0)}_{ij}$, $R = R(g^{(0)})$, $\nabla_i \equiv \nabla^{(0)}_i$, $\Box = \nabla^{(0)i}\nabla^{(0)}_i$, etc.)

$$\ell^2 \hat{R} = -d(d+1) + \rho\left[R + 2(d-1)\operatorname{tr}(g^{(1)})\right]$$
$$+ \rho^2\left[-g^{(1)}_{ij}R^{ij} - 4(3-d)\operatorname{tr}(g^{(2)}) - (2d-5)\operatorname{tr}((g^{(1)})^2)\right.$$
$$\left. - (\operatorname{tr}(g^{(1)}))^2 - \Box\operatorname{tr}(g^{(1)}) + \nabla^i\nabla^j g^{(1)}_{ij}\right] + \mathcal{O}(\rho^3),$$

$$\ell^4 \hat{R}_{MNPQ}\hat{R}^{MNPQ} = 2d(d+1) - 4\left[R + 2(d-1)\operatorname{tr}(g^{(1)})\right]\rho$$
$$+ \left[R_{ijkl}R^{ijkl} + 12g^{(1)}_{ij}R^{ij} + 4(\Box\operatorname{tr}(g^{(1)}) - \nabla^i\nabla^j g^{(1)}_{ij}) + 8(\operatorname{tr}(g^{(1)}))^2\right.$$
$$\left. - 16(d-3)\operatorname{tr}(g^{(2)}) + (12d-28)\operatorname{tr}((g^{(1)})^2)\right]\rho^2 + \mathcal{O}(\rho^3),$$

$$\ell^4 \hat{R}_{MN} \hat{R}^{MN} = d^2(d+1) - 2d\left[R + 2(d-1)\text{tr}(g^{(1)})\right]\rho$$
$$+ \left[R_{ij}R^{ij} + 4(d-1)g^{(1)}_{ij}R^{ij} + 2R\,\text{tr}(g^{(1)}) + 2d\left(\Box\,\text{tr}(g^{(1)}) - \nabla^i\nabla^j g^{(1)}_{ij}\right)\right.$$
$$+ (5d^2 - 14d + 4)\text{tr}\left((g^{(1)})^2\right) + (5d-4)(\text{tr}(g^{(1)}))^2 - 8d(d-3)\text{tr}\,g^{(2)}\Big]\rho^2$$
$$+ \mathcal{O}(\rho^3). \tag{18.248}$$

We also need
$$\sqrt{G} = \frac{\ell^{d+1}}{2\rho^{1+d/2}}\sqrt{g}\bigg\{1 + \frac{\rho}{2}\text{tr}(g^{(1)})$$
$$+ \rho^2\left[\frac{1}{2}\text{tr}(g^{(2)}) - \frac{1}{4}\text{tr}\left((g^{(1)})^2\right) + \frac{1}{8}\left(\text{tr}(g^{(1)})\right)^2\right]\bigg\} + \mathcal{O}(\rho^3). \tag{18.249}$$

These expressions simplify considerably if one uses the explicit expression for $g^{(1)}_{ij}$.

Under $g_{ij} \to g_{ij} + h_{ij}$ and $g^{ij} \to g^{ij} - h^{ij}$, the first order changes of various quantities are

$$\delta\sqrt{g} = \frac{1}{2}\sqrt{g}\,h, \qquad \delta\Gamma^k_{ij} = \frac{1}{2}\left(\nabla_i h_j{}^k + \nabla_j h_i{}^k - \nabla^k h_{ij}\right),$$

$$\delta R_{ijkl} = -\frac{1}{2}\left\{\nabla_k\nabla_i h_{jl} + \nabla_l\nabla_j h_{ik} - \nabla_k\nabla_j h_{il} - \nabla_l\nabla_i h_{jk}\right\} + \frac{1}{2}R_{kli}{}^m h_{mj} - \frac{1}{2}R_{klj}{}^m h_{mi},$$

$$\delta R_{ij} = -\frac{1}{2}\left\{\Box h_{ij} + \nabla_i\nabla_j h - \nabla_k\nabla_i h^k{}_j - \nabla_k\nabla_j h^k{}_i\right\},$$

$$\delta R = -\Box h + \nabla^i\nabla^j h_{ij} - R^{ij}h_{ij},$$

$$\delta(\Box R) = -h^{ij}\Box R_{ij} - h^{ij}\nabla_i\nabla_j R - \Box^2 h + \Box\nabla_i\nabla_j h^{ij}$$
$$- R^{ij}\Box h_{ij} - 2\nabla^k R_{ij}\nabla_k h^{ij} - \nabla_i R\nabla_j h^{ij} + \frac{1}{2}\nabla_i R\nabla^i h. \tag{18.250}$$

All indices are raised with g^{ij}, e.g. $h^{ij} = g^{ik}g^{jl}h_{kl}$ and $h = g^{ij}h_{ij}$.

Using $h_{ij} = 2\sigma g_{ij}$, one finds for the first order Weyl variations

$$\delta_\sigma R = -2\sigma R - 2(d-1)\Box\sigma,$$
$$\delta_\sigma R_{ij} = -2(d-2)\nabla_i\nabla_j\sigma R - g_{ij}\Box\sigma,$$
$$\delta_\sigma R_{ijkl} = 2\sigma R_{ijkl} - g_{ik}\nabla_j\nabla_\sigma\sigma - g_{jo}\nabla_i\nabla_k\sigma + g_{il}\nabla_{jk}\sigma + g_{jk}\nabla_i\nabla_l\sigma. \tag{18.251}$$

18.9 AdS/CFT Correspondence

The equations of motion derived from the action

$$S = \int d^D x \sqrt{g}\mathcal{L} = \int d^D x \sqrt{g}\left(R - \Lambda + aR^2 + bR^{ij}R_{ij} + cR^{ijkl}R_{ijkl}\right) \tag{18.252}$$

are

$$\frac{1}{2}\mathcal{L}g_{ij} - R_{ij} + 2a\left(\nabla_i\nabla_j R - \Box R g_{ij} - RR_{ij}\right)$$
$$+ b\left(-2R_{ik}R^k{}_j - \Box R_{ij} - \frac{1}{2}\Box R g_{ij} + (\nabla_k\nabla_i R^k{}_j + \nabla_k\nabla_j R^k{}_i)\right)$$
$$+ c\left(-2R_{iklm}R_j{}^{klm} - 4\nabla_l\nabla_k R_i{}^l{}_j{}^k\right) = 0. \tag{18.253}$$

They can be rewritten with the help of the identities

$$\nabla_k\nabla_i R^k{}_j + \nabla_k\nabla_j R^k{}_i = \nabla_i\nabla_j R + 2R_{ik}R^k{}_j - 2R^{kl}R_{ikjl},$$
$$\nabla_k\nabla_l R_i{}^k{}_j{}^l = \Box R_{ij} - \frac{1}{2}\nabla_i\nabla_j R - R_{ik}R^k{}_j + R_{ikjl}R^{kl}. \tag{18.254}$$

The condition on the dimensionless parameters $\alpha = a/L^2$, etc. where $\Lambda = -(D-1)(D-2)/L^2$ which has to be satisfied for the equations to have a symmetric space solution with

$$R_{ijkl} = k(g_{ik}g_{jl} - g_{il}g_{jk}), \qquad k = \frac{1}{\ell^2} \tag{18.255}$$

is ($q = L/\ell$)

$$-(D-2) + \frac{(D-2)}{q^2} + (D-4)\left[D(D-1)\alpha + (D-1)\beta + 2\gamma\right]q^2 = 0. \tag{18.256}$$

The derivative terms in (18.253) do not contribute because the metric is covariantly constant.

Further Reading

Two of the seminal papers on string dualities are

- C.M. Hull, P.K. Townsend, Unity of superstring dualities. Nucl. Phys. **B438**, 109–137 (1995) [arXiv:hep-th/9410167]
- E. Witten, String theory dynamics in various dimensions. Nucl. Phys. **B443**, 85–126 (1995) [arXiv:hep-th/9503124]

Early papers on S-duality in string theory include

- A. Font, L.E. Ibañez, D. Lüst, F. Quevedo, Strong – weak coupling duality and nonperturbative effects in string theory. Phys. Lett. B **249**, 35 (1990)
- A. Sen, *Electric magnetic duality in string theory*. Nucl. Phys. B **404**, 109 (1993) [arXiv:hep-th/9207053]

Reviews of string dualities include

- A. Sen, *An introduction to non-perturbative string theory* Cambridge 1997, Proceedings: *Duality and Supersymmetric Theories* (D.I. Olive and P.C. West, eds.), Cambridge University Press 1999, p. 297 [arXiv:hep-th/9802051]
- J.H. Schwarz, *Lectures on superstring and M theory dualities: Given at ICTP Spring School and at TASI Summer School*. Nucl. Phys. Proc. Suppl. **55B**, 1–32 (1997) [arXiv:hep-th/9607201]
- B. de Wit, J. Louis, *Supersymmetry and dualities in various dimensions* [arXiv: hep-th/9801132]
- N.A. Obers, B. Pioline, U duality and M theory. Phys. Rep. **318**, 113–225 (1999) [arXiv:hep-th/9809039]

Extremal and non-extremal brane solutions of supergravity were first systematically constructed in

- G.T. Horowitz, A. Strominger, Black strings and p-branes. Nucl. Phys. B **360**, 197 (1991)

Comprehensive reviews are

- M.J. Duff, R.R. Khuri, J.X. Lu, String solitons. Phys. Rep. **259**, 213 (1995) [arXiv:hep-th/9412184]
- K.S. Stelle, *BPS branes in supergravity* [arXiv:hep-th/9803116]

The type I – heterotic duality is discussed in

- J. Polchinski, E. Witten, Evidence for heterotic – type I string duality. Nucl. Phys. **B460**, 525–540 (1996) [arXiv:hep-th/9510169]

The seven-brane solution was constructed in

- B.R. Greene, A.D. Shapere, C. Vafa, S.-T. Yau, Stringy cosmic strings and noncompact Calabi-Yau manifolds. Nucl. Phys. **B337**, 1 (1990)

F-theory was proposed in

- C. Vafa, Evidence for F theory. Nucl. Phys. **B469**, 403–418 (1996) [arXiv:hep-th/9602022]

Two reviews are

- F. Denef, *Les Houches Lectures on Constructing String Vacua* [arXiv:0803.1194 [hep-th]]
- T. Weigand, *Lectures on F-theory compactifications and model building*. Class. Quant. Grav. **27**, 214004 (2010) [arXiv:1009.3497 [hep-th]]

18.9 AdS/CFT Correspondence

The fate of the $U(1)$ factor in F-theory was clarified in

- T.W. Grimm, T. Weigand, On Abelian Gauge symmetries and proton decay in global F-theory GUTs. Phys. Rev. **D82**, 086009 (2010) [arXiv:1006.0226 [hep-th]]

The absorption cross section of D-branes was computed in

- I.R. Klebanov, World-volume approach to absorption by non-dilatonic branes. Nucl. Phys. B **496**, 231–242 (1997) [arXiv:hep-th/9702076]
- S.S. Gubser, I.R. Klebanov, A.A. Tseytlin, String theory and classical absorption by Threebranes. Nucl. Phys. **B499**, 217–240 (1997) [arXiv:hep-th/97030404]

The original references for the Maldacena conjecture are

- J.M. Maldacena, The large N limit of superconformal field theories and supergravity. Adv. Theor. Math. Phys. **2**, 231 (1998) [arXiv:hep-th/9711200]
- S.S. Gubser, I.R. Klebanov, A.M. Polyakov, Gauge theory correlators from noncritical string theory. Phys. Lett. **B428**, 105 (1998) [arXiv:hep-th/9802109]
- E. Witten, Anti-de Sitter space and holography. Adv. Theor. Math. Phys. **2**, 253 (1998) [arXiv:hep-th/9802150]

Useful reviews and introductory lecture notes are

- O. Aharony, S.S. Gubser, J.M. Maldacena, H. Ooguri, Y. Oz, Large N field theories, string theory and gravity. Phys. Rep. **323**, 183–386 (2000) [arXiv:hep-th/9905111]
- E. D'Hoker, D.Z. Freedman, *Supersymmetric gauge theories and the AdS / CFT correspondence* [arXiv:hep-th/0201253]
- J.L. Petersen, Introduction to the Maldacena conjecture on AdS / CFT. Int. J. Mod. Phys. **A14**, 3597 (1999) [arXiv:hep-th/9902131]
- H. Nastase, *Introduction to AdS-CFT* [arXiv:0712.0689 [hep-th]]
- E. Papantonopoulos (ed.), From gravity to thermal gauge theories: The AdS/CFT correspondence. Lect. Notes Phys. **828** (2011)

In the last reference various applications of the AdS/CFT corespondence to QCD, condensed matter systems, etc. are also discussed.

The holographic Weyl anomaly was first computed in

- M. Henningson, K. Skenderis, The Holographic Weyl anomaly. JHEP **9807**, 023 (1998) [arXiv:hep-th/9806087]

Our presentation follows

- C. Imbimbo, A. Schwimmer, S. Theisen, S. Yankielowicz, Diffeomorphisms and holographic anomalies. Class. Quant. Grav. **17**, 1129 (2000) [arXiv:hep-th/9910267]
- A. Schwimmer, S. Theisen, Entanglement entropy, trace anomalies and holography. Nucl. Phys. **B801**, 1 (2008) [arXiv:0802.1017 [hep-th]]

The holographic reconstruction of space-time was developed in

- S. de Haro, S.N. Solodukhin, K. Skenderis, Holographic reconstruction of spacetime and renormalization in the AdS / CFT correspondence. Comm. Math. Phys. **217**, 595–622 (2001) [arXiv:hep-th/0002230]

The holographic result for η/s was first derived in

- G. Policastro, D.T. Son, A.O. Starinets, The shear viscosity of strongly coupled N = 4 supersymmetric Yang-Mills plasma. Phys. Rev. Lett. **87**, 081601 (2001) [arXiv:hep-th/0104066]

Our derivation follows

- S. Bhattacharyya, V. Hubeny, S. Minwalla, M. Rangamani, Nonlinear fluid dynamics from gravity. JHEP **0802**, 045 (2008) [arXiv:0712.2456 [hep-th]]

Aspects of integrability are reviewed in

- N. Beisert, C. Ahn, L.F. Alday, Z. Bajnok, J.M. Drummond, L. Freyhult, N. Gromov, R.A. Janik et al., Review of AdS/CFT integrability: An overview Lett. Math. Phys. Vol. 99 (2012) (special volume) [arXiv:1012.3982 [hep-th]]

Index

Abelian differential, 139
A-D-E classification, 351
  singularities, 716
ADM tension, 694
AdS/CFT correspondence
  strong, 748
  weak, 748
Almost complex structure, 454
Anomaly
  gauge, 652
  gravitational, 250
  holographic, 754
Anthropic principle, 3, 663
Anti-de-Sitter, 741
  global coordinates, 742
  metric, 735
  Poincaré coordinates, 742
A-roof genus, 633
Atlas, 450
Axion, 629

Background charge, 395
Beltrami
  differential, 129
  equation, 129
Betti numbers, 448, 461
  in Kähler manifolds, 468
Bianchi identity, 498, 625, 657
Bosonic string
  map, 561, 567, 568
  propagators, 37
Bosonization, 341, 391
Boundary changing operator, 98
Boundary condition
  A-type, 380
  B-type, 380

Dirichlet, 12, 19, 188
  mixed, 29, 189, 315
  Neumann, 12, 19, 188
Boundary state, 101, 164
  fermionic, 232
  formalism, 93
  fractional D-brane, 540
BPS, 241
  bound, 696
  state, 560, 684
Brane world scenario, 641
Breitenlohner-Freedman bound, 765
BRST
  charge, 385, 406
  current, 113, 387
  quantization, 110
Bulk, 641
Bundle
  canonical, 474
  line, 469
  stable, 503
  vector, 453
Buscher rules, 434

Calabi-Yau
  four-fold, 478
  Hodge numbers, 476
  holomorphic n-form, 475
  manifold, 385, 444, 473
  rigid, 532
  warped, 662
Calibrated submanifold, 648
Canonical quantization, 35
Cardy condition, 165, 170
Cardy formula, 93

Cartan
  matrix, 329
  subalgebra, 328
Cartan-Weyl basis, 328
Casimir, 325
Center of mass position, 24, 29
Central charge, 39, 323
Central extension, 39, 323, 720
Chan-Paton factor, 49, 244
Character, 153, 344
Charge conjugation, 153, 211, 348
Chern
  character, 473
  class, 471
  connection, 457, 470
Chern-Simons
  3-form, 628
  term, 624
Chiral field, 66
  $N = 2$ superconformal, 377
Chirality matrix, 212
Chirality operator, 211
Chiral ring, 383, 577
Christoffel symbols, 509
Clebsch-Gordan coefficients, 219
Clifford algebra, 198, 210
Closed-open string duality, 99
Closed string, 19
  classical solution, 32
  fermionic mode expansion, 189
  fermionic spectrum, 207
  mode expansion, 23
  partition function, 148
  spectrum, 50
Cocycle factor, 339
Coherent state, 101
Cohomology, 461
  de Rham, 461
  Dolbeault, 463
Compactification, 264, 440
  brane world, 641
  flux, 658
  heterotic on Calabi-Yau, 497
  supersymmetric, 441
  type II on Calabi-Yau, 487
Complex
  manifold, 130
  structure, 130, 450
  moduli, 485
Conformal
  bootstrap, 83
  coordinates, 129
  family, 76
  gauge, 16
  group, 63, 737
  Killing spinor, 182
  Killing vector, 18, 129, 182, 737
  spin, 66
  structure, 129
  Ward identity, 84
  weight, 66
Conformal field theory, 63
  boundary, 93
  rational, 79, 164, 167, 344
  super, 355
Conifold, 482
Conjugacy class, 331, 392
Constraint
  first class, 9
  primary, 8, 12
  secondary, 9
  second class, 9, 186, 279
Contorsion, 667
Coordinates
  homogeneous, 451
  inhomogeneous, 452
Coset, 331
  lattice, 328
Covariant
  derivative, 510
  lattice, 416, 578
Coxeter element, 524
Critical dimension
  bosonic string, 47
  fermionic string, 203
Critical string, 431
Crosscap, 124
  gluing conditions, 167
  Ishibashi states, 167
  state, 166, 248
Crossing symmetry, 83
Cross ratio, 82
Current algebra, 324
  families, 325
Cylinder amplitude, 157

D-brane, 30
  fractional, 540
  intersecting, 310, 646
  magnetized, 317
  R-R charge, 240
  SUGRA solution, 701
  tension, 163, 171, 240
Dedekind eta-function, 147
Dehn twist, 135

Descendant
  field, 69, 741
  state, 75
Dilation, 64, 740
Dimensional reduction, 435, 445
  double, 719
Dirac-Born-Infeld action, 436, 630
Dirac bracket, 186, 279
Dirac matrix, 210
Dirichlet boundary condition, 12, 19, 188
Discrete torsion, 300
Discriminant, 725
Disk, 94, 101, 145
Dolbeault cohomology, 384, 463
Doubling trick, 28, 96, 191
Dual
  Coxeter number, 325
  lattice, 327
  resonance model, 601
Duality, 688
  heterotic-type IIA in 6D, 713
  S- heterotic-type I in 10D, 710
  strong-weak, 676
  S- type IIB in 10D, 705
  T-, 269, 432
  target-space, 678
  U-, 712

Eddington-Finkelstein coordinates, 762
Effective action, 615
  D-brane, 630
  supergravity, 622
Effective field theory, 586
Einstein frame, 172, 619, 626, 690
  modified, 690
Einstein spaces, 736
Elliptic curve, 724
Elliptic fibration, 724
Energy-momentum tensor, 13, 67
  holographic, 757
Euler Beta-function, 607, 636
Euler number, 431, 468, 478

Faddeev-Popov ghosts, 53
Fefferman-Graham
  expansion, 752
  gauge, 751
First order system, 393
Fischler-Susskind mechanism, 319
Fixed point, 528
Flux, 657
  compactification, 658
  imaginary self-dual, 660
  metric, 658
  NS-NS three-form, 657
  R-R p-form, 658
Fractional D-brane, 540
Frenkel-Kač-Segal construction, 327
F-theory, 704, 724
Fubini-Study metric, 458
Fundamental form, 456
Fundamental region of torus, 137
Fundamental string
  SUGRA solution, 699
  tension, 10, 240
Fusion rules, 78, 344
  $N = 2$ superconformal, 381
  $\widehat{su}(2)_k$, 351

Gauge hierarchy problem, 439
Gauss-Bonnet term, 15
Gepner model, 567
  heterotic, 576
  partition function, 573
  type II, 573
Ghost, 36, 142
  action, 107
  field, 107
  number operator, 111
  picture
    canonical, 404
    charge, 403
  system, 394
Gluing
  automorphism, 164
  conditions, 102, 164, 233
Graviphoton, 489
Gravitino, 489, 546
  two-dimensional, 177
Green-Schwarz mechanism, 499, 629, 651
Green-Schwarz superstring, 176
GSO projection, 203, 230, 551

Hagedorn temperature, 93
Harmonic form, 466
Hawking temperature, 759
Hermitian
  conjugate, 36, 70
  connection, 457
  metric, 456
  Yang Mills, 470, 503
Heterotic string, 284
  compactification, 497
  covariant lattice, 422, 578
  Gepner model, 576

non-supersymmetric, 425
orbifold, 525
S-duality, 710
Hirzebruch L-polynomial, 634
Hodge
  decomposition, 384, 462, 463, 467
  diamond, 464, 477
  numbers, 463
  $*$ operator, 465
Holography, 749
Holonomy group, 298, 443, 460
  G_2, 445
  $SU(2)$, 445
  $SU(3)$, 444
Homology
  class, 461
  group, 461
Hořava-Witten theory, 723
Hydrodynamics, 760
Hypermultiplet, 489
  universal, 554, 574
Hypersurface, 452

Instanton
  D-, 706
  world-sheet, 487
  Yang-Mills, 678
Intersecting D-brane
  boundary conditions, 310
  models, 544, 646
Intersection number, 313, 463
  triple, 496
Ishibashi state, 164, 233, 284
Isometry, 432

Jacobi
  theta-function, 227
  triple identity, 228
Jacobian, 128
j-function, 703

Kähler
  cone, 484
    complexification, 486
  form, 457
  manifold, 457
  metric, 457
  moduli, 484
  potential, 457

Kač-Moody algebra, 321, 323
  representation, 343
  $SO(2n)$, 341
  $\widehat{su}(2)_k$, 349, 381
Kaluza-Klein
  Ansatz, 435
  modes, 264
  monopole, 720
Kawai-Lewellen-Tye relations, 614
Killing
  group, 132
  spinor, 442, 522
    equations, 664
  vector, 432
    conformal, 18, 129, 737
Klein bottle, 167, 242
  amplitude, 168, 242
K-theory, 251, 648
K3 surface, 445, 523, 530, 713

Lagrangian submanifold, 648
Laplacian, 448, 467
Large extra dimension scenario, 642
Large-N expansion, 748
Lattice, 327
  covariant, 416, 578
  D_5, 392
  $D_{5,1}$, 416
  $E_{5,1}$, 421
  even, 273, 328
  Lorentzian, 272
  Narain, 273, 293, 715
  odd, 328
  rational, 273
  self-dual, 287, 328, 423
  unimodular, 328
Leech lattice, 335
Lefschetz decomposition, 466
Level
  Kač-Moody algebra, 323
  matching, 40
Lichnerowicz operator, 483
Lie algebra, 322
  E_6, 330
  E_7, 330
  E_8, 330
  lattice, 328, 332
  simply laced, 329
  $SO(2n)$, 329
  $SU(n)$, 330
Lie derivative, 432
Lie group, 321
  simply laced, 326

Index

Light-cone
  coordinates
    space-time, 42
    world-sheet, 16
  gauge, 42, 200
Line bundle, 469, 726
Little group, 46, 178
Loop group, 321
Loop-channel, 159, 165, 236
  annulus, 245, 539
  Klein-bottle, 242, 537
  Möbius strip, 246, 542
Loop-channel – tree-channel equivalence, 159, 238, 242, 540

M-theory, 688, 719
M2-brane, 719
M5-brane, 720, 721
Möbius
  strip, 169, 246
    loop-channel, 170, 246
    tree-channel, 169, 247
  transformations, 72
Magnetic monopole, 240
Maldacena limit, 745
Mandelstam variables, 600
Manifold
  Calabi-Yau, 473
  complex, 450
  Hermitian, 457
  Kähler, 457
  Kähler-Einstein, 459
  orientable, 451
  Ricci-flat, 443
Mapping class group, 140
Marginal operator, 560
Massless spectrum
  Gepner model, 574, 577
  orientifold, 537
  $T^4/\mathbb{Z}_2$ orbifold, 529
  $T^4/\mathbb{Z}_2$ orientifold, 543
  $T^6/\mathbb{Z}_3$ orbifold, 531
Membrane, 719
Mirror
  map, 493
  symmetry, 277, 385, 449, 491, 560
  transformation, 560
Mittag-Leffler theorem, 609
Mode expansion
  closed string, 24
  fermionic closed string, 189
  fermionic open string, 191
  open string, 27

Modular
  group, 136
  invariance, 138, 148
  S-transformation, 137, 153
  T-transformation, 137, 153
Moduli, 269, 449
  bundle, 534
  Calabi-Yau, 483–487
  complex structure, 276, 485
  Kähler, 276, 484
  space Riemann surface, 140
  stabilization, 643
  2-torus, 276, 449
Momentum
  angular, 22
  canonical, 12
  conjugate, 8
  space-time, 24, 27
Monodromy charge, 155

Nambu-Goto action, 10
Narain lattice, 273, 293, 715
Navier-Stokes equation, 762
Near-horizon
  geometry, 736
  limit, 735
Neumann boundary condition, 12, 19, 188
Neveu-Schwarz sector, 188, 225
Newton's constant, 434
Niemeier lattices, 424
Nijenhuis tensor, 455
Non-BPS D-branes, 251
Non-linear sigma model, 315, 428
No-scale structure, 661
NS5-brane, 689
  SUGRA solution, 700

Open string, 19
  classical solution, 31
  fermionic mode expansion, 191
  fermionic spectrum, 203
  mode expansion, 27
  partition function, 157
  spectrum, 46
Operator product expansion, 69
Orbifold, 138
  circle S^1, 294
  heterotic, 525
  supersymmetric, 523
  $T^4/\mathbb{Z}_2$, 528
  $T^6/\mathbb{Z}_3$, 530

Orientifold, 51, 317
  compactification, 534
  construction, 241
  plane, 52, 169, 243, 317
  type IIA on CY, 646
  type IIB on $T^4/\mathbb{Z}_2$, 535
  type IIB on CY, 645

P-transformation, 170, 247, 252
Partition function, 80
  torus, 146
  $T^4/\mathbb{Z}_2$, 529
  $T^6/\mathbb{Z}_3$, 531
Peccei-Quinn shift symmetry, 490, 629
Penrose-Brown-Henneaux (PBH) transformations, 751
Period, 462
  matrix, 139
Physical state, 90, 115
  conditions, 40
Picard-Lefschetz theorem, 728
Picture
  changing operation, 405
  changing operator, 408
  ghost, 404
Planar limit, 748
Planck length, 3
Planck mass, 1, 52, 642
Plasma, 759
Poincaré
  duality, 461
  polynomial, 384
Point group, 298
Poisson
  bracket, 8, 21
  resummation formula, 229, 257, 286
Polyakov action, 12, 19, 176
Prepotential, 494
Primary field, 65, 69, 324, 741
Primitive p-form, 466
Projective space
  complex, 451, 458
  weighted, 480
Pure spinor formulation, 176, 429, 615

Quasi-primary field, 72
Quintic, 480, 486, 576

Radial
  ordering, 64
  quantization, 64, 741

Ramond
  ground state, 198, 379
  sector, 188, 225
Reflection coefficients, 165
Regge
  slope, 11
  trajectory, 32
Resolution of singularity, 523
Ricci
  form, 458
  scalar, 509
  tensor, 430, 458, 509
Riemann
  identity, 228, 257
    generalized, 257
  sphere, 72
  surface, 122
  tensor, 509
  zeta-function, 44
Riemannian
  connection, 457
  geometry, 508
  structure, 129
Riemann-Roch theorem, 132, 232, 397
RNS-superstring, 176
Root, 328
  lattice, 328, 329
  positive, 329
  simple, 329
R-symmetry, 371, 744

Scalar potential, 643, 659
Scattering amplitude
  closed string
    1-closed string, 611
    3-gauge bosons, 596
    3-gravitons, 598
    3-tachyon, 596
    4-tachyons, 600
  one loop, 123
  open string
    3-gluon, 606
    4-gauge bosons, 609
    2-open string, 610
    3-tachyon, 606
    4-tachyons, 608
  tree level, 121, 144
Schur's lemma, 685
Schwarzian derivative, 73
Schwinger term, 324
S-duality
  heterotic-type I, 710
  type IIB, 705

Secondary field, 66
Sen-limit, 730
Shapiro-Virasoro model
    extended, 51
    restricted, 51
Siegel's upper-half plane, 139
Simple current, 154, 570
    extension, 156
    orbit, 156
Singularity, 452
Soliton, 265, 678, 689, 701
Space
    anti-de-Sitter, 736
    conformally flat, 736
    Einstein, 736
    group, 297
    maximally symmetric, 736
Special
    coordinates, 494
    holonomy, 444
    Kähler geometry, 493
    Kähler manifolds, 493
    Lagrangian, 497, 648
Spectral flow, 379, 552
Spin
    connection, 509, 697
    field, 364, 412
    manifolds, 223
    structure, 223
Spinor
    covariantly constant, 442
    Dirac, 213
    Majorana, 216
    Majorana-Weyl, 217
    representation, 213
    Weyl, 217
Stückelberg
    mass, 499
    mechanism, 656
Standard embedding, 499
State-operator correspondence, 74, 741
String
    closed, 14, 23
    frame, 172, 620, 690
    functions, 350, 381
    landscape, 662
    length scale, 11
    mass scale, 11
    open, 14, 27
    scale, 449
    tension, 11
Structure constants, 322
Submanifold
    complex, 452

    transverse, 452
Sugawara construction, 325
Sugimoto model, 319
Super
    charge, 411
    current, 183
    derivative, 356
    interval, 356
    $N = 2$, 371
    Kač-Moody algebra, 556
    operator product expansion, 359
    Virasoro algebra, 191, 197, 358
Superconformal
    algebra, 358
    gauge, 180
    Hilbert space, 361
    transformation, 356
Superfield
    chiral, 356, 372, 377
    primary, 357
    quasi-primary, 361
Supergravity, 622
    democratic formulation, 627
    eleven-dimensional, 623, 717
    gauged $\mathcal{N} = 8$, 750
    heterotic, 629
    type I, 627
    type IIA/B, 624
Supermultiplet, 489, 531, 537
    short, 685
Superspace, 356
    $N = 2$, 371

Tachyon, 47, 90, 162, 251
Tadpole, 95, 101, 243
    cancellation, 247, 541, 647, 732
    local, 317
Tangent bundle, 452
    holomorphic, 453
T-duality, 269, 432
Teichmüller
    parameter, 134
    space, 134
Tension
    ADM, 694
    D-brane, 163, 171, 240
    M2-brane, 719
    M5-brane, 719
    NS5-brane, 695
    string, 11
Theta-function, 227, 254
't Hooft coupling, 748
Three-algebra, 721

Throat, 735
Topological field theory, 385
Torelli group, 141
Toric variety, 482
Toroidal compactification
   D-dimensional, 270
   spectrum, 266
Torsion, 455, 504
   intrinsic, 666
Trace anomaly, 84
Tree-channel, 159, 165
   annulus, 245, 540
   Klein-bottle, 243, 538
   Möbius strip, 247, 542
Twisted sector, 265, 296, 300
Type I superstring, 208, 241
   S-duality, 710
Type I' superstring, 317
Type IIA supergravity
   action, 624
   10D-field content, 488
   4D on CY, 488
Type IIA superstring, 207
   orientifold on CY, 646
Type IIB supergravity
   action, 624
   10D-field content, 488
   4D on CY, 490
Type IIB superstring, 206
   orientifold on CY, 645

Uhlenbeck-Yau theorem, 503
Unit cell, 327
Unitarity, 78, 343, 601
Unitary representation
   Kač-Moody algebra, 343
   $N = 2$ super Virasoro algebra, 380
   $\widehat{su}(2)_k$, 349
   super Virasoro algebra, 362
   Virasoro algebra, 79
Upstairs geometry, 318, 634

Vector bundle
   Hermitian, 469
   holomorphic, 453, 468
   stable, 503
   tangent, 452
Vectormultiplet, 489
Veneziano amplitude, 608
Verlinde formula, 154, 166
Verma module, 75
Vertex operator, 90, 392
   gaugino, 594
   gravitino/dilatino, 594
   graviton, 592
   open string gauge boson, 604
   open string tachyon, 604
   R-R p-form, 595
Virasoro algebra, 39, 59, 70, 323
   centerless, 26
   $N = 1$ supersymmetric, 360
   $N = 2$ supersymmetric, 375, 551
Virasoro constraints, 40
Virasoro-Shapiro amplitude, 601
Volume form, 475

W-algebra, 71
Ward identity
   boundary, 94
   conformal, 83
Warp factor, 440, 662
Weierstrass form, 725
Weight vector, 331
Weyl
   anomaly, 53, 429, 754
   fermion, 178
   invariance, 16
   rescaling, 15
   spinor, 217
   transformation, 179
Weyl-Kač character formula, 350
Wick
   rotation, 64
   theorem, 88
Wilson line, 290, 310, 527
   continuous, 284
   discrete, 504
   moduli, 715
Wilson loop, 766
Winding
   number, 265
   state, 265
Witt algebra, 26
World-line, 7
World-sheet, 10
   parity, 50, 207, 241
   T-duality, 317

Yau's theorem, 473

Zero mode, 18, 24
   dimensional reduction, 447
   fermionic, 196
Zeta-function
   regularization, 44, 202, 259
   Riemann, 44

Printed by Printforce, United Kingdom